Lewin's Essential GENES

Second Edition

Jones and Bartlett Titles in Biological Science

AIDS: Science and Society, Fifth Edition
Hung Fan, Ross F. Conner, & Luis P. Villarreal

AIDS: The Biological Basis, Fifth Edition
Benjamin S. Weeks & I. Edward Alcamo

Alcamo's Fundamentals of Microbiology, Eighth Edition
Jeffrey C. Pommerville

Alcamo's Fundamentals of Microbiology: Body Systems Edition
Jeffrey C. Pommerville

Aquatic Entomology
W. Patrick McCafferty & Arwin V. Provonsha

Atlas of the Visible Human Male: Reverse Engineering of the Human Body
Victor M. Spitzer & David G. Whitlock

Bioethics: An Introduction to the History, Methods, and Practice, Second Edition
Nancy S. Jecker, Albert R. Jonsen, & Robert A. Pearlman

Bioimaging: Current Concepts in Light and Electron Microscopy
Douglas E. Chandler & Robert W. Roberson

Biomedical Graduate School: A Planning Guide to the Admissions Process
David J. McKean & Ted Johnson

Biomedical Informatics
Jules J. Berman

Botany: An Introduction to Plant Biology, Fourth Edition
James D. Mauseth

Case Studies for Understanding the Human Body
Stanton Braude, Deena Goran, & Shelley Maxfield

CELLS
Benjamin Lewin, Lynne Cassimeris, Vishwanath R. Lingappa, & George Plopper

Defending Evolution: A Guide to the Evolution/Creation Controversy
Brian J. Alters

Diversity of Life: The Illustrated Guide to Five Kingdoms, Second Edition
Lynn Margulis, Karlene Schwartz, & Michael Dolan

DNA Sequencing: Optimizing the Process and Analysis
Jan Kieleczawa

DNA Sequencing II: Optimizing Preparation and Cleanup
Jan Kieleczawa

DNA Sequencing III: Dealing with Difficult Templates
Jan Kieleczawa

Early Life: Evolution on the PreCambrian Earth, Second Edition
Lynn Margulis & Michael Dolan

Electron Microscopy, Second Edition
John J. Bozzola & Lonnie D. Russell

Encounters in Microbiology, Volume 1, Second Edition
Jeffrey C. Pommerville

Encounters in Microbiology, Volume 2
Jeffrey C. Pommerville

Essential Genetics: A Genomics Perspective, Fourth Edition
Daniel L. Hartl & Elizabeth W. Jones (late)

Essentials of Molecular Biology, Fourth Edition
George M. Malacinski

Exploring Bioinformatics: A Project-Based Approach
Caroline St. Clair & Jonathan E. Visick

Exploring the Way Life Works: The Science of Biology
Mahlon Hoagland, Bert Dodson, & Judy Hauck

GENES IX
Benjamin Lewin

Genetics: Analysis of Genes and Genomes, Seventh Edition
Daniel L. Hartl & Elizabeth W. Jones (late)

Genetics of Populations, Third Edition
Philip W. Hedrick

Guide to Infectious Diseases by Body System
Jeffrey C. Pommerville

How the Human Genome Works
Edwin H. McConkey

Human Anatomy and Physiology Coloring Workbook, Third Edition
Paul D. Anderson

Human Biology, Sixth Edition
Daniel D. Chiras

Human Biology Laboratory Manual
Charles Welsh

Introduction to the Biology of Marine Life, Ninth Edition
John Morrissey & James L. Sumich

Laboratory and Field Investigations in Marine Life, Ninth Edition
James L. Sumich

Laboratory Fundamentals of Microbiology, Eighth Edition
Jeffrey C. Pommerville

Laboratory Investigations in Molecular Biology
Steven A. Williams, Barton E. Slatko, & John R. McCarrey

Laboratory Textbook of Anatomy and Physiology: Cat Version, Ninth Edition
Anne B. Donnersberger

Laboratory Textbook of Anatomy and Physiology: Fetal Pig, Eighth Edition
Anne B. Donnersberger & Anne E. Lesak Scott

Microbes and Society, Second Edition
Benjamin S. Weeks & I. Edward Alcamo

The Microbial Challenge: Human-Microbe Interactions
Robert I. Krasner

Microbial Genetics, Second Edition
Stanley R. Maloy, John E. Cronan, Jr., & David Freifelder

Microbiology Pearls of Wisdom, Second Edition
S. James Booth

Missing Links: Evolutionary Concepts & Transitions Through Time
Robert A. Martin

Molecular Biology: Genes to Proteins, Third Edition
Burton E. Tropp

Organic Chemistry and Biochemistry Structure Visualization Workbook
B. A. Luceigh

Perl Programming for Medicine and Biology
Jules J. Berman

Plant Cell Biology
Brian E. S. Gunning & Martin W. Steer

Plants, Genes, and Crop Biotechnology, Second Edition
Maarten J. Chrispeels & David E. Sadava

Plant Structure: A Color Guide, Second Edition
Bryan G. Bowes & James D. Mauseth

Principles of Modern Microbiology
Mark Wheelis

Protein Microarrays
Mark Schena

Protoctista Glossary
Lynn Margulis

Python for Bioinformatics
Jason Kinser

Ruby Programming for Medicine and Biology
Jules J. Berman

Strickberger's Evolution, Fourth Edition
Brian K. Hall & Benedikt Hallgrímsson

Teaching Biological Evolution in Higher Education
Brian J. Alters

Tropical Forests
Bernard A. Marcus

20th Century Microbe Hunters
Robert I. Krasner

Understanding Viruses
Teri Shors

Lewin's Essential GENES

Second Edition

Jocelyn E. Krebs
University of Alaska, Anchorage

Elliott S. Goldstein
Arizona State University

Stephen T. Kilpatrick
University of Pittsburgh at Johnstown

JONES AND BARTLETT PUBLISHERS
Sudbury, Massachusetts
BOSTON TORONTO LONDON SINGAPORE

World Headquarters
Jones and Bartlett Publishers
40 Tall Pine Drive
Sudbury, MA 01776
978-443-5000
info@jbpub.com
www.jbpub.com

Jones and Bartlett Publishers Canada
6339 Ormindale Way
Mississauga, Ontario L5V 1J2
Canada

Jones and Bartlett Publishers International
Barb House, Barb Mews
London W6 7PA
United Kingdom

Jones and Bartlett's books and products are available through most bookstores and online booksellers. To contact Jones and Bartlett Publishers directly, call 800-832-0034, fax 978-443-8000, or visit our website, www.jbpub.com.

Substantial discounts on bulk quantities of Jones and Bartlett's publications are available to corporations, professional associations, and other qualified organizations. For details and specific discount information, contact the special sales department at Jones and Bartlett via the above contact information or send an email to specialsales@jbpub.com.

Production Credits
Chief Executive Officer: Clayton Jones
Chief Operating Officer: Don W. Jones, Jr.
President, Higher Education and Professional Publishing: Robert W. Holland, Jr.
V.P., Sales: William J. Kane
V.P., Design and Production: Anne Spencer
V.P., Manufacturing and Inventory Control: Therese Connell
Publisher, Higher Education: Cathleen Sether
Acquisitions Editor: Molly Steinbach
Associate Editor: Megan R. Turner
Editorial Assistant: Caroline Perry
Production Manager: Louis C. Bruno, Jr.
Associate Production Editor: Leah Corrigan
Senior Marketing Manager: Andrea DeFronzo
Text Design: Anne Spencer
Cover Design: Kristin E. Parker
Illustrations: Imagineering Media Services, Inc. and Shepherd, Inc.
Photo Research Manager and Photographer: Kimberly Potvin
Freelance Photo Researcher: Asuka Ohsawa
Composition: Shepherd, Inc.
Printing and Binding: Courier Kendallville
Cover Printing: Courier Kendallville
Cover Image: © Carin L. Cain, The Art of Science: Medical and Scientific Visualization (www.carincain.com)

About the cover: In this picture, eukaryotic gene expression is depicted as a video game. mRNA (purple) is transcribed from chromosomes (blue), processed (scissors), and transported from the nucleus (green circle) where it is translated by ribosomes in the cytoplasm to polypeptides. Antisense RNAs (pink) aid in the regulation of mRNA stability.

Library of Congress Cataloging-in-Publication Data
Lewin's essential genes / by Jocelyn E. Krebs, Elliott S. Goldstein, and Stephen T. Kilpatrick. — 2nd ed.
p. ; cm.
Rev. ed. of: Essential genes / Benjamin Lewin. c2006.
Condensed and updated ed. of: Genes IX / Benjamin Lewin. c2008.
Includes index.
ISBN 978-0-7637-5915-5 (alk. paper)
1. Genetics. 2. Genes. I. Krebs, Jocelyn E. II. Goldstein, Elliott S. III. Kilpatrick, Stephen T. IV. Lewin Benjamin. Essential genes. V. Lewin, Benjamin. Genes IX. VI. Title Essential genes.
[DNLM: 1. Genes. 2. DNA Replication. 3. Eukaryotic Cells—physiology. 4. Gene Expression. 5. Proteins—genetics. 6. Recombination, Genetic. QU 470 E78 2009]
QH430.L4 2009
576.5—dc22 2008047274

6048

Printed in the United States of America
13 12 11 10 09 10 9 8 7 6 5 4 3 2 1

Dedication

To Benjamin Lewin, for setting the bar high.

To my mother Ellen Baker, for raising me with a love of science; to the memory of my stepfather Barry Kiefer, for convincing me science would stay fun; and to my partner Susannah Morgan, for always pretending my biology jokes are funny. Finally, to my son, Rhys, who arrived just in time for this dedication.

Jocelyn Krebs

To my family: my wife, Suzanne, whose patience, understanding and confidence in me are amazing; my children, Andy, Hyla, and Gary, who have taught me so much about using the computer; and my grandchildren, Seth and Elena, whose smiles and giggles inspire me. And to the memory of my mentor and dear friend, Lee A. Snyder, whose professionalism, guidance, and insight demonstrated the skills necessary to be a scientist and teacher. I have tried to live up to his expectations. This is for you, Doc.

Elliott Goldstein

To my wife, Lori; my parents, David and Sandra;
and my children, Jennifer, Andrew, and Sarah.

Stephen Kilpatrick

Brief Table of Contents

Contents

PART II. PROTEINS 154

Chapter 7. Messenger RNA 155

Chapter 8. Translation 183

Chapter 9. Using the Genetic Code 217

PART IV. DNA REPLICATION AND RECOMBINATION 368

Preface

Of the diverse ways to study the living world, molecular biology has been most remarkable in the speed and breadth of its expansion. New data are acquired daily, and new insights into well-studied processes come on a scale measured in weeks or months rather than years. It's difficult to believe that the first complete organismal genome sequence was obtained less than 15 years ago. The structure and function of genes and genomes and their associated cellular processes are sometimes elegantly and deceptively simple but frequently amazingly complex, and no single book can do justice to the realities and diversities of natural genetic systems. The purpose of this book is to provide a clear and concise overview of the field for the undergraduate student; it may also be appropriate for some medical school courses in the subject. Compared to the full edition, there is a redirected focus on essential topics and (in some areas) more background and introductory material.

This edition is generally updated and reorganized for a more logical flow of topics. In particular, discussion of chromatin organization and nucleosome structure now precedes the discussion of eukaryotic transcription, because chromosome organization is critical to all DNA transactions in the cell, and current research in the field of transcriptional regulation is heavily biased toward the study of the role of chromatin in this process. The discussion of transcriptional activation and chromatin remodeling has accordingly been combined into one chapter (Chapter 26). Two chapters on transposons and retrotransposons have been combined into one (Chapter 21). Chapter 30, "Genetic Engineering," has been expanded. Throughout the book, there is additional background material and some expanded coverage of evolutionary genetic concepts. Many new figures are included in this book, some reflecting new developments in the field.

This book is organized into five parts. **Part I (Genes)** comprises Chapters 1 through 6. Chapters 1 and 2 serve as an introduction to the structure and function of DNA and contain basic coverage of DNA replication and gene expression. Chapter 3 introduces the interrupted structures of eukaryotic genes, and Chapters 4 through 6 discuss genome structure and evolution.

Part II (Proteins) comprises Chapters 7 through 10. Chapters 7 to 9 provide general introductions to gene expression: transcription, translation, and the genetic code. Chapter 10 covers the transport of proteins within cells.

Part III (Prokaryotic Gene Expression) includes Chapters 11 through 14. Chapter 11 provides more in-depth coverage of bacterial transcription. In Chapters 12 and 13, the regulation of bacterial gene expression via operons and regulatory RNAs, including RNAi, are discussed. Chapter 14 covers the regulation of expression of genes during phage development as they infect bacterial cells.

Part IV (DNA Replication and Recombination) comprises Chapters 15 through 23. Chapters 15 to 18 provide detailed discussions of DNA replication in plasmids, viruses, and prokaryotic and eukaryotic cells. Chapters 19 through 22 cover recombination and its roles in DNA repair and the human immune system, with Chapter 21 focusing on different types of transposable elements. Chapter 23 discusses the structure of eukaryotic chromosomes.

Part V (Eukaryotic Gene Expression) includes Chapters 24 through 30. In Chapter 24, the structure of nucleosomes and chromatin is discussed; chromatin remodeling and other chromosomal alterations are covered in Chapters 26 and 27. Chapters 25 and 26 describe eukaryotic transcription and its regulation, with Chapter 28 detailing posttranscriptional RNA processing. Chapter 29 provides a discussion of ribozymes. Chapter 30 introduces basic molecular techniques for the study of genes and gene expression.

Although the chapter on genetic engineering has been placed at the end of the book, instructors may choose to use the information from this chapter at any point appropriate for their courses, such as following the introductory chapters, 1 and 2.

For instructors who prefer to order topics with the essentials of DNA replication and gene expression followed by more advanced topics, the following chapter sequence is suggested:

Introduction: Chapters 1–2

Gene and Genome Structure: Chapters 3–6

DNA Replication: Chapters 15–18

Transcription: Chapters 7, 11, 25, and 28

Translation: Chapters 8–9

Regulation of Gene Expression: Chapters 12–14, 22–23, and 26

Other chapters can be covered at the instructor's discretion.

To the Instructor

This edition contains many new pedagogical components to help the instructor engage students in the topic. The text has been revised to be more accessible for students at introductory levels. Each chapter section concludes with *Concept and Reasoning Checks*: one or two questions for review, conceptual synthesis, hypothesizing, or application of learning. Each chapter includes a set of *End of Chapter Questions* with answers to half of the questions provided to the students; the other questions could be used as homework assignments or quizzes. There are additional instructional tools available on the Instructor's ToolKit CD-ROM and accompanying Web site (see below).

To the Student

There are a number of features in the book to help you learn as you read. Each section is summarized with a bulleted list of *Key Concepts*. *Key Terms* are highlighted in boldface in the text and defined in the margin for easy reference as well as compiled into the *Glossary* at the end of the book. Each chapter includes a set of *End of Chapter Questions* intended for self-assessment. Most chapters contain at least one feature box with additional background material or more in-depth details on an issue relevant to the chapter's focus. Boxes fall into one of four categories: *Essential Ideas*, *Historical Perspectives*, *Methods and Techniques*, and *Medical Applications*. In many cases these represent areas of ongoing research in the field. Finally, each chapter concludes with suggested *Further Reading*, a brief list of current reviews and pivotal papers to supplement and reinforce the chapter content.

Ancillaries

Jones and Bartlett Publishers offers an impressive variety of traditional and interactive multimedia supplements to assist instructors and aid students in mastering molecular biology. Additional information and review copies of any of the following items are available through your Jones and Bartlett sales representative or by going to http://www.jbpub.com/biology.

Online Student Study Guide

Jones and Bartlett Publishers and Brent Nielsen of Brigham Young University have developed an interactive, electronic study guide dedicated exclusively to this title. Students will find a variety of study aids and resources at http://biology.jbpub.com/lewin/essentialgenes, all designed to explore the concepts of molecular biology in more depth and to help students master the material in the book. A variety of activities are available to help students review class material, such as an interactive summary, Web-based learning exercises, study quizzes, a searchable glossary, and links to animations, videos, and podcasts, all to help students master important terms and concepts.

Instructor's ToolKit CD-ROM

The *Instructor's ToolKit CD-ROM* contains a suite of files to help professors teach their courses. The materials are cross-platform for Windows and Macintosh systems. All the files on the CD are ready for online courses using the **WebCT** or **Blackboard** formats.

- The **PowerPoint® Image Bank** provides the illustrations, photographs, and tables (to which Jones and Bartlett Publishers holds the copyright or has permission to reprint digitally) inserted into PowerPoint slides. With the Microsoft PowerPoint program, you can quickly and easily copy individual image slides into your existing lecture slides.
- The **PowerPoint Lecture Outline Slides** presentation package provides images and lecture notes, created by Hao Nguyen, of California State University, Sacramento, for each chapter of *Lewin's Essential Genes, Second Edition*. A PowerPoint viewer is provided on the CD. Instructors with the Microsoft PowerPoint software can customize the outlines, figures, and order of presentation.
- The **Test Bank**, also created by Hao Nguyen, is provided as a text file with over 700 questions in a variety of formats.

Acknowledgments

The authors would like to thank the following individuals for their assistance in the preparation of this book:

The editorial, production, marketing, and sales teams at Jones & Bartlett have been exemplary in all aspects of this project. Cathy Sether, Molly Steinbach, and Lou Bruno deserve special mention. Cathy brought us together on this project and in doing so launched an efficient and amiable partnership. She has provided able leadership and has been an excellent resource as we ventured into new territories. Molly and Lou have handled the daily responsibilities of the writing and production phases with friendly professionalism, helpful guidance, and appropriate doses of humor.

Thanks for the creation of many of the End of Chapter Questions go to Brent Nielsen. We also thank the authors of the special topics boxes found throughout the text:

Loree Burns
Jamie Kass, New York Academy of Sciences
Brent Nielsen, Brigham Young University
Teri Shors, University of Wisconsin, Oshkosh
Esther Siegfried, University of Pittsburgh at Johnstown

We thank Hao Nguyen of California State University, Sacramento, for writing the test bank and lecture outlines. Finally, we would like express our gratitude to the reviewers of this edition, whose feedback helped to shape the text in many ways:

Salem Al-Maloul, Hashemite University
James Botsford, New Mexico State University
David Bourgaize, Whittier College
John Boyle, University of Mississippi
Mary Connell, Appalachian State University
Robert Dotson, Tulane University
Julia Frugoli, Clemson University
Daniel Herman, University of Wisconsin, Eau Claire
Stan Ivey, Delaware State University
Christi Magrath, Troy University
Mitch McVey, Tufts University
Hao Nguyen, California State University, Sacramento
Stacy Darling Novak, University of La Verne
Eva Sapi, University of New Haven
Ben Stark, Illinois Institute of Technology
Takashi Ueda, Florida Gulf Coast University
Ramakrishna Wusirika, Michigan Technological University
Anastasia Zimmerman, College of Charleston

Jocelyn E. Krebs
Elliott S. Goldstein
Stephen T. Kilpatrick

About the Authors

Benjamin Lewin founded the journal *Cell* in 1974 and was Editor until 1999. He founded the Cell Press journals *Neuron*, *Immunity*, and *Molecular Cell*. In 2000, he founded Virtual Text, which was acquired by Jones & Bartlett Publishers in 2005. He is also the author of *GENES* and *CELLS*.

Jocelyn E. Krebs received a B.A. in Biology from Bard College, Annandale-on-Hudson, NY, and a Ph.D. in Molecular and Cell Biology from the University of California, Berkeley. For her Ph.D. thesis, she studied the roles of DNA topology and insulator elements in transcriptional regulation. She performed her postdoctoral training as an American Cancer Society Fellow at the University of Massachusetts Medical School in the laboratory of Dr. Craig Peterson, where she focused on the roles of histone acetylation and chromatin remodeling in transcription. In 2000, Dr. Krebs joined the faculty in the Department of Biological Sciences at the University of Alaska, Anchorage, where she is now an Associate Professor. She directs a research group studying chromatin structure and function in transcription and DNA repair in the yeast *Saccharomyces cerevisiae* and the role of chromatin remodeling in embryonic development in the frog *Xenopus*. She teaches courses in molecular biology for undergraduates, graduate students, and first-year medical students. She also teaches a Molecular Biology of Cancer course and has taught Genetics and Introductory Biology. She lives in Eagle River, AK, with her partner and a house full of dogs and cats. Her non-work passions include hiking, camping, and snowshoeing.

Elliott S. Goldstein earned his B.S. in Biology from the University of Hartford (Connecticut) and his Ph.D. in Genetics from the University of Minnesota, Department of Genetics and Cell Biology. Following this, he was awarded an N.I.H. Postdoctoral Fellowship to work with Dr. Sheldon Penman at the Massachusetts Institute of Technology. After leaving Boston, he joined the faculty at Arizona State University in Tempe, where he is an Associate Professor in the Cellular, Molecular, and Biosciences program in the School of Life Sciences and in the Honors Disciplinary Program. His research interests are in the area of molecular and developmental genetics of early embryogenesis in *Drosophila melanogaster*. In recent years, he has focused on the *Drosophila* counterparts of the human proto-oncogenes *jun* and *fos*. His primary teaching responsibilities are in the undergraduate General Genetics course as well as the graduate level Molecular Genetics course. Dr. Goldstein lives in Tempe with his wife, his high school sweetheart. They have three children and two grandchildren. He is a bookworm who loves reading as well as underwater photography. His pictures can be found at http://www.public.asu.edu/~elliotg/.

Stephen T. Kilpatrick received a B.S. in Biology from Eastern College (now Eastern University) in St. Davids, PA, and a Ph.D. from the Program in Ecology and Evolutionary Biology at Brown University. His thesis research was an investigation of the population genetics of interactions between the mitochondrial and nuclear genomes of *Drosophila melanogaster*. Since 1995, Dr. Kilpatrick has taught at the University of Pittsburgh at Johnstown (UPJ) in Johnstown, PA. His regular teaching duties include undergraduate courses in nonmajors biology, introductory majors biology, and advanced undergraduate courses in genetics, evolution, molecular genetics, and biostatistics. He has also supervised a number of undergraduate research projects in evolutionary genetics. Dr. Kilpatrick's major professional focus has been in biology education. He has participated in the development and authoring of ancillary materials for several introductory biology, genetics, and molecular genetics texts as well as writing articles for educational reference publications. For his classes at UPJ, Dr. Kilpatrick has developed many active learning exercises in introductory biology, genetics, and evolution. Dr. Kilpatrick resides in Johnstown, PA, with his wife and three children. Outside of scientific interests, he enjoys music, literature, and theater and occasionally performs in local community theater groups.

I

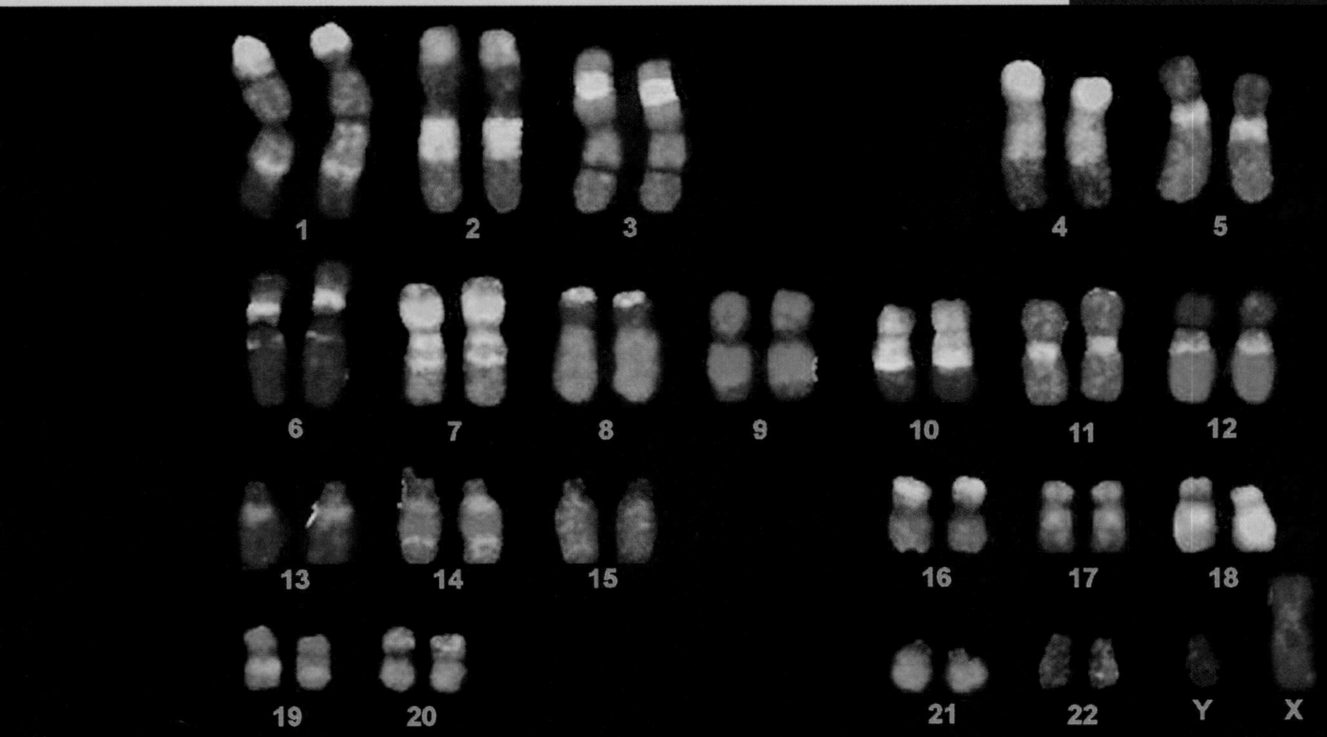

The human male karyotype, or complete set of chromosomes, shown using chromosome painting. The "paints" are created by hybridizing fluorescent molecular probes to specific chromosomal regions. Photo courtesy of Steve M. Carr, Memorial University. Adapted from a photograph by Genetix Ltd. Used with permission.

Genes

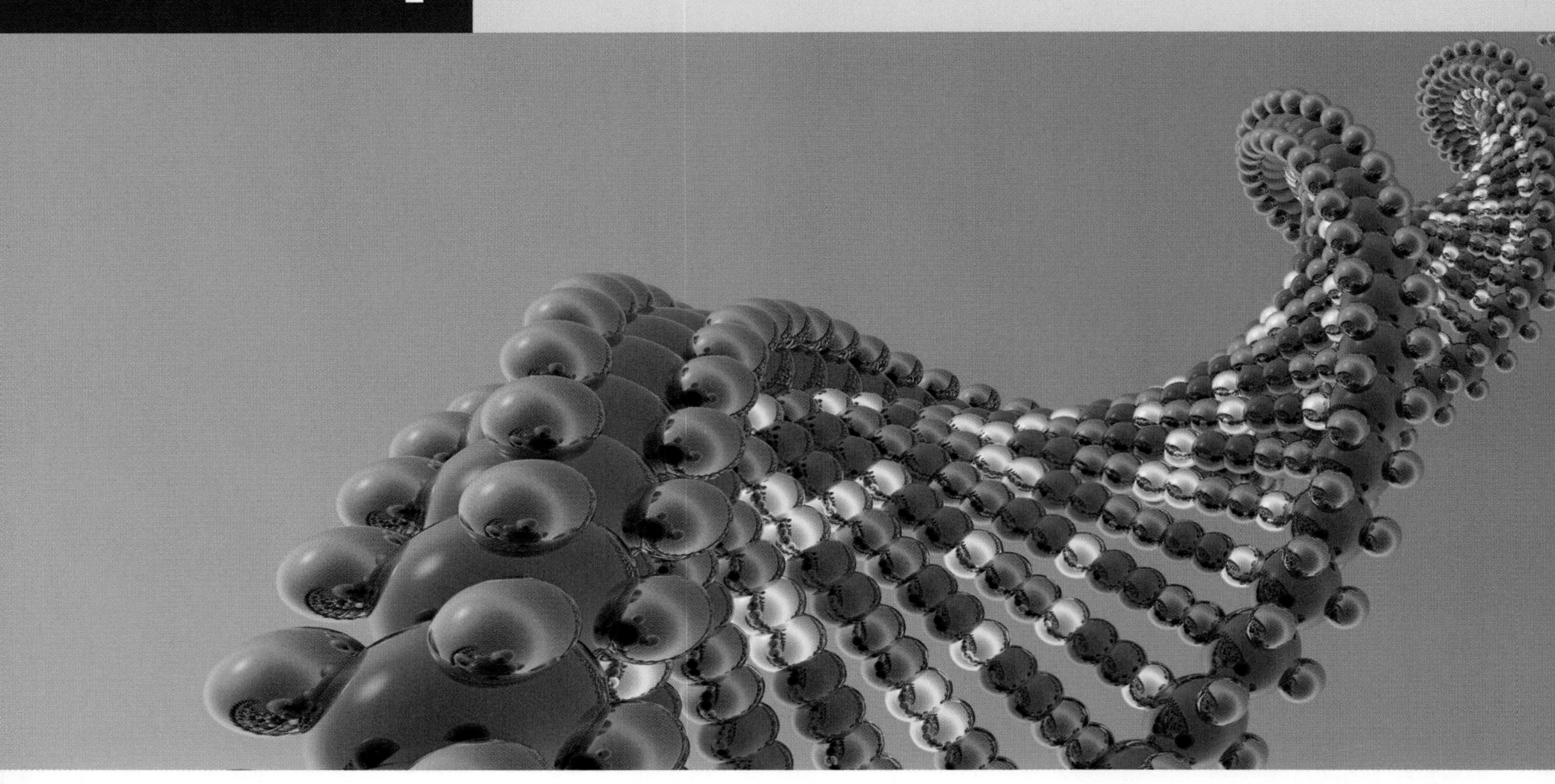

A strand of DNA in blue and red. DNA is the genetic material of eukaryotic cells, bacteria, and many viruses. © Artsilensecome/ShutterStock, Inc.

Genes Are DNA

CHAPTER OUTLINE

1.1 Introduction

The hereditary basis of every living organism is its **genome**, a long sequence of DNA that provides the complete set of hereditary information carried by the organism. The genome includes chromosomal DNA as well as DNA in plasmids and (in eukaryotes) organellar DNA as found in mitochondria and chloroplasts. We use the term *information* because the genome does not itself perform an active role in the development of the organism. It is the sequence of the individual subunits, or bases, of the DNA that determines development. By a complex series of interactions, the DNA sequence produces all of the proteins of the organism at the appropriate time and place. Proteins serve a diverse series of roles in the development and functioning of an organism: they can form part of the structure of the organism, they have the capacity to build the structures, they perform the metabolic reactions necessary for life, and they participate in regulation as transcription factors, receptors, key players in signal transduction pathways, and other molecules.

▸ **genome** The complete set of sequences in the genetic material of an organism. It includes the sequence of each chromosome plus any DNA in organelles.

Physically, the genome may be divided into a number of different DNA molecules, or **chromosomes**. The ultimate definition of a genome is the sequence of the DNA of each chromosome. Functionally, the genome is divided into genes. Each gene is a sequence of DNA that codes for a single type of RNA or polypeptide. Each of the discrete chromosomes comprising the genome may contain a large number of genes. Genomes for living organisms may contain as few as ~500 genes (for a mycoplasma, a type of bacterium) or as many as ~20,000 to 25,000 for a human being.

▸ **chromosome** A discrete unit of the genome carrying many genes. Each consists of a very long molecule of duplex DNA and an approximately equal mass of proteins. It is visible as a morphological entity only during cell division.

In this chapter, we explore the gene in terms of its basic molecular construction. **FIGURE 1.1** summarizes the stages in the transition from the historical concept of the gene to the modern definition of the genome.

The first definition of the gene as a functional unit followed from the discovery that individual genes are responsible for the production of specific proteins. The chemical differences between the DNA of the gene and its protein product led to the suggestion that a gene codes for a protein. This in turn led to the discovery of the complex apparatus by which the DNA sequence of a gene determines the amino acid sequence of a polypeptide.

Understanding the process by which a gene is expressed allows us to make a more rigorous definition of its nature. **FIGURE 1.2** shows the basic theme of this book. A gene is a sequence of DNA that directly produces a single strand of another nucleic acid, RNA, with a sequence that is identical to one of the two polynucleotide strands of DNA. In most cases, the RNA is in turn used to direct production of a polypeptide, while in other cases (such as rRNA and tRNA genes), the RNA transcribed from the gene is the functional end product. Thus a gene is a sequence of DNA that codes for an RNA, and in protein-coding (or **structural**) genes, the RNA in turn codes for a polypeptide.

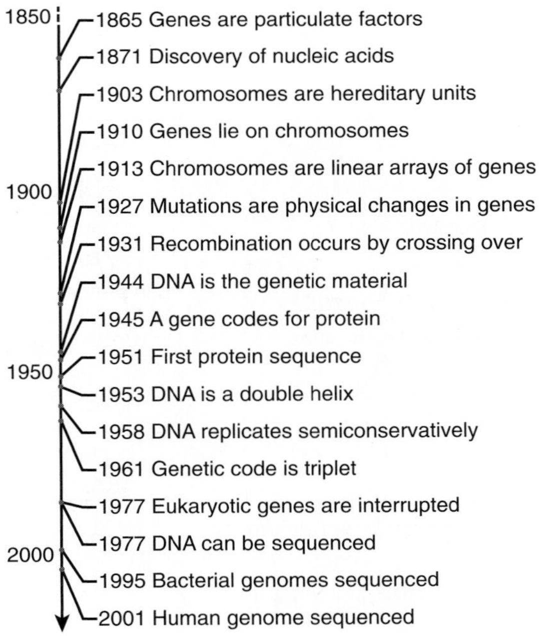

From the demonstration that a gene consists of DNA, and that a chromosome consists of a long stretch

1850
1900
1950
2000
1865 Genes are particulate factors
1871 Discovery of nucleic acids
1903 Chromosomes are hereditary units
1910 Genes lie on chromosomes
1913 Chromosomes are linear arrays of genes
1927 Mutations are physical changes in genes
1931 Recombination occurs by crossing over
1944 DNA is the genetic material
1945 A gene codes for protein
1951 First protein sequence
1953 DNA is a double helix
1958 DNA replicates semiconservatively
1961 Genetic code is triplet
1977 Eukaryotic genes are interrupted
1977 DNA can be sequenced
1995 Bacterial genomes sequenced
2001 Human genome sequenced

FIGURE 1.1 A brief history of genetics.

▸ **structural gene** A gene that codes for any RNA or polypeptide product other than a regulator.

FIGURE 1.2 A gene codes for an RNA, which may code for polypeptide.

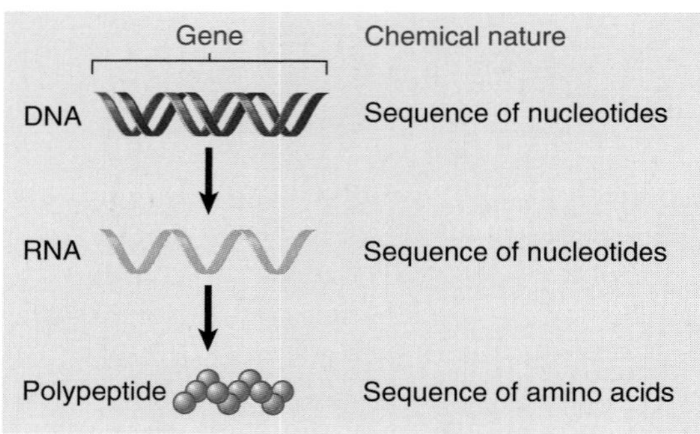

Gene
Chemical nature
DNA
Sequence of nucleotides
RNA
Sequence of nucleotides
Polypeptide
Sequence of amino acids

of DNA representing many genes, we will move to the overall organization of the genome. In *Chapter 3, The Interrupted Gene*, we take up in more detail the organization of the gene and its representation in proteins. In *Chapter 4, The Content of the Genome*, we consider the total number of genes, and in *Chapter 6, Clusters and Repeats*, we discuss other components of the genome and the maintenance of its organization.

CONCEPT AND REASONING CHECK

Why is it accurate to say that a genome has information for the development of an organism but does not directly participate in development?

1.2 DNA Is the Genetic Material of Bacteria, Viruses, and Eukaryotic Cells

▸ **transformation** In bacteria, it is the acquisition of new genetic material by incorporation of added DNA.

The idea that the genetic material is DNA has its roots in the discovery of **transformation** by Frederick Griffith in 1928 (see *Historical Perspectives: Determining That DNA Is the Genetic Material*). Purification of the transforming principle from bacterial cells in 1944 by Avery, MacLeod, and McCarty showed that it is deoxyribonucleic acid (DNA).

Having shown that DNA is the genetic material of bacteria, the next step was to demonstrate that DNA is the genetic material in a quite different system. Phage T2 is a virus that infects the bacterium *Escherichia coli*. When phage particles are added to bacteria, they attach to the outside surface, some material enters the cell, and then ~20 minutes later each cell bursts open, or lyses, to release a large number of progeny phage.

FIGURE 1.3 illustrates the results of an experiment in 1952 by Alfred Hershey and Martha Chase in which bacteria were infected with T2 phages that had been radioactively labeled either in their DNA component (with ^{32}P) or in their protein component (with ^{35}S). The infected bacteria were agitated in a blender, and two fractions were separated by centrifugation. One fraction contained the empty phage "ghosts" that were released from the surface of the bacteria, and the other consisted of the infected bacteria themselves. Previously, it had been shown that phage replication occurs intracellularly, so that the genetic material of the phage would have to enter the cell during infection.

FIGURE 1.3 The genetic material of phage T2 is DNA.

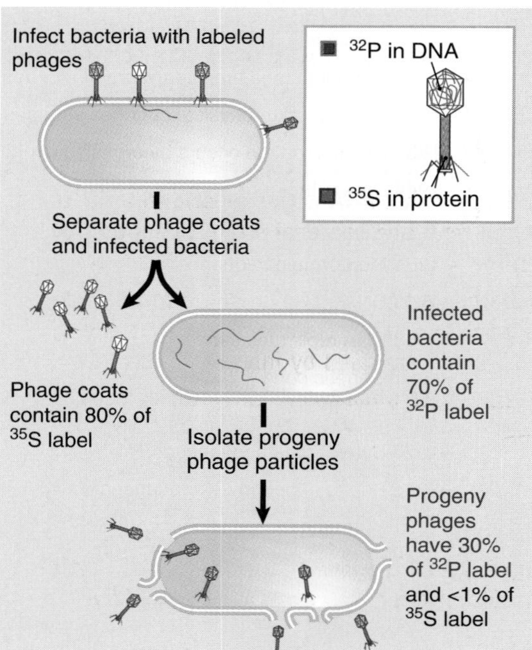

Infect bacteria with labeled phages
^{32}P in DNA
^{35}S in protein
Separate phage coats and infected bacteria
Infected bacteria contain 70% of ^{32}P label
Phage coats contain 80% of ^{35}S label
Isolate progeny phage particles
Progeny phages have 30% of ^{32}P label and <1% of ^{35}S label

Most of the ^{32}P label was present in the fraction containing infected bacteria. The progeny phage particles produced by the infection contained ~30% of the original ^{32}P label. The progeny received less than 1% of the protein contained in the original phage population. The phage ghosts consist of protein and therefore carried the ^{35}S radioactive label. This experiment directly showed that only the DNA of the parent phages enters the bacteria and becomes part of the progeny phages, which is exactly the pattern expected of genetic material.

A phage reproduces by commandeering the machinery of an infected host cell to manufacture more copies of itself. The phage possesses genetic material with properties analogous to those of cellular genomes: its traits are faithfully expressed and are subject to the same rules that govern inheritance of cellular traits. The case of T2 reinforces the general conclusion that DNA is genetic material of the genome of a cell or a virus.

Cells that lack *TK* gene cannot produce thymidine kinase and die in absence of thymidine
Add TK^+ DNA
Dead cells
Live cells
Colony of TK^+ cells
Some cells take up *TK* gene; descendants of transfected cell pile up into a colony

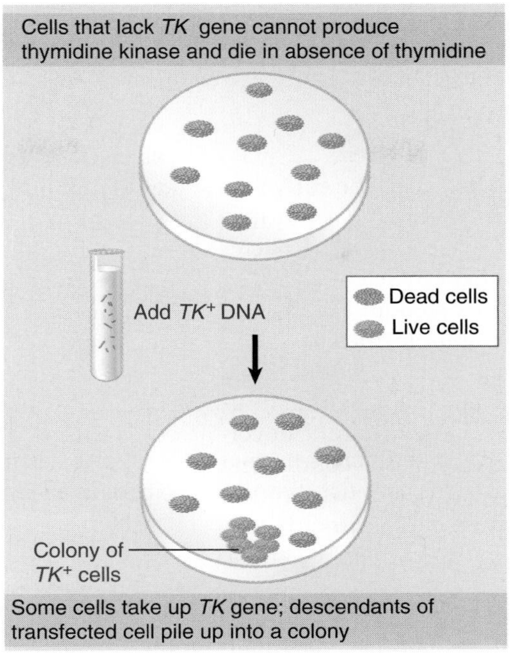

FIGURE 1.4 Eukaryotic cells can acquire a new phenotype as the result of transfection by added DNA.

When DNA is added to eukaryotic cells growing in culture, it enters the cells, and in some of them this results in the production of new proteins. When an isolated gene is used, its incorporation leads to the production of a particular protein, as depicted in **FIGURE 1.4**. Although for historical reasons these experiments are described as **transfection** when performed with animal cells, they are a direct counterpart to bacterial transformation. The DNA that is introduced into the recipient cell becomes part of its genome and is inherited with it, and expression of the new DNA results in a new trait. At first, these experiments were successful only with individual cells growing in culture, but in later experiments DNA was introduced into mouse eggs by microinjection and became a stable part of the genome of the mouse. Such experiments show directly that DNA is the genetic material in eukaryotes, and that it can be transferred between different species and remain functional.

▸ **transfection** In eukaryotic cells, it is the acquisition of new genetic markers by incorporation of added DNA.

The genetic material of all known organisms and many viruses is DNA. However, some viruses use RNA as the genetic material. Therefore, the general nature of the genetic material is that it is always nucleic acid; specifically, it is DNA, except in the RNA viruses.

KEY CONCEPTS

- Bacterial transformation provided the first support that DNA is the genetic material of bacteria. Genetic properties can be transferred from one bacterial strain to another by extracting DNA from the first strain and adding it to the second strain.
- Phage infection showed that DNA is the genetic material of viruses. When the DNA and protein components of bacteriophages are labeled with different radioactive isotopes, only the DNA is transmitted to the progeny phages produced by infecting bacteria.
- DNA can be used to introduce new genetic traits into animal cells or whole animals.
- In some viruses, the genetic material is RNA.

CONCEPT AND REASONING CHECK

If Hershey and Chase had observed that nearly none of the DNA but a substantial fraction of the protein of phage T2 enters *E. coli* cells during infection, what would they have concluded? Protein is the genetic material (of the phage T2).

HISTORICAL PERSPECTIVES

Determining That DNA Is the Genetic Material

Pneumonia was a leading cause of death in the early part of the twentieth century, and much effort was put into understanding the structure and function of the bacteria known to cause it: the pneumococci. Scientists knew that several pneumococcal types existed and that these types could be distinguished by the molecules (capsular polysaccharides) displayed on their surface, but they did not know whether the different types represented individual and stable strains of bacteria or different developmental stages of a single strain. While conducting experiments to distinguish between these possibilities, Fred Griffith set in motion a series of experiments that ultimately led to the identification of DNA as the genetic material of all cells.

Griffith studied S forms and R forms of pneumococcal bacteria (so-called for the smooth and rough appearance of the bacteria when grown in laboratory culture; see *Section 1.2, DNA Is the Genetic Material of Bacteria, Viruses, and Eukaryotic Cells*). The R form, which was generated in the laboratory from the S form, produced no capsular polysaccharides and did not cause pneumonia when injected into mice. S form was lethal to mice, but could be inactivated by exposing the bacteria to high heat before injection. Surprisingly, Griffith found that when R form was injected into mice alongside heat-killed S bacteria, the mice contracted pneumonia and died (**FIGURE B1.1**). What was more, live S type bacteria—virulent and coated with capsular proteins—could be isolated from the dead mice. R form bacteria had clearly been transformed into a stable S form, and the transformation process required some factor from the heat-killed S sample. Griffith suspected this factor, the so-called transforming principle, was a protein . . . and he was not alone.

Proteins were by far the most abundant macromolecule in living cells. Their basic building blocks, the amino acids, came in twenty varieties and could be

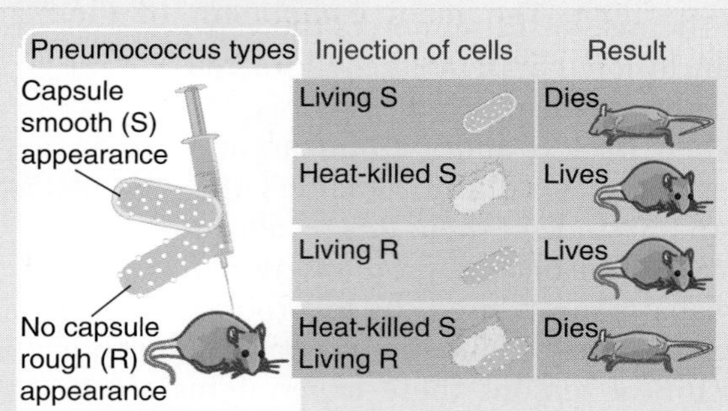

FIGURE B1.1 Neither heat-killed S-type nor live R-type bacteria can kill mice, but simultaneous injection of both can kill mice just as effectively as the live S-type.

1.3 Polynucleotide Chains Have Nitrogenous Bases Linked to a Sugar–Phosphate Backbone

- **purine** A double-ringed nitrogenous base, such as adenine or guanine.
- **pyrimidine** A single-ringed nitrogenous base, such as cytosine, thymine, or uracil.
- **nucleoside** A molecule consisting of a purine or pyrimidine base linked to the 1′ carbon of a pentose sugar.

The basic building block of nucleic acids (DNA and RNA) is the nucleotide, which has three components:

- a nitrogenous base,
- a sugar, and
- one or more phosphates.

The nitrogenous base is a **purine** or **pyrimidine** ring. The base is linked to the 1′ ("one prime") carbon on a pentose sugar by a glycosidic bond from the N_1 of pyrimidines or the N_9 of purines. The pentose sugar linked to a nitrogenous base is called a **nucleoside**. Nucleic acids are named for the type of sugar: DNA has 2′-deoxyribose, whereas RNA has ribose. The difference is that the sugar in RNA

joined in seemingly endless combinations to create molecules of infinite size, shape, and function. By contrast, DNA was a relatively minor component of cells and was known to comprise only four nucleotide types. These nucleotides were thought to be arranged in a specific, repetitive pattern, which contributed to the notion that deoxyribonucleic acids were molecules of "monotonous uniformity" that were unlikely to have any biological specificity. And, so, it was universally assumed that the transforming principle was a protein.

In 1944, Oswald Avery, Colin MacLeod, and Maclyn McCarty published the first report on the chemical nature of the transforming principle, and their results were startling. The main component of their active sample of purified transforming principle was not protein at all; it was DNA. To prove that the transforming activity of this sample was due to the DNA and not to some minor but active contaminant, the authors treated it with enzymes that degrade protein, RNA, and carbohydrate; in all cases the sample retained the ability to stably transform R bacteria into S bacteria. The authors concluded that the transforming principle "consists principally, if not solely, of a highly polymerized, viscous form of deoxyribonucleic acid."

Avery and colleagues strengthened their claim that DNA was the transforming (or genetic) material by later showing that DNase I, an enzyme that specifically degrades DNA, completely eliminated the transforming activity of their purified sample, but the scientific community remained skeptical. It was not until 1952, when A.D. Hershey and Martha Chase studied the independent roles of protein and DNA in bacteriophage T2 infection, that the notion of DNA as the genetic material finally gained favor.

Bacteriophage T2 is a virus that infects bacterial cells in discrete steps: the phage particle attaches itself to the bacterial cell wall, phage material is injected into the cell, the host cell is induced to produce new phage particles, and, finally, the bacterial cell bursts, releasing hundreds of new phage particles. To follow the movement of phage protein and phage DNA during this process, Hershey and Chase radioactively labeled each macromolecule separately: the protein with a radioactive form of sulfur and the DNA with a radioactive form of phosphorous. In their experiment, the majority of the phosphorous label entered the bacterial cell during infection, whereas the majority of the sulfur label stayed outside the cell. The authors went on to show that the new phage particles produced by the infected bacterial cell contained a large percentage of the labeled DNA but virtually none of the labeled protein (see Figure 1.3). These results indicated that it was DNA from the infecting phage, not protein, that entered the bacterial cell and induced genetic changes.

has a hydroxyl (–OH) group on the 2′ carbon of the pentose ring. The sugar can be linked by its 5′ or 3′ carbon to a phosphate group. A nucleoside linked to a phosphate is a **nucleotide**.

A **polynucleotide** is a long chain of nucleotides. **FIGURE 1.5** shows that the backbone of the polynucleotide chain consists of an alternating series of pentose (sugar) and phosphate residues. The chain is formed by linking the 5′ carbon of one pentose ring to the 3′ carbon of the next pentose ring via a phosphate group, so the sugar–phosphate backbone is said to consist of 5′–3′ phosphodiester linkages. The nitrogenous bases "stick out" from the backbone.

Each nucleic acid contains four types of nitrogenous base. The same two purines, adenine (A) and guanine (G), are present in both DNA and RNA. The two pyrimidines in DNA are cytosine (C) and thymine (T); in RNA uracil (U) is found instead of thymine. The only difference between uracil and thymine is the presence of a methyl group at position C_5.

‣ **nucleotide** A molecule consisting of a purine or pyrimidine base linked to the 1′ carbon of a pentose sugar and a phosphate group linked to either the 5′ or 3′ carbon of the sugar.

‣ **polynucleotide** A chain of nucleotides, such as DNA or RNA.

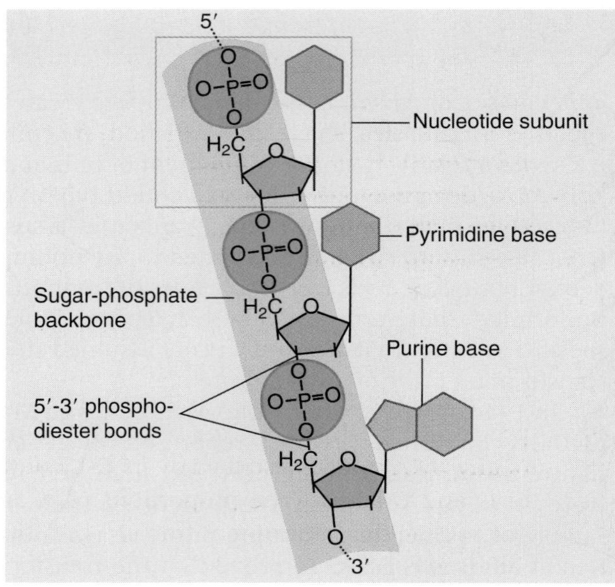

FIGURE 1.5 A polynucleotide chain consists of a series of 5′–3′ sugar-phosphate links that form a backbone from which the bases protrude.

The terminal nucleotide at one end of the chain has a free 5′ phosphate group, whereas the terminal nucleotide at the other end has a free 3′ hydroxyl group. It is conventional to write nucleic acid sequences in the 5′ to 3′ direction—that is, from the 5′ terminus at the left to the 3′ terminus at the right.

KEY CONCEPTS

- A nucleoside consists of a purine or pyrimidine base linked to the 1′ carbon of a pentose sugar.
- The difference between DNA and RNA is in the group at the 2′ position of the sugar. DNA has a deoxyribose sugar (2′–H); RNA has a ribose sugar (2′–OH).
- A nucleotide consists of a nucleoside linked to a phosphate group on either the 5′ or 3′ carbon of the (deoxy)ribose.
- Successive (deoxy)ribose residues of a polynucleotide chain are joined by a phosphate group between the 3′ carbon of one sugar and the 5′ carbon of the next sugar.
- One end of the chain (conventionally written on the left) has a free 5′ end and the other end of the chain has a free 3′ end.
- DNA contains the four bases adenine, guanine, cytosine, and thymine; RNA has uracil instead of thymine.

CONCEPT AND REASONING CHECK

List the structural differences between DNA and RNA nucleotides.

1.4 DNA Is a Double Helix

By the 1950s, the observation by Erwin Chargaff that the bases are present in different amounts in the DNAs of different species led to the concept that the sequence of bases is the form in which genetic information is carried. Given this concept, there were two remaining challenges: working out the structure of DNA, and explaining how a sequence of bases in DNA could determine the sequence of amino acids in a protein.

Three pieces of evidence contributed to the construction of the double helix model for DNA by James Watson and Francis Crick in 1953:

- X-ray diffraction data collected by Rosalind Franklin and Maurice Wilkins showed that the B-form of DNA (found in aqueous solution) is a regular helix, making a complete turn every 34 Å (3.4 nm), with a diameter of ~20 Å (2 nm). Since the distance between adjacent nucleotides is 3.4 Å (0.34 nm), there must be 10 nucleotides per turn.
- The density of DNA suggests that the helix must contain two polynucleotide chains. The constant diameter of the helix can be explained if the bases in each chain face inward and are restricted so that a purine is always paired with a pyrimidine, avoiding partnerships of purine–purine (which would be too wide) or pyrimidine–pyrimidine (which would be too narrow).
- Chargaff also observed that regardless of the absolute amounts of each base, the proportion of G is always the same as the proportion of C in DNA, and the proportion of A is always the same as that of T. Consequently, the composition of any DNA can be described by its G-C content, or the sum of the proportions of G and C bases. (The proportions of A and T bases can be determined by subtracting the G-C content from 1.) G-C content ranges from 0.26 to 0.74 for different species.

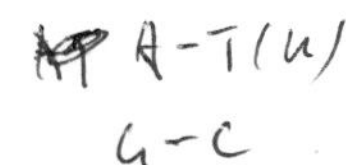

Watson and Crick proposed that the two polynucleotide chains in the double helix associate by hydrogen bonding between the nitrogenous bases. Normally, G can hydrogen bond specifically only with C, whereas A can bond specifically only with T. This hydrogen bonding between bases is described as *base pairing*, and the paired bases (G forming three hydrogen bonds with C, or A forming two hydrogen bonds with T) are said to be **complementary**. Complementary base pairing occurs because of complementary shapes of the complementary bases at the interfaces of where they pair, along with the location of just the right functional groups in just the right geometry along those interfaces so that hydrogen bonds can form.

The Watson-Crick model has the two polynucleotide chains running in opposite directions, so they are said to be **antiparallel**, as illustrated in **FIGURE 1.6**. Looking in one direction along the helix, one strand runs in the 5′ to 3′ direction, whereas its complement runs 3′ to 5′.

- **complementary** Base pairs that match up in the pairing reactions in double helical nucleic acids (A with T in DNA or with U in RNA, and C with G).
- **antiparallel** Strands of the double helix are organized in opposite orientation, so that the 5′ end of one strand is aligned with the 3′ end of the other strand.

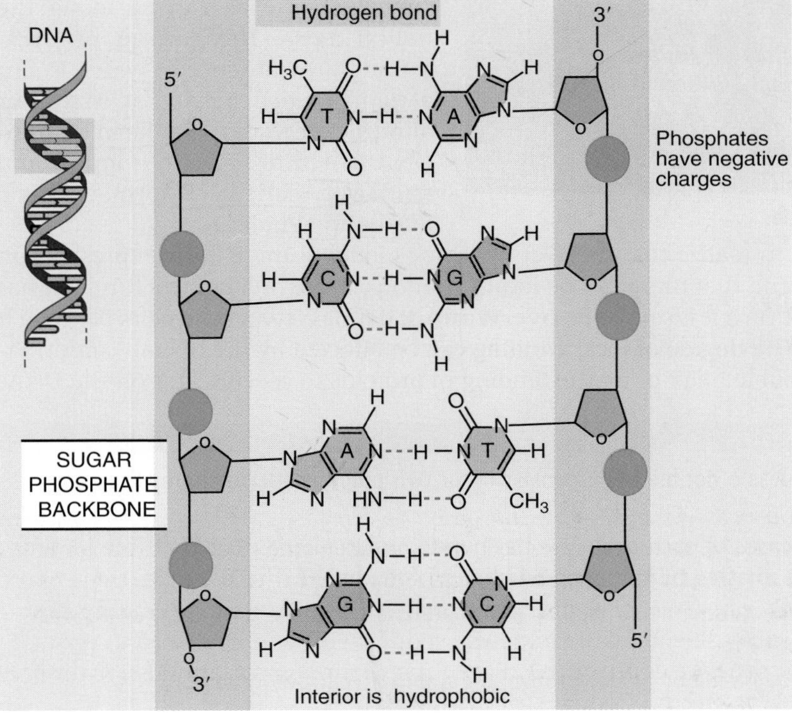

FIGURE 1.6 The double helix maintains a constant width because purines always face pyrimidines in the complementary A-T and G-C base pairs. The sequence in the figure is T-A, C-G, A-T, G-C.

FIGURE 1.7 Flat base pairs lie perpendicular to the sugar-phosphate backbone.

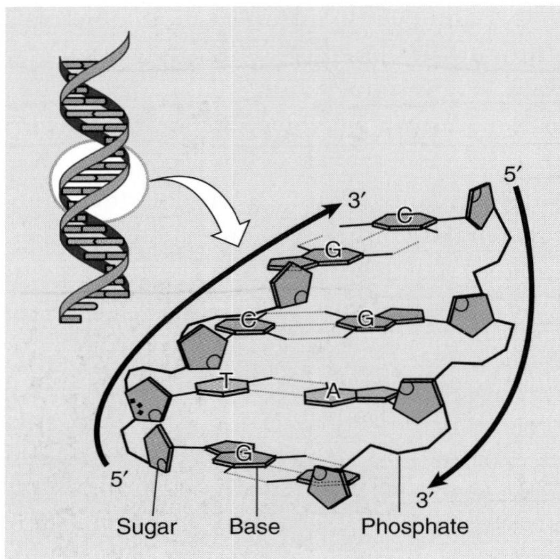

FIGURE 1.8 The two strands of DNA form a double helix. Photo © Photodisc.

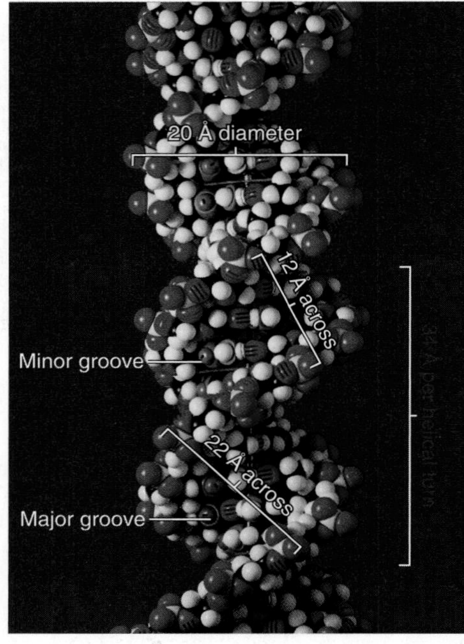

▸ **minor groove** A fissure running the length of the DNA double helix that is 12 Å across.

▸ **major groove** A fissure running the length of the DNA double helix that is 22 Å across.

The sugar–phosphate backbones are on the outside of the double helix and carry negative charges on the phosphate groups. When DNA is in solution *in vitro*, the charges are neutralized by the binding of metal ions, typically sodium (Na^+). In the cell, positively charged proteins provide some of the neutralizing force. These proteins play important roles in determining the organization of DNA in the cell.

The base pairs are on the inside of the double helix. They are flat and lie perpendicular to the axis of the helix. Using the analogy of the double helix as a spiral staircase, the base pairs form the steps, as illustrated schematically in **FIGURE 1.7**. Proceeding up the helix, bases are stacked above one another like a pile of plates.

Each base pair is rotated ~36° around the axis of the helix relative to the next base pair, so ~10 base pairs make a complete turn of 360°. The twisting of the two strands around one another forms a double helix with a **minor groove** that is ~12 Å (1.2 nm) across and a **major groove** that is ~22 Å (2.2 nm) across, as can be seen from the scale model of **FIGURE 1.8**. In B-DNA, the double helix is said to be "right-handed"; the turns run clockwise as viewed along the helical axis. (The A-form of DNA, found in the absence of water, is also a right-handed helix that is shorter and thicker than the B-form. A third DNA structure, Z-DNA, is longer and narrower than the B-form, and is a left-handed helix.)

It is important to realize that the Watson-Crick model of the B-form represents an average structure, and that there can be local variations in the precise structure. If it has more base pairs per turn it is said to be **overwound**; if it has fewer base pairs per turn it is **underwound**. The degree of local winding can be affected by the overall conformation of the DNA double helix or by the binding of proteins to specific sites on the DNA.

▸ **overwound** B-form DNA that has more than 10.5 base pairs per turn of the helix.

▸ **underwound** B-form DNA that has fewer than 10.5 base pairs per turn of the helix.

KEY CONCEPTS

- The B-form of DNA is a double helix consisting of two polynucleotide chains that run antiparallel.
- The nitrogenous bases of each chain are flat purine or pyrimidine rings that face inward and pair with one another by hydrogen bonding to form only A-T or G-C pairs.
- The diameter of the double helix is 20 Å, and there is a complete turn every 34 Å, with ten base pairs per turn.
- The double helix has a major (wide) groove and a minor (narrow) groove.

CONCEPT AND REASONING CHECKS

1. Summarize the evidence that Watson and Crick used in proposing their model of the B-form of DNA.
2. If the G-C content of a DNA duplex is 0.44, what are the proportions of the four bases?

1.5 Supercoiling Affects the Structure of DNA

The two strands of DNA are wound around each other to form the double helical structure; the double helix can also wind around itself to change the overall conformation, or *topology*, of the DNA molecule in space. This is called **supercoiling**. The effect can be imagined like a rubber band twisted around itself. Supercoiling creates tension in the DNA, and therefore can only occur if the DNA has no free ends (otherwise the free ends can rotate to relieve the tension) or in linear DNA if it is anchored to a protein scaffold, as in eukaryotic chromosomes. The simplest example of a DNA with no free ends is a circular molecule. The effect of supercoiling can be seen by comparing the nonsupercoiled circular DNA lying flat in **FIGURE 1.9** (center) with the supercoiled circular molecule that forms a twisted (and therefore more condensed) shape (**FIGURE 1.9**, bottom).

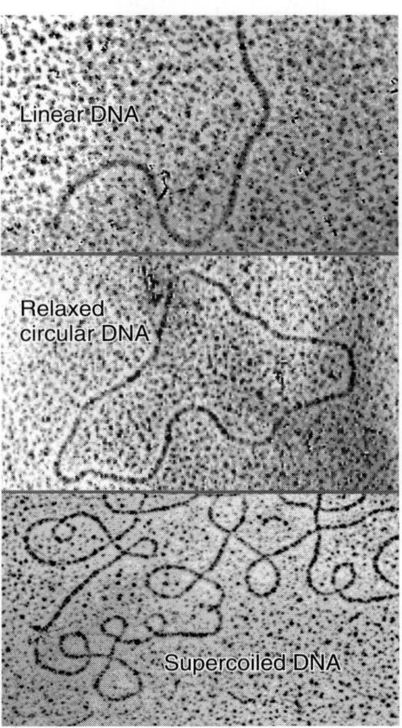

FIGURE 1.9 Linear DNA is extended (top); a circular DNA remains extended if it is relaxed (nonsupercoiled) (center); but a supercoiled DNA has a twisted and condensed form (bottom). Photos courtesy of Nirupam Roy Choudhury, International Centre for Genetic Engineering and Biotechnology (ICGEB).

▸ **supercoiling** The coiling of a closed duplex DNA in space so that it crosses over its own axis.

The consequences of supercoiling depend on whether the DNA is twisted around itself in the same direction as the two strands within the double helix (clockwise) or in the opposite direction. Twisting in the same direction produces *positive supercoiling*, which overwinds the DNA so that there are more base pairs per turn. Twisting in the opposite direction produces *negative supercoiling*, or underwinding, so there are fewer base pairs per turn. Both types of supercoiling of the double helix in space are tensions in the DNA (which is why DNA molecules with no supercoiling are called "relaxed"). Negative supercoiling can be thought of as creating tension in the DNA that is relieved by the unwinding of the double helix. The effect of severe negative supercoiling is to generate a region in which the two strands of DNA have separated (technically, zero base pairs per turn).

Topological manipulation of DNA is a central aspect of all its functional activities (recombination, replication, and transcription) as well as of the organization of its higher-order structure. All synthetic activities involving double-stranded DNA require the strands to separate. The strands do not simply lie side by side, though; they are intertwined. Their separation therefore requires the strands to rotate about each other in space. Some possibilities for the unwinding reaction are illustrated in **FIGURE 1.10**.

Unwinding a short linear DNA presents no problems, as the DNA ends are free to spin around the axis of the double helix to relieve any tension. However, DNA in a typical chromosome is not only extremely long, but is also coated with proteins that serve to anchor the DNA at numerous points. Therefore, even a linear eukaryotic chromosome does not functionally possess free ends.

Consider the effects of separating the two strands in a molecule whose ends are not free to rotate. When two intertwined strands are pulled apart from one end, the result is to *increase* their winding about each other farther along the molecule, resulting in positive supercoiling elsewhere in the molecule to balance the underwinding generated in the single-stranded region. The problem can be overcome by introducing a transient nick in one strand. An internal free end allows the nicked strand to rotate about the intact strand, after which the nick can be sealed. Each repetition of the

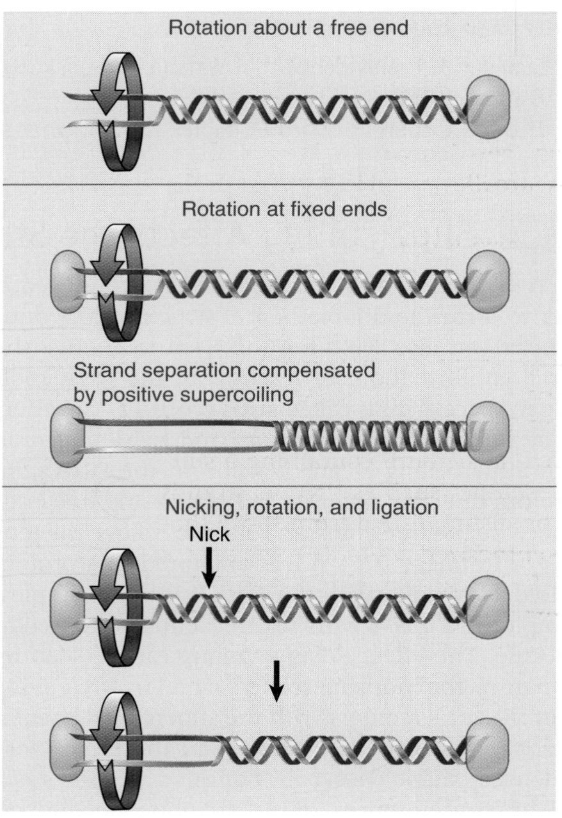

FIGURE 1.10 Separation of the strands of a DNA double helix could be achieved in several ways.

nicking and sealing reaction releases one superhelical turn. The topoisomerase enzymes that perform these reactions to control supercoiling in the cell will be discussed in *Section 19.7, Topoisomerases Relax or Introduce Supercoils in DNA*.

KEY CONCEPTS

- Supercoiling occurs only in "closed" DNA with no free ends.
- Closed DNA is either circular DNA or linear DNA in which the ends are anchored so that they are not free to rotate.

CONCEPT AND REASONING CHECK

Why does negative supercoiling facilitate unwinding of DNA, but positive supercoiling inhibits unwinding?

1.6 DNA Replication Is Semiconservative

It is crucial that DNA be reproduced accurately. The two polynucleotide strands are joined only by hydrogen bonds, so they are able to separate without the breakage of covalent bonds. The specificity of base pairing suggests that both of the separated parental strands could act as template strands for the synthesis of complementary daughter strands. **FIGURE 1.11** shows the principle that a new daughter strand is assembled from each parental strand. The sequence of the daughter strand is determined by the parental strand: an A in the parental strand causes a T to be placed in the daughter strand, a parental G directs incorporation of a daughter C, and so on.

The top part of Figure 1.11 shows an unreplicated parental duplex with the original two parental strands. The lower part shows the two daughter duplexes produced by complementary base pairing. Each of the daughter duplexes is identical in sequence to the original parent duplex, containing one parental strand and one newly synthesized strand. The structure of DNA carries the information needed for its own replication. The consequences of this mode of replication, called **semiconservative replication**, are illustrated in **FIGURE 1.12**. The unit conserved from one generation to the next is one of the two individual strands comprising the parental duplex.

Figure 1.12 illustrates a prediction of this model. If the parental DNA carries a "heavy" density label because the organism has been grown in medium containing a suitable isotope (such as ^{15}N), its strands can be distinguished from those that are synthesized when the organism is transferred to a medium containing "light" isotopes. The parental DNA is a duplex of two "heavy" strands (red). After one generation of growth in "light" medium, the duplex DNA is *hybrid* in density—it consists of one heavy parental strand (red) and one light daughter strand (blue). After a second generation, the two strands of each hybrid duplex have separated. Each strand gains a light partner, so that now one half of the duplex DNA remains hybrid and the other half is entirely light (both strands are blue).

The individual strands of these duplexes are entirely heavy or entirely light. This pattern was confirmed experimentally by Matthew Meselson and Franklin Stahl in 1958. Meselson and Stahl followed the semiconservative replication of DNA through three generations of growth of *E. coli*. When DNA was extracted from bacteria and separated in a density gradient by centrifugation, the DNA formed bands corresponding to its density—heavy for parental, hybrid for the first generation, and half hybrid bands and half light bands in the second generation, indicating that a single parental strand is retained in the daughter molecule. (See Box 15-1 for more detail on this experiment.)

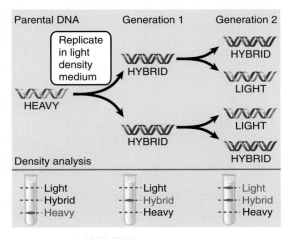

FIGURE 1.11 Base pairing provides the mechanism for replicating DNA.

▸ **semiconservative replication** DNA replication accomplished by separation of the strands of a parental duplex, each strand then acting as a template for synthesis of a complementary strand.

FIGURE 1.12 Replication of DNA is semiconservative.

KEY CONCEPTS

- The Meselson–Stahl experiment used "heavy" isotope labeling to show that the single polynucleotide strand is the unit of DNA that is conserved during replication.
- Each strand of a DNA duplex acts as a template for synthesis of a daughter strand.
- The sequences of the daughter strands are determined by complementary base pairing with the separated parental strands.

CONCEPT AND REASONING CHECK

What is the expected result of an experiment similar to that of Meselson and Stahl, but beginning with "light" DNA and culturing cells in a "heavy" medium?

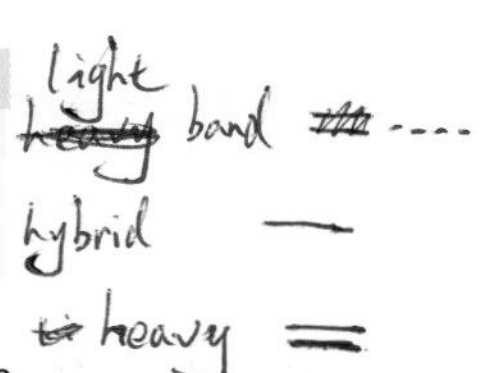

1.7 Polymerases Act on Separated DNA Strands at the Replication Fork

- **denaturation** A molecule's conversion from the physiological conformation to some other (inactive) conformation. In DNA, this involves the separation of the two strands due to breaking of hydrogen bonds between bases.
- **renaturation** The reassociation of denatured complementary single strands of a DNA double helix.
- **replication fork** The point at which strands of parental duplex DNA are separated so that replication can proceed. A complex of proteins including DNA polymerase is found there.
- **DNA polymerase** An enzyme that synthesizes a daughter strand(s) of DNA (under direction from a DNA template). Any particular enzyme may be involved in repair or replication (or both).
- **DNase** An enzyme that degrades DNA.
- **RNase** An enzyme that degrades RNA.
- **exonuclease** An enzyme that cleaves nucleotides one at a time from the end of a polynucleotide chain; it may be specific for either the 5′ or 3′ end of DNA or RNA.
- **endonuclease** An enzyme that cleaves bonds within a nucleic acid chain; it may be specific for RNA or for single-stranded or double-stranded DNA.

Replication requires the two strands of the parental duplex to separate, or **denature**. However, the disruption of the duplex is only transient and is reversed, or **renatured**, as the daughter duplex is formed. Only a small stretch of the duplex DNA is denatured at any moment during replication.

The helical structure of a molecule of DNA during replication is illustrated in FIGURE 1.13. The unreplicated region consists of the parental duplex, opening into the replicated region where the two daughter duplexes have formed. The duplex is disrupted at the junction between the two regions, which is called the **replication fork**. Replication involves movement of the replication fork along the parental DNA, so that there is continuous denaturation of the parental strands and formation of daughter duplexes.

The synthesis of DNA is aided by specific enzymes, **DNA polymerases**, that recognize the template strand and catalyze the addition of nucleotide subunits to the polynucleotide chain that is being synthesized. They are accompanied in DNA replication by ancillary enzymes such as helicases that unwind the DNA duplex, a primase that synthesizes an RNA primer required by DNA polymerase, and ligase that connects discontinuous DNA strands. Degradation of nucleic acids also requires specific enzymes: deoxyribonucleases (**DNases**) degrade DNA, and ribonucleases (**RNases**) degrade RNA. The nucleases fall into the general classes of **exonucleases** and **endonucleases**:

- Endonucleases break individual phosphodiester linkages within RNA or DNA molecules, generating discrete fragments. Some DNases cleave both strands of a duplex DNA at the target site, whereas others cleave only one of the two strands. Endonucleases are involved in cutting reactions, as shown in FIGURE 1.14.
- Exonucleases remove nucleotide residues one at a time from the end of the molecule, generating mononucleotides. They always function on a single nucleic acid strand, and each exonuclease proceeds in a specific direction, that is, starting either at a 5′ or at a 3′ end and proceeding toward the other end. They are involved in trimming reactions, as shown in FIGURE 1.15.

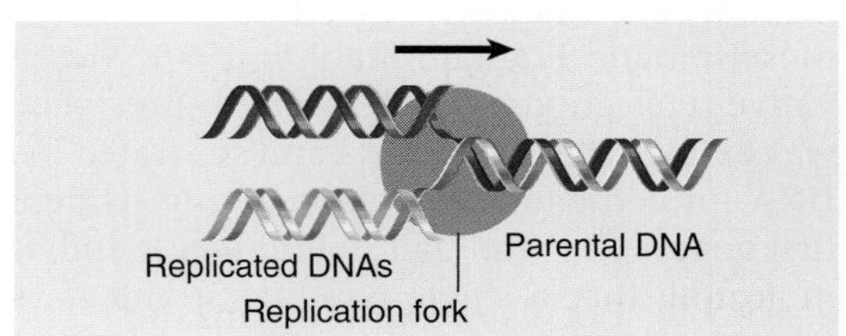

FIGURE 1.13 The replication fork is the region of DNA in which there is a transition from the unwound parental duplex to the newly replicated daughter duplexes.

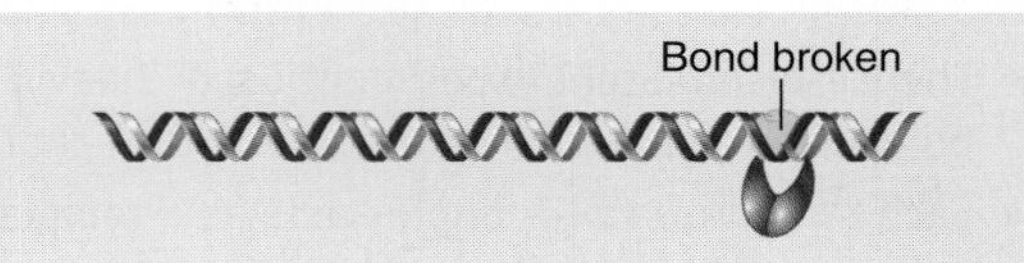

FIGURE 1.14 An endonuclease cleaves a bond within a nucleic acid. This example shows an enzyme that attacks one strand of a DNA duplex.

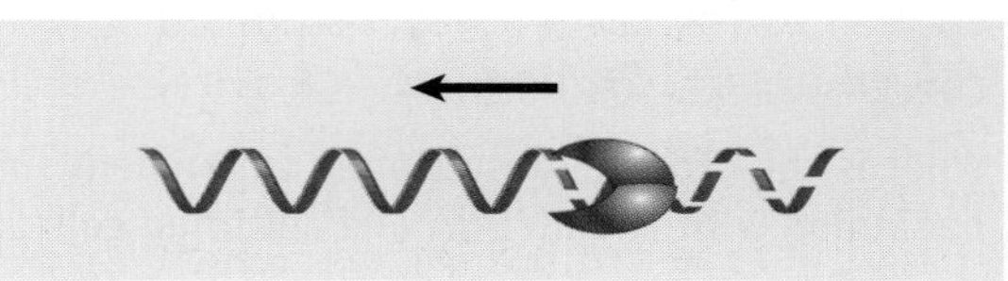

FIGURE 1.15 An exonuclease removes bases one at a time by cleaving the last bond in a polynucleotide chain.

KEY CONCEPTS

- Replication of DNA is undertaken by a complex of enzymes that separate the parental strands and synthesize the daughter strands.
- The replication fork is the point at which the parental strands are separated.
- The enzymes that synthesize DNA are called DNA polymerases.
- Nucleases are enzymes that degrade nucleic acids; they include DNases and RNases and can be categorized as endonucleases or exonucleases.

CONCEPT AND REASONING CHECK

What are the functions of DNA polymerases, DNases, and RNases in living cells?

1.8 Genetic Information Can Be Provided by DNA or RNA

The **central dogma** is the dominant paradigm of molecular biology. Structural genes exist as sequences of nucleic acid, but function by being expressed in the form of polypeptides. Replication makes possible the inheritance of genetic information, while transcription and translation are responsible for its expression to another form.

▸ **central dogma** Information cannot be transferred from protein to protein or protein to nucleic acid, but can be transferred between nucleic acids and from nucleic acid to protein.

▸ **RNA polymerase** An enzyme that synthesizes RNA using a DNA template (formally described as DNA-dependent RNA polymerases).

FIGURE 1.16 illustrates the roles of replication, transcription, and translation in the context of the central dogma:

- Transcription of DNA by a DNA-dependent **RNA polymerase** generates RNA molecules. Messenger RNAs (mRNAs) are translated to polypeptides. Other types of RNA, such as rRNAs and tRNAs, are functional themselves and are not translated.
- A genetic system may involve either DNA or RNA as the genetic material. Cells use only DNA. Some viruses use RNA, and replication of viral RNA by an RNA-dependent RNA polymerase occurs in the infected cell.
- The expression of cellular genetic information is usually unidirectional. Transcription of DNA generates RNA molecules; the exception is the reverse transcription of retroviral RNA to DNA that occurs when retroviruses infect cells (see below). Generally polypeptides cannot be retrieved for use as genetic information; translation of RNA into polypeptide is always irreversible.

These mechanisms are equally effective for the cellular genetic information of prokaryotes or eukaryotes and for the information carried by viruses. The genomes of all living organisms consist of duplex DNA. Viruses have genomes that consist of DNA or RNA, and there are examples of each type that are double-stranded (dsDNA or dsRNA) or single-stranded (ssDNA or ssRNA). Details of the mechanism used to replicate the nucleic acid vary among viruses, but the principle of replication via synthesis of complementary strands remains the same, as illustrated in **FIGURE 1.17**.

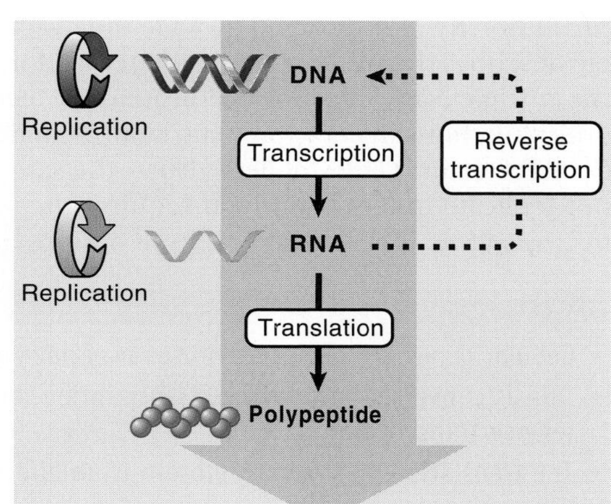

FIGURE 1.16 The central dogma states that information in nucleic acid can be perpetuated or transferred, but the transfer of information into a polypeptide (protein) is irreversible.

The restriction of a unidirectional transfer of information from DNA to RNA in cells is not absolute. It is broken by the retroviruses, which have genomes consisting of a single-stranded RNA molecule. During the retroviral cycle of infection, the RNA is converted into a single-stranded DNA by the process of **reverse transcription**, which

▸ **reverse transcription** Synthesis of DNA on a template of RNA. It is accomplished by the enzyme reverse transcriptase.

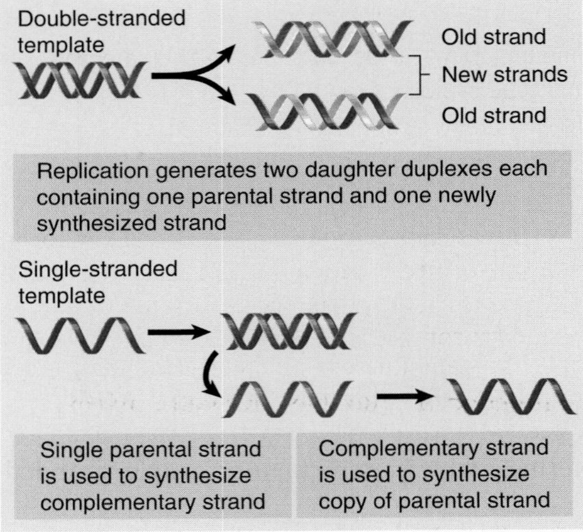
Double-stranded template
Old strand
New strands
Old strand
Replication generates two daughter duplexes each containing one parental strand and one newly synthesized strand
Single-stranded template
Single parental strand is used to synthesize complementary strand
Complementary strand is used to synthesize copy of parental strand

FIGURE 1.17 Double-stranded and single-stranded nucleic acids both replicate by synthesis of complementary strands governed by the rules of base pairing.

is accomplished by the enzyme *reverse transcriptase,* an RNA-dependent DNA polymerase. The resulting ssDNA is in turn converted into a double-stranded DNA. This duplex DNA becomes part of the genome of the host cell and is inherited like any other gene. So reverse transcription allows a sequence of RNA to be retrieved and used as DNA in a cell.

The existence of RNA replication and reverse transcription establishes the general principle that information in the form of either type of nucleic acid sequence can be converted into the other type. In the usual course of events, however, the cell relies on the processes of DNA replication, transcription, and translation. But on rare occasions (possibly mediated by an RNA virus), information from a cellular RNA is converted into DNA and inserted into the genome. Although retroviral reverse transcription is not necessary for the regular operations of the cell, it becomes a mechanism of potential importance when we consider the evolution of the genome.

Genome	Gene Number	Base Pairs
Organisms		
Plants	<50,000	$<10^{11}$
Mammals	30,000	$\sim 3 \times 10^{9}$
Worms	14,000	$\sim 10^{8}$
Flies	12,000	1.6×10^{8}
Fungi	6,000	1.3×10^{7}
Bacteria	2–4,000	$<10^{7}$
Mycoplasma	500	$<10^{6}$
dsDNA Viruses		
Vaccinia	<300	187,000
Papova (SV40)	~6	5,226
Phage T4	~200	165,000
ssDNA Viruses		
Parvovirus	5	5,000
Phage fX174	11	5,387
dsRNA Viruses		
Reovirus	22	23,000
ssRNA Viruses		
Coronavirus	7	20,000
Influenza	12	13,500
TMV	4	6,400
Phage MS2	4	3,569
STNV	1	1,300
Viroids		
PSTV RNA	0	359

FIGURE 1.18 The amount of nucleic acid in the genome varies over an enormous range.

The same principles for the perpetuation of genetic information apply to the massive genomes of plants or amphibians as well as the tiny genomes of mycoplasma and the even smaller genomes of DNA or RNA viruses. **FIGURE 1.18** presents some examples that illustrate the range of genome types and sizes. The reasons for such variation in genome size and gene number will be explored in Chapters 4 and 5.

Among the various living organisms, with genomes varying in size over a 100,000-fold range, a common principle prevails: the DNA codes for all the proteins that the cell(s) of the organism must synthesize, and the proteins in turn (directly or indirectly) provide the functions needed for survival. A similar principle describes the function of the genetic information of viruses, whether DNA or RNA: the nucleic acid codes for the protein(s) needed to package the genome and for any other functions in addition to those provided by the host cell that are needed to reproduce the virus. (The smallest virus—the satellite tobacco necrosis virus [STNV]—cannot replicate independently. It requires the presence of a "helper" virus—the tobacco necrosis virus [TNV], which is itself a normally infectious virus.)

KEY CONCEPTS

- Cellular genes are DNA, but viruses may have genomes of RNA.
- DNA is converted into RNA by transcription, and RNA may be converted into DNA by reverse transcription.
- The translation of RNA into protein is unidirectional.

CONCEPT AND REASONING CHECK

What types of enzymes would be necessary to replicate ssDNA, ssRNA, dsDNA, and dsRNA genomes to produce exact copies of the same type of nucleic acid?

polymerase.

1.9 Nucleic Acids Hybridize by Base Pairing

A crucial property of the double helix is the capacity to separate the two strands without disrupting the covalent bonds that form the polynucleotides and at the (very rapid) rates needed to sustain genetic functions. The specificity of the processes of denaturation and renaturation is determined by complementary base pairing.

The concept of base pairing is central to all processes involving nucleic acids. Disruption of the base pairs is crucial to the function of a double-stranded nucleic acid, whereas the ability to form base pairs is essential for the activity of a single-stranded nucleic acid. **FIGURE 1.19** shows that base pairing enables complementary single-stranded nucleic acids to form a duplex.

- An intramolecular duplex region can form by base pairing between two complementary sequences that are part of a single-stranded nucleic acid.
- A single-stranded nucleic acid may base pair with an independent, complementary single-stranded nucleic acid to form an intermolecular duplex.

Formation of duplex regions from single-stranded nucleic acids is most prevalent in RNA, but is also important for single-stranded viral DNA genomes. Base pairing between independent complementary single strands is not restricted to DNA–DNA or RNA–RNA, but can also occur between DNA and RNA.

The lack of covalent bonds between complementary strands makes it possible to manipulate DNA *in vitro*. The hydrogen bonds that stabilize the double helix are disrupted by heating or low salt concentration. The two strands of a double helix separate entirely when all the hydrogen bonds between them are broken.

Denaturation of DNA occurs over a narrow temperature range and results in striking changes in many of its physical properties. The midpoint of the temperature range over which the strands of DNA separate is called the **melting temperature** (T_m), and depends on the G-C content of the duplex. Because each G-C base pair has three hydrogen bonds, it is more stable than an A-T base pair, which has only two hydrogen bonds. The more G-C base pairs in a DNA, the greater the energy that is needed to separate the two strands. In solution under physiological conditions, a DNA that is 40% G-C (a value typical of mammalian genomes) denatures with a T_m of about 87°C, so duplex DNA is stable at the temperature of the cell.

▸ **melting temperature** The midpoint of the temperature range over which the strands of DNA separate.

The denaturation of DNA is reversible under appropriate conditions. Renaturation depends on specific base pairing between the complementary strands. **FIGURE 1.20** shows that the reaction takes place in two stages. First, single strands of DNA in the solution encounter one another by chance; if their sequences are complementary, the two strands base pair to generate a short double-stranded region. This region of base pairing then extends along the molecule, much like a zipper, to form a lengthy duplex.

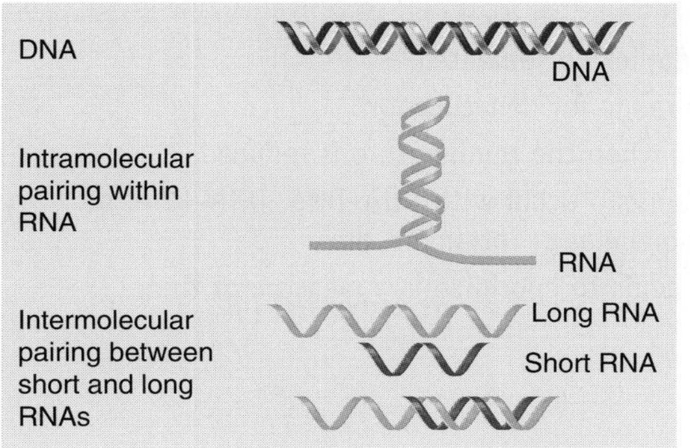

FIGURE 1.19 Base pairing occurs in duplex DNA and also in intra- and intermolecular interactions in single-stranded RNA (or DNA).

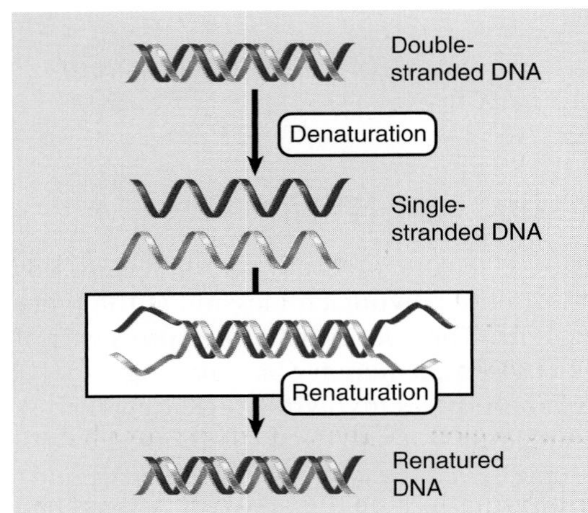

FIGURE 1.20 Denatured single strands of DNA can renature to give the duplex form.

- **annealing** The renaturation of a duplex structure from single strands that were obtained by denaturing duplex DNA.
- **hybridization** The pairing of complementary RNA and DNA strands to give an RNA-DNA hybrid.

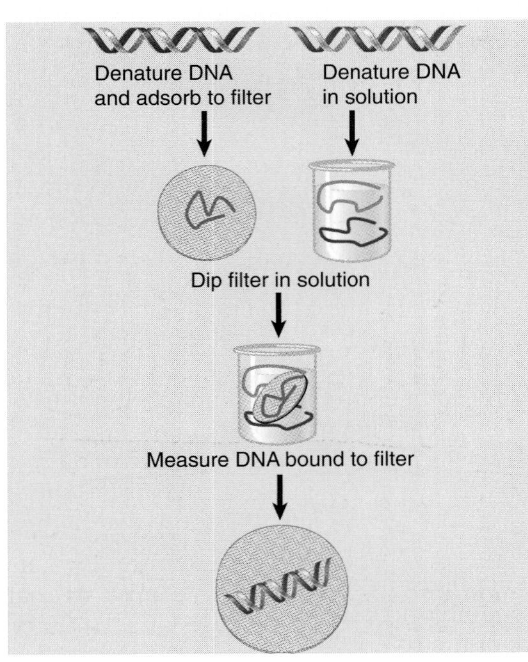

FIGURE 1.21 Filter hybridization establishes whether a solution of denatured DNA (or RNA) contains sequences complementary to the strands immobilized on the filter.

Complete renaturation restores the properties of the original double helix. The property of renaturation applies to any two complementary nucleic acid sequences. This is sometimes called **annealing**, but the reaction is more generally called **hybridization** whenever nucleic acids from different sources are involved, as in the case when DNA hybridizes to RNA. The ability of two nucleic acids to hybridize constitutes a precise test for their complementarity because only complementary sequences can form a duplex.

The principle of the hybridization reaction is to combine two single-stranded nucleic acids in solution and then to measure the amount of double-stranded material that forms. **FIGURE 1.21** illustrates a procedure in which a DNA preparation is denatured and the single strands are attached to a filter. Then a second denatured DNA (or RNA) preparation is added. The filter is treated so that the second preparation can attach to it only if it is able to base pair with the DNA that was originally attached. Usually the second preparation is labeled so that the hybridization reaction can be measured as the amount of label retained by the filter. Alternatively, hybridization in solution can be measured as the change in UV-absorbance of a nucleic acid solution at 260 nm as detected via spectrophotometry. As DNA denatures to single strands with increasing temperature, UV-absorbance of the DNA solution increases; UV-absorbance consequently decreases as ssDNA hybridizes to complementary DNA or RNA with decreasing temperature.

Two sequences need not be perfectly complementary to hybridize. If they are similar but not identical, an imperfect duplex is formed in which base pairing is interrupted at positions where the two single strands are not complementary.

KEY CONCEPTS

- Heating causes the two strands of a DNA duplex to separate.
- The T_m is the midpoint of the temperature range for denaturation.
- Complementary single strands can renature when the temperature is reduced.
- Denaturation and renaturation/hybridization can occur with DNA–DNA, DNA–RNA, or RNA–RNA combinations and can be intermolecular or intramolecular.
- The ability of two single-stranded nucleic acids to hybridize is a measure of their complementarity.

CONCEPT AND REASONING CHECK

Describe how measures of hybridization between DNA from different species can be used to estimate the evolutionary relationships between those species.

1.10 Mutations Change the Sequence of DNA

Mutations provide decisive evidence that DNA is the genetic material. When a change in the sequence of DNA causes an alteration in a protein, we may conclude that the DNA codes for that protein. Furthermore, a corresponding change in the phenotype of the organism may allow us to identify the function of that protein. The existence of many mutations in a gene may allow many variant forms of a protein to be compared, and a detailed analysis can be used to identify regions of the protein responsible for individual enzymatic or other functions.

All organisms suffer a certain number of mutations as the result of normal cellular operations or random interactions with the environment. These are called **spontaneous mutations**, and the rate at which they occur (the "background level") is characteristic for any particular organism. Mutations are rare events, and of course those that have deleterious effects are selected against during evolution. It is therefore difficult to observe large numbers of spontaneous mutants from natural populations.

The occurrence of mutations can be increased by treatment with certain compounds. These are called **mutagens**, and the changes they cause are called **induced mutations**. Most mutagens either modify a particular base of DNA or become incorporated into the nucleic acid. The potency of a mutagen is judged by how much it increases the rate of mutation above background. By using mutagens, it becomes possible to induce many changes in any gene.

Mutation rates can be measured at several levels of resolution: mutation across the whole genome (as the rate per genome per generation), mutation in a gene (as the rate per locus per generation), or mutation at a specific nucleotide site (as the rate per base pair per generation). These rates correspondingly decrease as smaller units are observed.

Spontaneous mutations that inactivate gene function occur in bacteriophages and bacteria at a relatively constant rate of $3–4 \times 10^{-3}$ per genome per generation. Given the large variation in genome sizes between bacteriophages and bacteria, this corresponds to great differences in the mutation rate per base pair. This suggests that the overall rate of mutation has been subject to selective forces that have balanced the deleterious effects of most mutations against the advantageous effects of some mutations. This conclusion is strengthened by the observation that an archaean that lives under harsh conditions of high temperature and acidity (which are expected to damage DNA) does not show an elevated mutation rate, but in fact has an overall mutation rate just below the average range. **FIGURE 1.22** shows that in bacteria, the mutation rate corresponds to ~10^{-6} events per locus per generation or to an average rate of change per base pair of 10^{-9} to 10^{-10} per generation. The rate at individual base pairs varies very widely, over a 10,000-fold range. We have no accu-

▸ **spontaneous mutations** Mutations occurring in the absence of any added reagent to increase the mutation rate, as the result of errors in replication (or other events involved in the reproduction of DNA) or by random changes to the chemical structure of bases.

▸ **mutagens** Substances that increase the rate of mutation by inducing changes in DNA sequence, directly or indirectly.

▸ **induced mutations** Mutations that result from the action of a mutagen. The mutagen may act directly on the bases in DNA or it may act indirectly to trigger a pathway that leads to a change in DNA sequence.

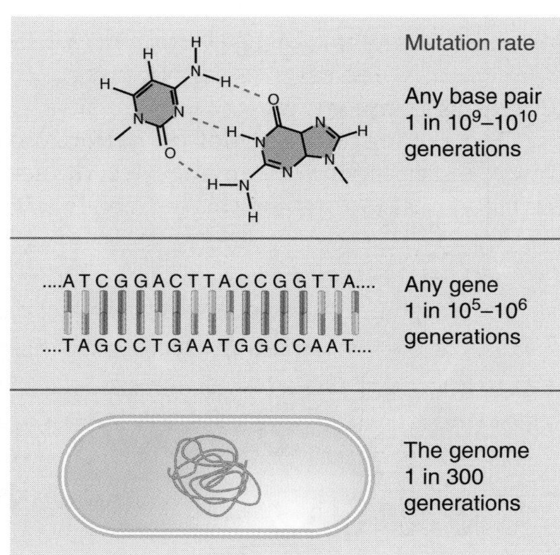

FIGURE 1.22 A base pair is mutated at a rate of 10^{-9}–10^{-10} per generation, a gene of 1000 bp is mutated at ~10^{-6} per generation, and a bacterial genome is mutated at 3×10^{-3} per generation.

rate measurement of the rate of mutation in eukaryotes, although usually it is thought to be somewhat similar to that of bacteria on a per-locus per-generation basis.

KEY CONCEPTS

- All mutations are changes in the sequence of DNA.
- Mutations may occur spontaneously or may be induced by mutagens.

CONCEPT AND REASONING CHECK

What are the advantages of maintaining a nonzero mutation rate? evolution force

1.11 Mutations May Affect Single Base Pairs or Longer Sequences

▸ **point mutation** A change in the sequence of DNA involving a single base pair.

▸ **transition** A mutation in which one pyrimidine is replaced by the other, or in which one purine is replaced by the other.

▸ **transversion** A mutation in which a purine is replaced by a pyrimidine or vice versa.

Any base pair of DNA can be mutated. A **point mutation** changes only a single base pair and can be caused by either of two types of event:

- Chemical modification of DNA directly changes one base into a different base.
- An error during the replication of DNA causes the wrong base to be inserted into a polynucleotide.

Point mutations can be divided into two types, depending on the nature of the base substitution:

- The most common class is the **transition**, resulting from the substitution of one pyrimidine by the other, or of one purine by the other. This replaces a G-C pair with an A-T pair or vice versa.
- The less common class is the **transversion**, in which a purine is replaced by a pyrimidine or vice versa, so that an A-T pair becomes a T-A or C-G pair.

As shown in **FIGURE 1.23**, the mutagen nitrous acid performs an oxidative deamination that converts cytosine into uracil, resulting in a transition. In the replication cycle following the transition, the U pairs with an A, instead of the G with which the original C would have paired. So the C-G pair is replaced by a T-A pair when the A pairs with the T in the next replication cycle. (Nitrous acid can also deaminate adenine, causing the reverse transition from A-T to G-C.)

Transitions are also caused by base mispairing, when noncomplementary bases pair instead of the usual Watson–Crick pairs. Base mispairing usually occurs as an aberration resulting from the incorporation into DNA of an abnormal base that has flexible pairing properties. **FIGURE 1.24** shows the example of the mutagen bromouracil (BrdU), an analog of thymine that contains a bromine atom in place of thymine's methyl group and can be incorporated into DNA in place of thymine. However, BrdU has flexible pairing properties, because the presence of the bromine atom allows a *tautomeric shift* from a keto (=O) form to an enol (–OH) form. The enol form of BrdU can pair with guanine, which after replication leads to substitution of the original A-T pair by a G-C pair. Tautomeric shifts can also occur when

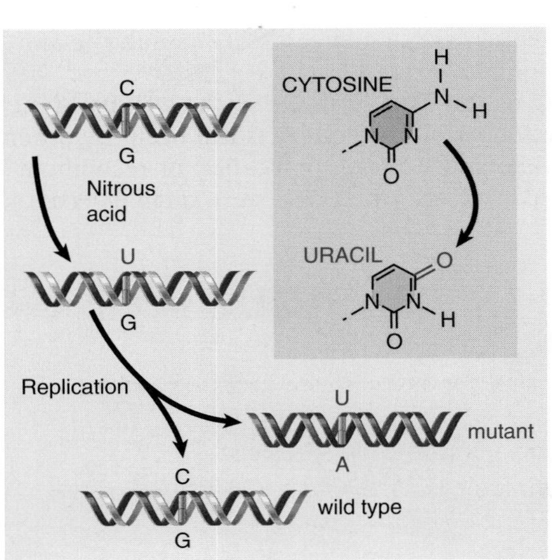

FIGURE 1.23 Mutations can be induced by chemical modification of a base.

a proton shifts position within a normal base and results in an anomalous, but more stable, base pairing. For example, while the common keto form of guanine pairs stably with cytosine, the rare enol form of guanine pairs stably with thymine.

Transversions are rarer than transitions, since they require a temporary purine-purine or pyrimidine-pyrimidine pairing that would alter the diameter of the DNA duplex. However, one cause of transversions is a proton shift followed by a 180° rotation of the base around the glycosidic bond. For example, a rotated syn-adenine (produced by a proton shift in adenine) will pair stably with a normal adenine.

Point mutations were thought for a long time to be the principal means of change in individual genes. However, we now know that insertions of short sequences are quite frequent. Some mutagens, such as intercalating agents, can cause the insertion or deletion of a single base pair. An intercalating agent will insert itself between two adjacent base pairs in a DNA duplex, so that the duplex is distorted and a DNA polymerase can either skip or add a base during DNA replication. If this occurs in the coding sequence of a gene, a frameshift mutation will result (see *Section 2.7, The Genetic Code Is Triplet*). Often, the insertions are the result of transposable elements, which are sequences of DNA with the ability to move from one site to another (see *Chapter 21, Transposons, Retroviruses, and Retroposons*). An insertion within a coding region usually abolishes the activity of the gene. However, both insertions and deletions of short sequences can occur by other mechanisms—for example, those involving errors during replication or recombination. In addition, a class of mutagens called acridines introduce very small insertions and deletions.

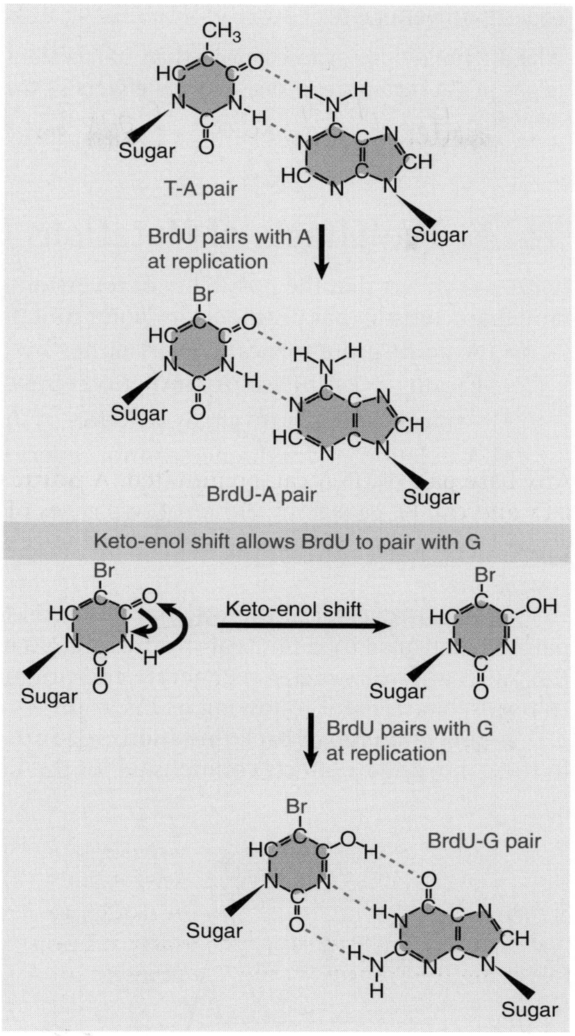

FIGURE 1.24 Mutations can be induced by the incorporation of base analogs into DNA.

KEY CONCEPTS

- A point mutation changes a single base pair.
- Point mutations can be caused by the chemical conversion of one base into another or by errors that occur during replication.
- A transition replaces a G-C base pair with an A-T base pair or vice versa.
- A transversion replaces a purine with a pyrimidine, such as changing A-T to T-A.
- Insertions can result from the movement of transposable elements.

CONCEPT AND REASONING CHECK

Why are transitions more common than tranversions? Consider how the DNA repair mechanisms might recognize errors, and the effects of these mutations on DNA structure in your answer. alter the diameter of DNA base & pair.

1.12 The Effects of Mutations Can Be Reversed

▸ **revertants** Reversions of a mutant cell or organism to the wild-type phenotype.

▸ **forward mutation** A mutation that inactivates a functional gene.

▸ **back mutation** A mutation that reverses the effect of a mutation that had inactivated a gene; thus it restores the original sequence or function of the gene product.

▸ **true reversion** A mutation that restores the original sequence of the DNA.

▸ **second-site reversion** A second mutation suppressing the effect of a first mutation.

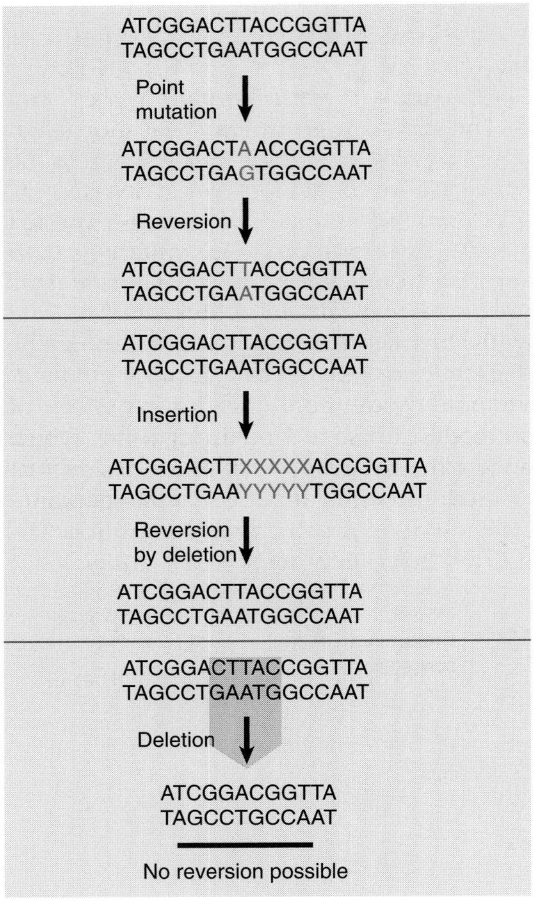

FIGURE 1.25 shows that the possibility of reversion mutations, or **revertants**, is an important characteristic that distinguishes point mutations and insertions from deletions.

- A point mutation can revert either by restoring the original sequence or by gaining a compensatory mutation elsewhere in the gene.
- An insertion can revert by deletion of the inserted sequence.
- A deletion of a sequence cannot revert in the absence of some mechanism to restore the lost sequence.

Mutations that inactivate a gene are called **forward mutations**. Their effects are reversed by **back mutations**, which are of two types: true reversions and second-site reversions.

An exact reversal of the original mutation is called **true reversion**. So if an A-T pair was replaced by a G-C pair in the original mutation, another mutation to restore the A-T pair will exactly regenerate the original sequence. The exact removal of a transposable element following its insertion is another example of a true reversion.

The second type of back mutation, **second-site reversion**, may occur elsewhere in the gene, and its effects compensate for the first mutation. For example, one amino acid change in a protein may abolish its function, but a second alteration may compensate for the first and restore protein activity.

FIGURE 1.25 Point mutations and insertions can revert, but deletions cannot revert.

A forward mutation results from any change that alters the function of a gene product, whereas a back mutation must restore the original function to the altered gene product. Therefore, the possibilities for back mutations are much more restricted than those for forward mutations. The rate of back mutations is correspondingly lower than that of forward mutations, typically by a factor of ~10.

▸ **suppression mutation** A second event eliminates the effects of a mutation without reversing the original change in DNA.

Mutations in other genes can also occur to circumvent the effects of mutation in the original gene. This effect is called **suppression**. A locus in which a mutation suppresses the effect of a mutation in another locus is called a *suppressor*. For example, a point mutation may cause an amino acid substitution in a polypeptide, while a second mutation in a tRNA gene may cause it to recognize the mutated codon, and as a result insert the original amino acid during translation. (Note that this suppresses the original mutation but causes errors during translation of other mRNAs.)

KEY CONCEPTS

- Forward mutations alter the function of a gene, and back mutations (or revertants) reverse their effects.
- Insertions can revert by deletion of the inserted material, but deletions cannot revert.
- Suppression occurs when a mutation in a second gene bypasses the effect of mutation in the first gene.

CONCEPT AND REASONING CHECK

Transposable elements were originally identified because the rate of reversion of mutations caused by them is much higher than the reversion rate for point mutations. Explain why the reversion rate is so high.

1.13 Mutations Are Concentrated at Hotspots

So far we have dealt with mutations in terms of individual changes in the sequence of DNA that influence the activity of the DNA in which they occur. When we consider mutations in terms of the alteration of function of the gene, most genes within a species show more or less similar rates of mutation relative to their size. This suggests that the gene can be regarded as a target for mutation, and that damage to any part of it can alter its function. As a result, susceptibility to mutation is roughly proportional to the size of the gene. But are all base pairs in a gene equally susceptible, or are some more likely to be mutated than others?

What happens when we isolate a large number of independent mutations in the same gene? Each is the result of an individual mutational event. Most mutations will occur at different sites, but some will occur at the same position. Two independently isolated mutations at the same site may constitute exactly the same change in DNA (in which case the same mutation has happened more than once), or they may constitute different changes (three different point mutations are possible at each base pair).

The histogram of **FIGURE 1.26** shows the frequency with which mutations are found at each base pair in the *lacI* gene of *E. coli*. The statistical probability that more than one mutation occurs at a particular site is given by random-hit kinetics (as seen in the Poisson distribution). Some sites will gain one, two, or three mutations, whereas others will not gain any. Some sites gain far more than the number of mutations expected from a random distribution; they may have 10× or even 100× more mutations than predicted by random hits. These sites are called **hotspots**. Spontaneous mutations may occur at hotspots, and different mutagens may have different hotspots.

▸ **hotspots** A site in the genome at which the frequency of mutation (or recombination) is very much increased, usually by at least an order of magnitude relative to neighboring sites.

A major cause of spontaneous mutation is the presence of an unusual base in the DNA. In addition to the four standard bases of DNA, *modified bases* are sometimes found. The name reflects their origin; they are produced by chemical modification of one of the four standard bases. The most common modified base is 5-methylcytosine, which is generated when a methylase enzyme adds a methyl group to cytosine residues at specific sites in the DNA. Sites containing 5-methylcytosine are hotspots for spontaneous point mutation in *E. coli*. In each case, the mutation is a G-C to A-T transition. The hotspots are not found in mutant strains of *E. coli* that cannot methylate cytosine.

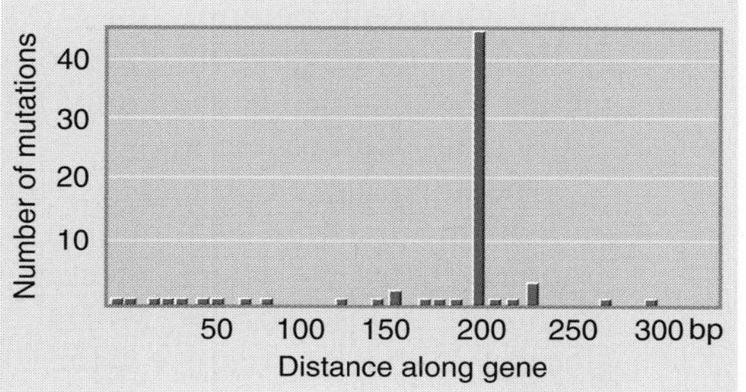

FIGURE 1.26 Spontaneous mutations occur throughout the *lacI* gene of *E. coli* but are concentrated at a hotspot.

The reason for the existence of these hotspots is that cytosine bases suffer a higher frequency of spontaneous deamination. In this reaction, the amino group is replaced by a keto group. Recall that deamination of cytosine generates uracil (see Figure 1.23). **FIGURE 1.27** compares this reaction with the deamination of 5-methylcytosine where deamination generates thymine. The effect is to generate the mismatched base pairs G-U and G-T, respectively.

FIGURE 1.28 shows that the consequences of deamination are different for 5-methylcytosine and cytosine. Deaminating the (rare) 5-methylcytosine causes a mutation, whereas deaminating cytosine does not have this effect. This happens because the DNA repair systems are much more effective in recognizing G-U than G-T, and always correct the U (which normally should be present only in RNA, not in DNA).

E. coli contain an enzyme, uracil-DNA-glycosidase, that removes uracil residues from DNA (see *Section 20.4, Base Excision Repair Systems Require Glycosylases*). This action leaves an unpaired G residue, and a repair system then inserts a complementary C base. The net result of these reactions is to restore the original sequence of the DNA. Thus, this system protects DNA against the consequences of spontaneous deamination of cytosine. (This system is not, however, efficient enough to prevent the effects of the increased deamination caused by nitrous acid; see Figure 1.23.)

Note that the deamination of 5-methylcytosine creates thymine and results in a mismatched base pair, G-T. If the mismatch is not corrected before the next replication cycle, a mutation results; the bases in the mispaired G-T separate, and then they pair with the correct complements to produce the original G-C and the mutant A-T.

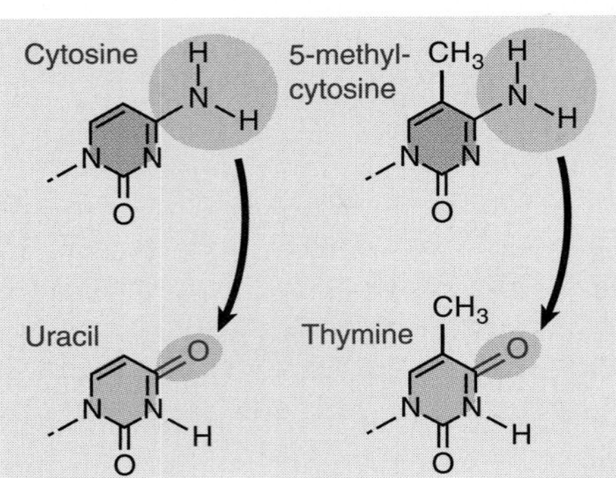

FIGURE 1.27 Deamination of cytosine produces uracil, whereas deamination of 5-methylcytosine produces thymine.

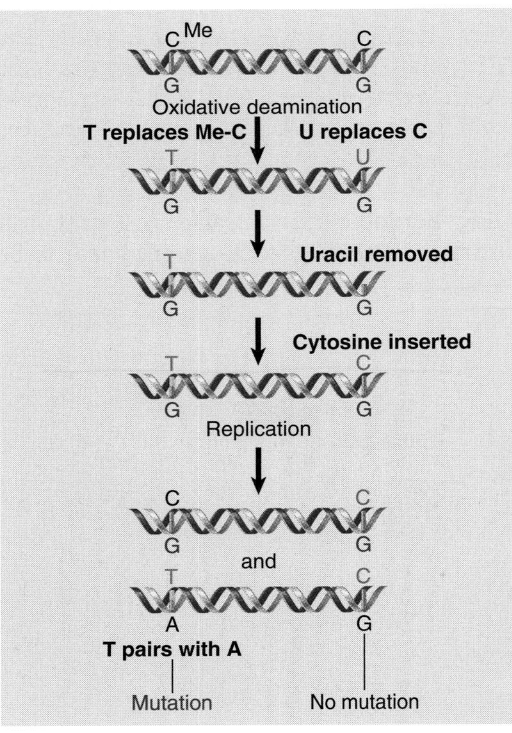

FIGURE 1.28 The deamination of 5-methylcytosine produces thymine (by C-G to T-A transitions), while the deamination of cytosine produces uracil (which usually is removed and then replaced by cytosine).

Deamination of 5-methylcytosine is the most common cause of mismatched G-T pairs in DNA. Repair systems that act on G-T mismatches have a bias toward replacing the T with a C (rather than the alternative of replacing the G with an A), which helps to reduce the rate of mutation (see *Section 20.6, Controlling the Direction of Mismatch Repair*). However, these systems are not as effective as those that remove U from G-U mismatches. As a result, deamination of 5-methylcytosine leads to mutation much more often than does deamination of cytosine.

5-methylcytosine also creates hotspots in eukaryotic DNA. It is common at CpG dinucleotides that are concentrated in regions called *CpG islands* (see *Section 27.6, CpG Islands Are Subject to Methylation*). Although 5-methylcytosine accounts for ~1% of the bases in human DNA, sites containing the modified base account for ~30% of all point mutations.

The importance of repair systems in reducing the rate of mutation is emphasized by the effects of eliminating the mouse enzyme MBD4, a glycosylase that can remove T (or U) from mismatches with G. The result is to increase the mutation rate at CpG sites by a factor of 3. (The reason the effect is not greater is that MBD4 is only one of several systems that act on G-T mismatches; probably the elimination of all the systems would increase the mutation rate much more.)

Another type of hotspot, though not often found in coding regions, is the "slippery sequence"—a homopolymer run, or region where a very short sequence (one or a few nucleotides) is repeated many times in tandem. During replication, a DNA polymerase may skip one repeat or replicate the same repeat twice, leading to a decrease or increase in repeat number.

KEY CONCEPTS

- The frequency of mutation at any particular base pair is statistically equivalent, except for hotspots, where the frequency is increased by at least an order of magnitude.
- A common cause of hotspots is the modified base 5-methylcytosine, which is spontaneously deaminated to thymine.
- A hotspot can result from the high frequency of change in copy number of a short tandemly repeated sequence.

CONCEPT AND REASONING CHECK

Suggest several possible reasons that a particular base pair can be a mutational hotspot.

1.14 Some Hereditary Agents Are Extremely Small

Viroids (or subviral pathogens) are infectious agents that cause diseases in higher plants. They are very small circular molecules of RNA. Unlike viruses—for which the infectious agent consists of a virion, a genome encapsulated in a protein coat—the viroid RNA is itself the infectious agent. The viroid consists solely of the RNA molecule, which is extensively folded by imperfect base pairing, forming a characteristic rod as shown in **FIGURE 1.29**. Mutations that interfere with the structure of this rod reduce the infectivity of the viroid.

▸ **viroid** A small infectious nucleic acid that does not have a protein coat.

A viroid RNA consists of a single molecule that is replicated autonomously and accurately in infected cells. Viroids are categorized into several groups. A given viroid is assigned to a group according to sequence similarity with other members of the group. For example, four viroids in the PSTV (potato spindle tuber viroid) group have 70%–83% sequence similarity with PSTV. Different isolates of a particular viroid strain vary from one another in sequence, which may result in phenotypic differences among infected cells. For example, the "mild" and "severe" strains of PSTV differ by three nucleotide substitutions.

Viroids are similar to viruses in having heritable nucleic acid genomes, but differ from viruses in both structure and function. Viroid RNA does not appear to be

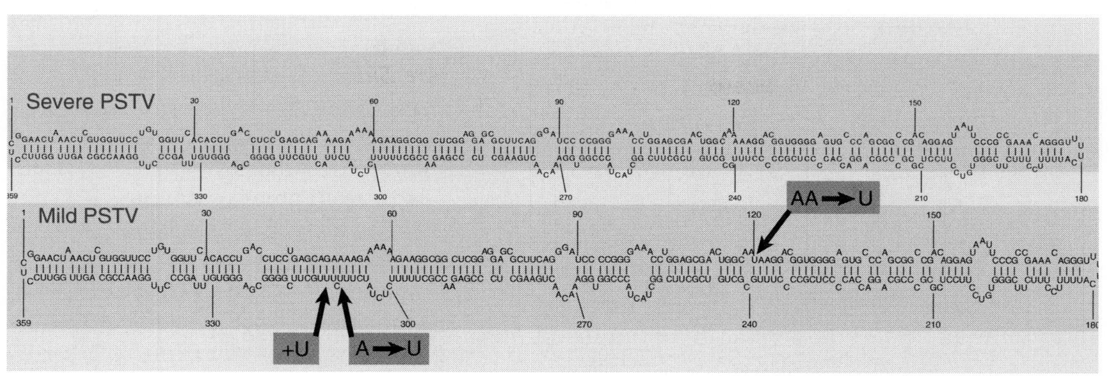

FIGURE 1.29 PSTV RNA is a circular molecule that forms an extensive double-stranded structure, interrupted by many interior loops. The severe and mild forms differ at three sites.

translated into polypeptide, so it cannot itself code for the functions needed for its survival. This situation poses two as yet unanswered questions: How does viroid RNA replicate, and how does it affect the phenotype of the infected plant cell?

Replication must be carried out by enzymes of the host cell. The heritability of the viroid sequence indicates that viroid RNA is the template for replication.

Viroids are presumably pathogenic because they interfere with normal cellular processes. They might do this in a relatively random way, for example, by taking control of an essential enzyme for their own replication or by interfering with the production of necessary cellular RNAs. Alternatively, they might behave as abnormal regulatory molecules, with particular effects upon the expression of individual genes.

▸ **prion** A proteinaceous infectious agent that behaves as an inheritable trait, although it contains no nucleic acid. One example is PrPSc, the agent of scrapie in sheep and bovine spongiform encephalopathy.

An even more unusual agent is the cause of scrapie, a degenerative neurological disease of sheep and goats. The disease is similar to the human diseases of kuru and Creutzfeldt–Jakob syndrome, which affect brain function. The infectious agent of scrapie does not contain nucleic acid. This extraordinary agent is called a **prion** (proteinaceous infectious agent). It is a 28-kD hydrophobic glycoprotein, PrP. PrP is encoded by a cellular gene (conserved among the mammals) that is expressed in normal brain cells. The protein exists in two forms: the version found in normal brain cells is called PrP^{c} and is entirely degraded by proteases during normal protein turnover. The version found in infected brains is called PrP^{sc} and is extremely resistant to degradation by proteases. PrP^{c} is converted to PrP^{sc} by a conformational change that confers protease resistance, and has yet to be fully defined.

As the infectious agent of scrapie, PrP^{sc} must in some way modify the synthesis of its normal cellular counterpart so that it becomes infectious instead of harmless (see *Section 27.9, Prions Cause Diseases in Mammals*). Mice that lack a PrP gene cannot develop scrapie, which demonstrates that PrP is essential for development of the disease.

KEY CONCEPT

- Some very small hereditary agents do not code for polypeptide but consist of RNA or protein with heritable properties.

CONCEPT AND REASONING CHECK

How would you distinguish whether a newly discovered infectious agent is an organism, a virus, a viroid, or a prion?

1.15 Summary

Two classic experiments provided strong evidence that DNA is generally the genetic material of bacteria, eukaryotic cells, and many viruses. DNA isolated from one strain of *Pneumococcus* bacteria can confer properties of that strain upon another strain. In addition, DNA is the only component that is inherited by progeny phages from parental phages. DNA can be used to transfect new properties into eukaryotic cells.

DNA is a double helix consisting of antiparallel strands in which the nucleotide units are linked by 5′ to 3′ phosphodiester bonds. The backbone is on the exterior; purine and pyrimidine bases are stacked in the interior in pairs in which A is complementary to T and G is complementary to C. In semiconservative replication, the two strands separate and daughter strands are assembled by complementary base pairing. Complementary base pairing is also used to transcribe an RNA from one strand of a DNA duplex.

A mutation consists of a change in the sequence of A-T and G-C base pairs in DNA. A mutation in a coding sequence may change the sequence of amino acids in the corresponding polypeptide. A point mutation changes only the amino acid represented by the codon in which the mutation occurs. Point mutations may be reverted by back mutation of the original mutation. Insertions may revert by loss of the inserted material, but deletions cannot revert. Mutations may also be suppressed indirectly when a mutation in a different gene counters the original defect.

The natural incidence of mutations is increased by mutagens. Mutations may be concentrated at hotspots. A type of hotspot responsible for some point mutations is caused by deamination of the modified base 5-methylcytosine. Forward mutations occur at a rate of ~10^{-6} per locus per generation; back mutations are rarer.

Although all genetic information in cells is carried by DNA, viruses have genomes of double-stranded or single-stranded DNA or RNA. Viroids are subviral pathogens that consist solely of small molecules of RNA with no protective packaging. The RNA does not code for protein and its mode of perpetuation and of pathogenesis is unknown. Scrapie results from a proteinaceous infectious agent, or prion.

CHAPTER QUESTIONS

1. Which strain of *S. pneumoniae* was found to be avirulent (non-lethal) when mice were infected with it?

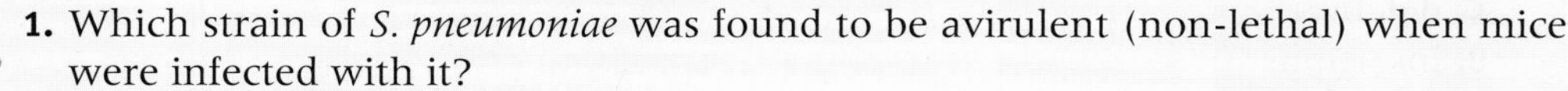

   A. the smooth strain
   B. the rough strain
   C. both strains
   D. none were virulent

2. What isotope can be used to specifically label protein?

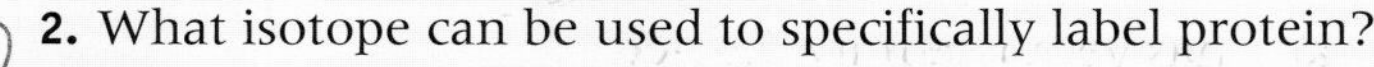

   A. ^{14}C
   B. ^{3}H
   C. ^{32}P
   D. ^{35}S

3. The difference between RNA and DNA is the presence of a:

   A. 2′ PO_4 group on the ribose sugar in RNA.
   B. 3′ PO_4 group on the ribose sugar in RNA.
   C. 2′ OH group on the ribose sugar in RNA.
   D. 3′ OH group on the ribose sugar in RNA.

4. Which pair of scientists determined that DNA replication is semiconservative?

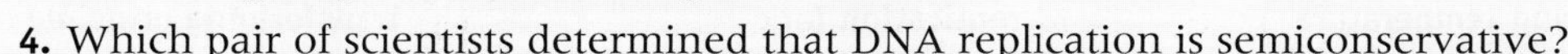

   A. Meselson and Stahl
   B. Watson and Crick
   C. Okazaki and Okazaki
   D. Griffith and Avery

5. In the density labeling experiment to study DNA replication, parental DNA in the cell was labeled with a high density isotope and, after one or more generations of growth, subjected to density gradient centrifugation. Which of the following populations was not detected after the second generation?
   A. the light density population
   B. the hybrid density population
   C. the heavy density population
   D. both the light and heavy density populations

6. Which class of molecules is inserted into the host genome as a double-stranded DNA segment?
   A. retroviruses
   B. double-stranded RNA viruses
   C. double-stranded DNA viruses
   D. viroids

7. The mutation rate in bacteria is about:
   A. 10^{-6} per locus per generation.
   B. 10^{-7} per locus per generation.

C. 10^{-8} per locus per generation.

D. 10^{-9} per locus per generation.

8. About 30% of human point mutations are associated with which of the following modified bases?

A. 5-methylguanine

B. 5-methyladenine

C. 5-methylthymine

D. 5-methylcytosine

9. What does the presence of the modified base in the previous question often lead to?

A. transitions

B. transversions

C. deletions

D. insertions

10. The reversal of an original base pair that was changed from A-T to G-C, then back to A-T is an example of a:

A. true reversion.

B. second-site reversion.

C. forward mutation.

D. suppression.

KEY TERMS

annealing
antiparallel
back mutation
central dogma
chromosome
complementary
denaturation
DNA polymerase
DNase
endonuclease
exonuclease
forward mutation
genome
hotspots
hybridization
induced mutations
major groove
melting temperature
minor groove
mutagens
nucleoside
nucleotide
overwound
point mutation
polynucleotide
prion
purine
pyrimidine
renaturation
replication fork
reverse transcription
revertants
RNA polymerase
RNase
second-site reversion
semiconservative replication
spontaneous mutations
structural gene
supercoiling
suppression mutation
transformation
transforming principle
transition
transversion
true reversion
underwound
viroid

FURTHER READING

Holmes, F. (2001). *Meselson, Stahl, and the Replication of DNA: A History of the Most Beautiful Experiment in Biology.* Yale University Press, New Haven, CT.

An account of Meselson and Stahl's scientific partnership with a unique look into the daily business of "doing science."

Maki, H. (2002). Origins of spontaneous mutations: specificity and directionality of base-substitution, frameshift, and sequence-substitution mutageneses. *Annu. Rev. Genet.* 36, 279–303.

A review of causes and effects of spontaneous mutations and mutational hotspots.

Prusiner, S. B. (1998). Prions. *Proc. Natl. Acad. Sci. USA* 95, 13363–13383.

An edited version of Prusiner's Nobel lecture, including an overview of the biology of prions and an account of their discovery.

Watson, J. D. (1981). *The Double Helix: A Personal Account of the Discovery of the Structure of DNA* (Norton Critical Editions). W. W. Norton, New York, NY.

Watson's 1968 best-selling personal account of the discovery of the double helix along with reprints of original publications and additional commentary.

2

A visual representation of gene expression analysis that allows researchers to see patterns in a network of expressed genes. Each dot represents a gene, and links between the dots occur where there are similar patterns of expression between the genes. © Anton Enright, The Sanger Institute/Wellcome Images.

Genes Code for Proteins

CHAPTER OUTLINE

2.1 Introduction

The gene is the functional unit of heredity. Each gene is a DNA sequence that functions by coding for a discrete product, which may be a polypeptide or an RNA. The basic pattern of inheritance of a gene was proposed by Mendel more than a century ago. Summarized in his two major principles of *segregation* and *independent assortment*, the gene was recognized as a "particulate factor" that passes unchanged from parent to progeny. A gene may exist in alternative forms, called **alleles**.

▸ **allele** One of several alternative forms of a gene occupying a given locus on a chromosome.

In diploid organisms, which have two sets of chromosomes, one copy of each chromosome is inherited from each parent. This is the same pattern of inheritance that is displayed by genes. One of the two copies of each gene is the paternal allele (inherited from the father), the other is the maternal allele (inherited from the mother). The shared pattern of inheritance of genes and chromosomes led to the discovery that chromosomes in fact carry the genes.

Each chromosome consists of a linear array of genes, and each gene resides at a particular location on the chromosome. The location is more formally called a genetic **locus**. The alleles of a gene are the different forms that are found at its locus. Although generally there are up to two alleles per locus in an individual, a population may have many alleles.

▸ **locus** The position on a chromosome at which the gene for a particular trait resides; it may be occupied by any one of the alleles for the gene.

▸ **linkage** The tendency of genes to be inherited together as a result of their location on the same chromosome; measured by percent recombination between loci.

The key to understanding the organization of genes into chromosomes was the discovery of genetic **linkage**—the tendency for genes on the same chromosome to remain together in the progeny instead of assorting independently as predicted by Mendel's principle. Once the unit of *recombination* (reassortment) was introduced as a measure of linkage, the construction of genetic maps became possible.

The resolution of the linkage map of a higher eukaryote is restricted by the small number of progeny that can be obtained from each mating. Recombination occurs so infrequently between nearby points that it is rarely observed between different variable sites in the same gene. As a result, classical linkage maps of eukaryotes can place the genes in order, but cannot resolve the locations of variable sites within a gene. By using a microbial system in which a very large number of progeny can be obtained from each genetic cross, researchers could demonstrate that recombination occurs within genes, and that it follows the same rules as those for recombination between genes.

Variable nucleotide sites among alleles of a gene can be mapped into a linear order, showing that the gene itself has the same linear construction as the array of genes on a chromosome. In other words, the genetic map is linear within, as well as between, loci, as an unbroken sequence of nucleotides. This conclusion leads naturally to the modern view summarized in **FIGURE 2.1** that the genetic material of a chromosome consists of an uninterrupted length of DNA representing many genes.

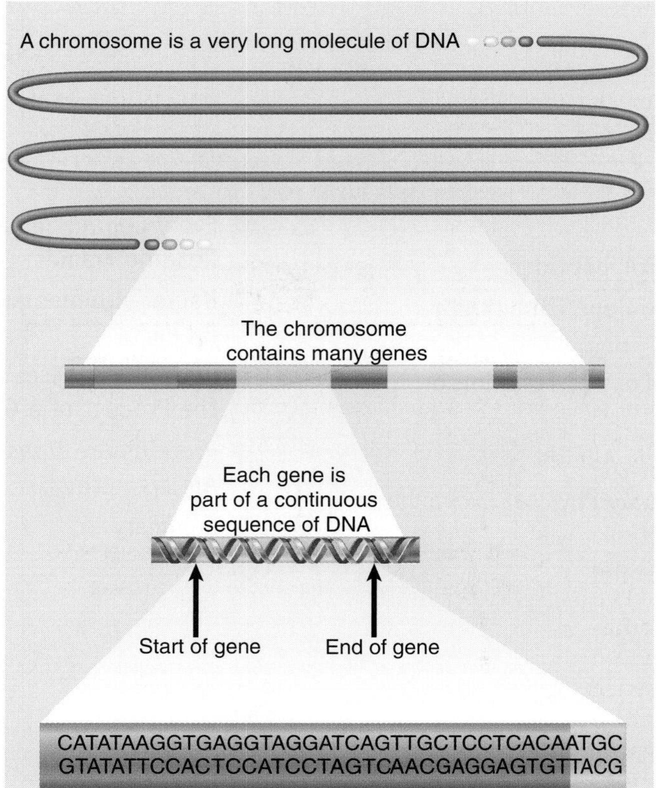
A chromosome is a very long molecule of DNA
The chromosome contains many genes
Each gene is part of a continuous sequence of DNA
Start of gene
End of gene
CATATAAGGTGAGGTAGGATCAGTTGCTCCTCACAATGC
GTATATTCCACTCCATCCTAGTCAACGAGGAGTGTTACG

FIGURE 2.1 Each chromosome has a single long molecule of DNA within which are the sequences of individual genes.

CONCEPT AND REASONING CHECK

Early work measuring recombination frequencies between genes led to the establishment of "linkage groups": sets of genes that do not assort independently. How does this support the concept that genes are carried on chromosomes?

2.2 Most Genes Code for Polypeptides

Sir Archibald Garrod was the first person to make a connection between a metabolic defect and a heritable factor, suggesting that gene action has chemical effects in cells. In studying a patient with alkaptonuria in 1902, he determined both that the disorder resulted from an enzymatic deficiency and that its pattern of inheritance in the patient's family was consistent with an autosomal recessive allele. Garrod later identified several other "inborn errors of metabolism" with autosomal recessive inheritance.

The work of George Beadle and Edward Tatum in the 1940s was the first systematic attempt to associate genes with enzymes. Beadle and Tatum showed that each step in a metabolic pathway is catalyzed by a single enzyme and can be blocked by mutation in a different gene. This led to the **one gene : one enzyme hypothesis**. A mutation in a gene alters the activity of the protein enzyme for which it codes.

A modification in the hypothesis is needed to apply to proteins that consist of more than one polypeptide subunit. If the subunits are all the same, the protein is a **homomultimer**, and is encoded by a single gene. If the subunits are different, the protein is a **heteromultimer**, and each different subunit is encoded by a different gene. Stated as a more general rule applicable to any heteromultimeric protein, the one gene : one enzyme hypothesis becomes more precisely expressed as **one gene : one polypeptide chain**. (Even this modification is not completely descriptive of the relationship between genes and proteins, as many genes code for alternate versions of a polypeptide; see *Section 28.11, Alternative Splicing Involves Differential Use of Splice Junctions.*)

Identifying the biochemical effects of a particular mutation can be a protracted task. The mutation responsible for creating Mendel's wrinkled-pea phenotype was identified only in 1990 as an alteration that inactivates the gene for a starch debranching enzyme!

It is important to remember that a gene does not directly generate a polypeptide. As shown previously in Figure 1.2, a gene codes for an RNA, which may in turn code for a polypeptide. Most genes are structural genes that code for polypeptides, but some genes code for RNAs that are not translated to polypeptides. These RNAs may be structural components of the protein synthesis machinery or may have roles in regulating gene expression. The basic principle is that *the gene is a sequence of DNA that specifies the sequence of an independent product.* The process of gene expression may terminate in a product that is either RNA or polypeptide.

A mutation in a coding region is generally a random event with regard to the structure and function of the gene (but see *Section 1.13, Mutations Are Concentrated at Hotspots*); mutations can have little or no effect (neutral mutations), or can damage or even abolish gene function. Most mutations that affect gene function are recessive; they result in an absence of function because the mutant allele does not produce its usual polypeptide. **FIGURE 2.2** illustrates the relationship between recessive and wild-type

▸ **one gene : one enzyme hypothesis** Beadle and Tatum's hypothesis that a gene is responsible for the production of a single enzyme.

▸ **homomultimer** A molecular complex (such as a protein) in which the subunits are identical.

▸ **heteromultimer** A molecular complex (such as a protein) composed of different subunits.

▸ **one gene : one polypeptide hypothesis** A modified version of the not generally correct one gene : one enzyme hypothesis; the hypothesis that a gene is responsible for the production of a single polypeptide.

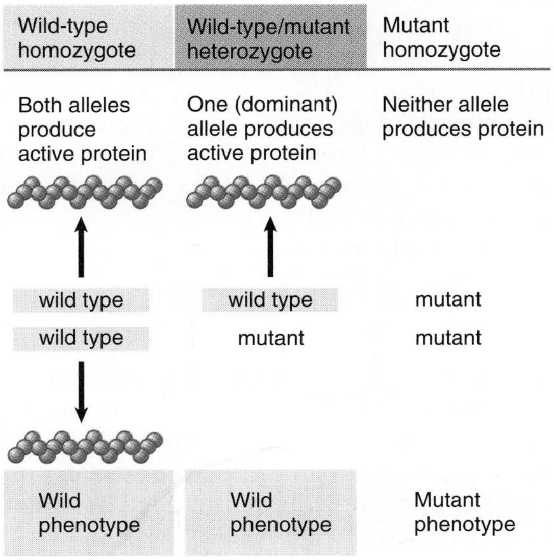

Wild-type homozygote	Wild-type/mutant heterozygote	Mutant homozygote
Both alleles produce active protein	One (dominant) allele produces active protein	Neither allele produces protein
wild type	wild type	mutant
wild type	mutant	mutant
Wild phenotype	Wild phenotype	Mutant phenotype

FIGURE 2.2 Genes code for proteins; dominance is explained by the properties of mutant proteins. A recessive allele does not contribute to the phenotype because it produces no protein (or protein that is nonfunctional).

HISTORICAL PERSPECTIVES

One Gene : One Enzyme—George W. Beadle and Edward L. Tatum, 1941

Genetic Control of Biochemical Reactions in *Neurospora*

How do genes control metabolic processes? The suggestion that genes control enzymes was made very early in the history of genetics, most notably by the British physician Archibald Garrod in his 1908 book *Inborn Errors of Metabolism*. But the precise relationship between genes and enzymes was still uncertain. Perhaps each enzyme is controlled by more than one gene, or perhaps each gene contributes to the control of several enzymes. The classic experiments of George Beadle and Edward Tatum showed that the relationship is often remarkably simple: one gene codes for one enzyme. The pioneering experiments united genetics and biochemistry, and for the "one gene, one enzyme" concept, Beadle and Tatum were awarded a Nobel Prize in 1958 (Joshua Lederberg shared the prize for his contributions to microbial genetics). Because we now know that some enzymes contain polypeptide chains encoded by two (or occasionally more) different genes, a more accurate statement of the principle is "one gene, one polypeptide." Beadle and Tatum's experiments also demonstrate the importance of choosing the right organism. *Neurospora* had been introduced as a genetic model organism only a few years earlier, and Beadle and Tatum realized that they could take advantage of the ability of this organism to grow on a simple medium composed of known substances.

> From the standpoint of physiological genetics the development and functioning of an organism consist essentially of an integrated system of chemical reactions controlled in some manner by genes.... In investigating the roles of genes, the physiological geneticist usually attempts to determine the physiological and biochemical bases of already known hereditary traits. . . . There are, however, a number of limitations inherent in this approach. Perhaps the most serious of these is that the investigator must in general confine himself to the study of nonlethal heritable characters. Such characters are likely to involve more or less nonessential so-called "terminal" reactions. . . . A second difficulty is that the standard approach to the problem implies the use of characters with visible manifestations. Many such characters involve morphological variations, and these are likely to be based on systems of biochemical reactions so complex as to make analysis exceedingly difficult. . . . Considerations such as those just outlined have led us to investigate the general problem of the genetic control of development and metabolic reactions by reversing the ordinary procedure and, instead of attempting to work out the chemical bases of known genetic characters, to set out to determine if and how genes control known biochemical reactions. The ascomycete *Neurospora* offers many advantages for such an approach and is well suited to genetic studies. Accordingly, our program has been built around this organism. The procedure is based on the assumption that x-ray treatment will induce mutations in genes concerned with the control of known specific chemical reactions. If the organism must be able to carry out a certain chemical reaction to survive on a given medium, a mutant unable to do this will obviously be lethal on this medium. Such a mutant can be maintained and studied, however, if it will grow on a medium to which has been added the essential product of the genetically blocked reaction. . . . Among approximately 2000 strains [derived from single cells after x-ray treatment], three mutants have been found that grow essentially normally on the complete medium and scarcely at all on the minimal medium. One of these strains proved to be unable to synthesize vitamin B6 (pyridoxine). A second strain turned out to be unable to synthesize vitamin B1 (thiamine). A third strain has been found to be unable to synthesize para-aminobenzoic acid. . . . These preliminary results appear to us to indicate that the approach may offer considerable promise as a method of learning more about how genes regulate development and function. For example, it should be possible, by finding a number of mutants unable to carry out a particular step in a given synthesis, to determine whether only one gene is ordinarily concerned with the immediate regulation of a given specific chemical reaction.

Source: Beadle, G. W. and Tatum, E. L., *Proc. Natl. Acad. Sci. USA* 27 (1941): 499–506.

alleles. When a heterozygote contains one wild-type allele and one mutant allele, the wild-type allele is able to direct production of the enzyme, and is therefore dominant. (This assumes that an adequate amount of protein is made by the single wild-type allele. When this is not true, the smaller amount made by one allele as compared to two alleles results in the intermediate phenotype of a partially dominant allele in a heterozygote.)

KEY CONCEPTS

- The one gene : one polypeptide hypothesis summarizes the basis of modern genetics: that a typical gene is a stretch of DNA coding for a single polypeptide chain.
- Genes can also code for RNA products that are not translated into proteins.
- Mutations can damage gene function.

CONCEPT AND REASONING CHECK

Propose a situation in which a mutant allele is dominant in a heterozygote with a wild-type allele.

2.3 Mutations in the Same Gene Cannot Complement

How do we determine whether two mutations that cause a similar phenotype lie in the same gene? If they map to positions that are very close together (*i.e.*, they recombine very rarely), they may be alleles. However, in the absence of information about their relative positions, they could also represent mutations in two different genes whose proteins are involved in the same function. The **complementation test** is used to determine whether two recessive mutations lie in the same gene or in different genes. The test consists of making a heterozygote for the two mutations (by mating parents homozygous for each mutation) and observing its phenotype.

▸ **complementation test** A test that determines whether two mutations are alleles of the same gene. It is accomplished by crossing two different recessive mutations that have the same phenotype and determining whether the wild-type phenotype can be produced. If so, the mutations are said to complement each other and are probably not mutations in the same gene.

If the mutations lie in the same gene, the parental genotypes can be represented as:

$$\frac{m_1}{m_1} \quad \text{and} \quad \frac{m_2}{m_2}$$

The first parent provides an m_1 mutant allele and the second parent provides an m_2 allele, so that the heterozygote progeny has the genotype:

$$\frac{m_1}{m_2}$$

No wild-type allele is present, so the heterozygote has a mutant phenotype.

If the mutations lie in different linked genes, the parental genotypes can be represented as:

$$\frac{m_1\,+}{m_1\,+} \quad \text{and} \quad \frac{+\,m_2}{+\,m_2}$$

Each chromosome has one wild-type allele at one locus (represented by the plus sign, +) and one mutant allele at the other locus. Then the heterozygote progeny has the genotype:

$$\frac{m_1\,+}{+\,m_2}$$

in which the two parents between them have provided a wild-type allele from each gene. The heterozygote has a wild-type phenotype, and the two genes are said to complement.

The complementation test is shown in more detail in **FIGURE 2.3**. The basic test consists of the comparison shown in the top part of the figure. If two mutations lie in the same gene, we see a difference in the phenotypes of the *trans* configuration (both mutations are not in the same allele) and the *cis* configuration (both mutations are in the same allele). The *trans* configuration is mutant, because each allele has a different mutation. However, the *cis* configuration is wild-type, because one allele has two mutations and the other allele has no mutations. The lower part of the figure shows that if the two mutations lie in different genes, we always see a wild phenotype. There is always one wild-type and one mutant allele of each gene in both the *cis* and *trans* configurations. "Failure to complement" means that the two mutations occurred in the same gene. The term **cistron** is used to describe the gene as defined by the complementation test, and

▸ **cistron** The genetic unit defined by the complementation test; it is equivalent to a gene.

FIGURE 2.3 The cistron is defined by the complementation test. Genes are represented by spirals; red stars identify sites of mutation.

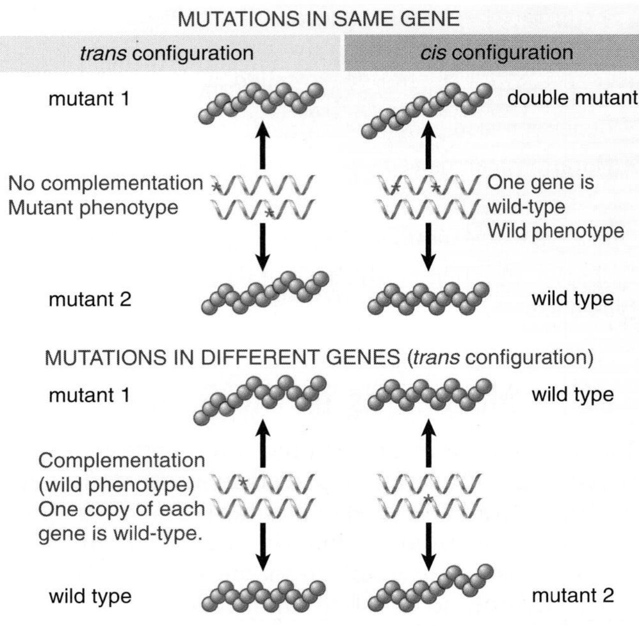

(like "gene") describes a stretch of DNA that functions as a unit to produce an RNA or polypeptide product. The properties of the gene with regard to complementation are explained by the fact that its product is a single molecule that behaves as a functional unit.

KEY CONCEPTS

- A mutation in a gene affects only the polypeptide encoded by the mutant copy of the gene and does not affect the polypeptide encoded by any other allele.
- Failure of two mutations to complement (produce wild-type phenotype when they are present in *trans* configuration in a heterozygote) means that they are part of the same gene.

CONCEPT AND REASONING CHECK

Would the complementation test work for mutations that are dominant? Why or why not?

2.4 Mutations May Cause Loss-of-Function or Gain-of-Function

The various possible effects of mutation in a gene are summarized in **FIGURE 2.4**.

In principle, when a gene has been identified, insight into its function can be gained by generating a mutant organism that entirely lacks the gene. A mutation that completely eliminates gene function—usually because the gene has been deleted—is called a **null mutation**. If a gene is essential to survival, a null mutation is lethal.

▸ **null mutation** A mutation that completely eliminates the function of a gene.

To determine how a gene affects the phenotype, it is necessary to characterize the effect of a null mutation. When a mutation fails to affect the phenotype, it is possible that it is a "leaky" mutation—enough active product is made to fulfill its function, even though the activity is quantitatively reduced or qualitatively different from the wild-type. However, if a null mutant fails to affect a phenotype, we may safely conclude that the gene function is not necessary.

▸ **loss-of-function mutation** A mutation that eliminates or reduces the activity of a gene. It is often, but not always, recessive.

Null mutations, or other mutations that impede gene function (but do not necessarily abolish it entirely), are called **loss-of-function mutations**. A loss-of-function mutation is generally recessive (as in the example of Figure 2.2). Sometimes a mutation has the opposite effect and causes a protein to acquire a new function; such a change is called a **gain-of-function mutation**. A gain-of-function mutation is generally dominant.

Not all mutations in genes lead to a detectable change in the phenotype. Mutations without apparent effect are called **silent mutations**. They fall into two categories: (1) base changes in a gene that do not cause any change in the amino acid in the resulting polypeptide; and (2) base changes in a gene that change the amino acid, but the replacement in the polypeptide does not affect its activity. Silent mutations in the second category are called **neutral substitutions**.

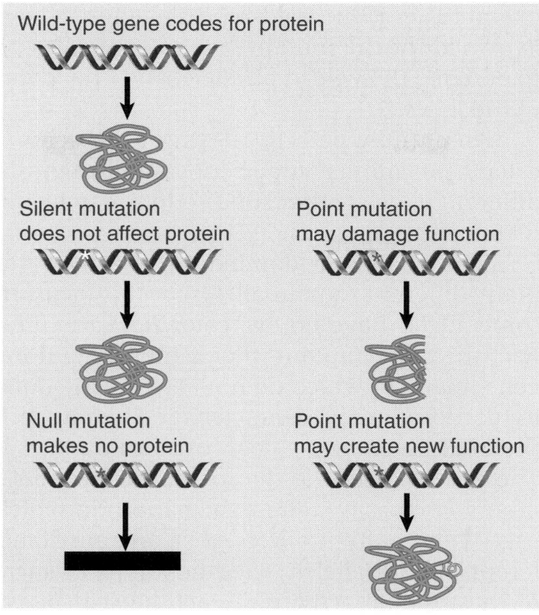

Wild-type gene codes for protein
Silent mutation
does not affect protein
Point mutation
may damage function
Null mutation
makes no protein
Point mutation
may create new function

FIGURE 2.4 Mutations that do not affect protein sequence or function are silent. Mutations that abolish all protein activity are null. Point mutations that cause loss-of-function are recessive; those that cause gain-of-function are dominant.

- **gain-of-function mutation** A mutation that causes an increase in the normal gene activity. It sometimes represents acquisition of certain abnormal properties. It is often, but not always, dominant.
- **silent mutation** A mutation that does not change the sequence of a polypeptide because it produces synonymous codons.
- **neutral substitutions** Substitutions in a protein that cause changes in amino acids that do not affect activity.

KEY CONCEPTS

- Recessive mutations are due to loss-of-function by the polypeptide product.
- Dominant mutations result from a gain-of-function.
- Testing whether a gene is essential requires a null mutation (one that completely eliminates its function).
- Silent mutations have no effect, either because the base change does not change the sequence or amount of polypeptide, or because the change in polypeptide sequence has no effect.
- "Leaky" mutations do affect the function of the gene product, but are not shown in the phenotype because sufficient activity remains.

CONCEPT AND REASONING CHECK

Explain why loss-of-function mutations are generally recessive and gain-of-function mutations are generally dominant.

2.5 A Locus May Have Many Alleles

If a recessive mutation is produced by every change in a gene that prevents the production of an active protein, there should be a large potential number of such mutations for any one gene. Many amino acid replacements may change the structure of the protein sufficiently to impede its function.

Different variants of the same gene are called *multiple alleles*, and their existence makes it possible to construct heterozygotes with two mutant alleles. The relationships between these multiple alleles can take various forms.

In the simplest case, a wild-type allele codes for a polypeptide product that is functional, whereas mutant allele(s) code for polypeptides that are nonfunctional. However, there are often cases in which a series of mutant alleles have different phenotypes. For example, wild-type function of the *white* locus of *Drosophila melanogaster* is required for development of the normal red color of the eye. The locus is named for the effect of null mutations which, in homozygotes, cause the fly to have white eyes.

In *Drosophila*, the name of the wild-type allele is indicated by a "plus" superscript after the name of the locus, which is usually named after a mutant phenotype. For example, w^+ is the wild-type allele for red eye color in *D. melanogaster*, and the locus is named for the mutant "white" eye color phenotype. Sometimes + is used by itself to describe the wild-type allele, and only the mutant alleles are indicated by the name of the locus.

An entirely defective form of the gene (or absence of phenotype) may be indicated by a "minus" superscript. To distinguish among a variety of mutant alleles with different effects, other superscripts may be introduced, such as w^i (ivory eye color) or w^a (apricot eye color).

The w^+ allele is dominant over any other allele in heterozygotes, and there are many different mutant alleles for this locus. **FIGURE 2.5** shows a small sample. Although some alleles have no eye color (*i.e.*, a white eye), many alleles produce some color. Each of these mutant alleles must therefore represent a different mutation of the gene, many of which do not eliminate its function entirely, but leave a residual activity that produces a characteristic phenotype. These alleles are named for the color of the eye in a homozygote. Most *w* mutations affect the quantity of pigment in the eye. The examples in the figure are arranged in roughly declining amount of color, but others, such as w^{sp}, affect the pattern in which pigment is deposited.

When multiple alleles exist, an animal may be a heterozygote that carries two different mutant alleles. The phenotype of such a heterozygote depends on the nature of the residual activity of each allele. The relationship between two mutant alleles is in principle no different from that between wild-type and mutant alleles: one allele may be dominant, there may be partial dominance, or there may be codominance.

Allele	Phenotype of homozygote
w^+	red eye (wild type)
w^{bl}	blood
w^{ch}	cherry
w^{bf}	buff
w^h	honey
w^a	apricot
w^e	eosin
w^i	ivory
w^z	zeste (lemon-yellow)
w^{sp}	mottled, color varies
w^1	white (no color)

FIGURE 2.5 The *w* locus has an extensive series of alleles whose phenotypes extend from wild-type (red) color to complete lack of pigment.

There is not necessarily a unique wild-type allele for any particular locus. Control of the *ABO* human blood group system provides an example. Lack of function is represented by the null, or *O*, allele. However, the functional alleles *A* and *B* are codominant with one another and dominant to the *O* allele. The basis for this relationship is illustrated in **FIGURE 2.6**.

The O (or H) antigen is generated in all individuals and consists of a particular carbohydrate group that is added to proteins. The *ABO* locus codes for a galactosyltransferase enzyme that adds a further sugar group to the O antigen. The specificity of this enzyme determines the blood group. The *A* allele produces an enzyme that uses the cofactor UDP-N-acetylgalactose, creating the A antigen. The *B* allele produces an enzyme that uses the cofactor UDP-galactose, creating the B antigen. The A and B versions of the transferase protein differ in four amino acids that presumably affect its recognition of the type of cofactor. The *O* allele has a small deletion that eliminates the activity of the transferase, so no modification of the O antigen occurs.

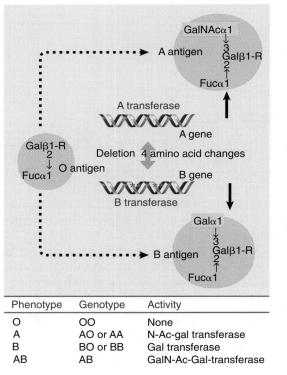

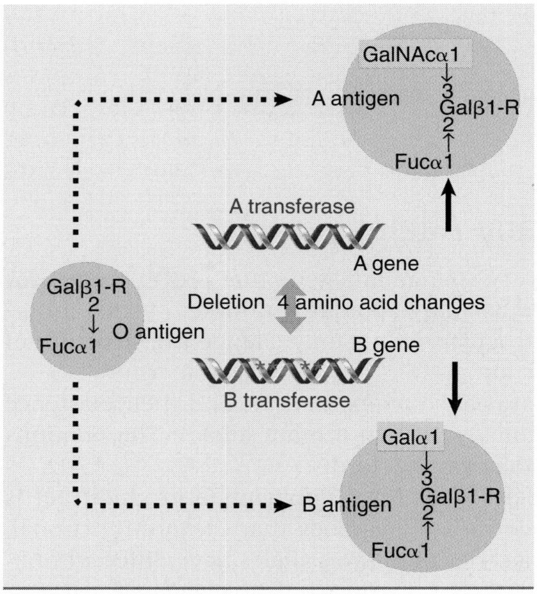

Phenotype	Genotype	Activity
O	OO	None
A	AO or AA	N-Ac-gal transferase
B	BO or BB	Gal transferase
AB	AB	GalN-Ac-Gal-transferase

FIGURE 2.6 The ABO blood group locus codes for a galactosyltransferase whose specificity determines the blood group.

This explains why *A* and *B* alleles are dominant in the *AO* and *BO* heterozygotes: the corresponding transferase activity creates the A or B antigen. The *A* and *B* alleles are codominant in *AB* heterozygotes, because both transferase activities are expressed. The *OO* homozygote is a null that has neither activity and therefore lacks both antigens.

Neither *A* nor *B* alleles can be regarded as uniquely wild-type, because they represent alternative activities rather than loss or gain of function. A situation such as this, in which there are multiple functional alleles in a population, is described as a **polymorphism** (see *Section 4.3, Individual Genomes Show Extensive Variation*).

▸ **polymorphism** The simultaneous occurrence in the population of alleles showing variations at a given position.

KEY CONCEPTS

- The existence of multiple alleles allows the possibility of heterozygotes representing any pairwise combination of alleles.
- A locus may have a polymorphic distribution of alleles with no individual allele that can be considered to be the sole wild-type.

CONCEPT AND REASONING CHECK

Explain why an individual fly with a new recessive mutation of the *white* locus will have a wild-type phenotype.

2.6 Recombination Occurs by Physical Exchange of DNA

The term **genetic recombination** describes the generation of new combinations of alleles at each generation in diploid organisms. Generally, each chromosome has a *homologous* partner with which it pairs during meiosis. The two homologous copies of each chromosome may have different alleles at some loci. By the exchange of corresponding segments between the homologs, recombinant chromosomes that are different from the parental chromosomes can be generated.

▸ **genetic recombination** A process by which separate DNA molecules are joined into a single molecule, due to such processes as crossing-over or transposition.

Recombination results from a physical exchange of chromosomal material. For example, recombination may result from the crossing-over that occurs during meiosis (the specialized division that produces haploid germ cells), specifically during Prophase I. Meiosis starts with a cell that has duplicated its chromosomes, so that it has four copies of each chromatid. (A *chromatid* is one of two identical copies of a chromosome following its duplication; since there are two homologs per cell, there are four chromatids of each type of chromosome.) Early in meiosis, all four copies are closely associated (synapsed) in a structure called a *bivalent*, and later a *tetrad*. At this point, pairwise exchanges of material between two nonsister (of the four total) chromatids may occur.

▸ **chiasma** A site at which two homologous chromosomes synapse during meiosis.

The point of synapsis between homologs is called a **chiasma** and is illustrated diagrammatically in FIGURE 2.7. A chiasma represents a site at which one strand in each of two nonsister chromatids in a tetrad has been broken and exchanged. If during the resolution of the chiasma the previously unbroken strands are also broken and exchanged, recombinant chromatids will be generated. Each recombinant chromatid consists of material derived from one chromatid on one side of the chiasma, with material from the other chromatid on the opposite side. The two recombinant chromatids have

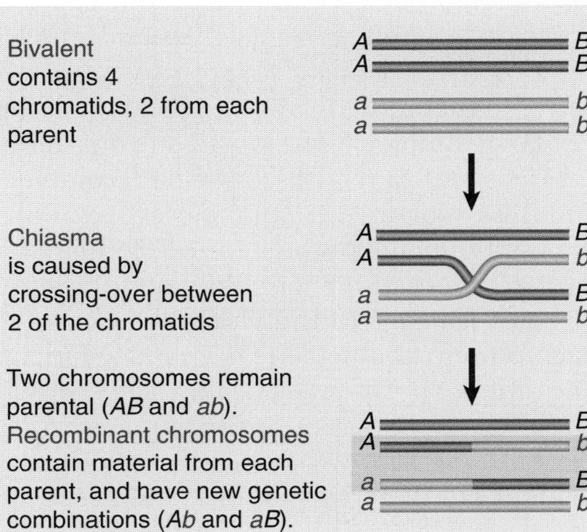

Bivalent contains 4 chromatids, 2 from each parent
A B
A B
a b
a b
Chiasma is caused by crossing-over between 2 of the chromatids
A B
A b
a B
a b
Two chromosomes remain parental (*AB* and *ab*). Recombinant chromosomes contain material from each parent, and have new genetic combinations (*Ab* and *aB*).
A B
A b
a B
a b

FIGURE 2.7 Chiasma formation is responsible for generating recombinants.

reciprocal structures. The event can be described as a breakage and reunion. Because each individual crossing-over event involves only two of the four associated chromatids, a single recombination event can produce only 50% recombinants.

The complementarity of the two strands of DNA is essential for the recombination process. Each of the chromatids shown in Figure 2.7 consists of a very long duplex of DNA. For them to be broken and reconnected without any addition or loss of material requires a mechanism to recognize exactly corresponding positions; this mechanism is complementary base pairing.

▸ **heteroduplex DNA** DNA that is generated by base pairing between complementary single strands derived from the different parental duplex molecules; it occurs during genetic recombination.

Recombination results from a process in which the single strands in the region of the crossover exchange their partners, resulting in a branch that may migrate for some distance in either direction. This creates a stretch of **heteroduplex DNA**, in which the single strand of one duplex is paired with its complement from the other duplex. The mechanism, of course, involves other stages in which strands must be broken and religated, which we discuss in more detail in *Chapter 19, Homologous and Site-Specific Recombination*, but the crucial feature that makes precise recombination possible is the complementarity of DNA strands. A stretch of heteroduplex DNA forms in the recombination intermediate when a single strand crosses over from one duplex to the other. Each recombinant consists of one parental duplex DNA which is connected by a stretch of heteroduplex DNA to the other parental duplex. Each duplex DNA corresponds to one of the chromatids involved in recombination in Figure 2.7.

The formation of heteroduplex DNA requires the sequences of the two recombining duplexes to be close enough to allow pairing between the complementary strands. If there are no differences between the two parental genomes in this region, formation of heteroduplex DNA will be perfect. However, the reaction can be tolerated even when there are small differences. In this case, the hybrid DNA has points of mismatch, at which a base in one strand is paired with a base in the other strand that is not complementary to it. The correction of such mismatches is another feature of genetic recombination (see *Chapter 20, Repair Systems*).

Over chromosomal distances, recombination events occur more or less at random, with a characteristic frequency. The probability that a crossover will occur within any specific region of the chromosome is more or less proportional to the length of the region, up to a saturation point. For example, a large human chromosome usually has three or four crossover events per meiosis, but a small chromosome has only one on average.

FIGURE 2.8 compares three situations: two genes on different chromosomes, two genes that are far apart on the same chromosome, and two genes that are close together on the same chromosome. Genes on different chromosomes segregate independently according to Mendel's laws, resulting in the production of 50% parental types and 50% recombinant types during meiosis. When genes are sufficiently far apart on the same chromosome, the probability of one or more recombination events in the region between them becomes so high that they behave in the same way as genes on different chromosomes and show 50% recombination.

But when genes are close together, the probability of a recombination event between them is reduced, and recombination occurs only in some proportion of meioses. For example, if it occurs in one quarter of the meioses, the overall rate of recombination is 12.5% (because a single recombination event produces 50% recombination, and this occurs in 25% of meioses). When genes are very close together, as shown in the bottom panel of the figure, recombination between them may never be observed in phenotypes of higher eukaryotes.

This leads us to the view that a chromosome contains an array of many genes. Each protein-coding gene is an independent unit of expression, and is represented in one or more polypeptide chains. The properties of a gene can be changed by mutation. The allelic combinations present on a chromosome can be changed by recombination. We can now ask, "what is the relationship between the sequence of a gene and the sequence of the polypeptide chain it represents?"

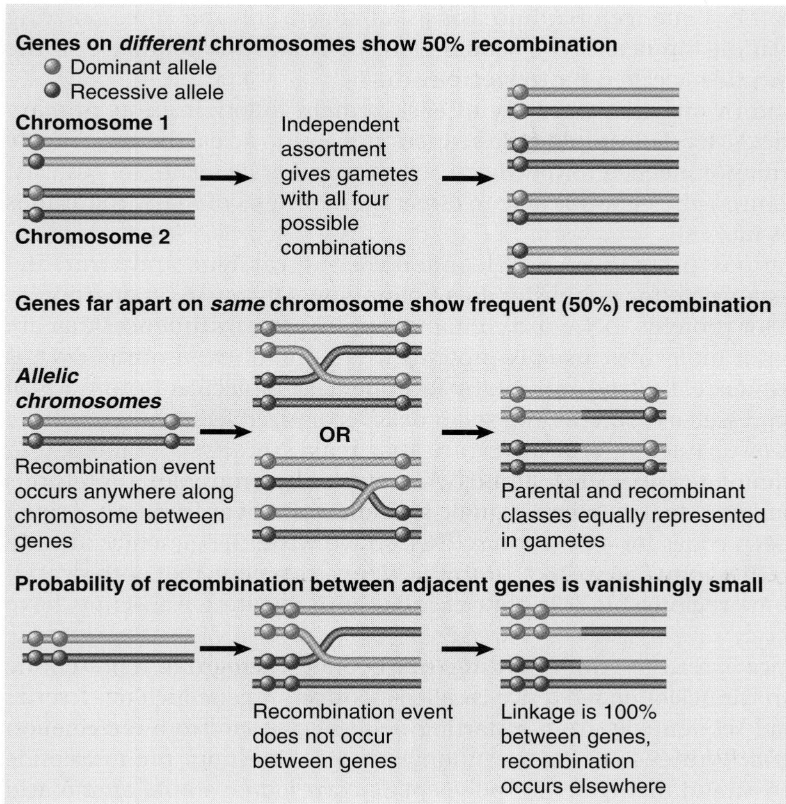

Genes on *different* chromosomes show 50% recombination
Dominant allele
Recessive allele
Chromosome 1
Chromosome 2
Independent assortment gives gametes with all four possible combinations
Genes far apart on *same* chromosome show frequent (50%) recombination
Allelic chromosomes
OR
Recombination event occurs anywhere along chromosome between genes
Parental and recombinant classes equally represented in gametes
Probability of recombination between adjacent genes is vanishingly small
Recombination event does not occur between genes
Linkage is 100% between genes, recombination occurs elsewhere

FIGURE 2.8 Genes on different chromosomes segregate independently so that all possible combinations of alleles are produced in equal proportions. Recombination occurs so frequently between genes that are far apart on the same chromosome that they effectively segregate independently. But recombination is reduced when genes are closer together, and for adjacent genes may hardly ever occur.

KEY CONCEPTS

- Recombination is the result of crossing-over that occurs at a chiasma and involves two of the four chromatids.
- Recombination occurs by a breakage and reunion that proceeds via an intermediate of heteroduplex DNA.
- Linkage between genes is determined by their distance apart.
- Genes that are close together are tightly linked because recombination between them is rare.
- Genes that are far apart may not be directly linked, because recombination is so frequent as to produce the same result as for genes on different chromosomes.

CONCEPT AND REASONING CHECK

Explain why the maximum frequency of recombination that can be measured between two loci is 50%.

2.7 The Genetic Code Is Triplet

Each structural gene codes for a particular polypeptide chain. The concept that each polypeptide consists of a particular sequence of amino acids dates from Sanger's characterization of insulin in the 1950s. The discovery that a gene consists of DNA presents us with the issue of how a sequence of nucleotides in DNA is used to construct a sequence of amino acids in protein.

A crucial feature of the general structure of DNA, the double helix, is that it is independent of the particular sequence of its component nucleotides. The sequence of nucleotides in a structural gene is important not because of its structure *per se*, but because it codes for the sequence of amino acids that constitutes the corresponding polypeptide. The relationship between a sequence of DNA and the sequence of the corresponding polypeptide is called the **genetic code**.

▸ **genetic code** The correspondence between triplets in DNA (or RNA) and amino acids in polypeptide.

The structure and/or enzymatic activity of each protein follow from its primary sequence of amino acids. By determining the sequence of amino acids, the gene is able to carry all the information needed to specify an active polypeptide chain. In this way, a single type of structure—the gene—is able to direct the synthesis of many thousands of polypeptide forms in a cell.

Together the various proteins of a cell undertake the catalytic and structural activities that are responsible for establishing its phenotype. Of course, in addition to sequences that code for proteins, DNA also contains certain control sequences that are recognized by regulator molecules, usually proteins. Here the function of the DNA is determined by its sequence directly, not via any intermediary molecule. Both types of sequence—genes expressed as proteins and sequences recognized as such—constitute genetic information.

The coding region of a gene is deciphered by a complex apparatus that interprets the nucleic acid sequence. In any given region, it is often the case that only one of the two strands of DNA codes for a functional RNA, so we write the genetic code as a sequence of bases (rather than base pairs). (Recent evidence suggests that both strands are transcribed in some regions, but it is not clear that both resulting transcripts have functional importance.)

A coding sequence is read in groups of three nucleotides, each group representing one amino acid. Each trinucleotide sequence is called a **codon**. A gene includes a series of codons that is read sequentially from a starting point at one end to a termination point at the other end. Written in the conventional 5′ to 3′ direction, the nucleotide sequence of the DNA strand that codes for polypeptide corresponds to the amino acid sequence of the polypeptide written in the direction from N-terminus to C-terminus.

▸ **codon** A triplet of nucleotides that codes for an amino acid, or a termination signal.

A coding sequence is read in nonoverlapping triplets from a fixed starting point:

- Nonoverlapping implies that each codon consists of three nucleotides and that successive codons are represented by successive trinucleotides. An individual nucleotide is part of only one codon.
- The use of a fixed starting point means that assembly of a protein must start at one end and proceed to the other, so that different parts of the coding sequence cannot be read independently.

The nature of the code predicts that two types of mutations will have different effects. If a particular sequence is read sequentially, such as:

UUU AAA GGG CCC (codons)

aa1 aa2 aa3 aa4 (amino acids)

then a nucleotide substitution will affect only one amino acid. For example, the substitution of an A by some other base (X) causes aa2 to be replaced by aa5:

UUU AAX GGG CCC

aa1 aa5 aa3 aa4

because only the second codon has been changed.

However, a mutation that inserts or deletes a single base will change the triplet sets for the entire subsequent sequence. A change of this sort is called a **frameshift**. An insertion might take the form:

UUU AAX AGG GCC C

aa1 aa5 aa6 aa7

Because the new sequence of triplets is completely different from the old one, the entire amino acid sequence of the protein is altered downstream from the site of mutation, so the function of the protein is likely to be lost completely.

Frameshift mutations are induced by the **acridines**, compounds that bind to DNA and distort the structure of the double helix, causing additional bases to be incorporated or omitted during replication. Each mutagenic event in the presence of an acridine results in the addition or removal of a single base pair.

▸ **frameshift** A mutation caused by deletions or insertions that are not a multiple of three base pairs. They change the frame in which triplets are translated into polypeptide.

▸ **acridines** Mutagens that act on DNA to cause the insertion or deletion of a single base pair. They were useful in defining the triplet nature of the genetic code.

If an acridine mutant is produced by, say, addition of a nucleotide, it should revert to wild type by deletion of the nucleotide. However, reversion also can be caused by deletion of a different base at a site close to the first. Combinations of such mutations provided revealing evidence about the nature of the genetic code.

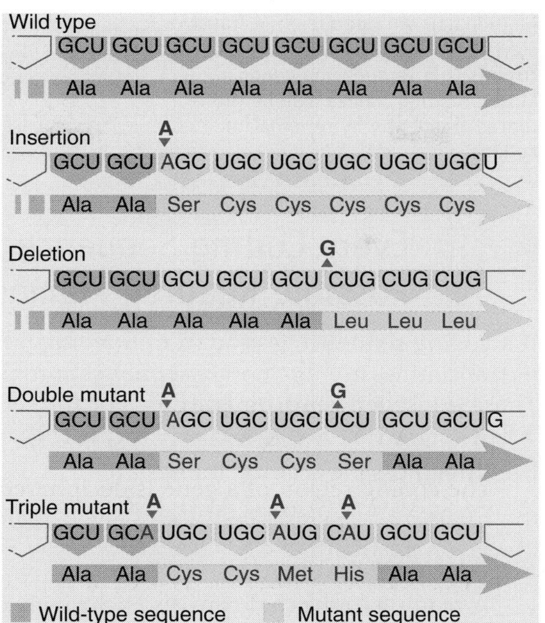

FIGURE 2.9 Frameshift mutations show that the genetic code is read in triplets from a fixed starting point.

FIGURE 2.9 illustrates the properties of frameshift mutations. An insertion or deletion changes the entire polypeptide sequence following the site of mutation. However, the combination of the insertion of a single nucleotide and the deletion of a single nucleotide causes the code to be read incorrectly only between the two sites of mutation; reading in the original frame resumes after the second site.

In 1961, genetic analysis of acridine mutations in the *rII* region of the phage T6 showed that all the mutations could be classified into one of two sets, described as (+) and (–). Either type of mutation by itself causes a frameshift: the (+) type by virtue of a base addition, and the (–) type by virtue of a base deletion. Double mutant combinations of the types (++) and (– –) continue to show mutant behavior. However, combinations of the types (+–) or (–+) suppress one another, so that one mutation is described as a frameshift suppressor of the other. (In the context of this work, "suppressor" is used in an unusual sense because the second mutation is in the same gene as the first; in fact, these are second-site reversions.)

These results show that the genetic code must be read as a sequence that is fixed by the starting point. Therefore, a single base addition and deletion compensate for each other, whereas double additions or double deletions remain frameshift mutants. However, these observations do not suggest how many nucleotides make up each codon.

When triple mutants are constructed, only (+++) and (– – –) combinations show the wild phenotype, whereas other combinations remain mutant. If we take three single base additions or three deletions to correspond respectively to the addition or omission overall of a single amino acid, this implies that the code is read in triplets. An incorrect amino acid sequence is found between the two outside sites of mutation and the sequence on either side remains wild-type, as indicated in Figure 2.9.

KEY CONCEPTS

- The genetic code is read in nucleotide triplets called *codons*.
- The triplets are nonoverlapping and are read from a fixed starting point.
- Mutations that insert or delete individual bases cause a shift in the triplet sets after the site of mutation; these are frameshift mutations.
- Combinations of mutations that together insert or delete three bases (or multiples of three) insert or delete amino acids, but do not change the reading of the triplets beyond the last site of mutation.

CONCEPT AND REASONING CHECK

Consider mutagens that insert or delete two adjacent nucleotides. Using "+" to mean an insertion of two nucleotides and "–" to mean a deletion of two nucleotides, what combinations of insertions and deletions would suppress one another?

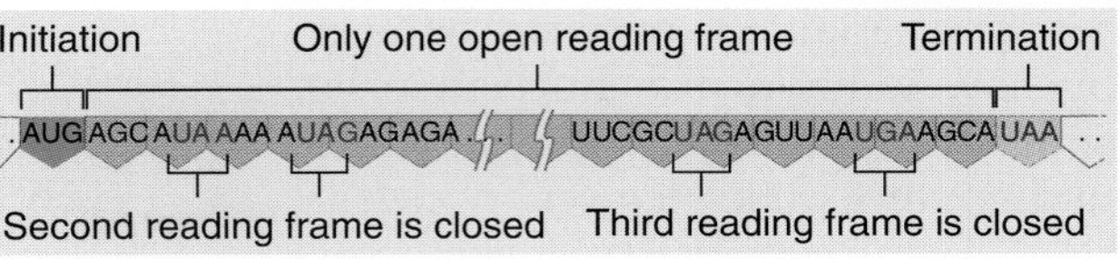

FIGURE 2.10 An open reading frame starts with AUG and continues in triplets to a termination codon. Closed reading frames may be interrupted frequently by termination codons.

2.8 Every Coding Sequence Has Three Possible Reading Frames

▸ **reading frame** One of three possible ways of reading a nucleotide sequence. Each divides the sequence into a series of successive triplets.

If a coding sequence is read in nonoverlapping triplets, there are three possible ways of translating any nucleotide sequence into protein, depending on the starting point. These are called **reading frames**. For the sequence

A C G A C G A C G A C G A C G A C G

the three possible reading frames are

ACG ACG ACG ACG ACG ACG
CGA CGA CGA CGA CGA CG
GAC GAC GAC GAC GAC G

▸ **open reading frame** A sequence of DNA consisting of triplets that can be translated into amino acids starting with an initiation codon and ending with a termination codon.

▸ **initiation codon** A special codon (usually AUG) used to start synthesis of a polypeptide.

▸ **termination codon** One of the three codons (UAA, UAG, UGA) that signal the termination of translation of a polypeptide.

▸ **closed (blocked) reading frame** A reading frame that cannot be translated into protein because of the occurrence of termination codons.

A reading frame that consists exclusively of triplets coding for amino acids is called an **open reading frame** or **ORF**. A sequence that is translated into polypeptide has a reading frame that starts with a special **initiation codon** (AUG) and then extends through a series of triplets coding for amino acids until it ends at one of three **termination codons** (see *Chapter 7, Messenger RNA*). ORFs can also be characterized by consensus promoter sequences at an appropriate distance upstream from the initiation codon, and (in eukaryotes) the presence of intron splicing junctions.

A reading frame that cannot be read into protein because termination codons occur frequently is said to be **closed** or **blocked**. If a sequence is closed in all three reading frames, it cannot have the function of coding for polypeptide.

When the sequence of a DNA region of unknown function is obtained, each possible reading frame can be analyzed to determine whether it is open or closed. Usually no more than one of the three possible reading frames is open in any single stretch of DNA. **FIGURE 2.10** shows an example of a sequence that can be read in only one reading frame because the alternative reading frames are blocked by frequent termination codons. A long open reading frame is unlikely to exist by chance; if it were not translated into polypeptide, there would have been no selective pressure to prevent the accumulation of termination codons. Therefore, the identification of a lengthy open reading frame is taken to be *prima facie* evidence that the sequence is (or until recently has been) translated into a polypeptide in that frame. An ORF for which no polypeptide product has been identified is sometimes called an **unidentified reading frame** (URF).

▸ **unidentified reading frame** An open reading frame with an as yet undetermined function.

KEY CONCEPT

- Usually only one of the three possible reading frames is translated and the other two are closed by frequent termination signals.

CONCEPT AND REASONING CHECK

Suppose you identify a stretch of DNA in which only one of the three possible reading frames is closed. What would you hypothesize from this observation?

2.9 Bacterial Genes Are Colinear with Their Products

▸ **colinearity** The relationship that describes the 1:1 correspondence of a sequence of triplet nucleotides to a sequence of amino acids.

By comparing the nucleotide sequence of a gene with the amino acid sequence of its polypeptide product, we can determine whether the gene and the polypeptide are **colinear**; that is, whether the sequence of nucleotides in the gene exactly corresponds to the sequence of amino acids in the polypeptide. In bacteria and their viruses, genes and their products are colinear. Each gene is a continuous stretch of DNA with a length that is three times the number of amino acids in the polypeptide for which it codes (due to the triplet nature of the genetic code). In other words, if a protein contains N amino acids, the gene encoding that protein contains 3N nucleotides.

The correspondence of the bacterial gene and its product means that a physical map of DNA will exactly match an amino acid map of the polypeptide. How well do these maps match the recombination map?

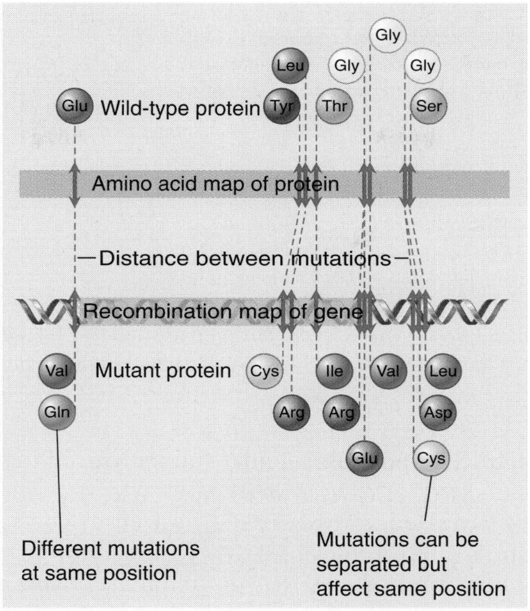

FIGURE 2.11 The recombination map of the tryptophan synthetase gene corresponds with the amino acid sequence of the protein.

The colinearity of gene and protein was originally investigated in the tryptophan synthetase gene of *E. coli*. Genetic distance was measured as the percent recombination between mutations; amino acid distance was measured as the number of amino acids separating sites of replacement. **FIGURE 2.11** compares the two maps. The order of seven sites of mutation is the same as the order of the corresponding sites of amino acid replacement, and the recombination distances are roughly similar to the actual distances in the protein. The recombination map expands the distances between some mutations, but otherwise there is little distortion of the recombination map relative to the physical map.

The recombination map leads to two further general points about the organization of the gene. Different mutations may cause a wild-type amino acid to be replaced with different alternatives. (See Figure 9.1 for the genetic table to see how base changes in specific codons can alter the resulting amino acid.) If two such mutations cannot recombine, they must involve different point mutations at the same position in DNA. If the mutations can be separated on the genetic map, but affect the same amino acid on the upper map (the connecting lines converge in the figure), they must involve point mutations at different positions in the same codon. This happens because the minimum size unit of genetic recombination (1 bp) is smaller than the unit coding for the amino acid (3 bp).

KEY CONCEPTS

- A bacterial gene consists of a continuous length of 3N nucleotides that codes for N amino acids.
- The gene is colinear with both its mRNA and polypeptide products.

CONCEPT AND REASONING CHECK

Although a recombination map and a physical map will show genetic markers in the same order, the distance between markers will usually be different when the two maps are compared. Why?

2.10 Several Processes Are Required to Express the Product of a Gene

In comparing a gene and its polypeptide product, we are restricted to dealing with the sequence of DNA that lies between the points corresponding to the N-terminus and C-terminus of the polypeptide. However, a gene is not directly translated into polypeptide, but is expressed via the production of a **messenger RNA** (abbreviated to **mRNA**),

messenger RNA (mRNA) The intermediate that represents one strand of a gene coding for polypeptide. Its coding region is related to the polypeptide sequence by the triplet genetic code.

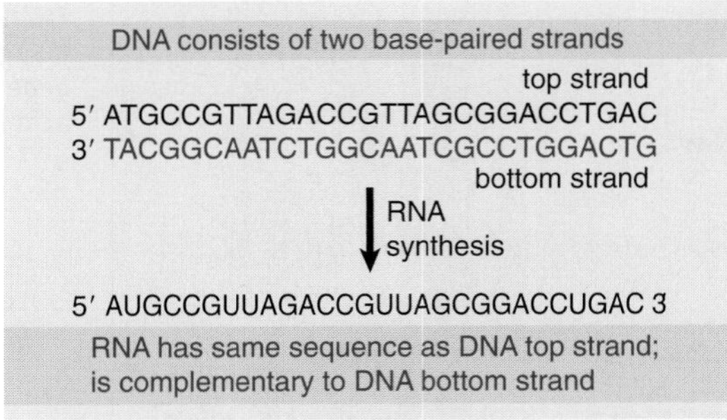

FIGURE 2.12 RNA is synthesized by using one strand of DNA as a template for complementary base pairing.

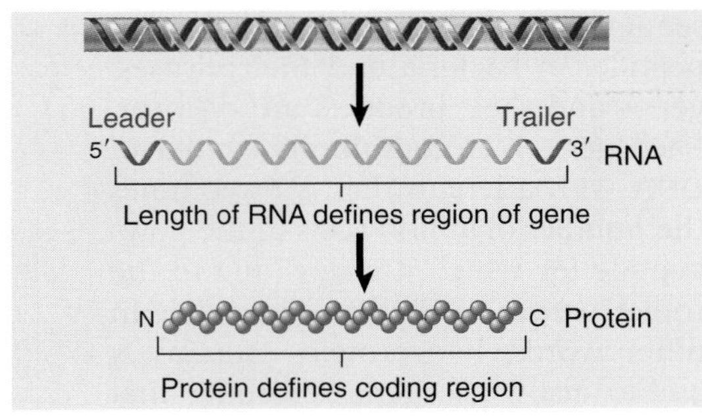

FIGURE 2.13 The gene may be longer than the sequence coding for protein.

a nucleic acid intermediate actually used to synthesize a protein (as we see in detail in *Chapter 7, Messenger RNA*).

Messenger RNA is synthesized by the same process of complementary base pairing used to replicate DNA, with the important difference that it corresponds to only one strand of the DNA double helix. **FIGURE 2.12** shows that the sequence of mRNA is complementary with the sequence of one strand of DNA (called the **template strand** or **antisense strand**) and is identical (apart from the replacement of T with U) with the other strand of DNA (called the **coding strand** or **sense strand**). The convention for writing DNA sequences is that the top strand is the coding strand and runs 5′→3′.

- **antisense (template) strand** The DNA strand that is complementary to the sense strand, and acts as the template for synthesis of mRNA.
- **coding (sense) strand** The DNA strand that has the same sequence as the mRNA and is related by the genetic code to the protein sequence that it represents.
- **gene expression** The process by which the information in a sequence of DNA in a gene is used to produce an RNA or polypeptide, involving transcription and (for polypeptides) translation.
- **transcription** Synthesis of RNA on a DNA template.
- **translation** Synthesis of protein on an mRNA template.
- **coding region** A part of a gene that codes for a polypeptide sequence.
- **leader (5′ UTR)** In mRNA, it is the untranslated sequence at the 5′ end that precedes the initiation codon.
- **trailer (3′ UTR)** An untranslated sequence at the 3′ end of an mRNA following the termination codon.

The process by which information from a gene is used to synthesize an RNA or polypeptide product is called **gene expression**. In bacteria, expression of a structural gene consists of two stages. The first stage is **transcription**, when an mRNA copy of the template strand of the DNA is produced. The second stage is **translation** of the mRNA into polypeptide. This is the process by which the sequence of an mRNA is read in triplets to give the series of amino acids that make the corresponding polypeptide.

An mRNA includes a sequence of nucleotides that corresponds with the sequence of amino acids in the polypeptide. This part of the nucleic acid is called the **coding region**. However, the mRNA includes additional sequences on either end that do not code for amino acids. The 5′ untranslated region is called the **leader** or **5′ UTR**, and the 3′ untranslated region is called the **trailer** or **3′ UTR**.

The gene includes the entire sequence represented in messenger RNA. Sometimes mutations impeding gene function are found in the additional, noncoding regions, confirming the view that these comprise a legitimate part of the genetic unit. **FIGURE 2.13** illustrates this situation, in which the gene is considered to comprise a continuous stretch of DNA needed to produce a particular polypeptide, including the leader, the coding region, and the trailer.

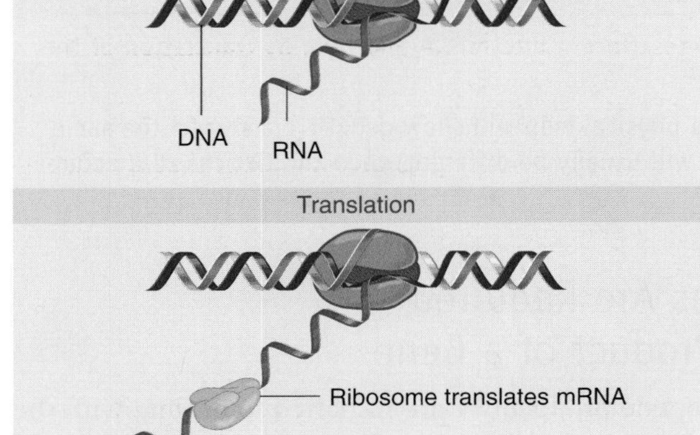

FIGURE 2.14 Transcription and translation take place in the same compartment in bacteria.

A bacterial cell has only a single compartment, so transcription and translation occur in the same place, as illustrated in **FIGURE 2.14**. In eukaryotes, transcription occurs in the nucleus, but the RNA product must be transported to the cytoplasm in order to be translated. This results in a spatial separation between transcription (in the nucleus) and translation (in the cytoplasm). For the simplest eukaryotic genes (as in bacteria), the transcript RNA is in fact the mRNA. However, for more complex genes, the primary transcript of the gene is a **pre-mRNA** that requires processing to generate the mature mRNA. The basic stages of gene expression in a eukaryote are outlined in **FIGURE 2.15**.

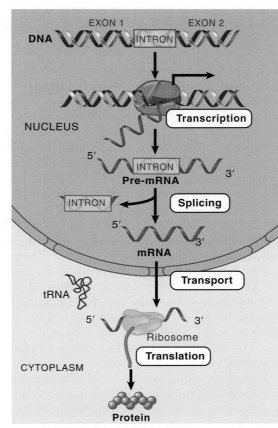

FIGURE 2.15 In eukaryotes, transcription occurs in the nucleus, and translation occurs in the cytoplasm.

- **pre-mRNA** The nuclear transcript that is processed by modification and splicing to give an mRNA.

The most important stage in **RNA processing** is **splicing**. Many genes in eukaryotes (and a majority in higher eukaryotes) contain internal regions called **introns** that do not carry coding information for the polypeptide products encoded by those genes. The process of splicing removes introns from the pre-mRNA to generate an RNA that has a continuous open reading frame (see Figure 3.1). Other processing events that occur at this stage involve the modification of the 5′ and 3′ ends of the pre-mRNA (see Figure 7.16).

Translation is accomplished by a complex apparatus that includes both protein and RNA components. The actual "machine" that undertakes the process is the **ribosome**, a large complex that includes some large RNAs (**ribosomal RNAs**, abbreviated to **rRNAs**) and many small proteins. The process of recognizing which amino acid corresponds to a particular nucleotide triplet requires an intermediate **transfer RNA** (abbreviated to **tRNA**); there is at least one tRNA species for every amino acid. Many ancillary proteins are involved. We describe translation in *Chapter 7, Messenger RNA*, but note for now that the ribosomes are the large structures in Figure 2.14 that translate the mRNA.

The important point to note at this stage is that the process of gene expression involves RNA not only as the essential substrate, but also in providing components of the apparatus. The rRNA and tRNA components are coded by genes and are generated by the process of transcription (like mRNA), but are not translated to polypeptide.

- **RNA processing** Modifications to RNA transcripts of genes. This may include alterations to the 3′ and 5′ ends and the removal of introns.
- **splicing** The process of excising introns from RNA and connecting the exons into a continuous mRNA.
- **intron** A segment of DNA that is transcribed, but later removed from within the transcript by splicing together the sequences (exons) on either side of it.
- **ribosome** A large assembly of RNA and proteins that synthesizes proteins under direction from an mRNA template.
- **ribosomal RNAs (rRNAs)** A major component of the ribosome.
- **transfer RNA (tRNA)** The intermediate in protein synthesis that interprets the genetic code. Each molecule can be linked to an amino acid. It has an anticodon sequence that is complementary to a triplet codon representing the amino acid.

KEY CONCEPTS

- A bacterial gene is expressed by transcription into mRNA and then by translation of the mRNA into polypeptide.
- In eukaryotes, a gene may contain introns that are not represented in polypeptide.
- Introns are removed from the RNA transcript by splicing to give an mRNA that is colinear with the polypeptide product.
- Each mRNA consists of an untranslated 5′ leader, a coding region, and an untranslated 3′ trailer.

CONCEPT AND REASONING CHECKS

1. Why is bacterial gene expression generally faster than eukaryotic gene expression?
2. Why do the recombination maps of eukaryotic genes usually not correspond to the amino acid maps of their products?

2.11 Proteins Are *trans*-Acting, but Sites on DNA Are *cis*-Acting

A crucial progression in the definition of the gene was the realization that all its parts must be present on one contiguous stretch of DNA. In genetic terminology, sites that are located on the same DNA are said to be in *cis*. Sites that are located on two different molecules of DNA are described as being in *trans*. So two mutations may be in *cis* (on the same DNA) or in *trans* (on different DNAs). The complementation test uses this concept to determine whether two mutations are in the same gene (see Figure 2.3 in *Section 2.3, Mutations in the Same Gene Cannot Complement*). We may now extend the concept of the difference between *cis* and *trans* effects from defining the coding region of a gene to describing the interaction between a gene and its regulatory elements.

Suppose that the ability of a gene to be expressed is controlled by a protein that binds to the DNA close to the coding region. In the example depicted in **FIGURE 2.16**, RNA can be synthesized only when the protein is bound to a control site on the DNA. Now suppose that a mutation occurs in the control site so that the protein can no longer bind to it. As a result, the gene can no longer be expressed.

So gene expression can be inactivated either by a mutation in a control site or by a mutation in a coding region. The mutations cannot be distinguished genetically, because both have the property of acting only on the DNA sequence of the single allele in which they occur. They have identical properties in the complementation test, so a mutation in a control region is defined as comprising part of the gene in the same way as a mutation in the coding region.

FIGURE 2.17 shows that a change in the control site affects only the coding region to which it is connected; it does not affect the ability of the homologous allele to be expressed. A mutation that acts solely by affecting the properties of the contiguous sequence of DNA is called ***cis*-acting**.

- ***cis*-acting sequence** A site that affects the activity only of sequences on its own molecule of DNA (or RNA); this property usually implies that the site does not code for protein.
- ***trans*-acting sequence** DNA sequence coding for a product that can function on any copy of its target DNA. This implies that it is a diffusible protein or RNA.

We may contrast the behavior of the *cis*-acting mutation shown in Figure 2.17 with the result of a mutation in the gene coding for the regulatory protein. **FIGURE 2.18** shows that the absence of regulatory protein would prevent both alleles from being expressed. A mutation of this sort is said to be ***trans*-acting**.

Reversing the argument, if a mutation is *trans*-acting, we know that its effects must be exerted through some diffusible product (typically a protein) that acts on multiple targets within a cell. However, if a mutation is *cis*-acting, it must function via affecting directly the properties of the contiguous DNA, which means that it is not expressed in the form of RNA or protein.

KEY CONCEPTS

- All gene products (RNA or polypeptides) are *trans*-acting. They can act on any copy of a gene in the cell.
- *cis*-acting mutations identify sequences of DNA that are targets for recognition by *trans*-acting products. They are not expressed as RNA or polypeptide and affect only the contiguous stretch of DNA.

FIGURE 2.16 Control sites in DNA provide binding sites for proteins; coding regions are expressed via the synthesis of RNA.

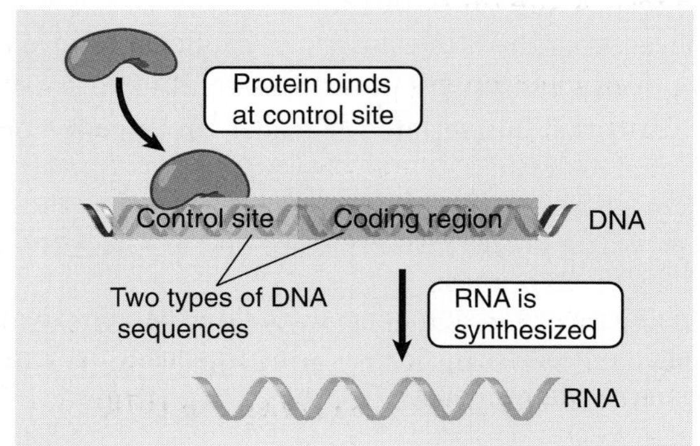

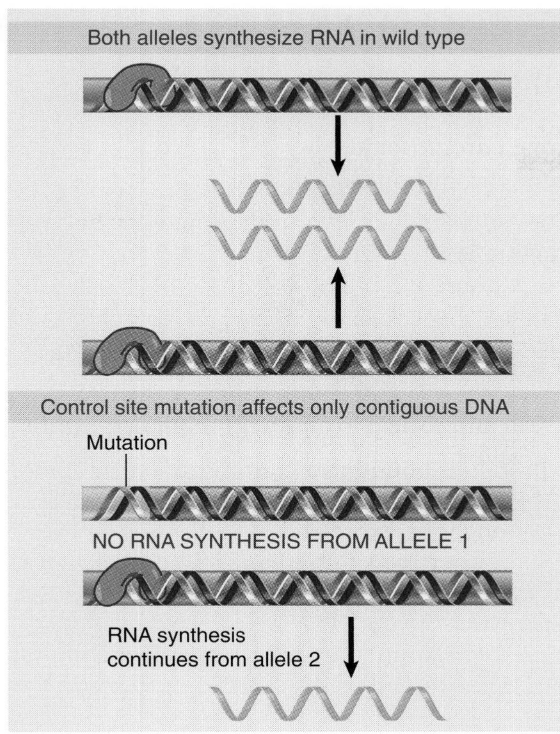
Both alleles synthesize RNA in wild type
Control site mutation affects only contiguous DNA
Mutation
NO RNA SYNTHESIS FROM ALLELE 1
RNA synthesis
continues from allele 2

FIGURE 2.17 A *cis*-acting site controls the adjacent DNA, but does not influence the other allele.

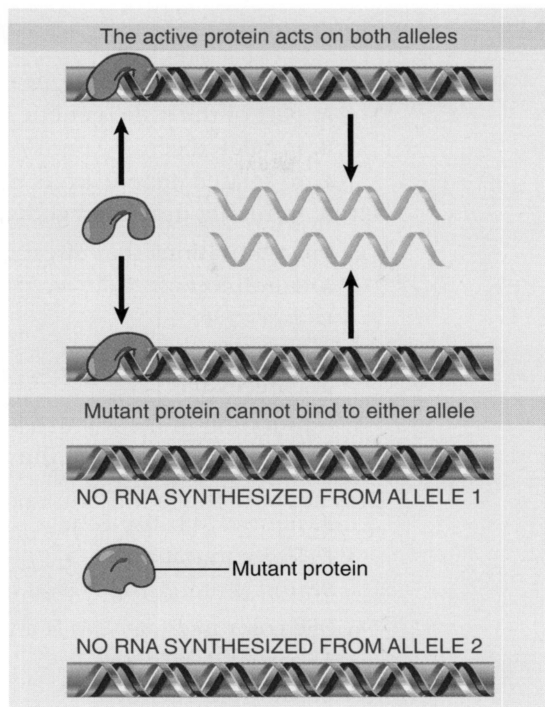
The active protein acts on both alleles
Mutant protein cannot bind to either allele
NO RNA SYNTHESIZED FROM ALLELE 1
Mutant protein
NO RNA SYNTHESIZED FROM ALLELE 2

FIGURE 2.18 A *trans*-acting mutation in a protein affects both alleles of a gene that it controls.

CONCEPT AND REASONING CHECK

Can a mutation in a control site alter the function of a gene's protein product without inactivating its expression? Explain. ~~Yes. cis-acting element can control the~~

No, control site doesn't affect the sequence of AA in polypeptides.

2.12 Summary

A chromosome consists of an uninterrupted length of duplex DNA that contains many genes. Each gene is transcribed into an RNA product, which in turn is translated into a polypeptide sequence if it is a structural gene. An RNA or protein product of a gene is said to be *trans*-acting. A gene is defined as a unit of a single stretch of DNA by the complementation test. A site on DNA that regulates the activity of an adjacent gene is said to be *cis*-acting.

When a gene codes for polypeptide, the relationship between the sequence of DNA and sequence of the polypeptide is given by the genetic code. Only one of the two strands of DNA codes for polypeptide. A codon consists of three nucleotides that represent a single amino acid. A coding sequence of DNA consists of a series of codons, read from a fixed starting point. Usually only one of the three possible reading frames can be translated into polypeptide.

A gene may have multiple alleles. Recessive alleles are caused by loss-of-function mutations that interfere with the function of the protein. A null allele has total loss-of-function. Dominant alleles are caused by gain-of-function mutations that create a new property in the protein.

CHAPTER QUESTIONS

1. Failure to complement means that two mutations are:
   A. part of the same genetic unit.
   B. in related genetic units on the same chromosome.
   C. in related genetic units on different chromosomes.
   D. in totally unrelated genetic units.
2. Most mutations that affect gene function are:
   A. dominant.
   B. recessive.
   C. codominant.
   D. corecessive.
3. A mutation in the DNA that changes an amino acid in a protein sequence without affecting the activity of the protein is called a:
   A. recessive mutation.
   B. neutral substitution.
   C. leaky mutation.
   D. null mutation.
4. Eye color in *Drosophila* is an example of:
   A. partial dominance.
   B. multiple alleles.
   C. codominance.
   D. leaky mutation.
5. Which blood group is the null phenotype?
   A. A
   B. B
   C. AB
   D. O
6. Any given segment of a genome could have how many possible reading frames in a single strand of the DNA?
   A. 1
   B. 2
   C. 3
   D. 4
7. What types of mutations do acridines cause?
   A. frameshift mutations
   B. point mutations
   C. long deletions
   D. long insertions
8. The change in base sequence from AAA AGC TTC GAC to AAA GCT TCG ACC is an example of a:
   A. point mutation.
   B. insertion.
   C. frameshift mutation.
   D. deamination.

9. The convention for writing one strand of the DNA sequence of a protein-coding gene is:
   A. 5′ → 3′ with the sequence being the same as for the mRNA (except for T instead of U).
   B. 3′ → 5′ with the sequence being the same as for the mRNA (except for T instead of U).
   C. 5′ → 3′ with the sequence being opposite that for the mRNA (except for T instead of U).
   D. 3′ → 5′ with the sequence being opposite that for the mRNA (except for T instead of U).

10. What type of mutation would prevent both alleles of a gene from being expressed?
   A. dominant
   B. recessive
   C. *trans*-acting
   D. *cis*-acting

KEY TERMS

acridines
allele
antisense (template) strand
chiasma
***cis*-acting sequence**
cistron
closed (blocked) reading frame
coding region
coding (sense) strand
codon
colinearity
complementation test
frameshift
gain-of-function mutation
gene expression
genetic code
genetic recombination
heteroduplex DNA
heteromultimer
homomultimer
initiation codon
intron
leader (5′ UTR)
linkage
locus
loss-of-function mutation
messenger RNA (mRNA)
neutral substitutions
null mutation
one gene : one enzyme hypothesis
one gene : one polypeptide hypothesis
open reading frame
polymorphism
pre-mRNA
reading frame
ribosomal RNAs (rRNAs)
ribosome
RNA processing
silent mutation
splicing
termination codon
trailer (3′ UTR)
***trans*-acting sequence**
transcription
transfer RNA (tRNA)
translation
unidentified reading frame

FURTHER READING

Carter Jr., C. W. (2008). Whence the genetic code?: thawing the 'frozen accident.' *Heredity* 100, 339–340.
A brief review of current hypotheses for the origin of the genetic code.

Ripley, L. S. (1990). Frameshift mutation: determinants of specificity. *Annu. Rev. Genet.* 24, 189–213.
A review of the mechanisms of frameshift mutations.

Yanofsky, C. (2001). Advancing our knowledge in biochemistry, genetics, and microbiology through studies on tryptophan metabolism. *Annu. Rev. Biochem.* 70, 1–37.
A personal account of Yanofsky's research, including establishing the colinearity of genes and their protein products and the regulation of *trp* expression via attenuation.

3

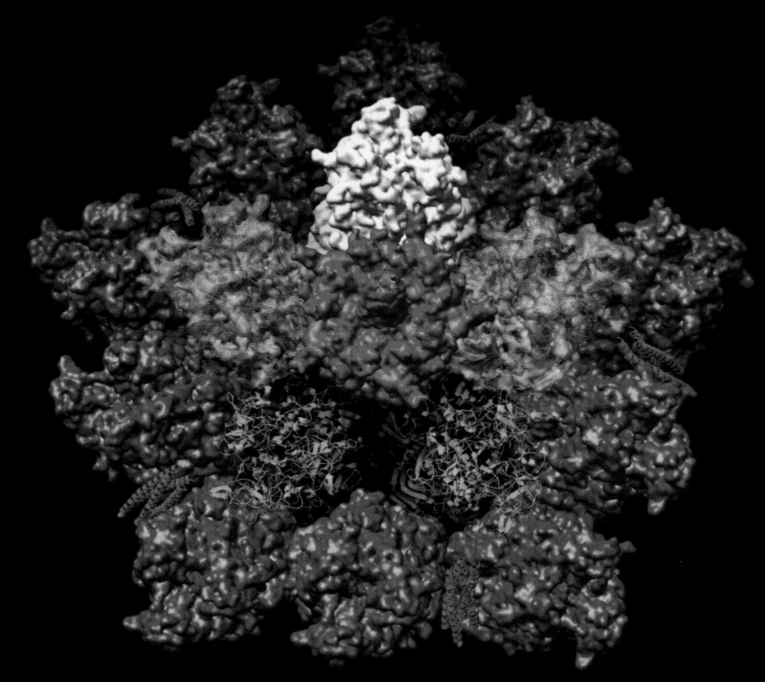

The structure of adenovirus at 6Å resolution. Viruses infecting eukaryotic cells, like their hosts, have interrupted genes. Reproduced from *J. Virol.*, 2006, vol. 80, p. 12049–12059, DOI and reproduced with permission from the American Society of Microbiology. Photo courtesy of Phoebe L. Stewart, Vanderbilt University Medical Center.

The Interrupted Gene

CHAPTER OUTLINE

3.1 Introduction

The simplest form of a gene is a length of DNA that directly corresponds to its protein product. Bacterial genes are almost always of this type, in which a continuous coding sequence of 3*N* base pairs codes for a protein of *N* amino acids. In eukaryotes, however, a gene may include additional sequences that lie within the coding region and interrupt the sequence that codes for the protein. These sequences are removed from the RNA product following transcription, generating an mRNA that includes a nucleotide sequence exactly corresponding to the protein product according to the rules of the genetic code.

The sequences of DNA comprising an **interrupted gene** are divided into the two categories depicted in **FIGURE 3.1**:

- **Exons** are the sequences retained in the final RNA product. By definition, a gene begins and ends with exons that correspond to the 5′ and 3′ ends of the RNA.
- **Introns** are the intervening sequences that are removed when the primary transcript is processed to give the final RNA product.

▸ **interrupted gene** A gene in which the coding sequence is not continuous due to the presence of introns.

▸ **exon** Any segment of an interrupted gene that is represented in the mature RNA product.

▸ **intron** A segment of DNA that is transcribed, but later removed from within the transcript by splicing together the sequences (exons) on either side of it.

The exon sequences are in the same order in the gene and in the RNA, but an interrupted gene is longer than its final RNA product because of the presence of the introns.

The processing of interrupted genes requires an additional step that is not needed for uninterrupted genes. The DNA of an interrupted gene is transcribed to an RNA copy (a transcript) that is exactly complementary to the original gene sequence. This RNA is only a precursor, though; it cannot yet be used to produce a protein. First, the introns must be removed from the RNA to give a messenger RNA that consists only of the series of exons. This process is called **RNA splicing** (see *Section 2.10, Several Processes Are Required to Express the Product of a Gene*) and involves precisely deleting the introns from the primary transcript and then joining the ends of the RNA on either side to form a covalently intact molecule (see *Chapter 28, RNA Splicing and Processing*).

▸ **RNA splicing** The process of excising introns from RNA and connecting the exons into a continuous mRNA.

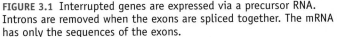

The original gene comprises the region in the genome between the points corresponding to the 5′ and 3′ terminal bases of the mature mRNA. We know that transcription starts at the DNA template corresponding to the 5′ end of the mRNA and usually extends beyond the complement to the 3′ end of the mature mRNA, which is generated by cleavage of the primary RNA transcript (see *Section 28.14, The 3′ Ends*

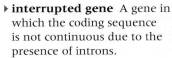

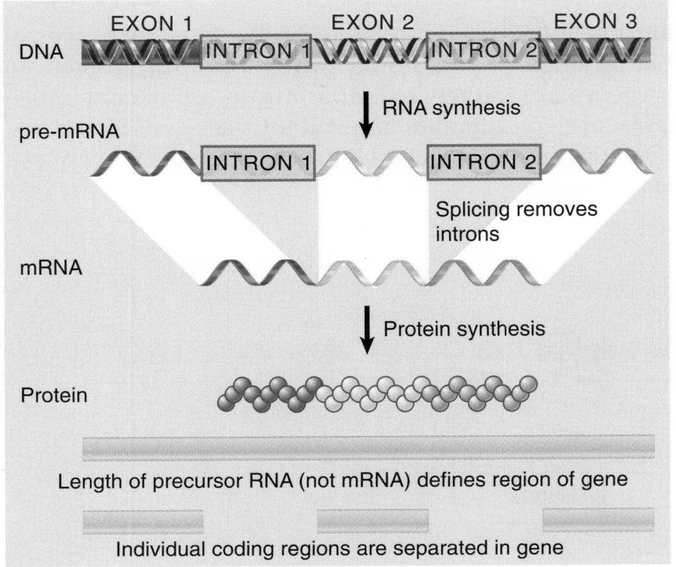

FIGURE 3.1 Interrupted genes are expressed via a precursor RNA. Introns are removed when the exons are spliced together. The mRNA has only the sequences of the exons.

of mRNAs Are Generated by Cleavage and Polyadenylation). The gene is also considered to include the regulatory regions on both sides of the gene that are required for initiating and (sometimes) terminating transcription.

CONCEPT AND REASONING CHECK

Why might it be difficult to express an intact human gene inserted into a bacterial cell?

3.2 An Interrupted Gene Consists of Exons and Introns

How does the existence of introns change our view of the gene? During splicing, the exons are always joined together in the same order they are found in the original DNA, so the correspondence between the gene and protein sequences is maintained. **FIGURE 3.2** shows that the order of exons in the gene remains the same as the order of exons in the processed mRNA, but the distances between sites in the gene (as determined by recombination analysis) do not correspond to the distances between sites in the processed mRNA. The length of the gene is defined by the length of the primary RNA transcript instead of the length of the processed mRNA.

All of the exons are present in the primary RNA transcript, and their splicing together occurs only as an intramolecular reaction. There is usually no joining of exons carried by different RNA transcripts, so the splicing mechanism excludes any splicing together of sequences from different alleles. (However, in a phenomenon known as *trans*-splicing, sequences from different mRNAs are ligated together into a single molecule for translation.) Mutations located in different exons of a gene cannot complement one another, so they continue to be defined as members of the same complementation group.

Mutations that directly affect the sequence of a protein must occur in exons. What are the effects of mutations in the introns? The introns are not part of the processed messenger RNA, so mutations in them cannot directly affect the polypeptide sequence. However, they may affect the processing of the messenger RNA by inhibiting the splicing together of exons. A mutation of this sort acts only on the allele that carries it. As a result, it fails to complement any other mutation in that allele and is part of the same complementation group as the exons.

Mutations that affect splicing are usually deleterious. The majority are single-base substitutions at the junctions between introns and exons. They may cause an exon to be left out of the product, cause an intron to be included, or make splicing occur at a different site. The most common outcome is a termination codon that truncates the protein sequence. About 15% of the point mutations that cause human diseases disrupt splicing.

Some eukaryotic genes are not interrupted, and, like prokaryotic genes, correspond directly with the protein product. In yeast, most genes are uninterrupted. In multicellular eukaryotes, most genes are interrupted, and introns are usually much longer than exons, so that genes are considerably larger than their coding regions. (See Figure 3.6 for the example of the mammalian β-globin genes, in which exons can be ~37% of the total length of the gene.)

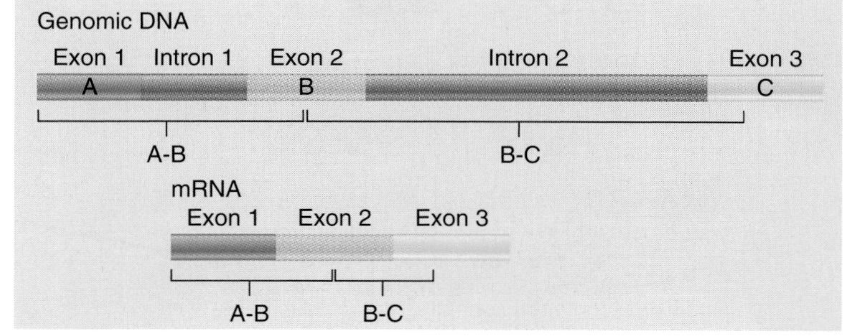

FIGURE 3.2 Exons remain in the same order in mRNA as in DNA, but distances along the gene do not correspond to distances along the mRNA or protein products. The distance from A–B in the gene is smaller than the distance from B–C; but the distance from A–B in the mRNA (and protein) is greater than the distance from B–C.

KEY CONCEPTS

- Introns are removed by the process of RNA splicing.
- Only mutations in exons can affect polypeptide sequence; however, mutations in introns can affect processing of the RNA and therefore prevent production of polypeptide.

CONCEPT AND REASONING CHECK

Describe the types of mutations that lead to abnormal splicing and their specific effects.

3.3 Organization of Interrupted Genes May Be Conserved

The characterization of eukaryotic genes was first made possible by the development of techniques for physically mapping DNA. When an mRNA is compared with the genomic sequence from which it was derived, the genomic sequence turns out to have extra regions that are not represented in the mRNA.

One technique for comparing mRNA with genomic DNA is to hybridize the mRNA with the complementary strand of the DNA. If the two sequences are colinear, a duplex is formed. **FIGURE 3.3** shows a typical result when an RNA made from an uninterrupted gene is hybridized with a DNA that includes the gene. The sequences on either side of the gene are not represented in the RNA, but the DNA sequence of the gene hybridizes with the RNA to form a continuous duplex region.

Suppose we now perform the same experiment with RNA made from an interrupted gene. The difference is that the sequences represented in mRNA lie on either side of a sequence that is not in the mRNA. **FIGURE 3.4** shows that the RNA-DNA hybrid forms a duplex, but the unhybridized DNA sequence in the middle remains

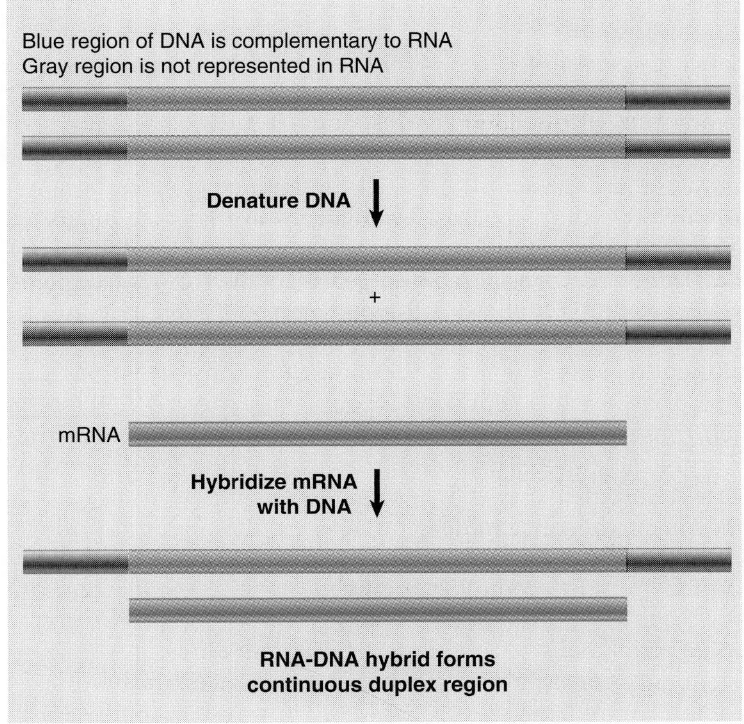
Blue region of DNA is complementary to RNA
Gray region is not represented in RNA
Denature DNA
+
mRNA
Hybridize mRNA with DNA
RNA-DNA hybrid forms continuous duplex region

FIGURE 3.3 Hybridizing an mRNA from an uninterrupted gene with the DNA of the gene generates a duplex region corresponding to the gene.

FIGURE 3.4 RNA hybridizing with the DNA made from an interrupted gene produces a duplex corresponding to the exons, with an intron excluded as a single-stranded loop between the exons.

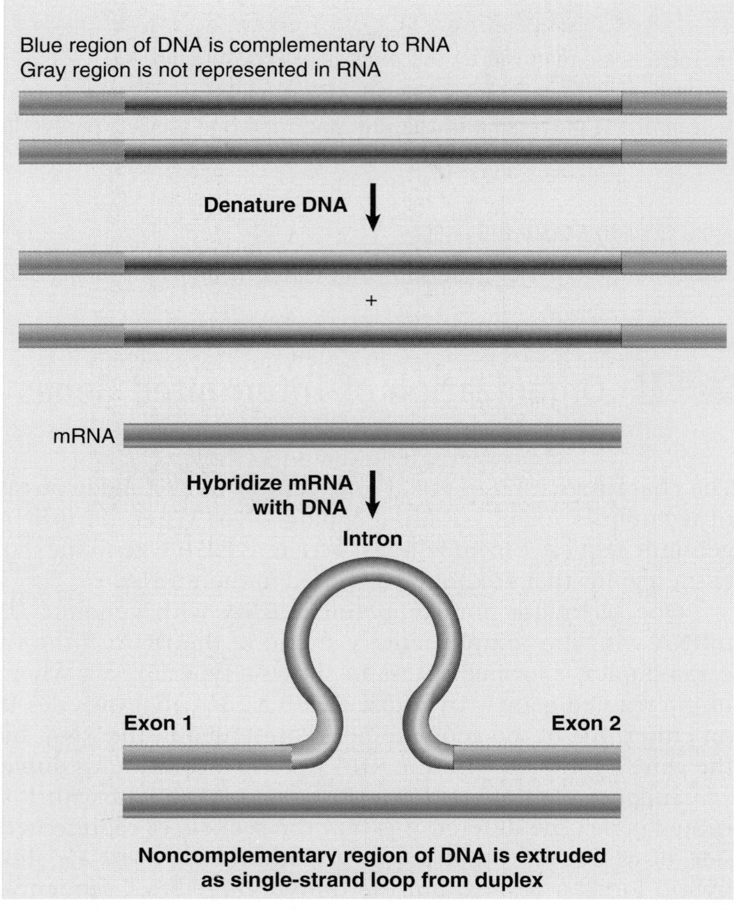

Blue region of DNA is complementary to RNA
Gray region is not represented in RNA
Denature DNA
+
mRNA
Hybridize mRNA with DNA
Intron
Exon 1
Exon 2
Noncomplementary region of DNA is extruded as single-strand loop from duplex

single-stranded, forming a loop that extrudes from the duplex. The hybridizing regions correspond to the exons, and the extruded loop corresponds to the intron.

The structure of the mRNA-DNA hybrid can be visualized by electron microscopy. One of the very first examples of the visualization of an interrupted gene is shown in FIGURE 3.5. Tracing the structure in the lower part of the figure shows that three introns are located close to the very beginning of the gene.

When a gene is uninterrupted, the restriction map of its DNA corresponds exactly with the map of its mRNA. When a gene has introns, the maps of the gene and its mRNA are different except at each end, corresponding to the first and last exon. The gene map will have additional regions due to the presence of introns. FIGURE 3.6 compares the restriction maps of a β-globin gene and a complementary DNA (**cDNA**) copy of its mRNA. The gene has two introns, each of which contains a series of restriction sites that are absent from the cDNA. The pattern of restriction sites in the exons is the same in both the cDNA and the gene.

▸ **cDNA** A single-stranded DNA complementary to an RNA, synthesized from it by reverse transcription *in vitro*.

Ultimately, a comparison of the nucleotide sequences of the gene and mRNA sequences precisely identifies the introns. As indicated in FIGURE 3.7, an intron usually has no open reading frame. An intact reading frame in the mRNA results from the removal of the introns.

The structures of eukaryotic genes vary extensively. Some genes are uninterrupted, so that the gene sequence corresponds to the mRNA sequence. Most multicellular eukaryotic genes are interrupted, but among genes the introns vary enormously

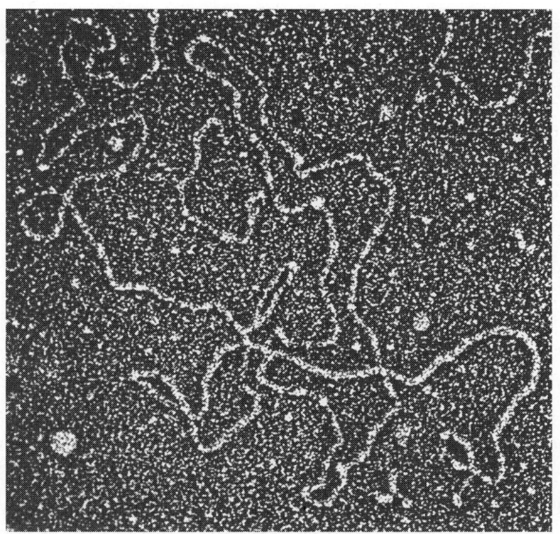

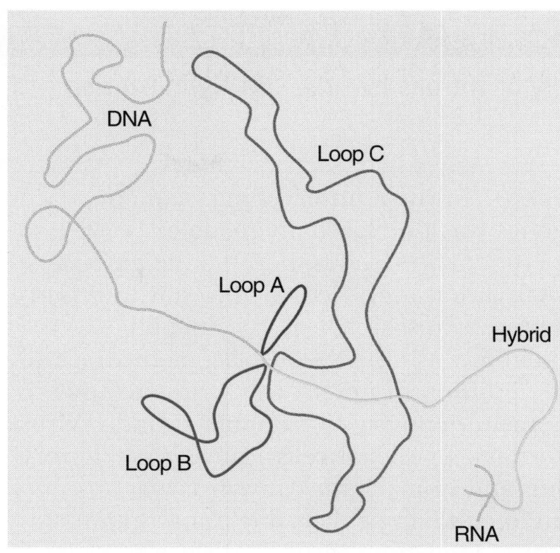

FIGURE 3.5 Hybridization between an adenovirus mRNA and its DNA identifies three loops corresponding to introns that are located at the beginning of the gene. Photo reproduced from Berget, S. M., Moore, C., and Sharp, P. A. *Proc. Natl. Acad. Sci. USA* 74 (1977): 3171–3175. Used with permission of Philip Sharp, Koch Institute for Integrative Cancer Research, Massachusetts Institute of Technology.

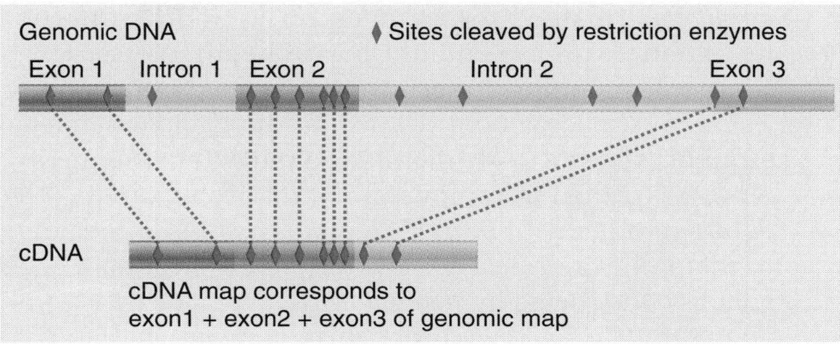

FIGURE 3.6 Comparison of the restriction maps of cDNA and genomic DNA for mouse β globin shows that the gene has two introns that are not present in the cDNA. The exons can be aligned exactly between cDNA and gene.

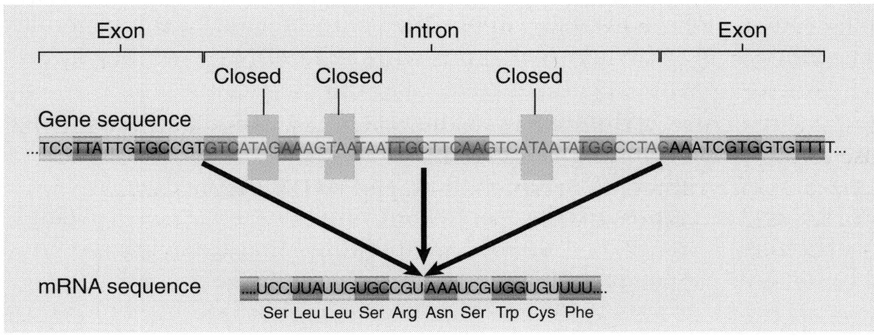

FIGURE 3.7 An intron is a sequence present in the gene but absent from the mRNA (here shown in terms of the cDNA sequence). The reading frame is indicated by the alternating open and shaded blocks; note that all three possible reading frames are blocked by termination codons in the intron.

in both number and size. (See Figure 3.10 for distributions of intron number variation among genes, and Figure 3.13 for distributions of intron size variation.)

Genes coding for proteins, rRNA, or tRNA may all have introns. Introns are also found in mitochondrial genes of plants, fungi, protists, and one metazoan (a sea anemone), and in chloroplast genes. Genes with introns have been found in every class of eukaryotes, archaea, bacteria, and bacteriophages, although they are extremely rare in prokaryotic genomes.

METHODS AND TECHNIQUES

The Discovery of Introns by DNA-RNA Hybridization

The discovery of introns (intervening sequences) in eukaryotic genes was unexpectedly made in 1977. Two different experimental methods, both taking advantage of DNA-RNA hybridization, led to this startling finding. One avenue of research was the mapping of the adenovirus genome using RNA displacement loops to determine the position of mRNA transcripts relative to the viral genome. Adenovirus is a double-stranded virus that infects human epithelial cells, causing respiratory cold symptoms and stomach flu. This virus was an early model system for studying eukaryotic transcription.

In 1977, two independent groups, the labs of Richard Roberts at Cold Spring Harbor Laboratories and Phillip Sharp at the Massachusetts Institute of Technology, used the method of *R loop mapping* to determine the regions of the adenovirus genome that are transcribed. Using this technique, double-stranded genomic DNA is mixed with individual mRNA transcripts under conditions that promote DNA-RNA hybrids between complementary sequences. Since the mRNA will anneal to only one strand of the DNA, the other strand of DNA is displaced, resulting in a loop (R loop) of single-stranded DNA visible under the electron microscope (see Figure 3.4). When mapping the R loops formed between adenovirus DNA and viral mRNA transcripts expressed late in the infection cycle, it was observed that the 5′ and 3′ ends of the RNA did not anneal with the DNA. This was not surprising for the 3′ end of the mRNA transcript, which was known to be modified following transcription by the addition of a polyadenylate tail; however, the discontinuity of the 150–200 nucleotides at the 5′ end was unexpected. Hybridization of the 5′ terminal mRNA and discrete single-stranded fragments of genomic DNA demonstrated that this mRNA sequence is actually composed of sequences derived from three distinct and dispersed regions of the adenovirus genome (see Figure 3.5). This was the first evidence of the post-transcriptional process of mRNA splicing, and Drs. Roberts and Sharp were recognized for their discovery of "interrupted genes" in 1993 with a Nobel Prize in Physiology or Medicine.

A second line of research, construction of physical maps of eukaryotic genes, revealed that interrupted genes are not unique to viral genomes. Prior to the advent of whole genome sequencing, physical maps of genomes were constructed by mapping restriction endonuclease cleavage sites in genomic DNA. Restriction endonucleases are bacterial enzymes that cleave double-stranded DNA in a site-specific fashion; different restriction endonucleases recognize and cleave distinct sequences (see *Section 30.2, Restriction Endonucleases Are a Key Tool in Manipulating DNA*). Whole genomic DNA can be fragmented by cleavage with restriction enzymes, and the resulting fragments are separated based on size through agarose gel electrophoresis. It is impossible to directly visualize the fragments for a specific gene because of the complexity and size of a typical eukaryotic genome; however, using the method of Southern blotting and DNA hybridization the fragments derived from a specific gene can be identified. The separated fragments of genomic DNA are denatured and transferred to a membrane in a process referred to as *Southern blotting* (named for its inventor Ed Southern). The membrane with the bound DNA is then immersed in a solution containing a specific radio-labeled DNA probe. The DNA probe will only anneal to the complementary sequences bound to the membrane, and when this membrane is washed to remove unbound probe, dried, and exposed to X-ray film, the fragments of DNA annealing to the probe are revealed. Typically, a physical map of a gene is constructed by using a cDNA (copied DNA from an mRNA transcript) to probe the genomic DNA. One of the first genes to be mapped in this fashion was the rabbit β-globin gene, performed by Jeffreys and Flavell in 1977. This led to the revelation that there is a sequence of 600 base pairs in the coding region of the genomic DNA that is not present in the cDNA. Another interrupted gene discovered at the same time is the chicken ovalbumin gene, which is also interrupted in the coding region. With time, it became clear that most eukaryotic genes are interrupted by intervening sequences that are transcribed and then removed by mRNA splicing. In 1978 Walter Gilbert proposed that these intervening sequences be referred to as *introns*.

Some interrupted genes have only one or a few introns. The globin genes provide an extensively studied example (see *Section 3.9, The Members of a Gene Family Have a Common Organization*). The two general classes of globin gene, α and β, share a common organization. They originated from an ancient gene duplication event and are described as **paralogous genes** or **paralogs**. The consistent structure of mammalian globin genes is evident from the "generic" globin gene presented in **FIGURE 3.8**.

▸ **paralogous genes (paralogs)** Genes that share a common ancestry due to gene duplication.

Introns are found at homologous positions (relative to the coding sequence) in all known active globin genes, including those of mammals, birds, and frogs. Although

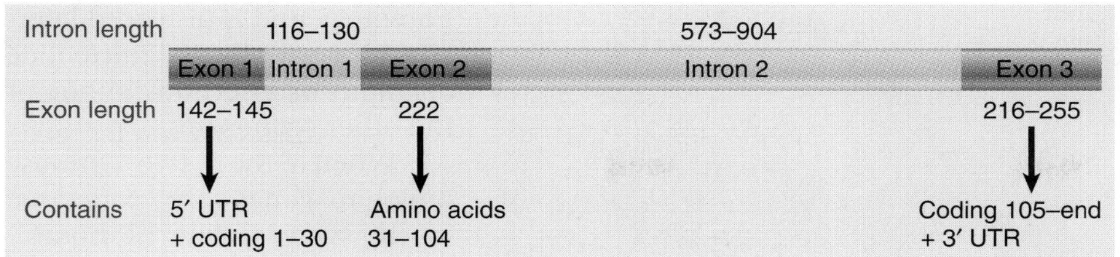

FIGURE 3.8 All functional globin genes have an interrupted structure with three exons. The lengths indicated in the figure apply to the mammalian β-globin genes.

intron lengths vary, the first intron is always fairly short, and the second usually is longer. Most of the variation in the lengths of different globin genes results from length variation in the second intron. For example, the second intron in the mouse α-globin gene is 150 bp of the total 850 bp of the gene, whereas the homologous intron in the mouse major β-globin gene is 585 bp of the total 1382 bp. The difference in length of the genes is much greater than that of their mRNAs (α-globin mRNA = 585 bases; β-globin mRNA = 620 bases).

The globin genes are examples of a general phenomenon: genes that share a common ancestry have similar organizations with conservation of the positions (of at least some) of the introns. Variations in the lengths of the genes are primarily due to intron length variation.

KEY CONCEPTS

- Introns can be detected by the presence of additional regions when genes are compared with their RNA products by restriction mapping or electron microscopy. The ultimate determination, though, is based on comparison of sequences.
- The positions of introns are usually conserved when homologous genes are compared between different organisms. The lengths of the corresponding introns may vary greatly.

CONCEPT AND REASONING CHECK

Why would the genes of viruses with DNA genomes that infect eukaryotic cells be expected to have introns?

3.4 Exon Sequences Are Conserved but Introns Vary

Is a single-copy structural gene completely unique among other genes in a genome? The answer depends on how "completely unique" is defined. Considered as a whole, the gene is unique, but its exons are often related to those of other genes. As a general rule, when two genes are related, the relationship between their exons is closer than the relationship between their introns. In an extreme case, the exons of two different genes may code for the same protein sequence whereas the introns are different. This situation can result from a duplication of a common ancestral gene followed by unique mutations in both copies, with mutations restricted in the exons by the need to code for a functional protein.

As we will see later when we consider the evolution of the gene, exons can be considered basic building blocks that may be assembled in various combinations. It is possible for a gene to have some exons related to those of another gene, with the remaining exons unrelated. Usually, in such cases, the introns are not related at all. Such homologies between genes may result from duplication and translocation of individual exons.

The homology between two genes can be plotted in the form of a dot matrix comparison, as in **FIGURE 3.9**. A dot is placed in each position that is identical in both genes. The dots form a solid line on the diagonal of the matrix if the two sequences are completely identical. If they are not identical, the line is broken by gaps that lack

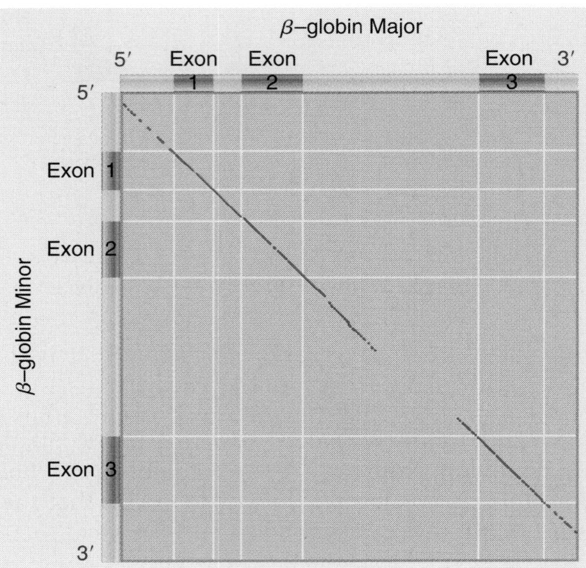

FIGURE 3.9 The sequences of the mouse β^{maj}- and β^{min}-globin genes are closely related in coding regions but differ in the flanking regions and long intron. Data provided by Philip Leder, Harvard Medical School.

homology and is displaced laterally or vertically by nucleotide deletions or insertions in one or the other sequence.

When the two mouse β-globin genes are compared in this way, a line of homology extends through the three exons and the small intron. The line disappears in the flanking regions and in the large intron. This is a typical pattern in related genes; the coding sequences and areas of introns adjacent to exons retain their similarity, but there is greater divergence in longer introns and the regions on either side of the gene.

The overall degree of divergence between two homologous exons in related genes corresponds to the differences between the proteins. It is mostly a result of base substitutions. In the translated regions, changes in exon sequences are constrained by selection against mutations that alter or destroy the function of the protein. Many of the preserved changes do not affect codon meanings because they change one codon into another for the same amino acid. Similarly, there are higher rates of changes in nontranslated regions of the gene (specifically, those that are transcribed to the 5′ leader and 3′ trailer of the mRNA).

In homologous introns, the pattern of divergence involves both changes in length (due to deletions and insertions) and base substitutions. Introns evolve much more rapidly than exons. When a gene is compared among different species, there are instances where its exons are homologous but its introns have diverged so much that very little homology is retained. Although mutations in certain intron sequences (branchpoint sequence, splicing junctions, and perhaps other sequences influencing splicing) will be subject to selection, most intron mutations are expected to be selectively neutral.

In general, mutations occur at the same rate in both exons and introns, but exon mutations are eliminated more effectively by selection. However, because of the low level of functional constraints, introns may more freely accumulate point substitutions and other changes. The empirical observation of faster evolution in introns implies that introns have fewer sequence-specific functions, but it is not clear whether their presence is required for normal gene function.

KEY CONCEPTS

- Comparisons of related genes in different species show that the sequences of the corresponding exons are usually conserved but the sequences of the introns are much less well related.
- Introns evolve much more rapidly than exons because of the lack of selective pressure to produce a protein with a useful sequence.

CONCEPT AND REASONING CHECK

In comparing two genes in different species, consider whether they are homologous given the following differences: (1) different exons; (2) the same exons, but completely different introns; (3) the same exons, and introns in the same positions but of different sizes.

3.5 Genes Show a Wide Distribution of Sizes Primarily Due to Intron Size and Number Variation

FIGURE 3.10 compares the organization of genes in a yeast, an insect, and mammals. In the yeast *Saccharomyces cerevisiae*, the majority of genes (~96%) are not interrupted, and those that have introns generally have three or fewer. There are virtually no *S. cerevisiae* genes with more than four exons.

In insects and mammals, the situation is reversed. Only a few genes have uninterrupted coding sequences (6% in mammals). Insect genes tend to have a fairly small number of exons—typically fewer than 10. Mammalian genes are split into more pieces, and some have more than 60 exons. About half of mammalian genes have more than 10 introns.

If we examine the effect of intron number variation on the total size of genes, we see in **FIGURE 3.11** that there is a striking difference between yeast and multicellular eukaryotes. The average yeast gene is 1.4 kb long, and very few are longer than 5 kb. The predominance of interrupted genes in high eukaryotes, however, means that the gene can be much larger than the sum total of the exon lengths. Only a small percentage of genes in flies or mammals are shorter than 2 kb, and most have lengths between 5 kb and 100 kb. The average human gene is 27 kb long (see Figure 5.11).

The switch from largely uninterrupted to largely interrupted genes seems to have occurred with the evolution of multicellular eukaryotes. In fungi other than yeasts, the majority of genes are interrupted, but they have a relatively small number of exons (<6) and are fairly short (<5 kb). In the fruit fly, genes are significantly larger. With this increase in the length of the gene due to the increased number of introns, the correlation between genome size and organism complexity becomes weak (see Figure 4.5).

Is there an evolutionary trend in intron length as well as intron number? Yes, organisms with larger genomes tend to have larger introns, whereas exon size does

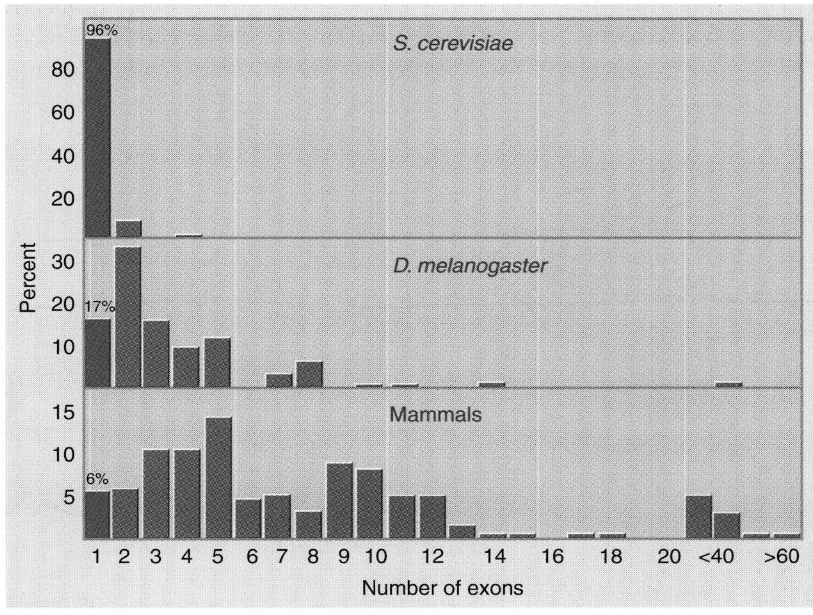

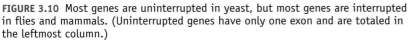

FIGURE 3.10 Most genes are uninterrupted in yeast, but most genes are interrupted in flies and mammals. (Uninterrupted genes have only one exon and are totaled in the leftmost column.)

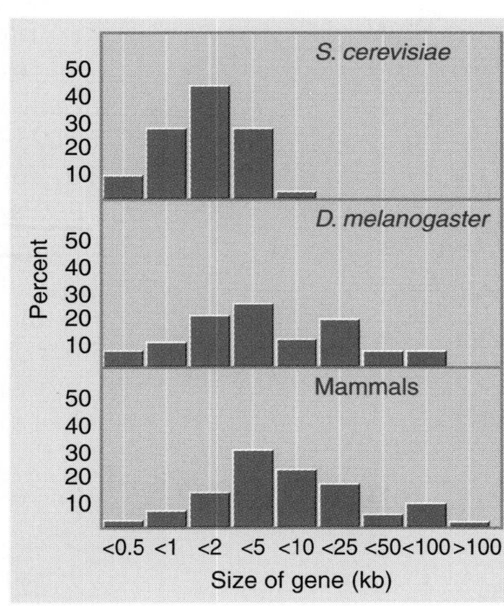

FIGURE 3.11 Yeast genes are short, but genes in flies and mammals have a dispersed distribution extending to very long sizes.

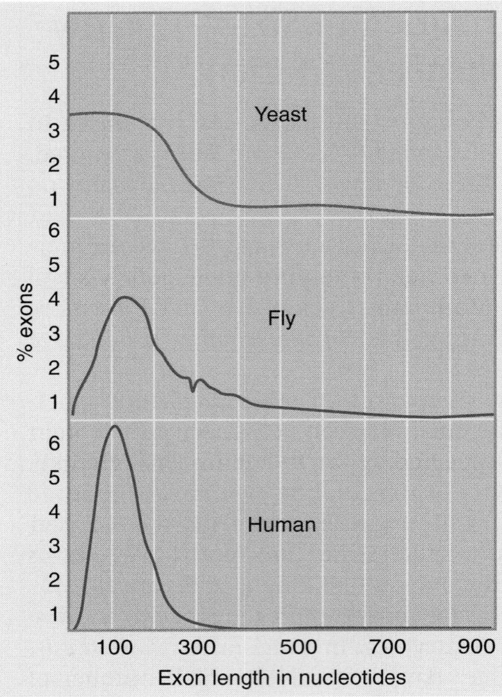

FIGURE 3.12 Exons coding for proteins usually are short.

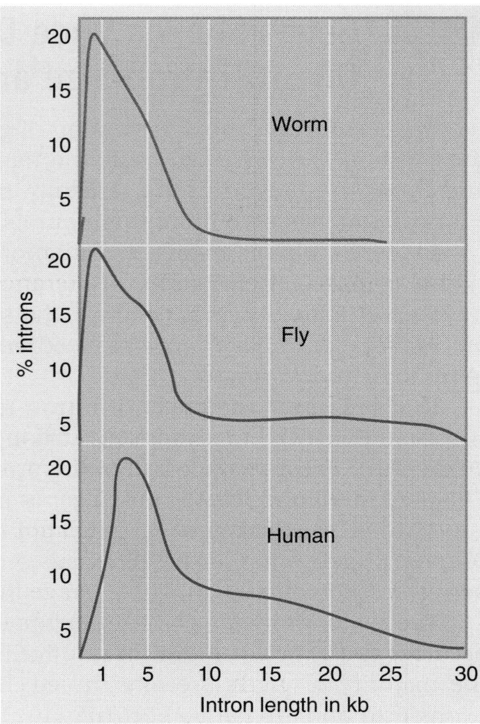

FIGURE 3.13 Introns range from very short to very long.

not tend to increase (**FIGURE 3.12**). In multicellular eukaryotes, the average exon codes for ~50 amino acids, and the general distribution is consistent with the hypothesis that genes have evolved by the gradual addition of exon units that code for short, functionally independent protein domains (see *Section 3.7, How Did Interrupted Genes Evolve?*). There is no significant difference in the average size of exons in different multicellular eukaryotes, although the size range is smaller in vertebrates for which there are few exons longer than 200 bp. In yeast, there are some longer exons that represent genes without introns.

FIGURE 3.13 shows that introns vary greatly in size among multicellular eukaryotes. In worms and flies, the average intron is not much longer than the average exon. There are no introns longer than about 10 kb in worms, but some fly introns can be up to 30 kb in length. In vertebrates, intron size range is much greater, from as small as about the same length as the average exon (<200 bp) to as much as 60 kb.

Very long genes are the result of very long introns, not the result of coding for longer products. There is no correlation between total gene size and total exon size in multicellular eukaryotes, nor is there a good correlation between gene size and number of exons. The size of a gene is, therefore, determined primarily by the lengths of its individual introns. In mammals and insects, the average gene length is approximately five times that of the total length of its exons.

KEY CONCEPTS

- Most genes are uninterrupted in yeasts, but are interrupted in higher eukaryotes.
- Exons are usually short, typically coding for ~100 amino acids.
- Introns are short in lower eukaryotes, but range up to several 10s of kb in length in higher eukaryotes.
- The overall length of a gene is determined largely by its introns.

CONCEPT AND REASONING CHECK

Would you expect there to be a general evolutionary trend for increase or decrease in the size of introns? Why or why not?

3.6 Some DNA Sequences Code for More Than One Polypeptide

Most structural genes consist of a sequence of DNA that codes for a single polypeptide, although the gene may include noncoding regions at both ends and introns within the coding region. However, there are some cases in which a single sequence of DNA codes for more than one polypeptide.

The simplest **overlapping gene** is one in which one gene is part of another. In other words, the first half (or second half) of a gene independently specifies a protein that is the first (or second) half of the protein specified by the full gene. This relationship is illustrated in **FIGURE 3.14**.

▸ **overlapping gene** A gene in which part of the sequence is found within part of the sequence of another gene.

A more complex form of overlapping gene occurs when the same sequence of DNA codes for two nonhomologous proteins because it has more than one reading frame. Usually, a coding DNA sequence is read in only one of the three potential reading frames. In some viral and mitochondrial genes, however, there is some overlap between two adjacent genes that are read in different reading frames, as illustrated in **FIGURE 3.15**. The length of overlap is usually short, so that most of the DNA sequence codes for a unique protein sequence.

In some genes, alternative proteins result from switches in the process of connecting the exons. A single gene may generate a variety of mRNA products that differ in their exon content. Often, this is because there are pairs of exons that are treated as mutually exclusive—one or the other is included in the **mature transcript**, but not

▸ **mature transcript** A modified RNA transcript. Modification may include the removal of intron sequences and alterations to the 5′ and 3′ ends.

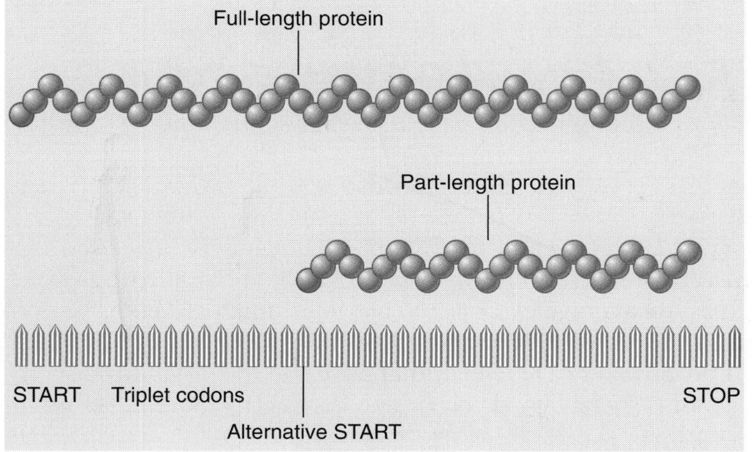

FIGURE 3.14 Two proteins can be generated from a single gene by starting (or terminating) expression at different points.

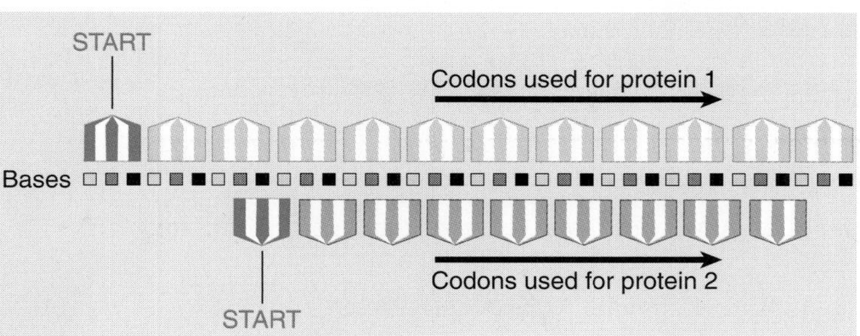

FIGURE 3.15 Two genes may share the same sequence by reading the DNA in different frames.

both. The alternative proteins have one part in common and one unique part. Such an example is presented in **FIGURE 3.16**. The 3′ half of the rat troponin T gene contains five exons, but only four are used to construct an individual mRNA. Three exons, W, X, and Z, are included in all mRNAs. However, in one **alternative splicing** pattern the α exon is included between X and Z, whereas in the other pattern it is replaced by the β exon. The α and β forms of troponin T therefore differ in the sequence of the amino acids between W and Z, depending on which of the alternative exons (α or β) is used.

▸ **alternative splicing** The production of different RNA products from a single product by changes in the usage of splicing junctions.

There are also cases of alternative splicing in which certain exons are optional; i.e., they may be included or spliced out, as in the example presented in **FIGURE 3.17**. There is a single **primary transcript** from the gene, but it can be spliced in either one of two ways. In the first, more standard way, the two introns are spliced out and the three exons are joined together. In the second way, the second exon is excluded along with the two introns as if a single large intron is spliced out. The two alternate proteins are the same at their ends, but one has an additional sequence in the middle. (Other types of combinations that are produced by alternative splicing are discussed in *Section 28.11, Alternative Splicing Involves Differential Use of Splice Junctions.*)

▸ **primary transcript** The original unmodified RNA product corresponding to a transcription unit.

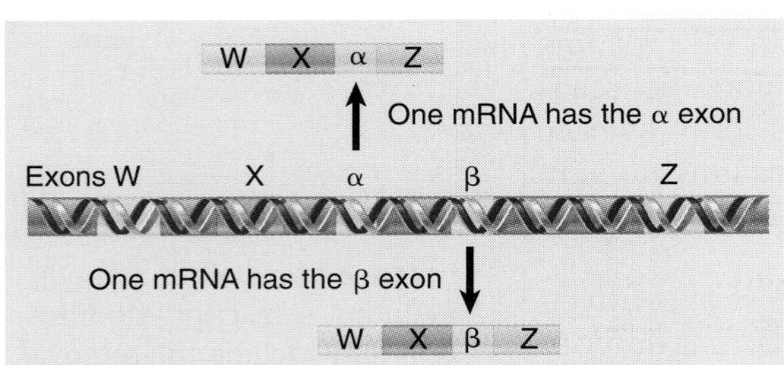

FIGURE 3.16 Alternative splicing generates the α and β variants of troponin T.

Sometimes the two alternative splicing patterns operate simultaneously, with a certain proportion of the primary mRNA transcripts being spliced in each way. However, in some genes the splicing patterns are alternatives that are expressed under different conditions, e.g., one in one cell type and one in another cell type.

So, alternative (or differential) splicing can generate different proteins with related sequences from a single stretch of DNA. It is curious that the multicellular eukaryotic genome is often extremely large, with long genes that are often widely dispersed along a chromosome, but at the same time there may be multiple products from

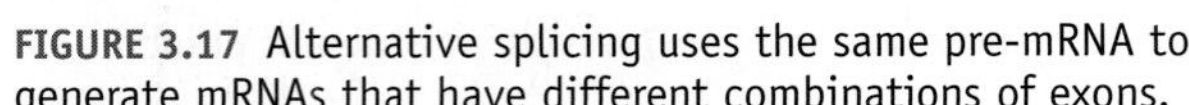

FIGURE 3.17 Alternative splicing uses the same pre-mRNA to generate mRNAs that have different combinations of exons.

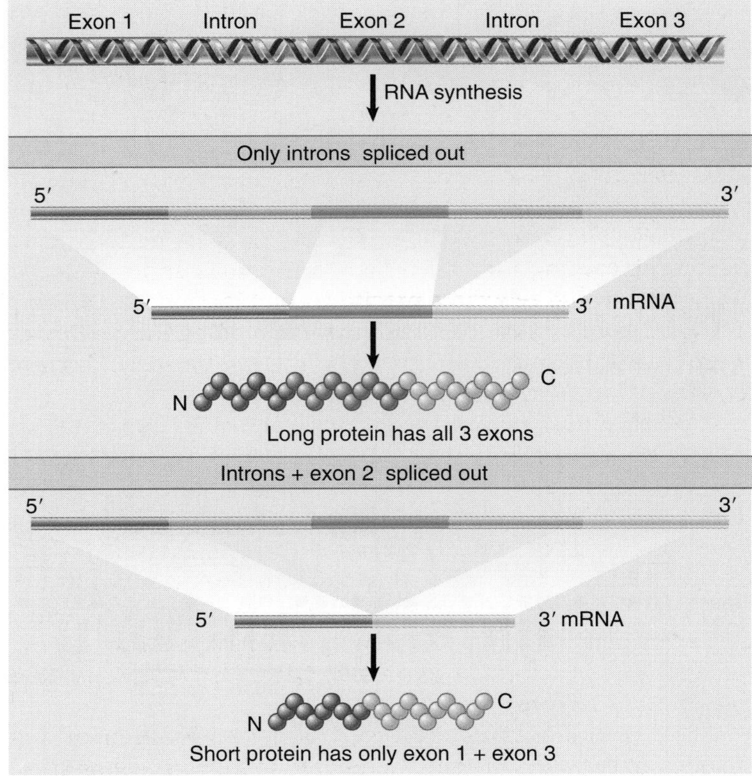

a single locus. Due to alternative splicing, there are ~15% more proteins than genes in flies and worms, but it is estimated that ~60% of human genes are alternatively spliced (see *Section 5.5, The Human Genome Has Fewer Genes Than Originally Expected*).

KEY CONCEPTS

- The use of alternative initiation or termination codons allows two proteins to be generated where one is equivalent to a fragment of the other.
- Nonhomologous protein sequences can be produced from the same sequence of DNA when it is read in different reading frames by two (overlapping) genes.
- Homologous proteins that differ by the presence or absence of certain regions can be generated by differential (alternative) splicing when certain exons are included or excluded. This may take the form of including or excluding individual exons or of choosing between alternative exons.

CONCEPT AND REASONING CHECKS

What are the advantages to having the same gene produce two or more related polypeptides? Why might different versions of the same polypeptide be necessary?

3.7 How Did Interrupted Genes Evolve?

The structure of many eukaryotic genes suggests a concept of the eukaryotic genome as a sea of mostly unique DNA sequences in which exon "islands" separated by intron "shallows" are strung out in individual gene "archipelagoes." What was the original form of genes that today are interrupted?

- The **"introns early" model** is the proposal that introns have always been an integral part of the gene. Genes originated as interrupted structures, and those without introns have lost them in the course of evolution.
- The **"introns late" model** is the proposal that the ancestral protein-coding sequences were uninterrupted and that introns were subsequently inserted into them.

"introns early" model The hypothesis that the earliest genes contained introns and some genes subsequently lost them.

"introns late" model The hypothesis that the earliest genes did not contain introns, and that introns were subsequently added to some genes.

To test these hypotheses, we must ask whether the difference between eukaryotic and prokaryotic genes can be accounted for by the acquisition of introns in the eukaryotes or by the loss of introns from the prokaryotes.

The "introns early" model suggests that the mosaic structure of genes is a remnant of an ancient approach of recombining gene segments to code for novel proteins. Suppose that an early cell had a number of separate protein-coding sequences: it is likely to have evolved by reshuffling different polypeptide units to construct new proteins. If a protein-coding unit must be a continuous series of codons, every such reshuffling event would require a precise recombination of DNA to place separate protein-coding units in sequence and in the same reading frame. However, if this combination doesn't produce a functional protein, the cell has been damaged because the original sequence of protein-coding units is lost.

The cell might survive, however, if some of the experimental recombination occurs in RNA transcripts, leaving the DNA intact. If a translocation event could place two protein-coding units in the same transcription unit, various RNA splicing experiments to combine the two proteins into a single polypeptide chain could be attempted. If some combinations are not successful, the original protein-coding units remain available for further trials. Also, this scenario does not require the two protein-coding units to be recombined precisely into a continuous coding sequence. There is evidence supporting this scenario: different genes have related exons, as if genes had been assembled by a process of **exon shuffling** (see *Section 3.8, Some Exons Can Be Equated with Protein Functions*).

exon shuffling The hypothesis that genes have evolved by the recombination of various exons coding for functional protein domains.

FIGURE 3.18 illustrates the result of a translocation of a random sequence that includes an exon into a gene. Compared to introns, exons are very small, so it is likely

FIGURE 3.18 An exon surrounded by flanking sequences that is translocated into an intron may be spliced into the RNA product.

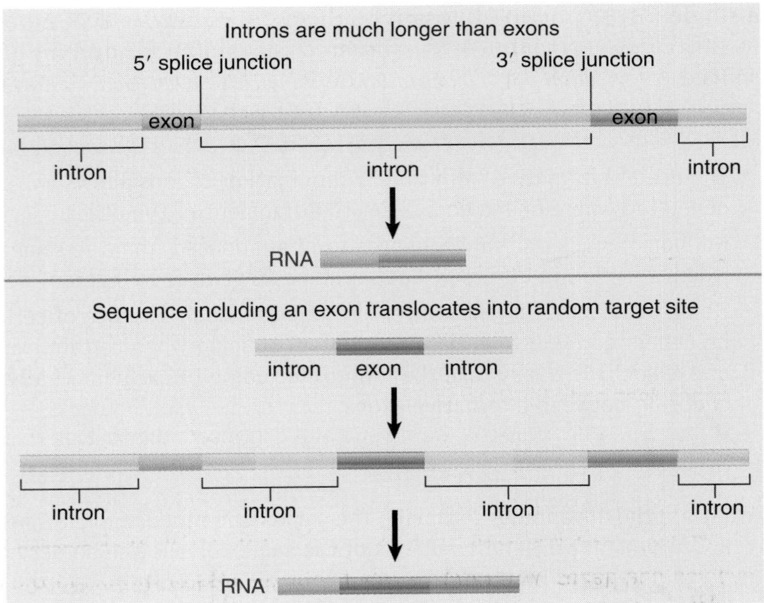

that the exon will insert within an intron and be flanked by functional 5′ and 3′ splice junctions. Splicing junctions are recognized in sequential pairs, so the splicing mechanism should recognize the 5′ splicing junction of the original intron and the 3′ splicing junction of the introduced exon, instead of the 3′ splice junction of the original intron. Similarly, the 5′ splicing junction of the new exon and the 3′ splicing junction of the original intron will be recognized as a pair, so the new exon will remain between the original two exons in the mature RNA transcript. As long as the new exon is in the same reading frame as the original exons, a new, longer protein will be produced. Exon shuffling events could have been responsible for generating new combinations of exons during evolution.

Some species have alternative forms of rRNA and tRNA genes, both with and without introns. For tRNAs, which all have the same general conformation, it seems unlikely that the two regions of the gene evolved independently because the two regions base pair to fold the molecule into a functional shape. In this case, the intron must have been inserted into a continuous gene.

There is also evidence that introns have been lost from some members of gene families. See *Section 3.9, The Members of a Gene Family Have a Common Organization*, for examples from the insulin and actin gene families. In the case of the actin gene family, it is sometimes not clear whether the presence of an intron in a member of the family indicates the ancestral state or an insertion event. Overall, current evidence suggests that genes originally had introns, but can evolve with both the loss and gain of introns.

Organelle genomes show the evolutionary connections between prokaryotes and eukaryotes. There are many general similarities between mitochondria or chloroplasts and certain bacteria, and thus it seems likely that the organelles originated by endosymbiosis in which a bacterial cell dwelled within the cytoplasm of a eukaryotic prototype. Although there are similarities to bacterial genetic processes—such as protein and RNA synthesis—some organelle genes possess introns and therefore resemble eukaryotic nuclear genes. Introns are found in several chloroplast genes, including some that are homologous to *E. coli* genes. This suggests that the endosymbiotic event occurred before introns were lost from the prokaryotic lineage.

Mitochondrial genome comparisons are particularly striking. The genes of yeast and mammalian mitochondria encode virtually identical proteins, in spite of a considerable difference in gene organization. Vertebrate mitochondrial genomes are very small and extremely compact, whereas yeast mitochondrial genomes are larger and have some complex interrupted genes. Which is the ancestral form? Yeast mitochondrial introns (and certain other introns) can be mobile—they are independent sequences that can splice out of the RNA and insert DNA copies elsewhere—which suggests that they may have arisen by insertions into the genome (see *Section 29.5, Some Group I Introns Code for Endonucleases That Sponsor Mobility* and *Section 29.6, Group II Introns Code for Reverse Transcriptases*).

KEY CONCEPTS

- A major evolutionary question is whether genes originated as sequences interrupted by introns or whether they were originally uninterrupted. More evidence supports the first ("introns early") hypothesis, although it appears that introns can be inserted into genes.
- Most protein-coding genes probably originated in an interrupted form, but interrupted genes that code for RNA may have originally been uninterrupted.
- A special class of introns are mobile and can insert themselves into genes.

CONCEPT AND REASONING CHECK

The "introns early" hypothesis is the proposal that cells that existed before eukaryotic cells evolved had genes with introns. Why might the loss of introns in the ancestral lineages of modern prokaryotes be advantageous, or at least not deleterious?

3.8 Some Exons Can Be Equated with Protein Functions

If proteins evolve by combining ancestral proteins that were originally separate, the accumulation of protein domains is likely to have occurred sequentially, with one exon added at a time. Are the current functions of these domains the same as their original functions? In other words, can we assign particular functions of current proteins to individual exons?

In some cases, there is a clear relationship between the structures of the gene and the protein. The example *par excellence* is provided by the immunoglobulin proteins, which are coded by genes in which every exon corresponds exactly to a known functional domain of the protein. **FIGURE 3.19** compares the structure of an immunoglobulin with its gene.

An immunoglobulin is a tetramer of two light chains and two heavy chains combined into a protein with several distinct domains. Light chains and heavy chains differ

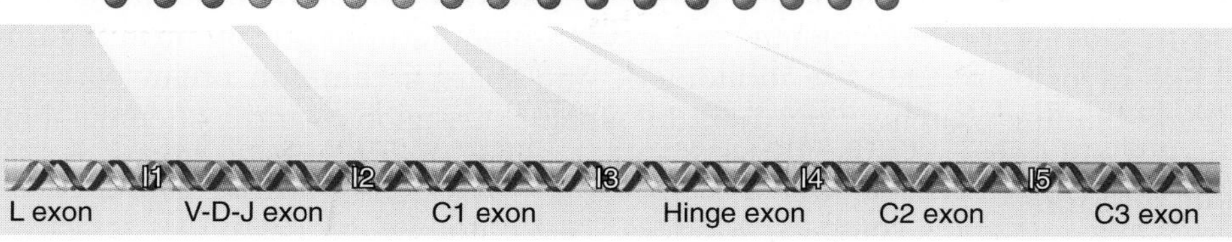

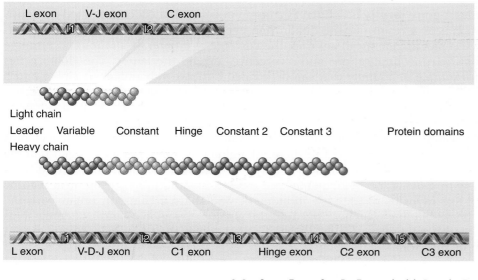

FIGURE 3.19 Immunoglobulin light chains and heavy chains are coded by genes whose structures (in their expressed forms) correspond with the distinct domains in the protein. Each protein domain corresponds to an exon; introns are numbered 1 to 5.

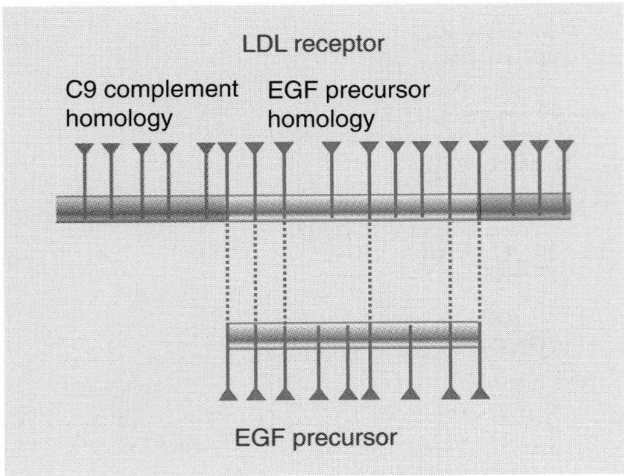

FIGURE 3.20 The LDL receptor gene consists of 18 exons, some of which are related to EGF precursor exons and some of which are related to the C9 blood complement gene. Triangles mark the positions of introns.

in structure, and there are several types of heavy chains. Each type of chain is produced from a gene that has a series of exons corresponding to the structural domains of the protein.

In many instances, some of the exons of a gene can be identified with particular functions. In secretory proteins, such as insulin, the first exon often specifies the signal sequence involved in membrane secretion.

The view that exons are the functional building blocks of genes is supported by cases in which two genes may share some related exons but also have unique exons. **FIGURE 3.20** summarizes the relationship between the human LDL (plasma low density lipoprotein) receptor and other proteins. The LDL receptor gene has a series of exons related to the exons of the EGF (epidermal growth factor) precursor gene, and another series of exons related to those of the blood protein complement factor C9. Apparently, the LDL receptor gene evolved by the assembly of modules suitable for its various functions. These modules are also used in different combinations in other proteins.

Exons tend to be fairly small (see Figure 5.11), around the size of the smallest polypeptide that can assume a stable folded structure (~20 to 40 residues). It may be that proteins were originally assembled from rather small modules. Each individual module need not correspond to a current function; several modules could have combined to generate a new functional unit. Larger genes tend to have more exons, which is consistent with the view that proteins acquire multiple functions by successively adding appropriate modules.

KEY CONCEPTS

- Many exons can be equated with coding for polypeptide sequences that have particular functions.
- Related exons are found in different genes.

CONCEPT AND REASONING CHECK

Would you expect two membrane-bound enzymes that catalyze different reactions to have (1) completely different exons, (2) some similar exons and some different exons, or (3) all similar exons? Why? (2).

3.9 The Members of a Gene Family Have a Common Organization

Many genes in a multicellular eukaryotic genome are related to others in the same genome. A **gene family** is defined as a group of genes that code for related or identical proteins as a result of gene duplication events. After the first duplication event, the two copies are identical, but then they diverge as different mutations accumulate in them. Further duplications and divergences extend the family. The globin genes are an example of a family that can be divided into two subfamilies (α globin and β globin), but all its members have the same basic structure and function. In some cases, we can find genes that are more distantly related, but still can be recognized as having common ancestry. Such a group of gene families is called a **superfamily**.

A fascinating case of evolutionary conservation is presented by the α and β globins and two other proteins related to them. Myoglobin is a monomeric oxygen-binding protein in animals. Its amino acid sequence suggests a common (though ancient) origin with globins. Leghemoglobins are oxygen-binding proteins found in legume plants; like myoglobin, they are monomeric and share a common origin with the other heme-binding proteins. Together, the globins, myoglobin, and leghemoglobins make up the globin superfamily—a set of gene families all descended from an ancient common ancestor.

- **gene family** A set of genes within a genome that code for related or identical proteins or RNAs. The members were derived by duplication of an ancestral gene followed by accumulation of changes in sequence between the copies. Most often the members are related but not identical.
- **superfamily** A set of genes all related by presumed descent from a common ancestor, but now showing considerable variation.

Both α- and β-globin genes have three exons and two introns at conserved positions (see Figure 3.8). The central exon represents the heme-binding domain of the globin chain. There is a single myoglobin gene in the human genome and its structure is essentially the same as that of the globin genes. The three-exon structure therefore predates the separation of the ancestral myoglobin and globin genes. Leghemoglobin genes contain three introns, the first and last of which are homologous to the two introns in the globin genes. This remarkable similarity suggests an exceedingly ancient origin for the interrupted structure of heme-binding proteins, as illustrated in **FIGURE 3.21**.

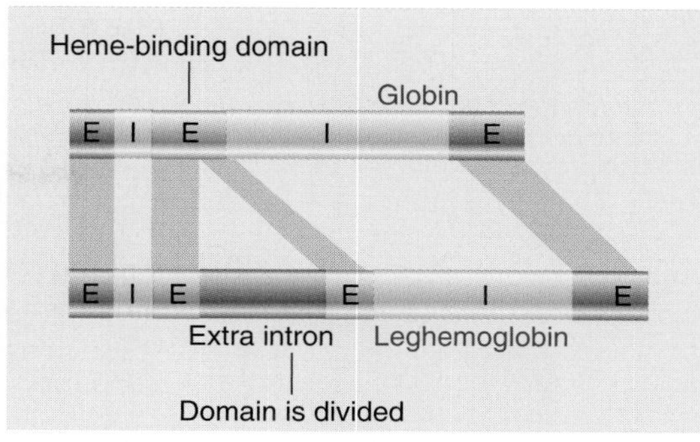

FIGURE 3.21 The exon structure of globin genes corresponds with protein function, but leghemoglobin has an extra intron in the central domain.

Orthologous genes, or **orthologs**, are genes that are **homologous** due to speciation; in other words, they are related genes in different species. Comparison of orthologs that differ in structure may provide information about their evolution. An example is insulin. Mammals and birds have only one gene for insulin, except for rodents, which have two. **FIGURE 3.22** illustrates the structures of these genes. We use the principle of parsimony in comparing the organization of orthologous genes by assuming that a common feature predates the evolutionary separation of the two species. In chickens, the single insulin gene has two introns; one of the two homologous rat genes has the same structure. The common structure implies that the ancestral insulin gene had two introns. However, since the second rat gene has only one intron, it must have evolved by a gene duplication in rodents that was followed by the precise removal of one intron from one of the homologs.

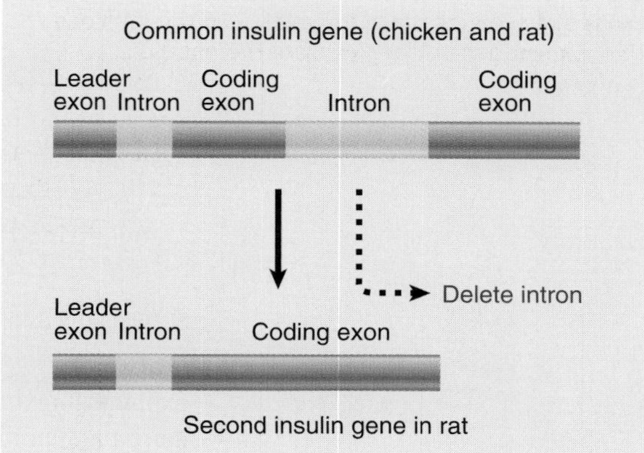

FIGURE 3.22 The rat insulin gene with one intron evolved by loss of an intron from an ancestor with two introns.

The organizations of some orthologs show extensive discrepancies between species. In these cases, there must have been removal or insertion of introns during evolution. A well-characterized example is that of the actin genes. The common features of actin genes are a nontranslated leader of <100 bases, a coding region of ~1200 bases, and a trailer of ~200 bases. Most actin genes have introns, and their positions can be aligned with regard to the coding sequence (except for a single intron sometimes found in the leader). **FIGURE 3.23** shows that almost every actin gene is different in its pattern of intron positions. Among all the genes being compared, introns occur at 19 different sites. However, the range of intron number per gene is zero to six. How did this situation arise? If we suppose that the primordial actin gene had introns, and that all current actin genes are related to it by loss of introns, different introns have been lost in each evolutionary branch. Probably some introns have been lost entirely, so the primordial gene could well have had

- **orthologous genes (orthologs)** Related genes in different species.
- **homologous genes (homologs)** Related genes in the same species, such as alleles on homologous chromosomes, or multiple genes in the same genome sharing common ancestry.

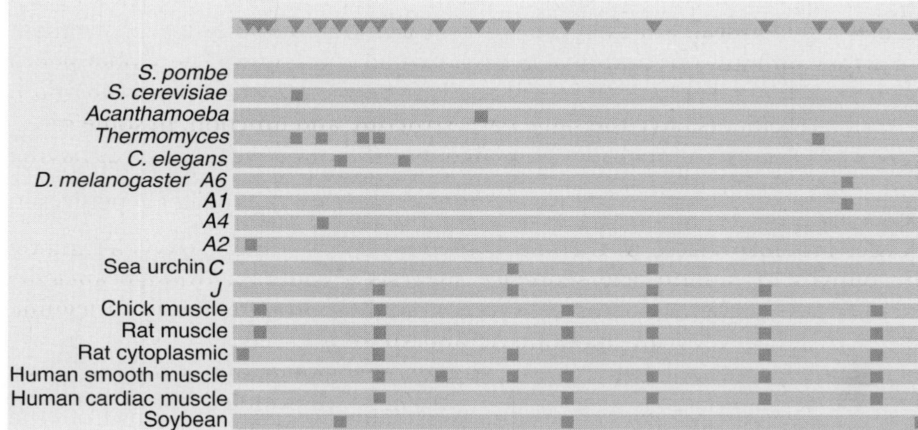

FIGURE 3.23 Actin genes vary widely in their organization. The sites of introns are indicated in dark green.

20 introns or more. The alternative is to suppose that a process of intron insertion continued independently in the different lines of evolution.

The relationship between individual exons and functional protein domains is somewhat erratic. In some cases there is a clear 1:1 relationship; in others no pattern can be discerned. One possibility is that the removal of introns has fused previously adjacent exons. This means that the intron must have been precisely removed, without changing the integrity of the coding region. An alternative is that some introns arose by insertion into an exon coding for a single domain. Together with the variations that we see in exon placement in cases such as the actin genes, the conclusion is that intron positions can evolve.

The correspondence of at least some exons with protein domains, and the presence of related exons in different proteins, leaves no doubt that the duplication and juxtaposition of exons have played important roles in evolution. It is possible that the number of ancestral exons—from which all proteins have been derived by duplication, variation, and recombination—could be relatively small, perhaps as little as a few thousand. The idea that exons are the building blocks of new genes is consistent with the "introns early" model.

KEY CONCEPTS

- A common feature in a set of genes is assumed to identify a property that preceded their separation in evolution.
- All globin genes have a common form of organization with three exons and two introns, suggesting that they are descended from a single ancestral gene.

CONCEPT AND REASONING CHECK

"Parsimony" is the least complex explanation for a given observation. For example, the explanation that modern actin genes evolved from an ancestral gene with many exons is more parsimonious than the explanation that modern actin genes gained introns independently. Evaluate the statement, "Selecting the most parsimonious explanation guarantees the choice of the correct explanation."

3.10 Summary

Virtually all eukaryotic genomes contain interrupted genes. The proportion of interrupted genes is low in yeasts, but few genes are uninterrupted in multicellular eukaryotes. Introns are found in all classes of eukaryotic genes. The structure of the interrupted gene is the same in all tissues: exons are spliced together in RNA in the same order as they are found in DNA, and the introns, which usually have no coding function, are removed from RNA by splicing.

Some genes are expressed by alternative splicing patterns, in which a particular sequence is removed as an intron in some situations, but retained as an exon in others. Often, when the organizations of orthologous genes are compared, the positions of introns are conserved. Intron sequences vary—and may even be unrelated—although exon sequences are clearly related. The conservation of exon sequence and position can be used to isolate related genes in different species.

The size of a gene is determined primarily by the lengths of its introns. Large introns probably first appeared in the multicellular eukaryotes, and there is an evolutionary trend toward increased intron (and consequently, gene) size. The range of gene sizes in mammals is generally from 1 to 100 kb, but it is possible to have even larger genes.

Some genes share only some of their exons with other genes, suggesting that they have been assembled by addition of exons representing functional "modular units" of the protein. Such modular exons may have been incorporated into a variety of different proteins. The hypothesis that genes have been assembled by accumulation of exons implies that introns were present in the genes of proto-eukaryotes. Some of the relationships between orthologous genes can be explained by loss of introns from the primordial genes, with different introns being lost in different lines of descent.

CHAPTER QUESTIONS

1. Mutations that affect splicing:
- **A.** are usually inconsequential.
- **B.** are usually deleterious.
- **C.** are always deleterious.
- **D.** increase gene expression.

2. In which group of organisms are most genes interrupted (have introns)?
- **A.** bacteria
- **B.** yeast
- **C.** animals
- **D.** more than one of the above

3. The length of genes in a gene family often varies; this variation is most often determined by:
- **A.** length of the 5′ untranslated region.
- **B.** number and size of exons.
- **C.** number and size of introns.
- **D.** length of the 3′ untranslated region.

4. In general, for two genes that are related, which shows the closest relationship?
- **A.** exons
- **B.** introns
- **C.** both introns and exons
- **D.** the promoter region and first exon

5. What percentage of genes is uninterrupted in animals?
- **A.** <5%
- **B.** <10%
- **C.** ~20%
- **D.** ~33%

6. In general, as genome size increases in different organisms:
- **A.** exons also increase in size.
- **B.** introns increase in size.
- **C.** both exons and introns increase in size.
- **D.** 5′ and 3′ untranslated regions increase in size.

7. Which group of organisms has the longest average intron size?
- **A.** bacteria
- **B.** flies
- **C.** worms
- **D.** insects

8. The presence of introns in a number of chloroplast genes that have homologies with *E. coli* genes suggests that:
- **A.** the endosymbiotic event occurred before introns were lost from the prokaryotic line.
- **B.** the endosymbiotic event occurred after introns were lost from the prokaryotic line.
- **C.** the endosymbiotic event occurred before introns were lost from the eukaryotic line.
- **D.** the endosymbiotic event occurred after introns were lost from the eukaryotic line.

9. Genes that code for related or identical proteins in an organism are part of:
   A. a gene family.
   B. a superfamily.
   C. homologous genes.
   D. orthologous genes.
10. A common feature of protein structure is that exon-intron boundaries often are:
   A. randomly distributed in the protein.
   B. located in active sites of the protein.
   C. located in the hydrophobic center of the protein.
   D. located at the surface of the protein.

KEY TERMS

alternative splicing
cDNA
exon
exon shuffling
gene family
homologous genes (homologs)
interrupted gene
intron
"introns early" model
"introns late" model
mature transcript
orthologous genes (orthologs)
overlapping gene
paralogous genes (paralogs)
primary transcript
RNA splicing
superfamily

FURTHER READING

Black, D. L. (2003). Mechanisms of alternative pre-messenger RNA splicing. *Annu. Rev. Biochem.* 72, 291–336.
An in-depth review of specific examples of alternative splicing, including the *Drosophila* sex determination system.

Faustino, N. A. and Cooper, T. A. (2003). Pre-mRNA splicing and human disease. *Genes Dev.* 17, 419–437.
A review of the mechanisms by which problems with alternative splicing of pre-mRNA can result in human diseases.

Ponting, C. P. and Russell, R. R. (2002). The natural history of protein domains. *Annu. Rev. Biophys. Biomol.* 31, 45–71.
A discussion of the origin and evolution of protein domains, including the suggestion that domains be classified in a hierarchical taxonomic system.

Reddy, A. S. N. (2007). Alternative splicing of pre-messenger RNAs in plants in the genomic era. *Annu. Rev. Plant Biol.* 58, 267–294.
A review of unexpected recent discoveries of alternative splicing of plant genes.

Rodríguez-Trelles, F., Tarrío, R., and Ayala, F. J. (2006). Origins and evolution of spliceosomal introns. *Annu. Rev. Genet.* 40, 47–76.
A review of classic and newer hypotheses about the origins of introns.

4

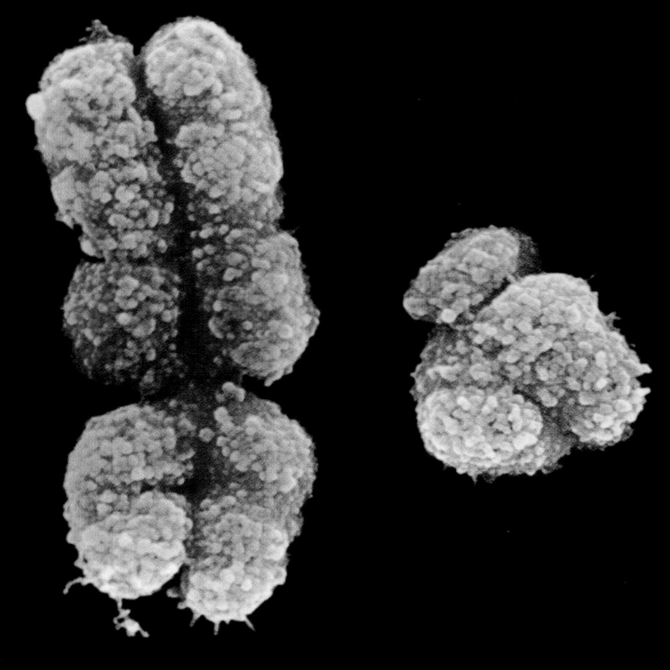

Scanning electron micrograph (SEM) of human X and Y chromosomes in metaphase (35,000×). Most of the content of the genome is found in chromosomal DNA, though some organelles contain their own genomes. © Biophoto Associates/Photo Researchers, Inc.

The Content of the Genome

CHAPTER OUTLINE

4.1 Introduction

One key question about the genome is how many genes it contains. An even more fundamental question, however, is "what is a gene?" Clearly, genes cannot solely be defined as a sequence of DNA that codes for polypeptide because many genes code for multiple polypeptides, and many code for RNAs that serve other functions. Given the variety of RNA functions and the complexities of gene expression, it seems prudent to focus on the gene as a unit of transcription. However, large areas of chromosomes previously thought to be devoid of genes now appear to be extensively transcribed, so at present the definition of a "gene" is a moving target.

We can attempt to characterize both the total number of genes and the number of protein-coding genes at four levels, which correspond to successive stages in gene expression:

- The **genome** is the complete set of genes of an organism. Ultimately it is defined by the complete DNA sequence, although as a practical matter it may not be possible to identify every gene unequivocally solely on the basis of sequence.
- The **transcriptome** is the complete set of genes expressed under particular conditions. It is defined in terms of the set of RNA molecules that is present and can refer to a single cell type or to any more complex assembly of cells, up to the complete organism. Because some genes generate multiple mRNAs, the transcriptome is likely to be larger than the number of genes defined directly in the genome. The transcriptome includes noncoding RNAs (such as tRNAs, rRNAs, and microRNAs or miRNAs—described in *Section 13.9, Eukaryotes Contain Regulator RNAs*) as well as mRNAs.
- The **proteome** is the complete set of polypeptides coded by the whole genome or produced in any particular cell or tissue. It should correspond to the mRNAs in the transcriptome, although there can be differences of detail reflecting changes in the relative abundance or stabilities of mRNAs and proteins. There may also be post-translational modifications to proteins that allow more than one protein to be produced from a single transcript (this is called *protein splicing*; see *Section 29.11, Protein Splicing Is Autocatalytic*).
- Proteins may function independently or as part of multiprotein or multimolecular complexes, such as holoenzymes and metabolic pathways where enzymes are clustered together. The RNA polymerase holoenzyme (see *Section 11.5, Bacterial RNA Polymerase Consists of the Core Enzyme and Sigma Factor*) and the spliceosome (see *Section 28.8, Five snRNPs Form the Spliceosome*) are two examples. If we could identify all protein–protein interactions, we could define the total number of independent complexes of proteins. This is sometimes referred to as the **interactome**.

▸ **genome** The complete set of sequences in the genetic material of an organism. It includes the sequence of each chromosome plus any DNA in organelles.

▸ **transcriptome** The complete set of RNAs present in a cell, tissue, or organism. Its complexity is due mostly to mRNAs, but it also includes noncoding RNAs.

▸ **proteome** The complete set of proteins that is expressed by the entire genome. Sometimes the term is used to describe the complement of proteins expressed by a cell at any one time.

▸ **interactome** The complete set of protein complexes/protein-protein interactions present in a cell, tissue, or organism.

The maximum number of protein-coding genes in the genome can be identified directly by characterizing open reading frames. Large-scale mapping of this nature is complicated by the fact that interrupted genes may consist of many separated open reading frames. We do not necessarily have information about the functions of the protein products—or indeed proof that they are expressed at all—so this approach is restricted to defining the *potential* of the genome. However, a strong presumption exists that any conserved open reading frame is likely to be expressed.

Another approach is to define the number of genes directly in terms of the transcriptome (by directly identifying all the RNAs) or proteome (by directly identifying all the polypeptides). This gives an assurance that we are dealing with *bona fide* genes that are expressed under known circumstances. It allows us to ask how many genes are expressed in a particular tissue or cell type, what variation exists in the relative levels of expression, and how many of the genes expressed in one particular cell are unique to that cell or are also expressed elsewhere. In addition, analysis of the transcriptome can reveal how many different mRNAs (e.g., mRNAs containing different combinations of exons) are generated from a given gene.

Concerning the types of genes, we may ask whether a particular gene is *essential*: what happens to a null mutant? If a null mutation is lethal, or the organism has a visible defect, we may conclude that the gene is essential or at least conveys a selective advantage. However, some genes can be deleted without apparent effect on the phenotype. Are these genes really dispensable, or does a selective disadvantage result from the absence of the gene, perhaps in other circumstances, or over longer periods of time? In some cases, the absence of these genes could be compensated for by a redundant mechanism, such as a gene duplication, providing a backup for an essential function.

CONCEPT AND REASONING CHECK

Explain why, in animals, the transcriptome and proteome of a cell are generally smaller than its genome, while the transcriptome and proteome of the whole organism are generally larger than its genome.

4.2 Genomes Can Be Mapped at Several Levels of Resolution

Defining the contents of a genome essentially means making a map. We can think about mapping genes and genomes at several levels of resolution:

- A **genetic** (or **linkage**) **map** identifies the distance between loci in terms of recombination frequencies. It is limited by its reliance on the occurrence of recombination of variable markers that are either visible (such as phenotypic traits) or can be visualized (such as by electrophoresis). For example, a **linkage map** can be constructed by measuring recombination between sites in genomic DNA that have sequence variations generating differences in the susceptibility to cleavage by certain restriction enzymes. Because such variations are common, such a map can be prepared for any organism irrespective of the occurrence of mutants. Because recombination frequencies can be distorted relative to the physical distance between sites, a linkage map does not accurately represent physical distances along the genetic material.
- A **restriction map** is constructed by cleaving DNA into fragments with restriction enzymes and measuring the physical distances, in terms of the length of DNA (determined by migration on an electrophoretic gel), between the sites of cleavage. A restriction map does not intrinsically identify sites of interest, such as a gene. For it to be related to the genetic map, mutations have to be characterized in terms of their effects upon the restriction sites. Large changes in the genome can be recognized because they affect the sizes or numbers of restriction fragments. Point mutations are more difficult to detect because they change only a single restriction site or lie between restriction sites and are undetectable.
- The ultimate genomic map is the sequence of the DNA. From the sequence, we can identify genes and the distances between them. By analyzing the protein-coding potential of a sequence of the DNA, we can hypothesize about its function. The basic assumption here is that natural selection prevents the accumulation of damaging mutations in sequences that code for proteins. Reversing the argument, we may assume that an intact coding sequence is likely to be used to generate a protein.

▸ **genetic map** *See* **linkage map**.

▸ **linkage map** A map of the positions of loci or other genetic markers on a chromosome obtained by measuring recombination frequencies between markers.

▸ **restriction map** A linear array of restriction sites on DNA, determined by clearing the DNA with various restriction endonucleases or by scanning a known sequence for restriction sites.

By comparing a wild-type DNA sequence with that of a mutant allele, we can determine the nature of a mutation and its exact site of occurrence. This provides a way to determine the relationship between the genetic map (based entirely on sites of mutation) and the physical map (based on, or even comprising, the sequence of DNA).

Similar techniques are used to identify and sequence genes and to map the genome, although there is of course a difference of scale. In each case, the principle is to characterize a series of overlapping fragments of DNA that can be connected into a continuous map. The crucial feature is that each segment is related to the next segment on the map by the overlap between them, so that we can be sure no segments are missing. This principle is applied both at the level of ordering large fragments into a map and in connecting the sequences that make up the fragments.

KEY CONCEPTS

- Linkage maps are based on the frequency of recombination between genetic markers; restriction maps are based on the physical distances between markers.
- Molecular characterization of mutations can be used to reconcile linkage maps with physical maps.

CONCEPT AND REASONING CHECK

If the same chromosome is sampled from different populations of the same organism, the physical map can be identical but the linkage map may be slightly different. Why?

4.3 Individual Genomes Show Extensive Variation

The original Mendelian view of the genome classified alleles as either wild-type or mutant. Subsequently we recognized the existence of multiple alleles, each with a different effect on the phenotype. In some cases it may not even be appropriate to define any one allele as "wild-type."

▸ **polymorphism** The simultaneous occurrence in the population of alleles showing variations at a given position.

The coexistence of multiple alleles at a locus is called genetic **polymorphism**. Any site at which multiple alleles exist as stable components of the population is by definition polymorphic. A locus is usually defined as polymorphic if two or more alleles are present at a frequency of >1% in the population.

Although not evident from the phenotype, the wild-type may itself be polymorphic. Multiple versions of the wild-type allele may be distinguished by differences in sequence that do not affect their function, and that therefore do not produce phenotypic variants. A population may have extensive polymorphism at the level of genotype. Many different sequence variants may exist at a given locus; some of them are evident because they affect the phenotype, but others are hidden because they have no visible effect.

So there may be a continuum of changes at a locus, including those that change DNA sequence but do not change protein sequence, those that change protein sequence without changing function, those that create proteins with different activities, and those that create mutant proteins that are nonfunctional.

▸ **single nucleotide polymorphism (SNP)** A polymorphism (variation in sequence between individuals) caused by a change in a single nucleotide. This is responsible for most of the genetic variation between individuals.

A change in a single nucleotide when alleles are compared is called a **single nucleotide polymorphism (SNP)**. On average, one occurs every ~1330 bases in the human genome. Defined by their SNPs, every human being is unique. SNPs can be detected by various means, ranging from direct comparisons of sequences to mass spectroscopy or biochemical methods that produce differences based on sequence variations in a defined region.

One aim of genetic mapping is to obtain a catalog of common variants. The observed frequency of SNPs per genome predicts that, over the human population as a whole (taking the sum of all human genomes of all living individuals), there should be >10 million SNPs that occur at a frequency of >1%. Already >1 million have been identified.

Some polymorphisms in the genome can be detected by comparing the restriction maps of different individuals. The criterion is a change in the pattern of fragments produced by cleavage with a restriction enzyme. **FIGURE 4.1** shows that when a target

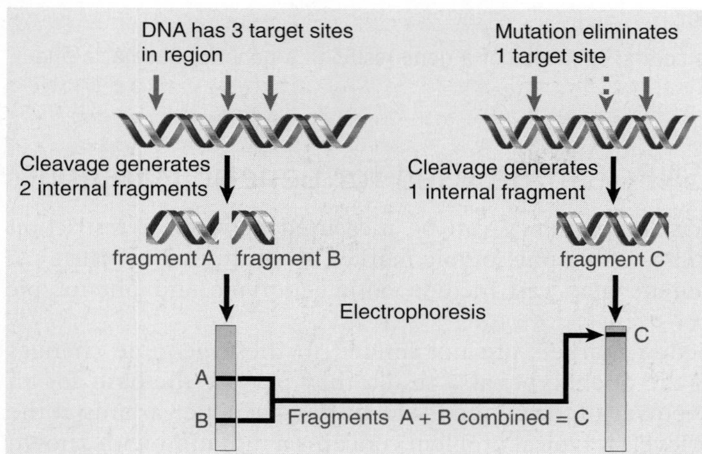
DNA has 3 target sites in region
Mutation eliminates 1 target site
Cleavage generates 2 internal fragments
Cleavage generates 1 internal fragment
fragment A
fragment B
fragment C
Electrophoresis
A
B
C
Fragments A + B combined = C

FIGURE 4.1 A point mutation that affects a restriction site is detected by a difference in restriction fragments.

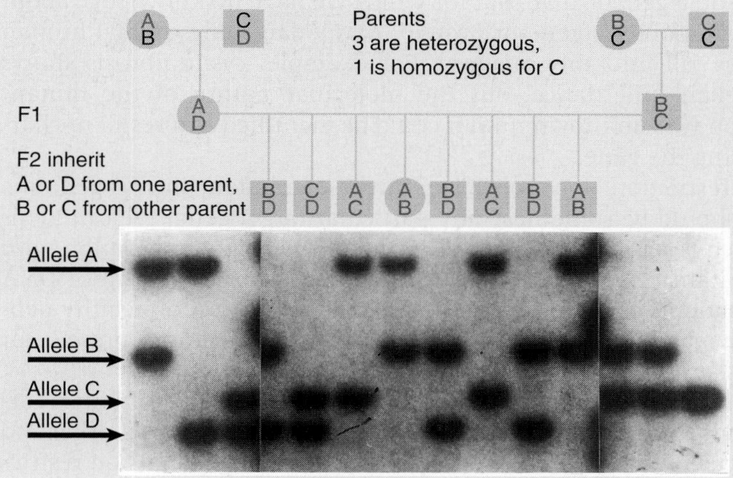
Parents
3 are heterozygous,
1 is homozygous for C
F1
F2 inherit
A or D from one parent,
B or C from other parent
Allele A
Allele B
Allele C
Allele D

FIGURE 4.2 Restriction site polymorphisms are inherited according to Mendelian rules. As visualized by Southern blotting and probe hybridization, four alleles for a restriction marker are found in all possible pairwise combinations and segregate independently at each generation. Photo courtesy of Ray White, Ernest Gallo Clinic and Research Center, University of California, San Francisco.

site is present in the genome of one individual and absent from another, the extra cleavage in the first genome will generate two fragments corresponding to the single fragment in the second genome. A difference in restriction maps between two individuals is called a **restriction fragment length polymorphism (RFLP)** or "riflip." Basically, an RFLP is an SNP that is located in the target site for a restriction enzyme. It can be used as a genetic marker in exactly the same way as any other marker. Instead of examining some feature of the phenotype, we directly assess the genotype, as revealed by the restriction map. FIGURE 4.2 shows a pedigree of a restriction polymorphism followed through three generations. It displays Mendelian segregation at the level of DNA marker fragments.

‣ **restriction fragment length polymorphism (RFLP)** Inherited differences in sites for restriction enzymes (for example, caused by base changes in the target site) that result in differences in the lengths of the fragments produced by cleavage with the relevant restriction enzyme. They are used for genetic mapping to link the genome directly to a conventional genetic marker.

The restriction map is independent of gene function; as a result, an RFLP at this level can be detected irrespective of whether the sequence change affects the phenotype. Probably very few of the RFLPs in a genome actually affect the phenotype. Most involve sequence changes that have no effect on the production of proteins (for example, because they lie between genes).

KEY CONCEPTS

- Polymorphism may be detected at the phenotypic level when a sequence affects gene function, at the restriction fragment level when it affects a restriction enzyme target site, and at the sequence level by direct analysis of DNA.
- The alleles of a gene show extensive polymorphism at the sequence level, but many sequence changes do not affect function.

CONCEPT AND REASONING CHECK

Why might a mutation in the coding sequence of a gene result in a genetic, but not a phenotypic, polymorphism?

4.4 RFLPs and SNPs Can Be Used for Genetic Mapping

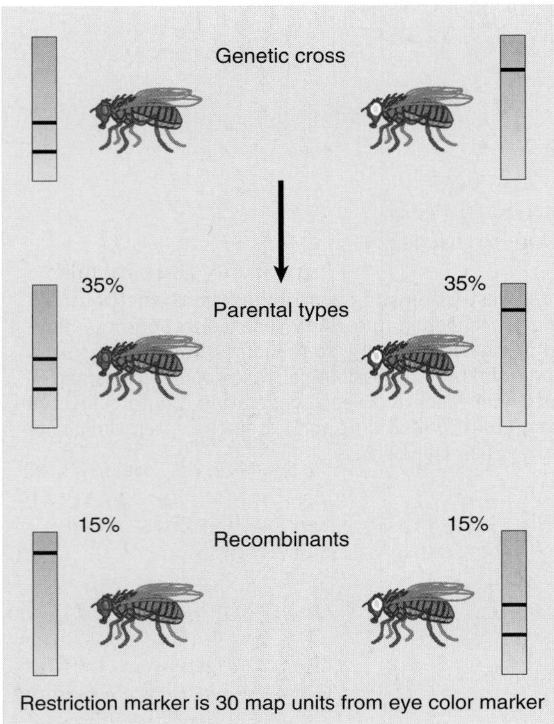

FIGURE 4.3 A restriction polymorphism can be used as a genetic marker to measure recombination distance from a phenotypic marker (such as eye color). The figure simplifies the situation by showing only the DNA bands corresponding to the allele of one genome in a diploid.

Recombination frequency can be measured between a restriction marker and a visible phenotypic marker, as illustrated in **FIGURE 4.3**. Thus a genetic map can include both genotypic and phenotypic markers.

Restriction markers are not limited to those genome changes that affect the phenotype; as a result, they provide the basis for an extremely powerful technique for identifying genetic variants at the molecular level. A typical problem concerns a mutation with known effects on the phenotype, where the relevant genetic locus can be placed on a genetic map, but for which we have no knowledge about the corresponding gene or protein. Many damaging or fatal human diseases fall into this category. For example, cystic fibrosis shows Mendelian inheritance, but the molecular nature of the mutant function was unknown until it could be identified as a result of characterizing the gene.

If restriction polymorphisms occur at random in the genome, there should be some near any particular target gene. We can identify such restriction markers by virtue of their tight association with the mutant phenotype. If we compare the restriction map of DNA from patients suffering from a disease with the DNA of healthy people, we may find that a particular restriction site is always present (or always absent) from the patients.

A hypothetical example is shown in **FIGURE 4.4**. This situation corresponds to finding 100% linkage between the restriction marker and the locus producing the phenotype. It would imply that the restriction marker lies so close to the mutant gene that it is never separated from it by recombination; it may in fact be the same mutation.

FIGURE 4.4 If a restriction marker is associated with a phenotypic characteristic, the restriction site must be located near the gene responsible for the phenotype. The mutation changing the band that is common in healthy people into the band that is common in patients is very closely linked to the disease gene.

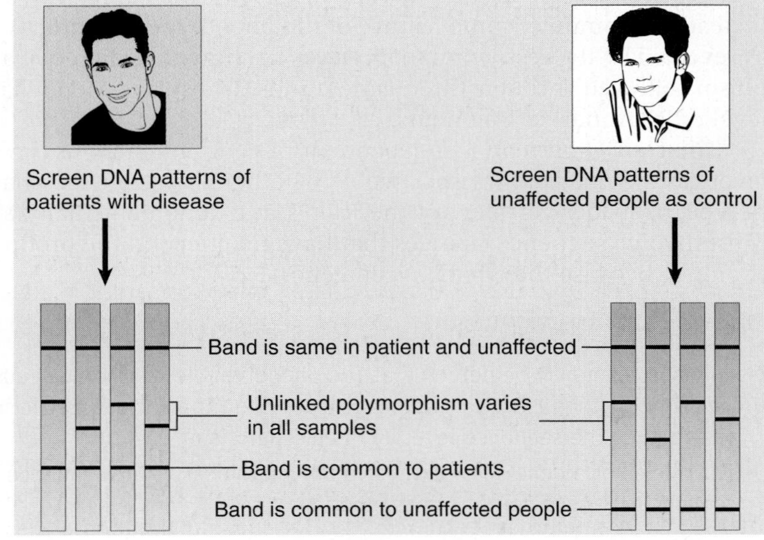

The identification of such a marker has two important consequences:

- It may offer a diagnostic procedure for detecting the disease. Some of the human diseases that have a known inheritance pattern but are ill defined in molecular terms cannot be easily diagnosed. If a restriction marker is closely linked to the phenotype, then its presence can be used to diagnose the probability of carrying the disease allele.
- It may lead to isolation of the gene. The restriction marker must lie relatively near the gene on the genetic map if the two loci rarely or never recombine. "Relatively near" in genetic terms can be a substantial distance in terms of base pairs of DNA; nonetheless, it provides a starting point from which we can proceed along the DNA to the gene itself.

The frequent occurrence of SNPs in the human genome makes them useful for genetic mapping. From the 1.5×10^6 SNPs that have already been identified, there is on average an SNP every 1 to 2 kb. This should allow rapid localization of new disease genes by locating them between the nearest SNPs.

On the same principle, RFLP mapping has been in use for some time. Once an RFLP has been assigned to a linkage group (i.e., a chromosome), it can be placed on the genetic map. RFLP mapping of both the human and mouse genomes has led to the construction of linkage maps for both. Any site with an unknown position can be tested for linkage to these sites, and by this means can be rapidly placed on the map. There are fewer RFLPs than SNPs, so the resolution of the RFLP map is in principle more limited.

The large proportion of polymorphic sites means that every individual has a unique constellation of SNPs and RFLPs. The particular combination of sites found in a specific region is called a **haplotype**, a genotype in miniature. Haplotype was originally introduced as a concept to describe the genetic constitution of the major histocompatibility locus, a region specifying proteins of importance in the immune system (see *Chapter 22, Immune Diversity*). The term now has been extended to describe the particular combination of alleles, restriction sites, or any other genetic markers present in some defined area of the genome. Using SNPs, a detailed haplotype map of the human genome has been made; this enables disease-causing genes to be mapped more easily.

The existence of RFLPs provides the basis for a technique to establish unequivocal parent–offspring relationships. In cases for which parentage is in doubt, a comparison of the RFLP map in a suitable chromosome region between potential parents and child allows absolute assignment of the relationship. The use of DNA restriction analysis to identify individuals has been called **DNA fingerprinting**. Analysis of especially variable "minisatellite" sequences is used in mapping the human genome (see *Section 6.12, Minisatellites Are Useful for Genetic Mapping*).

▸ **haplotype** The particular combination of alleles in a defined region of some chromosome—in effect, the genotype in miniature. Originally used to describe combinations of major histocompatibility complex (MHC) alleles, it now may be used to describe particular combinations of RFLPs, SNPs, or other markers.

▸ **DNA fingerprinting** A technique for analyzing the differences between individuals of the fragments generated by using restriction enzymes to cleave regions that contain short repeated sequences or by PCR. The lengths of the repeated regions are unique to every individual, and as a result the presence of a particular subset in any two individuals can be used to define their common inheritance (e.g., a parent–child relationship).

KEY CONCEPT

- RFLPs and SNPs can be the basis for linkage maps and are useful for establishing parent–offspring relationships.

CONCEPT AND REASONING CHECK

Describe how the position of a disease allele can be narrowed down to a specific chromosomal region by linkage analysis using RFLPs from members of a large family in which the disease allele is segregating.

4.5 Why Are Some Genomes So Large?

The total amount of DNA in the (haploid) genome is a characteristic of each living species known as its **C-value**. There is enormous variation in the range of C-values, from $<10^6$ bp for a mycoplasma to $>10^{11}$ bp for some plants and amphibians.

▸ **C-value** The total amount of DNA in the genome (per haploid set of chromosomes).

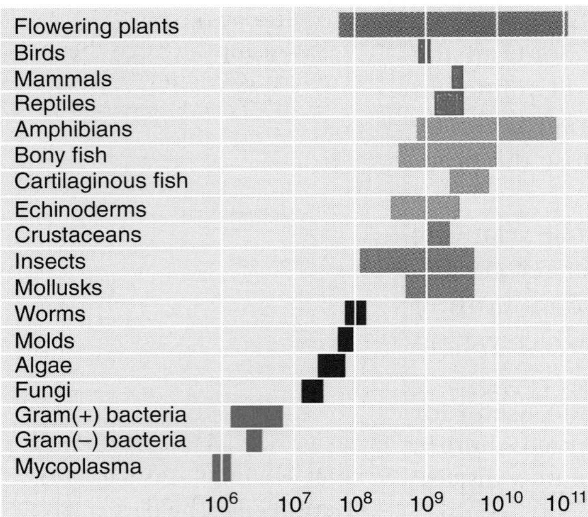

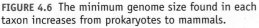

FIGURE 4.5 DNA content of the haploid genome increases with morphological complexity of lower eukaryotes, but varies extensively within some groups of higher eukaryotes. The range of DNA values within each group is indicated by the shaded area.

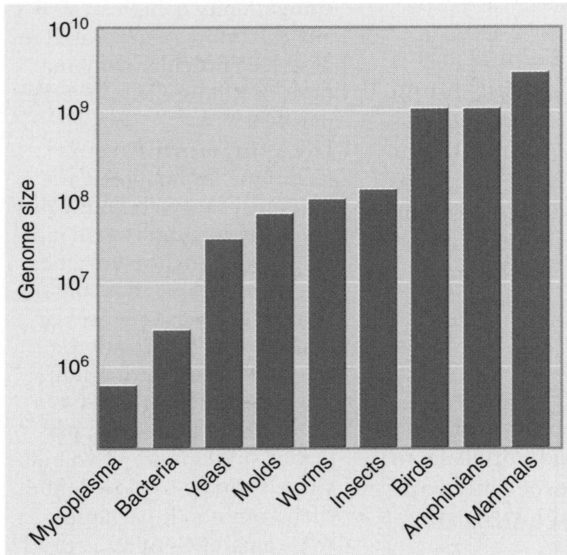

FIGURE 4.6 The minimum genome size found in each taxon increases from prokaryotes to mammals.

Phylum	Species	Genome (bp)
Algae	*Pyrenomas salina*	6.6×10^5
Mycoplasma	*M. pneumoniae*	1.0×10^6
Bacterium	*E. coli*	4.2×10^6
Yeast	*S. cerevisiae*	1.3×10^7
Slime mold	*D. discoideum*	5.4×10^7
Nematode	*C. elegans*	8.0×10^7
Insect	*D. melanogaster*	1.8×10^8
Bird	*G. domesticus*	1.2×10^9
Amphibian	*X. laevis*	3.1×10^9
Mammal	*H. sapiens*	3.3×10^9

FIGURE 4.7 The genome sizes of some common experimental organisms.

FIGURE 4.5 summarizes the range of C-values found in different taxa (groups of organisms classified together). There is an increase in the minimum genome size found in each group as the complexity increases. Although C-values are greater in the multicellular eukaryotes, we do see some wide variations in the genome sizes within some taxa.

Plotting the minimum amount of DNA required for a member of each group suggests in **FIGURE 4.6** that an increase in genome size is required to make more complex prokaryotes and lower eukaryotes.

Mycoplasma are the smallest prokaryotes and have genomes only ~3× the size of a large bacteriophage. More typical bacterial genome sizes start at ~2×10^6 bp. Unicellular eukaryotes (whose lifestyles may resemble those of prokaryotes) get by with genomes that are small, too, although they are larger than those of the bacteria. Being eukaryotic *per se* does not imply a vast increase in genome size; a yeast may have a genome size of ~1.3×10^7 bp, which is only about twice the size of an average bacterial genome.

A further twofold increase in genome size is adequate to support the slime mold *Dictyostelium discoideum*, which is able to live in either unicellular or multicellular modes. Another increase in complexity is necessary to produce the first fully multicellular organisms; the nematode worm *Caenorhabditis elegans* has a DNA content of 8×10^7 bp.

We also can see the steady increase in genome size with complexity in the listing in **FIGURE 4.7** of some of the most commonly analyzed organisms. It is necessary to increase the genome size in order to make insects, birds or amphibians, and mammals. After this point, though, there is no clear relationship between genome size and morphological complexity of the organism.

We know that genes are much larger than the sequences needed to code for polypeptides, because exons may comprise only a small part of the total length of a gene. This explains why there is much more DNA than is needed to provide reading frames for all the proteins of the organism. Large parts of an interrupted gene may not code for polypeptide. In addition, there also may be significant lengths of DNA between genes. So it is not possible to deduce from the overall size of the genome anything about the number of genes.

The **C-value paradox** refers to the lack of correlation between genome size and genetic complexity. There are some extremely curious observations about relative genome size, such as that the toad *Xenopus* and humans have genomes of essentially the same size. In some taxa there are extremely large variations in DNA content between organisms that do not vary much in complexity (see Figure 4.5). (This is especially marked in insects, amphibians, and plants, but does not occur in birds, reptiles, and mammals, which all show little variation within the group—with an ~2× range of genome sizes.) A cricket has a genome 11× the size of that of a fruit fly. In amphibians, the smallest genomes are $<10^9$ bp, whereas the largest are ~10^{11} bp. There is unlikely to be a large difference in the number of genes needed to specify these amphibians. The extent to which this variation is selectively neutral or subject to natural selection is not yet fully understood.

KEY CONCEPTS

- There is no clear correlation between genome size and genetic complexity.
- There is an increase in the minimum genome size required to make organisms of increasing complexity.
- There are wide variations in the genome sizes of organisms within many taxa.

CONCEPT AND REASONING CHECK

What are some advantages and disadvantages of a large genome? A small genome?

▸ **C-value paradox** The lack of relationship between the DNA content (C-value) of an organism and its coding potential.

4.6 Eukaryotic Genomes Contain Both Nonrepetitive and Repetitive DNA Sequences

The general nature of the eukaryotic genome can be assessed by the kinetics of reassociation of denatured DNA. This technique was used extensively before large-scale DNA sequencing became possible.

Reassociation kinetics identifies two general types of genomic sequences:

- **Nonrepetitive DNA** consists of sequences that are unique: there is only one copy in a haploid genome.
- **Repetitive DNA** consists of sequences that are present in more than one copy in each genome.

Repetitive DNA often is divided into two general types:

- Moderately repetitive DNA consists of relatively short sequences that are repeated typically 10–1000× in the genome. The sequences are dispersed throughout the genome and are responsible for the high degree of secondary structure formation in pre-mRNA, when inverted repeats in the introns pair to form duplex regions.
- Highly repetitive DNA consists of very short sequences (typically <100 bp) that are present many thousands of times in the genome, often organized as long regions of tandem repeats (see *Section 6.9, Satellite DNAs Often Lie in Heterochromatin*). Neither class is found in exons.

The proportion of the genome occupied by nonrepetitive DNA varies widely among taxa. **FIGURE 4.8** summarizes the genome organization of some representative organisms. Prokaryotes contain primarily nonrepetitive DNA. For lower eukaryotes, most of the DNA is nonrepetitive; <20% falls into one or more moderately repetitive components. In animal cells, up to half of the DNA often is occupied by moderately and highly repetitive components. In plants and amphibians, the moderately and highly repetitive components may account for up to 80% of the genome, so that the nonrepetitive DNA is reduced to a minority component.

A significant part of the moderately repetitive DNA consists of **transposons**, short sequences of DNA (~1 kb) that have the ability to move to new locations in the genome and/or to make additional copies of themselves (see *Chapter 21, Transposons, Retroviruses, and Retroposons*). In some higher eukaryotic genomes they may even occupy more than half of the genome (see *Section 5.5, The Human Genome Has Fewer Genes Than Expected*).

Transposons are sometimes viewed as **selfish DNA**, which is defined as sequences that propagate themselves within a genome without contributing to the development and functioning of the

▸ **nonrepetitive DNA** DNA that is unique (present only once) in a genome.

▸ **repetitive DNA** DNA that is present in many (related or identical) copies in a genome.

▸ **transposon** A DNA sequence able to insert itself (or a copy of itself) at a new location in the genome without having any sequence relationship with the target locus.

▸ **selfish DNA** DNA sequences that do not contribute to the phenotype of the organism but have self-perpetuation within the genome as their sole function.

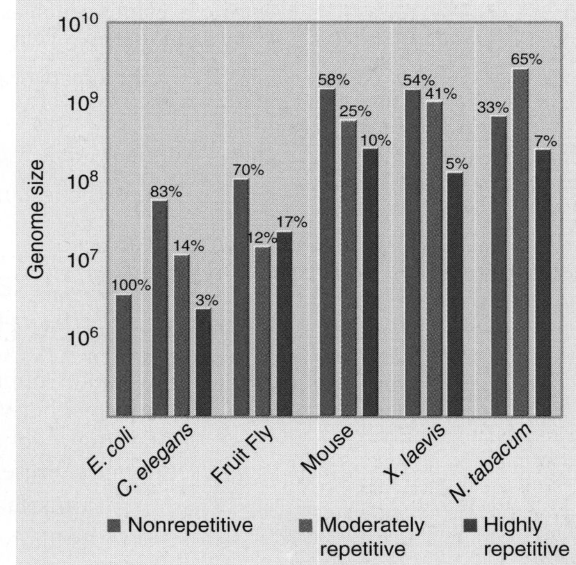

FIGURE 4.8 The proportions of different sequence components vary in eukaryotic genomes. The absolute content of nonrepetitive DNA increases with genome size but reaches a plateau at ~2×10^9 bp.

organism. Transposons may cause genome rearrangements, and these could confer selective advantages. It is fair to say, though, that we do not really understand why selective forces do not act against transposons becoming such a large proportion of the genome. It may be that they are selectively neutral as long as they do not interrupt or delete coding or regulatory regions. However, many organisms actively suppress transposition, as in some cases deleterious chromosome breakages result (see Figure 21.18). Another term that is used to describe the apparent excess of DNA in some genomes is "junk" DNA, meaning genomic sequences without any apparent function. Of course, it is likely that there is a balance in the genome between the generation of new sequences and the elimination of unwanted sequences, and some proportion of DNA that apparently lacks function may be in the process of being eliminated.

The length of the nonrepetitive DNA component tends to increase with overall genome size as we proceed up to a total genome size ~3×10^9 (characteristic of mammals). Further increases in genome size, however, generally reflect an increase in the amount and proportion of the repetitive components, so that it is rare for an organism to have a nonrepetitive DNA component >2×10^9. The nonrepetitive DNA content of genomes therefore accords better with our sense of the relative complexity of the organism. *E. coli* has 4.2×10^6 bp of nonrepetitive DNA, *C. elegans* has an order of magnitude more (6.6×10^7 bp), *D. melanogaster* has ~10^8 bp, and mammals yet another order of magnitude more at ~2×10^9 bp.

What type of DNA corresponds to polypeptide-coding genes? Reassociation kinetics typically shows that mRNA is derived from nonrepetitive DNA. The amount of nonrepetitive DNA is therefore a better indication of the coding potential than is the C-value. (However, more detailed analysis based on genomic sequences shows that many exons have related sequences in other exons [see *Section 3.5, Exon Sequences Are Conserved but Introns Vary*]. Such exons evolve by a duplication to give copies that initially are identical, but that then diverge in sequence during evolution.)

KEY CONCEPTS

- The kinetics of DNA reassociation after a genome has been denatured distinguish sequences by their frequency of repetition in the genome.
- Polypeptides are generally coded by sequences in nonrepetitive DNA.
- Larger genomes within a taxon do not contain more genes, but have large amounts of repetitive DNA.
- A large part of moderately repetitive DNA may be made up of transposons.

CONCEPT AND REASONING CHECK

What is the resolution of the C-value paradox?

4.7 Eukaryotic Protein-Coding Genes Can Be Identified by the Conservation of Exons

Some major approaches to identifying eukaryotic protein-coding genes are based on the contrast between the conservation of exons and the variation of introns. In a region containing a gene whose function has been conserved among a range of species, the sequence representing the polypeptide should have two distinctive properties:

- it must have an open reading frame, and
- it is likely to have a related (orthologous) sequence in other species.

These features can be used to isolate genes.

Suppose we know by linkage analysis that a gene influencing a particular trait is located in a given chromosomal region. If we lack knowledge about the nature of the gene product, how are we to identify the gene in a region that may be, for example, >1 Mb in size?

An approach that has proved successful with some genes of medical importance is to screen relatively short fragments from the region for the two properties expected of a conserved gene. First we seek to identify fragments that cross-hybridize with the genomes of other species, and then we examine these fragments for open reading frames.

The first criterion is applied by performing a **zoo blot**. We use short fragments from the region as labeled probes to test for homologous DNA from a variety of species by Southern blotting (a technique for transferring DNA fragments from an electrophoretic gel to a filter membrane, followed by hybridization of a probe to detect the complementary or near-complementary sequence). If we find hybridizing fragments in several species related to that of the probe (which is usually prepared from human DNA), the probe becomes a candidate for an exon of the gene.

▸ **zoo blot** The use of Southern blotting to test the ability of a DNA probe from one species to hybridize with the DNA from the genomes of a variety of other species.

The candidates are sequenced, and if they contain open reading frames they are used to isolate surrounding genomic regions. If these appear to be part of an exon, they can then be used to identify the entire gene, to isolate the corresponding cDNA (DNA reverse transcribed from the mRNA) or mRNA itself, and ultimately to identify the protein.

When a human disease is caused by a change in a known protein, the gene that is responsible can be identified because it codes for the protein, and its responsibility for the disease can be confirmed by showing that it has mutations in the DNA of patients but not in DNA of unaffected individuals. However, in many cases we do not know the cause of a disease at the molecular level, and it is necessary to identify the gene without any information about its protein product.

▸ **chromosomal walk** A technique for locating a gene by using the most closely linked markers as a probe for a genetic library.

The basic criterion for identifying a gene involved in a human disease is to show that in every patient with the disease the gene has a mutation that is not present in normal DNA. However, the extensive polymorphism between individual genomes means that we may find many changes when we compare patient DNA with normal DNA. Before the sequencing of the human genome, genetic linkage could be used to identify a region containing a disease gene, but the region could contain many candidate genes. For a very large gene, with introns spread over a long distance of the genome, it was difficult to identify the critical mutations in patients. The availability of high-resolution SNP maps and of the genome sequence now makes it much easier to pinpoint a smaller region containing the gene in which sequences of normal and patient DNA can be directly compared.

An example of the process by which a disease gene can be tracked down is provided by the gene responsible for Duchenne muscular dystrophy (DMD), a degenerative disorder of muscle that is X-linked and affects 1 in 3500 human males. The steps in identifying the gene are summarized in **FIGURE 4.9**.

Linkage analysis localized the DMD locus to chromosomal band Xp21. Patients with the disease often have chromosomal rearrangements involving this band. By comparing the ability of X-linked DNA probes to hybridize with DNA from patients with normal DNA, cloned fragments were obtained that correspond to the region that was rearranged or deleted in patients' DNA.

Once some DNA in the general vicinity of the target gene has been obtained, it is possible to "walk" along the chromosome until the gene is reached. A **chromosomal walk** was used to construct a restriction map of the region on either side of the probe, which covered a region of >100 kb. Analysis of the DNA from a series of patients identified large deletions in this region that extended in either direction. The most telling deletion is one that is contained entirely within the region, because this delineates a segment that must be important in gene function and indicates that the gene—or at least part of it—lies in this region.

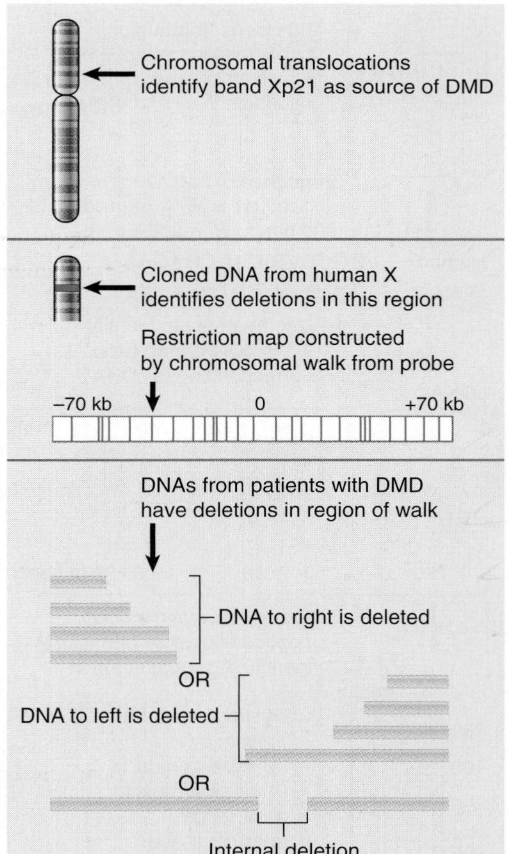

FIGURE 4.9 The gene involved in Duchenne muscular dystrophy was tracked down by chromosome mapping and "walking" to a region in which deletions can be identified with the occurrence of the disease.

After identifying the region of the gene, its exons and introns needed to be identified. A zoo blot identified fragments that cross-hybridize with the mouse X chromosome and with other mammalian DNAs. As summarized in **FIGURE 4.10**, these were scrutinized for open reading frames and the sequences typical of exon–intron junctions. Fragments that met these criteria were used as probes to identify homologous sequences in a cDNA library prepared from muscle mRNA.

The cDNA corresponding to the gene identifies an unusually large (14 kb) mRNA. Hybridization back to the genome shows that the mRNA is coded by >60 exons, which are spread over ~2000 kb of DNA. This makes DMD the longest gene identified.

The gene codes for a protein of ~500 kD called *dystrophin*, which is a component of muscle and is present in rather low amounts. All patients with the disease have deletions at this locus and lack (or have defective) dystrophin.

Muscle also has the distinction of having the largest known protein, titin, with almost 27,000 amino acids. The *titin* gene has the largest number of exons (178) and the longest single exon in the human genome (17,000 bp).

▸ **exon trapping** Inserting a genomic fragment into a vector whose function depends on the provision of splicing junctions by the fragment.

Another technique that allows genomic fragments to be scanned rapidly for the presence of exons is called **exon trapping**. **FIGURE 4.11** shows that it starts with a vector that contains a strong promoter and has a single intron between two exons. When this vector is transfected into cells, its transcription generates large amounts of an RNA containing the sequences of the two exons. A restriction site lies within the intron and is used to insert genomic fragments from a region of interest. If a fragment does not contain an exon, there is no change in the splicing pattern, and the RNA contains only the same sequences as the parental vector. If the genomic fragment contains an exon flanked by two partial intron sequences, though, the splicing sites on either side of this exon are recognized and the sequence of the exon is inserted into the RNA between the two exons

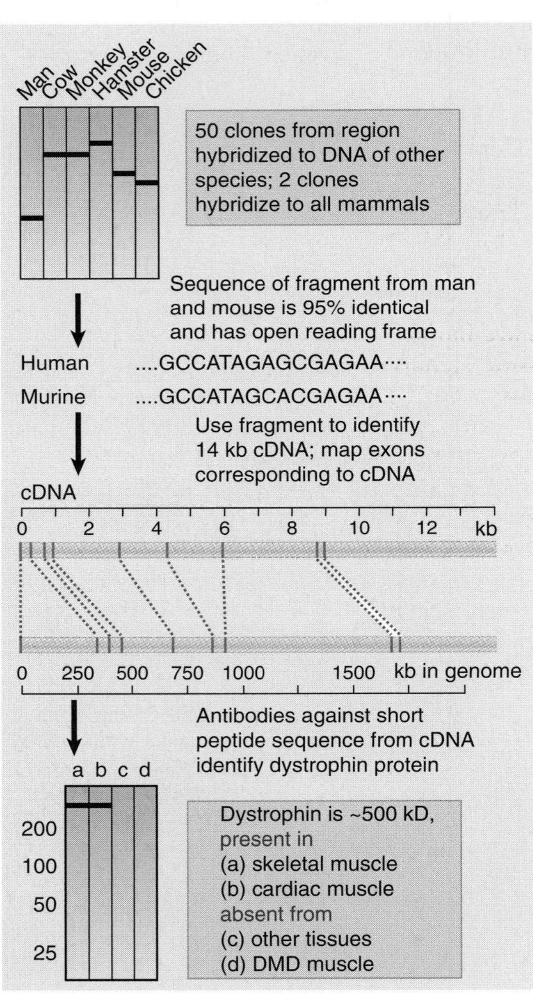

FIGURE 4.10 The Duchenne muscular dystrophy gene was characterized by zoo blotting, cDNA hybridization, genomic hybridization, and identification of the protein.

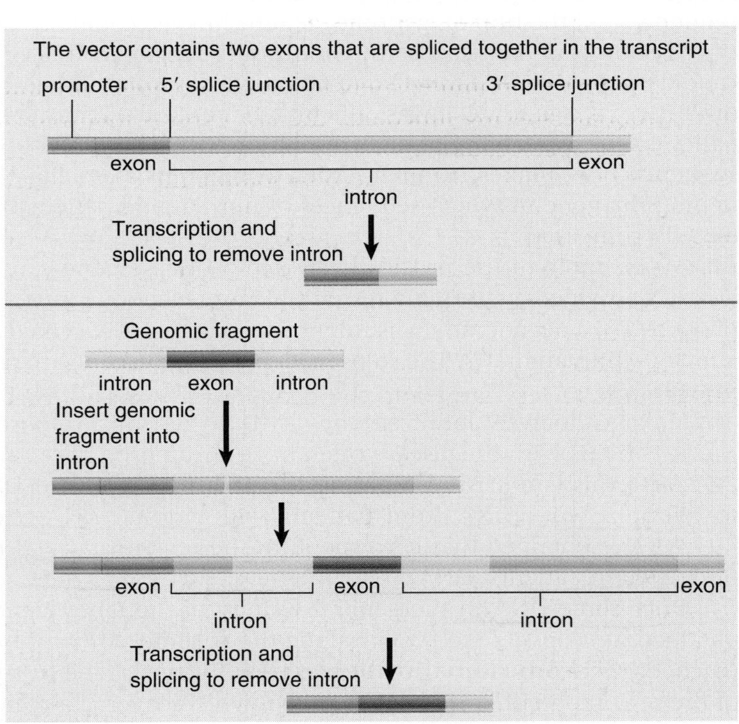

FIGURE 4.11 A special splicing vector is used for exon trapping. If an exon is present in the genomic fragment, its sequence will be recovered in the cytoplasmic RNA. If the genomic fragment consists solely of sequences from within an intron, though, splicing does not occur, and the mRNA is not exported to the cytoplasm.

of the vector. This can be detected readily by reverse transcribing the cytoplasmic RNA into cDNA and using PCR (called *RT-PCR*—see next section) to amplify the sequences between the two exons of the vector. So the appearance in the amplified population of sequences from the genomic fragment indicates that an exon has been "trapped." In mammalian protein-coding genes, introns are usually large and exons are small, so there is a high probability that a random piece of genomic DNA will contain the required structure of an exon surrounded by partial introns. In fact, exon trapping may mimic the events that have occurred naturally during evolution of genes (see *Section 3.7, How Did Interrupted Genes Evolve?*).

KEY CONCEPTS

- Conservation of exons can be used as the basis for identifying coding regions by identifying fragments whose sequences are present in multiple organisms.
- Human disease genes are identified by mapping and sequencing DNA of patients to find differences from normal DNA that are genetically linked to the disease.

CONCEPT AND REASONING CHECK

Probing for a particular exon may identify multiple sites in the genome. Why?

4.8 The Conservation of Genome Organization Helps to Identify Genes

Once we have determined the sequence of a genome, we still have to identify the genes within it. Coding sequences represent a very small fraction of the total genome. Potential exons can be identified as uninterrupted open reading frames flanked by appropriate sequences. What criteria need to be satisfied to identify a functional (intact) gene from a series of exons?

FIGURE 4.12 shows that a functional gene should consist of a series of exons for which the first exon immediately follows a promoter, the internal exons are flanked by appropriate splicing junctions, the last exon is followed by 3′ processing signals, and a single open reading frame starting with an initiation codon and ending with a termination codon can be deduced by joining the exons together. Internal exons can be identified as open reading frames flanked by splicing junctions. In the simplest cases, the first and last exons contain the start and end of the coding region, respectively (as well as the 5′ and 3′ untranslated regions). In more complex cases, the first or last exons may have only untranslated regions and may therefore be more difficult to identify.

The algorithms that are used to connect exons are not completely effective when the genome is very large and the exons may be separated by very large distances.

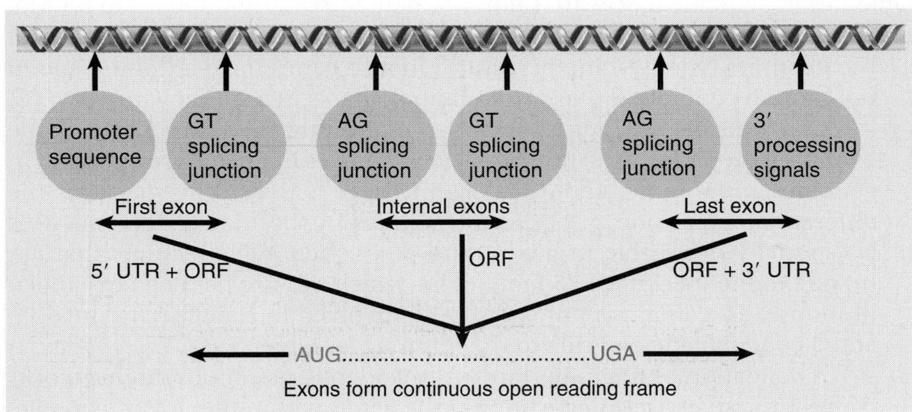

FIGURE 4.12 Exons of protein-coding genes are identified as coding sequences flanked by appropriate signals (with untranslated regions at both ends). The series of exons must generate an open reading frame with appropriate initiation and termination codons.

For example, the initial analysis of the human genome mapped 170,000 exons into 32,000 genes. This is unlikely to be correct because it gives an average of 5.3 exons per gene, whereas the average of individual genes that have been fully characterized is 10.2. Either we have missed many exons, or they should be connected differently into a smaller number of genes in the whole genome sequence.

Even when the organization of a gene is correctly identified, there is the problem of distinguishing functional genes from pseudogenes. Many pseudogenes can be recognized by obvious defects in the form of multiple mutations that create a nonfunctional coding sequence. Pseudogenes that have arisen more recently have not accumulated so many mutations and thus may be more difficult to recognize. In an extreme example, the mouse has only one functional *Gapdh* gene (coding for glyceraldehyde phosphate dehydrogenase), but has ~400 pseudogenes. Approximately 100 of these pseudogenes initially appeared to be functional in the mouse genome sequence, and individual examination was necessary to exclude them from the list of functional genes.

How can putative protein-coding genes be verified? If it can be shown that a DNA sequence is transcribed and processed into a translatable mRNA, it is assumed that it is functional. One technique for doing this is **reverse transcription polymerase chain reaction (RT-PCR)** in which RNA isolated from cells is reverse transcribed to DNA and subsequently amplified to many copies using the polymerase chain reaction. The amplified DNA products can then be sequenced or otherwise analyzed to see if they have the appropriate structural features of a mature transcript. RT-PCR can also be used as a quantitative assessment of gene expression.

▸ **reverse transcription polymerase chain reaction (RT-PCR)** A technique for the detection and quantification of expression of a gene by reverse transcription and amplification of RNAs from a cell sample.

Confidence that a gene is functional can be increased by comparing regions of the genomes of different species. There has been extensive overall reorganization of sequences between the mouse and human genomes, as seen in the simple fact that there are 23 chromosomes in the human haploid genome and 20 chromosomes in the mouse haploid genome. However, at the local level the order of genes is generally the same: when pairs of human and mouse homologs are compared, the genes located on either side also tend to be homologues. This relationship is called **synteny**.

▸ **synteny** A relationship between chromosomal regions of different species where homologous genes occur in the same order.

FIGURE 4.13 shows the relationship between mouse chromosome 1 and the human chromosomal set. We can recognize 21 segments in this mouse chromosome that have syntenic counterparts in human chromosomes. The extent of reshuffling that has occurred between the genomes is shown by the fact that the segments are spread among six different human chromosomes. The same types of relationships are found in all mouse chromosomes except for the X chromosome, which is syntenic only with the human X chromosome. This is explained by the fact that the X is a special case, subject to dosage compensation to adjust for the difference between the one copy of males and the two copies of females (see *Section 27.5, X Chromosomes Undergo Global Changes*). This restriction may apply selective pressure against the translocation of genes to and from the X chromosome.

Comparison of the mouse and human genome sequences shows that >90% of each genome lies in syntenic blocks that range widely in size from 300 kb to 65 Mb. There are a total of 342 syntenic segments, with an average length of 7 Mb (0.3% of the genome). Ninety-nine percent of mouse genes have a homologue in the human genome; for 96% that homologue is in a syntenic region.

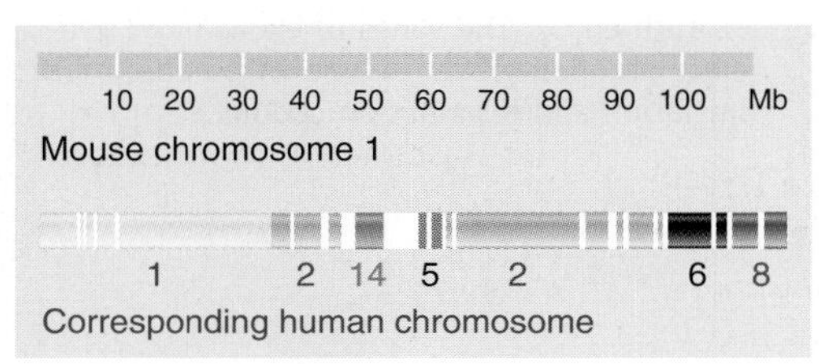

FIGURE 4.13 Mouse chromosome 1 has 21 segments of 1 to 25 Mb that are syntenic with regions corresponding to parts of six human chromosomes.

Comparison of genomes provides interesting information about the evolution of species. The number of gene families in the mouse and human genomes is the same, and a major difference between the species is the differential expansion of particular families in the mouse genome. This is especially noticeable in genes that affect phenotypic features that are unique to the species. Of 25 families for which the size has been expanded in mouse, 14 contain genes specifically involved in rodent reproduction, and 5 contain genes specific to the immune system.

A validation of the importance of the identification of syntenic blocks comes from pairwise comparisons of the genes within them. For example, a gene that is not in a syntenic location (that is, its context is different in the two species being compared) is twice as likely to be a pseudogene. Put another way, translocation away from the original locus tends to be associated with the formation of pseudogenes. The lack of a related gene in a syntenic position is therefore grounds for suspecting that an apparent gene may really be a pseudogene. Overall, >10% of the genes that are initially identified by analysis of the genome are likely to turn out to be pseudogenes.

As a general rule, comparisons between genomes add significantly to the effectiveness of gene prediction. When sequence features indicating active genes are conserved—for example, between human and mouse genomes—there is an increased probability that they identify active orthologs.

Identifying genes coding for RNAs other than mRNA is more difficult because we cannot use the criterion of the open reading frame. It is certainly true that the comparative genome analysis described above has increased the rigor of the analysis. For example, analysis of either the human or the mouse genome alone identifies ~500 genes coding for tRNA, but comparison of features suggests that <350 of these genes are in fact functional in each genome.

An active gene can be located through the use of an **expressed sequence tag (EST)**, a short portion of a transcribed sequence usually obtained from sequencing one or both ends of a cloned fragment from a cDNA library. An EST can confirm that a suspected gene is actually transcribed or help identify genes that influence particular disorders. Through the use of a physical mapping technique such as *in situ* hybridization (see *Section 6.9, Satellite DNAs Often Lie in Heterochromatin*), the chromosomal location of an EST can be determined.

▸ **expressed sequence tag (EST)** A short sequenced fragment of a cDNA sequence that can be used to identify an actively expressed gene.

KEY CONCEPTS

- Methods for identifying active genes are not perfect and many corrections must be made to preliminary estimates.
- Pseudogenes must be distinguished from active genes.
- There are extensive syntenic relationships between the mouse and human genomes, and most active genes are in a syntenic region.

CONCEPT AND REASONING CHECK

In the absence of definitive data about a sequence's expression, such as the presence of a functional gene product, what features of a eukaryotic sequence would strongly suggest that it is an active polypeptide-coding gene?

4.9 Some Organelles Have DNA

The first evidence for the presence of genes outside the nucleus was provided by **non-Mendelian inheritance** in plants (observed in the early years of the twentieth century, just after the rediscovery of Mendelian inheritance). Non-Mendelian inheritance is defined by the failure of the progeny of a mating to display Mendelian segregation for parental characters and is, therefore, taken to indicate the presence of genes that reside outside the nucleus and do not utilize segregation on the meiotic and mitotic spindles to distribute copies to gametes or to daughter cells, respectively. **FIGURE 4.14** shows that this happens when the mitochondria inherited from the male and female parents have different alleles, and a daughter cell receives an unbalanced distribution of mitochondria from only one parent (see *Section 17.11, How Do Mitochondria Replicate and Segregate?*). This is also true of chloroplasts in plants; both mitochondria and chloroplasts contain genomes with functional genes (see below).

▸ **non-Mendelian inheritance** A pattern of inheritance that does not follow that expected by Mendelian principles (each parent contributing a single allele to offspring). Extranuclear genes show a non-Mendelian inheritance pattern.

The extreme form of non-Mendelian inheritance is uniparental inheritance, which occurs when the genotype of only one parent is inherited and that of the other parent is not passed on to the offspring. In less extreme examples, one parental

METHODS AND TECHNIQUES

Using mtDNA to Reconstruct Human Phylogenies

Since the discovery in the early 1960s that mitochondria contain DNA, this mtDNA has been given much attention. In animals, mtDNA is relatively small, only about 16.6 kb, making it an ideal simple genome for analysis. This small genome encodes several genes that are absolutely essential for mitochondrial function, but the genes contain no introns and there is very little intergenic sequence. The only region that has significant variability is the D-loop, or control region, which is involved in regulation of transcription and DNA replication, and makes up about 1122 bp of the genome (with some variation in length). The control region does not code for genes and is highly variable and rapidly evolving relative to the rest of the mitochondrial genome. Human mtDNA, and animal mtDNA in general, has a much higher mutation rate than the average rate of nuclear genes, whereas the organization (order of genes in the genome) is very highly conserved. mtDNA is, therefore, a good target to identify differences among individuals even within populations.

Given the small size and highly conserved nature of animal mtDNA, it is relatively straightforward to sequence either portions or the entire molecule, and analysis of mtDNA has been an important element of human phylogenetic studies for nearly three decades. mtDNA has been a logical focus given the uniparental inheritance and relative lack of recombination among variants of this organelle genome, which simplifies genetic analysis. In contrast, analysis of autosomal genes lacks the clarity of uniparental markers (the mitochondrial genome from the mother and the Y chromosome from the father).

The first analysis of human mtDNA variation utilized restriction enzyme digestion and characterization of the entire mtDNA molecule (published in 1981 by Anderson et al.). This study reported on the ability of restriction analysis to trace human genetic history from a group of 21 people from diverse ethnic and geographic backgrounds, based on the observation of restriction fragment length polymorphisms (RFLPs). This led to the estimation of about 180,000 years for global mtDNA variation—the length of time for all variation to have appeared from a common starting point—which is remarkably consistent with more recent studies.

By the 1990s, the analysis of RFLPs was coupled with mtDNA sequence analysis of hypervariable segments (HVS) of the mtDNA control region to provide additional information for phylogenetic analysis. The HVS-I portion (about 350 bp) of the control region is highly variable but is susceptible to recurrent mutations that can make analysis difficult. Small regions such as these can be routinely amplified by PCR and sequenced to obtain data for construction of phylogenetic trees. In practice, only a few small regions of mtDNA are targeted for amplification and sequence analysis, mostly within the HVS segments. The rate of divergence of the mtDNA control region hypervariable segments has been estimated to be between 11.5% and 17.3% per one million years. By coupling sequence analysis with RFLP analysis, a fairly refined human phylogenetic tree has been produced, with major branches for different geographic regions of the world. As more sequence information becomes available, the tree becomes more refined.

As a result of analysis of the mtDNA control region many haplogroups have been characterized; individuals from different parts of the world have distinct mtDNA types, and thus are placed in different haplogroups. Each haplogroup then is associated with a particular region of the world. The major haplogroups are identified in broad global terms as Sub-Saharan Africa, East Asia, South Asia, Oceania, Europe, and Americas. People who migrate from one region to another bring their mtDNA type with them, providing a tag for genealogical and geographic analysis. This has provided the basis for determination of where the last common human ancestor of the modern mtDNA types lived.

mtDNA analysis has led to the generation of a relatively complex phylogenetic tree with origins in Africa approximately 200,000 years ago. Since the mitochondrial genome is inherited maternally, some have referred to the root, or most recent common ancestor of modern mtDNA haplotypes, as the Mitochondrial Eve. Many haplogroups have descended over time, with many nodes diverging ~80,000 to 90,000 years ago, which may correspond with changing climate after a glacial interstadial phase. Other nodes can be associated with migration and colonization in various areas of the world. This approach has been used to determine that Native Americans originated from Asia. In other studies it has been shown that native Europeans share essentially the same haplogroups with Near Eastern people but not with Africans or East Asians. Many, but not all, of the findings show strong correlation with analysis of genetic markers on the paternally transmitted Y chromosome.

As DNA sequencing technology improves and the cost to obtain DNA sequence information drops, it has become more feasible to sequence entire mtDNA molecules rather than only small segments. This has allowed much more complete and accurate analyses, with the result being an mtDNA tree with a much higher phylogeographic resolution, providing further refinement of geographical origins. For example, complete mtDNA sequence analysis has led to the conclusion that the Central Asian and Native American haplogroup pools are subsets of the East Asian mtDNA variation. This distinction would not be possible by comparing HVS sequences alone. However, complete mtDNA sequence analysis creates some challenge when comparing sequence information from short stretches of mtDNA obtained from ancient DNA samples from bones, teeth or other samples with complete mitochondrial genome information from modern samples. The ancient DNA samples often lack enough integrity to allow complete mtDNA sequence analysis.

One uncertainty that may affect interpretations based on mtDNA sequence information is the question of whether mtDNA undergoes homologous recombination. mtDNA recombination occurs readily in yeast and other fungi, and in plants. However, mtDNA recombination has been thought not to occur in animals, at least partly due to the extreme small size of the genome and the uniparental inheritance from the mother. The absence of recombination simplifies analysis by allowing maximum molecular resolution simply as a function of sequence length examined, rather than needing to take into consideration the variation of gene structure as a result of recombination, which is common for nuclear autosomal genes. Multiple investigators have reported that mtDNA recombination may occur at low levels and in some specific organs in humans and other animals. However, this is somewhat controversial and requires further experiments to determine the extent of recombination in animal mtDNA. If mtDNA recombination does occur, then this must be taken into consideration and may require some alteration of the phylogenetic trees.

In addition to phylogenetic analyses, mtDNA sequence information has been helpful in characterizing inherited and spontaneous mitochondrial disease. Over the past two decades a number of mtDNA diseases have been characterized, including mitochondrial myopathies leading to muscle deterioration and hereditary optic neuropathy that leads to blindness. These diseases are due to mutations in genes encoded in the mtDNA. Since the human mitochondrial genome is so compact, with very little noncoding sequence, nearly any mutation can lead to a defect if it causes a change in amino acid incorporated into the corresponding protein. Many of these diseases result from *deletions* of coding sequence. In fact, that's why they often result in "genetic anticipation"; the defective variant arises as one copy among many but can gradually increase in frequency within a cell because it is shorter and replicates faster. As a result, the disease can increase in severity within an individual over time (as mentioned), and (if in germ cells) can occur earlier in subsequent generations. However, since multiple mitochondria are present in a cell, it is only when molecules with a particular mutation are multiplied and accumulate that the defect becomes apparent. Thus, many of these diseases are progressive rather than having a sudden onset. Many other disorders have been associated with mtDNA defects, including mitochondrial encephalomyopathy, dystonia, some cases of diabetes and Alzheimer's disease, and also aging. Some of these diseases are inherited, whereas others appear to be caused by environmental factors. For inherited mitochondrial diseases, the same advantages of working with a small molecule with little or no recombination allow analysis of inheritance patterns.

Although mtDNA sequence analysis has been very successful for studying human mtDNA phylogenies, it is not suitable for analysis of fungal or plant mtDNA. In fungi and plants the mitochondrial genomes are larger, there is as yet no identified control region (given the vast difference in gene organization), and introns and long intergenic spaces are common. In addition, plant mtDNA has been shown to have a much lower mutation rate as compared to animal mtDNA, but a much higher rate of rearrangement. It is still possible and in fact is quite common to use sequence analysis of genes or noncoding regions of mtDNA from these organisms for phylogenetic studies.

Sources and Further Reading

Kraytsberg et al. (2004). Recombination of human mitochondrial DNA. *Science* 304:981.

Torroni et al. (2006). Harvesting the fruit of the human mtDNA tree. *Trends in Genetics* 22:339–345.

Underhill and Kivisild (2007). Use of Y chromosome and mitochondrial DNA population structure in tracing human migrations. *Annual Review of Genetics* 41:539–564.

Vigilant et al. (1991). African populations and the evolution of human mitochondrial DNA. *Science* 253:1503–1507.

Wallace (1997). Mitochondrial DNA in aging and disease. *Scientific American*, August:40–47.

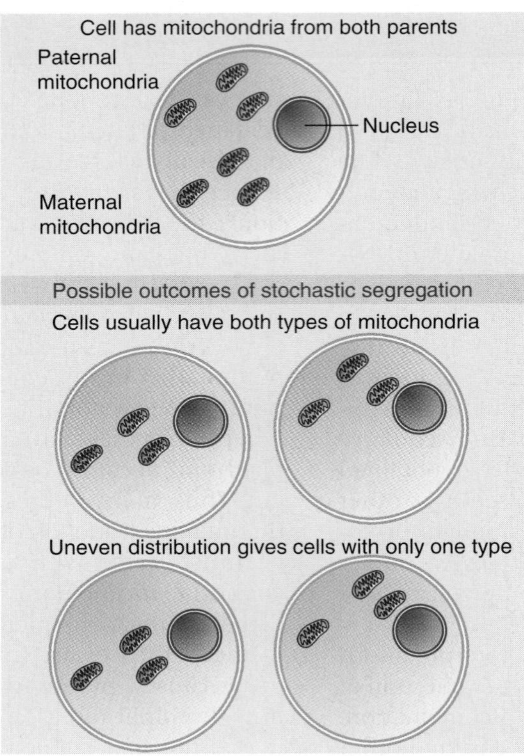

Cell has mitochondria from both parents
Paternal mitochondria
Nucleus
Maternal mitochondria
Possible outcomes of stochastic segregation
Cells usually have both types of mitochondria
Uneven distribution gives cells with only one type

FIGURE 4.14 When paternal and maternal mitochondrial alleles differ, a cell has two sets of mitochondrial DNAs. Mitosis usually generates daughter cells with both sets. Somatic variation may result if unequal segregation generates daughter cells with only one set.

▸ **maternal inheritance** The preferential survival in the progeny of genetic markers provided by one parent.

genotype exceeds the other genotype in the progeny. In animals and most plants it is the mother whose genotype is preferentially (or solely) inherited. This effect is sometimes described as **maternal inheritance**. The important point is that the genotype contributed by the parent of one particular sex predominates, as seen in abnormal segregation ratios when a cross is made between mutant and wild type. This contrasts with the behavior of Mendelian genetics, which occurs when reciprocal crosses show the contributions of both parents to be equally inherited.

The bias in parental genotypes is established at, or soon after, the formation of a zygote. There are various possible causes. The contribution of maternal or paternal information to the organelles of the zygote may be unequal; in the most extreme case, only one parent contributes. In other cases the contributions are equal, but the information provided by one parent does not survive. Combinations of both effects are possible. Whatever the cause, the unequal representation of the information from the two parents contrasts with nuclear genetic information, which derives equally from each parent.

Some non-Mendelian inheritance results from the presence in mitochondria and chloroplasts of DNA genomes that are inherited independently of nuclear genes. In effect, the organelle genome comprises a length of DNA that has been physically sequestered in a defined part of the cell and is subject to its own form of expression and regulation. An organelle genome can code for some or all of the tRNAs and rRNAs, but codes for only some of the polypeptides needed to perpetuate the organelle. The other polypeptides are coded in the nucleus, expressed via the cytoplasmic protein synthetic apparatus, and imported into the organelle.

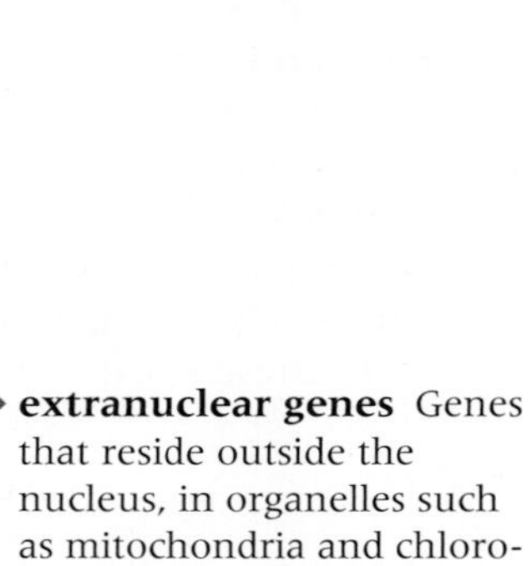

▸ **extranuclear genes** Genes that reside outside the nucleus, in organelles such as mitochondria and chloroplasts.

Genes not residing within the nucleus are generally described as **extranuclear genes**; they are transcribed and translated in the same organelle compartment (mitochondrion or chloroplast) in which they reside. By contrast, nuclear genes are expressed by means of cytoplasmic protein synthesis. (The term "cytoplasmic inheritance" sometimes is used to describe the behavior of genes in organelles. We shall not use this description, though, because it is important to be able to distinguish between events in the general cytosol and those in specific organelles.)

Animals show maternal inheritance of mitochondria, which can be explained if the mitochondria are contributed entirely by the ovum and not at all by the sperm. **FIGURE 4.15** shows that the sperm contributes only a copy of the nuclear DNA. Thus the mitochondrial genes are derived exclusively from the mother, and in males they are discarded each generation. Chloroplasts are generally also maternally inherited, though some plant taxa show paternal or biparental inheritance of chloroplasts.

The chemical environment of organelles is different from that of the nucleus, and organelle DNA therefore evolves at its own distinct rate. If inheritance is uniparental, there can be no recombination between parental genomes. In fact, recombination usually does not occur in those cases for which organelle genomes are inherited from both parents. Organelle DNA has a different replication system from that of the nucleus; as a result, the error rate during replication may be different. Mitochondrial DNA accumulates mutations more rapidly than nuclear DNA in mammals, but in plants the accumulation of mutations in the mitochondrion is slower than in the nucleus; chloroplast DNA has an intermediate mutation rate.

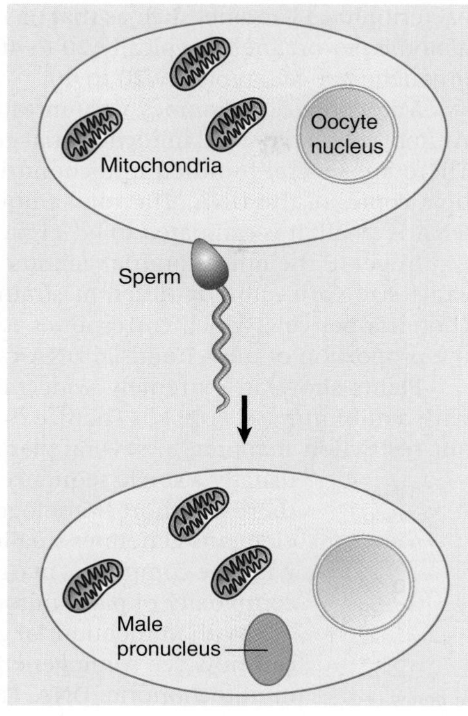

FIGURE 4.15 DNA from the sperm enters the oocyte to form the male pronucleus in the fertilized egg, but all the mitochondria are provided by the oocyte.

One consequence of maternal inheritance is that the sequence of mitochondrial DNA is more sensitive than nuclear DNA to reductions in the size of the breeding population. Comparisons of mitochondrial DNA sequences in a range of human populations allow an evolutionary tree to be constructed. The divergence among human mitochondrial DNAs spans 0.57%. A tree can be constructed in which the mitochondrial variants diverged from a common (African) ancestor. The rate at which mammalian mitochondrial DNA accumulates mutations is 2% to 4% per million years, which is >10× faster than the rate for globin. Such a rate would generate the observed divergence over an evolutionary period of 140,000 to 280,000 years. This implies that human mitochondrial DNA is descended from a single population that lived in Africa ~200,000 years ago. This cannot be interpreted as evidence that there was only a single population at that time, however; there may have been many populations, and some or all of them may have contributed to modern human *nuclear* genetic variation.

KEY CONCEPTS

- Mitochondria and chloroplasts have genomes that show non-Mendelian inheritance. Typically they are maternally inherited.
- Organelle genomes may undergo somatic segregation in plants.
- Comparisons of mitochondrial DNA suggest that it is descended from a single population that lived 200,000 years ago in Africa.

CONCEPT AND REASONING CHECK

Why might the mutation rate of mitochondrial DNA in mammals be an order of magnitude higher than that of nuclear DNA? (Hint: What compounds are found in mitochondria and not in the nucleus?)

4.10 Organelle Genomes Are Circular DNAs That Code for Organelle Proteins

Most organelle genomes take the form of a single circular molecule of DNA of unique sequence (denoted **mtDNA** in the mitochondrion and **ctDNA** or **cpDNA** in the chloroplast). There are a few exceptions in unicellular eukaryotes for which mitochondrial DNA is a linear molecule.

▸ **mtDNA** Mitochondrial DNA.
▸ **ctDNA (cpDNA)** Chloroplast DNA.

Usually there are several copies of the genome in the individual organelle. There are also multiple organelles per cell; therefore, there are many organelle genomes per cell, so the organelle genome can be considered a repetitive sequence.

Chloroplast genomes are relatively large, usually ~140 kb in higher plants and <200 kb in unicellular eukaryotes. This is comparable to the size of a large bacteriophage genome, such as that of T4 at ~165 kb. There are multiple copies of the genome per organelle, typically 20 to 40 in a higher plant, and multiple copies of the organelle per cell, typically 20 to 40.

Mitochondrial genomes vary in total size by more than an order of magnitude. Animal cells have small mitochondrial genomes (approximately 16.5 kb in mammals). There are several hundred mitochondria per cell, and each mitochondrion has multiple copies of the DNA. The total amount of mitochondrial DNA relative to nuclear DNA is small; it is estimated to be <1%.

In yeast, the mitochondrial genome is much larger. In *Saccharomyces cerevisiae*, the exact size varies among different strains but averages ~80 kb. There are ~22 mitochondria per cell, which corresponds to ~4 genomes per organelle. In growing cells, the proportion of mitochondrial DNA can be as high as 18%.

Plants show an extremely wide range of variation in mitochondrial DNA size, with a minimum of ~100 kb. The size of the genome makes it difficult to isolate intact, but restriction mapping in several plants suggests that the mitochondrial genome is usually a single sequence that is organized as a circle. Within this circle there are short homologous sequences. Recombination between these elements generates smaller, subgenomic circular molecules that coexist with the complete, "master" genome—a good example of the apparent complexity of plant mitochondrial DNAs.

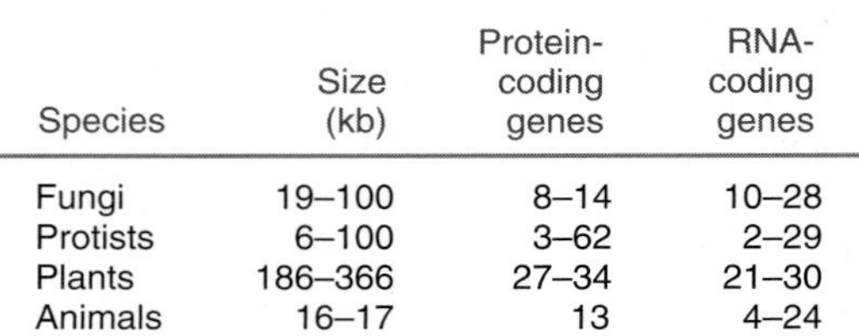

Species	Size (kb)	Protein-coding genes	RNA-coding genes
Fungi	19–100	8–14	10–28
Protists	6–100	3–62	2–29
Plants	186–366	27–34	21–30
Animals	16–17	13	4–24

FIGURE 4.16 Mitochondrial genomes have genes coding for (mostly complex I–IV) proteins, rRNAs, and tRNAs.

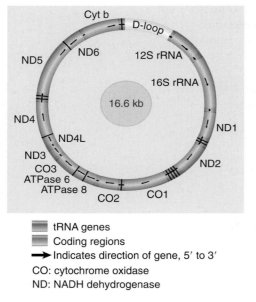

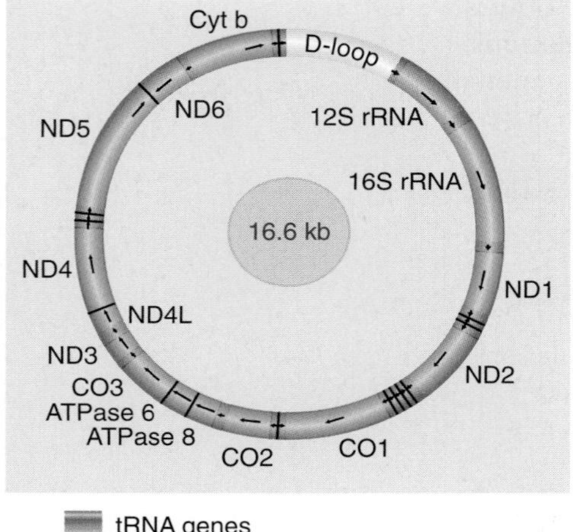

FIGURE 4.17 Human mitochondrial DNA has 22 tRNA genes, 2 rRNA genes, and 13 protein-coding regions. Fourteen of the 15 protein-coding or rRNA-coding regions are transcribed in the same direction. Fourteen of the tRNA genes are expressed in the clockwise direction and 8 are read counter-clockwise.

With mitochondrial genomes sequenced from many organisms, we can now see some general patterns in the representation of functions in mitochondrial DNA. **FIGURE 4.16** summarizes the distribution of genes in mitochondrial genomes. The total number of protein-coding genes is rather small and does not correlate with the size of the genome. The 16 kb mammalian mitochondrial genomes code for 13 proteins, whereas the 60 to 80 kb yeast mitochondrial genomes code for as few as 8 proteins. The much larger plant mitochondrial genomes code for more proteins. Introns are found in most mitochondrial genomes, although not in the very small mammalian genomes.

The two major rRNAs are always coded by the mitochondrial genome. The number of tRNAs coded by the mitochondrial genome varies from none to the full complement (25 to 26 in mitochondria). This accounts for the variation in Figure 4.16.

The major part of the protein-coding activity is devoted to the components of the multisubunit assemblies of respiration complexes I–IV. Many ribosomal proteins are encoded in protist and plant mitochondrial genomes, but there are few or none in fungi and animal genomes. There are genes coding for proteins involved in import in many protist mitochondrial genomes.

Animal mitochondrial DNA is extremely compact. There are extensive differences in the detailed gene organization found in different animal taxa, but the general principle of a small genome coding for a restricted number of functions is maintained. In mammalian mitochondria, the genome is extremely compact. There are no introns, some genes actually overlap, and almost every base pair can be assigned to a gene. With the exception of the **D loop**, a region involved with the initiation of DNA replication, no more than 87 of the 16,569 bp of the human mitochondrial genome lie in intercistronic regions.

The complete nucleotide sequences of animal mitochondrial genomes show extensive homology in organization. The map of the human mitochondrial genome is summarized in **FIGURE 4.17**. There are 13 protein-coding regions. All of the proteins are components of the electron transfer system of cellular respiration. These include cytochrome b, three subunits of cytochrome oxidase, one of the subunits of ATPase, and seven subunits (or associated proteins) of NADH dehydrogenase.

▸ **D loop** A region of the animal mitochondrial DNA molecule that is variable in size and sequence and contains the origin of replication.

The fivefold discrepancy in size between the *S. cerevisiae* (84 kb) and mammalian (16 kb) mitochondrial genomes alone alerts us to the fact that there must be a great difference in their genetic organization in spite of their common function. The number of endogenously synthesized products concerned with mitochondrial enzymatic functions appears to be similar. Does the additional genetic material in yeast mitochondria represent other proteins, perhaps concerned with regulation, or is it unexpressed?

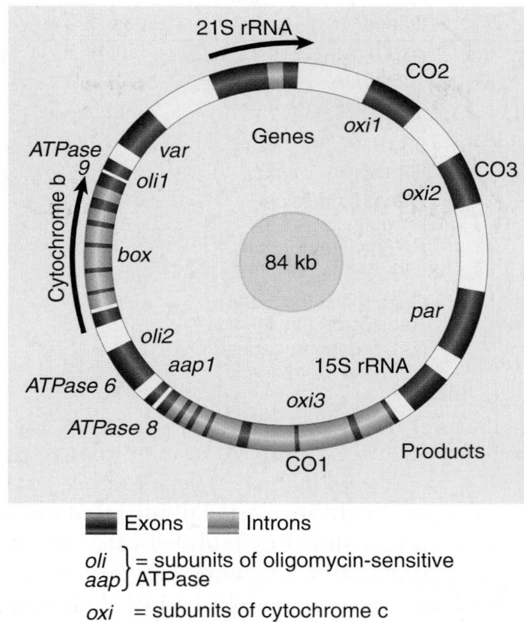

FIGURE 4.18 The mitochondrial genome of *S. cerevisiae* contains both interrupted and uninterrupted protein-coding genes, rRNA genes, and tRNA genes (positions not indicated). Arrows indicate direction of transcription.

The map shown in **FIGURE 4.18** accounts for the major RNA and protein products of the yeast mitochondrion. The most notable feature is the dispersion of loci on the map.

The two most prominent loci are the interrupted genes *box* (coding for cytochrome b) and *oxi3* (coding for subunit 1 of cytochrome oxidase). Together these two genes are almost as long as the entire mitochondrial genome in mammals! Many of the long introns in these genes have open reading frames in register with the preceding exon (see *Section 29.5, Some Group I Introns Code for Endonucleases That Sponsor Mobility*). This adds several proteins, all synthesized in low amounts, to the complement of the yeast mitochondrion.

The remaining genes are uninterrupted. They correspond to the other two subunits of cytochrome oxidase coded by the mitochondrion, to the subunit(s) of the ATPase, and (in the case of *var1*) to a mitochondrial ribosomal protein. The total number of yeast mitochondrial genes is unlikely to exceed ~25.

KEY CONCEPTS

- Organelle genomes are usually (but not always) circular molecules of DNA.
- Organelle genomes code for some, but not all, of the proteins found in the organelle.
- Animal cell mitochondrial DNA is extremely compact and typically codes for 13 proteins, 2 rRNAs, and 22 tRNAs.
- Yeast mitochondrial DNA is 5× longer than animal cell mtDNA because of the presence of long introns.

CONCEPT AND REASONING CHECK

What accounts for the variation in size of mitochondrial genomes among different eukaryotes?

4.11 The Chloroplast Genome Codes for Many Proteins and RNAs

What genes are carried by chloroplasts? Chloroplast DNAs vary in length from ~120 to 217 kb (the largest in geranium). The sequenced chloroplast genomes (>100 in total) have 87 to 183 genes. **FIGURE 4.19** summarizes the functions coded by the chloroplast genome in land plants. There is more variation in the chloroplast genomes of algae.

FIGURE 4.19 The chloroplast genome in land plants codes for 4 rRNAs, 30 tRNAs, and ~60 proteins.

Genes	Types
RNA-coding	
16S rRNA	1
23S rRNA	1
4.5S rRNA	1
5S rRNA	1
tRNA	30–32
Gene Expression	
r-proteins	20–21
RNA polymerase	3
Others	2
Chloroplast functions	
Rubisco and thylakoids	31–32
NADH dehydrogenase	11
Total	105–113

The chloroplast genome is generally similar to that of mitochondria, except that more genes are involved. The chloroplast genome codes for all the rRNA and tRNA species needed for protein synthesis. The ribosome includes two small rRNAs in addition to the major species. The tRNA set may include all of the necessary genes. The chloroplast genome codes for ~50 proteins, including RNA polymerase and ribosomal proteins. Again, the rule is that organelle genes are transcribed and translated within the organelle. About half of the chloroplast genes code for proteins involved in protein synthesis.

Introns in chloroplasts fall into two general classes. Those in tRNA genes are usually (although not inevitably) located in the anticodon loop, like the introns found in yeast nuclear tRNA genes (see *Section 28.13, Yeast tRNA Splicing Involves Cutting and Rejoining*). Those in protein-coding genes resemble the introns of mitochondrial genes (see *Chapter 29, Catalytic RNA*). This places the endosymbiotic event at a time in evolution before the separation of prokaryotes with uninterrupted genes.

The chloroplast is the site of photosynthesis. Many of its genes code for proteins of photosynthetic complexes located in the thylakoid membranes. The constitution of these complexes shows a different balance from that of mitochondrial complexes. Although some complexes are like mitochondrial complexes in that they have some subunits coded by the organelle genome and some by the nuclear genome, other chloroplast complexes are encoded entirely by one genome. For example, the gene for the large subunit of ribulose bisphosphate carboxylase (RuBisCO, which catalyzes the carbon fixation reaction of the Calvin cycle), *rbcL*, is contained in the chloroplast genome; variation in this gene is frequently used as a basis for reconstructing plant phylogenies. However, the gene for the small rubisco subunit, *rbcS*, is usually carried in the nuclear genome. On the other hand, genes for photosystem protein complexes are found on the chloroplast genome, whereas those for the LHC (light-harvesting complex) proteins are nuclear-encoded.

KEY CONCEPT

- Chloroplast genomes vary in size, but are large enough to code for 50 to 100 proteins as well as the rRNAs and tRNAs.

CONCEPT AND REASONING CHECK

What advantage is there to chloroplast genomes coding for RNA polymerase, tRNAs, rRNAs, and ribosomal proteins?

4.12 Mitochondria and Chloroplasts Evolved by Endosymbiosis

How is it that an organelle evolved so that it contains genetic information for some of its functions, whereas the information for other functions is encoded in the nucleus? **FIGURE 4.20** shows the endosymbiotic hypothesis for mitochondrial evolution, in which primitive cells captured bacteria that provided the function of cellular respiration and over time evolved into mitochondria. At this point, the proto-organelle must have contained all of the genes needed to specify its functions. A similar mechanism has been proposed for the origin of chloroplasts.

Sequence homologies suggest that mitochondria and chloroplasts evolved separately, from lineages that are common with different eubacteria, with mitochondria sharing an origin with α-purple bacteria and chloroplasts sharing an origin with cyanobacteria. The closest known relative of mitochondria among the bacteria is *Rickettsia* (the causative agent of typhus), which is an obligate intracellular parasite that is probably descended from free-living bacteria. This reinforces the idea that mitochondria originated in an endosymbiotic event involving an ancestor that is also common to *Rickettsia*.

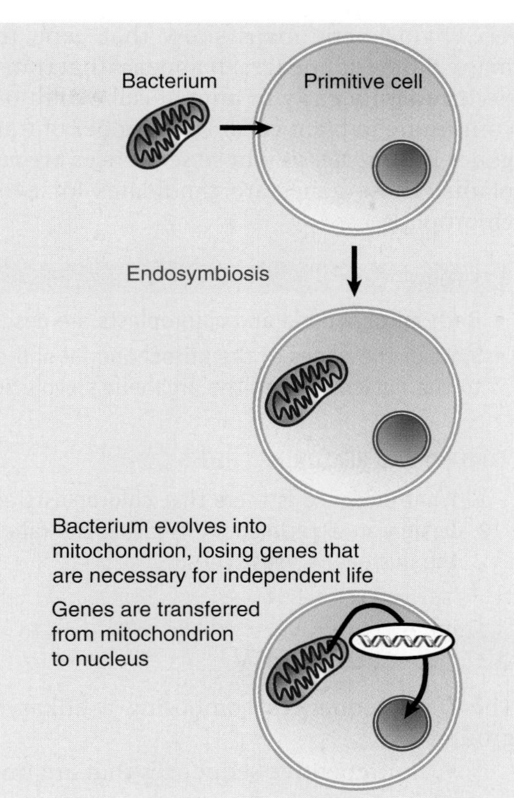
Bacterium
Primitive cell
Endosymbiosis
Bacterium evolves into mitochondrion, losing genes that are necessary for independent life
Genes are transferred from mitochondrion to nucleus

FIGURE 4.20 Mitochondria originated by an endosymbiotic event when a bacterium was captured by a eukaryotic cell.

The endosymbiotic origin of the chloroplast is emphasized by the relationships between its genes and their counterparts in bacteria. The organization of the rRNA genes in particular is closely related to that of a cyanobacterium, which pins down more precisely the last common ancestor between chloroplasts and bacteria. Not surprisingly, cyanobacteria are photosynthetic.

Two changes must have occurred as the bacterium became integrated into the recipient cell and evolved into the mitochondrion (or chloroplast). The organelles have far fewer genes than an independent bacterium and have lost many of the gene functions that are necessary for independent life (such as metabolic pathways). The majority of genes coding for organelle functions are in fact now located in the nucleus; thus these genes must have been transferred there from the organelle.

Transfer of DNA between an organelle and the nucleus has occurred over evolutionary history and still continues. The rate of transfer can be measured directly by introducing into an organelle a gene that can function only in the nucleus, for example, because it contains a nuclear intron, or because the protein must function in the cytosol. In terms of providing the material for evolution, the transfer rates from organelle to nucleus are roughly equivalent to the rate of single gene mutation. DNA introduced into mitochondria is transferred to the nucleus at a rate of 2×10^{-5} per generation. Experiments to measure transfer in the reverse direction, from nucleus to mitochondrion, suggest that the rate is much lower, $<10^{-10}$. When a nuclear-specific antibiotic resistance gene is introduced into chloroplasts, its transfer to the nucleus and successful expression can be followed by screening seedlings for resistance to the antibiotic. This shows that transfer occurs at a rate of 1 in 16,000 seedlings, or 6×10^{-5} per generation.

Transfer of a gene from an organelle to the nucleus requires physical movement of the DNA, of course, but successful expression also requires changes in the coding sequence. Organelle proteins that are coded by nuclear genes have special sequences that allow them to be imported into the organelle after they have been synthesized in the cytoplasm (see *Section 10.6, Posttranslational Membrane Insertion Depends on Signal Sequences*). These sequences are not required by proteins that are synthesized within the organelle. Perhaps the process of effective gene transfer occurred at a period when compartments were less rigidly defined, so that it was easier both for the DNA to be relocated and for the proteins to be incorporated into the organelle irrespective of the site of synthesis.

Phylogenetic maps show that gene transfers have occurred independently in many different lineages. It appears that transfers of mitochondrial genes to the nucleus occurred only early in animal cell evolution, but it is possible that the process is still continuing in plant cells. The number of transfers can be large; there are >800 nuclear genes in *Arabidopsis* whose sequences are related to genes in the chloroplasts of other plants. These genes are candidates for evolution from genes that originated in the chloroplast.

KEY CONCEPTS

- Both mitochondria and chloroplasts are descended from bacterial ancestors.
- Most of the genes of the mitochondrial and chloroplast genomes have been transferred to the nucleus during the organelle's evolution.

CONCEPT AND REASONING CHECKS

1. What evidence is there that chloroplasts are descended from prokaryotic ancestors?
2. Design an experiment to measure the rate of transfer of genes from mitochondria to the nucleus in yeast cells.

4.13 Summary

The DNA sequences composing a eukaryotic genome can be classified into three groups:

- nonrepetitive sequences that are unique;
- moderately repetitive sequences that are dispersed and repeated a small number of times, with some copies not being identical; and
- highly repetitive sequences that are short and usually repeated as tandem arrays.

The proportions of these types of sequences are characteristic for each genome, although larger genomes tend to have a smaller proportion of nonrepetitive DNA. Almost 50% of the human genome consists of repetitive sequences, the vast majority corresponding to transposon sequences. Most structural genes are located in nonrepetitive DNA. The amount of nonrepetitive DNA is a better reflection of the complexity of the organism than the total genome size; the greatest amount of nonrepetitive DNA in genomes is ~2×10^9 bp.

Non-Mendelian inheritance is explained by the presence of DNA in organelles in the cytoplasm. Mitochondria and chloroplasts are membrane-bounded systems in which some proteins are synthesized within the organelle, while others are imported. The organelle genome is usually a circular DNA that codes for all the RNAs and some of the proteins required by the organelle.

Mitochondrial genomes vary greatly in size from the small 16 kb mammalian genome to the 570 kb genome of higher plants. The larger genomes may code for additional functions. Chloroplast genomes range from ~120 to 217 kb. Those that have been sequenced have similar organizations and coding functions. In both mitochondria and chloroplasts, many of the major proteins contain some subunits synthesized in the organelle and some subunits imported from the cytosol. Transfers of DNA have occurred between chloroplasts or mitochondria and nuclear genomes.

CHAPTER QUESTIONS

1. Single nucleotide polymorphisms (SNPs) occur approximately how often in the human genome?
   A. one per 560 bases
   B. one per 950 bases
   C. one per 1330 bases
   D. one per 2040 bases
2. If a genetic cross between two flies, one with dark eye pigment and a single DNA fragment hybridizing with a probe and one lacking pigment and having two fragments hybridizing with the probe, results in progeny with 40% of one parental type and 40% of the other parental type, and 10% of each of the recombinant phenotypes, what is the genetic map distance between the restriction marker and the eye color marker?
   A. 10%
   B. 20%
   C. 40%
   D. 80%
3. Which of the following groups of organisms shows the least genome complexity?
   A. insects
   B. mollusks
   C. fungi
   D. algae
4. Which species shows the greatest percentage of repetitive DNA?
   A. *C. elegans*
   B. *X. laevis*
   C. *D. melanogaster*
   D. *H. sapiens*
5. Prokaryotic genomes generally contain:
   A. nearly 100% nonrepetitive DNA.
   B. mostly nonrepetitive DNA with about 20% moderately repetitive DNA.
   C. mostly moderately or highly repetitive DNA with about 20% nonrepetitive DNA.
   D. about equal amounts of moderately repetitive and nonrepetitive DNA.
6. Plant and amphibian genomes contain:
   A. nearly 100% nonrepetitive DNA.
   B. mostly nonrepetitive DNA with about 20% moderately repetitive DNA.
   C. mostly moderately or highly repetitive DNA with about 20% nonrepetitive DNA.
   D. about equal amounts of moderately repetitive and nonrepetitive DNA.
7. Duchenne muscular dystrophy (DMD) is caused by a _______________ in the gene for dystrophin:
   A. large deletion
   B. large insertion
   C. single base change
   D. small insertion
8. In most animals, mitochondrial DNA is inherited:
   A. equally from both parents.
   B. predominantly from the male parent.
   C. from the female parent only.
   D. from the male parent only.

9. Over the course of evolution, genes from the mitochondrial and chloroplast genomes have:

A. been transferred to nuclear chromosomes.
B. been transferred from one organelle to the other.
C. been deleted from the cell.
D. evolved to other functions.

10. The closest known relative of mitochondria among bacteria is:

A. cyanobacteria.
B. *Rickettsia*.
C. *Salmonella*.
D. *Bacillus*.

KEY TERMS

ctDNA (cpDNA)
C-value
C-value paradox
chromosomal walk
D loop
DNA fingerprinting
exon trapping
expressed sequence tag (EST)
extranuclear genes
genetic map
genome
haplotype
interactome
linkage map
maternal inheritance
mtDNA
non-Mendelian inheritance
nonrepetitive DNA
polymorphism
proteome
repetitive DNA
restriction fragment length polymorphism (RFLP)
restriction map
reverse transcription polymerase chain reaction (RT-PCR)
selfish DNA
single nucleotide polymorphism (SNP)
synteny
transcriptome
transposon
zoo blot

FURTHER READING

Altshuler, D., Brooks, L. D., Chakravarti, A., Collins, F. S., Daly, M. J., and Donnelly, P. (2005). A haplotype map of the human genome. *Nature* 437, 1299–1320.
A report on HapMap, the public database of SNPs (single nucleotide polymorphisms) that represents a large-scale attempt to document human genetic variation.

Gregory, T. R. (2001). Coincidence, coevolution, or causation? DNA content, cell size, and the C-value enigma. *Biol. Rev. Camb. Philos. Soc.* 76, 65–101.
A review of the C-value enigma (paradox) and the hypotheses to account for it, and a discussion of the positive correlation between C value and cell size.

Hinds, D. A., Stuve, L. L., Nilsen, G. B., Halperin, E., Eskin, E., Ballinger, D. G., Frazer, K. A., and Cox, D. R. (2005). Whole-genome patterns of common DNA variation in three human populations. *Science* 307, 1072–1079.
A report on over 1.5 million SNPs from 71 human individuals descended from three different populations as a sample of human genetic variation.

Lang, B. F., Gray, M. W., and Burger, G. (1999). Mitochondrial genome evolution and the origin of eukaryotes. *Annu. Rev. Genet.* 33, 351–397.
Evidence suggests that mitochondria evolved from an endosymbiotic α-proteobacterium at about the same time that the nucleus evolved.

Sharp, A. J., Cheng, Z., and Eichler, E. E. (2006). Structural variation of the human genome. *Annu. Rev. Genom. Hum. G.* 7, 407–442.
A review of the structural rearrangements of the human genome and their effects on phenotypic variation.

Sugiura, M., Hirose, T., and Sugita, M. (1998). Evolution and mechanism of translation in chloroplasts. *Annu. Rev. Genet.* 32, 437–459.
Comparison of multiple chloroplast genomes reveals multiple mechanisms for translation within these organelles.

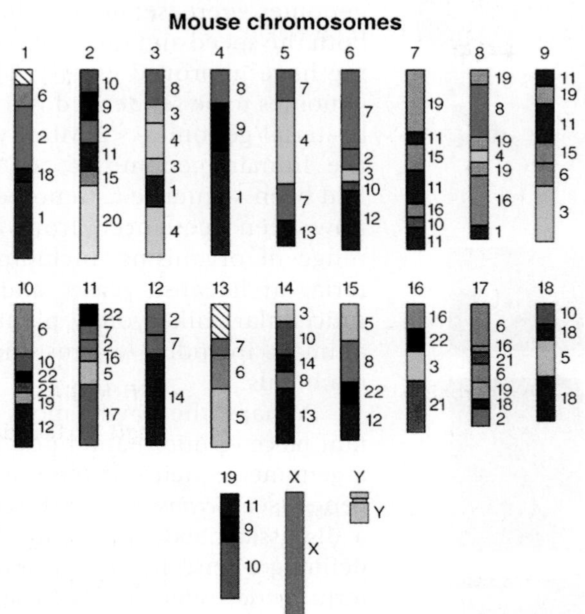

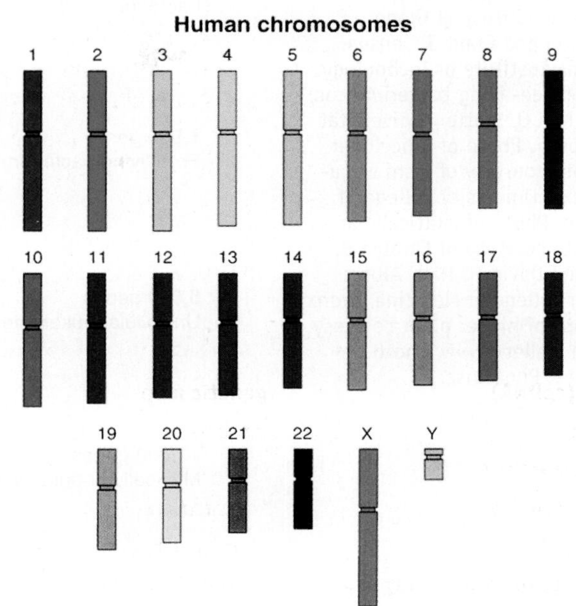

Mouse and human chromosomes. Both species' genomes have undergone chromosomal rearrangements from the genome of their most recent common ancestor. The colors and corresponding numbers on the mouse chromosomes indicate the human chromosomes containing homologous segments. Photo courtesy of U.S. Department of Energy. Used with permission of Lisa J. Stubbs, University of Illinois at Urbana–Champagne.

Genome Sequences and Gene Numbers

CHAPTER OUTLINE

FIGURE 5.1 The minimum gene number required for any type of organism increases with its complexity. Photo of intracellular bacterium courtesy of Gregory P. Henderson and Grant J. Jensen, California Institute of Technology. Photo of free-living bacterium courtesy of Karl O. Stetter, Universität Regensburg. Photo of unicellular eukaryote courtesy of Eishi Noguchi, Drexel University College of Medicine. Photo of multicellular eukaryote courtesy of Carolyn B. Marks and David H. Hall, Albert Einstein College of Medicine, Bronx, NY. Photo of higher plant courtesy of Keith Weller/USDA. Photo of mammal © Photodisc.

500 genes
Intracellular (parasitic) bacterium

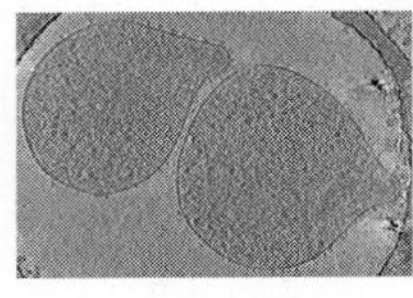

1,500 genes
Free-living bacterium

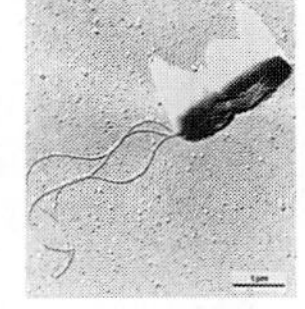

5,000 genes
Unicellular eukaryote

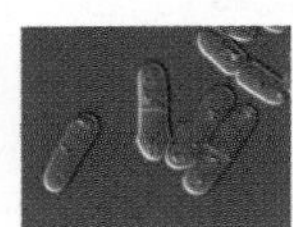

13,000 genes
Multicellular eukaryote

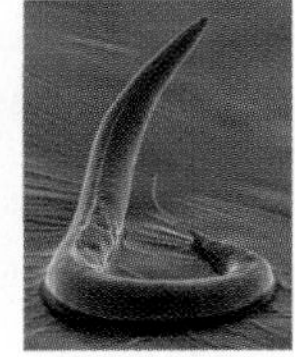

25,000 genes
Higher plants

25,000 genes
Mammals

5.1 Introduction

Since the first complete organismal genomes were sequenced in 1995, both the speed and range of sequencing have improved greatly. The first genomes to be sequenced were small bacterial genomes, <2 Mb. By 2002, the human genome of >3000 Mb had been sequenced. Genomes have now been sequenced from a wide range of organisms, including bacteria, archaeans, yeasts and other unicellular eukaryotes, plants, and animals, including worms, flies, and mammals.

Perhaps the single most important piece of information provided by a genome sequence is the number of genes (see *Section 4.1, Introduction* for a discussion about the difficulties of defining a gene; for our purposes, the term "gene" refers to a DNA sequence transcribed to mRNA, tRNA, or rRNA). *Mycoplasma genitalium,* a free-living parasitic bacterium, has the smallest known genome of any organism, with only ~470 genes. The genomes of free-living bacteria have from 1700 genes to 7500 genes. Archaean genomes have a similar range. Unicellular eukaryotic genomes start with about 5300 genes. Worms and flies have roughly 18,500 and 13,500 genes, respectively, but the number rises only to ~25,000 for the mouse and human genomes.

FIGURE 5.1 summarizes the minimum number of genes found in six groups of organisms. A cell requires ~500 genes, a free-living cell requires ~1500 genes, a cell with a nucleus requires >5000 genes, a multicellular organism requires >10,000 genes, and an organism with a nervous system requires >13,000 genes. Many species may have more than the minimum number of genes required, so the number of genes can vary widely even among closely related species.

Within bacteria and unicellular eukaryotes, most genes are unique. Within multicellular eukaryotic genomes, however, some genes are arranged into families of related members. Of course, some genes are unique (meaning the family has only one member), but many belong to families with ten or more members. The number of different families may be a better indication of the overall complexity of the organism than the number of genes.

Some of the most insightful information comes from comparing genome sequences. With the sequences now available for both the human and chimpanzee genomes, it is possible to begin to address some of the questions about what makes humans unique.

5.2 Prokaryotic Gene Numbers Range Over an Order of Magnitude

Large-scale efforts have now led to the sequencing of many genomes. The range of known genome sizes is summarized in **FIGURE 5.2**. They extend from the 0.6×10^6 bp of a mycoplasma to the 3.3×10^9 bp of the human genome and include several important experimental animals, such as yeasts, the fruit fly, and a nematode worm.

The sequences of the genomes of prokaryotes show that virtually all of the DNA (typically 85%–90%) codes for RNA or polypeptide. **FIGURE 5.3** shows that the range of genome sizes is about an order of magnitude, and that the genome size is proportional to the number of genes. The typical gene averages about 1000 bp in length.

All of the prokaryotes with genome sizes below 1.5 Mb are parasites—they can live within a eukaryotic host that provides them with small molecules. Their genome sizes suggest the minimum number of functions required for a cellular organism. All classes of genes are reduced in number compared to prokaryotes with larger genomes, but the most significant reduction is in loci that code for enzymes concerned with metabolic functions (which are largely provided by the host cell) and with regulation of gene expression. *Mycoplasma genitalium* has the smallest genome, with ~470 genes.

Archaeans have biological properties that are intermediate between those of other prokaryotes and those of eukaryotes, but their genome sizes and gene numbers fall in the same range as those of bacteria. Their genome sizes vary from 1.5 to 3 Mb, corresponding to 1500 to 2700 genes. *Methanococcus jannaschii* is a methane-producing species that lives under high pressure and temperature. Its total gene number is similar to that of *Haemophilus influenzae*, but fewer of its genes can be identified on the basis of comparison with genes known in other organisms. Its apparatus for gene expression resembles that of eukaryotes more than that of prokaryotes, but its apparatus for cell division better resembles that of prokaryotes.

The genomes of archaea and the smallest free-living bacteria suggest the minimum number of genes required to make a cell able to function independently in its environment. The smallest archaeal genome has ~1500 genes. The free-living nonparasitic bacterium with the smallest known genome is the thermophile *Aquifex aeolicus*, with a 1.5 Mb genome and 1512 genes. A "typical" gram-negative bacterium, *H. influenzae*, has 1743 genes, each of which are ~900 bp. So we can conclude that ~1500 genes are required by an exclusively free-living organism.

Prokaryotic genome sizes extend over about an order of magnitude, from 0.6 Mb to <8 Mb. As expected, the larger genomes have more genes. The prokaryotes with the largest genomes, *Sinorhizobium meliloti* and *Mesorhizobium loti*, are nitrogen-fixing bacteria that live on plant roots. Their genome sizes (~7 Mb) and total gene numbers (>7500) are similar to those of yeasts.

Species	Genomes (Mb)	Genes	Lethal loci
Mycoplasma genitalium	0.58	470	~300
Rickettsia prowazekii	1.11	834	
Haemophilus influenzae	1.83	1,743	
Methanococcus jannaschi	1.66	1,738	
B. subtilis	4.2	4,100	
E. coli	4.6	4,288	1,800
S. cerevisiae	13.5	6,034	1,090
S. pombe	12.5	4,929	
A. thaliana	119	25,498	
O. sativa (rice)	466	~30,000	
D. melanogaster	165	13,601	3,100
C. elegans	97	18,424	
H. sapiens	3,300	~25,000	

FIGURE 5.2 Genome sizes and gene numbers are known from complete sequences for several organisms. Lethal loci are estimated from genetic data.

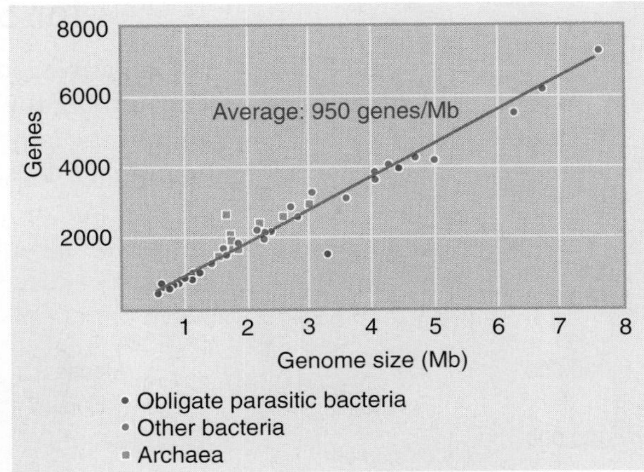

FIGURE 5.3 The number of genes in bacterial and archaeal genomes is proportional to genome size.

The size of the genome of *E. coli* is in the middle of the range for prokaryotes. The common laboratory strain has 4288 genes, with an average length of ~950 bp, and an average separation between genes of 118 bp. There can be quite significant differences between strains, though. The known extremes among strains of *E. coli* are from 4.6 Mb with 4249 genes to 5.5 Mb with 5361 genes.

We still do not know the functions of all of these genes. In most of these genomes, ~60% of the genes can be identified on the basis of homology with known genes in other species. These genes fall approximately equally into classes whose products are related to function in metabolism, cell structure or transport of components, and gene expression and its regulation. In virtually every genome, >25% of the genes cannot yet be ascribed any function. Many of these genes can be found in related organisms, which implies that they have a conserved function.

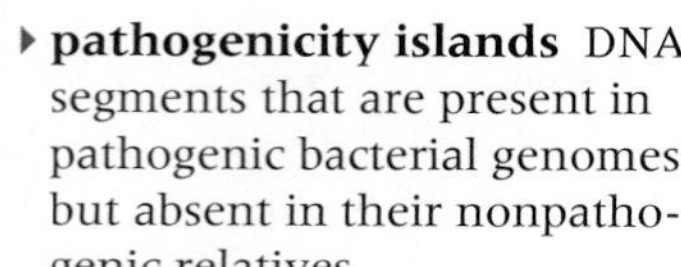

▸ **pathogenicity islands** DNA segments that are present in pathogenic bacterial genomes but absent in their nonpathogenic relatives.

▸ **horizontal transfer** The transfer of DNA from one cell to another by a process other than cell division, such as bacterial conjugation.

There has been some emphasis on sequencing the genomes of pathogenic bacteria, given their medical significance. An important insight into the nature of pathogenicity has been provided by the demonstration that **pathogenicity islands** are a characteristic feature of their genomes. These are large regions, ~10 to 200 kb, which are present in the genomes of pathogenic species but absent from the genomes of nonpathogenic variants of the same or related species. Their G-C content often differs from that of the rest of the genome, and it is likely that they migrate between bacteria by a process of **horizontal transfer**. For example, the bacterium that causes anthrax (*Bacillus anthracis*) has two large plasmids (extra chromosomal DNA), one of which has a pathogenicity island that includes the gene coding for the anthrax toxin.

KEY CONCEPT

- The minimum number of genes for a parasitic prokaryote is about 500; for a free-living nonparasitic prokaryote, about 1500.

CONCEPT AND REASONING CHECK

Why do some prokaryotes have a rather large number of genes while others thrive with many fewer genes?

5.3 Total Gene Number Is Known for Several Eukaryotes

As soon as we look at eukaryotic genomes, the relationship between genome size and gene number is weakened. The genomes of unicellular eukaryotes fall in the same size range as the largest bacterial genomes. Multicellular eukaryotes have more genes, but the number does not correlate with genome size, as can be seen in **FIGURE 5.4**.

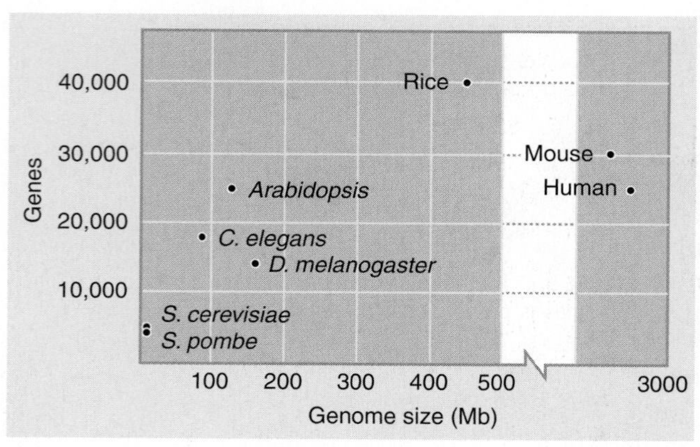

FIGURE 5.4 The number of genes in a eukaryote varies from 6000 to 40,000 but does not correlate with the genome size or the complexity of the organism.

The most extensive data for unicellular eukaryotes are available from the sequences of the genomes of the yeasts *Saccharomyces cerevisiae* and *Schizosaccharomyces pombe*. **FIGURE 5.5** summarizes the most important features. The yeast genomes of 12.5 Mb and 13.5 Mb have ~6000 and ~5000 genes, respectively. The average open reading frame (ORF) is ~1.4 kb, so that ~70% of the genome is occupied by coding regions. The major difference between them is that only 5% of *S. cerevisiae* genes have introns, compared to 43% in *S. pombe*. The density of genes is high; organization is generally similar, although the spaces between genes are a bit shorter in *S. cerevisiae*. About half of the genes identified by sequence were either known previously or related to known genes. The remainder were previously unknown, which gives some indication of the number of new types of genes that may be discovered.

The identification of long reading frames on the basis of sequence is quite accurate. However, ORFs coding for <100 amino acids cannot be identified solely by sequence because of the high occurrence of false positives. Analysis of gene expression suggests that only ~300 of 600 such ORFs in *S. cerevisiae* are likely to be active genes.

A powerful way to validate gene structure is to compare sequences in closely related species: if a gene is active, it is likely to be conserved. Comparisons between the sequences of four closely related yeast species suggest that 503 of the genes originally identified in *S. cerevisiae* do not have counterparts in the other species and therefore should be deleted from the catalog. This reduces the total estimated gene number for *S. cerevisiae* to 5726.

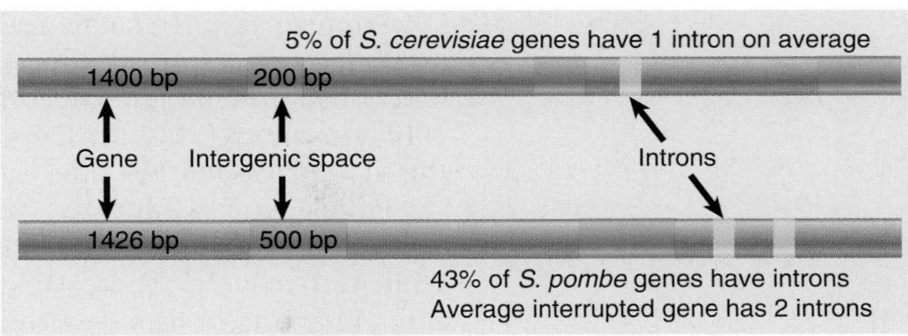

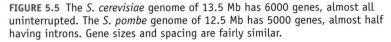

FIGURE 5.5 The *S. cerevisiae* genome of 13.5 Mb has 6000 genes, almost all uninterrupted. The *S. pombe* genome of 12.5 Mb has 5000 genes, almost half having introns. Gene sizes and spacing are fairly similar.

The genome of *Caenorhabditis elegans* DNA varies between regions rich in genes and regions in which genes are more sparsely distributed. The total sequence contains ~18,500 genes. Only ~42% of the genes have putative counterparts outside Nematoda.

The fly genome is larger than the worm genome, but there are fewer genes in some species (~14,000 in *D. melanogaster*) and more in others (e.g., ~23,000 in *D. persimilis*). The number of different transcripts is somewhat larger as the result of alternative splicing. We do not understand why *C. elegans*—arguably, a less complex organism—has 30% more genes than the fly, but it may be because *C. elegans* has a larger average number of genes per gene family than does *D. melanogaster*, so the number of *unique* genes of the two species is more similar. A comparison of twelve *Drosophila* genomes reveals that there can be a fairly large range of gene number among closely related species. In some cases, there are several thousand genes that are species-specific. This emphasizes forcefully the lack of an exact relationship between gene number and complexity of the organism.

The plant *Arabidopsis thaliana* has a genome size intermediate between the worm and the fly, but has a larger gene number (25,000) than either. This again shows the lack of a clear relationship and also emphasizes the special quality of plants, which may have more genes (due to ancestral duplications) than animal cells. A majority of the *Arabidopsis* genome is found in duplicated segments, suggesting that there was an ancient doubling of the genome (to result in a tetraploid). Only 35% of *Arabidopsis* genes are present as single copies.

The genome of rice (*Oryza sativa*) is ~4× larger than that of *Arabidopsis*, but the number of genes is only ~50% larger, probably ~40,000. Repetitive DNA occupies 42% to 45% of the genome. More than 80% of the genes found in *Arabidopsis* are also found in rice. Of these common genes, ~8000 are found in *Arabidopsis* and rice but not in any of the bacterial or animal genomes that have been sequenced. This is probably the set of genes that codes for plant-specific functions, such as photosynthesis.

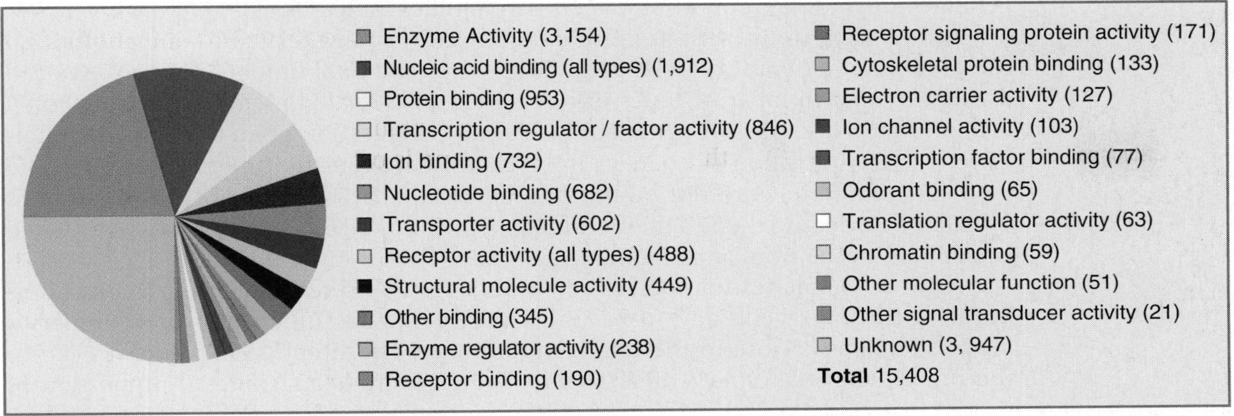

FIGURE 5.6 Functions of *Drosophila* genes based on comparative genomics of twelve species. The functions of about a quarter of the genes of *Drosophila* are unknown. Adapted from *Drosophila* 12 Genomes Consortium, "Evolution of genes and genomes on the *Drosophila* phylogeny," *Nature* 450 (2007): 203–218.

From the twelve sequenced *Drosophila* genomes, we can form an impression of how many genes are devoted to each type of function. **FIGURE 5.6** breaks down the functions into different categories. Among the genes that are identified, we find over 3000 enzymes, ~900 transcription factors, and ~700 transporters and ion channels. About a quarter of the genes encode products of unknown function.

Polypeptide size increases from prokaryotes to eukaryotes. The archaean *M. jannaschi* and bacterium *E. coli* have average polypeptide lengths of 287 and 317 amino acids, respectively, whereas *S. cerevisiae* and *C. elegans* have average lengths of 484 and 442 amino acids, respectively. Large polypeptides (>500 amino acids) are rare in bacteria, but comprise a significant component (~1/3) in eukaryotes. The increase in length is due to the addition of extra domains, with each domain typically constituting 100 to 300 amino acids. The increase in polypeptide size, however, is responsible for only a very small part of the increase in genome size.

Another insight into gene number is obtained by counting the number of expressed protein-coding genes. If we relied upon the estimates of the number of different mRNA species that can be counted in a cell, we would conclude that the average vertebrate cell expresses ~10,000 to 20,000 genes. The existence of significant overlaps between the mRNA populations in different cell types would suggest that the total expressed gene number for the organism should be within the same order of magnitude. The estimate for the total human gene number of 20,000 to 25,000 (see *Section 5.5, The Human Genome Has Fewer Genes Than Expected*) would imply that a significant proportion of the total gene number is actually expressed in any given cell.

▸ **monocistronic mRNA** mRNA that codes for one polypeptide.

▸ **polycistronic mRNA** mRNA that includes coding regions representing more than one gene.

Eukaryotic genes are transcribed individually, with each gene producing a **monocistronic** messenger. There is only one general exception to this rule: in the genome of *C. elegans*, ~15% of the genes are organized into **polycistronic** units (which are associated with the use of *trans*-splicing to allow expression of the downstream genes in these units; see *Section 28.13, trans-Splicing Reactions Use Small RNAs*).

KEY CONCEPT

- There are 6000 genes in yeast; 18,500 in a worm; 13,600 in a fly; 25,000 in the small plant *Arabidopsis*; and probably 20,000 to 25,000 in mice and humans.

CONCEPT AND REASONING CHECK

Why is it that in multicellular eukaryotes the genome size is not a good indication of the number of genes?

5.4 How Many Different Types of Genes Are There?

Some genes are unique; others belong to families in which the other members are related (but not usually identical). The proportion of unique genes declines with genome size and the proportion of genes in families increases.

Some genes are present in more than one copy or are related to one another, so the number of different types of genes is less than the total number of genes. We can divide the total number of genes into sets that have related members, as defined by comparing their exons. (A gene family arises by duplication of an ancestral gene followed by accumulation of changes in sequence between the copies. Most often the members of a family are related but not identical.) The number of types of genes is calculated by adding the number of unique genes (for which there is no other related gene at all) to the numbers of families that have two or more members.

FIGURE 5.7 compares the total number of genes with the number of distinct families in each of six genomes. In bacteria, most genes are unique, so the number of distinct families is close to the total gene number. The situation is different even in the unicellular eukaryote *S. cerevisiae*, for which there is a significant proportion of repeated genes. The most striking effect is that the number of genes increases quite sharply in the higher eukaryotes, but the number of gene families does not change much.

FIGURE 5.8 shows that the proportion of unique genes drops sharply with genome size. When genes are present in families, the number of members in a family is small in bacteria and unicellular eukaryotes, but is large in multicellular eukaryotes. Much of the extra genome size of *Arabidopsis* is accounted for by families with >4 members.

If every gene is expressed, the total number of genes will account for the total number of polypeptides required to make the organism (the proteome). There are two conditions, however, that cause the proteome to be different from the total gene number. First, genes can be duplicated, and as a result some of them code for the same polypeptide (although it may be expressed in a different time or place) and others may code for related polypeptides that again play the same role in different times or places. Second, the proteome can be larger than the number of genes because some genes can produce more than one polypeptide by means of alternative splicing.

What is the core proteome—the basic number of the different types of polypeptides in the organism? Although difficult to estimate because of the possibility of alternative splicing, a minimum estimate is given by the number of gene families, ranging from 1400 in the bacterium, to ~4000 in the yeast, and 11,000 to 14,000 for the fly and the worm.

What is the distribution of the proteome by type of protein? The 6000 proteins of the yeast proteome include 5000 soluble proteins and 1000 transmembrane proteins. About half of the proteins are cytoplasmic, a quarter are in the nucleus, and the remainder are split between the mitochondrion and the endoplasmic reticulum (ER)/Golgi system.

How many genes are common to all organisms (or to groups such as bacteria or multicellular eukaryotes), and how many are taxon-specific? **FIGURE 5.9** shows the comparison of fly genes to those of the worm (another multicellular eukaryote) and yeast (a unicellular eukaryote). Genes that code for corresponding polypeptides in different organisms are called **orthologs** (see *Section 3.9, The Members of a Gene Family Have a Common Organization*). Operationally, we usually consider that two genes in different organisms are orthologs if their sequences are similar over >80% of the length. By this criterion, ~20% of the fly genes have orthologs in both yeast and worm. These genes are probably required by all eukaryotes. The proportion increases to 30% when fly and worm are compared, probably representing the addition of gene functions that are common to multicellular eukaryotes. This still leaves a major proportion of genes as coding for proteins that are required specifically by either flies or worms, respectively.

A minimum estimate of the size of an organismal proteome can be deduced from the number and structures of genes, and a cellular or organismal proteome size can also be directly measured by analyzing the total polypeptide content of the cell or organism. By such approaches, some proteins have been identified that were not suspected on the basis of genome analysis;

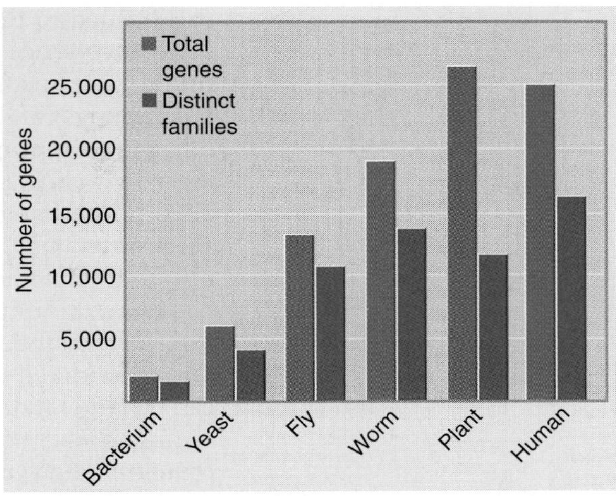

FIGURE 5.7 Many genes are duplicated, and as a result the number of different gene families is much less than the total number of genes. The histogram compares the total number of genes with the number of distinct gene families.

	Unique genes	Families with 2–4 members	Families with >4 members
H. influenzae	89%	10%	1%
S. cerevisiae	72%	19%	9%
D. melanogaster	72%	14%	14%
C. elegans	55%	20%	26%
A. thaliana	35%	24%	41%

FIGURE 5.8 The proportion of genes that are present in multiple copies increases with genome size in multicellular eukaryotes.

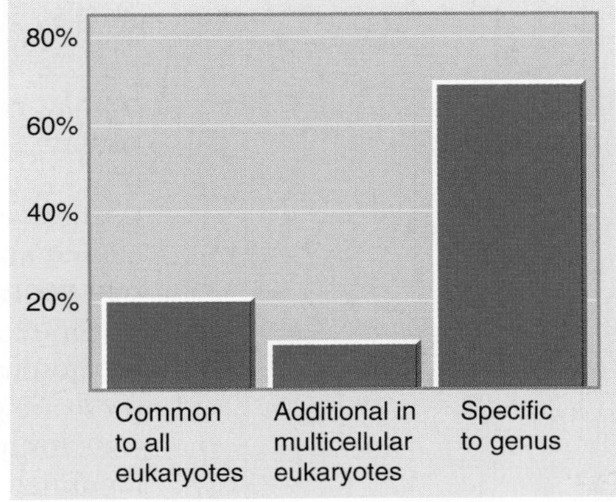

FIGURE 5.9 The fly genome can be divided into genes that are (probably) present in all eukaryotes, additional genes that are (probably) present in all multicellular eukaryotes, and genes that are more specific to subgroups of species that include flies.

▸ **orthologous genes (orthologs)** Related genes in different species.

this has led to the identification of new genes. Several methods are used for large-scale analysis of proteins. Mass spectrometry can be used for separating and identifying proteins in a mixture obtained directly from cells or tissues. Hybrid proteins bearing tags can be obtained by expression of cDNAs made by linking the sequences of ORFs to appropriate expression vectors that incorporate the sequences for affinity tags. This allows array analysis to be used to analyze the products. These methods also can be effective in comparing the proteins of two tissues—for example, a tissue from a healthy individual and one from a patient with disease—to pinpoint the differences.

In addition to functional genes, there are also copies of genes that have become nonfunctional (identified as such by interruptions in their protein-coding sequences). These are called *pseudogenes* (see *Section 6.1, Introduction*). The number of pseudogenes can be large. In the mouse and human genomes, the number of pseudogenes is ~10% of the number of (potentially) active genes (see *Section 4.8, The Conservation of Genome Organization Helps to Identify Genes*).

KEY CONCEPTS

- The sum of the number of unique genes and the number of gene families is an estimate of the number of types of genes.
- The minimum size of the proteome can be estimated from the number of types of genes.

CONCEPT AND REASONING CHECK

Why isn't the size of the proteome an accurate estimate of the number of types of genes?

5.5 The Human Genome Has Fewer Genes Than Originally Expected

The human genome was the first vertebrate genome to be sequenced. This massive task has revealed a wealth of information about the genetic makeup of our species and about the evolution of genomes in general. Our understanding is deepened further by the ability to compare the human genome sequence with other sequenced vertebrate genomes.

Mammal genomes generally fall into a narrow size range, ~3 $\times$ 10^9 bp (see *Section 4.5, Why Are Some Genomes So Large?*). The mouse genome is ~14% smaller than the human genome, probably because it has had a higher rate of deletion. The genomes contain similar gene families and genes, with most genes having an ortholog in the other genome, but with differences in the number of members of a family, especially in those cases for which the functions are specific to the species (see *Section 4.8, The Conservation of Genome Organization Helps to Identify Genes*). Originally estimated to have ~30,000 genes, the mouse genome is now thought to have about the same number of genes as the human genome, 20,000 to 25,000. The 23,000 protein-coding genes are accompanied by ~3000 genes representing RNAs that do not code for proteins; these are generally small (aside from the ribosomal RNAs). Almost half of these genes code for transfer RNAs. In addition to the active genes, ~1200 pseudogenes have been identified.

The human (haploid) genome contains 22 autosomes plus the X and Y chromosomes. The chromosomes range in size from 45 to 279 Mb of DNA, making a total genome size of 3286 Mb (~3.3 $\times$ 10^9 bp). On the basis of chromosome structure, the genome can be divided into regions of euchromatin (containing many active genes) and heterochromatin, with a much lower density of active genes (see *Section 23.5, Chromatin Is Divided into Euchromatin and Heterochromatin*). The euchromatin comprises the majority of the genome, ~2.9 $\times$ 10^9 bp. The identified genome sequence represents ~90% of the euchromatin. In addition to providing information on the genetic

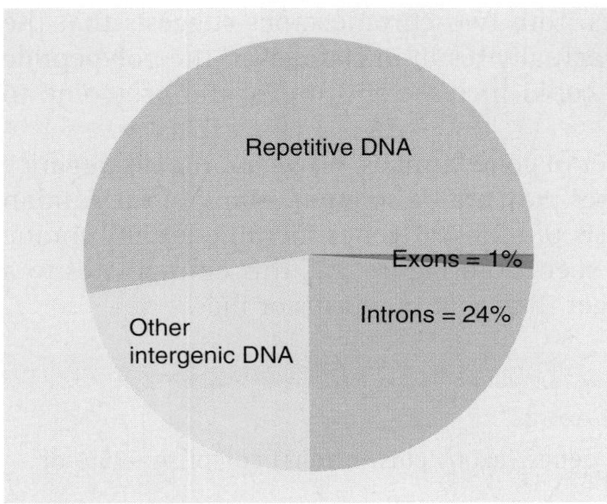

FIGURE 5.10 Genes occupy 25% of the human genome, but protein-coding sequences are only a tiny part of this fraction.

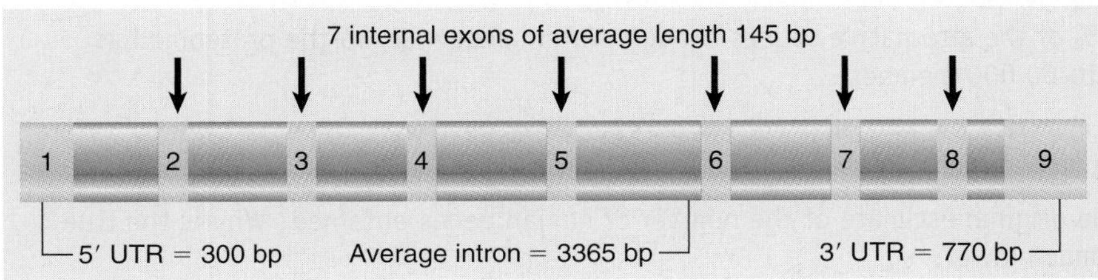

FIGURE 5.11 The average human gene is 27 kb long and has nine exons, usually comprising two longer exons at each end and seven internal exons. The UTRs in the terminal exons are the untranslated (noncoding) regions at each end of the gene. (This is based on the average. Some genes are extremely long, which makes the median length 14 kb with seven exons.)

content of the genome, the sequence also identifies features that may be of structural importance (see *Section 23.6, Chromosomes Have Banding Patterns*).

FIGURE 5.10 shows that a tiny proportion (~1%) of the human genome is accounted for by the exons that actually code for polypeptides. The introns that constitute the remaining sequences of protein-coding genes bring the total of DNA concerned with producing proteins to ~25%. As shown in **FIGURE 5.11**, the average human gene is 27 kb long, with nine exons that include a total coding sequence of 1340 bp. The average coding sequence is therefore only 5% of the length of an average protein-coding gene.

By any measure, the total human gene number of 20,000 to 25,000 is much less than was originally expected—most estimates before the genome was sequenced were ~100,000. It shows a relatively small increase over flies and worms (~13,600 and ~18,500, respectively), not to mention the plant *Arabidopsis thaliana* (~25,000) (see Figure 5.2). However, we should not be particularly surprised by the notion that it does not take a great number of additional genes to make a more complex organism. The difference in DNA sequences between the human and chimpanzee genomes is extremely small (there is >99% similarity), so it is clear that the functions and interactions between a similar set of genes can produce very different results. The functions of specific groups of genes may be especially important, because detailed comparisons of orthologous genes in humans and chimpanzees suggest that there has been rapid evolution of certain classes of genes, including some involved in early development, olfaction, and hearing—all functions that are relatively specialized in these species.

The number of protein-coding genes is less than the number of potential polypeptides because of mechanisms such as alternative splicing, alternate promoter selection, and alternate poly(A) site selection that can result in several polypeptides from the same gene (see *Section 28.11, Alternative Splicing Involves Differential Use of Splice Junctions*). The extent of alternative splicing is greater in humans than in flies or worms; it may affect as many as 60% of the genes, so the increase in size of the human proteome relative to that of the other eukaryotes may be larger than the increase in the number of genes. A sample of genes from two chromosomes suggests that the proportion of the alternative splices that actually result in changes in the polypeptide sequence may be as high as 80%. This could increase the size of the proteome to 50,000 to 60,000 members.

In terms of the diversity of the number of gene families, however, the discrepancy between humans and the other eukaryotes may not be so great. Many of the human genes belong to gene families. An analysis of ~25,000 genes identified 3500 unique genes and 10,300 gene pairs. As can be seen from Figure 5.7, this extrapolates to a number of gene families only slightly larger than that of worms or flies.

KEY CONCEPTS

- Only 1% of the human genome consists of exons.
- The exons comprise ~5% of each gene, so genes (exons plus introns) comprise ~25% of the genome.
- The human genome has 20,000 to 25,000 genes.
- ~60% of human genes are alternatively spliced.
- Up to 80% of the alternative splices change protein sequence, so the proteome has ~50,000 to 60,000 members.

CONCEPT AND REASONING CHECK

How was the original estimate of the number of human genes obtained? Why is the true number so much lower?

5.6 How Are Genes and Other Sequences Distributed in the Genome?

Are genes uniformly distributed in the genome? Some chromosomes are relatively poor in genes and have >25% of their sequences as "deserts"—regions longer than 500 kb where there are no ORFs. Even the most gene-rich chromosomes have >10% of their sequences as deserts. So overall, ~20% of the human genome consists of deserts that have no protein-coding genes.

Repetitive sequences account for ~50% of the human genome, as seen in **FIGURE 5.12**. The repetitive sequences fall into five classes:

- Transposons (either active or inactive) account for the vast majority (45% of the genome). All transposons are found in multiple copies.
- Processed pseudogenes (~3000 in all, account for ~0.1% of total DNA). (These are sequences that arise by insertion of a reverse transcribed DNA copy of an mRNA sequence into the genome; see *Section 6.1, Introduction*.)
- Simple sequence repeats (highly repetitive DNA such as $(CA)_n$ account for ~3%).
- Segmental duplications (blocks of 10 to 300 kb that have been duplicated into a new region) account for ~5%. Only a minority of these duplications are found on the same chromosome; in the other cases, the duplicates are on different chromosomes.
- Tandem repeats form blocks of one type of sequence (especially found at centromeres and telomeres).

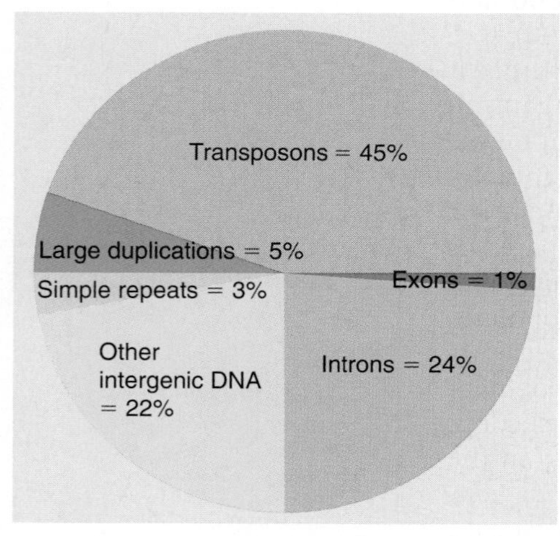

FIGURE 5.12 The largest component of the human genome consists of transposons. Other repetitive sequences include large duplications and simple repeats.

The sequence of the human genome emphasizes the importance of transposons. (Many transposons have the capacity to replicate themselves and insert into new locations. They may function exclusively as DNA elements or may have an active form that is RNA [see *Chapter 21, Transposons, Retroviruses, and Retroposons*]. Their distribution in the human genome is summarized in

Figure 21.35.) Most of the transposons in the human genome are nonfunctional; very few are currently active. However, the high proportion of the genome occupied by these elements indicates that they have played an active role in shaping the genome. One interesting feature is that some present genes originated as transposons and evolved into their present condition after losing the ability to transpose. Almost 50 genes appear to have originated in this manner.

Segmental duplication at its simplest involves the tandem duplication of some region within a chromosome (typically because of an aberrant recombination event at meiosis; see *Section 6.6, Unequal Crossing-over Rearranges Gene Clusters*). In many cases, however, the duplicated regions are on different chromosomes, implying that either there was originally a tandem duplication followed by a translocation of one copy to a new site, or that the duplication arose by some different mechanism altogether. The extreme case of a segmental duplication is when a whole genome is duplicated, in which case the diploid genome initially becomes tetraploid. As the duplicated copies develop differences from one another, the genome may gradually become effectively a diploid again, although homologies between the diverged copies leave evidence of the event. This is especially common in plant genomes. The present state of analysis of the human genome identifies many individual duplicated regions, but does not indicate whether there was a whole-genome duplication in the vertebrate lineage.

One curious feature of the human genome is the presence of sequences that do not appear to have coding functions, but that nonetheless show an evolutionary conservation higher than the background level. As detected by comparison with other genomes (such as the mouse genome), these represent about 5% of the total genome. Are these sequences associated with protein-coding sequences in some functional way? Their density on chromosome 18 is the same as elsewhere in the genome, although chromosome 18 has a significantly lower concentration of protein-coding genes. This suggests indirectly that their function is not connected with structure or expression of protein-coding genes.

KEY CONCEPTS

- Repeated sequences (present in more than one copy) account for >50% of the human genome.
- The great bulk of repeated sequences consist of copies of nonfunctional transposons.
- There are many duplications of large chromosome regions.

CONCEPT AND REASONING CHECK

What mechanisms may result in tandem duplications of chromosomal segments? What mechanism may account for a displaced duplication found on a different chromosome?

5.7 The Y Chromosome Has Several Male-Specific Genes

The sequence of the human genome has significantly extended our understanding of the role of the sex chromosomes. It is generally thought that the X and Y chromosomes have descended from a common (very ancient) autosome pair. Their evolution has involved a process in which the X chromosome has retained most of the original genes, whereas the Y chromosome has lost most of them.

The X chromosome is like the autosomes insofar as females have two copies and recombination can take place between them. The density of genes on the X chromosome is comparable to the density of genes on other chromosomes.

The Y chromosome is much smaller than the X chromosome and has many fewer genes. Its unique role results from the fact that only males have the Y chromosome, of which there is only one copy, so Y-linked loci are effectively haploid instead of diploid like all other human genes.

METHODS AND TECHNIQUES

Tracing Human History through the Y Chromosome

Because the Y chromosome does not undergo recombination along most of its length, genetic markers in the Y chromosome are completely linked and remain together as the chromosome is transmitted from generation to generation. The genetic relation between Y chromosomes can, therefore, be traced, because chromosomes that are closely related will share more alleles along their length than will more distantly related chromosomes. The set of alleles at two or more loci present in a particular chromosome is called a **haplotype**. For many genealogical studies of the Y chromosome, simple sequence repeat (SSR) polymorphisms are convenient because of their relatively high rate of mutation due to replication errors and the large number of alleles. The logic is that Y chromosomes with haplotypes that share alleles at each of 20–30 SSRs across the chromosome must have descended from the same ancestral Y chromosome in the very recent past. For haplotypes differing at a single locus the genetic relationship is less close, for those differing at two loci it is still less close, and so forth. This simple logic is the basis of tracing population history through Y-chromosome polymorphisms. Haplotypes that share many alleles have a more recent common ancestral Y chromosome than haplotypes that share fewer alleles. Furthermore, because the rate of SSR mutation can be estimated, the time at which the ancestral chromosome existed can be deduced. This reasoning forms the basis of the estimate that the most recent common ancestor of all extant human Y chromosomes existed 50,000 years ago. Such estimates are not highly precise, and there are many assumptions that must be made. Other studies using different markers yield estimates of 150,000 years. The reasons for the discrepancy are still unclear. Nevertheless, much can be learned about human population history through studies of the Y chromosome. For example, the legacy of Genghis Khan can be investigated by tracing the lineage of the Y chromosome.

At its maximum range, stretching from China to Russia through to the Middle East and then into Eastern Europe, the Mongol Empire of the 13th century comprised the largest land empire that history has known. The founder was born with the name Temujin around 1162. As a young man he organized a confederation of tribes, who around the year 1200 took to their small Mongolian ponies equipped with high wooden saddles and stirrups, and armed with bow and arrow began to conquer their neighbors. Soon thereafter, Temujin adopted the name Genghis Khan, which means Universal Ruler. He was often merciless, exterminating the men and boys of rebellious cities and kidnapping the women and girls. In answer to a question about the source of happiness, he is reputed to have said, "The greatest happiness is to vanquish your enemies, to chase them before you, to rob them of their wealth, to see those dear to them bathed in tears, to clasp to your bosom their wives and daughters." Through their multiple wives, concubines, and innumerable unrecorded sexual conquests, Genghis Khan and his descendants were very prolific. His eldest son Tushi had 40 acknowledged sons, and his grandson Kubilai Khan (under whom the Mongol Empire reached its greatest scope) had 22 acknowledged sons.

Although the legacy of Genghis Khan is well recorded in history, it was hardly expected that it would show up in studies of the Y chromosome. But genotyping studies of 32 markers along the Y chromosome of 2123 men sampled from throughout a large region of Asia yielded the remarkable result in **FIGURE B5.1**. Each circle represents a population sample, with its area proportional to the sample size. The red sectors denote the relative frequency of a group of nearly identical Y-chromosome haplotypes, whereas the white sectors represent the relative frequency of other haplotypes that are genetically much more diverse. The most recent

▸ **haplotype** The particular combination of alleles in a defined region of some chromosome-in effect, the genotype in miniature. Originally used to describe combinations of MHC alleles, it now may be used to describe particular combinations of RFLPs, SNPs, or other markers.

For many years, the Y chromosome was thought to carry almost no genes except for one or a few genes that determine maleness. The vast majority of the Y chromosome (>95% of its sequence) does not undergo crossing-over with the X chromosome, which led to the view that it could not contain active genes because there would be no means to prevent the accumulation of deleterious mutations. This region is flanked by short pseudoautosomal regions that exchange frequently with the X chromosome during male meiosis. It was originally called the *nonrecombining region*, but now has been renamed the *male-specific region*.

Detailed sequencing of the Y chromosome shows that the male-specific region contains three types of sequences, as illustrated in **FIGURE 5.13**.

- The *X-transposed sequences* consist of a total of 3.4 Mb comprising some large blocks resulting from a transposition from band q21 in the X chromosome

common ancestor of the closely related haplotypes is estimated to have existed 1000 ± 300 years ago. Furthermore, the geographical region in which the closely related haplotypes cluster is included largely within the Mongol Empire (shading). The sole exception is population 10, composed of the ethnic Hazara of Pakistan. This provides a clue to the origin of the closely related Y chromosomes, because the Hazara consider themselves to be of Mongol origin, and many claim to be direct male-line descendants of Genghis Khan. Whatever their origin, the closely related Y chromosomes are found in about 8% of the males throughout a large region of Asia (populations 1–16). Direct proof of the connection with Genghis Khan in principle could be obtained by determining the haplotype of the Y chromosome in material recovered from his grave. He died in 1227 from injuries sustained in a fall from a horse, and he is said to have been buried in secret at a site near his birthplace.

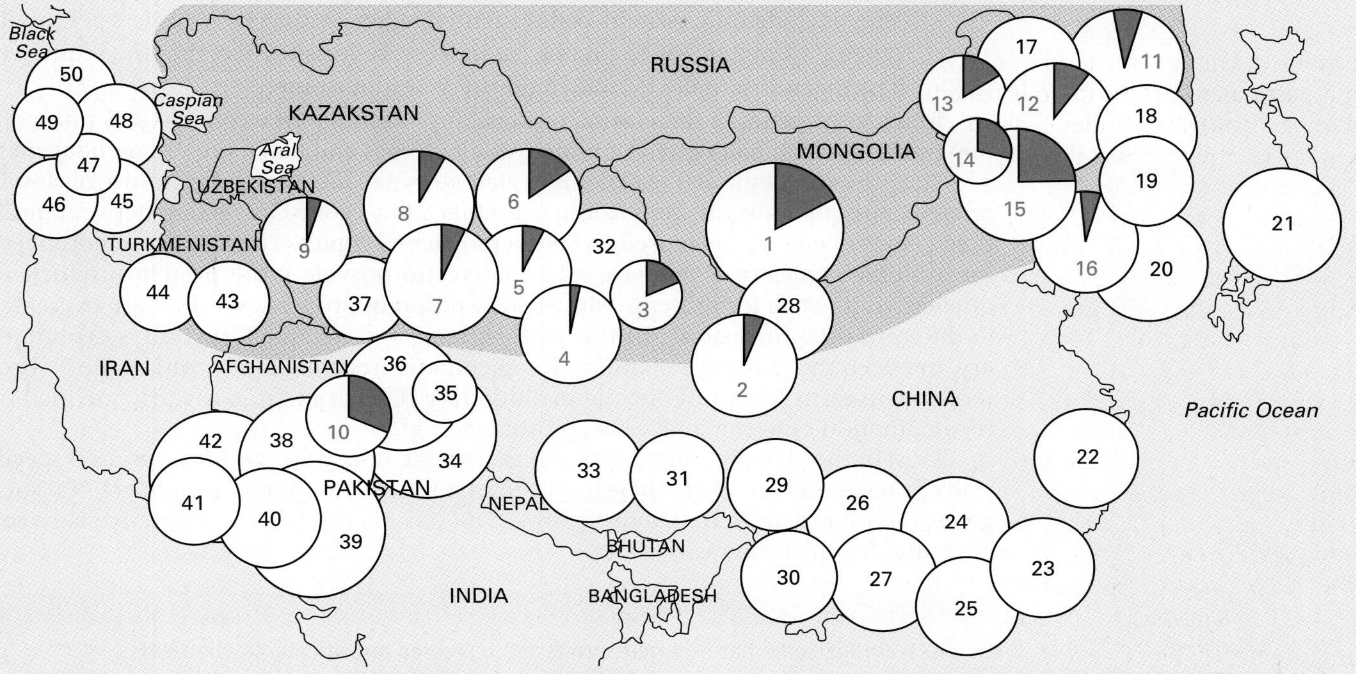

FIGURE B5.1 Distribution of Y-chromosome haplotypes (red), presumed to have descended from Genghis Khan or his close male relatives, among populations near and bordering the ancient Mongol Empire. The specific population groups are: (1) Mongolian, (2) Han [Gansu], (3) Chinese Kazak, (4) Han [Xinjiang], (5) Xibe, (6) Uyghur, (7) Kyrgyz, (8) Kazak, (9) Uzbek, (10) Hazara, (11) Hezhe, (12) Daur, (13) Ewenki, (14) Han [Inner Mongolian], (15) Inner Mongolian, (16) Manchu, (17) Oroqen, (18) Han [Heilongjiang], (19) Chinese Korean, (20) Korean, (21) Japanese, (22) Shezu, (23) Han [Guangdong], (24) Yaozu [Liannan], (25) Lizu, (26) Buyi, (27) Yaozu [Bama], (28) Huizu, (29) Han [Sichuan], (30) Hani, (31) Qiangzu, (32) Chinese Uyghur, (33) Tibetan, (34) Burusho, (35) Balti, (36) Kalash, (37) Tajik, (38) Baloch, (39) Parsi, (40) Makrani Negroid, (41) Makrani Baloch, (42) Brahui, (43) Turkmen, (44) Kurd, (45) Azeni, (46) Armenian, (47) Lezgi, (48) Georgian, (49) Ossetian, (50) Svan. [Adapted from Zerjal, T. *Am. J. Hum. Genet.* 72 (2003): 717–721.]

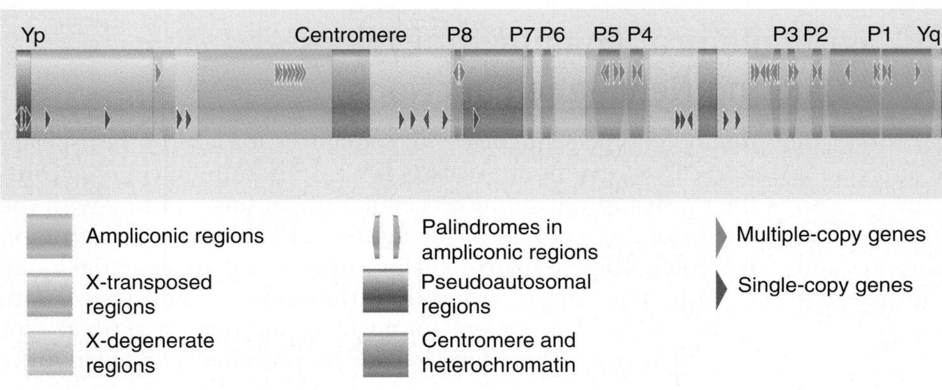

FIGURE 5.13 The Y chromosome consists of X-transposed regions, X-degenerate regions, and amplicons. The X-transposed X-degenerate regions have two and fourteen single-copy genes, respectively. The amplicons have eight large palindromes (P1–P8), which contain nine gene families. Each family contains at least two copies.

about 3 or 4 million years ago. This is specific to the human lineage. These sequences do not recombine with the X chromosome and have become largely inactive. They now contain only two active genes.

- The *X-degenerate segments* of the Y are sequences that have a common origin with the X chromosome (going back to the common autosome from which both X and Y have descended) and contain genes or pseudogenes related to X-linked genes. There are 14 active genes and 13 pseudogenes. The active genes have, in a sense, thus far defied the trend for genes to be eliminated from chromosomal regions that cannot recombine at meiosis.
- The *ampliconic segments* have a total length of 10.2 Mb and are internally repeated on the Y chromosome. There are eight large palindromic blocks. They include nine protein-coding gene families, with copy numbers per family ranging from 2 to 35. The name "amplicon" reflects the fact that the sequences have been internally amplified on the Y chromosome.

Totaling the genes in these three regions, the Y chromosome contains 156 transcription units, of which half represent protein-coding genes and half represent pseudogenes.

The presence of the active genes is explained by the fact that the existence of closely related gene copies in the ampliconic segments allows gene conversion between multiple copies of a gene to be used to regenerate active copies. The most common needs for multiple copies of a gene are quantitative (to provide more protein product) or qualitative (to code for proteins with slightly different properties or that are expressed in different times or places). In this case, though, the essential function is evolutionary. In effect, the existence of multiple copies allows recombination within the Y chromosome itself to substitute for the evolutionary diversity that is usually provided by recombination between allelic chromosomes.

Most of the protein-coding genes in the ampliconic segments are expressed specifically in testis and are likely to be involved in male development. If there are ~60 such genes out of a total human gene set of ~25,000, then the genetic difference between male and female humans is ~0.2%.

KEY CONCEPTS

- The Y chromosome has ~60 genes that are expressed specifically in the testis.
- The male-specific genes are present in multiple copies in repeated chromosomal segments.
- Gene conversion between multiple copies allows the active genes to be maintained during evolution.

CONCEPT AND REASONING CHECK

Why was it originally assumed that the human Y chromosome would have very few active genes?

5.8 Morphological Complexity Evolves by Adding New Gene Functions

Comparison of the human genome sequence with sequences found in other species is revealing about the process of evolution. **FIGURE 5.14** analyzes human genes according to the breadth of their distribution among all cellular organisms. Starting with the most generally distributed (top right corner of the figure), 21% of genes are common to eukaryotes and prokaryotes. These tend to code for proteins that are essential for all living forms—typically basic metabolism, replication, transcription, and translation. Moving clockwise, another 32% of genes are found in eukaryotes in general—for example, they may be found in yeast. These tend to code for proteins involved in functions that are general to eukaryotic cells but not to bacteria—for example, they may be concerned with specifying organelles or cytoskeletal components. Another 24% of genes are generally found in animals. These include genes necessary for multicellularity and for development of different tissue types. Twenty-two percent of genes are unique to vertebrates. These mostly code for proteins of the immune and nervous systems; they code for very few enzymes, consistent with the idea that enzymes have ancient origins, and that metabolic pathways originated early in evolution. We see, therefore, that the evolution of more complex morphology and specialization requires addition of groups of genes representing the necessary new functions.

One way to define essential proteins is to identify the proteins present in all proteomes. Comparing the human proteome in more detail with the proteomes of other organisms, 46% of the yeast proteome, 43% of the worm proteome, and 61% of the fly proteome is represented in the human proteome. A key group of ~1300 proteins is present in all four proteomes. The common proteins are basic "housekeeping" proteins required for essential functions, falling into the types summarized in **FIGURE 5.15**. The main functions are concerned with transcription and translation (35%), metabolism (22%), transport (12%), DNA replication and modification (10%), protein folding and degradation (8%), and cellular processes (6%).

One of the striking features of the human proteome is that it has many unique proteins compared with other eukaryotes, but has relatively few unique protein domains (portions of proteins having a specific function). Most protein domains appear to be common to the animal kingdom. There are many unique protein architectures, however, defined as unique combinations of domains. **FIGURE 5.16** shows that the greatest proportion of unique proteins are transmembrane and extracellular proteins. In yeast, the vast majority of architectures are concerned with intracellular proteins. About twice as many intracellular architectures are found in flies (or worms), but there is a very striking higher proportion of transmembrane and extracellular proteins, as might be expected from the additional functions required for the interactions between the cells of a multicellular organism. The additions in intracellular architectures required in a vertebrate (human) is relatively small, but there is again a higher proportion of transmembrane and extracellular architectures.

It has long been known that the genetic difference between humans and chimpanzees (our nearest relative) is very small, with ~99% identity between genomes. The sequence of the chimpanzee genome now allows us to investigate the 1% of differences in more detail to see whether features responsible for "humanity" can be identified. The comparison shows 35×10^6 nucleotide substitutions (1.2% sequence difference overall), 5×10^6 deletions or insertions (making ~1.5% of the euchromatic sequence specific to each species), and many chromosomal rearrangements. Homologous proteins are usually very similar; 29% are identical, and in most cases there are only one or two amino acid differences in the protein

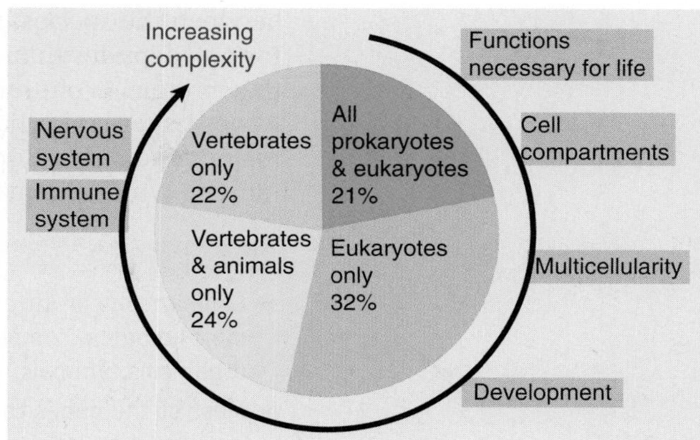

FIGURE 5.14 Human genes can be classified according to how widely their homologues are distributed in other species.

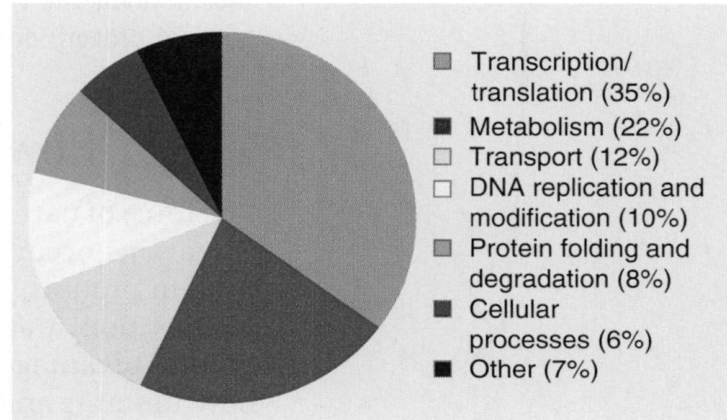

FIGURE 5.15 Common eukaryotic proteins are concerned with essential cellular functions.

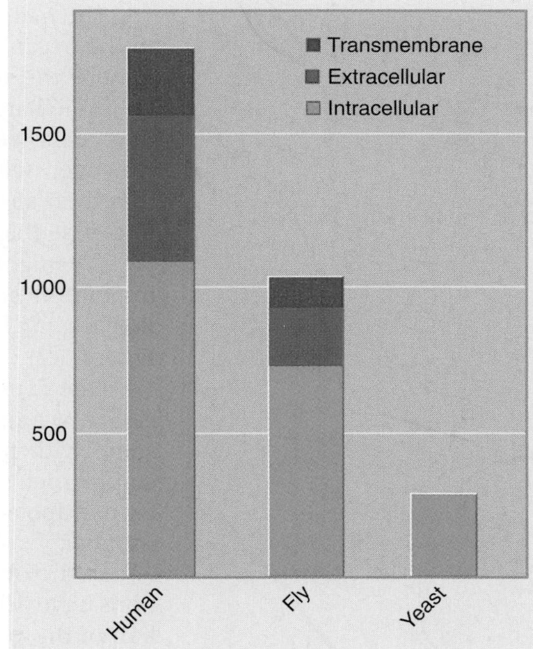

FIGURE 5.16 Increasing complexity in eukaryotes is accompanied by accumulation of new proteins for transmembrane and extracellular functions.

between the species. In fact, nucleotide substitutions occur less often in genes coding for polypeptides than are likely to be involved in specifically human traits, suggesting that protein evolution is not a major factor in human–chimpanzee differences. This leaves larger-scale changes in gene structure and/or changes in gene regulation as the major candidates. Some 25% of nucleotide substitutions occur in CpG dinucleotides (among which are many potential regulator sites).

KEY CONCEPTS

- Comparisons of different genomes show a positive correlation between gene number and morphological complexity as additional genes are needed in eukaryotes, multicellular organisms, animals, and vertebrates.
- Most of the genes that are unique to vertebrates are concerned with the immune or nervous systems.

CONCEPT AND REASONING CHECK

Account for the observation that the human proteome has many unique proteins but few unique protein domains.

5.9 How Many Genes Are Essential?

The force of natural selection ensures that functional genes are retained in the genome. Mutations occur at random, and a common effect in an ORF will be to damage the protein product. An organism with a damaging mutation will be at a disadvantage in competition, and ultimately the mutation may be eliminated. However, the frequency of a disadvantageous allele in the population is balanced between the generation of new mutants and the elimination of the allele by selection. Reversing this argument, whenever we see an intact, expressed ORF in the genome, we assume that its product plays a useful role in the organism. Natural selection must have prevented mutations from accumulating in the gene. The ultimate fate of a gene that ceases to be functional is to accumulate mutations until it is no longer recognizable.

The maintenance of a gene implies that it does not confer a selective disadvantage to the organism. In the course of evolution, though, even a small relative advantage may be the subject of natural selection, and a phenotypic defect may not necessarily be immediately detectable as the result of a mutation. Also, in diploid organisms, a new recessive mutation may be "hidden" in heterozygous form for many generations. However, we should like to know how many genes are actually essential, meaning that their absence is lethal to the organism. In the case of diploid organisms, it means of course that the homozygous null mutation is lethal.

We might assume that the proportion of essential genes will decline with increase in genome size, given that larger genomes may have multiple related copies of particular gene functions. So far this expectation has not been borne out by the data (see Figure 5.2).

One approach to the issue of gene number is to determine the number of essential genes by mutational analysis. If we saturate some specified region of the chromosome with mutations that are lethal, the mutations should map into a number of complementation groups that correspond to the number of lethal loci in that region. By extrapolating to the genome as a whole, we may calculate the total essential gene number.

In the organism with the smallest known genome (*M. genitalium*), random insertions have detectable effects only in about two thirds of the genes. Similarly, fewer than half of the genes of *E. coli* appear to be essential. The proportion is even lower in the yeast *S. cerevisiae*. When insertions were introduced at random into the genome in one early analysis, only 12% were lethal, and another 14% impeded growth. The majority

(70%) of the insertions had no effect. A more systematic survey based on completely deleting each of 5916 genes (>96% of the identified genes) shows that only 18.7% are essential for growth on a rich medium (that is, when nutrients are fully provided). **FIGURE 5.17** shows that these include genes in all categories. The only notable concentration of defects is in genes coding for products involved in protein synthesis, where ~50% are essential. Of course, this approach underestimates the number of genes that are essential for the yeast to live in the wild, when it is not so well provided with nutrients.

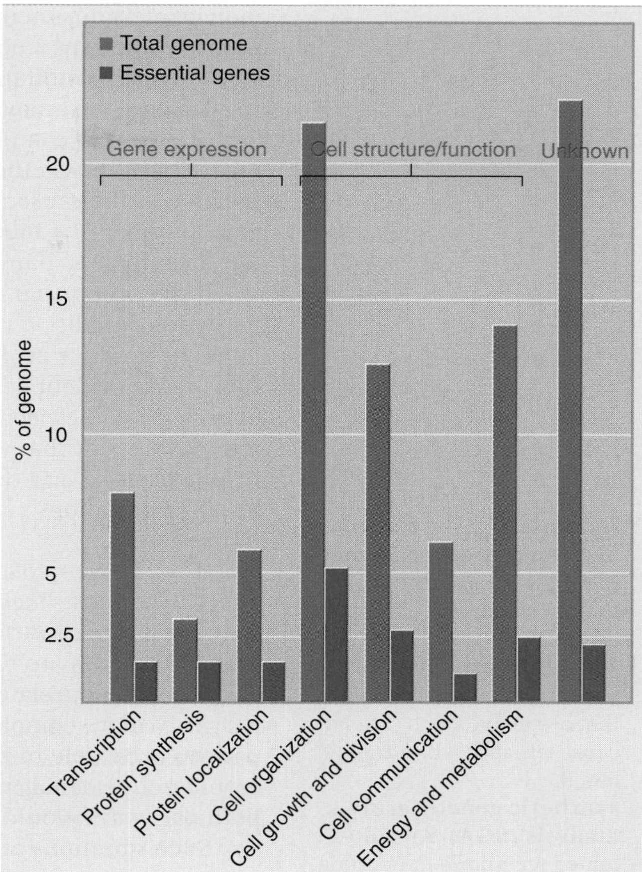

FIGURE 5.17 Essential yeast genes are found in all classes. Blue bars show total proportion of each class of genes, red bars show those that are essential.

FIGURE 5.18 summarizes the results of a systematic analysis of the effects of loss of gene function in the worm *C. elegans*. The sequences of individual genes were predicted from the genome sequence, and by targeting an inhibitory RNA against these sequences (see *Section 13.8, Eukaryotes Contain Regulator RNAs*), a large collection of worms was made in which one (predicted) gene was prevented from functioning in each worm. Detectable effects on the phenotype were only observed for 10% of these knockdowns, suggesting that most genes do not play essential roles.

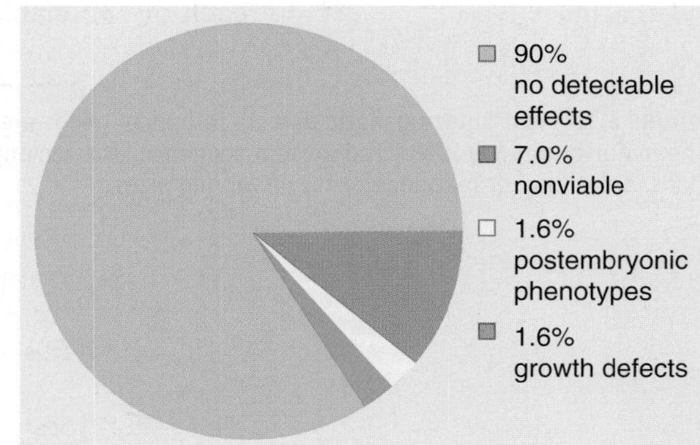

FIGURE 5.18 A systematic analysis of loss of function for 86% of worm genes shows that only 10% have detectable effects on the phenotype.

There is a greater proportion of essential genes (21%) among those worm genes that have counterparts in other eukaryotes, suggesting that widely conserved genes tend to have more basic functions. There is also an increased proportion of essential genes among those that are present in only one copy per haploid genome, compared with those where there are multiple copies of related or identical genes. This suggests that many of the multiple genes might be relatively recent duplications that can substitute for one another's functions.

Extensive analyses of essential gene number in a multicellular eukaryote have been made in *Drosophila* through attempts to correlate visible aspects of chromosome structure with the number of functional genetic units. The notion that this might be possible originated from the presence of bands in the polytene chromosomes of *D. melanogaster*. (These chromosomes are found at certain developmental stages and represent an unusually extended physical form, in which a series of bands [more formally called chromomeres] are evident; see *Section 23.10, Polytene Chromosomes Form Bands That Expand at Sites of Gene Expression*.) From the time of the early concept that the bands might represent a linear order of genes, there has been an attempt to correlate the organization of genes with the organization of bands. There are ~5000 bands in the *D. melanogaster* haploid set; they vary in size over an order of magnitude, but on average there is ~20 kb of DNA per band.

The basic approach is to saturate a chromosomal region with mutations. Usually the mutations are simply collected as lethals, without analyzing the cause of the lethality. Any mutation that is lethal is taken to identify a locus that is essential for the organism. Sometimes mutations cause visible deleterious effects short of lethality, in which case we also define them as essential loci. When the mutations are placed into complementation groups, the number can be compared with the number of bands in the region, or individual complementation groups may even be assigned to individual bands. The purpose of these experiments has been to determine whether there is a consistent relationship between bands and genes. For example, does every band contain a single gene?

Totaling the analyses that have been carried out over the past 35 years, the number of essential complementation groups is ~70% of the number of bands. It is an open question whether there is any functional significance to this relationship. Irrespective of the cause, the equivalence gives us a reasonable estimate for the essential gene number of ~3600. By any measure, the number of essential loci in *Drosophila* is significantly less than the total number of genes.

If the proportion of essential human genes is similar to that of other eukaryotes, we would predict a range of ~4000 to 8000 genes, in which mutations would be lethal or produce evidently damaging effects. At present, 1300 genes in which mutations cause evident defects have been identified. This is a substantial proportion of the expected total, especially in view of the fact that many lethal genes may act so early in development that we never see their effects. This sort of bias may also explain the results in **FIGURE 5.19**, which show that the majority of known genetic defects are due to point mutations (where there is more likely to be at least some residual function of the gene).

- **redundancy** The concept that two or more genes may fulfill the same function, so that no single one of them is essential.
- **synthetic lethality** This occurs when two mutations that are viable by themselves cause lethality when combined.
- **synthetic genetic array analysis (SGA)** An automated technique in budding yeast whereby a mutant is crossed to an array of approximately 5000 deletion mutants to determine if the mutations interact to cause a synthetic lethal phenotype.

How do we explain the persistence of genes whose deletion appears to have no effect? The most likely explanation is that the organism has alternative ways of fulfilling the same function. The simplest possibility is that there is **redundancy**, and that some genes are present in multiple copies. This is certainly true in some cases, in which multiple (related) genes must be knocked out in order to produce an effect. In a slightly more complex scenario, an organism might have two separate biochemical pathways capable of providing some activity. Inactivation of either pathway by itself would not be damaging, but the simultaneous occurrence of mutations in genes from both pathways would be deleterious.

Such situations can be tested by combining mutations. In this approach, deletions in two genes, neither of which is lethal by itself, are introduced into the same strain. If the double mutant dies, the strain is called a **synthetic lethal**. This technique has been used to great effect with yeast, where the isolation of double mutants can be automated. The procedure is called **synthetic genetic array analysis (SGA)**. **FIGURE 5.20** summarizes the results of an analysis in which an SGA screen was made for each of 132 viable deletions by testing whether it could survive in combination with

FIGURE 5.19 Most known genetic defects in human genes are due to point mutations. The majority directly affect the protein sequence. The remainder are due to insertions, deletions, or rearrangements of varying sizes.

Missense/nonsense	58%
Splicing	10%
Regulatory	<1%
Small deletions	16%
Small insertions	6%
Large deletions	5%
Large rearrangements	2%

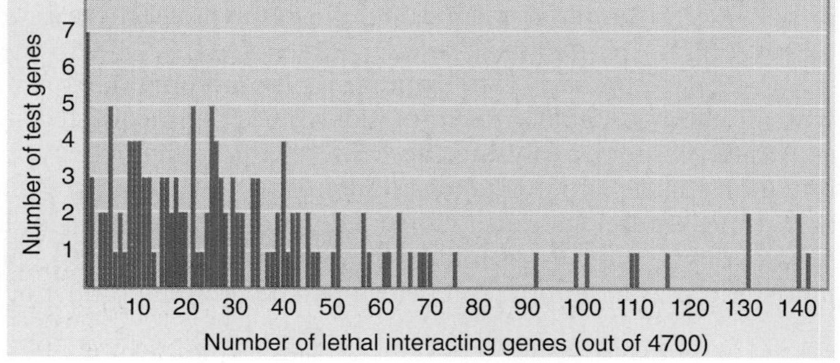

FIGURE 5.20 All 132 mutant test genes have some combinations that are lethal when they are combined with each of 4700 nonlethal mutations. The chart shows how many lethal interacting genes there are for each test gene.

any one of 4700 viable deletions. Every one of the tested genes had at least one partner with which the combination was lethal, and most of the tested genes had many such partners; the median is ~25 partners, and the greatest number is shown by one tested gene that had 146 lethal partners. A small proportion (~10%) of the interacting mutant pairs code for polypeptides that interact physically.

This result goes some way toward explaining the apparent lack of effect of so many deletions. Natural selection will act against these deletions when they are found in lethal pairwise combinations. To some degree, the organism is protected against the damaging effects of mutations by built-in redundancy. However, there is a price in the form of accumulating the "genetic load" of mutations that are not deleterious in themselves, but that may cause serious problems when combined with other such mutations in future generations. Presumably, the loss of the individual genes in such circumstances produces a sufficient disadvantage to maintain the active gene during the course of evolution.

KEY CONCEPTS

- Not all genes are essential. In yeast and flies, deletions of <50% of the genes have detectable effects.
- When two or more genes are redundant, a mutation in any one of them may not have detectable effects.
- We do not fully understand the persistence of genes that are apparently dispensable in the genome.

CONCEPT AND REASONING CHECK

What are some of the problems in measuring the number of essential genes by mutational analysis?

5.10 About 10,000 Genes Are Expressed at Widely Differing Levels in a Eukaryotic Cell

The proportion of DNA containing protein-coding genes being expressed in a specific cell at a specific time can be determined by the amount of the DNA that can hybridize with the mRNAs isolated from that cell. Such a saturation analysis conducted for many cell types at various times typically identifies ~1% of the DNA being expressed as mRNA. From this we can calculate the number of protein-coding genes, so long as we know the average length of an mRNA. For a unicellular eukaryote such as yeast, the total number of expressed protein-coding genes is ~4000. For somatic tissues of multicellular eukaryotes, including both plants and vertebrates, the number usually is 10,000 to 15,000. (The only consistent exception to this type of value is presented by mammalian brain cells, for which much larger numbers of genes appear to be expressed, although the exact number is not certain.)

The average number of molecules of each mRNA per cell is called its **abundance**. It can be calculated quite simply if the total mass of a specific mRNA species in the cell is known. For example, in chick oviduct cells, the total mRNA can be accounted for as 100,000 copies of ovalbumin mRNA, 4000 copies of each of 7 or 8 other mRNAs, and only ~5 copies of each of the 13,000 remaining mRNAs.

▸ **abundance** The average number of mRNA molecules per cell.

We can divide the mRNA population into two general classes, according to their abundance:

- The oviduct is an extreme case, with so much of the mRNA represented by only one species, but most cells do contain a small number of RNAs present in many copies each. This **abundant mRNA** component typically consists of <100 different mRNAs present in 1000 to 10,000 copies per cell. It often corresponds to a major part of the mass, approaching 50% of the total mRNA.

▸ **abundant mRNA** Consists of a small number of individual species, each present in a large number of copies per cell.

- About half of the mass of the mRNA consists of a large number of sequences, of the order of 10,000, each represented by only a small number of copies in the mRNA—say, <10. This is the **scarce mRNA** or **complex mRNA** class.

▸ **scarce mRNA** mRNA that consists of a large number of individual mRNA species, each present in very few copies per cell. This accounts for most of the sequence complexity in RNA.

▸ **complex mRNA** *See* **scarce mRNA**.

Many somatic tissues of higher eukaryotes have an expressed gene number in the range of 10,000 to 20,000. How much overlap is there between the genes expressed in different tissues? For example, the expressed gene number of chick liver is ~11,000 to 17,000, compared with the value for oviduct of ~13,000 to 15,000. How many of these two sets of genes are identical? How many are specific for each tissue? These questions are usually addressed by analyzing the transcriptome—the set of sequences represented in RNA.

We see immediately that there are likely to be substantial differences among the genes expressed in the abundant class. Ovalbumin, for example, is synthesized only in the oviduct, and not at all in the liver. This means that 50% of the mass of mRNA in the oviduct is specific to that tissue.

The abundant mRNAs represent only a small proportion of the number of expressed genes, though. In terms of the total number of genes of the organism, and of the number of changes in transcription that must be made between different cell types, we need to know the extent of overlap between the genes represented in the scarce mRNA classes of different cell phenotypes.

Comparisons between different tissues show that, for example, ~75% of the sequences expressed in liver and oviduct are the same. In other words, ~12,000 genes are expressed in both liver and oviduct, ~5000 additional genes are expressed only in liver, and ~3000 additional genes are expressed only in oviduct.

The scarce mRNAs overlap extensively. Between mouse liver and kidney, ~90% of the scarce mRNAs are identical, leaving a difference between the tissues of only 1000 to 2000 in terms of the number of expressed genes. The general result obtained in several comparisons of this sort is that only ~10% of the mRNA sequences of a cell are unique to it. The majority of sequences are common to many—perhaps even all—cell types.

This suggests that the common set of expressed gene functions, numbering perhaps ~10,000 in mammals, comprise functions that are needed in all cell types. Sometimes this type of function is referred to as a **housekeeping gene** or **constitutive gene**. It contrasts with the activities represented by specialized functions (such as ovalbumin or globin) needed only for particular cell phenotypes. These are sometimes called **luxury genes**.

▸ **housekeeping gene** A gene that is (theoretically) expressed in all cells because it provides basic functions needed for sustenance of all cell types.

▸ **constitutive gene** *See* **housekeeping gene**.

▸ **luxury gene** A gene coding for a specialized function, synthesized (usually) in large amounts in particular cell types.

KEY CONCEPTS

- In any given cell, most genes are expressed at a low level.
- Only a small number of genes, whose products are specialized for the cell type, are highly expressed.
- mRNAs expressed at low levels overlap extensively when different cell types are compared.
- The abundantly expressed mRNAs are usually specific for the cell type.
- ~10,000 expressed genes may be common to most cell types of a higher eukaryote.

CONCEPT AND REASONING CHECK

In a particular cell, why are some genes heavily expressed and others rarely expressed?

5.11 Expressed Gene Number Can Be Measured *en Masse*

Recent technology allows more systematic and accurate estimates of the number of expressed protein-coding genes. One approach (serial analysis of gene expression, or SAGE) allows a unique sequence tag to be used to identify each mRNA. The technology then allows the abundance of each tag to be measured. This approach identifies

4665 expressed genes in *S. cerevisiae* growing under normal conditions, with abundances varying from 0.3 to >200 transcripts/cell. This means that ~75% of the total gene number (~6000) is expressed under these conditions. **FIGURE 5.21** summarizes the number of different mRNAs found at each different abundance level.

The most powerful new technology uses chips that contain **microarrays**, arrays of many tiny DNA oligonucleotide samples. Their construction is made possible by knowledge of the sequence of the entire genome. In the case of *S. cerevisiae*, each of 6181 ORFs is represented on the microarray by twenty 25-mer oligonucleotides that perfectly match the sequence of the message and twenty mismatch oligonucleotides that differ at one base position. The expression level of any gene is calculated by subtracting the average signal of a mismatch from its perfect match partner. The entire yeast genome can be represented on four chips. This technology is sensitive enough to detect transcripts of 5460 genes (~90% of the genome), and shows that many genes are expressed at low levels, with abundances of 0.1 to 0.2 transcript/cell. An abundance of <1 transcript/cell means that not all cells have a copy of the transcript at any given moment.

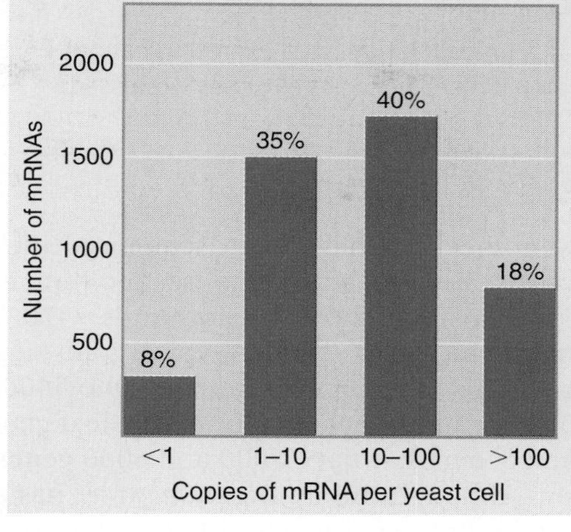

FIGURE 5.21 The abundances of yeast mRNAs vary from <1 per cell (meaning that not every cell has a copy of the mRNA) to >100 per cell (coding for the more abundant proteins).

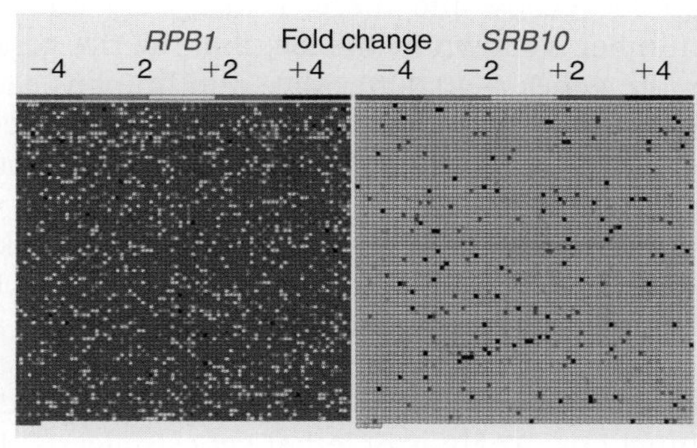

FIGURE 5.22 DNA microarray analysis allows change in expression of each gene to be measured. Each square represents one gene (top left is first gene on chromosome I, bottom right is last gene on chromosome XVI). Change in expression relative to wild type is indicated by red (reduction), white (no change), or blue (increase). Photos courtesy of Rick A. Young, Whitehead Institute, Massachusetts Institute of Technology.

The technology allows not only measurement of levels of gene expression, but also detection of differences in expression in mutant cells compared to wild-type cells growing under different growth conditions, and so on. The results of comparing two states are expressed in the form of a grid, in which each square represents a particular gene and the relative change in expression is indicated by color. The left part of **FIGURE 5.22** shows the effect of a mutation in the largest subunit RNA polymerase II (*RPB1*), the enzyme that produces mRNA, which, as might be expected, causes the expression of most genes to be heavily reduced. By contrast, the right part shows that a mutation in an ancillary component of the transcription apparatus (*SRB10*) has much more restricted effects, causing increases in expression of some genes.

▸ **microarray** An arrayed series of thousands of tiny DNA oligonucleotide samples imprinted on a small chip. mRNAs can be hybridized to microarrays to assess the amount and level of gene expression.

The extension of this technology to animal cells will allow the general descriptions based on RNA hybridization analysis to be replaced by exact descriptions of the genes that are expressed, and the abundances of their products, in any given cell type. A gene expression map of *D. melanogaster* detects transcriptional activity in some stage of the life cycle in almost all (93%) of predicted genes and shows that 40% have alternatively spliced forms.

KEY CONCEPTS

- DNA microarray technology allows a snapshot to be taken of the expression of the entire genome in a yeast cell.
- ~75% (~4500 genes) of the yeast genome is expressed under normal growth conditions.
- DNA microarray technology allows detailed comparisons of related animal cells to determine (for example) the differences in expression between a normal cell and a cancer cell.

CONCEPT AND REASONING CHECK

If a unicellular yeast is expressing about 75% of its genes under normal conditions, why aren't the other 25% of genes expressed?

5.12 Summary

Genomes that have been sequenced include those of many bacteria and archaea, yeasts, a worm, a fly, a mouse, and human. The minimum number of genes required to make a living cell (a parasite) is ~470. The minimum number required to make a free-living cell is ~1700. A typical gram-negative bacterium has ~1500 genes. Genomes of strains of *E. coli* vary from 4300 to 5400 genes. The average bacterial gene is ~1000 bp long and is separated from the next gene by a space of ~100 bp. The yeasts *S. pombe* and *S. cerevisiae* have 5000 and 6000 genes, respectively.

Although the fly *D. melanogaster* has a larger genome than the worm *C. elegans*, the fly has fewer genes (13,600) than the worm (18,500). The plant *Arabidopsis* has 25,000 genes, and the lack of a clear relationship between genome size and gene number is shown by the fact that the rice genome is 4× larger but contains only 50% more genes (~40,000). Mice and humans each have 20,000 to 25,000 genes, which is much less than had been originally expected. The complexity of development of an organism may depend on the nature of the interactions between genes as well as their total number.

About 8000 genes are common to prokaryotes and eukaryotes and are likely to be involved in basic functions. A further 12,000 genes are found in multicellular organisms. Another 8000 genes are found in animals, and an additional 8000 (largely involved with the immune and nervous systems) are found in vertebrates. In each organismal genome that has been sequenced, only ~50% of the genes have defined functions. Analysis of lethal genes suggests that only a minority of genes is essential in each organism.

The sequences comprising a eukaryotic genome can be classified in three groups: nonrepetitive sequences are unique; moderately repetitive sequences are dispersed and repeated a small number of times in the form of related, but not identical, copies; and highly repetitive sequences are short and usually repeated as tandem arrays. The proportions of the types of sequence are characteristic for each genome, although larger genomes tend to have a smaller proportion of nonrepetitive DNA. Almost 50% of the human genome consists of repetitive sequences, the vast majority corresponding to transposon sequences. Most structural genes are located in nonrepetitive DNA. The complexity of nonrepetitive DNA is a better reflection of the complexity of the organism than the total genome complexity.

Genes are expressed at widely varying levels. There may be 10^5 copies of mRNA for an abundant gene whose protein is the principal product of the cell, 10^3 copies of each mRNA for <10 moderately abundant messages, and <10 copies of each mRNA for >10,000 scarcely expressed genes. Overlaps between the mRNA populations of cells of different phenotypes are extensive; the majority of mRNAs are present in most cells.

CHAPTER QUESTIONS

1. What percent of the human genome codes for polypeptide (i.e., exons, not introns)?
   A. 1%
   B. 10%
   C. 25%
   D. 40%
2. What proportion of the total number of human genes is thought to be alternatively spliced?
   A. 24%
   B. 45%
   C. 60%
   D. 72%
3. Which of the following species has the highest proportion of gene families?
   A. *S. cerevisiae*
   B. *D. melanogaster*
   C. *C. elegans*
   D. *A. thaliana*
4. What is the estimated gene family number found in a multicellular eukaryote?
   A. ~1,200
   B. ~4,500
   C. ~12,000
   D. ~18,000
5. What percent of the *D. melanogaster* genome is unique to this genus (i.e., not shared with other eukaryotes)?
   A. 20%
   B. 45%
   C. 70%
   D. 90%
6. The minimal number of genes required in a free-living cell is:
   A. about 470 genes.
   B. about 1500 genes.
   C. about 5400 genes.
   D. about 20,500 genes.
7. The first vertebrate genome to be sequenced was that of the:
   A. mouse.
   B. rat.
   C. human.
   D. chimpanzee.
8. What is the approximate genetic identity between humans and chimpanzees?
   A. 67%
   B. 85%
   C. 92%
   D. 99%
9. What percentage of yeast genes has been determined to be essential (by analysis of deletion mutants)?
   A. 12%
   B. 19%
   C. 27%
   D. 43%

10. What is the approximate number of *expressed* genes that is thought to be common to most cell types in higher eukaryotes?
   A. 4500
   B. 10,000
   C. 15,000
   D. 21,000

KEY TERMS

abundance
abundant mRNA
complex mRNA
constitutive gene
haplotype
horizontal transfer
housekeeping gene
luxury gene
microarray
monocistronic mRNA
orthologous genes (orthologs)
pathogenicity islands
polycistronic mRNA
redundancy
scarce mRNA
synthetic genetic array analysis (SGA)
synthetic lethality

FURTHER READING

Adams, M. D. *et al.* (2000). The genome sequence of *D. melanogaster*. *Science* 287, 2185–2195.
The publication of most of the genome sequence of the long-time genetic model organism, the common fruit fly, with some initial analysis of genome structure and gene number.

Arabidopsis Initiative (2000). Analysis of the genome sequence of the flowering plant *Arabidopsis thaliana*. *Nature* 408, 796–815.
The publication of about 92% of the genome sequence of the genetic model organism for plants, including some comparison to the genome sequences of the other (at the time) recently published genome sequences of *D. melanogaster* and *C. elegans*.

Bentley, S. D. and Parkhill, J. (2004). Comparative genomic structure of prokaryotes. *Annu. Rev. Genet.* 38, 771–792.
A broad-ranging review of comparison of bacterial genome structures.

Blattner, F. R. *et al.* (1997). The complete genome sequence of *Escherichia coli* K-12. *Science* 277, 1453–1474.
The publication of the complete 4.6 Mb genome sequence of *E. coli*, the stalwart of molecular biology, with comparisons to several other prokaryotic genome sequences available at the time.

C. elegans Sequencing Consortium (1998). Genome sequence of the nematode *C. elegans*: a platform for investigating biology. *Science* 282, 2012–2022.
The publication of the nearly complete 97 Mb genome sequence of this model genetic organism, often referenced in this book as "the worm." Includes an analysis of genome structure and of gene homology with other species.

International Human Genome Sequencing Consortium (2004). Finishing the euchromatic sequence of the human genome. *Nature* 431, 931–945.
A "final draft" of the human genome sequence with greater coverage and accuracy than the 2001 "draft sequence" publication. At this point the number of human genes was estimated to be 20–25,000.

Phizicky, E., Bastiaens, P. I., Zhu, H., Snyder, M., and Fields, S. (2003). Protein analysis on a proteomic scale. *Nature* 422, 208–215.
A review of the approaches to proteomic characterization and analysis.

Tong, A. H. *et al.* (2004). Global mapping of the yeast genetic interaction network. *Science* 303, 808–813.
A large-scale study of interactions between genes conducted by crossing new mutations with known mutations and scoring effects. The study determined genetic components of biochemical pathways and their genomic locations.

Wood, V. *et al.* (2002). The genome sequence of *Schizosaccharomyces pombe*. *Nature* 415, 871–880.
The annotated genome sequence of "fission yeast," with comparison to the genome of "budding yeast" (*S. cerevisiae*). This species has a small number (~4800) of protein-coding genes for a eukaryote.

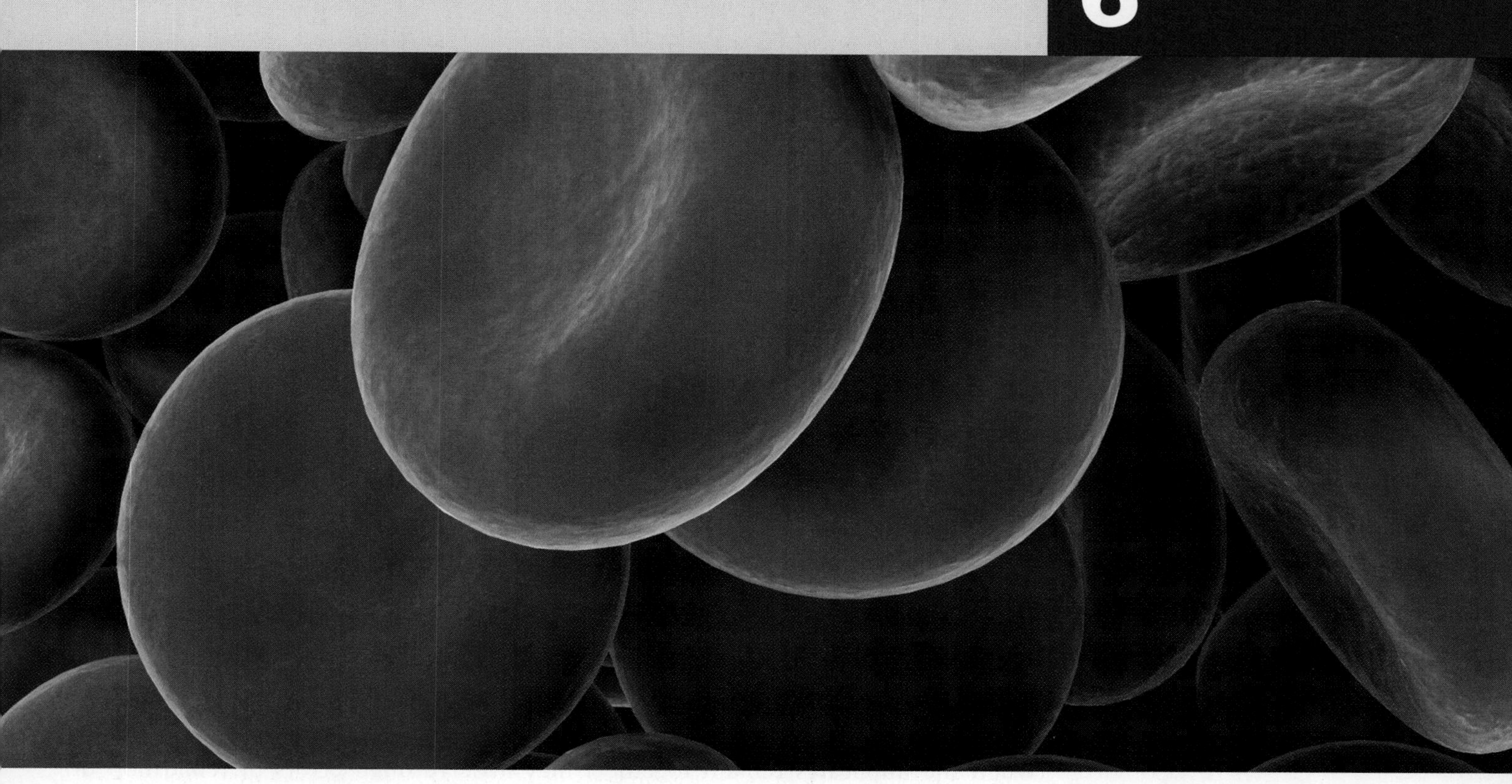

Scanning electron micrograph (SEM) of red blood cells. Genes that encode the protein subunits of hemoglobin, the oxygen-carrying complex found in red blood cells, are members of a large gene family that have evolved by duplication and divergence. © Sebastian Kaulitzki/ ShutterStock. Inc.

Clusters and Repeats

CHAPTER OUTLINE

6.1 Introduction

‣ **gene family** A set of genes within a genome that code for related or identical proteins or RNAs. The members were derived by duplication of an ancestral gene followed by accumulation of changes in sequence between the copies. Most often the members are related but not identical.

‣ **pseudogenes** Inactive but stable components of the genome derived by mutation of an ancestral active gene. Usually they are inactive because of mutations that block transcription or translation or both.

‣ **translocation** The reciprocal or nonrecriprocal exchange of chromosomal material between nonhomologous chromosomes.

‣ **gene cluster** A group of adjacent genes that are identical or related.

A set of genes descended by duplication and variation from a single ancestral gene is called a **gene family**. Its members may be clustered together or dispersed on different chromosomes (or a combination of both). Genome analysis to identify paralogous sequences shows that many genes belong to families; the 20,000 to 25,000 genes identified in the human genome fall into ~15,000 families, so the average gene has ~2 relatives in the genome (see Figure 5.7). Gene families vary enormously in the degree of relatedness between members, from those consisting of multiple identical members to those for which the relationship is quite distant. Genes are usually related only by their exons, with introns having diverged (see *Section 3.4, Exon Sequences Are Conserved but Introns Vary*). Genes may also be related by only some of their exons, whereas others are unique (see *Section 3.8, Some Exons Can Be Equated with Protein Functions*).

Some members of the gene family may evolve to become **pseudogenes**. Pseudogenes (ψ) are defined by their possession of sequences that are related to those of the functional genes, but that cannot be translated into a functional polypeptide.

Some pseudogenes have the same general structure as functional genes, with sequences corresponding to exons and introns in the usual locations. They may have been rendered inactive by mutations that prevent any or all of the stages of gene expression. The changes can take the form of abolishing the signals for initiating transcription, preventing splicing at the exon–intron junctions, or prematurely terminating translation.

The initial event that allows related exons or genes to develop is a duplication, when a copy of some sequence is generated within the genome. *Tandem duplication* (when the duplicates remain together) may arise through errors in replication or recombination. Separation of the duplicates can occur by a **translocation** that transfers material from one chromosome to another. A duplicate at a new location may also be produced directly by a transposition event that is associated with copying a region of DNA from the vicinity of the transposon. Duplications of intact genes, collections of exons, or even individual exons may occur. When an intact gene is involved, duplication generates two copies of a gene whose activities are indistinguishable, but then usually the copies diverge as each accumulates different substitutions.

The members of a structural gene family usually have related or even identical functions, although they may be expressed at different times or in different cell types. For example, different globin proteins are expressed in embryonic and adult red blood cells, whereas different actins are utilized in muscle and nonmuscle cells. When genes have diverged significantly, or when only some exons are related, the proteins may have different functions.

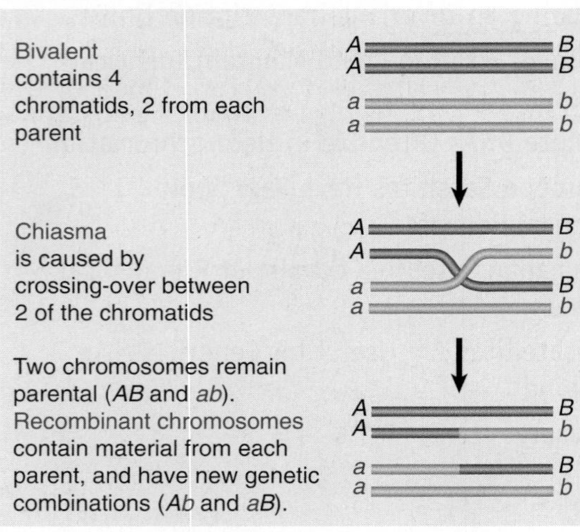

Bivalent contains 4 chromatids, 2 from each parent
A
A
a
a
B
B
b
b
Chiasma is caused by crossing-over between 2 of the chromatids
A
A
a
a
B
b
B
b
Two chromosomes remain parental (AB and ab). Recombinant chromosomes contain material from each parent, and have new genetic combinations (Ab and aB).
A
A
a
a
B
b
B
b

FIGURE 6.1 Chiasma formation represents the generation of recombinants.

Some gene families consist of identical members. Clustering is a prerequisite for maintaining identity between genes, although clustered genes are not necessarily identical. **Gene clusters** range from extremes in which a duplication has generated two adjacent related genes to cases where hundreds of identical genes lie in a tandem array. Extensive tandem repetition of a gene may occur when the product is needed in unusually large amounts. Examples are the genes for rRNA or histone proteins. This creates a special situation with regard to the maintenance of identity and the effects of selective pressure.

Gene clusters offer us an opportunity to examine the forces involved in the evolution of the genome over larger regions than single genes. Duplicated sequences, especially those that remain in the same vicinity, provide the substrate for further evolution by recombination. A population evolves by the classical homologous recombination illustrated in **FIGURES 6.1** and **6.2**, in which an exact crossing-over occurs. The recombinant chromosomes have the same organization as the parental chromosome; they contain precisely the same loci in the same order, but include different combinations of alleles, providing the raw material for natural selection. However, the existence of duplicated sequences allows aberrant events to occur occasionally, which changes the number of copies of genes and not just the combination of alleles.

Unequal crossing-over (also known as **nonreciprocal recombination**) describes a recombination event occurring between two sites that are similar or identical, but not precisely homologous in position. The feature that makes such events possible is the existence of repeated sequences. **FIGURE 6.3** shows that this allows one copy of a repeat in one chromosome to misalign for recombination with a different copy of the repeat in the homologous chromosome, instead of with the corresponding copy. When recombination occurs, this increases the number of repeats in one chromosome and decreases it in the other. In effect, one recombinant chromosome has a deletion and the other has an insertion. This mechanism is responsible for the evolution of clusters of related sequences. We can trace its operation in expanding or contracting the size of an array in both gene clusters and regions of highly repeated DNA.

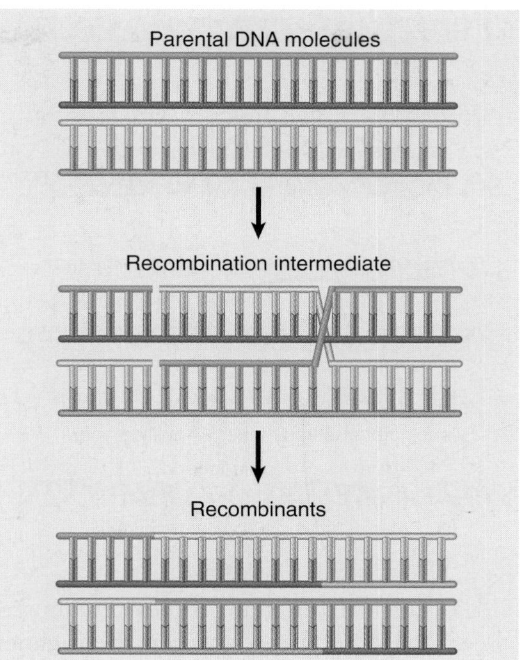

FIGURE 6.2 Recombination involves pairing between complementary strands of the two parental duplex DNAs.

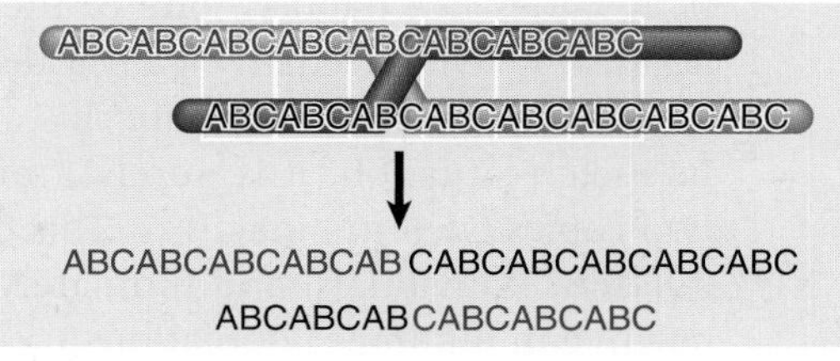

FIGURE 6.3 Unequal crossing-over results from pairing between nonequivalent repeats in regions of DNA consisting of repeating units. Here the repeating unit is the sequence ABC, and the third repeat of the blue chromosome has aligned with the first repeat of the black chromosome. Throughout the region of pairing, ABC units of one chromosome are aligned with ABC units of the other chromosome. Crossing-over generates chromosomes with ten and six repeats each, instead of the eight repeats of each parent.

The highly repetitive fraction of the genome consists of multiple tandem copies of very short repeating units. These often have unusual properties. One is that they may be identified as a separate peak on a density gradient analysis of DNA, which gave rise to the name **satellite DNA**. They often are associated with heterochromatic regions of the chromosomes and in particular with centromeres (which contain the points of attachment for segregation on a mitotic or meiotic spindle). As a result of their repetitive organization, they show some of the same behavior with regard to evolution as the tandem gene clusters. In addition to the satellite sequences, there are shorter stretches of DNA called **minisatellites**, tandem repeats in which each repeat is less than 10 base pairs in length, and they have similar properties. They are useful in showing a high degree of divergence between individual genomes that can be used for mapping or identification purposes.

All of these events that change the constitution of the genome are rare, but they are significant over the course of evolution.

6.2 Gene Duplication Is a Major Force in Evolution

Exons behave like modules for building genes that are tried out in the course of evolution in various combinations (see *Section 3.8, Some Exons Can Be Equated with Protein Functions*). At one extreme, an individual exon from one gene may be copied and used in another gene. At the other extreme, an entire gene, including both exons and introns, may be duplicated. In such a case, mutations can accumulate in one copy without elimination by natural selection as long as the other copy is under selection to remain functional. The selectively neutral copy may then evolve to a new function, become expressed in a different time or place from the first copy, or become a nonfunctional pseudogene.

- **unequal crossing-over (nonreciprocal recombination)** It results from an error in pairing and crossing over in which nonequivalent sites are involved in a recombination event. It produces one recombinant with a deletion of material and one with a duplication.
- **satellite DNA** DNA that consists of many tandem repeats (identical or related) of a short basic repeating unit.
- **minisatellite** DNAs consisting of ~10 copies of a short repeating sequence. The length of the repeating unit is measured in 10s of base pairs. The number of repeats varies between individual genomes.

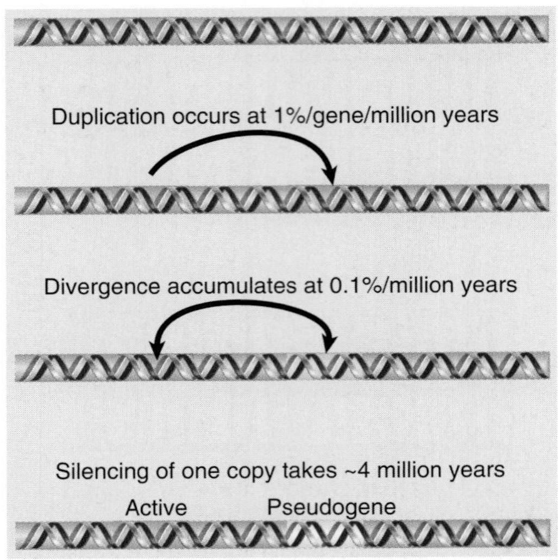
Duplication occurs at 1%/gene/million years
Divergence accumulates at 0.1%/million years
Silencing of one copy takes ~4 million years
Active
Pseudogene

FIGURE 6.4 After a gene has been duplicated, differences may accumulate between the copies. The genes may acquire different functions or one of the copies may become inactive.

FIGURE 6.4 summarizes our present view of the rates at which these processes occur. There is ~1% probability that a given gene will be included in a duplication in a period of one million years. After the gene has duplicated, differences develop as the result of the occurrence of different mutations in each copy. These accumulate at a rate of ~0.1% per million years (see *Section 6.4, Sequence Divergence Is the Basis for the Molecular Clock*).

Unless the gene encodes a product that is required in high concentration in the cell, the organism is not likely to need to retain two identical copies of the gene. As differences develop between the duplicated genes, one of two types of event is likely to occur:

- Both of the gene copies remain necessary. This can happen either because the differences between them generate proteins with different functions, or because they are expressed specifically in different times or places.
- If this does not happen, one of the genes is likely to become a pseudogene because it will by chance gain a deleterious mutation, and there will be no adverse selection to eliminate this copy. Typically this takes ~4 million years. In such a situation, it is purely a matter of chance in terms of which of the two copies becomes inactive. (This can contribute to incompatibility between different individuals, and ultimately to speciation, if different copies become inactive in different populations.)

Analysis of the human genome sequence shows that ~5% comprises duplications of identifiable segments ranging in length from 10 to 300 kb. These duplications have arisen relatively recently; that is, there has not been sufficient time for divergence between them for their homology to become obscured. They include a proportional share (~6%) of the expressed exons, which shows that the duplications are occurring more or less irrespective of genetic content. The genes in these duplications may be especially interesting because of the implication that they have evolved recently and therefore could be important for recent evolutionary developments (such as the separation of the human lineage from that of other primates).

KEY CONCEPT

- Duplicated genes may diverge to generate different genes, or one copy may become an inactive pseudogene.

CONCEPT AND REASONING CHECK

A single duplication event is likely to increase the rate of additional duplication events involving those two gene copies. Why?

6.3 Globin Clusters Are Formed by Duplication and Divergence

The most common type of duplication generates a second copy of the gene close to the first copy. In some cases, the copies remain associated and further duplication may generate a cluster of related genes. The best characterized example of a gene cluster is presented by the globin genes, which constitute an ancient gene family concerned with a function that is central to animals: the transport of oxygen.

The major constituent of the red blood cell is the globin tetramer, which is associated with its heme (iron-binding) group in the form of hemoglobin. Functional globin genes in all species have the same general structure: they are divided into three exons, as shown previously in Figure 3.8. We conclude that all globin genes are derived from a single ancestral gene, and by tracing the history of individual globin genes within and between species we may learn about the mechanisms involved in the evolution of gene families.

In blood cells of adult mammals, the globin tetramer consists of two identical chains and two identical β chains. Embryonic blood cells contain hemoglobin tetramers that are different from the adult form. Each tetramer contains two identical α-like chains and two identical β-like chains, each of which is related to the adult polypeptide and is later replaced by it in the adult form of the protein. This is an example of developmental control, in which different genes are successively switched on and off to provide alternative products that fulfill the same function at different times.

The division of globin chains into α-like and β-like reflects the organization of the genes. Each type of globin is encoded by genes organized into a single cluster. The structures of the two clusters in the primate genome are illustrated in **FIGURE 6.5**. Pseudogenes are indicated by the symbol ψ.

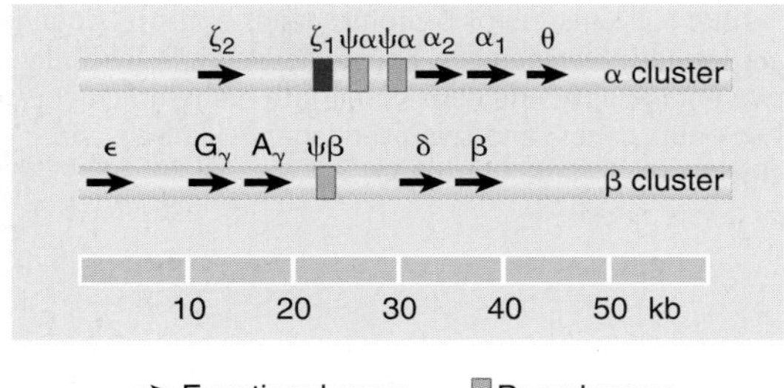

FIGURE 6.5 Each of the α-like and β-like globin gene families is organized into a single cluster that includes functional genes and pseudogenes (ψ).

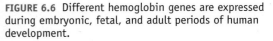

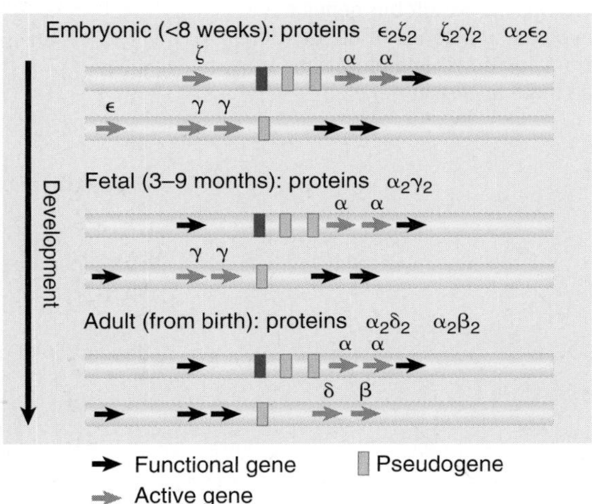

FIGURE 6.6 Different hemoglobin genes are expressed during embryonic, fetal, and adult periods of human development.

Stretching over 50 kb, the β cluster contains five functional genes (ε, two γ, δ, and β) and one nonfunctional pseudogene (ψβ). The two γ genes differ in their coding sequence in only one amino acid; the G variant has glycine at position 136, whereas the A variant has alanine.

The more compact α cluster extends over 28 kb and includes one active ζ gene, one nonfunctional ζ pseudogene, two α genes, two α nonfunctional pseudogenes, and the θ gene of unknown function. The two α genes code for the same protein. Two (or more) identical genes present on the same chromosome are described as **nonallelic** copies.

▸ **nonallelic genes** Two (or more) copies of the same gene that are present at different locations in the genome (contrasted with alleles, which are copies of the same gene derived from different parents and present at the same location on the homologous chromosomes).

The details of the relationship between embryonic and adult hemoglobins vary with the species. The human pathway has three stages: embryonic, fetal, and adult. The distinction between embryonic and adult is common to mammals, but the number of pre-adult stages varies. In humans, zeta and alpha are the two α-like chains. Epsilon, gamma, delta, and beta are the β-like chains. **FIGURE 6.6** shows how the chains are expressed at different stages of development. There is also tissue-specific expression associated with the developmental expression: embryonic hemoglobin genes are expressed in the yolk sac, fetal genes are expressed in the liver, and adult genes are expressed in bone marrow.

In the human pathway, ζ is the first α-like chain to be expressed, but it is soon replaced by α. In the β-pathway, ε and γ are expressed first, with δ and β replacing them later. In adults, the $\alpha_2\beta_2$ form provides 97% of the hemoglobin, $\alpha_2\delta_2$ provides ~2%, and ~1% is provided by persistence of the fetal form $\alpha_2\gamma_2$.

What is the significance of the differences between embryonic and adult globins? The embryonic and fetal forms have a higher affinity for oxygen, which is necessary in order to obtain oxygen from the mother's blood. This explains why there is no direct equivalent (although there is temporal expression of globins) in, for example, the chicken, for which the embryonic stages occur outside the mother's body (that is, within the egg).

Functional genes are defined by their expression to RNA and ultimately by the polypeptides they encode. Pseudogenes are defined as such by their inability to encode

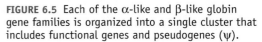

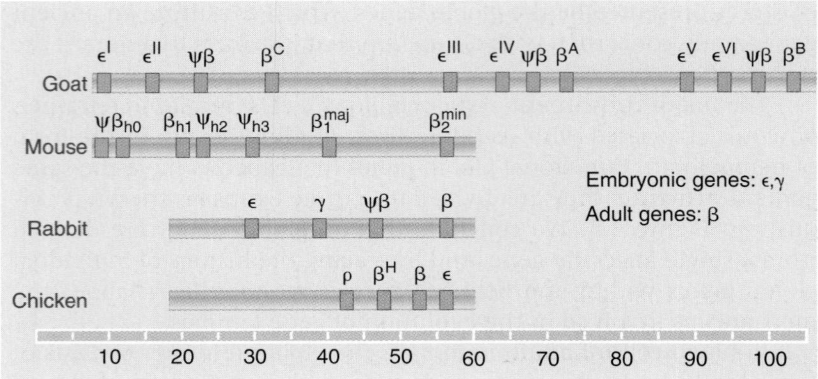

FIGURE 6.7 Clusters of β-globin genes and pseudogenes are found in vertebrates. Seven mouse genes include two early embryonic genes, one late embryonic gene, two adult genes, and two pseudogenes. Rabbit and chicken each have four genes.

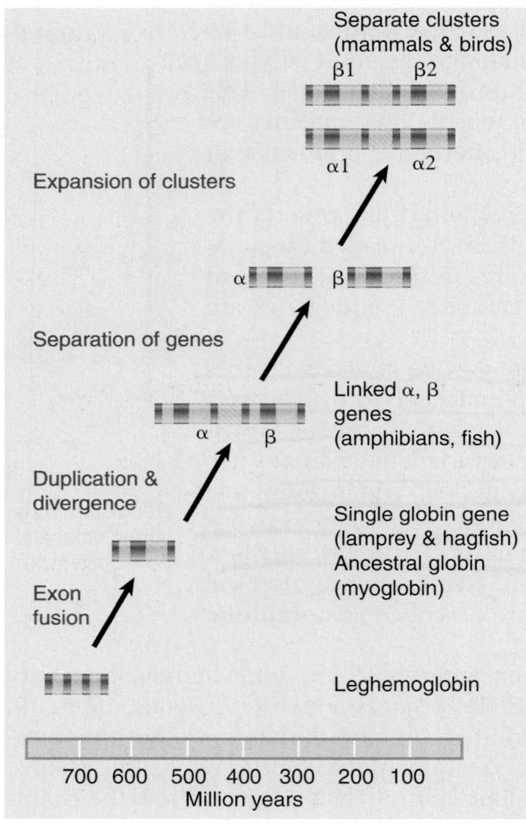

FIGURE 6.8 All globin genes have evolved by a series of duplications, transpositions, and mutations from a single ancestral gene.

polypeptides; the reasons for their inactivity vary, and the deficiencies may be in transcription or translation (or both). A similar general organization is found in other vertebrate globin gene clusters, but details of the types, numbers, and order of genes all vary, as illustrated in **FIGURE 6.7**. Each cluster contains both embryonic and adult genes. The total lengths of the clusters vary widely. The longest known cluster is found in the goat genome, where a basic cluster of four genes has been duplicated twice. The distribution of active genes and pseudogenes differs in each case, illustrating the random nature of the conversion of one copy of a duplicated gene into the inactive state.

The characterization of these gene clusters makes an important general point. There may be more members of a gene family, both functional and nonfunctional, than we would suspect on the basis of protein analysis. The extra functional genes may represent duplicates that code for identical polypeptides, or they may be related to—but different from—known proteins (and presumably expressed only briefly or in low amounts).

With regard to the question of how much DNA is needed to code for a particular function, we see that coding for the β-like globins requires a range of 20 to 120 kb in different mammals. This is much greater than we would expect just from scrutinizing the known β-globin proteins or from even considering the individual genes. However, clusters of this type are not common; most genes are found as individual loci.

From the organization of globin genes in a variety of species, we should be able to trace the evolution of present globin gene clusters from a single ancestral globin gene. Our present view of the evolutionary history is pictured in **FIGURE 6.8**.

The leghemoglobin gene of plants, which is related to the globin genes, may most closely represent the ancestral form. (Leghemoglobin is an oxygen carrier found in the nitrogen-fixing root nodules of legumes.) The furthest back that we can trace a globin gene in modern form is to the sequence of the single chain of mammalian myoglobin, which diverged from the globin line of descent ~800 million years ago. The myoglobin gene has the same organization as globin genes, so we may take the three-exon structure to represent their common ancestor.

Some members of the class *Chondrichthyes* (cartilaginous fish) have only a single type of globin chain, so they must have diverged from the line of evolution before the ancestral globin gene was duplicated to give rise to the α and β variants. This appears to have occurred ~500 million years ago, during the evolution of the *Osteichthyes* (bony fish).

The next stage of evolution is represented by the state of the globin genes in the amphibian *Xenopus laevis,* which has two globin clusters. Each cluster, though, contains both α and β genes, of both larval and adult types. The cluster must therefore have evolved by duplication of a linked α–β pair, followed by divergence between the individual copies. Later the entire cluster was duplicated.

The amphibians separated from the mammalian/avian line ~350 million years ago, so the separation of the α- and β-globin genes must have resulted from a transposition in the mammalian/avian forerunner after this time. This probably occurred in the period of early tetrapod evolution. There are separate clusters for α and β globins in both birds and mammals; thus the α and β genes must have been physically separated before the mammals and birds diverged from their common ancestor, an event that occurred probably ~270 million years ago.

Changes have taken place within the separate α and β clusters in more recent times, as we'll see from the description of the divergence of the individual genes in the following section.

KEY CONCEPTS

- All globin genes are descended by duplication and mutation from an ancestral gene that had three exons.
- The ancestral gene gave rise to myoglobin, leghemoglobin, and α and β globins.
- The α- and β-globin genes separated in the period of early vertebrate evolution, after which duplications generated the individual clusters of separate α- and β-like genes.
- Once a gene has been inactivated by mutation, it may accumulate further mutations and become a pseudogene, which is homologous to the active gene(s) but has no functional role.

CONCEPT AND REASONING CHECK

Although the plant leghemoglobin gene sequence is closer to that of the ancestral sequence than to those of the globin genes, it is very unlikely to be the precise ancestral sequence. Why?

6.4 Sequence Divergence Is the Basis for the Molecular Clock

Most changes in gene sequences occur by small mutations that accumulate slowly over time. Point mutations and small insertions and deletions occur by chance, probably with more or less equal probability in all regions of the genome. The exceptions to this are *hotspots,* where mutations occur much more frequently. Most mutations that change the amino acid sequence (*nonsynonymous substitutions*) are deleterious and will be eliminated by natural selection.

Few mutations are advantageous. When a rare advantageous mutation occurs, it is likely to spread through the population, eventually replacing the original sequence. When a new variant replaces the previous version of the gene, it is said to have become *fixed* in the population.

A historically contentious issue is what proportion of mutational changes in a protein-coding gene sequence are selectively **neutral**—that is, without any effect on the function of the encoded polypeptide—and able, therefore, to accrue as the result of random **genetic drift** and **fixation**.

▸ **genetic drift** The chance fluctuation (without selective pressure) of the frequencies of alleles in a population.

▸ **fixation** The process by which a new allele replaces the allele that was previously predominant in a population.

The rate at which mutational changes accumulate is a characteristic of each gene, presumably depending at least in part on its functional flexibility with regard to change. Within a species, a gene evolves by mutation, followed by elimination or fixation within the single population. Remember that when we scrutinize the gene pool of a species, we see only the variants that have been maintained, whether by selection or genetic drift. When multiple variants are present they may be stable (because none has any selective advantage), or they may in fact be transient because they are in the process of being displaced.

When a single species separates into two new species, each of the resulting species now constitutes an independent pool for evolution. By comparing orthologous genes in two species, we see the differences that have accumulated between them since the time when their ancestors ceased to interbreed. Some genes are highly conserved, showing little or no change from species to species. This indicates that almost any change is deleterious and therefore eliminated.

▸ **divergence** The corrected percent difference in nucleotide sequence between two related DNA sequences or in amino acid sequences between two proteins.

The difference between two genes is expressed as their **divergence**, the percent of positions at which the nucleotides are different, corrected for the possibility of convergent mutations (the same mutation at the same site in two separate lineages) and true revertants. There is usually a difference in the rate of evolution among the three codon positions within genes, since mutations at the third base position often are *synonymous* (have no effect on the encoded amino acid).

▸ **replacement sites** Sites in a coding region at which mutations have altered the amino acid that is encoded.

▸ **silent sites** Sites in a coding region at which mutations have not changed the amino acid that is encoded.

We may divide the nucleotide sequence of a coding region into potential **replacement sites** and **silent sites**:

- At replacement sites, a mutation has altered the amino acid that is encoded. The effect of the mutation (deleterious, neutral, or advantageous) depends on the functional result of the amino acid replacement.
- At silent sites, mutation has only substituted one synonymous codon for another, so there is no change in the resulting polypeptide.

In addition to the coding sequence, a gene contains untranslated regions. Here again, most mutations are potentially neutral, apart from their effects on either secondary structure or (usually rather short) regulatory signals.

Although silent mutations are expected to be neutral with regard to the polypeptide, they could affect gene expression via the sequence change in RNA. For example, a change in secondary structure might influence transcription, processing, or translation. Another possibility is that a change in synonymous codons calls for a different tRNA to respond, influencing the efficiency of translation. Species generally show a **codon bias**; when there are multiple codons for the amino acid, one codon is found in protein-coding genes in a high percentage, whereas the remaining codons are found in low percentages. There is a corresponding percentage difference in the tRNA species that recognize these codons. Consequently, a change from a common to a rare synonymous codon may reduce the rate of translation due to a lower concentration of appropriate tRNAs.

▸ **codon bias** A higher usage of one codon in genes to encode amino acids for which there are several synonymous codons.

To take the example of the human β- and δ-globin chains, there are 10 differences in 146 residues, a divergence of 6.9%. The DNA sequence has 31 changes in 441 residues. However, these changes are distributed very differently in the replacement and silent sites. There are 11 changes in the 330 replacement sites, but 20 changes in only 111 silent sites. This gives corrected rates of divergence of 3.7% in the replacement sites and 32% in the silent sites, an order of magnitude in difference.

The striking difference in the divergence of replacement and silent sites demonstrates the existence of much greater constraints on nucleotide changes that influence protein constitution relative to those that do not. Many fewer amino acid changes are neutral.

Suppose we take the rate of silent substitutions to indicate the underlying rate of mutational fixation (assuming there is no selection at all at the silent sites). Then over the period since the β and δ genes diverged, there should have been changes at 32% of the 330 replacement sites, for a total of 105. All but 11 of them have been eliminated, which means that ~90% of the mutations were not maintained.

The divergence between any pair of globin sequences is (more or less) proportional to the time since they separated. This provides a **molecular clock** that measures the accumulation of mutations at an approximately constant rate during the evolution of a given protein-coding gene.

▸ **molecular clock** An approximately constant rate of evolution that occurs in DNA sequences, such as by the genetic drift of neutral mutations.

The rate of divergence can be measured as the percent difference per million years, or as its reciprocal, the unit evolutionary period (UEP), the time in millions of years that it takes for 1% divergence to accrue. Once the rate of the molecular clock has been established by pairwise comparisons between species (remembering the practical difficulties in establishing the actual time of speciation), it can be applied to paralogous genes within a species. From their divergence, we can calculate how much time has passed since the duplication that generated them.

By comparing the sequences of orthologous genes in different species, the rate of divergence at both replacement and silent sites can be determined, as plotted in **FIGURE 6.9**.

In pairwise comparisons, there is an average divergence of 10% in the replacement sites of either the α- or β-globin genes of mammal lineages that have been separated since the mammalian radiation occurred ~85 million years ago. This corresponds to a replacement divergence rate of 0.12% per million years.

The rate is approximately constant when the comparison is extended to genes that diverged in the more distant past. For example, the average replacement divergence between orthologous mammalian and chicken globin genes is 23%. Relative to a common ancestor at ~270 million years ago, this gives a rate of 0.09% per million years.

Going further back, we can compare the α- with the β-globin genes within a species. They have been diverging since the original duplication event 500 million years ago (see Figure 6.8). They have an average replacement divergence of ~50%, which gives a rate of 0.1% per million years.

The summary of these data in Figure 6.9 shows that replacement divergence in the globin genes has an average rate of ~0.096% per million years (for a UEP of 10.4). Considering the uncertainties in estimating the times at which the species diverged, the results lend good support to the idea that there is a constant molecular clock.

The data on silent site divergence are much less clear. In every case, it is evident that the silent site divergence is much greater than the replacement site divergence, by a factor that varies from 2 to 10. The range of silent site divergences in pairwise comparisons, though, is too great to establish a molecular clock, so we must base temporal comparisons on the replacement sites.

From Figure 6.9, it is clear that the rate at silent sites is not constant over time. If we assume that there must be zero divergence at zero years of separation, we see that the rate of silent site divergence is much greater for the first ~100 million years of separation. One interpretation is that a fraction of roughly half of the silent sites is rapidly (within 100 million years) saturated by mutations; this fraction behaves as neutral sites. The other fraction accumulates mutations more slowly, at a rate approximately the same as that of

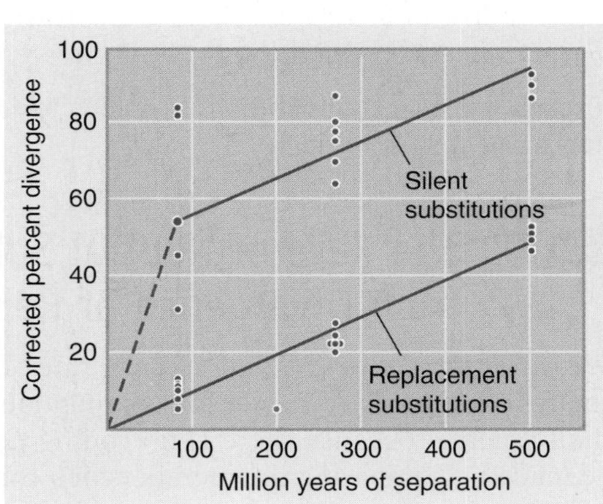

FIGURE 6.9 Divergence of DNA sequences depends on evolutionary separation. Each point on the graph represents a pairwise comparison.

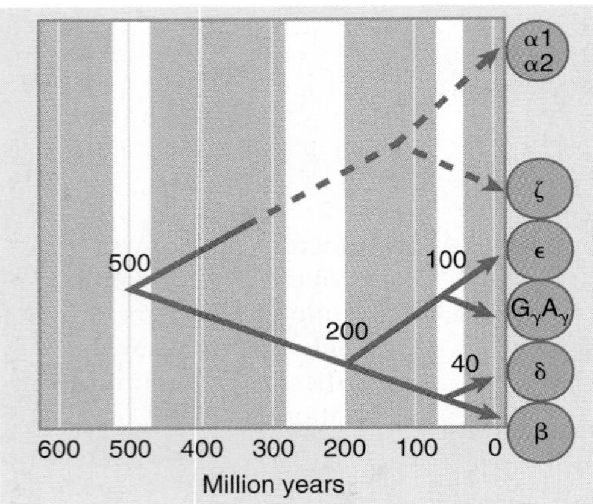

FIGURE 6.10 Replacement site divergences between pairs of β-globin genes allow the history of the human cluster to be reconstructed. This tree accounts for the separation of classes of globin genes.

the replacement sites; this fraction represents sites that are silent with regard to the polypeptide, but that are under selective constraint for some other reason.

Now we can reverse the calculation of divergence rates to estimate the times since paralogous genes were duplicated. The difference between the human β and δ genes is 3.7% for replacement sites. At a UEP of 10.4, these genes must have diverged $10.4 \times 3.7 = 40$ million years ago—about the time of the separation of the major primate lineages: New World monkeys, Old World monkeys, and great apes (including humans). All of these taxa have both β and δ genes, which suggests that the gene divergence commenced just before this point in evolution.

Proceeding further back, the divergence between the replacement sites of γ and ε genes is 10%, which corresponds to a duplication event ~100 million years ago. The separation between embryonic and fetal globin genes therefore may have just preceded or accompanied the mammalian radiation.

An evolutionary tree for the human globin genes is presented in **FIGURE 6.10**. Paralogous groups that evolved before the mammalian radiation—such as the separation of β/δ from γ—should be found in all mammals. Paralogous groups that evolved afterward—such as the separation of β- and δ-globin genes—should be found in individual lineages of mammals.

In each species, there have been comparatively recent changes in the structures of the clusters. We know this because we see differences in gene number (one adult β-globin gene in humans, two in the mouse) or in type (most often concerning whether there are separate embryonic and fetal genes).

When sufficient data have been collected on the sequences of a particular gene or gene family, the analysis can be reversed, and comparisons between orthologous genes can be used to assess taxonomic relationships. If a molecular clock has been established, the time to common ancestry between the previously analyzed species and a species newly introduced to the analysis can be estimated.

KEY CONCEPTS

- The sequences of orthologous genes in different species vary at replacement sites (where mutations have caused amino acid substitutions) and silent sites (where mutation has not affected the amino acid sequence).
- Silent substitutions accumulate ~10× faster than replacement substitutions.
- The evolutionary divergence between two DNA sequences is measured by the corrected percent of positions at which the corresponding nucleotides differ.
- Mutations may accumulate at a more or less constant rate after genes separate, so that the divergence between any pair of globin sequences is proportional to the time since they shared common ancestry.

CONCEPT AND REASONING CHECK

Explain why silent substitutions in coding regions occur at a much higher rate than replacement substitutions.

6.5 The Rate of Neutral Substitution Can Be Measured from Divergence of Repeated Sequences

We can make the best estimate of the rate of substitution at neutral sites by examining sequences that do not code for polypeptide. (We use the term "neutral" here rather than "silent," because there is no coding potential.) An informative comparison can be made by comparing the members of a common repetitive family in the human and mouse genomes.

The principle of the analysis is summarized in **FIGURE 6.11**. We start with a family of related sequences that have evolved by duplication and substitution from an original ancestral sequence. We assume that the ancestral sequence can be deduced by taking the base that is most common at each position. Then we can calculate the divergence of each individual family member as the proportion of bases that differ from the deduced ancestral sequence. In this example, individual members vary from 0.13 to 0.18 divergence and the average is 0.16.

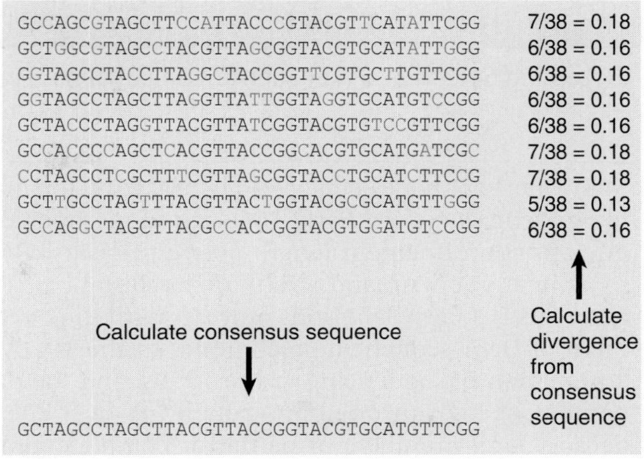

GCCAGCGTAGCTTCCATTACCCGTACGTTCATATTCGG 7/38 = 0.18
GCTGGCGTAGCCTACGTTAGCGGTACGTGCATATTGGG 6/38 = 0.16
GGTAGCCTACCTTAGGCTACCGGTTCGTGCTTGTTCGG 6/38 = 0.16
GGTAGCCTAGCTTAGGTTATTGGTAGGTGCATGTCCGG 6/38 = 0.16
GCTACCCTAGGTTACGTTATCGGTACGTGTCCGTTCGG 6/38 = 0.16
GCCACCCCAGCTCACGTTACCGGCACGTGCATGATCGC 7/38 = 0.18
CCTAGCCTCGCTTTCGTTAGCGGTACCTGCATCTTCCG 7/38 = 0.18
GCTTGCCTAGTTTACGTTACTGGTACGCGCATGTTGGG 5/38 = 0.13
GCCAGGCTAGCTTACGCCACCGGTACGTGGATGTCCGG 6/38 = 0.16

Calculate consensus sequence

Calculate divergence from consensus sequence

GCTAGCCTAGCTTACGTTACCGGTACGTGCATGTTCGG

FIGURE 6.11 An ancestral consensus sequence for a family is calculated by taking the most common base at each position. The divergence of each existing current member of the family is calculated as the proportion of bases at which it differs from the ancestral sequence.

One family used for this analysis in the human and mouse genomes derives from a sequence that is thought to have ceased to be active at about the time of the common ancestor between humans and rodents (the LINEs family; see *Section 21.10, Retroposons Fall into Three Classes*). This means that it has been diverging under limited selective pressure for the same length of time in both species. Its average divergence in humans is ~0.17 substitutions per site, corresponding to a rate of 2.2×10^{-9} substitutions per base per year over the 75 million years since the separation. In the mouse genome, however, neutral substitutions have occurred at twice this rate, corresponding to 0.34 substitutions per site in the family, or a rate of 4.5×10^{-9}. Note, however, that if we calculated the rate per generation instead of per year, it would be greater in humans than in the mouse ($\sim 2.2 \times 10^{-8}$ as opposed to $\sim 10^{-9}$).

These figures probably underestimate the rate of substitution in the mouse; at the time of divergence the rates in both lineages would have been the same, and the difference must have evolved since then. The current rate of neutral substitution per year in the mouse is probably 2–3× greater than the historical average. These rates reflect the balance between the occurrence of mutations (which may be higher in species with higher metabolic rates, like the mouse) and the loss of them due to genetic drift, which is largely a function of population size; genetic drift is a type of "sampling error" where allele frequencies fluctuate more widely in smaller populations. In addition to eliminating neutral alleles more quickly, smaller population sizes also allow faster fixation of neutral alleles. The faster rate of neutral substitution is likely to be a result of both the small effective population sizes and the shorter generation times (allowing more opportunities for substitutions per year) of rodent species.

Comparing the mouse and human genomes allows us to assess whether syntenic (homologous) regions show signs of conservation or have differed at the rate predicted from accumulation of neutral substitutions. The proportion of sites that show signs of selection is ~5%. This is much higher than the proportion found in exons (~1%). This observation implies that the genome includes many more stretches whose sequence is important for functions other than coding for RNA. Known regulatory elements are likely to comprise only a small part of this proportion. This number also suggests that most (i.e., the rest) of the genome sequences do not have any function that depends on the exact sequence.

KEY CONCEPT

- The rate of substitution per year at neutral sites is greater in the mouse than in the human genome.

CONCEPT AND REASONING CHECK

In comparing homologous sequences from related species, describe how both the ancestral sequence and the rates of evolution in the descendant lineages can be estimated.

METHODS AND TECHNIQUES

Detecting Selection via Sequence Comparisons

Many methods have been used over the years for analyzing genetic selection. Early DNA-based methods relied on restriction fragment length polymorphisms (RFLPs) to estimate genetic and evolutionary distances. However, this is a fairly laborious process. With the development of DNA sequencing techniques in the 1970s, the automation of sequencing in the 1990s, and the development of high throughput sequencing over the past decade, large numbers of partial or complete genome sequences are becoming available. Coupled with the polymerase chain reaction (PCR) to amplify specific genomic regions, DNA sequence analysis has become a valuable tool in many applications, including the study of genetic selection.

There is now an abundance of DNA sequence data from a wide range of organisms in various publicly available databases. Homologous gene sequences have been obtained from many species as well as from different individuals of the same species. This allows for determination of genetic changes across species lines as compared to changes within a species. These comparisons have led to the observation that there are high levels of DNA sequence polymorphism within species, most likely as a result of **neutral mutations** and random genetic drift within populations. By conducting both interspecific and intraspecific DNA sequence analysis, the level of divergence due to species differences can be determined.

Neutral mutations may change the sequence of an amino acid codon in the DNA to an alternate codon for the same amino acid, especially if the mutation occurs in the third (wobble) position of the codon. These are synonymous mutations, but not all synonymous mutations are neutral. While this may at first seem contradictory, the levels of individual tRNAs for a given amino acid are not the same in a cell. Some tRNAs for a given amino acid are more abundant than others, and a specific codon may lack sufficient tRNAs, whereas a different codon for the same amino acid may have a sufficient number. In the case of a codon that requires a rare tRNA in that organism, ribosomal frameshifting or other alterations in translation may occur. Alternatively, the codon may be changed to code for a different amino acid with the same general characteristics (called a *nonsynonymous mutation*), with little or no effect on the folding and activity of the protein. In either case neutral sequence changes have little effect on the organism. However, when a codon is changed to encode an amino acid with different properties, this is often deleterious to the function of the protein and thus to the organism. Depending on the location of the amino acid in the protein, such a change may cause only slight disruption of protein folding and activity. Only in rare cases is an amino acid change advantageous; in this case the mutational change may become subjected to positive selection and ultimately lead to fixation of the change in the population.

One common approach for determining selection is to use codon-based sequence information to study the evolutionary history of a given gene. This can be done by counting the number of synonymous (K_s) and nonsynonymous (K_a) amino acid substitutions in a gene from two different (but related) species, and determining the K_a/K_s ratio. This ratio is indicative of the selection constraints on the gene. A K_a/K_s ratio of 1 is expected for those genes under neutral selection, which neither favors nor disfavors amino acid sequence changes. In this case the changes that occur do not usually affect the activity of the protein, and this serves as a suitable control. A K_a/K_s ratio <1 is most commonly observed, and indicates negative selection where amino acid replacements are disfavored because they affect the activity of the protein. Thus there is pressure to retain the correct amino acid at these sites in order to maintain proper protein function. Positive selection occurs when the K_a/K_s ratio is >1, but is rarely observed. This indicates that the amino acid changes are favored and may become fixed in the gene. Some examples of this are antigenic proteins of some pathogens, such as viral coat proteins, which are under strong selection pressure to evade the immune response of the host, and some reproductive proteins that are under sexual selection. Some caution needs to be used with codon-based detection of selection,

▸ **neutral mutation** A mutation that has no significant effect on evolutionary fitness and usually has no effect on the phenotype.

6.6 Unequal Crossing-over Rearranges Gene Clusters

There are frequent opportunities for rearrangement in a cluster of related or identical genes. We can see the results by comparing the mammalian β-globin clusters included in Figure 6.7. Although all the clusters serve the same function, and all have the same general organization, each is different in size, there is variation in the total number and types of β-globin genes, and the numbers and structures of pseudogenes are different. All of these changes must have occurred since the mammalian radiation ~85 million years ago (the time of the common ancestor to all the mammals).

as there have been reports of unusually highly mutable regions of some proteins with a high proportion of polar amino acids that may affect the conclusions.

Intraspecific DNA sequence analysis can be used to detect positive selection by comparing the nucleotide sequence between two alleles, or two individuals of the same species. Nucleotide sequences are subject to neutral selection at a certain rate; variation in this rate is identified as nucleotide heterozygosity. If a variant sequence is favored the site will show a reduction in nucleotide heterozygosity, and the variant will increase in frequency and eventually become fixed in the population. Nearby linked neutral variants may also become fixed, a phenomenon termed *genetic hitchhiking*. These regions are characterized by having a lower level of DNA sequence polymorphism.

In practice it is more reliable to carry out both interspecific and intraspecific DNA sequence comparisons to determine alterations from neutral selection. By including sequence information from at least one closely related species, species-specific DNA polymorphisms can be distinguished from ancestral polymorphisms, and more accurate information can be obtained regarding the link between the polymorphisms and between-locus differences. With this combined analysis the degree of nonsynonymous changes between species can be determined; an excess of these changes can be evidence for positive selection on these amino acids.

Relative rate tests can also be used for determining selection. This involves at a minimum three related species, two closely related and one outgroup representative. The substitution rate is compared between the first two as compared to the outgroup member to see if the substitution rate is similar. This removes the dependence of the analysis on time, as long as the phylogenetic relationship between the species is certain. If the rate of substitutions compared to the outgroup member is different, this may be an indication of selection on the sequence. This method must take into account that some genes accumulate nucleotide or amino acid substitutions more rapidly (fast-clock) in some species than in others, possibly due to differences in metabolic rate, generation time, DNA replication, or DNA repair. To deal with this difference, additional related organisms need to be examined in order to identify and eliminate fast-clock effects. The reliability of this approach is improved if larger numbers of distantly related species are included. However, it is difficult to make accurate comparisons between taxonomic groups due to the inherent rate differences. As more work in this area has been done, corrections have been developed to adjust for differences in substitution rates.

Another method for detecting selection utilizes estimates of polymorphism at specific genetic loci. For example, sequence analysis of the *Teosinte branched 1* (*Tb-1*) locus, an important gene in domesticated maize, has been used to characterize the nucleotide substitution rate in domesticated and native maize varieties, with an estimate of 2.9×10^{-8} to 3.3×10^{-8} base substitutions per year. Estimations have also been used in mapping complex disease genes, and millions of single nucleotide polymorphisms (SNPs) are being characterized in humans, nonhuman animals, and plants, along with other species. This approach involves looking at individual base changes associated with different phenotypes, and can be a powerful tool in determining selection at a genetic locus. The presence of a certain pattern of SNPs for an organism can be associated with a specific phenotype, and by analyzing a sufficient number of generational samples a link between the SNP and selection can be determined. An example of this is the *Drosophila* Tb^1 locus involved in development. However, with such a large number of potential haplotypes, the number of SNPs for a given gene can be extremely high. Complex computer programs have been developed to assist in the analysis of the large number of SNPs.

Sources and Further Reading

Clark et al. (2005). Estimating a nucleotide substitution rate for maize from polymorphism at a major domestication locus. *Molecular Biology and Evolution* 22:2304–2312.

Geetha et al. (1999). Comparing protein sequence-based and predicted secondary structure-based methods for identification of remote homologs. *Protein Engineering* 12:527–534.

Robinson et al. (1998). Sensitivity of the relative-rate test to taxonomic sampling. *Molecular Biology and Evolution* 15:1091–1098.

http://www.ncbi.nlm.nih.gov/books/bv.fcgi?rid=eurekah.section.29338.

The comparison makes the general point that gene duplication, rearrangement, and variation is as important a factor in evolution as the slow accumulation of point mutations in individual genes. What types of mechanisms are responsible for gene reorganization?

As described in the introduction to this chapter, unequal crossing-over can occur as the result of pairing between two sites that are not homologous in position. Usually, recombination involves corresponding sequences of DNA held in exact alignment between the two homologous chromosomes. However, when there are two copies of a gene on each chromosome, an occasional misalignment allows pairing between

FIGURE 6.12 Gene number can be changed by unequal crossing-over. If gene 1 of one chromosome pairs with gene 2 of the other chromosome, the other gene copies are excluded from pairing. Recombination between the mispaired genes produces one chromosome with a single (recombinant) copy of the gene and one chromosome with three copies of the gene (one from each parent and one recombinant).

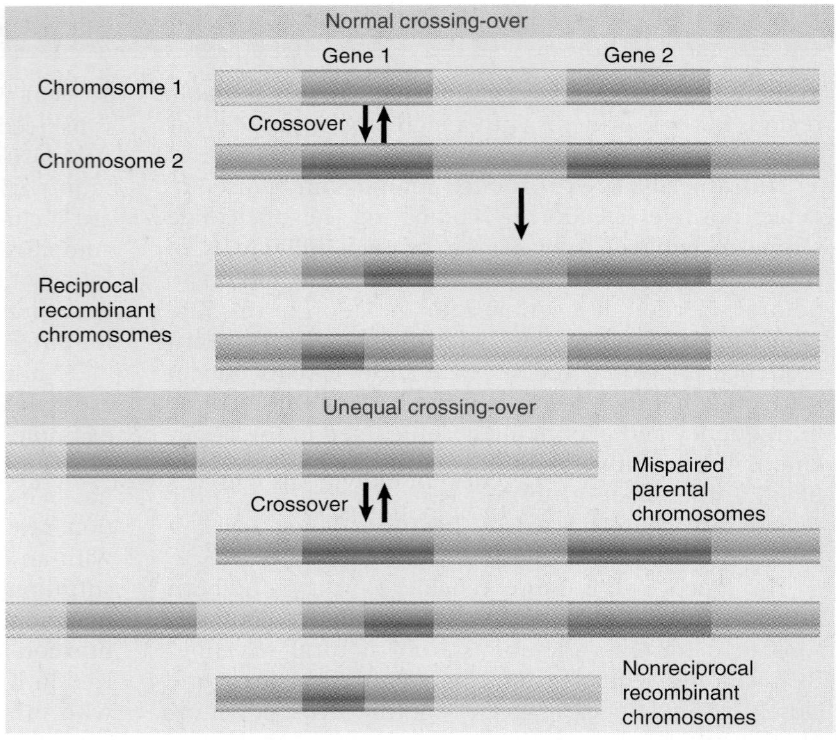

them. (This requires some of the adjacent regions to go unpaired.) This can happen in a region of short repeats (see Figure 6.3) or in a gene cluster. **FIGURE 6.12** shows that unequal crossing-over in a gene cluster can have two consequences—quantitative and qualitative:

- The number of repeats increases in one chromosome and decreases in the other. In effect, one recombinant chromosome has a deletion and the other has an insertion. This happens irrespective of the exact location of the crossover. In the figure, the first recombinant has an increase in the number of gene copies from two to three, whereas the second has a decrease from two to one.
- If the recombination event occurs within a gene (as opposed to between genes), the result depends on whether the recombining genes are identical or only related. If the nonhomologous gene copies 1 and 2 are identical in sequence, there is no change in the sequence of either gene. However, unequal crossing-over also can occur when the adjacent genes are well related (although the probability is less than when they are identical). In this case, each of the recombinant genes has a sequence that is different from either parent.

The determination of whether the chromosome has a selective advantage or disadvantage will depend on the consequence of any change in the sequence of the gene product, as well as on the change in the number of gene copies.

An obstacle to unequal crossing-over is presented by the interrupted structure of the genes. In a case such as the globins, the corresponding exons of adjacent gene copies are likely to be well enough related to support pairing; however, the sequences of the introns have diverged appreciably. The restriction of pairing to the exons considerably reduces the continuous length of DNA that can be involved. This lowers the chance of unequal crossing-over. So divergence between introns could enhance the stability of gene clusters by hindering the occurrence of unequal crossing-over.

▸ **thalassemia** A disease of red blood cells resulting from lack of either α- or β-globin.

Thalassemias result from mutations that reduce or prevent synthesis of either α- or β-globin. The occurrence of unequal crossing-over in the human globin gene clusters is revealed by the nature of certain thalassemias.

Many of the most severe thalassemias result from deletions of part of a cluster. In at least some cases, the ends of the deletion lie in regions that are homologous, which is exactly what would be expected if it had been generated by unequal crossing-over.

FIGURE 6.13 summarizes the deletions that cause the α-thalassemias. α-thal-1 deletions are long, varying in the location of the left end, with the positions of the right ends located beyond the known genes. They eliminate both the α genes. The α-thal-2 deletions are short and eliminate only one of the two α genes. The L deletion removes 4.2 kb of DNA, including the α2 gene. It probably results from unequal crossing-over, because the ends of the deletion lie in homologous regions, just to the right of the ψα and α2 genes, respectively. The R deletion results from the removal of exactly 3.7 kb of DNA, the precise distance between the α1 and α2 genes. It appears to have been generated by unequal crossing-over between the α1 and α2 genes themselves. This is precisely the situation depicted in Figure 6.12.

Depending on the diploid combination of thalassemic alleles, an affected individual may have any number of α chains, from zero to three. There are few differences from the wild type (four α genes) in individuals with three or two α genes. If an individual has only one α gene, though, the excess β chains form the unusual tetramer β4, which causes hemoglobin H **(HbH) disease**. The complete absence of α genes results in **hydrops fetalis**, which is fatal at or before birth.

- **HbH disease** A condition in which there is a disproportionate amount of the abnormal tetramer β4 relative to the amount of normal hemoglobin (α2β2).
- **hydrops fetalis** A fatal disease resulting from the absence of the hemoglobin α gene.

The same unequal crossing-over that generated the thalassemic chromosome should also have generated a chromosome with three α genes. Individuals with such chromosomes have been identified in several populations. In some populations, the frequency of the triple α locus is about the same as that of the single α locus; in others, the triple α genes are much less common than single α genes. This suggests that (unknown) selective factors operate in different populations to adjust the gene numbers.

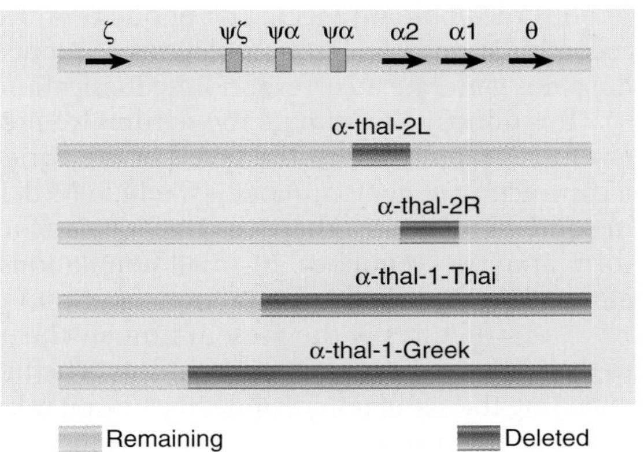

FIGURE 6.13 α thalassemias result from various deletions in the α-globin gene cluster.

Variations in the number of α genes are found relatively frequently, which suggests that unequal crossing-over in the cluster must be fairly common. It occurs more often in the α cluster than in the β cluster, possibly because the introns in α genes are much shorter and therefore present less of an impediment to mispairing between nonhomologous loci.

The deletions that cause β-thalassemias are summarized in **FIGURE 6.14**. In some (rare) cases, only the β gene is affected. These have a deletion of 600 bp, extending from the second intron through the 3′ flanking regions. In the other cases, more than one gene of the cluster is affected. Many of the deletions are very long, extending from the 5′ end indicated on the map for >50 kb toward the right.

The **Hb Lepore** type provided the classic evidence that deletion can result from unequal crossing-over

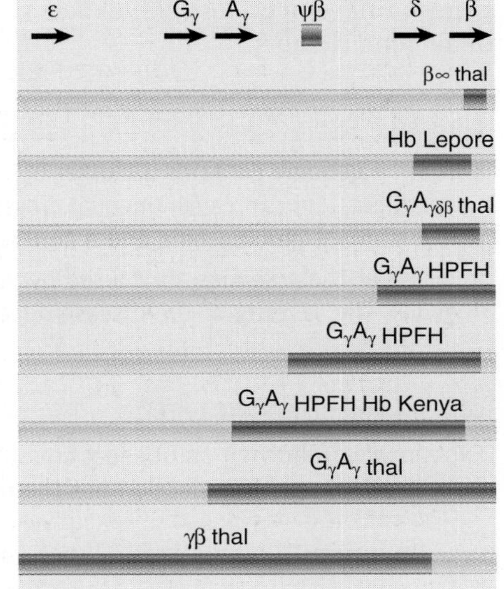

FIGURE 6.14 Deletions in the β-globin gene cluster cause several types of thalassemia.

- **Hb Lepore** An unusual globin protein that results from unequal crossing-over between the β and δ genes. The genes become fused together to produce a single β-like chain that consists of the N-terminal sequence of δ joined to the C-terminal sequence of β.

between linked genes. The β and δ genes differ by only ~7% in sequence. Unequal crossing-over deletes the material between the genes, thus fusing them together (see Figure 6.12). The fused gene produces a single β-like chain that consists of the N-terminal sequence of δ joined to the C-terminal sequence of β.

Several types of Hb Lepore are known, the difference between them lying in the point of transition from δ to β sequences. Thus when the δ and β genes pair for unequal crossing-over, the exact point of recombination determines the position at which the switch from δ to β sequence occurs in the amino acid chain.

▸ **Hb anti-Lepore** A fusion gene produced by unequal crossing-over that has the N-terminal part of β globin and the C-terminal part of δ globin.

▸ **Hb Kenya** A fusion gene produced by unequal crossing-over between the $^{A}\gamma$ and β globin genes.

The reciprocal of this event has been found in the form of **Hb anti-Lepore**, which is produced by a gene that has the N-terminal part of β and the C-terminal part of δ. The fusion gene lies between normal δ and β genes. Although heterozygotes for this mutation are phenotypically normal, those that also carry a β deletion in *trans* show a mild β-thalassemia.

Evidence that unequal crossing-over can occur between more distantly related genes is provided by the identification of **Hb Kenya**, another fused hemoglobin. This contains the N-terminal sequence of the $^{A}\gamma$ gene and the C-terminal sequence of the β gene. The fusion must have resulted from unequal crossing-over between $^{A}\gamma$ and α, which differ ~20% in sequence.

From the differences between the globin gene clusters of various mammals, we see that duplication followed (sometimes) by variation has been an important feature in the evolution of each cluster. The human thalassemic deletions demonstrate that unequal crossing-over continues to occur in both globin gene clusters. Each such event generates a duplication as well as the deletion, and we must account for the fate of both recombinant loci in the population. Deletions can also occur (in principle) by recombination between homologous sequences lying on the same chromosome. This does not generate a corresponding duplication.

It is difficult to estimate the natural frequency of these events, because evolutionary forces rapidly adjust the levels of the variant clusters in the population. Generally a contraction in gene number is likely to be deleterious and selected against. However, in some populations, there may be a balancing advantage that maintains the deleted form at a low frequency. In small populations, genetic drift is likely to play a role in eliminating effectively neutral new duplications.

The structures of the present human clusters show several duplications that attest to the importance of such mechanisms. The functional sequences include two α genes encoding the same polypeptide, fairly well related β and δ genes, and two almost identical γ genes. These comparatively recent independent duplications have persisted in the population, not to mention the more distant duplications that originally generated the various types of globin genes. Other duplications may have given rise to pseudogenes or have been lost. We expect continual duplication and deletion to be a feature of all gene clusters.

KEY CONCEPTS

- When a genome contains a cluster of genes with related sequences, mispairing between nonallelic loci can cause unequal crossing-over. This produces a deletion in one recombinant chromosome and a corresponding duplication in the other.
- Different thalassemias are caused by various deletions that eliminate α- or β-globin genes. The severity of the disease depends on the individual deletion.

CONCEPT AND REASONING CHECK

Explain why, following an unequal crossing-over event, the recombinant with a deletion may be eliminated by selection but the recombinant with the duplication may persist.

6.7 Genes for rRNA Form Tandem Repeats Including an Invariant Transcription Unit

In the cases we have discussed thus far, there are differences between the individual members of a gene cluster that allow selective pressure to act differently upon each gene. A contrast is provided by two cases of large gene clusters that contain many identical copies of the same gene or genes. Most organisms contain multiple copies of the genes for the histone proteins that are a major component of the chromosomes, and there are almost always multiple copies of the genes that code for the ribosomal RNAs. These situations pose some interesting evolutionary questions.

Ribosomal RNA is the predominant product of transcription, constituting some 80% to 90% of the total mass of cellular RNA in both eukaryotes and prokaryotes. The number of major rRNA genes varies from seven in *E. coli*, to 100 to 200 in unicellular eukaryotes, to several hundred in multicellular eukaryotes. The genes for the large and small rRNA (found in the large and small subunits of the ribosome, respectively) usually form a tandem pair. (The sole exception is the yeast mitochondrion.)

The lack of any detectable variation in the sequences of the rRNA molecules implies that all the copies of each gene must be identical, or at least must have differences below the level of detection in rRNA (~1%). A point of major interest is what mechanism(s) are used to prevent variations from accruing in the individual sequences.

In bacteria, the multiple rRNA genes are dispersed. In most eukaryotic genomes, the rRNA genes are contained in a tandem cluster or clusters. Sometimes these regions are called **rDNA**. (In some cases, the proportion of rDNA in the total DNA, together with its atypical base composition, is great enough to allow its isolation as a separate fraction directly from sheared genomic DNA.) An important diagnostic feature of a tandem cluster is that it generates a circular restriction map, as shown in **FIGURE 6.15**.

▸ **rDNA** Genes encoding ribosomal RNA (rRNA).

▸ **nucleolus** A discrete region of the nucleus where ribosomes are produced.

▸ **nucleolar organizer** The region of a chromosome carrying genes coding for rRNA.

Suppose that each repeat unit has three restriction sites. When we map these fragments by conventional means, we find that A is next to B, which is next to C, which is next to A, generating the circular map. If the cluster is large, the internal fragments (A, B, and C) will be present in much greater quantities than the terminal fragments (X and Y), which connect the cluster to adjacent DNA. In a cluster of 100 repeats, X and Y would be present at 1% of the level of A, B, and C. This can make it difficult to obtain the ends of a gene cluster for mapping purposes.

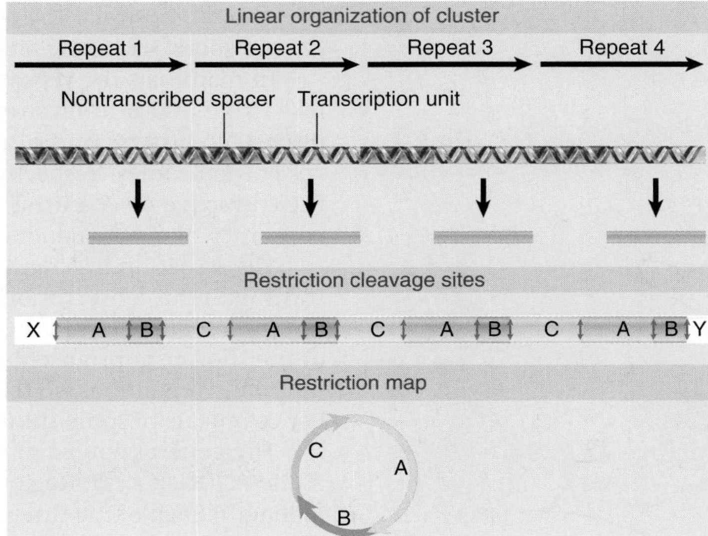

FIGURE 6.15 A tandem gene cluster has an alternation of transcription unit and nontranscribed spacer and generates a circular restriction map.

The region of the nucleus where 18S and 28S rRNA synthesis occurs has a characteristic appearance, with a fibrillar core surrounded by a granular cortex. The fibrillar core is where the rRNA is transcribed from the DNA template, and the granular cortex is formed by the ribonucleoprotein particles into which the rRNA is assembled. The whole area is called the **nucleolus**. Its characteristic morphology is evident in **FIGURE 6.16**.

The particular chromosomal regions associated with a nucleolus are called **nucleolar organizers**. Each nucleolar organizer corresponds to a cluster of tandemly repeated 18/28S rRNA genes on one chromosome. The concentration of the tandemly repeated rRNA genes, together with their very intensive transcription, is responsible for creating the characteristic morphology of the nucleoli.

The pair of major rRNAs is transcribed as a single precursor in both bacteria (where 5S and 16/23S rRNAs are cotranscribed) and the eukaryotic nucleolus (where

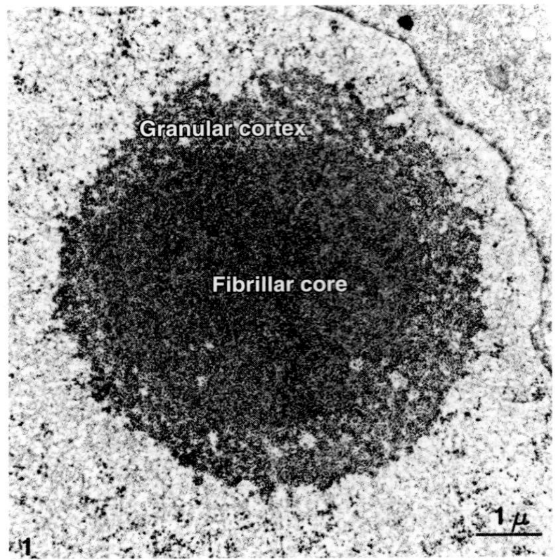

FIGURE 6.16 The nucleolar core identifies rDNA under transcription, and the surrounding granular cortex consists of assembling ribosomal subunits. This thin section shows the nucleolus of the newt *Notopthalmus viridescens*. Photo courtesy of Oscar Miller.

▸ **nontranscribed spacer** The region between transcription units in a tandem gene cluster.

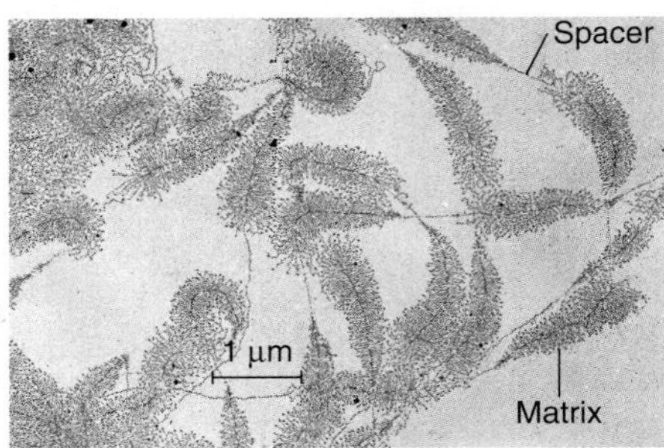

FIGURE 6.17 Transcription of rDNA clusters generates a series of matrices, each corresponding to one transcription unit and separated from the next by the nontranscribed spacer. Photo courtesy of Oscar Miller.

the 18S and 28S rRNAs are transcribed). In eukaryotes, 5S genes are also typically found in tandem clusters transcribed as a precursor with transcribed spacers. Following transcription, the precursor is cleaved to release the individual rRNA molecules. The transcription unit is shortest in bacteria and is longest in mammals (where it is known as 45S RNA, according to its rate of sedimentation). An rDNA cluster contains many transcription units, each separated from the next by a nontranscribed spacer. The alternation of transcription unit and **nontranscribed spacer** can be seen directly in electron micrographs. The example shown in FIGURE 6.17 is taken from the newt *Notopthalmus viridescens*, in which each transcription unit is intensively expressed, so that many RNA polymerases are simultaneously engaged in transcription on one repeating unit. The polymerases are so closely packed that the RNA transcripts form a characteristic matrix displaying increasing length along the transcription unit.

The length of the nontranscribed spacer varies a great deal between and (sometimes) within species. In yeast there is a short nontranscribed spacer that is relatively constant in length. In *D. melanogaster* there is almost a twofold variation in the length of the nontranscribed spacer between different copies of the repeating unit. A similar situation is seen in *X. laevis*. In each of these cases, all of the repeating units are present as a single tandem cluster on one particular chromosome. (In the example of *D. melanogaster*, this happens to be the sex chromosome. The cluster on the X chromosome is larger than the one on the Y chromosome, so female flies have more copies of the rRNA genes than male flies do.)

In mammals the repeating unit is much larger, comprising the transcription unit of ~13 kb and a nontranscribed spacer of ~30 kb. Usually, the genes lie in several dispersed clusters; in the cases of human and mouse the clusters reside on five and six chromosomes, respectively. One interesting (but unanswered) question is how the corrective mechanisms that presumably function within a single cluster to ensure constancy of rRNA sequence are able to work when there are several clusters.

The variation in length of the nontranscribed spacer in a single gene cluster contrasts with the conservation of sequence of the transcription unit. In spite of this variation, the sequences of longer nontranscribed spacers remain homologous with those of the shorter nontranscribed spacers. This implies that each nontranscribed spacer is internally repetitious, so that the variation in length results from changes in the number of repeats of some subunit.

The general nature of the nontranscribed spacer is illustrated by the example of *X. laevis* (FIGURE 6.18). Regions that are fixed in length alternate with regions that vary in length. Each of the three repetitious regions comprises a variable number of repeats of a rather short sequence. One type of repetitious region has repeats of a 97 bp sequence; the other, which occurs in two locations, has a repeating unit found in two forms, 60 bp and 81 bp long. The variation in the number of repeating units in the repetitious regions accounts for the overall variation in spacer length. The repeti-

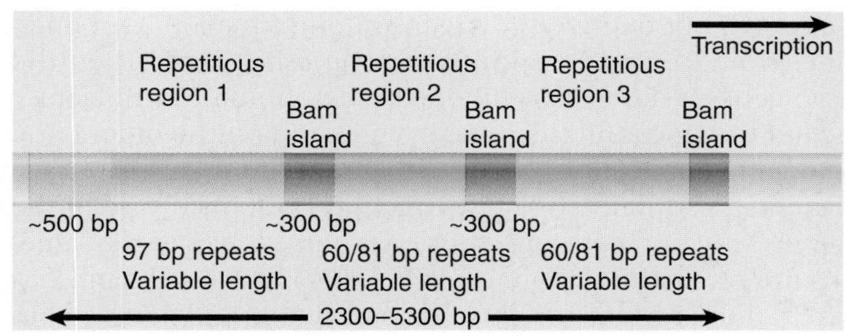

FIGURE 6.18 The nontranscribed spacer of *X. laevis* rDNA has an internally repetitious structure that is responsible for its variation in length. The Bam islands are short constant sequences that separate the repetitious regions.

tious regions are separated by shorter constant sequences called **Bam islands**. (This description takes its name from their isolation via the use of the *Bam*HI restriction enzyme.) From this type of organization, we see that the cluster has evolved by duplications involving the promoter region.

Bam islands A series of short, repeated sequences found in the nontranscribed spacer of *Xenopus* rDNA genes.

We need to explain the lack of variation in the expressed copies of the repeated genes. One hypothesis would suppose that there is a quantitative demand for a certain number of "good" sequences. This would, however, enable mutated sequences to accumulate up to a point at which their proportion of the cluster is great enough for selection to act against them. We can exclude this hypothesis because of the lack of such variation in the cluster.

The lack of variation implies that there is purifying selection against individual variations. One hypothesis would suppose that the entire cluster is regenerated periodically from one or a very few members. As a practical matter, any mechanism would need to involve regeneration every generation. We can exclude this hypothesis because a regenerated cluster would not show variation in the nontranscribed regions of the individual repeats.

We are left with a dilemma. Variation in the nontranscribed regions suggests that there is frequent unequal crossing-over. This will change the size of the cluster, but will not otherwise change the properties of the individual repeats. So how are mutations prevented from accumulating? We'll see in the next section that continuous contraction and expansion of a cluster may provide a mechanism for homogenizing its copies.

KEY CONCEPTS

- Ribosomal RNA is coded by a large number of identical genes that are tandemly repeated to form one or more clusters.
- Each rDNA cluster is organized so that transcription units giving a joint precursor to the major rRNAs alternate with nontranscribed spacers.
- The genes in an rDNA cluster all have an identical sequence.
- The nontranscribed spacers consist of shorter repeating units whose number varies so that the lengths of individual spacers are different.

CONCEPT AND REASONING CHECK

What are the advantages to having a large number of identical tandem repeats of important genes like rDNA?

6.8 Crossover Fixation Could Maintain Identical Repeats

Not all duplicated copies of genes become pseudogenes. How can selection prevent the accumulation of deleterious mutations?

The duplication of a gene is likely to result in an immediate relaxation of the selection pressure on its sequence. Now that there are two identical copies, a change in the sequence of either one will not deprive the organism of a functional product, because the original product continues to be encoded by the other copy. Then the selective pressure on the two genes is diffused, until one of them mutates sufficiently away from its original function to refocus all the selective pressure on the other.

Immediately following a gene duplication, changes might accumulate more rapidly in one of the copies, leading eventually to a new function (or to its disuse in the form of a pseudogene). If a new function develops, the gene then evolves at the same, slower rate characteristic of the original function. Probably this is the sort of mechanism responsible for the separation of functions between embryonic and adult globin genes.

Yet there are instances in which duplicated genes retain the same function, coding for identical or nearly identical products. Identical polypeptides are encoded by the two human α-globin genes, and there is only a single amino acid difference between the two γ-globin polypeptides. How does selection maintain their sequence identities?

The most obvious possibility is that the two genes do not actually have identical functions, but instead differ in some (undetected) property, such as time or place of expression. Another possibility is that the need for two copies is quantitative, because neither by itself produces a sufficient amount of product.

In more extreme cases of repetition, however, it is impossible to avoid the conclusion that no single copy of the gene is essential. When there are many copies of a gene, the immediate effects of mutation in any one copy must be very slight. The consequences of an individual mutation are diluted by the large number of copies of the gene that retain the wild-type sequence. Many mutant copies could accumulate before a lethal effect is generated.

Lethality becomes quantitative, a conclusion reinforced by the observation that half of the units of the rDNA cluster of *X. laevis* or *D. melanogaster* can be deleted without ill effect. So how are these units prevented from gradually accumulating deleterious mutations? What chance is there for the rare favorable mutation to display its advantages in the cluster?

The basic principle of hypotheses to explain the maintenance of identity among repeated copies is to suppose that nonallelic genes are continually regenerated from *one* of the copies of a preceding generation. In the simplest case of two identical genes, when a mutation occurs in one copy, either it is by chance eliminated (because the sequence of the other copy takes over), or it is spread to both duplicates. Spreading exposes a mutation to selection. The result is that the two genes evolve together as though only a single locus existed. This is called **concerted evolution** or **coincidental evolution**. It can be applied to a pair of identical genes or (with further assumptions) to a cluster containing many genes.

▸ **concerted evolution (coincidental evolution)** The ability of two or more related genes to evolve together as though constituting a single locus.

One mechanism supposes that the sequences of the nonallelic genes are directly compared with one another and homogenized by enzymes that recognize any differences. This can be done by exchanging single strands between them to form genes, one of whose strands derives from one copy, and one from the other copy. Any differences are revealed as improperly paired bases, which attract attention from enzymes able to excise and replace a base, so that only A-T and G-C pairs survive. This type of event is called **gene conversion** and is associated with genetic recombination. We should be able to ascertain the scope of such events by comparing the sequences of duplicate genes. If they are subject to concerted evolution, we should not see the accumulation of silent substitutions between them (because the homogenization process applies to these as well as to the replacement sites). We know that the extent of the maintenance mechanism need not extend beyond the gene itself, as there are cases of duplicate genes whose flanking sequences are entirely different. Indeed, we may see abrupt boundaries that mark the ends of the sequences that were homogenized.

▸ **gene conversion** The alteration of one strand of a heteroduplex DNA to make it complementary with the other strand at any position(s) where there were mispaired bases.

We must remember that the existence of such mechanisms can invalidate the determination of the history of such genes via their divergence, because the divergence reflects only the time since the last homogenization/regeneration event, not the original duplication.

The **crossover fixation** model supposes that an entire cluster is subject to continual rearrangement by the mechanism of unequal crossing-over. Such events can explain the concerted evolution of multiple genes if unequal crossing-over causes all the copies to be regenerated physically from one copy.

▸ **crossover fixation** A possible consequence of unequal crossing-over that allows a mutation in one member of a tandem cluster to spread through the whole cluster (or to be eliminated).

Following the sort of event depicted in Figure 6.12, for example, the chromosome carrying a triple locus could suffer deletion of one of the genes. Of the two remaining genes, 1½ represent the sequence of one of the original copies; only ½ of the sequence of the other original copy has survived. Any mutation in the first region now exists in both genes and is subject to selection.

Tandem clustering provides frequent opportunities for "mispairing" of loci whose sequences are the same, but that lie in different positions in their clusters. By continually expanding and contracting the number of units via unequal crossing-over, it is possible for all the units in one cluster to be derived from rather a small proportion of those in an ancestral cluster. The variable lengths of the spacers are consistent with the idea that unequal crossing-over events take place in spacers that are internally mispaired. This can explain the homogeneity of the genes compared with the variability of the spacers. The genes are exposed to selection when individual repeating units are amplified within the cluster; however, the spacers are functionally irrelevant and can accumulate changes.

In a region of nonrepetitive DNA, recombination occurs between precisely matching points on the two homologous chromosomes, thus generating reciprocal recombinants. The basis for this precision is the ability of two duplex DNA sequences to align exactly. We know that unequal recombination can occur when there are multiple copies of genes whose exons are related, even though their flanking and intervening sequences may differ. This happens because of the mispairing between corresponding exons in nonallelic genes.

Imagine how much more frequently misalignment must occur in a tandem cluster of identical or nearly identical repeats. Except at the very ends of the cluster, the close relationship between successive repeats makes it impossible even to define the exactly corresponding repeats! This has two consequences: there is continual adjustment of the size of the cluster; and there is homogenization of the repeating unit.

Consider a sequence consisting of a repeating unit "ab" with ends "x" and "y." If we represent one chromosome in black and the other in red, the exact alignment between "allelic" sequences would be:

xababababababababababababababy
xababababababababababababababy

It is likely, however, that *any* sequence *ab* in one chromosome could pair with *any* sequence *ab* in the other chromosome. In a misalignment such as:

  xababababababababababababababy
xababababababababababababababy,

the region of pairing is no less stable than in the perfectly aligned pair, although it is shorter. We do not know very much about how pairing is initiated prior to recombination, but very likely it starts between short corresponding regions and then spreads. If it starts within highly repetitive satellite DNA, it more likely than not involves repeating units that do not have exactly corresponding locations in their clusters.

Now suppose that a recombination event occurs within the unevenly paired region. The recombinants will have different numbers of repeating units. In one case, the cluster has become longer; in the other, it has become shorter,

xababababababababababababababy
×
xababababababababababababababy
↓
xababababababababababababababababy
+
xabababababababababababy,

where "x" indicates the site of the crossover.

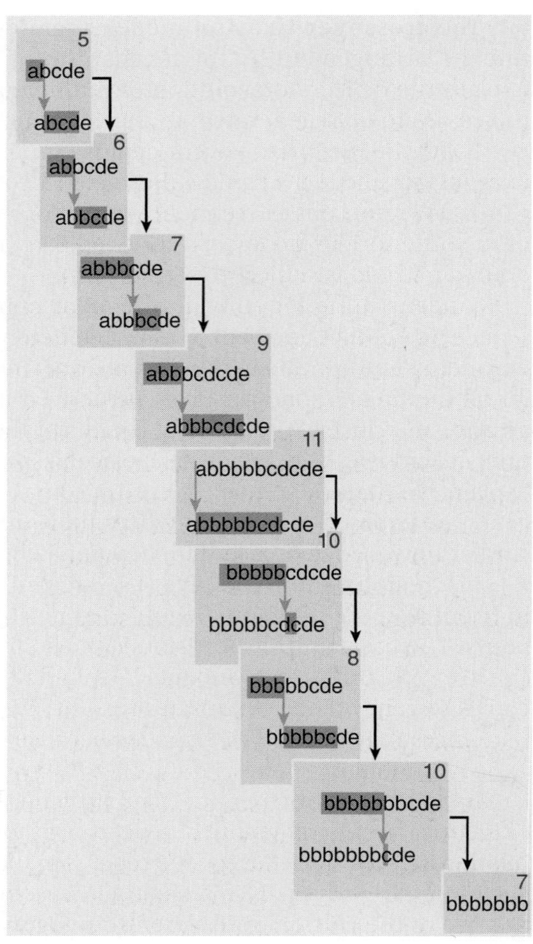

FIGURE 6.19 Unequal recombination allows one particular repeating unit to occupy the entire cluster. The numbers indicate the length of the repeating unit at each stage.

If this type of event is common, clusters of tandem repeats will undergo continual expansion and contraction. This can cause a particular repeating unit to spread through the cluster, as illustrated in **FIGURE 6.19**. Suppose that the cluster consists initially of a sequence *abcde*, where each letter represents a repeating unit. The different repeating units are closely enough related to one another to mispair for recombination. Then by a series of unequal recombination events, the size of the repetitive region increases or decreases, and one unit spreads to replace all the others.

The crossover fixation model predicts that *any sequence of DNA that is not under selective pressure will be taken over by a series of identical tandem repeats generated in this way*. The critical assumption is that the process of crossover fixation is fairly rapid relative to mutation, so that new mutations either are eliminated (their repeats are lost) or come to take over the entire cluster. In the case of the rDNA cluster, of course, a further factor is imposed by selection for a functional transcribed sequence.

KEY CONCEPTS

- Unequal crossing-over changes the size of a cluster of tandem repeats.
- Individual repeating units can be eliminated or can spread through the cluster.

CONCEPT AND REASONING CHECK

Describe two mechanisms by which concerted evolution of repeated sequences can take place.

6.9 Satellite DNAs Often Lie in Heterochromatin

Repetitive DNA is characterized by its (relatively) rapid rate of renaturation. The component that renatures most rapidly in a eukaryotic genome is called *highly repetitive* DNA and consists of very short sequences repeated many times in tandem in large clusters. As a result of its short repeating unit, it is sometimes described as **simple sequence DNA**. This type of component is present in almost all higher eukaryotic genomes, but its overall amount is extremely variable. In mammalian genomes it is typically <10%, but in (for example) *Drosophila virilis*, it amounts to ~50%. In addition to the large clusters in which this type of sequence was originally discovered, there are smaller clusters interspersed with nonrepetitive DNA. It typically consists of short sequences that are repeated in identical or related copies in the genome.

▸ **simple sequence DNA** short repeating units of DNA sequence.

The tandem repetition of a short sequence often creates a fraction with distinctive physical properties that can be used to isolate it. In some cases, the repetitive sequence has a base composition distinct from the genome average, which allows it to form a separate fraction by virtue of its distinct buoyant density. A fraction of this sort is called **satellite DNA**. The term *satellite DNA* is essentially synonymous with simple sequence DNA. Consistent with its simple sequence, this DNA may or may not be transcribed, but it is not translated. (In some species there is evidence that short RNAs are required for heterochromatin formation, suggesting that there is transcription of sequences in heterochromatic regions of chromosomes, which contain satellite DNA.)

Tandemly repeated sequences are especially liable to undergo misalignments during chromosome pairing, and thus the sizes of tandem clusters tend to be highly polymorphic, with wide variations between individuals. In fact, the smaller clusters of such sequences can be used to characterize individual genomes in the technique of "DNA fingerprinting" (see *Section 6.12, Minisatellites Are Useful for Genetic Mapping*).

The buoyant density of a duplex DNA depends on its G-C content and usually is determined by centrifuging DNA through a density gradient of CsCl. The DNA forms a band at the position corresponding to its own density. Fractions of DNA differing in G-C content by >5% can usually be separated on a density gradient.

When eukaryotic DNA is centrifuged on a density gradient, two types of material may be distinguished:

- Most of the genome forms a continuum of fragments that appear as a rather broad peak centered on the buoyant density corresponding to the average G-C content of the genome. This is called the *main band.*
- Sometimes an additional, smaller peak (or peaks) is seen at a different value. This material is the satellite DNA.

Satellites are present in many eukaryotic genomes. They may be either heavier or lighter than the main band, but it is uncommon for them to represent >5% of the total DNA. A clear example is provided by mouse DNA, shown in **FIGURE 6.20**. The graph is a quantitative scan of the bands formed when mouse DNA is centrifuged through a CsCl density gradient. The main band contains 92% of the genome and is centered on a buoyant density of 1.701 g-cm^{-3} (corresponding to its average G-C of 42%, typical for a mammal). The smaller peak represents 8% of the genome and has a distinct buoyant density of 1.690 g-cm^{-3}. It contains the mouse satellite DNA, whose G-C content (30%) is much lower than any other part of the genome.

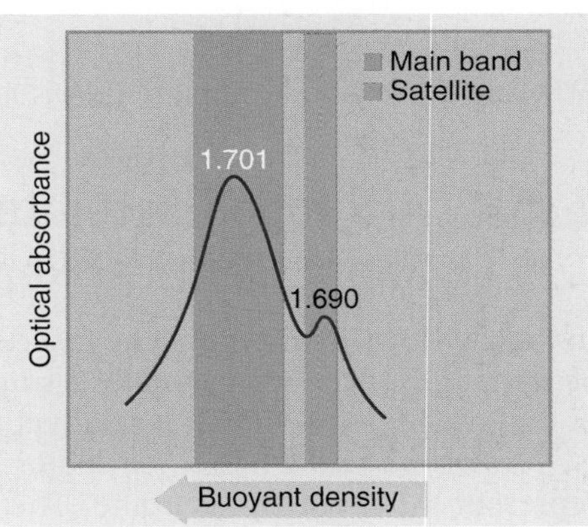

FIGURE 6.20 Mouse DNA is separated into a main band and a satellite by centrifugation through a density gradient of CsCl.

The behavior of satellite DNA on density gradients is often anomalous. When the actual base composition of a satellite is determined, it is different from the predic-

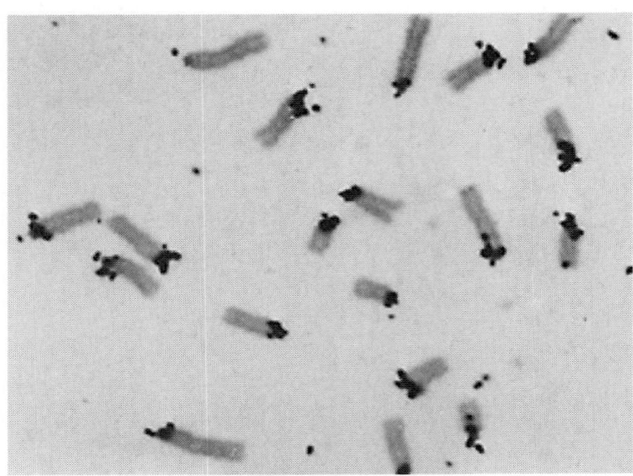

FIGURE 6.21 Cytological hybridization shows that mouse satellite DNA is located at the centromeres. Photo courtesy of Mary Lou Pardue and Joseph G. Gall, Carnegie Institution.

tion based on its buoyant density. The reason is that ρ is a function not just of base composition, but of the constitution in terms of nearest neighbor pairs. For simple sequences, these are likely to deviate from the random pairwise relationships needed to obey the equation for buoyant density. In addition, satellite DNA may be methylated, which changes its density.

Often, most of the highly repetitive DNA of a genome can be isolated in the form of satellites. When a highly repetitive DNA component does not separate as a satellite, on isolation its properties often prove to be similar to those of satellite DNA. That is to say, highly repetitive DNA consists of multiple tandem repeats with anomalous centrifugation. Material isolated in this manner is sometimes referred to as a **cryptic satellite**. Together the cryptic and apparent satellites usually account for all the large tandemly repeated blocks of highly repetitive DNA. When a genome has more than one type of highly repetitive DNA, each exists in its own satellite block (although sometimes different blocks are adjacent).

Where in the genome are the blocks of highly repetitive DNA located? An extension of nucleic acid hybridization techniques allows the location of satellite sequences to be determined directly in the chromosome complement. In the technique of ***in situ* hybridization**, the chromosomal DNA is denatured by treating cells that have been squashed on a cover slip. Next, a solution containing a radioactively labeled DNA or RNA probe is added. The probe hybridizes with its complements in the denatured genome. The location of the sites of hybridization can be determined by autoradiography (see Figure 23.15).

Satellite DNAs are found in regions of **heterochromatin**. Heterochromatin is the term used to describe regions of chromosomes that are permanently tightly coiled up and inert, in contrast with the **euchromatin** that represents most of the genome (see *Section 23.5, Chromatin Is Divided into Euchromatin and Heterochromatin*). Heterochromatin is commonly found at centromeres (the regions where the kinetochores are formed at mitosis and meiosis for controlling chromosome movement). The centromeric location of satellite DNA suggests that it has some structural function in the chromosome. This function could be connected with the process of chromosome segregation.

An example of the localization of satellite DNA for the mouse chromosomal complement is shown in **FIGURE 6.21**. In this case, one end of each chromosome is labeled, because this is where the centromeres are located in *Mus musculus* chromosomes.

- **cryptic satellite** A satellite DNA sequence not identified as such by a separate peak on a density gradient; that is, it remains present in main band DNA.
- ***in situ* hybridization** Hybridization performed by denaturing the DNA of cells squashed on a microscope slide so that reaction is possible with an added single-stranded RNA or DNA; the added preparation is radioactively labeled and its hybridization is followed by autoradiography.
- **heterochromatin** Regions of the genome that are highly condensed, are not transcribed, and are late-replicating. It is divided into two types: constitutive and facultative.
- **euchromatin** Regions that comprise most of the genome in the interphase nucleus, are less tightly coiled than heterochromatin, and contain most of the active or potentially active single copy genes.

KEY CONCEPTS

- Highly repetitive DNA has a very short repeating sequence and no coding function.
- It occurs in large blocks that can have distinct physical properties.
- It is often the major constituent of centromeric heterochromatin.

CONCEPT AND REASONING CHECK

Why is it assumed that satellite DNA is nonfunctional?

6.10 Arthropod Satellites Have Very Short Identical Repeats

In the arthropods, as typified by insects and crustaceans, each satellite DNA appears to be rather homogeneous. Usually, a single, very short repeating unit accounts for >90% of the satellite. This makes it relatively straightforward to determine the sequence.

Drosophila virilis has three major satellites and a cryptic satellite; together they represent >40% of the genome. The sequences of the satellites are summarized in

FIGURE 6.22. The three major satellites have closely related sequences. A single base substitution is sufficient to generate either satellite II or III from the sequence of satellite I.

Satellite	Predominant Sequence	Total Length	Genome Proportion
I	ACAAACT TGTTTGA	1.1×10^7	25%
II	ATAAACT TATTTGA	3.6×10^6	8%
III	ACAAATT TGTTTAA	3.6×10^6	8%
Cryptic	AATATAG TTATATC		

FIGURE 6.22 Satellite DNAs of *D. virilis* are related. More than 95% of each satellite consists of a tandem repetition of the predominant sequence.

The satellite I sequence is present in other species of *Drosophila* related to *virilis* and so may have preceded speciation. The sequences of satellites II and III seem to be specific to *D. virilis*, and so may have evolved from satellite I after speciation.

The main feature of these satellites is their very short repeating unit: only 7 bp. Similar satellites are found in other species. *D. melanogaster* has a variety of satellites, several of which have very short repeating units (5, 7, 10, or 12 bp). Comparable satellites are found in crustaceans.

The close sequence relationship found among the *D. virilis* satellites is not necessarily a feature of other genomes, for which the satellites may have unrelated sequences. *Each satellite has arisen by a lateral amplification of a very short sequence.* This sequence may represent a variant of a previously existing satellite (as in *D. virilis*), or it could have some other origin.

Satellites are continually generated and lost from genomes. This makes it difficult to ascertain evolutionary relationships, because a current satellite could have evolved from some previous satellite that has since been lost. The important feature of these satellites is that *they represent very long stretches of DNA of very low sequence complexity, within which constancy of sequence can be maintained.*

One feature of many of these satellites is a pronounced asymmetry in the orientation of base pairs on the two strands. In the example of the *D. virilis* satellites shown in Figure 6.22, in each of the major satellites one of the strands is much richer in T and G bases. This increases its buoyant density, so that upon denaturation this heavy strand (H) can be separated from the complementary light strand (L). This can be useful in sequencing the satellite.

KEY CONCEPT

- The repeating units of arthropod satellite DNAs are only a few nucleotides long. Most of the copies of the sequence are identical.

CONCEPT AND REASONING CHECK

Why is it difficult to assess evolutionary relationships using satellite sequences?

6.11 Mammalian Satellites Consist of Hierarchical Repeats

In the mammals, as typified by various rodents, the sequences comprising each satellite show appreciable divergence between tandem repeats. Common short sequences can be recognized by their preponderance among the oligonucleotide fragments released by chemical or enzymatic treatment. However, the predominant short sequence usually accounts for only a small minority of the copies. The other short sequences are related to the predominant sequence by a variety of substitutions, deletions, and insertions.

But a series of these variants of the short unit can constitute a longer repeating unit that is itself repeated in tandem with some variation. Thus mammalian satellite DNAs are constructed from a hierarchy of repeating units.

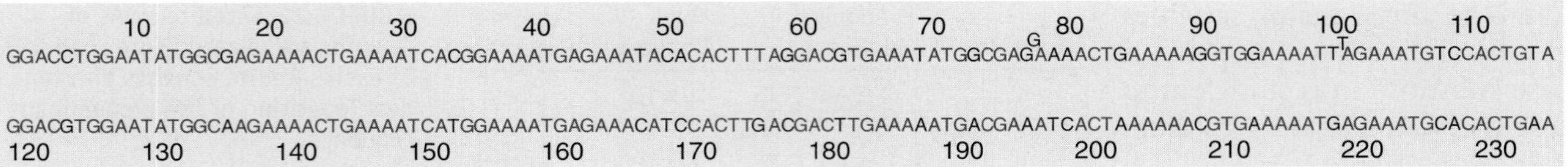

FIGURE 6.23 The repeating unit of mouse satellite DNA contains two half-repeats, which are aligned to show the identities (in blue).

FIGURE 6.24 The alignment of quarter-repeats identifies homologies between the first and second half of each half-repeat. Positions that are the same in all four quarter-repeats are shown in gray; identities that extend only through three quarter-repeats are indicated by black letters in the green area.

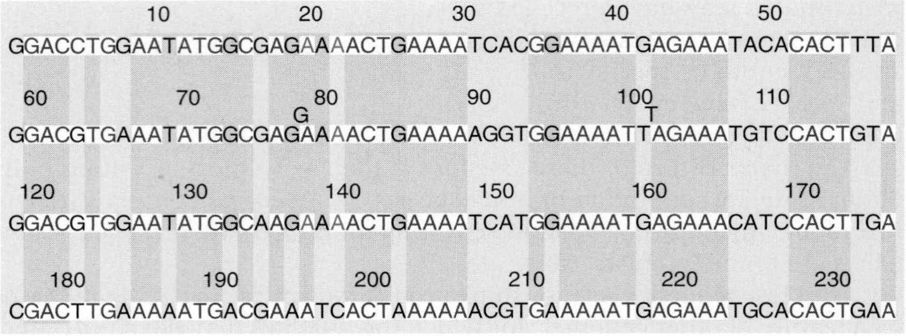

10 20 30 40 50
GGACCTGGAATATGGCGAGAAAACTGAAAATCACGGAAAATGAGAAATACACACTTTA

60 70 80 90 100 110
G T
GGACGTGAAATATGGCGAGAAAACTGAAAAAGGTGGAAAATTAGAAATGTCCACTGTA

120 130 140 150 160 170
GGACGTGGAATATGGCAAGAAAACTGAAAATCATGGAAAATGAGAAACATCCACTTGA

180 190 200 210 220 230
CGACTTGAAAAATGACGAAATCACTAAAAAACGTGAAAAATGAGAAATGCACACTGAA

When any satellite DNA is digested with an enzyme that has a recognition site in its repeating unit, one fragment will be obtained for every repeating unit in which the site occurs. In fact, when the DNA of a eukaryotic genome is digested with a restriction enzyme, most of it gives a general smear, due to the random distribution of cleavage sites. But satellite DNA generates sharp bands, because a large number of fragments of identical or almost identical size are created by cleavage at restriction sites that lie a regular distance apart.

The satellite DNA of the mouse *M. musculus* is cleaved by the enzyme EcoRII into a series of bands, including a predominant monomeric fragment of 234 bp. This sequence must be repeated with few variations throughout the 60% to 70% of the satellite that is cleaved into the monomeric band. We may analyze this sequence in terms of its successively smaller constituent repeating units.

FIGURE 6.23 depicts the sequence in terms of two half-repeats. By writing the 234 bp sequence so that the first 117 bp are aligned with the second 117 bp, we see that the two halves are quite closely related. They differ at 22 positions, corresponding to 19% divergence. This means that the current 234 bp repeating unit must have been generated at some time in the past by duplicating a 117 bp repeating unit, after which differences accumulated between the duplicates.

Within the 117 bp unit we can recognize two further subunits. Each of these is a quarter-repeat relative to the whole satellite. The four quarter-repeats are aligned in **FIGURE 6.24**. The upper two lines represent the first half-repeat of Figure 6.23; the lower two lines represent the second half-repeat. We see that the divergence between the four quarter-repeats has increased to 23 out of 58 positions, or 40%. The first three quarter-repeats are somewhat better related, and a large proportion of the divergence is due to changes in the fourth quarter-repeat.

Looking within the quarter-repeats, we find that each consists of two related subunits (one-eighth-repeats), shown as the α and β sequences in **FIGURE 6.25**. The α sequences all have an insertion of a C, and the β sequences all have an insertion of a trinucleotide, relative to a common consensus sequence. This suggests that the quarter-repeat originated by the duplication of a sequence like the consensus sequence, after which changes occurred to generate the components we now see as α and β. Further changes then took place between tandemly repeated $\alpha\beta$ sequences to generate the individual quarter- and half-repeats that exist today. Among the one-eighth-repeats, the present divergence is 19/31 = 61%.

The consensus sequence is analyzed directly in Figure 6.26, which demonstrates that the current satellite sequence can be treated as derivatives of a 9 bp sequence.

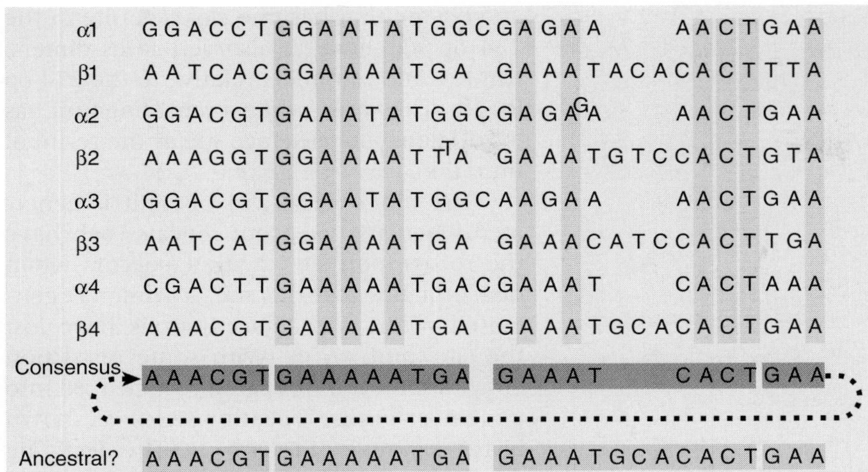

FIGURE 6.25 The alignment of eighth-repeats shows that each quarter-repeat consists of an α half and a β half. The consensus sequence gives the most common base at each position. The "ancestral" sequence shows a sequence very closely related to the consensus sequence, which could have been the predecessor to the α and β units. (The satellite sequence is continuous, so that for the purposes of deducing the consensus sequence we can treat it as a circular permutation, as indicated by joining the last GAA triplet to the first 6 bp.)

We can recognize three variants of this sequence in the satellite, as indicated at the bottom of **FIGURE 6.26**. If in one of the repeats we take the next most frequent base at two positions instead of the most frequent, we obtain three closely related 9 bp sequences:

GAAAAACGT
GAAAAATGA
GAAAAAACT

The origin of the satellite could well lie in an amplification of one of these three nonamers (9 bp units). The overall consensus sequence of the present satellite is GAAAAA$^{AG}_{TC}$T, which is effectively an amalgam of the three 9 bp repeats.

The average sequence of the monomeric fragment of the mouse satellite DNA explains its properties. The longest repeating unit of 234 bp is identified by the restriction cleavage. The unit of reassociation between single strands of denatured satellite DNA is probably the 117 bp half-repeat, because the 234 bp fragments can anneal both in register and in half-register (in the latter case, the first half-repeat of one strand renatures with the second half-repeat of the other).

G G A C C T
G G A A T A T G G C
G A G A A A A C T
G A A A A T C A C
G G A A A A T G A
G A A A T C A C T
T T A G G A C G T
G A A A T A T G G C
G A G A^G A A A C T
G A A A A A G G T
G G A A A A T^T T A
G A A A T* C A C T
G T A G G A C G T
G G A A T A T G G C
A A G A A A A C T
G A A A A T C A T
G G A A A A T G A
G A A A C* C A C T
T G A C G A C T T
G A A A A A T G A C
G A A A T C A C T
A A A A A A C G T
G A A A A A T G A
G A A A T* C A C T
G A A
G_{20} A_{16} A_{21} A_{20} A_{12} A_{17} T_8 G_{11} A_5
T_7 C_5 A_8 C_9 T_{15}
C_7

* indicates inserted triplet in β sequence
C in position 10 is extra base in α sequence

FIGURE 6.26 The existence of an overall consensus sequence is shown by writing the satellite sequence in terms of a 9 bp repeat.

So far, we have treated the present satellite as though it consisted of identical copies of the 234 bp repeating unit. Although this unit accounts for the majority of the satellite, variants of it also are present. Some of them are scattered at random throughout the satellite; others are clustered.

The existence of variants is implied by our description of the starting material for the sequence analysis as the "monomeric" fragment. When the satellite is digested by

FIGURE 6.27 Digestion of mouse satellite DNA with the restriction enzyme EcoRII identifies a series of repeating units (1, 2, 3) that are multimers of 234 bp and also a minor series (½, 1½, 2½) that includes half-repeats (see text this page). The band at the far left is a fraction resistant to digestion.

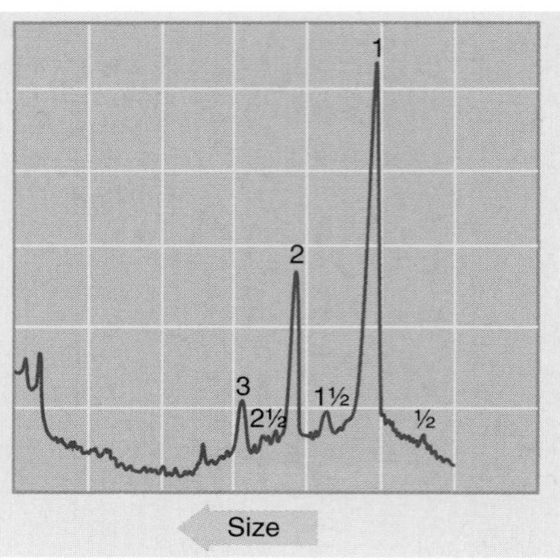

an enzyme that has one cleavage site in the 234 bp sequence, it also generates dimers, trimers, and tetramers relative to the 234 bp length. They arise when a repeating unit has lost the enzyme cleavage site as the result of mutation.

The monomeric 234 bp unit is generated when two adjacent repeats each have the recognition site. A dimer occurs when one unit has lost the site, a trimer is generated when two adjacent units have lost the site, and so on. With some restriction enzymes, most of the satellite is cleaved into a member of this repeating series, as shown in the example of **FIGURE 6.27**. The declining number of dimers, trimers, and so forth shows that there is a random distribution of the repeats in which the enzyme's recognition site has been eliminated by mutation.

KEY CONCEPT

- Mouse satellite DNA has evolved by duplication and mutation of a short repeating unit to give a basic repeating unit of 234 bp in which the original half, quarter, and eighth repeats can be recognized.

CONCEPT AND REASONING CHECK

Why is direct sequencing of satellite regions problematic?

6.12 Minisatellites Are Useful for Genetic Mapping

Sequences that resemble satellites—in that they consist of tandem repeats of a short unit, but overall are much shorter, for example, with 5 to 50 repeats—are common in mammalian genomes. They were discovered by chance as fragments whose size is extremely variable in genomic libraries of human DNA. The variability is seen when a population contains fragments of many different sizes that represent the same genomic region; when individuals are examined, it turns out that there is extensive polymorphism, and that many different alleles can be found.

▸ **microsatellite** DNAs consisting of repetitions of extremely short (typically <10 bp) units.

▸ **variable number tandem repeat (VNTR)** Very short repeated sequences, including microsatellites and minisatellites.

The name **microsatellite** is usually used when the length of the repeating unit is <10 bp, and the name *minisatellite* is used when the length of the repeating unit is ~10 to 100 bp. The terminology is not, however, precisely defined. These types of sequences are also called **variable number tandem repeat (VNTR)** regions.

The cause of the variation between individual genomes at microsatellites or minisatellites is that individual alleles have different numbers of the repeating unit. For example, one minisatellite has a repeat length of 64 bp and is found in the population with the following approximate distribution:

7% 18 repeats
11% 16 repeats
43% 14 repeats
36% 13 repeats
4% 10 repeats

The rate of genetic exchange at minisatellite sequences is high, ~10^{-4} per kb of DNA. (The frequency of exchanges per actual locus is assumed to be proportional to the length of the minisatellite.) This rate is ~10× greater than the rate of homologous recombination at meiosis, that is, in any random DNA sequence.

The high variability of minisatellites makes them especially useful for genomic mapping, because there is a high probability that individuals will vary in their alleles at such a locus. An example of mapping by minisatellites is illustrated in **FIGURE 6.28**.

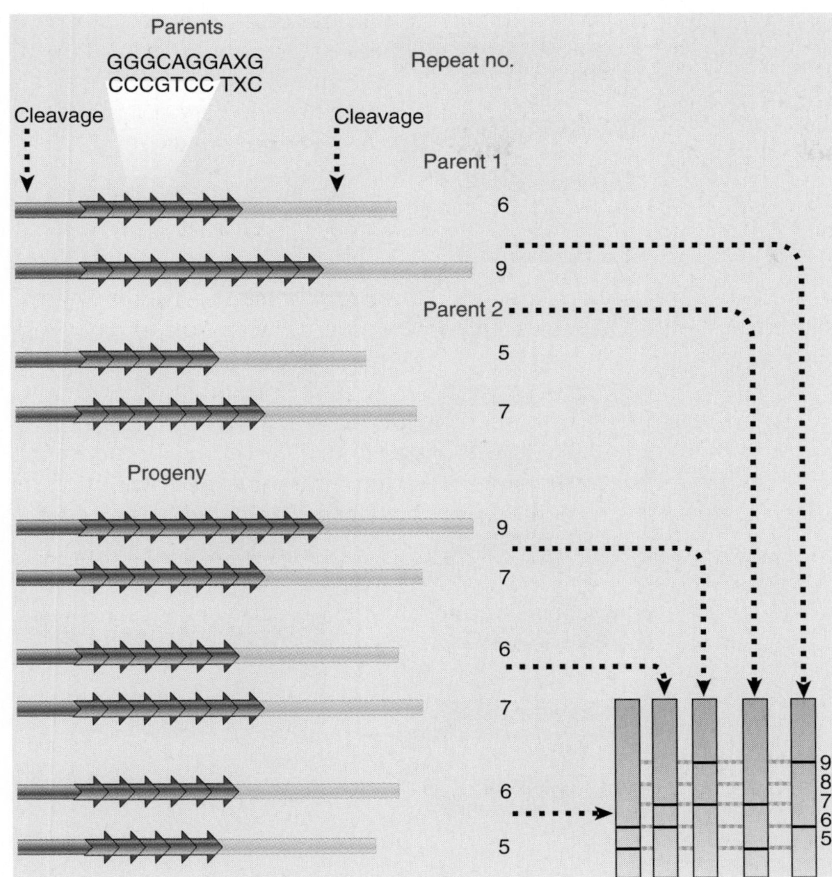

FIGURE 6.28 Alleles may differ in the number of repeats at a minisatellite locus, so that cleavage on either side generates restriction fragments that differ in length. By using a minisatellite with alleles that differ between parents, the pattern of inheritance can be followed.

This shows an extreme case in which two individuals both are heterozygous at a minisatellite locus, and in fact all four alleles are different. All progeny gain one allele from each parent in the usual way, and it is possible unambiguously to determine the source of every allele in the progeny. In the terminology of human genetics, the meioses described in this figure are highly informative because of the variation between alleles.

One family of minisatellites in the human genome shares a common "core" sequence. The core is a G-C-rich sequence of 10 to 15 bp, showing an asymmetry of purine/pyrimidine distribution on the two strands. Each individual minisatellite has a variant of the core sequence, but ~1000 minisatellites can be detected on Southern blot (see *Section 4.7, Eukaryotic Protein-Coding Genes Can Be Identified by the Conservation of Exons*) by a probe consisting of the core sequence.

Consider the situation shown in Figure 6.28, but multiplied many times by the existence of many such sequences. The effect of the variation at individual loci is to create a unique pattern for every individual. This makes it possible to assign heredity unambiguously between parents and progeny by showing that 50% of the bands in any individual are derived from a particular parent. This is the basis of the technique known as **DNA fingerprinting**.

▸ **DNA fingerprinting** Analysis of the differences between individuals of the fragments generated by using restriction enzymes to cleave regions that contain short repeated sequences or by PCR. The lengths of the repeated regions are unique to every individual, and, as a result, the presence of a particular subset in any two individuals can be used to define their common inheritance (e.g., a parent-child relationship).

Both microsatellites and minisatellites are unstable, although for different reasons. Microsatellites undergo intrastrand mispairing, when slippage during replication leads to expansion of the repeat, as shown in **FIGURE 6.29**. Systems that repair damage to DNA—in particular those that recognize mismatched base pairs—are important in reversing such changes, as shown by a large increase in frequency when repair genes are inactivated. Mutations in repair systems are an important contributory factor in the development of cancer; thus tumor cells often display variations in microsatellite sequences. Minisatellites undergo the same sort of unequal crossing-over between

FIGURE 6.29 Replication slippage occurs when the daughter strand slips back one repeating unit in pairing with the template strand. Each slippage event adds one repeating unit to the daughter strand. The extra repeats are extruded as a single-strand loop. Replication of this daughter strand in the next cycle generates a duplex DNA with an increased number of repeats.

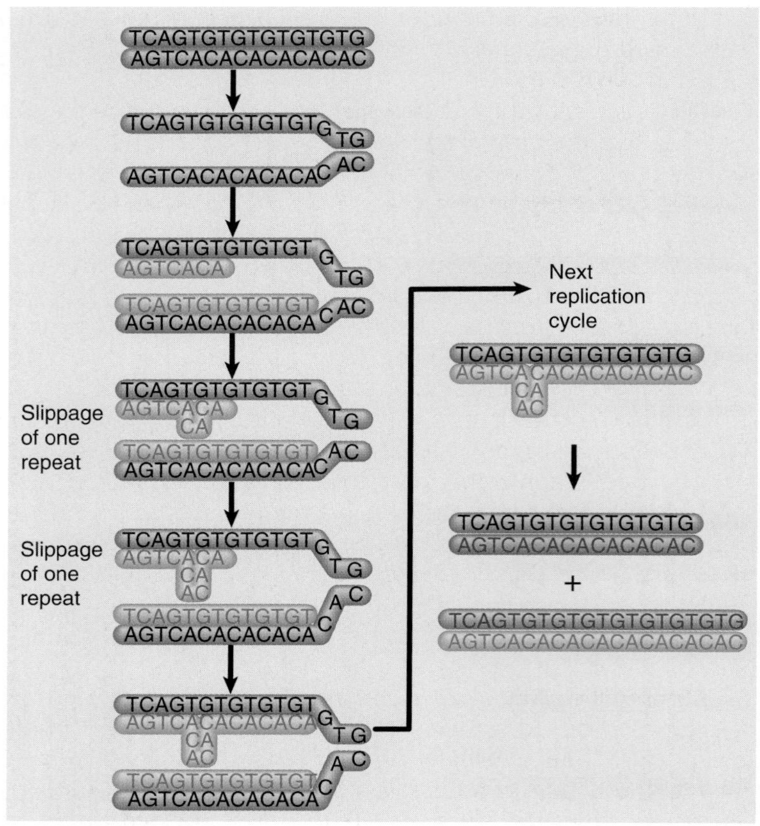

repeats that we have discussed for satellites (see Figure 6.3). One telling case is that increased variation is associated with a recombination hotspot. The recombination event is not usually associated with recombination between flanking markers, but has a complex form in which the new mutant allele gains information from both the sister chromatid and the other (homologous) chromosome.

It is not clear at what repeating length the cause of the variation shifts from replication slippage to unequal crossing-over.

KEY CONCEPT

- The variation between microsatellites or minisatellites in individual genomes can be used to identify heredity unequivocally by showing that 50% of the bands in an individual are derived from a particular parent.

CONCEPT AND REASONING CHECK

Why are VNTRs useful for DNA fingerprinting? Describe how the technique may be used to determine paternity.

6.13 Summary

Most genes belong to families, which are defined by the possession of related sequences in the exons of individual members. Families evolve by the duplication of a gene (or genes), followed by divergence between the copies. Some copies suffer inactivating mutations and become pseudogenes that no longer have any function.

An evolving set of genes may remain together in a cluster or may be dispersed to new locations by chromosomal rearrangement. The organization of existing clusters can sometimes be used to infer the series of events that has occurred. These events act with regard to sequence rather than function, and therefore include pseudogenes as well as active genes.

Silent substitutions accumulate more rapidly than replacement substitutions (which affect the amino acid sequence). The rate of divergence at replacement sites can sometimes be used to establish a molecular clock, which can be calibrated in percent divergence per million years. The clock can then be used to calculate the time of divergence between any two members of the family.

A tandem cluster consists of many copies of a repeating unit that includes the transcribed sequence(s) and a nontranscribed spacer(s). rRNA gene clusters code only for a single rRNA precursor. Maintenance of active genes in clusters depends on mechanisms such as gene conversion or unequal crossing-over, which cause mutations to spread through the cluster so that they become exposed to evolutionary pressure.

Satellite DNA consists of very short sequences repeated many times in tandem. Its distinct centrifugation properties reflect its biased base composition. Satellite DNA is concentrated in centromeric heterochromatin, but its function (if any) is unknown. The individual repeating units of arthropod satellites are identical. Those of mammalian satellites are related and can be organized into a hierarchy reflecting the evolution of the satellite by the amplification and divergence of randomly chosen sequences.

Unequal crossing-over appears to have been a major determinant of satellite DNA organization. Crossover fixation explains the ability of variants to spread through a cluster.

Minisatellites and microsatellites consist of even shorter repeating sequences than satellites: <10 bp for microsatellites and 10 to 50 bp for minisatellites. The number of repeating units is usually 5 to 50. There is high variation in the repeat number between individual genomes. A microsatellite repeat number varies as the result of slippage during replication; the frequency is affected by systems that recognize and repair damage in DNA. Minisatellite repeat number varies as the result of recombination-like events. Variations in repeat number can be used to determine hereditary relationships by the technique known as DNA fingerprinting.

CHAPTER QUESTIONS

1. Genes that have become nonfunctional and are no longer able to be transcribed to a functional protein are called:

A. allelic genes.
B. nonallelic genes.
C. pseudogenes.
D. faux genes.

2. The most common type of duplication generates a:

A. second copy of the gene in close proximity to the first copy.
B. second copy of the gene in a distant location of the same chromosome.
C. second copy of the gene on a different chromosome.
D. Any of the above—they occur with equal probability.

3. Most mutations that change the amino acid sequence of a protein are:

A. deleterious and will be eliminated by natural selection.
B. deleterious but cannot be eliminated.
C. neutral due to similar properties of many amino acids.
D. beneficial.

4. A silent substitution:
   - **A.** creates a premature stop codon, thus silencing expression of the gene.
   - **B.** substitutes one amino acid for a different amino acid with similar properties, so it does not affect protein function.
   - **C.** changes only one DNA base in a codon without changing the amino acid sequence.
   - **D.** occurs outside the protein coding region, in the 5′ or 3′ untranslated regions, and thus does not affect protein function.
5. Mutations that reduce or prevent synthesis of either α or β globin result in:
   - **A.** hydrops fetalis.
   - **B.** thalassemias.
   - **C.** HbH disease.
   - **D.** Hb anti-Lepore.
6. Which of the following show the highest number of multiple identical gene copies in eukaryotes?
   - **A.** RNA polymerase and histones
   - **B.** RNA polymerase and tRNA
   - **C.** histones and rRNA
   - **D.** rRNA and tRNA
7. In what region of a eukaryotic cell does rRNA synthesis occur?
   - **A.** nucleus
   - **B.** nucleolus
   - **C.** ribosome
   - **D.** cytoplasm
8. Satellite DNAs are typically not found within which of the following parts of chromosomes?
   - **A.** heterochromatin
   - **B.** euchromatin
   - **C.** telomeres
   - **D.** centromeres
9. Simple sequence DNA consists of:
   - **A.** single copy DNA sequences.
   - **B.** minimally repetitive DNA of short sequences.
   - **C.** moderately repetitive DNA of short sequences repeated in tandem in a cluster.
   - **D.** highly repetitive DNA of short sequences repeated in tandem in a cluster.
10. The crossover fixation model predicts that:
    - **A.** any sequence of DNA that is not under selective pressure will be taken over by a series of identical tandem repeats.
    - **B.** any sequence of DNA that is under selective pressure will be taken over by a series of identical tandem repeats.
    - **C.** any sequence of DNA that is not under selective pressure will be taken over by a series of non-identical tandem repeats.
    - **D.** any sequence of DNA that is under selective pressure will be taken over by a series of non-identical tandem repeats.

KEY TERMS

Bam islands
codon bias
concerted evolution (coincidental evolution)
crossover fixation
cryptic satellite
divergence
DNA fingerprinting
euchromatin
fixation
gene cluster
gene conversion
gene family
genetic drift
Hb anti-Lepore
Hb Kenya
Hb Lepore
HbH disease
heterochromatin
hydrops fetalis
***in situ* hybridization**
microsatellite
minisatellite
molecular clock
neutral mutation
nonallelic genes
nontranscribed spacer
nucleolar organizer
nucleolus
pseudogenes
rDNA
replacement sites
satellite DNA
silent sites
simple sequence DNA
thalassemia
translocation
unequal crossing-over (nonreciprocal recombination)
variable number tandem repeat (VNTR)

FURTHER READING

Bailey, J. A., Gu, Z., Clark, R. A., Reinert, K., Samonte, R. V., Schwartz, S., Adams, M. D., Myers, E. W., Li, P. W., and Eichler, E. E. (2002). Recent segmental duplications in the human genome. *Science* 297, 1003–1007.

An identification of regions of the human genome with duplicated segments. A total of 169 regions were identified, of which 24 have been associated with genetic diseases.

Hardison, R. (1998). Hemoglobins from bacteria to man: evolution of different patterns of gene expression. *J. Exp. Biol.* 201, 1099–1117.

A review of the evolutionary history of hemoglobin genes in bacteria and eukaryotes, including the duplications and divergences in sequence and function.

Waterston, R. H. *et al.* (2002). Initial sequencing and comparative analysis of the mouse genome. *Nature* 420, 520–562.

The publication of the "draft sequence" of the mouse (*Mus musculus*) genome with comparison to the human genome structure.

II

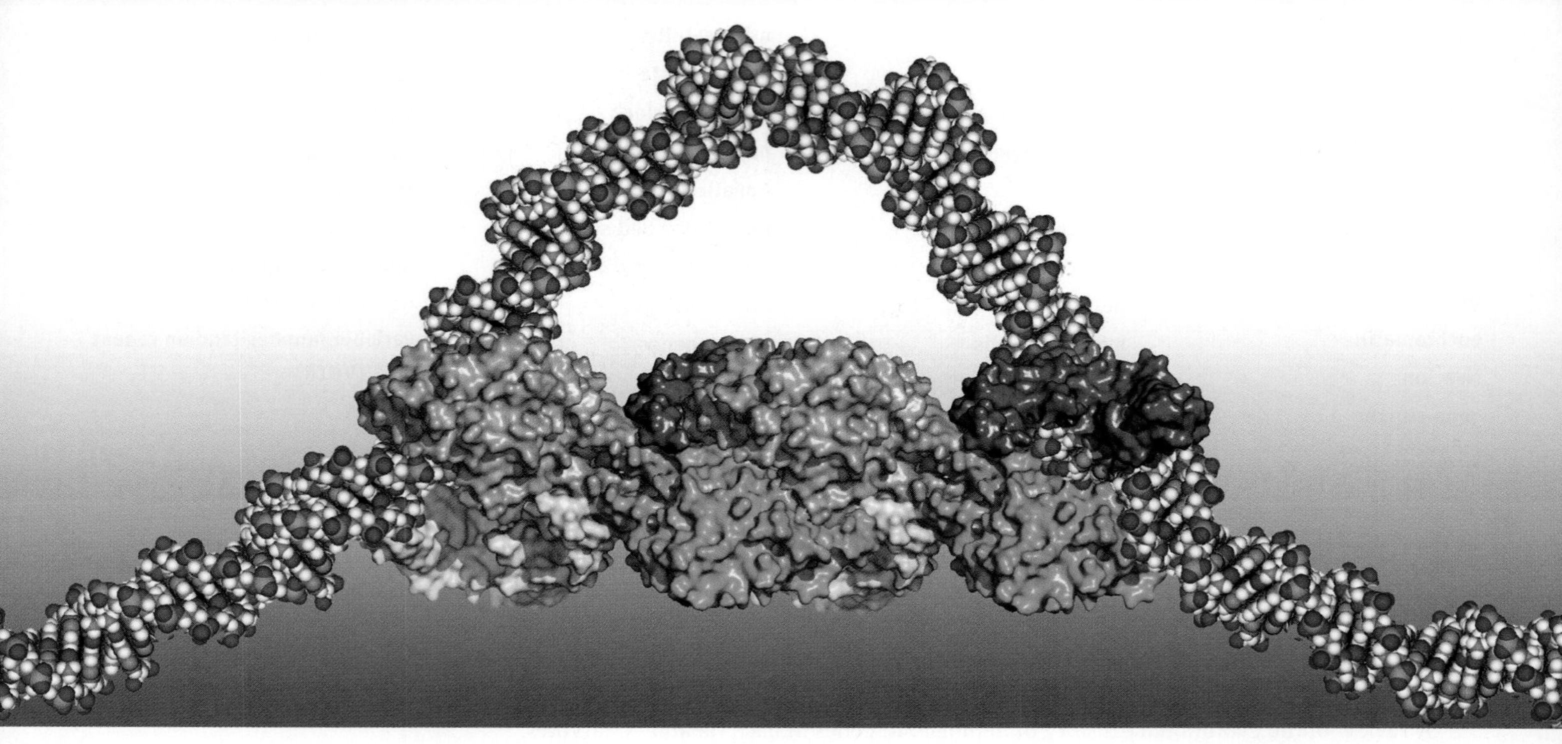

Bound proteins forming a loop in a stretch of DNA. In general, DNA-binding proteins influence the expression of genes. Photo courtesy of Mike White and Doug Smith, University of California, San Diego. Structure assembled using a model by Eva Vanamee and Aneel Aggarwal, Mount Sinai School of Medicine.

Proteins

7

A large ribosomal RNA, with the sugar-phosphate backbone shown in dark blue and individual bases shown in pastel colors, revealing the complex 3D structure of rRNA. Graphic produced using the UCSF Chimera package from the Resource for Biocomputing, Visualization, and Informatics at the University of California, San Francisco (supported by NIH P41 RR-01081).

Messenger RNA

CHAPTER OUTLINE

7.1 Introduction

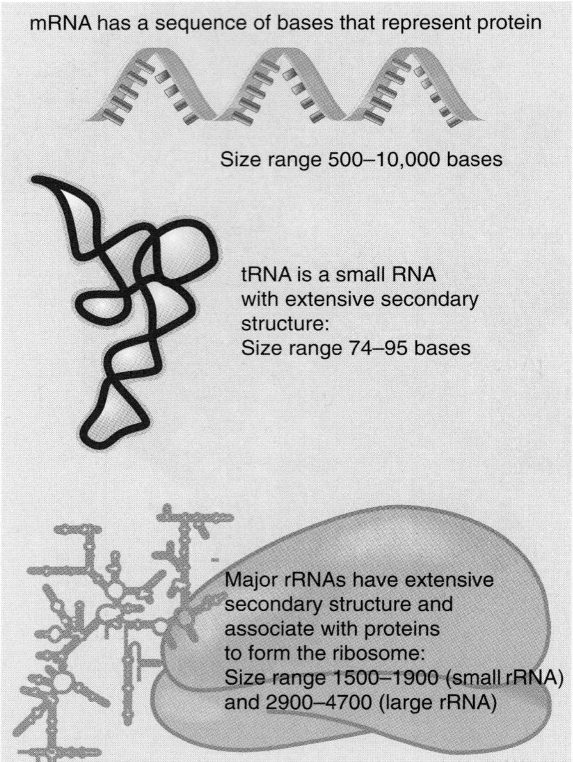

mRNA has a sequence of bases that represent protein
Size range 500–10,000 bases
tRNA is a small RNA
with extensive secondary
structure:
Size range 74–95 bases
Major rRNAs have extensive
secondary structure and
associate with proteins
to form the ribosome:
Size range 1500–1900 (small rRNA)
and 2900–4700 (large rRNA)

FIGURE 7.1 The three types of RNA universally required for gene expression are mRNA (carries the coding sequence), tRNA (provides the amino acid corresponding to each codon), and rRNA (a major component of the ribosome that provides the environment for protein synthesis).

RNA is a central player in gene expression. It was first characterized as an intermediate in protein synthesis, but since then many other RNAs have been discovered that play structural or functional roles at other stages of gene expression. The involvement of RNA in many functions concerned with gene expression supports the general view that the entire process may have evolved in an "RNA world" in which RNA was originally the active component in maintaining and expressing genetic information. Many of these functions were subsequently assisted or taken over by proteins, with a consequent increase in versatility and probably efficiency.

As summarized in **FIGURE 7.1**, three major classes of RNA are directly involved in the production of proteins:

- Messenger RNA (mRNA) provides an intermediate that carries the copy of a DNA sequence that represents protein.
- **Transfer RNAs (tRNA)** are small RNAs that are used to provide amino acids corresponding to each particular codon in mRNA.
- **Ribosomal RNAs (rRNA)** are components of the ribosome, a large ribonucleoprotein complex that contains many proteins as well as its RNA components, and provides the apparatus for actually polymerizing amino acids into a polypeptide chain.

▸ **transfer RNA (tRNA)** The intermediate in protein synthesis that interprets the genetic code. Each molecule can be linked to an amino acid. It has an anticodon sequence that is complementary to a triplet codon representing the amino acid.

▸ **ribosomal RNA (rRNA)** A major component of the ribosome.

The type of role that RNA plays in each of these cases is distinct. For messenger RNA, its sequence is the important feature: each nucleotide triplet within the coding region of the mRNA represents an amino acid in the corresponding protein. However, the structure of the mRNA—in particular the sequences on either side of the coding region—can play an important role in controlling its activity, and therefore the amount of polypeptide that is produced from it.

In tRNA, we see two of the common themes governing the use of RNA: its three-dimensional structure is important, and it has the ability to base pair with another RNA (mRNA). The three-dimensional structure is recognized first by an enzyme as providing a target that is appropriate for linkage to a specific amino acid. The linkage creates an aminoacyl-tRNA, which is recognized as the structure that is used for protein synthesis. The specificity with which an aminoacyl-tRNA is used is controlled by base pairing when a short triplet sequence (the anticodon) pairs with the nucleotide triplet representing its amino acid.

With rRNA, we see another type of activity. One role of RNA is structural, in providing a framework to which ribosomal proteins attach. It also participates directly in the activities of the ribosome. One of the crucial activities of the ribosome is the ability to catalyze the formation of a peptide bond by which an amino acid is incorporated into protein. This activity resides in one of the rRNAs.

The important thing about this background is that, as we consider the role of RNA in protein synthesis, we have to view it as a component that plays an active role and that can be a target for regulation by either proteins or other RNAs, and we should remember that the RNAs may have been the basis for the original apparatus. The theme that runs through all of the activities of RNA, in both protein synthesis and elsewhere, is that its functions depend critically upon base pairing, both to form its secondary structure and to interact specifically with other RNA molecules. The coding function of mRNA is unique, but tRNA and rRNA are examples of a much broader class of noncoding RNAs with a variety of functions in gene expression.

7.2 mRNA Is Produced by Transcription and Is Translated

Gene expression occurs by a two-stage process.

- Transcription generates a single-stranded RNA identical in sequence (with U in place of T) with one of the strands of the duplex DNA. (Uracil base pairs with adenine, so it is the functional analog of thymine in RNA both in protein synthesis and in the formation of secondary RNA structures.)
- In protein-coding genes for which mRNA is the product of transcription, translation converts the nucleotide sequence of mRNA into the sequence of amino acids comprising a polypeptide. The entire length of an mRNA is not translated, but each mRNA contains at least one coding region that is related to a polypeptide sequence by the genetic code: each nucleotide triplet (codon) of the coding region represents one amino acid.

For a single transcription event, only one strand of a DNA duplex is transcribed into a messenger RNA; the polymerase does not switch strands in the midst of transcription. (However, both strands in a certain region may be transcribed in separate events when genes overlap.) We distinguish the two strands of DNA as depicted in **FIGURE 7.2**:

- The strand of DNA that directs synthesis of the mRNA via complementary base pairing is called the **antisense strand** or **template strand**. (*Antisense* is used as a general term to describe a sequence of DNA or RNA that is complementary to mRNA.)
- The other DNA strand bears the same sequence as the mRNA (except for possessing T instead of U) and is called the **coding strand** or **sense strand**.

antisense strand (template strand) The DNA strand that is complementary to the sense strand, and acts as the template for synthesis of mRNA.

coding strand (sense strand) The DNA strand that has the same sequence as the mRNA and is related by the genetic code to the polypeptide sequence that it represents.

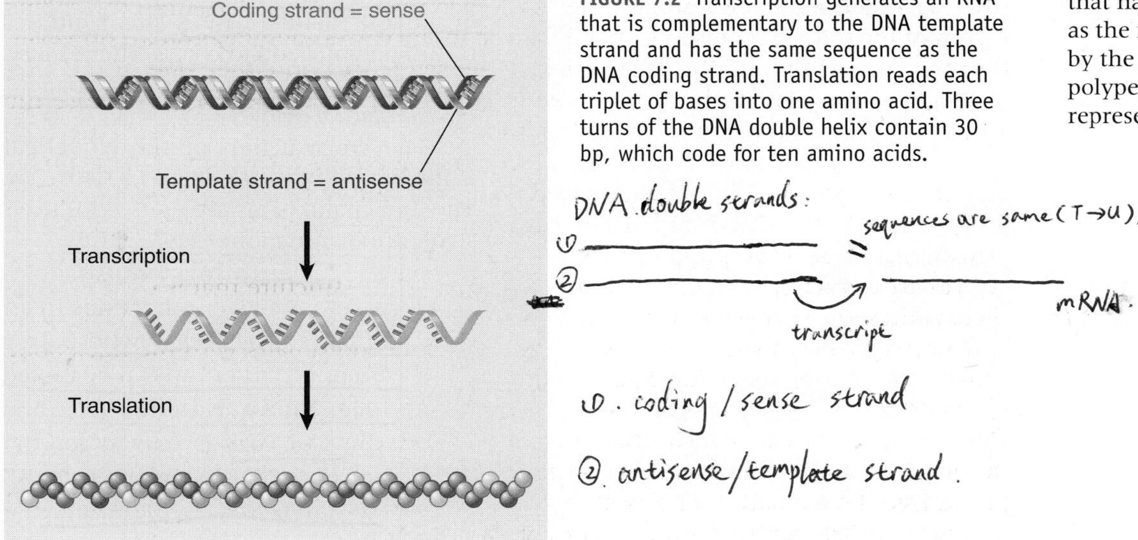

FIGURE 7.2 Transcription generates an RNA that is complementary to the DNA template strand and has the same sequence as the DNA coding strand. Translation reads each triplet of bases into one amino acid. Three turns of the DNA double helix contain 30 bp, which code for ten amino acids.

In this chapter we discuss mRNA and its use as a template for protein synthesis. In *Chapter 8, Translation,* we discuss the process by which a protein is synthesized. In *Chapter 9, Using the Genetic Code,* we discuss the way the genetic code is used to interpret the meaning of a sequence of mRNA. In *Chapter 10, Protein Localization,* we turn to the question of how a protein finds its proper location in the cell when or after it is synthesized.

KEY CONCEPT

- Within a gene, only one of the two strands of DNA is transcribed into RNA.

CONCEPT AND REASONING CHECK

In research publications containing DNA sequences, usually only the coding strand of a coding region is presented. Why?

▸ **anticodon** A trinucleotide sequence in tRNA that is complementary to the codon in mRNA and enables the tRNA to place the appropriate amino acid in response to the codon.

▸ **cloverleaf** The structure of tRNA drawn in two dimensions, forming four distinct arm-loops.

▸ **stem** The base-paired segment of a hairpin structure in RNA.

▸ **loop** A single-stranded region at the end of a hairpin in RNA (or single-stranded DNA); it corresponds to the sequence between inverted repeats in duplex DNA.

▸ **arm** One of the four (or in some cases five) stem-loop structures that make up the secondary structure of tRNA.

7.3 The Secondary Structure of Transfer RNA Is a Cloverleaf

Messenger RNA can be distinguished from the apparatus responsible for its translation by the use of *in vitro* cell-free systems to synthesize proteins. *A protein-synthesizing system from one cell type can translate the mRNA from another, demonstrating that both the genetic code and the translation apparatus are universal.*

Each nucleotide triplet in the mRNA corresponds to an amino acid. The incongruity of structure between trinucleotide and amino acid immediately raises the question of how each codon is matched to its particular amino acid. The "adaptor" is transfer RNA (tRNA). A tRNA has two crucial properties:

- It represents a single amino acid, to which it is *covalently linked.*
- It contains a trinucleotide sequence, the **anticodon**, which is *complementary to the codon representing its amino acid.* The anticodon enables the tRNA to recognize the codon via complementary base pairing.

All tRNAs have common secondary and tertiary structures. The tRNA secondary structure can be written in the form of a **cloverleaf**, illustrated in **FIGURE 7.3**, in which complementary base pairing forms **stems** for single-stranded **loops**. The stem–loop structures are called the **arms** of tRNA. Their sequences include nonstandard bases that are generated by modification of the four standard bases after synthesis of the polynucleotide chain. The tertiary, three-dimensional structure is "L-shaped" (see *Section 7.4* below).

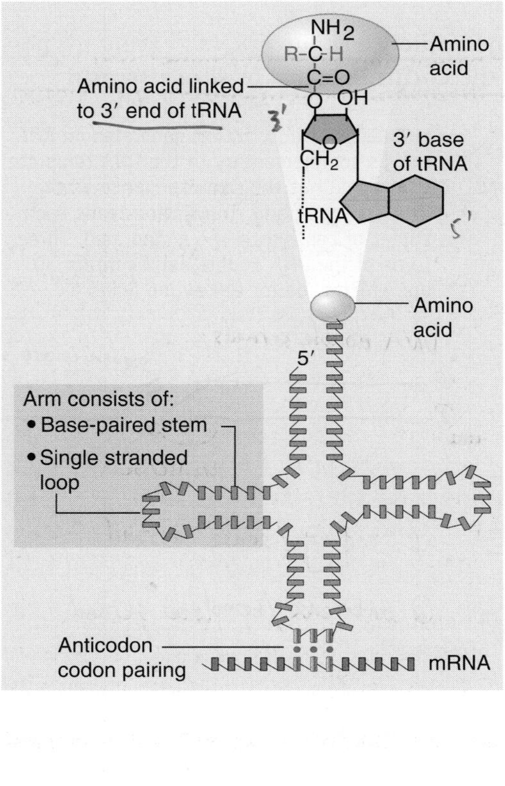

FIGURE 7.3 A tRNA has the dual properties of an adaptor that recognizes both the amino acid and codon. The 3′ adenosine is covalently linked to an amino acid. The anticodon pairs with the codon on mRNA.

The construction of the cloverleaf is illustrated in more detail in **FIGURE 7.4**. The four major arms are named for their structure or function:

▸ **acceptor arm** A short duplex on tRNA that terminates in the CCA sequence to which an amino acid is linked.

- The **acceptor arm** consists of a base-paired stem that ends in an unpaired sequence whose free 2′- or 3′-OH group can be linked to an amino acid.
- The *TψC arm* is named for the presence of this triplet sequence. (ψ stands for pseudouridine, a modified base.)

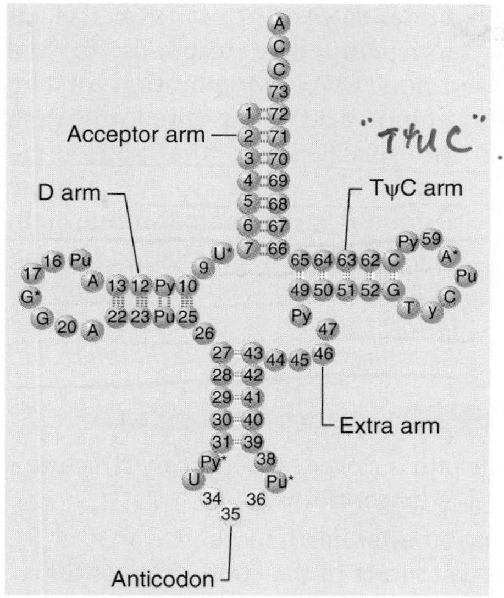

FIGURE 7.4 The tRNA cloverleaf has invariant and semi-invariant bases and a conserved set of base pairing interactions.

- The **anticodon arm** always contains the anticodon triplet in the center of the loop.
- The **D arm** is so named because it contains the base dihydrouridine (another of the modified bases in tRNA).
- The **extra arm** lies between the TψC and anticodon arms and varies from 3 to 21 bases in length.

The numbering system for tRNA illustrates the constancy of the structure. Positions are numbered from 5′ to 3′ according to the most common tRNA structure, which has 76 bases. The overall range of tRNA lengths is 74 to 95 bases. The variation in length is caused by differences in the D arm and extra arm.

The base pairing that maintains the secondary structure is shown in Figure 7.4. Within a given tRNA, most of the base pairings are conventional partnerships of A-U and G-C, but occasional G-U, G-ψ, or A-ψ pairs are found. The additional types of base pairs are less stable than the regular pairs, but still allow some double-helical structures to form in tRNAs.

When the sequences of tRNAs are compared, the bases found at some positions are **invariant** (or **conserved**); a particular base is almost always found at the position. Some positions are described as **semi-invariant** (or **semiconserved**) because they are restricted to one type of base (purine or pyrimidine), but either base of that type may be present.

When a tRNA is *charged* with the amino acid corresponding to its anticodon, it is called an **aminoacyl-tRNA**. The amino acid is linked by an ester bond from its carboxyl group to the 2′ or 3′ hydroxyl group of the ribose of the 3′ terminal base of the tRNA (which is always adenine). The process of charging a tRNA is catalyzed by a specific class of enzymes called **aminoacyl-tRNA synthetases**, of which there are (at least) 20. Each recognizes a single amino acid as well as all the tRNAs onto which it can legitimately be attached.

There is at least one tRNA (but usually more) for each amino acid. A tRNA is named by using the three-letter abbreviation for the amino acid as a superscript. If there is more than one tRNA for the same amino acid, subscript numerals are used to distinguish them. So two tRNAs for tyrosine would be described as $tRNA_1^{Tyr}$ and $tRNA_2^{Tyr}$. A tRNA carrying an amino acid—that is, an aminoacyl-tRNA—is indicated by a prefix that identifies the amino acid. Ala-tRNA describes $tRNA^{Ala}$ carrying its amino acid.

Does the anticodon sequence alone allow aminoacyl-tRNA to recognize the correct codon? A classic experiment to test this question is illustrated in **FIGURE 7.5**. Reductive desulfuration converts the amino acid of cysteinyl-tRNA into alanine, generating

- **anticodon arm** A stem-loop structure in tRNA that exposes the anticodon triplet at one end.
- **D arm** The arm of tRNA that has a high content of the base dihydrouridine.
- **extra arm** The arm of tRNA that shows length variation among different tRNAs.
- **invariant sequences** Base positions in tRNA that have the same nucleotide in virtually all (>95%) tRNAs.
- **conserved sequences** Sequences in which many examples of a particular nucleic acid or protein are compared and the same individual bases or amino acids are always found at particular locations.
- **semiconserved (semi-invariant)** A position where comparison of many individual sequences finds the same type of base (pyrimidine or purine) always present.
- **aminoacyl-tRNA** A tRNA linked to an amino acid. The COOH group of the amino acid is linked to the 3′- or 2′-OH group of the terminal base of the tRNA.
- **aminoacyl-tRNA synthetases** Enzymes responsible for covalently linking amino acids to the 2′- or 3′-OH position of tRNA.

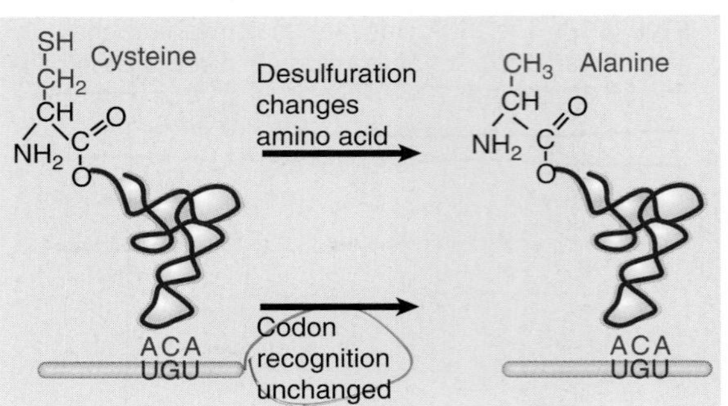

FIGURE 7.5 The meaning of tRNA is determined by its anticodon and not by its amino acid.

alanyl-tRNA^{Cys}. The tRNA has an anticodon that responds to the codon UGU. Modification of the amino acid does not influence the specificity of the anticodon–codon interaction, so the alanine residue is incorporated into protein in place of cysteine. *Once a tRNA has been charged, the amino acid plays no further role in its specificity, which is determined exclusively by the anticodon.*

KEY CONCEPTS

- A tRNA has a sequence of 74 to 95 bases that folds into a cloverleaf secondary structure with four constant arms (and an additional arm in the longer tRNAs).
- tRNA is charged to form aminoacyl-tRNA by forming an ester link from the 2′- or 3′-OH group of the adenylic acid at the end of the acceptor arm to the COOH group of the amino acid.
- The sequence of the anticodon is solely responsible for the specificity of the aminoacyl-tRNA during translation.

CONCEPT AND REASONING CHECK

Why does an aminoacyl-tRNA synthetase have to be specific for both its amino acid and tRNA substrates?

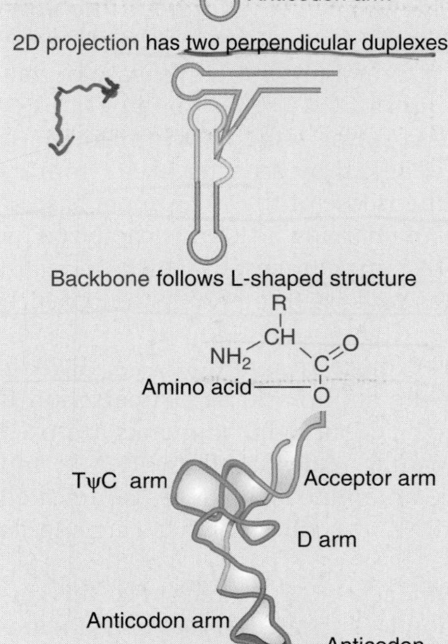

FIGURE 7.6 Transfer RNA folds into a compact L-shaped tertiary structure with the amino acid at one end and the anticodon at the other end.

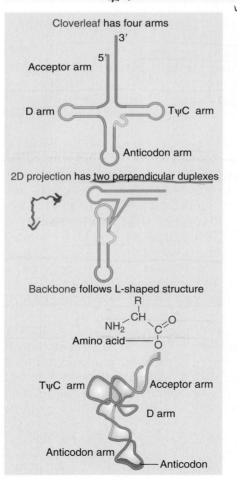

7.4 The Acceptor Arm and Anticodon Are at Opposite Ends of the tRNA Tertiary Structure

The secondary structure of each tRNA folds into a compact L-shaped tertiary structure in which the 3′ end that binds the amino acid is opposite from the anticodon that binds the mRNA. All tRNAs have the same general tertiary structure, although they are distinguished by individual variations.

The base paired double-helical stems of the secondary structure are maintained in the tertiary structure, but their arrangement in three dimensions essentially creates two double helices at right angles to each other, as illustrated in **FIGURE 7.6**. The acceptor arm and the TψC arm form one continuous double helix with a single gap; the D arm and anticodon arm form another continuous double helix, also with a gap. The region between the double helices, where the turn in the L-shape is made, contains the TψC loop and the D loop. So, the amino acid resides at the extremity of one arm of the L-shape and the anticodon loop forms the other end.

The tertiary structure is stabilized by hydrogen bonding, mostly involving bases that are unpaired in the secondary structure. Many of the invariant and semi-invariant bases are involved in these H bonds, which explains their conservation. Not every one of these interactions is universal, but probably they all identify the *general* pattern for establishing tRNA structure.

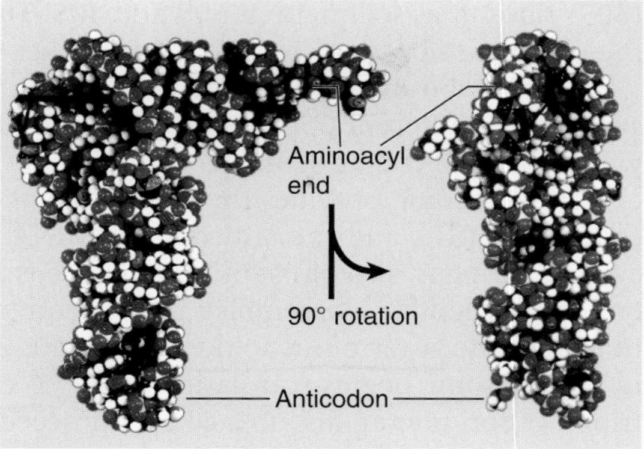

A molecular model of the structure of yeast $tRNA^{Phe}$ is shown in **FIGURE 7.7**. The left view corresponds with the bottom panel in Figure 7.6. Differences in the structure are found in other tRNAs, thus accommodating the dilemma that all tRNAs must have a similar shape, yet it must be possible to recognize differences between them. For example, in $tRNA^{Asp}$, the angle between the two axes is slightly greater, so the molecule has a slightly more open conformation.

FIGURE 7.7 A space-filling model shows that the $tRNA^{Phe}$ tertiary structure is compact. The two views of tRNA are rotated by 90°. Photo courtesy of Sung-Hou Kim, University of California, Berkeley.

The structure suggests a general conclusion about the function of tRNA: its sites for exercising particular functions are maximally separated. The amino acid is as far from the anticodon as possible, which is consistent with their roles in protein synthesis.

KEY CONCEPT

- The cloverleaf forms an L-shaped tertiary structure with the acceptor arm at one end and the anticodon arm at the other end.

CONCEPT AND REASONING CHECK

The tRNA tertiary structure has two functional ends. How is this consistent with the overall function of tRNA?

7.5 Messenger RNA Is Translated by Ribosomes

Translation of an mRNA into a polypeptide chain is catalyzed by the **ribosome**. Ribosomes are traditionally described in terms of their (approximate) rate of sedimentation (measured in Svedbergs, in which a higher S value indicates a greater rate of sedimentation and a larger mass). Bacterial ribosomes generally sediment at ~70S. The ribosomes of the cytoplasm of higher eukaryotic cells are larger, usually sedimenting at ~80S.

- **ribosome** A large assembly of RNA and proteins that synthesizes proteins under direction from an mRNA template.
- **ribonucleoprotein** A complex of RNA with proteins.

The ribosome is a compact **ribonucleoprotein** particle consisting of two subunits. Each subunit has an RNA component, including one very large RNA molecule and many proteins. The relationship between a ribosome and its subunits is depicted in **FIGURE 7.8**. The two subunits dissociate *in vitro* when the concentration of Mg^{2+} ions is reduced. In each case, the **large subunit** is about twice the mass of the **small subunit**. Bacterial (70S) ribosomes have subunits that sediment at 50S and 30S. The subunits of eukaryotic cytoplasmic

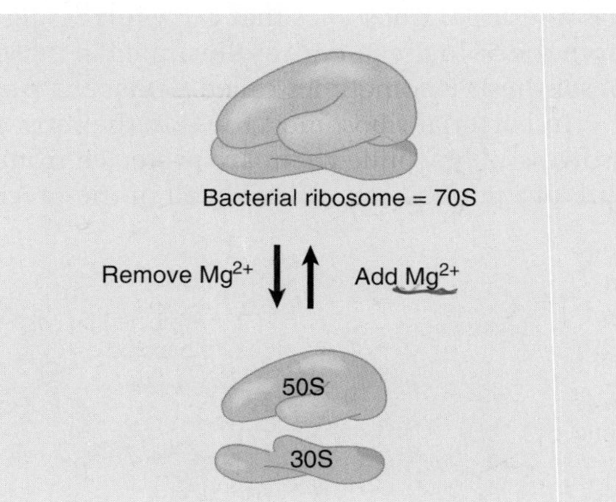

FIGURE 7.8 A ribosome consists of two subunits.

- **large subunit** The subunit of the ribosome (50S in bacteria, 60S in eukaryotes) that has the peptidyl transferase active site that synthesizes the peptide bond.
- **small subunit** The subunit of the ribosome (30S in bacteria, 40S in eukaryotes) that binds the mRNA.

(80S) ribosomes sediment at 60S and 40S. The two subunits work together as part of the complete ribosome, but each undertakes distinct reactions in protein synthesis.

All the ribosomes of a given cell compartment are identical. They undertake the synthesis of different polypeptides by associating with the different mRNAs that provide the actual coding sequences.

The ribosome provides the environment that controls the recognition between a codon of mRNA and the anticodon of tRNA. Reading the genetic code as a series of adjacent triplets, protein synthesis proceeds from the start of a coding region to the end. A polypeptide is assembled by the sequential addition of amino acids in the direction from the N-terminus to the C-terminus as a ribosome moves along the mRNA.

A ribosome begins translation at the 5′ end of a coding region; it translates each triplet codon into an amino acid as it proceeds toward the 3′ end. At each codon, the appropriate aminoacyl-tRNA associates with the ribosome, donating its amino acid to the polypeptide chain. At any given moment, the ribosome can accommodate the two aminoacyl-tRNAs corresponding to successive codons, making it possible for a peptide bond to form between the two corresponding amino acids. At each step, the growing polypeptide chain becomes longer by one amino acid.

KEY CONCEPTS

- Ribosomes are characterized by their rate of sedimentation (70S for bacterial ribosomes and 80S for eukaryotic ribosomes).
- A ribosome consists of a large subunit (50S or 60S for bacteria and eukaryotes) and a small subunit (30S or 40S).
- The ribosome provides the environment in which aminoacyl-tRNAs add amino acids to the growing polypeptide chain in response to the corresponding triplet codons.
- A ribosome moves along an mRNA from 5′ to 3′.

CONCEPT AND REASONING CHECK

Describe in general how a ribosome assembles a polypeptide.

7.6 Many Ribosomes Can Bind to One mRNA

When active ribosomes are isolated as the fraction associated with newly synthesized polypeptides, they are found in the form of a complex consisting of an mRNA associated with several ribosomes. This is the **polyribosome** or **polysome**. The small subunit of each ribosome is associated with the mRNA, and the large subunit carries the newly synthesized protein. tRNAs span both subunits.

▸ **polyribosome (polysome)** An mRNA that is simultaneously being translated by several ribosomes.

Each ribosome in the polysome independently synthesizes a single polypeptide during its traverse of the mRNA sequence. Essentially the mRNA is pulled through the ribosome, and each triplet nucleotide is translated into an amino acid. Thus the mRNA has a series of ribosomes that carry increasing lengths of the protein product, moving from the 5′ to the 3′ end, as illustrated in **FIGURE 7.9**. A polypeptide chain in the process of synthesis is sometimes called a **nascent polypeptide**.

▸ **nascent polypeptide** A protein that has not yet completed its synthesis; the polypeptide chain is still attached to the ribosome via a tRNA.

In bacterial ribosomes, roughly the most recent 30 to 35 amino acids added to a growing polypeptide chain are protected from the cellular environment by the structure of the ribosome. Probably all of the preceding part of the polypeptide protrudes

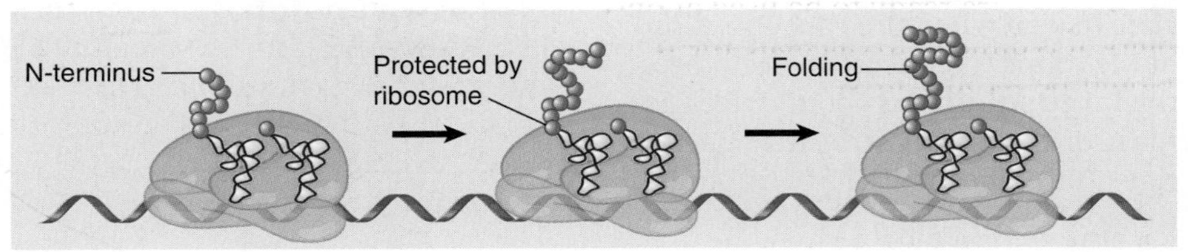

FIGURE 7.9 A polyribosome consists of an mRNA being translated simultaneously by several ribosomes moving in the direction from 5′–3′. Each ribosome has two tRNA molecules, one carrying the nascent polypeptide, the second carrying the next amino acid to be added.

and is free to start folding into its proper conformation. Thus proteins can display parts of the mature conformation even before synthesis has been completed.

A classic characterization of polysomes is shown in the electron micrograph of **FIGURE 7.10**. Globin protein is synthesized by a set of five ribosomes attached to each mRNA (pentasomes). The ribosomes appear as squashed spherical objects of ~7 nm (70 Å) in diameter, connected by a thread of mRNA. The ribosomes are located at various positions along the messenger. Those at one end have just started protein synthesis; those at the other end are about to complete production of a polypeptide chain.

The size of the polysome depends on several variables. In bacteria, it is very large, with tens of ribosomes simultaneously engaged in translation. The size is due in part to the length of the mRNA (which usually is *polycistronic,* coding for several polypeptides); it is also due in part to the high efficiency with which the ribosomes attach to the mRNA.

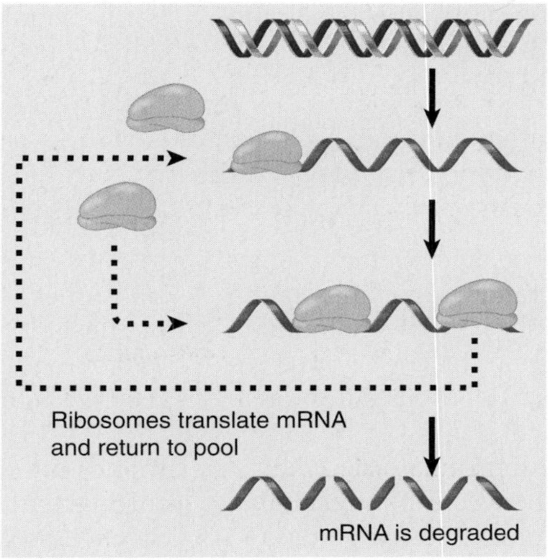

Polysomes in the cytoplasm of a eukaryotic cell are likely to be smaller than those in bacteria; again, their size is a function both of the length of the mRNA (usually representing only a single polypeptide in eukaryotes) and of the characteristic frequency with which ribosomes attach. An average eukaryotic mRNA has ~8 ribosomes attached at any one time.

FIGURE 7.11 illustrates the cycle of the ribosome. Ribosomes are drawn from a pool consisting of ribosomal subunits, used to translate an mRNA, and then returned to the pool for further cycles. The number of ribosomes on each mRNA molecule synthesizing a particular protein is not precisely determined in either bacteria or eukaryotes, but is a matter of statistical fluctuation, determined by the variables of mRNA size and efficiency.

An overall view of the attention devoted to protein synthesis in the intact bacterium is given in **FIGURE 7.12**. The 20,000 or so ribosomes account for a quarter of the cell mass. There are >3000 copies of each tRNA, and altogether the tRNA molecules outnumber the ribosomes by almost tenfold; most of them are present as aminoacyl-tRNAs and are ready to be used at once in protein synthesis. As a result of their instability, it is difficult to calculate the number of mRNA molecules, but a reasonable guess would be ~1500, in varying states of synthesis and decomposition. There are ~600 different types of mRNA in a bacterium. This suggests that there are usually only two to three copies of each mRNA per bacterium. On average, each codes for ~3 proteins. If there are 1850 different soluble proteins, there must be on average >1000 copies of each protein in a bacterium.

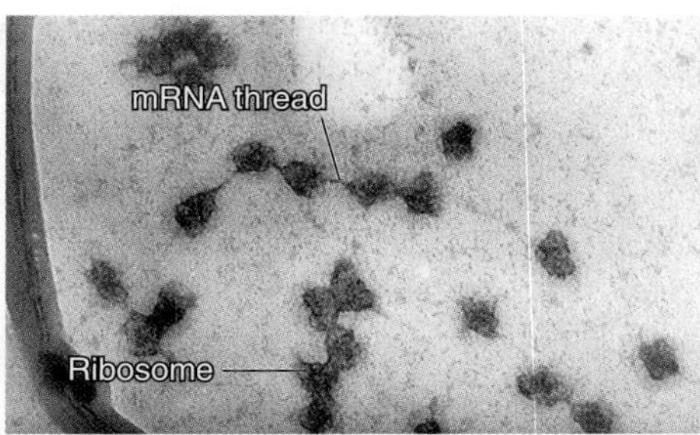

FIGURE 7.10 Protein synthesis occurs on polysomes. Photo courtesy of Alexander Rich, Massachusetts Institute of Technology.

FIGURE 7.11 Messenger RNA is translated by ribosomes that cycle through a pool.

Component	Dry cell mass (%)	Molecules/cell	Different types	Copies of each type
Wall	10	1	1	1
Membrane	10	2	2	1
DNA	1.5	1	1	1
mRNA	1	1,500	600	2–3
tRNA	3	200,000	60	>3,000
rRNA	16	38,000	2	19,000
Ribosomal proteins	9	10^6	52	19,000
Soluble proteins	46	2.0×10^6	1,850	>1,000
Small molecules	3	7.5×10^6	800	

FIGURE 7.12 Considering *E. coli* in terms of its macromolecular components.

METHODS AND TECHNIQUES

Demonstrating That Ribosomal Subunits Are Recycled

Elucidating the structure and assembly of ribosomes could not be achieved until Claude Hogeboom, W. C. Schneider (U.S.) and C. de Duve and his colleagues in Belgium worked out cell fractionation procedures in the 1950s. Cell fractionation requires cell disruption, then centrifugation steps to separate organelles and subcellular molecules into fractions that can be further purified and then collected and analyzed. Researchers began to isolate high molecular-weight complexes called *nucleoprotein particles* whose name was changed to *ribosomes* at a Biophysical Society meeting in 1958. At this time, it was proposed that ribosomes consisted of one-third to one-half RNA and two-thirds to one-half protein. It was also speculated that ribosomes were about 10–15 μm in diameter and with sedimentation values in the range of 20–100S. Ribosomes were found in all cell types and thought to be responsible for protein synthesis. By the mid-1960s the role of ribosomes in protein synthesis was fairly well established.

Ribosomes have two main functions: to decode the mRNA and to form peptide bonds between amino acids that make up the newly synthesized proteins. Both the large and small ribosomal subunits contain molecules needed for both of these activities. During initiation of translation, the small ribosomal subunits bind to mRNA and then are joined by the large ribosomal subunit. During elongation, the mRNA moves through the ribosome and is translated with each nucleotide triplet encoding an amino acid. Even though we usually talk about the ribosome moving along the mRNA, it is more realistic to think in terms of the mRNA being pulled through the ribosome. At termination the protein is released, mRNA is released, and the individual ribosomal subunits dissociate in order to be used again.

Ribosome recycling has been an overlooked cellular process. There is a long neglected *fourth step of translation* referred to as *recycling* of the post-termination complexes. Many textbooks give the wrong impression that the release of newly synthesized proteins during termination causes mRNA and tRNAs to be released from the ribosomes. Using *in vitro* assays, it was determined that *ribosome recycling factor (RRF)* is an *E. coli* protein involved in the dissociation of ribosomes from mRNA after the termination of translation. In bacteria, transcription and translation are coupled. The mRNAs contain bound polysomes during translation. In the final steps of protein synthesis, once the ribosome reaches a termination codon (UAA, UGA, or UGG), the nascent protein is released and the resulting complex of ribosome, mRNA, and tRNA must be recycled for the next round of protein synthesis. RRF specifically releases ribosomes from mRNA by working together with *GTP* and *elongation factor-G (EF-G)* to dissociate this complex into tRNA, ribosome, and mRNA.

The assay that worked out ribosome recycling involved adding purified RRF, along with EF-G and GTP to a preparation of polysomes. The polysomes are prepared by growing *E. coli* to log phase in a liquid medium containing tetracycline (an antibiotic that inhibits protein synthesis). The cells are quick-chilled on ice and pelleted by centrifugation. Subsequently the cells are lysed using a lysozyme solution. The cell lysate is treated with several chemicals and DNase I. The extracts were ultracentrifuged and the pellet containing polysomes on mRNAs was used for the recycling assays.

In this assay, EF-G bound on the ribosome hydrolyzes GTP. The assays were monitored by the use of sucrose gradients. Sucrose gradients are centrifugation tubes that contain layers of sucrose ranging from 5% to 20% sucrose in 5% increments in order to separate ribosomes from the extracts. A sample is placed on top of the gradient and centrifuged at forces in excess of 150,000 g. Particles travel through the gradient until it reaches a density that matches that of the surrounding sucrose. In this assay, polysomes were separated from monosomes because they migrated at different rates through sucrose gradients that are easily fractionated.

RRF is structurally similar to tRNA. Chemical analysis and cryo-electron microscopy has shown that RRF binds to the position of tRNA in the A-site of the ribosome. RRF and EF-G cooperate to split 70S ribosomes into subunits of 50S and 30S. *E. coli initiation factor-3 (IF-3)* also acts to keep the ribosome subunits dissociated during initiation of protein synthesis, but in the presence of high concentrations of magnesium (Mg^{2+}) the subunits are free to rejoin as 70S. Many experts in this field agree that ribosome recycling is a direct bridge from termination to initiation of protein synthesis (**FIGURE B7.1**).

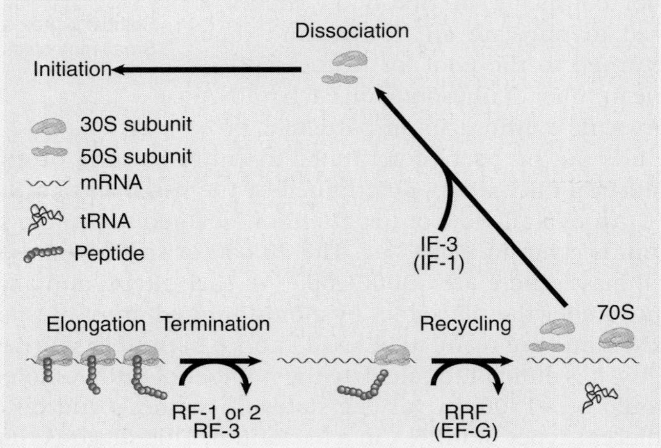

FIGURE B7.1 Model for ribosome recycling in *E. coli*. The ribosome cycle occurs in 4 stages: initiation, elongation, termination, and recycling. Adapted from Yoshida, H., et al., *J. Biochem.* 132 (2002): 983–989.

KEY CONCEPT

- An mRNA is simultaneously translated by several ribosomes. Each ribosome is at a different stage of progression along the mRNA.

CONCEPT AND REASONING CHECK

What factors determine the size of a polysome?

7.7 The Cycle of Bacterial Messenger RNA

Messenger RNA has the same function in all cells, but there are important differences in the details of the synthesis and structure of prokaryotic and eukaryotic mRNA.

A major difference in the production of mRNA depends on the locations where transcription and translation occur:

- In bacteria, mRNA is transcribed and translated in the single cellular compartment; the two processes are so closely linked that they occur simultaneously. Ribosomes attach to bacterial mRNA even before its transcription has been completed, so the polysome is likely still to be attached to DNA. Bacterial mRNA usually is unstable, and is therefore translated into polypeptides for only a few minutes.
- In a eukaryotic cell, synthesis and maturation of mRNA occur exclusively in the nucleus. Only after these events are completed is the mRNA exported to the cytoplasm, where it is translated by ribosomes. A typical eukaryotic mRNA is relatively stable and continues to be translated for several hours, although there is a great deal of variation in the stability of specific mRNAs.

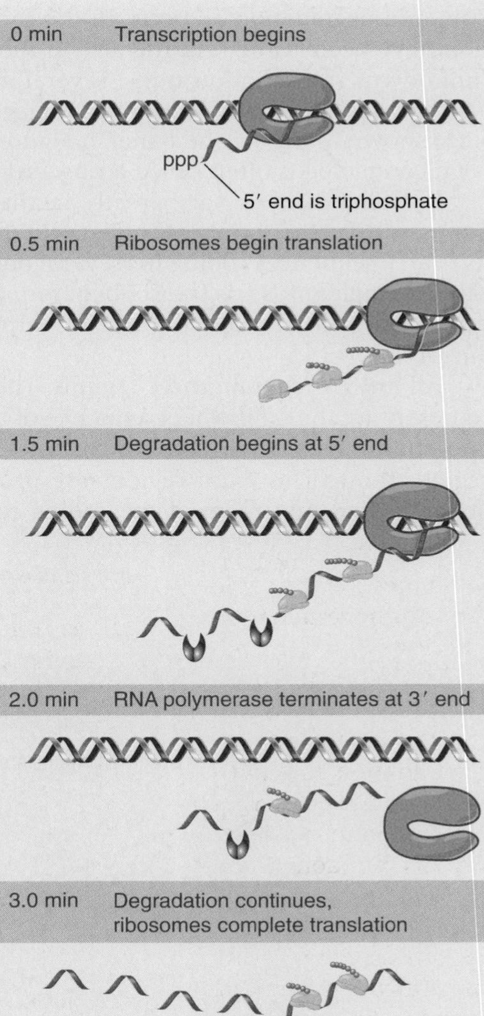
0 min
Transcription begins
ppp
5′ end is triphosphate
0.5 min
Ribosomes begin translation
1.5 min
Degradation begins at 5′ end
2.0 min
RNA polymerase terminates at 3′ end
3.0 min
Degradation continues, ribosomes complete translation

FIGURE 7.13 Overview: mRNA is transcribed, translated, and degraded simultaneously in bacteria.

FIGURE 7.13 shows that transcription and translation are intimately related in bacteria. Transcription begins when the enzyme RNA polymerase binds to DNA and then moves along, making a copy of one strand. As soon as transcription begins, ribosomes attach to the 5′ end of the mRNA and start translation, even before the rest of the message has been synthesized. A bunch of ribosomes move along the mRNA while it is being synthesized. The 3′ end of the mRNA is generated when transcription terminates. Ribosomes continue to translate the mRNA while it survives, but it is degraded in the overall 5′→3′ direction quite rapidly. The mRNA is synthesized, translated by the ribosomes, and degraded, all in rapid succession. An individual molecule of mRNA survives for only a matter of minutes at most.

Bacterial transcription and translation take place at similar rates. At 37° C, transcription of mRNA occurs at ~40 nucleotides/second. This is very close to the rate of protein synthesis, which is roughly 15 amino acids/second. It therefore takes ~2 minutes to transcribe and translate an mRNA of 5000 bp, corresponding to 180 kD of polypeptide. When expression of a new gene is initiated, its mRNA typically will appear in the cell within ~2.5 minutes. The corresponding polypeptide will appear within perhaps another 0.5 minute.

Bacterial translation is very efficient, and most mRNAs are translated by a large number of tightly packed ribosomes. In one example (*trp* mRNA), about 15 initiations of transcription occur every minute, and each of the 15 mRNAs probably is translated by ~30 ribosomes in the interval between its transcription and degradation.

The instability of most bacterial mRNAs is striking. Degradation of mRNA closely follows its translation and likely begins within one minute of the start of transcription. The 5′ end of the mRNA starts to decay before the 3′ end has been synthesized or translated. Degradation seems to follow the last ribosome of the convoy along the mRNA. Degradation proceeds more slowly, though—probably at about half the speed of transcription or translation.

The stability of mRNA has a major influence on the amount of polypeptide that is produced. It is usually expressed in terms of the half-life. The mRNA representing any particular gene has a characteristic half-life, but the average is ~2 minutes in bacteria.

This series of events is only possible, of course, because transcription, translation, and degradation all occur in the same direction. The dynamics of gene expression have been "caught in the act" in the electron micrograph of **FIGURE 7.14**. In these (unknown) transcription units, several mRNAs are under synthesis simultaneously, and each carries many ribosomes engaged in translation. (This corresponds to the stage shown in the second panel in Figure 7.13.) An RNA whose synthesis has not yet been completed is often called a **nascent RNA**.

- **nascent RNA** A ribonucleotide chain that is still being synthesized, so that its 3′ end is paired with DNA where RNA polymerase is elongating.
- **monocistronic mRNA** mRNA that codes for one polypeptide.
- **polycistronic mRNA** mRNA that includes coding regions representing more than one gene.
- **5′ UTR** The untranslated sequence upstream from the coding region of an mRNA.

Bacterial mRNAs vary greatly in the number of proteins for which they code. Some mRNAs carry only a single ORF; they are **monocistronic**. Others (the majority) carry sequences coding for several polypeptides; they are **polycistronic**. In these cases, a single mRNA is transcribed from a group of adjacent genes. (Such a cluster of genes constitutes an operon that is controlled as a single genetic unit; see *Chapter 12, The Operon*.)

All mRNAs contain three regions. The coding region consists of a series of codons representing the amino acid sequence of the polypeptide, starting (usually) with AUG and ending with a termination codon. The mRNA is always longer than the coding region, though, as extra regions are present at both ends. An additional sequence at the 5′ end, upstream of the coding region, is described as the leader or **5′ UTR**

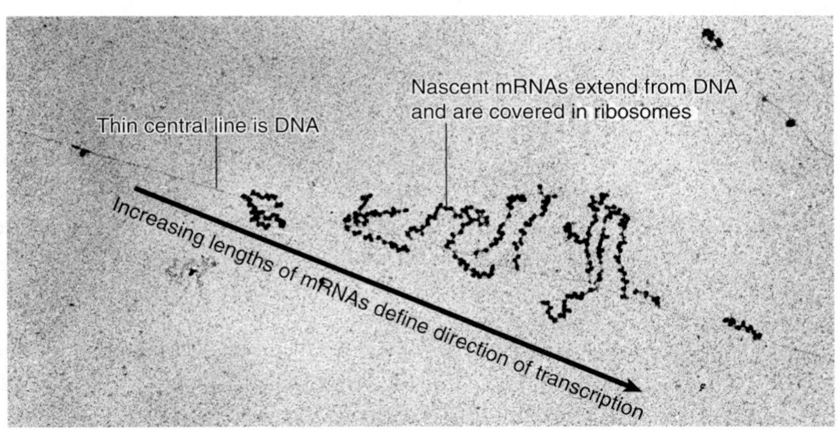
Nascent mRNAs extend from DNA and are covered in ribosomes
Thin central line is DNA
Increasing lengths of mRNAs define direction of transcription

FIGURE 7.14 Transcription units can be visualized in bacteria. Photo courtesy of Oscar Miller.

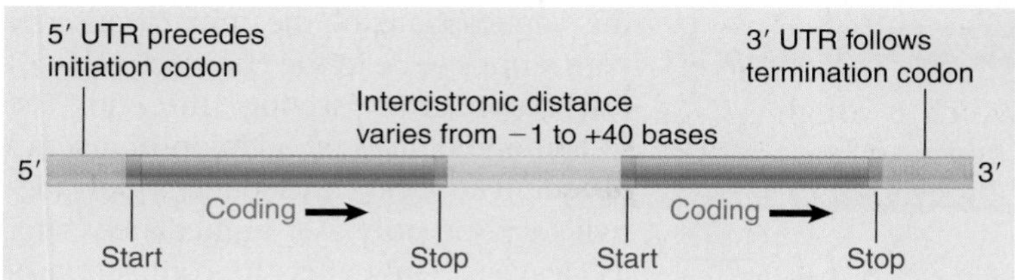

FIGURE 7.15 Bacterial mRNA includes untranslated as well as translated regions. Each coding region has its own initiation and termination signals. A typical mRNA may have several coding regions.

(untranslated region). An additional sequence downstream from the termination signal, forming the 3′ end, is called the trailer or **3′ UTR**. Although they are part of the transcription unit, these sequences do not encode polypeptide.

A polycistronic mRNA also contains **intercistronic regions**, as illustrated in **FIGURE 7.15**. They vary greatly in size. They may be as long as 30 nucleotides in bacterial mRNAs (and even longer in phage RNAs), or they may be very short, with as few as one or two nucleotides separating the termination codon for one polypeptide from the initiation codon for the next. In an extreme case, two genes actually overlap, so that the last base of one coding region is also the first base of the next coding region.

The number of ribosomes engaged in translating a particular cistron depends on the efficiency of its initiation site. The initiation site for the first cistron becomes available as soon as the 5′ end of the mRNA is synthesized. How are subsequent cistrons translated? Are the several coding regions in a polycistronic mRNA translated independently or is their expression connected? Is the mechanism of initiation the same for all cistrons, or is it different for the first cistron and the internal cistrons?

Translation of a bacterial mRNA proceeds sequentially through its cistrons. At the time when ribosomes attach to the first coding region, the subsequent coding regions have not yet even been transcribed. By the time the second ribosome site is available, translation is well under way through the first cistron. Typically ribosomes terminate translation at the end of the first cistron (and dissociate into subunits), and a new ribosome assembles independently at the start of the next coding region. (We discuss the processes of initiation and termination in *Chapter 8, Translation*.)

- **3′ UTR** The untranslated sequence downstream from the coding region of an mRNA.
- **intercistronic region** In a polycistronic mRNA, the distance between the termination codon of one gene and the initiation codon of the next gene.

KEY CONCEPTS

- Transcription and translation occur simultaneously in bacteria, as ribosomes begin translating an mRNA before its synthesis has been completed.
- Bacterial mRNA is unstable and has a half-life of only a few minutes.
- A bacterial mRNA may be polycistronic in having several coding regions that represent different genes.

CONCEPT AND REASONING CHECK

In bacteria, a polysome can form before transcription of the mRNA is complete. Why can't this happen in eukaryotes?

7.8 Eukaryotic mRNA Is Modified During or After Its Transcription

The production of eukaryotic mRNA involves additional stages after transcription. Transcription occurs in the usual way, initiating a transcript with a 5′ triphosphate end. However, the 3′ end is generated by cleaving the transcript, rather than by terminating transcription at a fixed site. Those RNAs that are derived from interrupted genes require splicing to remove the introns, generating a smaller mRNA that contains a continuous coding sequence.

FIGURE 7.16 shows that both ends of the transcript are modified by additions of further nucleotides (involving additional enzyme systems). The 5′ end of the RNA is modified by the addition of a "cap" virtually as soon as it is synthesized. This replaces

FIGURE 7.16 Eukaryotic mRNA is modified by the addition of a cap to the 5′ end and poly(A) to the 3′ end.

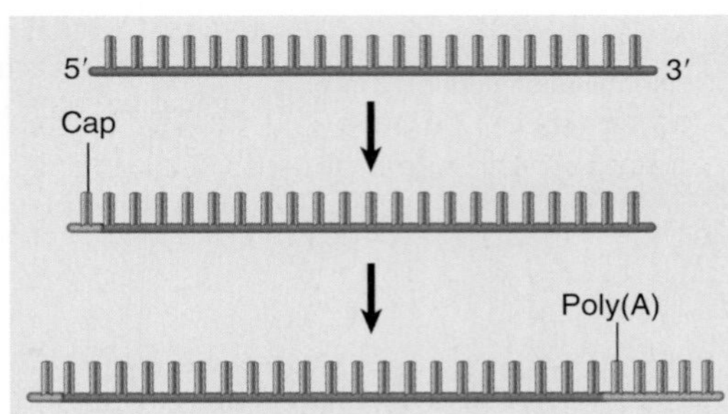

▸ **poly(A)** A stretch of ~200 bases of adenylic acid that is added to the 3′ end of mRNA following its synthesis.

FIGURE 7.17 Overview: Expression of mRNA in eukaryotes requires transcription, modification, processing, nucleocytoplasmic transport, and translation.

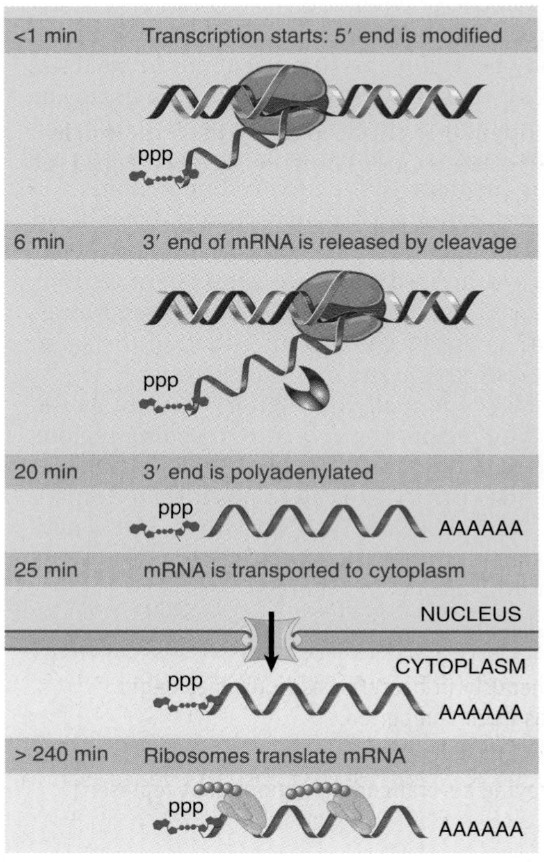

the triphosphate of the initial transcript with a nucleotide in the "reverse" (3′→5′) orientation, thus "sealing" the end. The 3′ end is modified by addition of a series of adenylic acid nucleotides [polyadenylic acid or **poly(A)**] immediately after its cleavage. Only after the completion of all modification and processing events can the mRNA be exported from the nucleus to the cytoplasm. The average delay in leaving for the cytoplasm is ~20 minutes. Once the mRNA has entered the cytoplasm, it is recognized by ribosomes and translated.

FIGURE 7.17 shows that the cycle of eukaryotic mRNA is longer than that of bacterial mRNA. Transcription in animal cells occurs at about the same speed as in bacteria, ~40 nucleotides per second. Many eukaryotic genes are large; a gene of 10,000 bp takes ~5 minutes to transcribe. Transcription of mRNA is not terminated by the release of enzyme from the DNA; instead the enzyme continues past the end of the gene. A coordinated series of events generates the 3′ end of the mRNA by cleavage, and adds a length of poly(A) to the newly generated 3′ end.

Eukaryotic mRNA constitutes only a small proportion of the total cellular RNA (~3% of the mass). In yeast, mRNA half-lives are relatively short, ranging from 1 to 60 minutes. There is a substantial increase in stability in multicellular eukaryotes; most animal cell mRNA is relatively stable, with half-lives ranging from 4 to 24 hours—there are exceptions of both some very short-lived and some very stable mRNAs.

Eukaryotic polysomes are reasonably stable. The modifications at both ends of the mRNA contribute to their stability.

KEY CONCEPTS

- A eukaryotic mRNA transcript is modified in the nucleus during or shortly after transcription.
- The modifications include the addition of a cap at the 5′ end and a sequence of poly(A) at the 3′ end.
- The mRNA is exported from the nucleus to the cytoplasm only after all modifications have been completed.

CONCEPT AND REASONING CHECK

What are the major purposes of the modifications to the 5′ and 3′ ends of eukaryotic mRNA?

7.9 The 5′ End of Eukaryotic mRNA Is Capped

Transcription starts with a nucleoside triphosphate (usually a purine, A or G). The first nucleotide retains its 5′ triphosphate group and makes the usual phosphodiester bond from its 3′ position to the 5′ position of the next nucleotide. The initial sequence of the transcript can be represented as:

5′ppp$^{A}/_{G}$pNpNpNp . . .

When the mature mRNA is treated *in vitro* with enzymes that should degrade it into individual nucleotides, however, the 5′ end does not give rise to the expected nucleoside triphosphate. Instead it contains two nucleotides, connected by a 5′–5′ triphosphate linkage and also bearing methyl groups. The terminal base is always a guanine that is added to the original RNA molecule after transcription.

Addition of the 5′ terminal G is catalyzed by a nuclear enzyme, guanylyl transferase. The reaction occurs so soon after transcription has started that it is not possible to detect more than trace amounts of the original 5′ triphosphate end in the nuclear RNA. The overall reaction can be represented as a condensation between GTP and the original 5′ triphosphate terminus of the RNA. Thus

5′ 5′
Gppp + pppApNpNp . . .
↓
5′—5′
GpppApNpNp . . . + pp + p

The new G residue added to the end of the RNA is in the reverse orientation from all the other nucleotides.

This structure is called a **cap**. It is a substrate for several methylation events. **FIGURE 7.18** shows the full structure of a cap after all possible methyl groups have been added. Types of caps are distinguished by how many of these methylations have occurred:

> **cap** The structure at the 5′ end of eukaryotic mRNA, and is introduced after transcription by linking the terminal phosphate of 5′ GTP to the terminal base of the mRNA.

> **cap 0** A cap at the 5′ end of mRNA that has a methyl group only on 7-guanine.

- The first methylation occurs in all eukaryotes, and consists of the addition of a methyl group to the 7 position of the terminal guanine. A cap that possesses this single methyl group is known as a **cap 0**. This is as far as the reaction proceeds in unicellular eukaryotes. The enzyme responsible for this modification is called guanine-7-methyltransferase.
- The next step is to add another methyl group, to the 2′-O position of the penultimate base (which was actually the original first base of the transcript before any modifications were made). This reaction is catalyzed by another

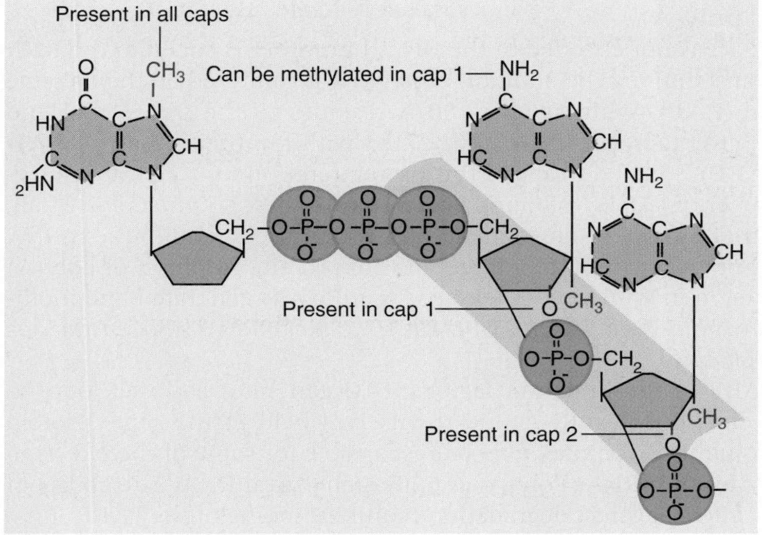

FIGURE 7.18 The cap blocks the 5′ end of mRNA and may be methylated at several positions.

enzyme (2′-O-methyl-transferase). A cap with the two methyl groups is called **cap 1**. This is the predominant type of cap in all eukaryotes except unicellular organisms.

- **cap 1** A cap at the 5′ end of mRNA that has methyl groups on the terminal 7-guanine and the 2′-O position of the next base.

- In a small minority of cases in higher eukaryotes, another methyl group is added to the second base. This happens only when the position is occupied by adenine; the reaction involves addition of a methyl group at the N^6 position. The enzyme responsible acts only on an adenosine substrate that already has the methyl group in the 2′-O position.
- In some species, a methyl group is added to the third base of the capped mRNA. The substrate for this reaction is the cap 1 mRNA that already possesses two methyl groups. The third-base modification is always a 2′-O ribose methylation. This creates the **cap 2** type. This cap usually represents less than 10% to 15% of the total capped population.

- **cap 2** A cap that has three methyl groups (7-guanine, 2′-O position of next base, and N6 adenine) at the 5′ end of mRNA.

In a population of eukaryotic mRNAs, every molecule is capped. The proportions of the different types of cap are characteristic for each particular organism. We do not know whether the structure of a particular mRNA is invariant or whether it can have more than one type of cap.

In addition to the methylation involved in capping, a low frequency of internal methylation occurs in the mRNA only of multicellular eukaryotes. This is accomplished by the generation of N^6 methyladenine residues at a frequency of about one modification per 1000 bases. There are one or two methyladenines in a typical higher eukaryotic mRNA, although their presence is not obligatory; some mRNAs do not have any.

KEY CONCEPT

- A 5′ cap is formed by adding a G to the terminal base of the transcript via a 5′–5′ link. One to three methyl groups are added to the base or ribose of the new terminal guanosine.

CONCEPT AND REASONING CHECK

What do you expect is the functional effect of the additional methylation of caps 1 and 2 as compared to cap 0?

7.10 The 3′ Terminus of Eukaryotic mRNA Is Polyadenylated

The 3′ terminal stretch of A residues is often described as the poly(A) tail; mRNA with this feature is denoted **poly(A)$^+$**.

- **poly(A)$^+$ mRNA** mRNA that has a 3′ terminal stretch of poly(A).
- **poly(A) polymerase (PAP)** The enzyme that adds the stretch of polyadenylic acid to the 3′ end of eukaryotic mRNA. It does not use a template.
- **poly(A) binding protein (PABP)** The protein that binds to the 3′ stretch of poly(A) on a eukaryotic mRNA.

The poly(A) sequence is not encoded in the DNA, but rather is added to the RNA in the nucleus after transcription. The addition of poly(A) is catalyzed by the enzyme **poly(A) polymerase (PAP)**, which adds ~200 A residues to the free 3′-OH end of the mRNA. The poly(A) tail of mRNA is associated with a protein, the **poly(A)-binding protein (PABP)**. Related forms of this protein are found in many eukaryotes. One PABP monomer of ~70 kD is bound every 10 to 20 bases of the poly(A) tail. Thus a common feature in many or most eukaryotes is that the 3′ end of the mRNA consists of a stretch of poly(A) bound to a large mass of protein. Addition of poly(A) occurs as part of a reaction in which the 3′ end of the mRNA is generated and modified by a complex of enzymes (see *Section 28.14, The 3′ Ends of mRNAs Are Generated by Cleavage and Polyadenylation*).

Binding of the PABP to the initiation factor eIF4G generates a closed loop, in which the 5′ and 3′ ends of the mRNA find themselves held in the same protein complex. The formation of this complex may be responsible for some of the effects of poly(A) on the properties of mRNA. Poly(A) usually stabilizes mRNA. The ability of the poly(A) to protect mRNA against degradation requires binding of the PABP.

Removal of poly(A) inhibits the initiation of translation *in vitro*, and depletion of PABP has the same effect in yeast *in vivo*. These effects could depend on the binding of PABP to the initiation complex at the 5′ end of mRNA. There are many examples in early embryonic development where polyadenylation of a particular mRNA is correlated with its translation. In some cases, mRNAs are stored in a nonpolyadenylated form, and poly(A) is added when their translation is required; in other cases, poly(A)$^+$ mRNAs are deadenylated, and their translation is reduced.

KEY CONCEPTS

- A length of poly(A) ~200 nucleotides long is added to a nuclear transcript after transcription.
- The poly(A) is bound by a specific protein (PABP).
- The poly(A) stabilizes the mRNA against degradation.

CONCEPT AND REASONING CHECK

What is one hypothesis about how the poly(A) tail stabilizes eukaryotic mRNA?

7.11 Bacterial mRNA Degradation Involves Multiple Enzymes

Bacterial mRNA is constantly degraded by a combination of endonucleases and exonucleases. Endonucleases cleave an RNA at an internal site. Exonucleases are involved in trimming reactions in which the extra residues are whittled away, base by base, from the end. Bacterial exonucleases that act on single-stranded RNA proceed along the nucleic acid chain from the 3′ end.

The way the two types of enzymes work together to degrade an mRNA is shown in **FIGURE 7.19**. Degradation of a bacterial mRNA is initiated by an endonucleolytic attack. Several 3′ ends may be generated by endonucleolytic cleavages within the mRNA. The overall direction of degradation (as measured by loss of ability to synthesize proteins) is 5′→3′. This probably results from a succession of endonucleolytic cleavages following the last ribosome. Degradation of the released fragments of mRNA into nucleotides then proceeds by exonucleolytic attack from the free 3′-OH end toward the 5′ terminus (that is, in the opposite direction from transcription). Endonucleolytic attack releases fragments that may have different susceptibilities to exonucleases. A region of secondary structure within the mRNA may provide an obstacle to the exonuclease, thus protecting the regions on its 5′ side. The stability of each mRNA is therefore determined by the susceptibility of its particular sequence to both endo- and exonucleolytic cleavages.

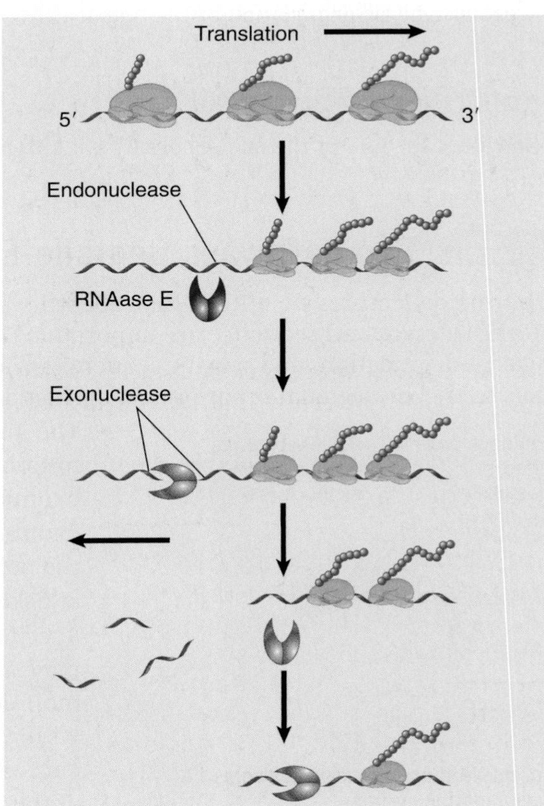

FIGURE 7.19 Degradation of bacterial mRNA is a two-stage process. Endonucleolytic cleavages proceed 5′–3′ behind the ribosomes. The released fragments are degraded by exonucleases that move 3′–5′.

There are ~12 ribonucleases in *E. coli*. Mutants in the endoribonucleases (except ribonuclease I, which is without effect) accumulate unprocessed precursors to rRNA and tRNA, but are viable. Mutants in the exonucleases often have apparently unaltered phenotypes, which suggests that one enzyme can substitute for another. Mutants lacking multiple enzymes sometimes are inviable.

Ribonuclease E (RNase E) is the key enzyme in initiating cleavage of mRNA. It may be the enzyme that makes the first cleavage for many mRNAs. Bacterial mutants that have a defective RNase E have increased stability (two- to threefold) of mRNA. However, this is not its only function. RNase E was originally discovered as the enzyme that is responsible for processing 5′ rRNA from the primary transcript by a specific endonucleolytic processing event.

▸ **degradosome** A complex of bacterial enzymes, including RNase and helicase activities, that may be involved in degrading mRNA.

The process of degradation may be catalyzed by a multienzyme complex (sometimes called the **degradosome**) that includes RNase E, PNPase, and a helicase. RNase E plays dual roles. Its N-terminal domain provides an endonuclease activity. The C-terminal domain provides a scaffold that holds together the other components. The helicase unwinds the substrate RNA to make it available to PNPase. According to this model, RNase E makes the initial cut and then passes the fragments to the other components of the complex for processing.

Polyadenylation may play a role in initiating degradation of some mRNAs in bacteria. PAP is associated with ribosomes in *E. coli*, and short (10 to 40 nucleotide) stretches of poly(A) are added to at least some mRNAs. Triple mutations that remove PAP, RNase E, and polynucleotide phosphorylase (PNPase is a 3′–5′ exonuclease) have a strong effect on stability. (Mutations in individual genes or pairs of genes have only a weak effect.) PAP may create a poly(A) tail that acts as a binding site for the nucleases. The role of poly(A) in bacteria would therefore be different from that in eukaryotic cells.

KEY CONCEPTS

- The overall direction of degradation of bacterial mRNA is 5′–3′.
- Degradation results from the combination of endonucleolytic cleavages followed by exonucleolytic degradation of the fragment from 3′→5′.

CONCEPT AND REASONING CHECK

How does the role of the mRNA poly(A) tail differ in prokaryotes and eukaryotes?

7.12 Two Pathways Degrade Eukaryotic mRNA

The major features of mRNA that affect its stability are summarized in **FIGURE 7.20**. Both structure and sequence are important. The 5′ and 3′ terminal structures protect against degradation, and specific sequences within the mRNA may either serve as targets to trigger degradation or protect against degradation:

- The modifications at the 5′ and 3′ ends of mRNA play an important role in preventing exonuclease attack. The cap prevents 5′–3′ exonucleases from attacking the 5′ end, and the poly(A) prevents 3′–5′ exonucleases from attacking the 3′ end.
- Specific sequence elements within the mRNA may stabilize or destabilize it. The most common location for destabilizing elements is within the 3′ untranslated region. The presence of such an element shortens the lifetime of the mRNA.

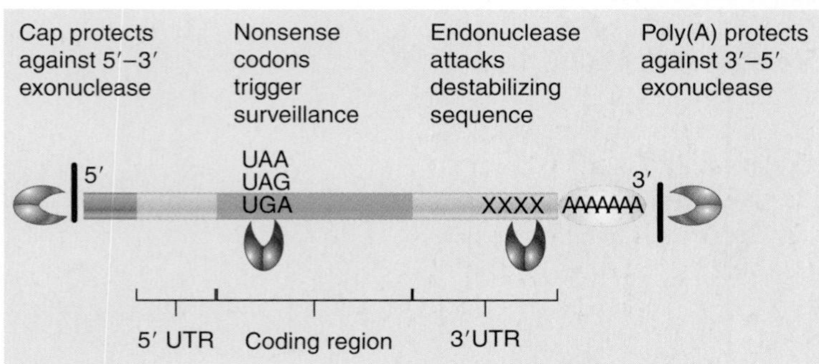

FIGURE 7.20 The terminal modifications of mRNA protect it against degradation. Internal sequences may activate degradation systems.

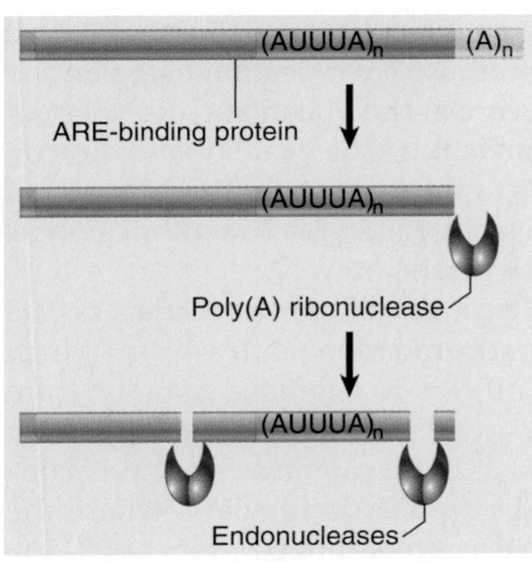

FIGURE 7.21 An ARE in a 3′ nontranslated region initiates degradation of mRNA.

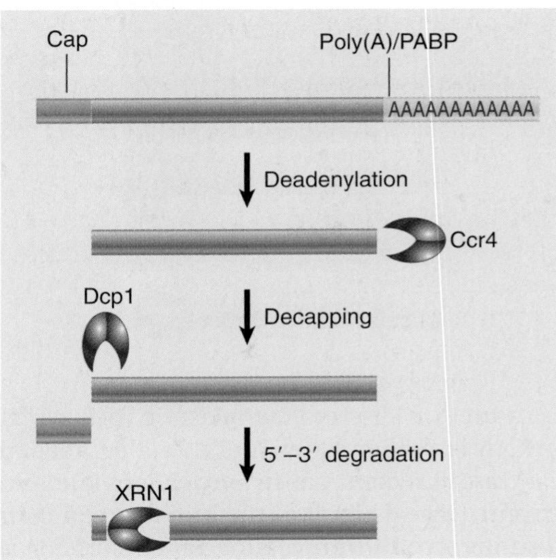

FIGURE 7.22 Deadenylation allows decapping to occur, which leads to endonucleolytic cleavage from the 5′ end.

- Within the coding region, mutations that create termination codons trigger a surveillance system that degrades the mRNA (see *Section 7.13, Nonsense Mutations Trigger a Surveillance System*).

Destabilizing elements have been found in several yeast mRNAs, although as yet we do not see any common sequences or know how they destabilize the mRNA. They do not necessarily act directly (by providing targets for endonucleases), but may function indirectly, perhaps by facilitating deadenylation. The criterion for defining a destabilizing sequence element is that its introduction into a new mRNA may cause it to be degraded. The removal of an element from an mRNA does not necessarily stabilize it, suggesting that an individual mRNA can have more than one destabilizing element.

A common feature in some unstable mRNAs is the presence of an AU-rich sequence of ~50 bases (called the ARE) that is found in the 3′ trailer region. The consensus sequence in the ARE is the pentanucleotide AUUUA, repeated several times. **FIGURE 7.21** shows that the ARE triggers destabilization by a two-stage process: first the mRNA is deadenylated, and then it decays. The deadenylation is probably needed because it causes loss of the poly(A)-binding protein, whose presence stabilizes the 3′ region (see below).

We know the most about the degradation of mRNA in yeast. There are basically two pathways. Both start with removal of the poly(A) tail. This is catalyzed by a specific deadenylase, which probably functions as part of a large protein complex. (The catalytic subunit is the exonuclease Ccr4 in yeast; it is also the exonuclease PARN in vertebrates, which is related to RNAase D.) The enzyme action is processive—once it has started to degrade a particular mRNA substrate, it continues to whittle away at that mRNA, base by base.

The major degradation pathway is summarized in **FIGURE 7.22**. Deadenylation at the 3′ end triggers decapping at the 5′ end. The basis for this relationship is that the presence of the PABP (poly(A)-binding protein) on the poly(A) prevents the decapping enzyme from binding to the 5′ end. PABP is released when the length of poly(A) falls below 10 to 15 residues. Decapping occurs by cleaving the methylated base off the 5′ end to leave a monophosphate in the reaction: $m^7GpppX \ldots \rightarrow m^7GDP + pX \ldots$. The enzyme requires the 7-methyl group.

Each end of the mRNA influences events that occur at the other end. This is explained by the fact that the two ends of the mRNA are held together by the factors involved in protein synthesis. The effect of PABP on decapping allows the 3′ end to

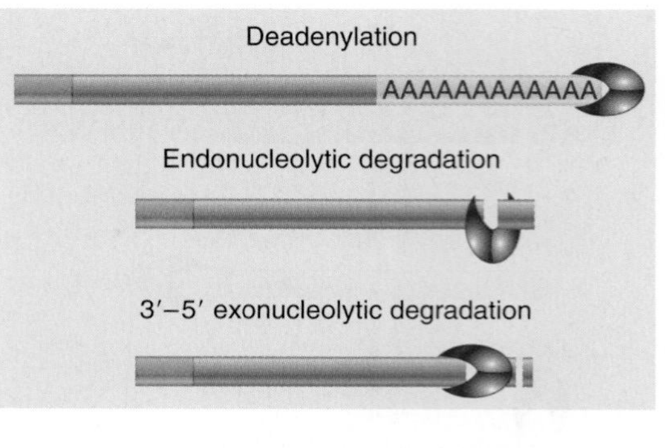

FIGURE 7.23 Deadenylation may lead directly to endonucleolytic cleavage and exonucleolytic cleavage from the 3′ end(s).

have an effect in stabilizing the 5′ end. There is also a connection between the structure at the 5′ end and degradation at the 3′ end. The deadenylase directly binds to the 5′ cap, and this interaction is in fact needed for its exonucleolytic attack on the poly(A).

Removal of the cap triggers the 5′–3′ degradation pathway in which the mRNA is degraded rapidly from the 5′ end, by the 5′–3′ exonuclease XRN1. The decapping enzyme is concentrated in discrete cytoplasmic foci, which may be "processing bodies" where the mRNA is deadenylated and then degraded after it has been decapped.

In the second pathway, deadenylated yeast mRNAs can be degraded by the 3′–5′ exonuclease activity of the **exosome**, a complex of >9 exonucleases. The exosome is also involved in processing precursors for rRNAs. The aggregation of the individual exonucleases into the exosome complex may enable 3′–5′ exonucleolytic activities to be coordinately controlled. The exosome may also degrade fragments of mRNA released by endonucleolytic cleavage. FIGURE 7.23 shows that the 3′–5′ degradation pathway may actually involve combinations of endonucleolytic and exonucleolytic action. The exosome is also found in the nucleus, where it degrades unspliced precursors to mRNA.

- **exosome** A complex of several exonucleases involved in degrading mRNA.

Yeast mutants lacking either exonucleolytic pathway degrade their mRNAs more slowly, but the loss of both pathways is lethal.

KEY CONCEPTS

- The modifications at both ends of mRNA protect it against degradation by exonucleases.
- Specific sequences within an mRNA may have stabilizing or destabilizing effects.
- Destabilization may be triggered by loss of poly(A).
- Degradation of yeast mRNA requires removal of the 5′ cap and the 3′ poly(A).
- One yeast pathway involves exonucleolytic degradation from 5′→3′.
- Another yeast pathway uses a complex of several exonucleases that work in the 3′→5′ direction.

CONCEPT AND REASONING CHECK

Summarize the two ways in which eukaryotic mRNA may be degraded.

7.13 Nonsense Mutations Trigger a Surveillance System

Another pathway for degradation is identified by **nonsense-mediated mRNA decay**. FIGURE 7.24 shows that the introduction of a nonsense mutation often leads to increased degradation of the mRNA. As may be expected from dependence on a termination codon, the degradation occurs in the cytoplasm. It may represent a quality control or **surveillance system** for removing nonfunctional mRNAs.

The surveillance system has been studied best in yeast and *C. elegans*, but may also be generally important in animal cells. For example, during the formation of immunoglobulins and T cell receptors in cells of the immune system, genes are modified by somatic recombination and mutation (see *Chapter 22, Immune Diversity*). This generates a significant number of nonfunctional genes whose RNA products are disposed of by a surveillance system.

- **nonsense-mediated mRNA decay** A pathway that degrades an mRNA that has a nonsense mutation prior to the last exon.
- **surveillance systems** Systems that check nucleic acids for errors. The term is used in several different contexts. One example is the system that degrades mRNAs that have nonsense mutations. Another is the set of systems that react to damage in the double helix. The common feature is that the system recognizes an invalid sequence or structure and triggers a response.

In yeast, the degradation requires sequence elements (called *DSE*) that are downstream of the nonsense mutation. The simplest possibility would be that these are destabilizing elements and that translation suppresses their use. When translation is blocked, however, the mRNA is stabilized. This suggests that the process of degradation is linked to translation of the mRNA or to the termination event in some direct way.

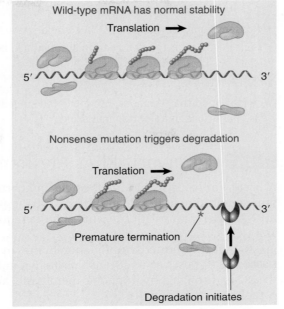

FIGURE 7.24 Nonsense mutations may cause mRNA to be degraded.

Genes that are required for the process have been identified in *S. cerevisiae* (*upf* loci) and *C. elegans* (*smg* loci) by identifying suppressors of nonsense-mediated degradation. Mutations in these genes stabilize aberrant mRNAs, but do not affect the stability of most wild-type transcripts. One of these genes is conserved in eukaryotes (*upf1/smg2*). It codes for an ATP-dependent helicase (an enzyme that unwinds double-stranded nucleic acids into single strands). This implies that recognition of the mRNA as an appropriate target for degradation requires a change in its structure.

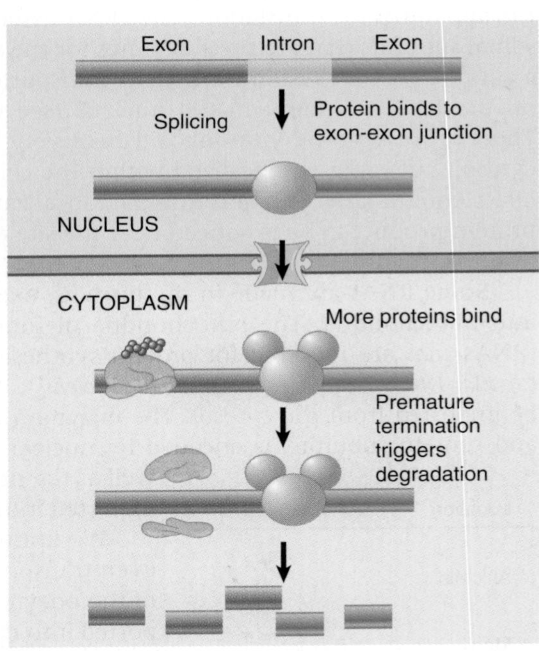

FIGURE 7.25 A surveillance system could have two types of components. Protein(s) must bind in the nucleus to mark the result of a splicing event. Other proteins could bind to the mark either in the nucleus or cytoplasm. They are triggered to act to degrade the mRNA when ribosomes terminate prematurely.

Upf1 interacts with the release factors (eRF1 and eRF3) that catalyze termination, which is probably how it recognizes the termination event. It may then "scan" the mRNA by moving toward the 3′ end to look for the downstream sequence elements.

In mammalian cells, the surveillance system appears to work only on mutations located prior to the last exon; in other words, there must be an intron after the site of mutation. This suggests that the system requires some event to occur in the nucleus, before the introns are removed by splicing. One possibility is that proteins attach to the mRNA in the nucleus at the exon–exon boundary when a splicing event occurs. **FIGURE 7.25** shows a general model for the operation of such a system. This is similar to the way in which an mRNA may be marked for export from the nucleus (see *Section 28.9, Splicing Is Connected to Export of mRNA*). Attachment of a protein to the exon–exon junction creates a mark of the event that persists into the cytoplasm. Human homologues of the yeast Upf2 and Upf3 proteins may be involved in such a system. They bind specifically to mRNA that has been spliced.

KEY CONCEPTS

- Nonsense mutations cause mRNA to be degraded.
- Genes coding for the degradation system have been found in yeast and worms.

CONCEPT AND REASONING CHECK

Why is it good that the nonsense-mediated surveillance system is only triggered by nonsense mutations in a nonterminal exon?

7.14 Eukaryotic RNAs Are Transported

A bacterium consists of only a single compartment, so all the RNAs function in the same environment in which they are synthesized. This is most striking in the case of mRNA, where translation occurs simultaneously with transcription (see *Section 7.7, The Cycle of Bacterial Messenger RNA*).

RNA is transported through membranes in the variety of instances summarized in **FIGURE 7.26**. It poses a significant thermodynamic problem to transport a highly negative RNA through a hydrophobic membrane, and the solution is to transport the RNA complexed with proteins.

In eukaryotic cells, RNAs are transcribed in the nucleus, but translation occurs in the cytoplasm. Each type of RNA must be transported into the cytoplasm to assemble the apparatus for translation. The rRNA assembles with ribosomal proteins into ribosomal subunits that are the substrates for the transport system. tRNA is transported by a specific protein system. mRNA is transported as a ribonucleoprotein, which forms on the RNA transcript in the nucleus (see *Chapter 28, RNA Splicing and Processing*). These processes are common to all eukaryotic cells. Many mRNAs are translated in the cytosol, but some are localized within the cell by means of attachment to a cytoskeletal element. One situation in which localization occurs is when it is important for a protein product to be produced near the site of its incorporation into some macromolecular structure.

Some RNAs are made in the nucleus, exported to the cytosol, and then imported into mitochondria. The mitochondria of some organisms do not code for all of the tRNAs that are required for protein synthesis (see *Section 4.10, Organelle Genomes Are Circular DNAs That Code for Organelle Proteins*). In these cases, the additional tRNAs must be imported from the cytosol. The enzyme ribonuclease P, which contains both RNA and protein subunits, is encoded by nuclear genes, but is found in mitochondria as well as the nucleus. This means that the RNA must be imported into the mitochondria.

RNA	Transport	Location
All RNA	Nucleus→cytoplasm	All cells
tRNA	Nucleus→mitochondrion	Many cells
mRNA	Nurse cell→oocyte	Fly embryogenesis
mRNA	Anterior→posterior oocyte	Fly embryogenesis
mRNA	Cell→cell	Plant phloem

FIGURE 7.26 RNAs are transported through membranes in a variety of systems.

We know of some situations in which mRNA is even transported between cells. During development of the oocyte in *Drosophila*, certain mRNAs are transported into the egg from the nurse cells that surround it. The nurse cells have specialized junctions with the oocyte that allow passage of material needed for early development. This material includes certain mRNAs. Once inside the egg, these mRNAs take up specific locations. Some simply diffuse from the anterior end where they enter, but others are transported the full length of the egg to the posterior end by a motor protein attached to microtubules.

The most striking case of transport of mRNA has been found in plants. Movement of individual nucleic acids over long distances was first discovered in plants, where viral movement proteins help propagate the viral infection by transporting an RNA virus genome through the plasmodesmata (connections between cells). Plants also have a defense system that causes cells to silence an infecting virus. This, too, may involve the spread of components (including RNA) over long distances between cells. It appears that similar systems may transport mRNAs between plant cells. Although the existence of the systems has been known for some time, it is only recently that their functional importance has been demonstrated. This was shown by grafting wild-type tomato plants onto plants that had the dominant mutation *Me* (which causes a change in the shape of the leaf). mRNA from the mutant stock was transported into the leaves of the wild-type graft, where it changed their shape.

KEY CONCEPTS

- RNA is transported through a membrane as a ribonucleoprotein particle.
- All eukaryotic RNAs that function in the cytoplasm must be exported from the nucleus.
- tRNAs and the RNA component of a ribonuclease are imported into mitochondria.
- mRNAs can travel long distances between plant cells.

CONCEPT AND REASONING CHECK

Why would mRNAs need to be transported to specific cellular locations?

7.15 mRNA Can Be Localized Within a Cell

An mRNA is synthesized in the nucleus but translated in the cytoplasm of a eukaryotic cell. It passes into the cytoplasm in the form of a ribonucleoprotein particle that is transported through a nuclear pore. Once in the cytosol, the mRNA may associate with ribosomes and be translated. The cytosol generally has a high concentration of proteins. It is not clear how freely a polysome can diffuse within the cytosol, and most mRNAs are probably translated in random locations, determined by their point of entry into the cytosol, and the distance that they may have moved away from it. However, some mRNAs are translated at specific sites. This may be accomplished by several mechanisms:

- An mRNA species may be specifically transported to a site where it is translated.
- An mRNA species may be universally distributed but degraded at all sites except the site of translation.
- An mRNA species may be freely diffusible but become trapped at the site of translation.

One of the best-characterized cases of localization within a cell is that of Ash1 in yeast. Ash1 represses expression of the HO endonuclease in the budding daughter cell, with the result that HO is expressed only in the mother cell. The consequence is that mating type is changed only in the mother cell. The cause of the restriction to the daughter cell is that all the *ASH1* mRNA is transported from the mother cell, where it is made, into the budding daughter cell.

Mutations in any one of five genes, called *SHE1-5*, prevent the specific localization and cause *ASH1* mRNA to be symmetrically distributed in both mother and daughter compartments. The proteins She1, She2, and She3 bind *ASH1* mRNA into a ribonucleoprotein particle that transports the mRNA into the daughter cell. **FIGURE 7.27** shows the functions of the proteins. She1 is a myosin (previously identified as Myo4), and She3 and She2 are proteins that connect the myosin to the mRNA. The myosin is

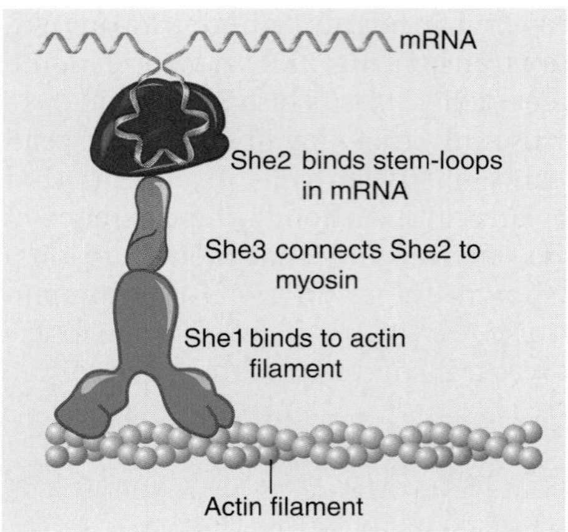
mRNA
She2 binds stem-loops in mRNA
She3 connects She2 to myosin
She1 binds to actin filament
Actin filament

FIGURE 7.27 *ASH1* mRNA forms a ribonucleoprotein containing a myosin motor that moves it along an actin filament.

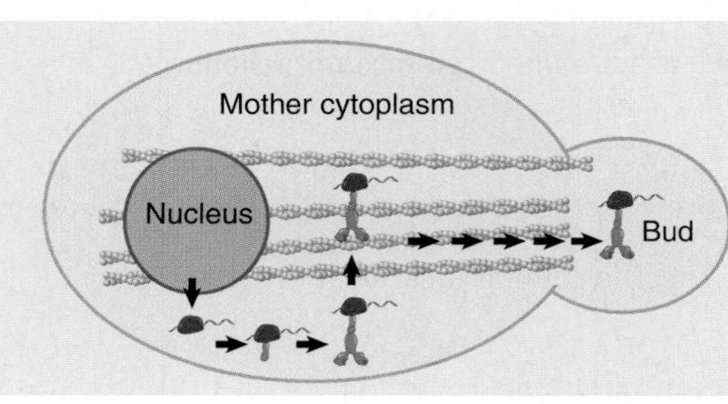
Mother cytoplasm
Nucleus
Bud

FIGURE 7.28 *ASH1* mRNA is exported from the nucleus into the cytoplasm, where it is assembled into a complex with the She proteins. The complex transports it along actin filaments to the bud.

a motor that moves the mRNA along actin filaments.

FIGURE 7.28 summarizes the overall process. *ASH1* mRNA is exported from the nucleus in the form of a ribonucleoprotein. In the cytoplasm it is first bound by She2, which recognizes some stem–loop secondary structures within the mRNA. Then She3 binds to She2, after which the myosin She1 binds. Next, the particle hooks onto an actin filament and moves to the bud. When *ASH1* mRNA reaches the bud, it is anchored there, probably by proteins that bind specifically to the mRNA.

Similar principles govern other cases where mRNAs are transported to specific sites. The mRNA is recognized by means of *cis*-acting sequences, which usually are regions of secondary structure in the 3′ untranslated region. (*ASH1* mRNA is unusual in that the *cis*-acting regions are in the coding frame.) The mRNA is packaged into a ribonucleoprotein particle. In some cases, the transported mRNA can be visualized in very large particles called *mRNA granules*. The particles are several times the size of a ribosome, large enough to contain many protein and RNA components.

A transported mRNP must be connected to a motor that moves it along a system of tracks. The tracks can be either actin filaments or microtubules. Whereas *ASH1* uses a myosin motor on actin tracks, *oscar* mRNA in the *Drosophila* egg uses a kinesin motor to move along microtubules. Once the mRNA reaches its destination, it needs to be anchored in order to prevent it from diffusing away. Less is known about this, but the process appears to be independent of transport. An mRNA that is transported along microtubules may anchor to actin filaments at its destination.

KEY CONCEPTS

- Yeast *ASH1* mRNA forms a ribonucleoprotein that binds to a myosin motor.
- A motor transports it along actin filaments into the daughter bud.
- It is anchored and translated in the bud, so that the protein is found only in the bud.

CONCEPT AND REASONING CHECK

Once mRNAs are transported to specific locations, how can they be prevented from diffusing away?

7.16 Summary

Protein-coding information carried by DNA is expressed in two stages: transcription of DNA into mRNA and translation of the mRNA into protein. Messenger RNA is transcribed from one strand of DNA. It is complementary to this (template or antisense) strand and identical with the other (coding or sense) strand. The sequence of mRNA, in triplet codons 5′–3′, is related to the amino acid sequence of polypeptide, N- to C-terminal.

The adaptor that interprets the meaning of a codon is transfer RNA, which has a compact L-shaped tertiary structure; one end of the tRNA has an anticodon that is complementary to the codon, and the other end can be covalently linked to the specific amino acid that corresponds to the target codon. A tRNA carrying an amino acid is called an aminoacyl-tRNA.

The ribosome provides the apparatus that allows aminoacyl-tRNAs to bind to codons on mRNA. The small subunit of the ribosome is bound to mRNA; the large subunit carries the nascent polypeptide. A ribosome moves along mRNA from an initiation codon in the 5′ region to a termination codon in the 3′ region, and the appropriate aminoacyl-tRNAs recognize their codons and deliver their amino acids, so that the growing polypeptide chain extends by one residue for each codon.

A typical mRNA contains both a nontranslated 5′ UTR (leader) and a 3′ UTR (trailer) as well as a coding region or regions. Bacterial mRNA is usually polycistronic, with untranslated regions between the cistrons. Each cistron is represented by a coding region that starts with a specific initiation codon and ends with a termination codon. Ribosome subunits associate at the initiation codon and dissociate at the termination codon of each coding region.

A single mRNA can be translated by many ribosomes simultaneously, generating a polyribosome (or polysome). Bacterial polysomes are large, typically with tens of ribosomes bound to a single mRNA. Eukaryotic polysomes are smaller, typically with fewer than ten ribosomes; each mRNA carries only a single coding sequence.

Bacterial mRNA has an extremely short half-life of only a few minutes. The 5′ end starts translation even while the downstream sequences are being transcribed. Degradation is initiated by endonucleases that cut at discrete sites, following the ribosomes in the 5′–3′ direction, after which exonucleases reduce the fragments to nucleotides by degrading them from the released 3′ end toward the 5′ end. Individual sequences may promote or retard degradation in bacterial mRNAs.

Eukaryotic mRNA must be processed in the nucleus before it is transported to the cytoplasm for translation. A methylated cap is added to the 5′ end. It consists of a nucleotide added to the original end by a 5′–5′ bond, after which methyl groups are added. Most eukaryotic mRNA has an ~200 base sequence of poly(A) added to its 3′ terminus in the nucleus after transcription, but mRNAs lacking poly(A) appear to be translated and degraded with the same kinetics as mRNAs with poly(A). Eukaryotic mRNA exists as a ribonucleoprotein particle; in some cases mRNPs are stored that fail to be translated. Eukaryotic mRNAs are usually stable for several hours. They may have multiple sequences that initiate degradation; examples are known in which the process is regulated.

Yeast mRNA is degraded by at least two pathways. Both start with the removal of poly(A) from the 3′ end, causing loss of poly(A)-binding protein, which in turn leads to removal of the methylated cap from the 5′ end. One pathway degrades the mRNA from the 5′ end by an exonuclease. Another pathway degrades from the 3′ end by the exosome, a complex containing several exonucleases.

Nonsense-mediated degradation leads to the destruction of mRNAs that have a termination (nonsense) codon prior to the last exon. The *upf* loci in yeast and the *smg* loci in worms are required for the process. They include a helicase activity to unwind mRNA and a protein that interacts with the factors that terminate protein synthesis. The features of the process in mammalian cells suggest that some of the proteins attach to the mRNA in the nucleus when RNA splicing occurs to remove introns.

mRNAs can be transported to specific locations within a cell (especially in embryonic development). In the Ash1 system in yeast, mRNA is transported from the mother cell into the daughter cell by a myosin motor that moves on actin filaments. In plants, mRNAs can be transported long distances between cells.

CHAPTER QUESTIONS

1. Eukaryotic mRNA usually is:
   A. relatively stable and translated into proteins for several hours.
   B. relatively stable and translated into proteins for up to an hour.
   C. relatively unstable and translated into proteins for several minutes.
   D. relatively unstable and translated into proteins for only a few minutes.
2. In bacteria, mRNA survives for about what length of time in the cell?
   A. seconds
   B. minutes
   C. hours
   D. days
3. Aminoacyl-tRNA synthetases attach the proper amino acid to the tRNA at what location in the tRNA?
   A. on the 2′ hydroxyl group of the 5′ terminal ribose of the tRNA
   B. on the 3′ hydroxyl group of the 3′ terminal ribose of the tRNA
   C. on either the 2′ or 3′ hydroxyl group of the 5′ terminal ribose of the tRNA
   D. on either the 2′ or 3′ hydroxyl group of the 3′ terminal ribose of the tRNA
4. What is the location where transcription and translation occur in bacterial cells?
   A. both occur in the nucleus
   B. both occur in the cytoplasm
   C. transcription occurs in the nucleus and translation in the cytoplasm
   D. transcription occurs in the cytoplasm and translation in the nucleoids
5. What is the sugar-sugar linkage in the 5′ cap on mRNA molecules?
   A. 5′–3′
   B. 3′–5′
   C. 3′–3′
   D. 5′–5′
6. Besides the 5′ cap, what modification is common to the 5′ end of many eukaryotic mRNAs?
   A. acetylation
   B. methylation
   C. adenylation
   D. all of the above
7. When and where in a eukaryotic cell does 5′ capping of mRNA occur:
   A. in the nucleus prior to completion of transcription of the mRNA
   B. in the nucleus after completion of transcription of the mRNA
   C. in the cytoplasm after completion of transcription of the mRNA
   D. 5′ capping does not occur in eukaryotes
8. What region of an mRNA is most commonly associated with transcript destabilization?
   A. the 5′ untranslated region
   B. the 3′ untranslated region
   C. the exonic coding regions
   D. the intronic regions

9. Removal of the 5′ cap triggers degradation of the mRNA by:

A. endonucleases.

B. 5′–3′ exonucleases.

C. 3′–5′ exonucleases.

D. random degradation from multiple sites in the transcript.

10. Introduction of a nonsense mutation in a eukaryotic protein coding gene often leads to:

A. retention of the mRNA in the nucleus so it is not translated.

B. increased degradation of the mRNA in the nucleus.

C. increased degradation of the mRNA in the cytoplasm.

D. decreased ribosome binding in the cytoplasm to the mRNA due to its shorter size.

KEY TERMS

3′ UTR
5′ UTR
acceptor arm
aminoacyl-tRNA
aminoacyl-tRNA synthetases
anticodon
anticodon arm
antisense strand (template strand)
arm
cap
cap 0
cap 1
cap 2
cloverleaf
coding strand (sense strand)
conserved sequences
D arm
degradosome
exosome
extra arm
intercistronic region
invariant sequences
large subunit
loop
monocistronic mRNA
nascent polypeptide
nascent RNA
nonsense-mediated mRNA decay
poly(A)
poly(A) binding protein (PABP)
poly(A) polymerase (PAP)
poly(A)$^+$ mRNA
polycistronic mRNA
polyribosome (polysome)
ribonucleoprotein
ribosomal RNA (rRNA)
ribosome
semiconserved (semi-invariant)
small subunit
stem
surveillance systems
transfer RNA (tRNA)

FURTHER READING

Coller, J. and Parker, R. (2004). Eukaryotic mRNA decapping. *Annu. Rev. Biochem.* 73, 861–890.
A review of the potential control steps of the process of eukaryotic mRNA decapping (which triggers mRNA decay): removal of the poly(A) tail, exit from translation, assembly of a decapping complex, and cytoplasmic sequestering.

Grunberg-Manago, M. (1999). mRNA stability and its role in control of gene expression in bacteria and phages. *Annu. Rev. Genet.* 33, 193–227.
A review of the mechanisms and control of mRNA decay in prokaryotes.

Hilleren, P. and Parker, R. (1999). Mechanisms of mRNA surveillance in eukaryotes. *Annu. Rev. Genet.* 33, 229–260.
A review of the process of the eukaryotic mRNA degradation system and a comparison to other systems of kinetic proofreading.

Jansen, R. P. (2001). mRNA localization: message on the move. *Nat. Rev. Mol. Cell Biol.* 2, 247–256.
A review of the processes and mechanisms involved in mRNA transport and localization.

Moore, M. J. (2005). From birth to death: the complex lives of eukaryotic mRNAs. *Science* 309, 1514–1518.
A review of the various roles of mRNAs in eukaryotic gene expression and its regulation.

Söll, D. and RajBhandary, U. L. (1995). *tRNA Structure, Biosynthesis, and Function*. Washington, DC: American Society for Microbiology.

A comprehensive treatment of the structures and functions of tRNAs.

Vance, V. and Vaucheret, H. (2001). RNA silencing in plants—defense and counterdefense. *Science* 292, 2277–2280.

A review of the role of gene regulation via RNA silencing in plants as antiviral defenses and the viral adaptations of suppressors of silencing.

8

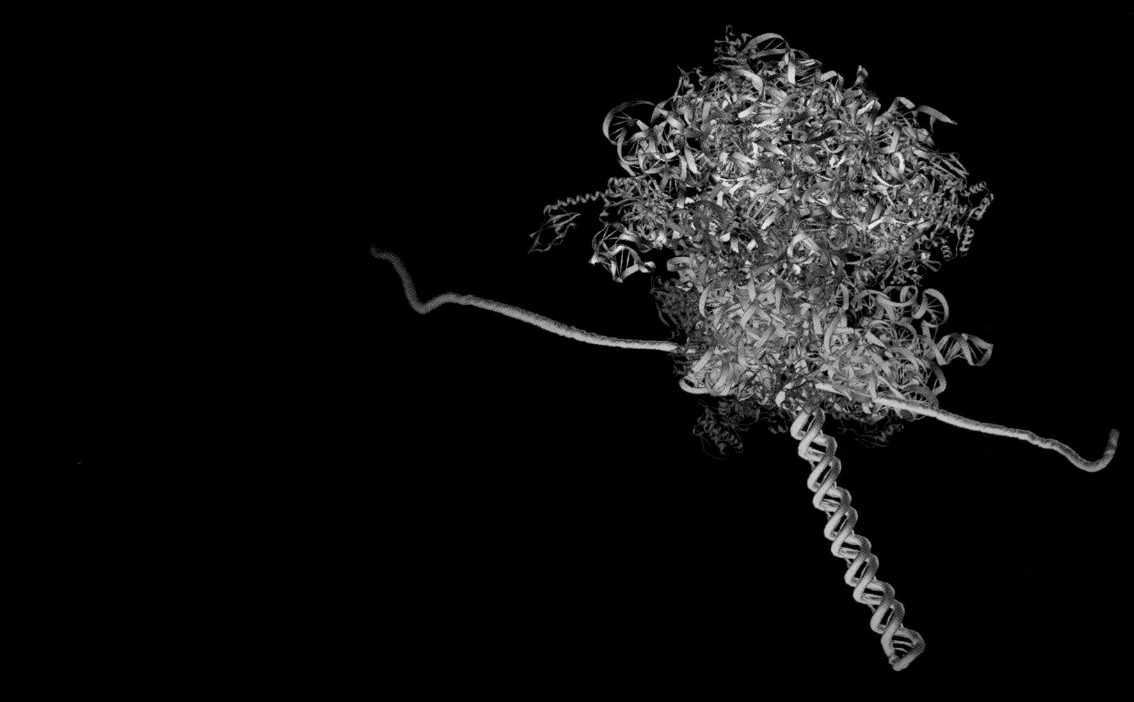

Translation of a single mRNA (in yellow) by an *E. coli* ribosome, one codon at a time. By fixing the ends of the mRNA, researchers can measure how long the ribosome spends at each codon and have learned that the ribosome often pauses after translocation cycles. Used with permission of Courtney Hodges, University of California, Berkeley, and Laura Lancaster, University of California, Santa Cruz. Photo courtesy of Carlos Bustamante, University of California, Berkeley.

Translation

CHAPTER OUTLINE

8.1 Introduction

An mRNA contains a series of codons that interact with the anticodons of aminoacyl-tRNAs so that a corresponding series of amino acids is incorporated into a polypeptide chain. The ribosome facilitates the interaction between mRNA and aminoacyl-tRNA. The ribosome behaves like a small migrating factory that travels along the template engaging in rapid cycles of peptide bond synthesis. Aminoacyl-tRNAs shoot in and out of the particle at a fearsome rate while depositing amino acids, and elongation factors cyclically associate with and dissociate from the ribosome. Together with its accessory factors, the ribosome provides the full range of activities required for all the steps of translation.

FIGURE 8.1 shows the relative dimensions of the components of the translation apparatus. The ribosome consists of two subunits that have specific roles in translation. Messenger RNA is associated with the small subunit; ~35 bases of the mRNA are bound at any time. The mRNA threads its way along the surface close to the junction of the subunits. Two tRNA molecules are active in translation at any moment, so polypeptide elongation involves reactions taking place at just two of the (roughly) ten codons covered by the ribosome. The two tRNAs are inserted into internal sites that stretch across the subunits. A third tRNA may remain on the ribosome after it has been used in protein synthesis, before being recycled.

The basic form of the ribosome has been evolutionarily conserved, but there are appreciable variations in the overall size and proportions of rRNAs and proteins in the ribosomes of bacteria, eukaryotic cytoplasm, and mitochondria and chloroplasts. **FIGURE 8.2** compares the components of bacterial and mammalian ribosomes. Both are ribonucleoprotein particles that contain more RNA than protein. The ribosomal proteins are known as *r-proteins*.

Each of the ribosome subunits contains a large rRNA and a number of small proteins. The large subunit may also contain smaller rRNA(s). In *E. coli*, the small (30S) subunit consists of the 16S rRNA and 21 r-proteins. The large (50S) subunit contains 23S rRNA, the small 5S RNA, and 31 proteins. With the exception of one protein that is present in four copies per ribosome, there is one copy of each protein. The large RNAs constitute the major part of the mass of the bacterial ribosome. Their presence is pervasive, and most or all of the ribosomal proteins actually contact rRNA (see Figure 8.34). So the large rRNAs form what is sometimes thought of as the "backbone" of each subunit—a continuous thread whose presence dominates the structure and determines the positions of the ribosomal proteins.

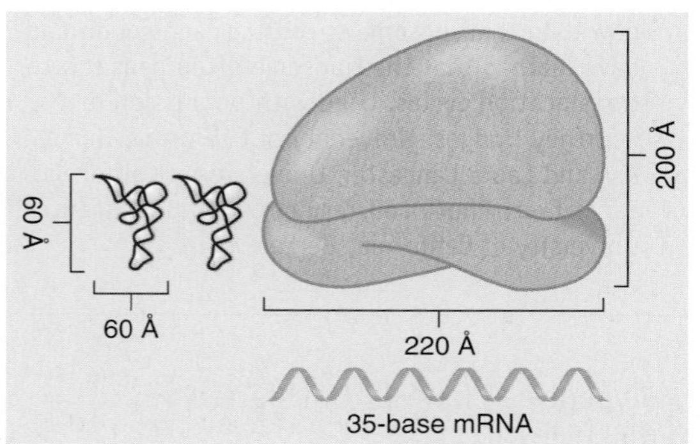

FIGURE 8.1 Size comparisons show that the ribosome is large enough to bind tRNAs and mRNA.

Ribosomes		rRNAs	r-proteins
Bacterial (70S) mass: 2.5 MDa 66% RNA	50S	23S = 2904 bases 5S = 120 bases	31
	30S	16S = 1542 bases	21
Mammalian (80S) mass: 4.2 MDa 60% RNA	60S	28S = 4718 bases 5.8S = 160 bases 5S = 120 bases	49
	40S	18S = 1874 bases	33

FIGURE 8.2 Ribosomes are large ribonucleoprotein particles that contain more RNA than protein and dissociate into large and small subunits.

The ribosomes in the cytoplasm of multicellular eukaryotes are larger than those of bacteria. The total content of both RNA and protein is greater; the large rRNA molecules are longer (called 18S and 28S rRNAs), and there are more proteins. RNA is still the predominant component by mass.

The ribosomes of mitochondria and chloroplasts are distinct from the ribosomes of the cytosol and they take varied forms. In some cases, they are almost the size of bacterial ribosomes and have 70% RNA; in other cases, they are only 60S and have <30% RNA.

The ribosome possesses several active centers, each of which is constructed from a group of proteins associated with a region of ribosomal RNA. The active centers require the direct participation of rRNA in a structural or even catalytic role (where the RNA functions as a ribozyme) with proteins supporting these functions in secondary roles. Some catalytic functions require individual proteins, but none of the activities can be reproduced by isolated proteins or groups of proteins; they function only in the context of the ribosome.

In analyzing the functions of structural components of the ribosome, there are two experimental approaches. First, the effects of mutations in genes for particular ribosomal proteins or at specific bases in rRNA genes shed light on the participation of these molecules in particular reactions. Second, structural analysis, including direct modification of components of the ribosome and comparisons to identify conserved features in rRNA, identifies the physical locations of components involved in particular functions.

8.2 Translation Occurs by Initiation, Elongation, and Termination

An amino acid is brought to the ribosome by an aminoacyl-tRNA. Its addition to the growing polypeptide chain occurs by an interaction with the tRNA that brought the previous amino acid. Each of these tRNAs lies in a distinct site on the ribosome. **FIGURE 8.3** shows that the two sites have different features:

- An incoming aminoacyl-tRNA binds to the **A site**. Prior to the entry of aminoacyl-tRNA, the site exposes the codon representing the next amino acid to be added to the chain.
- The codon representing the most recent amino acid to have been added to the nascent polypeptide chain lies in the **P site**. This site is occupied by **peptidyl-tRNA**, a tRNA carrying the nascent polypeptide chain.

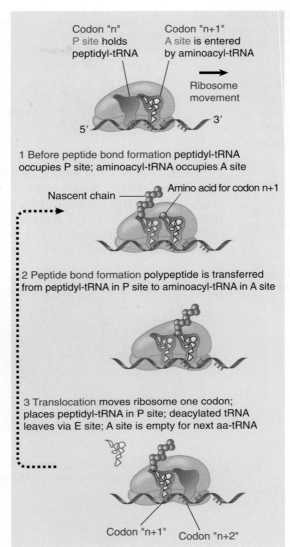

Codon "n"
P site holds peptidyl-tRNA
Codon "n+1"
A site is entered by aminoacyl-tRNA
Ribosome movement
5′
3′
1 Before peptide bond formation peptidyl-tRNA occupies P site; aminoacyl-tRNA occupies A site
Nascent chain
Amino acid for codon n+1
2 Peptide bond formation polypeptide is transferred from peptidyl-tRNA in P site to aminoacyl-tRNA in A site
3 Translocation moves ribosome one codon; places peptidyl-tRNA in P site; deacylated tRNA leaves via E site; A site is empty for next aa-tRNA
Codon "n+1"
Codon "n+2"

FIGURE 8.3 The ribosome has two sites for binding charged tRNA.

FIGURE 8.4 shows that the aminoacyl end of the tRNA is located on the large subunit, whereas the anticodon at the other end interacts with the mRNA bound by the small subunit. So the P and A sites each extend across both ribosomal subunits.

For a ribosome to synthesize a peptide bond, it must be in the state shown in step 1 in Figure 8.3, when peptidyl-tRNA is in the P site and aminoacyl-tRNA is in the A site. Peptide bond formation occurs when the polypeptide carried by the peptidyl-tRNA is transferred to the amino acid carried by the aminoacyl-tRNA. This reaction is catalyzed by the large subunit of the ribosome.

- **A site** The site of the ribosome that an aminoacyl-tRNA enters to base pair with the codon.
- **P site** The site in the ribosome that is occupied by peptidyl-tRNA, the tRNA carrying the nascent polypeptide chain, still paired with the codon to which it bound in the A site.
- **peptidyl-tRNA** The tRNA to which the nascent polypeptide chain has been transferred following peptide bond synthesis during polypeptide translation.

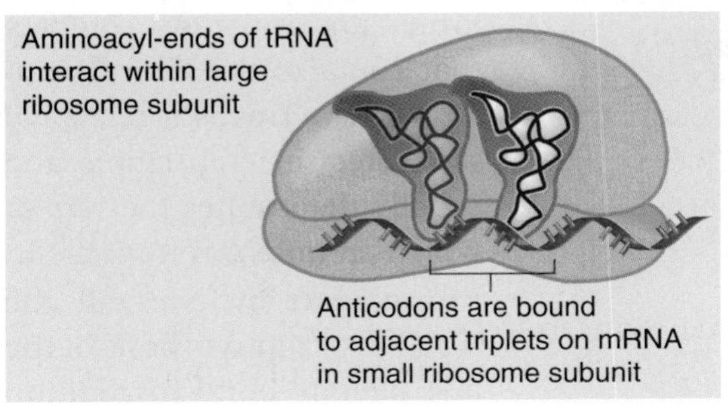
Aminoacyl-ends of tRNA interact within large ribosome subunit
Anticodons are bound to adjacent triplets on mRNA in small ribosome subunit

FIGURE 8.4 The P and A sites position the two interacting tRNAs across both ribosome subunits.

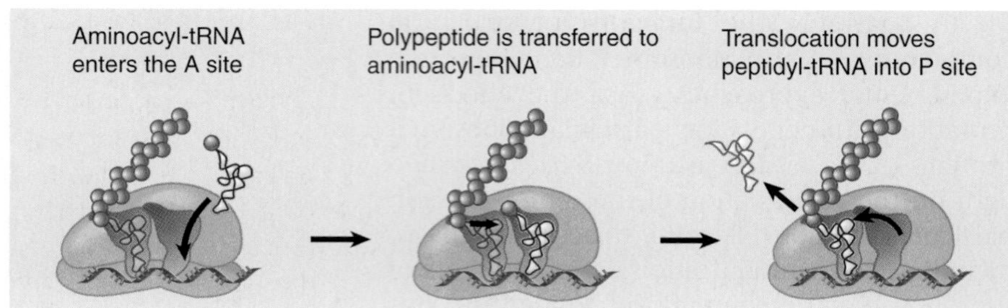

FIGURE 8.5 Aminoacyl-tRNA enters the A site, receives the polypeptide chain from peptidyl-tRNA, and is transferred into the P site for the next cycle of elongation.

▸ **deacylated tRNA** tRNA that has no amino acid or polypeptide chain attached because it has completed its role in protein synthesis and is ready to be released from the ribosome.

▸ **translocation** The movement of the ribosome one codon along mRNA after the addition of each amino acid to the polypeptide chain.

Transfer of the polypeptide generates the ribosome shown in step 2, in which the **deacylated tRNA**, lacking any amino acid, lies in the P site and a new peptidyl-tRNA is in the A site. This peptidyl-tRNA is one amino acid residue longer than the peptidyl-tRNA that had been in the P site in step 1.

The ribosome now moves one triplet along the messenger RNA. This stage is called **translocation**. The movement transfers the deacylated tRNA out of the P site and moves the peptidyl-tRNA into the P site (see step 3 in the figure). The next codon to be translated now lies in the A site, ready for a new aminoacyl-tRNA to enter, when the cycle will be repeated. **FIGURE 8.5** summarizes the interaction between tRNAs and the ribosome.

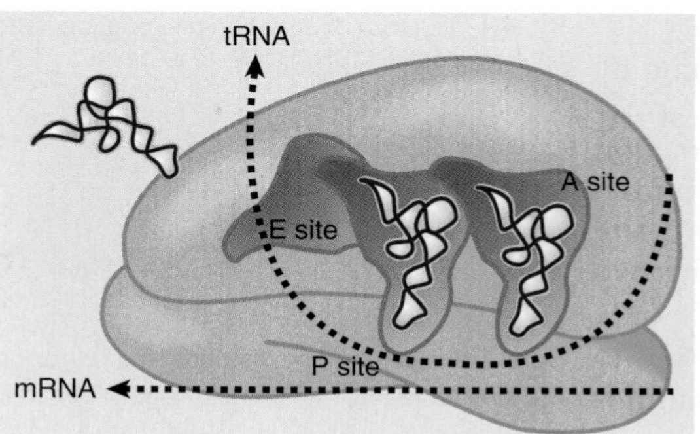

FIGURE 8.6 tRNA and mRNA move through the ribosome in the same direction.

The deacylated tRNA leaves the ribosome via another tRNA-binding site, the E site. This site is transiently occupied by the tRNA *en route* between leaving the P site and being released from the ribosome into the cytosol. Thus the flow of tRNA is into the A site, through the P site, and out through the E site (see also Figure 8.21 in *Section 8.10*). **FIGURE 8.6** compares the movement of tRNA and mRNA, which may be thought of as a sort of ratchet in which the reaction is driven by the codon–anticodon interaction.

FIGURE 8.7 Translation falls into three stages.

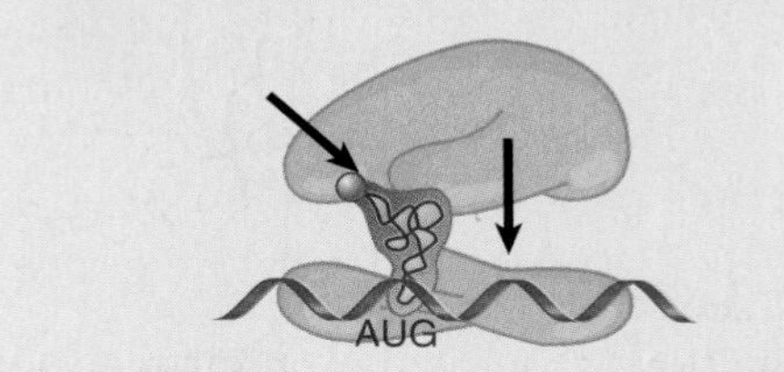

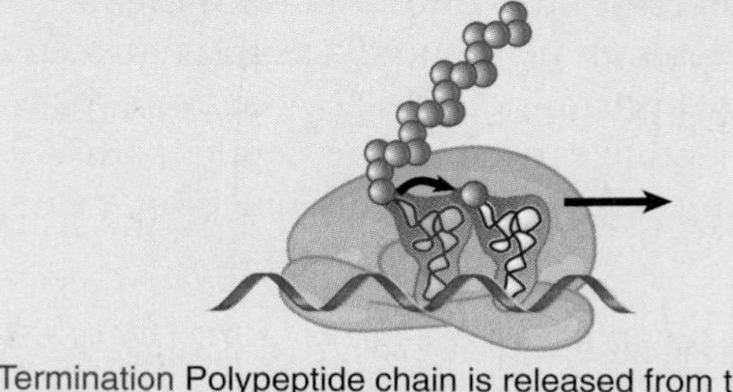

Elongation Ribosome moves along mRNA, extending protein by transfer from peptidyl-tRNA to aminoacyl-tRNA

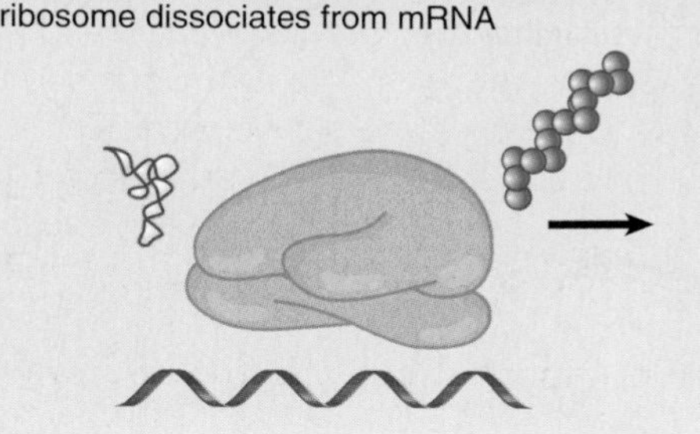

▸ **initiation** The stages of translation up to synthesis of the first peptide bond of the polypeptide.

▸ **elongation** The stage in a macromolecular synthesis reaction (replication, transcription, or translation) in which the nucleotide or polypeptide chain is extended by the addition of individual subunits.

Translation falls into the three stages shown in **FIGURE 8.7**:

- **Initiation** involves the reactions that precede formation of the peptide bond between the first two amino acids of the protein. It requires the ribosome to bind to the mRNA, which forms an initiation complex that contains the first aminoacyl-tRNA. This is a relatively slow step in translation and usually determines the rate at which an mRNA is translated.
- **Elongation** includes all the reactions from synthesis of the first peptide bond to addition of the last amino acid. Amino

acids are added to the chain one at a time; the addition of an amino acid is the most rapid step in translation.

- **Termination** encompasses the steps that are needed to release the completed polypeptide chain; at the same time, the ribosome dissociates from the mRNA.

Different sets of accessory factors assist the ribosome at each stage. Energy is provided at various stages by the hydrolysis of guanine triphosphate (GTP).

During initiation, the small ribosomal subunit binds to mRNA and then is joined by the large subunit. During elongation, the mRNA moves through the ribosome and is translated in nucleotide triplets. (Although we usually talk about the ribosome moving along mRNA, it is more realistic to think in terms of the mRNA moving through the ribosome.) At termination, the polypeptide is released, mRNA is released, and the individual ribosomal subunits dissociate and can be used again.

▸ **termination** A separate reaction that ends a macromolecular synthesis reaction (replication, transcription, or translation), by stopping the addition of subunits, and (typically) causing disassembly of the synthetic apparatus.

KEY CONCEPTS

- The ribosome has three tRNA-binding sites.
- An aminoacyl-tRNA enters the A site.
- Peptidyl-tRNA is bound in the P site.
- Deacylated tRNA exits via the E site.
- An amino acid is added to the polypeptide chain by transferring the polypeptide from peptidyl-tRNA in the P site to aminoacyl-tRNA in the A site.

CONCEPT AND REASONING CHECK

Why would you expect the initiation stage of translation to be the slowest stage?

8.3 Special Mechanisms Control the Accuracy of Translation

We know that translation is generally accurate, because of the consistency that is found when we determine the amino acid sequence of a polypeptide. There are few detailed measurements of the error rate *in vivo*, but it is generally thought to lie in the range of one error for every 10^4 to 10^5 amino acids incorporated. Considering that most polypeptides are produced in large quantities, this means that the error rate is too low to have much effect on the phenotype of the cell.

It is not immediately obvious how such a low error rate is achieved. In fact, the nature of discriminatory events is a general issue raised by several steps in gene expression:

- How do synthetases recognize just the corresponding tRNAs and amino acids?
- How does a ribosome recognize only the tRNA corresponding to the codon in the A site?
- How do the enzymes that synthesize DNA or RNA recognize only the base complementary to the template?

Each case poses a similar problem: how to distinguish one particular member from the entire set, all of which share the same general features.

Probably any substrate initially can contact the active center by a random-hit process, but then the wrong substrates are rejected and only the appropriate one is accepted. The appropriate member is always in a minority (one of twenty amino acids, one of ~30 to 50 tRNAs, one of four bases), so the criteria for discrimination must be strict. The point is that the enzyme must have some mechanism for increasing discrimination from the level that would be achieved merely by making contacts with the available surfaces of the substrates.

FIGURE 8.8 summarizes the error rates at the steps that can affect the accuracy of protein synthesis.

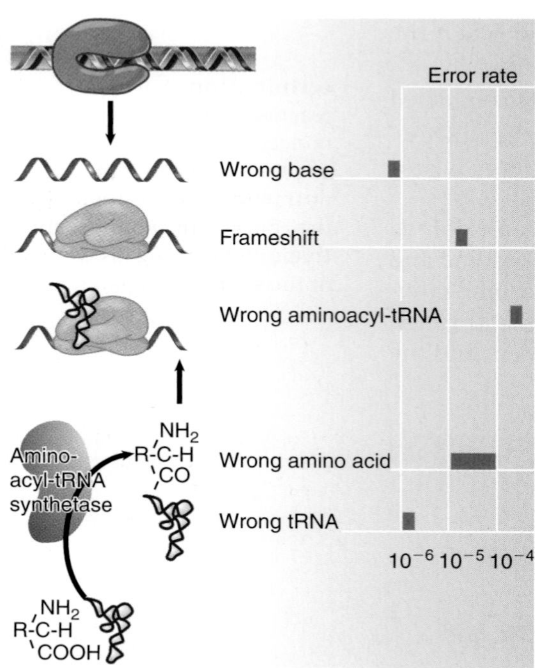

FIGURE 8.8 Errors occur at rates from 10^{-6} to 5×10^{-4} at different stages of translation.

Errors in transcribing mRNA are rare—probably $<10^{-6}$. This is an important stage for accuracy, because a single mRNA molecule is translated into many polypeptide copies. We do not know very much about the mechanisms that ensure transcriptional accuracy.

The ribosome can make two types of errors in translation. First, it may cause a frameshift by skipping a base when it reads the mRNA (or in the reverse direction by reading a base twice—once as the last base of one codon and then again as the first base of the next codon). These errors are rare, occurring at $\sim 10^{-5}$. Or, it may allow an incorrect aminoacyl-tRNA to (mis)pair with a codon, so that the wrong amino acid is incorporated. This is probably the most common error in protein synthesis, occurring at $\sim 5 \times 10^{-4}$. It is primarily controlled by ribosome structure.

A tRNA synthetase can make two types of errors: it can place the wrong amino acid on its tRNA, or it can charge its amino acid with the wrong tRNA (see *Section 9.8, tRNAs Are Charged with Amino Acids by Synthetases*). The incorporation of the wrong amino acid is more common, probably because the tRNA offers a larger surface with which the enzyme can make many more contacts to ensure specificity. Aminoacyl-tRNA synthetases have specific mechanisms to correct errors before a mischarged tRNA is released (see *Section 9.10, Synthetases Use Proofreading to Improve Accuracy*).

KEY CONCEPT

- The accuracy of protein synthesis is controlled by specific mechanisms at each stage.

CONCEPT AND REASONING CHECK

Why are translation errors less serious than DNA replication errors?

8.4 Initiation in Bacteria Needs 30S Subunits and Accessory Factors

Bacterial ribosomes engaged in elongating a polypeptide chain exist as 70S particles. At termination, they are released from the mRNA as free ribosomes or ribosomal subunits. In growing bacteria, the majority of ribosomes are synthesizing polypeptides; the free pool is likely to contain ~20% of the ribosomes.

Ribosomes in the free pool can dissociate into separate subunits; this means that 70S ribosomes are in dynamic equilibrium with 30S and 50S subunits. Initiation of translation is not a function of intact ribosomes, but is undertaken by the separate subunits, which reassociate during the initiation reaction. **FIGURE 8.9** summarizes the ribosomal subunit cycle during translation in bacteria.

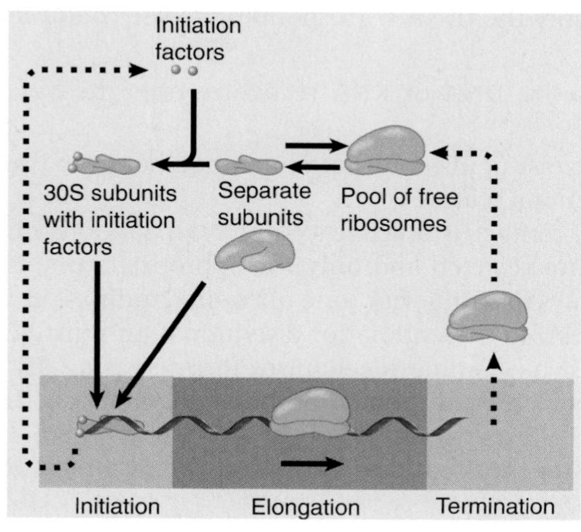

FIGURE 8.9 Initiation requires free ribosome subunits. When ribosomes are released at termination, the 30S subunits bind initiation factors and dissociate to generate free subunits. When subunits reassociate to give a functional ribosome at initiation, they release the factors.

Initiation occurs at a special sequence on mRNA called the **ribosome-binding site** (including the *Shine-Dalgarno sequence*). This is a short sequence of bases that precedes the coding region (see Figure 8.1) and is

ribosome-binding site A sequence on bacterial mRNA that includes an initiation codon that is bound by a 30S subunit in the initiation phase of polypeptide translation.

complementary to a portion of the 16S rRNA (see *Section 8.16, Two rRNAs Play Active Roles in Translation*). The small and large subunits associate at the ribosome-binding site to form an intact ribosome. The reaction occurs in two steps:

- Recognition of mRNA occurs when a small subunit binds to form an initiation complex at the ribosome-binding site.
- A large subunit then joins the complex to generate a complete ribosome.

Although the 30S subunit is involved in initiation, it is not sufficient by itself to undertake the reactions of binding mRNA and tRNA. It requires additional proteins called **initiation factors (IFs)**. These factors are found only on 30S subunits, and they are released when the 30S subunits associate with 50S subunits to generate 70S ribosomes. This action distinguishes initiation factors from the structural proteins of the ribosome. The initiation factors are concerned solely with formation of the initiation complex, they are absent from 70S ribosomes, and they play no part in the stages of elongation. FIGURE 8.10 summarizes the stages of initiation.

▸ **initiation factors (IFs)** Proteins that associate with the small subunit of the ribosome specifically at the stage of initiation of polypeptide translation.

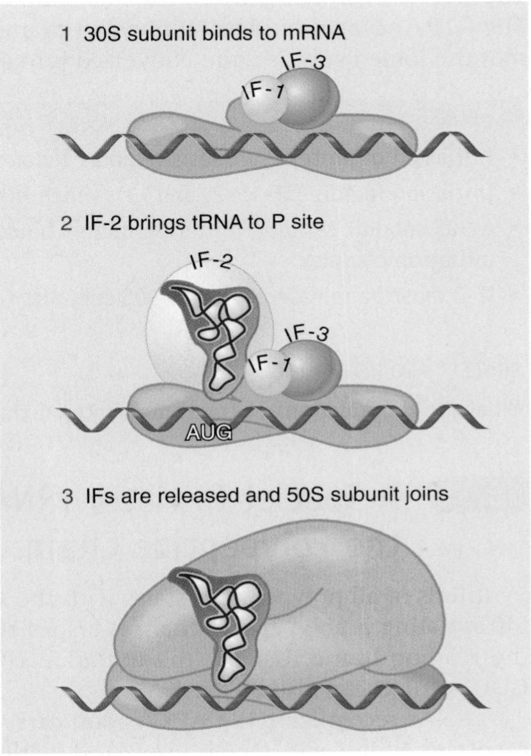

FIGURE 8.10 Initiation factors stabilize free 30S subunits and bind initiator tRNA to the 30S-mRNA complex.

Bacteria use three initiation factors, numbered **IF-1**, **IF-2**, and **IF-3**. They are needed for both mRNA and tRNA to enter the initiation complex:

- IF-3 is needed for 30S subunits to bind specifically to initiation sites in mRNA, and to inhibit the premature binding of 50S subunits.
- IF-2 binds a special initiator tRNA and controls its entry into the ribosome.
- IF-1 binds to 30S subunits only as a part of the complete initiation complex. It binds to the A site and prevents aminoacyl-tRNA from entering. Its location also may impede the 30S subunit from binding to the 50S subunit.

▸ **IF-1** A bacterial initiation factor that stabilizes the initiation complex for polypeptide translation.

▸ **IF-2** A bacterial initiation factor that binds the initiator tRNA to the initiation complex for polypeptide translation.

▸ **IF-3** A bacterial initiation factor required for 30S ribosomal subunits to bind to initiation sites in mRNA. It also prevents 30S subunits from binding to 50S ribosomal subunits.

IF-3 has multiple functions: it is needed first to stabilize (free) 30S subunits; then it enables them to bind to mRNA; and as part of the 30S-mRNA complex, it checks the accuracy of recognition of the first aminoacyl-tRNA (see *Section 8.6, mRNA Binds a 30S Subunit to Create the Binding Site for a Complex of IF-2 and fMet-tRNA$_f$*).

The first function of IF-3 controls the equilibrium between ribosomal states, as shown in FIGURE 8.11. IF-3 binds to free 30S subunits that are released from the pool of 70S ribosomes. The presence of IF-3 prevents the 30S subunit from reassociating with a 50S subunit; one molecule of

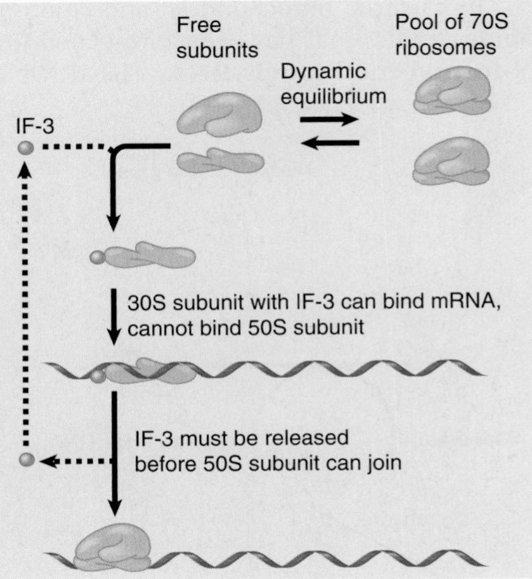

FIGURE 8.11 Initiation requires 30S subunits that carry IF-3.

IF-3 binds per subunit. IF-3 binds to the surface of the 30S subunit in the vicinity of the A site. There is significant overlap between the bases in 16S rRNA protected by IF-3 and those protected by binding of the 50S subunit, suggesting that it physically prevents junction of the subunits. There is a relatively small amount of IF-3, so its availability determines the number of free 30S subunits.

The second function of IF-3 is to control the ability of 30S subunits to bind to mRNA. Small subunits must have IF-3 in order to form initiation complexes with mRNA. IF-3 must be released from the 30S-mRNA complex in order to enable the 50S subunit to join. On its release, IF-3 immediately recycles by finding another 30S subunit.

IF-2 has a ribosome-dependent GTPase activity: it sponsors the hydrolysis of GTP in the presence of ribosomes, releasing the energy stored in the high-energy bond. The GTP is hydrolyzed when the 50S subunit joins to generate a complete ribosome. The GTP cleavage could be involved in changing the conformation of the ribosome, so that the joined subunits are converted into an active 70S ribosome.

KEY CONCEPTS

- Initiation of protein synthesis requires separate 30S and 50S ribosome subunits.
- Initiation factors (IF-1, -2, and -3), which bind to 30S subunits, are also required.
- A 30S subunit carrying initiation factors binds to an initiation site on mRNA to form an initiation complex.
- IF-3 must be released to allow 50S subunits to join the 30S-mRNA complex.

CONCEPT AND REASONING CHECK

What are the functions of the three bacterial translation initiation factors?

8.5 A Special Initiator tRNA Starts the Polypeptide Chain

▸ **N-formyl-methionyl-tRNA** The aminoacyl-tRNA that initiates bacterial polypeptide translation. The amino group of the methionine is formylated.

▸ **tRNA$_f^{Met}$** The special RNA used to initiate polypeptide translation in bacteria. It mostly uses AUG, but can also respond to GUG and CUG.

Synthesis of all polypeptides starts with the same amino acid: methionine. The signal for initiating a polypeptide chain is a special initiation codon that marks the start of the reading frame. Usually the initiation codon is the triplet AUG, but in bacteria, GUG or UUG can also be used.

tRNAs recognizing the AUG codon carry methionine, and two types of tRNA can carry this amino acid. One is used for initiation, the other for recognizing AUG codons during elongation.

In bacteria, mitochondria, and chloroplasts, the initiator tRNA carries a methionine residue that has been formylated on its amino group, forming a molecule of **N-formyl-methionyl-tRNA**. The tRNA is known as **tRNA$_f^{Met}$**. The name of the aminoacyl-tRNA is usually abbreviated to fMet-tRNA$_f$.

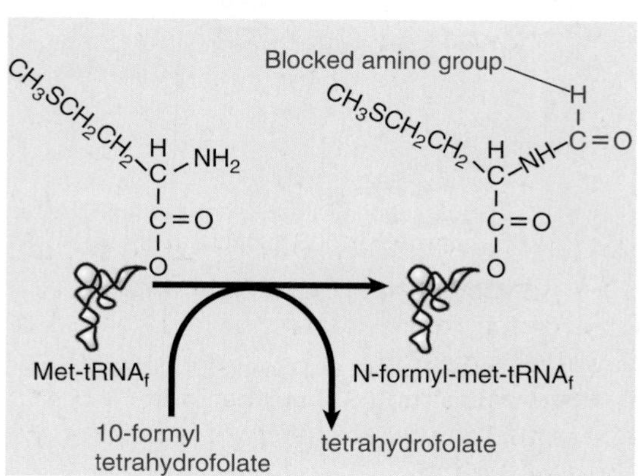

FIGURE 8.12 The initiator N-formyl-methionyl-tRNA (fMet-tRNA$_f$) is generated by formylation of methionyl-tRNA, using formyl-tetrahydrofolate as cofactor.

The initiator tRNA gains its modified amino acid in a two-stage reaction. First, it is charged with the amino acid to generate Met-tRNA$_f$; and then the formylation reaction shown in **FIGURE 8.12** blocks the free NH_2 group. Although the blocked amino acid group would prevent the initiator from participating in chain elongation, it does not interfere with the ability to initiate a polypeptide.

This tRNA is used only for initiation. It recognizes the codons AUG or GUG (occasionally UUG). The codons are not recognized equally well: the extent of initiation declines by about half when AUG is replaced by GUG, and declines by about half again when UUG is employed.

The tRNA species responsible for recognizing AUG codons in internal locations is **tRNA$_m$^{Met}**. This tRNA responds only to internal AUG codons. Its methionine is not formylated.

▸ **tRNA$_m$^{Met}** The bacterial tRNA that inserts methionine at internal AUG codons.

What features distinguish the fMet-tRNA$_f$ initiator and the Met-tRNA$_m$ elongator? Some characteristic features of the tRNA sequence are important, as summarized in **FIGURE 8.13**. Some of these features are needed to prevent the initiator from being used in elongation, whereas others are necessary for it to function in initiation:

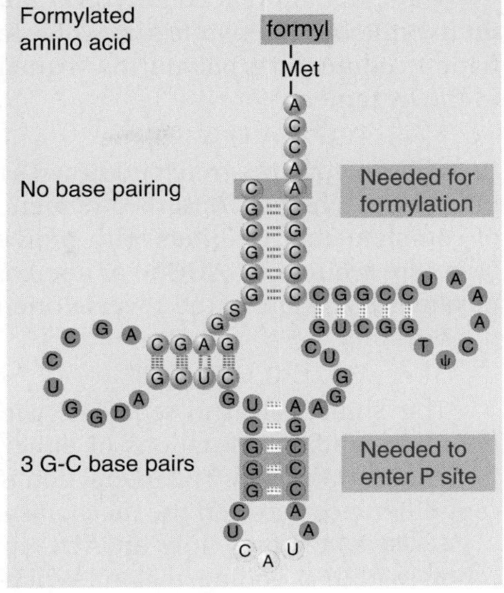

FIGURE 8.13 fMet-tRNA$_f$ has unique features that distinguish it as the initiator tRNA.

- Formylation is not strictly necessary, because nonformylated Met-tRNA$_f$ can function as an initiator. Formylation improves the efficiency with which the Met-tRNA$_f$ is used, though, because it is one of the features recognized by the factor IF-2 that binds the initiator tRNA.
- The bases that face one another at the last position of the stem to which the amino acid is connected are paired in all tRNAs except tRNA$_f$^{Met}. Mutations that create a base pair in this position of tRNA$_f$^{Met} allow it to function in elongation. The absence of this pair is therefore important in preventing tRNA$_f$^{Met} from being used in elongation. It is also needed for the formylation reaction.
- A series of three G-C pairs in the stem that precedes the loop containing the anticodon is unique to tRNA$_f$^{Met}. These base pairs are required to allow the fMet-tRNA$_f$ to be inserted directly into the P site.

KEY CONCEPTS

- Translation starts with a methionine amino acid usually coded by AUG.
- In bacteria, different methionine tRNAs are involved in initiation and elongation.
- The initiator tRNA has unique structural features that distinguish it from all other tRNAs.

CONCEPT AND REASONING CHECK

What advantage is there for bacteria to have two species of tRNA that recognize the AUG codon?

8.6 mRNA Binds a 30S Subunit to Create the Binding Site for a Complex of IF-2 and fMet-tRNA$_f$

In bacterial translation, the meaning of the AUG and GUG codons depends on their **context**. When the AUG codon is used for initiation, a formyl-methionine begins the polypeptide; when it is used within the coding region, methionine is added to the polypeptide. The meaning of the GUG codon is even more dependent on its location. When present as the first codon, formyl-methionine is added, but when present within a gene, it is bound by Val-tRNA, one of the regular members of the tRNA set, to provide valine as specified by the genetic code.

▸ **context** The fact that neighboring sequences may change the efficiency with which a codon is recognized by its aminoacyl-tRNA or is used to terminate polypeptide translation.

The initiation reaction involves binding of a 30S subunit to a *ribosome-binding site* on the mRNA. The two features of a bacterial ribosome-binding site are the AUG initiation codon and a polypurine sequence preceding it by ~10 bases that corresponds to the hexamer:

5′ . . . A G G A G G . . . 3′

▸ Shine–Dalgarno sequence The polypurine sequence AGGAGG centered about 10 bp before the AUG initiation codon on bacterial mRNA. It is complementary to the sequence at the 3′ end of 16S rRNA.

This polypurine stretch is known as the **Shine–Dalgarno sequence**. It is complementary to a highly conserved sequence close to the 3′ end of 16S rRNA. (The extent of complementarity differs with individual mRNAs, and may extend from the four-base core sequence GAGG to a nine-base sequence extending beyond each end of the hexamer.) Written in the reverse orientation, the complementary rRNA sequence is the hexamer:

3′ . . . U C C U C C . . . 5′

The Shine-Dalgarno sequence pairs with its rRNA complement during mRNA-ribosome binding. Mutations of either sequence in this reaction prevent an mRNA from being translated. The interaction is specific for bacterial ribosomes; this is a significant difference between the mechanisms of initiation for prokaryotes and eukaryotes.

FIGURE 8.14 shows how an AUG initiation codon is distinguished from an AUG codon within a coding region. When an initiation complex forms at a ribosome-binding site, the initiation codon lies within the part of the P site carried by the small subunit. The only aminoacyl-tRNA that can become part of the initiation complex is the initiator, which has the unique property of being able to enter directly into the partial P site to bind to its complementary codon.

When the large subunit joins the complex, the partial tRNA-binding sites are converted into the intact P and A sites. The initiator fMet-$tRNA_f$ occupies the P site, and the A site is available for entry of the aminoacyl-tRNA that is complementary to the second codon of the gene. The first peptide bond forms between the initiator and the next aminoacyl-tRNA.

Initiation occurs when an AUG (or GUG) codon lies within a ribosome-binding site, because only the initiator tRNA can enter the partial P site in the 30S subunit. AUG (or GUG) codons within the message, which are encountered by a ribosome that is continuing to translate an mRNA, are not recognized by the initiator tRNA because only the regular aminoacyl-tRNAs can enter the (complete) A site in the 70S ribosome.

Accessory factors are critical in controlling the usage of aminoacyl-tRNAs. All aminoacyl-tRNAs associate with the ribosome by binding to an accessory factor. The factor used in initiation is IF-2 (see *Section 8.4, Initiation in Bacteria Needs 30S Subunits and Accessory Factors*). The accessory factor used at elongation is discussed in *Section 8.8, Elongation Factor Tu Loads Aminoacyl-tRNA into the A Site*.

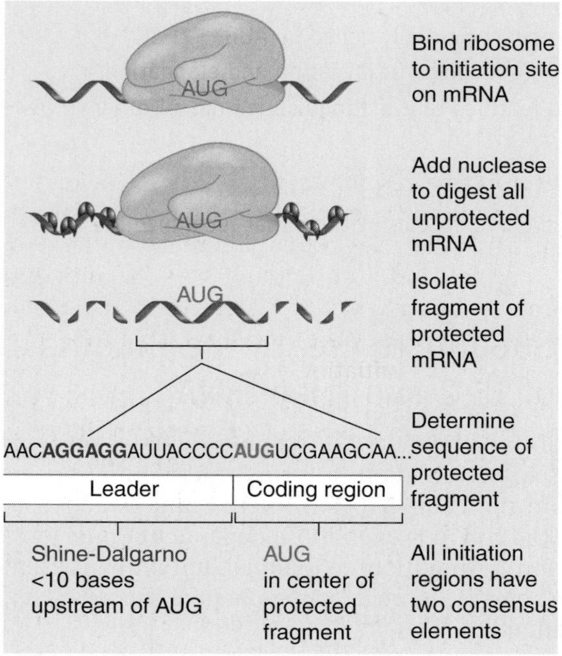

FIGURE 8.14 Ribosome-binding sites on mRNA can be recovered from initiation complexes. They include the upstream Shine–Dalgarno sequence and the initiation codon.

The initiation factor IF-2 places the initiator tRNA into the P site. By forming a complex specifically with fMet-$tRNA_f$, IF-2 ensures that only the initiator tRNA, and none of the regular aminoacyl-tRNAs, participates in the initiation reaction.

Conversely, the accessory factor that places aminoacyl-tRNAs in the A site cannot bind fMet-$tRNA_f$, which is therefore excluded from use during elongation.

The accuracy of initiation is also assisted by IF-3, which stabilizes binding of the initiator tRNA by recognizing correct base pairing with the second and third bases of the AUG initiation codon.

FIGURE 8.15 details the series of events by which IF-2 places the fMet-tRNA$_f$ initiator in the P site. IF-2, bound to GTP, associates with the P site of the 30S subunit. At this point, the 30S subunit carries all the initiation factors. fMet-tRNA$_f$ then binds to the IF-2 on the 30S subunit. IF-2 then transfers the tRNA into the partial P site.

IF-2 has a ribosome-dependent GTPase activity that is triggered when the 50S subunit joins to generate a complete ribosome. This probably provides the energy required for conformational changes associated with the joining of the subunits.

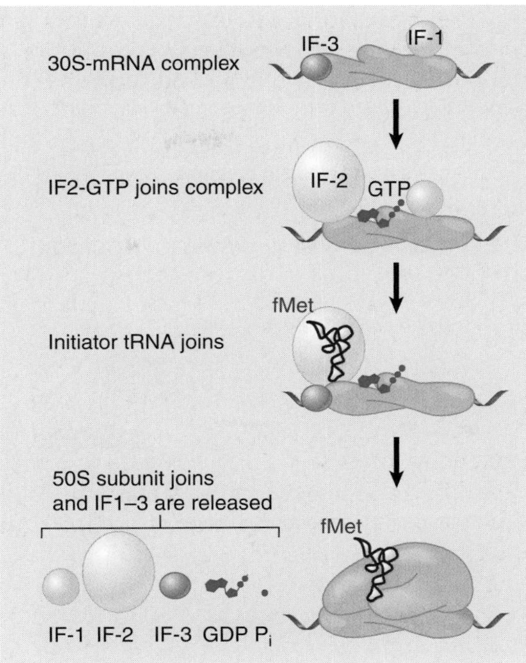

FIGURE 8.15 IF-2 is needed to bind fMet-tRNA$_f$ to the 30S-mRNA complex. After 50S binding, all IF factors are released and GTP is cleaved.

KEY CONCEPTS

- An initiation site on bacterial mRNA consists of the AUG initiation codon preceded with a gap of ~10 bases by the Shine–Dalgarno polypurine hexamer.
- The rRNA of the 30S bacterial ribosomal subunit has a complementary sequence that base pairs with the Shine–Dalgarno sequence during initiation.
- IF-2 binds the initiator fMet-tRNA$_f$ and allows it to enter the partial P site on the 30S subunit.

CONCEPT AND REASONING CHECK

During bacterial translation, how are AUG and GUG initiation codons distinguished from AUG and GUG codons later in an ORF?

8.7 Small Eukaryotic Subunits Scan for Initiation Sites on mRNA

Initiation of translation in eukaryotic cytoplasm resembles the process that occurs in bacteria, but the order of events is different and the number of accessory factors is greater. Some of the differences in initiation are related to a difference in the way that bacterial 30S and eukaryotic 40S subunits find their binding sites for initiating translation on mRNA. In eukaryotes, small subunits first recognize the 5′ end of the mRNA and then move to the initiation site, where they are joined by large subunits. (In prokaryotes, small subunits bind directly to the initiation site.)

Virtually all eukaryotic mRNAs are monocistronic, but each mRNA usually is substantially longer than necessary just to code for its protein. The average mRNA in eukaryotic cytoplasm is 1000 to 2000 bases long, has a methylated cap at the 5′ terminus, and carries 100 to 200 bases of poly(A) at the 3′ terminus. The nontranslated 5′ leader is relatively short, usually <100 bases. The length of the coding region is determined by the size of the polypeptide product. The nontranslated 3′ trailer is often rather long, at times reaching lengths of up to ~1000 bases.

The first feature to be recognized during translation of a eukaryotic mRNA is the methylated cap that marks the 5′ end. Messenger RNAs whose caps have been

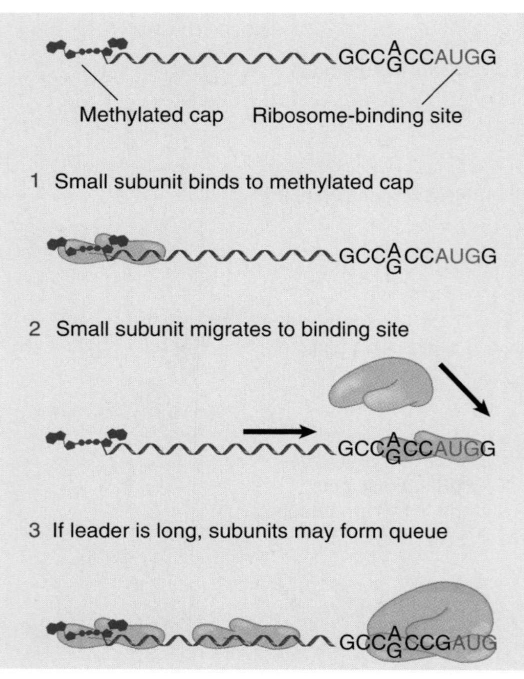
GCC A/G CCAUGG
Methylated cap
Ribosome-binding site
1 Small subunit binds to methylated cap
GCC A/G CCAUGG
2 Small subunit migrates to binding site
GCC A/G CCAUGG
3 If leader is long, subunits may form queue
GCC A/G CCGAUG

FIGURE 8.16 Eukaryotic ribosomes migrate from the 5′ end of mRNA to the ribosome binding site, which includes an AUG initiation codon.

removed are not translated efficiently *in vitro*. Binding of 40S subunits to mRNA requires several initiation factors, including proteins that recognize the structure of the cap. In some mRNAs, the AUG initiation codon lies within 40 bases of the 5′ terminus of the mRNA, so that both the cap and AUG lie within the span of ribosome binding. In many mRNAs, however, the cap and AUG are farther apart; in extreme cases, they can be as much as 1000 bases away from each other. Yet the presence of the cap still is necessary for a stable complex to be formed at the initiation codon. How can the ribosome rely on two sites so far apart?

FIGURE 8.16 illustrates the "scanning" model, which has the 40S subunit initially recognizing the 5′ cap and then "migrating" along the mRNA. Scanning from the 5′ end is a linear process. When 40S subunits scan the leader region, they can melt secondary structure hairpins with stabilities ≤30 kcal, but hairpins of greater stability impede or prevent migration.

Migration stops when the 40S subunit encounters the AUG initiation codon. Usually, although not always, the first AUG triplet sequence to be encountered will be the initiation codon. However, the AUG triplet by itself is not sufficient to halt migration; it is recognized efficiently as an initiation codon only when it is in the right context. The most important determinants of context are the bases in positions –4 and +1. An initiation codon may be recognized in the sequence NNNPuNN*AUG*G. The purine (A or G) 3 bases before the AUG codon, and the G immediately following it, can increase the efficiency of translation by 10×. When the leader sequence is long, further 40S subunits can recognize the 5′ end before the first has left the initiation site, creating a queue of subunits proceeding along the leader to the initiation site.

▸ **IRES (internal ribosome entry site)** A eukaryotic messenger RNA sequence that allows a ribosome to initiate polypeptide translation without migrating from the 5′ end.

The vast majority of eukaryotic initiation events involve scanning from the 5′ cap, but there is an alternative means of initiation, used especially by certain viral RNAs, in which a 40S subunit associates directly with an internal site called an **IRES** (internal ribosome entry site). (This entirely bypasses any AUG codons that may be in the 5′ nontranslated region.) There are few sequence homologies between known IRES elements. The most common type of IRES element includes the AUG initiation codon at its left boundary. The 40S subunit binds directly to it, using a subset of the same factors that are required for initiation at 5′ ends.

Modification at the 5' end occurs to almost all cellular and viral mRNAs, and is essential for their translation in eukaryotic cytosol. The sole exception to this rule is provided by a few viral mRNAs (such as poliovirus) that are not capped. They use the IRES pathway. This is especially important in picornavirus infection, where it was first discovered, because the virus inhibits host translation by destroying cap structures and inhibiting the initiation factors that bind them. This prevents the translation of host mRNAs. Viral mRNAs can be translated because they use the IRES.

Eukaryotic cells have more initiation factors than bacteria: the current list includes 12 factors that are directly or indirectly required for initiation. The factors are named similarly to those in bacteria, sometimes by analogy with the bacterial factors, and are given the prefix "e" to indicate their eukaryotic origin. They act at all stages of the process, including:

- forming an initiation complex with the 5′ end of mRNA;
- forming a complex with Met-tRNA$_i$;
- binding the mRNA-factor complex to the Met-tRNA$_i$-factor complex;
- enabling the ribosome to scan mRNA from the 5′ end to the first AUG;
- detecting binding of initiator tRNA to AUG at the start site; and
- mediating joining of the 60S subunit.

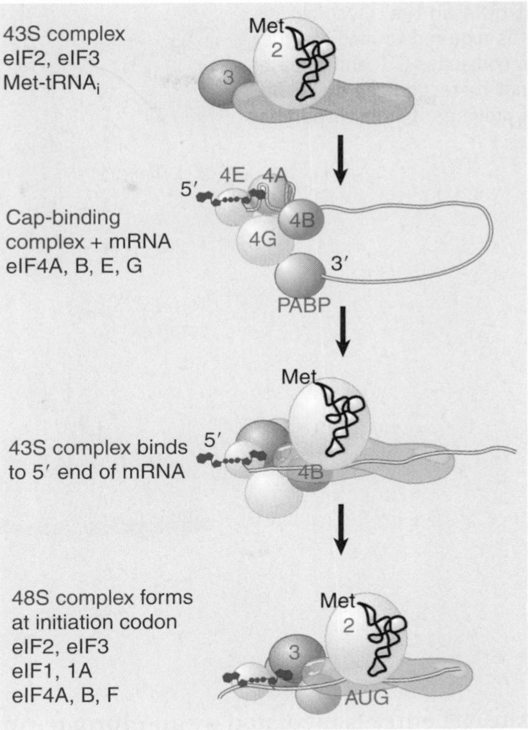

FIGURE 8.17 Some initiation factors bind to the 40S ribosome subunit to form the 43S complex; others bind to mRNA. When the 43S complex binds to mRNA, it scans for the initiation codon and can be isolated as the 48S complex.

FIGURE 8.17 summarizes the stages of initiation and shows which initiation factors are involved at each stage. eIF2 and eIF3 form a ternary complex with the initiator tRNA that binds to the 40S ribosome subunit to form a 43S complex. Then the ternary complex scans mRNA looking for an initiation codon. One role of the factors during scanning is to help unwind any base-paired regions in the mRNA. eIF4A, eIF4B, eIF4F, and eIF4G bind to the 5′ mRNA cap. eIF1 and eIF1A bind to the ribosome subunit-mRNA complex.

Note the circular arrangement of the mRNA associated with the cap-binding complex in Figure 8.17 (second line), when the interaction between the PABP and eIF4G brings the 5′ and 3′ ends of the mRNA into proximity. This enhances formation of initiation complexes, so that, in effect, PABP acts as an initiation factor.

KEY CONCEPTS

- Eukaryotic 40S ribosomal subunits bind to the 5′ end of mRNA and scan the mRNA until they reach an initiation site.
- A eukaryotic initiation site consists of a ten-nucleotide sequence that includes an AUG codon.
- Initiation factors are required for all stages of initiation, including binding of the initiator tRNA, 40S subunit attachment to mRNA, movement along the mRNA, and joining of the 60S subunit.
- eIF2 and eIF3 bind the initiator Met-tRNA$_i$ and GTP, and the complex binds to the 40S subunit before it associates with mRNA.

CONCEPT AND REASONING CHECK

How do eukaryotic 40S ribosomal subunits find the initiation codon of an mRNA?

8.8 Elongation Factor Tu Loads Aminoacyl-tRNA into the A Site

Once the complete ribosome is formed at the initiation codon, the stage is set for a cycle in which aminoacyl-tRNA enters the A site of a ribosome whose P site is occupied by peptidyl-tRNA. Any aminoacyl-tRNA except the initiator can enter the A

FIGURE 8.18 EF-Tu-GTP places aminoacyl-tRNA on the ribosome and then is released as EF-Tu-GDP. EF-Ts is required to mediate the replacement of GDP by GTP. The reaction consumes GTP and releases GDP. The only aminoacyl-tRNA that cannot be recognized by EF-Tu-GTP is fMet-$tRNA_f$, whose failure to bind prevents it from responding to internal AUG or GUG codons.

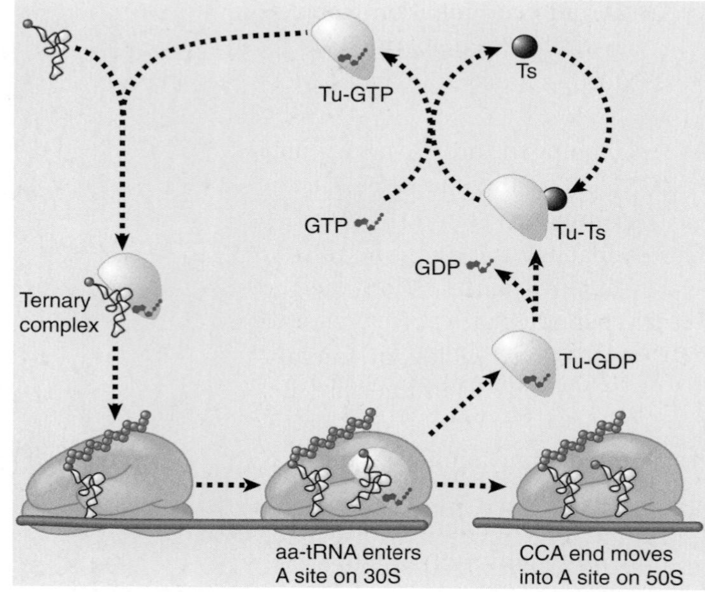

- **elongation factors** Proteins that associate with ribosomes cyclically during the addition of each amino acid to the polypeptide chain.
- **EF-Tu** The elongation factor that binds aminoacyl-tRNA and places it into the A site of a bacterial ribosome.

site. Its entry is mediated by an **elongation factor** (**EF-Tu** in bacteria). The process is similar in eukaryotes. EF-Tu is a highly conserved protein throughout bacteria and mitochondria and is homologous to its eukaryotic counterpart.

Just like its counterpart in initiation (IF-2), EF-Tu is associated with the ribosome only during the process of aminoacyl-tRNA entry. Once the aminoacyl-tRNA is in place, EF-Tu leaves the ribosome, to work again with another aminoacyl-tRNA.

FIGURE 8.18 shows the role of EF-Tu in bringing aminoacyl-tRNA to the A site. The binary complex of EF-Tu-GTP binds aminoacyl-tRNA to form a ternary complex of aminoacyl-tRNA-EF-Tu-GTP. The ternary complex binds only to the A site of ribosomes whose P site is already occupied by peptidyl-tRNA. This is the critical reaction in ensuring that the aminoacyl-tRNA and peptidyl-tRNA are correctly positioned for peptide bond formation.

Aminoacyl-tRNA is loaded into the A site in two stages. First, the anticodon end binds to the A site of the 30S subunit. Then, codon–anticodon recognition triggers a change in the conformation of the ribosome. This stabilizes tRNA binding and causes EF-Tu to hydrolyze its GTP. The CCA end of the tRNA now moves into the A site on the 50S subunit. The binary complex EF-Tu-GDP is released. This form of EF-Tu is inactive and does not bind aminoacyl-tRNA effectively. Another factor, EF-Ts, mediates the regeneration of the used form, EF-Tu-GDP, into the active form EF-Tu-GTP.

The presence of EF-Tu prevents the aminoacyl end of aminoacyl-tRNA from entering the A site on the 50S subunit (see Figure 8.22). So the release of EF-Tu-GDP is needed for the ribosome to undertake peptide bond formation. The same principle is seen at other stages of translation: one reaction must be completed properly before the next can proceed.

In eukaryotes, the factor eEF1α is responsible for bringing aminoacyl-tRNA to the ribosome, again in a reaction that involves cleavage of a high-energy bond in GTP. Like its prokaryotic homolog (EF-Tu), it is an abundant protein. After hydrolysis of GTP, the active form is regenerated by the factor eEF1βγ, a counterpart to EF-Ts.

KEY CONCEPTS

- EF-Tu is a monomeric G protein whose active form (bound to GTP) binds aminoacyl-tRNA.
- The EF-Tu-GTP-aminoacyl-tRNA complex binds to the ribosome A site.

CONCEPT AND REASONING CHECK

What are the roles of bacterial elongation factors EF-Tu and EF-Ts?

8.9 The Polypeptide Chain Is Transferred to Aminoacyl-tRNA

The ribosome remains in place while the polypeptide chain is elongated by transferring the polypeptide attached to the tRNA in the P site to the aminoacyl-tRNA in the A site. The reaction is shown in **FIGURE 8.19**. The activity responsible for synthesis of the peptide bond is called **peptidyl transferase**. It is a function of the large (50S or 60S) ribosomal subunit. The reaction is triggered when the aminoacyl end of the tRNA in the A site swings into a location close to the end of the peptidyl-tRNA following the release of EF-Tu. This site has a peptidyl transferase activity that essentially ensures a rapid transfer of the peptide chain to the aminoacyl-tRNA. Both rRNA and 50S subunit proteins are necessary for this activity, but the actual act of catalysis is a property of the ribosomal RNA of the large subunit (see *Section 8.16, Two rRNAs Play Active Roles in Translation*).

The nature of the transfer reaction is revealed by the ability of the antibiotic puromycin to inhibit protein synthesis. Puromycin resembles an amino acid attached to the terminal adenosine of tRNA. **FIGURE 8.20** shows that puromycin has an N instead of the O that joins an amino acid to tRNA. The antibiotic is treated by the ribosome as though it were an incoming aminoacyl-tRNA, after which the polypeptide attached to peptidyl-tRNA is transferred to the NH_2 group of the puromycin.

The puromycin moiety is not anchored to the A site of the ribosome, and as a result the polypeptidyl-puromycin complex is released from the ribosome in the form of polypeptidyl-puromycin. This premature termination of translation is responsible for the lethal action of the antibiotic.

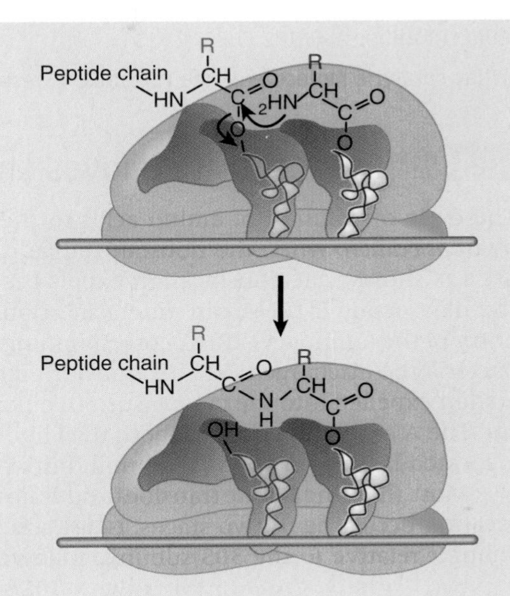

FIGURE 8.19 Peptide bond formation takes place by reaction between the polypeptide of peptidyl-tRNA in the P site and the amino acid of aminoacyl-tRNA in the A site.

peptidyl transferase The activity of the large ribosomal subunit that synthesizes a peptide bond when an amino acid is added to a growing polypeptide chain. The actual catalytic activity is a property of the rRNA.

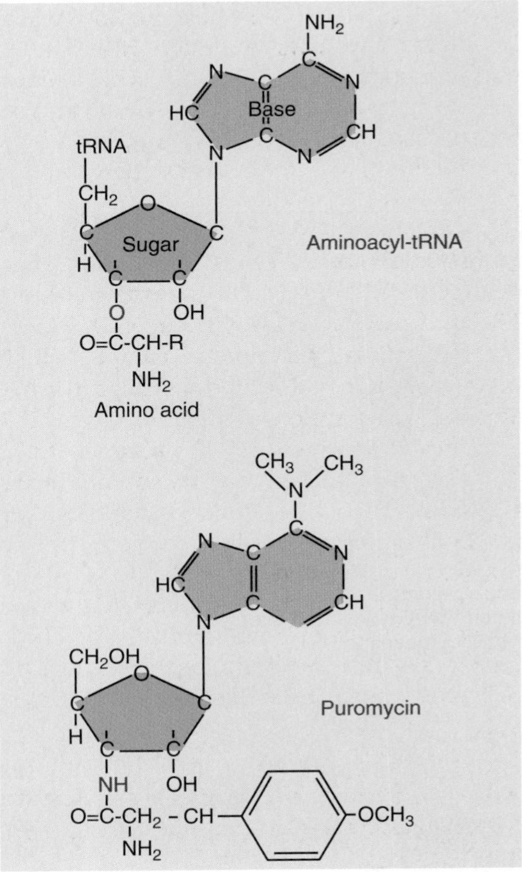

FIGURE 8.20 Puromycin mimics aminoacyl-tRNA because it resembles an aromatic amino acid linked to a sugar-base moiety.

KEY CONCEPTS

- The 50S subunit has peptidyl transferase activity.
- The nascent polypeptide chain is transferred from peptidyl-tRNA in the P site to aminoacyl-tRNA in the A site.
- Peptide bond synthesis generates deacylated tRNA in the P site and peptidyl-tRNA in the A site.

CONCEPT AND REASONING CHECK

What catalytic molecule of the ribosome is responsible for formation of the peptide bond?

8.10 Translocation Moves the Ribosome

The cycle of addition of amino acids to the growing polypeptide chain is completed by *translocation*, when the ribosome advances three nucleotides along the mRNA. **FIGURE 8.21** shows that translocation expels the uncharged tRNA from the P site, so that the new peptidyl-tRNA can enter. The ribosome then has an empty A site ready for entry of the aminoacyl-tRNA corresponding to the next codon. As the figure shows, in bacteria the discharged tRNA is transferred from the P site to the E site (from which it is then expelled into the cytoplasm). In eukaryotes it is expelled directly into the cytosol. The A and P sites straddle both the large and small subunits; the E site (in bacteria) is located largely on the 50S subunit, but has some contacts in the 30S subunit.

Most thinking about translocation follows the *hybrid state model*, which has translocation occurring in two stages. **FIGURE 8.22** shows that first there is a shift of the 50S subunit relative to the 30S subunit, followed by a second shift that occurs when the 30S subunit moves along mRNA to restore the original conformation. The basis for this model was the observation that the pattern of contacts

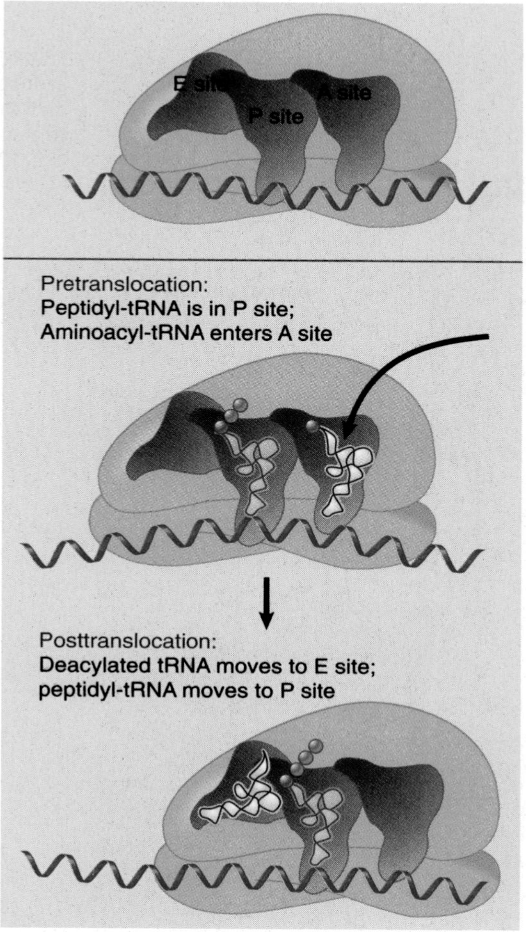

FIGURE 8.21 A bacterial ribosome has three tRNA-binding sites. Aminoacyl-tRNA enters the A site of a ribosome that has peptidyl-tRNA in the P site. Peptide bond synthesis deacylates the P site tRNA and generates peptidyl-tRNA in the A site. Translocation moves the deacylated tRNA into the E site and moves peptidyl-tRNA into the P site.

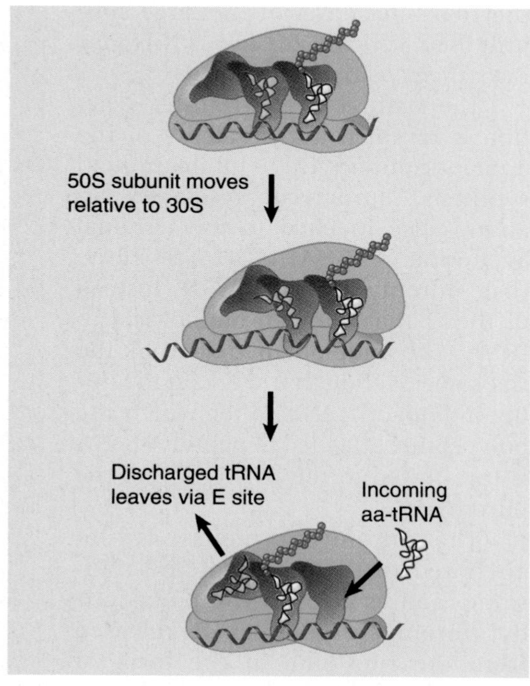

FIGURE 8.22 Models for translocation involve two stages. First, at peptide bond formation the aminoacyl end of the tRNA in the A site becomes relocated in the P site. Second, the anticodon end of the tRNA becomes relocated in the P site.

that tRNA makes with the ribosome (measured by chemical footprinting) changes in two stages. When puromycin is added to a ribosome that has an aminoacylated tRNA in the P site, the contacts of tRNA on the 50S subunit change from the P site to the E site, but the contacts on the 30S subunit do not change. This suggests that the 50S subunit has moved to a posttransfer state, but the 30S subunit has not changed.

The interpretation of these results is that first the aminoacyl ends of the tRNAs (located in the 50S subunit) move into the new sites (while the anticodon ends remain bound to their anticodons in the 30S subunit). At this stage, the tRNAs are effectively bound in hybrid sites, consisting of the 50SE/30SP and the 50SP/30SA sites. Then movement is extended to the 30S subunits, so that the anticodon–codon pairing region finds itself in the right site. The most likely means of creating the hybrid state is by a movement of one ribosomal subunit relative to the other, so that translocation in effect involves two stages, with the normal structure of the ribosome being restored by the second stage.

KEY CONCEPTS

- Ribosomal translocation moves the mRNA through the ribosome by three bases.
- Translocation moves deacylated tRNA into the E site and peptidyl-tRNA into the P site, and empties the A site.
- The hybrid state model proposes that translocation occurs in two stages, in which the 50S moves relative to the 30S, and then the 30S moves along mRNA to restore the original conformation.

CONCEPT AND REASONING CHECK

Ribosomal translocation occurs in a way that preserves codon-anticodon pairing. Why is this important?

8.11 Elongation Factors Bind Alternately to the Ribosome

Translocation requires GTP and another elongation factor, EF-G. This factor is a major constituent of the cell: it is present at a level of ~1 copy per ribosome (20,000 molecules per cell).

Ribosomes cannot bind EF-Tu and EF-G simultaneously, so translation follows the cycle illustrated in **FIGURE 8.23**, in which the factors are alternately bound to, and released from, the ribosome. Thus EF-Tu-GDP must be released before EF-G can bind; and then EF-G must be released before aminoacyl-tRNA-EF-Tu-GTP can bind.

Does the ability of each elongation factor to exclude the other rely on an allosteric effect on the overall conformation of the ribosome or on direct competition for overlapping binding sites? **FIGURE 8.24** shows an extraordinary similarity between the structures of the ternary complex of aminoacyl-tRNA-EF-Tu-GDP and

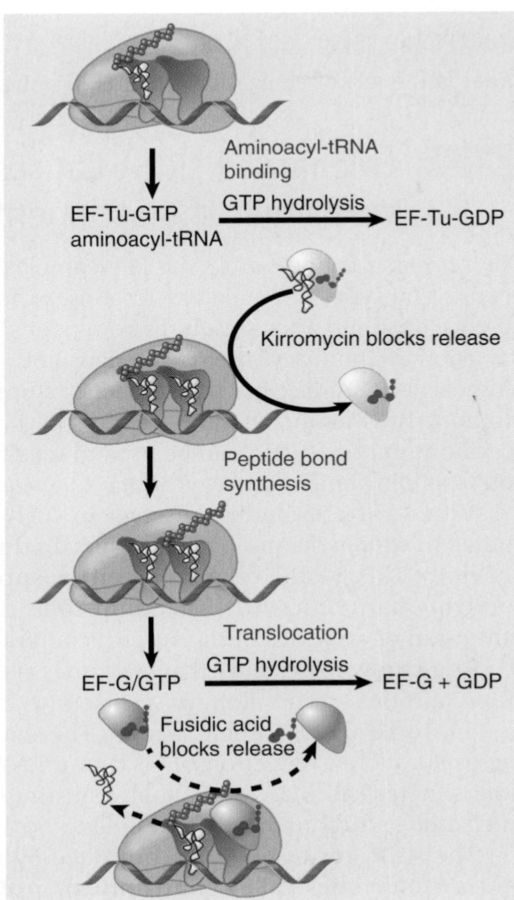

FIGURE 8.23 Binding of factors EF-Tu and EF-G alternates as ribosomes accept new aminoacyl-tRNA, form peptide bonds, and translocate.

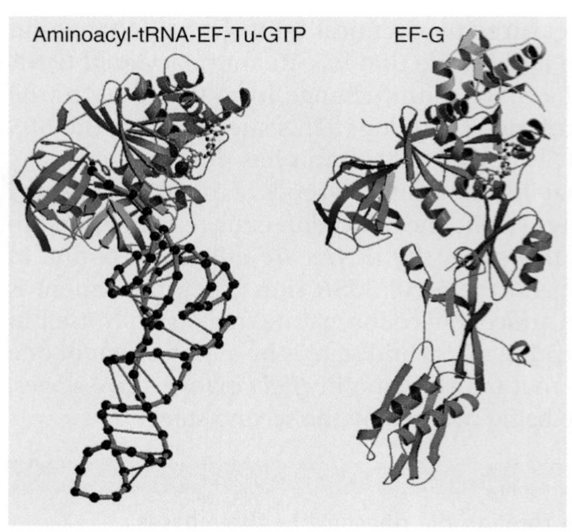

FIGURE 8.24 The structure of the ternary complex of aminoacyl-tRNA-EF-Tu-GTP (left) resembles the structure of EF-G (right). Structurally conserved domains of EF-Tu and EF-G are in red and green; the tRNA and the domain resembling it in EF-G are in purple. Photo courtesy of Poul Nissen, University of Aarhua, Denmark.

EF-G. The structure of EF-G mimics the overall structure of EF-Tu bound to the amino acceptor stem of aminoacyl-tRNA. This creates the immediate assumption that they compete for the same binding site (presumably in the vicinity of the A site). The need for each factor to be released before the other can bind ensures that the events of translation proceed in an orderly manner.

Both elongation factors are monomeric GTP-binding proteins that are active when bound to GTP, but inactive when bound to GDP. The triphosphate form is required for binding to the ribosome, which ensures that each factor obtains access to the ribosome only when complexed with the GTP that it needs to fulfill its function.

KEY CONCEPTS

- Translocation requires EF-G, whose structure resembles the aminoacyl-tRNA-EF-Tu-GTP complex.
- Binding of EF-Tu and EF-G to the ribosome is mutually exclusive.
- Translocation requires GTP hydrolysis, which triggers a change in EF-G, which in turn triggers a change in ribosome structure.

CONCEPT AND REASONING CHECK

What is the role of elongation factor EF-G in bacterial translation?

8.12 Uncharged tRNA Causes the Ribosome to Trigger the Stringent Response

The ribosome is not merely the key complex that synthesizes polypeptides; it also triggers several types of regulatory response. The initial stimulus for these responses is the absence of amino acids (resulting from poor growth conditions), which in turn leads to a depletion of aminoacyl-tRNA. The ribosome is in a key position to detect a deficiency in aminoacyl-tRNA, because this deficiency stops it from functioning in translation. Lack of aminoacyl-tRNAs in general is used to trigger an alarm response; also, the absence of a specific aminoacyl-tRNA can be used to regulate the metabolic systems that produce the corresponding amino acid (see *Section 13.5, Attenuation Can Be Controlled by Translation*).

When bacteria find themselves in such poor growth conditions that they lack a sufficient supply of amino acids to sustain translation, they shut down a wide range of activities. This is called the **stringent response**. We can view it as a mechanism for surviving hard times: the bacterium conserves its resources by engaging in only the minimum of activities until nutrient conditions improve.

The stringent response causes a massive (10–20×) reduction in the synthesis of rRNA and tRNA. This alone is sufficient to reduce the total amount of RNA synthesis to ~5% to 10% of its previous level. The synthesis of certain mRNAs is reduced, leading to an overall reduction of ~3× in mRNA synthesis. The rate of protein degradation is increased. Many metabolic adjustments occur, as seen in reduced synthesis of nucleotides, carbohydrates, and lipids.

The stringent response is controlled by the accumulation of two unusual nucleotides (sometimes called **alarmones**). **ppGpp** is guanosine tetraphosphate, with

- **stringent response** The ability of a bacterium to shut down synthesis of tRNA and ribosomes in a poor-growth medium.
- **alarmone** A small molecule in bacteria that is produced as a result of stress and that acts to alter the state of gene expression. The unusual nucleotides ppGpp and pppGpp are examples.
- **ppGpp** Guanosine tetraphosphate, with diphosphate groups attached to both the 5′ and 3′ positions, is an alarmone.

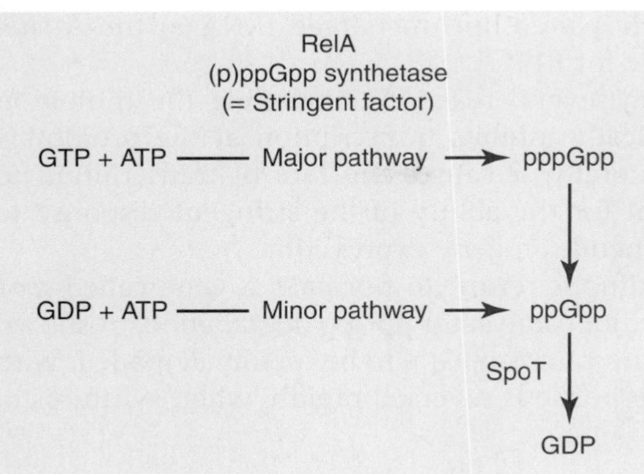

FIGURE 8.25 Stringent factor catalyzes the synthesis of pppGpp and ppGpp; ribosomal proteins can dephosphorylate pppGpp to ppGpp. ppGpp is degraded when it is no longer needed.

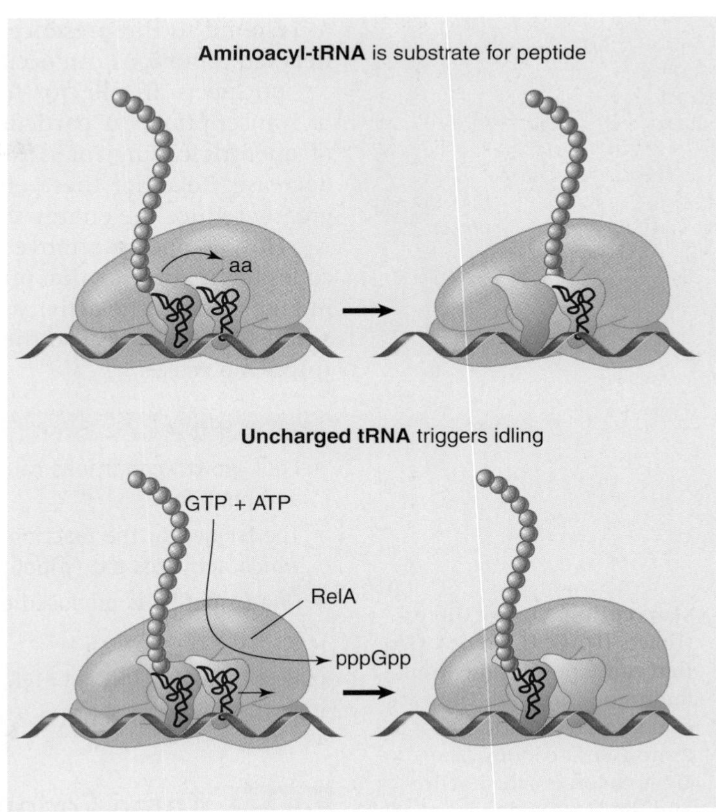

FIGURE 8.26 In normal translation, the presence of aminoacyl-tRNA in the A site is a signal for peptidyl transferase to transfer the polypeptide chain, followed by movement catalyzed by EF-G; but under stringent conditions, the presence of uncharged tRNA causes RelA protein to synthesize (p)ppGpp and to expel the tRNA.

diphosphates attached to both 5′ and 3′ positions. **pppGpp** is guanosine pentaphosphate, with a 5′ triphosphate group and a 3′ diphosphate. These nucleotides are typical small-molecule effectors that function by binding to target proteins to alter their activities. Sometimes they are known collectively as *(p)ppGpp*.

Deprivation of any one amino acid or a mutation that inactivates any aminoacyl-tRNA synthetase is sufficient to initiate the stringent response. The trigger that sets the entire series of events in motion is *the presence of uncharged tRNA in the A site of the ribosome.* Under normal conditions, of course, only aminoacyl-tRNA is placed in the A site by EF-Tu (see *Section 8.8, Elongation Factor Tu Loads Aminoacyl-tRNA into the A Site*). But when there is no aminoacyl-tRNA available to respond to a particular codon, the uncharged tRNA becomes able to gain entry.

Bacterial mutants that cannot produce the stringent response are called **relaxed mutants**. The most common site of relaxed mutation lies in the gene *relA*, which codes for a protein called the **stringent factor**. This factor is associated with ribosomes, although the amount is rather low—say, <1 molecule for every 200 ribosomes. So perhaps only a minority of the ribosomes are able to produce the stringent response.

The presence of uncharged tRNA in the A site blocks translation, and it triggers an **idling reaction** by wild-type ribosomes. Provided that the A site is occupied by an uncharged tRNA *specifically responding to the codon, the RelA protein catalyzes a reaction* in which ATP donates a pyrophosphate group to the 3′ position of either GTP or GDP. The formal name for this activity is *(p)ppGpp synthetase*.

FIGURE 8.25 shows the pathways for synthesis of (p)ppGpp. The RelA enzyme uses GTP as substrate more frequently, so that pppGpp is the predominant product. However, pppGpp is converted to ppGpp by several enzymes; among those able to perform this dephosphorylation are the translation factors EF-Tu and EF-G. The production of ppGpp via pppGpp is the most common route, *and ppGpp is the usual effector of the stringent response*.

The response of the ribosome to entry of uncharged tRNA is compared with normal translation in **FIGURE 8.26**. When EF-Tu places aminoacyl-tRNA in the A site, peptide bond synthesis is followed by ribosomal translocation. But when uncharged tRNA is paired with the codon in the A site, the ribosome remains stationary and engages in the idling reaction. The connection that activates RelA may be a ribosomal protein, L11 of the 50S subunit, which is located in the vicinity of the A and P sites, in a position

- **pppGpp** Guanosine pentaphosphate, with a triphosphate group attached to the 5′ position and a diphosphate group attached to the 3′ position, is an alarmone.
- **relaxed mutants** In *E. coli*, these do not display the stringent response to starvation for amino acids (or other nutritional deprivation).
- **stringent factor** The protein RelA, which is associated with ribosomes. It synthesizes ppGpp and pppGpp when an uncharged tRNA enters the A site.
- **idling reaction** This results in the production of pppGpp and ppGpp by ribosomes when an uncharged tRNA is present in the A site; this triggers the stringent response.

to respond to the presence of a properly paired but uncharged tRNA in the A site. Relaxed mutations can occur in the gene for L11.

ppGpp is an effector for controlling several reactions, including the inhibition of transcription. In particular, it specifically inhibits transcription at the promoters of operons coding for rRNA. More generally, it causes the rate of transcription to decrease. Together these effects account for the ability of the stringent response to greatly reduce the energy that the cell spends on gene expression.

How is ppGpp removed when conditions return to normal? A gene called *spoT* codes for an enzyme that provides the major catalyst for ppGpp degradation, as shown in Figure 8.25. The activity of this enzyme causes ppGpp to be rapidly degraded, with a half-life of ~20 sec; so the stringent response is reversed rapidly when synthesis of (p)ppGpp ceases.

KEY CONCEPTS

- Poor growth conditions cause bacteria to produce the small molecule regulators ppGpp and pppGpp.
- The trigger for the reaction is the entry of uncharged tRNA into the ribosomal A site, which activates the (p)ppGpp synthetase of the stringent factor RelA.
- One (p)ppGpp is produced every time an uncharged tRNA enters the A site.

CONCEPT AND REASONING CHECK

What is the function of the stringent response?

8.13 Three Codons Terminate Translation and Are Recognized by Protein Factors

Only 61 triplet codons specify amino acids. The other three triplets are termination codons (or **stop codons**), which end protein synthesis. They have informal names from the history of their discovery (see the accompanying Historical Context box). The UAG triplet is called the **amber** codon, UAA is the **ochre** codon, and UGA is sometimes called the **opal codon**.

The UAG, UAA, and UGA triplet sequences are necessary and sufficient to signal the end of translation, whether it occurs naturally at the end of an ORF or is created by mutation within a coding sequence.

(Sometimes the term *nonsense codon* is used to describe the termination triplets. "Nonsense" is really a misnomer, given that the codons do have meaning—albeit a disruptive one in a mutant gene. A better term is *stop codon*.)

In bacterial genes, UAA is the most commonly used termination codon. UGA is used more often than UAG, although there appear to be more errors reading UGA. (An error in reading a termination codon—when an aminoacyl-tRNA improperly responds to it—results in the continuation of translation until another termination codon is encountered.)

Two stages are involved in ending translation. The termination reaction itself involves release of the polypeptide chain from the last tRNA. The post-termination reaction involves release of the tRNA and mRNA and dissociation of the ribosome into its subunits.

None of the termination codons is recognized by a tRNA. They function in an entirely different manner from other codons and are recognized directly by protein factors. (The reaction does not depend on codon–anticodon recognition, so there seems to be no particular reason why it should require a triplet sequence.)

Termination codons are recognized by class 1 **release factors (RF)**. In *E. coli*, two class 1 release factors are specific for different sequences. **RF1** recognizes UAA and UAG; **RF2** recognizes UGA and UAA. The factors act at the ribosomal A site and require polypeptidyl-tRNA in the P site. The RFs are present at much lower levels than

- **stop codon** One of three triplets (UAG, UAA, or UGA) that cause polypeptide translation to terminate. They are also known historically as nonsense codons. The UAA codon is called ochre and the UAG codon is called amber, after the names of the nonsense mutations by which they were originally identified.
- **amber codon** The triplet UAG, one of the three termination codons that end polypeptide translation.
- **ochre codon** The triplet UAA, one of the three termination codons that end polypeptide translation.
- **opal codon** The triplet UGA, one of the three termination codons that end polypeptide translation. It has evolved to code for an amino acid in a small number of organisms or organelles.
- **release factor (RF)** A protein required to terminate polypeptide translation to cause release of the completed polypeptide chain and the ribosome from mRNA.
- **RF1** The bacterial release factor that recognizes UAA and UAG as signals to terminate polypeptide translation.
- **RF2** The bacterial release factor that recognizes UAA and UGA as signals to terminate polypeptide translation.

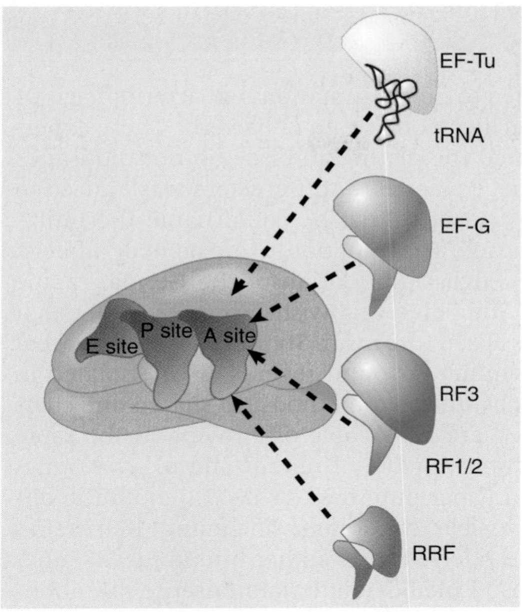

FIGURE 8.27 Molecular mimicry enables the elongation factor Tu-tRNA complex, the translocation factor EF-G, and the release factors RF1/2-RF3 to bind to the same ribosomal site. RRF is the ribosome recycling factor (see Figure 8.29).

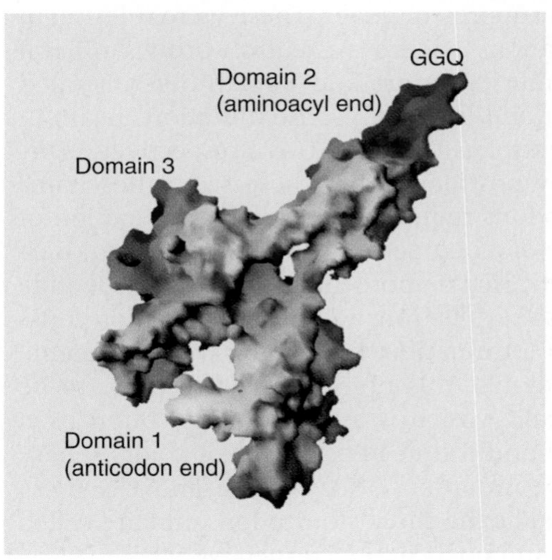

FIGURE 8.28 The eukaryotic termination factor eRF1 has a structure that mimics tRNA. The motif GGQ at the tip of domain 2 is essential for hydrolyzing the polypeptide chain from tRNA. Photo courtesy of David Barford, The Institute of Cancer Research.

initiation or elongation factors; there are ~600 molecules of each per cell, equivalent to one RF per ten ribosomes. In eukaryotes, there is only a single class 1 release factor, called eRF.

The class 1 release factors are assisted by class 2 release factors, which are not codon-specific. The class 2 factors are GTP-binding proteins. In *E. coli*, the role of the class 2 factor, **RF3**, is to release the class 1 factor from the ribosome. RF3 is a GTP-binding protein that is related to the elongation factors.

RF3 A polypeptide translation termination factor related to the elongation factor EF-G. It functions to release the factors RF1 or RF2 from the ribosome when they act to terminate polypeptide translation.

RF3 resembles the GTP-binding domains of EF-Tu and EF-G, and RF1 and RF2 resemble the C-terminal domain of EF-G, which mimics tRNA. This suggests that the release factors utilize the same site that is used by the elongation factors. **FIGURE 8.27** illustrates the basic idea that these factors all have the same general shape and bind to the ribosome successively at the same site (basically the A site or a region extensively overlapping with it).

The eukaryotic class 1 release factor, eRF1, is a single protein that recognizes all three termination codons. Its sequence is unrelated to the bacterial factors. It can terminate protein synthesis *in vitro* without the class 2 factor, eRF2, although eRF2 is essential in yeast *in vivo*. The structure of eRF1 follows a familiar theme: **FIGURE 8.28** shows that it consists of three domains that mimic the structure of tRNA.

HISTORICAL PERSPECTIVES

The Naming of the Amber, Ochre, and Opal Codons

Amber, ochre, and opal codons are the names for the three stop codons (UAG, UAA, and UGA, respectively) used to terminate translation *in vivo*. These codons do not specify any amino acid, but instead recruit release factors to terminate translation. The naming of these codons and the discovery of the nucleotide triplets that specify the stop codons date back to genetic studies of the bacteriophage T4. In the late 1950s and early 1960s, bacterial geneticists in the United States and England were isolating suppressible mutations in bacteriophage, phage capable of growth in some bacterial strains (permissive strains), but not able to grow in others (nonpermissive strains). Furthermore, these mutations were known to result from so-called nonsense codons, nucleotide triplets that do not code for any amino acid. The presence of these nonsense mutations in coding regions interrupts the reading frame of the transcripts and results in truncated polypeptides. This premature termination of translation is suppressed when the mutant phage are grown in a permissive bacterial strain. The differences between the permissive and nonpermissive bacterial strains are mutations in suppressor genes capable of suppressing the nonsense mutations. Suppressor genes are required to generate suppressor tRNAs that have altered anticodon sequences, such that they recognize and anneal to a specific nonsense mutation and insert an amino acid.

The term *amber mutation* is attributed to Harris Bernstein, at the time a graduate student at Caltech in Pasadena, California. Bernstein was recruited by two other biologists at Caltech, Dick Epstein and Charley Steinberg, to settle a scientific dispute with a series of experiments designed to isolate mutations of bacteriophage T4 that were specific to one strain of bacteria. These experiments required the plating and isolation of thousands of bacteriophage plaques, and Bernstein was enticed to participate by an offer that he could name the mutations after himself. The mutations were named "amber," which is the English translation of the German name "Bernstein." Amber is a yellowish-brown gemstone that is actually fossilized tree resin. Soon it became clear that these amber mutations, and several others isolated in different labs by different methods, all shared one characteristic: they are all capable of growing in the same permissive strains. Sydney Brenner and his colleagues proposed in a paper published in 1965 that all the different suppressible mutations, including Bernstein's mutations, be referred to as amber mutations. Brenner and others also isolated additional nonsense mutations that appeared to be distinct; they were suppressible by a different set of suppressor genes. These were referred to as "ochre mutations"; ochre is a reddish-brown mineral pigment. Brenner, Stretton, and Kaplan demonstrated that the amber and ochre nonsense mutations result in the nucleotide triplets UAG and UAA, respectively. Furthermore, they postulated that these were the chain-terminating codons required for normal termination of the translation of a polypeptide. In 1967, Brenner, Barnett, Katz, and Crick demonstrated that the nucleotide triplet UGA also results from a nonsense mutation causing translation termination. In keeping with the theme of colored minerals, this nonsense codon came to be known as "opal," which is a type of silica often used as a gemstone and found in a variety of colors. These three nonsense mutations, UAG, UAA, and UGA, are now recognized as the three stop codons and are called amber, ochre, and opal codons, respectively.

The termination reaction releases the completed polypeptide, but leaves a deacylated tRNA and the mRNA still associated with the ribosome. **FIGURE 8.29** shows that the dissociation of the remaining components (tRNA, mRNA, 30S, and 50S subunits) requires *ribosome recycling factor (RRF)*. RRF acts together with EF-G in a reaction that uses hydrolysis of GTP. As for the other factors involved in release, RRF has a structure that mimics tRNA, except that it lacks an equivalent for the 3′ amino acid-binding region. IF-3 is also required. RRF acts on the 50S subunit, and IF-3 acts to remove deacylated tRNA from the 30S subunit. Once the subunits have separated, IF-3 remains necessary, of course, to prevent their reassociation.

FIGURE 8.30 compares the functional and sequence homologies of the prokaryotic and eukaryotic translation factors.

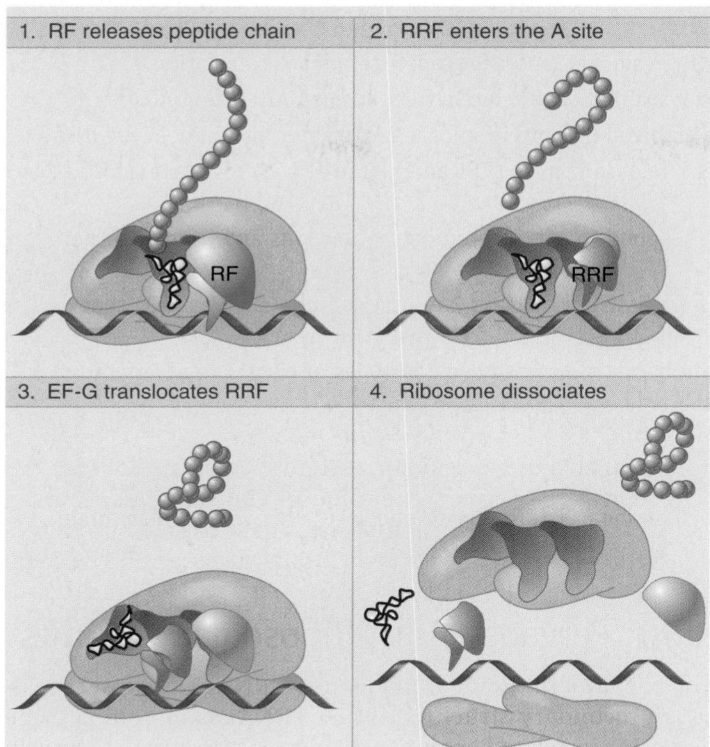
1. RF releases peptide chain
2. RRF enters the A site
RF
RRF
3. EF-G translocates RRF
4. Ribosome dissociates

FIGURE 8.29 The RF (release factor) terminates translation by releasing the polypeptide chain. The RRF (ribosome recycling factor) releases the last tRNA, and EF-G releases RRF, causing the ribosome to dissociate.

Initiation Factors			
Prokaryotic	Eukaryotic	General Function	Notes
IF-1	eIF1A	Blocks A site	eIF1A assists eIF2 in promoting Met-tRNA$_i^{Met}$ to binding to 40S; also promotes subunit dissociation
IF-2*†	eIF2, eIF3, eIF5B*	Entry of initiator tRNA	eIF2 is a GTPase eIF3 stimulates formation of the ternary complex, its binding to 40S, and binding and scanning of mRNA eIF5B is involved in initiator tRNA entry and is a GTPase
IF-3	eIF1, eIF4 complex, eIF3	Small subunit binding to mRNA	eIF4 complex functions in cap binding
Elongation Factors			
Prokaryotic	Eukaryotic	General Function	
EF-Tu†‡, EF-G† EF-Ts EF-G§	eEF1α‡ eEF1β, eEF1γ eEF2§	GTP-binding GDP-exchanging Ribosome translocation	
Release Factors			
Prokaryotic	Eukaryotic	General Function	
RF1 RF2 RF3†	eRF1 eRF1 eRF3	UAA/UAG recognition UAA/UGA recognition Stimulation of other RF(s)	

* IF-2 and eIF5B have sequence homology.
† IF-2, EF-Tu, EF-G, and RF3 have sequence homology.
‡ EF-Tu and eEF1α have sequence homology.
§ EF-G and eEF2 have sequence homology.

FIGURE 8.30 Functional homologies of prokaryotic and eukaryotic translation factors.

KEY CONCEPTS

- The codons UAA (ochre), UAG (amber), and UGA (opal) terminate translation.
- In bacteria they are used most often with relative frequencies UAA>UGA>UAG.
- Termination codons are recognized by protein release factors, not by aminoacyl-tRNAs.
- The structures of the class 1 release factors (RF1 and RF2 in *E. coli*) resemble aminoacyl-tRNA·EF-Tu and EF-G.
- The class 1 release factors respond to specific termination codons and hydrolyze the polypeptide-tRNA linkage.
- The class 1 release factors are assisted by class 2 release factors (such as RF3) that depend on GTP.
- The mechanism is similar in bacteria (which have two types of class 1 release factors) and eukaryotes (which have only one class 1 release factor).

CONCEPT AND REASONING CHECK

Why do class 1 release factors, aminoacyl-tRNA·EF-Tu, and EF-G all have a similar three-dimensional conformation?

8.14 Ribosomal RNA Pervades Both Ribosomal Subunits

Two-thirds of the mass of the bacterial ribosome is made up of rRNA. The most revealing approach to analyzing the secondary structure of large RNAs is to compare the sequences of corresponding rRNAs in related organisms. Those regions that are important in the secondary structure retain the ability to interact by base pairing. Thus if a base pair is required, it can form at the same relative position in each rRNA. This approach has enabled detailed models of both 16S and 23S rRNA to be constructed.

Each of the major rRNAs has a secondary structure with several discrete domains. Four general domains are formed by 16S rRNA, in which just under half of the sequence is base paired (see Figure 8.40). Six general domains are formed by 23S rRNA. The individual double-helical regions tend to be short (<8 bp). Often the duplex regions are not perfect and contain bulges of unpaired bases. Comparable models of mitochondrial rRNAs (which are shorter and have fewer domains) and eukaryotic cytosolic rRNAs (which are longer and have more domains) have been constructed. The greater length of eukaryotic rRNAs is due largely to the acquisition of sequences for additional domains. The crystal structure of the ribosome shows that in each subunit the domains of the major rRNA fold independently and have discrete locations.

The 70S ribosome has an asymmetric structure. **FIGURE 8.31** shows a schematic of the structure of the 30S subunit, which is divided into four regions: the head, neck, body, and platform. **FIGURE 8.32** shows a similar representation of the 50S subunit, where two prominent features are the central protuberance (where 5S rRNA is located) and the stalk (made of multiple copies of protein L7). **FIGURE 8.33** shows that the platform of the

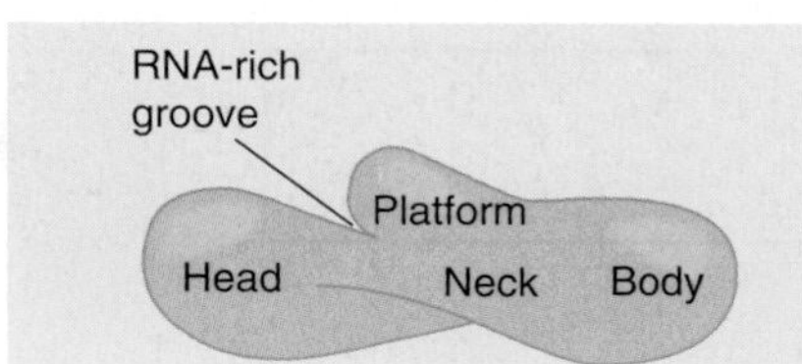

FIGURE 8.31 The 30S subunit has a head separated by a neck from the body, with a protruding platform.

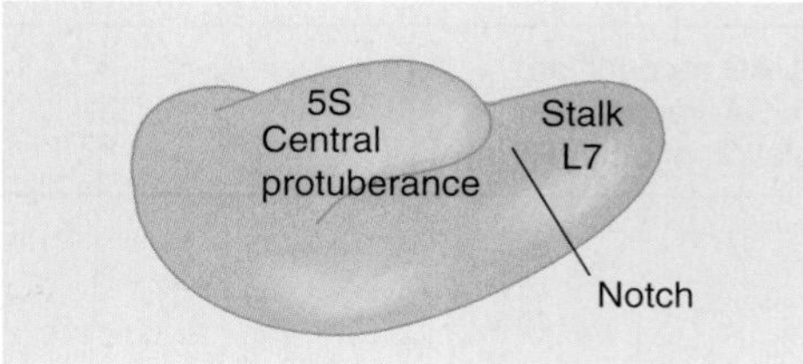

FIGURE 8.32 The 50S subunit has a central protuberance where 5S rRNA is located, separated by a notch from a stalk made of copies of the protein L7.

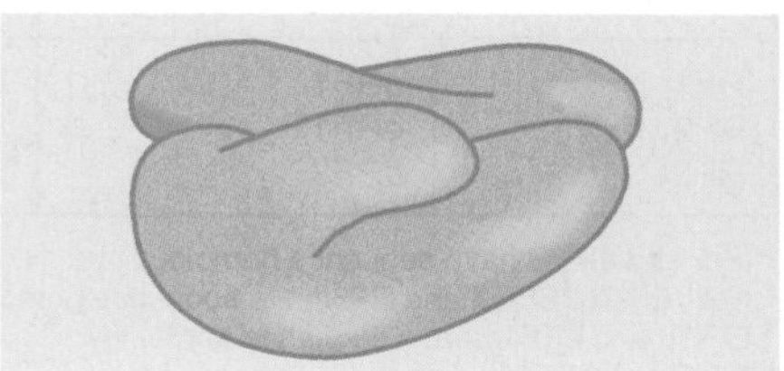

FIGURE 8.33 The platform of the 30S subunit fits into the notch of the 50S subunit to form the 70S ribosome.

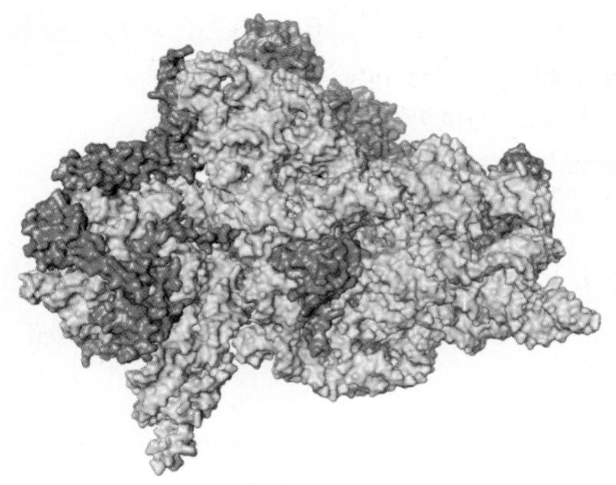

FIGURE 8.34 The 30S ribosomal subunit is a ribonucleoprotein particle. Ribosomal proteins are brown, and rRNA is yellow. Photo courtesy of V. Ramakrishnan, Medical Research Council (UK).

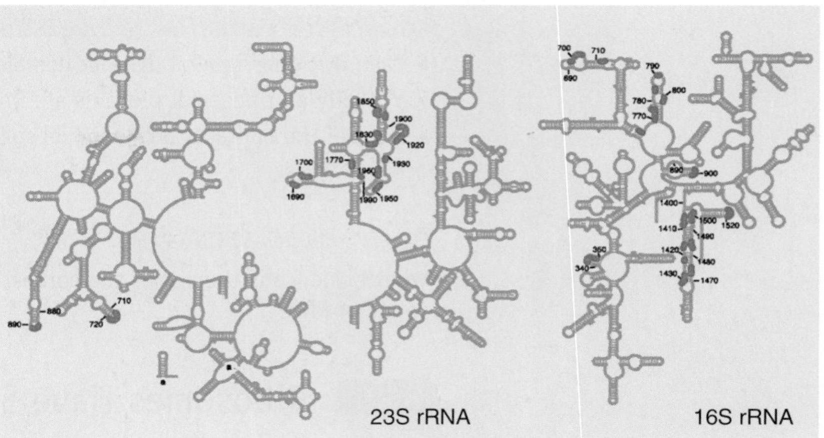

FIGURE 8.35 Contact points between the rRNAs are located in two domains of 16S rRNA and one domain of 23S rRNA. Reproduced from Yusupov, M. M., *Science* 292 (2001): 883–896. © 2001 AAAS. Photo courtesy of Harry Noller, University of California, Santa Cruz.

small subunit fits into the notch of the large subunit. There is a cavity (resembling a doughnut, but not visible in Figure 8.33) between the subunits that contains some of the important sites.

The structure of the 30S subunit follows the organization of 16S rRNA, with each structural feature corresponding to a domain of the rRNA. The body is based on the 5′ domain, the platform on the central domain, and the head on the 3′ region. **FIGURE 8.34** shows that the 30S subunit has an asymmetrical distribution of RNA and protein. One important feature is that the platform of the 30S subunit that provides the interface with the 50S subunit is composed almost entirely of RNA. Only two proteins (a small part of S7 and possibly part of S12) lie near the interface. This means that the association and dissociation of ribosomal subunits must depend on interactions with the 16S rRNA. This observation supports the idea that the evolutionary origin of the ribosome may have been as a particle consisting of RNA rather than protein.

The 50S subunit has a more even distribution of components than the 30S subunit, with long rods of double-stranded RNA crisscrossing the structure. The RNA forms a mass of tightly packed helices. The exterior surface largely consists of protein, except for the peptidyl transferase center (see *Section 8.16, Two rRNAs Play Active Roles in Translation*). Almost all segments of the 23S rRNA interact with protein, but many of the proteins are relatively unstructured.

The junction of subunits in the 70S ribosome involves contacts between 16S rRNA (many in the platform region) and 23S rRNA. There are also some interactions between rRNAs of each subunit and proteins in the other, and a few protein–protein contacts. **FIGURE 8.35** identifies the contact points on the rRNA structures. **FIGURE 8.36** opens out the structure (imagine the 50S subunit rotated counterclockwise and the 30S subunit rotated clockwise around the axis shown in the figure) to show the locations of the contact points on the face of each subunit.

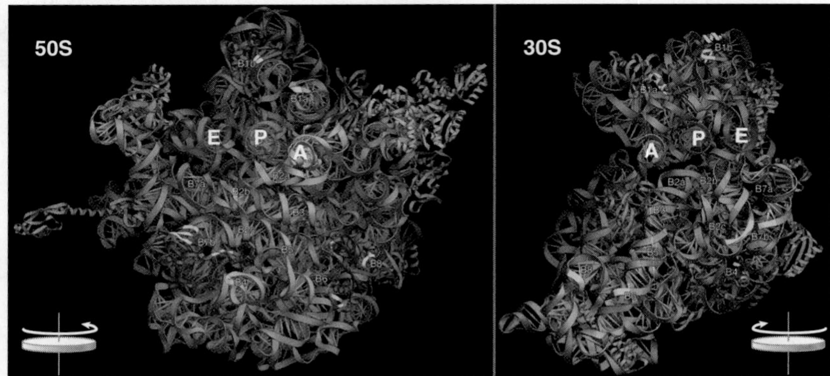

FIGURE 8.36 Contacts between the ribosomal subunits are mostly made by RNA (shown in purple). Contacts involving proteins are shown in yellow. The two subunits are rotated away from one another to show the faces where contacts are made; from a plane of contact perpendicular to the screen, the 50S subunit is rotated 90° counterclockwise, and the 30S is rotated 90° clockwise (this shows it in the reverse of the usual orientation). Photo courtesy of Harry Noller, University of California, Santa Cruz.

KEY CONCEPTS

- Each rRNA has several distinct domains that fold independently.
- Virtually all ribosomal proteins are in contact with rRNA.
- Most of the contacts between ribosomal subunits are made between the 16S and 23S rRNAs.

CONCEPT AND REASONING CHECK

What evidence is there that the earliest version of the ribosome may have been composed solely of RNA?

8.15 Ribosomes Have Several Active Centers

The basic ribosomal feature to remember is that it is a cooperative structure that depends on changes in the relationships among its active sites during translation. The active sites are not small, discrete regions like the active centers of enzymes. They are large regions whose construction and activities may depend just as much on the rRNA as on the ribosomal proteins. The crystal structures of the individual subunits and bacterial ribosomes give us a good impression of the overall organization and emphasize the role of the rRNA. At 5.5 Å resolution, the structure clearly identifies the locations of the tRNAs and the functional sites. We can now account for many ribosomal functions in terms of its structure.

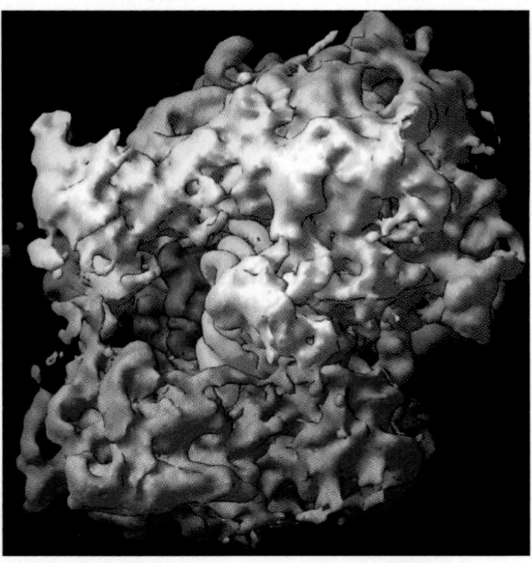

FIGURE 8.37 The 70S ribosome consists of the 50S subunit (white) and the 30S subunit (purple) with three tRNAs located superficially: yellow in the A site, blue in the P site, and green in the E site. Photo courtesy of Harry Noller, University of California, Santa Cruz.

Ribosomal functions are centered around the interactions with tRNAs. **FIGURE 8.37** shows the 70S ribosome with the positions of tRNAs in the three binding sites. The tRNAs in the A and P sites are nearly parallel to one another. All three tRNAs are aligned with their anticodon loops bound to the mRNA in the groove on the 30S subunit. The rest of each tRNA is bound to the 50S subunit. The environment surrounding each tRNA is mostly provided by rRNA. In each site, the rRNA contacts the tRNA at parts of the structure that are universally conserved.

It has always been a struggle to understand how two bulky tRNAs can fit next to one another in reading adjacent codons. The crystal structure shows a 45° kink in the mRNA between the P and A sites, which allows the tRNAs to fit as shown in the expansion of **FIGURE 8.38**. The tRNAs in the P and A sites are angled at 26° relative to each other at their anticodons. The closest approach between the backbones of the tRNAs occurs at the 3′ ends, where they converge to within 5 Å (perpendicular to the plane of the page). This allows the peptide chain to be transferred from the peptidyl-tRNA in the P site to the aminoacyl-tRNA in the A site.

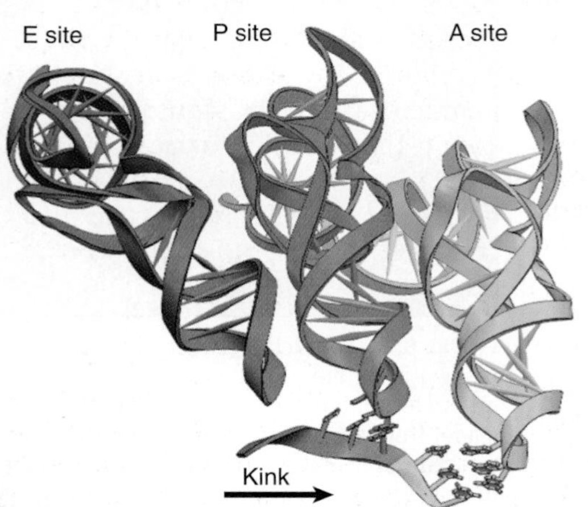

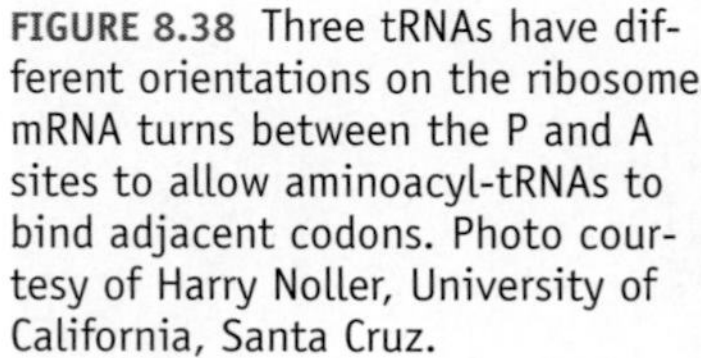

FIGURE 8.38 Three tRNAs have different orientations on the ribosome. mRNA turns between the P and A sites to allow aminoacyl-tRNAs to bind adjacent codons. Photo courtesy of Harry Noller, University of California, Santa Cruz.

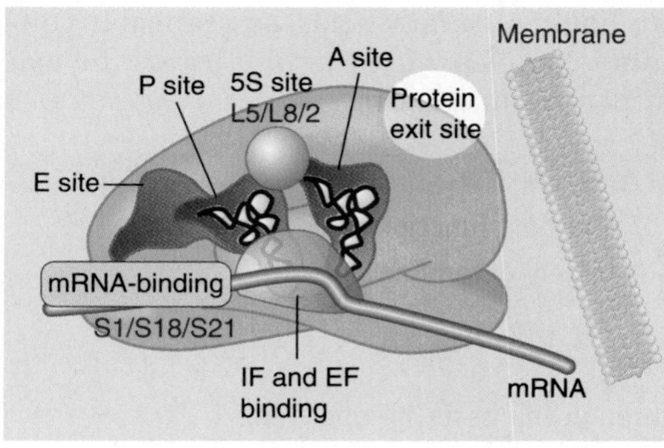

FIGURE 8.39 The ribosome has several active centers. It may be associated with a membrane. mRNA takes a turn as it passes through the A and P sites, which are angled with regard to each other. The E site lies beyond the P site. The peptidyl transferase site (not shown) stretches across the tops of the A and P sites. Part of the site bound by EF-Tu/G lies at the base of the A and P sites.

Translocation involves large movements in the positions of the tRNAs within the ribosome. The anticodon end of tRNA moves ~28 Å from the A site to the P site, and then a further 20 Å from the P site to the E site. As a result of the angle of each tRNA relative to the anticodon, the bulk of the tRNA moves much larger distances: 40 Å from A site to P site and 55 Å from P site to E site. This suggests that translocation requires a major reorganization of structure.

Much of the structure of the bacterial ribosome is occupied by its active centers. The schematic view of the ribosomal sites in **FIGURE 8.39** shows they comprise about two thirds of the ribosomal structure. A tRNA enters the A site, is transferred by translocation into the P site, and then leaves the ribosome by the E site. The A and P sites extend across both ribosome subunits; tRNA is paired with mRNA in the 30S subunit, but peptide transfer takes place in the 50S subunit. The A and P sites are adjacent, enabling translocation to move the tRNA from one site into the other. The E site is located near the P site (representing a position *en route* to the surface of the 50S subunit). The peptidyl transferase center is located on the 50S subunit, close to the aminoacyl ends of the tRNAs in the A and P sites (see *Section 8.16, Two rRNAs Play Active Roles in Translation*).

All of the GTP-binding proteins that function in translation (EF-Tu, EF-G, IF-2, and RF1, RF2, and RF3) bind to the same factor-binding site (sometimes called the GTPase center), which probably triggers their hydrolysis of GTP. This site is located at the base of the stalk of the large subunit, which consists of the proteins L7 and L12. (L7 is a modification of L12 and has an acetyl group on the N-terminus.) In addition to this region, the complex of protein L11 with a 58-base stretch of 23S rRNA provides the binding site for some antibiotics that affect GTPase activity. Neither of these ribosomal structures actually possesses GTPase activity, but they are both necessary for it. The role of the ribosome is to trigger GTP hydrolysis by factors bound in the factor-binding site.

Initial binding of 30S subunits to mRNA requires protein S1, which has a strong affinity for single-stranded nucleic acid. It is responsible for maintaining the single-stranded state in mRNA that is bound to the 30S subunit. This action is necessary to prevent the mRNA from taking up a base-paired conformation that would be unsuitable for translation. S1 has an extremely elongated structure and associates with S18 and S21. The three proteins constitute a domain that is involved in the initial binding of mRNA and in binding initiator tRNA. This locates the mRNA-binding site in the vicinity of the cleft of the small subunit (see Figure 8.31). The 3′ end of rRNA, which pairs with the mRNA initiation site, is located in this region.

A nascent polypeptide extends through the ribosome, away from the active sites, into the region in which ribosomes may be attached to membranes (see *Chapter 10, Protein Localization*). A polypeptide chain emerges from the ribosome through an exit channel, which leads from the peptidyl transferase site to the surface of the 50S subunit. The tunnel is composed mostly of rRNA. It is quite narrow—only 1 to 2 nm wide—and is ~10 nm long. The nascent polypeptide emerges from the ribosome ~15 Å away from the peptidyl transferase site. The tunnel can hold ~50 amino acids, and probably constrains the polypeptide chain so that it cannot completely fold until it leaves the exit domain, though some limited secondary structures may form.

KEY CONCEPTS

- Interactions involving rRNA are a key part of ribosome function.
- The environment of the tRNA-binding sites is largely determined by rRNA.

CONCEPT AND REASONING CHECK

What are the paths of tRNAs and the mRNA through the bacterial ribosome?

8.16 Two rRNAs Play Active Roles in Translation

The ribosome was originally viewed as a collection of proteins with various catalytic activities held together by protein–protein and RNA–protein interactions. The discovery of RNA molecules with catalytic activities (see *Chapter 28, RNA Splicing and Processing*) immediately suggests, however, that rRNA might play a more active role in ribosome function. There is now evidence that rRNA interacts with mRNA or tRNA at each stage of translation, and that the proteins are necessary to maintain the rRNA in a structure in which it can perform the catalytic functions. Several interactions involve specific regions of rRNA:

- The 3′ terminus of the rRNA interacts directly with mRNA at initiation.
- Specific regions of 16S rRNA interact directly with the anticodon regions of tRNAs in both the A and P sites. Similarly, 23S rRNA interacts with the CCA terminus of peptidyl-tRNA in both the P site and A site.
- Subunit interaction involves interactions between 16S and 23S rRNAs (see *Section 8.14, Ribosomal RNA Pervades Both Ribosomal Subunits*).

The functions of rRNA have been investigated by two types of approach. Structural studies show that particular regions of rRNA are located in important sites of the ribosome and that chemical modifications of these bases impede particular ribosomal functions. In addition, mutations identify bases in rRNA that are required for particular ribosomal functions. **FIGURE 8.40** summarizes the sites in 16S rRNA that have been identified by these means.

The tRNA makes contacts with the 23S rRNA in both the P and A sites. The classic criterion for demonstrating the importance of the interaction is satisfied: a mutation in tRNA can be compensated by a mutation in rRNA, so there is a required role for rRNA in both the tRNA-binding sites. Indeed, we are moving toward describing the movements of tRNA between the A and P sites in terms of making and breaking contacts with rRNA.

What is the nature of the site on the 50S subunit that provides peptidyl transferase function? A long search for ribosomal proteins that might possess the catalytic activity was unsuccessful and led to the discovery that 23S rRNA can catalyze the formation of a peptide bond between peptidyl-tRNA and aminoacyl-tRNA. But since the rRNA has the basic catalytic activity, the role of the proteins must be indirect, serving to fold the rRNA properly or to present the substrates to it. Activity is abolished by mutations in domain V, which lies in the P site. The crystal structure of an archaeal 50S subunit shows that the peptidyl transferase site basically consists of 23S rRNA. There is no protein within 18 Å of the active site where the transfer reaction occurs between peptidyl-tRNA and aminoacyl-tRNA!

Peptide bond synthesis requires an attack by the amino group of one amino acid on the carboxyl group of another amino acid. Catalysis requires a basic residue to accept the hydrogen atom that is released from the amino group, as shown in **FIGURE 8.41**. If rRNA is the catalyst, it must provide this residue, but we do not know how this happens. The purine and pyrimidine bases are not basic at physiological pH. A

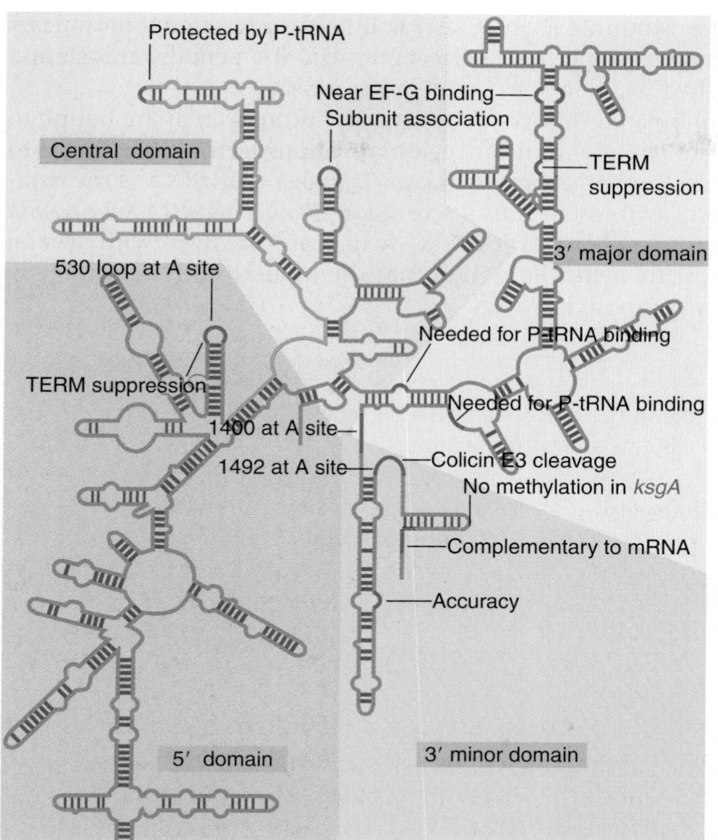

FIGURE 8.40 Some sites in 16S rRNA are protected from chemical probes when 50S subunits join 30S subunits or when aminoacyl-tRNA binds to the A site. Others are the sites of mutations that affect protein synthesis. TERM suppression sites may affect termination at some or several termination codons. The large colored blocks indicate the four domains of the rRNA.

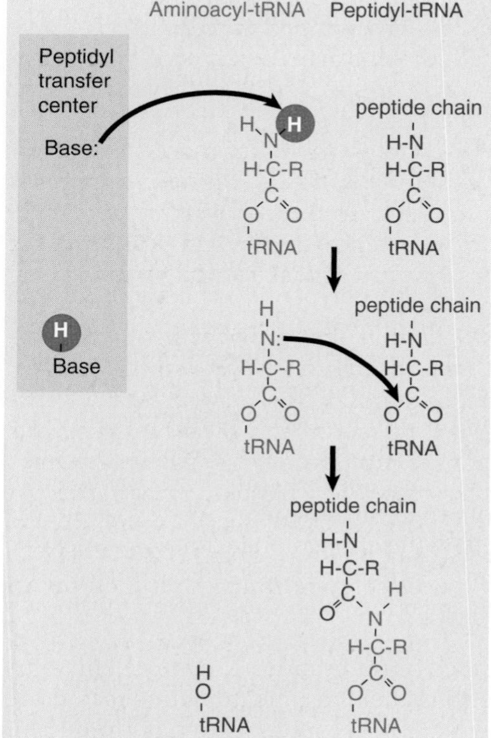

FIGURE 8.41 Peptide bond formation requires acid-base catalysis in which an H atom is transferred to a basic residue.

highly conserved base (at position 2451 in *E. coli*) had been implicated, but appears now neither to have the right properties nor to be crucial for peptidyl transferase activity.

The catalytic activity of isolated rRNA is quite low, and proteins that are bound to the 23S rRNA outside of the peptidyl transfer region are almost certainly required to enable the rRNA to form the proper structure *in vivo*. The idea that rRNA is the catalytic component is consistent with the results discussed in *Chapter 28, RNA Splicing and Processing,* which identify catalytic properties in RNA that are involved with several RNA processing reactions. It fits with the notion that the modern ribosome evolved from a prototype originally composed of RNA.

KEY CONCEPTS

- 16S rRNA plays an active role in the functions of the 30S subunit. It interacts directly with mRNA, with the 50S subunit, and with the anticodons of tRNAs in the P and A sites.
- Peptidyl transferase activity resides exclusively in the 23S rRNA.

CONCEPT AND REASONING CHECK

What are the functional roles of bacterial 16S and 23S rRNAs?

8.17 Summary

A codon in mRNA is recognized by an aminoacyl-tRNA, which has an anticodon complementary to the codon and carries the amino acid corresponding to the codon. A special initiator tRNA (fMet-$tRNA_f$ in prokaryotes or Met-$tRNA_i$ in eukaryotes) recognizes the AUG codon, which is used to start all coding sequences. In prokaryotes, GUG is also used. Only the termination (stop) codons, UAA, UAG, and UGA, are not recognized by aminoacyl-tRNAs.

Ribosomes are released from translation to enter a pool of free ribosomes that are in equilibrium with separate small and large subunits. Small subunits bind to mRNA and then are joined by large subunits to generate an intact ribosome that undertakes translation. Recognition of a prokaryotic initiation site involves binding of a sequence at the 3′ end of rRNA to the Shine–Dalgarno motif, which precedes the AUG (or GUG) codon in the mRNA. Recognition of a eukaryotic mRNA involves binding to the 5′ cap; the small subunit then migrates to the initiation site by scanning for AUG codons. When it recognizes an appropriate AUG codon (usually, but not always, the first it encounters), it is joined by a large subunit.

A ribosome can carry at least two aminoacyl-tRNAs simultaneously: its P site is occupied by a polypeptidyl-tRNA, which carries the polypeptide chain synthesized so far, whereas the A site is used for entry by an aminoacyl-tRNA carrying the next amino acid to be added to the chain. Bacterial ribosomes also have an E site, through which deacylated tRNA passes before it is released after being used in translation. The polypeptide chain in the P site is transferred to the aminoacyl-tRNA in the A site, creating a deacylated tRNA in the P site and a peptidyl-tRNA in the A site.

Following peptide bond synthesis, the bacterial ribosome translocates one codon along the mRNA, moving deacylated tRNA into the E site and peptidyl tRNA from the A site into the P site. Translocation is catalyzed by the elongation factor EF-G and, like several other stages of ribosome function, requires hydrolysis of GTP. During translocation, the ribosome passes through a hybrid stage in which the 50S subunit moves relative to the 30S subunit.

Additional factors are required at each stage of translation. They are defined by their cyclic association with, and dissociation from, the ribosome. Initiation factors are involved in prokaryotic initiation. IF-3 is needed for 30S subunits to bind to mRNA, and also is responsible for maintaining the 30S subunit in a free form. IF-2 is needed for fMet-$tRNA_f$ to bind to the 30S subunit and is responsible for excluding other aminoacyl-tRNAs from the initiation reaction. GTP is hydrolyzed after the initiator tRNA has been bound to the initiation complex. The initiation factors must be released in order to allow a large subunit to join the initiation complex.

Eukaryotic initiation involves a greater number of protein factors. Some of them are involved in the initial binding of the 40S subunit to the capped 5′ end of the mRNA, at which point the initiator tRNA is bound by another group of factors. After this initial binding, the small subunit scans the mRNA until it recognizes the correct AUG codon. At this point, initiation factors are released and the 60S subunit joins the complex.

Prokaryotic elongation factors are involved in elongation. EF-Tu binds aminoacyl-tRNA to the 70S ribosome. GTP is hydrolyzed when EF-Tu is released, and EF-Ts is required to regenerate the active form of EF-Tu. EF-G is required for translocation. Binding of the EF-Tu and EF-G factors to ribosomes is mutually exclusive, which ensures that each step must be completed before the next can be started.

The level of translation itself provides an important coordinating signal. Deficiency in aminoacyl-tRNAs causes an idling reaction on the ribosome, which leads to the synthesis of the unusual nucleotide ppGpp. This is an effector that inhibits initiation of transcription at certain promoters; it also has a general effect in inhibiting elongation on all templates.

Termination occurs at any one of the three special codons, UAA, UAG, and UGA. Class 1 release factors that specifically recognize the termination codons activate the ribosome to hydrolyze the peptidyl-tRNA. A class 2 RF is required to release the class 1 RF from the ribosome. The GTP-binding factors IF-2, EF-Tu, EF-G, and RF3 all have similar structures, with the latter two mimicking the RNA-protein structure of the first two when they are bound to tRNA. They all bind to the same ribosomal site, the G-factor binding site.

Ribosomes are ribonucleoprotein particles in which a majority of the mass is provided by rRNA. The shapes of all ribosomes are generally similar, but only those of bacteria (70S) have been characterized in detail. The small (30S) subunit has a squashed shape, with a "body" containing about two-thirds of the mass divided from the "head" by a cleft. The large (50S) subunit is more spherical, with a prominent "stalk" on the right and a "central protuberance." Approximate locations of all proteins in the small subunit are known.

Each subunit contains a single major rRNA, 16S and 23S in prokaryotes. Both major rRNAs have extensive base pairing, mostly in the form of short, imperfectly paired duplex stems with single-stranded loops. Conserved features in the rRNA can be identified by comparing sequences and the secondary structures that can be drawn for rRNA of a variety of organisms. The 16S rRNA has four distinct domains; the 23S rRNA has six distinct domains. Eukaryotic rRNAs have additional domains.

The crystal structure shows that the 30S subunit has an asymmetrical distribution of RNA and protein. RNA is concentrated at the interface with the 50S subunit. The 50S subunit has a surface of protein, with long rods of double-stranded RNA crisscrossing the structure. 30S-to-50S joining involves contacts between 16S rRNA and 23S rRNA.

Each subunit has several active centers, which are concentrated in the translational domain of the ribosome where polypeptides are synthesized. Polypeptides leave the ribosome through the exit domain, which can associate with a membrane. The major active sites are the P and A sites, the E site, the EF-Tu and EF-G binding sites, peptidyl transferase, and the mRNA-binding site. Ribosome conformation may change at stages during translation; differences in the accessibility of particular regions of the major rRNAs have been detected.

The tRNAs in the A and P sites are parallel to one another. The anticodon loops are bound to mRNA in a groove on the 30S subunit. The rest of each tRNA is bound to the 50S subunit. A conformational shift of tRNA within the A site is required to bring its aminoacyl end into juxtaposition with the end of the peptidyl-tRNA in the P site.

The peptidyl transferase site that links the P- and A-binding sites is a domain of the 23S rRNA, which has the peptidyl transferase catalytic activity, although proteins are probably needed to acquire the right structure.

An active role for the rRNAs in translation is indicated by mutations that affect ribosomal function, interactions with mRNA or tRNA that can be detected by chemical crosslinking, and the requirement to maintain individual base pairing interactions with the tRNA or mRNA. The 3′ terminal region of the rRNA base pairs with mRNA at initiation. Internal regions make individual contacts with the tRNAs in both the P and A sites. Ribosomal RNA is the target for some antibiotics or other agents that inhibit translation.

CHAPTER QUESTIONS

1. How many binding sites for tRNAs are present in each bacterial ribosome?
   A. 1
   B. 2
   C. 3
   D. 4
2. In translating ribosomes, new amino acids are added to the growing peptide chain from the:
   A. P site to the A site.
   B. A site to the P site.
   C. P site to the E site.
   D. A site to the E site.
3. Which of the following is not a termination codon for translation?
   A. UGG
   B. UAG
   C. UGA
   D. UAA
4. What step in transcription/translation generally has the highest error rate?
   A. insertion of the incorrect aminoacyl tRNA into the ribosome
   B. charging of the wrong amino acid onto a tRNA
   C. insertion of the wrong amino acid into a growing protein
   D. insertion of the wrong base in mRNA during transcription
5. What is the usual start codon for translation in both prokaryotes and eukaryotes?
   A. UAG
   B. UGA
   C. AUG
   D. AGU
6. The Shine-Dalgarno sequence occurs in what location in a bacterial mRNA molecule?
   A. about 35 bases upstream of the AUG initiation codon
   B. about 10 bases upstream of the AUG initiation codon
   C. about 15 bases downstream of the AUG initiation codon
   D. about 10 bases downstream of the termination codon
7. What is the basis of interaction between the ribosome and an mRNA molecule in bacteria?
   A. specific protein-RNA interactions between ribosomal proteins and the mRNA
   B. specific base pairing between the 3′ end of the 16S rRNA in the 30S subunit and a conserved sequence in the mRNA

C. specific base pairing between the 5′ end of the 16S rRNA in the 30S subunit and a conserved sequence in the mRNA

D. specific interactions between ribosomal proteins and a binding protein that associates with the 5′ end of the mRNA

8. What is the modification to the initiator methionine tRNA in bacteria?
   A. methylation
   B. phosphorylation
   C. acetylation
   D. formylation
9. In normally growing bacteria with sufficient nutrients, termination of translation and dissociation of the ribosome from the mRNA occurs when:
   A. tRNA molecules become limiting.
   B. a stop codon is encountered by the ribosome.
   C. release factors enter the A site of the ribosome.
   D. the mRNA is degraded.
10. The 23S rRNA of the large subunit of bacterial ribosomes interacts with:
   A. large subunit ribosomal proteins only.
   B. the 5′ untranslated region of the mRNA.
   C. the 3′ untranslated region of the mRNA.
   D. the 3′ CCA terminus of peptidyl tRNA in the P and A sites.

KEY TERMS

A site
alarmone
amber codon
context
deacylated tRNA
EF-Tu
elongation
elongation factors
idling reaction
IF-1
IF-2
IF-3
initiation
initiation factors (IFs)
IRES (internal ribosome entry site)
N-formyl-methionyl-tRNA
ochre codon
opal codon
P site
peptidyl transferase
peptidyl-tRNA
ppGpp
pppGpp
relaxed mutants
release factor (RF)
RF1
RF2
RF3
ribosome-binding site
Shine-Dalgarno sequence
stop codon
stringent factor
stringent response
termination
translocation
$tRNA_f^{Met}$
$tRNA_m^{Met}$

FURTHER READING

Hellen, C. U. and Sarnow, P. (2001). Internal ribosome entry sites in eukaryotic mRNA molecules. *Genes Dev.* 15, 1593–1612.

A review of the known mechanisms of eukaryotic translation initiation.

Moore, P. B. and Steitz, T. A. (2003). The structural basis of large ribosomal subunit function. *Annu. Rev. Biochem.* 72, 813–850.

A review of the process that led to the detailed crystal structure of the bacterial large ribosomal subunit, the structure itself, and the functions associated with that structure.

Noller, H. F. (2005). RNA structure: reading the ribosome. *Science* 309, 1508–1514.

A review of the roles of ribosomal RNA tertiary structures in tRNA selection, peptidyl transferase activity, and other ribosomal processes.

Ramakrishnan, V. (2002). Ribosome structure and the mechanism of translation. *Cell* 108, 557–572.

A review of the known and as-yet-unknown (at the time of publication) details of bacterial protein translation, based on the recently published ribosome crystal structures.

Schuwirth, B. S., Borovinskaya, M. A., Hau, C. W., Zhang, W., Vila-Sanjurjo, A., Holton, J. M., and Doudna Cate, J. H. (2005). Structures of the bacterial ribosome at 3.5 Å resolution. *Science* 310, 827–834.

A detailed examination of the structure of the complete functional bacterial ribosome.

Selmer, M., Dunham, C. M., Murphy, IV, F. V., Weixlbaumer, A., Petry, S., Kelley, A. C., Weir, J. R., and V. Ramakrishnan, V. (2006). Structure of the 70*S* ribosome complexed with mRNA and tRNA. *Science* 313, 1935–1942.

The bacterial ribosomal structure at 2.8 Å resolution.

Stark, H., Rodnina, M. V., Wieden, H. J., van Heel, M., and Wintermeyer, W. (2000). Large-scale movement of elongation factor G and extensive conformational change of the ribosome during translocation. *Cell* 100, 301–309.

The structure of the EF-G-ribosome complex both before and after ribosome translocation, and a proposal for the mechanism of translocation.

Yonath, A. (2005). Antibiotics targeting ribosomes: resistance, selectivity, synergism and cellular regulation. *Annu. Rev. Biochem.* 74, 649–679.

A review of the structural features of the bacterial ribosome targeted by various antibiotics and the molecular bases for resistances to them.

9

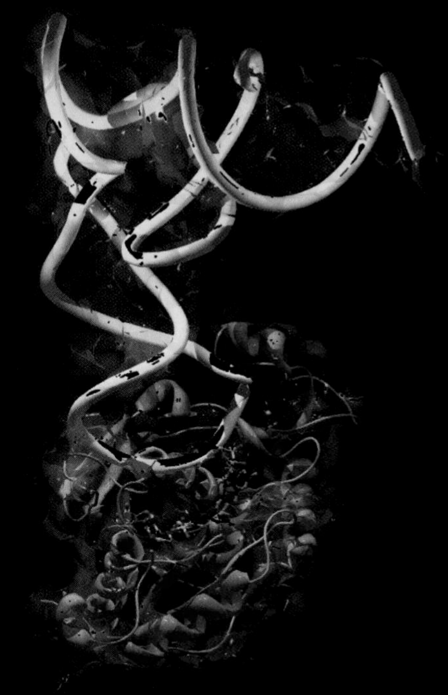

An aminoacyl-tRNA synthetase is shown complexed to a tRNA. Reproduced from Blaise, M., et al., *Nucleic Acids Res.* 32 (2004): 2768–2775. Used with permission of Oxford University Press. Photo courtesy of Pr. Daniel Kern, Université Louis Pasteur, Institut de Biologie Moléculaire et Cellulaire, UPR 9002 du SNRS, Strasbourg.

Using the Genetic Code

CHAPTER OUTLINE

9.1 Introduction

The sequence of a coding strand of DNA, read in the direction from 5′ to 3′, consists of nucleotide triplets (codons) corresponding to the amino acid sequence of a polypeptide read from N-terminus to C-terminus. Sequencing of DNA and proteins makes it possible to compare corresponding nucleotide and amino acid sequences directly. There are sixty-four codons (each of four possible nucleotides can occupy each of the three positions of the codon, making $4^3 = 64$ possible trinucleotide sequences).

The determination of the genetic code originally showed that genetic information is stored in the form of nucleotide triplets, but did not reveal how each codon specifies its corresponding amino acid. Before the advent of DNA sequencing, codon assignments were deduced on the basis of two types of *in vitro* studies. A system involving the translation of synthetic polynucleotides was introduced in 1961, when Nirenberg showed that polyuridylic acid [poly(U)] directs the assembly of phenylalanine into polyphenylalanine. This result means that UUU must be a codon for phenylalanine. A second system—in which a trinucleotide was used to mimic a codon, thus causing the corresponding aminoacyl-tRNA to bind to a ribosome—was later used. By identifying the amino acid component of the aminoacyl-tRNA, the meaning of the codon could be found. The two techniques together assigned meaning to all of the codons that represent amino acids.

Sixty-one of the sixty-four codons represent amino acids. The other three cause termination of translation. The assignment of amino acids to codons is not random, but shows relationships in which the third (3′) base has less effect on codon meaning. In addition, related amino acids are often represented by related codons. The meaning of a codon that represents an amino acid is determined by the tRNA that corresponds to it; the meaning of the termination codons is determined directly by protein factors.

9.2 Related Codons Represent Related Amino Acids

- **synonym codons** Codons that have the same meaning (specifying the same amino acid, or specifying termination of translation) in the genetic code.
- **third-base degeneracy** The lesser effect on codon meaning of the nucleotide present in the third (3′) codon position.

The code is summarized in **FIGURE 9.1**. There are more codons than there are amino acids, and as a result almost all amino acids are represented by more than one codon. The only exceptions are methionine and tryptophan. Codons that have the same meaning are called **synonyms**.

Codons representing the same or similar amino acids tend to be similar in sequence. Often the base in the third position of a codon is not significant, because the four codons differing only in the third base represent the same amino acid. Sometimes a distinction is made only between a purine versus a pyrimidine in this position. The reduced specificity at the last position is known as **third-base degeneracy**.

First base	Second base: U	Second base: C	Second base: A	Second base: G
U	UUU, UUC – Phe UUA, UUG – Leu	UCU, UCC, UCA, UCG – Ser	UAU, UAC – Tyr UAA, UAG – STOP	UGU, UGC – Cys UGA STOP UGG Trp
C	CUU, CUC, CUA, CUG – Leu	CCU, CCC, CCA, CCG – Pro	CAU, CAC – His CAA, CAG – Gln	CGU, CGC, CGA, CGG – Arg
A	AUU, AUC, AUA – Ile AUG Met	ACU, ACC, ACA, ACG – Thr	AAU, AAC – Asn AAA, AAG – Lys	AGU, AGC – Ser AGA, AGG – Arg
G	GUU, GUC, GUA, GUG – Val	GCU, GCC, GCA, GCG – Ala	GAU, GAC – Asp GAA, GAG – Glu	GGU, GGC, GGA, GGG – Gly

FIGURE 9.1 All the triplet codons have meaning: Sixty-one represent amino acids, and three cause termination (STOP).

The tendency for similar amino acids to be represented by related codons minimizes the effects of mutations. It increases the probability that a single random base change will result in no amino acid substitution or in one involving amino acids of similar character. For example, a mutation of CUC to CUG does not change the resulting polypeptide, because both codons represent leucine. A mutation of CUU to AUU results in replacement of leucine with isoleucine, a closely related amino acid.

FIGURE 9.2 plots the number of codons representing each amino acid against the frequency with which the amino acid is used in proteins (in *E. coli*). There is only a slight tendency for amino acids that are more common to be represented by more codons, and therefore it does not seem that the genetic code has been optimized with regard to the utilization of amino acids.

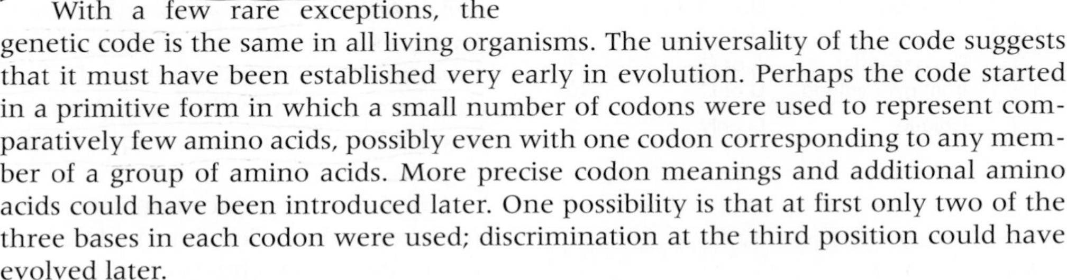

FIGURE 9.2 The number of codons for each amino acid does not correlate closely with its frequency of use in proteins.

The three codons (UAA, UAG, and UGA) that do not represent amino acids are used specifically to terminate translation. One of these **stop codons** marks the end of every open reading frame.

▸ **stop codon** One of three triplets (UAG, UAA, or UGA) that causes translation to terminate. They are also known historically as *nonsense codons*. The UAA codon is called *ochre* and the UAG codon is called *amber*, after the names of the nonsense mutations by which they were originally identified.

With a few rare exceptions, the genetic code is the same in all living organisms. The universality of the code suggests that it must have been established very early in evolution. Perhaps the code started in a primitive form in which a small number of codons were used to represent comparatively few amino acids, possibly even with one codon corresponding to any member of a group of amino acids. More precise codon meanings and additional amino acids could have been introduced later. One possibility is that at first only two of the three bases in each codon were used; discrimination at the third position could have evolved later.

Evolution of the code could have become "frozen" at a point at which the system had become so complex that any changes in codon meaning would disrupt functional proteins by substituting unacceptable amino acids. Its universality implies that this must have happened at such an early stage that all living organisms are descended from a single pool of primitive cells in which this occurred.

KEY CONCEPTS

- Sixty-one of the sixty-four possible triplets code for twenty amino acids.
- Three codons do not represent amino acids and cause termination of translation.
- The genetic code was frozen at an early stage of evolution and is universal.
- Most amino acids are represented by more than one codon.
- The multiple codons for an amino acid are usually related.
- Related amino acids often have related codons, minimizing the effects of mutation.

CONCEPT AND REASONING CHECK

How does degeneracy of the genetic code minimize the effects of mutations?

9.3 Codon–Anticodon Recognition Involves Wobbling

The function of tRNA in translation is fulfilled when it recognizes the mRNA codon in the ribosomal A site. The interaction between anticodon and codon takes place by base pairing, but under rules that extend pairing beyond the usual G-C and A-U partnerships. We can deduce the rules governing the interaction from the sequences of the anticodons that correspond to particular codons.

The genetic code itself yields some important clues about the process of codon recognition. The pattern of third-base degeneracy is clear in **FIGURE 9.3**, which shows that in almost all cases either the third base is irrelevant or a distinction is made only between purines and pyrimidines.

FIGURE 9.3 Third bases have the least influence on codon meanings. Boxes indicate groups of codons within which third-base degeneracy ensures that the meaning is the same.

UUU	UCU	UAU	UGU
UUC	UCC	UAC	UGC
UUA	UCA	UAA	UGA
UUG	UCG	UAG	UGG
CUU	CCU	CAU	CGU
CUC	CCC	CAC	CGC
CUA	CCA	CAA	CGA
CUG	CCG	CAG	CGG
AUU	ACU	AAU	AGU
AUC	ACC	AAC	AGC
AUA	ACA	AAA	AGA
AUG	ACG	AAG	AGG
GUU	GCU	GAU	GGU
GUC	GCC	GAC	GGC
GUA	GCA	GAA	GGA
GUG	GCG	GAG	GGG

Third-base relationship		Third bases with same meaning	Codon Number
Third base irrelevant		U, C, A, G	32
		U, C, A	3
Purines differ from pyrimidines		A or G	14
		U or C	10
Unique		G only	2

FIGURE 9.4 Wobble in base pairing allows G-U pairs to form between the third base of the codon and the first base of the anticodon.

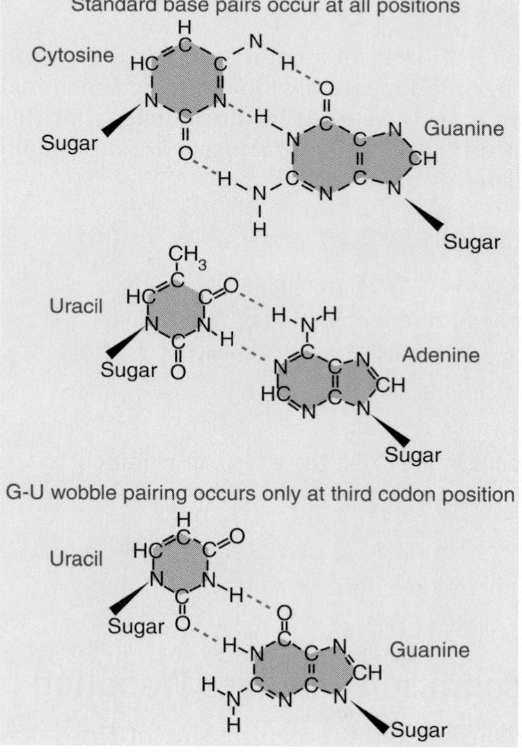

There are eight codon families in which all four codons sharing the same first two bases have the same meaning, so that the third base has no role at all in specifying the amino acid. There are seven codon pairs in which the meaning is the same regardless of which pyrimidine is present at the third position, and there are five codon pairs in which either purine may be present without changing the amino acid that is coded.

There are only three cases in which a unique meaning is conferred by the presence of a particular base at the third position: AUG (for methionine), UGG (for tryptophan), and UGA (termination). So C and U never have a unique meaning in the third position, and A never signifies a unique amino acid.

Because the anticodon is complementary to the codon, it is the first base in the anticodon sequence written conventionally in the direction from 5′ to 3′ that pairs with the third base in the codon sequence written by the same convention. So the combination

Codon 5′ A C G 3′
Anticodon 3′ U G C 5′

is usually written as codon ACG/anticodon CGU, where the anticodon sequence must be read backward for complementarity with the codon. To avoid confusion, anticodon sequences will be written as 3′-NNN-5′.

Does each triplet codon require its own tRNA with a precisely complementary anticodon? Or can a single tRNA respond to both members of a codon pair and to all (or at least some) of the four members of a codon family?

Often one tRNA can recognize more than one codon. This means that the base in the first position of the anticodon must be able to partner alternative bases in the corresponding third position of the codon. Base pairing at this position cannot be limited to the usual G-C and A-U partnerships.

wobble hypothesis The ability of a tRNA to recognize more than one codon by unusual (non-G-C, non-A-T) pairing with the third base of a codon.

The rules governing the recognition patterns are summarized in the **wobble hypothesis**, which states that the pairing between codon and anticodon at the first two codon positions always follows the usual rules, but that exceptional "wobbles" occur at the third position. Wobbling occurs because the conformation of the tRNA anticodon loop permits flexibility at the first base of the anticodon. **FIGURE 9.4** shows that G-U pairs can form in addition to the usual pairs.

This single addition creates a pattern of base pairing in which A can no longer have a unique meaning in the codon (because the U that recognizes it must also recognize G). Similarly, C also no longer has a unique meaning (because the G that recognizes it also must recognize U). **FIGURE 9.5** summarizes the pattern of recognition.

Base in first position of anticodon	Base(s) recognized in third position of codon
U	A or G
C	G only
A	U only
G	C or U

FIGURE 9.5 Codon–anticodon pairing involves wobbling at the third position.

Wobbling can also result from flexible pairing with modified bases in the tRNA anticodon, such as inosine (I) (see *Section 9.5, Modified Bases Affect Anticodon–Codon Pairing*).

KEY CONCEPTS

- Multiple codons that represent the same amino acid most often differ at the third base position.
- The wobble in pairing between the first base of the anticodon and the third base of the codon results from the structure of the anticodon loop.

CONCEPT AND REASONING CHECK

What is the wobble hypothesis, and how is it consistent with the third-base degeneracy of the genetic code?

▸ **modification** All changes made to the nucleotides of DNA or RNA after their initial incorporation into the polynucleotide chain.

9.4 tRNA Contains Modified Bases

Transfer RNA is unique among nucleic acids in its content of bases that are produced by **modification** of one of the four usual bases after it has been incorporated into the polyribonucleotide chain.

All classes of RNA display some degree of modification, but in all cases except tRNA this is confined to rather simple events, such as the addition of methyl groups. In tRNA, there is a vast range of modifications, ranging from simple methylation to wholesale restructuring of the purine ring. Modifications occur in all parts of the tRNA molecule. There are >50 different types of modified bases in tRNA.

FIGURE 9.6 shows some of the more common modified bases. Modifications of pyrimidines (C and U) are less complex than those of purines (A and G). In addition to the modifications of the bases themselves, methylation at the 2′-O position of the ribose ring also occurs.

Modifications of uridine are common. Methylation at position 5 creates ribothymidine (T). The base is the same commonly found in DNA, but here it is attached to ribose rather than deoxyribose. In RNA, thymine constitutes an unusual base that originates by modification of U.

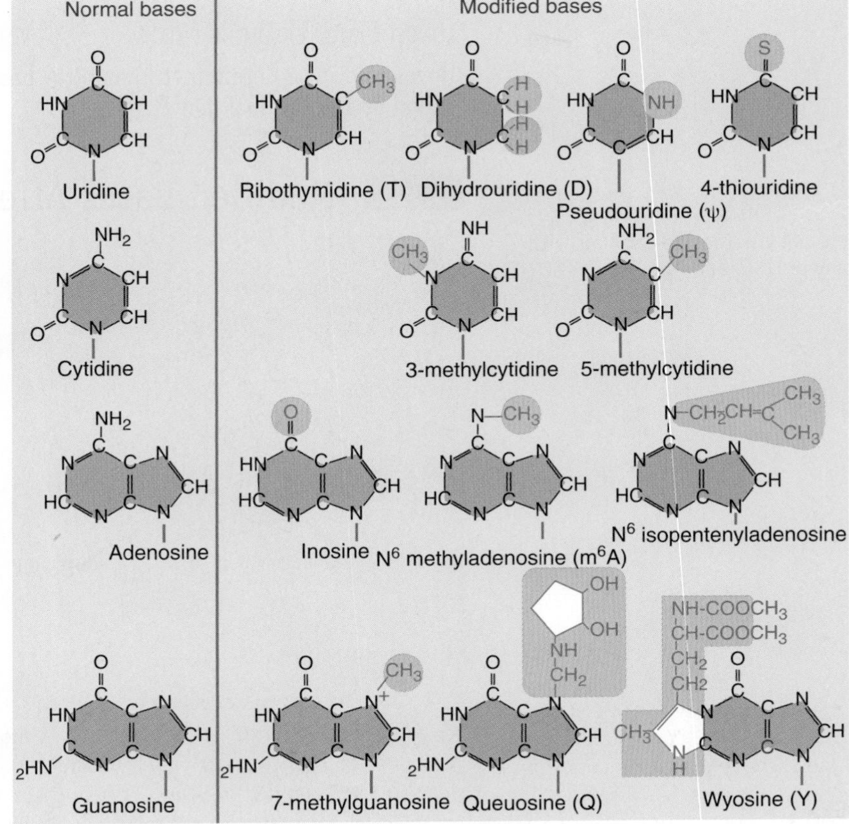

FIGURE 9.6 All of the four bases in tRNA can be modified.

Dihydrouridine (D) is generated by the saturation of a double bond, which changes the ring structure. Pseudouridine (ψ) interchanges the positions of N and C atoms (see Figure 28.31). In 4-thiouridine, sulfur is substituted for oxygen.

The nucleoside inosine (I) is found normally in the cell as an intermediate in the purine biosynthetic pathway. It is not, however, incorporated directly into RNA. Instead, its existence depends on modification of A to create I. Other modifications of A include the addition of complex groups.

Some modifications are constant features of all tRNA molecules—for example, the D residues that give rise to the name of the D arm and the ψ found in the TψC sequence. On the 3′ side of the anticodon there is always a modified purine, although the modification varies widely.

Other modifications are specific for particular tRNAs or groups of tRNAs. For example, wyosine bases are characteristic of $tRNA^{Phe}$ in bacteria, yeast, and mammals. There are also some species-specific patterns.

The many tRNA-modifying enzymes (~60 in yeast) vary greatly in specificity. In some cases, a single enzyme acts to make a particular modification at a single position. In other cases, an enzyme can modify bases at several different target positions. Some enzymes undertake single reactions with individual tRNAs; others have a range of substrate molecules. The features recognized by the tRNA-modifying enzymes are unknown, but probably involve recognition of structural features surrounding the site of modification. Some modifications require the successive actions of more than one enzyme.

KEY CONCEPTS

- tRNAs contain >50 modified bases.
- Modification usually involves direct alteration of the primary bases in tRNA, but there are some exceptions in which a base is removed and replaced by another base.

CONCEPT AND REASONING CHECK

How could it be determined that tRNA bases are modified posttranscriptionally and not coded by the original gene?

9.5 Modified Bases Affect Anticodon–Codon Pairing

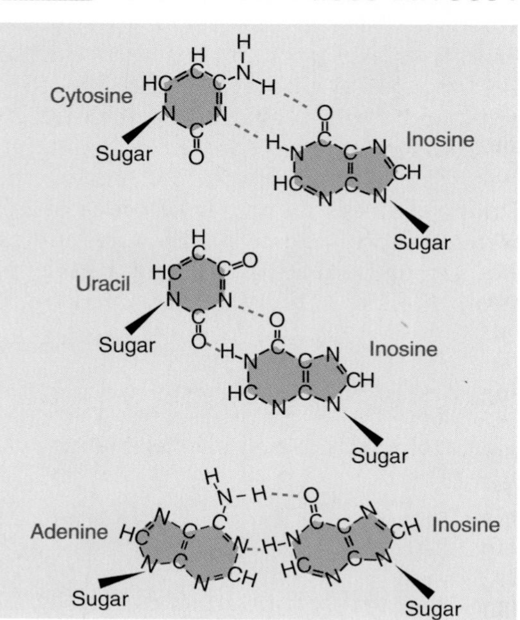

FIGURE 9.7 Inosine can pair with any of U, C, and A.

The most direct effect of modification is seen in the anticodon, where the change of sequence influences the ability to pair with the codon and determine its meaning. Modifications elsewhere in the vicinity of the anticodon also influence its pairing.

When bases in the anticodon are modified, further pairing patterns—in addition to those predicted by the regular and wobble pairing involving A, C, U, and G—become possible. **FIGURE 9.7** shows the use of inosine (I), which is often present at the first position of the anticodon. Inosine can pair with any one of the three bases U, C, and A.

This flexible pairing ability is especially important in the isoleucine codons, where AUA codes for isoleucine, whereas AUG codes for methionine. It is not possible with the usual bases to recognize A alone in the third position, and as a result any tRNA with U starting its anticodon would have to recognize AUG as well as AUA. Thus AUA must be read together with AUU and AUC, a problem that is solved by the existence of tRNA with I in the anticodon's first position.

Some modifications allow preferential readings of some codons with respect to others. Anticodons with uridine-5-oxyacetic acid and 5-methoxyuridine in the first position recognize A and G efficiently as third bases of the codon, but recognize U less efficiently. Another case in which multiple pairings can occur—but with some pairings preferred to others—is provided by the series of queuosine and its derivatives. These modified G bases continue to recognize both C and U, but pair with U more readily.

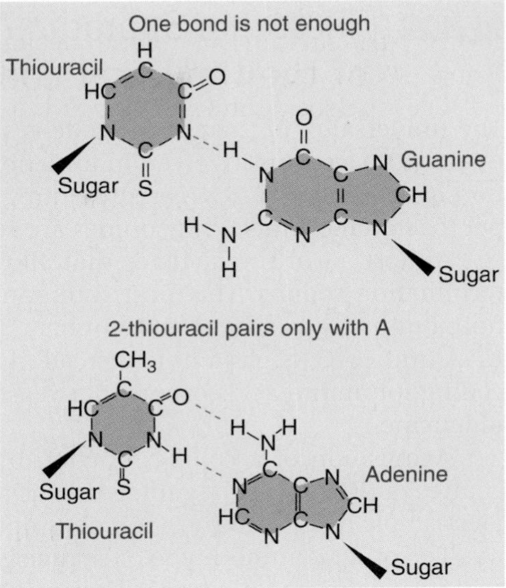

FIGURE 9.8 Modification to 2-thiouridine restricts pairing to A alone because only one H-bond can form with G.

A restriction not allowed by the usual rules can be achieved by the employment of 2-thiouridine in the anticodon. **FIGURE 9.8** shows that its modification allows the base to continue to pair with A, but prevents it from participating in wobble pairing with G.

These and other pairing relationships make the general point that there are multiple ways to construct a set of tRNAs able to recognize all the sixty-one codons representing amino acids. No particular pattern predominates in any given organism, although the absence of a certain pathway for modification can prevent the use of some recognition patterns. Thus a particular codon family is read by tRNAs with different anticodons in different organisms.

Often the tRNAs will have overlapping responses, so that a particular codon is read by more than one tRNA. In such cases there may be differences in the efficiencies of the alternative recognition reactions. (As a general rule, codons that are commonly used tend to be more efficiently read.) In addition to the construction of a set of tRNAs that are able to recognize all the codons, there may be multiple tRNAs that respond to the same codons.

The predictions of wobble pairing accord very well with the observed abilities of almost all tRNAs. There are, however, exceptions in which the codons recognized by a tRNA differ from those predicted by the wobble rules. Such effects probably result from the influence of neighboring bases and/or the conformation of the anticodon loop in the overall tertiary structure of the tRNA. Indeed, the importance of the structure of the anticodon loop is inherent in the idea of the wobble hypothesis itself. Further support for the influence of the surrounding structure is provided by the isolation of occasional mutants in which a change in a base in some other region of the molecule alters the ability of the anticodon to recognize codons.

KEY CONCEPT

- Modifications in the anticodon affect the pattern of wobble pairing and therefore are important in determining tRNA specificity.

CONCEPT AND REASONING CHECK

Why is it that some species have as few as forty tRNA species and others have as many as sixty?

9.6 There Are Sporadic Alterations of the Universal Code

The universality of the genetic code is striking, but some exceptions exist. They tend to affect the codons involved in initiation or termination and result from the production (or absence) of tRNAs representing certain codons. The changes found in principal (bacterial or nuclear) genomes are summarized in **FIGURE 9.9**.

Almost all of the changes that allow a codon to represent an amino acid affect termination codons. The most common change is that UGA is not used for termination, but instead codes for tryptophan (Trp). Some ciliates (unicellular protozoa) read UAA and UAG as glutamine instead of termination signals. The only substitution in coding for amino acids occurs in a yeast (*Candida*), where CUG means serine instead of leucine.

Acquisition of a coding function by a termination codon requires two types of change: a tRNA must be mutated so as to recognize the codon, and the class 1 release factor must be altered so that it does not terminate at this codon.

The other common type of change is loss of the tRNA that responds to a codon, so that the codon no longer specifies any amino acid. What happens at such a codon will depend on whether a release factor evolves to recognize it.

All of these changes appear to have occurred independently in the evolution of specific lineages. They may be concentrated on termination codons because these changes do not involve substitution of one amino acid for another. Once a specific genetic code was established early in the evolutionary history of cellular organisms, or even protocells, any general change in the meaning of a codon would cause a substitution in all the polypeptides that contain that amino acid. It seems likely that the change would be deleterious in at least some of these polypeptides, with the result that it would be strongly selected against. The divergent uses of the termination codons could represent their "capture" for normal coding purposes. If some termination codons were used only rarely, they could be recruited for coding purposes by changes that allowed tRNAs to recognize them.

Exceptions to the universal genetic code also occur in the mitochondria of several species. **FIGURE 9.10** presents a phylogeny for the changes. It suggests that there was a universal code that was changed at various points in mitochondrial evolution. The earliest change was the employment of UGA to code for tryptophan, which is common to all (nonplant) mitochondria.

Why have changes been able to evolve in the mitochondrial code? The mitochondrion synthesizes only a small number of proteins (~10), and as a result the

UUU Phe UUC Phe UUA Leu UUG Leu	UCU Ser UCC Ser UCA Ser UCG Ser	UAU Tyr UAC Tyr UAA STOP→Gln UAG STOP→Gln	UGU Cys UGC Cys UGA STOP→Trp, Cys, Sel UGG Trp
CUU Leu CUC Leu CUA Leu CUG Leu→Ser	CCU Pro CCC Pro CCA Pro CCG Pro	CAU His CAC His CAA Gln CAG Gln	CGU Arg CGC Arg CGA Arg CGG Arg→NONE
AUU Ile AUC Ile AUA Ile→NONE AUG Met	ACU Thr ACC Thr ACA Thr ACG Thr	AAU Asn AAC Asn AAA Lys AAG Lys	AGU Ser AGC Ser AGA Arg→NONE AGG Arg
GUU Val GUC Val GUA Val GUG Val	GCU Ala GCC Ala GCA Ala GCG Ala	GAU Asp GAC Asp GAA Glu GAG Glu	GGU Gly GGC Gly GGA Gly GGG Gly

FIGURE 9.9 Changes in the genetic code in bacterial or eukaryotic nuclear genomes usually assign amino acids to stop codons or change a codon so that it no longer specifies an amino acid. A change in meaning from one amino acid to another is unusual.

problem of disruption by changes in meaning is much less severe. It is likely that the codons that are altered were not used extensively in locations where amino acid substitutions would have been deleterious. The variety of changes found in mitochondria of different species suggests that they have evolved separately, rather than by common descent, from an ancestral mitochondrial code.

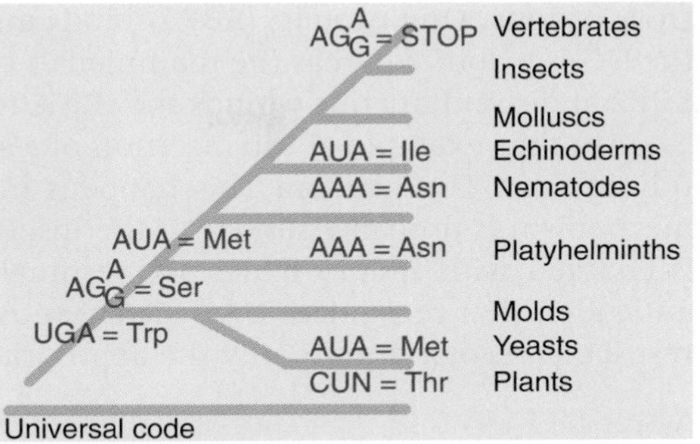

FIGURE 9.10 Changes in the genetic code in mitochondria can be traced in phylogeny. The minimum number of independent changes is generated by supposing that the AUA=Met and the AAA=Asn changes each occurred independently twice, and that the early AUA=Met change was reversed in echinoderms.

KEY CONCEPTS

- Changes in the universal genetic code have occurred in some species.
- These changes are more common in mitochondrial genomes, where a phylogenetic tree can be constructed for the changes.
- In nuclear genomes, the changes are sporadic and usually affect only termination codons.

CONCEPT AND REASONING CHECK

What is the evidence that the few variations in the universal genetic code are derived changes and were not present in the common ancestor of current species?

9.7 Novel Amino Acids Can Be Inserted at Certain Stop Codons

Specific changes in reading the code occur in individual genes. The specificity of such changes implies that the reading of the particular codon must be influenced by the surrounding bases.

A striking example is the incorporation of the modified amino acid seleno-cysteine at certain UGA codons within the genes that code for selenoproteins in both prokaryotes and eukaryotes. Usually these proteins catalyze oxidation-reduction reactions, and contain a single seleno-cysteine residue, which forms part of the active site. The most is known about the use of the UGA codons in three *E. coli* genes coding for formate dehydrogenase isozymes. The internal UGA codon is read by a seleno-Cys-tRNA.

The system in *E. coli* was identified by mutations in four genes that create a deficiency in selenoprotein synthesis. *selC* codes for tRNA (with the anticodon 3′-ACU-5′) that is charged with serine. *selA* and *selD* are required to modify the serine to seleno-cysteine. SelB is an alternative elongation factor. It is a guanine nucleotide-binding protein that acts as a specific translation factor for entry of seleno-Cys-tRNA into the A site; it thus provides (for this single tRNA) a replacement for factor EF-Tu. The sequence of SelB is related to both EF-Tu and IF-2.

Why is seleno-Cys-tRNA inserted only at certain UGA codons? These codons are followed by a stem-loop structure in the mRNA. **FIGURE 9.11** shows that the stem of this structure is recognized by an additional domain in SelB (one that is not present in EF-Tu or IF-2). A similar mechanism interprets some UGA codons in mammalian cells, except that two proteins are required to identify the appropriate

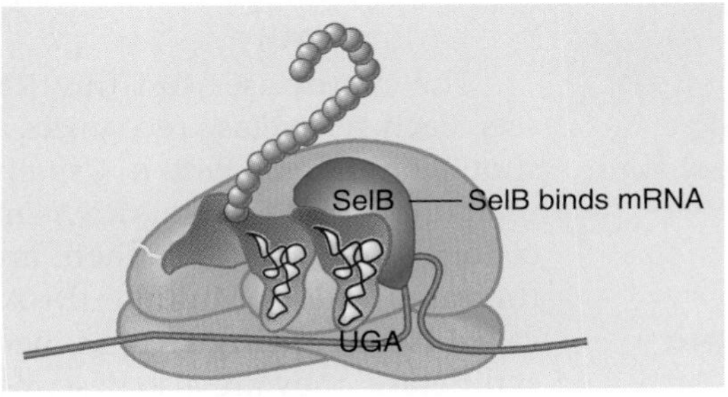

FIGURE 9.11 SelB is an elongation factor that specifically binds Seleno-Cys-tRNA to a UGA codon that is followed by a stem-loop structure in mRNA.

UGA codons. One protein (SBP2) binds a stem-loop structure far downstream from the UGA codon, whereas the mammalian counterpart of SelB (called SECIS) binds to SBP2 and simultaneously binds the tRNA to the UGA codon.

Another example of the insertion of a special amino acid is the placement of pyrrolysine at a UAG codon. This happens in both an archaean and a bacterium. The mechanism is probably similar to the insertion of seleno-cysteine. An unusual tRNA is charged with lysine, which is presumably then modified. The tRNA has a CUA anticodon that recognizes UAG. There must be other components of the system that restrict its recognition to only the appropriate UAG codons.

KEY CONCEPTS

- Changes in the reading of specific codons can occur in individual genes.
- The insertion of seleno-Cys-tRNA at certain UGA codons requires several proteins to modify the Cys-tRNA and insert it into the ribosome.
- Pyrrolysine can be inserted at certain UAG codons.

CONCEPT AND REASONING CHECK

Why is it assumed that changes in the reading of specific codons require a specific sequence context?

- **isoaccepting tRNAs** *See* **cognate tRNAs**.
- **cognate tRNAs** tRNAs recognized by a particular aminoacyl-tRNA synthetase. All are charged with the same amino acid.

9.8 tRNAs Are Charged with Amino Acids by Synthetases

It is necessary for tRNAs to have certain characteristics in common, but yet be distinguished by other characteristics. The crucial feature that confers this capacity is the ability of tRNA to fold into a specific tertiary structure. Changes in the details of this structure, such as the angle of the two arms of the "L" or the protrusion of individual bases, may distinguish the individual tRNAs.

All tRNAs can fit in the P and A sites of the ribosome. At one end they are associated with mRNA via codon–anticodon pairing, and at the other end the polypeptide is being synthesized and transferred. Similarly, all tRNAs (except the initiator) share the ability to be recognized by elongation factors (EF-Tu or eEF1) for binding to the ribosome. The initiator tRNA is recognized instead by IF-2 or eIF2. Thus the tRNA set must possess common features for interaction with elongation factors, but the initiator tRNA can be distinguished.

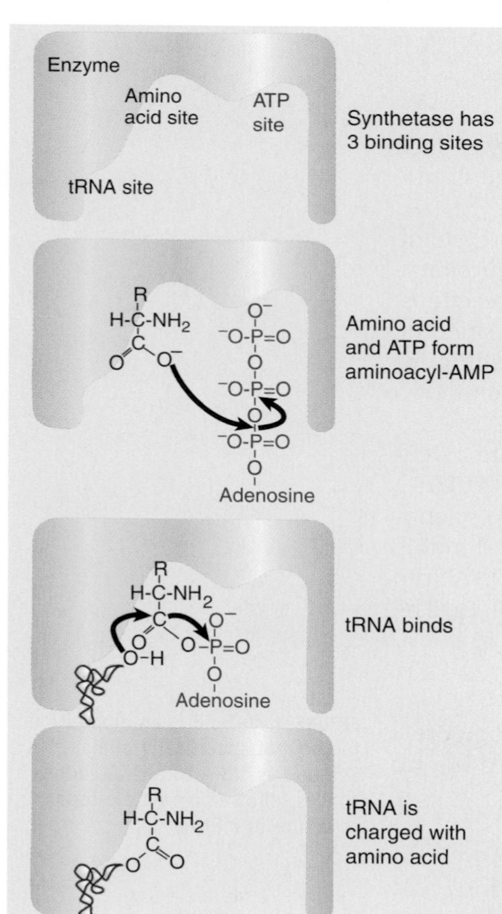

FIGURE 9.12 An aminoacyl-tRNA synthetase charges tRNA with an amino acid.

Amino acids enter the translation pathway through the aminoacyl-tRNA synthetases, which provide the interface for connection with nucleic acid. All synthetases function by the two-step mechanism depicted in **FIGURE 9.12**:

- First, the amino acid reacts with ATP to form aminoacyl-adenylate, releasing pyrophosphate. Energy for the reaction is provided by cleaving the high energy bond of the ATP.
- Second, the activated amino acid is transferred to the tRNA, releasing AMP.

The synthetases sort the tRNAs and amino acids into corresponding sets. Each synthetase recognizes a single amino acid and all the tRNAs that should be charged with it. Usually, each amino acid is represented by more than one tRNA. Several tRNAs may be needed to respond to synonymous codons, and sometimes there are multiple species of tRNA reacting with the same codon. Multiple tRNAs representing the same amino acid are called **isoaccepting tRNAs**; because they are all recognized by the same synthetase, they are also described as its **cognate tRNAs**.

A group of isoaccepting tRNAs must be charged only by the single aminoacyl-tRNA synthetase specific for their amino acid, so isoaccepting tRNAs must share some common feature(s) that enables the enzyme to distinguish them from the other tRNAs. The entire complement of tRNAs is divided into twenty isoaccepting groups, and each group is able to identify itself to its particular synthetase.

Many attempts to deduce similarities in sequence between cognate tRNAs, or to induce chemical alterations that affect their charging, have shown that the basis for recognition is not the same for different tRNAs, and it does not necessarily lie in some feature of primary or secondary structure alone. tRNAs are identified by their synthetases by contacts that recognize a small number of bases, typically from one to five. Three types of features are commonly used:

- Usually (but not always), at least one base of the anticodon is recognized. Sometimes all the positions of the anticodon are important.
- Often, one of the last three base pairs in the acceptor stem is recognized. An extreme case is represented by alanine tRNA, which is identified by a single unique base pair in the acceptor stem.
- The so-called discriminator base, which lies between the acceptor stem and the CCA terminus, is always invariant among isoacceptor tRNAs.

No one of these features constitutes a unique means of distinguishing twenty sets of tRNAs, or provides sufficient specificity, so it appears that recognition of tRNAs is idiosyncratic, with each following its own rules. Recognition depends on an interaction between a few points of contact in the tRNA, concentrated at the extremities, and a few amino acids constituting the active site in the protein. The relative importance of the roles played by the acceptor stem and anticodon is different for each tRNA-synthetase interaction.

KEY CONCEPTS

- Aminoacyl-tRNA synthetases are enzymes that charge tRNA with an amino acid to generate aminoacyl-tRNA in a two-stage reaction that uses energy from ATP.
- There are twenty aminoacyl-tRNA synthetases in each cell. Each charges all the tRNAs that represent a particular amino acid.
- Recognition of a tRNA is based on a small number of points of contact in the tRNA sequence.

CONCEPT AND REASONING CHECK

What features of a tRNA allow its unique recognition by an aminoacyl-tRNA synthetase?

9.9 Aminoacyl-tRNA Synthetases Fall into Two Groups

In spite of their common function, synthetases are a rather diverse group of proteins. The individual subunits vary in molecular weight from 40 to 110 kD, and the enzymes may be monomeric, dimeric, or tetrameric. Synthetases have been divided into two general groups, each containing ten enzymes, on the basis of the structure of the domain that contains the active site. The two groups show no relationship. They may have evolved independently of one another. This makes it seem possible that an early form of life with proteins that were made up of just the ten amino acids coded by one type of synthetase or the other could have existed.

A general model for synthetase-tRNA binding suggests that the protein binds the tRNA along the "side" of the L-shaped molecule. The same general principle applies for all synthetase-tRNA binding: the tRNA is bound principally at its two extremities, and most of the tRNA sequence is not involved in recognition by a synthetase. However, the detailed nature of the interaction is different between class I and class II enzymes, as can be seen from the models of **FIGURE 9.13**, which are based on crystal structures.

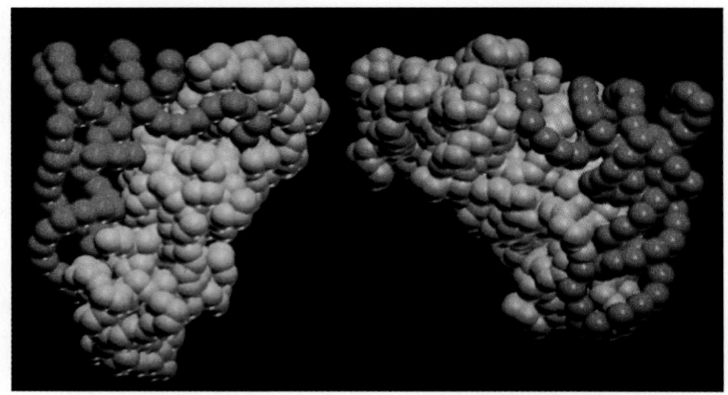

FIGURE 9.13 Crystal structures show that class I and class II aminoacyl-tRNA synthetases bind the opposite faces of their tRNA substrates. The tRNA is shown in red and the protein in blue. Photo courtesy of Dino Moras, Institute of Genetics and Molecular and Cellular Biology.

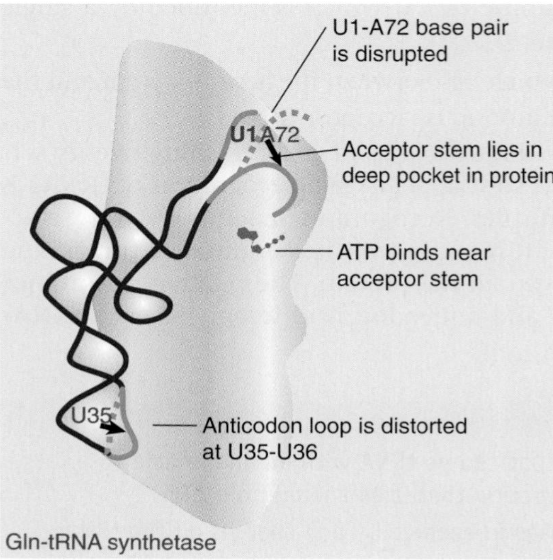
U1-A72 base pair is disrupted
U1A72
Acceptor stem lies in deep pocket in protein
ATP binds near acceptor stem
U35
Anticodon loop is distorted at U35-U36
Gln-tRNA synthetase

FIGURE 9.14 A class I tRNA synthetase contacts tRNA at the minor groove of the acceptor stem and at the anticodon.

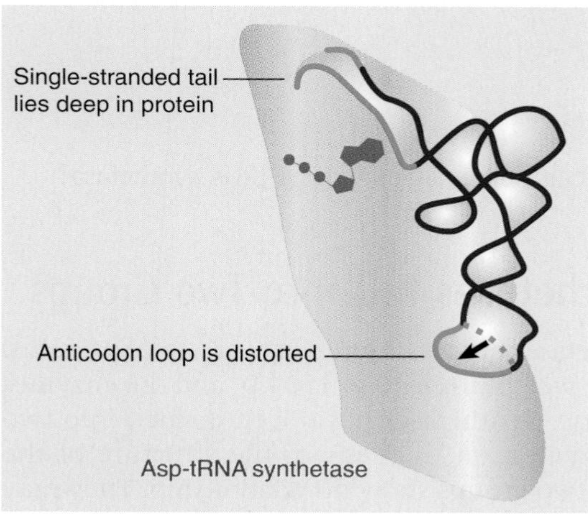
Single-stranded tail lies deep in protein
Anticodon loop is distorted
Asp-tRNA synthetase

FIGURE 9.15 A class II aminoacyl-tRNA synthetase contacts tRNA at the major groove of the acceptor helix and at the anticodon loop.

The two types of enzyme approach the tRNA from opposite sides, with the result that the tRNA-protein complexes look almost like mirror images of one another.

A class I enzyme (for example, Gln-tRNA synthetase) approaches the D-loop side of the tRNA. It recognizes the minor groove of the acceptor stem at one end of the binding site, and interacts with the anticodon loop at the other end. FIGURE 9.14 is a diagrammatic representation of the crystal structure of the $tRNA^{Gln}$-synthetase complex. A revealing feature of the structure is that contacts with the enzyme change the structure of the tRNA at two important points. These can be seen by comparing the dotted and solid lines in the anticodon loop and acceptor stem.

- Bases U35 and U36 in the anticodon loop are pulled farther out of the tRNA into the protein.
- The end of the acceptor stem is seriously distorted, with the result that base pairing between U1 and A72 is disrupted. The single-stranded end of the stem pokes into a deep pocket in the synthetase protein, which also contains the binding site for ATP.

This structure shows why changes in U35, G73, or the U1-A72 base pair affect the recognition of the tRNA by its synthetase. At all of these positions, hydrogen bonding occurs between the protein and tRNA.

A class II enzyme (for example, Asp-tRNA synthetase) approaches the tRNA from the other side; it recognizes both the variable loop and the major groove of the acceptor stem, as shown in FIGURE 9.15. The acceptor stem remains in its regular helical conformation. ATP is probably bound near the terminal adenine. At the other end of the binding site, there is a tight contact with the anticodon loop, which has a change in conformation that allows the anticodon to be in close contact with the protein.

KEY CONCEPT

- Aminoacyl-tRNA synthetases are divided into the class I and class II groups by sequence and structural similarities.

CONCEPT AND REASONING CHECK

What are the differences between the tRNA-protein interactions of class I and class II synthetases?

9.10 Synthetases Use Proofreading to Improve Accuracy

Aminoacyl-tRNA synthetases have a difficult job. Each synthetase must distinguish one out of twenty amino acids, and must differentiate cognate tRNAs (typically one to three) from the total set.

Many amino acids are closely related to one another, and all amino acids are related to the metabolic intermediates in their particular synthetic pathway. It is especially difficult to distinguish between two amino acids that differ only in the length of the carbon backbone (that is, by one —CH_2— group). Intrinsic discrimination based on relative energies of binding two such amino acids would be only ~1/5. The synthetase enzymes improve this ratio ~1000-fold.

Intrinsic discrimination between tRNAs is better, because the tRNA offers a larger surface with which to make more contacts. It is still true, however, that all tRNAs conform to the same general structure, and there may be a quite limited set of features that distinguish the cognate tRNAs from the noncognate tRNAs.

Synthetases use proofreading mechanisms to control the recognition of both types of substrates. They improve significantly on the intrinsic differences among amino acids or among tRNAs, but, consistent with the intrinsic differences in each group, make more mistakes in selecting amino acids (error rates are 10^{-4} to 10^{-5}) than in selecting tRNAs (for which error rates are ~10^{-6}) (see Figure 8.8).

Synthetases use **kinetic proofreading** to improve their discrimination of tRNAs. This relies on the greater intrinsic affinity of a cognate tRNA for its synthetase. A correctly bound tRNA makes contacts with the enzyme surface that allow the enzyme to aminoacylate it rapidly. An incorrectly bound tRNA makes fewer contacts, so the aminoacylation is not triggered so quickly, and this allows more time for the tRNA to dissociate from the enzyme before it is trapped by the reaction.

Specificity for amino acids varies among the synthetases. Some are highly specific for initially binding a single amino acid, whereas others can also activate amino acids closely related to the proper substrate. The analog amino acid can sometimes be converted to the adenylate form, but in none of these cases is an incorrectly activated amino acid actually used to form a stable aminoacyl-tRNA.

There are two stages at which proofreading of an incorrect aminoacyl-adenylate may occur during formation of aminoacyl-tRNA. FIGURE 9.16 shows that both use **chemical proofreading**, in which the catalytic reaction is reversed. The extent to which one pathway or the

- **kinetic proofreading** A proofreading mechanism that depends on incorrect events proceeding more slowly than correct events, so that incorrect events are reversed before a subunit is added to a polymeric chain.
- **chemical proofreading** A proofreading mechanism in which the correction event occurs after the addition of an incorrect subunit to a polymeric chain, by means of reversing the addition reaction.

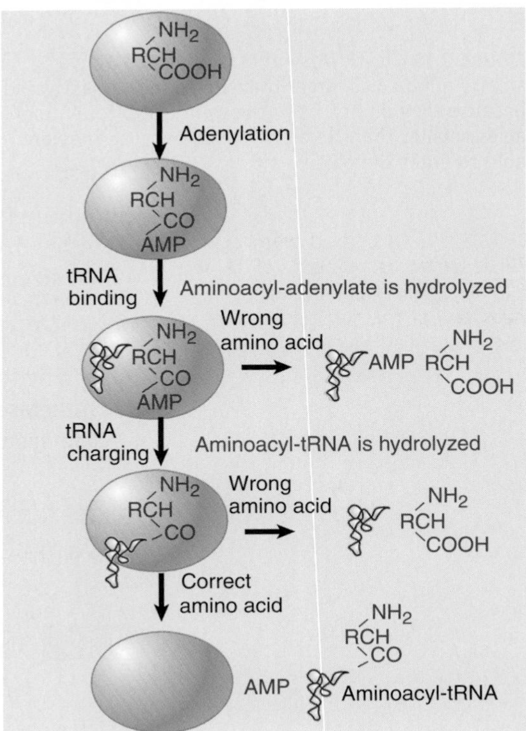

FIGURE 9.16 When a synthetase binds the incorrect amino acid, proofreading requires binding of the cognate tRNA. It may take place either by a conformation change that causes hydrolysis of the incorrect aminoacyl-adenylate or by transfer of the amino acid to tRNA, followed by hydrolysis.

Step	Frequency of Error
Activation of valine to Val-AMP^{Ile}	1/225
Release of Val–tRNA	1/270
Overall rate of error	1/225 × 1/270 = 1/60,000

FIGURE 9.17 The accuracy of charging tRNA^{Ile} by its synthetase depends on error control at two stages.

other predominates varies with the individual synthetase. The presence of the cognate tRNA usually is needed to trigger proofreading:

- The noncognate aminoacyl-adenylate may be hydrolyzed when the cognate tRNA binds. This mechanism is used predominantly by several synthetases, including those for methionine, isoleucine, and valine.
- Some synthetases use chemical proofreading at a later stage. The wrong amino acid is actually transferred to tRNA, is then recognized as incorrect by its structure in the tRNA binding site, and so is hydrolyzed and released. The process requires a continual cycle of linkage and hydrolysis until the correct amino acid is transferred to the tRNA.

A classic example in which discrimination between amino acids depends on the presence of tRNA is provided by the Ile-tRNA synthetase of *E. coli*. The enzyme can charge valine with AMP, but hydrolyzes the valyl-adenylate when tRNA^{Ile} is added. The overall error rate depends on the specificities of the individual steps, as summarized in **FIGURE 9.17**. The overall error rate of 1.5×10^{-5} is less than the measured rate at which valine is substituted for isoleucine (in rabbit globin), which ranges from 2 to 5×10^{-4}. So mischarging probably provides only a small fraction of the errors that actually occur in protein synthesis.

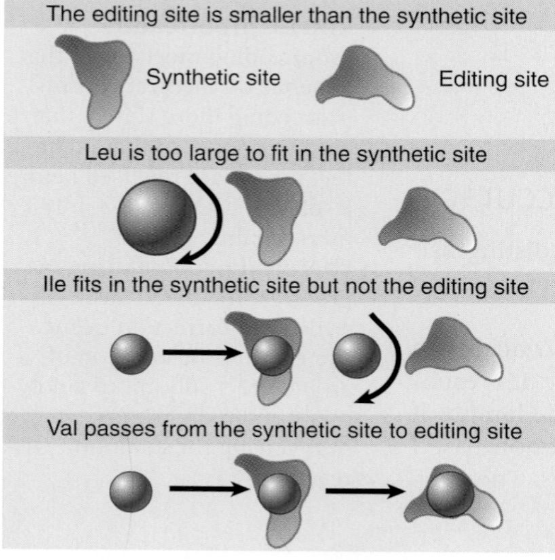

FIGURE 9.18 Ile-tRNA synthetase has two active sites. Amino acids larger than Ile cannot be activated because they do not fit in the synthetic site. Amino acids smaller than Ile are removed because they are able to enter the editing site.

Ile-tRNA synthetase uses size as a basis for discrimination among amino acids. **FIGURE 9.18** shows that it has two active sites: the synthetic (or activation) site and the editing (or hydrolytic) site. The crystal structure of the enzyme shows that the synthetic site is too small to allow leucine (a close analog of isoleucine) to enter. All amino acids larger than isoleucine are excluded from activation because they cannot enter the synthetic site. An amino acid that can enter the synthetic site is placed on tRNA. Then the enzyme tries to transfer it to the editing site. Isoleucine is safe from editing because it is too large to enter the editing site. However, valine can enter this site, and as a result an incorrect Val-tRNA^{Ile} is hydrolyzed. Essentially the enzyme provides a double molecular sieve, in which size of the amino acid is used to discriminate between closely related species.

KEY CONCEPT

- Specificity of recognition of both amino acid and tRNA is controlled by aminoacyl-tRNA synthetases by proofreading reactions that reverse the catalytic reaction if the wrong component has been incorporated.

CONCEPT AND REASONING CHECK

How is the error rate of an aminoacyl-tRNA synthetase minimized?

9.11 Suppressor tRNAs Have Mutated Anticodons That Read New Codons

Isolation of mutant tRNAs has been one of the most potent tools for analyzing the ability of a tRNA to recognize its codon(s) in mRNA, and for determining the effects that changes in different parts of the tRNA molecule have on codon–anticodon recognition.

Mutant tRNAs are isolated by virtue of their ability to overcome the effects of mutations in genes coding for polypeptides. In genetic terminology, a mutation that is able to overcome the effects of another mutation is called a **suppressor**.

▸ **suppressor** A second mutation that compensates for or alters the effects of a primary mutation.

In tRNA suppressor systems, the primary mutation changes a codon in an mRNA so that the polypeptide product is altered. The secondary suppressor mutation changes the anticodon of a tRNA, so that it recognizes the mutant codon instead of (or as well as) its original target codon. The amino acid that is now inserted restores the original polypeptide. The suppressors are described as **nonsense suppressors** or **missense suppressors**, depending on the nature of the original mutation.

In a wild-type cell, a nonsense mutation is recognized only by a release factor, which terminates translation. The suppressor mutation creates an aminoacyl-tRNA that can recognize the termination codon; by inserting an amino acid, it allows translation to continue beyond the site of the nonsense mutation. This new capacity of the translation system allows a full-length polypeptide to be synthesized, as illustrated in **FIGURE 9.19**. If the amino acid inserted by suppression is different from the amino acid that was originally present at this site in the wild-type protein, the activity of the polypeptide may be altered. Each type of nonsense suppressor is specific for one of the three nonsense codons; that is, its anticodon recognizes one of the codons UAA, UAG, or UGA.

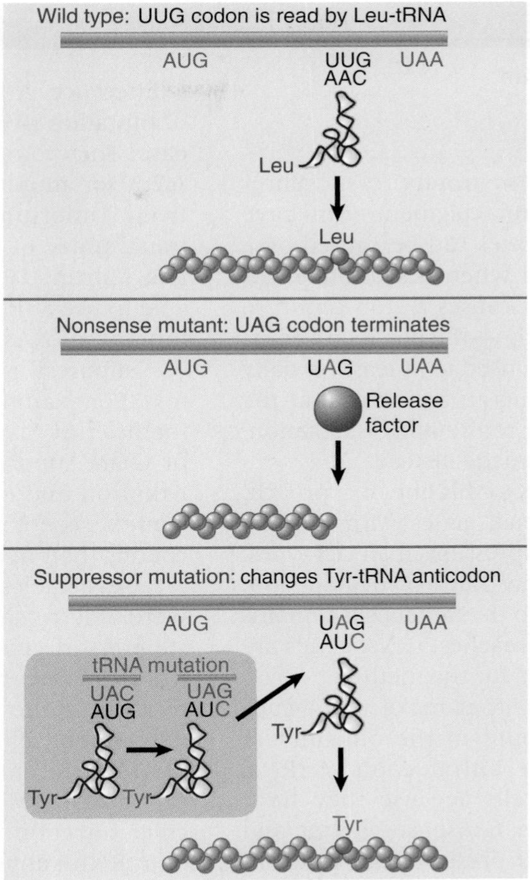

FIGURE 9.19 Nonsense mutations can be suppressed by a tRNA with a mutant anticodon, which inserts an amino acid at the mutant codon, producing a full-length protein in which the original Leu residue has been replaced by Tyr.

- **nonsense suppressor** A gene coding for a mutant tRNA that is able to respond to one or more of the termination codons and insert an amino acid at that site.
- **missense suppressor** A suppressor that codes for a tRNA that has been mutated to recognize a different codon. By inserting a different amino acid at a mutant codon, the tRNA suppresses the effect of the original mutation.

A nonsense suppressor is isolated by its ability to recognize a mutant nonsense codon. The same triplet sequence, however, constitutes one of the normal termination signals of the cell! The mutant tRNA that suppresses the nonsense mutation must, in principle, be able to suppress natural termination at the end of any gene that uses this codon. **FIGURE 9.20** shows that this **readthrough** results in the synthesis of a longer polypeptide, with additional C-terminal sequence. The extended polypeptide will end at the next termination triplet sequence found in the reading frame. Any extensive suppression of termination is likely to be deleterious to the cell by producing extended polypeptides whose functions are thereby altered.

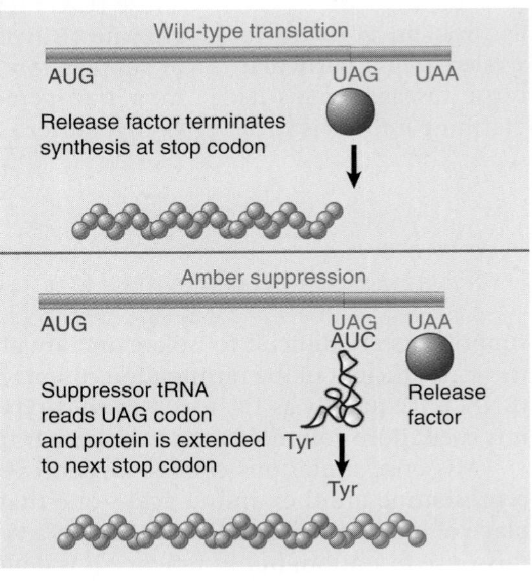

FIGURE 9.20 Nonsense suppressors also read through natural termination codons, synthesizing polypeptides that are longer than wild-type.

- **readthrough** This occurs at transcription or translation when RNA polymerase or the ribosome, respectively, ignores a termination signal because of a mutation of the template or the behavior of an accessory factor.

Amber codons are used relatively infrequently to terminate translation in *E. coli*, and as a result amber suppressors tend to be relatively efficient (10–50%). The ochre codon is used most frequently as a natural termination signal, and as a result ochre

MEDICAL APPLICATIONS

Therapies for Nonsense Mutations

What do Duchenne muscular dystrophy, cystic fibrosis, β-thalassamia, and xeroderma pigmentosum have in common? Each of these diseases can be caused by a mutation in a single gene. And when that mutation is a nonsense mutation—one that causes a stop codon to occur prematurely in the transcript—the possibility exists that a suppressor tRNA could be used to at least partially correct the defect. By introducing an amino acid at the position of the stop codon that is terminating translation, the full-length protein could be made instead.

There are several advantages to this line of approach. tRNAs are encoded by very small genes with internal promoters that robustly drive expression in all cell types. Once transcribed, they have low rates of degradation. Although the delivery of genes to the target cells remains a problem for gene therapy approaches, tRNA genes are in principle very good candidates for the method.

But the very characteristics that make tRNA suppression attractive also contribute to the difficulty of carrying it out in practice. The introduction of tRNA suppressors is deleterious to cells because they have no way to distinguish between nonsense codons and natural stop codons. Thus, their presence can result in readthrough of natural stop codons and the production of many abnormal proteins.

One way to get around this problem is to try to exploit any difference in the prevalence of a particular stop codon in mutations that cause disease versus its normal prevalence in cells. To determine if there is such a difference, Atkinson and Martin looked at 179 events of mutation to nonsense codons that cause human disease. They found that there is a ratio of approximately 1:2:3 for mutation to UAA, UAG, and UGA, respectively. Unfortunately, this ratio is similar to the proportional usage of those codons in human cells (Atkinson and Martin, 1994). Therefore, researchers will not be able to use a tRNA suppressor that preferentially affects mutant mRNAs.

Suppressor efficiency also depends on 3′ codon context. For example, UAG codons flanked 3′ by A are very inefficiently suppressed, whereas those followed by C or G are suppressed with much higher efficiency. But Atkinson and Martin found that the distribution of 3′ contexts is not significantly different in premature stop codons than in naturally occurring stop codons.

Nevertheless, as early as 1982, Temple and others were able to show that coinjection of a human tRNA suppressor gene and the poly-A^+ reticulocyte RNA from a patient with β-thalassamia into *Xenopus* oocytes resulted in the production of full length β-globin protein (Temple et al., 1982).

More recently, studies in mice have also shown some promise. Buvoli and others have shown that intramuscular injection of a plasmid containing several copies of a tRNA suppressor gene resulted in the readthrough of an ochre codon in a chloramphenicol acetyl transferase (CAT) reporter gene and expression of the active enzyme (Buvoli et al., 2000). However, the efficiency of suppression was very low, reaching 0.28% at most. In a transgenic mouse expressing the same mutant CAT reporter gene in the heart, injection of the tRNA

suppressors are difficult to isolate and are always much less efficient (<10%). UGA is the least efficient of the termination codons in its natural function; it is misread by Trp-tRNA as frequently as 1% to 3% in wild-type cells. In spite of this deficiency, however, it is used more commonly than the amber triplet to terminate bacterial genes.

Missense mutations change a codon representing one amino acid into a codon representing another amino acid—one that may not function in the polypeptide in place of the original residue. (Formally, any substitution of amino acids constitutes a missense mutation, but in practice it is detected only if it changes the activity of the polypeptide.) The mutation can be suppressed by the insertion either of the original amino acid or of some other amino acid that can function in the protein.

FIGURE 9.21 demonstrates that missense suppression can be accomplished in the same way as nonsense suppression, by mutating the anticodon of a tRNA carrying an acceptable amino acid so that it responds to the mutant codon. So missense suppression involves a change in the meaning of the codon from one amino acid to another.

suppressor gene construct into the heart resulted in 1% to 2% CAT activity. The long-term effects of the treatment were not reported.

Another approach to readthrough is the use of aminoglycoside antibiotics, which bind to the decoding site of the small subunit rRNA and decrease the accuracy requirements of codon-anticodon pairing (Moazed et al., 1986). This results in a full-length protein with a missense mutation. Gentamicin is one such drug that has been used in this manner, but it has low efficiency and toxic side effects, and it suffers from the same lack of discrimination between premature stop codons and natural ones as other readthrough approaches.

Recently, a small molecule has been identified in screens for compounds that can read through premature stop UGA codons. Called PTC124, this molecule is remarkable for its ability to discriminate between premature stop codons and naturally occurring stop codons (Welch et al., 2007). Normally, premature stop codons are recognized by the ribosome and lead to nonsense-mediated decay of the transcript. Normal termination is mechanistically different and does not lead to decay of the transcript. Presumably, PTC124 is sensitive to this difference in some way.

Finally, the success of most nonsense-suppression approaches will depend in part upon the amount of normal protein required to correct a defect. Studies have suggested that to correct the muscle defect of Duchenne and Becker muscular dystrophy, the level of expression must reach at least 30% to 40% of normal protein (Hoffman et al., 1988). By contrast, in canine hemophilia B, 1% of normal factor IX levels results in partial correction of the coagulation defect (Snyder et al., 1999).

Nonsense mutations in monogenic disorders seem like good candidates for new therapies because the cause of the defect is very well defined. Although intervening in cellular activities without causing unwanted side effects is always challenging, there is reason to hope that a thorough understanding of the mechanism of translation and mRNA decay will lead to effective treatments in the future.

References

Atkinson J, Martin R. (1994). Mutations to nonsense codons in human genetic disease: implications for gene therapy by nonsense suppressor tRNAs. *Nucleic Acids Research* 22:1327–1334.

Buvoli M, Buvoli A, Leinwand, LA. (2000). Suppression of nonsense mutations in cell culture and mice by multimerized suppressor tRNA genes. *Molecular and Cellular Biology* 20:3116–3124.

Hoffman EP, Fischbeck KH, Brown RH, et al. (1988). Characterization of dystrophin in muscle-biopsy specimens from patients with Duchenne's or Becker's muscular dystrophy. *New England Journal of Medicine* 318:1363–1368.

Moazed D, Noller HF. (1986).Transfer RNA shields specific nucleotides in 16S ribosomal RNA from attack by chemical probes. *Cell* 47:985–994.

Snyder RO, Miao C, Meuse L, et al. (1999). Correction of hemophilia B in canine and murine models using recombinant adeno-associated viral vectors. *Nature Medicine* 5:64–70.

Temple GF, Dozy AM, Roy KL, Kan YW. (1982). Construction of a functional human suppressor tRNA gene: an approach to gene therapy for β-thalassaemia. *Nature* 296:537–540.

Welch EM, Bart ER, Zhuo J, et al. (2007). PTC124 targets genetic disorders caused by nonsense mutations. *Nature* 447:87–91.

As with nonsense suppressors, an event that suppresses a missense mutation at one site may replace the wild-type amino acid at another site with a new amino acid. The change may inhibit normal polypeptide function. The absence of any strong missense suppressors is therefore explained by the damaging effects that would be caused by a general and efficient substitution of amino acids.

Suppression is most often considered in the context of a mutation that changes the reading of a codon. There are, however, some situations in which a stop codon is read as an amino acid at a low frequency in the wild-type situation. The first example to be discovered was the coat protein gene of the RNA phage Qβ. The formation of infective Qβ particles requires that the stop codon at the end of this gene be suppressed at a low frequency to generate a small proportion of coat proteins with a C-terminal extension. In effect, this stop codon is leaky. The reason is that Trp-tRNA recognizes the codon at a low frequency.

FIGURE 9.21 Missense suppression occurs when the anticodon of tRNA is mutated so that it responds to the wrong codon. The suppression is only partial because both the wild-type tRNA and the suppressor tRNA can recognize AGA.

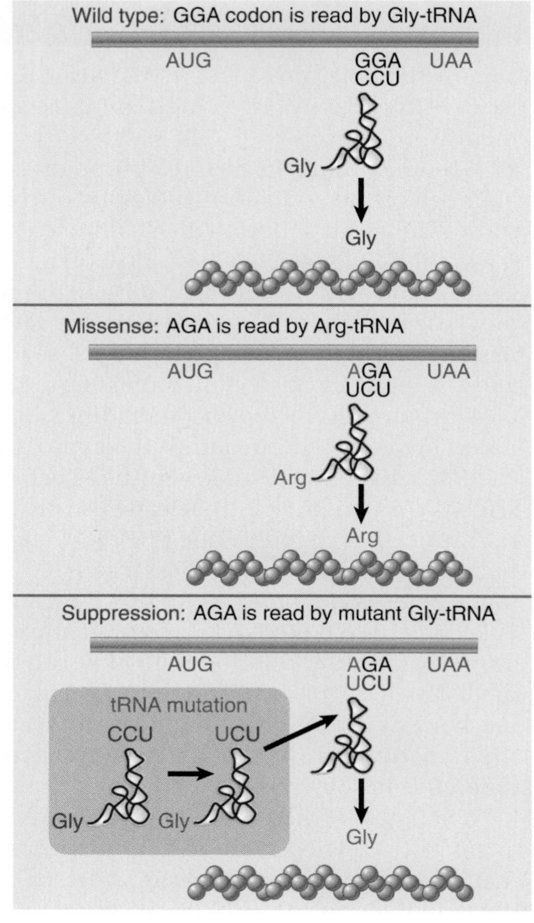

Readthrough past stop codons also occurs in eukaryotes, where it is employed most often by RNA viruses. This may involve the suppression of UAG/UAA by Tyr-tRNA, Gln-tRNA, or Leu-tRNA, or the suppression of UGA by Trp-tRNA or Arg-tRNA. The extent of partial suppression is dictated by the context surrounding the codon.

KEY CONCEPTS

- A suppressor tRNA typically has a mutation in the anticodon that changes the codons to which it responds.
- Each type of nonsense codon is suppressed by tRNAs with mutant anticodons.
- When the new anticodon corresponds to a termination codon, an amino acid is inserted and the polypeptide chain is extended beyond the termination codon. This results in nonsense suppression at a site of nonsense mutation, or in readthrough at a natural termination codon.
- Suppressor tRNAs compete with wild-type tRNAs that have the same anticodon to read the corresponding codon(s).
- Efficient suppression is deleterious because it results in readthrough past normal termination codons.
- Missense suppression occurs when the tRNA recognizes a different codon from usual, so that one amino acid is substituted for another.

CONCEPT AND REASONING CHECK

Which type of mutation is more likely to be deleterious, nonsense suppressors or missense suppressors? Why?

9.12 Recoding Changes Codon Meanings

The reading frame of a messenger RNA is usually invariant. Translation starts at an AUG codon and continues in triplets to a termination codon. Reading does not depend on the sense of the message: insertion or deletion of a base causes a frameshift mutation, in which the reading frame is changed beyond the site of mutation. Ribosomes and tRNAs inevitably continue reading in triplets, synthesizing an entirely different series of amino acids.

Suppression is caused by mutated anticodon

NNNNNUGANNNNNNNNNN

Special factor + tRNA recognizes codon

Sel-Cys

NN NN NN NN

NNNNNUGANNNNNNNNNN

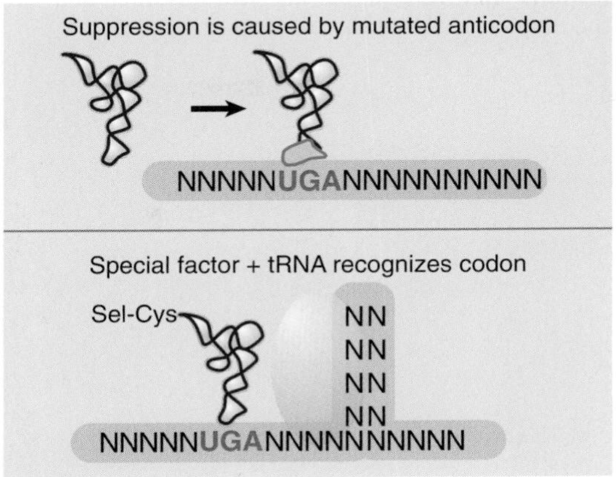

FIGURE 9.22 A mutation in an individual tRNA (usually in the anticodon) can suppress the usual meaning of that codon. In a special case, a specific tRNA is bound by an unusual elongation factor to recognize a termination codon adjacent to a hairpin loop.

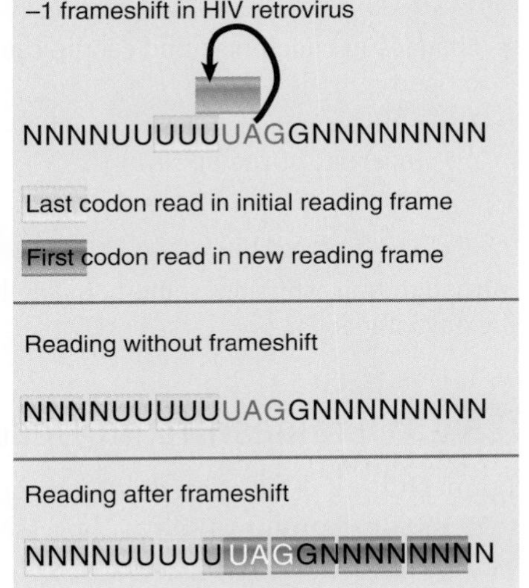

FIGURE 9.23 A tRNA that slips one base in pairing with a codon causes a frameshift that can suppress termination. The efficiency is usually ~5%.

There are some exceptions to the usual pattern of translation that enable a reading frame with an interruption of some sort—such as a nonsense codon or frameshift—to be translated into a full-length polypeptide. **Recoding** events are responsible for making exceptions to the usual rules.

In one type of recoding, changing the meaning of a single codon allows one amino acid to be substituted in place of another, or for an amino acid to be inserted at a termination codon. **FIGURE 9.22** shows that these changes rely on the properties of an individual tRNA that responds to the codon:

- Suppression involves recognition of a codon by a (mutant) tRNA that usually would respond to a different codon (see *Section 9.11, Suppressor tRNAs Have Mutated Anticodons That Read New Codons*).
- Redefinition of the meaning of a codon occurs when an aminoacyl-tRNA is modified (see *Section 9.7, Novel Amino Acids Can Be Inserted at Certain Stop Codons*).

Changing the reading frame, another type of recoding, occurs in two types of situations:

- Frameshifting typically involves changing the reading frame when aminoacyl-tRNA slips by one base, either +1 forward or –1 backward (see *Section 9.13, Frameshifting Occurs at Slippery Sequences*). The result shown in **FIGURE 9.23** is that translation continues past a termination codon.
- Bypassing involves a movement of the ribosome to change the codon that is paired with the peptidyl-tRNA in the P site. The sequence between the two codons fails to be represented in the resulting polypeptide. As shown in **FIGURE 9.24**, this allows translation to continue past any termination codons in the intervening region.

▸ **recoding** Events that occur when the meaning of a codon or series of codons is changed from that predicted by the genetic code. It may involve altered interactions between aminoacyl-tRNA and mRNA that are influenced by the ribosome.

FIGURE 9.24 Bypassing occurs when the ribosome moves along mRNA so that the peptidyl-tRNA in the P site is released from pairing with its codon and then repairs with another codon farther along.

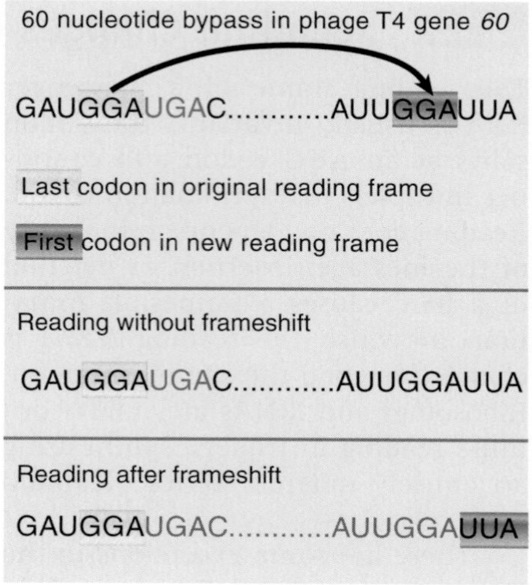

KEY CONCEPTS

- Changes in codon meaning can be caused by mutant tRNAs or by tRNAs with special properties.
- The reading frame can be changed by frameshifting or bypassing, both of which depend on properties of the mRNA.

CONCEPT AND REASONING CHECK

Although frameshifting would normally be deleterious, under what circumstances would it be advantageous?

9.13 Frameshifting Occurs at Slippery Sequences

Frameshifting is associated with specific tRNAs in two circumstances:

- Some mutant tRNA suppressors recognize a "codon" for four bases instead of the usual three bases.
- Certain "slippery" sequences allow a tRNA to move a base up or down mRNA in the A site.

The simplest type of external frameshift suppressor corrects the reading frame when a mutation has been caused by inserting an additional base within a stretch of identical residues. For example, a G may be inserted in a run of several contiguous G bases. The frameshift suppressor is a $tRNA^{Gly}$ that has an extra base inserted in its anticodon loop, converting the anticodon from the usual triplet sequence 3′-CCC-5′ to the quadruplet sequence 3′-CCCC-5′. The suppressor tRNA recognizes a 4-base "codon."

Situations in which frameshifting is a normal event occur in phages and viruses. Such events may affect the continuation or termination of translation, and may result from the intrinsic properties of the mRNA.

In retroviruses, translation of the first gene is terminated by a nonsense codon in phase with the reading frame. The second gene lies in a different reading frame, and (in some viruses) is translated by a frameshift that changes into the second reading frame and therefore bypasses the termination codon (see Figure 9.23). The efficiency of the frameshift is low, typically ~5%. In fact, this is important in the biology of the virus; an increase in efficiency can be damaging. **FIGURE 9.25** illustrates the similar situation of the yeast Ty element, in which the termination codon of *tya* must be bypassed by a frameshift in order to read the subsequent *tyb* gene.

Such situations make the important point that the rare (but predictable) occurrence of "misreading" events can be relied on as a necessary step in natural translation. This is called **programmed frameshifting**. It occurs at particular sites at frequencies that are 100 to 1000× greater than the rate at which errors are made at nonprogrammed sites (~3×10^{-5} per codon).

There are two common features in this type of frameshifting:

- A "slippery" sequence allows an aminoacyl-tRNA to pair with its codon and then to move +1 (rarely) or –1 base (more commonly) to pair with an overlapping triplet sequence that can also pair with its anticodon.
- The ribosome is delayed at the frameshifting site to allow time for the aminoacyl-tRNA to rearrange its pairing. The cause of the delay can be an adjacent codon that requires a scarce aminoacyl-tRNA, a termination codon that is recognized slowly by its release factor, or a structural impediment in mRNA (for example, a "pseudoknot," a particular conformation of RNA) that impedes the ribosome.

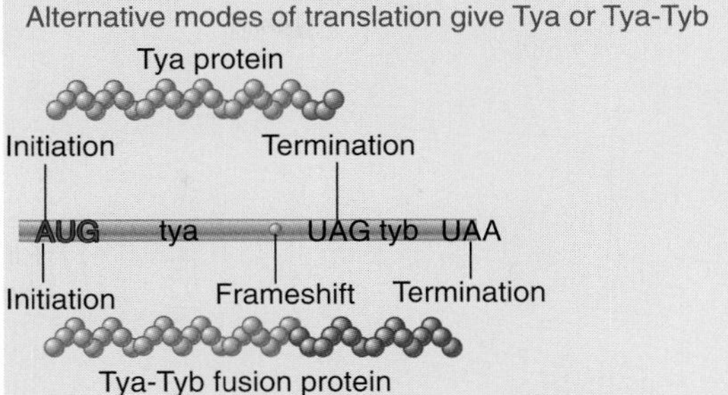

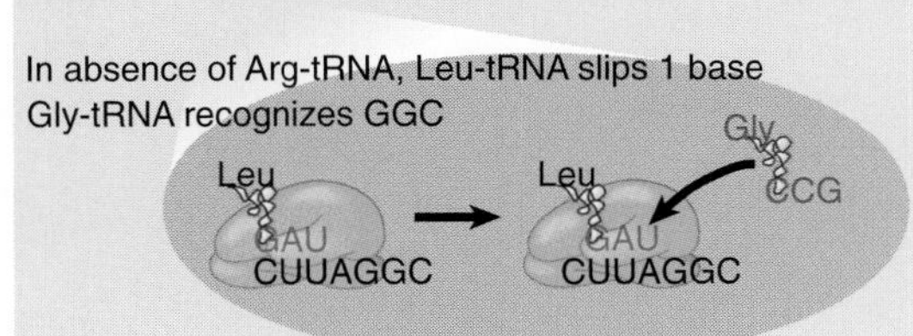

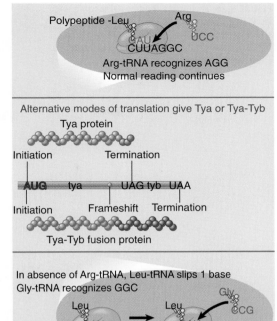

FIGURE 9.25 A +1 frameshift is required for expression of the *tyb* gene of the yeast Ty element. The shift occurs at a 7-base sequence at which two Leu codon(s) are followed by a scarce Arg codon.

▸ **programmed frameshifting** Frameshifting that is required for expression of the polypeptide sequences encoded beyond a specific site at which a +1 or –1 frameshift occurs at some typical frequency.

Slippery events can involve movement in either direction; a –1 frameshift is caused when the tRNA moves backward, and a +1 frameshift is caused when it moves forward. In either case, the result is to expose an out-of-phase triplet in the A site for the next aminoacyl-tRNA. The frameshifting event occurs before peptide bond synthesis. In the most common event, when it is triggered by a slippery sequence in conjunction with a downstream hairpin in mRNA, the surrounding sequences influence its efficiency.

The frameshifting in Figure 9.25 shows the behavior of a typical slippery sequence. The seven-nucleotide sequence CUUAGGC is usually recognized by Leu-tRNA at CUU, followed by Arg-tRNA at AGC. However, the Arg-tRNA is scarce, and when its scarcity results in a delay, the Leu-tRNA slips from the CUU codon to the overlapping UUA triplet. This causes a frameshift, because the next triplet in phase with the new pairing (GGC) is read by Gly-tRNA. Slippage usually occurs in the P site (when the Leu-tRNA actually has become peptidyl-tRNA, carrying the nascent chain).

KEY CONCEPTS

- The reading frame may be influenced by the sequence of mRNA and the ribosomal environment.
- Slippery sequences allow a tRNA to shift by one base after it has paired with its anticodon, thereby changing the reading frame.
- Translation of some genes depends upon the regular occurrence of programmed frameshifting.

CONCEPT AND REASONING CHECK

What types of coding sequences are likely to be "slippery," allowing programmed frameshifting?

FIGURE 9.26 In bypass mode, a ribosome with its P site occupied can stop translation. It slides along mRNA to a site where peptidyl-tRNA pairs with a new codon in the P site. Then translation is resumed.

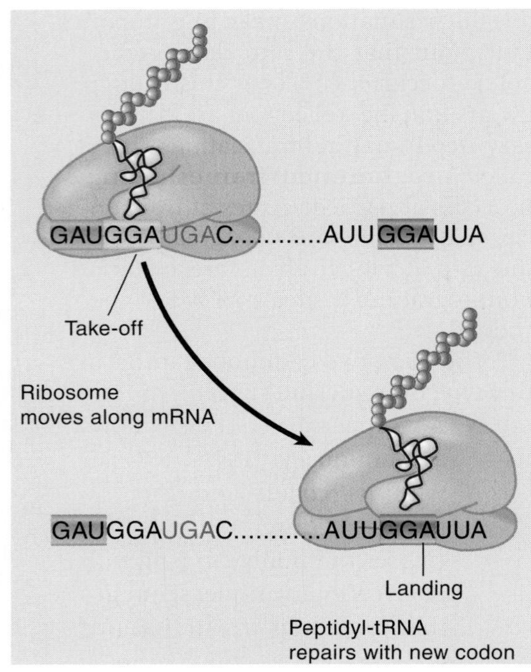

9.14 Bypassing Involves Ribosome Movement

Certain sequences trigger a bypass event, in which a ribosome stops translation, slides along mRNA with peptidyl-tRNA remaining in the P site, and then resumes translation (see Figure 9.24). This is a very rare phenomenon, with only one authenticated example: in gene *60* of phage T4, the ribosome moves sixty nucleotides along the mRNA, as shown in Figure 9.24.

The key to the bypass system is that there are identical (or synonymous) codons at either end of the sequence that is skipped. They are sometimes referred to as the "take-off" and "landing" sites. Before bypass, the ribosome is positioned with a peptidyl-tRNA paired with the take-off codon in the P site, with an empty A site waiting for an aminoacyl-tRNA to enter. **FIGURE 9.26** shows that the ribosome slides along mRNA in this condition until the peptidyl-tRNA can become paired with the codon in the landing site. A remarkable feature of the system is its high efficiency, ~50%.

The sequence of the mRNA triggers the bypass. The important features are the two GGA codons for take-off and landing, the spacing between them, a stem-loop structure that includes the take-off codon, and the stop codon adjacent to the take-off codon. The polypeptide under synthesis is also involved.

The take-off stage requires that the peptidyl-tRNA unpair from its codon. This is followed by a movement of the mRNA that prevents it from re-pairing. Then the ribosome scans the mRNA until the peptidyl-tRNA can repair with the codon in the landing reaction. This is followed by the resumption of translation when aminoacyl-tRNA enters the A site in the usual way.

Like frameshifting, the bypass reaction depends on a pause by the ribosome. The probability that peptidyl-tRNA will dissociate from its codon in the P site is increased by delays in the entry of aminoacyl-tRNA into the A site. Starvation for an amino acid can trigger bypassing in bacterial genes because of the delay that occurs when there is no aminoacyl-tRNA available to enter the A site. In phage T4 gene *60*, one role of mRNA structure may be to reduce the efficiency of termination, thus creating the delay that is needed for the take-off reaction.

KEY CONCEPT

- When a ribosome encounters a GGA codon adjacent to a stop codon in a specific stem-loop structure, it moves directly to a specific GGA downstream without adding amino acids to the polypeptide.

CONCEPT AND REASONING CHECK

Explain how bypassing may allow alternate gene expression in times of bacterial starvation.

9.15 Summary

The sequence of mRNA read in triplets 5′ → 3′ is related by the genetic code to the amino acid sequence of a polypeptide read from N- to C-terminus. Of the sixty-four triplets, sixty-one code for amino acids and three provide termination signals. Synonymous codons that represent the same amino acids are related, often by a difference in the third base of the codon. This third-base degeneracy, coupled with a pattern in which related amino acids tend to be coded by related codons, minimizes the effects of mutations. The genetic code is nearly universal and must have been established very early in evolution. Variations of the code in nuclear genomes are rare, but some changes have occurred during mitochondrial evolution.

Multiple tRNAs may recognize a particular codon. The set of tRNAs responding to the various codons for each amino acid is distinctive for each organism. Codon–anticodon recognition involves wobbling at the first position of the anticodon (third position of the codon), which allows some tRNAs to recognize multiple codons. All tRNAs have modified bases, introduced by enzymes that recognize target bases in the tRNA structure. Codon–anticodon pairing is influenced by modifications of the anticodon itself and also by the context of adjacent bases, especially on the 3′ side of the anticodon.

Each amino acid is recognized by a particular aminoacyl-tRNA synthetase, which also recognizes all of the tRNAs coding for that amino acid.

Aminoacyl-tRNA synthetases vary widely, but fall into two general groups according to the structure of the catalytic domain. Synthetases of each group bind the tRNA from the side, making contacts principally with the extremities of the acceptor stem and the anticodon stem-loop; the two types of synthetases bind tRNA from opposite sides. The relative importances of the acceptor stem and the anticodon region for specific recognition vary with the individual tRNA. Aminoacyl-tRNA synthetases have proofreading functions that scrutinize the aminoacyl-tRNA products and hydrolyze incorrectly joined aminoacyl-tRNAs.

Mutations may allow a tRNA to read different codons; the most common form of such mutations occurs in the anticodon itself. Alteration of its specificity may allow a tRNA to suppress a mutation in a protein-coding gene. A tRNA that recognizes a termination codon provides a nonsense suppressor; one that changes the amino acid responding to a codon is a missense suppressor. Suppressors of UAG and UGA codons are more efficient than those of UAA codons, which is explained by the fact that UAA is the most commonly used natural termination codon. The efficiency of all suppressors, however, depends on the context of the individual target codon.

Frameshifts of the +1 type may be caused by aberrant tRNAs that read "codons" of four bases. Frameshifts of either +1 or –1 may be caused by slippery sequences in mRNA that allow a peptidyl-tRNA to slip from its codon to an overlapping sequence that can also pair with its anticodon. This frameshifting also requires another sequence that causes the ribosome to delay. Frameshifts determined by the mRNA sequence may be required for expression of natural genes. Bypassing occurs when a ribosome stops translation and moves along mRNA with its peptidyl-tRNA in the P site until the peptidyl-tRNA pairs with an appropriate codon; then translation resumes.

CHAPTER QUESTIONS

1. Which two amino acids are encoded by a single codon each in the universal genetic code?
   - A. arginine and tryptophan
   - B. cysteine and isoleucine
   - C. methionine and tryptophan
   - D. histidine and praline
2. Which codon position has the least meaning in determining the amino acid to be inserted during translation of an mRNA?
   - A. 1
   - B. 2
   - C. 3
   - D. All positions have equal meaning.
3. The 5′ ends of mature tRNA molecules are generated by:
   - A. the first base of transcribed RNA.
   - B. digestion by RNase P.
   - C. digestion by RNase E.
   - D. digestion by RNase A.
4. What is the usual sequence at the 3′ ends of tRNAs?
   - A. random sequence
   - B. CAA
   - C. CCA
   - D. AAC
5. Inosine can form base pairs with all but which one of the four usual bases in RNA?
   - A. A
   - B. G
   - C. C
   - D. T
6. How many different types of modified bases can be found in tRNA molecules?
   - A. less than 12
   - B. about 25
   - C. about 36
   - D. more than 50
7. What is the actual minimum number of tRNAs required for responding to the regular codons in mitochondria?
   - A. 20
   - B. 22
   - C. 23
   - D. 31
8. Isoaccepting tRNAs have which one of the following properties?
   - A. They recognize two or more different triplet codons in a transcript.
   - B. They recognize a single triplet codon in a transcript.
   - C. They each represent the same amino acid but are recognized by different synthetases.
   - D. They each represent the same amino acid and are recognized by the same synthetases.

9. A mutation that is able to overcome the effects of another mutation is commonly called a:
   - **A.** depressor.
   - **B.** suppressor.
   - **C.** compensatory mutation.
   - **D.** sense mutation.
10. A missense suppressor for a particular gene may also be a:
    - **A.** terminator for another gene.
    - **B.** mutator for another gene.
    - **C.** suppressor for another gene.
    - **D.** more than one of the above.

KEY TERMS

chemical proofreading
cognate tRNAs
isoaccepting tRNAs
kinetic proofreading
missense suppressor
modification
nonsense suppressor
programmed frameshifting
readthrough
recoding
stop codon
suppressor
synonym codons
third-base degeneracy
wobble hypothesis

FURTHER READING

Farabaugh, P. J. (1996). Programmed translational frameshifting. *Annu. Rev. Genet.* 30, 507–528.
A review of the examples of this phenomenon.

Herr, A. J., Atkins, J. F., and Gesteland, R. F. (2000). Coupling of open reading frames by translational bypassing. *Annu. Rev. Biochem.* 69, 343–372.
A review of the bypassing mechanism of take-off, scanning, and landing.

Hopper, A. K. and Phizicky, E. M. (2003). tRNA transfers to the limelight. *Genes Dev.* 17, 162–180.
A review of the processes involved in tRNA processing in yeast.

Ibba, M. and Söll, D. (2004). Aminoacyl-tRNAs: setting the limits of the genetic code. *Genes Dev.* 18, 731–738.
A brief review of the molecules involved in charging tRNAs and their activities.

Silvian, L. F., Wang, J., and Steitz, T. A. (1999). Insights into editing from an Ile-tRNA synthetase structure with $tRNA^{Ile}$ and mupirocin. *Science* 285, 1074–1077.
The 2.2 Å resolution crystal structure of a bacterial aminoacyl-tRNA synthetase and what it suggests about the editing process during the charging of a tRNA.

Srinivasan, G., James, C. M., and Krzycki, J. A. (2002). Pyrrolysine encoded by UAG in Archaea: charging of a UAG-decoding specialized tRNA. *Science* 296, 1459–1462.
A report of the amino acid pyrrolysine being encoded in the genome of a *Methanosarcina* species using the UAG codon (a termination codon in the standard genetic code). This species also produces a unique tRNA and aminoacyl-tRNA synthetase for incorporation of this amino acid in proteins.

10

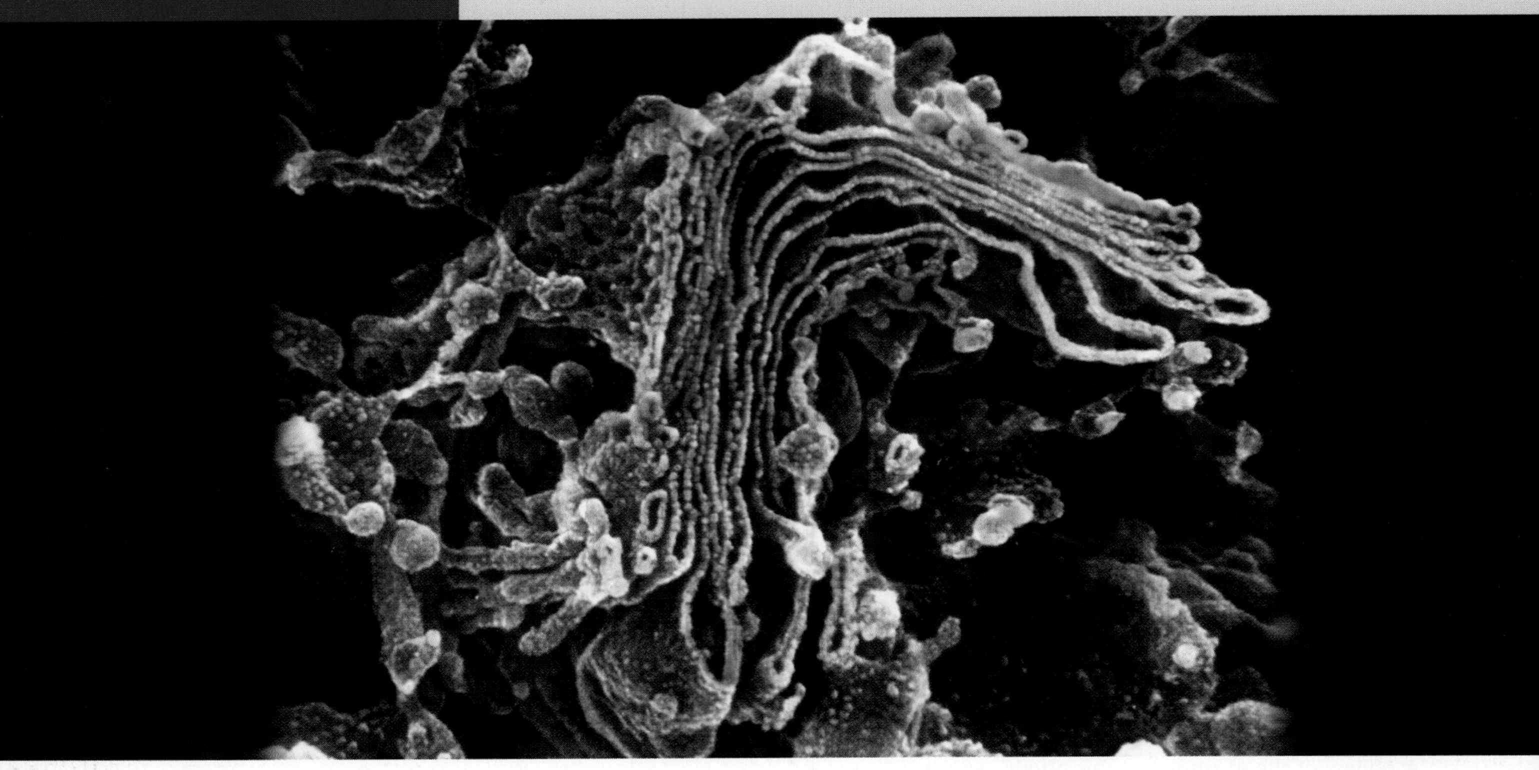

Colored, high-resolution scanning electron micrograph (SEM) of the Golgi apparatus, a stack of flattened interconnecting membranous sacs. Vesicles (pink and purple spheres) pinch off the Golgi apparatus and transport materials to other parts of the cell or cell membrane.

Protein Localization

CHAPTER OUTLINE

10.1 Introduction

All proteins are synthesized on ribosomes either in the cytosol or within organelles. The vast majority of proteins is synthesized in the cytosol, but a small minority is synthesized within organelles on ribosomes specific for those organelles.

Proteins synthesized in the cytosol can be divided into two general classes with regard to localization: those that are not associated with membranes and those that are associated with membranes. **FIGURE 10.1** shows a "sorting map" of the cell that illustrates the possible ultimate destinations for a newly synthesized protein and the systems that transport it. Proteins can be synthesized on "free ribosomes," which are not associated with any membrane. These proteins are released into the cytosol after synthesis is complete. At this point, these proteins can be divided into three classes depending on how they are localized after release from the ribosome:

1. Cytosolic (or "soluble") proteins remain free in the cytosol.
2. Some proteins associate with macromolecular cytosolic structures such as the cytoskeleton or centrioles (associated with the regions that become the poles of the mitotic spindle).
3. Some proteins are transported into the nucleus or into other organelles such as mitochondria or plastids (e.g., chloroplasts). All nuclear proteins and most of the proteins in cytoplasmic organelles are synthesized on free ribosomes in the cytosol.

Alternatively, proteins can be synthesized on "membrane-associated ribosomes." In eukaryotic cells, a complex series of membranes called the **endoplasmic reticulum (ER)** extends from the outer membrane of the nucleus. Close to the ER are the membrane-enclosed stacks of the **Golgi apparatus**, which extend toward the plasma membrane. All proteins that associate with the ER, Golgi, or plasma membrane are synthesized by membrane-associated ribosomes that associate with the ER. These

- **endoplasmic reticulum (ER)** A network of membranes extending from the nuclear membrane responsible for the synthesis, modification, and transport of proteins.
- **Golgi apparatus** A stack of flattened, membrane-enclosed compartments involved in the modification and sorting of lipids and proteins.

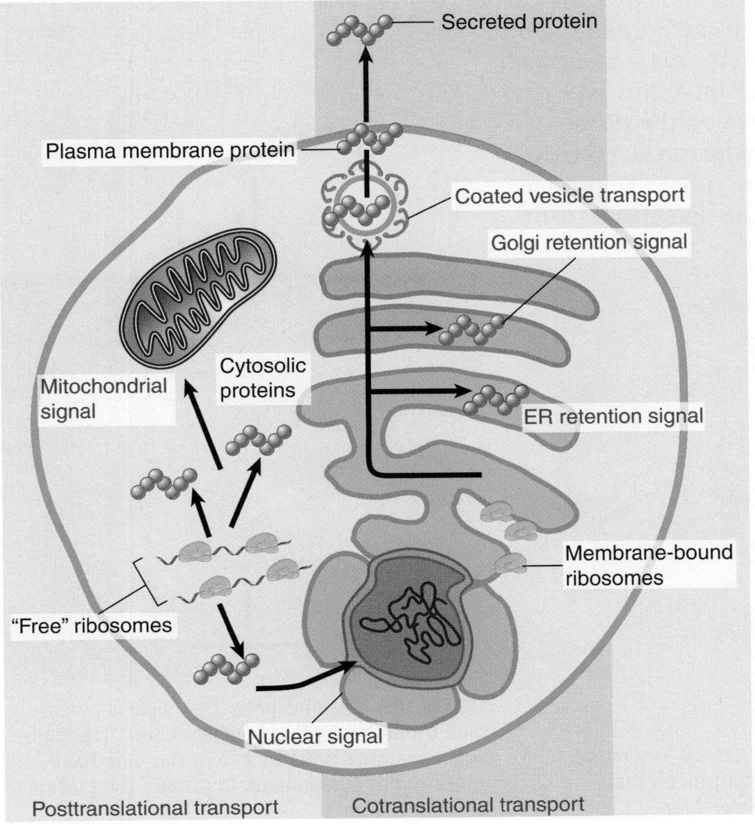

FIGURE 10.1 Overview: Proteins that are localized posttranslationally are released into the cytosol after synthesis on free ribosomes. Some have signals for targeting to organelles such as the nucleus or mitochondria. Proteins that are localized cotranslationally associate with the ER membrane during synthesis, so their ribosomes are "membrane bound." The proteins pass into the endoplasmic reticulum, along to the Golgi apparatus, and then through the plasma membrane, unless they have signals that cause retention at one of the steps on the pathway. They may also be directed to other organelles, such as endosomes or lysosomes.

- **protein translocation** The movement of a protein across a membrane. This occurs across the membranes of organelles in eukaryotes, or across the plasma membrane in bacteria. Each membrane across which proteins are moved has a channel specialized for the purpose.
- **carrier protein** General term for a protein that carries other molecules within the cell, between cellular compartments, or around the body.
- **plastids** A group of related organelles found in plants and algae that contain multiple copies of a circular plastid genome. Different plastids have specialized functions, such as chloroplasts, which contain the machinery of photosynthesis, or chromoplasts, which are used for pigment storage. All plastid types develop from undifferentiated protoplastids.
- **translocon** An integral membrane protein that provides a channel for displacement of polypeptide segments across the membrane; it is usually required in movement of a protein across a lipid bilayer.
- **peroxisome** An organelle in the cytoplasm enclosed by a single membrane. It contains oxidizing enzymes and is involved in fatty acid metabolism.
- **nuclear pore** Pore in the nuclear membrane that allows transport of RNA, proteins, and other molecules in and out of the nucleus.
- **cotranslational translocation** The movement of a protein across a membrane as the protein is being synthesized. The term is usually restricted to cases in which the ribosome binds to the channel. It may be restricted to the endoplasmic reticulum.

proteins are inserted directly into ER membranes and then are directed to their particular locations by the transport system of the Golgi apparatus. The Golgi has numerous functions, including posttranslational modification of proteins, packaging of proteins for delivery to specific membranes or for secretion, and transport of non-protein cargos (such as lipids).

10.2 Protein Translocation May Be Posttranslational or Cotranslational

The process of inserting a protein into or passing a protein through a membrane is called **protein translocation**. There is a dilemma that must be solved every time a protein passes through a membra ne. The protein presents a hydrophilic surface, but the membrane is hydrophobic. Like oil and water, the two have difficulty mixing. The solution is to create a special structure in the membrane through which the protein can pass. Proteins can either interact with these structures directly, or be assisted by specific **carrier proteins** that facilitate translocation.

The endoplasmic reticulum, mitochondria, and **plastids** all contain proteinaceous structures embedded in their membranes that allow proteins to pass through without contacting the surrounding hydrophobic lipids. These structures are generically referred to as **translocons**. FIGURE 10.2 shows that a substrate protein binds directly to the structure, is transported by it to the other side, and then is released, without the need for a carrier protein. **Peroxisomes** also have such structures in their membranes, but the substrate proteins do not bind directly to them. Instead they bind to carrier proteins in the cytosol, the carrier protein is transported through the channel into the peroxisome, and then the substrate protein is released.

For transport into the nucleus, a much larger and more complex structure, known as the **nuclear pore**, is employed. FIGURE 10.3 shows that although the pore provides the environment that allows a substrate to enter (or to leave) the nucleus, it does not actually provide the apparatus that binds to the substrate proteins and moves them through. Included in this apparatus are carrier proteins that bind to the substrates and transport them through the pore to the other side.

The nascent protein may associate with the translocation apparatus while it is still being synthesized on the ribosome. This is called **cotranslational translocation** and it is what causes ribosomes to become membrane-associated. This

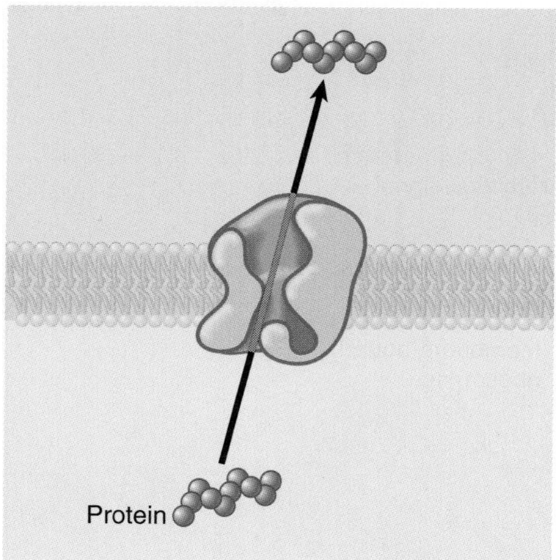

FIGURE 10.2 Proteins enter the ER or a mitochondrion by binding to a translocon that transports them across the membrane.

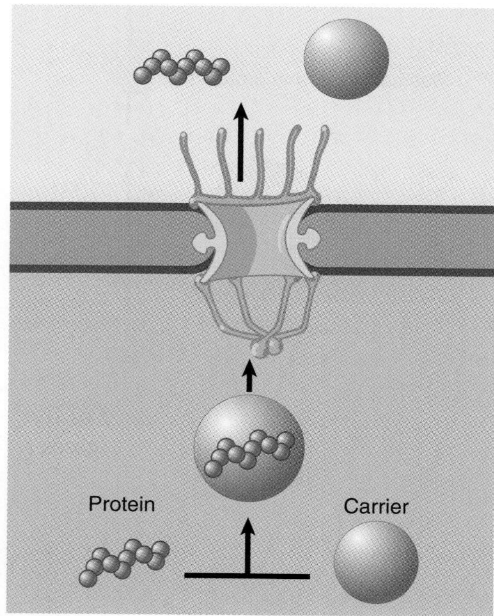

FIGURE 10.3 Proteins enter the nucleus by passage through very large nuclear pores. The transport apparatus is distinct from the pore itself and includes components that carry the protein through the pore.

mechanism is used for proteins that enter the ER; the consequence is that the ribosome is localized to the surface of the ER by the signal sequence of the protein being translated. Alternatively, the protein may be released from a ribosome after translation has been completed, then subsequently diffuse to the appropriate membrane and associate with the translocation apparatus. This is called **posttranslational translocation**.

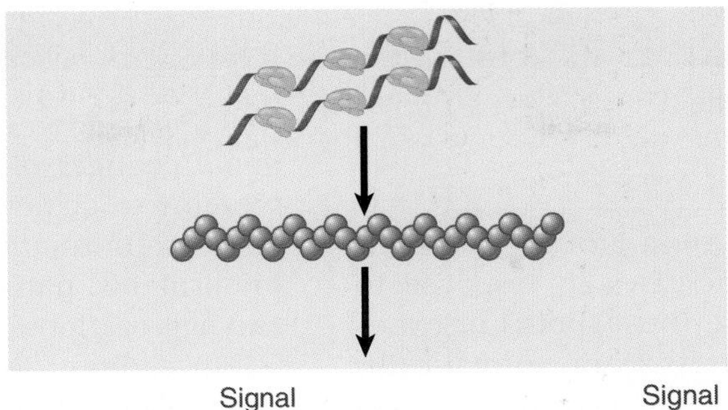

Organelle	Signal location	Type	Signal length
Mitochondrion	N-terminal	Amphipathic helix	12–30
Chloroplast	N-terminal	Charged	>25
Nucleus	Internal	Basic or bipartite	4–9
Peroxisome	C-terminal	Short peptide	3–4

FIGURE 10.4 Proteins synthesized on free ribosomes in the cytosol are directed after their release to specific destinations by short signal motifs.

▸ **posttranslational translocation** The movement of a protein across a membrane after the synthesis of the protein is completed and it has been released from the ribosome.

In order to be transported into the nucleus or an organelle, or to be associated with any cellular membrane, a protein requires a specific **signal sequence** (also known as a *signal peptide*). This applies to proteins synthesized either by free ribosomes (where the signal sequences are sometimes known as "transit peptides") or by membrane-bound ribosomes. The signal for interacting with the target membrane is provided by a short sequence of amino acids in the protein that is recognized by a receptor associated with the organelle.

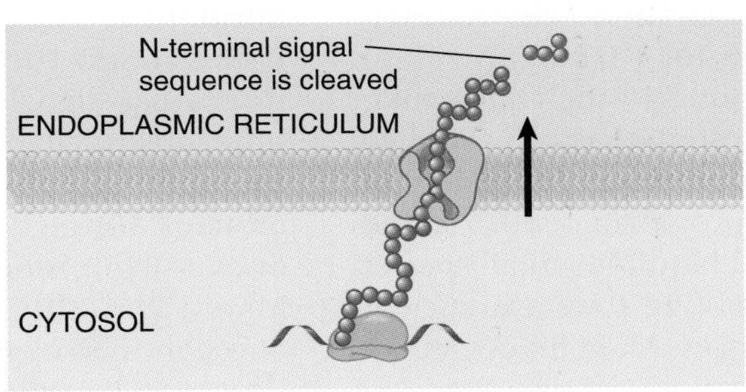

FIGURE 10.5 Proteins can enter the ER-Golgi pathway only by associating with the endoplasmic reticulum while they are being synthesized.

▸ **signal sequence** A short region of a protein that directs it to the endoplasmic reticulum for cotranslational translocation.

FIGURE 10.4 summarizes some signals used by proteins released from cytosolic ribosomes. Import into the nucleus results from the presence of a variety of rather short sequences within proteins. These **nuclear localization signals (NLS)** enable the proteins to pass through nuclear pores. One type of signal that determines transport to the peroxisome is a very short C-terminal sequence. Mitochondrial and chloroplast proteins are synthesized on free ribosomes; after their release into the cytosol they associate with the organelle membranes by means of N-terminal sequences of ~25 amino acids in length that are recognized by receptors on the organelle envelope.

▸ **nuclear localization signal (NLS)** Short sequence of amino acids that targets a polypeptide to the nucleus via the nuclear pore.

Proteins destined for the ER, Golgi, endosomes, or lysosomes enter the ER while they are being synthesized. Cotranslational translocation into the ER is depicted in **FIGURE 10.5**. An important feature of this system is that the nascent protein is responsible for recognizing the translocation apparatus. This requires the signal for cotranslational translocation to be part of the protein that is first synthesized, and, in fact, it is usually located at the N-terminus. Usually this is a sequence of 15 to 30 N-terminal amino acids that is cleaved from the protein during translocation. At or close to the N-terminus are several polar residues, and within the leader is a hydrophobic core consisting exclusively or very largely of hydrophobic amino acids. There is no other conservation of sequence. **FIGURE 10.6** gives an example.

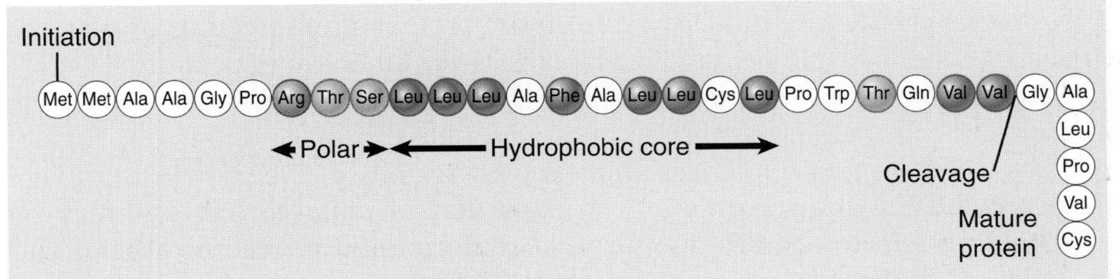

FIGURE 10.6 The signal sequence of bovine growth hormone consists of the N-terminal 29 amino acids and has a central highly hydrophobic region, preceded or flanked by regions containing polar amino acids.

ESSENTIAL IDEAS

Conservation and Function of the Hsp Family of Chaperones

Heat shock proteins (Hsps) are a group of proteins (see table) whose synthesis increases in response to sudden increases in temperature (heat shock) or exposure to other environmental stressors in normal cells. Examples of stressors include chemical and physical stressors such as infection, inflammation, starvation, oxygen deprivation, water or nitrogen deficiency, radiation, and exposure to toxins (e.g., heavy metals, arsenic, and ultraviolet light, among many others). The Hsps are sometimes referred to as stress proteins because their upregulation at a transcriptional level is triggered by a stress response. *HSP* gene transcription requires activated heat shock transcription factor 1 (HSF1).

The Hsps are named according to their molecular weights. For example, the most extensively studied (Hsp60, Hsp70, and Hsp90) are members of families of heat shock proteins that are approximately 60, 70, and 90 kilodaltons in size. Hsps are ubiquitous in all species from bacteria to humans, suggesting they are ancient in evolutionary terms and possess a high degree of sequence conservation. The function of Hsps is similar in almost all living organisms. A table of the families of Hsps and their function is provided below.

Hsps function as molecular chaperones and play a critical role in protein folding of newly synthesized and unfolded proteins and peptides, signal transduction, aggregation prevention, stress management, degradation, and protein transport. Hsp60 and Hsp70 chaperones (sometimes called *chaperonins*) recognize and bind to hydrophobic segments in extended or globular proteins as they assist protein-folding processes in almost all cellular compartments. These chaperonins utilize ATP binding and hydrolysis to drive ordered conformational changes of newly synthesized and translocated proteins to achieve their native conformational form. These conformational changes lead to stabilization of the proteins' native states and prevent unwanted protein aggregation.

In addition to general protective chaperone properties, Hsp90 preferentially acts with specific subsets of proteins of the **proteome**. The proteome is the full set of proteins expressed by a cell, tissue, organ, or whole organism found at a given time or under any particular set of conditions. Hsp90 is associated with cytoskeletal proteins such as actin and tubulin. It also plays a key role in regulating the stability and function of steroid hormone receptors, tyrosine kinases, and serine-threonine kinases. These are receptors or enzymes involved in cellular signaling pathways that regulate key cell functions such as growth, differentiation, and anti-apoptotic signaling. Unregulated activation of these proteins, through mechanisms such as point mutations or over-expression, can lead to various forms of cancer. More than 70% of the known oncogenes and proto-oncogenes involved in cancer code for tyrosine kinases. Hsps also become overexpressed in a wide variety of cancer cells. The exact mechanism of this overexpression is unknown. Research suggests that Hsp90 helps cancer cells survive.

The expression of Hsp27 is associated with poor recovery or response to chemotherapy treatment of patients with gastric, liver, prostate, and bone cancers. Hsp70 is correlated with poor prognosis in patients who have breast, endometrial, uterine, cervical, and bladder cancers. Transcription factor HSF1 is also overexpressed in cancer and plays a role in the invasion and metastasis of cancer. Metastasis is the transfer or spread of

- **proteome** The complete set of proteins that is expressed by the entire genome. Sometimes the term is used to describe the complement of proteins expressed at any one time (or set of conditions) in a given cell, tissue or organism.
- **leader** In a protein, it is a short N-terminal sequence responsible for initiating passage into or through a membrane; in mRNA, it is the untranslated sequence at the 5′ end that precedes the initiation codon.
- **preprotein** A protein to be imported into an organelle or secreted from bacteria before its signal sequence has been removed.

The N-terminal sequences that allow proteins to be transported cotranslationally to the ER or posttranslationally to mitochondria or chloroplasts are rarely retained following transport. These N-terminal sequences comprise **leaders** that are not part of the mature proteins. The protein carrying this leader is called a **preprotein** and is a transient precursor to the mature protein. The leader is cleaved from the protein during protein translocation.

Some proteins contain **signal patches** rather than N- or C-terminal sequences. These patches consist of amino acids that may be quite distant in the primary sequence of the protein, but are clustered together in the folded protein to form a functional targeting signal. These signal patches are not cleaved from the protein after targeting, as they are not discrete sequences that can be removed like the N-terminal leader peptides.

When a cytosolic protein is initially synthesized—that is to say, when it exits the ribosome to enter the cytosol—it appears in an unfolded form. Spontaneous folding then occurs as the emerging sequence interacts with regions of the protein that were synthesized previously. **Chaperones** are proteins that mediate correct assembly by causing a target protein to acquire one possible conformation instead of others. This is accomplished by binding to reactive surfaces in the target protein that are exposed

Examples of the Families of Eukaryotic/Mammalian Heat Shock Proteins

Hsp Family	Cellular Location	Function
Hsp100	Cytosol Mitochondria	ATP-dependent unfolding of protein aggregates formed by extreme heat stress so that they can be refolded by other chaperones
Hsp90	Cytosol Endoplasmic Reticulum	Conformational folding of steroid receptors, signal transduction kinases, and transcription factors
Hsp70	Cytosol Mitochondria Endoplasmic Reticulum	ATP-dependent protein folding of newly synthesized and unfolded proteins and peptides
Hsp60	Cytosol Mitochondria	ATP-dependent protein folding of newly synthesized and unfolded proteins and peptides
Hsp40	Cytosol	Co-chaperone of Hsp70: facilitates ATPase activity of Hsp70
Hsp27	Cytosol	Marker of epidermal differentiation; changes in expression have been linked with aging; increased levels of Hsp70 found in sun-protected aged skin

malignant cells from a tumor from one part of the body to another. Hsp expression appears to be an important factor in predicting cancer patient survivorship.

Clinical trials are under way to test a new class of drugs that inhibit Hsp90. *Tanespimycin* is the first anti-Hsp90 drug. Tanespimycin has been shown to induce apoptosis of multiple myeloma cell lines (multiple myeloma is a cancer of plasma cells in the bone marrow). It also inhibits expression of various cell surface cytokines, such as insulin-like growth factor 1 receptor (IGF-1R) and interleukin 6 receptor (IL-6R), which are involved in growth, survival, and drug resistance of multiple myeloma cells. Destabilization of proteins that interact with Hsp90 (e.g., p53, kinases, and steroid hormone receptors) with tanespimycin while blocking their degradation with another drug called bortezomib promotes the accumulation of cytotoxic proteins, leading to cancer cell death. At the time of this writing, tanespimycin treatment was showing antitumor activity during clinical trials that involved patients with multiple myelomas.

References

Bukau B, Horwich AL. (1998). The Hsp 70 and Hsp 60 chaperone machines. *Cell* 92:351–366.

Davenport EL, Morgan GJ, and Davies FE. (2008). Untangling the unfolded protein responses. *Cell Cycle* 7(7):865–869.

Vastag B. (2006). Hsp 90 Inhibitors promise to complement cancer therapies. *Nature Biotechnology* 24:1307.

during the assembly process and preventing those surfaces from interacting with other regions of the protein to form an incorrect conformation. Chaperones function by preventing formation of incorrect structures rather than by promoting formation of correct structures, or by providing an environment that promotes a specific folding pattern.

Chaperones are also required to assist during the transport of proteins through membranes. A persistent theme in membrane passage is that control (or delay) of protein folding is an important feature. **FIGURE 10.7** shows that it may be necessary to maintain a protein in an unfolded state before it enters the membrane because of the geometry of passage: the mature protein could simply be too large to fit into the available channel. Chaperones may prevent a protein from acquiring a conformation that would prevent passage through the membrane; in this capacity, their role is

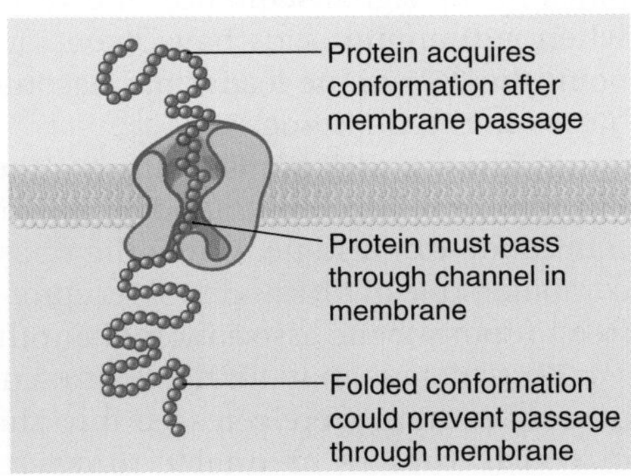
Protein acquires conformation after membrane passage
Protein must pass through channel in membrane
Folded conformation could prevent passage through membrane

FIGURE 10.7 A protein is constrained to a narrow passage as it crosses a membrane.

▸ **signal patch** A protein targeting signal consisting of amino acids that can be distant in the primary sequence of the protein but are clustered together in the folded protein to form a functional signal.

▸ **chaperone** A class of proteins that bind to incompletely folded or assembled proteins in order to assist their folding or prevent them from aggregating.

basically to maintain the protein in an unfolded, flexible state. Once the protein has passed through the membrane, it may require another chaperone to assist with folding to its mature conformation in much the same way that a cytosolic protein requires assistance from a chaperone as it emerges from the ribosome. The state of the protein as it emerges from a membrane is probably similar to that as it emerges from the ribosome—basically extended in a more or less linear condition.

KEY CONCEPTS

- Proteins pass across membranes through specialized protein structures embedded in the membrane.
- A large and complex apparatus is required for transport into the nucleus.
- Proteins that are imported into cytoplasmic organelles are synthesized on free ribosomes in the cytosol.
- Proteins that are imported into the ER-Golgi system are synthesized on ribosomes that are associated with the ER.
- Proteins associate with membranes by means of specific amino acid sequences called *signal sequences*.
- Signal sequences are most often leaders that are located at the N-terminus.
- N-terminal signal sequences are usually cleaved off the protein during the insertion process.
- Chaperones act on newly synthesized proteins, proteins that are passing through membranes, or proteins that have been denatured.

CONCEPT AND REASONING CHECK

Why are most translocation signal sequences N-terminal? Peroxisome signal sequences are C-terminal; could a C-terminal sequence function for ER targeting? Why or why not?

10.3 The Signal Sequence Interacts with the SRP

Protein translocation into the ER can be divided into two general stages: first, ribosomes carrying nascent polypeptides associate with the membranes, and then the nascent chain is transferred to the channel and translocates through it.

The attachment of ribosomes to membranes requires the **signal recognition particle (SRP)**. The SRP is located in the cytosol and has two key functions:

1. it binds to the signal sequence of a nascent protein targeted for the ER; and
2. it binds to a protein (the **SRP receptor** located in the ER membrane.

The SRP and SRP receptor function catalytically to transfer a ribosome carrying a nascent protein to the membrane. The first step is the recognition of the signal sequence by the SRP. The SRP then binds to the SRP receptor and the ribosome binds to the membrane. The stages of translation of membrane proteins are summarized in **FIGURE 10.8**.

The role of the SRP receptor in protein translocation is transient. When the SRP binds to the signal sequence, it arrests translation elongation. This usually happens when ~70 amino acids have been incorporated into the polypeptide chain (at this point the 25-residue leader has become exposed, with the next ~40 amino acids still buried in the ribosome).

When the SRP binds to the SRP receptor, the SRP releases the signal sequence. The ribosome becomes bound by another component of the membrane. At this point, translation can resume. When the ribosome has been passed on to the membrane, the combined role of SRP and SRP receptor has been completed. They now recycle and are free to sponsor the association of another nascent polypeptide with the membrane.

This process may be needed to control the conformation of the protein. If the nascent protein were released into the cytoplasm, it could take up a conformation in which it might be unable to traverse the membrane. The ability of the SRP to

- **signal recognition particle (SRP)** A ribonucleoprotein complex that recognizes signal sequences during translation and guides the ribosome to the translocation channel. Particles from different organisms may have different compositions, but all contain related proteins and RNAs.
- **SRP receptor** A dimeric protein in the ER membrane that recognizes the signal recognition particle (SRP) and is involved in the insertion of the nascent polypeptide into the ER translocon.

Inner face of ER
Cytosolic face
Leader
Ribosome initiates protein synthesis on "free" mRNA
SRPreceptor
SRP
SRP attaches to leader sequence; translation halts
SRP is bound by SRP receptor; ribosome attaches to membrane; translation resumes
Leader sequence enters membrane
Protein passes through membrane; leader is cleaved; translation continues
Protein is secreted through membrane; ribosome subunits are released from mRNA

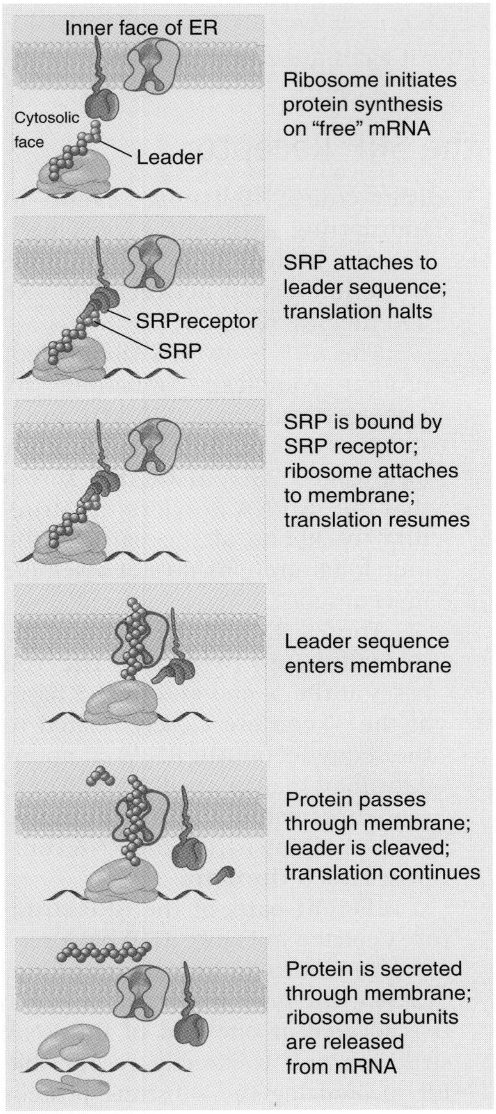

FIGURE 10.8 Ribosomes synthesizing secretory proteins are attached to the membrane via the signal sequence on the nascent polypeptide.

inhibit translation while the ribosome is being handed over to the membrane is therefore important in preventing the protein from being released into the aqueous environment.

The signal peptide is cleaved from a translocating protein by a complex of five proteins called the **signal peptidase**. The complex is several times more abundant than the SRP and SRP receptor. Its amount is roughly equivalent to the amount of bound ribosomes, suggesting that it functions in a structural capacity. It is located on the lumenal (inner) face of the ER membrane, which implies that the entire signal sequence must cross the membrane before cleavage occurs. Homologous signal peptidases can be recognized in bacteria, archaea, and eukaryotes.

▸ **signal peptidase** An enzyme within the membrane of the ER that specifically removes the signal sequences from proteins as they are translocated. Analogous activities are present in bacteria, archaea, and in each organelle in a eukaryotic cell into which proteins are targeted and translocated by means of removable targeting sequences. It is one component of a larger protein complex.

KEY CONCEPTS

- The signal sequence binds to the SRP (signal recognition particle).
- Signal-SRP binding causes protein synthesis to pause.
- Protein synthesis resumes when the SRP binds to the SRP receptor in the membrane.
- The signal sequence is cleaved from the translocating protein by the signal peptidase located on the "inside" face of the membrane.

CONCEPT AND REASONING CHECK

Why does the SRP only bind to the SRP receptor after it binds to a nascent protein?

10.4 The SRP Interacts with the SRP Receptor

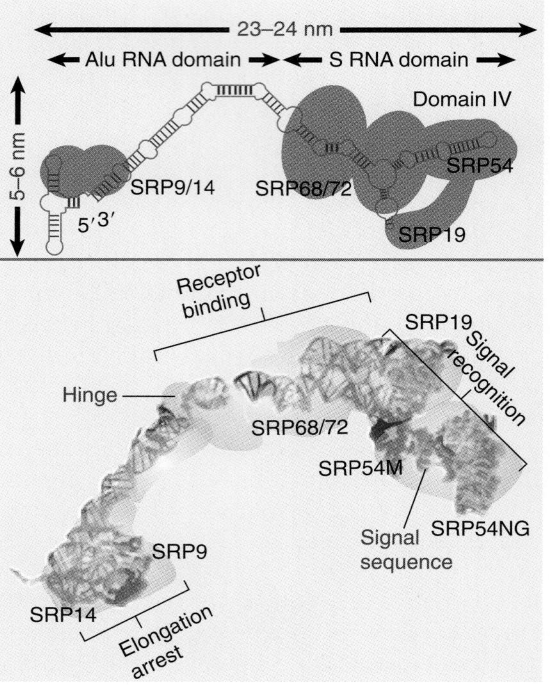

FIGURE 10.9 7S RNA of the SRP has two domains. Proteins bind as shown on the two-dimensional diagram above to form the crystal structure shown below. Each function of the SRP is associated with a discrete part of the structure.

▸ **7S RNA** A small RNA component of the signal recognition particle (SRP), transcribed by RNA polymerase III. Also known as 7SL RNA

▸ **Alu domain** The parts of the 7S RNA of the SRP that are related to Alu RNA.

▸ **S domain** The sequence of 7S RNA of the SRP that is not related to Alu RNA.

The critical initiating event in transferring a ribosome carrying a nascent protein to the membrane is the interaction between the SRP and the SRP receptor.

The SRP is an 11S ribonucleoprotein complex, containing six proteins (total mass 240 kD) and a small (305 base, 100 kD) **7S RNA** (also called 7SL). FIGURE 10.9 shows that the 7S RNA provides the structural backbone of the particle; the individual proteins do not assemble in its absence.

The 7S RNA of the SRP particle is divided into two parts. The 100 bases at the 5′ end and the 45 bases at the 3′ end are closely related to the sequence of Alu RNA, an abundant mammalian small RNA. These sequences define the **Alu domain**. The remaining part of the RNA comprises the **S domain**.

Different parts of the SRP structure depicted in Figure 10.9 have separate functions in protein targeting. SRP54 is the most important subunit. It is located at one end of the RNA structure and is directly responsible for recognizing the substrate protein by binding to the signal sequence. It also binds to the SRP receptor in conjunction with the SRP68-SRP72 dimer that is located at the central region of the RNA. The SRP9-SRP14 dimer is located at the other end of the molecule; it is responsible for elongation arrest.

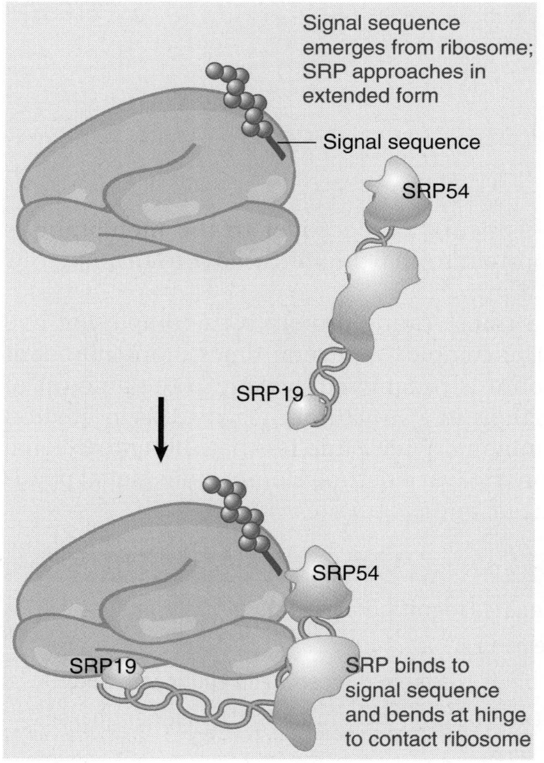

FIGURE 10.10 SRP binds to a signal sequence as it emerges from the ribosome. The binding causes the SRP to change conformation by bending at a "hinge," allowing SRP54 to contact the ribosome at the protein exit site while SRP19 makes a second set of contacts.

The SRP is a flexible structure. In its unengaged form (not bound to signal sequence), it is quite extended, as can be seen from the crystal structure in Figure 10.9. FIGURE 10.10 shows that binding to a signal sequence triggers a dramatic change of conformation. The protein bends at a hinge to allow the SRP54 end to contact the ribosome at the protein exit site, while the SRP19 swings around to contact the ribosome at the elongation factor binding site. This enables it to cause the elongation arrest that gives time for targeting to the translocation site on the membrane.

The SRP receptor is a dimer containing subunits SRα (72 kD) and SRβ (30 kD). The β subunit is an **integral membrane protein**. The amino-terminal end of the large α subunit is anchored by the β subunit. The bulk of the α protein protrudes into the cytosol. A large part of the sequence of the cytoplasmic region of the protein resembles a nucleic acid-binding protein with many positive residues. This suggests the possibility that the SRP receptor recognizes the 7S RNA in the SRP.

▸ **integral membrane protein** A protein that requires disruption of the lipid bilayer in order to be released from the membrane.

There is a counterpart to SRP in bacteria, although it contains fewer components. *E. coli* contains a 4.5S RNA that associates with ribosomes and is homologous to the 7S RNA of the SRP. It associates with two proteins: Ffh is homologous to SRP54 and FtsY is homologous to the α subunit of the SRP receptor. In fact, FtsY replaces the functions of both the α and β SRP subunits; its N-terminal domain substitutes for SRPβ in membrane targeting, and the C-terminal domain interacts with the target protein. The role of this complex is more limited than that of SRP–SRP receptor. It is probably required to keep some (but not all) secreted proteins in a conformation that enables them to interact with the secretory apparatus. This could be the original connection between protein synthesis and secretion; in eukaryotes the SRP has acquired the additional roles of causing translational arrest and targeting to the membrane.

Why should the SRP have an RNA component? The answer must lie in the evolution of the SRP: it must have originated very early in evolution, in an RNA-dominated world, possibly in conjunction with a ribosome whose functions were mostly carried out by RNA. The crystal structure of the complex between the protein-binding domain of 4.5S RNA and the RNA-binding domain of Ffh, shown in **FIGURE 10.11**, suggests that RNA continues to play a role in the function of SRP.

The 4.5S RNA has a region (domain IV) that is very similar to domain IV in 7S RNA (see Figure 10.9). Ffh consists of three domains (N, G, and M). N refers to the N-terminal region, and G refers to the GTPase domain. The M domain (named for a high content of methionines) performs the key binding functions. It has a hydrophobic pocket that binds the signal sequence of a target protein. This pocket is formed by a cleft in the protein structure lined by the hydrophobic side chains of methionine residues. Next to the pocket is a helix-turn-helix motif that is typical of DNA-binding proteins (see *Section 14.10, Repressor Uses a Helix-Turn-Helix Motif to Bind DNA*).

The crystal structure shows that the helix-loop-helix of the M domain binds to a duplex region of the 4.5S RNA in domain IV. The negatively charged backbone of the RNA is adjacent to the hydrophobic pocket. This raises the possibility that a signal sequence actually binds to both the protein and RNA components of the SRP. The signal sequence contains positively charged residues that could interact with the RNA and a hydrophobic region that could sit in the pocket.

GTP hydrolysis plays an important role in inserting the signal sequence into the membrane. Both the SRP and the SRP receptor have GTPase capability. The signal-binding subunit of the SRP, SRP54, is a GTPase, and both subunits of the SRP receptor are GTPases. All of the GTPase activities are necessary for a nascent protein to be transferred to the membrane. **FIGURE 10.12** shows that the SRP is associated with GDP when it initially binds to the signal sequence. The ribosome then stimulates replacement of the GDP with GTP.

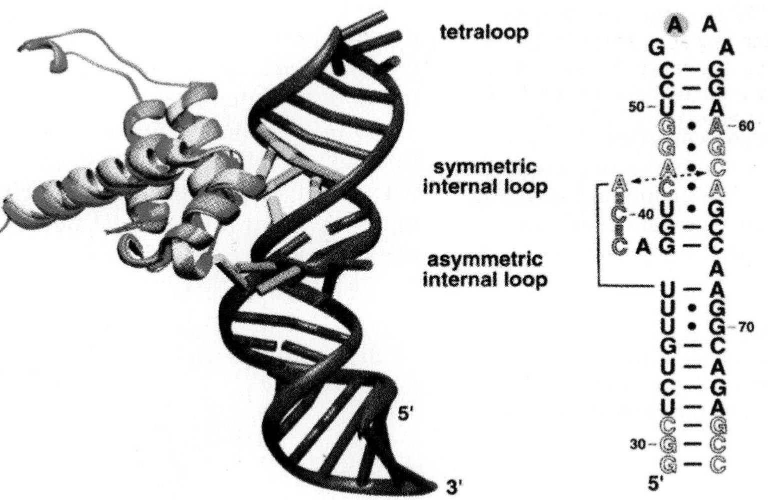

FIGURE 10.11 Crystal structure of the universally conserved ribonucleoprotein core of the *E. coli* signal recognition particle. The M domain is represented as a light blue ribbon. Nucleotide bases within the 4.5S RNA (blue) are shown as rods, with universally conserved and highly conserved nucleotides depicted as yellow and green, respectively. To the right is the secondary structure of the 4.5S RNA. Reproduced with permission from Batey, R. T., et al, *Science* 287 (2000): 1232–1239. © 2000 AAAS. Photo courtesy of Jennifer A. Doudna, University of California, Berkeley.

FIGURE 10.12 The SRP carries GDP when it binds the signal sequence. The ribosome causes the GDP to be replaced with GTP.

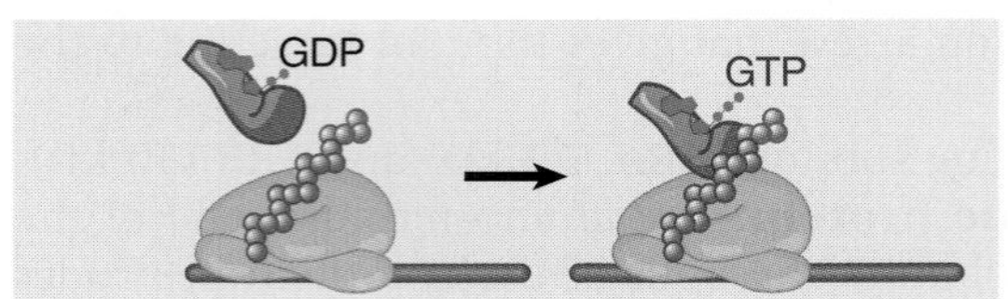

FIGURE 10.13 The SRP and SRP receptor both hydrolyze GTP when the signal sequence is transferred to the membrane.

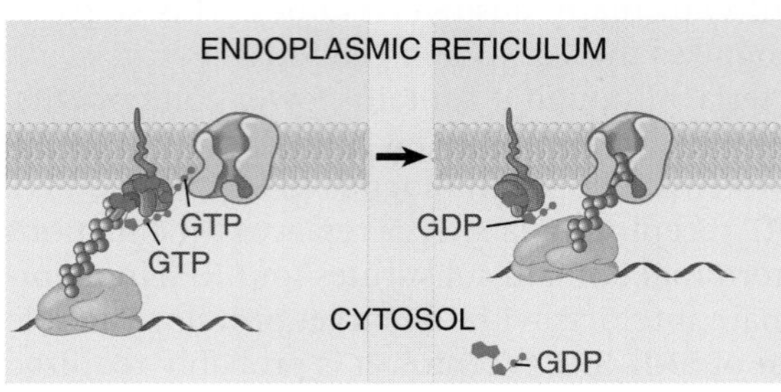

The signal sequence inhibits hydrolysis of the GTP. This ensures that the complex has GTP bound when it encounters the SRP receptor.

For the nascent protein to be transferred from the SRP to the membrane, the SRP must be released from the SRP receptor. **FIGURE 10.13** shows that this requires hydrolysis of the GTPs of both the SRP and the SRP receptor. The reaction has been characterized in the bacterial system, which has the unusual feature that Ffh activates hydrolysis by FtsY, and FtsY reciprocally activates hydrolysis by Ffh.

KEY CONCEPTS

- The SRP is a complex of 7S RNA with six proteins.
- The bacterial equivalent to the SRP is a complex of 4.5S RNA with two proteins.
- The SRP receptor is a dimer.
- GTP hydrolysis releases the SRP from the SRP receptor after their interaction.

CONCEPT AND REASONING CHECK

Why would GTP hydrolysis be important for transfer of proteins from the SRP to the membrane? Why do both SRP and the SRP receptor have to hydrolyze GTP?

10.5 The Translocon Forms a Pore

The basic problem in passing a (largely) hydrophilic protein through a hydrophobic membrane is that the interaction between the charged protein and the hydrophobic lipids is highly unfavorable. However, a protein in the process of translocation across the ER membrane can be extracted by denaturants that are effective in an aqueous environment. The same denaturants do not extract proteins that are resident components of the membrane. This suggests the model for translocation illustrated in **FIGURE 10.14**, in which proteins embedded in the ER membrane form an aqueous channel through the bilayer. A translocating protein moves through this channel, interacting with the resident proteins of the channel rather than with the lipid bilayer. The channel is *sealed* on the lumenal side to prevent free transfer of ions between the ER and the cytosol.

‣ **Sec61 complex** A heterotrimeric complex of proteins that forms a membrane-spanning pore, the translocon. Known as SecY in prokaryotes.

FIGURE 10.14 The translocon is a trimer of Sec61 that forms a channel through the membrane. It is sealed on the lumenal (ER) side.

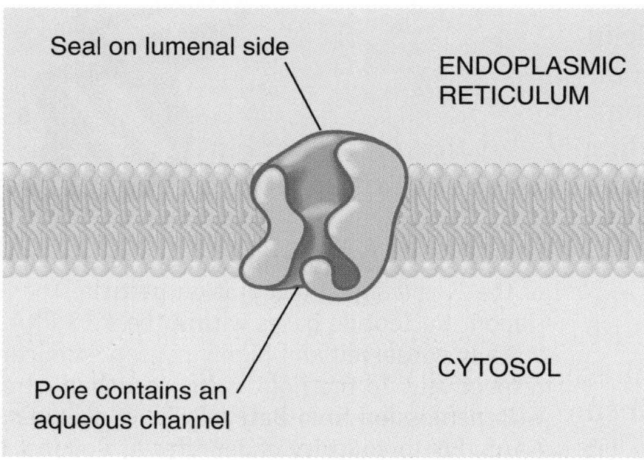

The channel through the membrane is called the *translocon*. It consists of a membrane-spanning cylinder formed from the heterotrimeric **Sec61 complex**. The Sec61 heterotrimer consists of three transmembrane proteins, named Sec61α, β, and γ in mammals. The Sec61 complex may also oligomerize to form larger aggregates, although whether this is necessary for its function is not clear.

A similar trimeric structure for the channel is found in all organisms, and the components are well-conserved in evolution. In bacteria and archaea it is called the SecY complex. The Sec61α subunit (or the corresponding SecY subunit in bacteria/archaea) provides the pore through which the protein passes and is the best conserved in sequence. Whereas a single Sec61/SecY heterotrimer can form a pore of ~15 to 20 Å, some structures have been observed in which the pore is created from dimers of the trimeric complex, organized back-to-back so that the α subunits are fused into a single channel.

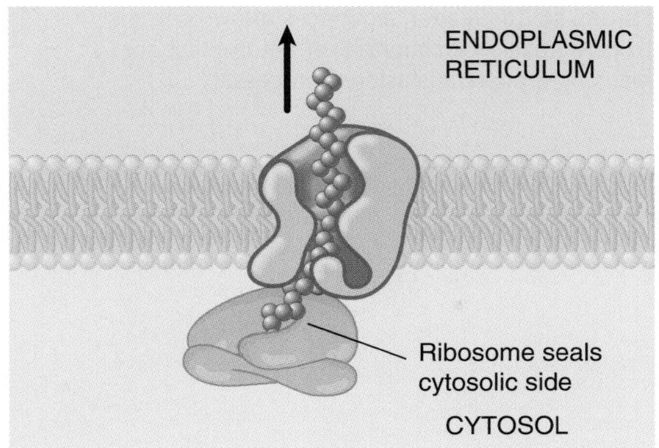

FIGURE 10.15 A nascent protein is transferred directly from the ribosome to the translocon. The ribosome seals the channel on the cytosolic side.

Access to the pore is controlled (or "gated") on both sides of the membrane. Before attachment of the ribosome, the pore is closed on the lumenal side. **FIGURE 10.15** shows that when the ribosome attaches, it seals the pore on the cytosolic side. When the nascent protein reaches a length of ~70 amino acids—most likely, when it extends fully across the channel—the pore opens on the lumenal side. The translocating protein fills the channel completely, so ions cannot pass through during translocation. Thus at all times, the pore is closed on one side or the other, maintaining the ionic integrities of the separate compartments.

In addition to its role in transferring nascent proteins into the ER lumen as discussed above, the translocon is versatile and can promote different outcomes for translocating proteins. Integral membrane proteins of the ER are also targeted to the translocon; delivery of these proteins to the membrane requires the channel to open or disaggregate in some unknown way so that the protein can move laterally into the lipid bilayer. Proteins can also be transferred from the ER back to the cytosol. This is known as **reverse translocation** and is generally used as part of a quality control system; these proteins are degraded once they reach the cytosol. The mechanism of reverse translocation is not known.

reverse translocation Transfer of proteins from the endoplasmic reticulum to the cytosol, usually for degradation.

KEY CONCEPTS

- The Sec61 trimeric complex provides the channel for proteins to pass through a membrane.
- A translocating protein passes directly from the ribosome to the translocon without exposure to the cytosol.

CONCEPT AND REASONING CHECKS

What would be the advantage of forming a pore out of two Sec61/SecY heterotrimers? Why is it necessary for the pore to be gated at all times?

10.6 Posttranslational Membrane Insertion Depends on Signal Sequences

Mitochondria and chloroplasts synthesize a small fraction of the proteins necessary for organelle function. Mitochondria synthesize only ~10 organelle proteins; chloroplasts synthesize ~50 proteins. The majority of organelle proteins are encoded by the nuclear genome and are synthesized in the cytosol by the same pool of free ribosomes that synthesize cytosolic proteins. They must then be imported into the organelle.

Many proteins that enter mitochondria or chloroplasts by a posttranslational process have signal sequences that are responsible for primary recognition of the outer membrane of the organelle. As shown in the simplified diagram in **FIGURE 10.16**, the signal sequence initiates the interaction between the precursor and the organelle

FIGURE 10.16 Leader sequences allow proteins to recognize mitochondrial or chloroplast surfaces by a posttranslational process.

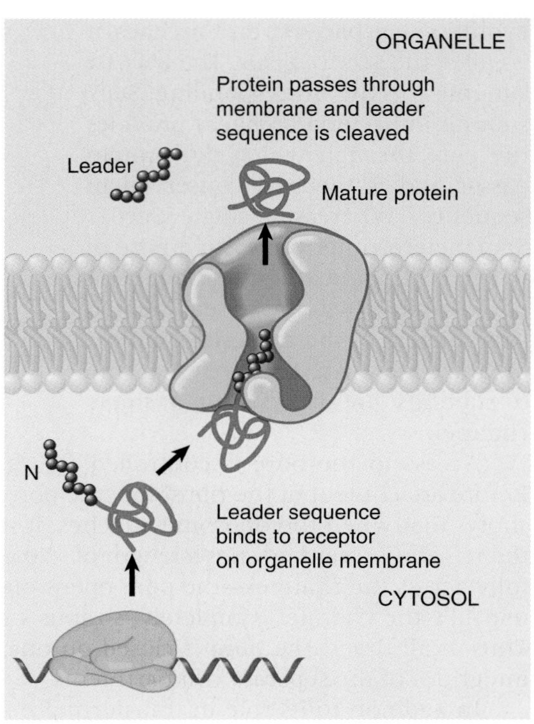

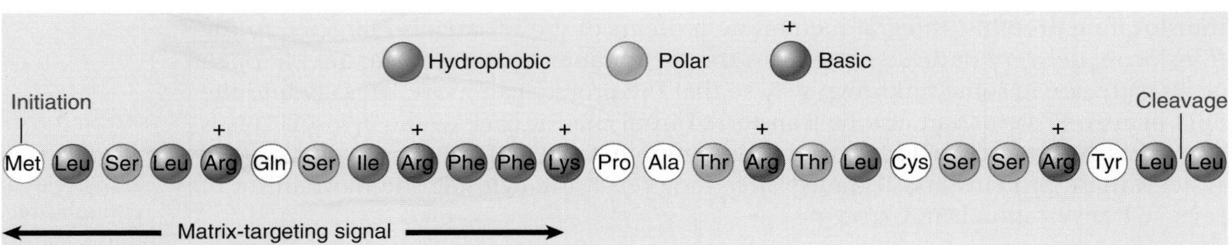

FIGURE 10.17 The leader sequence of yeast cytochrome c oxidase subunit IV consists of twenty-five neutral and basic amino acids. The first twelve amino acids are sufficient to transport any attached polypeptide into the mitochondrial matrix.

membrane. The protein passes through the membrane, and the leader is cleaved by a protease on the organelle side to produce the mature protein.

The signal sequences of proteins imported into mitochondria and chloroplasts usually have both hydrophobic and basic amino acids. They consist of stretches of uncharged amino acids interrupted by basic amino acids, and they lack acidic amino acids. There is little other homology. An example is given in **FIGURE 10.17**. Recognition of the leader does not depend on its exact sequence, but rather on its ability to form an *amphipathic helix*, in which one face consists of hydrophobic amino acids and the other face presents the basic amino acids.

The mitochondrion is surrounded by an envelope consisting of two membranes. Proteins imported into mitochondria may be targets to any of four different destinations: the outer membrane, the intermembrane space, the inner membrane, or the matrix (the internal lumen of the mitochondrion). Additionally, a protein that is a component of one of the membranes may be oriented so that it faces one side or the other. Proteins targeted to chloroplasts face similar alternatives, although the targeting in this case is even more complex, since a chloroplast consists of three membranes.

What is responsible for directing a mitochondrial or chloroplast protein to the appropriate compartment? The leader sequence actually contains all the information needed to properly localize an organelle protein. The "default" pathway for a protein imported into a mitochondrion is to move through both membranes into the matrix. A protein that is localized within the intermembrane space or in the inner membrane

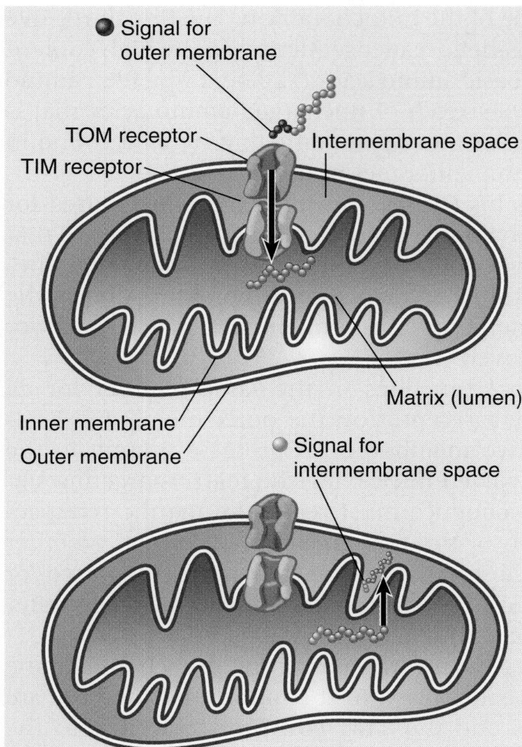

FIGURE 10.18 Mitochondria have receptors for protein transport in the outer and inner membranes. Recognition at the outer membrane may lead to transport through both receptors into the matrix, where the leader is cleaved. If it has a membrane-targeting signal, it may be re-exported.

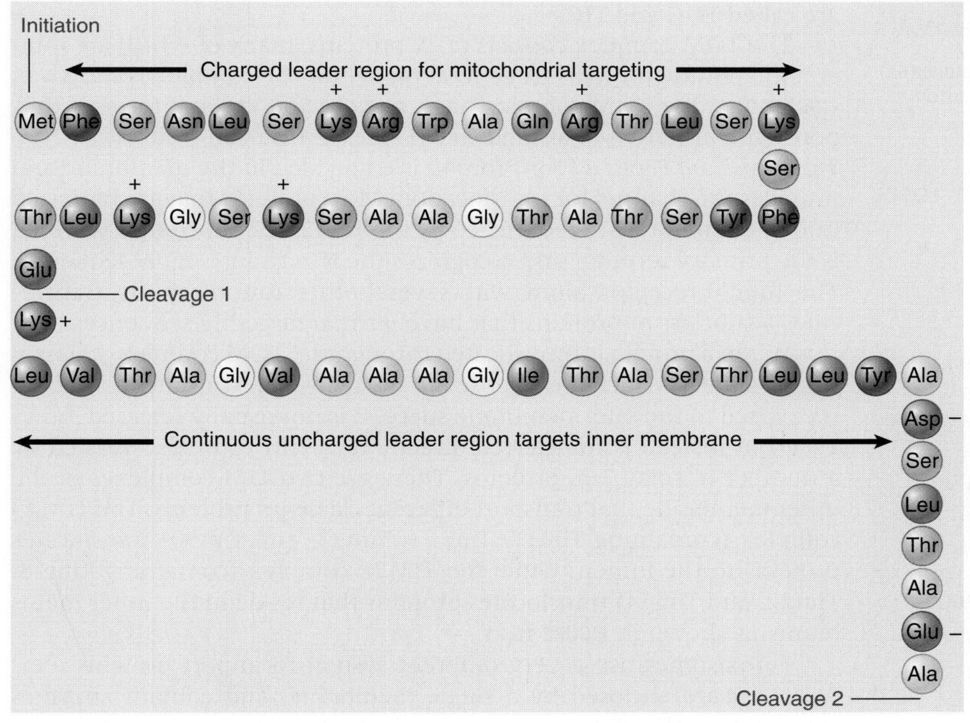

FIGURE 10.19 The leader of yeast cytochrome c1 contains an N-terminal region that targets the protein to the mitochondrion, followed by a region that targets the (cleaved) protein to the inner membrane. The leader is removed by two cleavage events.

itself requires an additional signal that specifies its destination within the organelle. A multipart leader contains signals that function in a hierarchical manner, as summarized in **FIGURE 10.18**. The first part of the leader targets the protein to the organelle, and the second part is required if its destination is elsewhere than the matrix. The two parts of the leader are removed by successive cleavages.

The two parts of a leader that contains both types of signal have different compositions. **FIGURE 10.19** shows an example of the yeast cytochrome c1 leader. The mature

- **TOM complex** A complex that resides in the outer membrane of the mitochondrion and is responsible for importing proteins from the cytosol into the space between the membranes.
- **TIM complex** A complex that resides in the inner membrane of mitochondria and is responsible for transporting proteins from the intermembrane space into the interior of the organelle.

c1 protein is located in the inner membrane of the mitochondrion. The first thirty-five N-terminal amino acids resemble other organelle leader sequences in the high content of uncharged amino acids, punctuated by basic amino acids. The next nineteen amino acids, however, comprise an uninterrupted stretch of uncharged amino acids that is long enough to span a lipid bilayer. A similar membrane-targeting signal is found in the leader sequences of proteins targeted to membranes of the ER.

Cleavage of the matrix-targeting signal is the sole processing event required for proteins that reside in the matrix. This signal must also be cleaved from proteins that reside in the intermembrane space; however, following this cleavage, the membrane-targeting signal (which is now the N-terminal sequence of the protein) directs the protein to its destination in the outer membrane, intermembrane space, or inner membrane. The signal then, in turn, is cleaved.

The N-terminal matrix-targeting signal functions in the same manner for all mitochondrial proteins. Its recognition by a receptor on the outer membrane leads to transport through the two membranes. A protease present in the mitochondrial matrix is involved in cleaving the matrix-targeting signal. Thus the N-terminal sequence must reach the matrix, irrespective of the final destination of the protein. It is not known whether entire proteins are imported into the matrix and then re-exported, or whether there is a mechanism to redirect proteins immediately after cleavage of the matrix signal sequence.

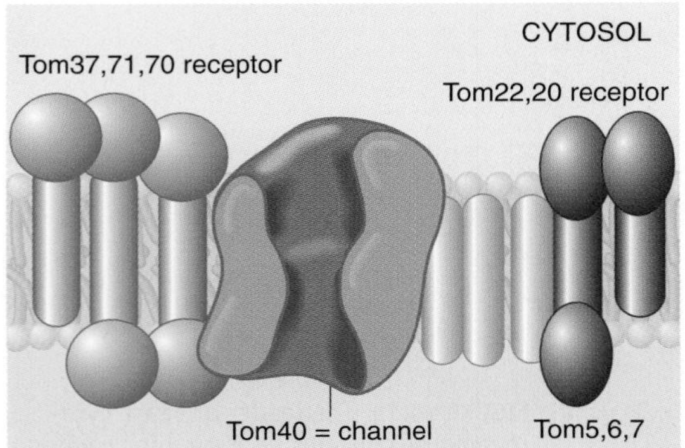

FIGURE 10.20 TOM proteins form receptor complex(es) that are needed for translocation across the mitochondrial outer membrane.

There are different receptors for transport through each membrane in the mitochondrion and chloroplast. In the mitochondrion they are called the **TOM complex** and the **TIM complex**, referring to the outer and inner membranes, respectively, and in the chloroplast they are called TOC and TIC.

The TOM complex consists of ~9 proteins, many of which are integral membrane proteins. A general model for the complex is shown in FIGURE 10.20. The TOM complex has a size of >500 kD, and forms a 20 Å pore, similar to that of a single Sec61/SecY translocon (see *Section 10.5, The Translocon Forms a Pore*). Tom40 is embedded in the membrane and provides the channel for translocation. Most proteins that are imported into mitochondria are recognized by the Tom20/22 subcomplex, which is the primary receptor and recognizes the N-terminal signal sequence. The Tom70 receptor, along with several other Tom receptors, translocates a subclass of proteins that have *internal* targeting sequences.

When a protein is translocated through the TOM complex, it passes from a state in which it is exposed to the cytosol into a state in which it is exposed to the intermembrane space. It is not usually released, however, but instead is transferred directly to a TIM complex, assisted by a number of small Tim proteins. There are two TIM complexes in the inner membrane that transport different classes of proteins. The TIM23 complex (containing Tim17, Tim 21, Tim23, and Tim50) translocates proteins to the lumen, while the TIM22 complex (containing Tim18, Tim22, and Tim54) translocates proteins that reside in the inner membrane, as shown in FIGURE 10.21.

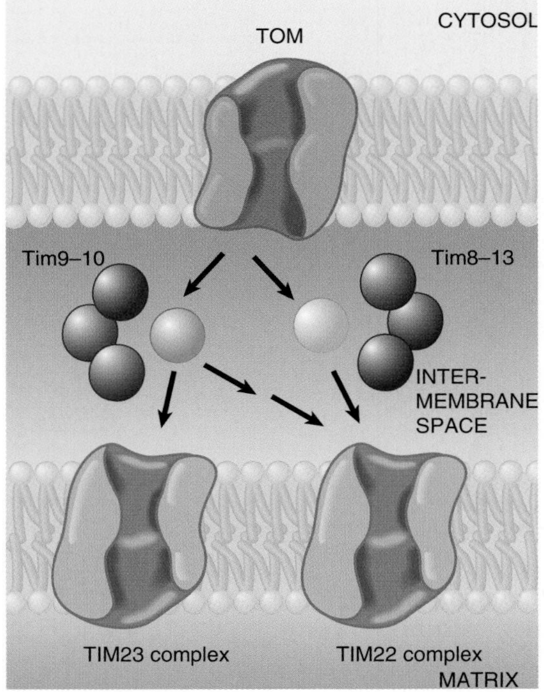

FIGURE 10.21 Proteins are transferred from TOM to either TIM complex, depending on the final destination of the protein.

Peroxisomes use a very different system to import proteins. Peroxisomes are enclosed by a single membrane, and contain enzymes concerned with oxygen utilization, which convert oxygen to hydrogen peroxide by removing hydrogen atoms from substrates. Catalase then uses the hydrogen peroxide to oxidize a variety of other substrates. Their activities are crucial for the cell. Since the fatal disease of Zellweger syndrome was found to be caused by lack of peroxisomes, >15 human diseases have been linked to disorders in peroxisome function.

All of the components of the peroxisome are imported from the cytosol, and transport of these proteins occurs posttranslationally.

Proteins that are imported into the matrix have one of two distinct short sequences, called peroxisomal targeting signal (PTS)1 and PTS2. The PTS1 signal is a tri- or tetrapeptide at the C-terminus (see Figure 10.4). The PTS2 signal is a sequence of nine amino acids that can be located near the N-terminus or internally. The peroxisomal receptors that bind the two types of signals are called Pex5p and Pex7p, respectively.

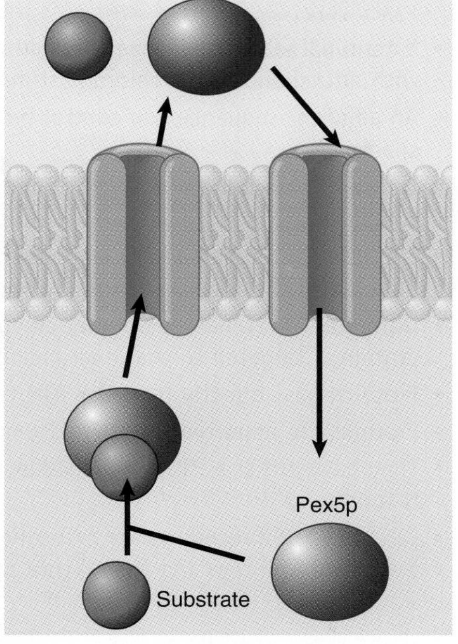

FIGURE 10.22 The Pex5p receptor binds a substrate protein in the cytosol, carries it across the membrane into the peroxisome, and then returns to the cytosol.

Transport into the peroxisome has unusual features that mark important differences from the system used for transport into other organelles.

Surprisingly, proteins can be imported into the peroxisome in their mature, fully folded state. This contrasts with the requirement to unfold a protein for passage into the ER or mitochondrion, where it passes through a channel in the membrane into the organelle in something akin to an unfolded thread of amino acids. In fact, intact multimeric protein complexes, or even proteins experimentally conjugated to large gold particles, can be imported into the peroxisome, verifying that proteins imported into the peroxisome are not restricted to a narrow channel as occurs in other organelles.

The Pex5p and Pex7p receptors are not integral membrane proteins, but rather are largely cytosolic, with only a small proportion associated with peroxisomes. They behave as carrier proteins, binding their protein cargo in the cytoplasm and delivering this cargo to the peroxisome, as shown in **FIGURE 10.22**. The import pathways converge at the peroxisomal membrane, where Pex5p and Pex7p both interact with the same membrane protein complex, which consists of Pex14p and Pex13p. Upon docking with the peroxisome, the receptor moves with its cargo through the membrane into the interior, and then returns to the cytosol to undertake another cycle. This *shuttling* behavior resembles the carrier system for import into the nucleus. All of the organelle transport systems discussed so far are summarized in **FIGURE 10.23**.

Organelle	Signal Type See Figure 10.4 for more detail	Transport Mechanism	Transport/Accessory Factors
Nucleus	Basic internal or bipartite peptide	Folded protein escorted by carrier	Nuclear Pore Complex (NPC, >30 proteins)
ER	N-terminal peptide	Unfolded protein cotranslationally fed into translocase	Signal Recognition Particle (SRP, 7S RNA + 6 proteins) SRP receptor (dimer) Translocon (Sec61 trimer)
Mitochondrion	N-terminal amphipathic helix	Delivered by cytoplasmic chaperone, unfolded protein fed through channel	Outer membrane translocon (TOM; 9 proteins) Inner membrane translocon (TIM; several complexes)
Chloroplast	N-terminal charged peptide	Delivered by cytoplasmic chaperone, unfolded protein fed through channel	Outer membrane translocon (TOC; >5 proteins) Inner membrane translocon (TIC; >7 proteins)
Peroxisome	C-terminal short peptide	Folded protein escorted by shuttling receptor	Signal receptors (Pex5, 7) PEX translocon (>4 proteins)

FIGURE 10.23 Summary of different transport mechanisms into organelles.

KEY CONCEPTS

- N-terminal leader sequences provide the information that allows proteins to associate with mitochondrion or chloroplast membranes.
- An adjacent sequence can control further targeting to a membrane or the intermembrane spaces.
- Transport through the outer and inner mitochondrial membranes uses different receptor complexes.
- The TOM (outer membrane) complex is a large complex in which substrate proteins are directed to the Tom40 channel by one of two subcomplexes.
- Different TIM (inner membrane) complexes are used depending on whether the substrate protein is targeted to the inner membrane or to the lumen.
- Proteins pass directly from the TOM to the TIM complex.
- Proteins are imported into peroxisomes in their fully folded state.
- They have either a PTS1 sequence at the C-terminus or a PTS2 sequence at the N-terminus.
- PTS1 and PTS2 receptors are cytosolic proteins that shuttle into the peroxisome carrying a substrate protein and then return to the cytosol.

CONCEPT AND REASONING CHECKS

Explain how transport systems utilize two-part, hierarchical signal sequences to sort proteins in complex organelles. How might these systems work if there was instead a single unique signal sequence for each possible destination? What is unusual about protein import into peroxisomes?

10.7 Bacteria Use Both Cotranslational and Posttranslational Translocation

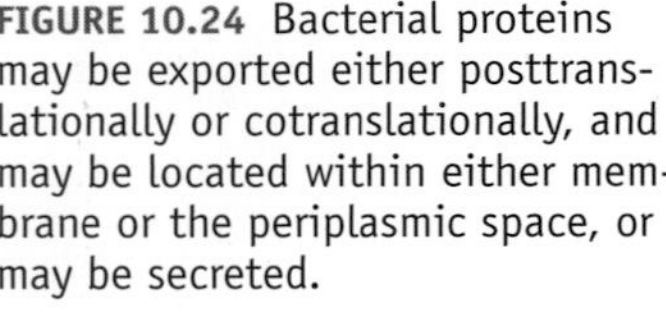

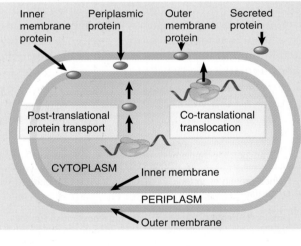

FIGURE 10.24 Bacterial proteins may be exported either posttranslationally or cotranslationally, and may be located within either membrane or the periplasmic space, or may be secreted.

▸ **periplasm (periplasmic space)** The region between the inner and outer membranes in the bacterial cell envelope.

The bacterial envelope consists of two membrane layers. The space between them is called the **periplasm.** Proteins are exported from the cytoplasm to reside in the envelope or to be secreted from the cell. The mechanisms of secretion from bacteria are similar to those characterized for eukaryotic cells, and we can recognize some related components. **FIGURE 10.24** shows that proteins that are exported from the cytoplasm have one of four fates:

- They can be inserted into the inner membrane.
- They can be translocated through the inner membrane to remain in the periplasm.
- They can be inserted into the outer membrane.
- They can be translocated through the outer membrane and secreted into the external environment.

Different protein complexes in the inner membrane are responsible for transport of proteins depending on whether their fate is to pass through or stay within the inner membrane. This resembles the situation in mitochondria, where different complexes in each of the inner and outer membranes handle different subsets of protein substrates depending on their destinations (see *Section 10.6, Posttranslational Membrane Insertion Depends on Signal Sequences*). A difference from import into organelles is that transfer in *E. coli* may be either co- or posttranslational. Some proteins are secreted both cotranslationally and posttranslationally, and the relative kinetics of translation versus secretion through the membrane could determine the balance.

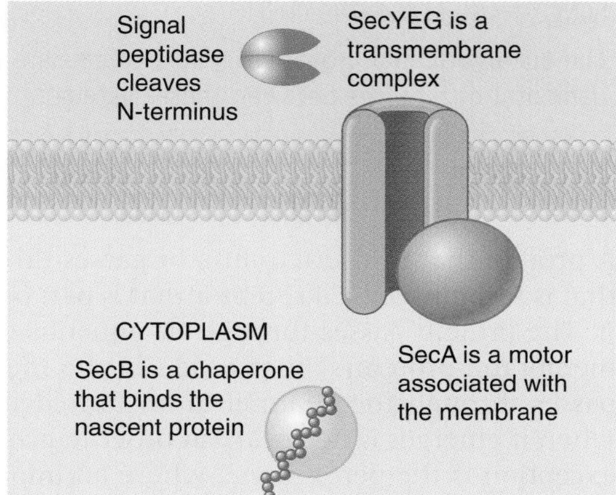

FIGURE 10.25 The Sec system has the SecYEG translocon embedded in the membrane, the SecA-associated protein that pushes proteins through the channel, the SecB chaperone that transfers nascent proteins to Sec A, and the signal peptidase that cleaves the N-terminal signal from the translocated protein.

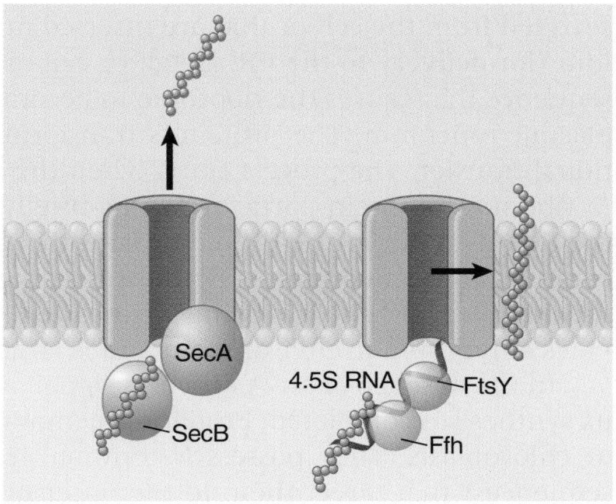

FIGURE 10.26 SecB/SecA transfer proteins to the translocon that pass through the membrane. 4.5S RNA transfers proteins that enter the membrane.

Exported bacterial proteins have N-terminal leader sequences with a hydrophilic N-terminus and an adjacent hydrophobic core, similar to the signal sequences used for protein targeting in eukaryotes. The leader is cleaved by a signal peptidase that recognizes precursor forms of several exported proteins. The signal peptidase is an integral membrane protein located in the inner membrane.

There are several systems for transport through the inner membrane. The best characterized is the Sec system, whose components are shown in **FIGURE 10.25**. The translocon that is embedded in the membrane consists of three subunits that are related to the components of mammalian/yeast Sec61 (see *Section 10.5, The Translocon Forms a Pore*). Each of the subunits is an integral transmembrane protein. The functional translocon is a trimer with one copy of each subunit. The major pathway for directing proteins to the translocon is mediated by the proteins SecB and SecA. SecB binds to the nascent protein and transfers the protein to SecA, which in turn transfers it to the translocon. Bacteria have two alternative routes to direct proteins to the same translocon as shown in **FIGURE 10.26**.

SecB behaves as a chaperone and binds to a nascent protein to retard folding. SecB also has an affinity for the protein SecA. This allows it to target a precursor protein to the membrane. SecA is an ATPase that provides the motor that pushes the substrate protein through the SecYEG translocon. The SecB/SecYEG pathway is used for translocation of proteins that are secreted into the periplasm.

The *E. coli* ribonucleoprotein complex of 4.5S RNA with Ffh and FtsY proteins is a counterpart to the eukaryotic SRP (see *Section 10.4, The SRP Interacts with the SRP Receptor*). It is needed for the secretion of some, but not all, proteins. As we see in Figure 10.24, its substrates are integral membrane proteins. The basis for differential selection of substrates is that the *E. coli* SRP recognizes an **anchor sequence** in the protein (anchor sequences by definition are present only in integral membrane proteins).

‣ **anchor sequence** A segment of a transmembrane protein that resides in the membrane.

KEY CONCEPT

- Bacterial proteins that are exported to or through membranes use both cotranslational and posttranslational mechanisms.

CONCEPT AND REASONING CHECK

The eukaryotic and prokaryotic SRP systems are related. What are the similarities and fundamental differences between these systems?

10.8 Summary

A protein that is inserted into, or passes through, a membrane has a signal sequence that is recognized by a receptor that is part of the membrane or that can associate with it. The protein passes through an aqueous channel (a pore) that is created by transmembrane protein(s) that reside in the membrane. In almost all cases, the protein passes through the channel in an unfolded form, and association with chaperones when it emerges is necessary in order to acquire the correct conformation. The major exception is the peroxisome, where an imported protein in its mature conformation binds to a cytosolic protein that carries it through the channel in the membrane.

Synthesis of proteins in the cytosol starts on "free" ribosomes. Proteins that are secreted from the cell or that are inserted into membranes of the endoplasmic reticulum (for delivery to the ER, Golgi, or cell membrane) start with an N-terminal signal sequence that causes the ribosome to become attached to the membrane of the endoplasmic reticulum. The protein is translocated through the membrane by cotranslational transfer. The process starts when the signal sequence is recognized by the SRP (a ribonucleoprotein particle), which briefly interrupts translation. The SRP binds to the SRP receptor in the ER membrane and transfers the signal sequence to the Sec61 complex in the membrane. Synthesis resumes, and the protein is translocated through the membrane while it is being synthesized. The channel through the membrane provides a hydrophilic environment and is largely made of the protein Sec61.

In the absence of any particular signal, a protein is released into the cytosol when its synthesis is completed. Proteins are imported posttranslationally into mitochondria or chloroplasts. They possess N-terminal leader sequences (or occasionally internal sequences) that target them to the outer membrane of the organelle envelope; they then are transported through the outer and inner membranes into the matrix. The N-terminal leader is cleaved by a protease within the organelle. Proteins that reside within the membranes or intermembrane space possess a signal (which becomes N-terminal when the first part of the leader is removed) that either causes export from the matrix to the appropriate location or halts transfer before all of the protein has entered the matrix.

Mitochondria (and chloroplasts) have separate receptor complexes that create channels through each of the outer and inner membranes. All imported proteins pass directly from the TOM complex in the outer membrane to a TIM complex in the inner membrane. Proteins that reside in the intermembrane space or in the outer membrane are re-exported from the TIM complex after entering the matrix. The TOM complex uses different receptors for imported proteins depending on whether they have N-terminal or internal signal sequences and directs both types into the Tom40 channel. There are two TIM receptors in the inner membrane: one is used for proteins whose ultimate destination is the inner matrix; the other is used for proteins that are re-exported to the intermembrane space or the outer membrane.

Bacteria have components for membrane translocation that are related to those of the cotranslational eukaryotic system, but translocation often occurs by a posttranslational mechanism. SecYEG forms the translocon, and SecA associates with the channel and is involved in inserting and propelling the substrate protein. SecB is a chaperone that brings the protein to the channel. Some integral membrane proteins are inserted into the channel by an interaction with an apparatus resembling the SRP, which consists of 4.5S RNA and the Ffh and FtsY proteins.

CHAPTER QUESTIONS

1. What three components are sufficient to insert a nascent protein into a membrane translocon?
2. List four possible destinations for proteins imported into mitochondria.
3. Match the following cellular destination for proteins with the proper characteristics of the targeting signal for the protein from the list at the right.

Nucleus	**A.** N-terminal amphipathic helix of 12–30 amino acids
Mitochondrion	**B.** C-terminal short peptide of 3–4 amino acids
Chloroplast	**C.** Internal basic or bipartite signal of 4–9 amino acids
Peroxisome	**D.** N-terminal charged signal of >25 amino acids

4. Place the following events in synthesis of secretory proteins in their proper order:
   A. Leader sequence enters membrane.
   B. SRP attaches to the leader sequence.
   C. Ribosome subunits are released from the mRNA.
   D. SRP is bound by SRP receptor and ribosome attaches to membrane.
   E. Ribosome initiates protein synthesis on free mRNA.
   F. Protein passes through membrane and leader is cleaved.
5. Cotranslational translocation is most commonly used for proteins that will:
   A. be exported from the cell.
   B. be transported into the nucleus.
   C. enter the endoplasmic reticulum for processing.
   D. become part of cellular structures such as the cytoskeleton.
6. Proteins that are imported into the endoplasmic reticulum typically are characterized by:
   A. a cleavable leader sequence of 15–30 amino acids with a highly hydrophobic region flanked by polar amino acids.
   B. a cleavable leader sequence of 15–30 amino acids with a highly hydrophilic region flanked by polar amino acids.
   C. a cleavable leader sequence of 15–30 amino acids with a highly hydrophobic region flanked by nonpolar amino acids.
   D. a cleavable leader sequence of 15–30 amino acids with a highly hydrophilic region flanked by nonpolar amino acids.
7. What is the default pathway for protein import into mitochondria (i.e., what is the first to happen)?
   A. import into the intermembrane space
   B. import into the mitochondrial outer membrane
   C. import into the mitochondrial inner membrane
   D. import into the mitochondrial matrix
8. Zellweger syndrome in humans (fatal) is caused by a lack of:
   A. mitochondria.
   B. endoplasmic reticulum.
   C. peroxisomes.
   D. liposomes.
9. What is unique about protein import into peroxisomes?
   A. They can be imported by the C-terminus first.
   B. They can be imported by the N-terminus first.
   C. They can be imported as a loop with the central region first.
   D. They can be imported in their fully folded state.
10. Explain the problem of passing a hydrophilic protein through a lipid bilayer membrane.

KEY TERMS

Alu domain
anchor sequence
carrier protein
chaperone
cotranslational translocation
endoplasmic reticulum
Golgi apparatus
integral membrane protein
leader
nuclear localization signal (NLS)
nuclear pore
periplasm (periplasmic space)
peroxisome
plastids
posttranslational translocation
preprotein
protein translocation
proteome
reverse translocation
S domain
Sec61 complex
signal patch
signal peptidase
signal recognition particle (SRP)
signal sequence
SRP receptor
TIM complex
TOM complex
translocon
7S RNA

FURTHER READING

Brown L. A. and Baker A. (2008). Shuttles and cycles: transport of proteins into the peroxisome matrix (review). *Mol. Memb. Biol.* 25(5): 363–375.

A review of the protein transport into peroxisomes in mammals, yeast and plants.

Driessen A. J. M. and Nouwen N. (2008). Protein translocation cross the bacterial cytoplasmic membrane. *Ann. Rev. Biochem.* 77: 643–667.

A review of protein translocation in prokaryotes that discusses the signal recognition particle (SRP) and the Sec translocase.

Neupert W. and Herrmann J. M. (2007). Translocation of proteins into mitochondria. *Ann. Rev. Biochem* 76: 723–749.

A detailed review of the current understanding of mitochondrial protein sorting and import.

Pool M. R. (2005). Signal recognition particles in chloroplasts, bacteria, yeast and mammals (review). *Mol. Memb. Biol.* 22(1–2): 3–15.

A review of signal recognition particles across many species.

III

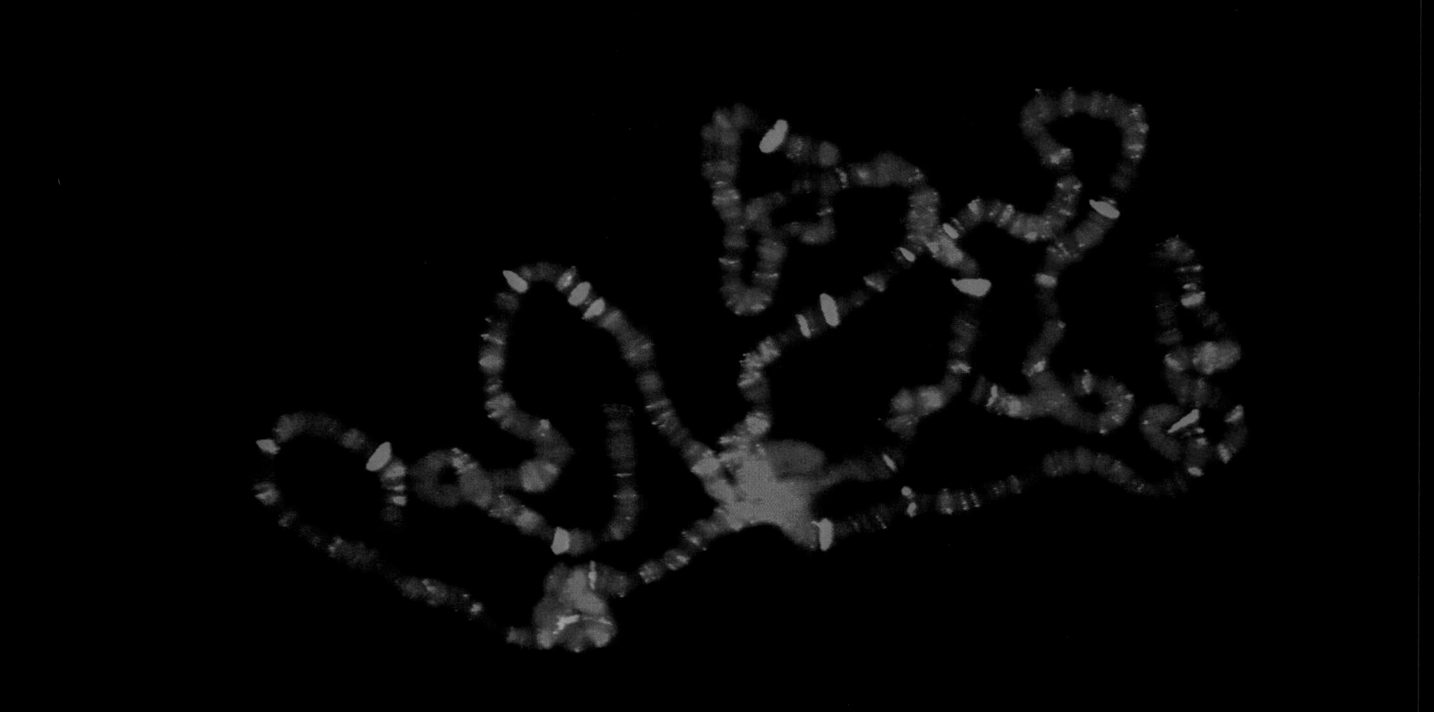

A *Drosophila* salivary gland chromosome stained so that methylated histone H4 appears red and RNA polymerase II appears green. Note that sites of active transcription do not overlap with this particular site of histone methylation (histone H4 K20). Reproduced from Karachentsev, D., et al., *Genes Dev.* 19 (2005): 431–435. Used with permission of Cold Spring Harbor Laboratory Press. Photo courtesy of Ruth Steward, Waksman Institute, Rutgers University.

Prokaryotic Gene Expression

11

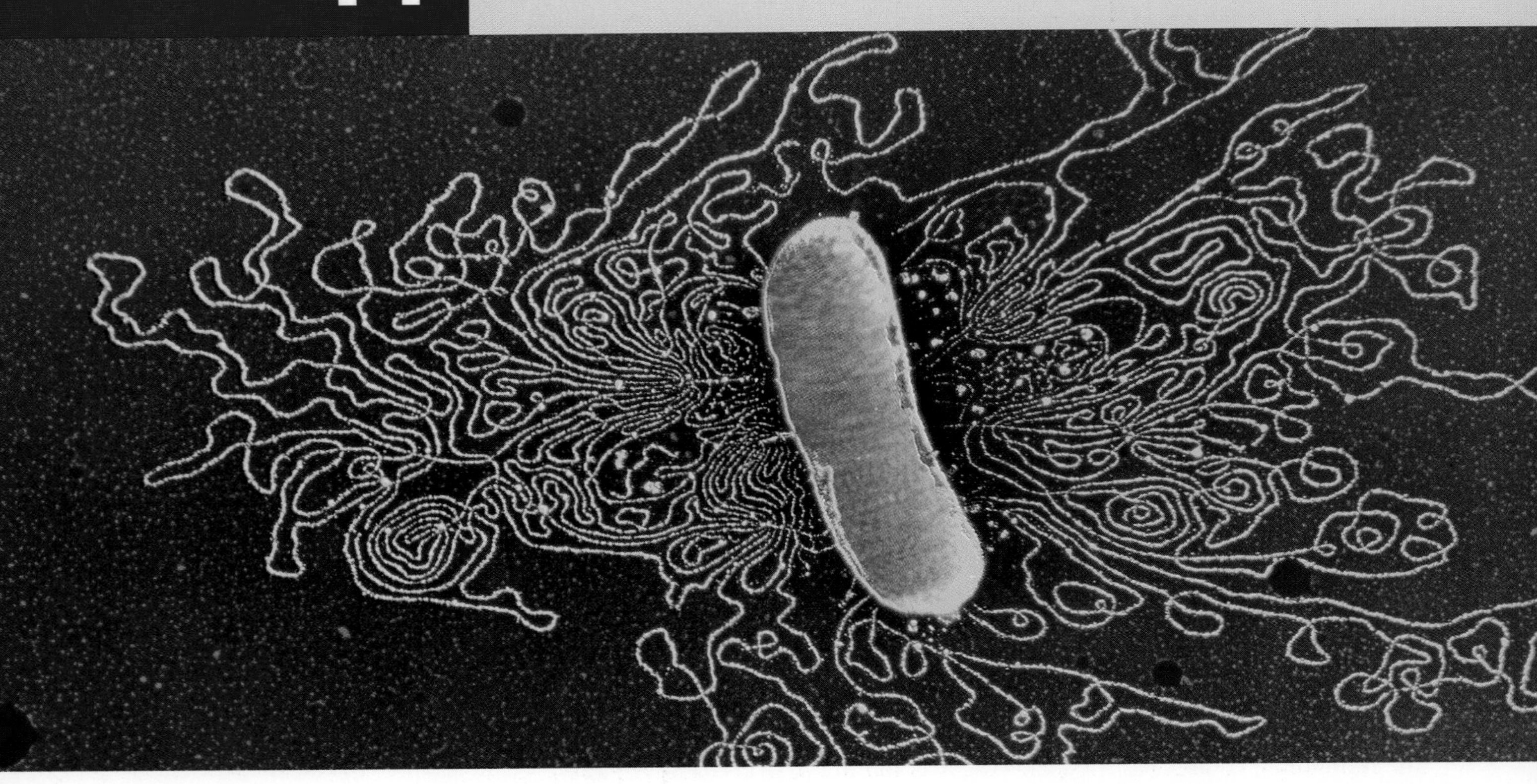

A transmission electron micrograph of the bacterium *E. coli* (after treatment to break its cell wall) surrounded by its DNA (colored gold). © Dr. Gopal Murti/ Photo Researchers, Inc.

Bacterial Transcription

CHAPTER OUTLINE

11.1 Introduction

Transcription produces an RNA chain representing a copy of one strand of a DNA duplex. The newly synthesized RNA transcript is made 5′ to 3′ from a **template strand** that is 3′ to 5′ (FIGURE 11.1). Note that the RNA transcript is *complementary* to the template and has the same sequence as the **coding strand**.

RNA synthesis is catalyzed by the enzyme **RNA polymerase**. Transcription starts when RNA polymerase binds to a special region, the **promoter**, at the start of the gene. From the promoter, RNA polymerase moves along the template, synthesizing RNA, until it reaches a **terminator** (t) sequence. This action defines a **transcription unit** that extends from the promoter to the terminator. The critical feature of the transcription unit, depicted in FIGURE 11.2, is that it constitutes a stretch of DNA used as a template for the production of a *single* RNA molecule. A transcription unit may include more than one gene.

Sequences prior to the **startpoint**, the first base that is transcribed into RNA, are described as **upstream** of it; those after the startpoint (within the transcribed sequence) are **downstream** of it. Sequences are conventionally written so that transcription proceeds from left (upstream) to right (downstream). This corresponds to writing the mRNA in the usual 5′ to 3′ direction.

The DNA sequence often is written to show only the coding strand, which has the same sequence as the RNA. Base positions are numbered in both directions away from the startpoint, which is assigned the value +1; numbers increase as they go downstream. The base before the startpoint is numbered –1, and the negative numbers increase going upstream. (There is no base assigned the number 0.)

The immediate product of transcription is called the **primary transcript**. It consists of an RNA extending from the promoter to the terminator and possesses the original 5′ and 3′ ends. The primary transcript is, however, almost always immediately modified. In prokaryotes, it is rapidly degraded (mRNA) or cleaved to give mature products (rRNA and tRNA). Transcription is the first stage in gene expression and the principal step at which it is controlled. Regulatory proteins determine whether a particular gene is available to be transcribed by RNA polymerase. The initial (and often the only) step in regulation is the decision of whether or not to transcribe a gene. Most regulatory events occur at the initiation of transcription, although subsequent stages in transcription (or other stages of gene expression) are sometimes regulated.

Within this context, there are two basic questions in gene expression:

- How does RNA polymerase find promoters on DNA? This is a particular example of a more general question: How do proteins read the DNA sequence to find their specific binding sites in DNA?

- **template strand** The DNA strand that is copied by the polymerase.
- **coding strand** The DNA strand that has the same sequence as the mRNA and is related by the genetic code to the protein sequence that it represents.
- **RNA polymerase** An enzyme that synthesizes RNA using a DNA template (formally described as a DNA-dependent RNA polymerase).
- **promoter** A region of DNA where RNA polymerase binds to initiate transcription.
- **terminator** A sequence of DNA that causes RNA polymerase to terminate transcription.
- **transcription unit** The sequence between sites of initiation and termination by RNA polymerase; it may include more than one gene.
- **startpoint** The position on DNA corresponding to the first base incorporated into RNA.
- **upstream** Sequences in the opposite direction from expression.
- **downstream** Sequences proceeding farther in the direction of expression within the transcription unit.
- **primary transcript** The original unmodified RNA product corresponding to a transcription unit.

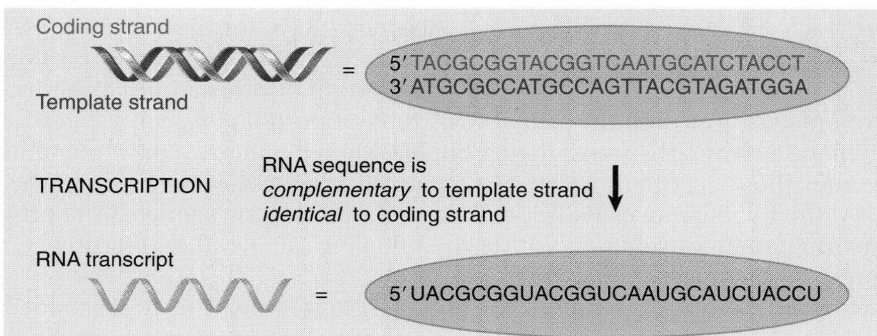

FIGURE 11.1 RNA is transcribed 5′ to 3′ from a 3′ to 5′ template.

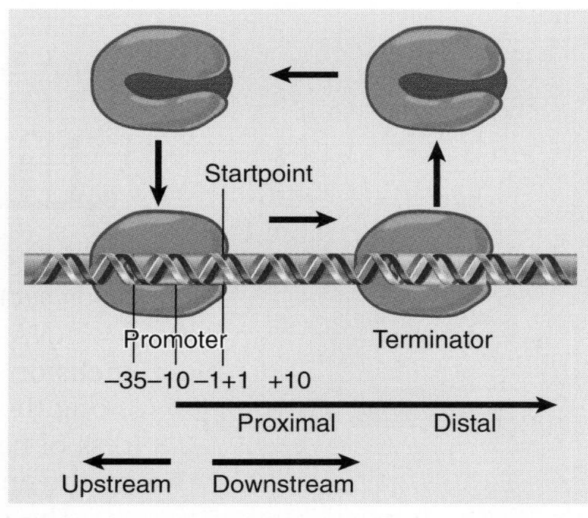

FIGURE 11.2 A transcription unit is a sequence of DNA transcribed into a single RNA, starting at the promoter and ending at the terminator.

- How do regulatory proteins interact with RNA polymerase (and with one another) to activate or to repress specific steps in the initiation, elongation, or termination of transcription?

In this chapter, we analyze the interactions of bacterial RNA polymerase with DNA from its initial contact with a gene, through the act of transcription, and then finally its release when the transcript has been completed. *Chapter 12, The Operon,* describes the various means by which regulatory proteins can assist or prevent bacterial RNA polymerase from recognizing a particular gene for transcription. *Chapter 13, Regulatory RNA,* discusses other means of regulation, including the use of small RNAs, and considers how these interactions can be connected into larger regulatory networks. In *Chapter 14, Phage Strategies,* we consider how individual regulatory interactions can be connected into more complex networks. In *Chapter 25, Eukaryotic Transcription* and *Chapter 26 Eukaryotic Transcription Regulation,* we consider the analogous reactions between eukaryotic RNA polymerases and their templates.

CONCEPT AND REASONING CHECK

RNA is synthesized from 5′ to 3′. Why is the template 3′ to 5′?

11.2 Transcription Occurs by Base Pairing in a "Bubble" of Unpaired DNA

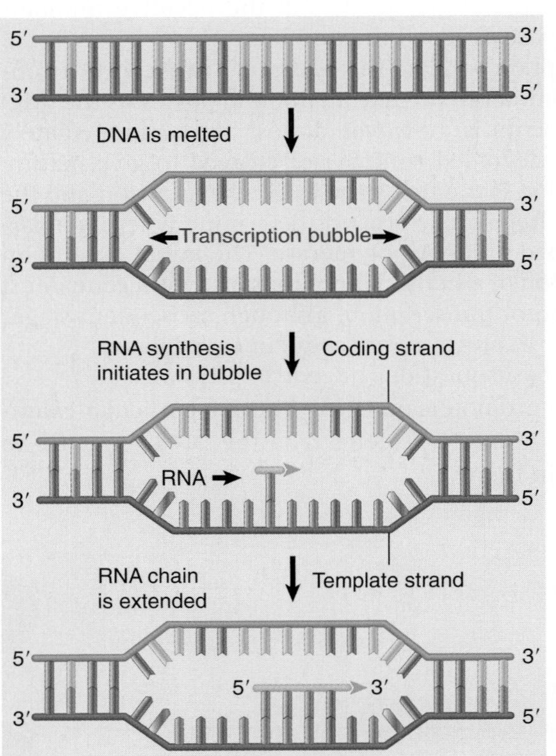

FIGURE 11.3 DNA strands separate to form a transcription bubble. RNA is synthesized 5′ to 3′ by complementary base pairing with the template strand.

Transcription takes place by the usual process of complementary base pairing. **FIGURE 11.3** illustrates the general principle of transcription. RNA synthesis takes place within a "transcription bubble," in which DNA is transiently separated into its single strands and the template strand is used to direct synthesis of the RNA strand.

The RNA chain is synthesized from the 5′ end toward the 3′ end from a template that runs from 3′ to 5′. The 3′—OH group of the last nucleotide added to the chain reacts with an incoming nucleoside 5′ triphosphate. The incoming nucleotide loses its terminal two phosphate groups (γ and β); its α group is used in the phosphodiester bond linking it to the growing chain. The overall reaction rate is ~40 nucleotides/second at 37°C for the bacterial RNA polymerase; this is about the same as the rate of translation (15 amino acids/sec), but much slower than the rate of DNA replication (800 bp/sec).

RNA polymerase creates the transcription bubble when it binds to a promoter and melts or separates the two strands. **FIGURE 11.4** shows that as RNA polymerase moves along the DNA, the bubble moves with it and the RNA chain grows longer. The process of base pairing and base addition within the bubble is catalyzed and scrutinized by the enzyme.

The structure of the bubble within RNA polymerase is shown in the expanded view of **FIGURE 11.5**. As RNA polymerase moves along the DNA template, it unwinds the duplex at the front of the bubble (the unwinding point), and rewinds the DNA at

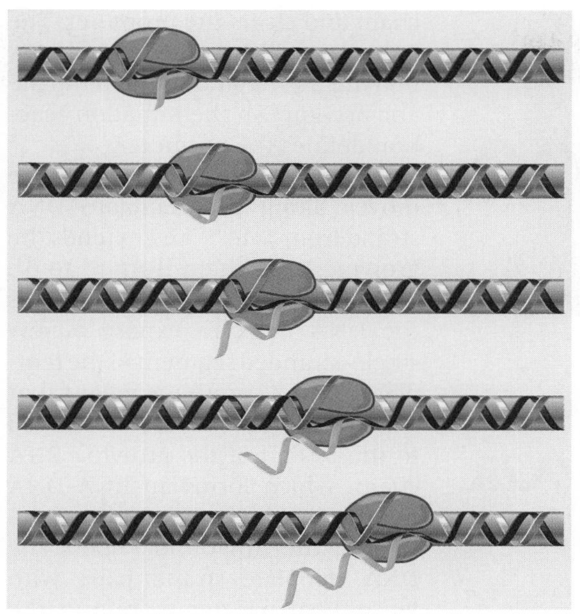

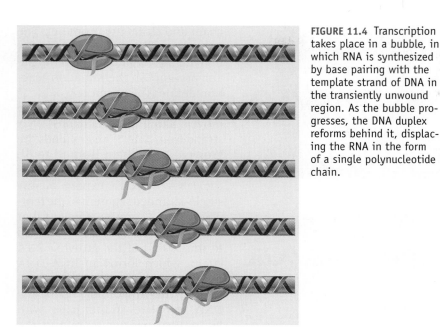

FIGURE 11.4 Transcription takes place in a bubble, in which RNA is synthesized by base pairing with the template strand of DNA in the transiently unwound region. As the bubble progresses, the DNA duplex reforms behind it, displacing the RNA in the form of a single polynucleotide chain.

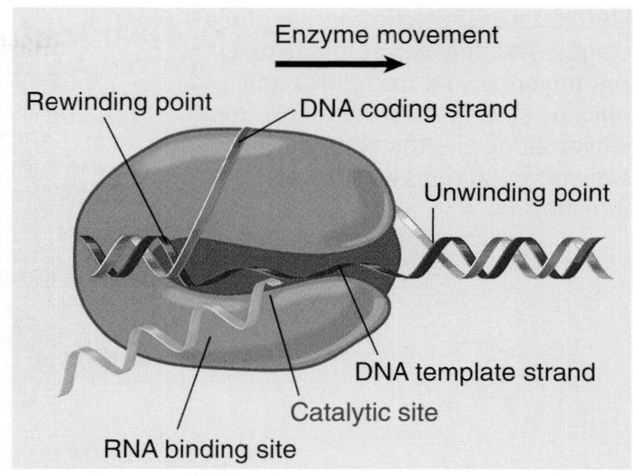

FIGURE 11.5 During transcription, the bubble is maintained within RNA polymerase, which unwinds and rewinds DNA and synthesizes RNA.

the back (the rewinding point). The length of the transcription bubble is ~12 to 14 bp, but the length of the RNA-DNA hybrid region within it is ~8 to 9 bp. As the enzyme moves on, the DNA duplex reforms, and the RNA is displaced as a free polynucleotide chain. Roughly the last twenty-five ribonucleotides added to a growing chain are complexed with DNA and/or enzyme at any moment.

KEY CONCEPTS

- RNA polymerase separates the two strands of DNA in a transient "bubble" and uses one strand that runs 3′ to 5′ as a template to direct synthesis of a complementary sequence of RNA running 5′ to 3′.
- The length of the bubble is ~12 to 14 bp, and the length of RNA-DNA hybrid within it is ~8 to 9 bp.

CONCEPT AND REASONING CHECK

If the RNA polymerase is moving 3′ to 5′ along the template strand, why are nucleotides being added to the 3′ end of the growing chain?

11.3 The Transcription Reaction Has Three Stages

The transcription reaction can be arbitrarily divided into the three stages illustrated in FIGURE 11.6: **initiation**, in which the promoter is recognized, a bubble is created, and RNA synthesis begins; **elongation**, as the bubble moves along the DNA; and finally **termination**, in which the RNA transcript is released and the bubble closes.

Initiation is the most complex part of transcription and consists of multiple steps. It begins with template recognition, the binding of RNA polymerase to the double-stranded DNA at a DNA sequence called the *promoter* to form a **closed complex**. The polymerase then separates the strands of DNA to form the **open complex** that makes the template strand available for base pairing with incoming ribonucleotides. The transcription bubble is created by local unwinding that begins at the promoter and is catalyzed by RNA polymerase. The next step is the synthesis of the first nucleotide bonds in RNA. The enzyme remains at the promoter while it synthesizes the first ~9 nucleotide bonds. The initiation phase is protracted by the occurrence of abortive events, in which the enzyme makes short transcripts, releases them, and then starts synthesis of RNA again. The initiation phase ends when the enzyme succeeds in extending the

- **initiation** The stages of transcription up to synthesis of the first bond in RNA. This includes binding of RNA polymerase to the promoter and melting a short region of DNA into single strands.
- **elongation** The stage in a macromolecular synthesis reaction (replication, transcription, or translation) when the nucleotide or polypeptide chain is extended by the addition of individual subunits.
- **termination** A separate reaction that ends a macromolecular synthesis reaction (replication, transcription, or translation), by stopping the addition of subunits, and (typically) causing disassembly of the synthetic apparatus.
- **closed complex** The stage of initiation of transcription before RNA polymerase causes the two strands of DNA to separate to form the "transcription bubble." The DNA is double stranded.
- **open complex** The stage of initiation of transcription when RNA polymerase causes the two strands of DNA to separate to form the "transcription bubble."

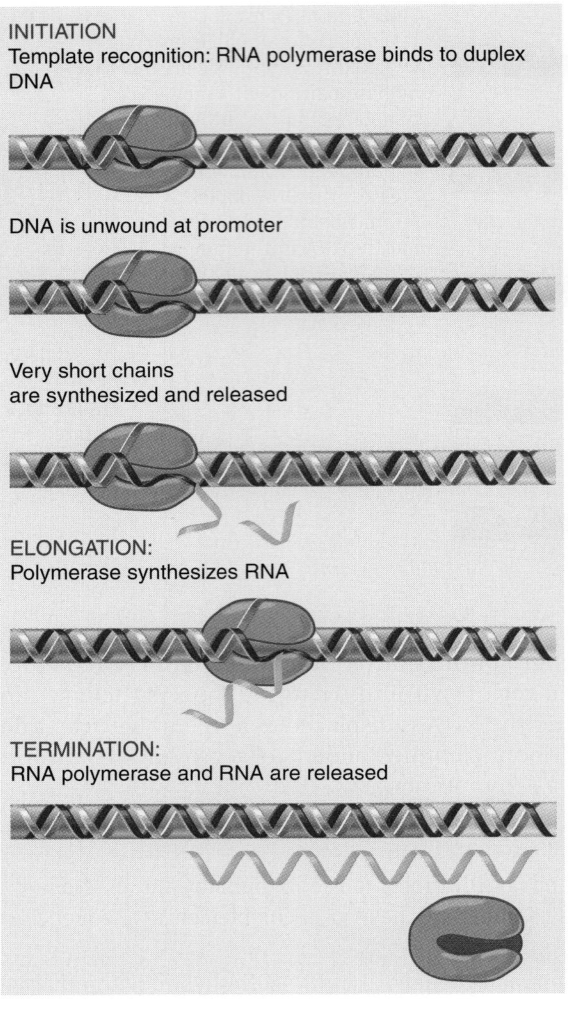
INITIATION
Template recognition: RNA polymerase binds to duplex DNA
DNA is unwound at promoter
Very short chains
are synthesized and released
ELONGATION:
Polymerase synthesizes RNA
TERMINATION:
RNA polymerase and RNA are released

FIGURE 11.6 Transcription has three stages: The polymerase binds to the promoter and melts DNA and remains stationary during *initiation*; moves along the template during *elongation*; and dissociates at *termination*.

chain and clears the promoter. The sequence of DNA needed for RNA polymerase to bind to the template and accomplish the initiation reaction defines the promoter.

During elongation the enzyme moves along the template DNA strand from 3′ to 5′ and extends the growing RNA chain from 5′ to 3′. As the enzyme moves, it unwinds the DNA helix to expose a new single-stranded segment of the template. 5′ to 3′ synthesis means that nucleotides are covalently added to the 3′ end of the growing RNA chain, which forms an RNA-DNA hybrid in the unwound region. Behind the unwound region, the DNA template strand pairs with its original partner to reform the double helix as the bubble moves. The RNA emerges as a free single strand. Elongation involves the movement of the transcription bubble by melting the DNA structure. The transiently unwound region is paired with the nascent RNA at the growing point.

Termination involves recognition of the point at which no further bases should be added to the chain; this point is, like the promoter, a specific DNA sequence. To terminate transcription, the formation of phosphodiester bonds must cease, and the transcription complex must come apart. When the last base is added to the RNA chain, the transcription bubble collapses as the RNA-DNA hybrid is disrupted, the DNA reforms in duplex state, and the enzyme and RNA are both released. The sequence of DNA required for these reactions defines the terminator.

The classical view of elongation was that it is a monotonic process, in which the enzyme moves forward at a constant rate. We now know that this is not correct. Certain DNA sequences can cause the polymerase to slow down or even pause indefinitely. As we will see in *Section 11.13, Bacterial Transcription Termination,* prolonged pausing can lead to transcription termination. In addition, the insertion of an incorrect base will alter the DNA structure and cause the polymerase to "backtrack." This is part of an error editing mechanism.

KEY CONCEPTS

- RNA polymerase binds to the promoter on the DNA to form a closed complex.
- RNA polymerase initiates transcription after opening the DNA duplex to form a transcription bubble.
- During elongation the transcription bubble moves along DNA and the RNA chain is extended in the 5′ to 3′ direction by adding nucleotides to the 3′ end of the growing chain.
- When transcription stops, the DNA duplex reforms and RNA polymerase dissociates at a terminator site.

CONCEPT AND REASONING CHECK

DNA polymerase requires a helicase to move the replication fork. How does RNA polymerase move the "bubble"?

11.4 A Model for Enzyme Movement Is Suggested by the Crystal Structure

We now have abundant information about the structure and function of RNA polymerase as the result of the crystal structures of bacterial enzymes. Bacterial RNA polymerase has overall dimensions of ~90 × 95 × 160 Å. Structural analysis shows a "channel" or groove on the surface ~25 Å wide that could be the path for DNA. An example of this channel is illustrated in **FIGURE 11.7**. This groove is long enough to hold ~16 bp, but it represents only part of the total length of DNA bound during transcription. The enzyme surface is largely negatively charged, but the groove is lined with positive charges, enabling it to interact with the negatively charged phosphate groups of DNA.

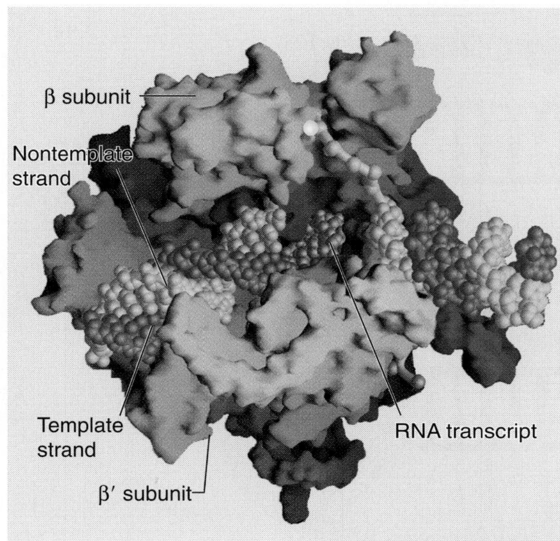

FIGURE 11.7 The β (light blue) and β′ subunit (pink) of RNA polymerase have a channel for the DNA template. Synthesis of an RNA transcript (copper) has just begun; the DNA template (red) and coding (yellow) strands are separated in a transcription bubble. Photo courtesy of Seth Darst, Rockefeller University.

Once DNA has been melted, the individual strands have a flexible structure in the transcription bubble. This enables DNA to take its turn in the active site. Before transcription starts, though, the DNA double helix is a relatively rigid straight structure. How does this structure enter the polymerase without being blocked by the wall? The answer is that a large conformational shift must occur in the enzyme. Adjacent to the wall is a clamp. In the free form of RNA polymerase, this clamp swings away from the wall to allow DNA to follow a straight path through the enzyme. After DNA has been melted to create the transcription bubble, the clamp must swing back into position against the wall.

One of the dilemmas for any nucleic acid polymerase is that the enzyme must make tight contacts with the nucleic acid substrate and product, but must break these contacts and remake them with each cycle of nucleotide addition. Consider the situation illustrated in **FIGURE 11.8**. A polymerase makes a series of specific contacts with the bases at particular positions. For example, contact "1" is made with the base at the end of the growing chain and contact "2" is made with the base in the template strand that is complementary to the next base to be added. Note, however, that the bases that occupy these locations in the nucleic acid chains change every time a nucleotide is added!

The top and bottom panels of the figure show the same situation: a base is about to be added to the growing chain. The difference is that the growing chain has been extended by one base in the bottom panel. The geometry of both complexes is exactly the same, but contacts "1" and "2" in the bottom panel are made to bases in the nucleic acid chains that are located one position farther along the chain. The middle panel shows that this must mean that after the base is added, and before the enzyme moves relative to the nucleic acid, the contacts made to specific positions must be broken so that they can be remade to bases that occupy those positions after the movement.

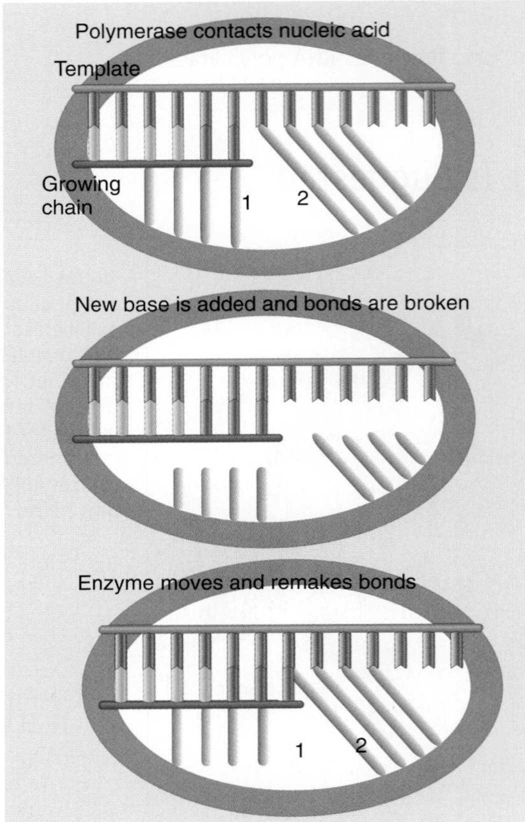

FIGURE 11.8 Movement of RNA polymerase requires breaking and remaking bonds to the nucleotides at fixed positions relative to the enzyme structure. The nucleotides in these positions change each time the enzyme moves a base along the template.

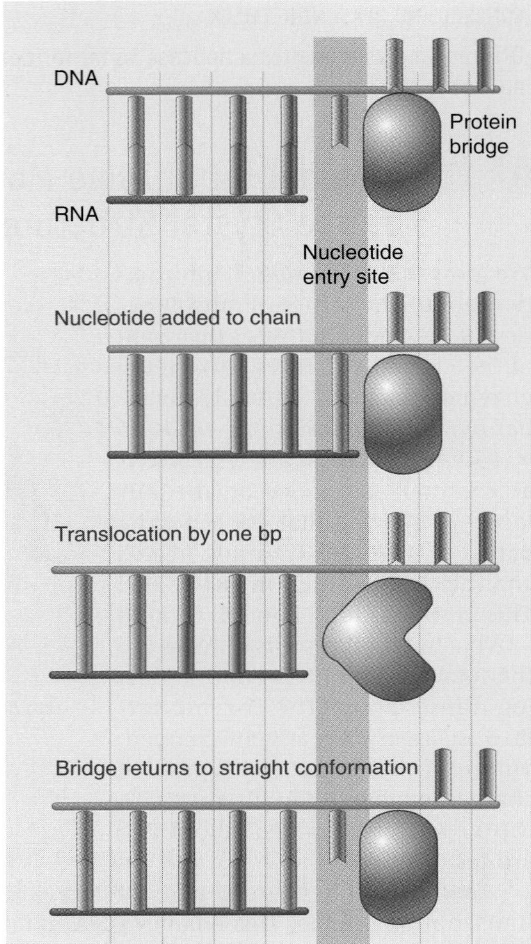

FIGURE 11. 9 The RNA polymerase elongation cycle starts with a straight bridge adjacent to the nucleotide entry site. After nucleotide addition, the enzyme moves one base pair and the bridge bends as it retains contact with the newly added nucleotide. When the bridge is released, the cycle can start again.

The RNA polymerase structure suggests an insight into how the enzyme retains contact with its substrate while breaking and remaking bonds. A structure in the protein called the *bridge* is adjacent to the active site. **FIGURE 11.9** suggests that the change in conformation of the bridge structure is closely related to translocation of the enzyme along the nucleic acid.

At the start of the cycle of translocation, the bridge has a straight conformation adjacent to the nucleotide entry site. This allows the next nucleotide to bind at the nucleotide entry site. The bridge is in contact with the newly added nucleotide. The protein then moves one base pair along the substrate. The bridge changes its conformation, bending to keep contact with the newly added nucleotide. In this conformation, the bridge obscures the nucleotide entry site. To end the cycle, the bridge returns to its straight conformation, allowing access again to the nucleotide entry site. The bridge acts as a ratchet that releases the DNA and RNA strands for translocation while holding on to the end of the growing chain.

KEY CONCEPTS

- RNA polymerase contains a groove on the surface for the path of the DNA.
- A protein bridge changes conformation to control the entry of nucleotides to the active site.

CONCEPT AND REASONING CHECK

Why is understanding structure necessary for understanding function?

11.5 Bacterial RNA Polymerase Consists of the Core Enzyme and Sigma Factor

The best-characterized RNA polymerases are those of eubacteria, for which *E. coli* is a typical case. A single type of RNA polymerase appears to be responsible for almost all synthesis of mRNA, and all synthesis of rRNA and tRNA, in a eubacterium. About 7000 RNA polymerase molecules are present in an *E. coli* cell. Many of them are engaged in transcription; probably 2000 to 5000 enzymes are synthesizing RNA at any one time, with the number depending on the growth conditions.

The complete enzyme or **holoenzyme** in *E. coli* consists of six subunits and has a molecular weight of ~465 kD. The holoenzyme ($\alpha_2\beta\beta'\sigma\omega$) can be separated into two components, the **core enzyme** ($\alpha_2\beta\beta'\omega$) and the **sigma factor** (the σ polypeptide), which is concerned specifically with promoter recognition. The β and β' subunits together make up the catalytic center. Their sequences are related to those of the largest subunits of eukaryotic RNA polymerases (see *Section 25.2, Eukaryotic RNA Polymerases Consist of Many Subunits*). The α subunit is required for assembly of the core enzyme, and it also plays a role in the interaction of RNA polymerase with some regulatory factors through its carboxy terminal domain (**C-terminal domain**). The ω subunit also plays a role in assembly and may also play a role in certain regulatory functions.

The σ subunit is concerned specifically with promoter recognition, and we have more information about its functions than for any other subunit. Only the holoenzyme can recognize a gene at its promoter and initiate transcription. Sigma factor ensures that bacterial RNA polymerase binds in a stable manner to DNA only at the promoter. The sigma factor is sometimes released when the RNA chain reaches eight to nine bases, leaving the core enzyme to undertake elongation. The core enzyme has the ability to synthesize RNA on a DNA template, but it cannot initiate transcription at the proper sites.

The core enzyme has a general affinity for DNA, in which electrostatic attraction between the basic protein and the acidic nucleic acid plays a major role. Any (random) sequence of DNA that is bound by core polymerase in this general binding reaction is described as a loose binding site. No change occurs in the DNA, which remains duplex. The complex at such a site is stable, with a half-life for dissociation of the enzyme from DNA of ~60 minutes. Core enzyme does not distinguish between promoters and other sequences of DNA.

FIGURE 11.10 shows that the sigma factor introduces a major change in the affinity of RNA polymerase for DNA. The holoenzyme has a drastically reduced ability to recognize loose binding sites—that is, to bind to any general sequence of DNA. The half-life of the holoenzyme-DNA complex is <1 second, so sigma factor destabilizes the general binding ability dramatically.

Note that sigma factor also confers the ability to recognize specific DNA

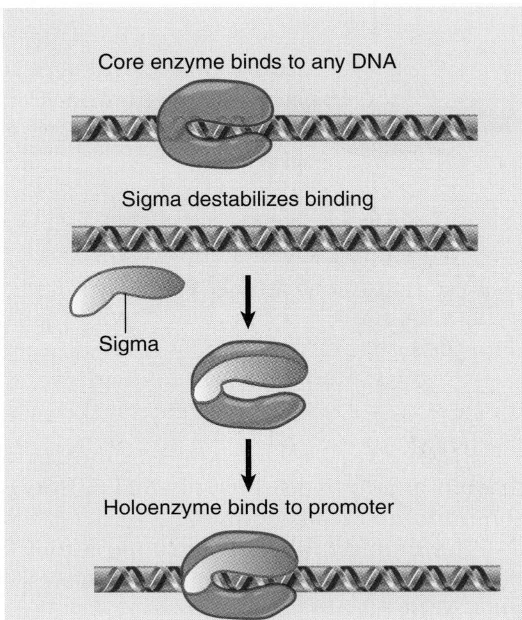

Core enzyme binds to any DNA
Sigma destabilizes binding
Sigma
Holoenzyme binds to promoter

FIGURE 11.10 Core enzyme binds indiscriminately to any DNA. Sigma factor reduces the affinity for sequence-independent binding and confers specificity for promoters.

- **holoenzyme** The RNA polymerase form that is competent to initiate transcription. It consists of the four subunits of the core enzyme ($\alpha_2\beta\beta'\omega$) and σ factor.
- **core enzyme** The complex of RNA polymerase subunits needed for elongation. It does not include additional subunits or factors that may be needed for initiation or termination.
- **sigma factor** The subunit of bacterial RNA polymerase needed for initiation; it is the major influence on selection of promoters.
- **C-terminal domain** The domain of RNA polymerase that is involved in stimulating transcription by contact with regulatory proteins.

binding sites. The holoenzyme binds to promoters very tightly, with an association constant increased from that of core enzyme by (on average) 1000 times, and with a half-life of several hours.

The specificity of holoenzyme for promoters compared to other sequences is ~10^7, but this is only an average because there is wide variation in the affinity with which the holoenzyme binds to different promoter sequences. This is an important parameter in determining the efficiency of an individual promoter in initiating transcription.

KEY CONCEPTS

- Bacterial RNA polymerase holoenzyme can be divided into an $\alpha_2\beta\beta'$ core enzyme that catalyzes transcription and a sigma (σ) subunit that is required only for initiation.
- Sigma factor changes the DNA-binding properties of RNA polymerase so that its affinity for general DNA is reduced and its affinity for promoters is increased.
- Binding constants of RNA polymerase for different promoters vary greatly, and correlate with the frequency with which transcription is initiated at each promoter.

CONCEPT AND REASONING CHECK

How does sigma factor provide promoter specificity?

11.6 How Does RNA Polymerase Find Promoter Sequences?

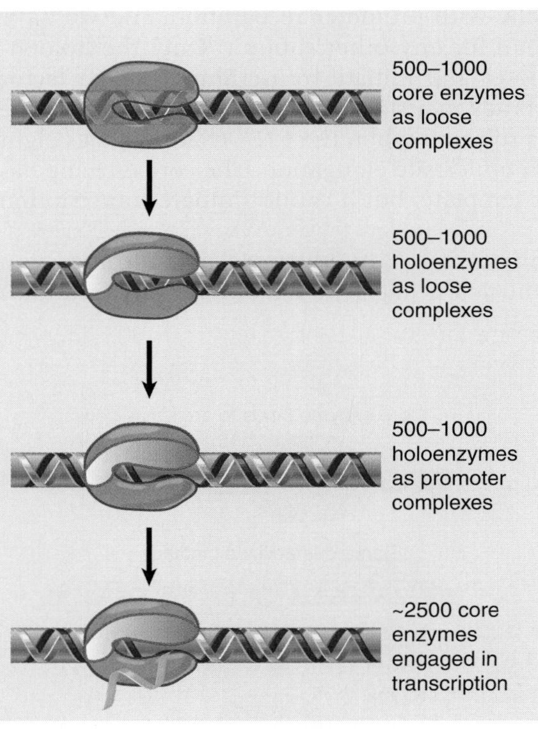

FIGURE 11.11 Core enzyme and holoenzyme are distributed on DNA, and very little RNA polymerase is free.

How is RNA polymerase distributed in the cell? Virtually all of it is bound to DNA, with no free core enzyme or holoenzyme. A (somewhat speculative) picture of the enzyme's binding profile is depicted in **FIGURE 11.11**:

- Excess core enzyme exists largely as closed loose complexes because the enzyme enters into them rapidly and leaves them slowly. There is very little, if any, free core enzyme.
- There is enough sigma factor for about one-third of the polymerases to exist as holoenzymes, and they are distributed between loose complexes at nonspecific sites and binary complexes (mostly closed) at promoters.
- About half of the RNA polymerases consist of core enzymes engaged in transcription.

RNA polymerase must find promoters within the context of the genome. Suppose that a promoter is a stretch of ~60 bp. How is it distinguished from the 4×10^6 bp that comprise the *E. coli* genome?

The simplest model is to suppose that RNA polymerase moves by random diffusion. Holoenzyme very rapidly associates with, and dissociates from, loose binding sites. Thus it could continue to make and break a series of closed complexes until (by chance) it encounters a promoter, and its recognition of the specific sequence would allow tight binding to occur by formation of an open complex. Movement from one site to another is limited by the speed of diffusion through the medium. However, the actual speed with which RNA polymerase finds some promoters is greater than would be possible by diffusion.

RNA polymerase must therefore use some other means to seek its binding sites. **FIGURE 11.12** shows that the process could be sped up if the initial target for RNA polymerase is the whole genome, not just a specific promoter sequence. When the target size is increased, the rate constant for diffusion to DNA is correspondingly increased and is no longer limiting.

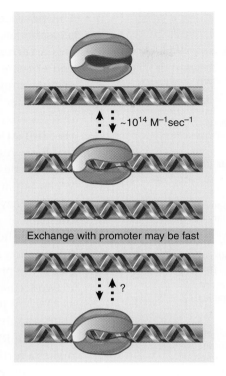

FIGURE 11.12 RNA polymerase binds very rapidly to random DNA sequences and could find a promoter by direct displacement of the bound DNA sequence.

If this hypothesis is correct, a free RNA polymerase binds DNA and then remains in contact with it. How does the enzyme move from a random (loose) binding site on DNA to a promoter? The most likely model is to suppose that the bound sequence is directly displaced by another sequence. Having taken hold of DNA, the enzyme very rapidly exchanges this sequence with another sequence and continues to exchange sequences until a promoter is found. The enzyme then forms a stable, open complex, after which initiation occurs. The search process becomes much faster because association and dissociation are virtually simultaneous and time is not spent commuting between sites. Direct displacement can give a "directed walk," in which the enzyme moves preferentially from a weak site to a stronger site.

KEY CONCEPTS

- The rate at which RNA polymerase binds to promoters is too fast to be accounted for by random diffusion.
- RNA polymerase probably binds to random sites on DNA and exchanges them with other sequences very rapidly until a promoter is found.

CONCEPT AND REASONING CHECK

How does sigma factor "read" the DNA sequence of the promoter?

11.7 Sigma Factor Controls Binding to Promoters

We can now describe the stages of transcription in terms of the interactions between different forms of RNA polymerase and the DNA template. The initiation reaction can be described by the parameters that are summarized in **FIGURE 11.13**:

- The holoenzyme-promoter reaction starts by forming a closed binary complex in the template recognition stage. "Closed" means that the DNA remains duplex.
- The closed complex is converted into an open complex by the "melting" of a short region of DNA within the sequence bound by the enzyme. The series of events leading to formation of an open complex is called **tight binding**. This reaction is fast. Sigma factor is involved in the melting reaction (see *Section 11.11, Substitution of Sigma Factors May Control Initiation*).

▸ **tight binding** The binding of RNA polymerase to DNA in the formation of an open complex (when the strands of DNA have separated).

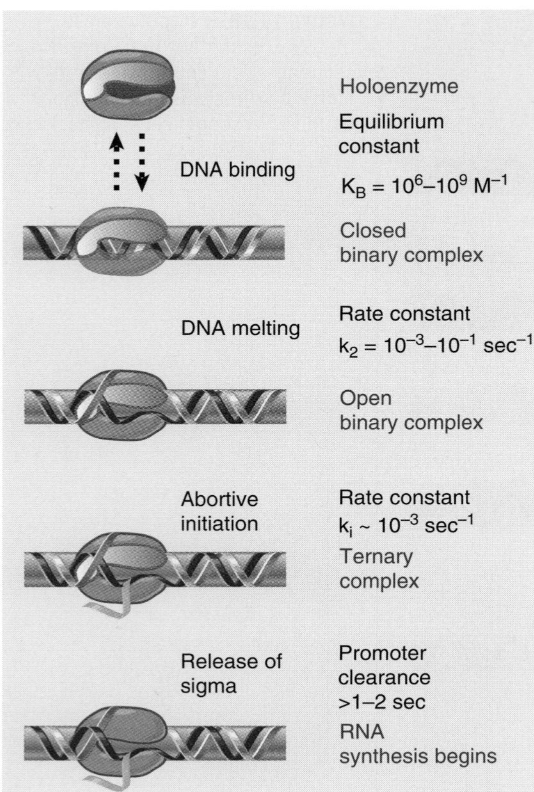

FIGURE 11.13 RNA polymerase passes through several steps prior to elongation. A closed binary complex is converted to an open form and then into a ternary complex.

- The next step is to incorporate the first two nucleotides; a phosphodiester bond then forms between them. This generates a **ternary complex** that contains RNA as well as DNA and enzyme. Further nucleotides can be added without any enzyme movement to generate an RNA chain of up to nine bases. After each base is added, there is a certain probability that the enzyme will release the chain. This comprises an **abortive initiation**, after which the enzyme begins again with the first base. A cycle of abortive initiations usually occurs to generate a series of very short oligonucleotides.
- When initiation succeeds, sigma factor is no longer necessary, and the enzyme makes the transition to the elongation ternary complex of core polymerase-DNA-nascent RNA. The critical parameter here is how long it takes for the polymerase to leave the promoter so another polymerase can initiate. This parameter is the promoter clearance time; its minimum value of one to two seconds establishes the maximum frequency of initiation as <1 event per second. The enzyme then moves along the template, and the RNA chain extends beyond ten bases.
- When the RNA chain extends to 15 to 20 bases, the enzyme makes a further structural transition, to form the complex that undertakes elongation.

Since soon after the discovery of sigma factor, it has been a tenet of transcription that sigma factor is released after initiation. This may not be strictly true. Direct measurements of elongating RNA polymerase complexes show that ~70% of them retain sigma factor. A third of elongating polymerases lack sigma; hence the original conclusion is certainly correct that it is not necessary for elongation. In those cases where it remains associated with core enzyme, the nature of the association has almost certainly changed (see *Section 11.11, Substitution of Sigma Factors May Control Initiation*).

- **ternary complex** The complex in initiation of transcription that consists of RNA polymerase and DNA as well as a dinucleotide that represents the first two bases in the RNA product.
- **abortive initiation** It describes a process in which RNA polymerase starts transcription but terminates before it has left the promoter. It then reinitiates. Several cycles may occur before the elongation stage begins.

RNA polymerase encounters a dilemma in reconciling its needs for initiation with those for elongation. Initiation requires tight binding only to particular sequences (promoters), whereas elongation requires close association with all sequences that the enzyme encounters during transcription. **FIGURE 11.14** illustrates how the dilemma is solved by the reversible association between sigma factor and core enzyme.

Sigma factor is involved only in initiation. It becomes unnecessary when abortive initiation is concluded and RNA synthesis has been successfully initiated. At this point, the core enzyme in the ternary complex is bound very tightly to DNA. It is essentially "locked in" until elongation has been completed. When transcription terminates, the core enzyme is released. It is then "stored" by binding to a loose site on DNA. Core enzyme has a high intrinsic affinity for DNA, which is increased by the presence of nascent RNA. Its affinity for loose binding sites is, however, too high to allow the enzyme to efficiently distinguish promoters from other sequences. By reducing the stability of the loose complexes, sigma factor allows the process to occur much more rapidly; and by stabilizing the association at tight binding sites, the factor drives the reaction irreversibly into the formation of open complexes. When the enzyme releases sigma factor (or changes its association with it), it reverts to a general affinity for all DNA, irrespective of sequence, that allows it to continue transcription.

What is responsible for the ability of holoenzyme to bind specifically to promoters? Sigma factor has domains that recognize (read) the promoter DNA sequence. As an independent polypeptide, sigma factor does not bind to DNA, but when holoenzyme forms a tight binding complex, sigma factor contacts the DNA in the region upstream of the startpoint. This difference is due to a change in the conformation of sigma factor

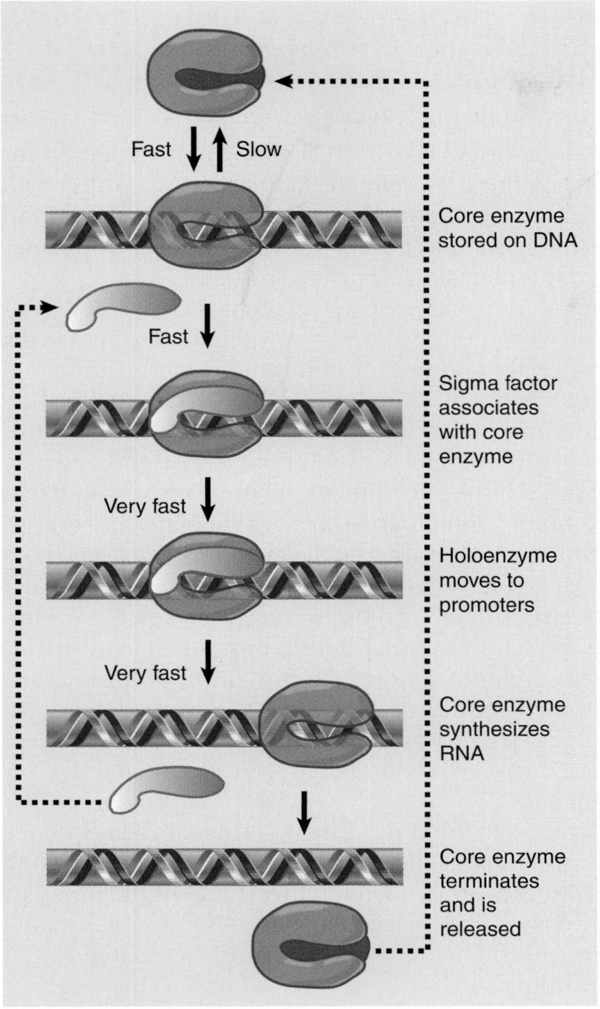

FIGURE 11.14 Sigma factor and core enzyme recycle at different points in transcription.

Fast
Slow
Core enzyme stored on DNA
Fast
Sigma factor associates with core enzyme
Very fast
Holoenzyme moves to promoters
Very fast
Core enzyme synthesizes RNA
Core enzyme terminates and is released

when it binds to core enzyme. The N-terminal region of free sigma factor suppresses the activity of the DNA-binding region; when sigma binds to core, this inhibition is released, and it becomes able to bind specifically to promoter sequences (see also Figure 11.23 in *Section 11.12, Sigma Factors Directly Contact DNA*). The inability of free sigma factor to recognize promoter sequences may be important: if sigma factor could freely bind to promoters, it might block holoenzyme from initiating transcription.

KEY CONCEPTS

- RNA polymerase binds to the promoter as a closed complex in which the DNA remains double-stranded.
- RNA polymerase then separates the DNA strands to form an open complex and a transcription bubble that incorporates up to nine nucleotides into RNA.
- There may be a cycle of abortive initiations before the enzyme moves to the next phase.
- Sigma factor may be released from RNA polymerase when the nascent RNA chain reaches eight to nine bases in length.
- A change in association between sigma factor and holoenzyme changes binding affinity for DNA so that core enzyme can move along DNA.

CONCEPT AND REASONING CHECK

What does it mean to convert a promoter from the closed complex to the open complex?

11.8 Promoter Recognition Depends on Consensus Sequences

As a sequence of DNA whose function is to be recognized by proteins, a promoter differs from sequences whose role is to be transcribed or translated. The information for promoter function is provided directly by the DNA sequence: its structure is the signal. This is a classic example of a *cis*-acting site, as defined previously in Figure 2.16 and Figure 2.17. By contrast, expressed regions gain their meaning only after the information is transferred into the form of some other nucleic acid or protein.

A key question in examining the interaction between a protein such as RNA polymerase and the DNA sequence we call a promoter is how the protein reads the specific DNA promoter sequence. In the bacterial genome, the minimum length that could provide an adequate signal is 12 bp. (Any shorter sequence is likely to occur—just by chance—a sufficient number of additional times to provide false signals. The minimum length required for unique recognition increases with the size of genome.) The 12 bp sequence need not be contiguous. If a specific number of base pairs separate two constant shorter sequences, their combined length could be less than 12 bp, because the distance of separation itself provides a part of the signal (even if the intermediate sequence is itself irrelevant).

Initial attempts to identify the features in DNA that are necessary for RNA polymerase binding involved comparing the sequences of different promoters. Any nucleotide sequence essential for polymerase binding should be present in *all* promoters. Such a sequence is said to be a **conserved sequence**. However, a conserved sequence need not necessarily be identical at every single position; some variation is permitted. How do we analyze a sequence of DNA to determine whether it is sufficiently conserved to constitute a recognizable signal?

▸ **conserved sequence** Sequences in which many examples of a particular nucleic acid or protein are compared and the same individual bases or amino acids are always found at particular locations.

▸ **consensus sequence** An idealized sequence in which each position represents the base most often found when many actual sequences are compared.

Putative DNA recognition sites can be defined in terms of an idealized sequence that represents the base most often present at each position. A **consensus sequence** is defined by aligning all known examples so as to maximize their homology. For a sequence to be accepted as a consensus, each particular base must be reasonably predominant at its position, and most of the actual examples must be related to the consensus by only one or two substitutions.

The most striking feature in the sequence of promoters in *E. coli* is the presence of two short 6 bp consensus sequences within the 60 bp bound by RNA polymerase. The sequence of much of the rest of the binding site is irrelevant. Conservation of only very short consensus sequences is a typical feature of regulatory sites (such as promoters) in both prokaryotic and eukaryotic genomes.

There are four (perhaps five) conserved features in a bacterial promoter: the startpoint, the –10 sequence, the –35 sequence, the separation between the –10 and –35 sequences, and (sometimes) the UP element:

- The startpoint is usually (>90% of the time) a purine. It is common for the startpoint to be the central base in the sequence CAT, but the conservation of this triplet is not great enough to regard it as an obligatory signal.
- Just upstream of the startpoint, a 6-bp region is recognizable in almost all promoters. The center of this hexamer is generally about 10 bp upstream of the startpoint; the distance varies in known promoters from position –18 to –9. Named for its location, this hexamer is called the *–10 sequence*, the *Pribnow box*, or the **TATA box**. Its consensus is TATAAT and can be summarized in the form

  $T_{80}\ A_{95}\ T_{45}\ A_{60}\ A_{50}\ T_{96}$

  where the subscript denotes the percent occurrence of the most frequently found base, which varies from 45% to 96%. (A position at which there is no discernible preference for any base is indicated by "N.") If the frequency of

▸ **TATA box** A conserved A-T-rich octamer found about 25 bp before the startpoint of each eukaryotic RNA polymerase II transcription unit; it is involved in positioning the enzyme for correct initiation.

occurrence indicates the likely importance in binding RNA polymerase, we would expect the initial highly conserved TA and the final almost completely conserved T in the –10 sequence to be the most important bases.

- Another conserved hexamer is centered ~35 bp upstream of the startpoint. This is called the *–35 sequence* or **–35 box** because you cannot pronounce it. The consensus is TTGACA; in more detailed form, the conservation is T_{82} T_{84} G_{78} A_{65} C_{54} A_{45}.
- The distance separating the –35 and –10 sites is between 16 and 18 bp in 90% of promoters; in the exceptions, it is as little as 15 bp or as great as 20 bp. Although the actual sequence in the intervening region is unimportant, the distance is critical in holding the two sites at the appropriate separation for the geometry of RNA polymerase.
- Some promoters have an A-T-rich sequence located farther upstream. This is called the *UP element*. Proteins that bind the element interact with the α subunit of the RNA polymerase. It is typically found in promoters for genes that are highly expressed, such as the promoters for rRNA genes.

–35 box The consensus sequence centered about 35 bp before the startpoint of a bacterial gene. It is involved in initial recognition by RNA polymerase.

The optimal promoter is a sequence consisting of the –35 hexamer, separated by 17 bp from the –10 hexamer, lying 7 bp upstream of the startpoint. The structure of a promoter, showing the permitted range of variation from this optimum, is illustrated in **FIGURE 11.15**. The implication that the consensus promoter sequence is the optimal promoter means, first, that there are many different promoter sequences, and second, that there are "good" promoters (with sequence similar to the consensus) and poor promoters (whose sequences vary from the consensus). There is a range of promoter strengths.

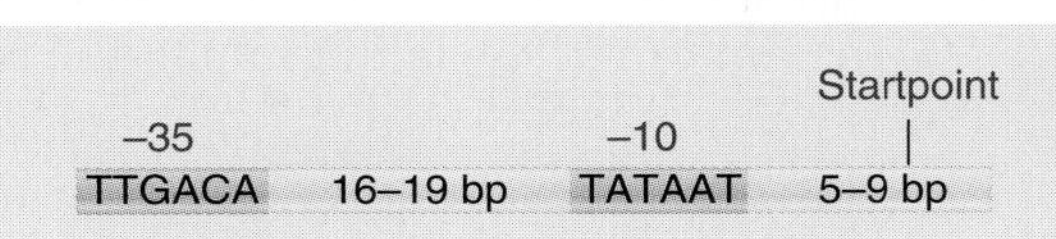

FIGURE 11.15 A typical promoter has three components, consisting of consensus sequences at –35, –10, and the startpoint.

RNA polymerase initially binds the promoter region from –50 to +20. **FIGURE 11.16** shows the contacts that RNA polymerase makes with the DNA at a typical promoter. The regions at –35 and –10 contain most of the contact points for the enzyme. Viewed in three dimensions, the points of contact upstream of the –10 sequence all lie on one face of the DNA double helix.

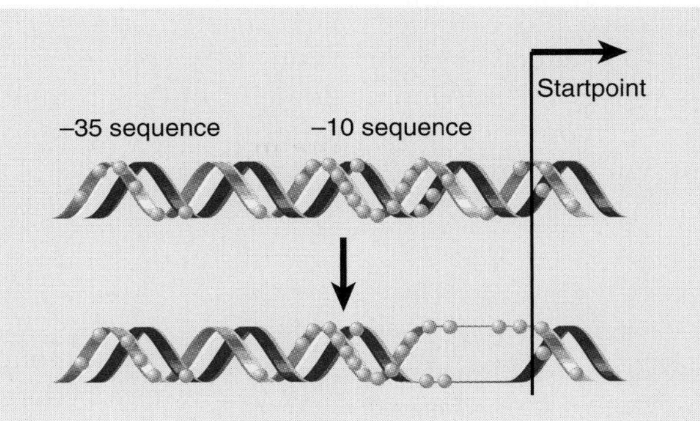

FIGURE 11.16 The consensus sequences at –35 and –10 provide most of the contact points for RNA polymerase in the promoter. Most points of contact lie on one face of the DNA.

KEY CONCEPTS

- A promoter is the site where RNA polymerase binds the DNA.
- The promoter consensus sequences consist of a purine at the startpoint, the hexamer TATAAT centered at –10, and another hexamer centered at –35.
- Individual promoters usually differ from the consensus at one or more positions.

CONCEPT AND REASONING CHECKS

- When RNA polymerase binds to a promoter, how does it choose which strand is the template strand?
- Why is the TATA box AT rich?

11.9 Promoter Efficiencies Can Be Increased or Decreased by Mutation

Mutations are a major source of information about promoter function. Mutations in promoters affect the level of expression of the gene(s) they control without altering the gene products themselves. Most are identified as bacterial mutants that have lost, or have very much reduced, transcription of the adjacent genes. They are known as **down mutations**. Less often, mutants are found in which there is increased transcription from the promoter. They are **up mutations**.

- **down mutation** A mutation in a promoter that decreases the rate of transcription.
- **up mutation** A mutation in a promoter that increases the rate of transcription.

Is the most effective promoter one that has the actual consensus sequences? This expectation is borne out by the simple rule that up mutations usually increase homology with one of the consensus sequences or bring the distance between them closer to 17 bp. Down mutations usually decrease the resemblance of either site with the consensus or make the distance between them more distant from 17 bp. Down mutations tend to be concentrated in the most highly conserved positions, which confirms their particular importance as the main determinant of promoter efficiency. There are, however, occasional exceptions to these rules.

There is ~100-fold variation in the rate at which RNA polymerase binds to different promoters *in vitro*, which correlates well with the frequencies of transcription when their genes are expressed *in vivo*. By measuring the kinetic constants for formation of a closed complex and its conversion to an open complex, we can dissect two stages of the initiation reaction. Mutations in the two regions have different effects:

- Down mutations in the –35 sequence reduce the rate of closed complex formation (but they do not inhibit the conversion to an open complex).
- Down mutations in the –10 sequence do not affect the initial formation of a closed complex, but they slow its conversion to the open form.

These results suggest the model shown in **FIGURE 11.17**. The function of the –35 sequence is to provide the signal for recognition by RNA polymerase, whereas the –10 sequence allows the complex to convert from closed to open form. We might view the –35 sequence as comprising a "recognition domain," whereas the –10 sequence comprises an "unwinding domain" of the promoter.

The consensus sequence of the –10 site consists exclusively of A-T base pairs, a configuration that assists the initial melting of DNA into single strands. The lower energy needed to disrupt A-T pairs compared with G-C pairs means that a stretch of A-T pairs demands the minimum amount of energy for strand separation.

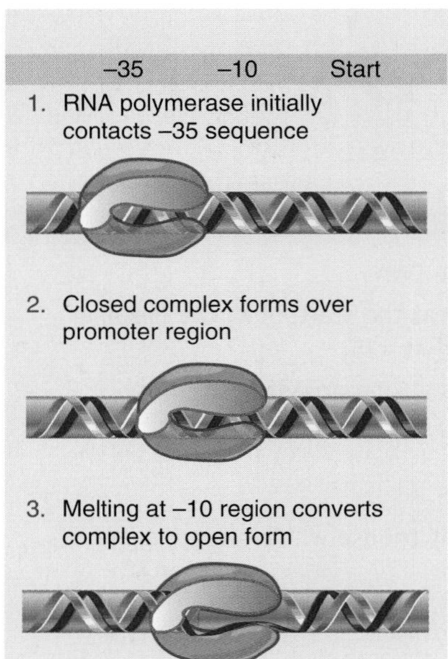

–35
–10
Start
1. RNA polymerase initially contacts –35 sequence
2. Closed complex forms over promoter region
3. Melting at –10 region converts complex to open form

FIGURE 11.17 The –35 sequence is used for initial recognition and the –10 sequence is used for the melting reaction that converts a closed complex to an open complex.

The sequence immediately around the startpoint influences the initiation event. The initial transcribed region (from +1 to +30) influences the rate at which RNA polymerase clears the promoter and therefore has an effect upon promoter strength. Thus the overall strength of a promoter cannot be predicted entirely from its –35 and –10 consensus sequences.

Notice that the promoter is asymmetrical; that is, the TATA box is closest to the startpoint, the –35 box furthest away. This allows the polymerase to know whether it should be facing to the right or to the left, or in other words, which strand is the template strand.

A "typical" promoter relies upon its –35 and –10 sequences to be recognized by RNA polymerase, but one or the other of these sequences can be absent from some (exceptional) promoters. In at least some of these cases, the promoter cannot be recognized by RNA polymerase alone; the reaction requires ancillary proteins, which overcome the deficiency in intrinsic interaction between RNA polymerase and the promoter.

KEY CONCEPTS

- Down mutations that decrease promoter efficiency usually decrease conformance to the consensus sequences, whereas up mutations have the opposite effect.
- Mutations in the –35 sequence usually affect initial binding of RNA polymerase.
- Mutations in the –10 sequence usually affect the melting reaction that converts a closed to an open complex.

CONCEPT AND REASONING CHECK

What is the DNA sequence difference between a "good" promoter and a "bad" promoter?

11.10 Supercoiling Is an Important Feature of Transcription

Supercoiling has an important influence on transcription at both initiation and elongation. When DNA is negatively supercoiled, the two strands are less tightly wound around one another; at high enough negative supercoiling density they may separate (see *Section 1.5, Supercoiling Affects the Structure of DNA*). Negative supercoiling may assist initiation by making it easier for the strands of DNA to be separated. The efficiency of some promoters is influenced by the degree of supercoiling. Why, though, should some promoters be influenced by the extent of supercoiling, whereas others are not? One possibility is that the dependence of a promoter on supercoiling is determined by its surrounding sequence, or **sequence context**. This would predict that some promoters have sequences that are easier to melt (and are therefore less dependent on supercoiling), whereas others have more difficult sequences (and have a greater need to be supercoiled). An alternative is that the location of the promoter might be important if different regions of the bacterial chromosome have different degrees of supercoiling.

sequence context The sequence surrounding a consensus sequence. It may modulate the activity of the consensus sequence.

Supercoiling also has a continuing involvement with transcription. As RNA polymerase transcribes DNA, unwinding and rewinding occurs. The consequences of the rotation of DNA are illustrated in **FIGURE 11.18**, in what is often referred to as the *twin domain model* for transcription. As RNA polymerase pushes forward along the double helix, it generates positive supercoils (more tightly wound DNA) ahead and leaves negative supercoils (partially unwound DNA) behind. For each helical turn traversed by RNA polymerase, +1 turn is generated ahead and –1 turn behind. Too much positive supercoiling can actually prevent the continued forward motion of the polymerase.

Transcription therefore has a significant effect on the local structure of DNA. As a result, the enzymes gyrase (introduces negative supercoils) and topoisomerase I (removes negative supercoils) are required to rectify the situation in front of and behind the polymerase, respectively. Blocking the activities of gyrase and topoisomerase causes major changes in the supercoiling of DNA. A possible implication of these results is that transcription is responsible for generating a significant proportion of the supercoiling that occurs in the cell.

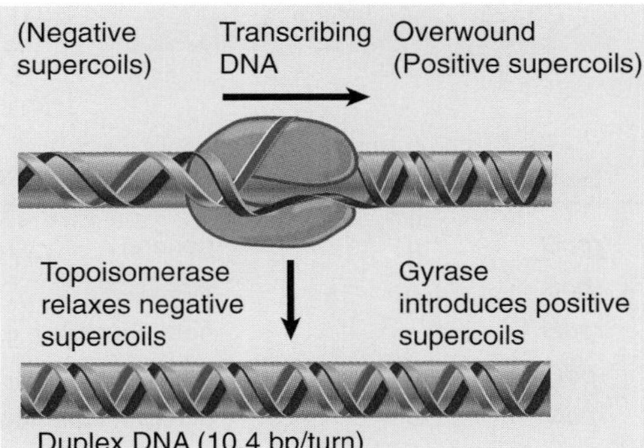

FIGURE 11.18 Transcription generates more tightly wound (positively supercoiled) DNA ahead of RNA polymerase, while the DNA behind becomes less tightly wound (negatively supercoiled).

KEY CONCEPTS

- Negative supercoiling increases the efficiency of some promoters by assisting the melting reaction.
- Transcription generates positive supercoils ahead of the enzyme and negative supercoils behind it, and these must be removed by gyrase and topoisomerase.

CONCEPT AND REASONING CHECK

Why is it more important to relieve positive supercoils rather than negative supercoils that are generated during transcription?

11.11 Substitution of Sigma Factors May Control Initiation

Because sigma factor determines the choice of the promoter for initiation of transcription, it is possible to control the choice of which gene or group of genes are to be transcribed by changing the sigma factor. **FIGURE 11.19** illustrates the basic idea. When the sigma factor is changed, RNA polymerase behaves in exactly the same way as before, except that it binds to a different promoter sequence. *E. coli* uses alternative sigma factors to respond to general environmental changes; they are listed in **FIGURE 11.20** and are named either by molecular weight of the product or for the gene. The general factor, which is responsible for transcription of most genes under normal conditions, is σ^{70}. The alternative sigma factors σ^{S}, σ^{32}, σ^{E}, and σ^{54} are activated in response to environmental changes; σ^{28} is used for expression of flagellar genes during normal growth, but its level of expression responds to changes in the environment. All the sigma factors except σ^{54} belong to the same protein family and function in the same general manner.

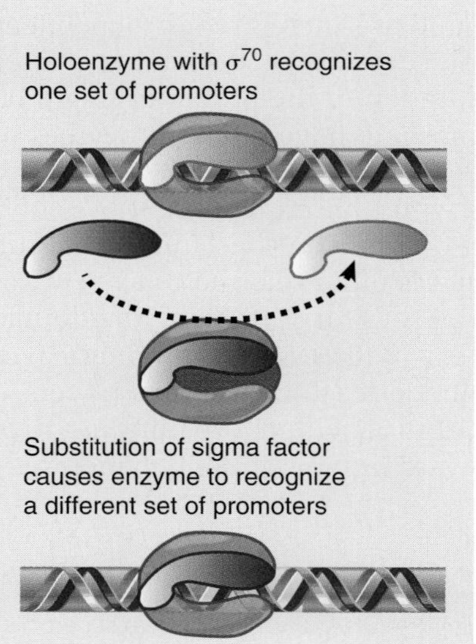
Holoenzyme with σ70 recognizes one set of promoters
Substitution of sigma factor causes enzyme to recognize a different set of promoters

FIGURE 11.19 The sigma factor associated with core enzyme determines the set of promoters at which transcription is initiated.

Temperature fluctuation is a common type of environmental challenge. Many organisms, both prokaryotic and eukaryotic, respond in a similar way. Upon an increase in temperature, synthesis of the proteins currently being made is turned off or down, and a new set of proteins is synthesized. The new proteins are the products of the heat shock genes, which play a role in protecting the cell against environmental stress. Heat shock genes are synthesized in response to conditions other than heat shock as well. Several of the heat shock proteins are chaperones. In *E. coli*, the expression of seventeen heat shock proteins is triggered by changes at transcription. The gene *rpoH* is a regulator needed to switch on the heat shock response. Its product is σ^{32}, which functions as an alternative sigma factor that causes transcription of the heat shock genes.

The heat shock response is accomplished by increasing the amount of σ^{32} when the temperature increases and decreasing its activity when the temperature change is reversed. The basic signal that induces production of σ^{32} is the accumulation of unfolded (partially denatured) proteins that results from increase in temperature. The σ^{32} protein is unstable, which is important in allowing its quantity to be increased or decreased

Gene	Factor	Use
rpoD	σ^{70}	general
rpoS	σ^{S}	stress
rpoH	σ^{32}	heat shock
rpoE	σ^{E}	heat shock
rpoN	σ^{54}	nitrogen starvation
fliA	σ^{28} (σ^{F})	flagellar synthesis

FIGURE 11.20 In addition to σ^{70}, *E. coli* has several sigma factors that are induced by particular environmental conditions. (A number in the name of a factor indicates its mass.)

rapidly. The proteins σ^{70} and σ^{32} can compete for the available core enzyme, so that the set of genes transcribed during heat shock depends on the balance between them. More complex regulatory circuits can be constructed by forming cascades of sigma factors; in such a cascade, a set of genes activated by one sigma factor includes a gene coding for another sigma factor that in turn activates another set of genes.

Sigma factors are used extensively to control initiation of transcription in the bacterium *B. subtilis*, for which ~10 different sigma factors are known. Some are present in vegetative cells; others are produced only in the special circumstances of phage infection or the change from vegetative growth to sporulation.

The major RNA polymerase found in *B. subtilis* cells engaged in normal vegetative growth has the same structure as that of *E. coli*, $\alpha_2\beta\beta'\sigma$. Its sigma factor (described as σ^{43} or σ^{A}) recognizes promoters with the same consensus sequences used by the *E. coli* enzyme under direction from σ^{70}. Several variants of the RNA polymerase that contain other sigma factors are found in much smaller amounts. The variant enzymes recognize different promoter DNA sequences on the basis of consensus sequences at −35 and −10.

- **sporulation** The generation of a spore by a bacterium (by morphological conversion) or by a yeast (as the product of meiosis).
- **vegetative phase** The period of normal growth and division of a bacterium. For a bacterium that can sporulate, this contrasts with the sporulation phase, when spores are being formed.

Perhaps the most dramatic example of switches in sigma factors is provided by **sporulation**, an alternative lifestyle available to some bacteria. At the end of the **vegetative phase** in a bacterial culture, logarithmic growth ceases because nutrients in the medium become depleted. This triggers sporulation. DNA is replicated, a genome is segregated to one end of the cell, and eventually the genome is surrounded by the tough spore coat. **FIGURE 11.21** shows how successive sigma factors play key regulatory roles in such a cascade.

Sporulation takes approximately eight hours. It can be viewed as a primitive sort of differentiation, in which a parent cell (the vegetative bacterium) gives rise to two different daughter cells with distinct fates: the mother cell is eventually lysed, and the spore that is released has an entirely different structure from the original bacterium.

Sporulation involves a drastic change in the biosynthetic activities of the bacterium, in which many genes are involved. The basic level of control lies at transcription. Some of the genes that functioned in the vegetative phase are turned off during sporulation, but most continue to be expressed. In addition, the genes specific for sporulation are expressed only during this period. At the end of sporulation, ~40% of the bacterial mRNA is sporulation-specific.

New forms of the RNA polymerase become active in sporulating cells; they contain the same core enzyme as vegetative cells, but have different sigma factors in place of the vegetative σ^{43}. The changes in transcriptional specificity occur in both the mother cell and the spore. The principle is that in each compartment the existing sigma factor is successively displaced by a new factor that causes transcription of a different set of genes.

Each sigma factor causes RNA polymerase to initiate at a different promoter with a different DNA sequence. The promoters have the same size and location relative to the startpoint; they differ in the −10 and −35 consensus sequences.

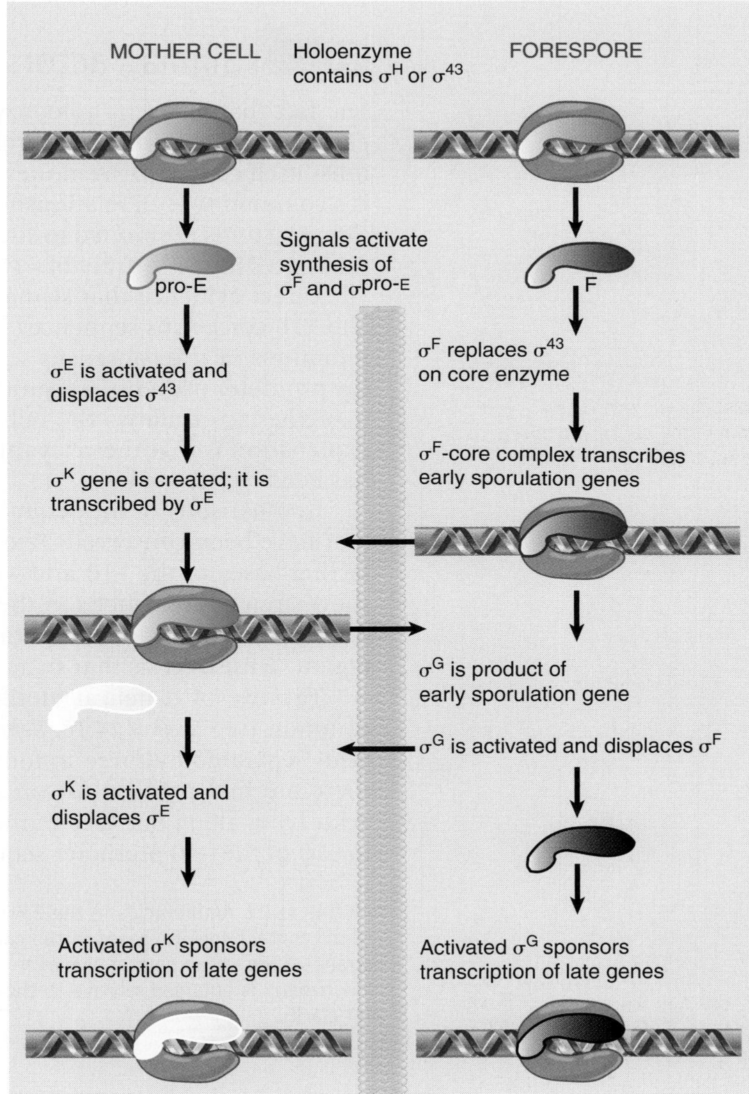

FIGURE 11.21 Sporulation involves successive changes in the sigma factors that control the initiation specificity of RNA polymerase. The cascades in the mother cell (left) and the forespore (right) are related by signals passed across the septum (indicated by horizontal arrows).

KEY CONCEPTS

- *E. coli* has several sigma factors, each of which causes RNA polymerase to initiate at a set of promoters defined by specific –35 and –10 sequences.
- σ^{70} is used for general transcription, and the other sigma factors are activated by special conditions.
- A cascade of sigma factors is created when one sigma factor is required to transcribe the gene coding for the next sigma factor.
- Sporulation divides a bacterium into two compartments: a mother cell that is lysed and a spore that is released.
- Each compartment advances to the next stage of development by synthesizing a new sigma factor that displaces the previous sigma factor.

CONCEPT AND REASONING CHECK

How do different sigma factors recognize different promoters?

11.12 Sigma Factors Directly Contact DNA

The fact that different holoenzymes with different sigma factors and a common core enzyme recognize different promoter sequences implies that the sigma factor subunit must itself contact DNA in these regions. This suggests the general principle that there is a common type of relationship between sigma factor and core enzyme, in which the sigma factor is positioned in such a way as to make critical contacts with the promoter sequences in the vicinity of –35 and –10.

Direct evidence that sigma factor contacts the promoter directly at both the –35 and –10 consensus sequences is provided by mutations in sigma factor that suppress mutations in the consensus sequences. When a mutation at a particular position in the promoter prevents recognition by RNA polymerase, and a compensating mutation in sigma factor allows the polymerase to use the mutant promoter, the most likely explanation is that the relevant base pair in DNA is contacted by the amino acid that has been substituted.

Comparisons of the sequences of several bacterial sigma factors identify regions that have been conserved. Two short regions named 2.4 and 4.2 are involved in contacting bases in the –10 and –35 elements, respectively. Both of these regions form short stretches of α-helix in the protein. They contact bases principally on the coding strand, and continue to hold these contacts after the DNA has been unwound in this region. This suggests that sigma factor could be important in the melting reaction.

The use of α-helical motifs in proteins to recognize duplex DNA sequences is common (see *Section 14.10, Repressor Uses a Helix-Turn-Helix Motif to Bind DNA*). Amino acids separated by three to four positions lie on the same face of an α-helix and are therefore in a position to contact adjacent base pairs. **FIGURE 11.22** shows that amino acids lying along one face of the 2.4 region α-helix contact the bases at positions –12 to –10 of the –10 promoter sequence.

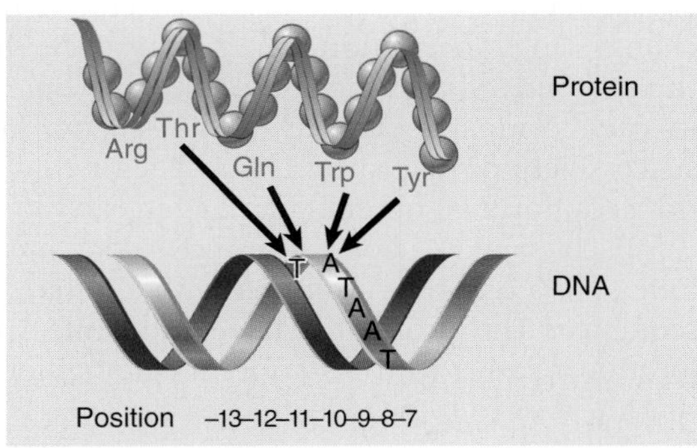

FIGURE 11.22 Amino acids in the 2.4 α-helix of σ^{70} contact specific bases in the coding strand of the –10 promoter sequence. The 2.4 α-helix is enlarged relative to the DNA for clarity.

The N-terminal region of σ^{70} has important regulatory functions. If it is removed, the shortened protein becomes able to bind specifically to promoter sequences in the absence of RNA polymerase core enzyme. This suggests that the N-terminal region behaves as an autoinhibition domain. It occludes the DNA-binding domains when σ^{70} is free. Association with core enzyme changes the conformation of sigma so that the inhibition is released, and the DNA-binding domains can contact DNA.

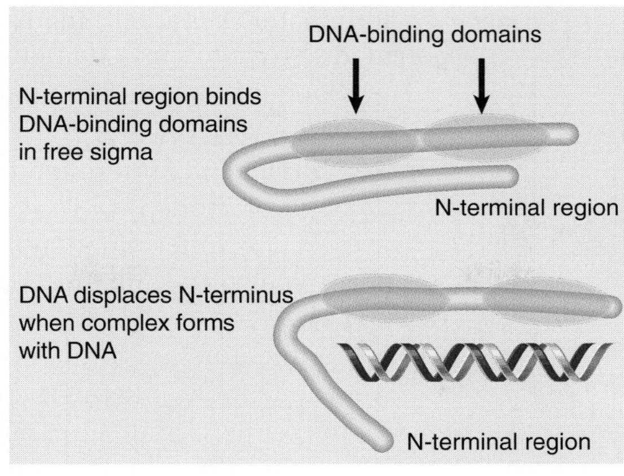

FIGURE 11.23 The N-terminus of sigma blocks the DNA-binding regions from binding to DNA. When an open complex forms, the N-terminus swings 20 Å away, and the two DNA-binding regions separate by 15 Å.

FIGURE 11.23 schematizes the conformational change in sigma factor at open complex formation. When sigma factor binds to the core polymerase, the N-terminal domain swings ~20 Å away from the DNA-binding domains, and the DNA-binding domains separate from one another by ~15 Å, presumably to acquire a more elongated conformation appropriate for contacting DNA. Mutations in either the –10 or –35 sequences prevent an (N-terminal-deleted) σ^{70} from binding to DNA, which suggests that σ^{70} contacts both sequences simultaneously. In the free holoenzyme, the N-terminal domain is located in the active site of the core enzyme components, essentially mimicking the location that DNA will occupy when a transcription complex is formed. When the holoenzyme forms an open complex on DNA, the N-terminal sigma domain is displaced from the active site. Its relationship with the rest of the protein is therefore very flexible; the relationship changes when sigma factor binds to core enzyme and again when the holoenzyme binds to DNA.

KEY CONCEPTS

- σ^{70} changes its structure to expose its DNA-binding regions when it associates with core enzyme.
- σ^{70} binds both the –35 and –10 sequences.

CONCEPT AND REASONING CHECK

How can changing sigma factors change which gene is expressed?

11.13 Bacterial Transcription Termination

Transcription termination at a site called the *terminator* occurs when the RNA polymerase stops adding nucleotides to the growing RNA chain, releases the completed product, and dissociates from the DNA template. Termination requires that all hydrogen bonds holding the RNA–DNA hybrid together be broken, after which the DNA duplex reforms. It is misleading, however, to consider that termination and elongation are separate processes. In fact, they are a continuum. As discussed in *Section 11.3, The Transcription Reaction Has Three Stages,* the rate of transcription by RNA polymerase is not constant. Transcription can proceed rapidly in some regions, slowly in other regions, or even pause, depending on the underlying DNA sequence context. Pausing is a prerequisite for transcription termination. All organisms have multiple elongation control regulators to prevent prolonged pausing at nonterminator sites.

It is sometimes difficult to define the termination point of an RNA molecule that has been synthesized in the living cell. It is always possible that the 3′ end of the molecule has been generated by cleavage of the primary transcript, and therefore does not represent the actual site at which RNA polymerase terminated.

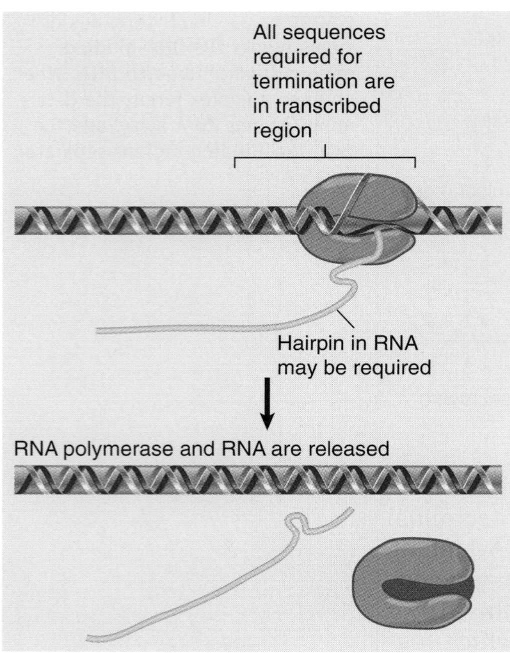
All sequences required for termination are in transcribed region
Hairpin in RNA may be required
RNA polymerase and RNA are released

FIGURE 11.24 The DNA sequences required for termination are located upstream of the terminator sequence. Formation of a hairpin in the RNA may be necessary.

Terminator sites are distinguished in *E. coli* according to whether RNA polymerase requires any additional factors to terminate transcription:

- Core enzyme can terminate *in vitro* at certain sites in the absence of any other factor. These sites are called *intrinsic terminators*.
- Rho-dependent terminators are defined by the need for addition of rho factor (ρ) *in vitro*, and mutations show that the factor is involved in termination *in vivo*.

Both types of terminators share the characteristics summarized in **FIGURE 11.24**.

- Terminators are located before the point at which the last base is added to the RNA. The responsibility for termination lies with the sequences already transcribed by RNA polymerase. Thus termination relies on scrutiny of the template or product that the polymerase is currently transcribing.
- Many terminators require a hairpin to form in the secondary structure of the RNA being transcribed. This indicates that termination depends on the RNA product and is not determined simply by scrutiny of the DNA sequence during transcription.

Terminators vary widely in their efficiency. There are "strong" terminators and "weak" terminators. At some terminators, the termination event can be prevented by specific ancillary factors that interact with RNA polymerase. **Antitermination** causes the enzyme to continue transcription past the terminator sequence, an event called *readthrough* (the same term used in *Section 9.11, Suppressor tRNAs Have Mutated Anticodons That Read New Codons*, to describe a ribosome's suppression of termination codons).

▸ **antitermination** A mechanism of transcriptional control in which termination is prevented at a specific terminator site, allowing RNA polymerase to read into the genes beyond it.

In approaching the termination event, we must regard it not simply as a mechanism for generating the 3′ end of the RNA molecule, but as an opportunity to control gene expression. Thus the stages when RNA polymerase associates with DNA (initiation) or dissociates from it (termination) both are subject to specific control. There are interesting parallels between the systems employed in initiation and termination. Both require breaking of hydrogen bonds (initial melting of DNA at initiation and RNA–DNA dissociation at termination), and both require additional proteins to interact with the core enzyme.

KEY CONCEPTS

- Termination may require both recognition of the terminator sequence in DNA and the formation of a hairpin structure in the RNA product.

CONCEPT AND REASONING CHECK

Why is pausing required for transcription termination?

11.14 Intrinsic Termination Requires a Hairpin and U-Rich Region

▸ **hairpin** An RNA sequence that can fold back on itself forming double stranded RNA.

Intrinsic terminators have the two structural features evident in **FIGURE 11.25**: a **hairpin** in the secondary structure, and a region at the very end of the transcription unit that is rich in U residues. Both features are needed for termination. The hairpin usually contains a G-C-rich region near the base of the stem. The typical distance between the hairpin and the U-rich region is seven to nine bases. About half of the genes in *E. coli* have intrinsic terminators.

Pausing creates an opportunity for termination to occur. Pausing occurs at sites that resemble terminators but have an increased separation (typically ten to eleven bases) between the hairpin and the U-run. If the pause site does not correspond to a terminator and does not last long enough, though, the enzyme usually moves on again to continue transcription. The length of the pause varies, but at a typical terminator lasts ~60 seconds.

A downstream U-rich region destabilizes the RNA–DNA hybrid when RNA polymerase pauses at the hairpin. The rU-dA RNA–DNA hybrid has an unusually weak base-paired structure; it requires the least energy of any RNA–DNA hybrid to break the association between the two strands. When the polymerase pauses, the RNA–DNA hybrid unravels from the weakly bonded rU-dA terminal region. Often, the actual termination event takes place at any one of several positions toward or at the end of the U-rich region, as though the enzyme "stutters" during termination. The U-rich region in RNA corresponds to an A-T-rich region in DNA, so we see that A-T-rich regions are important in intrinsic termination as well as initiation.

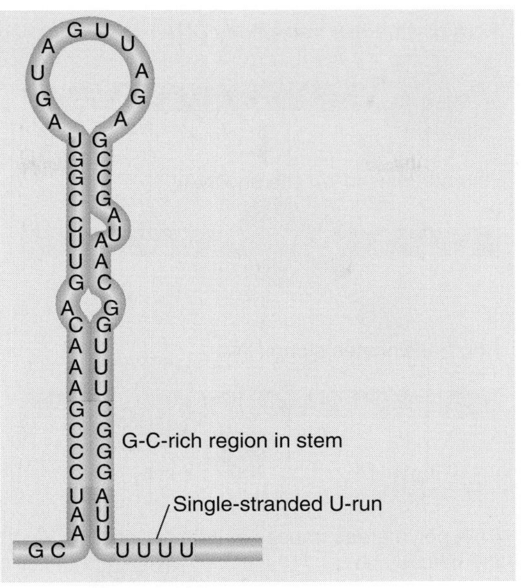

FIGURE 11.25 Intrinsic terminators include palindromic regions that form hairpins varying in length from 7 to 20 bp. The stem-loop structure includes a G-C-rich region and is followed by a run of U residues.

Both the sequence of the hairpin and the length of the U-run influence the efficiency of termination. Termination efficiency *in vitro*, however, varies from 2% to 90% and does not correlate in any simple way with the constitution of the hairpin or the number of U residues in the U-rich region. The hairpin and U-region are therefore necessary, but not sufficient, and additional parameters influence the interaction with RNA polymerase. In particular, the sequences both upstream and downstream of the intrinsic terminator influence its efficiency.

KEY CONCEPT

- Intrinsic terminators consist of a G-C-rich hairpin in the RNA product followed by a U-rich region in which termination occurs.

CONCEPT AND REASONING CHECK

What are the two kinds of transcription terminator sites and how do they differ?

11.15 Rho Factor Is a Site-Specific Terminator Protein

Rho factor is an essential protein in *E. coli* that functions solely at the stage of termination. It acts at **rho-dependent terminators**, which account for about half of *E. coli* terminators.

FIGURE 11.26 shows how rho functions. First it binds to a sequence within the transcript upstream of the site of termination. This sequence is called a ***rut*** site (an acronym for rho utilization). Rho then tracks along the RNA until it catches up to RNA polymerase. When the RNA polymerase reaches the termination site, rho acts on the RNA–DNA hybrid in the enzyme to cause release of the RNA. Pausing by the polymerase at the site of termination allows time for rho factor to translocate to the hybrid stretch, and is an important feature of termination.

We see an important general principle here. When we know the site on DNA at which some protein exercises its effect, we cannot assume that this coincides with the DNA sequence that it initially recognizes. They can be separate, and there need not be a fixed relationship between them. In fact, *rut* sites in different transcription units

- **rho factor** A protein involved in assisting *E. coli* RNA polymerase to terminate transcription at certain terminators (called rho-dependent terminators).
- **rho dependent termination** Transcriptional termination by bacterial RNA polymerase in the presence of the rho factor.
- ***rut*** An acronym for rho utilization site, the sequence of RNA that is recognized by the rho termination factor.

are found at varying distances preceding the sites of termination. A similar distinction is made by antitermination factors (see the next section).

The common feature of *rut* sites is that the sequence is rich in C residues and poor in G residues and has no secondary structure. An example is given in **FIGURE 11.27**. C is by far the most common base (41%) and G is the least common base (14%). Rho is a member of the family of hexameric ATP-dependent helicases. Each subunit has an RNA-binding domain and an ATP hydrolysis domain. The hexamer functions by passing nucleic acid through a hole in the middle of the assembly formed from the RNA-binding domains of the subunits.

FIGURE 11.28 shows that rho binds the RNA along the surfaces of the N-terminal domains from one end, pushing the other end of the bound RNA into the interior of the hexamer, where it binds to the C-terminal domains. When it reaches the RNA-DNA hybrid region at the point of transcription, rho uses its helicase activity to unwind the duplex structure and to release the RNA.

Some rho mutations can be suppressed by mutations in other genes. Studying the mutants is an excellent way to identify proteins that interact with rho, and these studies have implicated the β subunit of RNA polymerase as interacting with the factor.

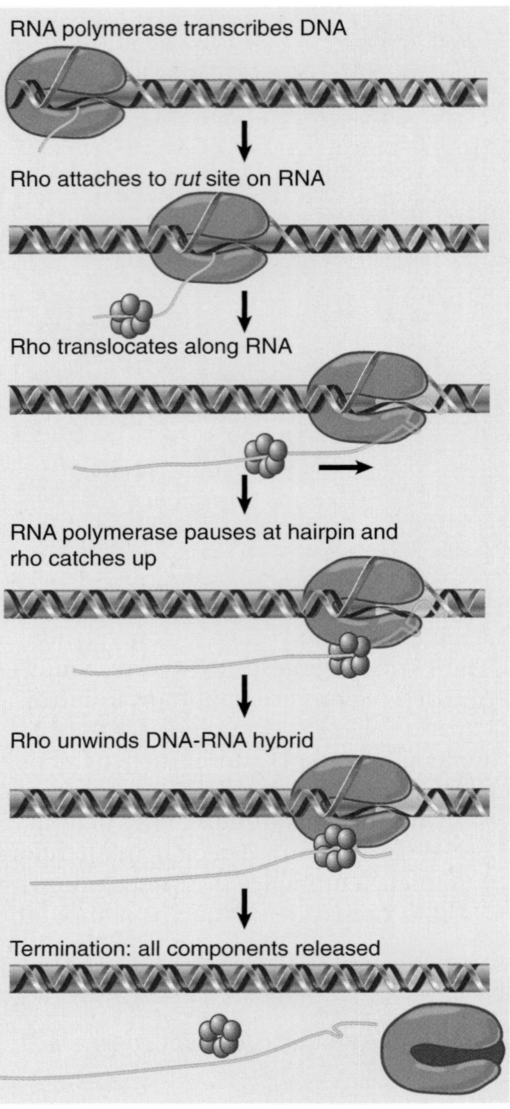

RNA polymerase transcribes DNA
Rho attaches to *rut* site on RNA
Rho translocates along RNA
RNA polymerase pauses at hairpin and rho catches up
Rho unwinds DNA-RNA hybrid
Termination: all components released

FIGURE 11.26 Rho factor binds to RNA at a *rut* site and translocates along RNA until it reaches the RNA–DNA hybrid in RNA polymerase, where it releases the RNA from the DNA.

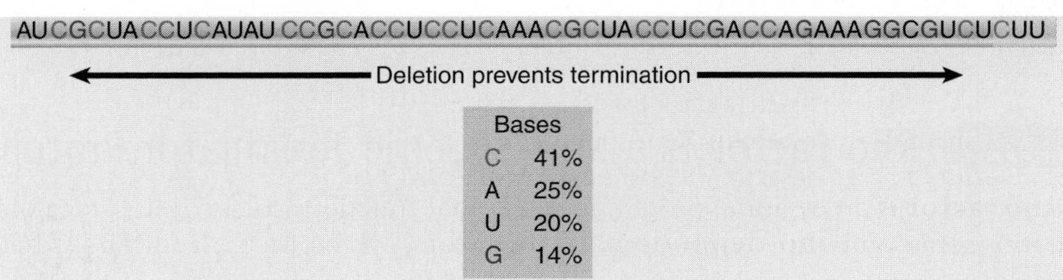

AUCGCUACCUCAUAUCCGCACCUCCUCAAACGCUACCUCGACCAGAAAGGCGUCUCUU
Deletion prevents termination
Bases
C 41%
A 25%
U 20%
G 14%

FIGURE 11.27 A *rut* site has a sequence rich in C and poor in G preceding the actual site(s) of termination. The sequence corresponds to the 3′ end of the RNA.

KEY CONCEPT

- Rho factor is a terminator protein that binds to a *rut* site on a nascent RNA and tracks along the RNA to release it from the RNA-DNA hybrid structure at the RNA polymerase.

CONCEPT AND REASONING CHECK

If transcription termination is a process that occurs on the DNA template, why does rho factor bind to the RNA transcript?

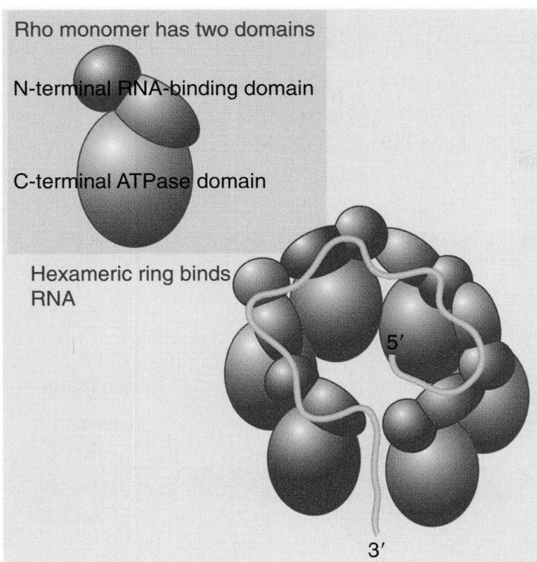
Rho monomer has two domains
N-terminal RNA-binding domain
C-terminal ATPase domain
Hexameric ring binds RNA
5′
3′

FIGURE 11.28 Rho has an N-terminal RNA-binding domain and a C-terminal ATPase domain. A hexamer in the form of a gapped ring binds RNA along the exterior of the N-terminal domains. The 5′ end of the RNA is bound by a secondary binding site in the interior of the hexamer.

11.16 Antitermination May Be a Regulated Event

Antitermination is used as a control mechanism in both phage regulatory circuits and bacterial operons. **FIGURE 11.29** shows that antitermination controls the ability of the enzyme to read past a terminator into downstream genes. In the example shown in the figure, the default pathway is for RNA polymerase to terminate at the end of region 1; however, antitermination allows it to continue transcription through region 2. The promoter does not change, so as a result both situations produce an RNA with the same 5′ sequences; the difference is that after antitermination the RNA is extended to include new sequences at the 3′ end.

antitermination protein Proteins that allow RNA polymerase to transcribe through certain terminator sites.

Antitermination was discovered in bacteriophage infections. A common feature in the control of phage infection is that very few of the phage genes (the "early" genes) can be transcribed by the bacterial host RNA polymerase. Among these genes, however, are regulator(s) whose product(s) allow the next set of phage genes to be expressed (see *Section 14.4, Two Types of Regulatory Events Control the Lytic Cascade*). One of these types of regulator is an **antitermination protein**. **FIGURE 11.30** shows that it enables RNA polymerase to read through a terminator, thus extending the RNA transcript. In the absence of the antitermination protein, RNA polymerase terminates at the terminator (top panel). When the antitermination protein is present, it continues past the terminator (middle panel).

The best-characterized example of antitermination is provided by phage lambda, in which the phenomenon was discovered. It is used at two stages of phage expression. The antitermination protein produced at each stage is specific for the particular transcription units that are expressed at that stage, as summarized in the bottom panel of Figure 11.30.

The host RNA polymerase initially transcribes two genes, which are called the *immediate early genes*. The transition to the next stage of expression is controlled

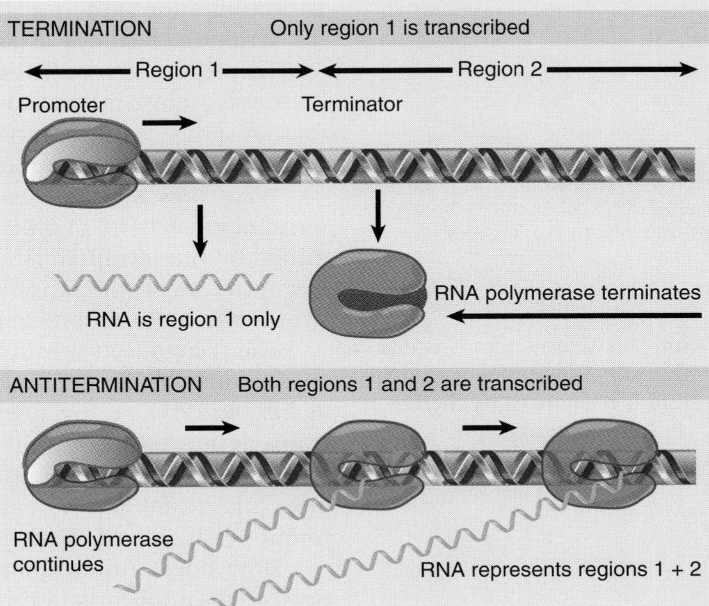
TERMINATION
Only region 1 is transcribed
Region 1
Region 2
Promoter
Terminator
RNA is region 1 only
RNA polymerase terminates
ANTITERMINATION
Both regions 1 and 2 are transcribed
RNA polymerase continues
RNA represents regions 1 + 2

FIGURE 11.29 Antitermination can control transcription by determining whether RNA polymerase terminates or reads through a particular terminator into the following region.

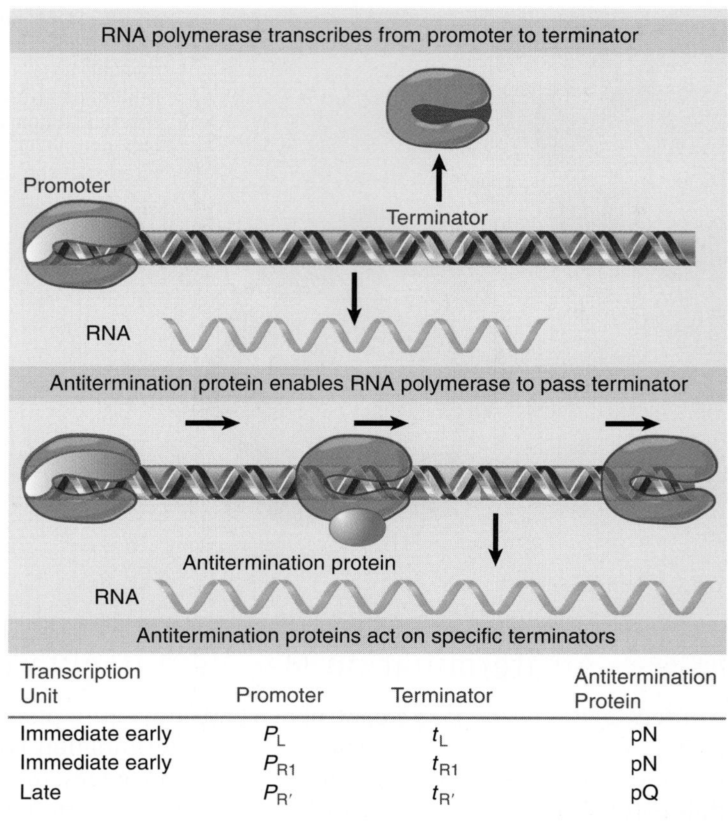

RNA polymerase transcribes from promoter to terminator
Promoter
Terminator
RNA
Antitermination protein enables RNA polymerase to pass terminator
Antitermination protein
RNA
Antitermination proteins act on specific terminators

Transcription Unit	Promoter	Terminator	Antitermination Protein
Immediate early	P_L	t_L	pN
Immediate early	P_{R1}	t_{R1}	pN
Late	$P_{R'}$	$t_{R'}$	pQ

FIGURE 11.30 An antitermination protein can act on RNA polymerase to enable it to read through a specific terminator.

by preventing termination at the ends of the immediate early genes, with the result that the *delayed early genes* are expressed. (We discuss the overall regulation of lambda development in *Chapter 14, Phage Strategies.*) The antitermination protein pN acts specifically on the immediate early transcription units. Later during infection, another antitermination protein pQ acts specifically on the late transcription unit, to allow its transcription to continue past a termination sequence.

The different specificities of pN and pQ establish an important general principle: RNA polymerase interacts with transcription units in such a way that an ancillary factor can sponsor antitermination specifically for some transcripts. Termination can be controlled with the same sort of precision as initiation.

What sites are involved in controlling the specificity of antitermination? The antitermination activity of pN is highly specific, but the antitermination event is not determined by the terminators t_{L1} and t_{R1}; the recognition site needed for antitermination lies upstream in the transcription unit, that is, at a different place from the terminator site at which the action eventually is accomplished.

▸ ***nut*** An acronym for N utilization site, the sequence of DNA that is recognized by the N antitermination factor.

The recognition sites required for pN action are called ***nut*** (for N utilization). The sites responsible for determining leftward and rightward antitermination are described as nut_L and nut_R, respectively. **FIGURE 11.31** shows that their locations relative to each transcription unit are quite different. nut_L lies between the startpoint of P_L and the beginning of the N coding region. By contrast, nut_R lies between the end of the cro gene and t_{R1}. So nut_L is near the promoter, but nut_R is near the terminator. (*qut* is different yet again, and lies within the promoter.)

How does antitermination occur? When pN recognizes the *nut* site, it forms a persistent antitermination complex in cooperation with a number of *E. coli* host proteins. These include three host Nus proteins, NusA, B, and C. NusA is an interesting protein. By itself in *E. coli*, it is part of the transcription termination system. However, when co-opted by N, it participates in antitermination. The complex must act on RNA polymerase to ensure that the enzyme can no longer respond to the terminator. The

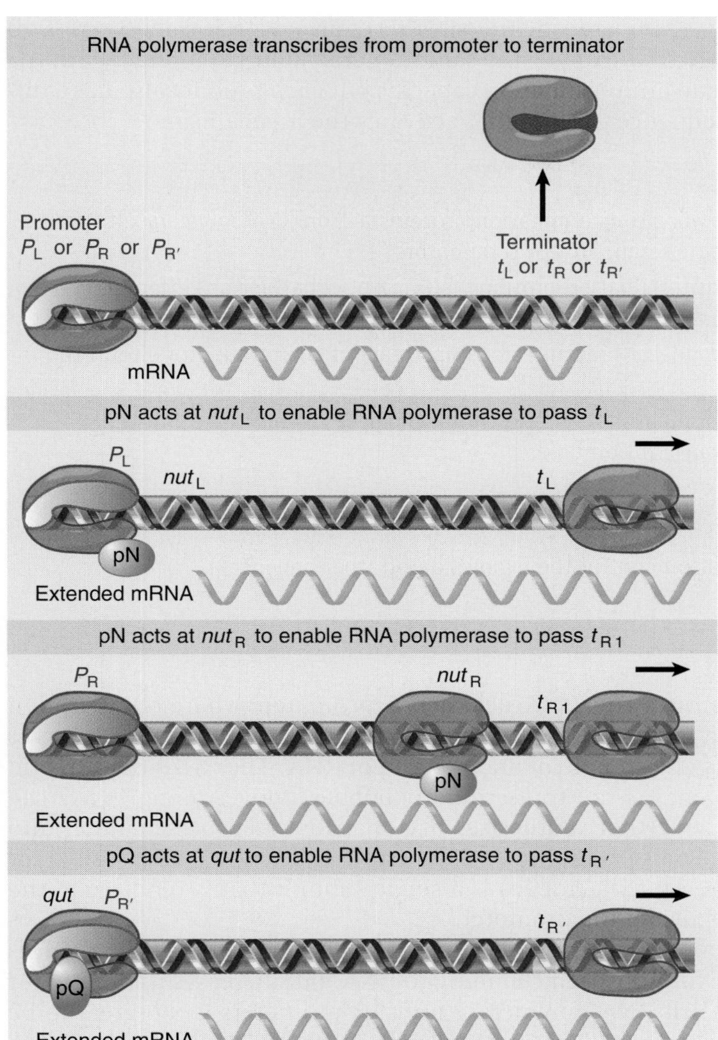

FIGURE 11.31 Host RNA polymerase transcribes lambda genes and terminates at *t* sites. pN allows it to read through terminators in the L and R1 units; pQ allows it to read through the R′ terminator. The sites at which pN acts (*nut*) and at which pQ acts (*qut*) are located at different relative positions in the transcription units.

variable locations of the *nut* sites indicate that this event is linked neither to initiation nor to termination, but can occur to RNA polymerase as it elongates the RNA chain past the *nut* site. As illustrated in **FIGURE 11.32**, the polymerase then becomes a juggernaut that continues past the terminator, heedless of its signal. Is the ability of pN to recognize a short sequence within the transcription unit an example of a more widely used mechanism for antitermination? Phages that are related to lambda have different N genes and different antitermination specificities. The region of the phage genome in which the *nut* sites lie has a different sequence in each of these phages, and each

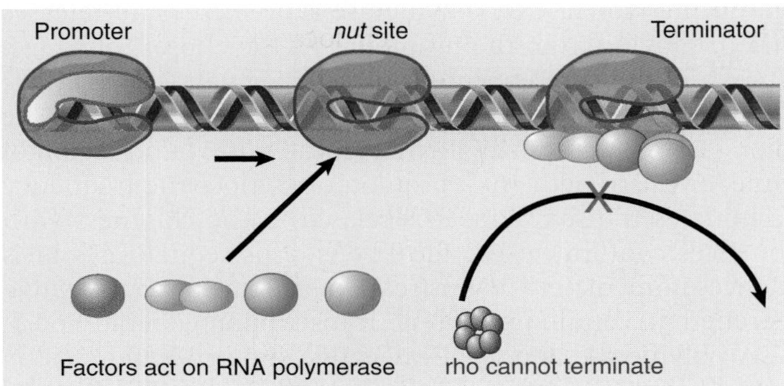

FIGURE 11.32 Ancillary factors bind to RNA polymerase as it passes the nut site. They prevent rho from causing termination when the polymerase reaches the terminator.

phage must therefore have characteristic *nut* sites recognized specifically by its own pN. Each of these pN products must have the same general ability to interact with the transcription apparatus in an antitermination capacity, but each product also has a different specificity for the sequence of DNA that activates the mechanism.

KEY CONCEPTS

- Termination is prevented when antitermination proteins act on RNA polymerase to cause it to read through a specific terminator or terminators.
- Phage lambda has two antitermination proteins, pN and pQ, that act on different transcription units.
- The site where an antiterminator protein acts is upstream of the terminator site in the transcription unit.
- The location of the antiterminator site varies in different cases and can be in the promoter or within the transcription unit.

CONCEPT AND REASONING CHECK

How can termination regulation control the expression of other genes?

11.17 Summary

A transcription unit comprises the DNA between a promoter, where transcription initiates, and a terminator, where it ends. One strand of the DNA in this region serves as a template for synthesis of a complementary strand of RNA. The RNA–DNA hybrid region is short and transient, as the transcription "bubble" moves along DNA. The RNA polymerase holoenzyme that synthesizes bacterial RNA can be separated into two components. Core enzyme is a multimer of structure $\alpha_2\beta\beta'$ that is responsible for elongating the RNA chain. Sigma factor (σ) is a single subunit that is required at the stage of initiation for recognizing the promoter.

Core enzyme has a general affinity for DNA. The addition of sigma factor reduces the affinity of the enzyme for nonspecific binding to DNA, but increases its affinity for promoters. The rate at which RNA polymerase finds its promoters is too great to be accounted for by diffusion and random contacts with DNA; direct exchange of DNA sequences held by the enzyme may be involved.

Bacterial promoters are identified by two short conserved sequences centered at –35 and –10 relative to the startpoint. Most promoters have sequences that are well related to the consensus sequences at these sites. The distance separating the consensus sequences is 16 to 18 bp. RNA polymerase initially "touches down" at the –35 sequence and then extends its contacts over the –10 region. The initial "closed" binary complex is converted to an "open" binary complex by melting of a sequence of ~12 bp that extends from the –10 region to the startpoint. The A-T-rich base pair composition of the –10 sequence may be important for the melting reaction.

The binary complex between RNA polymerase and DNA is converted to a ternary complex by the incorporation of ribonucleotide precursors. There are multiple cycles of abortive initiation, during which RNA polymerase synthesizes and releases very short RNA chains without moving from the promoter. At the end of this stage, there is a change in structure, and the core enzyme contracts. Sigma factor is either released (30% of cases) or it changes its form of association with the core enzyme. The core enzyme then moves along DNA, synthesizing RNA. A locally unwound region of DNA moves with the enzyme. The "strength" of a promoter describes the frequency at which RNA polymerase initiates transcription; it is related to the closeness with which its –35 and –10 sequences conform to the ideal consensus sequences, but is influenced also by the sequences immediately downstream of the startpoint. Negative supercoiling increases the strength of certain promoters. Transcription generates positive supercoils ahead of RNA polymerase and leaves negative supercoils behind the enzyme. The supercoiling must be resolved by topoisomerases.

The core enzyme can be directed to recognize promoters with different consensus sequences by alternative sigma factors. In *E. coli,* these sigma factors are activated by adverse conditions, such as heat shock or nitrogen starvation. *B. subtilis* contains a single major sigma factor with the same specificity as the *E. coli* sigma factor and also contains a variety of minor sigma factors. A specific sequence of sigma factors is activated when sporulation is initiated; sporulation is regulated by two cascades in which sigma factor replacements occur in the forespore (a precursor to the final spore) and mother cell. The geometry of RNA polymerase-promoter recognition is similar for holoenzymes containing all sigma factors. Each sigma factor causes RNA polymerase to initiate transcription at a promoter that conforms to a particular consensus at –35 and –10. Direct contacts between sigma and DNA at these sites have been demonstrated for *E. coli* σ^{70}. The σ^{70} factor of *E. coli* has an N-terminal autoinhibitory domain that prevents the DNA-binding regions from recognizing DNA. The autoinhibitory domain is displaced by DNA when the holoenzyme forms an open complex.

Bacterial RNA polymerase terminates transcription at two types of sites. Intrinsic terminators contain a G-C-rich hairpin followed by a U-rich region. They are recognized *in vitro* by core enzyme alone. Rho-dependent terminators require rho factor both *in vitro* and *in vivo*; rho binds to *rut* sites that are rich in C and poor in G residues and that precede the actual site of termination. Rho is a hexameric ATP-dependent helicase that translocates along the RNA until it reaches the RNA–DNA hybrid region in the transcription bubble of RNA polymerase, where it dissociates the RNA from DNA. In both types of termination, pausing by RNA polymerase is important in order to allow time for the actual termination event to occur.

Antitermination is used by some phages to regulate progression from one stage of gene expression to the next. The lambda gene N codes for an antitermination protein (pN) that is necessary to allow RNA polymerase to read through the terminators located at the ends of the immediate early genes. Another antitermination protein, pQ, is required later in phage infection. pN and pQ act on RNA polymerase as it passes specific sites (*nut* and *qut,* respectively). These sites are located at different relative positions in their respective transcription units.

CHAPTER QUESTIONS

1. Abortive initiation usually occurs when the new transcript is less than:
- **A.** ~5 bases.
- **B.** ~9 bases.
- **C.** ~12 bases.
- **D.** ~16 bases.

2. When RNA polymerase initially binds to DNA, it covers a region of about:
- **A.** 40 to 45 base pairs.
- **B.** 55 to 60 base pairs.
- **C.** 75 to 80 base pairs.
- **D.** 85 to 90 base pairs.

3. In a bacterial genome, what is the minimal length of DNA that is required to provide a specific signal for protein binding (such as RNA polymerase)?
- **A.** 6 base pairs
- **B.** 12 base pairs
- **C.** 21 base pairs
- **D.** 35 base pairs

4. In bacteria, the first base of a transcription unit is almost always (>90%):
- **A.** a purine.
- **B.** a pyrimidine.
- **C.** G.
- **D.** C.

5. What is the most common, conserved distance between the –35 and –10 regions of a bacterial gene promoter?
   A. 8–10 base pairs
   B. 12–15 base pairs
   C. 16–18 base pairs
   D. 20–23 base pairs
6. A mutation at the –10 region of a bacterial gene from TATAGT to TCTAGT would be expected to result in:
   A. a significant increase in expression.
   B. elimination of expression.
   C. no change in expression.
   D. a decrease in expression.
7. Down mutations in the –35 sequence of a bacterial gene promoter usually affect:
   A. initial binding of RNA polymerase to form a closed complex.
   B. transition of a closed promoter complex to an open complex.
   C. initiation of transcription.
   D. clearance of the promoter and transition to an elongation complex.
8. Down mutations in the –10 sequence of a bacterial gene promoter usually affect:
   A. initial binding of RNA polymerase.
   B. transition of a closed promoter complex to an open complex.
   C. initiation of transcription.
   D. clearance of the promoter and transition to an elongation complex.
9. All transcription termination in bacteria is most often associated with:
   A. a hairpin structure caused by internal RNA pairing.
   B. a specific termination sequence.
   C. an AU-rich run of sequences.
   D. a GC-rich run of sequences.
10. The bacteriophage lambda N protein functions in transcription by binding:
   A. directly to the DNA template.
   B. to the mRNA transcript.
   C. directly to the RNA polymerase to alter its structure.
   D. to other transcription factors to alter specificity.

KEY TERMS

–35 box
abortive initiation
antitermination
antitermination protein
C-terminal domain
closed complex
coding strand
consensus sequence
conserved sequence
core enzyme
down mutation
downstream
elongation
hairpin
holoenzyme
initiation
nut
open complex
primary transcript
promoter
rho dependent termination
rho factor
RNA polymerase
rut
sequence context
sigma factor
sporulation
startpoint
TATA box
template strand
termination
terminator
ternary complex
tight binding
transcription unit
upstream
up mutation
vegetative phase

FURTHER READING

Greenblat, J. F. (2008). Transcription termination: pulling out all the stops. *Cell* 132, 917–919.
A short review of transcription termination, and see the article on page 971 of the same issue.

Herbert, K. M., Greenleaf, W. J., and Block, S. M. (2008). Single-molecule studies of RNA polymerase: motoring along. *Annu. Rev. Biochem. 77,* 149–176.
An elegant review demonstrating how studies of individual molecules can provide insight into mechanism of action.

Shilatifard, A., Conway, R. C., and Conway, J. W. (2003). The RNA polymerase II elongation complex. *Annu. Rev. Biochem.* 72, 693–715.
A review of the RNA polymerase structure.

Vassylyev, D. G., Vassylyev, M. N., Perederina, A., Tahirov, T. H., and Artsimovitch, I. (2007). Structural basis for transcription elongation by bacterial RNA poymerase. *Nature* 448, 157–163.
The correlation between structure and function during transcription elongation.

Wang, Q., Tullius, T. D., and Levin, J. R. (2007). Effects of discontinuities in the DNA template on abortive initiation and promoter escape by *E. coli* RNA polymerase. *J. Biol. Chem.* 282, 26917–26927.
A report on the steps required to complete the initiation process by RNA polymerase.

12

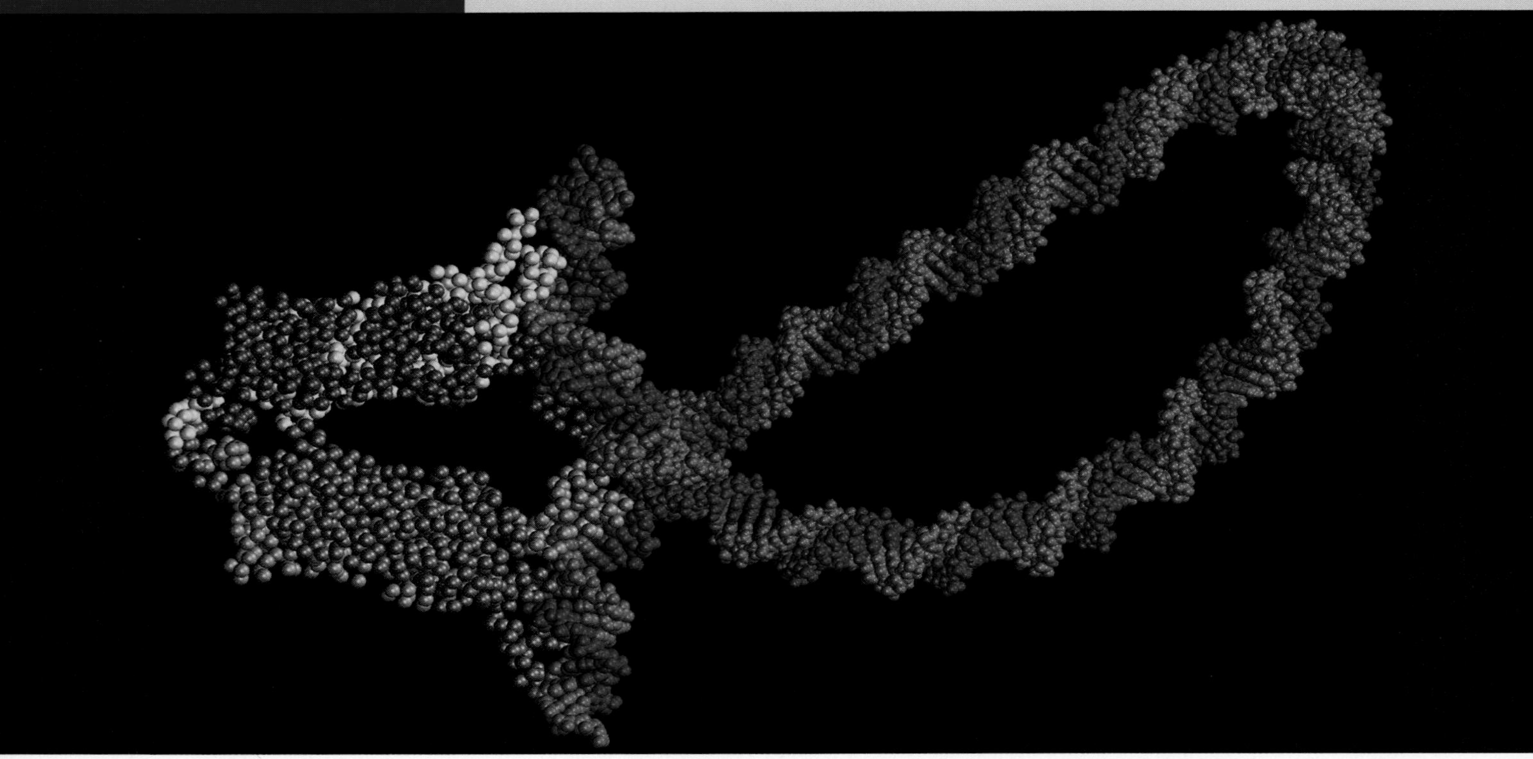

A DNA loop formed by the binding of the *lac* repressor protein to two operator sites of the *E. coli lac* operon. Photo courtesy of Noel Perkins, University of Michigan. More information at Goyal, S., *Biophys. J.* 83 (2007): 4342–4359. Graphics created with VMD (Humphrey, D., Dalke, A., and Schulten, K., "VMD—Visual Molecular Dynamics," *J. Molec. Graphics*, 1996, vol. 14, pp. 33–38 [http://www.ks.uiuc.edu/Research/vmd/]).

The Operon

CHAPTER OUTLINE

12.1 Introduction

Gene expression can be controlled at any of several stages, which we divide broadly into transcription, processing, and translation:

- Transcription often is controlled at the stage of initiation. Transcription is not usually controlled at elongation but may be controlled at termination to determine whether RNA polymerase is allowed to proceed past a terminator to the gene(s) beyond.
- In bacteria, an mRNA is typically available for translation while it is being synthesized; this is called **coupled transcription/translation**. (In eukaryotic cells, processing of the RNA product may be regulated at the stages of modification, splicing, transport, or stability.)
- Translation in bacteria may also be directly regulated, but more commonly it is passively modulated. The coding portion or open reading frame of a gene can be assembled either with common codons or rare codons, which correspond to common or rare tRNAs. mRNAs containing a number of rare codons are more difficult to translate.

▸ **coupled transcription/translation** The phenomena in bacteria where translation of the mRNA occurs simultaneously with its transcription.

The basic concept for the way transcription is controlled in bacteria is called the **operon** model and was proposed by François Jacob and Jacques Monod in 1961. They distinguished between two types of sequences in DNA: sequences that code for ***trans*-acting** products (usually proteins) and ***cis*-acting** sequences that function exclusively within the DNA. Gene activity is regulated by the specific interactions of the *trans*-acting products with the *cis*-acting sequences (see *Section 2.11, Proteins Are* Trans-*acting, but Sites on DNA Are* cis-*acting*). In more formal terms:

- A gene is a sequence of DNA that codes for a diffusible product. *The crucial feature is that the product diffuses away from its site of synthesis to act elsewhere.*
- The description *cis*-acting applies to any sequence of DNA that functions exclusively as a DNA sequence, affecting only the DNA to which it is physically linked.

To help distinguish between the components of regulatory circuits and the genes that they regulate, we sometimes use the terms *structural gene* and *regulator gene*. A **structural gene** is simply any gene that codes for a protein (or RNA) product. Protein structural genes represent an enormous variety of structures and functions, including structural proteins, enzymes with catalytic activities, and regulatory proteins. A type of structural gene is a **regulator gene**, which simply describes a gene that codes for a protein or an RNA involved in regulating the expression of other genes.

▸ **operon** A unit of bacterial gene expression and regulation, including structural genes and control elements in DNA recognized by regulator gene product(s).

▸ ***trans*-acting** A product that can function on any copy of its target DNA. This implies that it is a diffusible protein or RNA.

▸ ***cis*-acting** A site that affects the activity only of sequences on its own molecule of DNA (or RNA); this property usually implies that the site does not code for protein.

▸ **structural gene** A gene that codes for any RNA or protein product other than a regulator.

▸ **regulator gene** A gene that codes for a product (typically protein) that controls the expression of other genes (usually at the level of transcription).

The simplest form of the regulatory model is illustrated in **FIGURE 12.1**: *a regulator gene codes for a protein that controls transcription by binding to particular site(s) on DNA.* This interaction can regulate a target gene in either a positive manner (the interaction turns the gene on) or a negative manner (the interaction turns the gene off). The sites on DNA are usually (but not exclusively) located just upstream of the target gene.

The sequences that mark the beginning and end of the transcription unit—the promoter and terminator—are examples of *cis*-acting sites. *A promoter serves to initiate transcription only of the gene or genes physically connected to it on the same stretch of DNA.* In the same way, a terminator can terminate transcription only by an RNA polymerase that has traversed the preceding gene(s). In their simplest forms, promoters and terminators are *cis*-acting elements that are recognized by the same *trans*-acting species, that is, by RNA polymerase (although other factors also participate at each site).

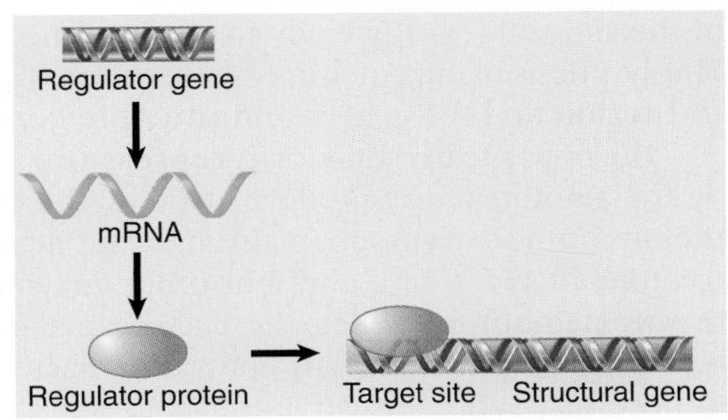

FIGURE 12.1 A regulator gene codes for a protein that acts at a target site on DNA.

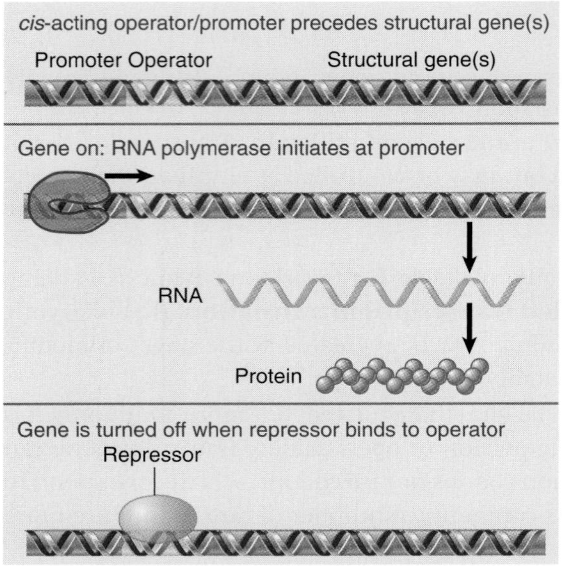

FIGURE 12.2 In negative control, a *trans*-acting repressor binds to the *cis*-acting operator to turn off transcription.

- **negative control** This describes a mechanism of gene regulation in which a regulator is required to turn the gene off.
- **operator** The site on DNA at which a repressor protein binds to prevent transcription from initiating at the adjacent promoter.

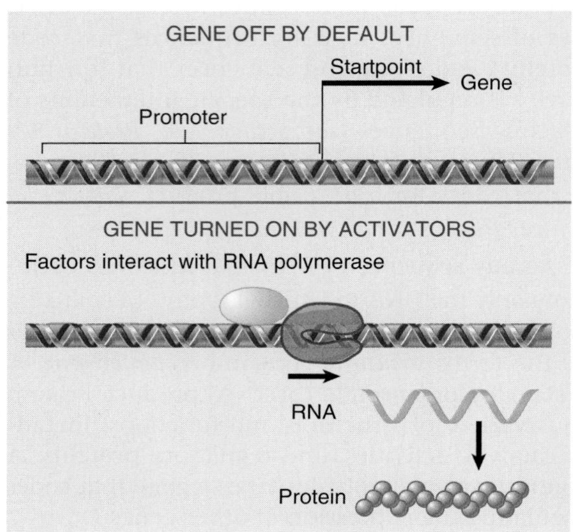

FIGURE 12.3 In positive control, a *trans*-acting factor must bind to *cis*-acting site in order for RNA polymerase to initiate transcription at the promoter.

- **positive control** This describes a system in which a gene is not expressed unless some action turns it on.

- **induction** The ability to synthesize certain enzymes only when their substrates are present; applied to gene expression, it refers to switching on transcription as a result of interaction of the inducer with the regulator protein.
- **inducible gene** A gene that is turned on by the presence of its substrate.
- **repression** The ability to prevent synthesis of certain enzymes when their products are present; more generally, it refers to inhibition of transcription (or translation) by binding of repressor protein to a specific site on DNA (or mRNA).
- **repressible gene** A gene that is turned off by its product.

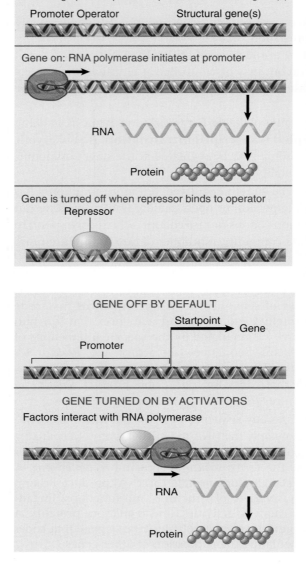

Additional *cis*-acting regulatory sites are often combined with the promoter. A bacterial promoter may have one or more such sites located close by, that is, in the immediate vicinity of the startpoint. A eukaryotic promoter is likely to have a greater number of sites that are spread out over a longer distance, as we will see in *Section 12.5, cis-Acting Constitutive Mutations Identify the Operator.*

A classic mode of transcription control in bacteria is ***negative control***: a repressor protein prevents a gene from being expressed. FIGURE 12.2 shows that in the absence of the negative regulator, the gene is expressed. Close to the promoter is another *cis*-acting site called the **operator**, which is the binding site for the repressor protein. When the repressor binds to the operator, RNA polymerase is prevented from initiating transcription, and *gene expression is, therefore, turned off.* An alternative mode of control is ***positive control***. This is used in bacteria (probably) with about equal frequency to negative control, and it is the most common mode of control in eukaryotes. A transcription factor is required to assist RNA polymerase in initiating at the promoter. FIGURE 12.3 shows that in the absence of the positive regulator, the gene is inactive: RNA polymerase cannot by itself initiate transcription at the promoter.

In addition to negative and positive control, there is another way to look at regulation of gene transcription for a gene that encodes an enzyme: whether it is controlled by the substrate or by the product of the enzyme. Bacteria need to respond swiftly to changes in their environment. Fluctuations in the supply of nutrients can occur at any time, and survival depends on the ability to switch from metabolizing one substrate to another. Yet economy is important, too: a bacterium that indulges in energetically expensive ways to meet the demands of the environment is likely to be at a disadvantage. Thus a bacterium avoids synthesizing the enzymes of a pathway in the absence of the substrate, but is ready to produce the enzymes if the substrate should appear. The synthesis of enzymes in response to the appearance of a specific substrate is called **induction** and the gene is an **inducible gene**.

The opposite of induction is **repression**, where the **repressible gene** is controlled by the amount of the product made by the enzyme. For example, *E. coli* synthesizes the amino acid tryptophan through the actions of an enzyme complex containing tryptophan synthetase and four other enzymes. If, however, tryptophan is provided in the medium on which the bacteria are growing, the production of the enzyme is immediately halted. This allows the bacterium to avoid devoting its resources to unnecessary synthetic activities.

Induction and repression represent similar phenomena. In one case the bacterium adjusts its ability to use a given substrate (such as the complex sugar lactose) for growth; in the other it adjusts its ability to synthesize a particular metabolic intermediate (such as an essential amino acid). The trigger for either type of adjustment is a small molecule that is the substrate, or related to the substrate for the enzyme, or the product of the enzyme activity, respectively. Small molecules that cause the production of enzymes that are able to metabolize them (or their analogues) are called **inducers**. Those that prevent the production of enzymes that are able to synthesize them are called **corepressors**.

These two ways of looking at regulation—negative versus positive control and inducible versus repressible control—are typically combined to give four different patterns of gene regulation: **negative inducible**, **negative repressible**, **positive inducible**, **positive repressible** as shown in FIGURE 12.4. This enables a bacterium to perform the ultimate in inventory control of its metabolism to allow survival in rapidly changing environments.

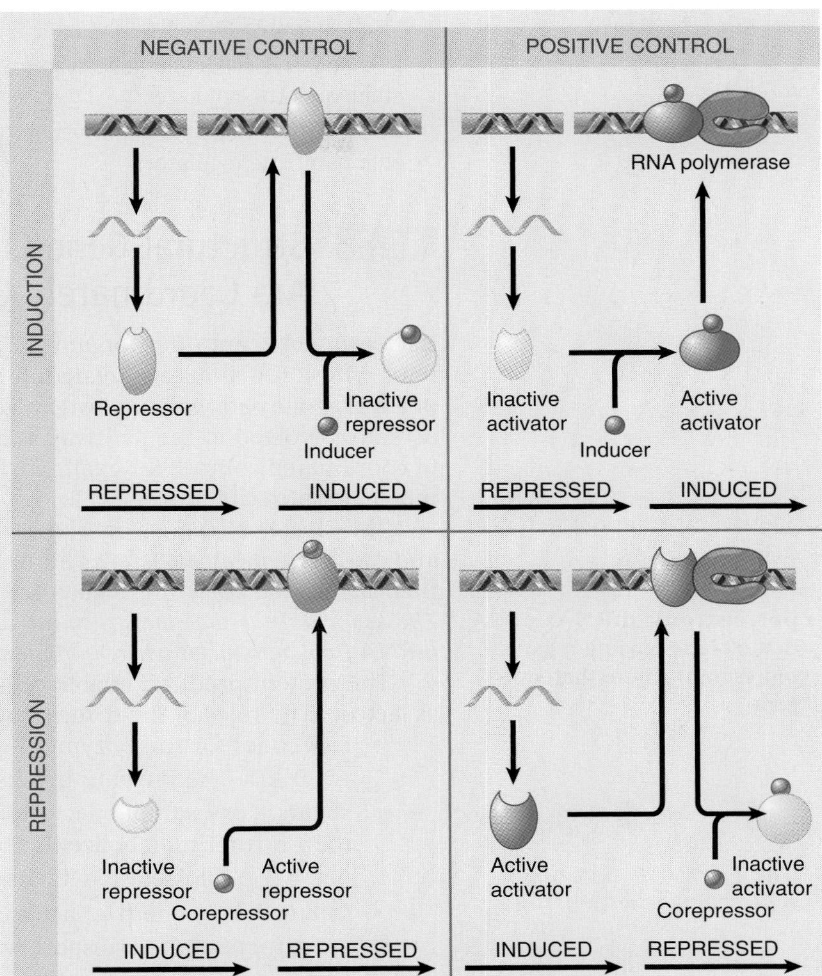

FIGURE 12.4 Regulatory circuits can be designed from all possible combinations of positive and negative control with inducible and repressible control.

The unifying theme is that regulatory proteins are *trans*-acting factors that recognize *cis*-acting elements (usually) upstream of the gene. The consequences of this recognition are either to activate or to repress the gene, depending on the individual type of regulatory protein. A typical feature is that the protein functions by recognizing a very short sequence in DNA, usually <10 bp in length, although the protein actually binds over a somewhat greater distance of DNA. The bacterial promoter is an example: RNA polymerase covers >70 bp of DNA at initiation, but the crucial sequences that it recognizes are the hexamers centered at −35 and −10.

A significant difference in gene organization between prokaryotes and eukaryotes is that structural genes in bacteria are organized in operons coordinately controlled by means of interactions at a single regulator, whereas genes in eukaryotes are controlled individually. As a result, in bacteria, an entire related set of genes is either transcribed or not transcribed. In this chapter, we discuss this mode of control and its use by bacteria. The means employed to coordinate control of dispersed eukaryotic genes are discussed in *Chapter 25, Eukaryotic Transcription*.

KEY CONCEPTS

- In negative regulation, a repressor protein binds to an operator to prevent a gene from being expressed.
- In positive regulation, a transcription factor is required to bind at the promoter in order to enable RNA polymerase to initiate transcription.
- In inducible regulation, the gene is regulated by the presence of its substrate.
- In repressible regulation, the gene is regulated by the product of its enzyme pathway.
- We can combine these in all four combinations: negative inducible, negative repressible, positive inducible, and positive repressible.

- **inducer** A small molecule that triggers gene transcription by binding to a regulator protein.
- **corepressor** A small molecule that triggers repression of transcription by binding to a regulator protein.
- **negative inducible** A control circuit in which an active repressor is inactivated by the substrate of the operon.
- **negative repressible** A control circuit in which an inactive repressor is activated by the product of the operon.
- **positive inducible** A control circuit in which an inactive positive regulator is converted into an active regulator by the substrate of the operon.
- **positive repressible** A control circuit in which an active positive regulator is inactivated by the product of the operon.

CONCEPT AND REASONING CHECKS

- In a negative inducible gene system, what would be the effect of a mutation that eliminates the regulator?
- In a positive inducible gene system, what would be the effect of a mutation that eliminates the regulator?

12.2 Structural Gene Clusters Are Coordinately Controlled

Bacterial genes are often organized into operons that include genes coding for proteins whose functions are related. It is common for the genes coding for the enzymes of a metabolic pathway to be organized into such a cluster. In addition to the enzymes actually involved in the pathway, other related activities may be included in the unit of coordinated control; for example, the protein responsible for transporting the small molecule substrate into the cell.

▸ **polycistronic mRNA** mRNA that includes coding regions representing more than one gene.

The cluster of the *lac* operon containing the three *lac* structural genes, *lacZ, lacY,* and *lacA,* is typical. **FIGURE 12.5** summarizes the organization of the structural genes, their associated *cis*-acting regulatory elements, and the *trans*-acting regulatory gene. *The key feature is that the structural gene cluster is transcribed into a single* ***polycistronic mRNA*** *from a promoter where initiation of transcription is regulated.*

The protein products enable cells to take up and metabolize β-galactosides, such as lactose. The roles of the three structural genes are:

- *lacZ* codes for the enzyme β-galactosidase, whose active form is a tetramer of ~500 kD. The enzyme breaks the complex β-galactoside into its component sugars. For example, lactose is cleaved into glucose and galactose (which are then further metabolized). This enzyme also produces an important by-product, β-1,6-allolactase, which we will see below has a role in regulation.
- *lacY* codes for the β-galactoside permease, a 30-kD membrane-bound protein constituent of the transport system. This transports β-galactosides into the cell.
- *lacA* codes for β-galactoside transacetylase, an enzyme that transfers an acetyl group from acetyl-CoA to β-galactosides.

Mutations in either *lacZ* or *lacY* can create the *lac* genotype, in which cells cannot utilize lactose. (The genotypic description *"lac"* without a qualifier indicates loss-of-function.) The *lacZ* mutations abolish enzyme activity, directly preventing metabolism of lactose. The *lacY* mutants cannot take up lactose efficiently from the medium. (No defect is identifiable in *lacA* cells, which is puzzling. It is possible that the acetylation reaction gives an advantage when the bacteria grow in the presence of certain analogs of β-galactosides that cannot be metabolized, because the modification results in detoxification and excretion.)

The entire system, including structural genes and the elements that control their expression, forms a common unit of regulation called an operon. The activity of the operon is controlled by regulator gene(s) whose protein products interact with the *cis*-acting control elements.

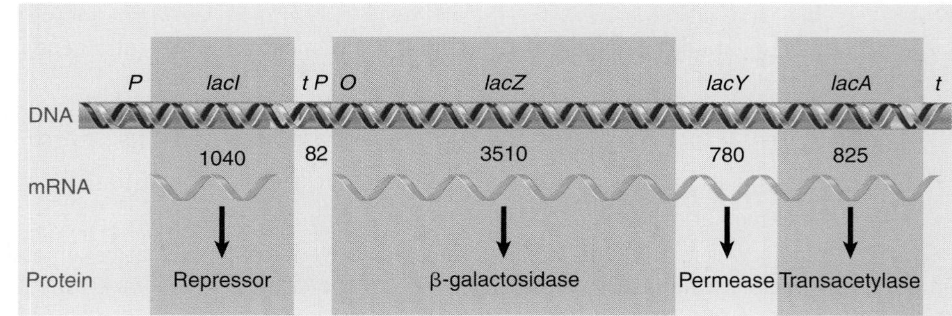

FIGURE 12.5 The *lac* operon occupies ~6000 bp of DNA. At the left the *lacI* gene has its own promoter and terminator. The end of the *lacI* region, *t*, is adjacent to the *lacZYA* promoter, *P*. Its operator, *O*, occupies the first 26 bp of the transcription unit. The long *lacZ* gene starts at base 39, and is followed by the *lacY* and *lacA* genes and a terminator, *t*.

HISTORICAL PERSPECTIVES

An Unstable Intermediate Carrying Information from Genes to Ribosomes for Protein Synthesis

Brenner and Jacob were guest investigators at the California Institute of Technology in 1960. At that time there was great interest in the mechanisms by which genes code for proteins. One possibility, which seemed reasonable at the time, was that each gene produced a different type of ribosome, differing in its RNA, which in turn produced a different type of protein. François Jacob and Jacques Monod had recently proposed an alternative, which was that the informational RNA ("messenger RNA") is actually an unstable molecule that breaks down rapidly. In this model, the ribosomes are nonspecific protein-synthesizing centers that synthesize different proteins according to specific instructions they receive from the genes through the messenger RNA. The key to the experiment is density-gradient centrifugation, which can separate macromolecules made "heavy" or "light" according to their content of 15N or 14N, respectively. (This technique is described in the Historical Perspectives Box in Chapter 15.) The experiment is a purely biochemical proof of an issue absolutely critical for genetics—that genes code for proteins through the intermediary of a relatively short-lived messenger RNA.

> A large amount of evidence suggests that genetic information for protein structure is encoded in deoxyribonucleic acid (DNA) while the actual assembling of amino acids into proteins occurs in cytoplasmic ribonucleoprotein particles called ribosomes. The fact that proteins are not synthesized directly on genes demands the existence of an intermediate information carrier. . . . Jacob and Monod have put forward the hypothesis that ribosomes are nonspecialized structures which receive genetic information from the gene in the form of an unstable intermediate or "messenger." We present here the results of experiments on phage-infected bacteria which give direct support to this hypothesis. . . . When growing bacteria are infected with T2 bacteriophage, synthesis of DNA stops immediately, to resume 7 minutes later, while protein synthesis continues at a constant rate; in all likelihood, the protein is genetically determined by the phage. . . . Phage-infected bacteria therefore provide a situation in which the synthesis of a protein is suddenly switched from bacterial to phage control. . . . It is possible to determine experimentally [whether an unstable messenger RNA is produced] in the following way: Bacteria are grown in heavy isotopes so that all cell constituents are uniformly labeled "heavy." They are infected with phage and transferred immediately to a medium containing light isotopes so that all constituents synthesized after infection are "light." The distribution of new RNA and new protein, labeled with radioactive isotopes, is then followed by density gradient centrifugation of purified ribosomes. . . . We may summarize our findings as follows: (1) After phage infection no new ribosomes can be detected. (2) A new RNA with a relatively rapid turnover is synthesized after phage infection. This RNA, which has a base composition corresponding to that of the phage DNA, is added to pre-existing ribosomes, from which it can be detached in a cesium chloride gradient by lowering the magnesium concentration. (3) Most, if not all, protein synthesis in the infected cell occurs in pre-existing ribosomes. . . . The results also suggest that the messenger RNA may be large enough to code for long polypeptide chains. . . . It is a prediction of the messenger RNA hypothesis that the messenger RNA should be a simple copy of the gene, and its nucleotide composition should therefore correspond to that of the DNA. This appears to be the case in phage-infected cells. . . . If this turns out to be universally true, interesting implications for the coding mechanisms will be raised.

(*Source*: Brenner, S., Jacob, F., and Meselson, M. An Unstable Intermediate Carrying Information from Genes to Ribosomes for Protein Synthesis. *Nature* 190, 576–581.)

KEY CONCEPT

- Genes coding for proteins that function in the same pathway may be located adjacent to one another and controlled as a single unit that is transcribed into a polycistronic mRNA.

CONCEPT AND REASONING CHECKS

In the *lacZYA* operon, what would be the effect of a promoter mutation on *lacZ*? *lacY*?

12.3 The *lac* Operon Is Negative Inducible

We can distinguish between structural genes and regulator genes by the effects of mutations. A mutation in a structural gene deprives the cell of the particular protein for which the gene codes. A mutation in a regulator gene, however, influences the expression of all the structural genes that it controls. The consequences of a regulatory mutation reveal the type of regulation.

Transcription of the *lacZYA* genes is controlled by a regulator protein encoded by the *lacI* gene. It happens that *lacI* is located adjacent to the structural genes, but it comprises an independent transcription unit with its own promoter and terminator. In principle, *lacI* need not be located near the structural genes because it specifies a diffusible product. It can function equally well if moved elsewhere, or can be carried on a separate DNA molecule (the classic test for a *trans*-acting regulator).

The *lacZYA* genes are negatively regulated: *they are transcribed unless turned off by the regulator protein.* A mutation that inactivates the regulator causes the structural genes to be continually expressed, a condition called **constitutive expression**. The product of *lacI* is called the ***lac* repressor**, because its function is to prevent the expression of the structural genes.

▸ **constitutive expression** This describes a gene system where the gene is expressed continuously.

▸ ***lac* repressor** A negative gene regulator encoded by the *lacI* gene that turns off the *lac* operon.

The repressor is a tetramer of identical subunits of 38 kD each. There are ~10 tetramers in a wild-type cell. The regulator gene is not controlled; it is an unregulated gene. It is transcribed into a monocistronic mRNA at a rate that appears to be governed simply by the affinity of its (poor) promoter for RNA polymerase. In addition, it is transcribed into a poor mRNA. This is a common way to restrict the amount of protein made. In this case, the mRNA has virtually no 5′ UTR, which restricts the ability of a ribosome to start translation. These two features account for the low abundance of Lac Repressor protein in the cell.

The repressor functions by binding to an operator (formally denoted O_{lac}) at the start of the *lacZYA* cluster. The operator lies between the promoter (P_{lac}) and the structural genes *(lacZYA). When the repressor binds at the operator, it prevents RNA polymerase from initiating transcription at the promoter.* **FIGURE 12.6** expands our view of the region at the start of the *lac* structural genes. The operator extends from position –5 just upstream of the mRNA startpoint to position +21 within the transcription unit; thus it overlaps the 3′, right end of the promoter. A mutation that inactivates the operator also causes constitutive expression.

When cells of *E. coli* are grown in the absence of a β-galactoside there is no need for β-galactosidase, and they contain very few molecules of the enzyme, about 5 per cell. When a suitable substrate is added, the enzyme activity appears very rapidly in the bacteria. Within two to three minutes some enzyme is present, and soon there are ~5000 molecules of enzyme per bacterium. (Under suitable conditions, β-galactosidase can account for 5% to 10% of the total soluble protein of the bacterium.) If the substrate is removed from the medium, the synthesis of enzyme stops as rapidly as it started.

FIGURE 12.7 summarizes the essential features of this induction. Control of transcription of the *lac* operon responds very rapidly to the inducer, as shown in the upper

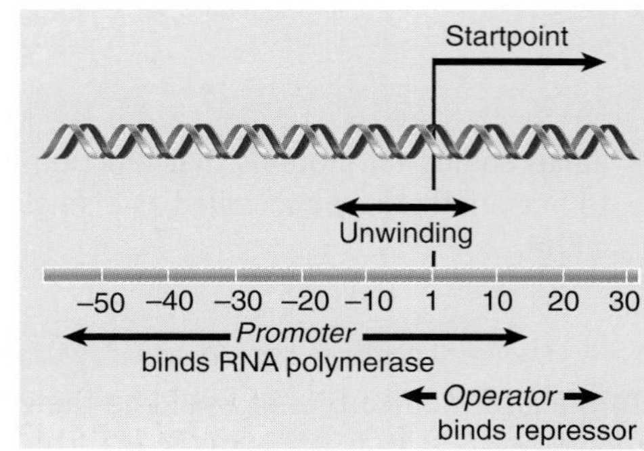

FIGURE 12.6 Repressor and RNA polymerase bind at sites that overlap around the transcription startpoint of the *lac* operon.

part of the figure. In the absence of inducer, the operon is transcribed at a very low basal level (this is an important concept, see the next section). Transcription is stimulated as soon as inducer is added; the amount of *lac* mRNA increases rapidly to an induced level that reflects a balance between synthesis and degradation of the mRNA.

The *lac* mRNA (as most mRNA is in bacteria) is extremely unstable and decays with a half-life of only ~3 minutes. This feature allows induction to be reversed rapidly by repressing transcription as soon as the inducer is removed. In a very short time all the *lac* mRNA is destroyed and enzyme synthesis ceases.

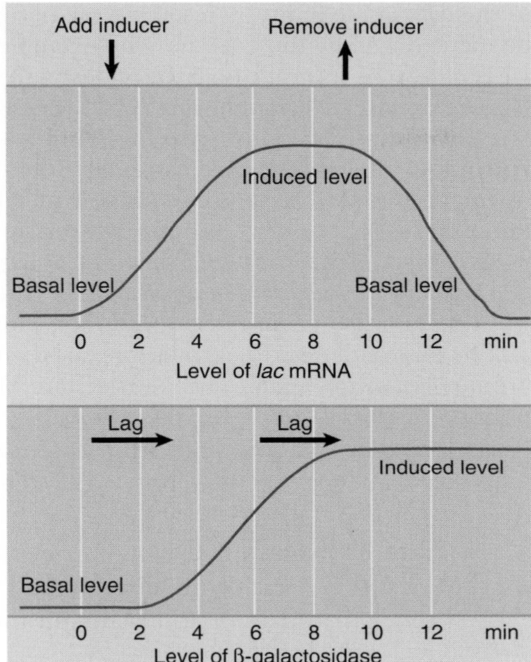

FIGURE 12.7 Addition of inducer results in rapid induction of *lac* mRNA, and is followed after a short lag by synthesis of the enzymes; removal of inducer is followed by rapid cessation of synthesis.

The production of protein is followed in the lower part of the figure. Translation of the *lac* mRNA produces β-galactosidase (and the products of the other *lac* genes). There is a short lag between the appearance of *lac* mRNA and appearance of the first completed enzyme molecules (it is ~2 minutes after the rise of mRNA from basal level before protein begins to increase). There is a similar lag between reaching maximal induced levels of mRNA and protein. When inducer is removed, synthesis of enzyme ceases almost immediately (as the mRNA is degraded), but the β-galactosidase in the cell is more stable than the mRNA, so the enzyme activity remains at the induced level for longer.

KEY CONCEPTS

- Transcription of the *lacZYA* operon is controlled by a repressor protein that binds to an operator that overlaps the promoter at the start of the cluster.
- The repressor protein is a tetramer of identical subunits coded by the *lacI* gene.
- β-galactosides, the substrates of the *lac* operon, are its inducer.
- In the absence of β-galactosides, the *lac* operon is expressed only at a very low (basal) level.
- Addition of specific β-galactosides induces transcription of all three genes of the *lac* operon.
- The *lac* mRNA is extremely unstable; as a result, induction can be rapidly reversed.

CONCEPT AND REASONING CHECK

In the *lac* operon, what would be the effect of a mutation in the operator?

12.4 *lac* Repressor Is Controlled by a Small-Molecule Inducer

The ability to act as inducer or corepressor is highly specific. Only the substrate/product or a closely related molecule can serve. In most cases, *the activity of the small molecule, however, does not depend on its interaction with the target enzyme.* For the *lac* system, however, the natural inducer is not lactose, but a by-product of the lacZ enzyme, **allolactose**. Allolactose is also a substrate of the lacZ enzyme, so it does not persist

▸ **allolactose** A by-product of the LacZ enzyme, the true inducer of the *lac* operon.

in the cell. Some inducers resemble the natural inducers of the *lac* operon but cannot be metabolized by the enzyme. The example *par excellence* is isopropylthiogalactoside (IPTG), one of several thiogalactosides with this property. IPTG is not recognized by β-galactosidase; even so, it is a very efficient inducer of the *lac* genes.

▸ **gratuitous inducer** Inducers that resemble authentic inducers of transcription, but are not substrates for the induced enzymes.

Molecules that induce enzyme synthesis but are not metabolized are called **gratuitous inducers**. They are extremely useful because they remain in the cell in their original form. (A real inducer would be metabolized, interfering with study of the system.) The existence of gratuitous inducers reveals an important point. *The system must possess some component, distinct from the target enzyme, that recognizes the appropriate substrate, and its ability to recognize related potential substrates is different from that of the enzyme.*

The component that responds to the inducer is the repressor protein encoded by *lacI*. Its target, the *lacZYA* structural genes, is transcribed into a single mRNA from the promoter just upstream of *lacZ*. The state of the repressor determines whether this promoter is turned off or on.

- **FIGURE 12.8** shows that in the absence of an inducer the genes are not transcribed, because the repressor protein is in an active form that is bound to the operator.
- **FIGURE 12.9** shows that when an inducer is added, the repressor is converted into either a form with lower affinity for operator or a lower affinity form that leaves the operator. Transcription then starts at the promoter and proceeds through the genes to a terminator located beyond the 3′ end of *lacA*.

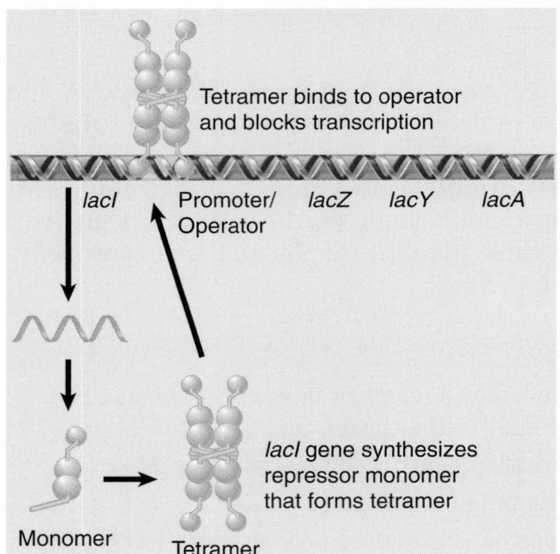

FIGURE 12.8 *lac* Repressor maintains the *lac* operon in the inactive condition by binding to the operator. The shape of the repressor is represented as a series of connected domains as revealed by its crystal structure (see Figure 12.12).

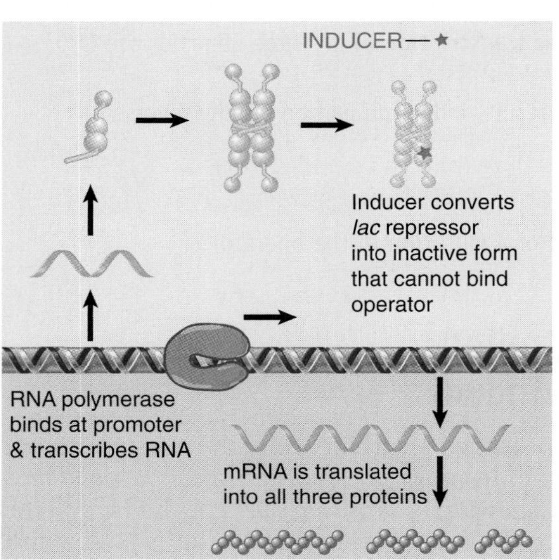

FIGURE 12.9 Addition of inducer converts *lac* repressor to an inactive form that cannot bind the operator. This allows RNA polymerase to initiate transcription.

The crucial features of the control circuit reside in the dual properties of the repressor: it can prevent transcription, and it can recognize the small-molecule inducer. The repressor has two binding sites: one for the operator DNA and one for the inducer. When the inducer binds at its site, it changes the conformation of the protein in such a way as to influence the activity of the operator-binding site. The ability of one site in the protein to control the activity of another is called **allosteric control**.

▸ **allosteric control** The ability of a protein to change its conformation (and therefore activity) at one site as the result of binding a small molecule to a second site located elsewhere on the protein.

Induction accomplishes a coordinate regulation: *all the genes are expressed (or not expressed) in unison.* The mRNA is translated sequentially from its 5′ end, which explains why induction always causes the appearance of β-galactosidase, β-galactoside permease, and β-galactoside transacetylase, in that order. Translation of a common mRNA explains why the relative amounts of the three enzymes always remain the same under varying conditions of induction. Usually, the most important enzyme is first in the operon.

There are several potential paradoxes in the constitution of the operon. First, the *lac* operon contains the structural gene *(lacZ)* coding for the β-galactosidase activity needed to metabolize the sugar; it also includes the gene *(lacY)* that codes for the protein needed to transport the substrate into the cell. If the operon is in a repressed state, though, how does the inducer enter the cell to start the process of induction?

Two features ensure that there is always a minimal amount of the protein present in the cell—enough to start the process off. There is a basal level of expression of the operon: even when it is not induced, it is expressed at a residual level (0.1% of the induced level). In addition, some substrate enters anyway via another uptake system.

The second paradox is that it requires β-galactosidase to make the inducer allolactose to induce the synthesis of β-galactosidase. That means that an operon with a mutant *lacZ* gene cannot be induced.

KEY CONCEPTS

- An inducer functions by converting the repressor protein into a form with lower operator affinity.
- *lac* repressor has two binding sites, one for the operator DNA and another for the inducer.
- *lac* repressor is inactivated by an allosteric interaction in which binding of inducer at its site changes the properties of the DNA-binding site.
- The true inducer is allolactose, not the actual substrate.

CONCEPT AND REASONING CHECK

If there are two complete copies of the *lac* operon in a cell and a mutation in one of the copies of *lacI*, what would be effect on the cell?

12.5 *cis*-Acting Constitutive Mutations Identify the Operator

▸ **uninducible** A mutant in which the affected gene(s) cannot be expressed.

Mutations in the regulatory circuit may either abolish expression of the operon or cause constitutive expression. Mutants that cannot be expressed at all are called **uninducible**. Mutants that are continuously expressed are called *constitutive mutants*.

Components of the regulatory circuit of the operon can be identified by mutations that *affect the expression of all the structural genes and map outside them.* They fall into two classes, *cis*-acting and *trans*-acting. The promoter and the operator are identified as targets for the regulatory proteins (RNA polymerase and repressor, respectively) by *cis*-acting mutations. The locus *lacI* is identified as the gene that codes for the repressor protein by mutations that eliminate the *trans*-acting product.

The operator was originally identified by constitutive mutations, denoted O^c, whose distinctive properties provided the first evidence for *an element that functions without being represented in a diffusible product.*

The structural genes contiguous with an O^c mutation are expressed constitutively because the mutation changes the operator so that the repressor no longer recognizes it and binds to it. Thus the repressor cannot prevent RNA polymerase from initiating transcription. The operon is transcribed constitutively, as illustrated in **FIGURE 12.10**.

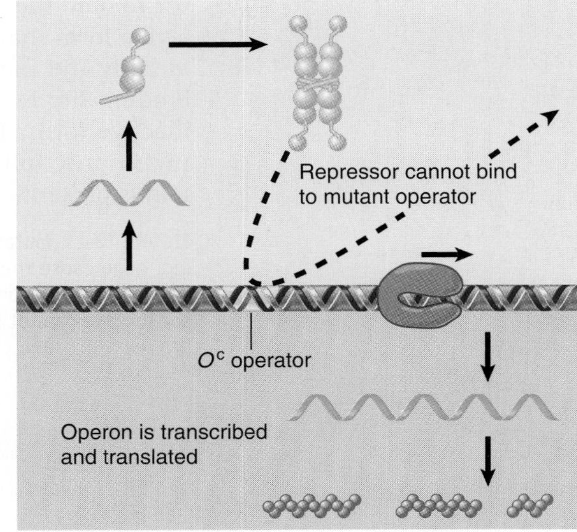
Repressor cannot bind to mutant operator
O^c operator
Operon is transcribed and translated

FIGURE 12.10 Operator mutations are constitutive because the operator is unable to bind repressor protein; this allows RNA polymerase to have unrestrained access to the promoter. The O^c mutations are *cis*-acting, because they affect only the contiguous set of structural genes.

The operator can control only the lac genes that are adjacent to it. If a second *lac* operon is introduced into the bacterium on an independent molecule of DNA, it has its own operator. Neither operator is influenced by the other. Thus if one operon has a wild-type operator it will be repressed under the usual conditions, whereas a second operon with an O^c mutation will be expressed in its characteristic fashion.

These properties define the operator as a typical *cis*-acting site, whose function depends upon recognition of its DNA sequence by some *trans*-acting factor. The operator controls the adjacent genes irrespective of the presence in the cell of other alleles of the site. A mutation in such a site—for example, the O^c mutation—is formally described as ***cis*-dominant**.

▸ *cis*-dominant A site or mutation that affects the properties only of its own molecule of DNA, often indicating that a site does not code for a diffusible product.

KEY CONCEPTS

- Mutations in the operator cause constitutive expression of all three *lac* structural genes.
- These mutations are *cis*-acting and affect only those genes on the contiguous stretch of DNA.

CONCEPT AND REASONING CHECK

If there are two complete copies of the *lac* operon in a cell and a mutation in only one of the operator sites controlling *lacZYA*, what would be the effect on the cell?

12.6 *trans*-Acting Mutations Identify the Regulator Gene

Mutations in the *lacI* gene can cause the operon to be either (1) uninducible (cannot be turned on) or (2) constitutive (always on). These two types of mutations identify different active sites in the protein. Constitutive transcription is caused by mutations of the *lacI*$^-$ type, which are caused by loss of function (including deletions of the gene). When the repressor is inactive or absent, transcription of the *lac* operon can initiate at the *lac* operon promoter. **FIGURE 12.11** shows that the *lacI*$^-$ mutants express the structural genes all the time (constitutively), *irrespective of whether the inducer is present or absent,* because the repressor is inactive. One important subset of *lacI*$^-$ mutations (called *lacI*$^{-d}$; see below) is localized in the DNA-binding site of the repressor. They abolish the ability to turn off the gene by damaging the site that the repressor uses to contact the operator. They are dominant mutations because they do not allow a tetramer with normal and mutant repressor subunits to bind the operator (see below).

Uninducible mutants are caused by mutations that abolish the ability of repressor to bind the inducer. They are described as *lacI*^{s}. The repressor is "locked in" to the active form that recognizes the operator and prevents transcription. These mutations identify the inducer-binding site. The addition of inducer has no effect because its binding site is absent, and therefore it is impossible to convert the repressor to the inactive form. The mutant repressor binds to all *lac* operators in the cell to prevent their transcription, and cannot be removed from the operator, even if wild-type protein is present.

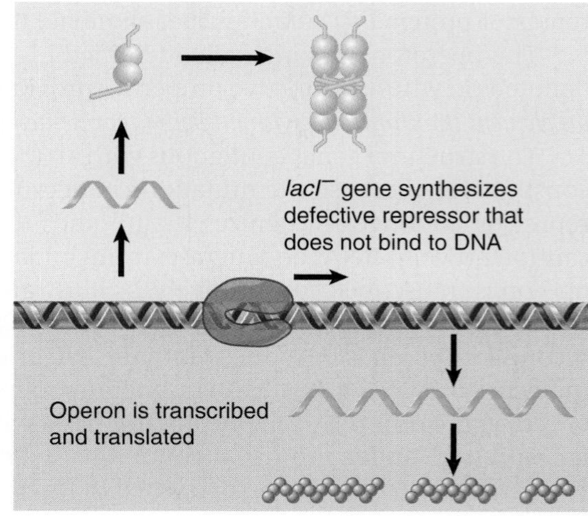
lacI$^-$ gene synthesizes defective repressor that does not bind to DNA
Operon is transcribed and translated

FIGURE 12.11 Mutations that inactivate the *lacI* gene cause the operon to be constitutively expressed, because the mutant repressor protein cannot bind to the operator.

An important feature of the repressor is that it is multimeric. Repressor subunits associate at random in the cell to form the active protein tetramer. When two different alleles of the *lacI* gene are present, the subunits made by each can associate to form a heterotetramer, whose properties differ from those of either homotetramer. This type of interaction between subunits is a characteristic feature of multimeric proteins and is described as **interallelic complementation**.

Combinations of certain repressor mutants display a form of interallelic complementation called **negative complementation**. The *lacI*$^{-d}$ mutation alone results in the production of a repressor that cannot bind the operator, and it is therefore constitutive like the *lacI*$^{-}$ alleles. Because the *lacI*$^{-}$ type of mutation inactivates the repressor, it is usually recessive to the wild type. However, the *–d* notation indicates that this variant of the negative type is dominant when paired with a wild-type allele. Such mutations are called **dominant negative**. The reason for their behavior is that one mutant subunit in a tetramer can antagonize the function of the wild-type subunits, as discussed in the next section.

- **interallelic complementation** The change in the properties of a heteromultimeric protein brought about by the interaction of subunits coded by two different mutant alleles; the mixed protein may be more or less active than the protein consisting of subunits of only one or the other type.
- **negative complementation** It occurs when interallelic complementation allows a mutant subunit to suppress the activity of a wild-type subunit in a multimeric protein.
- **dominant negative** A mutation that results in a mutant gene product that prevents the function of the wild-type gene product, causing loss or reduction of gene activity in cells containing both the mutant and wild-type alleles. The most common cause is that the gene codes for a homomultimeric protein whose function is lost if only one of the subunits is a mutant.

KEY CONCEPTS

- Mutations in the *lacI* gene are *trans*-acting and affect expression of all *lacZYA* clusters in the bacterium.
- Mutations that eliminate *lacI* function cause constitutive expression and are recessive.
- Mutations in the DNA-binding site of the repressor are constitutive because the repressor cannot bind the operator.
- Mutations in the inducer-binding site of the repressor prevent it from being inactivated and cause uninducibility.
- Mutations in the promoter are uninducible and *cis*-acting.
- When mutant and wild-type subunits are present, a single *lacI*$^{-d}$ mutant subunit can inactivate a tetramer whose other subunits are wild type.
- *lacI*$^{-d}$ mutations occur in the DNA-binding site. Their effect is explained by the fact that repressor activity requires all DNA-binding sites in the tetramer to be active.

CONCEPT AND REASONING CHECK

If there are two complete copies of the *lac* operon in a cell and a mutation in one copy of *lacZ*, what would be the effect on the cell?

12.7 *lac* Repressor Is a Tetramer Made of Two Dimers

The repressor has several domains, as shown in the crystal structure illustrated in **FIGURE 12.12**. A major feature is that the DNA-binding domain is separate from the rest of the protein.

The DNA-binding domain occupies residues 1–59. It consists of two α-helices separated by a turn. This is a common DNA-binding motif known as the HTH (helix-turn-helix); the two α-helices fit into the major groove of DNA, where they make contacts with specific bases (see *Section 14.10, Repressor Uses a Helix-Turn-Helix Motif to Bind DNA*). This region is connected by a hinge to the main body of the protein. In the DNA-binding form of repressor, the hinge forms a small α-helix (as shown in Figure 12.12); but when the repressor is not bound to DNA, this region is disordered. The HTH and hinge together correspond to the *headpiece*.

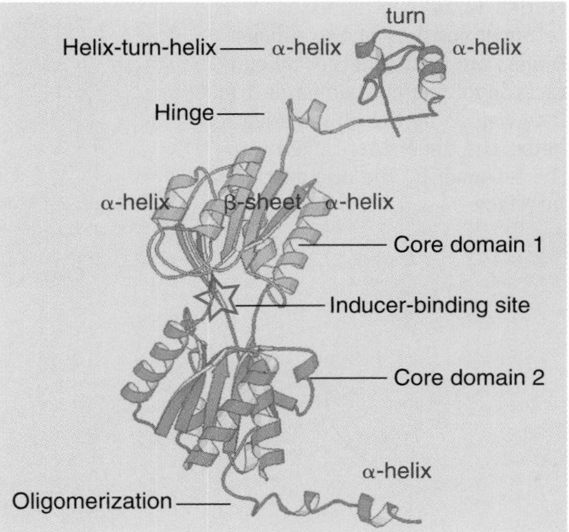

FIGURE 12.12 The structure of a monomer of *lac* repressor identifies several independent domains. Structure rendered from Protein Data Bank 1lbg by Hangli Zhan and provided by Kathleen S. Matthews, Rice University.

FIGURE 12.13 The crystal structure of the core region of *lac* repressor identifies the interactions between monomers in the tetramer. Each monomer is identified by a different color. Photos courtesy of Benjamin Wieder and Ponzy Lu, University of Pennsylvania.

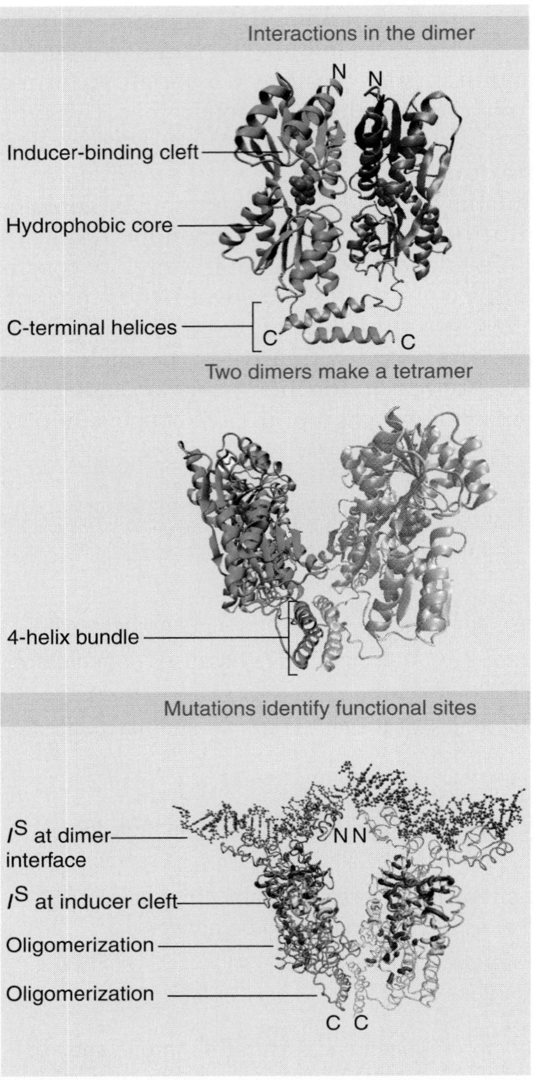

The bulk of the core consists of two regions with similar structures (core domains 1 and 2). Each has a six-stranded parallel β-sheet sandwiched between two α-helices on either side. The inducer binds in a cleft between the two regions.

At the C-terminus, there is an α-helix that contains two leucine heptad repeats. This is the oligomerization domain. The oligomerization helices of four monomers associate to maintain the tetrameric structure.

FIGURE 12.13 shows the structure of the tetrameric core (lacking the headpieces). It consists, in effect, of two dimers. The body of each dimer contains a loose interface between the N-terminal regions of the core monomers, a cleft at which inducer binds, and a hydrophobic core (top). The C-terminal regions of each monomer protrude as parallel helices. (The headpiece would join with the N-terminal regions at the top.) Together the dimers interact to form a tetramer (center) that is held together by a C-terminal bundle of four helices.

Sites of mutations are shown by color-coded beads on the structure at the bottom. *lacI*^{s} mutations make the repressor unresponsive to the inducer, so that the operon is uninducible. They map in two groups: gray shows those in the inducer-binding cleft, and yellow shows those that affect the core domain 1 dimer interface. The first group abolishes the inducer-binding site; the second group prevents the effects of inducer binding from being transmitted to the DNA-binding site. *lacI*$^-$ mutations that affect oligomerization map in two groups. White shows mutations in core domain 2 that prevent dimer formation. Purple shows those in the oligomerization helix that prevent the dimers from forming tetramers.

FIGURE 12.14 The repressor tetramer consists of two dimers. Dimers are held together by contacts involving core domains 1 and 2 as well as by the oligomerization helix. The dimers are linked into the tetramer by the oligomerization interface.

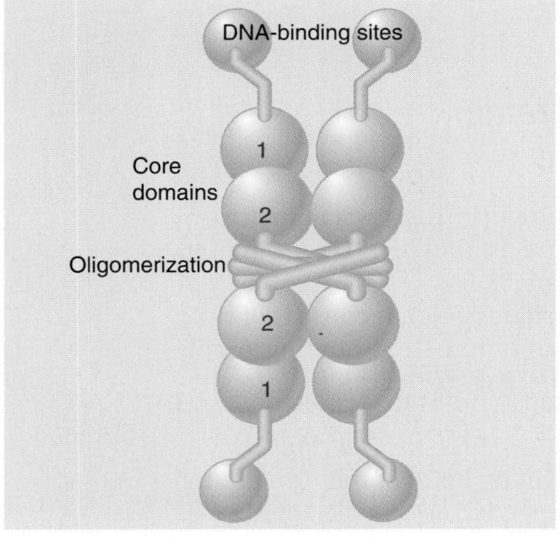

From these data we can derive the schematic of **FIGURE 12.14**, which shows how the monomers are organized into the tetramer. Two monomers form a dimer by means of contacts at core domain 2 and in the oligomerization helix. The dimer has two DNA-binding domains at one end of the structure and the oligomerization helices at the other end. Two dimers then form a tetramer by interactions at the oligomerization interface. We can map the types of mutations onto this structure as summarized in the scheme of **FIGURE 12.15**.

The special class of dominant-negative *lacI*$^{-d}$ mutations lie in the DNA-binding site of the repressor subunit. This explains their ability to prevent mixed tetramers from binding to the operator; a reduction in the number of binding sites reduces the specific affinity for the operator. The role of the N-terminal region in specifically binding DNA is also shown by its location as the site of occurrence of "tight binding" mutations. These increase the affinity of the repressor for the operator, sometimes so much that it cannot be released by inducer. They are rare.

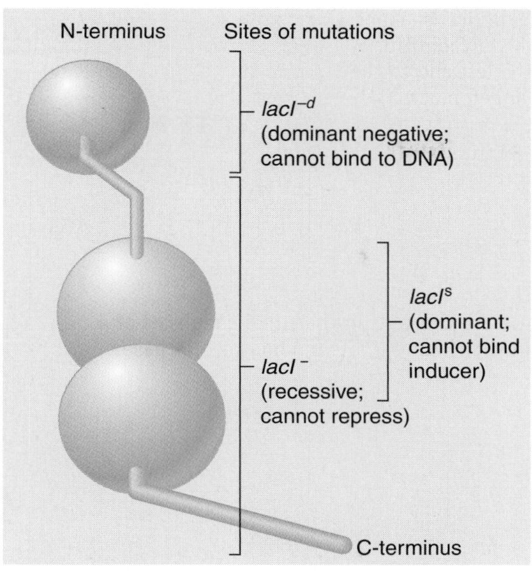

FIGURE 12.15 The locations of three type of mutations in lactose repressor are mapped on the domain structure of the protein. Recessive *lacI*$^{-}$ mutants that cannot repress can map anywhere in the protein. Dominant negative *lacI*$^{-d}$ mutants that cannot repress map to the DNA-binding domain. Dominant *lacI*^{s} mutants that cannot induce because they do not bind inducer or cannot undergo the allosteric change map to core domain 1.

Uninducible *lacI*^{s} mutations map largely in a region of the core domain 1 extending from the inducer-binding site to the hinge. One group lies in amino acids that contact the inducer, and these mutations function by preventing binding of inducer. The remaining mutations lie at sites that must be involved in transmitting the allosteric change in conformation to the hinge when inducer binds.

KEY CONCEPTS

- A single repressor subunit can be divided into the N-terminal DNA-binding domain, a hinge, and the core of the protein.
- The DNA-binding domain contains two short α-helical regions that bind the major groove of DNA.
- The inducer-binding site and the regions responsible for multimerization are located in the core.
- Monomers form a dimer by making contacts between core domains 1 and 2.
- Dimers form a tetramer by interactions between the oligomerization helices.
- Different types of mutations occur in different domains of the repressor protein.

CONCEPT AND REASONING CHECK

If there are two complete copies of the *lac* operon in a cell and a mutation in one of the operator sites controlling *lacZYA* as well as a second mutation in the *lacZ* gene in that operon, what would be the effect on the cell?

12.8 *lac* Repressor Binding to the Operator Is Regulated by an Allosteric Change in Conformation

How does the repressor recognize the specific sequence of operator DNA? The operator has a feature common to many recognition sites for regulator proteins: it is a type of **palindrome** known as an inverted repeat. The inverted repeats are highlighted in **FIGURE 12.16**. Each repeat can be regarded as a half site of the operator.

▸ **palindrome** A symmetrical sequence that reads the same forward and backward.

The importance of particular bases within the operator sequence can be determined by identifying those that contact the repressor protein or in which mutations change the binding of repressor. The region of DNA contacted by protein extends for 26 bp, and within this region are eight sites at which constitutive mutations occur. This emphasizes the same point made by promoter mutations: *A small number of*

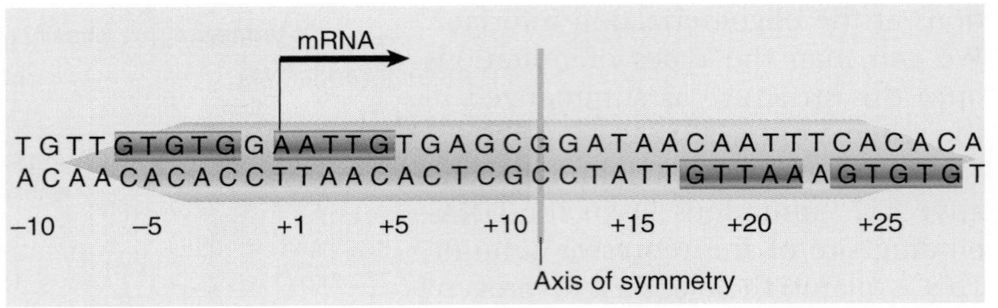

FIGURE 12.16 The lac operator has a symmetrical sequence. The sequence is numbered relative to the startpoint for transcription at +1. The pink arrows to the left and to the right identify the two dyad repeats. The green blocks indicate the positions of symmetry.

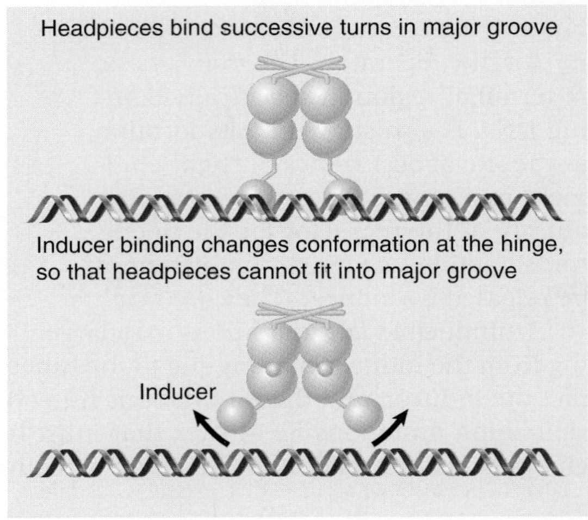

FIGURE 12.17 Inducer changes the structure of the core so that the headpieces of a repressor dimer are no longer in an orientation that permits binding to DNA bases.

essential specific contacts within a larger region can be responsible for sequence-specific association of a protein binding to DNA.

The symmetry of the DNA sequence reflects the symmetry in the protein. Each of the identical subunits in a repressor tetramer has a DNA-binding site. Two of these sites contact the operator in such a way that each inverted repeat or half site of the operator makes the same pattern of contacts with a repressor monomer. This is shown by symmetry in the contacts that repressor makes with the operator (the pattern between +1 and +6 is identical with that between +21 and +16) and by matching constitutive mutations in each inverted repeat.

Early work suggested a model in which the headpiece is relatively independent of the core. It can bind to operator DNA by making the same pattern of contacts with a half-site as intact repressor. Its affinity for DNA, however, is many orders of magnitude less than that of intact repressor. The reason for the difference is that the dimeric form of intact repressor allows two headpieces to contact the operator simultaneously, each binding to one half-site. **FIGURE 12.17** shows that the two DNA-binding domains in a dimeric unit contact DNA by inserting into successive turns of the major groove. This enormously increases affinity for the operator. A key element of binding is the insertion of the short hinge helix into the minor groove of operator DNA, binding the DNA by ~45°. This bend orients the major groove for HTH binding.

Binding of inducer causes an immediate conformational change in the repressor protein. Binding of two molecules of inducer to the repressor tetramer is adequate to release repression. Binding of inducer disrupts the hinge helices and changes the orientation of the headpieces relative to the core, with the result that the two headpieces in a dimer can no longer bind DNA simultaneously. This eliminates the advantage of the multimeric repressor and reduces the affinity for the operator.

An inducer reduces the affinity of the repressor for the operator below the threshold that is needed to ensure that operators are bound. When the inducer enters the cell, it binds to the repressor. After repressor binding, any repressor that is bound to an operator is released; any repressor that is free has too low an affinity for the operator to bind to one.

KEY CONCEPTS

- *lac* repressor protein binds to the double-stranded DNA sequence of the operator.
- The operator is a palindromic sequence of 26 bp.
- Each inverted repeat of the operator binds to the DNA-binding site of one repressor subunit.
- Inducer binding causes a change in repressor conformation that reduces its affinity for DNA and releases it from the operator.

CONCEPT AND REASONING CHECK

How does changing the structure of a protein (like the *lac* repressor) change the way it recognizes the DNA?

12.9 *lac* Repressor Binds to Three Operators and Interacts with RNA Polymerase

The allosteric transition that results from binding of inducer occurs in the repressor dimer. Why, then, is a tetramer required to establish full repression?

Each dimer can bind an operator sequence. This enables the intact repressor to bind to two operator sites simultaneously. In fact, there are two further operator sites in the initial region of the *lac* operon. The original operator, *O1,* is located just at the start of the *lacZ* gene. It has the strongest affinity for repressor. Weaker operator sequences are located on either side; *O2* is 410 bp downstream of the startpoint in *lacZ* and *O3* is 88 bp upstream of it in *lacI*.

FIGURE 12.18 shows what happens when a DNA-binding protein can bind simultaneously to two separated sites on DNA. The DNA between the two sites forms a loop from a base where the protein has bound the two sites. The length of the loop depends on the distance between the two binding sites. When *lac* repressor binds simultaneously to *O1* and to one of the other operators, it causes the DNA between them to form a rather short loop, significantly constraining the DNA structure. A scale model for binding of tetrameric repressor to two operators is shown in **FIGURE 12.19**.

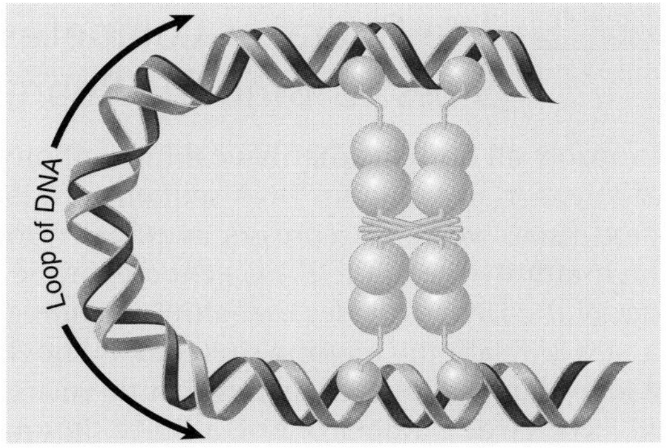

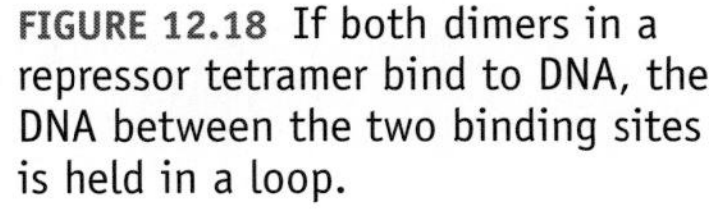

FIGURE 12.18 If both dimers in a repressor tetramer bind to DNA, the DNA between the two binding sites is held in a loop.

Binding at the additional operators affects the level of repression. Elimination of either the downstream operator *(O2)* or the upstream operator *(O3)* reduces the efficiency of repression by 2× to 4×. If, however, *both O2* and *O3* are eliminated, repression is reduced 100×. *This suggests that the ability of the repressor to bind to one of the two other operators as well as to O1 is important for establishing strong repression.* We do not know how and why this simultaneous binding increases repression. We know about the direct effects of binding of repressor to the operator *(O1)*. It

FIGURE 12.19 When a repressor tetramer binds to two operators, the stretch of DNA between them is forced into a tight loop. (The blue structure in the center of the looped DNA represents CRP, which is another regulator protein that binds in this region.) Reproduced with permission from Lewis, M. et al., *Science* 271 (1996): cover. © 1996 AAAS. Photo courtesy of Ponzy Lu, University of Pennsylvania.

was originally thought that repressor binding would occlude RNA polymerase from binding to the promoter. We now know that the two proteins may be bound to DNA simultaneously, and that *the binding of repressor actually enhances the binding of RNA polymerase!* But the bound enzyme is prevented from initiating transcription. The repressor in effect causes RNA polymerase to be stored at the promoter. When inducer is added, the repressor is released, and RNA polymerase can initiate transcription immediately. The overall effect of repressor is to speed up the induction process.

Does this model apply to other systems? The interaction between RNA polymerase, repressor, and the promoter/operator region is distinct in each system, because the operator does not always overlap with the same region of the promoter (see Figure 12.23). For example, in phage lambda, the operator lies in the upstream region of the promoter, and binding of repressor occludes the binding of RNA polymerase (see *Chapter 14, Phage Strategies*). Thus a bound repressor does not interact with RNA polymerase in the same way in all systems.

KEY CONCEPTS

- Each dimer in a repressor tetramer can bind an operator, so that the tetramer can bind two operators simultaneously.
- Full repression requires the repressor to bind to an additional operator downstream or upstream as well as to the operator at the *lacZ* promoter.
- Binding of repressor at the operator stimulates binding of RNA polymerase at the promoter but precludes transcription.

CONCEPT AND REASONING CHECK

Why does the *lac* operon need three operators?

12.10 The Operator Competes with Low-Affinity Sites to Bind *lac* Repressor

Probably all proteins that have a high affinity for a specific sequence also possess a low affinity for any random DNA sequence. A large number of low-affinity sites will compete just as well for a repressor as a small number of high-affinity sites. There is only one high-affinity site in the *E. coli* genome for the *lac* repressor: the *lac* operator. The remainder of the DNA provides low-affinity binding sites. Every base pair in the genome starts a new low-affinity binding site. Simply moving one base pair from the operator creates a low-affinity site! That means that there are 4.2×10^6 low-affinity sites.

The large number of low-affinity sites means that even in the absence of a specific binding site, all or virtually all of the repressors are bound to DNA; virtually no repressors are free in solution. All but 0.01% of repressors are bound to random DNA. Since there are only about 10 molecules of repressor tetramer per cell, this says that there is no free repressor protein. This has an important implication for the interaction of repressor for the operator: it means that the critical factor is the partitioning of the repressor on DNA, in which the single high-affinity site of the operator competes with a large number of low-affinity sites.

DNA	Repressor	Repressor + Inducer
Operator	2×10^{13}	2×10^{10}
Other DNA	2×10^{6}	2×10^{6}
Specificity	10^{7}	10^{4}
Operators bound	96%	3%
Operon is:	repressed	induced

FIGURE 12.20 *lac* repressor binds strongly and specifically to its operator but is released by inducer. All equilibrium constants are in M^{-1}.

The efficiency of repression therefore depends on the relative affinity of the repressor for its operator compared with other random DNA sequences. The affinity must be great enough to overcome the large number of random sites. We can see how this works by comparing the equilibrium constants for *lac* repressor/operator binding with repressor/general DNA binding. **FIGURE 12.20** shows that the ratio is 10^7 for an active repressor, enough to ensure that the operator is bound by repressor 96% of the time so that transcription is effectively—but not completely, repressed. (Remember that because allolactose is the inducer and not lactose, we always need a little β-galactosidase in the cell.) When inducer is added, the ratio is reduced to 10^4. At this level, only 3% of the operators are bound and the operon is effectively induced.

The consequence of these affinities is that in an uninduced cell, one tetramer of repressor usually is bound to the operator. All, or almost all, of the remaining tetramers are bound at random to other regions of DNA, as illustrated in **FIGURE 12.21**. There are likely to be very few or no repressor tetramers free within the cell.

The addition of inducer abolishes the ability of repressor to bind specifically at the operator. Those repressors bound at the operator are released and bind to random (low-affinity) sites. Thus in an induced cell, the repressor tetramers are "stored" on random DNA sites. In a noninduced cell, a tetramer is bound at the operator, whereas the remaining repressor molecules are bound to nonspecific sites. The effect of induction is therefore to change the distribution of repressor on DNA, rather than to generate free repressor. In the same way that RNA polymerase probably moves between promoters and other DNA by swapping one sequence for another, the repressor also may directly displace one bound DNA sequence with another to move between sites. We can define the parameters that influence the ability of a regulator protein to saturate its target site by comparing the equilibrium equations for specific and nonspecific binding. As might be expected, the important parameters are:

- The size of the genome dilutes the ability of a protein to bind specific target sites (remember how large eukaryote genomes are).
- The specificity of a protein counters the effect of the mass of the DNA.
- The amount of the protein that is required increases with the total amount of DNA in the genome and decreases the specificity of DNA binding.
- The amount of the protein also must be in reasonable excess of the total number of specific target sites, so we expect regulators with many targets to be found in greater quantities than regulators with fewer targets.

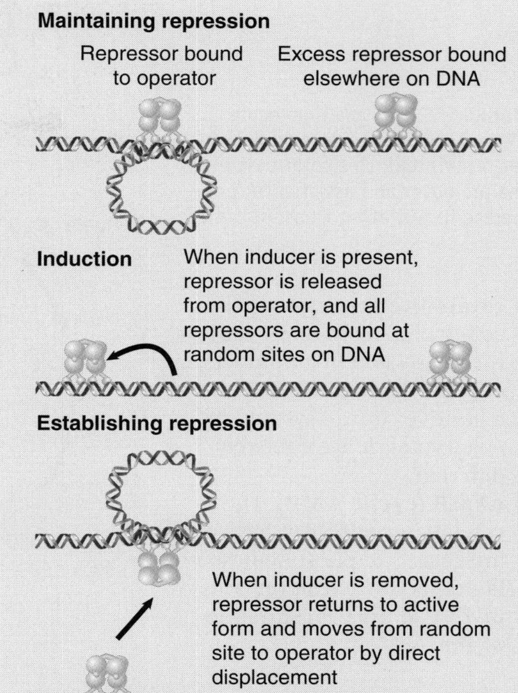

Maintaining repression
Repressor bound to operator
Excess repressor bound elsewhere on DNA
Induction
When inducer is present, repressor is released from operator, and all repressors are bound at random sites on DNA
Establishing repression
When inducer is removed, repressor returns to active form and moves from random site to operator by direct displacement

FIGURE 12.21 Virtually all the repressor in the cell is bound to DNA.

KEY CONCEPTS

- Proteins that have a high affinity for a specific DNA sequence also have a low affinity for other DNA sequences.
- Every base pair in the bacterial genome is the start of a low-affinity binding site for repressor.
- The large number of low-affinity sites ensures that all repressor protein is bound to DNA.
- *lac* repressor binds to the operator by moving from a low-affinity site rather than by equilibrating from solution.
- In the absence of inducer, the operator has an affinity for repressor that is $10^7\times$ that of a low-affinity site.
- The level of 10 repressor tetramers per cell ensures that the operator is bound by repressor 96% of the time.
- Induction reduces the affinity for the operator to $10^4\times$ that of low-affinity sites, so that only 3% of operators are bound.
- Induction causes repressor to move from the operator to a low-affinity site by direct displacement.

CONCEPT AND REASONING CHECK

Why can't a mutant *lac* operon be induced without a functional *lacZ* gene?

12.11 The *lac* Operon Has a Second Layer of Control: Catabolite Repression

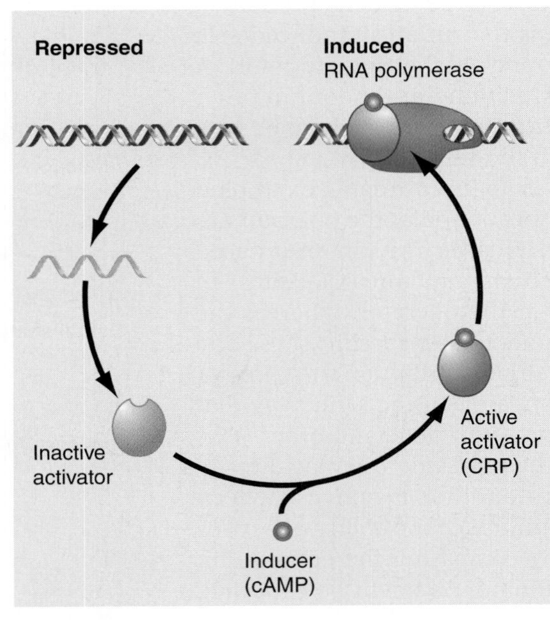

FIGURE 12.22 A small molecule inducer, cAMP, converts an activator protein CRP to a form that binds the promoter and assists RNA polymerase in initiating transcription.

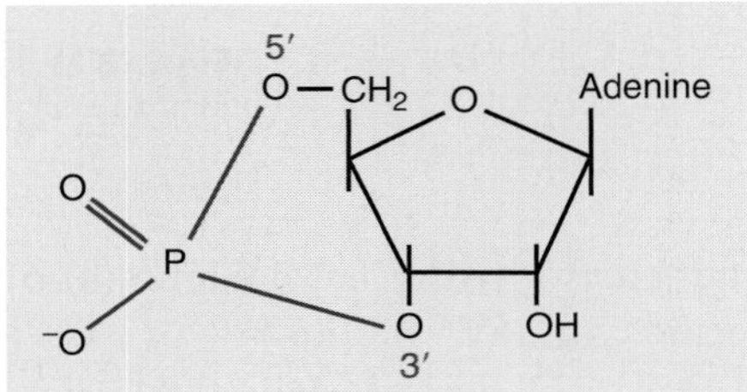

FIGURE 12.23 Cyclic AMP has a single phosphate group connected to both the 3′ and 5′ positions of the sugar ring.

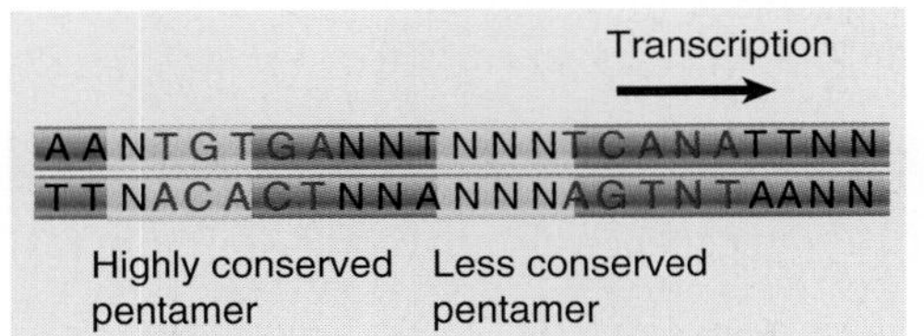

FIGURE 12.24 The consensus sequence for CRP contains the well-conserved pentamer TGTGA and (sometimes) an inversion of this sequence (TCANA).

- **catabolite repression** The ability of glucose to prevent the expression of a number of genes. In bacteria this is a positive control system; in eukaryotes, it is completely different.
- **cAMP (cyclic AMP)** The coregulator of CRP, it has an internal 3′–5′ phosphodiester bond. Its concentration is inverse to the concentration of glucose.
- **Catabolite repressor protein (CRP)** A positive regulator protein activated by cyclic AMP. It is needed for RNA polymerase to initiate transcription of many operons of *E. coli*.

The *E. coli lac* operon is negative inducible. Transcription is turned on by the presence of lactose by removing the Lac Repressor. However, this operon is also under a second layer of control. It cannot be turned on by lactose if there is a sufficient supply of glucose. The rationale for this is that glucose is a better energy source than lactose, so there is no need to turn on the operon if there is glucose available. This system is part of a global network called **catabolite repression** that affects about 20 genes in *E. coli*. Catabolite repression is exerted through a second messenger called **cyclic AMP (cAMP)** and the positive regulator protein called the **catabolite repressor protein** or **CRP** (also sometimes called *catabolite activator protein*, CAP). The *lac* operon is, therefore, under dual control.

Thus far we have dealt with the promoter as a DNA sequence that is competent to bind RNA polymerase, which then initiates transcription. There are, however, some promoters at which RNA polymerase cannot initiate transcription without assistance from an ancillary protein. Such proteins are positive regulators, because their presence is necessary to switch on the transcription unit. Typically, the activator overcomes a deficiency in the promoter, for example, a poor consensus sequence at –35 or –10.

One of the most widely acting activators is CRP. This protein is a positive regulator whose presence is necessary to initiate transcription at dependent promoters. CRP is active *only in the presence of cAMP*, which behaves as a classic small-molecule inducer for positive control (**FIGURE 12.22**).

cAMP is synthesized by the enzyme **adenylate cyclase**. The reaction uses ATP as substrate and introduces an internal 3′–5′ link via a phosphodiester bond, which generates the structure drawn in **FIGURE 12.23**. The level of cAMP is inversely related to the level of glucose. High levels of glucose repress adenylate cyclase. Only in low levels of glucose is the enzyme active and able to synthesize cAMP.

CRP binds to DNA only when complexed with cAMP. The factor is a dimer of two identical subunits of 22.5 kD, which can be activated by a single molecule of cAMP. A CRP monomer contains a DNA-binding region and a transcription-activating region.

A CRP dimer binds to a site of ~22 bp at a responsive promoter. The binding sites include variations of the consensus sequence given in **FIGURE 12.24**. Mutations preventing CRP action usually are located within the well-conserved pentamer, which appears to be the essential element in recognition. CRP binds most strongly to sites that contain two (inverted) versions of the pentamer, because this enables both subunits of the dimer to bind to the DNA.

CRP introduces a large bend when it binds DNA. In the *lac* promoter, this point lies at the center of dyad symmetry. No CRP-dependent promoter has a good –35 sequence and some also lack good –10 sequences. In fact, we might argue that effective control by CRP would be difficult if the promoter had effective –35 and –10 regions that interacted independently with RNA polymerase.

CRP is bound to the same face of DNA as RNA polymerase. When the α subunit of RNA polymerase has a deletion in the C-terminal domain, transcription appears normal except for the loss of ability to be activated by CRP. CRP has an "activation domain" that is required. This activating region, which consists of an exposed loop of ~10 amino acids, is a small patch that interacts directly with the C-terminal domain of the α subunit of RNA polymerase to stimulate the polymerase.

KEY CONCEPTS

- CRP is an activator protein that binds to a target sequence at a promoter.
- A dimer of CRP is activated by a single molecule of cAMP.
- cAMP is controlled by the level of glucose in the cell; a low glucose level allows cAMP to be made.
- CRP interacts with the C-terminal domain of the α subunit of RNA polymerase to activate it.

CONCEPT AND REASONING CHECK

What are the two conditions that must be met for transcription of the *lac* operon to occur?

12.12 The *trp* Operon Is a Repressible Operon with Three Transcription Units

The *lac* repressor acts only on the operator of the *lacZYA* cluster. Some repressors, however, control dispersed structural genes by binding at more than one operator. An example is the *trp* repressor, which controls three unlinked sets of genes:

- An operator at the cluster of structural genes *trpEDCBA* controls coordinate synthesis of the enzymes that synthesize tryptophan. This is an example of a *repressible operon,* one that is controlled by the product of the operon: tryptophan.
- The *trpR* regulator gene is repressed by its own product, the *trp* repressor. Thus the repressor protein acts to reduce its own synthesis: it is **autoregulated**. (Remember, the *lacI* regulator gene is unregulated.) Such circuits are quite common in regulatory genes and may be either negative or positive (see *Section 12.13, Translation Can Be Regulated,* and *Section 14.12, Lambda Repressor Maintains an Autoregulatory Circuit*).
- An operator at another locus controls the *aroH* gene, which codes for one of the three enzymes that catalyzes the initial reaction in the common pathway of aromatic amino acid biosynthesis.

▸ **autoregulation** A site or mutation that affects the properties only of its own molecule of DNA, often indicating that a site does not code for a diffusible product.

A related 21 bp operator sequence is present at each of the three loci at which the *trp* repressor acts. The conservation of sequence is indicated in **FIGURE 12.25**. Each operator contains appreciable (but not identical) dyad symmetry. The features conserved at all three operators include the important points of contact for *trp* repressor. This explains how one repressor protein acts on several loci: *each locus has a copy of a specific DNA-binding sequence recognized by the repressor* (just as each promoter shares consensus sequences with other promoters).

FIGURE 12.26 summarizes the variety of relationships between operators and promoters. A notable feature of the dispersed operators recognized by TrpR is their presence at

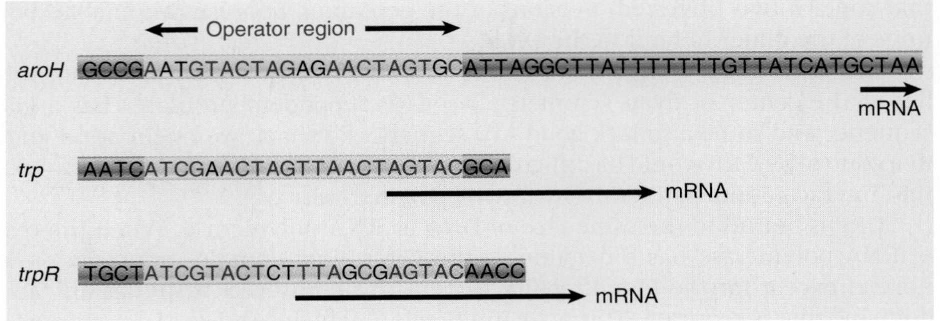

FIGURE 12.25 The *trp* repressor recognizes operators at three loci. Operators are highlighted in red; conserved bases within operators are shown in blue. The location of the startpoint and mRNA varies, as indicated by the black arrows.

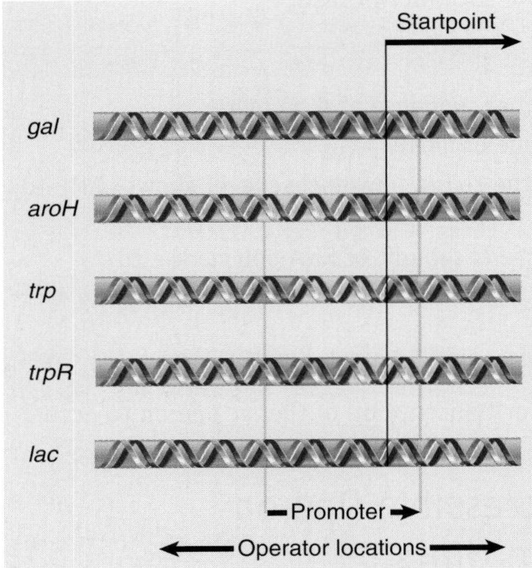

FIGURE 12.26 Operators may lie at various positions relative to the promoter.

different locations within the promoter in each locus. In *trpR* the operator lies between positions –12 and +9, whereas in the *trp* operon it occupies positions –23 to –3. In another gene system, the *aroH* locus, it lies farther upstream, between –49 and –29. In other cases, the operator lies downstream from the promoter (as in *lac*), or apparently just upstream of the promoter (as in *gal*, for which the nature of the repressive effect is not quite clear). The ability of the repressors to act at operators whose positions are different in each target promoter suggests that there could be differences in the exact mode of repression, the common feature being that RNA polymerase is prevented from initiating transcription at the promoter.

As we will see in the next chapter, the *trp* operon is also under dual control, like the *lac* operon above, but the second level is quite different.

KEY CONCEPTS

- The *trp* operon is negatively controlled by the level of its product, the amino acid tryptophan.
- The amino acid tryptophan activates an inactive repressor encoded by *trpR*.
- A repressor (or activator) will act on all loci that have a copy of its target operator sequence.

CONCEPT AND REASONING CHECK

What would the effect on a cell be if the *trpR* gene were mutant?

12.13 Translation Can Be Regulated

Control over which protein is made and how much is exerted first at the level of transcription control (as we have just discussed), then through RNA processing control (rare in bacteria, but common in eukaryotes), and then finally translation level control, which we will examine here.

The *lac* repressor is encoded by the *lacI* gene; this is an unregulated gene that is continuously transcribed, but from a poor promoter. Also, the coding region of the *lac* repressor is in a very poor mRNA. This simply means that the 5′ UTR (untranslated region) of the mRNA has a poor sequence context that does not allow rapid ribosome binding or movement onto the ORF. Just as we have seen that promoters can be "good" or "poor," so can mRNAs. Together, this means that ribosomes do not translate the small amount of mRNA at the same level as the *lacZYA* polycistronic mRNA. We find very little *lac* repressor in a cell, only about 10 tetramers.

A second way that translation can be modulated is by **codon usage**. Multiple codons exist for most of the amino acids. These codons are not decoded equally by tRNAs. Some have abundant tRNAs and some do not. If an ORF is constructed from codons with abundant tRNAs, it can be rapidly translated, whereas another ORF that contains codons with less abundant tRNAs will be translated much more slowly.

▸ **codon usage** A description of the relative abundance of tRNAs for each codon.

Additional, more active mechanisms exist for translation-level control. One mechanism for controlling gene expression at the level of translation is an exact parallel to the use of a repressor to prevent transcription. Repressor function is provided by a protein that binds to a target region on mRNA to prevent ribosomes from recognizing the initiation region. One example is found in operons coding for components of the protein synthetic apparatus. The operon provides an arrangement for *coordinate* regulation of a group of structural genes. *Repressor function is provided by a protein that binds to a target region on mRNA to prevent ribosomes from recognizing the initiation region.* Formally this is equivalent to a repressor protein binding to DNA to prevent RNA polymerase from utilizing a promoter. **FIGURE 12.27** illustrates the most common form of this interaction, in which the regulator protein binds directly to a sequence that includes the AUG initiation codon, thereby preventing the ribosome from binding.

Some examples of translational repressors and their targets are summarized in **FIGURE 12.28**. A classic example of how the product of translation can directly control the translation of its mRNA is the coat protein of the RNA phage R17; it binds to a hairpin that encompasses the ribosome-binding site in the phage mRNA. Similarly, the phage T4 RegA protein binds to a consensus sequence that includes the AUG initiation codon in several T4 early mRNAs, and T4 DNA polymerase binds to a sequence in its own mRNA that includes the Shine–Dalgarno element needed for ribosome binding. Autoregulation occurs whenever a protein (or RNA) regulates its own production. The usual type of effect is that accumulation of the protein prevents production of more protein at the level of either transcription or translation. Sometimes this applies to an individual product, such as T4 p32, where the protein prevents translation of its own RNA. In other cases, accumulation of one product of an operon may turn off the synthesis of all products of the operon, as in the inhibition of translation of the operons coding for ribosomal proteins. The genes coding for the ribosomal proteins (r-proteins) are grouped into several operons. In each of these operons, accumulation of one protein inhibits further synthesis of both itself and some of the other gene products. Each of the regulators is a ribosomal protein that binds directly to rRNA. Its effect on translation is a result of its ability to bind also to its own mRNA. The sites on mRNA at which these proteins bind either overlap the sequence where translation is initiated or lie nearby and probably influence the accessibility of the initiation site by inducing conformational change.

The use of r-proteins that bind rRNA to establish autogenous regulation immediately suggests that this provides a mechanism to link r-protein synthesis to rRNA synthesis. A generalized model is depicted

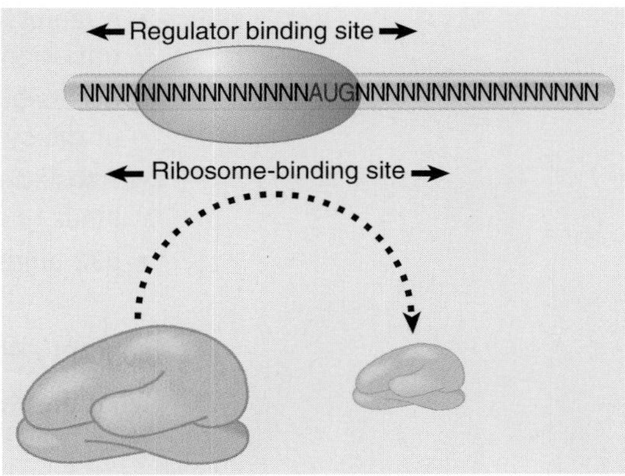

FIGURE 12.27 A regulator protein may block translation by binding to a site on mRNA that overlaps the ribosome-binding site at the initiation codon.

Repressor	Target Gene	Site of Action
R17 coat protein	R17 replicase	hairpin that includes ribosome binding site
T4 RegA	early T4 mRNAs	various sequences including initiation codon
T4 DNA polymerase	T4 DNA polymerase	Shine-Dalgarno sequence
T4 p32	gene 32	single-stranded 5′ leader

FIGURE 12.28 Proteins that bind to sequences within the initiation regions of mRNAs may function as translational repressors.

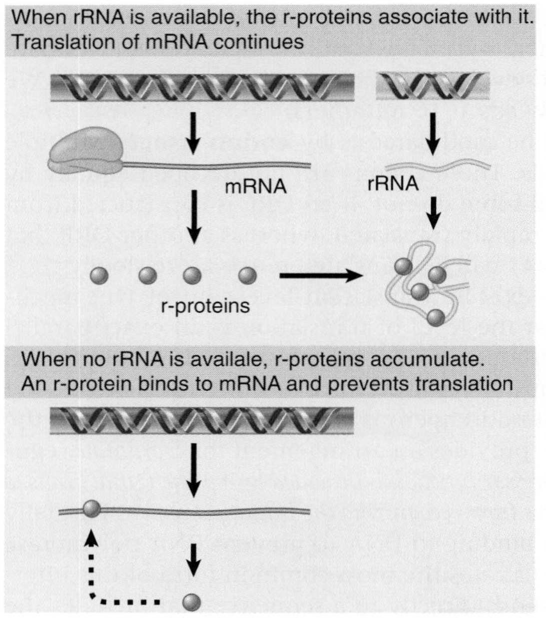
When rRNA is available, the r-proteins associate with it. Translation of mRNA continues
mRNA
rRNA
r-proteins
When no rRNA is availale, r-proteins accumulate. An r-protein binds to mRNA and prevents translation

FIGURE 12.29 Translation of the r-protein operons is autogenously controlled and responds to the level of rRNA.

in **FIGURE 12.29**. Suppose that the binding sites for the autogenous regulator r-proteins on rRNA are much stronger than those on the mRNAs. As long as any free rRNA is available, the newly synthesized r-proteins will associate with it to start ribosome assembly. There will be no free r-protein available to bind to the mRNA, so its translation will continue. As soon as the synthesis of rRNA slows or stops, though, free r-proteins begin to accumulate. They are then available to bind their mRNAs and thus repress further translation. This circuit ensures that each r-protein operon responds in the same way to the level of rRNA: as soon as there is an excess of r-protein relative to rRNA, synthesis of the protein is repressed.

Autoregulation is a common type of control among proteins that are incorporated into macromolecular assemblies. The assembled particle itself may be unsuitable as a regulator because it is too large, too numerous, or too restricted in its location. The need for synthesis of its components, though, may be reflected in the pool of free precursor subunits. If the assembly pathway is blocked for any reason, free subunits accumulate and shut off the unnecessary synthesis of further components.

KEY CONCEPTS

- Translation can be modulated by the 5′ untranslated region of an mRNA or by codon usage.
- A repressor protein can regulate translation by preventing a ribosome from binding to an initiation codon.
- Accessibility of initiation codons in a polycistronic mRNA can be controlled by changes in the structure of the mRNA that occur as the result of translation.
- Translation of an r-protein operon can be controlled by a product of the operon that binds to a site on the polycistronic mRNA.
- p32 binds to its own mRNA to prevent initiation of translation.

CONCEPT AND REASONING CHECK

What are the many parts that make up an mRNA molecule?

12.14 Summary

Transcription is regulated by the interaction between *trans*-acting factors and *cis*-acting sites. A *trans*-acting factor is the product of a regulator gene. It is usually protein but also can be RNA. It diffuses in the cell, and as a result it can act on any appropriate target gene. A *cis*-acting site in DNA (or RNA) is a sequence that functions by being recognized *in situ*. It has no coding function and can regulate only those sequences with which it is physically contiguous. Bacterial genes coding for proteins whose functions are related, such as successive enzymes in a pathway, may be organized in a cluster that is transcribed into a polycistronic mRNA from a single promoter. Control of this promoter regulates expression of the entire pathway. The unit of regulation, which contains structural genes and *cis*-acting elements, is called the operon.

Initiation of transcription is regulated by interactions that occur in the vicinity of the promoter. The ability of RNA polymerase to initiate at the promoter is prevented or activated by other proteins. Genes that are active unless they are turned off by binding the regulator are said to be under negative control. Genes that are active only when the regulator is bound to them are said to be under positive control. The type of control can be determined by the dominance relationships between wild-type genes and mutants that are constitutive/derepressed (permanently on) or uninducible/ super-repressed (permanently off). A second way of looking at regulation of genes that code for enzymes is whether they are controlled by the enzyme substrate (inducible) or by the enzyme product (repressible).

A repressor protein prevents RNA polymerase from either binding to the promoter or activating transcription. The repressor binds to a target sequence, the operator, which is usually located around or upstream of the startpoint. Operator sequences are short and often are palindromic. The repressor is often a homomultimer whose symmetry reflects that of its target.

The ability of the repressor protein to bind to its operator is often regulated by small molecules. In a negative inducible gene, the substrate, an inducer, prevents a repressor from binding. In a negative repressible gene, the product or corepressor activates an inactive regulator to turn off the gene. Binding of the inducer or corepressor to its site on the regulator protein produces a change in the structure of the DNA-binding site of the protein. This allosteric reaction occurs both in free repressor proteins and directly in repressor proteins already bound to DNA.

The lactose pathway in *E. coli* operates by negative induction. When an inducer, the substrate β-galactoside, prevents the repressor from binding its operator, transcription and translation of the *lacZ* gene then produce β-galactosidase, the enzyme that metabolizes β-galactosides.

A protein with a high affinity for a particular target sequence in DNA has a lower affinity for all DNA. The ratio defines the specificity of the protein. There are many more nonspecific sites (any DNA sequence) than specific target sites in a genome; as a result, a DNA-binding protein such as a repressor or RNA polymerase is "stored" on DNA. (It is likely that none, or very little, is free.) The specificity for the target sequence must be great enough to counterbalance the excess of nonspecific sites over specific sites. The balance for bacterial proteins is adjusted so that the amount of protein and its specificity allow specific recognition of the target in "on" conditions but allow almost complete release of the target in "off" conditions.

Some promoters cannot be recognized by RNA polymerase, or are recognized only poorly unless a specific activator protein, a positive regulator, is present. Activator proteins may also be regulated by small molecules. The CRP activator is only able to bind to target sequences when complexed with cAMP, which only happens in conditions of low glucose. All promoters that are controlled by catabolite repression have at least one copy of the CRP-binding site. Direct contact between CRP and RNA polymerase occurs through the C-terminal domain of the α subunits.

The tryptophan pathway operates by negative repression. The corepressor tryptophan, the product of the pathway, activates the repressor protein so that it binds to the operator and prevents expression of the genes that code for the enzymes that synthesize tryptophan. A repressor or activator can control multiple targets that have copies of an operator, or its consensus sequence.

Gene expression may also be modulated at the level of translation by the ability of an mRNA to attract a ribosome and by the abundance of specific tRNAs that recognize different codons. More active mechanisms that regulate at the level of translation are also found. Translation may be regulated by a protein that can bind to the mRNA to prevent the ribosome from binding. Most proteins that repress translation possess this capacity in addition to other functional roles; in particular, translation is controlled in some cases of autoregulation when a gene product regulates translation of the mRNA containing its own open reading frame.

CHAPTER QUESTIONS

1. Gene expression at the transcription level is commonly regulated at all but which one of the following?
   A. initiation
   B. elongation
   C. termination
   D. transcript stability
2. The default level of expression of a gene under negative control is
   A. at high level in the absence of a negative regulatory protein.
   B. at high level in the presence of a negative regulatory protein.
   C. turned off in the absence of a negative regulatory protein.
   D. turned off in the presence of a negative regulatory protein.
3. Any gene product that is free to diffuse to find its target is described as a
   A. structural gene.
   B. regulator gene.
   C. *cis*-acting element.
   D. *trans*-acting element.
4. The *lac* mRNA is.
   A. extremely unstable with a half-life of ~1 minute.
   B. extremely unstable with a half-life of ~3 minutes.
   C. very stable with a half-life of ~15 minutes.
   D. very stable with a half-life of ~25 minutes.
5. Promoter mutations are
   A. *trans*-dominant.
   B. *cis*-dominant.
   C. *trans*-recessive.
   D. *cis*-recessive.
6. The *trp* repressor regulates expression of its own gene. This is an example of
   A. allosteric control.
   B. positive regulation.
   C. autogenous control.
   D. coordinate regulation.
7. When glucose levels drop in a bacterial cell
   A. cAMP synthesis is induced.
   B. cAMP synthesis is repressed.
   C. cGMP synthesis is induced.
   D. cGMP synthesis is repressed.
8. The positive regulator of the *lac* operon is
   A. CRP.
   B. Allolactose.
   C. lactose.
   D. IPTG.
9. The *trp* repressor functions to control its own expression as an example of
   A. induction by negative control.
   B. induction by positive control.
   C. repression by negative control.
   D. repression by positive control.

10. CRP controls expression of

A. the *lac* operon.
B. many operons involved in the metabolism of sugars.
C. the *trp* operon.
D. many operons involved in biosynthesis of amino acids.

KEY TERMS

allolactose
allosteric control
autoregulation
cAMP (cyclic AMP)
catabolite regulation
***cis*-acting**
***cis*-dominant**
codon usage
constitutive expression
corepressor
coupled transcription/translation
CRP
dominant negative
gratuitous inducer
inducer
inducible gene
induction
interallelic complementation
***lac* repressor**
negative complementation
negative control
negative inducible
negative repressible
operon
operator
palindrome
polycistronic mRNA
positive control
positive inducible
positive repressible
regulator gene
repressible gene
repression
structural gene
***trans*-acting**
uninducible

FURTHER READING

Elf, J., Li, G.-W., and Xie, X. S. (2007). Probing transcription factor dynamics at the single-molecule level in a living cell. *Science* 316, 1191–1194.
Using single molecules to probe the interaction between the *lac* repressor and its operator site.

Swigon, D., Coleman, B. D., and Olson, W. K. (2006). Modeling the *lac* repressor-operator assembly: the influence of DNA looping on *lac* repressor conformation. *Proc. Natl. Acad. Sci.* 103, 9879–9884.
A report on the dynamics of the assembly of *lac* repressor complexes on operator sites.

Yu, H. and Gertstein, M. (2006). Genomic analysis of the hierarchical structure of regulatory networks. *Proc. Natl. Acad. Sci.* 103, 14724–14731.
A review which describes how individual gene regulatory networks fit into the larger metabolic pathways.

Wilson, C. J., Zahn, H., Swint-Kruse, L., and Matthews, K. S. (2006). The lactose repressor system: paradigms for regulation, allosteric behavior and protein folding. *Cell. Mol. Life Sci.* 64, 3–16.
A nice review on the *lac* system.

13

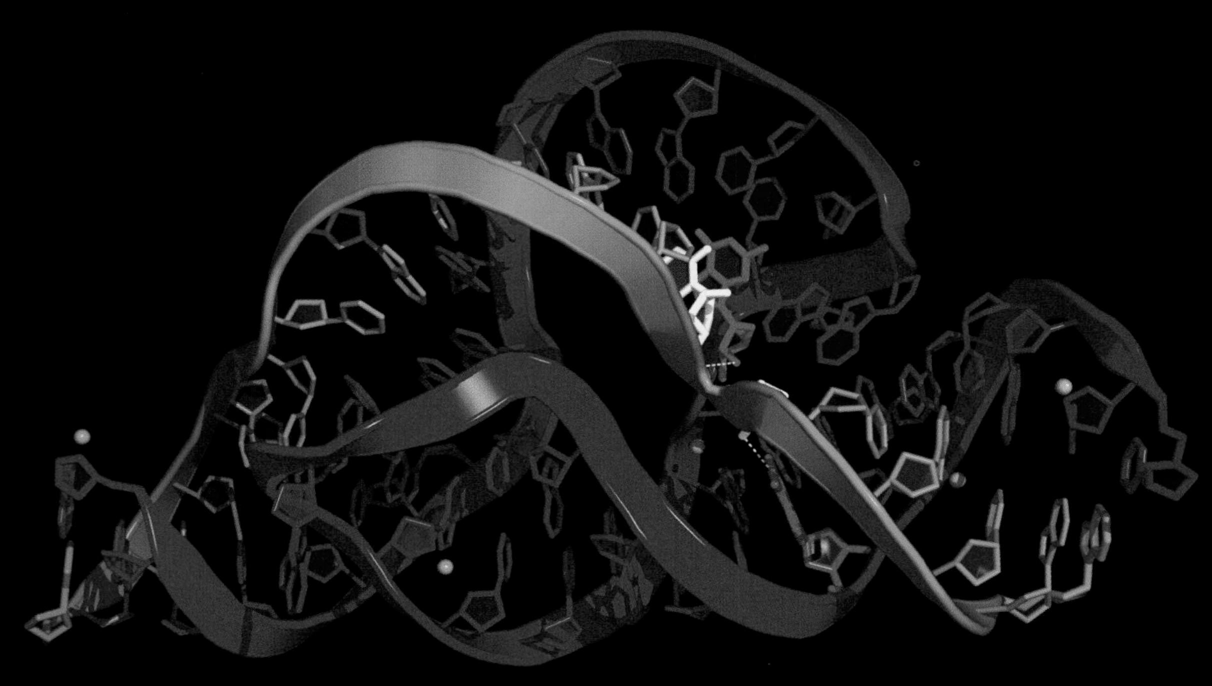

A hammerhead ribozyme that catalyzes a self-cleaving reaction. Many plant viroids are hammerhead RNAs. Reproduced from *Chem. Bio.*, vol. 15, Martick, M., et al., *Solvent Structure and Hammerhead Ribozyme Catalysis*, pp. 332–342. Copyright 2008, with permission from Elsevier [http://www.sciencedirect.com/science/journal/10745521]. Photo courtesy of William Scott, University of California, Santa Cruz.

Regulatory RNA

CHAPTER OUTLINE

13.1 Introduction

The basic principle of regulation in bacteria is that gene expression is controlled by a regulator that interacts with a specific sequence or structure in DNA or mRNA at some stage prior to the synthesis of protein. The stage of expression that is controlled can be transcription, when the target for regulation is DNA, or it can be at translation, when the target for regulation is RNA. Control during transcription can be at initiation, elongation, or termination. The regulator can be a protein or an RNA. "Controlled" can mean that the regulator turns off (represses) or turns on (activates) the target. Expression of many genes can be coordinately controlled by a single regulator gene on the principle that each target contains a copy of the sequence or structure that the regulator recognizes. Regulators may themselves be regulated, most typically in response to small molecules whose supply responds to environmental conditions. Regulators may be controlled by other regulators to make complex circuits. Let's compare the ways that different types of regulators work.

Many protein regulators work on the principle of allosteric changes. The protein has two binding sites—one for a nucleic acid target, the other for a small molecule. Binding of the small molecule to its site changes the conformation in such a way as to alter the affinity of the other site for the nucleic acid. The way in which this happens is known in detail for the Lac Repressor in *E. coli*. Protein regulators are often multimeric, with a symmetrical organization that allows two subunits to contact a palindromic target on DNA. This can generate cooperative binding effects that create a more sensitive response to regulation.

Regulation via RNA uses changes in secondary structure base pairing as the guiding principle. The ability of an RNA to shift between different conformations with regulatory consequences is the nucleic acid's alternative to the allosteric changes of protein conformation. The changes in structure may result from either intramolecular or intermolecular interactions.

The most common role for intramolecular changes is for an RNA molecule to assume alternative secondary structures by utilizing different schemes for base pairing. The properties of the alternative conformations may be different. Changes in secondary structure of an mRNA can result in a change in its ability to be translated. Secondary structure also is used to regulate the termination of transcription, when the alternative structures differ in whether they permit termination.

In intermolecular interactions, an RNA regulator recognizes its target by the familiar principle of complementary base pairing. **FIGURE 13.1** shows that the regulator is usually a small RNA molecule with extensive secondary structure, but with

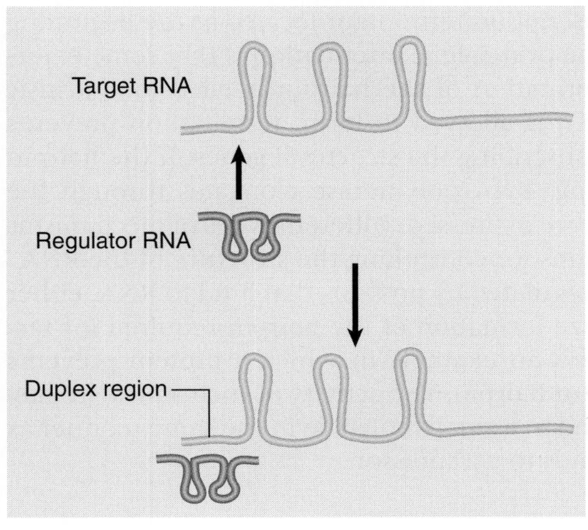

FIGURE 13.1 A regulator RNA is a small RNA with a single-stranded region that can pair with a single-stranded region in a target RNA.

a single-stranded region(s) that is complementary to a single-stranded region in its target. The formation of a double helical region between regulator and target can have two types of consequence:

- Formation of the double helical structure may itself be sufficient for regulatory purposes. In some cases, a protein(s) can bind only to the single-stranded form of the target sequence and is therefore prevented from acting by duplex formation. In other cases, the duplex region becomes a target for binding, for example, by nucleases that degrade the RNA and therefore prevent its expression.
- Duplex formation may be important because it sequesters a region of the target RNA that would otherwise participate in some alternative secondary structure.

KEY CONCEPT

- RNA functions as a regulator by forming a region of secondary structure (either inter- or intramolecular) that changes the properties of a target sequence.

CONCEPT AND REASONING CHECK

How could mRNA sequences that are not coding be important in the mRNA?

13.2 Attenuation: Alternative RNA Secondary Structure Control

▸ **attenuation** The regulation of bacterial operons by controlling termination of transcription at a site located before the first structural gene.

▸ **attenuator** A terminator sequence at which attenuation occurs.

RNA structure provides an opportunity for regulation in both prokaryotes and eukaryotes. Its most common role occurs when an RNA molecule can take up alternative secondary structures by utilizing different schemes for intramolecular base pairing. The properties of the alternative conformations may be different. This type of mechanism can be used to regulate the termination of transcription, when the alternative structures differ in whether they permit termination. Another means of controlling conformation (and thereby function) is provided by the cleavage of an RNA; by removing one segment of an RNA, the conformation of the rest may be altered. It is possible also for a (small) RNA molecule to control the activity of a target RNA by base pairing with it; the role of the small RNA is directly analogous to that of a regulator protein (see *Section 13.7, Bacteria Contain Regulator RNAs*). Both these mechanisms allow an interaction at one site in the molecule to affect the structure of another site.

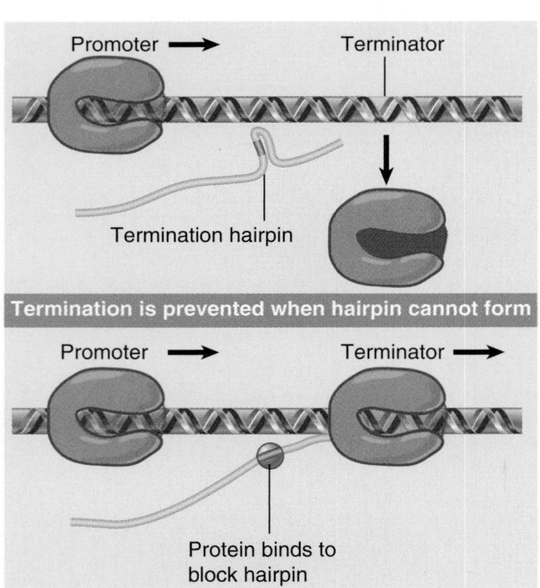
Promoter
Terminator
Termination hairpin
Termination is prevented when hairpin cannot form
Promoter
Terminator
Protein binds to block hairpin

FIGURE 13.2 Attenuation occurs when a terminator hairpin in RNA is prevented from forming.

Several operons are regulated by **attenuation**, a mechanism that controls the ability of RNA polymerase to read past an **attenuator**, which is an intrinsic transcription terminator located at the beginning of a transcription unit. The principle of attenuation is that some external event controls the formation of the hairpin needed for intrinsic termination. If the hairpin is allowed to form, termination prevents RNA polymerase from transcribing the structural genes. If the hairpin is prevented from forming, RNA polymerase elongates through the terminator and the genes are expressed. Different types of mechanisms are used in different systems for controlling the structure of the RNA.

Attenuation may be regulated by proteins that bind to RNA, either to stabilize or to destabilize formation of the hairpin required for termination. **FIGURE 13.2** shows an example in which a protein prevents formation of the terminator hairpin. The activity of such a protein may be intrinsic or may respond to a small molecule in the same manner as a repressor protein responds to corepressor.

KEY CONCEPTS

- Termination of transcription can be attenuated by controlling formation of the necessary hairpin structure in RNA.
- The most direct mechanisms for attenuation involve proteins that either stabilize or destabilize the hairpin.

CONCEPT AND REASONING CHECK

Based on what you know about transcription termination, provide a model for how the messenger RNA molecule itself can determine whether or not the RNA polymerase will terminate transcription.

13.3 Termination of *Bacillus subtilis trp* Genes Is Controlled by Tryptophan and by $tRNA^{Trp}$

The circuitry that controls transcription via termination can use both direct and indirect means to respond to the level of small molecule products or substrates. The basic principle is that when the product of a pathway is available, it causes transcription to be terminated in order to stop production of the enzymes responsible for producing it. Such systems have been extensively characterized for the control of tryptophan synthesis.

In *B. subtilis*, a protein called tryptophan RNA-binding attenuation protein (TRAP) (formerly called MtrB) is activated by tryptophan to bind to a sequence in the leader of the nascent transcript. TRAP forms a multimer of eleven subunits. Each subunit binds a single tryptophan amino acid and a trinucleotide (GAG or UAG) of RNA. The RNA is wound in a circle around the protein. **FIGURE 13.3** shows that the result is to ensure the availability of the regions that are required to form the termination hairpin. The termination of transcription then prevents production of the tryptophan biosynthetic enzymes. In effect, TRAP is a terminator protein that responds to the level of tryptophan. In the absence of TRAP, an alternative secondary structure precludes the formation of the termination hairpin.

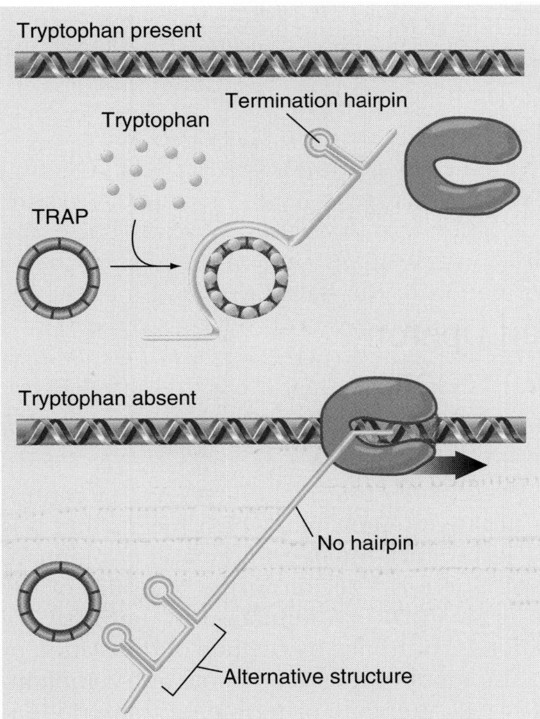

FIGURE 13.3 TRAP is activated by tryptophan and binds to trp mRNA. This allows the termination hairpin to form, with the result that RNA polymerase terminates and the genes are not expressed. In the absence of tryptophan, TRAP does not bind, and the mRNA adopts a structure that prevents the terminator hairpin from forming.

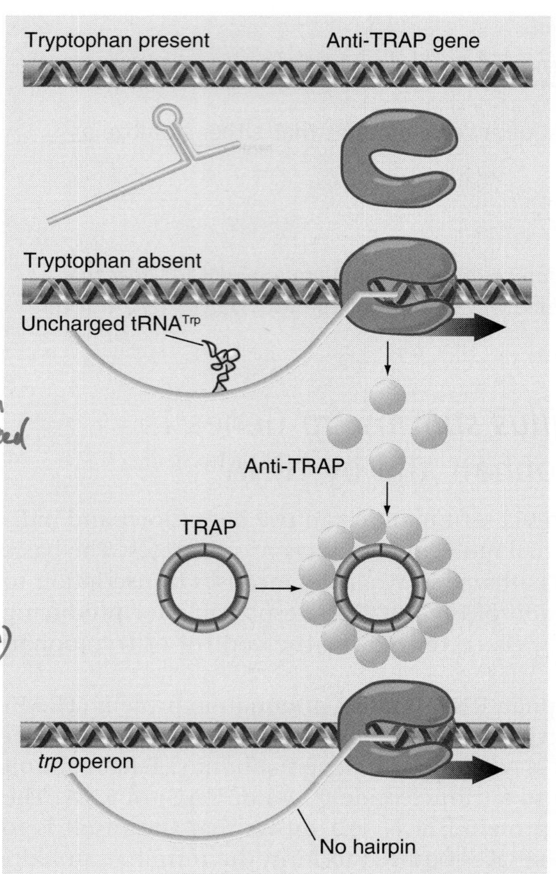

FIGURE 13.4 Under normal conditions (in the presence of tryptophan) transcription terminates before the anti-TRAP gene. When tryptophan is absent, uncharged tRNA^{Trp} base pairs with the anti-TRAP mRNA, preventing formation of the terminator hairpin, thus causing expression of anti-TRAP.

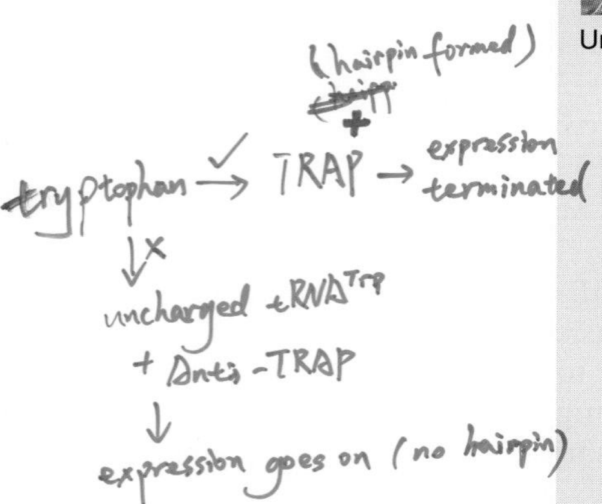

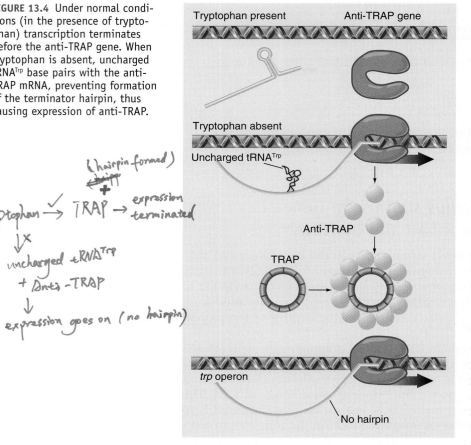

The TRAP, however, is also controlled by tRNA^{Trp}. **FIGURE 13.4** shows that uncharged tRNA^{Trp} binds to the mRNA for a protein called antiTRAP (AT). This suppresses formation of a termination hairpin in the mRNA. The uncharged tRNA^{Trp} also increases the translation of the mRNA. The combined result of these two actions increases synthesis of antiTRAP, which binds to TRAP and prevents it from repressing the tryptophan operon. By this complex series of events, the absence of tryptophan generates the uncharged tRNA, which causes synthesis of antiTRAP. This in turn prevents function of TRAP, which causes expression of tryptophan genes.

Expression of the *B. subtilis trp* genes is therefore controlled by both tryptophan and tRNA^{Trp}. When tryptophan is present, there is no need for it to be synthesized. This is accomplished when tryptophan activates TRAP and therefore inhibits expression of the enzymes that synthesize tryptophan. The presence of uncharged tRNA^{Trp} indicates that there is a shortage of tryptophan. The uncharged tRNA activates the antiTRAP, and thereby activates transcription of the *trp* genes.

KEY CONCEPTS

- A terminator protein called TRAP is activated by tryptophan to prevent transcription of trp genes.
- Activity of TRAP is (indirectly) inhibited by uncharged tRNA^{Trp}.

CONCEPT AND REASONING CHECK

Why is it necessary for the cell to use both tryptophan and tRNA^{Trp} to control the *trp* operon?

13.4 The *E. coli* Tryptophan Operon Is Controlled by Attenuation

A complex regulatory system is also used in the *E. coli trp* operon (where attenuation was originally discovered). As discussed in the last chapter (see *Section 12.12, The* trp *Operon Is a Repressible Operon with Three Transcription Units*), the first level of control of gene expression is that the operon is negative repressible, which means that it is prevented from initiating transcription by its product, the free amino acid tryptophan. Attenuation is the second level of control. There is a region in the 5′ leader of the mRNA called the *attenuator* that contains a small ORF. Attenuation in the *E. coli trp* operon means that transcription termination is controlled by the rate of translation of the attenuator ORF. This allows *E. coli* to also monitor the second pool of tryptophan, that of Trp-tRNA. High levels of Trp-tRNA will attenuate or terminate transcription,

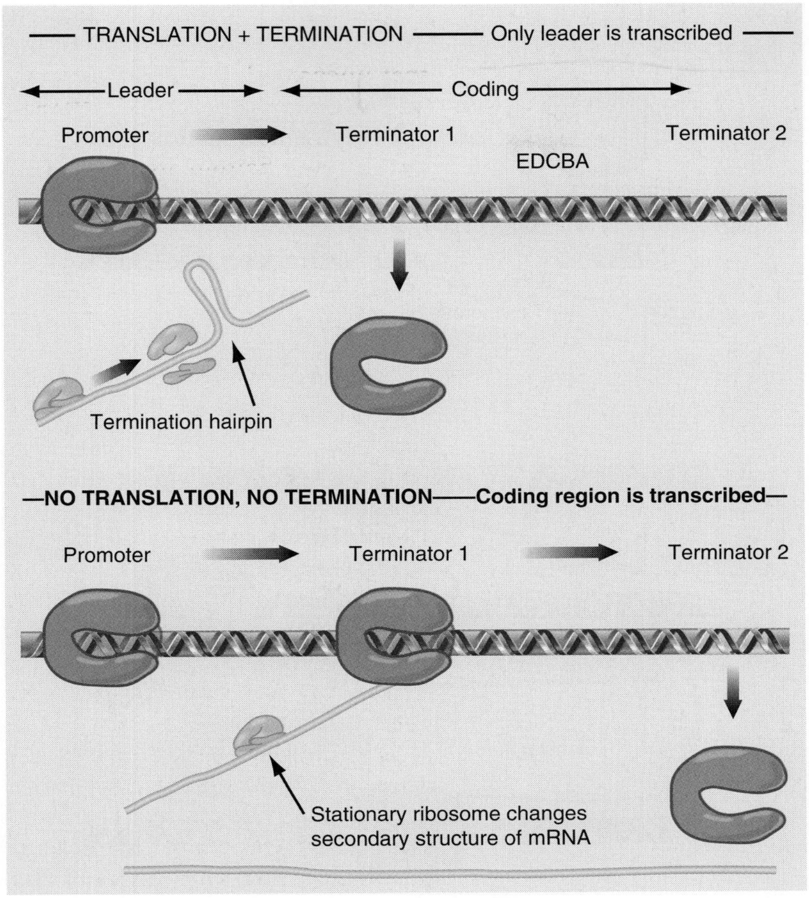

FIGURE 13.5 Termination can be controlled via changes in RNA secondary structure that are determined by ribosome movement.

whereas low levels will allow the *trpEDCBA* operon to be transcribed. This is accomplished by changes in secondary structure of the attenuator RNA that are determined by the position of the ribosome on mRNA. **FIGURE 13.5** shows that termination requires that the ribosome can translate the attenuator. When the ribosome translates the leader region, a termination hairpin forms at terminator 1. When the ribosome is prevented from translating the leader, though, the termination hairpin does not form, and RNA polymerase transcribes the coding region. This mechanism of antitermination therefore depends upon the level of Trp-tRNA to influence the rate of ribosome movement in the leader region.

Attenuation was first revealed by the observation that deleting a sequence between the operator and the *trpE* coding region can increase the expression of the structural genes. This effect is independent of repression: both the basal and derepressed levels of transcription are increased. Thus this site influences events that occur after RNA polymerase has set out from the promoter (irrespective of the conditions prevailing at initiation).

Termination at the attenuator responds to the level of Trp-tRNA, as illustrated in **FIGURE 13.6**. In the presence of adequate amounts of Trp-tRNA, termination is efficient. With low levels of Trp-tRNA, however, RNA polymerase can continue into the structural genes.

Repression and attenuation respond in the same way to the levels of the two pools of tryptophan. When free amino acid tryptophan is present, the operon is repressed. When tryptophan is removed, RNA polymerase has free access to the promoter, and can start transcribing the operon. When Trp-tRNA is present, the operon is attenuated and transcription terminates. When the pool of tryptophan bound to its tRNA is depleted, the RNA polymerase can continue to transcribe the operon. Note the pool

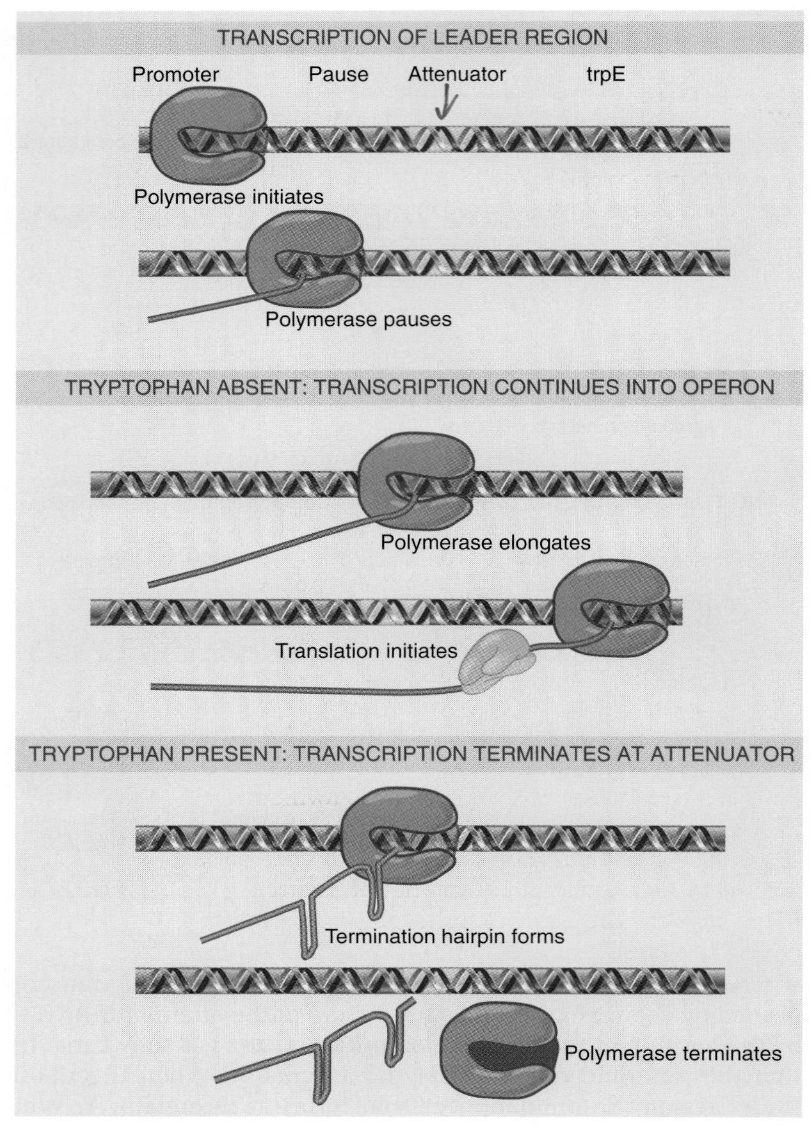

FIGURE 13.6 An attenuator controls the progression of RNA polymerase into the *trp* genes. RNA polymerase initiates at the promoter and then proceeds to position 90, where it pauses before proceeding to the attenuator at position 140. In the absence of tryptophan, the polymerase continues into the structural genes (trp starts at 163). In the presence of tryptophan there is 90% probability of termination to release the 140-base leader RNA.

of free tryptophan may be low and allow transcription to begin, but if the Trp-tRNA is fully changed, transcription will terminate.

Attenuation has ~10× effect on transcription. When tryptophan is present termination is effective, and the attenuator allows only ~10% of the RNA polymerases to proceed. In the absence of tryptophan, attenuation allows virtually all of the polymerases to proceed. Together with the ~70× increase in initiation of transcription that results from the release of repression, this allows an ~700-fold range of regulation of the operon.

KEY CONCEPTS

- An attenuator (intrinsic terminator) is located between the promoter and the first gene of the *trp* cluster.
- The absence of Trp-tRNA suppresses termination and results in a 10x increase in transcription.

CONCEPT AND REASONING CHECK

The *E. coli trp* operon is under dual control. What are the two conditions that allow transcription?

13.5 Attenuation Can Be Controlled by Translation

How can termination of transcription at the attenuator respond to the level of Trp-tRNA? The sequence of the leader region suggests a mechanism. It has a short coding sequence that could represent a leader peptide of fourteen amino acids. **FIGURE 13.7** shows that it contains a ribosome binding site whose AUG codon is followed by a short coding region that contains two successive codons for tryptophan. When the cell has a low level of Trp-tRNA, ribosomes initiate translation of the leader peptide but stop when they reach the Trp codons. The sequence of the mRNA suggests that this ribosome stalling influences termination at the attenuator.

The leader sequence can be written in alternative base-paired structures. The ability of the ribosome to proceed through the leader region controls transitions between these structures. The structure determines whether the mRNA can provide the features needed for termination.

FIGURE 13.8 shows these structures. In the first, region 1 pairs with region 2 and region 3 pairs with region 4. The pairing of regions 3 and 4 generates the hairpin that precedes the U_8 sequence: this is the essential signal for intrinsic termination. It is likely that the RNA would form this structure automatically.

A different structure is formed if region 1 is prevented from pairing with region 2. In this case, region 2 is free to pair with region 3. Region 4 then has no available pairing partner, so it is compelled to remain single-stranded. Thus the terminator hairpin cannot be formed.

FIGURE 13.9 shows that the position of the ribosome can determine which structure is formed in such a way that termination is attenuated only in the absence of tryptophan. The crucial feature is the position of the Trp codons in the leader peptide coding sequence.

When Trp-tRNA is abundant, ribosomes are able to synthesize the leader peptide. They continue along the leader section of the mRNA to the UGA codon, which lies between regions 1 and 2. As shown in the lower part of the figure,

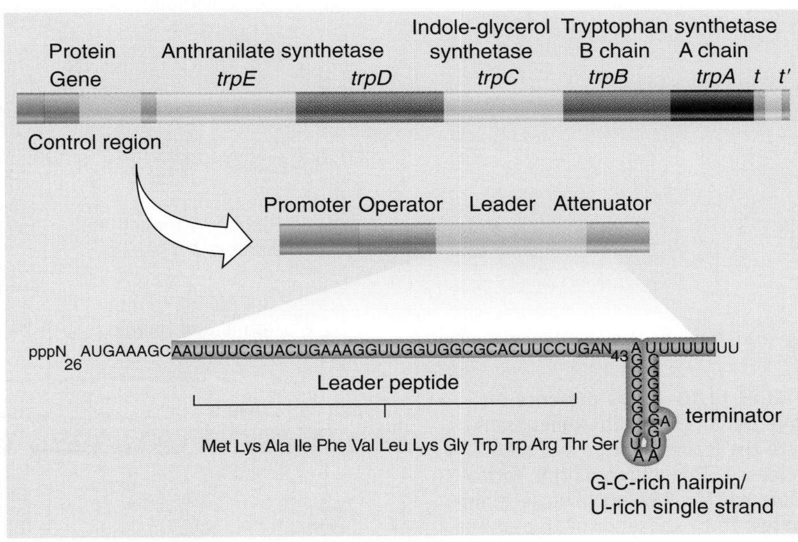

FIGURE 13.7 The *trp* operon consists of five contiguous structural genes preceded by a control region that includes a promoter, operator, leader peptide coding region, and attenuator.

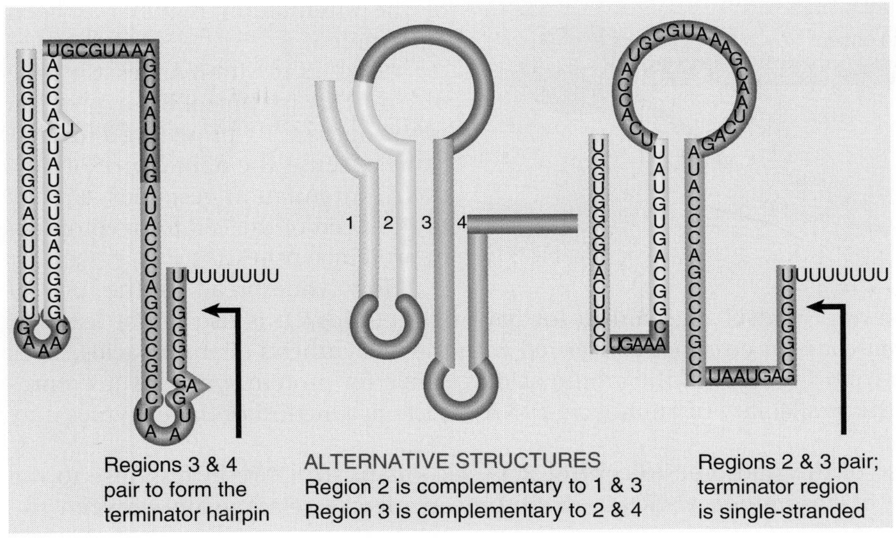

FIGURE 13.8 The *trp* leader region can exist in alternative base-paired conformations. The center shows the four regions that can base pair. Region 1 is complementary to region 2, which is complementary to region 3, which is complementary to region 4. On the left is the conformation produced when region 1 pairs with region 2 and region 3 pairs with region 4. On the right is the conformation when region 2 pairs with region 3, leaving regions 1 and 4 unpaired.

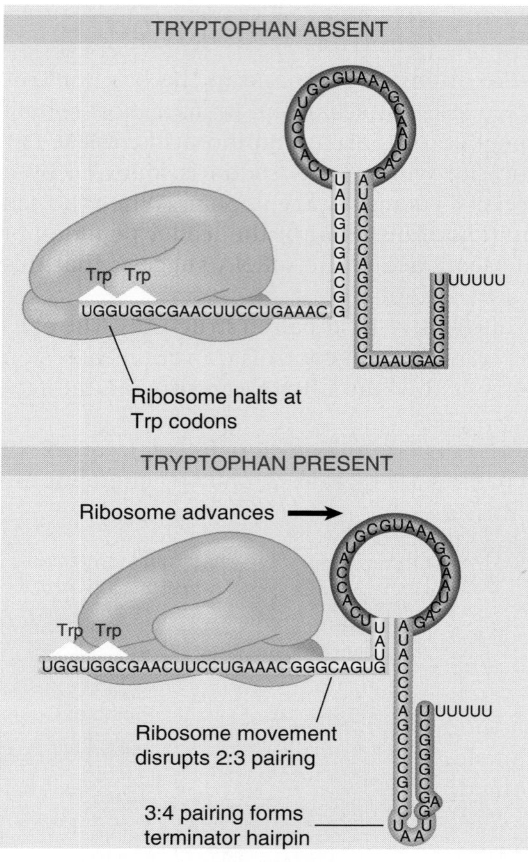

FIGURE 13.9 The alternatives for RNA polymerase at the attenuator depend on the location of the ribosome, which determines whether regions 3 and 4 can pair to form the terminator hairpin.

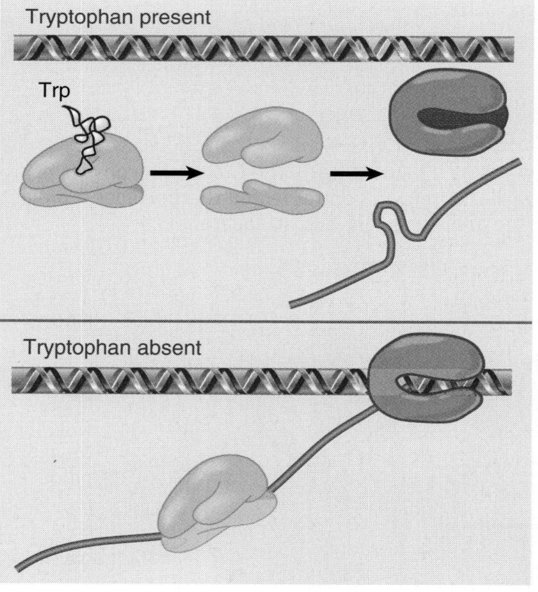

FIGURE 13.10 In the presence of tryptophan tRNA, ribosomes translate the leader peptide and are released. This allows hairpin formation, so that RNA polymerase terminates. In the absence of tryptophan tRNA, the ribosome is blocked, the termination hairpin cannot form, and RNA polymerase continues.

by progressing to this point, the ribosomes extend over region 2 and prevent it from base pairing. The result is that region 3 is available to base pair with region 4, which generates the terminator hairpin. Under these conditions, therefore, RNA polymerase terminates at the attenuator.

When Trp-tRNA is not abundant, ribosomes stall at the Trp codons, which are part of region 1, as shown in the upper part of the figure. Thus region 1 is sequestered within the ribosome and cannot base pair with region 2. This means that regions 2 and 3 become base paired before region 4 has been transcribed. This compels region 4 to remain in a single-stranded form. In the absence of the terminator hairpin, RNA polymerase continues transcription past the attenuator.

Control by attenuation requires a precise timing of events. For ribosome movement to determine formation of alternative secondary structures that control termination, translation of the leader must occur at the same time when RNA polymerase approaches the terminator site. A critical event in controlling the timing is the presence of a site that causes the RNA polymerase to pause at base 90 along the leader. The RNA polymerase remains paused until a ribosome translates the leader peptide. The polymerase is then released and moves off toward the attenuation site. By the time it arrives there, the secondary structure of the attenuation region has been determined.

FIGURE 13.10 summarizes the role of Trp-tRNA in controlling expression of the operon. By providing a mechanism to sense the abundance of Trp-tRNA, attenuation responds directly to the need of the cell for tryptophan in protein synthesis.

How widespread is the use of attenuation as a control mechanism for bacterial operons? It is used in at least six operons that code for enzymes concerned with the biosynthesis of amino acids. Thus a feedback from the level of the amino acid available for protein synthesis (as represented by the availability of aminoacyl-tRNA) to the production of the enzymes may be common.

The use of the ribosome to control RNA secondary structure in response to the availability of an aminoacyl-tRNA establishes an inverse relationship between the presence of aminoacyl-tRNA and the transcription of the operon, which is equivalent to a situation in which aminoacyl-tRNA functions as a corepressor of transcription. The regulatory mechanism is mediated by changes in the formation of duplex regions; thus attenuation provides a striking example of the importance of secondary structure in the termination event and of its use in regulation.

E. coli and *B. subtilis*, therefore, use the same types of mechanisms, which involve control of mRNA structure in response to the presence or absence of a tRNA, but they have combined the individual interactions in different ways. The end result is the same: to inhibit production of the enzymes when there is an excess supply of the amino acid, and to activate production when a shortage is indicated by the accumulation of uncharged $tRNA^{Trp}$.

KEY CONCEPTS

- The leader region of the trp operon has a fourteen-codon open reading frame that includes two codons for tryptophan.
- The structure of RNA at the attenuator depends on whether this reading frame is translated.
- In the presence of Trp-tRNA, the leader is translated, and the attenuator is able to form the hairpin that causes termination.
- In the absence of Trp-tRNA, the ribosome stalls at the tryptophan codons and an alternative secondary structure prevents formation of the hairpin, so that transcription continues.

CONCEPT AND REASONING CHECK

How does a hairpin structure in the RNA cause transcription termination on the DNA?

- **riboswitch** A catalytic RNA whose activity responds to a metabolite product or another small ligand.
- **ribozyme** An RNA that has catalytic activity.

13.6 A Riboswitch in the 5′ UTR Region Can Control Translation of the mRNA

As we have seen in the above two sections and in *Section 12.13, Translation Can Be Regulated,* an mRNA is more than simply an open reading frame. We have seen that regions in the 5′ UTR (5′ untranslated region) contain elements that, due to coupled transcription/translation, can control transcription termination. We have also seen that the 5′ UTR sequence itself can make an mRNA into a "good" message, which supports a high level of translation, or a "poor" message, which does not. What we will see now is another type of element in the 5′ UTR that can control expression of the mRNA with a different mechanism called a *riboswitch.*

A **riboswitch** is an element in the 5′ UTR of mRNAs that can assume alternate base pairing configurations that can control translation of the mRNA. **FIGURE 13.11** summarizes the regulation of the system that produces the metabolite GlcN6P. The gene *glmS* codes for an enzyme that synthesizes GlcN6P (Glucosamine-6-phosphate) from fructose-6-phosphate and glutamine. GlcN6P is a fundamental intermediate in bacterial cell wall biosynthesis. The mRNA contains a long 5′ UTR before the coding region of the mRNA. Within the 5′ UTR is a **ribozyme**—a sequence of RNA that has catalytic activity (see *Section 29.4, Ribozymes Have Various Catalytic Activities*). In this case, the catalytic activity is an endonuclease that cleaves its own RNA. It is activated by binding of the metabolite product, GlcN6P, to the ribozyme. The consequence is that accumulation of GlcN6P activates the ribozyme,

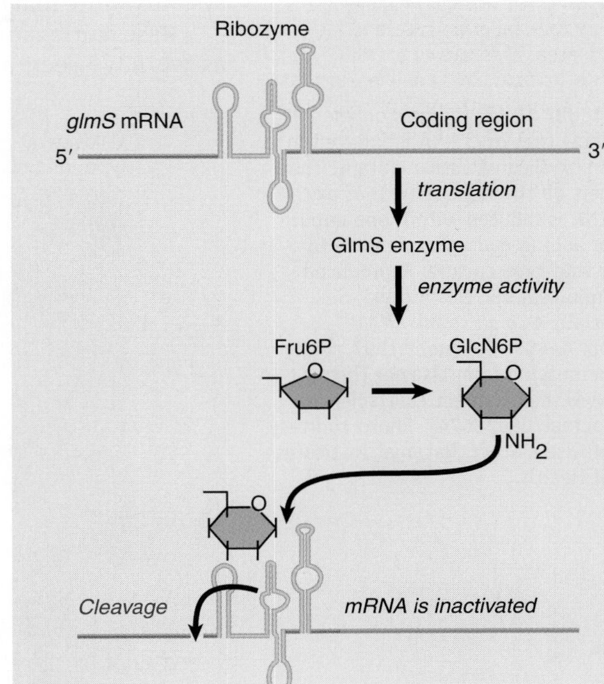

FIGURE 13.11 The 5′ untranslated region of the mRNA for the enzyme that synthesizes GlcN6P contains a ribozyme that is activated by the metabolic product. The ribozyme inactivates the mRNA by cleaving it.

which cleaves the mRNA, which in turn prevents further translation. This is an exact parallel to allosteric control of a repressor protein by the end product of a metabolic pathway. There are several examples of such riboswitches in bacteria.

Not all riboswitches encode a ribozyme that controls the mRNA stability. Other riboswitches have alternate configurations of the RNA that allow or prevent expression of the mRNA by affecting ribosome binding. Riboswitches are found predominantly in bacteria and rarely in eukaryotes.

KEY CONCEPTS

- A riboswitch is an RNA whose activity is controlled by the metabolite product or another small ligand. (A ligand is any molecule that binds to another.)
- A riboswitch may be a ribozyme.

CONCEPT AND REASONING CHECK

An mRNA consists of how many different elements?

13.7 Bacteria Contain Regulator sRNAs

▸ **sRNA** A small bacterial RNA that functions as a regulator of gene expression.

▸ **antisense RNA** RNA that has a complementary sequence to an RNA that is its target.

Bacteria contain many, up to hundreds, of genes that code for regulator RNAs. These are short RNA molecules ranging from about 50 nucleotides to about 200 nucleotides that are collectively known as **sRNA**s. Some of the sRNAs are general regulators that affect many target genes; others are specific for a single transcript. These sRNAs typically function as imperfect (meaning that only small regions within the sRNA are complementary to the target) **antisense RNA**s; that is, their sequences are complementary to their target RNAs.

At what level does the antisense RNA inhibit expression? It could in principle (1) prevent transcription of the gene, (2) affect processing of its RNA product, (3) affect translation of the messenger, or (4) affect stability of the RNA. Results with different systems show that inhibition depends on formation of RNA–RNA duplex molecules.

Base pairing offers a powerful means for one RNA to control the activity of another, as we have seen above. There are many cases in both prokaryotes and eukaryotes where a (usually rather short) single-stranded RNA base pairs with a complementary region of an mRNA, and as a result it prevents expression of the mRNA.

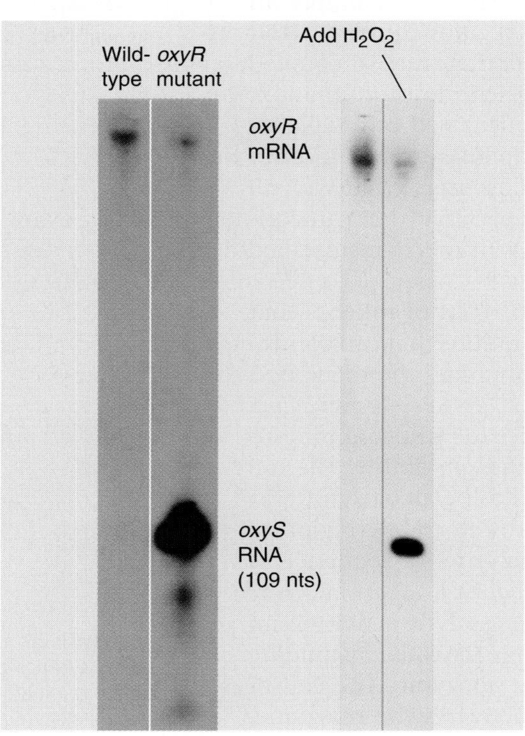

FIGURE 13.12 The gels on the left show that *oxyS* RNA is induced in an *oxyR* constitutive mutant. The gels on the right show that *oxyS* RNA is induced within one minute of adding hydrogen peroxide to a wild-type culture. Reproduced from *Cell*, vol. 90, Altuvia, S., et al., *A small stable RNA . . .*, pp. 44–53. Copyright 1997, with permission from Elsevier [http://www.sciencedirect.com/science/journal/00928674]. Photo courtesy of Gisela Storz, National Institutes of Health.

Oxidative stress in *E. coli* provides an interesting example of a general control system in which an sRNA is the regulator. When exposed to reactive oxygen species, bacteria respond by inducing antioxidant defense genes. Hydrogen peroxide activates the transcription activator OxyR, which controls the expression of several inducible genes. One of these genes is *oxyS*, which codes for a small RNA.

FIGURE 13.12 shows two salient features of the control of *oxyS* expression. In a wild-type bacterium under normal conditions, it is not expressed. The pair of gels on the left side of the figure show that it is expressed at high levels in a mutant bacterium with a constitutively active *oxyR* gene. This identi-

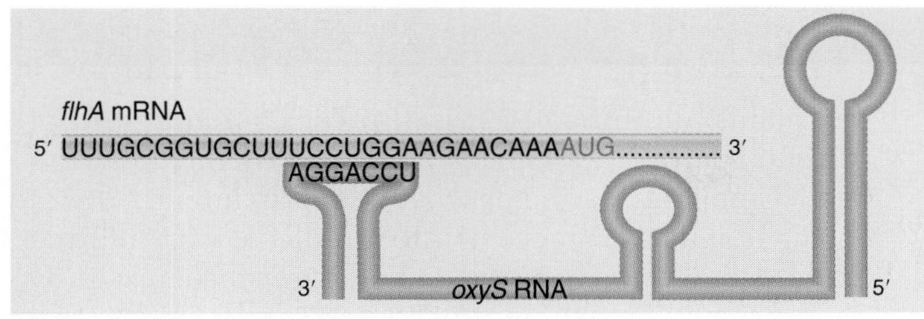

FIGURE 13.13 *oxyS* RNA inhibits translation of *flhA* mRNA by base pairing with a sequence just upstream of the AUG initiation codon.

fies *oxyS* as a target for activation by *oxyR*. The pair of gels on the right side of the figure show that *oxyS* RNA is transcribed within one minute of exposure to hydrogen peroxide.

The *oxyS* RNA is a short sequence (109 nucleotides) that does not code for protein. It is a *trans*-acting regulator that affects gene expression at the level of translation. It has >10 target mRNAs; at some of them, it activates expression, and at others it represses expression. **FIGURE 13.13** shows the mechanism of repression of one target, the *flhA* mRNA. Three stem-loop double-stranded RNA structures protrude in the secondary structure of *oxyS* mRNA, and the loop closest to the 3′ terminus is complementary to a sequence just preceding the initiation codon of *flhA* mRNA. Base pairing between *oxyS* RNA and *flhA* RNA prevents the ribosome from binding to the initiation codon and therefore represses translation. There is also a second pairing interaction that involves a sequence within the coding region of *flhA*.

Another target for *oxyS* is *rpoS*, the gene coding for an alternative sigma factor (which activates a general stress response). By inhibiting production of the sigma factor, *oxyS* ensures that the specific response to oxidative stress does not trigger the response that is appropriate for other stress conditions. The *rpoS* gene is also regulated by two other sRNAs (*dsrA* and *rprA*), which activate it. These three sRNAs appear to be global regulators that coordinate responses to various environmental conditions.

The actions of all three sRNAs are assisted by an RNA-binding protein called Hfq. The Hfq protein was originally identified as a bacterial host factor needed for replication of the RNA bacteriophage Qβ. It is related to the Sm proteins of eukaryotes that bind to many of the snRNAs (small nuclear RNAs) that have regulatory roles in gene expression (see *Section 28.5, snRNAs Are Required for Splicing*). Mutations in its gene have many effects; this identifies it as a pleiotropic protein. Hfq binds to many of the sRNAs of *E. coli*, and it increases the effectiveness of *oxyS* RNA by enhancing its ability to bind to its target mRNAs. The effect of Hfq is probably mediated by causing a small change in the secondary structure of *oxyS* RNA that improves the exposure of the single-stranded sequences that pair with the target mRNAs.

This technique offers a powerful approach for turning off genes at will; for example, the function of a regulatory gene can be investigated by introducing an antisense version. An extension of this technique is to place the antisense gene under the control of a promoter that is itself subject to regulation. The target gene can then be turned off and on by regulating the production of antisense RNA. This technique allows investigation of the importance of the timing of expression of the target gene.

KEY CONCEPTS

- Bacterial regulator RNAs are called sRNAs.
- Several of the sRNAs are bound by the protein Hfq, which increases their effectiveness.
- The *oxyS* sRNA activates or represses expression of >10 loci at the posttranscriptional level.

CONCEPT AND REASONING CHECK

How could an sRNA activate translation?

MEDICAL APPLICATIONS

Artificial Antisense Genes Can Be Used to Turn Off Viruses and Cancer Genes

The therapeutic application of antisense technology is a turning point in molecular biology and medicine. Artificial or laboratory synthesized antisense DNA or RNA oligonucleotides can be designed and synthesized to target and consequently turn off, or *silence*, the function of virtually any gene unique to cells or viruses. Short, single-stranded DNA oligonucleotides bind between the groove of double-stranded DNA gene sequences, producing a triple helix structure that prevents its transcription into mRNA (**FIGURE B13.1**). Short RNA antisense oligonucleotides target the mRNA transcripts (sense strand) of viral or cellular genes, preventing their translation (**FIGURE B13.2**).

Compared to DNA antisense oligonucleotides, RNAi is less toxic to cells and much more potent. RNA antisense technology takes advantage of a natural phenomenon called *RNA interference* or RNAi (also referred to as *gene silencing*). The RNA interference mechanism was originally discovered in the worm *Caenorhabditis elegans* but is known to play a pivotal role in regulating gene expression in a wide variety of organisms, including plants and humans. It is also considered a form of antiviral defense, especially in plants. When laboratory-made small interfering dsRNAs of approximately 21–23 bp in length are injected into uninfected or infected mammalian cells, the molecules are recognized by the enzymatic machinery of RNAi. Through the RNAi pathway, the end result is homology-dependent inhibition of expression of a target mRNA. Exploiting this native gene-silencing pathway is a way to regulate gene expression. Applications of RNAi have the potential to give rise to a cornucopia of drugs that silence disease-causing genes or disarm viral pathogens.

Theoretically, viral and bacterial infections should be amenable to antisense-mediated silencing. In addition, diseases caused by dominant mutations should also be amenable. The early phases of clinical trials using RNAi therapeutics are in progress to treat certain viral infections and cancers. Known and emerging viruses for which there are no effective vaccines or antiviral drugs available are an increasingly serious threat to public health. Drug designers of RNAi-based antivirals have prioritized their efforts toward treating viral infections caused by cytomegaloviruses, HIV-1, respiratory syncytial viruses (RSV), hepatitis B and C viruses,

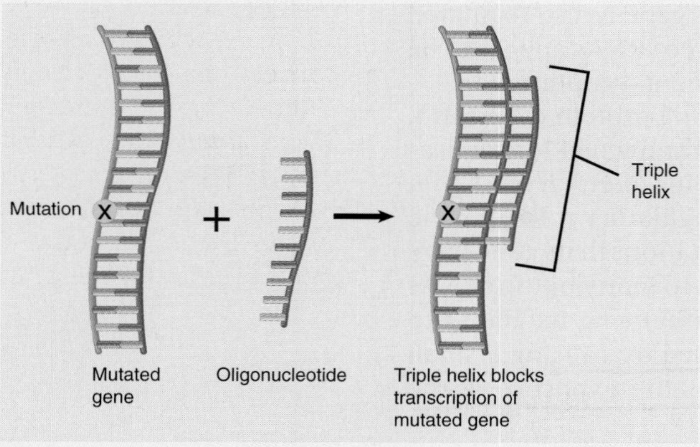

FIGURE B13.1 Preventing transcription of a mutated gene using triple-helix-forming oligonucleotides. Reproduced from *New Approaches to Gene Therapy* from the Genetics Science Learning Center. Used under license from the University of Utah [http://learn.genetics.utah.edu].

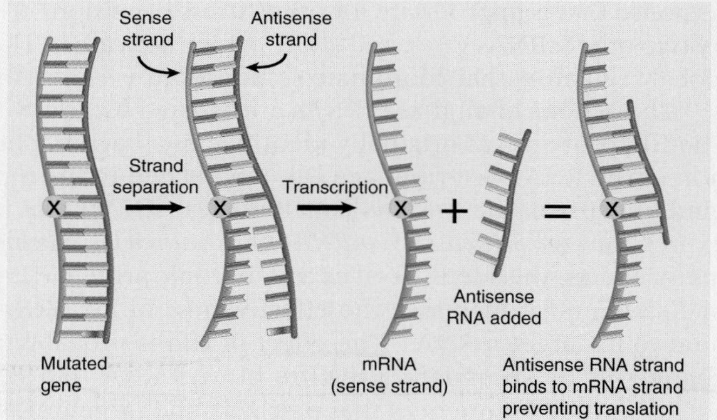

FIGURE B13.2 Preventing translation using antisense technology. Reproduced from *New Approaches to Gene Therapy* from the Genetics Science Learning Center. Used under license from the University of Utah [http://learn.genetics.utah.edu].

13.8 Eukaryotes Contain Regulator RNAs

Eukaryotes, like bacteria, use RNAs to regulate gene expression. This section will be divided into two general areas: (1) control of expression in the nucleus at the level of the DNA gene and (2) control in the cytoplasm at the level of the mRNA. As we will see, the eukaryote mechanisms, while related to the bacterial mechanisms, are very different.

Like bacteria, eukaryotes use RNAs to regulate transcription, but at the level of initiation as opposed to termination. Attenuation is not possible in eukaryotes as it is and human papilloma viruses. Critical to the design of RNAi drugs is ensuring that the antisense molecules target conserved sequences of the viral RNA genome or mRNAs that are essential viral factors. This approach may allow drug development to keep pace with viruses that mutate rapidly.

HIV-1 was the first virus targeted by RNAi treatment because the life cycle and gene expression pattern of the virus was well understood. Synthetic antisense RNAs were used to target HIV-1-encoded RNAs such as the TAR element, or *tat, rev, gag, env, vif, nef,* and reverse transcriptase mRNAs. The high mutation rate of HIV represents a substantial challenge. For this reason, an alternate approach is being tried. This targets conserved cellular cofactors, such as CCR5, required for HIV infection. This is possible because CCR5 is nonessential for normal immune system function, but central to HIV entry into cells. HIV-1 binds to the CD4 antigen and CCR5 co-receptor present on macrophages. Binding to the CCR5 co-receptor causes a conformational change in HIV's gp41 protein, allowing the virus to fuse with the plasma membrane of the host cell and subsequent entry. People living with a genetic variation of the CCR5 gene are resistant to HIV infection and still maintain a healthy immune system. CCR5 plays a role in the inflammatory response to infection; however, when this gene is deleted or not functional in the host, other cellular chemokines compensate for this loss of function.

Hepatitis C virus (HCV) infects an estimated 3% of the world's population. It is a major cause of chronic hepatitis. About 80% of HCV-infected people suffer from liver cirrhosis and hepatocellular cancer. It is the leading cause of liver transplantation in the United States. The only therapy available to treat patients is a combination of interferon and the drug ribavirin. The majority of HCV-infected individuals do not respond to this therapy because it often produces very toxic side effects in patients. At the time of this writing, RNAi studies *in vitro* were able to show that Huh-7.5 hepatoma cells containing persistently replicating HCV replicons could be "cured." To date, experimental RNAi directed at HCV RNA targets have not moved past Phase II clinical trials.

RNAi may also be used to silence errant genes containing dominant mutations that cause certain cancers. Even though there have been recent advances in surgery, radiotherapy, and chemotherapy, the prognosis and survival rate for patients with brain tumors remains poor. New therapies that specifically target and eradicate tumor cells are needed to improve the life expectancy of these patients. The human epidermal growth factor receptor (EGFR) gene represents a potential target for treating cancers in general. In normal cells, epidermal growth factor (EGF) binds to EGFR, causing the induction of cell proliferation or differentiation in mammalian cells. In cancer cells, the EGFR gene is overexpressed or mutated, causing the induction of uncontrolled cell growth and a malignant phenotype (solid tumors). Studies are under way that use RNAi treatment to target EGFR in an experimental human brain tumor in a SCID (severe combined immunodeficiency) mouse model.

There are still a number of hurdles and safety concerns before RNAi can be successfully applied as a therapeutic. The major hurdle of RNAi therapies is the delivery of these macromolecules to the desired cell type, tissue, or organ. RNA molecules do not readily cross membranes because of their negative charges and sizes. Liposome-based carriers, nanoparticles or other delivery schemes may eventually overcome this challenge. Other hurdles include stability and specificity of RNAi. Despite these obstacles, RNAi is a powerful tool to study gene function. RNAi discovery enables the development of a new class of human therapeutics in the pipeline that has the potential to treat a wide array of important diseases.

References

Aagaard L and Rossi JJ (2007). RNAi Therapeutics: Principles, Prospects and Challenges. *Adv. Drug Deliv. Rev.* 59(2–3):75–86.

Boado RJ (2005). RNA Interference and Nonviral Targeted Gene Therapy of Experimental Brain Cancer. *NeuroRX* 25:139–150.

Haasnoot J, et al. (2007). RNA Interference Against Viruses: Strike and Counterstrike. *Nature Biotech.* 25: 1435–1443.

in *E. coli* because the nuclear membrane separates the processes of transcription and translation. Because eukaryote mRNA is so much more stable than bacterial mRNA, with an average half-life of hours as opposed to minutes, much more translation-level control is used in eukaryotes, both at the level of translation initiation and mRNA stability control itself.

The classical view of eukaryotic transcription was that most control occurs at the level of gene regulation. Classical gene transcription units comprise about 25% of the DNA in higher eukaryotes (but only about 2% is actually coding exons). However, whole genome tiling array experiments (see Methods and Techniques box Microarrays and Tiling Experiments) have revealed that 90% or more of the eukaryote genome is expressed. In addition to the genes, intergenic regions are transcribed and vast amounts of antisense RNA are produced; that is, gene regions are transcribed extensively on both strands. Even simple sequence heterochromatic DNA, long thought to be transcriptionally silent, is expressed. These RNAs are called **ncRNA** or noncoding RNA. We know the function of some of this ncRNA, as we will see below, but much of it is still mysterious and remains to be discovered.

- **ncRNA** RNA which does not contain an open reading frame.
- **nested genes** A gene located within an intron of another gene.
- **CUT** Cryptic unstable antisense transcripts generated by promoters located at the 3′ end of genes.
- **miRNA** Very short RNAs that may regulate gene expression.
- **RNA interference (RNAi)** A process by which short 21 to 23 nucleotide antisense RNAs, derived from longer double-stranded RNAs, can modulate expression of mRNA by translation inhibition or degradation.
- **stRNA** Short temporal RNA, a form of miRNA in eukaryotes that modulates mRNA expression during development.
- **siRNA** Short interfering RNA, a miRNA that prevents gene expression.
- **Dicer** An endonuclease that processes double-stranded RNA to 21 to 23 nucleotide RNAi molecules.

Eukaryotic antisense RNA has been known for some time, since the first genome sequencing projects. Genes located within the introns of other genes, **nested genes**, are not as rare as once first thought, comprising as much as 5% to 10% of genes. If the nested gene is transcribed from the opposite strand, then antisense RNA is produced. This head to head arrangement of a nested gene will also lead to transcription interference since both genes cannot be transcribed simultaneously.

A more direct role for antisense RNA in transcription control has recently been discovered. In the yeast *S. cerevisiae,* the gene *PHO84* is regulated in part by a form of antisense RNA called cryptic unstable transcripts or **CUT**. As shown in **FIGURE 13.14**, in addition to the promoter at the 5′ end of the gene, there is another promoter (which is unregulated) on the opposite strand. Transcription from this promoter on the opposite strand produces an antisense RNA. Under normal conditions, this RNA is rapidly degraded as it is produced. In the absence of degradation or in aging cells, the antisense RNA persists. This antisense RNA, or CUT, is involved with the recruitment of enzymes which remove acetate groups from histones, thereby causing the chromatin over the gene region be remodeled and condensed so that the gene can no longer be transcribed (see *Section 26.9, Histone Acetylation Is Associated with Transcription Activation*). This is gene-specific remodeling directed by the antisense RNA and does not extend to the neighboring genes.

Very small RNAs or microRNAs (**miRNA**s) are gene expression regulators found in most, if not all eukaryotes. These bear some resemblance to their bacterial sRNA counterparts, but as we will see, they are smaller and their mechanism of action is different. The human genome has an estimated 300 genes that code for miRNAs that participate in **RNA interference (RNAi)**. This is a general mechanism to repress gene expression, usually (but not always) at the level of translation. These miRNAs go by a number of names and are sometimes called **stRNA**, short temporal RNA (because they are involved in development). Some miRNAs have also been shown to affect transcription initiation by binding to the gene's promoter. It is estimated that hundreds of miRNAs control thousands of mRNAs. It may be that a large fraction of mRNAs are at some point targeted by miRNAs at all stages of development. Another type of very small RNA is **siRNA**, short interfering RNA, typically produced during a virus infection.

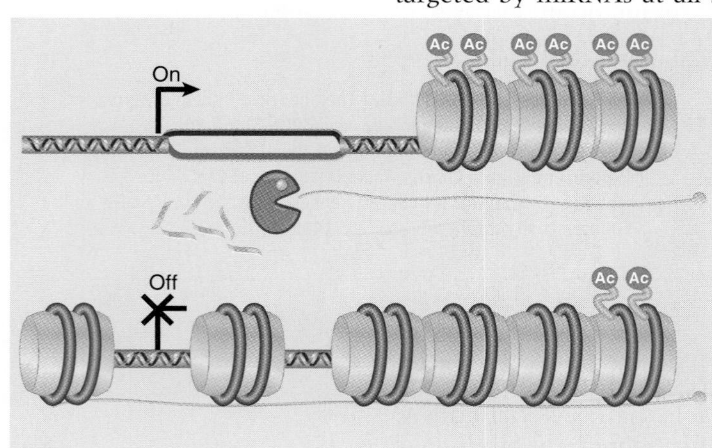

FIGURE 13.14 *PHO84* antisense RNA stabilization is paralleled by histone deacetylase recruitment, histone deacetylation and *PHO84* transcription repression. Under normal conditions, the RNA is rapidly degraded. In aging cells, antisense transcripts are stabilized and recruit the histone deacetylase to repress transcription. Adapted from Camblong, J., et al., *Cell* 131 (2007): 706–717.

One of the earliest known examples of RNAi in animals was discovered in the nematode *Caenorhabditis elegans* as the result of the interaction between the regulator gene *lin4* and its target gene, *lin14*. **FIGURE 13.15** illustrates the behavior of this regulatory system. The *lin14* target gene regulates larval development. Expression of *lin14* is controlled by *lin4,* which codes for a small transcript of twenty-two nucleotides. The *lin4* transcripts are complementary to a ten-base sequence that is repeated seven times in the 3′ nontranslated region (3′ UT) of the *lin14* mRNA.

The *lin4* gene is transcribed into a transcript that contains a double-stranded RNA hairpin that is a target for an endonuclease called **Dicer**. Dicer has an N-terminal helicase activity, which enables it to unwind the double-stranded region, and two nuclease domains that are related to the bacterial ribonuclease III. Related enzymes are nearly universal in eukaryotes. Cleavage

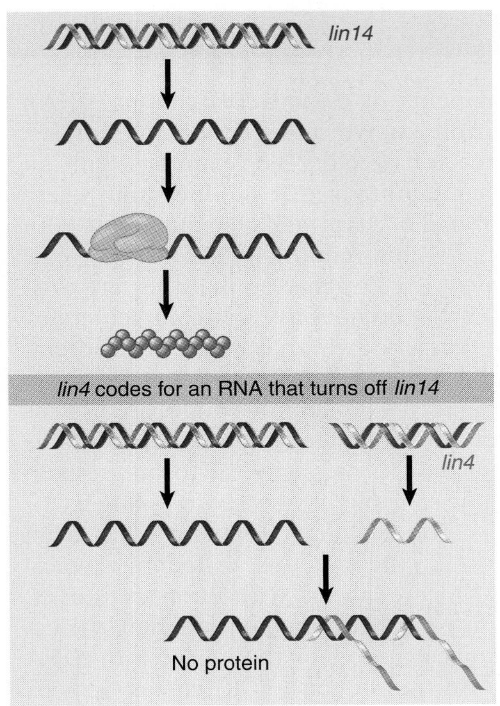

FIGURE 13.15 *lin4* RNA regulates expression of *lin14* by binding to the 3′ nontranslated region.

dsRNA >26 nucleotides
PKR
2′,5′AS
eIF2a
P
RNase L
X
Protein synthesis cannot initiate
Degradation of all mRNA
siRNA targets complementary mRNA
mRNA degraded

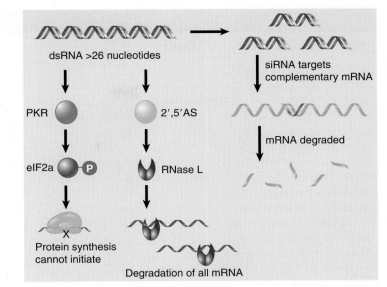

FIGURE 13.16 Long dsRNA inhibits protein synthesis and triggers degradation of all mRNA in mammalian cells, as well as having sequence-specific effects. Short dsRNA (>26 nt) leads to degradation only of complementary mRNAs.

of the initial transcript generates double-stranded sections of RNA of twenty-one to twenty-three base pairs. These are delivered to a complex called **RISC** (RNA-induced silencing complex). Proteins in the Argonaute family are components of this complex and are required for the final processing of a single strand to be delivered to the 3′ UT region of its target mRNA.

▸ **RISC** RNA-induced silencing complex, a ribonucleoprotein particle composed of a short single-stranded siRNA and a nuclease that cleaves mRNAs complementary to the siRNA. It receives siRNA from Dicer and delivers it to the mRNA.

The consequence of delivering an siRNA or miRNA to an mRNA differs somewhat in plants and animals. In plants, it is more likely that the miRNA has a higher percentage of base pairing, and thus tighter binding. Tight binding leads to mRNA-targeted degradation. In animals, imperfect binding is more common and this leads to translation repression rather than degradation.

Many of the *C. elegans* miRNA genes have homologs in mammals, so the mechanism may be widespread. They are also found in plants. Of sixteen miRNAs in *Arabidopsis*, eight are completely conserved in rice, suggesting widespread conservation of this regulatory mechanism. The mechanism of production of the miRNAs is also widely conserved.

RNAi has become a powerful technique for ablating the expression of a specific target gene in invertebrate cells. The technique, however, has been limited in mammalian cells, which have the more generalized response to dsRNA of shutting down protein synthesis and degrading mRNA. **FIGURE 13.16** shows that this happens because of two reactions. The dsRNA activates the enzyme PKR, which inactivates the translation initiation factor eIF2a by phosphorylating it. It also activates 2′5′ oligoadenylate synthetase, whose product activates RNase L, which degrades all RNAs in the cell. It turns out, however, that these reactions require dsRNA that is longer than twenty-six nucleotides. If shorter dsRNA (twenty-one to twenty-three nucleotides) is introduced into mammalian cells, it triggers the specific degradation of complementary RNAs just as with the RNAi technique in worms and flies. With this advance, RNAi has become the mechanism of choice for turning off the expression of a specific gene.

▸ **RNA silencing** The ability of an RNA, especially ncRNA, to alter chromatin structure in order to prevent gene transcription.

RNA interference is related to natural processes in which gene expression is silenced. Plants and fungi show **RNA silencing** (sometimes called *posttranscriptional gene silencing*), in which dsRNA inhibits expression of a gene. The most common

METHODS AND TECHNIQUES

Microarrays and Tiling Experiments

Scientific progress often depends on technological breakthroughs, such as those witnessed after cloning and PCR, and now microarrays. The development of genomic microarray technology itself was made possible because of breakthroughs in the technology of inexpensive oligonucleotide synthesis and DNA sequencing technologies. Microarray experiments are simply classical nucleic acid hybridization experiments, but done on a massive scale.

A DNA microarray, sometimes called "DNA on a chip," is basically a very simple device. A support platform can be a microscope slide. Target nucleic acid sequences, typically DNA, are spotted onto the slide. The earliest microarrays had been made by manually spotting the DNA. By the 1990s, automated machines were developed that allowed precise and high-density arrays to be made. Schematics to build these devices circulated on the Internet, allowing many laboratories to assemble their own devices.

The array density can range from 10 to 1000 spots per square mm (millimeter). Commercially produced arrays have become very popular for a number of reasons, which are discussed below. Standardized arrays for many model organisms are readily available from numerous vendors, and custom arrays for individual projects can also be synthesized.

Design of the target sequence probes to be spotted is a complex task. The DNA can be genomic PCR fragments, PCR fragments of cloned cDNA (copy DNA), copies of mRNA from known genes, or oligonucleotides, typically of 25 to 60 base pairs. For example, one can make an array containing a PCR product from every gene in an organism. For a typical higher eukaryote with 20,000 genes, that would require 40,000 PCR primers. The products have to be designed so that they are each unique sequences that do not have self-complementarity. The melting temperatures, that is, the GC content, have to be within a given range for mass production and use. The same is true if oligonucleotides are used as the target sequence. Computer programs are available to assist in target design. An array set to query exon usage and alternate splicing may be 10 times larger.

Once the decision on target sequences has been made, the next issue is the RNA to be tested. In a typical experiment, mRNAs are labeled with fluorescent dyes. The mRNA pool must first be amplified and then labeled. Amplification can be achieved by the synthesis of cDNA or RT-PCR (reverse transcriptase-PCR) copies. Experiments of this nature can ask a number of different questions, including how gene expression changes during development, or with altered metabolic challenges such as starvation. It can also compare the gene expression profile in a normal cell to that of a cancer cell.

In these types of experiments, where two cell types are being compared, the amplified mRNAs each have a different color of fluorescent tag attached, usually red and green. The two pools of amplified tagged RNAs are then mixed and applied to the microarray. Classical

sources of the RNA are a replicating virus or a transposable element. This mechanism may have evolved as a defense against these elements. When a virus infects a plant cell, the formation of dsRNA triggers the suppression of expression from the plant genome. Similarly, transposable elements also produce dsRNA. RNA silencing has the further remarkable feature that it is not limited to the cell in which the viral infection occurs: it can spread throughout the plant systemically. Presumably the propagation of the signal involves passage of RNA or fragments of RNA. It may require some of the same features that are involved in movement of the virus itself. RNA silencing in plants involves an amplification of the signal by an RNA-dependent RNA polymerase, which uses the siRNA as a primer to synthesize more RNA on a template of complementary RNA.

KEY CONCEPTS

- Eukaryote genomes produce antisense RNAs.
- Antisense RNAs regulate gene expression at the level of transcription and translation.
- Eukaryote genomes code for many short (~22 base) RNA molecules called microRNAs.
- MicroRNAs regulate gene expression by base pairing with complementary sequences in target mRNAs.
- RNA interference triggers degradation or translation inhibition of mRNAs complementary to miRNA or siRNA.
- dsRNA may cause silencing of host genes.

hybridization kinetics dictate that the most abundant RNA will hybridize to the probe more readily than the least abundant RNA.

The third step in the process is data acquisition and interpretation. Special scanners need to be employed in order to read each dot on the microarray at different fluorescent wavelengths. As might be expected, a massive data set can come from a single experiment. Bioinformatic software packages are required to sort and organize the data into a usable format. In the experiments described above, this entails the grouping together of genes that are expressed together. The choice of red and green labels results in data that range from a red spot indicating that the red labeled pool of RNA predominated in the mixture, to a yellow spot which indicates that both pools are equally represented, to a green spot indicating that the second pool predominates. A black spot indicates that the gene was not expressed in either pool (**FIGURE B13.3**).

A genomic tiling array is an extension of the expression microarray in that the target covers the entire genome, chromosome, or region of interest on both strands. The oligonucleotides can be of different lengths and spacing to obtain uniform coverage. The same issues of target probe design apply here as in the expression arrays described above. Repeat and self-complementary sequences must be avoided, and the melting temperatures must be within a critical range. The question asked here is also somewhat different. Instead of querying known genes at different times, now the question becomes, "is this region ever expressed?"

For further reading see *Annual Review of Biochemistry 74*, 53–82, 2005.

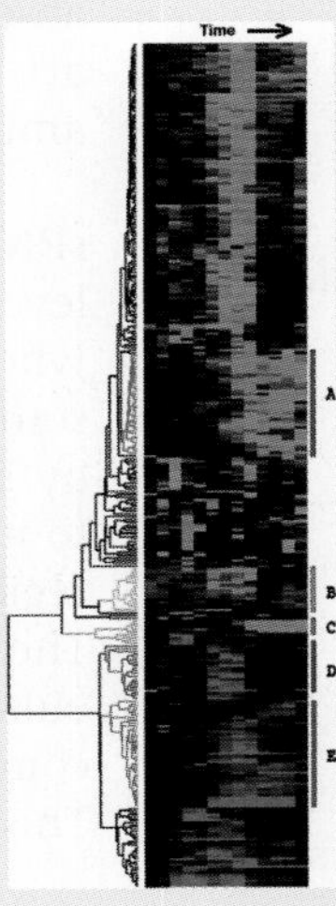

FIGURE B13.3 An example of analysis of gene expression in which similarly expressed genes have been grouped together. This experiment identified genes expressed over time during serum stimulation in human fibroblast cells. Green represents genes whose expression is reduced compared to unstimulated cells, while red indicates genes whose expression increases after stimulation. Groups A–E are genes with related functions that also show similar expression patterns. Reproduced from Eisen, M. B. et al., *Proc. Natl. Acad. Sci. USA* 95 (1998): 14863–14868. Copyright 1998 National Academy of Sciences, USA. Photo courtesy of Michael Eisen, University of California, Berkeley.

CONCEPT AND REASONING CHECK

Predict the consequence of a mutation that alters a miRNA.

13.9 Summary

Gene expression can be regulated positively by factors that activate a gene or negatively by factors that repress a gene. The first and most common level of control is at the initiation of transcription, but termination of transcription may also be controlled. Translation may be controlled by regulators that interact with mRNA. The regulatory products may be proteins, which often are controlled by allosteric interactions in response to the environment, or RNAs, which function by base pairing with the target RNA to change its secondary structure. Regulatory networks can be created by linking regulators so that the production or activity of one regulator is controlled by another.

Attenuation is a mechanism that relies on regulation of termination to control transcription through bacterial operons. It is commonly used in operons that code for enzymes involved in biosynthesis of an amino acid. The polycistronic mRNA of the operon starts with a sequence that can form alternative secondary structures. One of the structures has a hairpin loop that provides an intrinsic terminator upstream of the structural genes; the alternative structure lacks the terminator hairpin. Various types of interaction determine whether the hairpin forms. One interaction occurs when a protein binds to the mRNA to prevent formation of the alternative structure. In the *trp* operon of *B. subtilis*, the TRAP protein has this function; it is controlled by the anti-TRAP protein, whose production in turn is controlled by the level of uncharged aminoacyl-tRNA^{Trp}. In the *trp* (and other) operons of *E. coli*, the choice of which structure forms is controlled by the progress of translation through a short leader sequence that includes codons for the amino acid(s) that are the product of the system. In the presence of aminoacyl-tRNA bearing such amino acid(s), ribosomes translate the leader peptide, which allows a secondary structure to form that supports termination. In the absence of this aminoacyl-tRNA the ribosome stalls, which results in a new secondary structure in which the hairpin needed for termination cannot form. The supply of aminoacyl-tRNA therefore (inversely) controls amino acid biosynthesis.

Small regulator RNAs are found in both bacteria and eukaryotes. *E. coli* has ~17 sRNA species. The *oxyS* sRNA controls about ten target loci at the posttranscriptional level; some of them are repressed, whereas others are activated. Repression is caused when the sRNA binds to a target mRNA to form a duplex region that includes the ribosome-binding site. MicroRNAs are ~22 bases long and are produced in most eukaryotes by Dicer cleavage of a longer transcript, which is then delivered to RISC for delivery to its target mRNA. They function by base pairing with target mRNAs to form duplex regions that are susceptible to cleavage by endonucleases or inhibition of translation. The technique of RNA interference is becoming the method of choice for inactivating eukaryotic genes. It uses the introduction of short dsRNA sequences with one strand complementary to the target RNA, and it works by inducing degradation of the targets. This may be related to a natural defense system in plants called RNA silencing.

CHAPTER QUESTIONS

1. The *trp* attenuator is located:
 A. in the promoter region of the *trp* operon prior to any coding region.
 B. between the promoter region and the *trpE* gene in the operon.
 C. between the *trpC* and *trpB* genes of the operon.
 D. in the 3′ untranslated region of the operon.

2. If the *trp* attenuator is deleted from the operon, what will be the effect on expression?
 A. Expression will be eliminated.
 B. The operon will be constitutively expressed.
 C. Full expression of the operon will only occur in the absence of tryptophan.
 D. Full expression of the operon will only occur in the presence of tryptophan.

3. In the presence of tryptophan, the *trp* attenuator functions as:
 A. a rho-dependent terminator.
 B. an intrinsic terminator.
 C. a positive regulatory element.
 D. an autogenous regulatory element.

4. Deleting a sequence between the *trp* operator and the first structural gene in *E. coli* results in:
 A. increased expression of the *trp* structural genes independent of repression.
 B. decreased expression of the *trp* structural genes independent of activation.
 C. increased expression of the *trp* structural genes only when repressor is absent.
 D. decreased expression of the *trp* structural genes only when activator is present.

5. In *E. coli*, the *trp* operon is regulated by:
 A. attenuation.
 B. repression.
 C. both attenuation and repression.
 D. none of the above.

6. Can an attenuation mechanism similar to the *trp* operon function in eukaryotic nuclear gene expression?
   A. Yes, folding of the attenuator can result in termination of transcription in response to tryptophan in the cell.
   B. Yes, even though the transcript must be transported from the nucleus to the cytoplasm to be translated, the attenuator can still form to regulate expression.
   C. No, transcription and translation must be coupled for attenuation to function in regulation of gene expression.
   D. No, eukaryotic genes would not have a leader sequence with a separate initiation codon as is present for the *trp* operon.
7. At what level does antisense RNA function to inhibit gene expression?
   A. It may prevent transcription of the original gene in the cell.
   B. It may affect processing of the RNA product of the gene.
   C. It may alter translation of the mRNA.
   D. All of the above.
8. The *lin*4 microRNA in *C. elegans* regulates expression of the *lin*14 gene by:
   A. binding to complementary sequences in an early exon of *lin*14 mRNA.
   B. binding to complementary sequences in the first intron of *lin*14 mRNA.
   C. binding to complementary sequences in the 5′ untranslated region of *lin*14 mRNA.
   D. binding to complementary sequences in the 3′ untranslated region of *lin*14 mRNA.
9. Dicer cleaves double-stranded RNA molecules or regions to generate oligonucleotides of:
   A. 18–20 bases.
   B. 21–23 bases.
   C. 24–26 bases.
   D. 27–29 bases.

KEY TERMS

antisense RNA	**Dicer**	**riboswitch**	**RISC**
attenuation	**miRNA**	**ribozyme**	**sRNA**
attenuator	**ncRNA**	**RNA interference (RNAi)**	**siRNA**
CUT	**nested genes**	**RNA silencing**	**stRNA**

FURTHER READING

A collection of current news and a number of reviews about microRNAs and other regulatory RNAs in the RNA World can be found in *Science* 319, 1781–1799, 2008.

Proudfoot, N. and Gullerova, M. (2007). Gene silencing cuts both ways. *Cell* 131, 649–651. An interesting article about how noncoding RNA can control gene expression.

Tijsterman, M.,Ketting, R.E., and Plasterk, R. H. (2002). The genetics of RNA silencing. *Annu. Rev. Genet.* 36, 485–490.

14

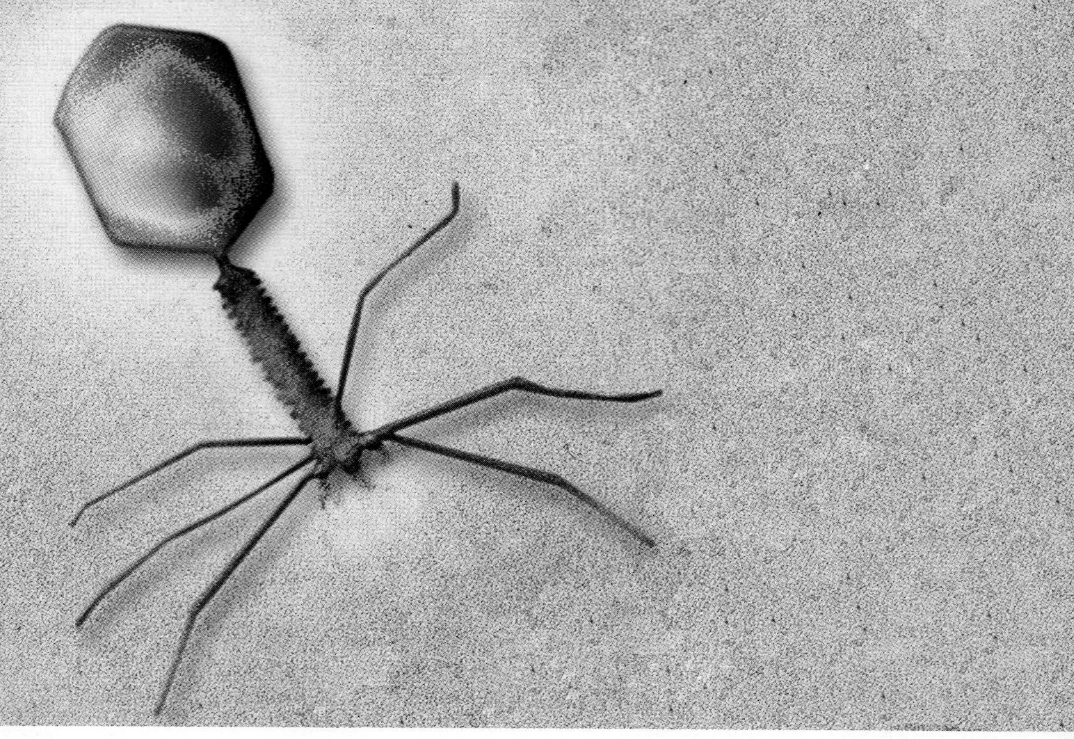

A transmission electron micrograph of a T4 phage. (Magnification: 450,000×.) © Dr. Harold Fisher/ Visuals Unlimited.

Phage Strategies

CHAPTER OUTLINE

14.1 Introduction

A virus consists of a nucleic acid genome contained in a protein coat. In order to reproduce, the virus must infect a host cell. The typical pattern of an infection is to subvert the functions of the host cell for the purpose of producing a large number of progeny viruses. Viruses that infect bacteria are generally called **bacteriophages**, often abbreviated to *phage* or simply ϕ. Usually a **phage** infection kills the bacterium. The process by which a phage infects a bacterium, reproduces itself, and then kills its host is called **lytic infection**. In the typical lytic cycle, the phage DNA (or RNA) enters the host bacterium, its genes are transcribed in a set order, the phage genetic material is replicated, and the protein components of the phage particle are produced. Finally, the host bacterium is broken open (*lysed*) to release the assembled progeny particles by the process of **lysis**. For some phages, called **virulent phages**, this is their only strategy for survival.

Other phages have a dual existence. They are able to perpetuate themselves via the same sort of lytic cycle in what amounts to an open strategy for producing as many copies of the phage as rapidly as possible. They also have an alternative form of existence, though, in which the phage genome is present in the bacterium in a latent form known as a **prophage**. This form of propagation is called **lysogeny** and the infected bacteria are known as *lysogens*. Phages that follow this pathway are called **temperate phages**.

In a lysogenic bacterium, the prophage is inserted into the bacterial genome and is inherited in the same way as bacterial genes. The process by which it is converted from an independent phage genome into a prophage that is a linear part of the bacterial genome is described as **integration**. By virtue of its possession of a prophage, a lysogenic bacterium has **immunity** against infection by other phage particles of the same type. Immunity is established by a single integrated prophage, so in general a bacterial genome contains only one copy of a prophage of any particular type.

There are transitions between the lysogenic and lytic modes of existence. FIGURE 14.1 shows that when a temperate phage produced by a lytic cycle enters a new bacterial host cell, it either repeats the lytic cycle or enters the lysogenic state. The outcome depends on the conditions of infection and the genotypes of phage and bacterium.

A prophage is freed from the restrictions of lysogeny by a process called **induction**. First the phage DNA is released from the bacterial chromosome by **excision**; then the free DNA proceeds through the lytic pathway.

The alternative forms in which these phages are propagated are determined by the regulation of transcription. Lysogeny is maintained by the interaction of a phage repressor with an operator. The lytic cycle requires a cascade of transcriptional controls. The transition between the two lifestyles is accomplished by the establishment of repression (lytic cycle to lysogeny) or by the relief of repression (induction of lysogen to lytic phage). These regulatory processes provide a wonderful example of how a series of relatively simple regulatory actions can be built up into complex developmental pathways.

14.2 Lytic Development Is Divided into Two Periods

Phage genomes by necessity are small. As with all viruses, they are restricted by the need to package the nucleic acid within the protein coat. This limitation dictates many of the viral strategies for reproduction. Typically a virus takes over the apparatus of the host cell, which then replicates and expresses phage genes instead of the bacterial genes.

Usually the phage has genes whose function is to ensure preferential replication of phage DNA. These genes are concerned with the initiation of replication and may even include a new DNA polymerase. Changes are introduced in the capacity of the host cell to engage in transcription. They involve replacing the RNA polymerase or modifying its capacity for initiation or termination. The result is always the

- **bacteriophage** A bacterial virus.
- **phage** An abbreviation of bacteriophage or bacterial virus.
- **lytic infection** Infection of a bacterium by a phage that ends in the destruction of the bacterium with release of progeny phage.
- **lysis** The death of bacteria at the end of a phage infective cycle when they burst open to release the progeny of an infecting phage (because phage enzymes disrupt the bacterium's cytoplasmic membrane or cell wall). The same term also applies to eukaryotic cells; for example, when infected cells are attacked by the immune system.
- **virulent phage** A bacteriophage that can only follow the lytic cycle.
- **prophage** A phage genome covalently integrated as a linear part of the bacterial chromosome.
- **lysogeny** The ability of a phage to survive in a bacterium as a stable prophage component of the bacterial genome.
- **temperate phage** A bacteriophage that can follow the lytic or lysogenic pathway.
- **integration** Insertion of a viral or another DNA sequence into a host genome as a region covalently linked on either side to the host sequences.
- **immunity** In phages, the ability of a prophage to prevent another phage of the same type from infecting a cell. In plasmids, the ability of a plasmid to prevent another of the same type from becoming established in a cell. It can also refer to the ability of certain transposons to prevent others of the same type from transposing to the same DNA molecule.
- **induction of phage** A phage's entry into the lytic (infective) cycle as a result of destruction of the lysogenic repressor, which leads to excision of free phage DNA from the bacterial chromosome.
- **excision** Release of phage from the host chromosome as an autonomous DNA molecule.

FIGURE 14.1 Lytic development involves the reproduction of phage particles with destruction of the host bacterium, but lysogenic existence allows the phage genome to be carried as part of the bacterial genetic information.

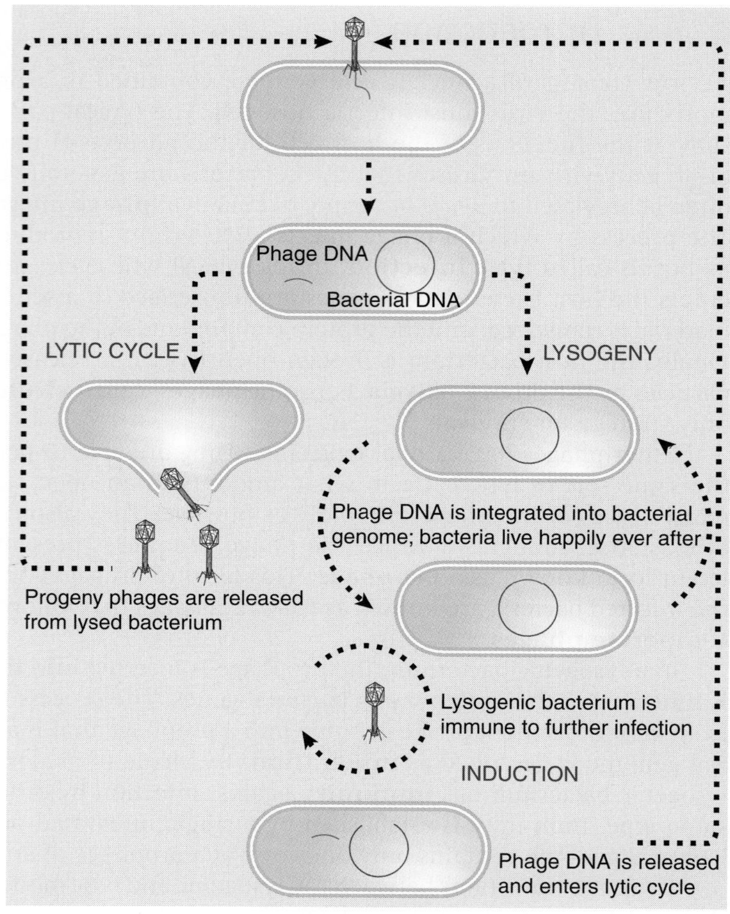

Phage DNA
Bacterial DNA
LYTIC CYCLE
LYSOGENY
Phage DNA is integrated into bacterial genome; bacteria live happily ever after
Progeny phages are released from lysed bacterium
Lysogenic bacterium is immune to further infection
INDUCTION
Phage DNA is released and enters lytic cycle

same: phage mRNAs are preferentially transcribed. As far as protein synthesis is concerned, the phage is, for the most part, content to use the host apparatus, redirecting its activities principally by replacing bacterial mRNA with phage mRNA.

Lytic development is accomplished by a pathway in which the phage genes are expressed in a particular order. This ensures that the right amount of each component is present at the appropriate time. The cycle can be divided into the two general parts illustrated in FIGURE 14.2:

- **Early infection** describes the period from entry of the DNA to the start of its replication.
- **Late infection** defines the period from the start of replication to the final step of lysing the bacterial cell to release progeny phage particles.

early infection The part of the phage lytic cycle between entry and replication of the phage DNA. During this time, the phage synthesizes the enzymes needed to replicate its DNA.

late infection The part of the phage lytic cycle from DNA replication to lysis of the cell. During this time, the DNA is replicated and structural components of the phage particle are synthesized.

The early phase is devoted to the production of enzymes involved in the reproduction of DNA. These include the enzymes concerned with DNA synthesis, recombination, and sometimes modification. Their activities cause a *pool* of phage genomes to accumulate. In this pool, genomes are continually replicating and recombining, so that *the events of a single lytic cycle concern a population of phage genomes.*

During the late phase, the protein components of the phage particle are synthesized. Often many different proteins are needed to make up head and tail structures, so the largest part of the phage genome consists of late functions. In addition to the structural proteins, "assembly proteins" are needed to help construct the particle, although they are not incorporated into it themselves. By the time the structural components are assembling into heads and tails, replication of DNA has reached its maximum rate. The genomes then are inserted into the empty protein heads, tails are added, and the host cell is lysed to allow release of new viral particles.

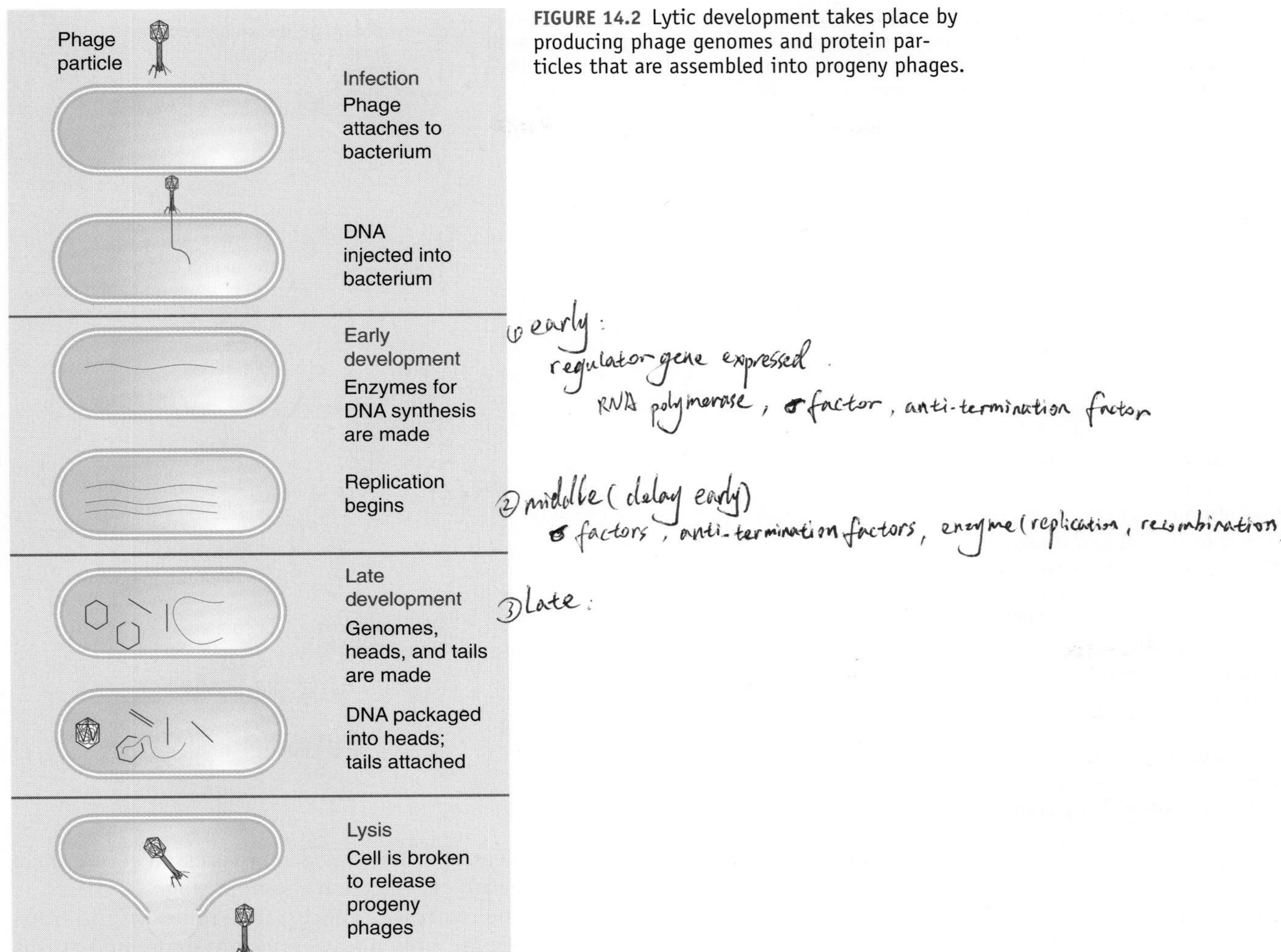
Phage particle
Infection
Phage attaches to bacterium
DNA injected into bacterium
Early development
Enzymes for DNA synthesis are made
Replication begins
Late development
Genomes, heads, and tails are made
DNA packaged into heads; tails attached
Lysis
Cell is broken to release progeny phages

FIGURE 14.2 Lytic development takes place by producing phage genomes and protein particles that are assembled into progeny phages.

KEY CONCEPTS

- A phage infective cycle is divided into the early period (before replication) and the late period (after the onset of replication).
- A phage infection generates a pool of progeny phage genomes that replicate and recombine.

CONCEPT AND REASONING CHECK

Most phages are virulent; therefore, why might it be advantageous to be temperate?

14.3 Lytic Development Is Controlled by a Cascade

The organization of the phage genetic map often reflects the sequence of lytic development. The concept of the operon is taken to somewhat of an extreme, in which the genes coding for proteins with related functions are clustered to allow their control with the maximum economy. This allows the pathway of lytic development to be controlled with a small number of regulatory switches.

The lytic cycle is under positive control, so that each group of phage genes can be expressed only when an appropriate signal is given. **FIGURE 14.3** is an overview showing

FIGURE 14.3 Phage lytic development proceeds by a regulatory cascade, in which a gene product at each stage is needed for expression of the genes at the next stage.

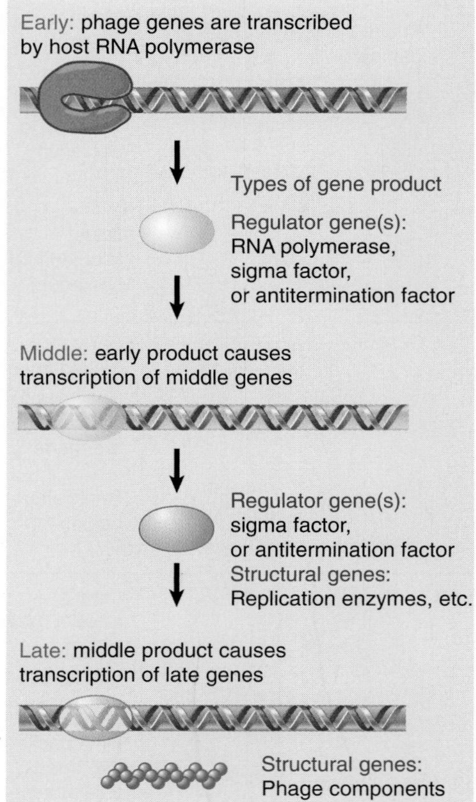

Early: phage genes are transcribed by host RNA polymerase
Types of gene product
Regulator gene(s): RNA polymerase, sigma factor, or antitermination factor
Middle: early product causes transcription of middle genes
Regulator gene(s): sigma factor, or antitermination factor
Structural genes: Replication enzymes, etc.
Late: middle product causes transcription of late genes
Structural genes: Phage components

- **cascade** A sequence of events, each of which is stimulated by the previous one. In transcriptional regulation, as seen in sporulation and phage lytic development, it means that regulation is divided into stages, and at each stage, one of the genes that is expressed codes for a regulator needed to express the genes of the next stage.
- **early genes** Genes that are transcribed before the replication of phage DNA. They code for regulators and other proteins needed for later stages of infection.
- **immediate early genes** Genes in phage lambda that are equivalent to the early class of other phages. They are transcribed immediately upon infection by the host RNA polymerase.
- **delayed early genes** Genes in phage lambda that are equivalent to the middle genes of other phages. They cannot be transcribed until regulator protein(s) coded by the immediate early genes have been synthesized.
- **middle genes** Phage genes that are regulated by the proteins coded by early genes. Some proteins coded by them catalyze replication of the phage DNA; others regulate the expression of a later set of genes.
- **late gene** A gene transcribed when phage DNA is being replicated. It code for components of the phage particle.

that the regulatory genes function in a **cascade**, in which a gene expressed at one stage is necessary for synthesis of the genes that are expressed at the next stage.

The early part of the first stage of gene expression necessarily relies on the transcription apparatus of the host cell. In general, only a few genes are expressed at this time. Their promoters are indistinguishable from those of host genes. The name of this class of genes depends on the phage. In most cases, they are known as the **early genes**. In phage lambda, they are given the evocative description of **immediate early genes**. Irrespective of the name, they constitute only a preliminary set of genes, representing just the initial part of the early period. Sometimes they are exclusively occupied with the transition to the next period. At all cases, *one of these genes always codes for a protein, a gene regulator that is necessary for transcription of the next class of genes.*

This next class of genes in the early stage is known variously as the **delayed early** or **middle gene** group. Its expression typically starts as soon as the regulator protein coded by the early gene(s) is available. Depending on the nature of the control circuit, the initial set of early genes may or may not continue to be expressed at this stage. Often, the expression of host genes is reduced. Together the two sets of early genes account for all necessary phage functions except those needed to assemble the particle coat itself and to lyse the cell.

When the replication of phage DNA begins, it is time for the **late genes** to be expressed. Their transcription at this stage usually is arranged by embedding an additional regulator gene within the previous (delayed early or middle) set of genes. This regulator may be another antitermination factor (as in lambda) or it may be another sigma factor (as in SPO1).

A lytic infection often falls into the stages described above, beginning with the early genes transcribed by host RNA polymerase (sometimes the regulators are the only products at this stage). This stage is followed by those genes transcribed under the direction of the regulator produced in the first stage (most of these genes code for enzymes needed for replication of phage DNA). The final stage consists of genes for phage components, which are transcribed under the direction of a regulator synthesized in the second stage.

The use of these successive controls, in which each set of genes contains a regulator that is necessary for expression of the next set, creates a cascade in which groups of genes are turned on (and sometimes off) at particular times. The means used to construct each phage cascade are different but the results are similar.

KEY CONCEPTS

- The early genes transcribed by host RNA polymerase following infection include, or comprise, regulators required for expression of the middle set of phage genes.
- The middle group of genes includes regulators to transcribe the late genes.
- This results in the ordered expression of groups of genes during phage infection.

CONCEPT AND REASONING CHECK

Why would a phage benefit from using a cascade of gene regulators instead of transcribing all of its genes at once? less gene, more effecient

14.4 Two Types of Regulatory Events Control the Lytic Cascade

At every stage of phage expression, one or more of the active genes is a regulator that is needed for the subsequent stage. The regulator may take the form of a new sigma factor that redirects the specificity of the host RNA polymerase (see *Section 11.12, Sigma Factors Directly Contact DNA*) or an antitermination factor that allows it to read a new group of genes (see *Section 11.16, Antitermination May Be a Regulated Event*). Now, let's compare the use of switching at initiation or termination to control gene expression.

FIGURE 14.4 shows that phages use two types of mechanisms for recognizing new phage promoters. One is to replace the sigma factor of the host enzyme with another factor that redirects its specificity in initiation. An alternative is to synthesize a new phage RNA polymerase. In either case, the critical feature that distinguishes the new set of genes is their possession of *different promoters from those originally recognized by host RNA polymerase.* **FIGURE 14.5** shows that the two sets of transcripts are independent; as a consequence, early gene expression can cease after the new sigma factor or polymerase has been produced.

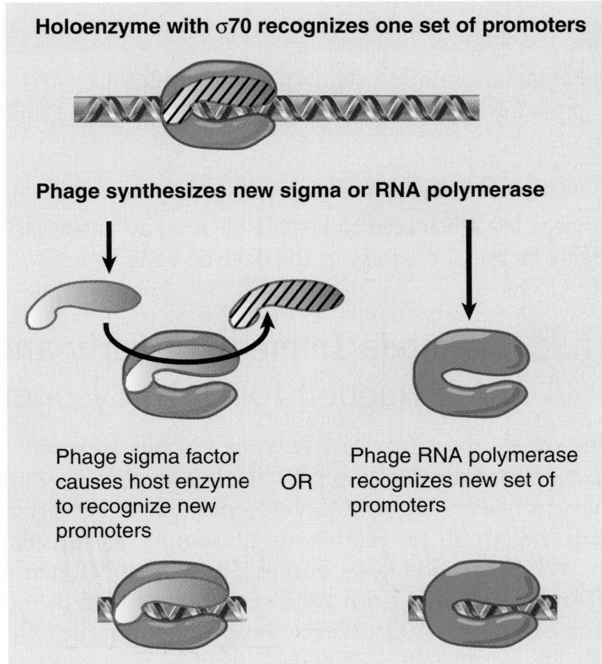
Holoenzyme with σ70 recognizes one set of promoters
Phage synthesizes new sigma or RNA polymerase
Phage sigma factor causes host enzyme to recognize new promoters
OR
Phage RNA polymerase recognizes new set of promoters

FIGURE 14.4 A phage may control transcription at initiation either by synthesizing a new sigma factor that replaces the host sigma factor or by synthesizing a new RNA polymerase.

Antitermination provides an alternative mechanism for phages to control the switch from early genes to the next stage of expression. The use of antitermination depends on a particular arrangement

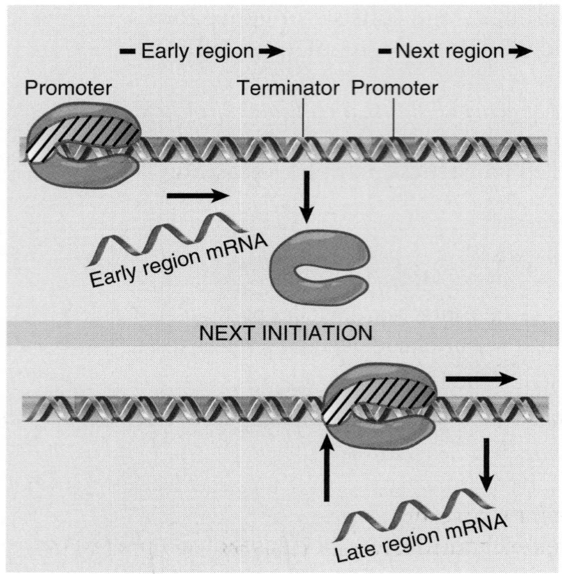

FIGURE 14.5 Control at initiation utilizes independent transcription units, each with its own promoter and terminator, which produce independent mRNAs. The transcription units need not be located near one another.

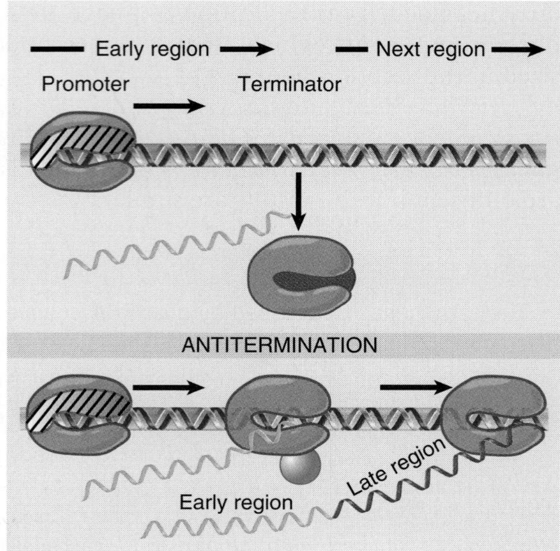

FIGURE 14.6 Control at termination requires adjacent units, so that transcription can read from the first gene into the next gene. This produces a single mRNA that contains both sets of genes.

of genes. FIGURE 14.6 shows that the early genes lie adjacent to the genes that are to be expressed next, but are separated from them by terminator sites. *If termination is prevented at these sites, the polymerase reads through into the genes on the other side.* So in antitermination, the *same promoters* continue to be recognized by RNA polymerase. The new genes are expressed only by extending the RNA chain to form molecules that contain the early gene sequences at the 5′ end and the new gene sequences at the 3′ end. Because the two types of sequences remain linked, early gene expression inevitably continues.

KEY CONCEPT

- Regulator proteins used in phage cascades may sponsor initiation at new (phage) promoters or cause the host polymerase to read through transcription terminators.

CONCEPT AND REASONING CHECK

What is the advantage to lambda of using an antitermination mechanism instead of a new sigma factor to progress to the delayed early genes?

14.5 Lambda Immediate Early and Delayed Early Genes Are Needed for Both Lysogeny and the Lytic Cycle

One of the most intricate cascade circuits is provided by phage lambda. Actually, the cascade for lytic development itself is straightforward, with two regulators controlling the successive stages of development. But the circuit for the lytic cycle is interlocked with the circuit for establishing lysogeny, as summarized in FIGURE 14.7.

When lambda DNA enters a new host cell, the lytic and lysogenic pathways start off the same way. Both require expression of the immediate early and delayed early genes, but then they diverge: lytic development follows if the late genes are expressed, and lysogeny ensues if synthesis of a gene regulator called the lambda repressor is established by turning on its gene, the *cI* gene.

Lambda has only two immediate early genes, transcribed independently by host RNA polymerase:

- The *N gene* codes for an antitermination factor whose action at the *nut* sites allows transcription to proceed into the delayed early genes (see *Section 11.16, Antitermination May Be a Regulated Event*).
- The *cro gene* has dual functions: it codes for a repressor that prevents expression of the *cI* gene coding for the lambda repressor (essentially de-repressing the late genes, a necessary action if the lytic cycle is to proceed), and it turns off expression of the immediate early genes (which are not needed later in the lytic cycle).

The delayed early genes, turned on by *N*, include two replication genes (needed for lytic infection), seven recombination genes (some involved in recombination during lytic infection, and two necessary to integrate lambda DNA into the bacterial chromosome for lysogeny), and three regulator genes. These regulator genes have opposing functions:

- The *cII-cIII* pair of regulator genes is needed to establish the synthesis of the lambda repressor.
- The *Q* regulator gene codes for an antitermination factor that allows host RNA polymerase to transcribe the late genes and is necessary for the lytic cycle.

Thus the delayed early genes serve two masters: some are needed for the phage to enter lysogeny, and the others are concerned with controlling the order of the lytic cycle. At this point, lambda is keeping open the option to choose either pathway.

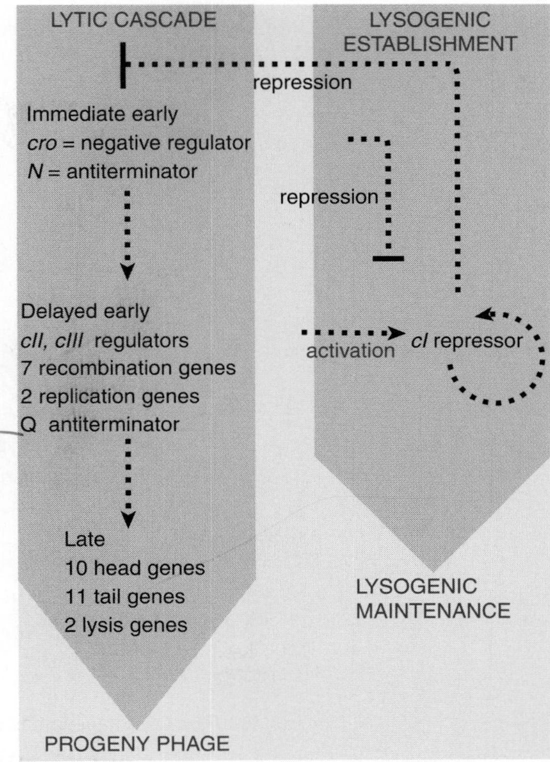

FIGURE 14.7 The lambda lytic cascade is interlocked with the circuitry for lysogeny.

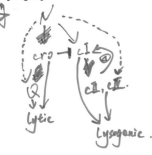

KEY CONCEPTS

- Lambda has two immediate early genes, *N* and *cro*, which are transcribed by host RNA polymerase.
- The *N* gene is required to express the delayed early genes.
- Three of the delayed early genes are regulators.
- Lysogeny requires the delayed early genes *cII-cIII*.
- The lytic cycle requires the immediate early gene *cro* and the delayed early gene *Q*.

CONCEPT AND REASONING CHECK

What would be the effect if the *N* gene were mutant?

14.6 The Lytic Cycle Depends on Antitermination by pN

To disentangle the lytic and lysogenic pathways, let's first consider just the lytic cycle. **FIGURE 14.8** gives the map of lambda phage DNA. A group of genes concerned with regulation is surrounded by genes needed for recombination and replication. The genes coding for structural components of the phage are clustered. All of the genes necessary for the lytic cycle are expressed in polycistronic transcripts from three promoters.

FIGURE 14.9 shows that the two immediate early genes, *N* and *cro*, are transcribed by host RNA polymerase. *N* is transcribed toward the left and *cro* toward the right. Each transcript is

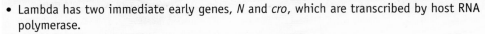

Promoters for the lytic cycle $P_L P_R$ $P_{R'}$

Head genes Tail genes Recombination Regulation Replication Lysis

AWBCNu3DEF$_I$F$_{II}$ZUVGTHMLKIJ att int xis αβγ cIII N cI cro cII O P QSR

Required for:	
lysogeny	*cIII* maintains *cII*
lysogeny and lysis	*N* turns on delayed early
lysogeny	*cI* is lysogenic repressor
lysis	*cro* turns off repressor
lysogeny	*cII* turns on repressor
lysis	*Q* turns on late

FIGURE 14.8 The lambda map shows clustering of related functions. The genome is 48,514 bp.

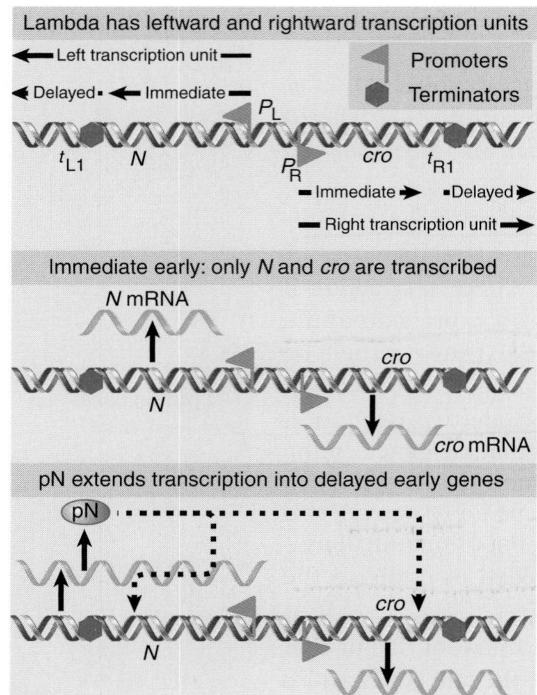

FIGURE 14.9 Phage lambda has two early transcription units. In the "leftward" unit, the "upper" strand is transcribed toward the left; in the "rightward" unit, the "lower" strand is transcribed toward the right. Genes *N* and *cro* are the immediate early functions and are separated from the delayed early genes by the terminators. Synthesis of N protein allows RNA polymerase to pass the terminators t_{L1} to the left and t_{R1} to the right.

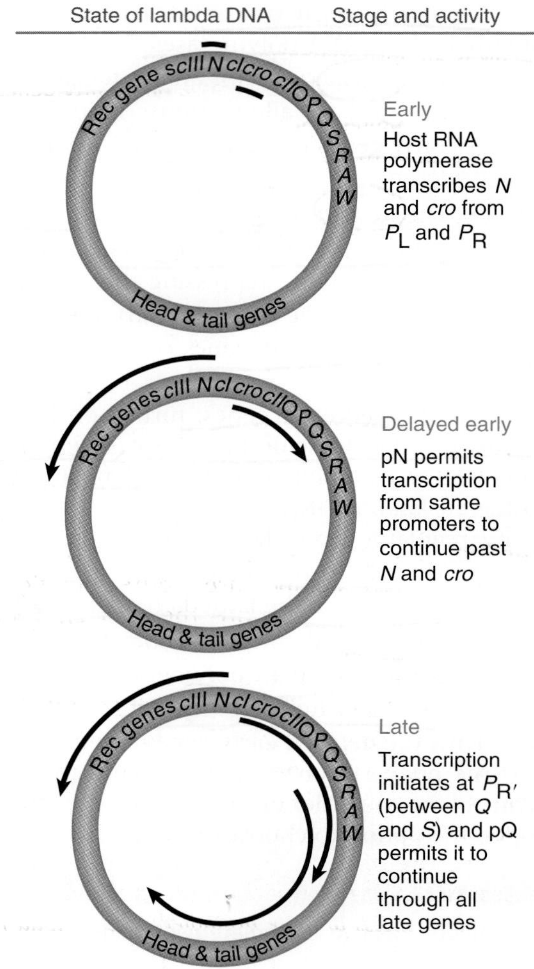

FIGURE 14.10 Lambda DNA circularizes during infection, so that the late gene cluster is intact in one transcription unit.

terminated at the end of the gene. The protein pN is the regulator, the antitermination factor that allows transcription to continue into the delayed early genes by suppressing use of the terminators t_L and t_R (see *Section 11.16, Antitermination May Be a Regulated Event*). In the presence of pN, transcription continues to the left of the *N* gene into the recombination genes and to the right of the *cro* gene into the replication genes.

The map in Figure 14.8 gives the organization of the lambda DNA as it exists in the phage particle. Shortly after infection, though, the ends of the DNA join to form a circle. **FIGURE 14.10** shows the true state of lambda DNA during infection. The late genes are welded into a single group, which contain the lysis genes *S-R* from the right end of the linear DNA and the head and tail genes *A-J* from the left end.

The late genes are expressed as a single transcription unit, starting from a promoter $P_{R'}$ that lies between *Q* and *S*. The late promoter is used constitutively. However, in the absence of the product of gene *Q* (which is the last gene in the rightward delayed early unit), late transcription terminates at a site t_{R3}. The transcript resulting from this termination event is 194 bases long; it is known as 6S RNA. When pQ becomes available, it suppresses termination at t_{R3} and the 6S RNA is extended, with the result that the late genes are expressed.

KEY CONCEPTS

- pN is an antitermination factor that allows RNA polymerase to continue transcription past the ends of the two immediate early genes.
- pQ is the product of a delayed early gene and is an antiterminator that allows RNA polymerase to transcribe the late genes.
- Lambda DNA circularizes after infection; as a result, the late genes form a single transcription unit.

CONCEPT AND REASONING CHECK

Predict what would happen to a cell infected by a lambda phage with a mutant *cI* gene.

14.7 Lysogeny Is Maintained by the Lambda Repressor Protein

Looking at the lambda lytic cascade, we see that the entire program is set in motion by the initiation of transcription at the two promoters P_L and P_R for the immediate early genes *N* and *cro*. Lambda uses antitermination to proceed to the next stage of (delayed early) expression; therefore, the same two promoters continue to be used throughout the early period.

The expanded map of the regulatory region drawn in **FIGURE 14.11** shows that the promoters P_L and P_R lie on either side of the *cI* gene. Associated with each promoter is an operator (O_L, O_R) at which repressor protein binds to prevent RNA polymerase from initiating transcription. The sequence of each operator overlaps with the promoter that it controls, and because this occurs so often these sequences are described as the P_L/O_L and P_R/O_R control regions.

Because of the sequential nature of the lytic cascade, the control regions provide a pressure point at which entry to the entire cycle can be controlled. *By denying RNA polymerase access to these promoters, the lambda repressor protein prevents the phage genome from entering the lytic cycle.* The lambda repressor functions in the same way as repressors of bacterial operons: it binds to specific operators.

The lambda repressor protein is encoded by the *cI* gene. Note in Figure 14.11 that the *cI* gene has two promoters, P_{RM} (promoter right maintenance) and P_{RE} (promoter right establishment). Mutants in this gene cannot maintain lysogeny, but always enter the lytic cycle. In the time since the original isolation of the lambda repressor protein, the characterization of the repressor protein has shown how it both maintains the lysogenic state and provides immunity for a lysogen against superinfection by new phage lambda genomes.

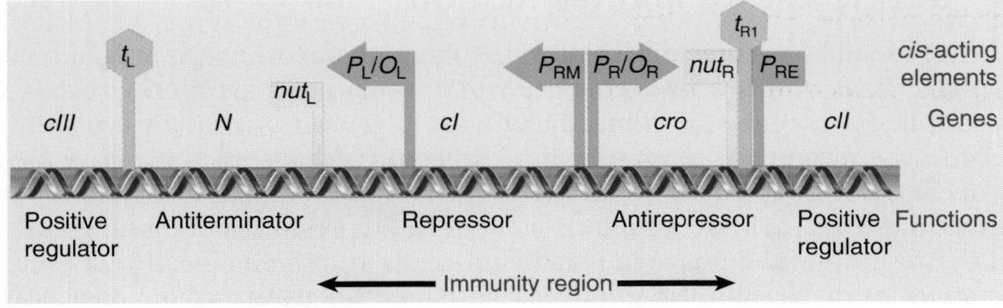

FIGURE 14.11 The lambda regulatory region contains a cluster of *trans*-acting functions and *cis*-acting elements.

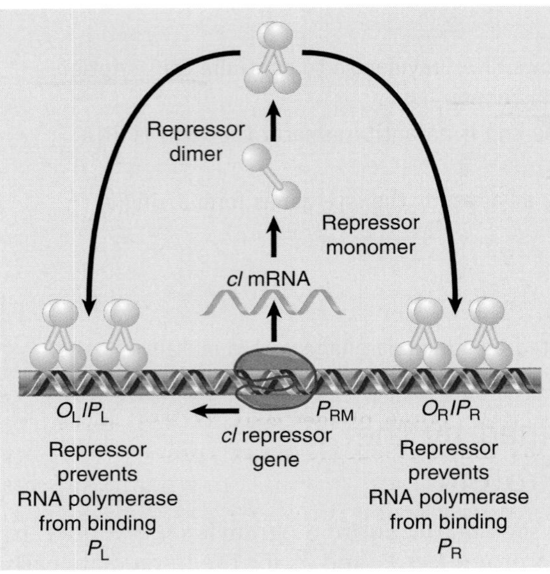

FIGURE 14.12 Repressor acts at the left operator and right operator to prevent transcription of the immediate early genes (*N* and *cro*). It also acts at the promoter P_{RM} to activate transcription by RNA polymerase of its own gene.

The lambda repressor binds independently to the two operators. Its ability to repress transcription at the associated promoters is illustrated in **FIGURE 14.12**.

At O_L the lambda repressor has the same sort of effect that we have already discussed for several other systems: it prevents RNA polymerase from initiating transcription at P_L. This stops the expression of gene *N*. Because P_L is used for all leftward early gene transcription, this action prevents expression of the entire leftward early transcription unit. So the lytic cycle is blocked before it can proceed beyond early stages.

At O_R, repressor binding prevents the use of P_R so *cro* and the other rightward early genes cannot be expressed. The lambda repressor protein binding at O_R also stimulates transcription of *cI*, its own gene from P_{RM}.

The nature of this control circuit explains the biological features of lysogenic existence. Lysogeny is stable because the control circuit ensures that, so long as the level of lambda repressor is adequate, there is continued expression of the *cI* gene. The result is that O_L and O_R remain occupied indefinitely. By repressing the entire lytic cascade, this action maintains the prophage in its inert form.

KEY CONCEPTS

- The lambda repressor, encoded by the *cI* gene, is required to maintain lysogeny.
- The lambda repressor acts at the O_L and O_R operators to block transcription of the immediate early genes.
- The immediate early genes trigger a regulatory cascade; as a result, their repression prevents the lytic cycle from proceeding.

CONCEPT AND REASONING CHECK

Predict what would happen to a cell infected by a lambda phage with a mutant O_R.

14.8 The Lambda Repressor and Its Operators Define the Immunity Region

The presence of lambda repressor explains the phenomenon of *immunity*. If a second lambda phage DNA enters a lysogenic cell, repressor protein synthesized from the resident prophage genome will immediately bind to O_L and O_R in the new genome. This prevents the second phage from entering the lytic cycle.

▸ **virulent mutations** Phage mutants that are unable to establish lysogeny.

The operators were originally identified as the targets for repressor action by **virulent mutations** (λvir). These mutations prevent the repressor from binding at O_L or O_R, with the result that the phage inevitably proceeds into the lytic pathway when it infects a new host bacterium. Note that λvir mutants can grow on lysogens because the virulent mutations in O_L and O_R allow the incoming phage to ignore the resident repressor and thus enter the lytic cycle. Virulent mutations in phages are the equivalent of operator-constitutive mutations in bacterial operons.

A prophage is induced to enter the lytic cycle when the lysogenic circuit is broken. This happens when the repressor is inactivated (see *Section 14.9, The DNA-Binding Form of the Lambda Repressor Is a Dimer*). The absence of repressor allows RNA polymerase to bind at P_L and P_R, starting the lytic cycle as shown in the lower part of **FIGURE 14.13**.

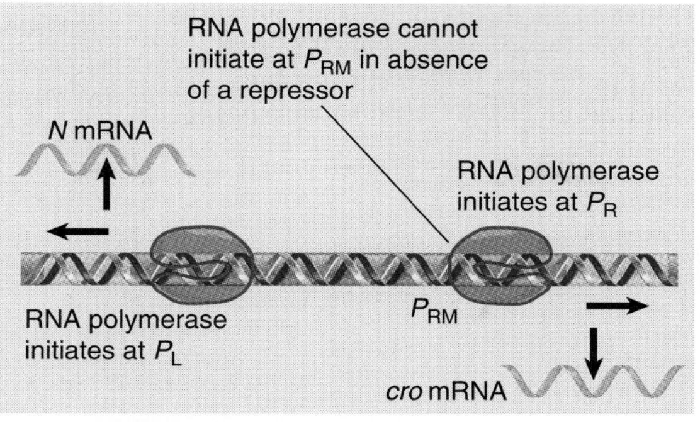

FIGURE 14.13 In the absence of repressor, RNA polymerase initiates at the left and right promoters. It cannot initiate at P_{RM} in the absence of repressor.

The region including the left and right operators, the *cI* gene, and the *cro* gene determines the immunity of the phage. Any phage that possesses this region has the same type of immunity, because *it specifies both the repressor protein and the sites on which the repressor acts.* Accordingly, this is called the **immunity region** (as marked in Figure 14.11). Each of the four lambdoid phages φ80, *21*, *434*, and λ has a unique immunity region. When we say that a lysogenic phage confers immunity to any other phage of the same type, we mean more precisely that the immunity is to any other phage that has the same immunity region (irrespective of differences in other regions).

immunity region A segment of the phage genome that enables a prophage to inhibit additional phage of the same type from infecting the bacterium. This region has a gene that encodes for the repressor, as well as the sites to which the repressor binds.

KEY CONCEPTS

- Several lambdoid phages have different immunity regions.
- A lysogenic phage confers immunity to further infection by any other phage with the same immunity region.

CONCEPT AND REASONING CHECK

Why can't another lambda phage infect a lysogenic bacterium?

14.9 The DNA-Binding Form of the Lambda Repressor Is a Dimer

The lambda repressor subunit is a polypeptide of 27 kD with the two distinct domains summarized in **FIGURE 14.14**.

- The N-terminal domain, residues 1–92, provides the operator-binding site.
- The C-terminal domain, residues 132–236, is responsible for dimerization.

Each domain can exercise its function independently of the other. The C-terminal fragment can form oligomers. The N-terminal fragment can bind the operators, although with a lower affinity than the intact lambda repressor. Thus the information for specifically contacting DNA is contained within the N-terminal domain, but the efficiency of the process is enhanced by the attachment of the C-terminal domain.

The dimeric structure of the lambda repressor is crucial in maintaining lysogeny. The induction of a lysogenic prophage to enter the lytic cycle is caused by cleavage of the repressor subunit in the connector region, between residues 111 and 113. (This is a counterpart to the allosteric change in conformation that results when a small-molecule inducer inactivates the repressor of a bacterial operon, a capacity that the lysogenic repressor does not have.) Induction occurs under certain adverse conditions, such as exposure of lysogenic bacteria to UV irradiation, which leads to proteolytic inactivation of the repressor.

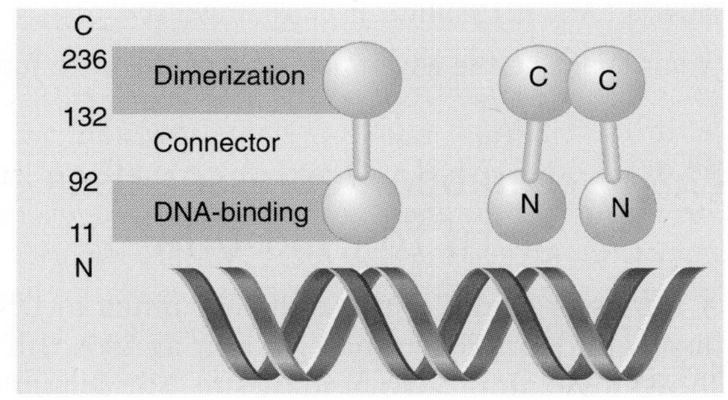

FIGURE 14.14 The N-terminal and C-terminal regions of repressor form separate domains. The C-terminal domains associate to form dimers; the N-terminal domains bind DNA.

In the intact state, dimerization of the C-terminal domains ensures that when the repressor binds to

FIGURE 14.15 Repressor dimers bind to the operator. The affinity of the N-terminal domains for DNA is controlled by the dimerization of the C-terminal domains.

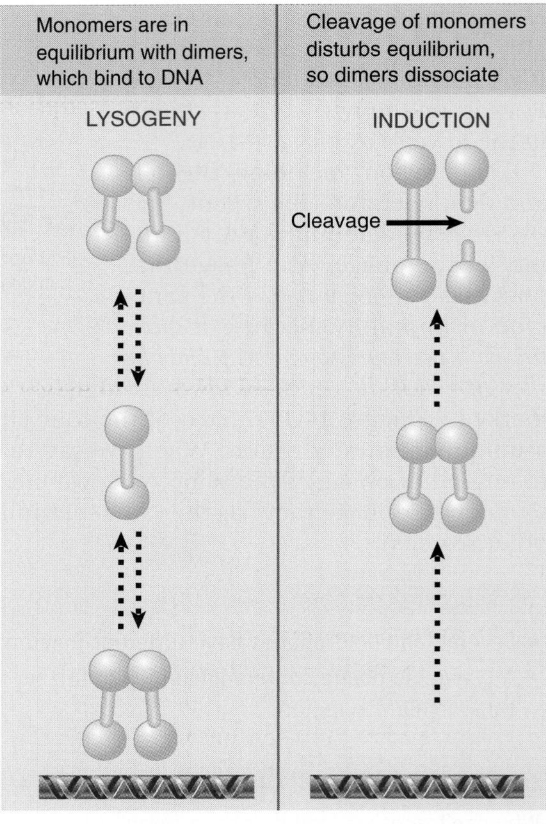
Monomers are in equilibrium with dimers, which bind to DNA
Cleavage of monomers disturbs equilibrium, so dimers dissociate
LYSOGENY
INDUCTION
Cleavage

DNA, its two N-terminal domains each contact DNA simultaneously. But cleavage releases the C-terminal domains from the N-terminal domains. As illustrated in FIGURE 14.15, this means that the N-terminal domains can no longer dimerize; as a result, they do not have sufficient affinity for the lambda repressor to remain bound to DNA. Also, two dimers usually cooperate to bind at an operator, and the cleavage destabilizes this interaction. The balance between lysogeny and the lytic cycle depends on the concentration of repressor.

KEY CONCEPTS

- A repressor monomer has two distinct domains.
- The N-terminal domain contains the DNA-binding site.
- The C-terminal domain dimerizes.
- Binding to the operator requires the dimeric form so that two DNA-binding domains can contact the operator simultaneously.
- Cleavage of the repressor between the two domains reduces the affinity for the operator and induces a lytic cycle.

CONCEPT AND REASONING CHECK

What might be the advantage for a regulator to function as a dimer?

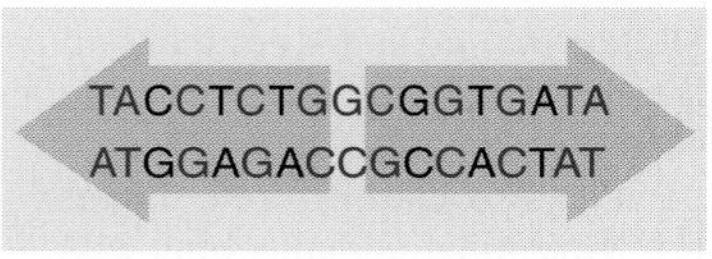

TACCTCTGGCGGTGATA
ATGGAGACCGCCACTAT

FIGURE 14.16 The operator is a 17-bp sequence with an axis of symmetry through the central base pair. Each half-site is marked in light blue. Base pairs that are identical in each operator half are in dark blue.

14.10 Lambda Repressor Uses a Helix-Turn-Helix Motif to Bind DNA

A repressor dimer is the unit that binds to DNA. It recognizes a sequence of 17 bp displaying partial symmetry about an axis through the central base pair. FIGURE 14.16 shows an example of a binding site. The sequence on each side of the central base pair is sometimes called a "half-site." Each individual N-terminal region contacts a half-site. Several DNA-binding proteins that regulate bacterial transcription share a similar mode of holding DNA, in which the active domain contains two short regions of α-helix that contact DNA. (Some transcription factors in eukaryotic cells use a similar motif. See *Section 26.5, There Are Many Types of DNA-Binding Domains.*)

The N-terminal domain of lambda repressor contains several stretches of α-helix, which are arranged as illustrated diagrammatically in **FIGURE 14.17**. Two of the helical regions are responsible for binding DNA. The **helix-turn-helix** model for contact is illustrated in **FIGURE 14.18**. Looking at a single monomer, α-helix-3 consists of nine amino acids, each of which lies at an angle to the preceding region of seven amino acids that forms α-helix-2. In the dimer, the two apposed helix-3 regions lie 34 Å apart, enabling them to fit into successive major grooves of DNA. The helix-2 regions lie at an angle that would place them across the groove. The symmetrical binding of dimer to the site means that each N-terminal domain of the dimer contacts a similar set of bases in its half-site.

▸ **helix-turn-helix** The motif that describes an arrangement of two α-helices that form a site that binds to DNA, one fitting into the major groove of DNA and the other lying across it.

Related forms of the α-helical motifs employed in the helix-turn-helix of the lambda repressor are found in several DNA-binding proteins, including catabolite repressor protein (CRP), the *lac* repressor, and several other phage repressors. By comparing the abilities of these proteins to bind DNA, we can define the roles of each helix:

- Contacts between helix-2 and helix-3 are maintained by interactions between hydrophobic amino acids.
- Contacts between helix-3 and DNA rely on hydrogen bonds between the amino acid side chains and the exposed positions of the base pairs. This helix is responsible for recognizing the specific target DNA sequence and is therefore also known as the **recognition helix**. By comparing the contact patterns summarized in **FIGURE 14.19**, we see that the lambda repressor and Cro

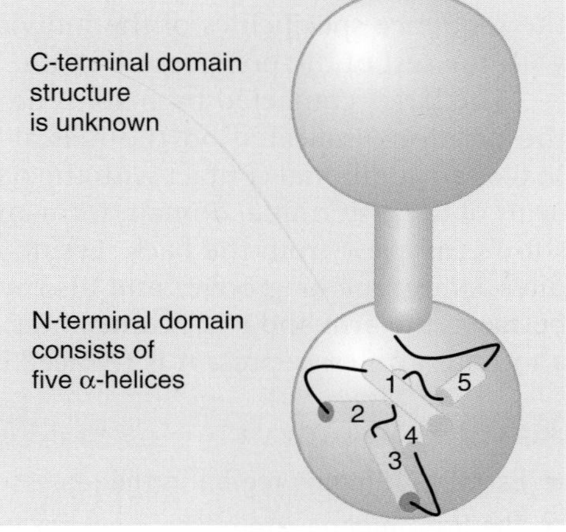

FIGURE 14.17 Lambda repressor's N-terminal domain contains five stretches of α-helix; helices 2 and 3 bind DNA.

▸ **recognition helix** One of the two helices of the helix-turn-helix motif that makes contacts with DNA that are specific for particular bases. This determines the specificity of the DNA sequence that is bound.

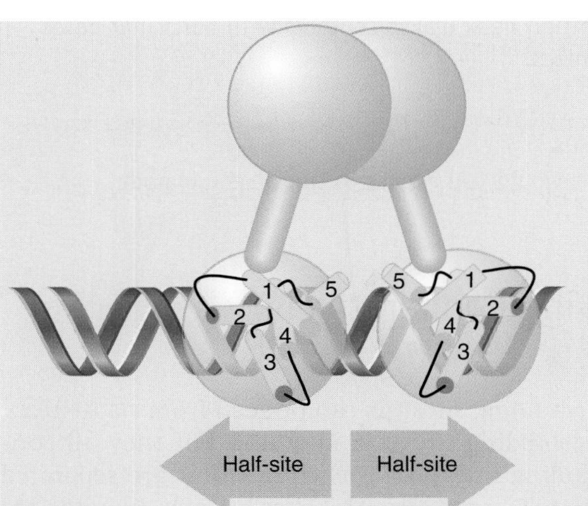

FIGURE 14.18 In the two-helix model for DNA binding, helix-3 of each monomer lies in the wide groove on the same face of DNA, and helix-2 lies across the groove.

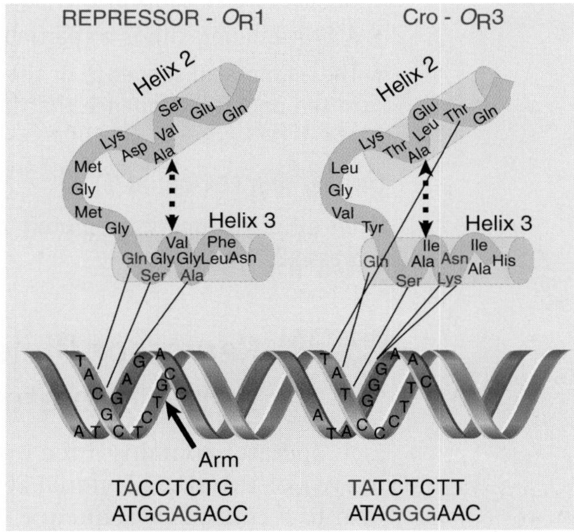

FIGURE 14.19 Two proteins that use the two-helix arrangement to contact DNA recognize lambda operators with affinities determined by the amino acid sequence of helix-3.

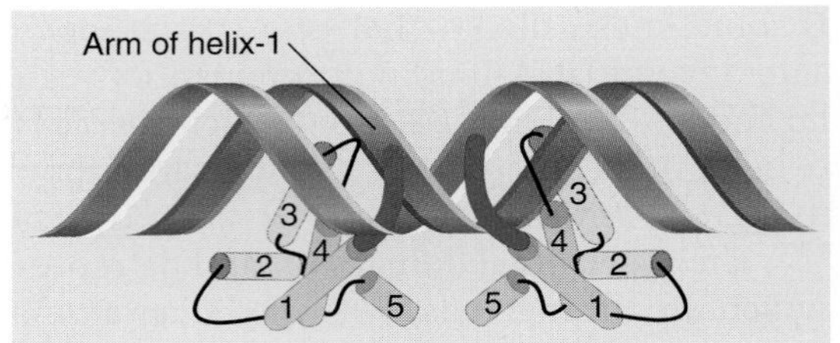

FIGURE 14.20 A view from the back shows that the bulk of the repressor contacts one face of DNA, but its N-terminal arms reach around to the other face.

select different sequences in the DNA as their most favored targets because they have different amino acids in the corresponding positions in helix-3.

- Contacts from helix-2 to the DNA take the form of hydrogen bonds connecting with the phosphate backbone. These interactions are necessary for binding, but do not control the specificity of target recognition. In addition to these contacts, a large part of the overall energy of interaction with DNA is provided by ionic interactions with the phosphate backbone.

What happens if we manipulate the coding sequence to construct a new protein by substituting the recognition helix in one repressor with the corresponding sequence from a closely related repressor? The specificity of the hybrid protein is that of its new recognition helix. The amino acid sequence of this short region determines the sequence specificities of the individual proteins and is able to act in conjunction with the rest of the polypeptide chain.

The bases contacted by helix-3 lie on one face of the DNA, as can be seen from the positions indicated on the helical diagram in Figure 14.19. However, repressor makes an additional contact with the other face of DNA. The last six N-terminal amino acids of the N-terminal domain form an "arm" extending around the back. **FIGURE 14.20** shows the view from the back. Lysine residues in the arm make contact with G residues in the major groove, and also with the phosphate backbone. The interaction between the arm and DNA contributes heavily to DNA binding; the binding affinity of a mutant armless repressor is reduced by ~1000 fold.

KEY CONCEPTS

- Each DNA-binding region in the repressor contacts a half-site in the DNA.
- The DNA-binding site of the repressor includes two short α-helical regions that fit into the successive turns of the major groove of DNA.
- A DNA-binding site is a (partially) palindromic sequence of 17 bp.
- The amino acid sequence of the recognition helix makes contacts with particular bases in the operator sequence that it recognizes.

CONCEPT AND REASONING CHECK

What different functional protein domains would you expect to find in the lambda repressor?

14.11 Repressor Dimers Bind Cooperatively to the Operator

Each operator contains three repressor-binding sites. As can be seen from **FIGURE 14.21**, no two of the six individual repressor-binding sites are identical, but they all conform to a consensus sequence. The binding sites within each operator are separated by spacers of 3 to 7 bp that are rich in A-T base pairs. The sites at each operator are numbered so that O_R consists of the series of binding sites O_R1-O_R2-O_R3, whereas O_L consists of the series O_L1-O_L2-O_L3. In each case, site 1 lies closest to the startpoint for transcription in the promoter, and sites 2 and 3 lie farther upstream.

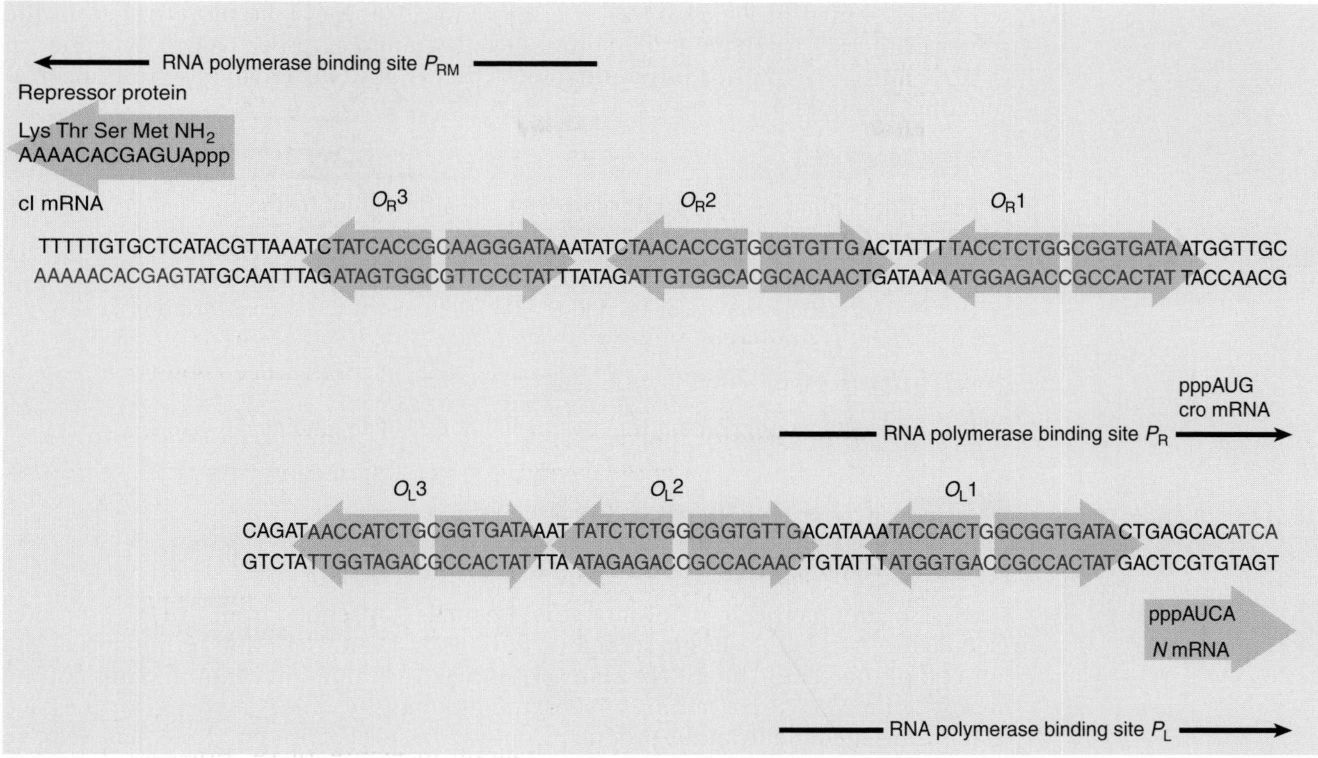

FIGURE 14.21 Each operator contains three repressor-binding sites and overlaps with the promoter at which RNA polymerase binds. The orientation of O_L has been reversed from usual to facilitate comparison with O_R.

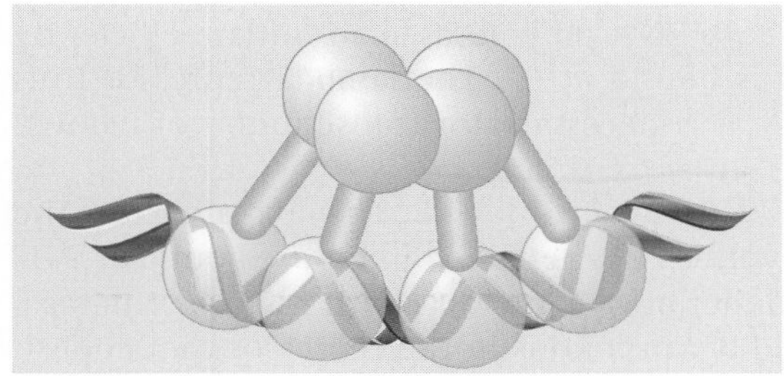

FIGURE 14.22 When two lambda repressor dimers bind cooperatively, each of the subunits of one dimer contacts a subunit in the other dimer.

Faced with the triplication of binding sites at each operator, how does the lambda repressor decide where to start binding? At each operator, site 1 has a greater affinity (roughly tenfold) than the other sites for the lambda repressor. Thus it always binds first to O_L1 and O_R1.

Lambda repressor binds to subsequent sites within each operator in a cooperative manner. The presence of a dimer at site 1 greatly increases the affinity with which a second dimer can bind to site 2. When both sites 1 and 2 are occupied, this interaction does *not* extend farther, to site 3. At the concentrations of the lambda repressor usually found in a lysogen, both sites 1 and 2 are filled at each operator, but site 3 is not occupied.

The C-terminal domain is responsible for the cooperative interaction between dimers, as well as for the dimer formation between subunits. **FIGURE 14.22** shows that it involves both subunits of each dimer; that is, each subunit contacts its counterpart in the other dimer, forming a tetrameric structure.

A result of cooperative binding is to increase the effective affinity of repressor for the operator at physiological concentrations. This enables a lower concentration of repressor to achieve occupancy of the operator. This is an important consideration in a system in which release of repression has irreversible consequences. From the sequences shown in Figure 14.21, we see that O_L1 and O_R1 lie more or less in the center of the RNA polymerase binding sites of P_L and P_R, respectively. Occupancy of O_L1-O_L2 and O_R1-O_R2 thus physically blocks access of RNA polymerase to the corresponding promoters.

KEY CONCEPTS

- Repressor binding to one operator increases the affinity for binding a second repressor dimer to the adjacent operator.
- The affinity is 10× greater for O_L1 and O_R1 than other operators, so they are bound first.
- Cooperativity allows repressor to bind the O_L2/O_R2 sites at lower concentrations.

CONCEPT AND REASONING CHECK

Why does each operator contain three parts and not a single binding site?

14.12 Lambda Repressor Maintains an Autoregulatory Circuit

Once lysogeny has been established, the *cI* gene is transcribed from the P_{RM} promoter (see Figure 14.11) that lies to its right, close to P_R/O_R. Transcription terminates at the left end of the gene. The mRNA starts with the AUG initiation codon; because of the absence of a 5′ UTR containing a ribosome binding site, this is a very poor message that is translated inefficiently, producing only a low level of protein. Note that we have not yet described how transcription for the *cI* gene is established (see *Section 14.16, The Cro Repressor Is Needed for Lytic Infection*).

The presence of the lambda repressor at O_R has dual effects as noted above (*Section 14.7, Lysogeny Is Maintained by the Lambda Repressor Protein*). It blocks expression from P_R, but it assists transcription from P_{RM}. RNA polymerase can initiate efficiently at P_{RM} only when the lambda repressor is bound at O_R. The lambda repressor thus behaves as a positive regulator protein that is necessary for transcription of its own gene, *cI*. This is the definition of an autoregulatory circuit.

The RNA polymerase binding site at P_{RM} is adjacent to O_R2. This explains how the lambda repressor autoregulates its own synthesis. When two dimers are bound at O_R1-O_R2, the amino terminal domain of the dimer at O_R2 interacts with RNA polymerase. The nature of the interaction is identified by mutations in the repressor that abolish positive control because they cannot stimulate RNA polymerase to transcribe from P_{RM}. They map within a small group of amino acids, located on the outside of helix-2 or in the turn between helix-2 and helix-3. The mutations reduce the negative charge of the region; conversely, mutations that increase the negative charge enhance the activation of RNA polymerase. This suggests that the group of amino acids constitutes an "acidic patch" that functions by an electrostatic interaction with a basic region on RNA polymerase to activate it.

The location of these "positive control mutations" in the repressor is indicated in **FIGURE 14.23**. They lie at a site on repressor that is close to a phosphate group on DNA, which is also close to RNA polymerase. Thus the group of amino acids on repressor that is involved in positive control is in a position to contact the polymerase. The important principle is that *protein–protein interactions can release energy that is used to help to initiate transcription.*

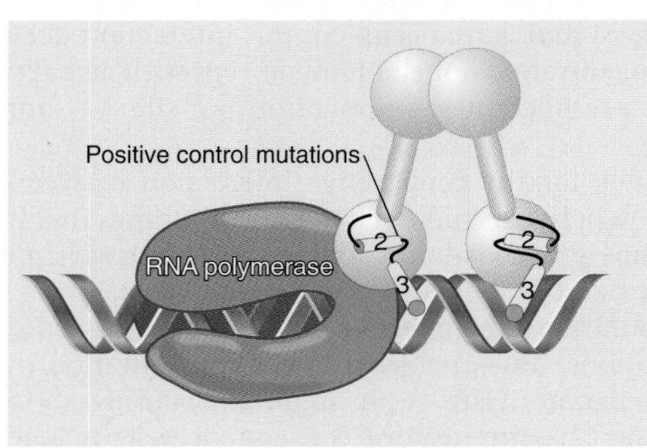

FIGURE 14.23 Positive control mutations identify a small region at helix-2 that interacts directly with RNA polymerase.

The target site on RNA polymerase that the repressor contacts is in the σ^{70} subunit, which is within the region that contacts the –35 region of the promoter. The interaction between repressor and polymerase is needed for the polymerase to make the transition from a closed complex to an open complex.

This explains how low levels of repressor positively regulate its own synthesis. As long as enough repressor is available to fill O_R2, RNA polymerase will continue to transcribe the *cI* gene from P_{RM}.

KEY CONCEPTS

- The DNA-binding region of repressor at O_R2 contacts RNA polymerase and stabilizes its binding to P_{RM}.
- This is the basis for the autoregulatory control of repressor maintenance.
- Repressor binding at O_L blocks transcription of gene *N* from P_L.
- Repressor binding at O_R blocks transcription of Cro, but also is required for transcription of *cI*.
- Repressor binding to the operators therefore simultaneously blocks entry to the lytic cycle and promotes its own synthesis.

CONCEPT AND REASONING CHECK

How is the battle for control of which developmental pathway lambda chooses determined by *cro* and the lambda repressor?

14.13 Cooperative Interactions Increase the Sensitivity of Regulation

Lambda repressor dimers interact cooperatively at both the left and right operators, so that their normal condition when occupied by repressor is to have dimers at both the 1 and 2 binding sites. In effect, each operator has a tetramer of repressor. However, this is not the end of the story. The two dimers interact with one another to form an octamer as depicted in **FIGURE 14.24**, which shows the distribution of repressors at the operator sites that are occupied in a lysogen. Repressors are occupying O_L1, O_L2, O_R1, and O_R2, and the repressor at the last of these sites is interacting with RNA polymerase, which is initiating transcription at P_{RM}.

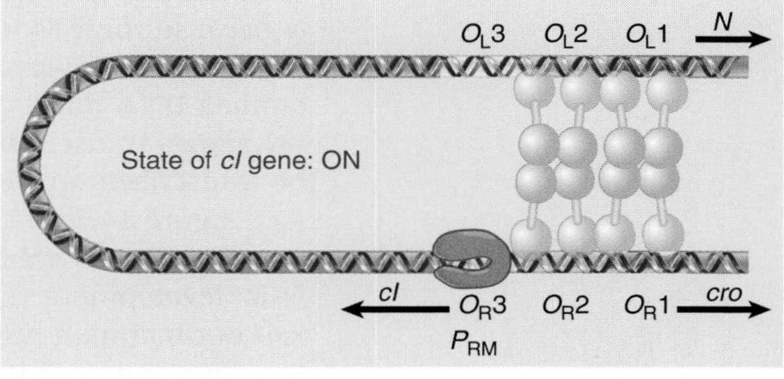

FIGURE 14.24 In the lysogenic state, the repressors bound at O_L1 and O_L2 interact with those bound at O_R1 and O_R2. RNA polymerase is bound at P_{RM} (which overlaps with O_R3) and interacts with the repressor bound at O_R2.

The interaction between the two operators has several consequences. It stabilizes repressor binding, thereby making it possible for repressor to occupy operators at lower concentrations. Binding at O_R2 stabilizes RNA polymerase binding at P_{RM}, which enables low concentrations of repressor to autogenously stimulate their own production.

The DNA between the O_L and O_R sites (that is, the gene *cI*) forms a large loop, which is held together by the repressor octamer. The octamer brings the sites O_L3 and O_R3 into proximity. As a result, two repressor dimers can bind to these sites and interact with one another, as shown in **FIGURE 14.25**. The occupation of O_R3 prevents RNA polymerase from binding to P_{RM} and therefore turns off expression of repressor.

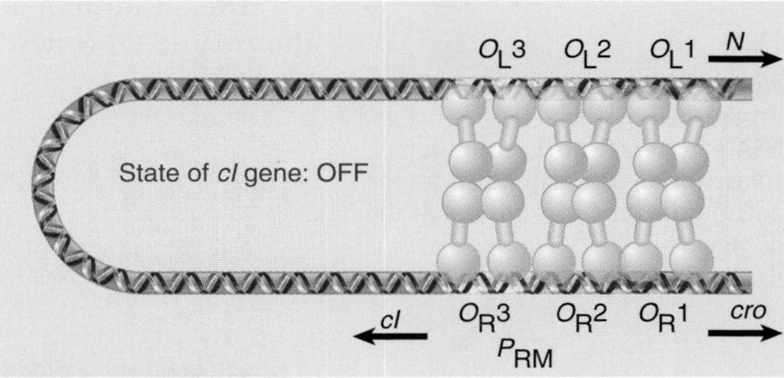

FIGURE 14.25 O_L3 and O_R3 are brought into proximity by formation of the repressor octamer, and an increase in repressor concentration allows dimers to bind at these sites and to interact.

This shows us how the expression of the *cI* gene becomes exquisitely sensitive to repressor concentration. At the lowest concentrations, it forms the octamer and activates RNA polymerase in a positive autogenous regulation. An increase in concentration allows binding to O_L3 and O_R3 and turns off transcription in a negative autogenous regulation. The threshold levels of repressor that are required for each of these events are reduced by the cooperative interactions, which makes the overall regulatory system much more sensitive. Any change in repressor level triggers the appropriate regulatory response to restore the lysogenic level.

Because the overall level of repressor has been reduced (about threefold from the level that would be required if there were no cooperative effects), there is less repressor that has to be eliminated when it becomes necessary to induce the phage. This increases the efficiency of induction.

KEY CONCEPTS

- Repressor dimers bound at O_L1 and O_L2 interact with dimers bound at O_R1 and O_R2 to form octamers.
- These cooperative interactions increase the sensitivity of regulation.

CONCEPT AND REASONING CHECK

What does it mean to say the *cI* gene regulates itself?

14.14 The *cII* and *cIII* Genes Are Needed to Establish Lysogeny

The control circuit for maintaining lysogeny presents a paradox. The presence of repressor protein is necessary for its own synthesis. This explains how the lysogenic condition is perpetuated. How, though, is the synthesis of repressor established in the first place?

When a lambda DNA enters a new host cell, RNA polymerase cannot transcribe *cI* because there is no repressor present to aid its binding at P_{RM}. This same absence of repressor, however, means that P_R and P_L are available. Thus the first event after lambda DNA infects a bacterium is when genes *N* and *cro* are transcribed. After this, pN allows transcription to be extended farther. This allows *cIII* (and other genes) to be transcribed on the left, whereas *cII* (and other genes) are transcribed on the right (see Figure 14.11).

The *cII* and *cIII* genes share with *cI* the property that mutations in them hinder lytic development. There is, however, a difference. The *cI* mutants can neither establish nor maintain lysogeny. The *cII* or *cIII* mutants have some difficulty in establishing lysogeny, but once it is established they are able to maintain it by the *cI* autoregulatory circuit.

This implicates the *cII* and *cIII* genes as positive regulators whose products are needed for an alternative system for repressor synthesis. The system is needed only to *initiate* the expression of *cI* to circumvent the inability of the autogenous circuit to engage in *de novo* synthesis. They are not needed for continued expression.

The cII protein acts directly on gene expression as a positive regulator. Between the *cro* and *cII* genes is the second *cI* promoter, called P_{RE}. This promoter can be recognized by RNA polymerase only in the presence of cII protein, whose action is illustrated in **FIGURE 14.26**. As is the case with most promoters that require a positive regulator to function, the P_{RE} promoter has a poor fit with the consensus sequence of the TATA box and lacks a –35 sequence. This explains its dependence on the cII protein, a classic positive regula-

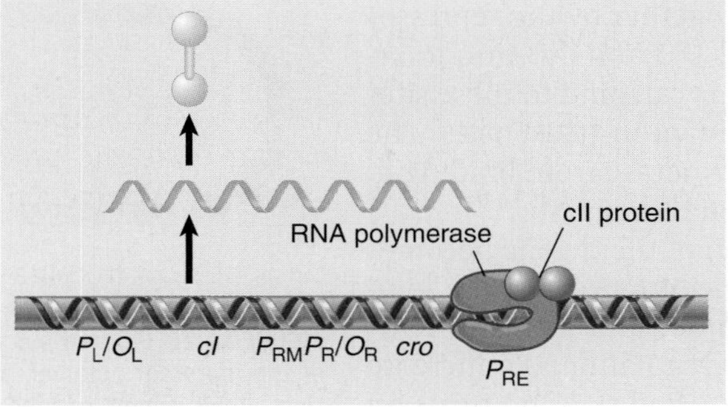

FIGURE 14.26 Repressor synthesis is established by the action of cII and RNA polymerase at P_{RE} to initiate transcription that extends from the antisense strand of *cro* through the *cI* gene.

tor: a protein that functions at the promoter to enable RNA polymerase to initiate transcription.

The cII protein is extremely unstable *in vivo,* because it is degraded as the result of the activity of a host protein called HflA—"*hfl*" stands for *h*igh *f*requency *l*ysogenization. The role of cIII is to protect cII against this degradation.

Transcription from P_{RE} promotes lysogeny in two ways. Its direct effect is that *cI* mRNA is translated into repressor protein. An indirect effect is that transcription proceeds through the *cro* gene in the "wrong" direction. Thus the 5′ part of the RNA corresponds to an antisense transcript of *cro;* in fact, it hybridizes to authentic *cro* mRNA, which inhibits its translation. This is important because *cro* expression is needed to enter the lytic cycle (see *Section 14.16, The cro Repressor Is Needed for Lytic Infection*).

The *cI* coding region on the P_{RE} transcript is very efficiently translated, in contrast with the weak translation of the P_{RM} transcript. In fact, repressor is synthesized approximately seven to eight times more effectively via expression from P_{RE} than from P_{RM}. This reflects the fact that the P_{RE} transcript has an efficient 5′UTR containing a strong ribosome-binding site, whereas the P_{RM} transcript has no ribosome-binding site and actually starts with the AUG initiation codon.

KEY CONCEPTS

- The delayed early gene products *cII* and *cIII* are necessary for RNA polymerase to initiate transcription at the promoter P_{RE}.
- *cII* acts directly at the promoter and *cIII* protects *cII* from degradation.
- Transcription from P_{RE} leads to synthesis of repressor and also blocks the transcription of *cro*.
- P_{RE} has atypical sequences at –10 and –35.
- RNA polymerase binds the promoter only in the presence of *cII*.
- *cII* binds to sequences close to the –35 region.

CONCEPT AND REASONING CHECK

How is the battle for control of which developmental pathway lambda chooses determined by *cII*?

14.15 Lysogeny Requires Several Events

Now we can see how lysogeny is established during an infection. **FIGURE 14.27** recapitulates the early stages and shows what happens as the result of expression of *cIII* and *cII*. The presence of *cII* allows P_{RE} to be used for transcription extending through *cI*. Lambda repressor protein is synthesized in high amounts from this transcript and immediately binds to O_L and O_R.

By directly inhibiting any further transcription from P_L and P_R, repressor binding turns off the expression of all phage genes. This halts the synthesis of cII and cIII proteins, which are unstable; they decay rapidly, with the result that P_{RE} can no longer be used. Thus the synthesis of repressor via the establishment circuit is brought to a halt.

But repressor is now present at O_R. It switches on the maintenance circuit for expression from P_{RM}. Repressor continues to be synthesized, although at the lower level typical of P_{RM} function. So the establishment circuit starts off repressor synthesis at a high level; then repressor turns off all other functions, while at the same time turning on the maintenance circuit, which functions at the low level adequate to sustain lysogeny.

We shall not at this point deal in detail with the other functions needed to establish lysogeny, but we can just briefly remark that the infecting lambda DNA must be inserted into the bacterial genome (see *Section 19.5, Specialized Enzymes Catalyze 5′ End Resection and Single-Strand Invasion*). The insertion requires the product of gene *int,*

FIGURE 14.27 A cascade is needed to establish lysogeny, but then this circuit is switched off and replaced by the autogenous repressor-maintenance circuit.

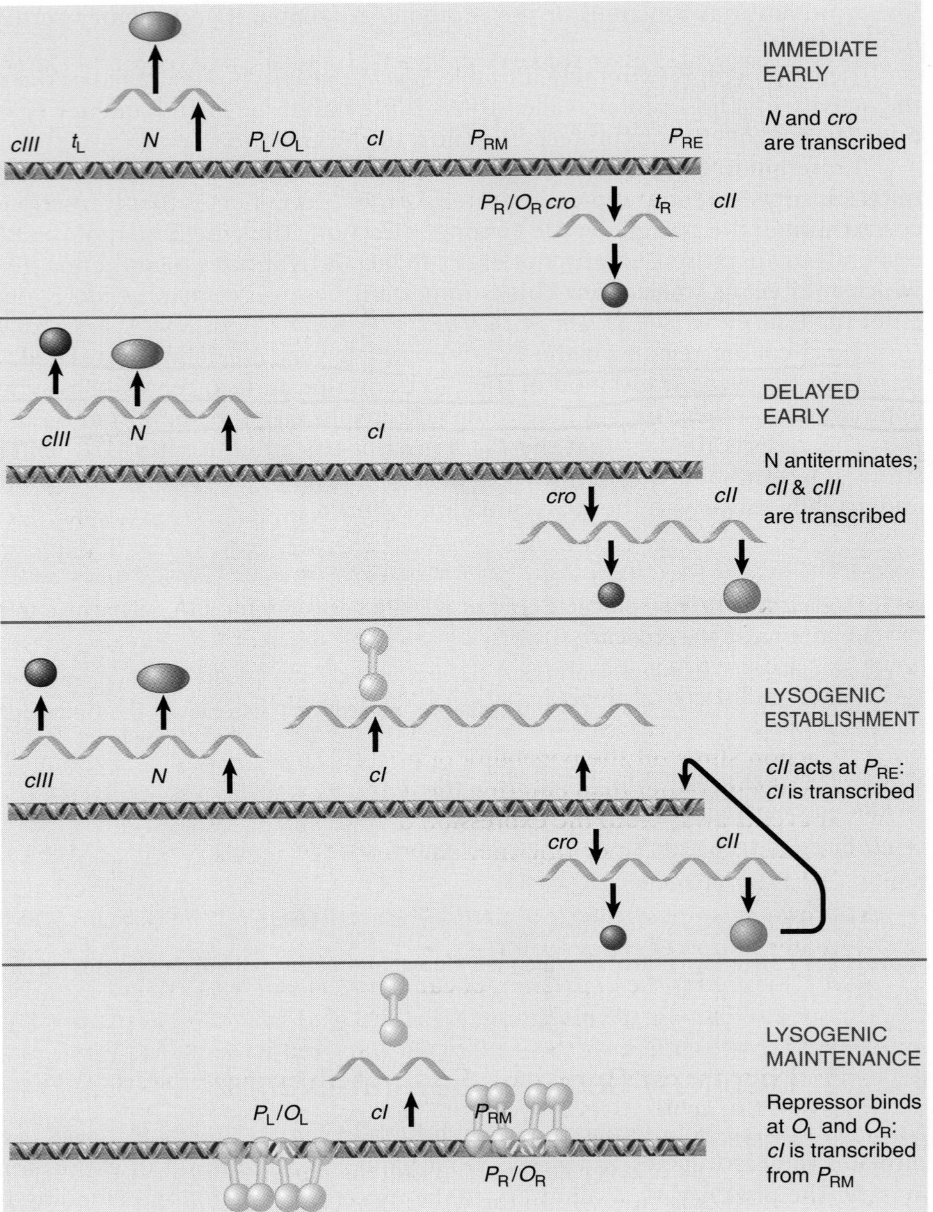

which is expressed from its own promoter P_I, at which the cII positive regulator also is necessary. The functions necessary for establishing the lysogenic control circuit are therefore under the same control as the function needed to integrate the phage DNA into the bacterial genome. Thus the establishment of lysogeny is under a control that ensures all the necessary events occur with the same timing.

Emphasizing the tricky quality of lambda's intricate cascade, we now know that *cII* promotes lysogeny in another, indirect manner. It sponsors transcription from a promoter called $P_{\text{anti-}Q}$, which is located within the *Q* gene. This transcript is an antisense version of the *Q* region, and it hybridizes with *Q* mRNA to prevent translation of Q protein, whose synthesis is essential for lytic development. Thus the same mechanisms that directly promote lysogeny by causing transcription of the *cI* repressor gene also indirectly help lysogeny by inhibiting the expression of *cro* (see above) and *Q*, the regulator genes needed for the antagonistic lytic pathway.

KEY CONCEPTS

- cII and cIII cause repressor synthesis to be established and also trigger inhibition of late gene transcription.
- Establishment of repressor turns off immediate and delayed early gene expression.
- Repressor turns on the maintenance circuit for its own synthesis.
- Lambda DNA is integrated into the bacterial genome at the final stage in establishing lysogeny.

CONCEPT AND REASONING CHECK

How does *cIII* determine the outcome of the battle between *cro* and *cI*?

14.16 The Cro Repressor Is Needed for Lytic Infection

Because lambda is a temperate virus, it has the alternatives of entering either the lysogenic pathway or the lytic pathway. Lysogeny is initiated by establishing an autoregulatory maintenance circuit that inhibits the entire lytic cascade through applying pressure at two points, P_LO_L and P_RO_R. The two pathways begin exactly the same—with the immediate early gene expression of the *N* gene and the *cro* gene, followed by the pN-directed delayed early transcription. We now face a problem. How does the phage enter the lytic cycle?

The key influence on the lytic cycle is the role of gene *cro*, which codes for another repressor protein. *Cro is responsible for preventing the synthesis of the lambda repressor protein;* this action shuts off the possibility of establishing lysogeny. *Cro* mutants usually establish lysogeny rather than entering the lytic pathway, because they lack the ability to switch events away from the expression of repressor.

Cro forms a small dimer (the monomer is 9 kD) that acts within the immunity region. It has two effects:

- It prevents the synthesis of the lambda repressor via the maintenance circuit; that is, it prevents transcription via P_{RM}.
- It also inhibits the expression of early genes from both P_L and P_R.

This means that when a phage enters the lytic pathway, Cro has responsibility both for preventing the synthesis of the lambda repressor and subsequently for turning down the expression of the early genes once there has been enough product made.

Note that Cro achieves its function by binding to the same operators as (*cI*) *lambda* repressor protein. How can two proteins have the same sites of action, yet have such opposite effects? The answer lies in the different affinities that each protein has for the individual binding sites within the operators. Let us just consider O_R, about which more is known, and where Cro exerts both its effects. The series of events is illustrated in **FIGURE 14.28**. (Note that the first two stages are identical to those of the lysogenic circuit shown in Figure 14.27.)

The affinity of Cro for O_R3 is greater than its affinity for O_R2 or O_R1. Thus it binds first to O_R3. This inhibits RNA polymerase from binding to P_{RM}. As a result, Cro's first action is to prevent the maintenance circuit for lysogeny from coming into play.

Cro then binds to O_R2 or O_R1. Its affinity for these sites is similar, and there is no cooperative effect. Its presence at either site is sufficient to prevent RNA polymerase from using P_R. This in turn stops the production of the early functions (including Cro itself). As a result of *cII*'s instability, any use of P_{RE} is brought to a halt. Thus the two actions of Cro together block *all* production of the lambda repressor.

As far as the lytic cycle is concerned, Cro turns down (although it does not completely eliminate) the expression of the early genes. Its incomplete effect is explained by its affinity for O_R1 and O_R2, which is about eight times lower than that of the lambda repressor. This effect of Cro does not occur until the early genes have become more or less superfluous, because the pQ protein is present; by this time, the phage has started late gene expression and is concentrating on the production of progeny phage particles.

FIGURE 14.28 The lytic cascade requires Cro protein, which directly prevents repressor maintenance via P_{RM}, as well as turning off delayed early gene expression, indirectly preventing repressor establishment.

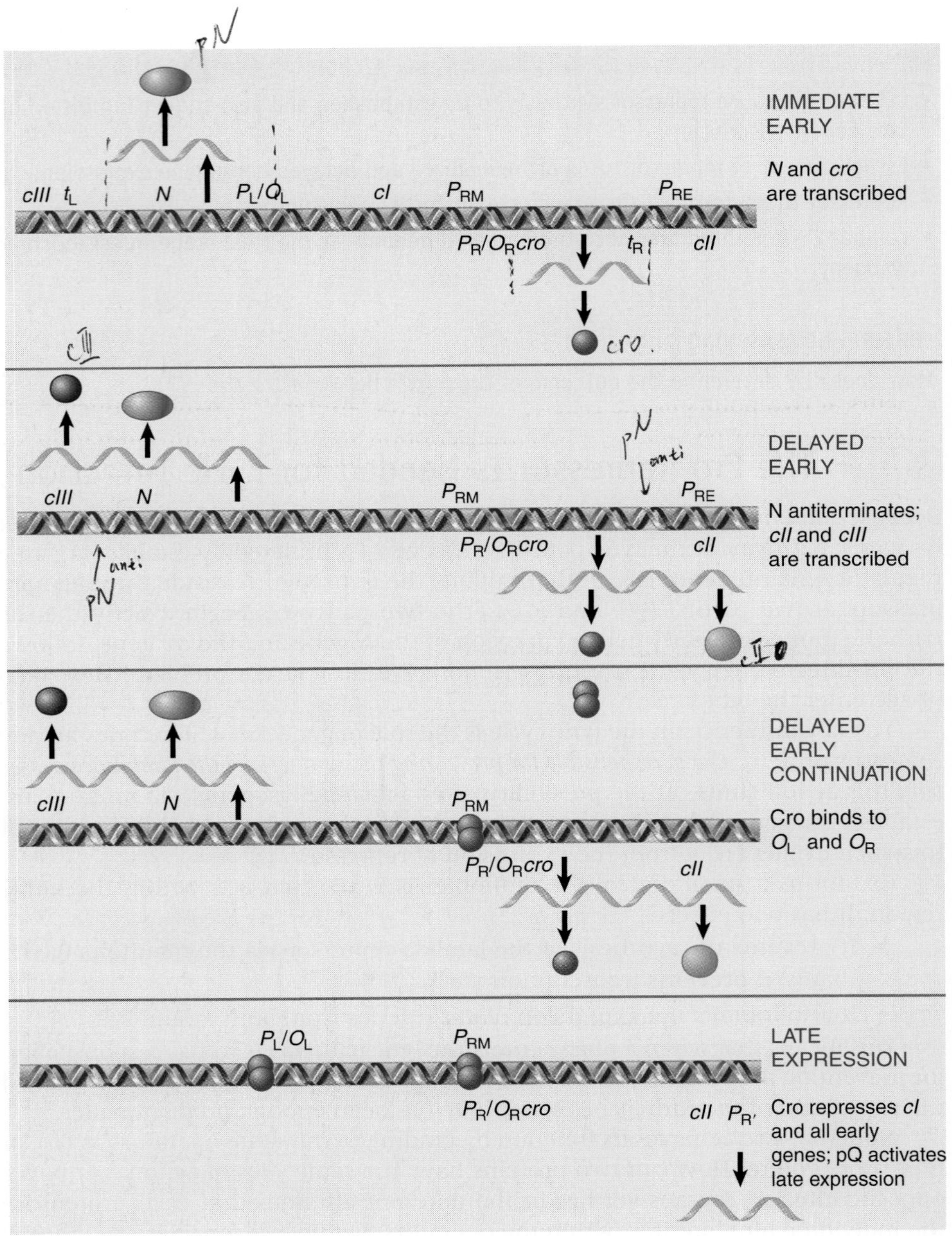

Note that in the early stages of the infection, Cro is given a head start over the lambda repressor, so it would seem that the lytic pathway is favored. Ultimately, the outcome will be determined by the concentration of the two proteins and their intrinsic DNA binding affinities.

KEY CONCEPTS

- Cro binds to the same operators as the lambda repressor, but with different affinities.
- When Cro binds to O_R3, it prevents RNA polymerase from binding to P_{RM} and blocks the maintenance of repressor promoter.
- When Cro binds to other operators at O_R or O_L, it prevents RNA polymerase from expressing immediate early genes, which (indirectly) blocks repressor establishment.

CONCEPT AND REASONING CHECK

If *cro* transcription begins much earlier than *cI* transcription, why doesn't *cro* always win?

ESSENTIAL IDEAS

The Mechanism of pQ Antitermination

One of the most fascinating aspects of biology is the wide variety of mechanisms that exist to regulate gene expression. Even within one approach to a problem—how to carry out transcription of genes that lie downstream of a terminator, for example—there may be several solutions. In bacteriophage λ, antitermination occurs at two points in the lytic cycle. First, the antiterminator protein pN acts to allow transcription to proceed beyond a terminator into the delayed early genes. Later on, the antiterminator protein pQ acts to allow transcription into the late genes.

Proteins often recognize sites within DNA and bind there to carry out some action. However, pN acts in a different manner. The phage DNA encodes sites called N utilization (*nut*) sites, but pN does not bind to the DNA site. Rather it binds to the corresponding region on the nascent RNA strand (called NUT to indicate it is an RNA site). Once bound, pN recruits additional proteins called NusA, NusB, NusD, and NusE to the RNA polymerase. The resulting complex remains on the RNA polymerase and is able to proceed through the terminators t_{L1} or t_{R1} (see Figure 11.32).

The antiterminator protein pQ uses an entirely different mechanism to enable RNA polymerase to proceed through terminators downstream of the late gene promoter $P_{R'}$. pQ binds directly to the q utilization (*qut*) site on the DNA. The *qut* site is located between the promoter –10 and –35 elements of $P_{R'}$, far from the terminator elements. In addition to the *qut* site, a pause-inducing element is required for pQ-mediated antitermination to occur. The pause site is located in the initial transcribed region and resembles a promoter –10 element. When RNA polymerase pauses at this site, pQ is able to associate with the polymerase. The protein remains associated and ultimately permits transcription to occur beyond the terminator site downstream and into the late genes. No additional proteins have been found to be required for this antitermination event.

Additional research is needed to determine the details of pQ action, but it is clear that its mechanism of action is completely different from that of pN. Although the bacteriophage λ genome is only 48.5 kb, it encompasses a complexity we are still working to understand.

14.17 What Determines the Balance Between Lysogeny and the Lytic Cycle?

The programs for the lysogenic and lytic pathways are so intimately related that it is impossible to predict the fate of an individual phage genome when it enters a new host bacterium. Will the antagonism between the lambda repressor and Cro be resolved by establishing the autoregulatory maintenance circuit shown in Figure 14.27, or by turning off lambda repressor synthesis and entering the late stage of development shown in Figure 14.28?

The same pathway is followed in both cases right up to the brink of decision. Both involve the expression of the immediate early genes and extension into the delayed early genes. The difference between them comes down to the question of whether the lambda repressor or Cro will obtain occupancy of the two operators O_L and P_L.

The early phase during which the decision is made is limited in duration in either case. No matter which pathway the phage follows, expression of all early genes will be prevented as P_L and P_R are repressed and, as a consequence of the disappearance of cII and cIII, production of repressor via P_{RE} will cease.

The critical question comes down to whether the cessation of transcription from P_{RE} is followed by activation of P_{RM} and the establishment of lysogeny, or whether P_{RM} fails to become active and the pQ regulator commits the phage to lytic development. **FIGURE 14.29** shows the critical stage at which both repressor and Cro are being synthesized. This will be determined by how much lambda repressor was made. This in turn will be determined by how much cII transcription factor was made. Finally, this in turn will be—at least partly—determined by how much cIII protein was made.

The initial event in establishing lysogeny is the binding of lambda repressor at O_L1 and O_R1. Binding at the first sites is rapidly succeeded by cooperative binding of further repressor dimers at O_L2 and O_R2. This shuts off the synthesis of Cro and starts up the synthesis of lambda repressor via P_{RM}.

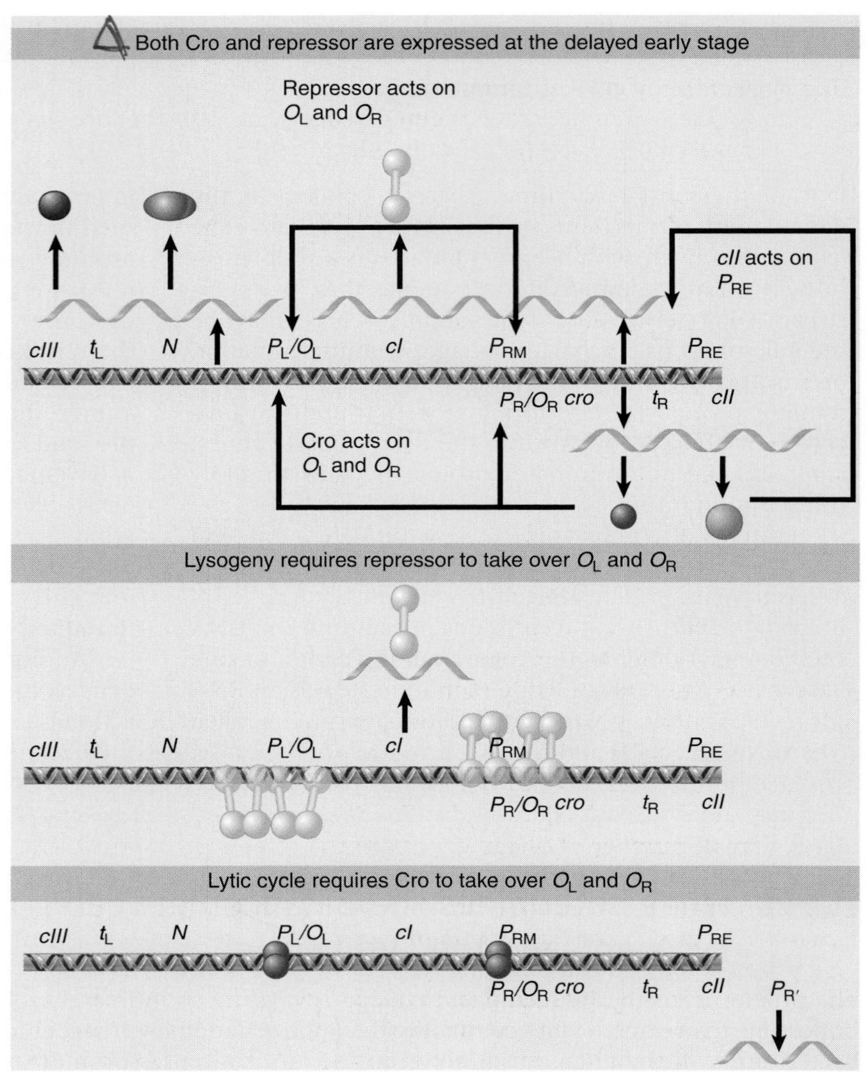

FIGURE 14.29 The critical stage in deciding between lysogeny and lysis is when delayed early genes are being expressed. If *cII* causes sufficient synthesis of repressor, lysogeny will result because repressor occupies the operators. Otherwise Cro occupies the operators, resulting in a lytic cycle.

The initial event in entering the lytic cycle is the binding of Cro at O_R3. This stops the lysogenic-maintenance circuit from starting up at P_{RM}. Cro must then bind to O_R1 or O_R2, and to O_L1 or O_L2, to turn down early gene expression. By halting production of cII and cIII, this action leads to the cessation of lambda repressor synthesis via P_{RE}. The shutoff of lambda repressor establishment occurs when the unstable cII and cIII proteins decay.

The critical influence over the switch between lysogeny and lysis is how much cII protein is made. If cII is abundant, synthesis of repressor via the establishment promoter is effective, and, as a result, repressor gains occupancy of the operators. If cII is not abundant, lambda repressor establishment fails, and Cro binds to the operators.

The level of cII protein under any particular set of circumstances determines the outcome of an infection. Mutations that increase the stability of cII increase the frequency of lysogenization. Such mutations occur in *cII* itself or in other genes. The cause of *cII*'s instability is its susceptibility to degradation by host proteases. Its level in the cell is influenced by *cIII* as well as by host functions.

The effect of the lambda protein cIII is secondary: it helps to protect cII against degradation. The presence of cIII does not guarantee the survival of cII; however, in the absence of cIII, cII is virtually always inactivated.

Host gene products act on this pathway. Mutations in the host genes *hflA* and *hflB* increase lysogeny. The mutations stabilize cII because they inactivate host protease(s) that degrade it.

The influence of the host cell on the level of cII provides a route for the bacterium to interfere with the decision-taking process. For example, host proteases that degrade cII are activated by growth on rich medium. Thus lambda tends to lyse cells that are growing well, but is more likely to enter lysogeny on cells that are starving (and that lack components necessary for efficient lytic growth).

KEY CONCEPTS

- The delayed early stage when both Cro and repressor are being expressed is common to lysogeny and the lytic cycle.
- The critical event is whether cII causes sufficient synthesis of repressor to overcome the action of Cro.

CONCEPT AND REASONING CHECK

In the end, how does *E. coli* itself determine whether lambda will enter the lytic or lysogenic path?

14.18 Summary

Virulent phages follow a lytic life cycle, in which infection of a host bacterium is followed by production of a large number of phage particles, lysis of the cell, and release of the viruses. Temperate phages can follow the lytic pathway or the lysogenic pathway, in which the phage genome is integrated into the bacterial chromosome and is inherited in this inert, latent form like any other bacterial gene.

In general, lytic infection can be described as falling into three phases. In the first phase a small number of phage genes are transcribed by the host RNA polymerase. One or more of these genes is a regulator that controls expression of the group of genes expressed in the second phase. The pattern is repeated in the second phase, when one or more genes is a regulator needed for expression of the genes of the third phase. Genes active during the first two phases code for enzymes needed to reproduce phage DNA; genes of the final phase code for structural components of the phage particle. It is common for the very early genes to be turned off during the later phases.

In phage lambda, the genes are organized into groups whose expression is controlled by individual regulatory events. The immediate early gene *N* codes for an antiterminator that allows transcription of the leftward and rightward groups of delayed early genes from the early promoters P_R and P_L. The delayed early gene *Q* has a similar antitermination function that allows transcription of all late genes from the promoter $P_{R'}$. The lytic cycle is repressed, and the lysogenic state maintained, by expression of the *cI* gene, whose product is a repressor protein, the lambda repressor, that acts at the operators O_R and O_L to prevent use of the promoters P_R and P_L, respectively. A lysogenic phage genome expresses only the *cI* gene from its promoter, P_{RM}. Transcription from this promoter involves positive autoregulation, in which repressor bound at O_R activates RNA polymerase at P_{RM}.

Each operator consists of three binding sites for the lambda repressor. Each site is palindromic, consisting of symmetrical half-sites. Lambda repressor functions as a dimer. Each half-binding site is contacted by a repressor monomer. The N-terminal domain of repressor contains a helix-turn-helix motif that contacts DNA. Helix-3 is the recognition helix and is responsible for making specific contacts with base pairs in the operator. Helix-2 is involved in positioning helix-3; it is also involved in contacting RNA polymerase at P_{RM}. The C-terminal domain is required for dimerization. Induction is caused by cleavage between the N- and C-terminal domains, which prevents the DNA-binding regions from functioning in dimeric form, thereby reducing their affinity for DNA and making it impossible to maintain lysogeny. Lambda repressor-operator binding is cooperative, so that once one dimer has bound to the first site, a second dimer binds more readily to the adjacent site.

The helix-turn-helix motif is used by other DNA-binding proteins, including lambda Cro. Cro binds to the same operators but has a different affinity for the individual operator sites, which are determined by the sequence of helix-3. Cro binds individually to operator sites, starting with O_R3, in a noncooperative manner. It is needed for progression through the lytic cycle. Its binding to O_R3 first prevents synthesis of repressor from P_{RM}, and then its binding to O_R2 and O_R1 prevents continued expression of early genes, an effect also seen in its binding to O_L1 and O_L2.

Establishment of lambda repressor synthesis requires use of the promoter P_{RE}, which is activated by the product of the *cII* gene. The product of *cIII* is required to stabilize the *cII* product against degradation. By turning off *cII* and *cIII* expression, Cro acts to prevent lysogeny. By turning off all transcription except that of its own gene, the repressor acts to prevent the lytic cycle. The choice between lysis and lysogeny depends on whether repressor or Cro gains occupancy of the operators in a particular infection. The stability of cII protein in the infected cell is a primary determinant of the outcome.

CHAPTER QUESTIONS

1. In a wild-type bacterial cell, how many copies of the lambda prophage can be present in the chromosome?

A. none
B. one
C. two
D. three or more

2. Lysogeny is maintained by the interaction of a(n) _______ with an operator.

A. positive regulator
B. repressor
C. antiterminator
D. antirepressor

3. The immediate early genes of bacteriophage lambda are:

A. *N* and *Q*.
B. *cI* and *Q*.
C. *N* and *cro*.
D. *cI* and *cro*.

4. The bacteriophage lambda cro protein functions as a(n):

A. antitermination factor.
B. repressor.
C. antirepressor.
D. inducer.

5. What factor is needed for transcription of the phage lambda late genes?

A. Cro protein
B. cI protein
C. N protein
D. Q protein

6. What effect would an N-mutation have on phage lambda infection?

A. no effect, infection would proceed as normal to either lytic or lysogenic pathway
B. lytic pathway blocked; only lysogeny possible
C. lysogenic pathway blocked; only lytic pathway possible
D. complete abolishment of infection; neither lytic nor lysogenic pathway possible

7. What effect would a *cII*-mutation have on phage lambda infection?
   A. no effect; infection would proceed as normal to either lytic or lysogenic pathway
   B. lytic pathway blocked; only lysogeny possible
   C. lysogenic pathway blocked; only lytic pathway possible
   D. complete abolishment of infection; neither lytic nor lysogenic pathway possible

8. Transcription from the phage lambda P_{RE} promoter requires which of the following factors?
   A. cI and pN
   B. cI and cII
   C. cII and cIII
   D. pN and cII

9. Which of the following is the order of gene expression for the bacteriophage lambda lytic pathway upon infection of a wild-type host?
   A. $N \rightarrow cI \rightarrow cII \rightarrow$ late genes
   B. $N \rightarrow cro \rightarrow cII \rightarrow$ late genes
   C. $N \rightarrow cII \rightarrow Q \rightarrow$ late genes
   D. $N \rightarrow Q \rightarrow$ late genes

10. The major function of the bacteriophage cro protein is to:
   A. prevent synthesis of the cI protein.
   B. prevent synthesis of the cII protein.
   C. prevent synthesis of the N protein.
   D. prevent synthesis of the Q protein.

KEY TERMS

bacteriophage
cascade
delayed early genes
early genes
early infection
excision
helix-turn-helix
immediate early genes
immunity
immunity region
induction of phage
integration
late gene
late infection
lysis
lysogeny
lytic infection
middle genes
phage
prophage
recognition helix
temperate phage
virulent mutations
virulent phage

FURTHER READING

Anderson and Yang (2008). DNA looping can enhance lysogenic cI transcription in phage lambda. *Proc. Natl. Acad. Sci. USA* 105, 5827–5832.

Oppenheim, A. B., Kobiler, O., Stavans, J., Court, D. L., and Adhya, S. (2005). Switches in bacteriophage lambda development. *Annu. Rev. Gen.* 39, 409–429.

Ptashne, M. (2004). *The Genetic Switch: Phage Lambda Revisited.* Cold Spring Harbor, NY: Cold Spring Harbor Press.

IV

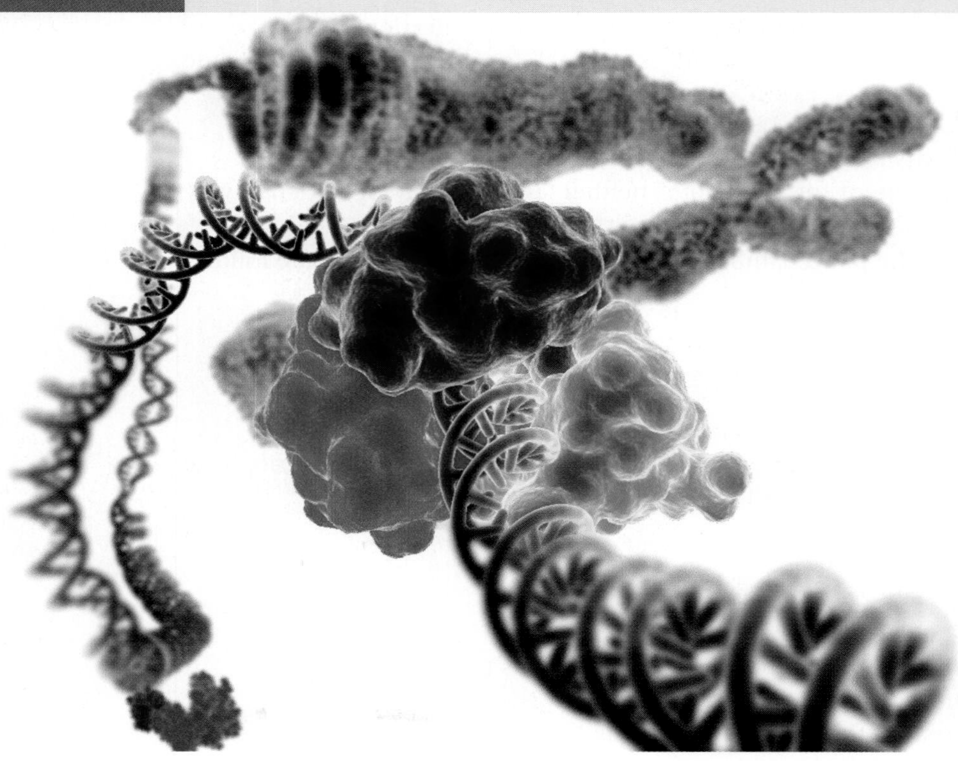

Breaks in the DNA double helix are repaired by protein-mediated mechanisms (in color). Photo courtesy of David Gohara, SciAna Film Works (http://www.scianafilms.com).

DNA Replication and Recombination

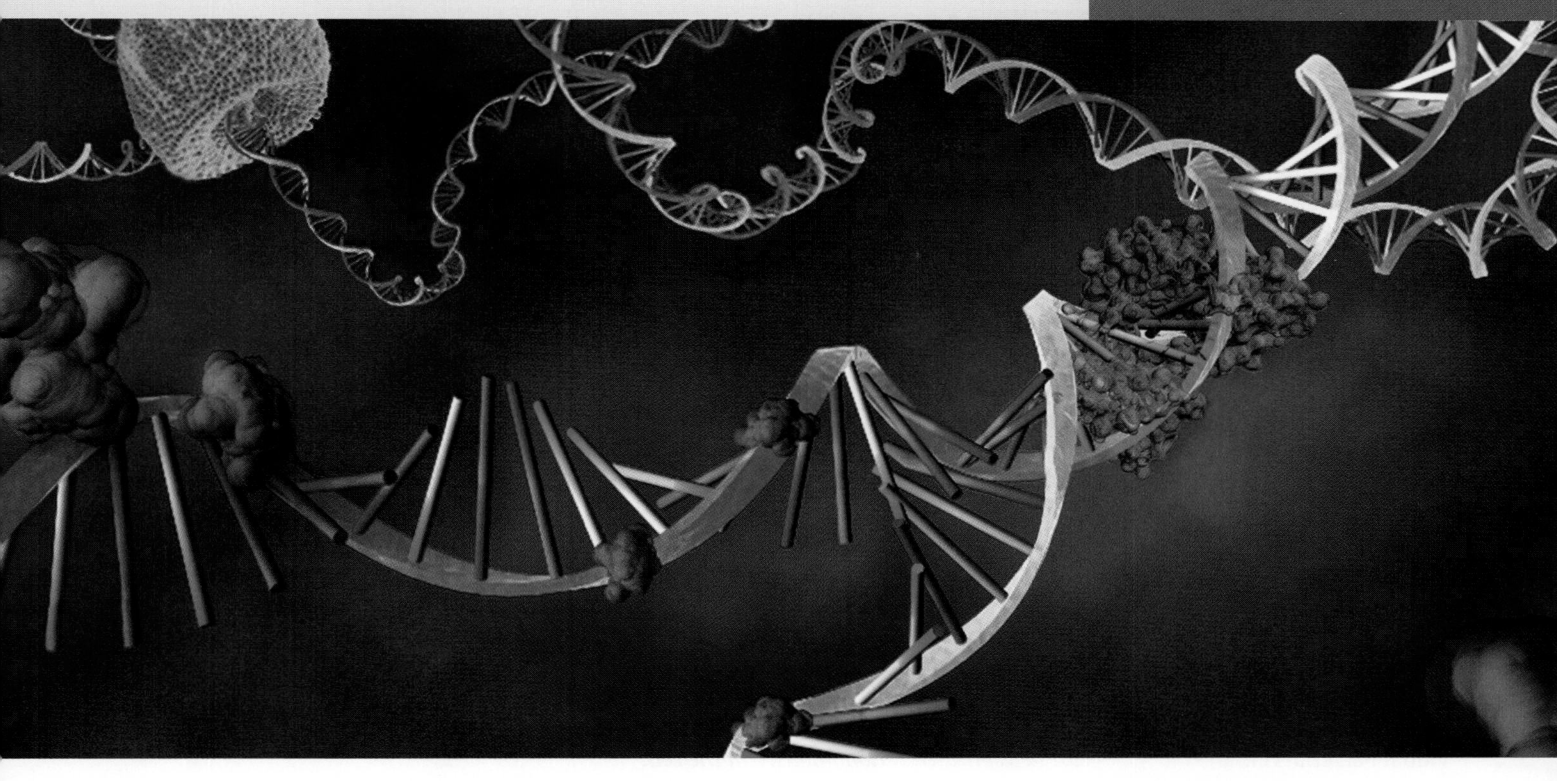

A helicase (in green) unwinding DNA at a replication fork. © Cindy Schroeder. Used with permission.

The Replicon

CHAPTER OUTLINE

15.1 Introduction

Replication of DNA is a key regulatory event in cell division. Every time that a cell divides, its entire set of DNA must be replicated once and only once. This is accomplished by controlling the initiation of replication, which occurs only at a unique site called the **origin** or *ori*.

▸ **origin** A sequence of DNA at which replication is initiated.

▸ **replicon** A unit of the genome in which DNA is replicated. Each contains an origin for initiation of replication.

The origin is the unit of DNA which controls the initiation of replication. The origin lies within the **replicon**, the DNA that is replicated whenever an initiation event occurs at the origin. **FIGURE 15.1** illustrates the general nature of the relationship between replicons and chromosomes or genomes in prokaryotes and eukaryotes.

A genome in a prokaryotic cell (usually a single circular molecule of DNA) constitutes a single replicon. This means that replication of the entire bacterial chromosome depends on a single initiation event that occurs at the unique origin. The initiation event occurs once for every cell division and is known as **single-copy replication control**. The frequency of initiation at the bacterial origin is controlled by its state of methylation (see *Section 15.4, Methylation of the Bacterial Origin Regulates Initiation*).

▸ **single-copy replication control** A control system in which there is only one copy of a replicon per unit bacterium. The bacterial chromosome and some plasmids have this type of regulation.

▸ **multicopy replication control** Replication occurs when the control system allows the plasmid to exist in more than one copy per individual bacterial cell.

Bacteria may contain additional genetic information in the form of plasmids. *A plasmid is an autonomous circular DNA that constitutes a separate replicon.* Plasmid replicons may show single copy control, which means that they replicate once every time the bacterial chromosome replicates, or they may be under **multicopy replication control**, when they are present in a greater number of copies than the bacterial chromosome. Each phage or virus DNA also constitutes a replicon, and thus is able to initiate many times during an infectious cycle. Perhaps a better way to view the prokaryotic replicon, therefore, is to reverse the definition: *any DNA molecule that contains an origin can be replicated autonomously in the cell.*

A major difference in the organization of bacterial and eukaryotic genomes is seen in their replication. Each eukaryotic chromosome (usually a very long linear molecule of DNA) contains a large number of replicons spaced unevenly throughout the chromosomes. We will see later why the origin is placed in the center of the replicon. Like the single replicon that constitutes the bacterial chromosome, each eukaryote origin "fires" once and only once in each cell cycle. There are exceptions to this rule for situations like regulated gene amplification and polytene chromosome formation. Eukaryote origin usage is controlled by the ability of regulator proteins to bind to it (see *Section 15.7, Licensing Factor Controls Eukaryotic Rereplication and Consists of MCM Proteins*). The replicons of a eukaryote genome are not active simultaneously, but are activated over a fairly protracted period, which is called the S phase. This implies the existence of additional levels of control to ensure that each replicon does fire once and does not fire a second time. There are additional rules that control when each origin is allowed to fire. Because many replicons are activated independently, another signal must exist to indicate when the entire process has been completed. In contrast to nuclear chromosomes, which have a single-copy type of control, the DNA of mitochondria and chloroplasts may be regulated more like plasmids that exist in multiple copies per bacterium. There are multiple copies of each organelle per cell and each organelle contains multiple copies of DNA. The control of organelle DNA replication must be coordinated with the cell cycle.

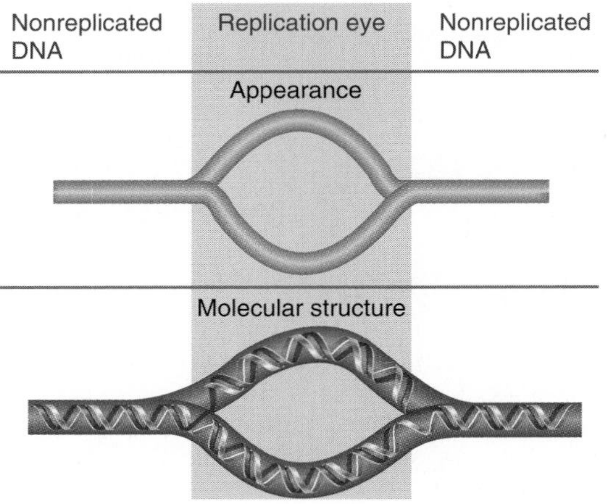

FIGURE 15.1 Replicated DNA is seen as a replication eye flanked by nonreplicated DNA.

In all these systems, the key question is to define the DNA sequences that function as origins and to determine how they are recognized by the appropriate proteins of the replication apparatus. We start by considering the basic construction of replicons and the various forms that they take in bacteria and eukaryotic cells. In *Chapter 16, Extrachromosomal Replicons,* we consider autonomously replicating units in bacteria. In *Chapter 17, Bacterial Replication Is Connected to the Cell Cycle,* we turn to the question of how replication of the genome is coordinated with bacterial division and what is responsible for segregating the genomes to daughter bacteria. In *Chapter 18, DNA Replication,* we examine the biochemistry of DNA synthesis.

15.2 An Origin Usually Initiates Bidirectional Replication

Replication starts at an origin by separating or melting the two strands of the DNA duplex. **FIGURE 15.2** shows that each of the parental strands then acts as a template to synthesize a complementary daughter strand. This model of replication, in which a parental duplex gives rise to two daughter duplexes, each containing one original parental strand and one new strand, is called **semiconservative replication** (see the Historical Perspectives box, *The Meselson-Stahl Experiment,* showing semiconservative replication).

A molecule of DNA engaged in replication has two types of regions. **FIGURE 15.3** shows that when replicating DNA is viewed by electron microscopy, the replicated region appears as a **replication eye** or bubble within the nonreplicated DNA. The nonreplicated region consists of the parental duplex; this opens into the replicated region where the two daughter duplexes have formed.

The point at which replication occurs is called the **replication fork** (sometimes also known as the **growing point**). *A replication fork moves sequentially along the DNA from its starting point at the origin.*

The form of replication used by both bacteria and eukaryote nuclear chromosomes is **bidirectional replication**, with two replication forks setting out from the origin in opposite directions as shown in **FIGURE 15.4**. As the replication forks continue to move apart, the replication eye increases in size, and eventually it becomes larger than the nonreplicated region.

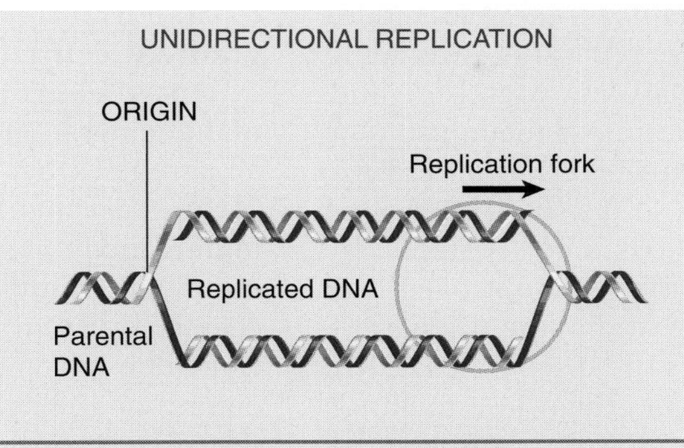

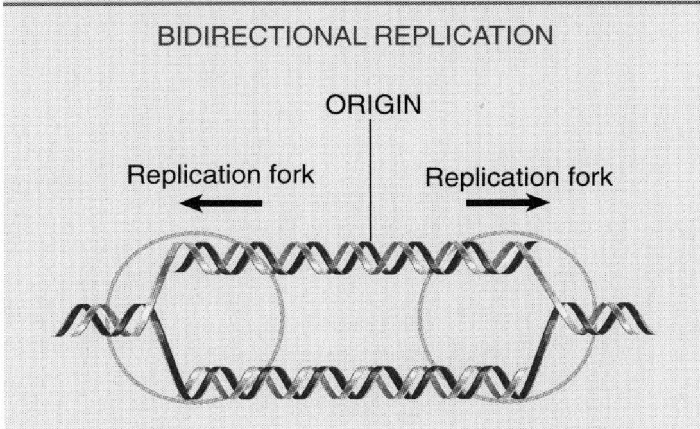

FIGURE 15.2 Replicons may be unidirectional or bidirectional, depending on whether one or two replication forks are formed at the origin.

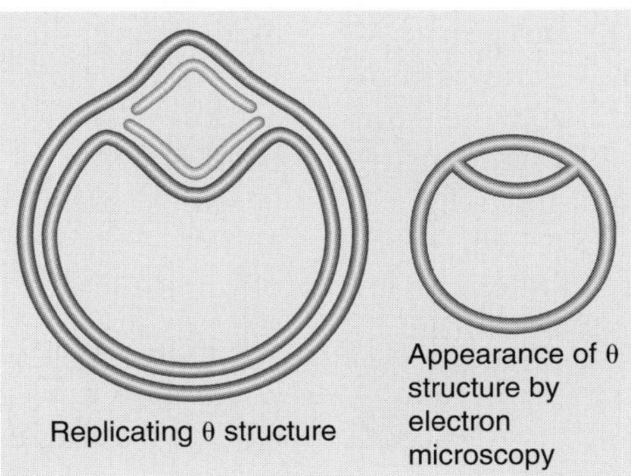

FIGURE 15.3 A replication eye forms a θ structure in circular DNA.

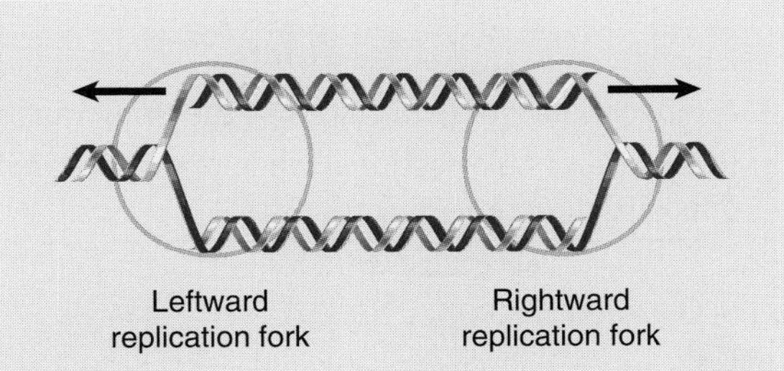

FIGURE 15.4 Replication is bidirectional when two replication forks are established at the origin, moving in opposite directions.

- **semiconservative replication** Replication accomplished by separation of the strands of a parental duplex, with each strand then acting as a template for synthesis of a complementary strand.
- **replication eye** A region in which DNA has been replicated within a longer, unreplicated region.
- **replication fork** The point at which strands of parental duplex DNA are separated so that replication can proceed. A complex of proteins including DNA polymerase is found there.
- **growing point** *See* replication fork.
- **bidirectional replication** A system in which an origin generates two replication forks that proceed away from the origin in opposite directions.

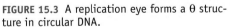

HISTORICAL PERSPECTIVES

The Meselson-Stahl Experiment

In theory, DNA could be replicated by a number of mechanisms other than the semiconservative mode, but the reality of semiconservative replication was demonstrated experimentally by Matthew Meselson and Franklin Stahl in 1958. The experiment made use of a newly developed high-speed centrifuge (an *ultracentrifuge*) that could spin a solution so fast that molecules differing only slightly in density could be separated. In their experiment, the heavy ^{15}N isotope of nitrogen was used to alter the density of DNA and allow the parental and daughter DNA molecules to be separated. DNA isolated from the bacterium *E. coli* grown in a medium containing ^{15}N as the only available source of nitrogen is denser than DNA from bacteria grown in medium with the normal ^{14}N isotope. These DNA molecules can be separated in an ultracentrifuge, because they have about the same density as a very concentrated solution of cesium chloride (CsCl).

When a CsCl solution containing DNA is centrifuged at high speed, the Cs^+ ions gradually sediment toward the bottom of the centrifuge tube. This movement is counteracted by diffusion (the random movement of molecules), which prevents complete sedimentation. At equilibrium, a linear gradient of CsCl concentration (and, therefore, density) is present, increasing in cesium concentration and density from the top of the centrifuge tube to the bottom. DNA in the tube moves upward or downward to a position in the gradient at which the local density of the solution is equal to its own density. At equilibrium, a mixture of ^{14}N-containing ("light") and ^{15}N-containing ("heavy") *E. coli* DNA will separate into two distinct zones in a density gradient, even though they differ only slightly in density. DNA from *E. coli* containing ^{14}N in the purine and pyrimidine rings has a density of 1.708 g/cm^3, whereas DNA with ^{15}N in the purine and pyrimidine rings has a density of 1.722 g/cm^3. These molecules can be separated because a solution of 6.6 molar CsCl has a density of 1.700 g/cm^3. When spun in a centrifuge, the CsCl solution forms a gradient of density that brackets the densities of the light and heavy DNA molecules. It is for this reason that the separation technique is called **equilibrium density-gradient centrifugation**.

Imagine an experiment using bacteria grown for many generations in a ^{15}N-containing medium so that all parental DNA strands are "heavy." At the beginning of the experiment, the cells are transferred to a ^{14}N-containing medium so that newly synthesized DNA strands are "light." Duplex DNA is isolated from samples of cells taken from the culture at intervals, and equilibrium density-gradient centrifugation is carried out to determine the density of the molecules. With semiconservative replication, the expected result of the experiment is as follows. After one round of replication, each duplex consists of one heavy and one light strand, so all daughter molecules are of intermediate density. After two rounds of replication, the duplexes containing an original parental strand are again intermediate in density, but now there are an equal number of duplexes consisting of two light strands, so two bands differing in density result from centrifugation.

The actual result of the Meselson-Stahl experiment is shown in **FIGURE B15.1**. The bottom part of the figure is that of the CsCl solution in centrifuge tubes oriented vertically. The positions of the DNA molecules in the density gradient are, therefore, indicated by dark bands.

At the start of the experiment ("Parental DNA"), all the DNA was heavy (^{15}N). After the transfer to ^{14}N, a band of lighter density began to appear, and it gradually became more prominent as the cells replicated their DNA and divided. After one generation of growth (one round of replication of the DNA molecules and a doubling of the number of cells), all the DNA had a "hybrid" density exactly intermediate between the densities of ^{15}N-DNA and ^{14}N-DNA. The observation of molecules with a hybrid density indicates that the replicated molecules contain equal amounts of the two nitrogen isotopes.

equilibrium density-gradient centrifugation A gradient method used to separate macromolecules on the basis of differences in their density. For DNA, it is prepared from a heavy soluble compound such as CsCl.

KEY CONCEPTS

- A replication fork is initiated at the origin and then moves sequentially along DNA.
- Replication is bidirectional when an origin creates two replication forks that move in opposite directions.

CONCEPT AND REASONING CHECK

What would the CsCl gradient pattern from the Meselson-Stahl experiment look like if replication were fully conservative, that is, a parental strand gave rise to one daughter with all original DNA and a second daughter with all new DNA?

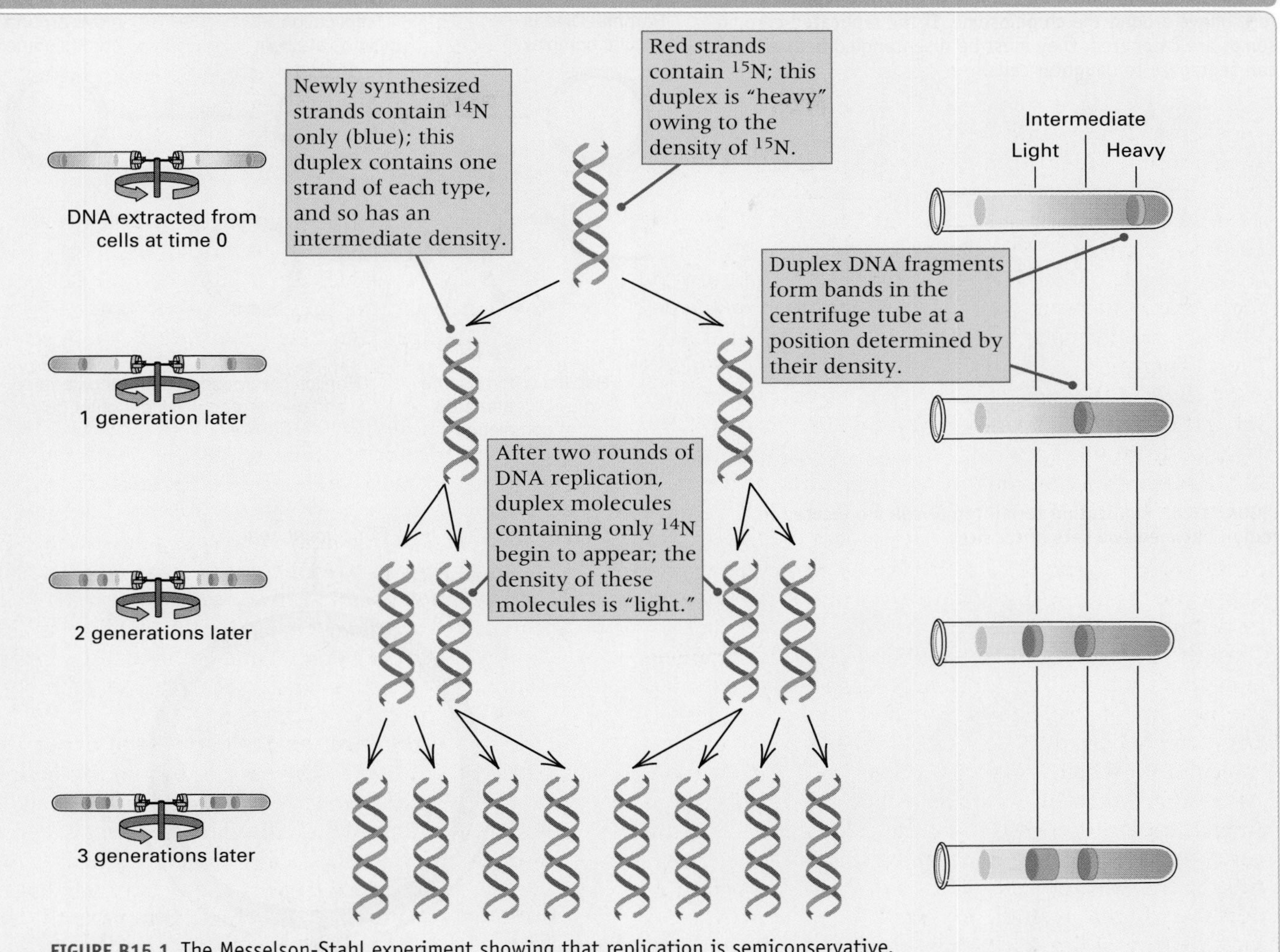
DNA extracted from cells at time 0
1 generation later
2 generations later
3 generations later
Newly synthesized strands contain ¹⁴N only (blue); this duplex contains one strand of each type, and so has an intermediate density.
Red strands contain ¹⁵N; this duplex is "heavy" owing to the density of ¹⁵N.
After two rounds of DNA replication, duplex molecules containing only ¹⁴N begin to appear; the density of these molecules is "light."
Duplex DNA fragments form bands in the centrifuge tube at a position determined by their density.
Intermediate
Light
Heavy

FIGURE B15.1 The Messelson-Stahl experiment showing that replication is semiconservative.

After two generations of replication in the ^{14}N medium, approximately half of the DNA had the density of DNA with ^{14}N in both strands ("light" DNA) and the other half had the hybrid density. This distribution of ^{15}N atoms is precisely the result predicted from semiconservative replication. Similar experiments with replicating DNA from numerous viruses and bacteria later confirmed the semiconservative mode of replication in these groups.

15.3 The Bacterial Genome Is a Single Circular Replicon

Prokaryote replicons are usually circular, so that the DNA forms a closed circle with no free ends. Circular structures include the bacterial chromosome itself, all plasmids, and many bacteriophages, and are also common in chloroplasts and mitochondrial DNAs. **FIGURE 15.5** summarizes the stages of replicating a circular chromosome. After replication has initiated at the origin, two replication forks proceed in opposite directions. The circular chromosome is sometimes described as a θ (theta) structure at this stage, because of its appearance. An important consequence of circularity is that the completion of the process can generate two chromosomes that are linked because one passes through the other (they are said to be catenated), and specific enzyme systems

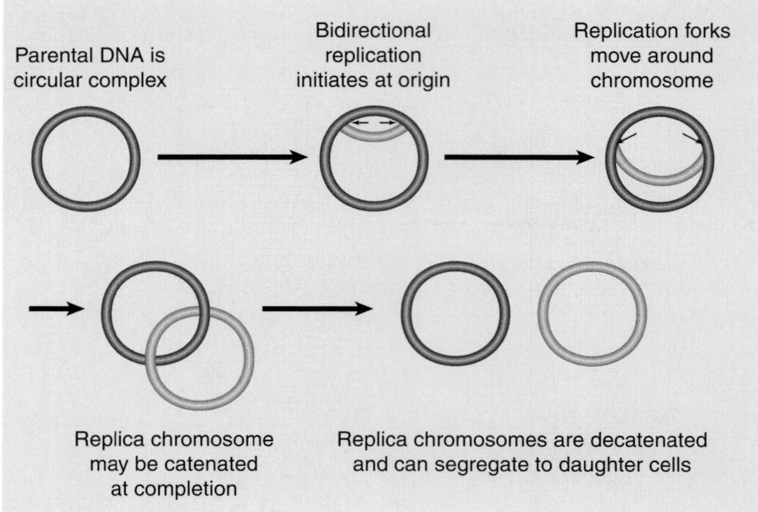
Parental DNA is
circular complex
Bidirectional
replication
initiates at origin
Replication forks
move around
chromosome
Replica chromosome
may be catenated
at completion
Replica chromosomes are decatenated
and can segregate to daughter cells

FIGURE 15.5 Bidirectional replication of a circular bacterial chromosome is initiated at a single origin. The replication forks move around the chromosome. If the replicated chromosomes are catenated, they must be disentangled before they can segregate to daughter cells.

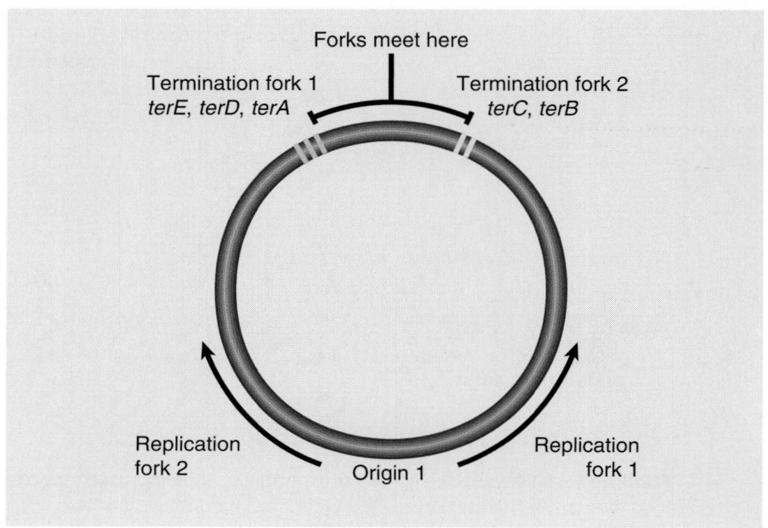
Forks meet here
Termination fork 1
terE, terD, terA
Termination fork 2
terC, terB
Replication
fork 2
Origin 1
Replication
fork 1

FIGURE 15.6 Replication termini in *E. coli* are located in a region between two sets of *ter* sites.

may be required to separate them (see *Section 17.7, Chromosomal Segregation May Require Site-Specific Recombination*).

The genome of *E. coli* is replicated bidirectionally from a single unique site called the origin, identified as the genetic locus *oriC*. Two replication forks initiate at *oriC* and move around the genome at approximately the same speed to a special termination region. One interesting question is what ensures that the DNA is replicated right across the region where the two forks meet.

Sequences that are involved with termination are called *ter* sites. A *ter* site contains a short, ~23 bp sequence. The termination sequences are unidirectional; that is, they function in only one orientation. The *ter* site is recognized by a protein called Tus in *E. coli* and RTP in *B. subtilis* that recognizes the consensus sequence and prevents the replication fork from proceeding. However, deletion of the *ter* sites does not prevent normal replication cycles from occurring, although it does affect segregation of the daughter chromosomes.

Termination in *E. coli* and *B. subtilis* has the interesting features shown in **FIGURE 15.6**. The two replication forks meet and halt in a region approximately halfway around the chromosome from the origin. In *E. coli,* two clusters of five *ter* sites each, including *terE,D,A* on one side and *terC* and *B* on the other side, are located ~100 kb on either side of this termination region. Each set of *ter* sites is specific for one direction of fork movement; that is, each set of *ter* sites allows a replication fork into the termination region, but does not allow it out the other side. For example, replication fork 1 can pass through *terC* and *terB* into the region, but cannot continue past *terE, D,* and *A*. This arrangement creates a "replication fork trap." If for some reason, one fork is delayed, so that the forks fail to meet in the middle, the faster fork will be trapped at the distal *ter* sites to wait for the slower fork.

What happens when a replication fork encounters a protein bound to DNA? We assume that repressors, for example, are displaced and then rebind. A particularly interesting question is what happens when a replication fork encounters an RNA polymerase engaged in transcription. A replication fork moves 10× faster than RNA polymerase. If they are proceeding in the same direction, either the replication fork must displace the RNA polymerase or it must slow down as it waits for the RNA polymerase to reach its terminator. It appears that a DNA polymerase moving in the same direction as an RNA polymerase can "bypass" it without disrupting transcription, but we do not understand how this happens.

A conflict arises when the replication fork meets an RNA polymerase traveling in the opposite direction, toward it. Can it displace the polymerase? Or do both replication and transcription come to a halt? An indication that these encounters cannot be easily resolved is provided by the gene organization on the *E. coli* chromosome. Almost all active transcription units are oriented so that they are expressed in the same direction as the replication fork that passes them. The exceptions all comprise small transcription units that are infrequently expressed. The difficulty of generating inversions containing highly expressed genes suggests that head-on encounters between a replication fork and a series of transcribing RNA polymerases may be lethal.

KEY CONCEPTS

- Bacterial replicons are usually circles that replicate bidirectionally from a single origin.
- The origin of *E. coli*, *oriC*, is 245 bp in length.
- The two replication forks usually meet halfway around the circle, but there are *ter* sites that cause termination if the replication forks go too far.

CONCEPT AND REASONING CHECK

Almost all active transcription units are oriented so that they are expressed in the same direction as the replication fork. Why do we think the other transcription units are in the opposite orientation and infrequently expressed?

15.4 Methylation of the Bacterial Origin Regulates Initiation

The bacterial origin contains sequences that are methylated and that are in different methylation states before and after replication. This difference is used as a mark to distinguish a replicated origin from a nonreplicated origin.

The *E. coli oriC* contains eleven copies of the sequence $\frac{\text{GATC}}{\text{CTAG}}$, which is a target for methylation at the N^6 position of adenine by the Dam methylase enzyme. These sites are also found throughout the genome. The reaction is illustrated in **FIGURE 15.7**.

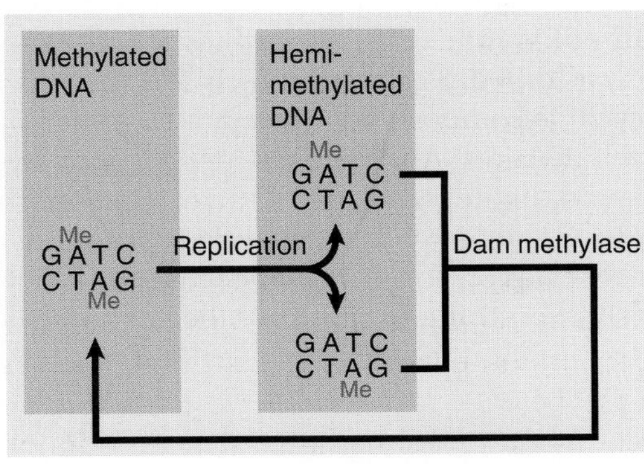

FIGURE 15.7 Replication of methylated DNA gives hemimethylated DNA, which maintains its state at GATC sites until the Dam methylase restores the fully methylated condition.

Before replication, the palindromic target site is methylated on the adenines of each strand. Replication inserts the normal (nonmodified) bases into the daughter strands. This generates

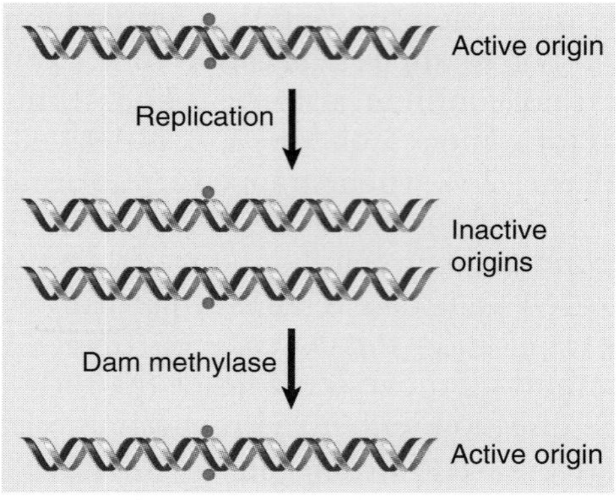

FIGURE 15.8 Only fully methylated origins can initiate replication; hemimethylated daughter origins cannot be used again until they have been restored to the fully methylated state.

hemimethylated DNA, in which one strand is methylated and one strand is unmethylated. Thus the replication event converts Dam target sites from fully methylated to hemimethylated condition.

What is the consequence for replication? The ability of a plasmid relying upon *oriC* to replicate in *dam*⁻ *E. coli* depends on its state of methylation. If the plasmid is methylated it undergoes a single round of replication, and then the hemimethylated products accumulate, as described in FIGURE 15.8. The hemimethylated plasmids then accumulate, rather than being replaced by unmethylated plasmids, suggesting that a hemimethylated origin cannot be used to initiate a replication cycle.

▸ **hemimethylated DNA** DNA that is methylated on one strand of a target sequence that has a cytosine on each strand.

This suggests two explanations. Initiation may require full methylation of the Dam target sites in the origin, or it may be inhibited by hemimethylation of these sites. The latter seems to be the case, because an origin of nonmethylated DNA can function effectively.

Thus hemimethylated origins cannot initiate again until the Dam methylase has converted them into fully methylated origins. The GATC sites at the origin remain hemimethylated for ~13 minutes after replication. This long period is unusual because at typical GATC sites elsewhere in the genome, remethylation begins immediately (<1.5 minutes) following replication.

DNA methylation in bacteria serves a second function as well. It allows the DNA mismatch recognition machinery to distinguish the old template strand from the new strand. If the DNA polymerase has made an error, such as creating an A-C base pair, the repair system will use the methylated strand as a template to replace the base on the nonmethylated strand. Without that methylation, the enzyme would have no way to determine which is the new strand.

KEY CONCEPTS

- *oriC* contains eleven GATC/CTAG repeats that are methylated on adenine on both strands.
- Replication generates hemimethylated DNA, which cannot initiate replication.
- There is a 13-minute delay before the GATC/CTAG repeats are remethylated.

CONCEPT AND REASONING CHECK

Why is it important for the cell to know whether or not a replication origin has replicated?

15.5 Each Eukaryotic Chromosome Contains Many Replicons

In eukaryotic cells, the replication of DNA is confined to the second part of the cell cycle called **S phase**, which follows the G1 phase (FIGURE 15.9). The eukaryote cell cycle is composed of alternating rounds of growth followed by DNA replication and cell division. After the cell divides into two daughter cells, each has to grow back to approximately the size of the original mother cell before cell division can occur again. The G1 phase of the cell cycle is primarily concerned with growth (although G1 is an abbreviation for *first gap* because the early cytologists could not see any activity). In G1, everything except DNA begins to be doubled: RNA, protein, lipids, and carbohy-

▸ **S phase** The restricted part of the eukaryotic cell cycle during which synthesis of DNA occurs.

drate. The progression from G1 into S is very tightly regulated and is controlled by a **checkpoint**. In order for a cell to be allowed to progress into S phase, there must be a certain minimum amount of growth that is biochemically measured. In addition, there must not be any damage to the DNA. Damaged DNA or too little growth prevents the cell from progressing into S phase. When S phase is completed, G2 phase commences. There is no control point and no sharp demarcation.

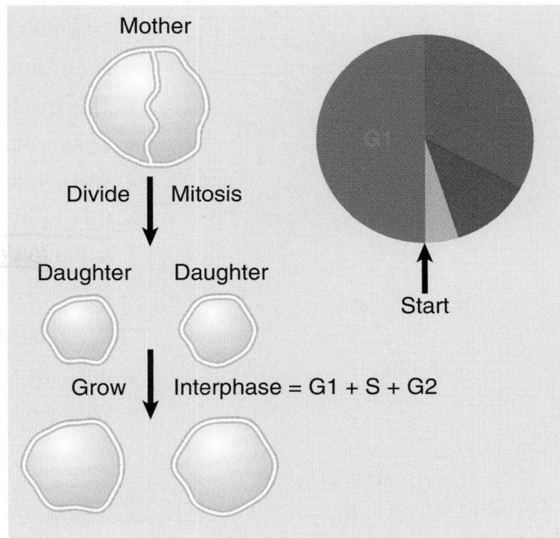

FIGURE 15.9 A growing cell alternates between cell division of a mother cell into two daughter cells and growth back to the original size.

Replication of the large amount of DNA contained in a eukaryotic chromosome is accomplished by dividing it into many individual replicons. Only some of these replicons are engaged in replication at any point in S phase. Presumably each replicon is activated at a specific time during S phase, although the evidence on this issue is not decisive.

The start of S phase is signaled by the activation of the first replicons. Over the next few hours, initiation events occur at other replicons in an ordered manner. Chromosomal replicons usually display bidirectional replication.

Individual replicons in eukaryotic genomes are relatively small, typically ~40 kb in yeast or fly and ~100 kb in animal cells. However, they can vary more than tenfold in length within a genome. The rate of replication is ~2000 bp/min, which is much slower than the 50,000 bp/min of bacterial replication fork movement, presumably because the chromosome is assembled into chromatin, not naked DNA.

checkpoint A biochemical control mechanism that prevents the cell from progressing from one stage to the next unless specific goals and requirements have been met.

From the speed of replication, it is evident that a mammalian genome could be replicated in ~1 hour if all replicons functioned simultaneously. But S phase actually lasts for >6 hours in a typical somatic cell, which implies that no more than 15% of the replicons are likely to be active at any given moment. There are some exceptional cases, such as the early embryonic divisions of *Drosophila* embryos, where the duration of S phase is compressed by the simultaneous functioning of a large number of replicons.

How are origins selected for initiation at different times during S phase? There is a general hierarchy to the order of replication. Replicons near active genes are replicated earliest and replicons in heterochromatin replicate last.

Available evidence suggests that most chromosomal replicons do not have a termination region like that of bacteria, at which the replication forks cease movement and (presumably) dissociate from the DNA. It seems more likely that a replication fork continues from its origin until it meets a fork proceeding toward it from the adjacent replicon. We have already mentioned the potential topological problem of joining the newly synthesized DNA at the junction of the replication forks.

The propensity of replicons located in the same vicinity to be active at the same time could be explained by "regional" controls, in which groups of replicons are initiated more or less coordinately, as opposed to a mechanism in which individual replicons are activated one by one in dispersed areas of the genome. Two structural features suggest the possibility of large-scale organization. Quite large regions of the chromosome can be characterized as "early replicating" or "late replicating," implying that there is little interspersion of replicons that fire at early or late times. Visualization of replicating forks by labeling with DNA precursors identifies 100 to 300 "foci" instead of uniform staining; each focus shown in **FIGURE 15.10** probably contains >300 replication forks. The foci could represent fixed structures through which replicating DNA must move.

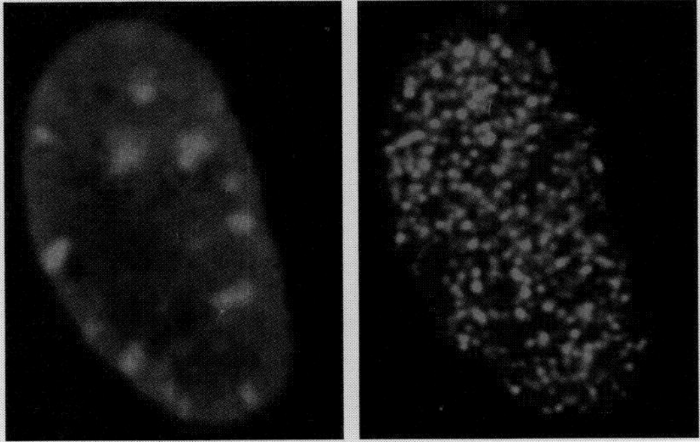

FIGURE 15.10 Replication forks are organized into foci in the nucleus. Cells were labeled with BrdU. The leftmost panel was stained with propidium iodide to identify bulk DNA. The right panel was stained using an antibody to BrdU to identify replicating DNA. Photo courtesy of Anthony D. Mills and Ron Laskey, Hutchinson/MRC Research Center, University of Cambridge.

KEY CONCEPTS

- A chromosome is divided into many replicons.
- The progression into S phase is tightly controlled.
- Eukaryotic replicons are 40 to 100 kb in length.
- Individual replicons are activated at characteristic times during S phase.
- Regional activation patterns suggest that replicons near one another are activated at the same time.

CONCEPT AND REASONING CHECKS

- Why do bacteria have only one *ori* whereas eukaryotes have many per chromosome?
- Why would organisms like *Drosophila* need such high rates of embryonic replication?

15.6 Replication Origins Bind the ORC

▸ ***ARS*** An origin for replication in yeast. The common feature among different examples of these sequences is a conserved 11 bp sequence called the A domain.

▸ **ORC** Origin Recognition Complex, found in eukaryotes, a multiprotein complex that binds to the replication origin, the ARS, and remains associated with it throughout the cell cycle.

Because a eukaryote chromosome contains many replicons, we cannot identify origins directly by mutations that prevent replication, as we can with a bacterial chromosome. However, any segment of DNA that has an origin should be able to replicate. This provided the first means to identify an origin by testing sequences for their ability to support replication of artificially created independent DNA molecules. A sequence that confers the ability to replicate efficiently in yeast is called an ***ARS*** (for Autonomously Replicating Sequence). *ARS* elements are derived from origins of replication.

An *ARS* element consists of an A-T-rich region that contains discrete sites in which mutations affect origin function. Base composition rather than sequence may be important in the rest of the region. **FIGURE 15.11** shows a systematic mutational analysis along the length of an origin. Origin function is abolished completely by mutations in a 14-bp "core" region, called the **A domain**, which contains an 11-bp consensus sequence consisting of A-T base pairs. This consensus sequence (sometimes called the *ACS* for ARS Consensus Sequence) is the only homology between known *ARS* elements.

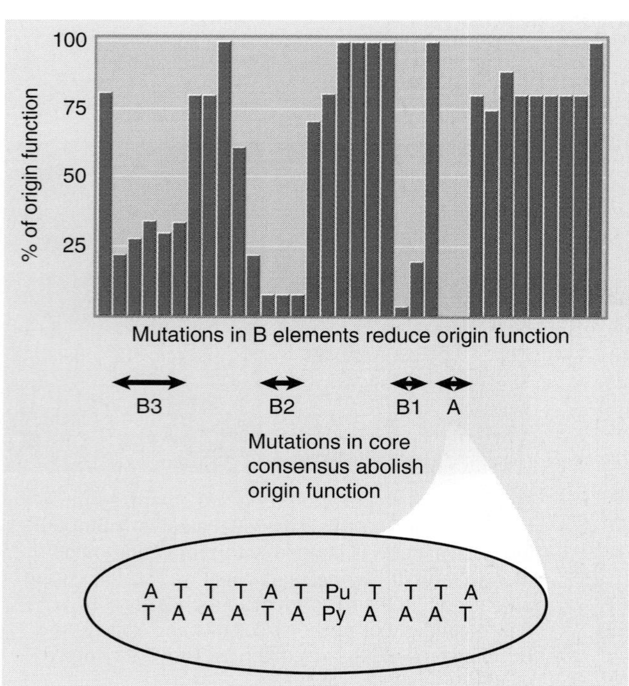

FIGURE 15.11 An *ARS* extends for ~50 bp and includes a consensus sequence (A) and additional elements (B1–B3).

Mutations in three adjacent elements, numbered B1 to B3, reduce origin function. An origin can function effectively with any two of the B elements, so long as a functional A element is present. (Imperfect copies of the core consensus, typically conforming at 9 out of 11 positions, are found close to, or overlapping with, each B element, but they do not appear to be necessary for origin function.)

The **ORC** (origin recognition complex) is a complex of six proteins with a mass of ~400 kD. ORC binds to the A and B1 elements on the A-T-rich strand and is associated with *ARS* elements throughout the cell cycle. This means that initiation depends on changes in its condition rather than *de novo* association with an origin (see *Section 15.7, Licensing Factor Controls Eukaryotic Rereplication and Consists of MCM Proteins*). By counting the number of sites to which ORC binds, we can estimate that there are about 400 origins of replication in the yeast genome. This means that the average length of a replicon is ~35,000 bp. Counterparts to ORC are found in multicellular eukaryotic cells.

ORC was first found in *S. cerevisiae* (where it is called scORC), but similar complexes have now been characterized in *Schizosaccharomyces pombe* (spORC), *Drosophila* (DmORC), and *Xenopus* (XlORC). All of the ORC complexes bind to DNA. Although none of the binding sites have been characterized in the same detail as in *S. cerevisiae,* in several cases they are at locations associated with the initiation of replication. It seems clear that ORC is an initiation complex whose binding identifies an origin of replication. Although the ORC is conserved, thus far it has not been possible to identify origins of replication in higher eukaryotic DNA that function like *ARS* elements.

KEY CONCEPTS

- Origins in *S. cerevisiae* are short A-T-rich sequences that have an essential 11-bp sequence.
- The ORC is a complex of six proteins that binds to an ARS. It remains bound through the cell cycle, but is activated only during S phase.
- Related ORC complexes are found in higher eukaryotes.

CONCEPT AND REASONING CHECK

Why do origins of replication contain a core of A-T rich DNA?

15.7 Licensing Factor Controls Eukaryotic Rereplication and Consists of MCM Proteins

A eukaryotic genome is divided into multiple replicons, and the origin in each replicon is activated once and only once in a single division cycle. This could be achieved by the provision of some rate-limiting component that functions only once at an origin or by the presence of a repressor that prevents rereplication at origins that have been used. The critical questions about the nature of this regulatory system are how the system determines whether any particular origin has been replicated and what protein components are involved.

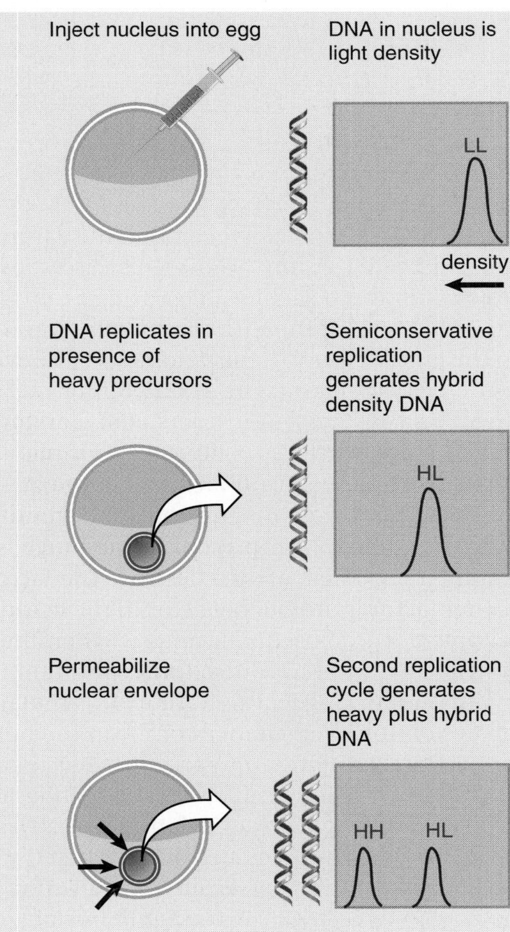

FIGURE 15.12 A nucleus injected into a Xenopus egg can replicate only once unless the nuclear membrane is permeabilized to allow subsequent replication cycles.

Insights into the nature of the protein components have been provided by using a system in which a substrate DNA undergoes only one cycle of replication. *Xenopus* eggs have all the components needed to replicate DNA—in the first few hours after fertilization they undertake eleven division cycles without new gene expression—and they can replicate the DNA in a nucleus that is injected into the egg. **FIGURE 15.12** summarizes the features of this system.

When a sperm or interphase nucleus is injected into the egg, its DNA is replicated only once (this can be followed by use of a density label, just like the original experiment that characterized semiconservative replication, shown in the Historical Perspectives box in this chapter). If protein synthesis is blocked in the egg, the membrane around the injected material remains intact and the DNA cannot replicate again. In the presence

FIGURE 15.13 Licensing factor in the nucleus is inactivated after replication. A new supply of licensing factor can enter only when the nuclear membrane breaks down at mitosis.

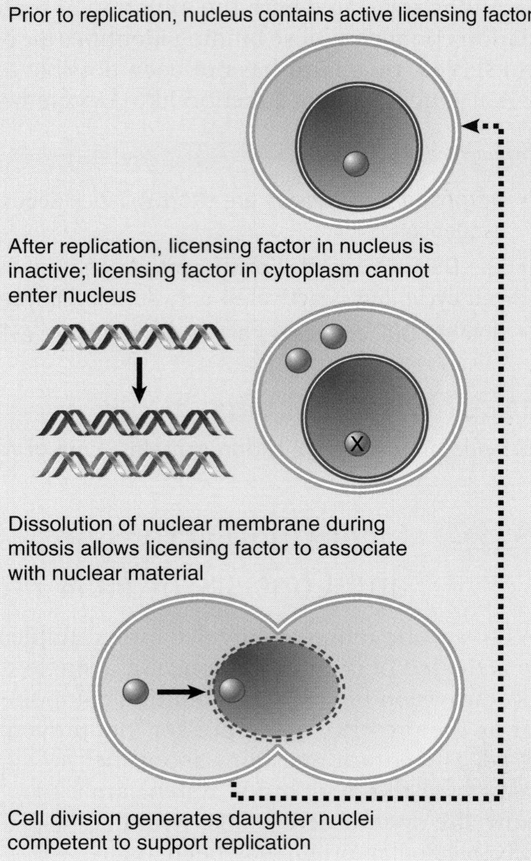
Prior to replication, nucleus contains active licensing factor
After replication, licensing factor in nucleus is inactive; licensing factor in cytoplasm cannot enter nucleus
X
Dissolution of nuclear membrane during mitosis allows licensing factor to associate with nuclear material
Cell division generates daughter nuclei competent to support replication

of protein synthesis, however, the nuclear membrane breaks down just as it would for a normal cell division, and in this case subsequent replication cycles can occur. The same result can be achieved by using agents that permeabilize the nuclear membrane. This suggests that the nucleus contains a protein(s) needed for replication that is used up in some way by a replication cycle, so even though more of the protein is present in the egg cytoplasm, it can only enter the nucleus if the nuclear membrane breaks down. The system can in principle be taken further by developing an *in vitro* extract that supports nuclear replication, thus allowing the components of the extract to be isolated and the relevant factors identified.

FIGURE 15.13 explains the control of reinitiation by proposing that this protein is a **licensing factor**. It is present in the nucleus prior to replication. One round of replication either inactivates or destroys the factor, and another round cannot occur until further factor is provided. Factor in the cytoplasm can gain access to the nuclear material only at the subsequent mitosis when the nuclear envelope breaks down.

▸ **licensing factor** A factor located in the nucleus and necessary for replication; it is inactivated or destroyed after one round of replication. New factors must be provided for further rounds of replication to occur.

The key event in controlling replication is the behavior of the ORC complex at the origin. The striking feature is that ORC remains bound at the origin through the entire cell cycle. However, changes occur in the binding of other proteins to the ORC-origin complex. FIGURE 15.14 summarizes the cycle of events at the origin.

At the end of the cell cycle, ORC is bound to A–B1 elements of the origin and protects the origin against degradation by DNAase, but there is a site that is hypersensitive to the enzyme in the center of B1. There is a change during G1, seen most strikingly by the loss of the hypersensitive site. This results from the binding of Cdc6 protein to the ORC. In yeast, Cdc6 is a highly unstable protein, with a half-life of <5 minutes. It is synthesized during G1 and typically binds to the ORC between the exit from mitosis and late G1. Its rapid degradation means that no protein is available later in the

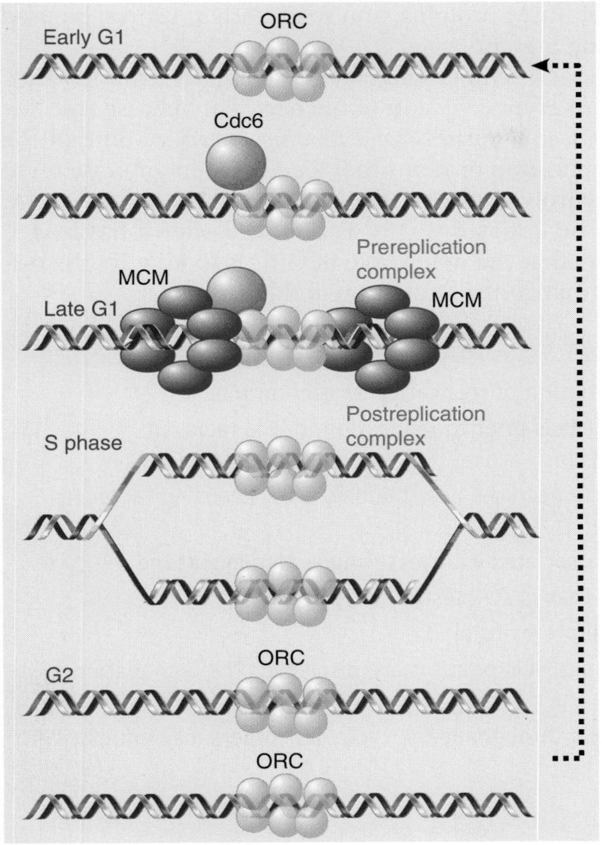

FIGURE 15.14 Proteins at the origin control susceptibility to initiation.

cycle. In mammalian cells Cdc6 is controlled differently; it is phosphorylated during S phase, and as a result it is exported from the nucleus. This feature makes Cdc6 the key licensing factor. Cdc6 also provides the connection between ORC and a complex of proteins that is involved in initiation. Cdc6 has an ATPase activity that is required for it to support initiation.

The licensing factor and the system that controls its availability in yeast is identified by two different types of mutations:

- The licensing factor is identified by mutations in *MCM2,3,5*, which prevent initiation of replication.
- Mutations that have the opposite effect, and allow the accumulation of excess quantities of DNA, are found in genes that code for components of the ubiquitination system that is responsible for programmed degradation of specific proteins. This suggests that licensing factor may be destroyed after the start of the replication cycle.

In yeast, free MCM2,3,5 enter the nucleus only during mitosis. Homologs are found in animal cells, where MCM3 is bound to chromosomal material before replication, but is released after replication. The animal cell MCM2,3,5 complex remains in the nucleus throughout the cell cycle, suggesting that it may be only one component of the licensing factor. Another component, able to enter only at mitosis, may be necessary for MCM2,3,5 to associate with chromosomal material.

The presence of Cdc6 at the yeast origin allows MCM proteins to bind to the complex. Their presence is necessary for initiation. The origin therefore enters S phase in the condition of a **prereplication complex**, which contains ORC, Cdc6, and MCM proteins. When initiation occurs, Cdc6 is displaced and MCM is converted to the active helicase for elongation, returning the origin to the state of the **postreplication complex**, which contains only ORC. Because Cdc6 is rapidly degraded during S phase, it is

- **prereplication complex** A protein-DNA complex at the origin in *S. cerevisiae* that is required for DNA replication. The complex contains the ORC complex, Cdc6, and the MCM proteins.
- **postreplication complex** A protein-DNA-complex in *S. cerevisiae* that consists of the ORC complex bound to the origin.

not available to support reloading of MCM proteins, and so the origin cannot be used for a second cycle of initiation during S phase.

The MCM2-7 proteins form a six-member ring-shaped complex around DNA. This complex is believed to be the eukaryotic version of the bacterial helicase (see *Section 18.7, Replication Requires a Helicase and Single-Strand Binding Protein*). Some of the ORC proteins have similarities to replication proteins that load DNA polymerase on to DNA. It is possible that ORC uses hydrolysis of ATP to load the MCM ring on to DNA. In *Xenopus* extracts, replication can be initiated if ORC is removed after it has loaded Cdc6 and MCM proteins. This shows that the major role of ORC is to identify the origin to the Cdc6 and MCM proteins that control initiation and licensing.

KEY CONCEPTS

- Licensing factor is necessary for initiation of replication at each origin.
- Licensing factor is present in the nucleus prior to replication, but is removed, inactivated, or destroyed by replication.
- Initiation of another replication cycle becomes possible only after licensing factor re-enters the nucleus after mitosis.
- The ORC is a protein complex that is associated with yeast origins throughout the cell cycle.
- Cdc6 protein is an unstable protein that is synthesized only in G1.
- Cdc6 binds to ORC and allows MCM proteins to bind.
- When replication is initiated, Cdc6 and MCM proteins are displaced. The degradation of Cdc6 prevents reinitiation.
- Some MCM proteins are in the nucleus throughout the cycle, but others may enter only after mitosis.

CONCEPT AND REASONING CHECK

What mechanism do you think could allow some proteins to have long half-lives and other proteins to have short half-lives?

15.8 Summary

Replicons in bacterial or eukaryotic chromosomes have a single unifying feature: replication is initiated at an origin, once and only once in each cell cycle. The origin is located within the replicon, and replication typically is bidirectional, with replication forks proceeding away from the origin in both directions. Replication is not usually terminated at specific sequences, but continues until DNA polymerase meets another DNA polymerase halfway around a circular replicon, or at the junction between two linear replicons.

An origin consists of a discrete sequence at which replication of DNA is initiated. Origins of replication tend to be rich in A-T base pairs. A bacterial chromosome contains a single origin, which is responsible for initiating replication once every cell cycle. The *oriC* in *E. coli* is a sequence of 245 bp. Any DNA molecule with this sequence can replicate in *E. coli*. Replication of the circular bacterial chromosome produces a θ structure, in which the replicated DNA starts out as a small replicating eye. Replication proceeds until the eye occupies the whole chromosome. The bacterial origin contains sequences that are methylated on both strands of DNA. Replication produces hemimethylated DNA, which cannot function as an origin. There is a delay before the hemimethylated origins are remethylated to convert them to a functional state, and this is responsible for preventing improper reinitiation.

A eukaryotic chromosome is divided into many individual replicons. Replication occurs during a discrete part of the cell cycle called S phase, but because not all replicons are active simultaneously, the process may take several hours. Eukaryotic replication is at least an order of magnitude slower than bacterial replication. Origins sponsor bidirectional replication and are probably used in a fixed order during S phase. Each replicon is activated only once in each cycle. Origins of replication were isolated as *ARS* sequences in yeast by virtue of their ability to support replication of any sequence attached to them. The core of an *ARS* is an 11 bp A-T rich sequence that is bound by the ORC protein complex, which remains bound throughout the cell cycle. Utilization of the origin is controlled by the MCM licensing factors that associate with the ORC.

CHAPTER QUESTIONS

1. For cell division to occur, DNA replication in the cell must be:
A. initiated and have cleared the origin.
B. at least halfway completed.
C. at least 80% completed.
D. 100% completed.

2. Replication origins are:
A. *cis*-acting elements.
B. *trans*-acting elements.
C. both *cis*- and *trans*-acting elements.
D. not considered either *cis*- or *trans*-acting elements.

3. Bacterial plasmids may be present in:
A. single copy
B. multicopy.
C. the host chromosome as an integrated copy.
D. any of the above.

4. How many copies of the mitochondrial genome are present in each mitochondrion?
A. less than one
B. one
C. two
D. more than two

5. What Greek letter is used to describe how DNA replicates?
A. θ
B. γ
C. β
D. ω

6. Differential labeling of replicating eukaryotic cell DNA showed that this genome replicates by:
A. bidirectional replication from multiple origins.
B. unidirectional replication from multiple origins.
C. bidirectional replication from a single origin.
D. unidirectional replication from a single origin.

7. In both *E. coli* and mammalian chromosomes, DNA replication is:
A. unidirectional and semiconservative.
B. bidirectional and semiconservative.
C. unidirectional and conservative.
D. bidirectional and conservative.

8. Replication origins in the yeast chromosome have a tendency to be:
A. AT-rich.
B. GC-rich.
C. AG-rich.
D. TC-rich.

9. In yeast, what protein binds to ORC to support initiation of DNA replication?
   A. SeqA
   B. Cdc2
   C. Cdc6
   D. DnaA
10. What proportion of replicons in mammalian cells are actively replicating at any given time?
   A. ~10%
   B. ~15%
   C. ~25%
   D. ~60%

KEY TERMS

ARS
bidirectional replication
checkpoint
equilibrium density-gradient centrifugation
growing point
hemimethylated DNA
licensing factor
multicopy replication control
ORC
origin
postreplication complex
prereplication complex
replication eye
replication fork
replicon
S phase
semiconservative replication
single-copy replication control

FURTHER READING

Bastia, D., Zzaman, S., Krings, G., Saxena, M., Peng, X., and Greenberg, M. M. (2008). Replication termination mechanism as revealed by Tus-mediated polar arrest of a sliding helicase. *Proc. Natl. Acad. Sci. USA* 105, 12831–12836.
A report describing bacterial transcription termination.

Bell, S. P. and Dutta, A. (2002). DNA replication in eukaryote cells. *Annu. Rev. Biochem*. 71, 333–374.

Gilbert, D. M. (2001). Making sense of eukaryote DNA replication origins. *Science* 294, 96–100.
A review describing how origins of replication differ between eukaryotes and prokaryotes.

16

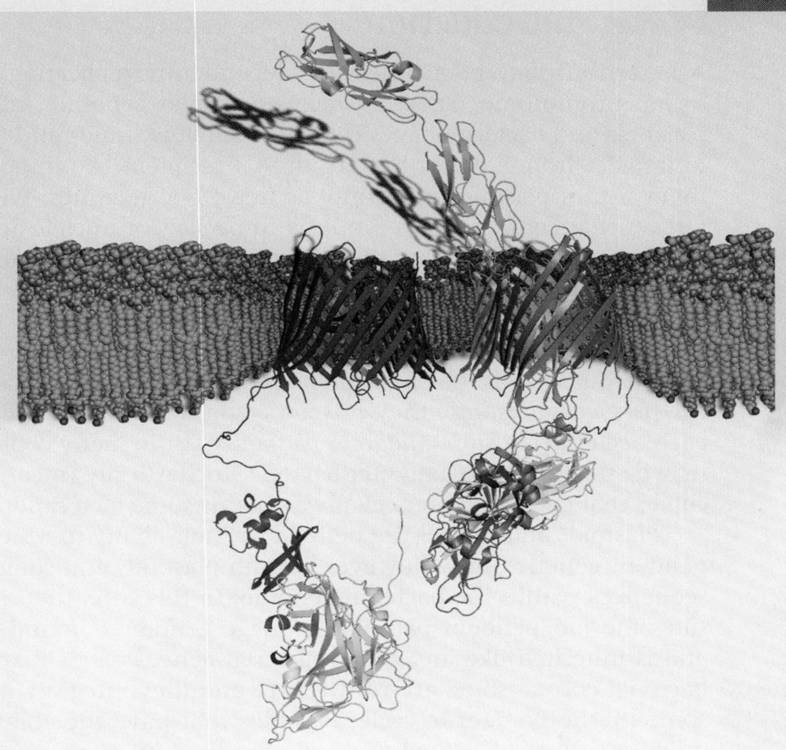

A model depicting pilus assembly in bacteria. The pilus is being built at the bacterial outer membrane on an assembly platform consisting of two "usher" complexes that include membrane channels through which the pilus components pass. Photo courtesy of Dr. Hans Remaut* and Gabriel Waksman, ISMB–UK. [*present address: Structural Biology Brussels, VUB/VIB, Belgium].

Extrachromosomal Replicons

CHAPTER OUTLINE

16.1 Introduction

A bacterium may be a host for independently replicating genetic units in addition to its chromosome. These extrachromosomal genomes fall into two general types: plasmids and bacteriophages (phages). Some plasmids and all phages have the ability to transfer from a donor bacterium to a recipient. An important distinction between them is that plasmids exist only as free DNA genomes, whereas bacteriophages are viruses that contain genes required to package a nucleic acid genome into a protein coat and are released from the bacterium at the end of an infective cycle.

▸ **plasmid** Circular, extrachromosomal DNA. It is autonomous and can replicate itself.

Plasmids are self-replicating circular molecules of DNA that are maintained in the cell in a stable and characteristic number of copies; that is, the number remains constant from generation to generation. *Single-copy plasmids* are maintained at the same relative quantity as the bacterial host chromosome, at one per unit bacterium. As with the host chromosome, they rely on a specific apparatus to be segregated equally at each bacterial division. *Multicopy plasmids* exist in many copies per unit bacterium and may be segregated to daughter bacteria stochastically (meaning that there are enough copies that each daughter cell always gains some by a random distribution).

Plasmids and phages are defined by their ability to reside in a bacterium as independent genetic units. However, certain plasmids and some phages can also exist as sequences within the bacterial genome. In this case, the same sequence that constitutes the independent plasmid or phage genome is found within the chromosome, and is inherited like any other bacterial gene. Phages that are found as part of the bacterial chromosome are **temperate** and they integrate into the bacterial chromosome by the **lysogenic** cycle. Plasmids that have the ability to behave like this are called **episomes**. Related processes are used by phages and episomes to insert into and excise from the bacterial chromosome.

▸ **temperate phage** A phage that can enter a lysogenic cycle within the host (can become a prophage integrated into the host genome).

▸ **lysogenic** The ability of a phage to survive in a bacterium as a stable prophage component of the bacterial genome.

▸ **episome** A plasmid able to integrate into bacterial DNA.

▸ **immunity** In phages, the ability of a prophage to prevent another phage of the same type from infecting a cell. In plasmids, the ability of a plasmid to prevent another of the same type from becoming established in a cell. It can also refer to the ability of certain transposons to prevent others of the same type from transposing to the same DNA molecule.

A parallel between lysogenic phages and plasmids and episomes is that they maintain a selfish possession of their bacterium and often make it impossible for another element of the same type to become established. This effect is called **immunity**, although the molecular basis for plasmid immunity is different from lysogenic immunity, and is a consequence of the replication control system.

FIGURE 16.1 summarizes the types of genetic units that can be propagated in bacteria as independent genomes. Virulent phages may have genomes of any type of nucleic acid, DNA or RNA, double stranded or single stranded; they transfer between cells by release of infective particles. Temperate phages have double-stranded DNA genomes, as do plasmids and episomes. Some plasmids transfer between cells by a conjugative process (with direct contact between donor and recipient cells). A feature of the transfer process in both cases is that on occasion some bacterial host genes are transferred with the phage or plasmid DNA, so these events play a role in allowing exchange of genetic information between bacteria.

The key feature in determining the behavior of each type of unit is how its origin is used. An origin in a bacterial or eukaryotic chromosome is used to initiate a single

FIGURE 16.1 Several types of independent genetic units exist in bacteria.

Type of unit	Genome structure	Mode of propagation	Consequences
Lytic phage	ds- or ssDNA or RNA; linear or circular	Infects susceptible host	Usually kills host
Lysogenic phage	dsDNA	Linear sequence in host chromosome	Immunity to infection
Plasmid	dsDNA circle	Replicates at defined copy number; may be transmittable	Immunity to plasmids in same group
Episome	dsDNA circle	Free circle or linear integration	May transfer host DNA

replication event that extends across the replicon. Replicons, however, can also be used to sponsor other forms of replication. The most common alternative is used by the small, independently replicating units of viruses. The objective of a viral replication cycle is to produce many copies of the viral genome before the host cell is lysed to release them. Some viruses replicate in the same way as a host genome, with an initiation event leading to production of duplicate copies, each of which then replicates again, and so on. Others use a mode of replication in which many copies are produced as a tandem array following a single initiation event. A similar type of event is triggered by episomes when an integrated plasmid DNA ceases to be inert and initiates a replication cycle.

Many prokaryotic replicons are circular, and this indeed is a necessary feature for replication modes that produce multiple tandem copies. Some extrachromosomal replicons are linear, though, and in such cases we have to account for the ability to replicate the end of the replicon. (Of course, eukaryotic chromosomes are linear, so the same problem applies to the replicons at each end. These replicons, however, have a special system for resolving the problem.)

16.2 The Ends of Linear DNA Are a Problem for Replication

None of the replicons that we have considered so far have a linear end: either they are circular (as in the *E. coli* or mitochondrial genomes) or they are part of longer segregation units (as in eukaryotic chromosomes). Linear replicons do occur, though—in some cases as single extrachromosomal units, and of course at the ends of eukaryotic chromosomes.

The end of a linear replicon poses a problem for DNA polymerases. First, they can only synthesize from 5′ to 3′ and second, there must be a primer. Consider the two parental strands depicted in **FIGURE 16.2**. The lower strand presents no problem. It can act as a template to synthesize a daughter strand that runs right up to the end, where presumably the polymerase falls off. To synthesize a complement at the end of the upper strand, however, synthesis must start right at the very last base, or else this strand would become shorter in successive cycles of replication.

No DNA replication polymerase can initiate replication. It can only extend what is there already. We usually think of a polymerase as binding at a site *surrounding* the position at which a base is to be incorporated. Thus a special mechanism must be employed for replication at the ends of linear replicons to provide a 3′–OH group. Several types of solution may be imagined to accommodate the need to copy a terminus:

- The problem may be circumvented by converting a linear replicon into a circular or multimeric molecule. Phages such as T4 or lambda use such mechanisms (see *Section 16.4, Rolling Circles Produce Multimers of a Replicon*).
- The DNA may form an unusual structure—for example, by creating a hairpin at the terminus, so that there is no free end. Formation of a crosslink is involved in replication of the linear mitochondrial DNA of *Paramecium.*
- Instead of being precisely determined, the end may be variable. Eukaryotic chromosomes may adopt this solution, in which the number of copies of a short repeating unit at the end of the DNA changes (see *Section 23.10, Telomeres Have Simple Repeating Sequences*). A mechanism to add or remove units makes it unnecessary to replicate right up to the very end.

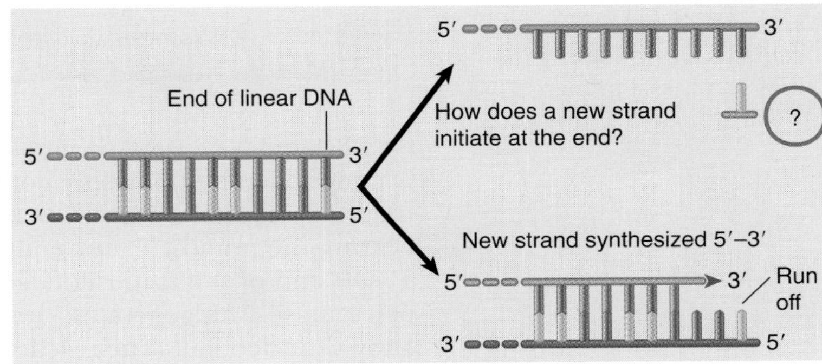

FIGURE 16.2 Replication can run off the end of the template strand containing a 5′ end, but can it initiate on the template strand with a 3′ end?

- A protein may intervene to make initiation possible at the actual terminus. Several linear viral nucleic acids have proteins that are *covalently linked to the 5′ terminal base*. The best characterized examples are adenovirus DNA, phage φ29 DNA, and poliovirus RNA.

KEY CONCEPT

- Special arrangements must be made to replicate the DNA strand with a 5′ end.

CONCEPT AND REASONING CHECK

What is the problem during replication at the end of a linear chromosome?

16.3 Terminal Proteins Enable Initiation at the Ends of Viral DNAs

Linear DNA
5′
3′
3′
5′
DNA synthesis initiates at left 5′ end
Fork proceeds
Single strand is displaced when fork reaches end
Termini base pair to form duplex origin
DNA synthesis proceeds

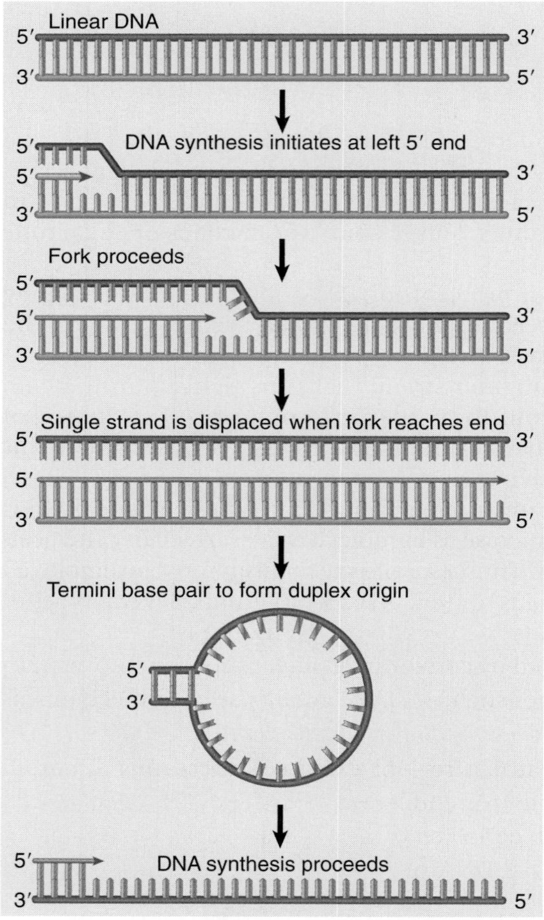

FIGURE 16.3 Adenovirus DNA replication is initiated separately at the two ends of the molecule and proceeds by strand displacement.

▸ **strand displacement** A mode of replication of some viruses in which a new DNA strand grows by displacing the previous (homologous) strand of the duplex.

▸ **terminal protein** A protein that allows replication of a linear phage genome to start at the very end. It attaches to the 5′ end of the genome through a covalent bond, is associated with a DNA polymerase, and contains a cytosine residue that serves as a primer.

An example of initiation at a linear end is provided by adenovirus and φ29 DNAs, which actually replicate from both ends using the mechanism of **strand displacement** illustrated in **FIGURE 16.3**. The same events can occur independently at either end. Synthesis of a new strand starts at one end, displacing the homologous strand that was previously paired in the duplex. When the replication fork reaches the other end of the molecule, the displaced strand is released as a free single strand. It is then replicated independently; this requires the formation of a duplex origin by base pairing between some short complementary sequences at the ends of the molecule.

In several viruses that use such mechanisms, a protein is found covalently attached to each 5′ end. In the case of adenovirus, a **terminal protein** is linked to the mature viral DNA via a phosphodiester bond to serine, as indicated in **FIGURE 16.4**.

How does the attachment of the protein overcome the initiation problem? The terminal protein has a dual role: it carries a cytidine nucleotide that provides the primer, and it is associated with DNA polymerase. In fact, linkage of terminal protein to a nucleotide is undertaken by DNA polymerase in the presence of adenovirus DNA. This suggests the model illustrated in **FIGURE 16.5**. The complex of polymerase and terminal protein, bearing the priming C nucleotide, binds to the end of the adenovirus DNA. The free 3′–OH end of the C nucleotide is used to prime the elongation reaction by the DNA polymerase. This generates a new strand whose 5′ end is covalently linked to the initiating C nucleotide. (The reaction actually involves displacement of protein from DNA rather than binding *de novo*. The 5′ end of adenovirus DNA is bound to the terminal

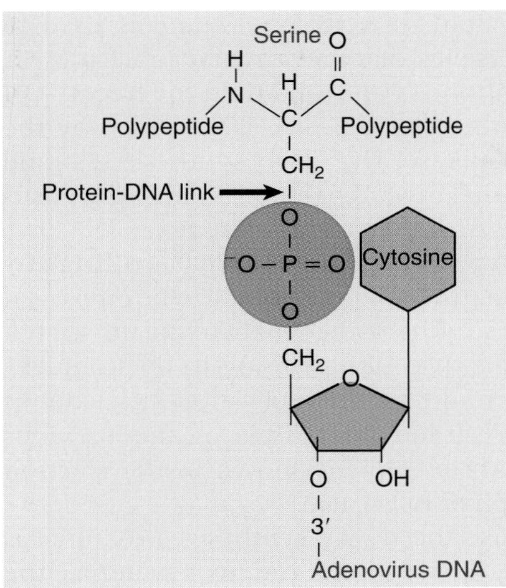

FIGURE 16.4 The 5′ terminal phosphate at each end of adenovirus DNA is covalently linked to serine in the 55 kD AD-binding protein.

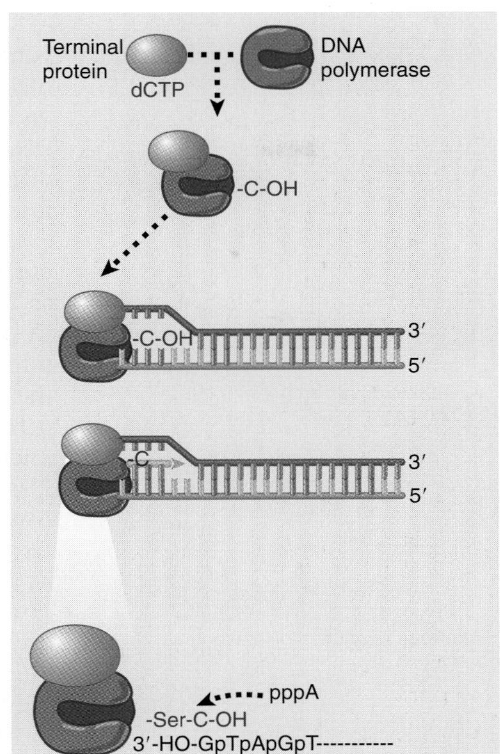

FIGURE 16.5 Adenovirus terminal protein binds to the 5′ end of DNA and provides a C–OH end to prime synthesis of a new DNA strand.

protein that was used in the previous replication cycle. The old terminal protein is displaced by the new terminal protein for each new replication cycle.)

Terminal protein binds to the region located between 9 and 18 bp from the end of the DNA. The adjacent region, between positions 17 and 48, is essential for the binding of a host protein, nuclear factor I, which is also required for the initiation reaction. The initiation complex may, therefore, form between positions 9 and 48, a fixed distance from the actual end of the DNA.

KEY CONCEPTS

- The dsDNA viruses adenovirus and φ29 have terminal proteins that initiate replication by generating a new 5′ end.
- The newly synthesized strand displaces the corresponding strand of the original duplex.
- The strand that is released base pairs at the ends to form a duplex origin that initiates synthesis of the complementary strand.

CONCEPT AND REASONING CHECK

How can a DNA polymerase use a protein as a primer for replication?

16.4 Rolling Circles Produce Multimers of a Replicon

The structures generated by replication depend on the relationship between the template and the replication fork. The critical features are whether the template is circular or linear, and whether the replication fork is engaged in synthesizing both strands of DNA or only one.

Replication of only one strand is used to generate copies of some circular molecules. A nick opens one strand, and then the free 3′–OH end generated by the nick is extended by the DNA polymerase. The newly synthesized strand displaces the original parental strand. The ensuing events are depicted in **FIGURE 16.6**.

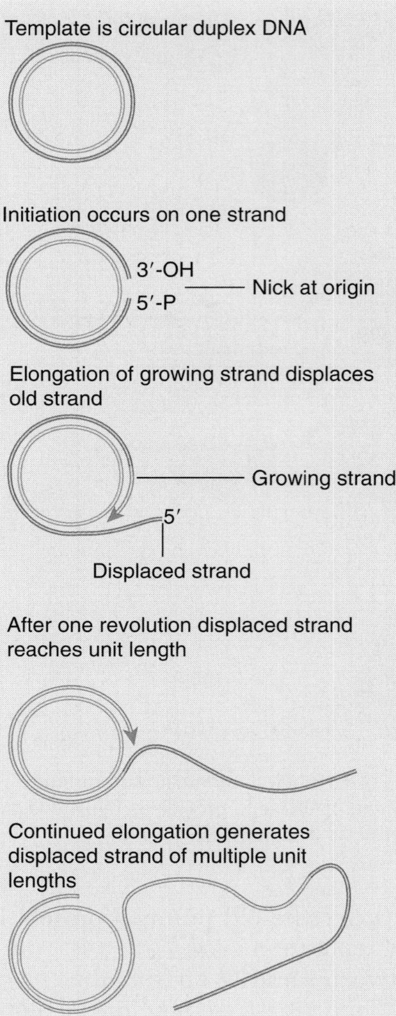
Template is circular duplex DNA
Initiation occurs on one strand
3′-OH
5′-P
Nick at origin
Elongation of growing strand displaces old strand
Growing strand
5′
Displaced strand
After one revolution displaced strand reaches unit length
Continued elongation generates displaced strand of multiple unit lengths

FIGURE 16.6 The rolling circle generates a multimeric single-stranded tail.

‣ **rolling circle** A mode of replication in which a replication fork proceeds around a circular template for an indefinite number of revolutions; the DNA strand newly synthesized in each revolution displaces the strand synthesized in the previous revolution, giving a tail containing a linear series of sequences complementary to the circular template strand.

This type of structure is called a **rolling circle**, because the growing point can be envisaged as rolling around the circular template strand. In principle, it could continue to do so indefinitely. As it moves, the replication fork extends the outer strand and displaces the previous partner. An example is shown in the electron micrograph of **FIGURE 16.7**.

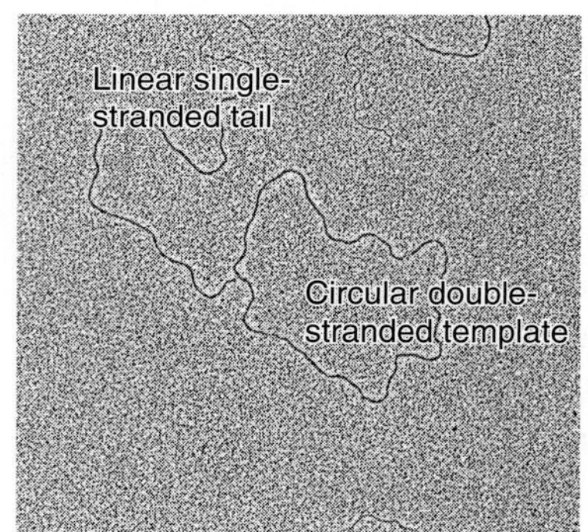
Linear single-stranded tail
Circular double-stranded template

FIGURE 16.7 A rolling circle appears as a circular molecule with a linear tail by electron microscopy. Photo courtesy of Ross B. Inman, Institute of Molecular Virology, Bock Laboratory and Department of Biochemistry, University of Wisconsin, Madison, Wisconsin, USA.

Because the newly synthesized material is covalently linked to the original material, the displaced strand has the original unit genome at its 5′ end. The original unit is followed by any number of unit genomes, synthesized by continuing revolutions of the template. Each revolution displaces the material synthesized in the previous cycle.

The rolling circle is used in several ways *in vivo*. Some pathways that are used to replicate DNA are depicted in **FIGURE 16.8**.

Cleavage of a unit length tail generates a copy of the original circular replicon in linear form. The linear form may be maintained as a single strand, or may be converted into a duplex by synthesis of the complementary strand (which is identical in sequence to the template strand of the original rolling circle).

The rolling circle provides a means for amplifying the original (unit) replicon. This mechanism is used to generate amplified ribosomal DNA (rDNA) in the *Xenopus* oocyte. The genes for ribosomal RNA (rRNA) are organized as a large number of contiguous repeats in the genome. A single repeating unit from the genome is converted into a rolling circle. The displaced tail, which contains many units, is converted into duplex DNA; later it is cleaved from the circle so that the two ends can be joined together to generate a large circle of amplified rDNA. The amplified material, therefore, consists of a large number of identical repeating units.

KEY CONCEPT

- A rolling circle generates single-stranded multimers of the original sequence.

CONCEPT AND REASONING CHECK

Why is a rolling circle model of replication advantageous for viral replication but not bacterial replication?

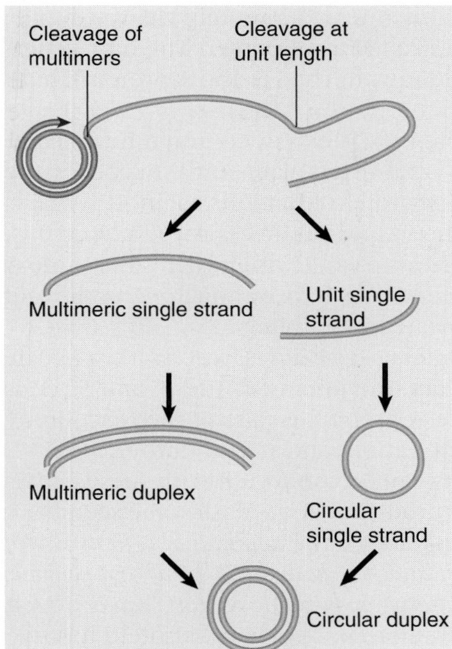

FIGURE 16.8 The fate of the displaced tail determines the types of products generated by rolling circles. Cleavage at unit length generates monomers, which can be converted to duplex and circular forms. Cleavage of multimers generates a series of tandemly repeated copies of the original unit. Note that the conversion to double-stranded form could occur earlier, before the tail is cleaved from the rolling circle.

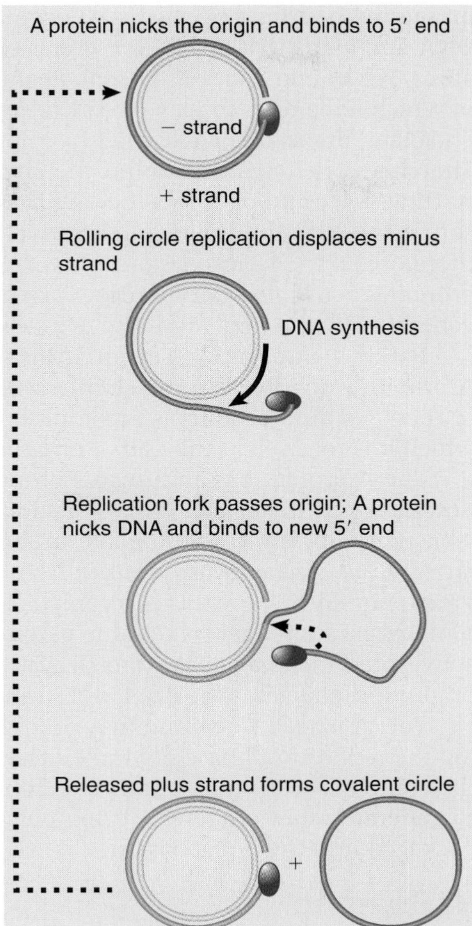

FIGURE 16.9 φX174 RF DNA is a template for synthesizing single-stranded viral circles. The A protein remains attached to the same genome through indefinite revolutions, each time nicking the origin on the viral (+) strand and transferring to the new 5′ end. At the same time, the released viral strand is circularized.

16.5 Rolling Circles Are Used to Replicate Phage Genomes

Replication by rolling circles is common among bacteriophages. Unit genomes can be cleaved from the displaced tail, generating monomers that can be packaged into phage particles or used for further replication cycles. Phage φX174 consists of a single-stranded circular DNA known as the plus (+) strand. A complementary strand, called the minus (–) strand, is synthesized as the first step in replication. This action generates the duplex circle shown at the top of **FIGURE 16.9**. Replication then proceeds by a rolling circle mechanism.

The duplex circle is converted to a covalently closed form, which becomes supercoiled. A protein coded by the phage genome, the A protein, nicks the (+) strand of the duplex DNA at a specific site that defines the origin for replication. After nicking the origin, the A protein remains connected to the 5′ end that it generates, while the 3′ end is extended by DNA polymerase.

The structure of the DNA plays an important role in this reaction, for the DNA can be nicked *only when it is negatively supercoiled* (i.e., wound about its axis in space in the opposite sense from the handedness of the double helix; see *Section 1.5, Supercoiling Affects the Structure of DNA*). The A protein is able to bind to a single-stranded decamer fragment of DNA that surrounds the site of the nick. This suggests that the supercoiling is needed to assist the formation of a single-stranded region that provides the A

▸ **relaxase** An enzyme that cuts one strand of DNA and binds to the free 5′ end.

protein with its binding site. (An enzymatic activity in which a protein cleaves duplex DNA and binds to a released 5′ end is sometimes called a **relaxase**.) The nick generates a 3′–OH end and a 5′–phosphate end (covalently attached to the A protein), both of which have roles to play in φX174 replication.

Using the rolling circle, the 3′–OH end of the nick is extended into a new chain. The chain is elongated around the circular (–) strand template until it reaches the starting point and displaces the origin. Now the A protein functions again. It remains connected with the rolling circle as well as to the 5′ end of the displaced tail, and is, therefore, in the vicinity as the growing point returns past the origin. Thus the same A protein is available again to recognize the origin and nick it, now attaching to the end generated by the new nick. The cycle can be repeated indefinitely.

Following this nicking event, the displaced single (+) strand is freed as a circle. The A protein is involved in the circularization. In fact, the joining of the 3′ and 5′ ends of the (+) strand product is accomplished by the A protein as part of the reaction by which it is released at the end of one cycle of replication, and starts another cycle.

The A protein has an unusual property that may be connected with these activities. It is *cis*-acting *in vivo.* (This behavior is not reproduced *in vitro,* as can be seen from its activity on any DNA template in a cell-free system.) *The implication is that in vivo the A protein synthesized by a particular genome can attach only to the DNA of that genome.* We do not know how this is accomplished. Its activity *in vitro,* however, shows how it remains associated with the same parental (–) strand template. The A protein has two active sites; this may allow it to cleave the "new" origin while still retaining the "old" origin; it then ligates the displaced strand into a circle.

The displaced (+) strand may follow either of two fates after circularization. During the replication phase of viral infection, it may be used as a template to synthesize the complementary (–) strand. The duplex circle may then be used as a rolling circle to generate more progeny. During phage morphogenesis, the displaced (+) strand is packaged into the phage virion.

KEY CONCEPT

- The φX A protein is a *cis*-acting relaxase that generates single-stranded circles from the tail produced by rolling circle replication.

CONCEPT AND REASONING CHECK

What happens to a virus that uses the rolling circle model of replication but has no means to cut monomers at discrete sites and cuts only unit lengths?

16.6 The F Plasmid Is Transferred by Conjugation between Bacteria

▸ **conjugation** A process in which two cells come in contact and transfer genetic material. In bacteria, DNA is transferred from a donor to a recipient cell. In protozoa, DNA passes from each cell to the other.

▸ **F plasmid** An episome that can be free or integrated in *E. coli,* and that can sponsor conjugation in either form.

An interesting example of a connection between replication and the propagation of a genetic unit is provided by bacterial **conjugation**, in which a plasmid genome or host chromosome is transferred from one bacterium to another.

Conjugation is mediated by the **F plasmid**, which is the classic example of an episome—an element that may exist as a free circular plasmid, or that may become integrated into the bacterial chromosome as a linear sequence like a temperate bacteriophage. The F plasmid is a large circular DNA ~100 kb in length, 5% of the *E. coli* genome size.

The F factor can integrate at multiple sites in the *E. coli* chromosome, often by a recombination event involving certain sequences (called IS sequences; see *Section 21.2, Insertion Sequences Are Simple Transposition Modules*) that are present on both the host chromosome and F plasmid. In its free (plasmid) form, the F plasmid utilizes its own replication origin (*oriV*) and control system, and is maintained at a level of one copy per bacterial chromosome. When it is integrated into the bacterial chromosome, this system is suppressed, and F DNA is replicated as a part of the chromosome.

The presence of the F plasmid, whether free or integrated, has important consequences for the host bacterium. Bacteria that are F^+, that is, that contain the episome, are able to conjugate (or mate) with bacteria that are F^-, that do not contain the episome. Conjugation involves contact between donor F^+ and recipient F^- bacteria; contact is followed by transfer of the F factor. If the F factor exists as a free plasmid in the donor bacterium, it is transferred as a plasmid, and the infective process converts the F^- recipient into an F^+ state. If the F factor is present in an integrated form in the donor, the transfer process may also cause some or all of the bacterial chromosome to be transferred. Many plasmids have conjugation systems that operate in a generally similar manner, but the F factor was the first to be discovered and remains the paradigm for this type of genetic transfer.

A large (~33 kb) region of the F plasmid called the **transfer region** is required for conjugation. It contains ~40 genes, named *tra* and *trb* loci, which are required for the transmission of DNA. Only four of the *tra* genes in the major transcription unit are concerned directly with the transfer of DNA; most are concerned with the properties of the bacterial cell surface and with maintaining contacts between mating bacteria.

F^+ bacteria possess surface appendages called **pili** (singular ***pilus***) that are encoded by the F factor. The gene *traA* codes for the single subunit protein, **pilin**, that is polymerized into the pilus. At least 12 *tra* genes are required for the modification and assembly of pilin into the pilus. The F-pili are hairlike structures, 2 to 3 μm long, that protrude from the bacterial surface. A typical F-positive cell has two to three pili. The pilin subunits are polymerized into a hollow cylinder, ~8 nm in diameter, with a 2 nm axial hole.

Mating is initiated when the tip of the F-pilus contacts the surface of the recipient F^- cell. FIGURE 16.10 shows an example of *E. coli* cells beginning to mate. An F^+ donor cell does not contact other F^+ cells, because the genes *traS* and *traT* code for "surface exclusion" proteins that make the cell a poor recipient in such contacts. This effectively restricts donor cells to mating with F^- cells. (The presence of F-pili has secondary consequences; they provide the sites to which RNA phages and some single-stranded DNA phages attach, so F^+ bacteria are susceptible to infection by these phages, whereas F^{--} bacteria are resistant.)

The initial contact between donor and recipient cells is easily broken, but other *tra* genes act to stabilize the association; this brings the mating cells closer together. The F-pili are essential for initiating pairing, but retract or disassemble as part of the process by which the mating cells are brought into close contact. There must be a channel through which DNA is transferred, but the pilus itself does not appear to provide it.

- **transfer region** A segment on the F plasmid that is required for bacterial conjugation.
- **pili** A surface appendage on a bacterium that allows the bacterium to attach to other bacterial cells. It appears as a short, thin, flexible rod. During conjugation, it is used to transfer DNA from one bacterium to another.
- **pilin** The subunit that is polymerized into the pilus in bacteria.

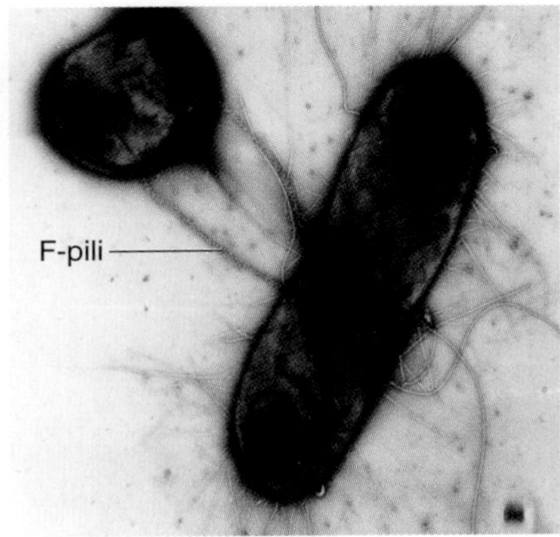

FIGURE 16.10 Mating bacteria are initially connected when donor F pili contact the recipient bacterium. Photo courtesy of Ron Skurray, School of Biological Sciences, University of Sydney.

KEY CONCEPTS

- A free F factor is a replicon that is maintained at the level of one plasmid per bacterial chromosome.
- An F factor can integrate into the bacterial chromosome, in which case its own replication system is suppressed.
- The F factor codes for specific pili that form on the surface of the bacterium.
- An F-pilus enables an F-positive bacterium to contact an F-negative bacterium and to initiate conjugation.

CONCEPT AND REASONING CHECK

How can an integrated F factor be used to map the bacterial chromosome?

16.7 Conjugation Transfers Single-Stranded DNA

Transfer of the F factor is initiated at a site called *oriT*, the origin of transfer, which is located at one end of the transfer region. The transfer process may be initiated when TraM recognizes that a mating pair has formed. TraY then binds near *oriT* and causes

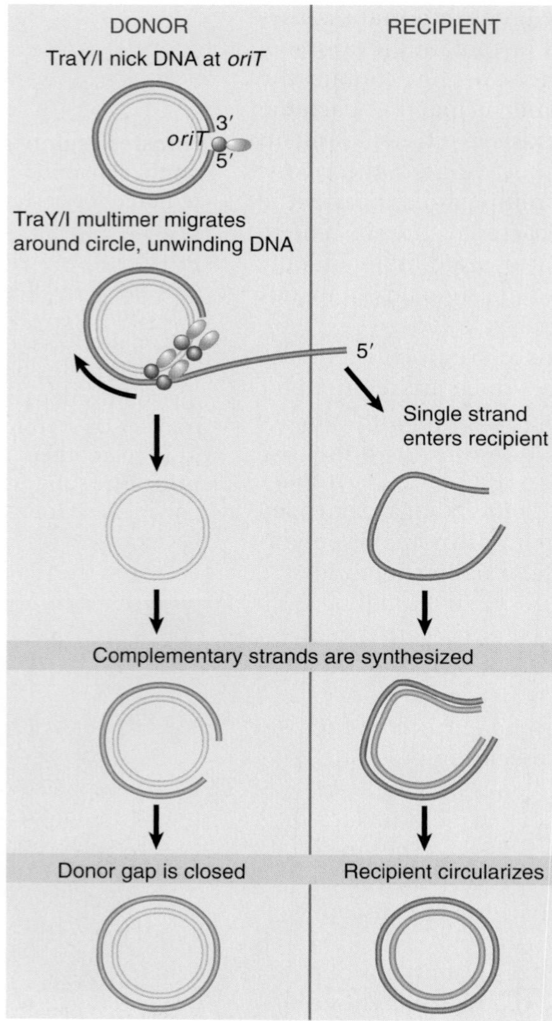

FIGURE 16.11 Transfer of DNA occurs when the F factor is nicked at *oriT* and a single strand is led by the 5′ end into the recipient. Only one unit length is transferred. Complementary strands are synthesized to the single strand remaining in the donor and to the strand transferred into the recipient.

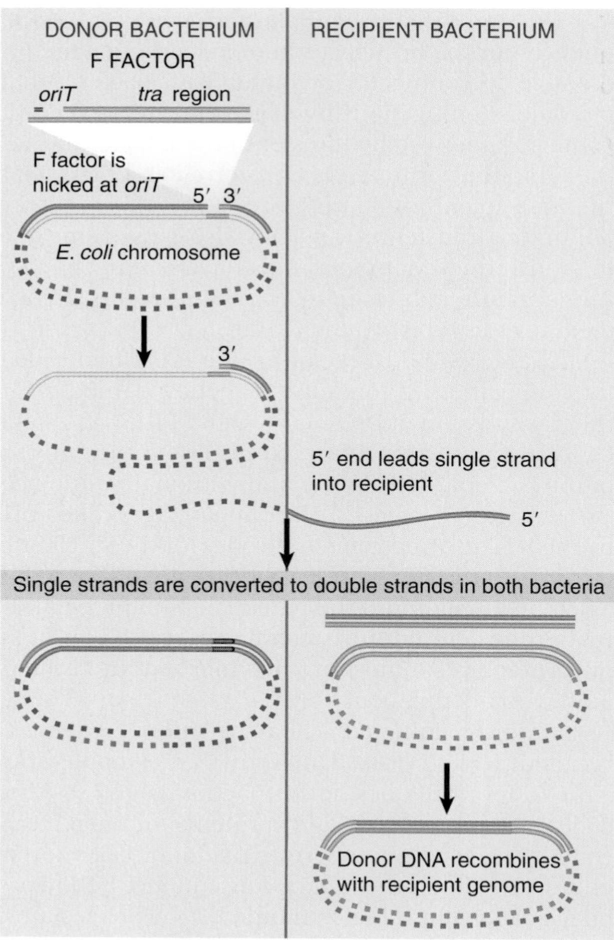

FIGURE 16.12 Transfer of chromosomal DNA occurs when an integrated F factor is nicked at *oriT*. Transfer of DNA starts with a short sequence of F DNA and continues until prevented by loss of contact between the bacteria.

TraI to bind. TraI is a relaxase, like φX174 A protein. TraI nicks *oriT* at a unique site (called *nic*), and then forms a covalent link to the 5′ end that has been generated. TraI also catalyzes the unwinding of ~200 bp of DNA (this is a helicase activity). **FIGURE 16.11** shows that the freed 5′ end leads the way into the recipient bacterium. A complement for the transferred single strand is synthesized in the recipient bacterium, which as a result is converted to the F-positive state.

A complementary strand must be synthesized in the donor bacterium to replace the strand that has been transferred. If this happens concomitantly with the transfer process, the state of the F plasmid will resemble the rolling circle of Figure 16.6. Conjugating DNA usually appears like a rolling circle, but replication as such is not necessary to provide the driving energy, and single-strand transfer is independent of DNA synthesis. Only a single unit length of the F factor is transferred to the recipient bacterium. This implies that some (unidentified) feature terminates the process after one revolution, after which the covalent integrity of the F plasmid is restored.

When an integrated F plasmid initiates conjugation, the orientation of transfer is directed away from the transfer region and into the bacterial chromosome. **FIGURE 16.12** shows that, following a short leading sequence of F DNA, bacterial DNA is transferred.

The process continues until it is interrupted by the breaking of contacts between the mating bacteria. It takes ~100 minutes to transfer the entire bacterial chromosome, and under standard conditions contact is often broken before the completion of transfer.

Donor DNA that enters a recipient bacterium is converted to double-stranded form and may recombine with the recipient chromosome. (Note that two recombination events are required to insert the donor DNA.) Thus conjugation affords a means to exchange genetic material between bacteria (a contrast with their usual asexual growth). A strain of *E. coli* with an integrated F factor supports such recombination at relatively high frequencies (compared to strains that lack integrated F factors); such strains are described as **Hfr** (for high frequency recombination). Each position of integration for the F factor gives rise to a different Hfr strain, with a characteristic pattern of transferring bacterial markers to a recipient chromosome.

▸ **Hfr** A bacterium that has an integrated F plasmid within its chromosome. Hfr stands for *high frequency recombination*, referring to the fact that chromosomal genes are transferred from an Hfr cell to an F^-cell much more frequently than from an F^+ cell.

KEY CONCEPTS

- Transfer of an F factor is initiated when rolling circle replication begins at oriT.
- The free 5′ end initiates transfer into the recipient bacterium.
- The transferred DNA is converted into double-stranded form in the recipient bacterium.
- When an F factor is free, conjugation "infects" the recipient bacterium with a copy of the F factor.
- When an F factor is integrated, conjugation causes transfer of the bacterial chromosome until the process is interrupted by (random) breakage of the contact between donor and recipient bacteria.

CONCEPT AND REASONING CHECK

Why do recipients from an Hfr mating rarely become Hfr?

16.8 The Bacterial Ti Plasmid Transfers Genes into Plant Cells

The interaction between bacteria and certain plants involves the transfer of DNA from the bacterial genome to the plant genome. **Crown gall disease**, shown in **FIGURE 16.13**, can be induced in most dicotyledonous plants by the soil bacterium *Agrobacterium tumefaciens*. The bacterium is a parasite that effects a genetic change in the eukaryotic host cell, with consequences for both parasite and host: it improves conditions for survival of the parasite, and it causes the plant cell to grow as a tumor.

▸ **crown gall disease** A tumor that can be induced in many plants by infection with the bacterium *Agrobacterium tumefaciens*.

▸ **Ti plasmid** An episome of the bacterium *Agrobacterium tumefaciens* that carries the genes responsible for the induction of crown gall disease in infected plants.

▸ **opine** A derivative of arginine that is synthesized by plant cells infected with crown gall disease.

Agrobacteria are necessary to induce tumor formation, but the tumor cells do not require the continued presence of bacteria. As with animal tumors, the plant cells have been transformed into a state in which new mechanisms govern growth and differentiation. Transformation is caused by the expression within the plant cell of genetic information transferred from the bacterium.

The ability of *Agrobacterium* to produce tumors resides in the **Ti plasmid**, which is perpetuated as an independent replicon within the bacterium. The plasmid carries genes involved in various bacterial and plant cell activities, including those required to generate the transformed state, and a set of genes concerned with synthesis or utilization of **opines** (novel derivatives of arginine). Ti plasmids (and thus the *Agrobacteria* in which they reside) can be divided into four groups, according to the types of opine that are made.

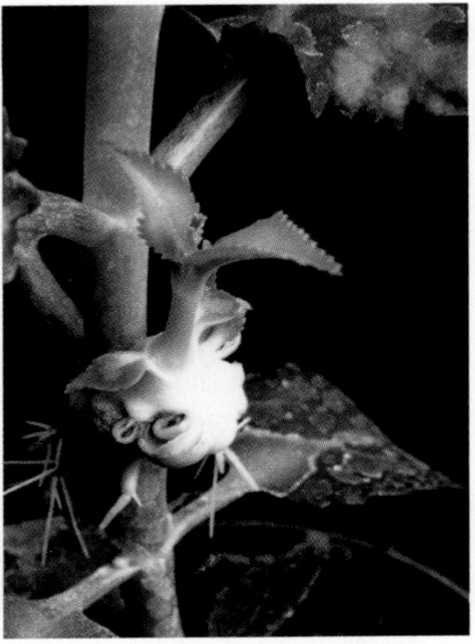

FIGURE 16.13 An *Agrobacterium* carrying a Ti plasmid of the nopaline type induces a teratoma, in which differentiated structures develop. Photo courtesy of the estate of Jeff Schell. Used with permission of the Max Planck Institute for Plant Breeding Research, Cologne.

FIGURE 16.14 T-DNA is transferred from *Agrobacterium* carrying a Ti plasmid into a plant cell, where it becomes integrated into the nuclear genome and expresses functions that transform the host cell.

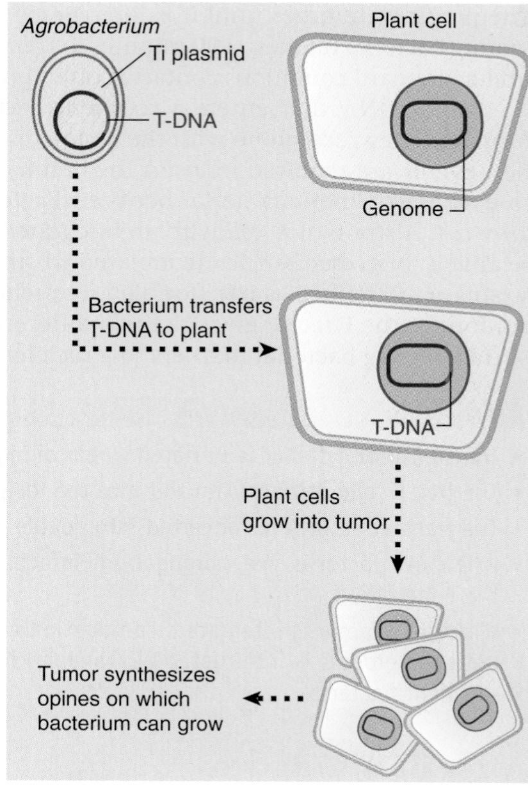

Agrobacterium
Plant cell
Ti plasmid
T-DNA
Genome
Bacterium transfers T-DNA to plant
T-DNA
Plant cells grow into tumor
Tumor synthesizes opines on which bacterium can grow

▸ **T-DNA** The ribonucleoprotein enzyme that creates repeating units of one strand at the telomere by adding individual bases to the DNA 3′ end, as directed by an RNA sequence in the RNA component of the enzyme.

The interaction between *Agrobacterium* and a plant cell is illustrated in **FIGURE 16.14**. The bacterium does not enter the plant cell, but rather transfers part of the Ti plasmid to the plant nucleus. The transferred part of the Ti genome is called **T-DNA**. It becomes integrated into the plant genome, where it expresses the functions needed to synthesize opines and to transform the plant cell.

Transformation of plant cells requires three types of functions carried in the *Agrobacterium:*

- Three loci on the *Agrobacterium* chromosome, *chvA, chvB,* and *pscA,* are required for the initial stage of binding the bacterium to the plant cell. They are responsible for synthesizing a polysaccharide on the bacterial cell surface.
- The *vir* region carried by the Ti plasmid outside the T-DNA region is required to release and initiate transfer of the T-DNA.
- The T-DNA is required to transform the plant cell.

The T-region occupies ~23 kb of the ~200 kb Ti genome, and codes for the proteins necessary to maintain the plant cell in a transformed state. However, functions affecting oncogenicity—the ability to form tumors—are not confined to the T-region. Those genes located outside the T-region must be concerned with establishing the tumorigenic state, but their products are not needed to perpetuate it. They may be concerned with transfer of T-DNA into the plant nucleus or perhaps with subsidiary functions such as the balance of plant hormones in the infected tissue.

KEY CONCEPTS

- Infection with the bacterium *A. tumefaciens* can transform plant cells into tumors.
- The infectious agent is a plasmid carried by the bacterium.
- The plasmid also carries genes for synthesizing and metabolizing opines (arginine derivatives) that are used by the tumor cell.
- Part of the DNA of the Ti plasmid is transferred to the plant cell nucleus, but the *vir* genes outside this region are required for the transfer process.

CONCEPT AND REASONING CHECK

How is crown gall disease similar to cancer in humans?

16.9 Transfer of T-DNA Resembles Bacterial Conjugation

The transfer of T-DNA from *Agrobacterium* into a plant cell is directed by six virulence loci, *virA-G*, which reside in a 40 kb region outside the T-DNA. Their organization is summarized in **FIGURE 16.15**. Each locus is transcribed as an individual unit; some are polycistronic.

We may divide the transforming process into (at least) two stages:

- *Agrobacterium* contacts a plant cell and the *vir* genes are induced.
- *vir* gene products cause T-DNA to be transferred to the plant cell nucleus, where it is integrated into the genome.

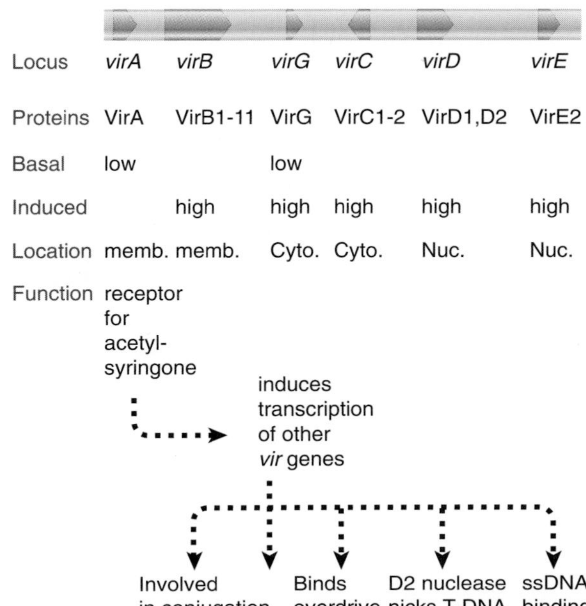

FIGURE 16.15 The *vir* region of the Ti plasmid has six loci that are responsible for transferring T-DNA to an infected plant.

The *vir* genes fall into two groups, corresponding to these stages. Genes *virA* and *virG* are regulators that respond to a change in the plant by inducing the other genes. So mutations in *virA* and *virG* are avirulent: the mutants carrying them cannot express the remaining *vir* genes. Genes *virB, C, D, E* code for proteins involved in the transfer of DNA. Mutations in *virB* and *virD* are avirulent in all plants, but the effects of mutations in *virC* and *virE* vary with the type of host plant.

The genes *virA* and *virG* are expressed constitutively at a low level. The signal to which they respond is provided by phenolic compounds generated by plants as a response to wounding. **FIGURE 16.16** presents an example. *Nicotiana tabacum* (tobacco) generates the molecules acetosyringone and α-hydroxyacetosyringone. Exposure to these compounds activates *virA*, which acts on *virG*, which in turn induces the expression of *virB, C, D, E*. This reaction explains why *Agrobacterium* infection succeeds only on wounded plants.

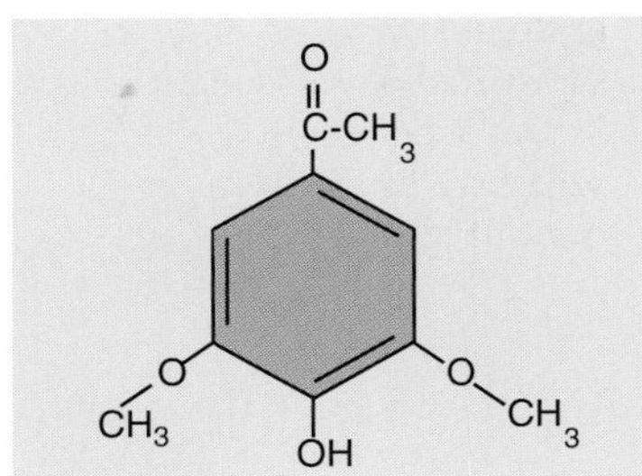

FIGURE 16.16 Acetosyringone (4-acetyl-2,6-dimethoxyphenol) is produced by *N. tabacum* upon wounding and induces transfer of T-DNA from *Agrobacterium*.

VirA and VirG are an example of a classic type of bacterial system in which stimulation of a sensor protein causes autophosphorylation and transfer of the phosphate to a second protein. VirA forms a homodimer that is located in the inner membrane; it may respond to the presence of the phenolic compounds in the periplasmic space. Exposure to these compounds causes VirA to become autophosphorylated on histidine. The phosphate group is then transferred to an aspartate residue in VirG. The phosphorylated VirG binds to promoters of the *virB, C, D,* and *E* genes to activate transcription. When *virG* is activated, its transcription is induced from a new startpoint, different from that used for constitutive expression, with the result that the amount of VirG protein is increased.

Of the other *vir* loci, *virD* is the best characterized. The *virD* locus has four open reading frames. Two of the proteins encoded by *virD*—VirD1 and VirD2—provide an endonuclease that initiates the transfer process by nicking T-DNA at a specific site.

The transfer process actually selects the T-region for entry into the plant. **FIGURE 16.17** shows that the T-DNA of a nopaline plasmid is demarcated from the flanking regions in the Ti plasmid by repeats of 25 bp, which differ at only two positions between the left and right ends. When T-DNA is integrated into a plant genome, it has a well-defined right junction, which retains 1 to 2 bp of the right repeat. The left junction is variable; the boundary of T-DNA in the plant genome may be located at the 25 bp repeat or at one of a series of sites extending over ~100 bp within the T-DNA. Sometimes multiple tandem copies of T-DNA are integrated at a single site.

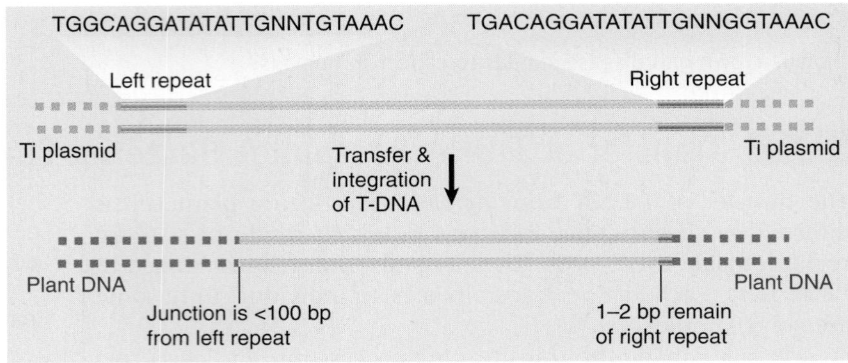

FIGURE 16.17 T-DNA has almost identical repeats of 25 bp at each end in the Ti plasmid. The right repeat is necessary for transfer and integration to a plant genome. T-DNA that is integrated in a plant genome has a precise junction that retains 1 to 2 bp of the right repeat, but the left junction varies and may be up to 100 bp short of the left repeat.

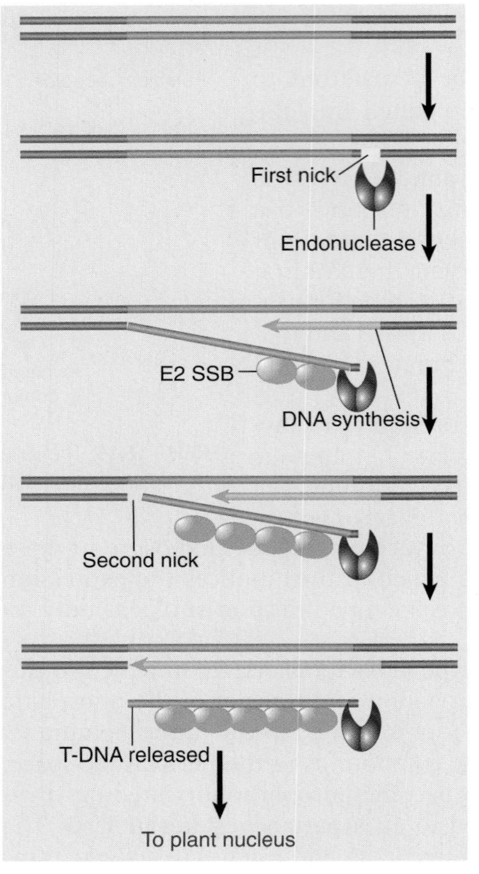

FIGURE 16.18 T-DNA is generated by displacement when DNA synthesis starts at a nick made at the right repeat. The reaction is terminated by a nick at the left repeat.

A model for transfer is illustrated in **FIGURE 16.18**. A nick is made at the right 25 bp repeat. It provides a priming end for synthesis of a DNA single strand. Synthesis of the new strand displaces the old strand, which is used in the transfer process. The transfer is terminated when the DNA synthesis reaches a nick at the left repeat. This model explains why the right repeat is essential, and it accounts for the polarity of the process. If the left repeat fails to be nicked, production of DNA for transfer could continue farther along the Ti plasmid.

The single molecule of single-stranded DNA produced for transfer in the infecting bacterium is transferred as a DNA-protein complex, sometimes called the T-complex. The DNA is covered by the VirE2 single-strand binding protein, which has a nuclear localization signal and is responsible for transporting T-DNA into the plant cell nucleus. A single molecule of the D2 subunit of the endonuclease remains bound at the 5′ end. The *virB* operon codes for 11 products that are involved in the transfer process.

This model for transfer of T-DNA closely resembles the events involved in bacterial conjugation, when the *E. coli* chromosome is transferred from one cell to another in single-stranded form. The genes of the *virB* operon are homologous to the *tra* genes of certain bacterial plasmids that are involved in conjugation (see *Section 16.7, Conjugation Transfers Single-Stranded DNA*). A difference is that the transfer of T-DNA is (usually) limited by the boundary of the left repeat, whereas transfer of bacterial DNA is indefinite.

We do not know how the transferred DNA is integrated into the plant genome. At some stage, the newly generated single strand must be converted into duplex DNA. Circles of T-DNA that are found in infected plant cells appear to be generated by recombination between the left and right 25 bp repeats, but we do not know if they are intermediates. The actual event is likely to involve a nonhomologous recombination, because there is no homology between the T-DNA and the sites of integration.

As a practical matter, the ability of *Agrobacterium* to transfer T-DNA to the plant genome makes it possible to introduce new genes into plants. The transfer/integration and oncogenic functions are separate; thus it is possible to engineer new Ti plasmids in which the oncogenic functions have been replaced by other genes whose effect on the plant we wish to test. The existence of a natural system for delivering genes to the plant genome should greatly facilitate genetic engineering of plants.

KEY CONCEPTS

- The *vir* genes are induced by phenolic compounds released by plants in response to wounding.
- The membrane protein VirA is autophosphorylated on histidine when it binds an inducer and activates VirG by transferring the phosphate to it.
- T-DNA is generated when a nick at the right boundary creates a primer for synthesis of a new DNA strand.
- The preexisting single strand that is displaced by the new synthesis is transferred to the plant cell nucleus.
- Transfer is terminated when DNA synthesis reaches a nick at the left boundary.
- The T-DNA is transferred as a complex of single-stranded DNA with the VirE2 single strand-binding protein.
- The single-stranded T-DNA is converted into double-stranded DNA and integrated into the plant genome.
- The mechanism of integration is not known. T-DNA can be used to transfer genes into a plant nucleus.

CONCEPT AND REASONING CHECK

What features of the T plasmid make it attractive to use as a system to transfer genes into plants?

16.10 Summary

The rolling circle is an alternative form of replication for circular DNA molecules in which an origin is nicked to provide a priming end. One strand of DNA is synthesized from this end; this displaces the original partner strand, which is extruded as a tail. Multiple genomes can be produced by continuing revolutions of the circle.

Rolling circles are used to replicate some phages. The A protein that nicks the φX174 origin has the unusual property of *cis*-action. It acts only on the DNA from which it was synthesized. It remains attached to the displaced strand until an entire strand has been synthesized, and then nicks the origin again; this releases the displaced strand and starts another cycle of replication.

Rolling circles also characterize bacterial conjugation, which occurs when an F plasmid is transferred from a donor to a recipient cell following the initiation of contact between the cells by means of the F-pili. A free F plasmid infects new cells by this means; an integrated F factor creates an Hfr strain that may transfer chromosomal DNA. In conjugation, replication is used to synthesize complements to the single strand remaining in the donor and to the single strand transferred to the recipient, but does not provide the motive power.

Agrobacteria induce tumor formation in wounded plant cells. The wounded cells secrete phenolic compounds that activate *vir* genes carried by the Ti plasmid of the bacterium. The *vir* gene products cause a single strand of DNA from the T-DNA region of the plasmid to be transferred to the plant cell nucleus. Transfer is initiated at one boundary of T-DNA, but ends at variable sites. The single strand is converted into a double strand and integrated into the plant genome. Genes within the T-DNA transform the plant cell and cause it to produce particular opines (derivatives of arginine). Genes in the Ti plasmid allow *Agrobacteria* to metabolize the opines produced by the transformed plant cell. T-DNA has been used to develop vectors for transferring genes into plant cells.

CHAPTER QUESTIONS

How does each of the linear genomes in the list at the left replicate its full genome? Choose the correct mechanism for each from the list at the right. Choices may be used more than once.

1. Lambda phage	**A.** A hairpin is created at the terminus so there is no free end.
2. Adenovirus	**B.** A protein covalently binds to the 5′ terminal base to provide an OH group for DNA synthesis.
3. T4 phage	**C.** The linear genome may convert to a circular or multimeric form for replication.

4. Lytic phages may have genomes composed of:
- **A.** single-stranded RNA.
- **B.** single-stranded DNA.
- **C.** double-stranded DNA.
- **D.** any of the above.

5. Bacterial plasmids may have genomes composed of:
- **A.** single-stranded RNA.
- **B.** single-stranded DNA.
- **C.** double-stranded DNA.
- **D.** any of the above.

6. Bacterial conjugation can result in transfer of:
- **A.** only mRNA transcripts.
- **B.** plasmid DNA sequences only.
- **C.** plasmid and host chromosome DNA sequences.
- **D.** host chromosome DNA sequences only.

7. What DNA replication mechanism is commonly used by lytic bacteriophages to enable rapid generation of new phage genomes?
- **A.** terminal-protein mediated replication of a linear genome
- **B.** bidirectional DNA replication from a specific origin
- **C.** unidirectional DNA replication from a specific origin
- **D.** rolling circle DNA replication

8. F-positive bacteria possess surface appendages called:
- **A.** flagella, hairlike structures used for mobility.
- **B.** flagella, hairlike structures used to conjugate with other cells.
- **C.** pili, hairlike structures used for mobility.
- **D.** pili, hairlike structures used to conjugate with other cells.

9. The F plasmid is present in how many copies per cell?
- **A.** one
- **B.** two
- **C.** variable, from one to several
- **D.** high copy number

10. T-DNA is transferred from the bacterial cell to the plant cell as a:
- **A.** single-stranded DNA beginning from a nick at the leftward T-DNA repeat.
- **B.** single-stranded DNA beginning from a nick at the rightward T-DNA repeat.
- **C.** double-stranded DNA beginning from a double-strand cut at the leftward T-DNA repeat.
- **D.** double-stranded DNA beginning from a double-strand cut at the rightward T-DNA repeat.

KEY TERMS

conjugation
crown gall disease
episome
F plasmid
Hfr
immunity
lysogenic
opine
pili
pilin
plasmid
relaxase
rolling circle
strand displacement
temperate phage
terminal protein
T-DNA
Ti plasmid
transfer region

FURTHER READING

Ulher, B., Li, Y., Logemann, E., Somssich, I. E., and Weisshaas, B. (2008). T-DNA mediated transfer of *Agrobacterium tumefaciens* chromosomal DNA into plants. *Nat. Biotechnol.* 26, 1015–1017.

Zzaman, S., Abhyanker, M. M., and Batistia, D. (2004). Reconstruction of F factor DNA replication in vitro with purified proteins. *J. Biolog. Chem.* 279, 17404–17410.

17

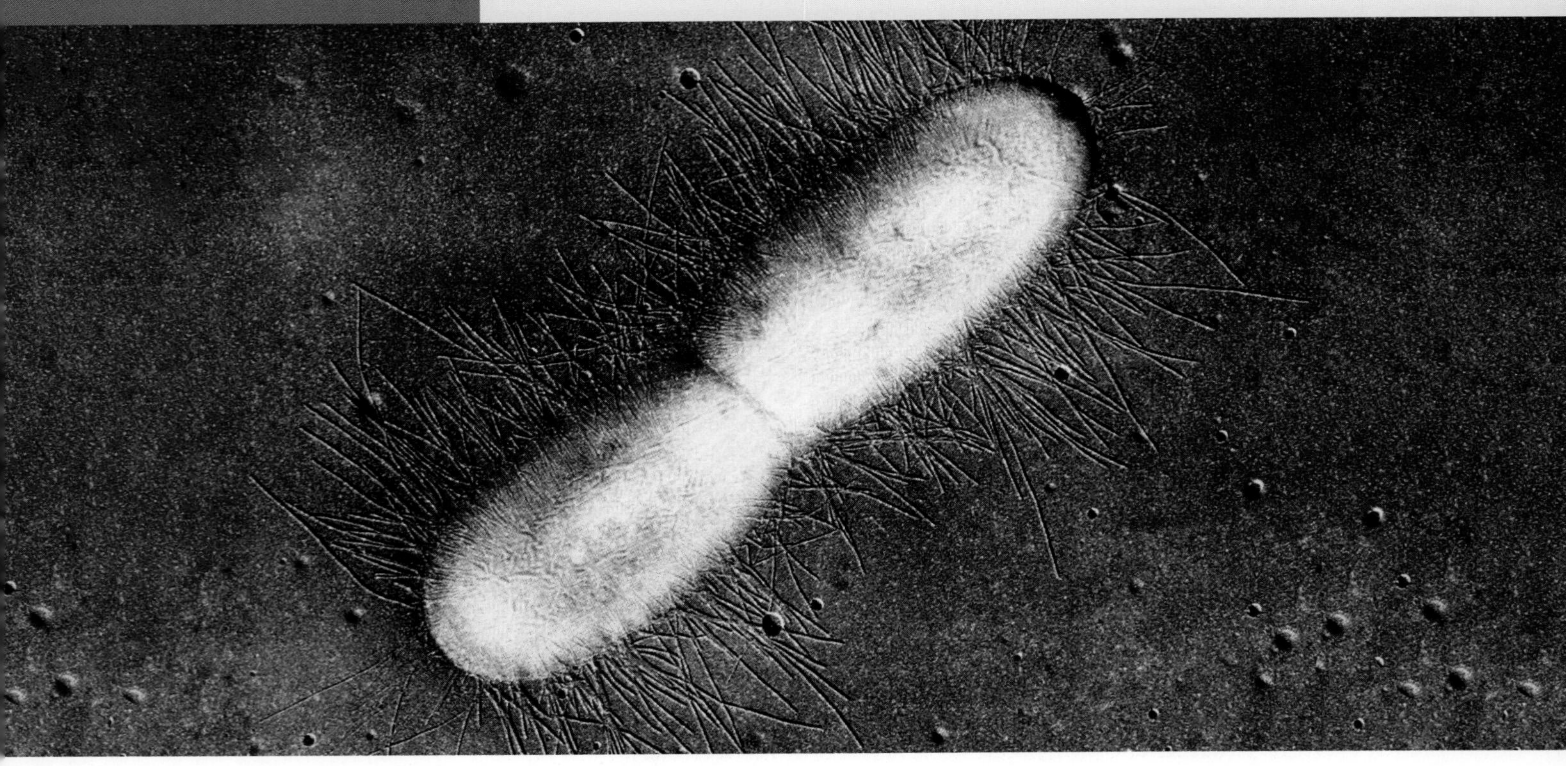

A transmission electron micrograph of an *E. coli* bacterium in the early stages of binary fission, the process by which the bacterium divides. The hair-like appendages around the bacterium are pili, structures used for bacterial conjugation. (Magnification: 17,500×.)

Bacterial Replication Is Connected to the Cell Cycle

CHAPTER OUTLINE

17.1 Introduction

A major difference between prokaryotes and eukaryotes is the way in which replication is controlled and linked to the cell cycle.

In eukaryotes, the following are true:

- chromosomes reside in the nucleus,
- each chromosome consists of many replicons,
- replication requires coordination of these replicons to reproduce DNA during a discrete period of the cell cycle,
- the decision about whether to replicate is determined by a complex pathway that regulates the cell cycle, and
- the duplicated chromosomes are segregated to daughter cells during mitosis by means of a special apparatus.

A unit cell has a circular chromosome
Replication initiates when cell passes critical size
Replication generates catenated daughter chromosomes
Daughter chromosomes are separated
Septum divides cell
Daughter cells separate

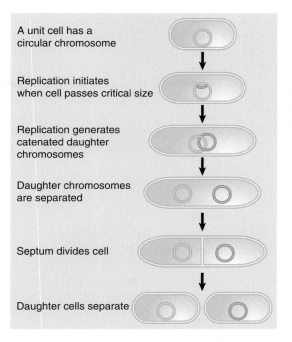

FIGURE 17.1 Replication initiates at the bacterial origin when a cell passes a critical threshold of size. Completion of replication produces daughter chromosomes that may be linked by recombination or that may be catenated. They are separated and moved to opposite sides of the septum before the bacterium is divided into two.

FIGURE 17.1 shows that in bacteria, replication is triggered at a single origin when the cell mass increases past a threshold level, and the segregation of the daughter chromosomes is accomplished by ensuring that they find themselves on opposite sides of the septum that grows to divide the bacterium into two.

How does the cell know when to initiate the replication cycle? The initiation event occurs at a constant ratio of cell mass to the number of chromosome origins. Cells growing more rapidly are larger and possess a greater number of origins. The growth of *E. coli* can be described in terms of the **unit cell**, an entity 1.7 μm long. A bacterium contains one origin per unit cell; a rapidly growing cell with two origins is 1.7 to 3.4 μm long.

▸ **unit cell** The state of an *E. coli* bacterium generated by a new division. It is 1.7 μm long and has a single replication origin.

How is cell mass titrated? An initiator protein could be synthesized continuously throughout the cell cycle; accumulation of a critical amount would trigger initiation. This explains why protein synthesis is needed for the initiation event. Another possibility is that an inhibitor protein might be synthesized at a fixed point, and diluted below an effective level by the increase in cell volume.

Some of the events in partitioning the daughter chromosomes are consequences of the circularity of the bacterial chromosome. Circular chromosomes are said to be *catenated* when one passes through another, connecting them. **Topoisomerases** are required to separate them. An alternative type of structure is formed when a recombination event occurs: a single recombination between two monomers converts them into a single dimer. This is resolved by a specialized recombination system that recreates the independent monomers. Essentially the partitioning process is handled by enzyme systems that act directly on discrete DNA sequences.

▸ **topoisomerase** An enzyme that changes the number of times the two strands in a closed DNA molecule cross each other. It does this by cutting the DNA, passing DNA through the break, and resealing the DNA.

17.2 Replication Is Connected to the Cell Cycle

Bacteria have two links between replication and cell growth:

- The frequency of initiation of cycles of replication is adjusted to fit the rate at which the cell is growing.
- The completion of a replication cycle is connected with division of the cell.

The rate of bacterial growth is assessed by the **doubling time**, the period required for the number of cells to double. The shorter the doubling time, the faster the growth

▸ **doubling time** The period (usually measured in minutes) that it takes for a bacterial cell to reproduce.

rate. *E. coli* cells can grow at rates ranging from doubling times as fast as 18 minutes to slower than 180 minutes. The bacterial chromosome is a single replicon; thus the frequency of replication cycles is controlled by the number of initiation events at the single origin. The replication cycle can be defined in terms of two constants:

- *C* is the fixed time of ~40 minutes required to replicate the entire bacterial chromosome. Its duration corresponds to a rate of replication fork movement of ~50,000 bp/minute. (The rate of DNA synthesis is more or less invariant at a constant temperature; it proceeds at the same speed unless and until the supply of precursors becomes limiting.)
- *D* is the fixed time of ~20 minutes that elapses between the completion of a round of replication and the cell division with which it is connected. This period may represent the time required to assemble the components needed for division.

(The constants *C* and *D* can be viewed as representing the maximum speed with which the bacterium is capable of completing these processes. They apply for all growth rates between doubling times of 18 and 60 minutes, but both constant phases become longer when the cell cycle occupies >60 minutes.)

A cycle of chromosome replication must be initiated at a fixed time of $C + D =$ 60 minutes before a cell division. For bacteria dividing more frequently than every 60 minutes, a cycle of replication must be initiated before the end of the preceding division cycle. You might say that a cell is "born already pregnant" with the next generation.

Consider the example of cells dividing every 35 minutes. The cycle of replication connected with a division must have been initiated 25 minutes before the preceding division. This situation is illustrated in **FIGURE 17.2**, which shows the chromosomal complement of a bacterial cell at 5-minute intervals throughout the cycle.

At division (35/0 minutes), the cell receives a partially replicated chromosome. The replication fork continues to advance. At 10 minutes, when this "old" replication fork has not yet reached the terminus, initiation occurs at both origins on the partially replicated chromosome. The start of these "new" replication forks creates a **multiforked chromosome**.

multiforked chromosome A bacterial chromosome that has more than one set of replication forks, because a second initiation has occurred before the first cycle of replication has been completed.

At 15 minutes—that is, at 20 minutes before the next division—the old replication fork reaches the terminus. Its arrival allows the two daughter chromosomes to separate; each of them has already been partially replicated by the new replication forks (which now are the only replication forks). These forks continue to advance.

At the point of division, the two partially replicated chromosomes segregate. This recreates the point at which we started. The single replication fork becomes "old," it terminates at 15 minutes, and 20 minutes later, there is a division. We see that the initiation event occurs $1^{25}/_{35}$ cell cycles before the division event with which it is associated.

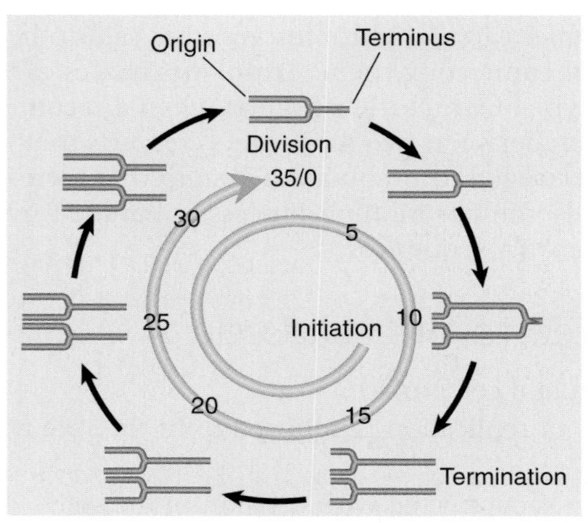

FIGURE 17.2 The fixed interval of 60 minutes between initiation of replication and cell division produces multiforked chromosomes in rapidly growing cells. Note that only the replication forks moving in one direction are shown; the chromosome actually is replicated symmetrically by two sets of forks moving in opposite directions on circular chromosomes.

The general principle of the link between initiation and the cell cycle is that as cells grow more rapidly (the cycle is shorter), the initiation event occurs an increasing number of cycles before the related division. There are correspondingly more chromosomes in the individual bacterium. This relationship can be viewed as the cell's response to its inability to reduce the periods of *C* and *D* to keep pace with the shorter cycle.

HISTORICAL PERSPECTIVES

John Cairns and the Bacterial Chromosome

The intact DNA molecules of most prokaryotes and many viruses are circular. The existence of circular cellular DNA molecules was not noticed for many years after the discovery of DNA because large DNA molecules usually break during isolation (eukaryote chromosomes do not break because they are held together as chromatin). Eventually, many people began to believe that the small fragments seen in bacterial DNA preparations were broken fragments. In 1963, while he was working in Australia, the British physician and molecular biologist John Cairns used autoradiography to obtain the first image of an intact bacterial DNA chromosome. This technique takes advantage of the fact that radioactive tritium-labeled DNA emits β-particles, which upon striking a photographic emulsion produce an image of the DNA. Cairns cultured *E. coli* in a medium containing [^{3}H]thymidine, a specific precursor for DNA, and then gently released labeled DNA from the bacteria by treating the cells with a combination of lysozyme (an egg white enzyme) to digest the bacterial cell wall and detergent to disrupt the cell membrane. After collecting the released DNA on a dialysis membrane (which can be used to separate molecules of different sizes), he coated the dried membrane with a photographic emulsion and stored the preparation in the dark for two months to allow sufficient time for the β-particles to produce an image. Analysis of the array of dark spots, which appeared after developing the emulsion, revealed that *E. coli* DNA is a double-stranded circular molecule with a contour length of approximately 1 mm, or about 1000 times longer than the bacteria itself.

Cairns's work was important for a number of reasons. First, it demonstrated that the *E. coli* chromosome is a single molecule. Second, it demonstrated a circular cellular chromosome for the first time. And, third, it confirmed the prediction made by Watson and Crick that semiconservative replication proceeds via a Y fork. In fact, Cairns observed two Y forks. What could not be determined from this experiment was the nature of the two Y forks. Are they both moving around the chromosome (giving bidirectional replication) or is one moving (giving unidirectional replication) and the second acting as a fixed swivel?

KEY CONCEPTS

- It requires 40 minutes to replicate the bacterial chromosome (at normal temperature).
- Completion of a replication cycle triggers a bacterial division 20 minutes later.
- If the doubling time is ~60 minutes, a replication cycle is initiated before the division resulting from the previous replication cycle.
- Fast rates of growth therefore produce multiforked chromosomes.

CONCEPT AND REASONING CHECK

Since it takes 40 minutes to replicate the *E. coli* chromosome, how can the cell have a cell cycle of 35 minutes?

17.3 The Septum Divides a Bacterium into Progeny That Each Contain a Chromosome

Chromosome segregation in bacteria is especially interesting because the DNA itself is involved in the mechanism for partition. (This contrasts with eukaryotic cells, in which segregation is achieved by the complex apparatus of mitosis.) The bacterial apparatus is quite accurate; however, **anucleate cells**, which lack a **nucleoid**, form <0.03% of a bacterial population.

The division of a bacterium into two daughter cells is accomplished by the formation of a **septum**, a structure that forms in the center of the cell as an invagination from the surrounding envelope. The septum forms an impenetrable barrier between

- **anucleate cell** Bacteria that lack a nucleoid, but are of similar shape to wild-type bacteria.
- **nucleoid** The structure in a prokaryotic cell that contains the genome. The DNA is bound to proteins and is not enclosed by a membrane.
- **septum** The structure that forms in the center of a dividing bacterium, providing the site at which the daughter bacteria will separate. The same term is used to describe the cell wall that forms between plant cells at the end of mitosis.

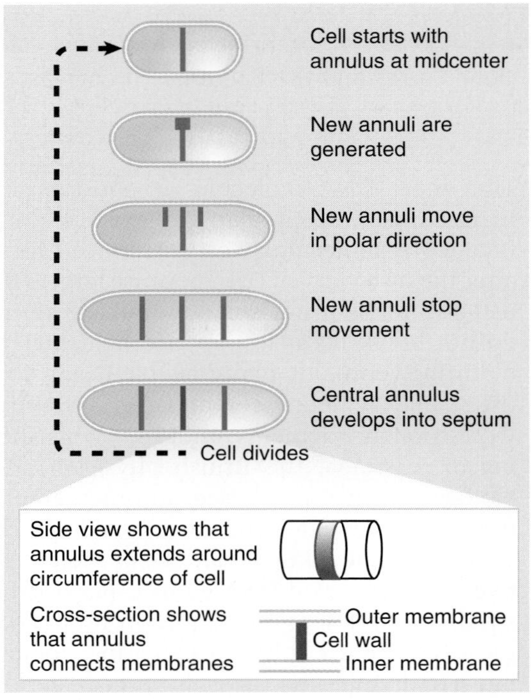
Cell starts with annulus at midcenter
New annuli are generated
New annuli move in polar direction
New annuli stop movement
Central annulus develops into septum
Cell divides
Side view shows that annulus extends around circumference of cell
Cross-section shows that annulus connects membranes
Outer membrane
Cell wall
Inner membrane

FIGURE 17.3 Duplication and displacement of the periseptal annulus give rise to the formation of a septum that divides the cell.

the two parts of the cell and provides the site at which the two daughter cells eventually separate entirely. Two related questions address the role of the septum in division: What determines the location at which it forms? and What ensures that the daughter chromosomes lie on opposite sides of it?

▸ **periseptal annulus** A ring-like area where inner and outer membranes appear fused. Formed around the circumference of the bacterium, it determines the location of the septum.

The formation of the septum is preceded by the organization of the **periseptal annulus**. This is observed as a zone in *E. coli* or *Salmonella typhimurium*, in which the structure of the envelope is altered so that the inner membrane is connected more closely to the cell wall and outer membrane layer. As its name suggests, the annulus extends around the cell. **FIGURE 17.3** summarizes its development.

The annulus starts at a central position in a new cell. As the cell grows, two events occur: a septum forms at the mid-cell position defined by the annulus, and new annuli form on either side of the initial annulus. These new annuli are displaced from the center and move along the cell to positions at ¼ and ¾ of the cell length. They will become the mid-cell positions after the next division. The displacement of the periseptal annulus to the correct position may be the crucial event that ensures the division of the cell into daughters of equal size. (The mechanism of movement is unknown.) Septation begins when the cell reaches a fixed length (2L), and the distance between the new annuli is always L. We do not know how the cell measures length, but the relevant parameter appears to be linear distance as such (not area or volume).

The septum consists of the same components as the cell envelope: there is a rigid layer of peptidoglycan in the periplasm, between the inner and outer membranes. The peptidoglycan is made by polymerization of tri- or pentapeptide-disaccharide units in a reaction involving connections between both types of subunit (transpeptidation and transglycosylation). The rodlike shape of the bacterium is maintained by a pair of activities, PBP2 and RodA. They are interacting proteins and are encoded by the same operon. RodA is a member of the SEDS family (SEDS stands for shape, elongation, division, and sporulation) present in all bacteria that have a peptidoglycan cell wall. Each SEDS protein functions together with a specific transpeptidase, which catalyzes the formation of the crosslinks in the peptidoglycan. PBP2 (penicillin-binding protein 2) is the transpeptidase that interacts with RodA. Mutations in the gene for either protein cause the bacterium to lose its extended shape and become round. This demonstrates the important principle that shape and rigidity can be determined by the simple extension of a polymeric structure. Another enzyme is responsible for generating the peptidoglycan in the septum (see *Section 17.5, FtsZ Is Necessary for Septum Formation*). The septum initially forms as a double layer of peptidoglycan, and the protein EnvA is required to split the covalent links between the layers so that the daughter cells may separate.

KEY CONCEPTS

- Septum formation is initiated at the annulus, which is a ring around the cell where the structure of the envelope is altered.
- New annuli are initiated at 50% of the distance from the septum to each end of the bacterium.
- When the bacterium divides, each daughter has an annulus at the mid-center position.
- Septation starts when the cell reaches a fixed length.
- The septum consists of the same peptidoglycans that comprise the bacterial envelope.

CONCEPT AND REASONING CHECKS

- What would happen to a bacterial chromosome during cell division if it were not connected to the membrane?
- How do eukaryotes divide their pairs of chromosomes into two separate cells?

17.4 Mutations in Division or Segregation Affect Cell Shape

A difficulty in isolating mutants that affect cell division is that mutations in the critical functions may be lethal and/or pleiotropic. For example, if formation of the annulus occurs at a site that is essential for overall growth of the envelope, it would be difficult to distinguish mutations that specifically interfere with annulus formation from those that inhibit envelope growth generally. Most mutations in the division apparatus have been identified as conditional mutants (whose division is affected under nonpermissive conditions; typically they are temperature sensitive). Mutations that affect cell division or chromosome segregation cause striking phenotypic changes. **FIGURE 17.4** and **FIGURE 17.5** illustrate the opposite consequences of failure in the division process and failure in segregation:

- Long *filaments* form when septum formation is inhibited, but chromosome replication is unaffected. The bacteria continue to grow—and even continue to segregate their daughter chromosomes—but septa do not form. Thus the cell consists of a very long filamentous structure, with the nucleoids (bacterial chromosomes) regularly distributed along the length of the cell. This phenotype is displayed by *fts* mutants (named for temperature-sensitive filamentation), which identify a defect or multiple defects that lie in the division process itself.
- **Minicells** form when septum formation occurs too frequently or in the wrong place, with the result that one of the new daughter cells lacks a chromosome. The minicell has a rather small size and lacks DNA, but otherwise appears morphologically normal. *Anucleate* cells form when segregation is aberrant; like minicells, they lack a chromosome, but because septum formation is normal, their size is unaltered. This phenotype is caused by *par* (partition) mutants (named because they are defective in chromosome segregation).

▸ **minicell** An anucleate bacterial (*E. coli*) cell produced by a division that generates a cytoplasm without a nucleus.

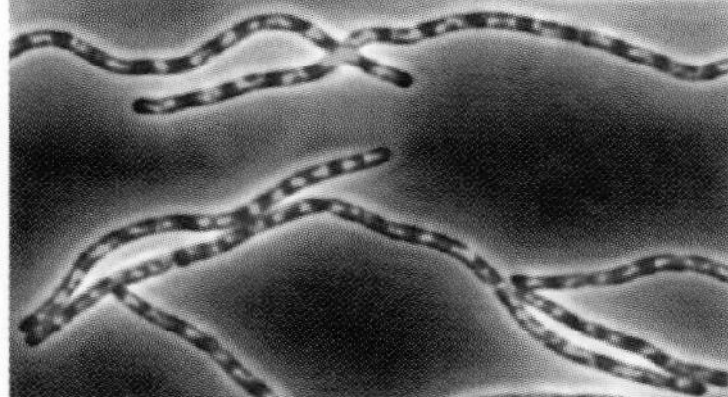

FIGURE 17.4 Failure of cell division under nonpermissive temperatures generates multinucleated filaments. Photo courtesy of Sota Hiraga, Kyoto University.

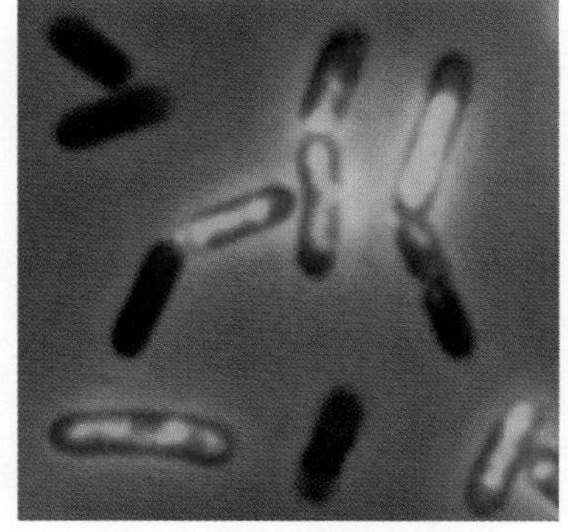

FIGURE 17.5 *E. coli* generate anucleate cells when chromosome segregation fails. Cells with chromosomes stain blue; daughter cells lacking chromosomes have no blue stain. This field shows cells of the mukB mutant; both normal and abnormal divisions can be seen. Photo courtesy of Sota Hiraga, Kyoto University.

KEY CONCEPTS

- *fts* mutants form long filaments because the septum that divides the daughter bacteria fails to form.
- Minicells form in mutants that produce too many septa; they are small and lack DNA.
- Anucleate cells of normal size are generated by partition mutants in which the duplicate chromosomes fail to separate.

CONCEPT AND REASONING CHECK

How could you isolate a loss of function mutation in an essential gene when mutating the gene can kill the cell?

17.5 FtsZ Is Necessary for Septum Formation

The gene *ftsZ* plays a central role in division. Mutations in *ftsZ* block septum formation and generate filaments. Overexpression induces minicells by causing an increased number of septation events per unit cell mass. *ftsZ* mutants act at stages varying from the displacement of the periseptal annuli to septal morphogenesis. FtsZ is therefore required for usage of preexisting sites for septum formation, but does not itself affect the formation of the periseptal annuli or their localization.

FtsZ functions at an early stage of septum formation. Early in the division cycle, FtsZ is localized throughout the cytoplasm. As the cell elongates and begins to constrict in the middle, FtsZ becomes localized in a ring around the circumference. The structure is called the **Z-ring**. **FIGURE 17.6** shows that it lies in the position of the mid-center annulus of Figure 17.3. The formation of the Z-ring is the rate-limiting step in septum formation. In a typical division cycle, it forms in the center of the cell 1 to 5 minutes after division, remains for 15 minutes, and then quickly constricts to pinch the cell into two.

▸ **Z-ring** *See* septal ring.

The structure of FtsZ resembles tubulin, suggesting that assembly of the ring could resemble the formation of microtubules in eukaryotic cells. FtsZ has GTPase activity, and GTP cleavage is used to support the oligomerization of FtsZ monomers into the ring structure. The Z-ring is a dynamic structure, in which there is continuous exchange of subunits with a cytoplasmic pool.

Two other proteins needed for division, ZipA and FtsA, interact directly and independently with FtsZ. ZipA is an integral membrane protein that is located in the inner bacterial membrane. It provides the means for linking FtsZ to the membrane. FtsA is a cytosolic protein, but is often found associated with the membrane. The Z-ring can form in the absence of either ZipA or FtsA, but it cannot form if both are absent. Both are needed for subsequent steps. This suggests that they have overlapping roles in stabilizing the Z-ring and perhaps in linking it to the membrane.

The products of several other *fts* genes join the Z-ring in a defined order after FtsA has been incorporated. They are all transmembrane proteins. The final structure is sometimes called the **septal ring**. It consists of a multiprotein complex that is presumed to have the ability to constrict the membrane. One of the last components to be incorporated into the septal ring is FtsW, which is a protein belonging to the SEDS family. *ftsW* is expressed as part of an operon with *ftsI*, which codes for a transpeptidase (also called PBP3 for penicillin-binding protein 3), a membrane-bound protein that has its catalytic site in the periplasm. FtsW is responsible for incorporating FtsI into the septal ring. This suggests a model for septum formation in which the transpeptidase activity then causes the peptidoglycan to grow inward, thus pushing the inner membrane and pulling the outer membrane.

▸ **septal ring** A complex of several proteins coded by *fts* genes of *E. coli* that forms at the mid-point of the cell. It gives rise to the septum at cell division. The first of the proteins to be incorporated is FtsZ, which gave rise to the original name of the Z-ring.

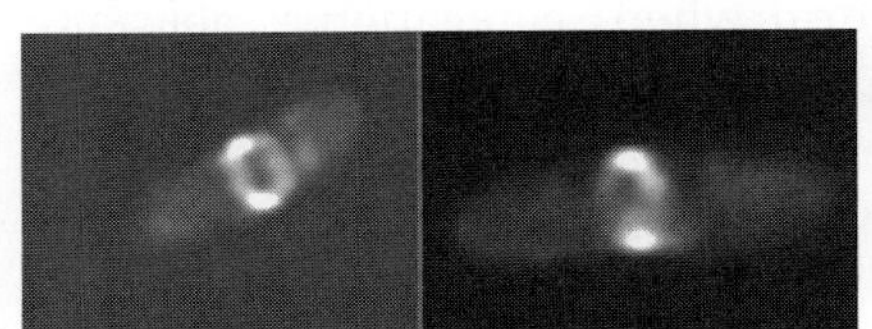

FIGURE 17.6 Immunofluorescence with an antibody against FtsZ shows that it is localized at the mid-cell. Photo courtesy of William Margolin, University of Texas Medical School at Houston.

FtsZ is the major cytoskeletal component of septation. It is common in bacteria, and also is found in chloroplasts. Mitochondria, which also share an evolutionary origin with bacteria, usually do not have FtsZ. Instead, they use a variant of the protein dynamin, which is involved in pinching off vesicles from membranes of eukaryotic cytoplasm. This functions from the outside of the organelle, squeezing the membrane to generate a constriction.

The common feature, then, in the division of bacteria, chloroplasts, and mitochondria is the use of a cytoskeletal protein that forms a ring around the organelle and either pulls or pushes the membrane to form a constriction.

KEY CONCEPTS

- The product of *ftsZ* is required for septum formation at preexisting sites.
- FtsZ is a GTPase that forms a ring on the inside of the bacterial envelope. It is connected to other cytoskeletal components.

CONCEPT AND REASONING CHECK

How is it possible that a single gene like *ftsZ* can have mutations that have different phenotypes?

17.6 *min* Genes Regulate the Location of the Septum

Information about the localization of the septum is provided by minicell mutants. The original minicell mutation lies in the locus *minB;* deletion of *minB* generates minicells by allowing septation to occur at the poles as well as (or instead of) at mid-cell. This suggests that the cell possesses the ability to initiate septum formation either at mid-cell or at the poles, and that the role of the wild-type *minB* locus is to suppress septation at the poles. In terms of the events depicted in Figure 17.3, this implies that a newborn cell has potential septation sites associated with both the annulus at mid-center and the poles. One pole was formed from the septum of the previous division; the other pole represents the septum from the division before that. Perhaps the poles retain remnants of the annuli from which they were derived and these remnants can nucleate septation.

The *minB* locus consists of three genes, *minC, -D,* and *-E.* Their roles are summarized in **FIGURE 17.7**. The products of *minC* and *minD* form a division inhibitor. (MinD is required to activate MinC, which prevents FtsZ from polymerizing into the Z-ring).

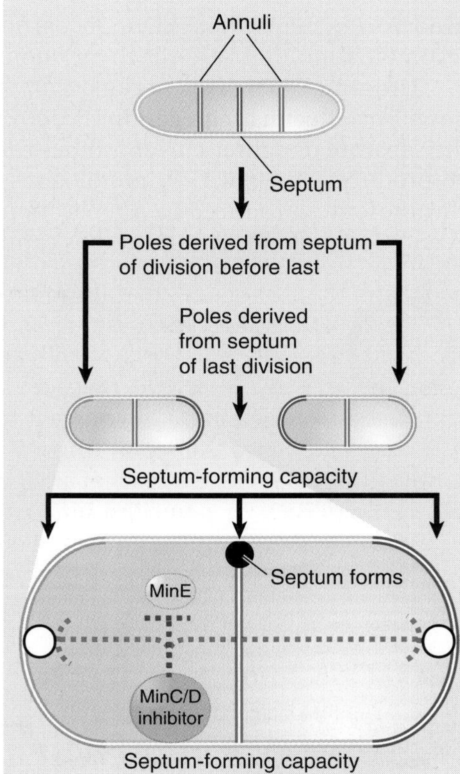

FIGURE 17.7 MinC/D is a division inhibitor whose action is confined to the polar sites by MinE.

Expression of MinCD in the absence of MinE, or overexpression even in the presence of MinE, causes a generalized inhibition of division. The resulting cells grow as long filaments without septa. Expression of MinE at levels comparable to MinCD confines the inhibition to the polar regions, so restoring normal growth. MinE protects the mid-cell sites from inhibition. Overexpression of MinE induces minicells, because the presence of excess MinE counteracts the inhibition at the poles as well as at mid-cell, allowing septa to form at both locations.

The determinant of septation at the proper (mid-cell) site is, therefore, the ratio of MinCD to MinE. The wild-type level prevents polar septation but permits mid-cell septation. The effects of MinC/D and MinE are inversely related; absence of MinCD, or too much MinE, causes indiscriminate septation, forming minicells; too much MinCD or absence of MinE inhibits mid-cell as well as polar sites, resulting in filamentation.

MinE forms a ring at the septal position. Its accumulation suppresses the action of MinCD in the vicinity, thus allowing formation of the septal ring (which includes FtsZ and ZipA).

KEY CONCEPTS

- The location of the septum is controlled by *minC*, *-D*, and *-E*.
- The number and location of septa is determined by the ratio of MinE/MinC,D.
- The septum forms where MinE is able to form a ring.
- At normal concentrations, MinC/D allows a mid-center ring, but prevents additional rings of MinE from forming at the poles.

CONCEPT AND REASONING CHECK

What would *minC* mutant bacteria look like?

17.7 Chromosomal Segregation May Require Site-Specific Recombination

After replication has created duplicate copies of a bacterial chromosome or plasmid, the copies can recombine. FIGURE 17.8 demonstrates the consequences. A single intermolecular recombination event between two circles generates a dimeric circle; further recombination can generate higher multimeric forms. Such an event reduces the number of physically segregating units. In the extreme case of a single-copy plasmid that has just replicated, formation of a dimer by recombination means that the cell only has one unit to segregate, and the plasmid therefore must inevitably be lost from one daughter cell. To counteract this effect, plasmids often have **site-specific recombination** systems that act upon particular sequences to sponsor an intramolecular recombination that restores the monomeric condition.

site-specific recombination Recombination that occurs between two specific sequences, as in phage integration/excision or resolution of cointegrated structures during transposition.

The same types of event can occur with the bacterial chromosome; FIGURE 17.9 shows how they affect its segregation. If no recombination occurs, there is no problem, and the separate daughter chromosomes can segregate to the daughter cells. A dimer will be produced, however, if homologous recombination occurs between the daughter chromosomes produced by a replication cycle. If there has been such a recombination event, the daughter chromosomes cannot separate. In this case, a second recombination is required to achieve resolution in the same way as a plasmid dimer.

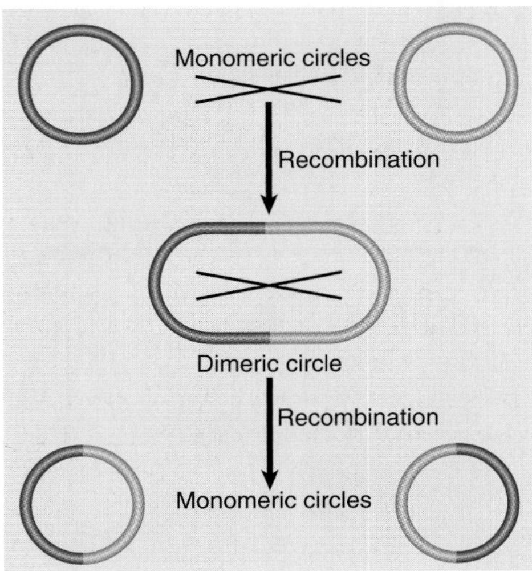

FIGURE 17.8 Intermolecular recombination merges monomers into dimers, and intramolecular recombination releases individual units from oligomers.

Most bacteria with circular chromosomes possess the Xer site-specific recombination system. In *E. coli*, this consists of two recombinases, XerC and XerD, which act on a 28-bp target site, called *dif*, which is located in the terminus region of the chromosome. The use of the Xer system is related to cell division in an interesting way. The relevant events are summarized in FIGURE 17.10. XerC can bind to a pair of *dif* sequences and form a Holliday junction between them. The complex may form soon after the replication fork passes over the *dif* sequence, which explains how the two copies of the target sequence can find

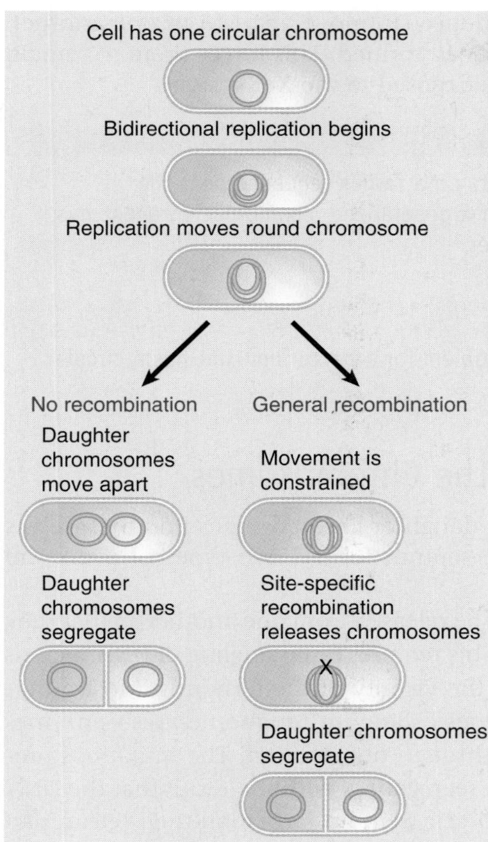

FIGURE 17.9 A circular chromosome replicates to produce two monomeric daughters that segregate to daughter cells. A generalized recombination event, however, generates a single dimeric molecule. This can be resolved into two monomers by a site-specific recombination.

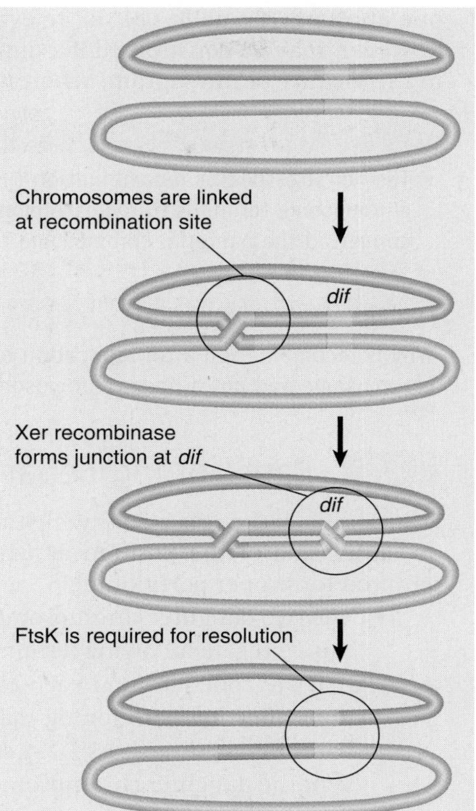

FIGURE 17.10 A recombination event creates two linked chromosomes. Xer creates a Holliday junction at the *dif* site, but can resolve it only in the presence of FtsK.

one another consistently. Resolution of the junction to give recombinants, however, occurs only in the presence of FtsK, a protein located in the septum that is required for chromosome segregation and cell division. In addition, the *dif* target sequence must be located in a region of ~30 kb; if it is moved outside of this region, it cannot support the reaction.

Resolution of the junction to give recombinants occurs only in the presence of FtsK, a protein located in the septum that is required for chromosome segregation and cell division. FtsK is a large transmembrane protein. Its N-terminal domain is associated with the membrane, and causes it to be localized to the septum. Its C-terminal domain has two functions. One is to cause Xer to resolve a dimer into two monomers. It also has an ATPase activity, which it can use to translocate along DNA. This could be used to pump DNA through the septum. Bacteria that have the Xer system always have an FtsK homolog, and vice versa, which suggests that the system has evolved so that resolution is connected to the septum.

So there is a site-specific recombination available when the terminus sequence of the chromosome is close to the septum. The bacterium, however, wants to have a recombination only when there has already been a general recombination event to generate a dimer. (Otherwise the site-specific recombination would create the dimer!) How does the system know whether the daughter chromosomes exist as independent monomers or have been recombined into a dimer?

The answer may be that segregation of chromosomes starts soon after replication. If there has been no recombination, the two chromosomes move apart from one another. The ability of the relevant sequences to move apart from one another, however, may be constrained if a dimer has been formed. This forces them to remain in the vicinity of the septum, where they are exposed to the Xer system.

KEY CONCEPT

- The Xer site-specific recombination system acts on a target sequence near the chromosome terminus to recreate monomers if a generalized recombination event has converted the bacterial chromosome to a dimer.

CONCEPT AND REASONING CHECK

Why is recombination after replication only a problem for a bacterium that has a circular chromosome and not a linear chromosome?

17.8 Partitioning Separates the Chromosomes

Partitioning is the process by which the two daughter chromosomes find themselves on either side of the position at which the septum forms. Two types of event are required for proper partitioning:

- The two daughter chromosomes must be released from one another so that they can segregate following termination. This requires disentangling of DNA regions that are coiled around each other in the vicinity of the terminus. Most mutations affecting partitioning map in genes coding for topoisomerases—enzymes with the ability to pass DNA strands through one another. The mutations prevent the daughter chromosomes from segregating, with the result that the DNA is located in a single large mass at mid-cell. Septum formation then releases an anucleate cell and a cell containing both daughter chromosomes. This tells us that the bacterium must be able to disentangle its chromosomes topologically in order to be able to segregate them into different daughter cells.
- Mutations that affect the partition process itself are rare. We expect to find two classes: (1) *cis*-acting mutations should occur in DNA sequences that are the targets for the partition process; and (2) *trans*-acting mutations should occur in genes that code for the protein(s) that cause segregation, which could include proteins that bind to DNA or activities that control the locations on the envelope to which DNA might be attached. Both types of mutation have been found in the systems responsible for partitioning plasmids, but only *trans*-acting functions have been found in the bacterial chromosome. In addition, mutations in plasmid site-specific recombination systems increase plasmid loss (because the dividing cell has only one dimer to partition instead of two monomers), and therefore have a phenotype that is similar to partition mutants.

Origins of replicating chromosomes attached to membrane
Daughter chromosomes attached to envelope
Septum grows between chromosomes
Septum divides cell
Chromosomes distributed to daughter cells

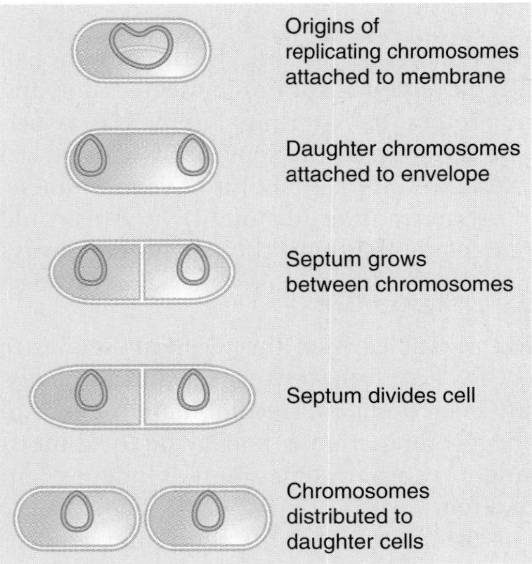

FIGURE 17.11 Attachment of bacterial DNA to the membrane could provide a mechanism for segregation.

Chromosome segregation requires that the daughter chromosomes find themselves on opposite sides of the septum when it divides the cell. **FIGURE 17.11** shows that the formation of a septum could segregate the chromosomes into the different daughter cells if the origins are connected to sites that lie on either side of the septum. The question now becomes what ensures that the daughter chromosomes are in such locations. Replicated chromosomes are capable of abrupt movements to their final positions at one quarter and three quarters of the cell length. If protein synthesis is inhibited before the termination of replication, the chromosomes fail to segregate and thus remain close to the mid-cell position. But when protein synthesis is allowed to resume, the chromosomes move to the quarter positions in the absence of any further envelope elongation. This suggests that an active process—one that requires protein synthesis—may move the chromosomes to specific locations.

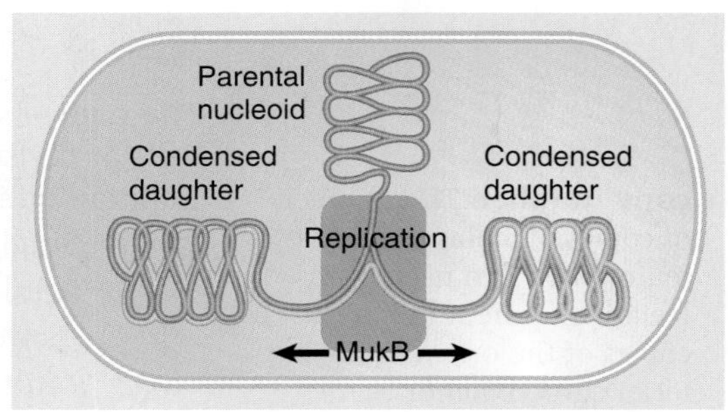

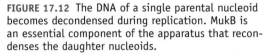

FIGURE 17.12 The DNA of a single parental nucleoid becomes decondensed during replication. MukB is an essential component of the apparatus that recondenses the daughter nucleoids.

Segregation is interrupted by mutations of the *muk* class, which give rise to anucleate progeny at a much increased frequency: both daughter chromosomes remain on the same side of the septum instead of segregating. Mutations in the *muk* genes are not lethal, and they may identify components of the apparatus that segregates the chromosomes. The gene *mukA* is identical to the gene for a known outer membrane protein (*tolC*) whose product could be involved with attaching the chromosome to the envelope. The gene *mukB* codes for a large (180 kD) globular protein, which has the same general type of organization as the two groups of structural maintenance of chromosomes (SMC) proteins that are involved in condensing and in holding together eukaryotic chromosomes. SMC-like proteins have also been found in other bacteria.

The insight into the role of MukB was the discovery that some mutations in *mukB* can be suppressed by mutations in *topA,* the gene that codes for topoisomerase I. MukB forms a complex with two other proteins, MukE and MukF, and the MukBEF complex is considered to be a condensin analogous to eukaryotic condensins. It uses a supercoiling mechanism to condense the chromosome. A defect in this function is the cause of failure to segregate properly. The defect can be compensated for by preventing topoisomerases from relaxing negative supercoils; the resulting increase in supercoil density helps to restore the proper state of condensation and thus allows segregation.

We still do not understand how genomes are positioned in the cell, but the process may be connected with condensation. **FIGURE 17.12** shows a current model. The parental genome is centrally positioned. It must be decondensed in order to pass through the replication apparatus. The daughter chromosomes emerge from replication, are disentangled by topoisomerases, and then passed in an uncondensed state to MukBEF, which causes them to form condensed masses at the positions that will become the centers of the daughter cells.

A physical link either directly or indirectly through chromosome-bound proteins exists between bacterial DNA and the membrane. Bacterial DNA can be found in membrane fractions, which tend to be enriched in genetic markers near the origin, the replication fork, and the terminus. The proteins present in these membrane fractions may be affected by mutations that interfere with the initiation of replication. The growth site could be a structure on the membrane to which the origin must be attached for initiation.

KEY CONCEPTS

- Replicon origins are attached to the inner bacterial membrane.
- Chromosomes make abrupt movements from the mid-center to the one quarter and three quarter positions.

CONCEPT AND REASONING CHECK

Why do bacteria have to attach their chromosomes to the membrane for effective separation, whereas eukaryotes do not?

17.9 Single-Copy Plasmids Have a Partitioning System

▸ **copy number** The number of copies of a plasmid that is maintained in a bacterium (relative to the number of copies of the origin of the bacterial chromosome).

The type of system that a plasmid uses to ensure that it is distributed to both daughter cells at division depends upon its type of replication system. Each type of plasmid is maintained in its bacterial host at a characteristic **copy number**:

- Single-copy control systems resemble that of the bacterial chromosome and result in one replication per cell division. A single-copy plasmid effectively maintains parity with the bacterial chromosome.
- Multicopy control systems allow multiple initiation events per cell cycle, with the result that there are several copies of the plasmid per bacterium. Multicopy plasmids exist in a characteristic number (typically 10 to 20) per bacterial chromosome.

Copy number is primarily a consequence of the type of replication control mechanism. The system responsible for initiating replication determines how many origins can be present in the bacterium. Each plasmid consists of a single replicon, and as a result the number of origins is the same as the number of plasmid molecules.

Single-copy plasmids have a system for replication control whose consequences are similar to those of the system for replication governing the bacterial chromosome. A single origin can be replicated once, and then the daughter origins are segregated to the different daughter cells.

Multicopy plasmids have a replication system that allows a pool of origins to exist. If the number is great enough (in practice, >10 per bacterium), an active segregation system becomes unnecessary, because even a statistical distribution of plasmids to daughter cells will result in the loss of plasmids at frequencies of $<10^{-6}$.

Plasmids are maintained in bacterial populations with very low rates of loss ($<10^{-7}$ per cell division is typical, even for a single-copy plasmid). The systems that control plasmid segregation can be identified by mutations that increase the frequency of loss, but that do not act upon replication itself. Several types of mechanisms are used to ensure the survival of a plasmid in a bacterial population. It is common for a plasmid to carry several systems, often of different types, all acting independently to ensure its survival. Single-copy plasmids require partitioning systems to ensure that the duplicate copies find themselves on opposite sides of the septum at cell division, and are therefore segregated to a different daughter cell. In fact, functions involved in partitioning were first identified in plasmids. The components of a common system are summarized in **FIGURE 17.13**. Typically there are two *trans*-acting loci (*parA* and *parB*) and a *cis*-acting element (*parS*) located just downstream of the two genes. ParA is an ATPase. It binds to ParB, which binds to the *parS* site on DNA. Deletions of any of the three loci prevent proper partition of the plasmid. Systems of this type have been characterized for the plasmids F, P1, and R1. In spite of their overall similarities, there are no significant sequence homologies between the corresponding genes or *cis*-acting sites.

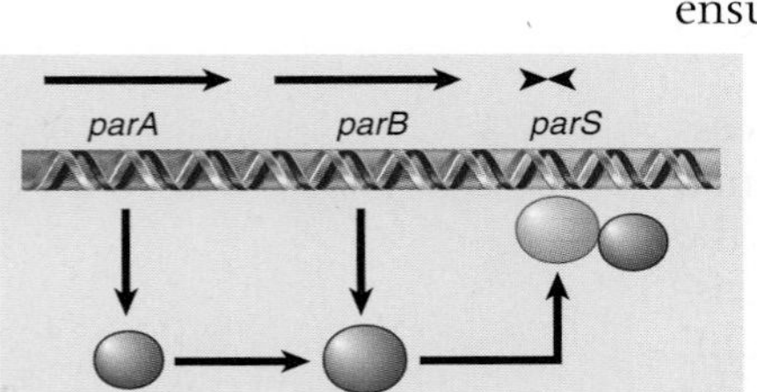

FIGURE 17.13 A common segregation system consists of genes *parA* and *parB* and the target site *parS*.

parS plays a role for the plasmid that is equivalent to the centromere in a eukaryotic cell. Binding of the ParB protein to it creates a structure that segregates the plasmid copies to opposite daughter cells. A bacterial protein, IHF, also binds at this site to form part of the structure. The complex of ParB and IHF with *parS* is called the *partition complex*. *parS* is a 34-bp sequence containing the IHF-binding site and is flanked on either side by sequences called *boxA* and *boxB* that are bound by ParB.

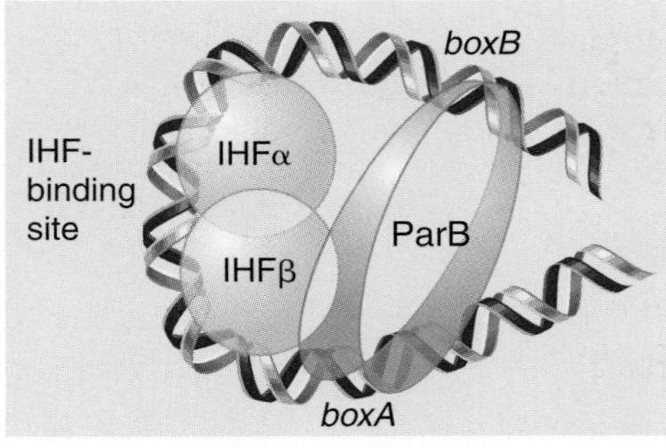

FIGURE 17.14 The partition complex is formed when IHF binds to DNA at parS and bends it so that ParB can bind to sites on either side. The complex is initiated by a heterodimer of IHF and a dimer of ParB, and then more ParB dimers bind.

IHF is the integration host factor. It is named for the role in which it was first discovered (forming a structure that is involved in the integration of phage lambda DNA into the host chromosome). IHF is a heterodimer that has the capacity to form a large structure in which DNA is wrapped on the surface. The role of IHF is to bend the DNA so that ParB can bind simultaneously to the separated *boxA* and *boxB* sites, as indicated in **FIGURE 17.14**.

Complex formation is initiated when *parS* is bound by a heterodimer of IHF together with a dimer of ParB. This enables further dimers of ParB to bind cooperatively. The interaction of ParA with the partition complex structure is essential but transient.

The protein-DNA complex that assembles on IHF during phage lambda integration binds two DNA molecules to enable them to recombine. The role of the partition complex is different: to ensure that two DNA molecules segregate from one another. We do not know yet how the formation of the individual complex accomplishes this task. One possibility is that it attaches the DNA to some physical site—for example, on the membrane—and then the sites of attachment are segregated by growth of the septum.

Proteins related to ParA and ParB are found in several bacteria. In *B. subtilis,* they are called Soj and SpoOJ, respectively. Mutations in these loci prevent sporulation because of a failure to segregate one daughter chromosome into the forespore. It is possible that SpoOJ binds both old and newly synthesized origins, maintaining a status equivalent to chromosome pairing until the chromosomes are segregated to the opposite poles. In *Caulobacter crescentus,* ParA and ParB localize to the poles of the bacterium and ParB binds sequences close to the origin, thus localizing the origin to the pole. These results suggest that a specific apparatus is responsible for localizing the origin to the pole. The next stage of the analysis will be to identify the cellular components with which this apparatus interacts.

KEY CONCEPTS

- Single-copy plasmids exist at one plasmid copy per bacterial chromosome origin.
- Multicopy plasmids exist at >1 plasmid copy per bacterial chromosome origin.
- Homologous recombination between circular plasmids generates dimers and higher multimers.
- Plasmids have site-specific recombination systems that undertake intramolecular recombination to regenerate monomers.
- Partition systems ensure that duplicate plasmids are segregated to different daughter cells produced by a division.

CONCEPT AND REASONING CHECK

Why don't multicopy plasmids use the same kind of partitioning system as single-copy plasmids?

17.10 Plasmid Incompatibility Is Determined by the Replicon

The phenomenon of plasmid incompatibility is related to the regulation of plasmid copy number and segregation. A **compatibility group** is defined as a set of plasmids whose members are unable to coexist in the same bacterial cell. The reason for their incompatibility is that they cannot be distinguished from one another at some stage that is essential for plasmid maintenance. DNA replication and segregation are stages at which this may apply.

▸ **compatibility group** A group of plasmids that contains members unable to coexist in the same bacterial cell.

The negative control model for plasmid incompatibility follows the idea that copy number control is achieved by synthesizing a repressor that measures the concentration of origins. (Formally, this is the same as the titration model for regulating replication of the bacterial chromosome.)

The introduction of a new origin in the form of a second plasmid of the same compatibility group mimics the result of replication of the resident plasmid; two origins now are present. Thus any further replication is prevented until after the two plasmids have been segregated to different cells to create the correct prereplication copy number, as illustrated in **FIGURE 17.15**.

FIGURE 17.15 Two plasmids are incompatible (they belong to the same compatibility group) if their origins cannot be distinguished at the stage of initiation. The same model could apply to segregation.

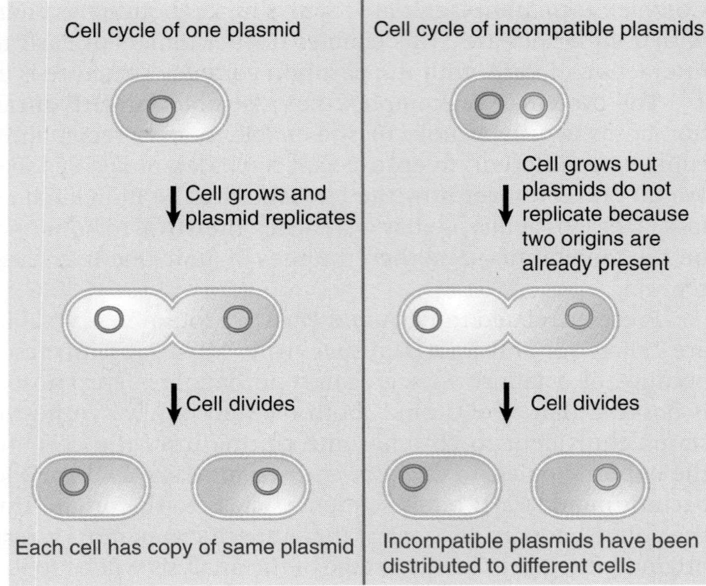

A similar effect would be produced if the system for segregating the products to daughter cells could not distinguish between two plasmids. For example, if two plasmids have the same *cis*-acting partition sites, competition between them would ensure that they would be segregated to different cells, and therefore could not survive in the same line.

The presence of a member of one compatibility group does not directly affect the survival of a plasmid belonging to a different group. Only one replicon of a given compatibility group (of a single-copy plasmid) can be maintained in the bacterium, but it does not interact with replicons of other compatibility groups.

KEY CONCEPT

- Plasmids in a single compatibility group have origins that are regulated by a common control system.

CONCEPT AND REASONING CHECK

How does one plasmid know whether or not another plasmid is in the same family?

17.11 How Do Mitochondria Replicate and Segregate?

Mitochondria must be duplicated during the cell cycle and segregated to the daughter cells. We understand some of the mechanics of this process, but not its regulation.

At each stage in the duplication of mitochondria—DNA replication, DNA segregation to duplicate mitochondria, and organelle segregation to daughter cells—the process appears to be stochastic, governed by a random distribution of each copy. The theory of distribution in this case is analogous to that of multicopy bacterial plasmids, with the same conclusion that >10 copies are required to ensure that each daughter gains at least one copy (see *Section 17.9, Single-Copy Plasmids Have a Partitioning System*). When there are mtDNAs with allelic variations in the same cell, called **heteroplasmy** (either because of inheritance from different parents or because of mutation), the stochastic distribution may generate cells that have only one of the alleles.

▸ **heteroplasmy** Having more than one mitochondrial allelic variant in a cell.

Replication of mtDNA may be stochastic because there is no control over which particular copies are replicated, so that in any cycle some mtDNA molecules may replicate more times than others. The total number of copies of the genome may be controlled by titrating mass in a way similar to bacteria (see *Section 17.2, Replication Is Connected to the Cell Cycle*).

A mitochondrion divides by developing a ring around the organelle that constricts to pinch it into two halves. The mechanism is similar in principle to that involved in bacterial division. The apparatus that is used in plant cell mitochondria is similar to that used in bacteria and uses a homolog of the bacterial protein FtsZ (see *Section 17.5, FtsZ Is Necessary for Septum Formation*). The molecular apparatus is different in animal cell mitochondria and uses the protein dynamin, which is involved in formation of membranous vesicles. An individual organelle may have more than one copy of its genome.

We do not know whether there is a partitioning mechanism for segregating mtDNA molecules within the mitochondrion, or whether they are simply inherited by daughter mitochondria according to which half of the mitochondrion they happen to lie in. **FIGURE 17.16** shows that the combination of replication and segregation mechanisms can result in a stochastic assignment of DNA to each of the copies, that is, so that the distribution of mitochondrial genomes to daughter mitochondria does not depend on their parental origins.

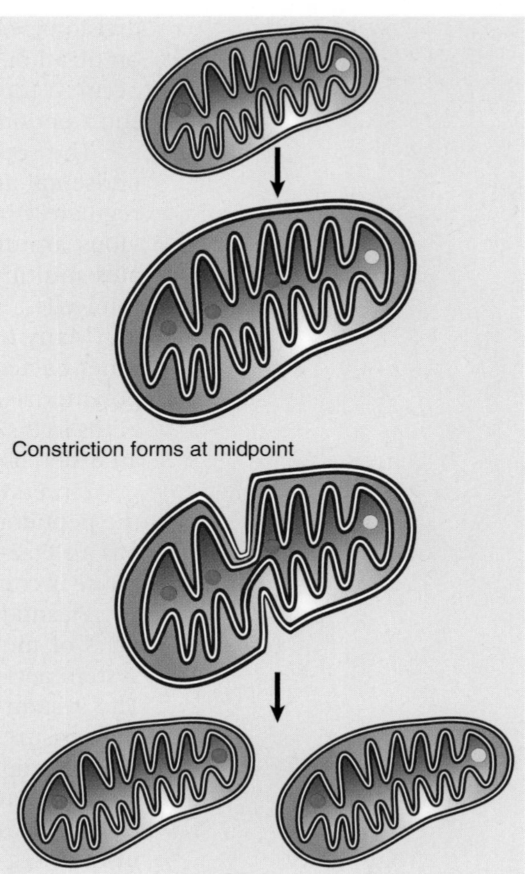

FIGURE 17.16 Mitochondrial DNA replicates by increasing the number of genomes in proportion to mitochondrial mass but without ensuring that each genome replicates the same number of times. This can lead to changes in the representation of alleles in the daughter mitochondria.

The assignment of mitochondria to daughter cells at mitosis also appears to be random. Indeed, it was the observation of somatic variation in plants that first suggested the existence of genes that could be lost from one of the daughter cells because they were not inherited according to Mendel's laws (see Figure 4.15).

In some situations a mitochondrion has both paternal and maternal alleles. This has two requirements: that both parents provide alleles to the zygote (which of course is not the case when there is maternal inheritance; see *Section 4.9, Some Organelles Have DNA*); and that the parental alleles are found in the same mitochondrion. For this to happen, parental mitochondria must have fused.

The size of the individual mitochondrion may not be precisely defined. Indeed, there is a continuing question as to whether an individual mitochondrion represents a unique and discrete copy of the organelle or whether it is in a dynamic flux in which it can fuse with other mitochondria. We know that mitochondria can fuse in yeast, because recombination between mtDNAs can occur after two haploid yeast strains have mated to produce a diploid strain. This implies that the two mtDNAs must have been exposed to one another in the same mitochondrial compartment. Attempts have been made to test for the occurrence of similar events in animal cells by looking for complementation between alleles after two cells have been fused, but the results are not clear.

KEY CONCEPTS

- mtDNA replication and segregation to daughter mitochondria is stochastic.
- Mitochondrial segregation to daughter cells is also stochastic.

CONCEPT AND REASONING CHECK

Why is the mitochondrial replication system not identical to eukaryote nuclear replication?

17.12 Summary

A fixed time of 40 minutes is required to replicate the *E. coli* chromosome and a further 20 minutes is required before the cell can divide. When cells divide more rapidly than every 60 minutes, a replication cycle is initiated before the end of the preceding division cycle. This generates multiforked chromosomes. The initiation event depends on titration of cell mass, probably by accumulating an initiator protein. Initiation may occur at the cell membrane because the origin is associated with the membrane for a short period after initiation.

The septum that divides the cell grows at a location defined by the preexisting periseptal annulus; a locus of three genes (*minC, -D,* and *-E*) codes for products that regulate whether the mid-cell periseptal annulus or the polar sites derived from previous annuli are used for septum formation. Absence of septum formation generates multinucleated filaments; an excess of septum formation generates anucleate minicells.

Many transmembrane proteins interact to form the septum. ZipA is located in the inner bacterial membrane and binds to FtsZ, which is a tubulin-like protein that can polymerize into a filamentous structure called a Z-ring. FtsA is a cytosolic protein that binds to FtsZ. Several other *fts* products, all transmembrane proteins, join the Z-ring in an ordered process that generates a septal ring. The last proteins to bind are the SEDS protein FtsW and the transpeptidase ftsI (PBP3), which together function to produce the peptidoglycans of the septum. Chloroplasts use a related division mechanism that has an FtsZ-like protein, but mitochondria use a different process in which the membrane is constricted by a dynamin-like protein.

Plasmids and bacteria have site-specific recombination systems that regenerate pairs of monomers by resolving dimers created by general recombination. The Xer system acts on a target sequence located in the terminus region of the chromosome. The system is active only in the presence of the FtsK protein of the septum, which may ensure that it acts only when a dimer needs to be resolved.

Plasmid partitioning involves the interaction of the ParB protein with the *parS* target site to build a structure that includes the IHF protein. This partition complex ensures that replica chromosomes segregate into different daughter cells. The mechanism of segregation may involve movement of DNA, possibly by the action of MukB in condensing chromosomes into masses at different locations as they emerge from replication.

Plasmids have a variety of systems that ensure or assist partition, and an individual plasmid may carry systems of several types. The copy number of a plasmid describes whether it is present at the same level as the bacterial chromosome (one per unit cell) or in greater numbers. Plasmid incompatibility can be a consequence of the mechanisms involved in either replication or partition (for single-copy plasmids). Two plasmids that share the same control system for replication are incompatible because the number of replication events ensures that there is only one plasmid for each bacterial genome.

CHAPTER QUESTIONS

1. How many replication forks would be present in the *E. coli* chromosome for cells that are growing under optimal conditions?

A. one
B. two
C. four
D. more than four

2. How long does it take to replicate the full bacterial chromosome at normal growth temperature?

A. 10 minutes
B. 20 minutes
C. 40 minutes
D. 60 minutes

3. What is the fixed interval of time required between initiation of DNA replication and cell division in *E. coli*?
   A. 10 minutes
   B. 20 minutes
   C. 40 minutes
   D. 60 minutes
4. The bacterial *ftsZ* gene is required for:
   A. septum formation.
   B. periseptal annulus formation and localization.
   C. DNA replication.
   D. partitioning of DNA.
5. The location of septum formation in a bacterial cell is controlled by the:
   A. *par* locus.
   B. *min* locus.
   C. *xer* locus.
   D. *fts* locus.
6. Current evidence suggests that the bacterial chromosome is:
   A. freely soluble in the cytoplasm of the cell.
   B. attached to a specific cytoskeleton structure in the cytoplasm.
   C. attached to random sites on the inner membrane of the cell.
   D. attached to one specific site on the inner membrane of the cell.
7. Occasionally two new daughter molecules can undergo homologous recombination to form a dimer of the bacterial genome. How does the bacterial genome deal with this to undergo cell division?
   A. It cannot; it is lethal and the cell dies.
   B. One daughter cell receives the dimer and survives; the other daughter cell lacks DNA.
   C. Site-specific DNA recombination separates the two monomers, one for each new cell.
   D. A reversal of the homologous recombination event occurs at random sites to separate the daughter molecules.
8. Two plasmids are of the same compatibility group if:
   A. they can coexist in the same bacterial cell.
   B. they cannot coexist in the same bacterial cell.
   C. they carry the same antibiotic resistance genes.
   D. they carry the same toxin genes.
9. Mitochondrial DNA replication and segregation is:
   A. tightly controlled with each new mitochondrion receiving one complete genome.
   B. tightly controlled with each new mitochondrion receiving at least 10 copies of the genome.
   C. stochastic with each new mitochondrion receiving one complete genome.
   D. stochastic with each new mitochondrion having at least 10 copies of the genome.
10. Mitochondria divide by:
   A. binary fission.
   B. budding.
   C. either binary fission or budding.
   D. neither binary fission nor budding.

KEY TERMS

anucleate cell
compatibility group
copy number
doubling time
heteroplasmy
minicell
multiforked chromosome
nucleoid
periseptal annulus
septal ring
septum
site-specific recombination
topoisomerase
unit cell
Z-ring

FURTHER READING

Falkenberg, M., Larsson, N.-G., and Gustafsson, C. M. (2007). DNA replication and transcription in mammalian mitochondria. *Annu. Rev. Biochem.* 76, 679–699.

Ghosh, S.K., Hajra, S., Paek, A., and Jayaram, M. (2006). Mechanisms for chromosomal and plasmid segregation. *Annu. Rev. Biochem.* 75, 211–241.

Lutkenhaus, J. (2007). Assembly dynamics of the bacterial MinCDE system and special resolution of the Z ring. *Annu. Rev. Biochem.* 76, 539–562.

18

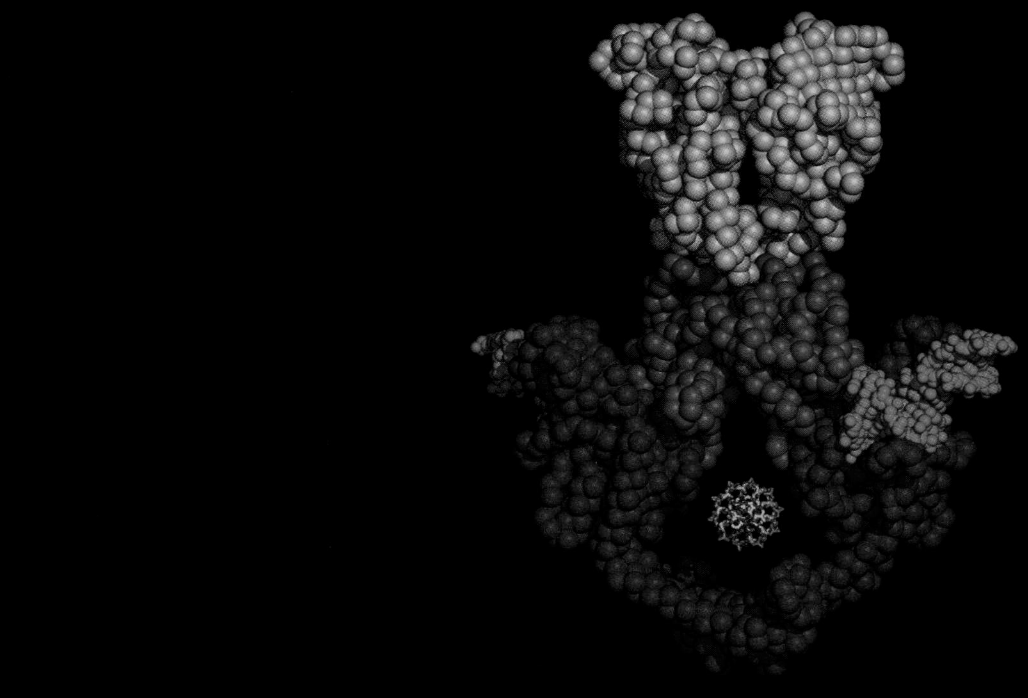

A type II topoisomerase acting on a knotted or supercoiled DNA substrate. The cleaved DNA is shown in green. The DNA that is being passed through the break is viewed end-on. Photo courtesy of James M. Berger, California Institute for Quantitative Biology, University of California, Berkeley.

DNA Replication

CHAPTER OUTLINE

18.1 Introduction

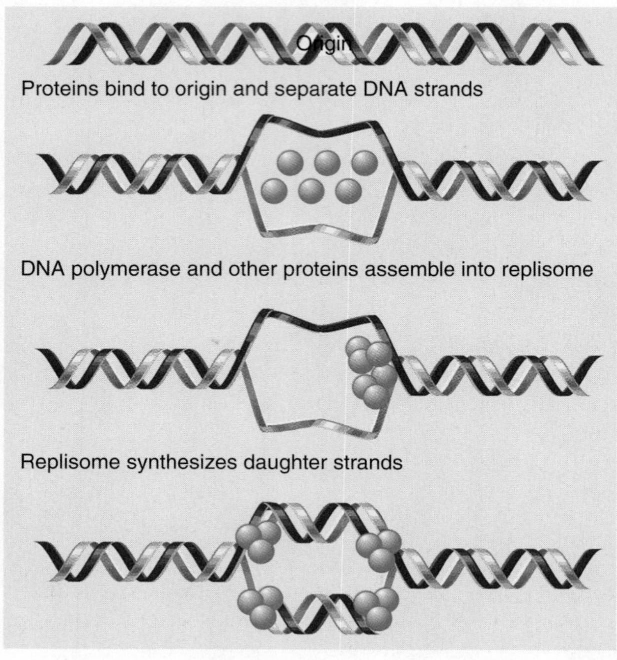

Origin
Proteins bind to origin and separate DNA strands
DNA polymerase and other proteins assemble into replisome
Replisome synthesizes daughter strands

FIGURE 18.1 Replication initiates when a protein complex binds to the origin and melts the DNA there. Then the components of the replisome, including DNA polymerase, assemble. The replisome moves along DNA, synthesizing both new strands.

Replication of duplex DNA is a complicated endeavor involving multiple enzyme complexes. Different activities are involved in the stages of initiation, elongation, and termination. Before initiation can occur, however, the supercoiled chromosome must be relaxed (see *Section 1.5, Supercoiling Affects the Structure of DNA*). This occurs in segments beginning with the replication origin region. This alteration to the structure of the chromosome is accomplished by the enzyme **topoisomerase**. Replication cannot occur on supercoiled DNA, only the relaxed form. FIGURE 18.1 shows an overview of the first stages of the process.

- *Initiation* involves recognition of an origin by a large protein complex. Before DNA synthesis can begin, the parental strands must be separated and transiently stabilized in the single-stranded state, creating a replication bubble. After this stage, synthesis of daughter strands can be initiated at the replication fork.
- *Elongation* is undertaken by another complex of proteins. The **replisome** exists only as a protein complex associated with the particular structure that DNA takes at the replication fork. It does not exist as an independent unit (for example, analogous to the ribosome), but assembles *de novo* at the origin for each replication cycle. As the replisome moves along DNA, the parental strands unwind and daughter strands are synthesized.
- At the end of the replicon, *joining* and/or *termination* reactions are necessary. Following termination, the duplicate chromosomes must be separated from one another, which requires manipulation of higher-order DNA structure.

Inability to replicate DNA is fatal for a growing cell. Mutants for replication must therefore be obtained as **conditional lethals**. These are able to accomplish replication under *permissive* conditions (provided by the normal temperature of incubation), but they are defective under *nonpermissive* conditions (provided by the higher temperature of 42°C). A comprehensive series of such temperature-sensitive mutants in *E. coli* identifies a set of loci called the *dna* genes.

▸ **topoisomerase** An enzyme that changes the number of times the two strands in a closed DNA molecule cross each other. It does this by cutting the DNA, passing DNA through the break, and resealing the DNA.

▸ **replisome** The multiprotein structure that assembles at the bacterial replication fork to undertake synthesis of DNA. It contains DNA polymerase and other enzymes.

▸ **conditional lethal** A mutation that is lethal under one set of conditions, but not lethal under a second set of conditions, such as temperature.

18.2 Initiation: Creating the Replication Forks at the Origin

Initiation of replication of duplex DNA in *E. coli* at the origin of replication, *oriC*, requires several successive activities:

- Protein synthesis is required to synthesize the origin recognition protein, DnaA. This is the *E. coli* **licensing factor** that must be made anew for each round of replication. Drugs that block protein synthesis block a new round of replication, but not continuation of replication.

▸ **licensing factor** A factor located in the nucleus and necessary for replication; it is inactivated or destroyed after one round of replication. New factors must be provided for further rounds of replication to occur.

- There is a requirement for transcription activation. This is not synthesis of the mRNA for DnaA, but rather either one of two genes that flank *oriC* must be transcribed. This transcription near the origin aids DnaA in twisting open the origin.
- There must be membrane/cell wall synthesis. Drugs (like penicillin) that inhibit cell wall synthesis block initiation of replication.

Most events that are required for initiation therefore occur uniquely at the origin; others recur with the initiation of each Okazaki fragment (see *Section 18.6, The Two New DNA Strands Have Different Modes of Synthesis*) during the elongation phase.

Initiation of replication at *oriC* starts with formation of a complex that ultimately requires six proteins: DnaA, DnaB, DnaC, HU, Gyrase, and SSB. Of the six proteins, DnaA draws our attention as the one uniquely involved in the initiation process. DnaB, an ATP hydrolysis-dependent 5′ to 3′ **helicase**, provides the "engine" of initiation after the origin has been opened (and the DNA is single-stranded), by its ability to further unwind the DNA. These events will only happen if the DNA at the origin is fully methylated on both strands.

▸ **helicase** An enzyme that uses energy provided by ATP hydrolysis to separate the strands of a nucleic acid duplex.

The first stage in initiation is binding of the DnaA-ATP protein complex to the fully methylated *oriC* sequence. This takes place in association with the inner membrane. DnaA is in the active form only when bound to ATP. DnaA has intrinsic ATPase activity that hydrolyzes ATP to ADP and thus inactivates itself when the initiation stage ends. This ATPase activity is stimulated by membrane phospholipids and single-stranded DNA. Single-stranded DNA forms once the origin is open. This mechanism is used to prevent reinitiation of replication. The origin of replication region remains attached to the membrane for about one third of the cell cycle as part of the mechanism to prevent reinitiation. While sequestered in the membrane, the newly synthesized strand of *oriC* cannot be methylated and so *oriC* remains hemimethylated until DnaA is degraded.

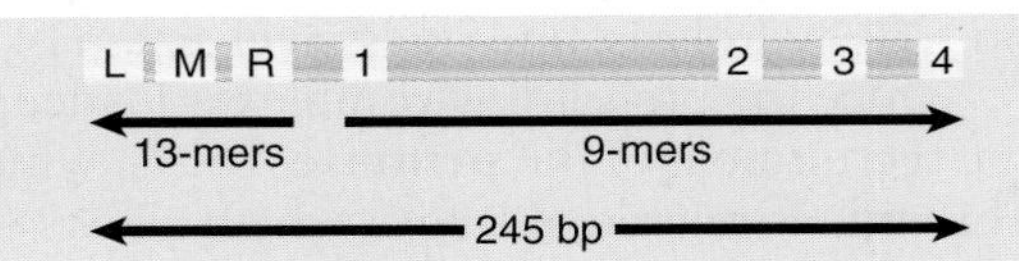
L M R 1 2 3 4
13-mers
9-mers
245 bp

FIGURE 18.2 The minimal origin is defined by the distance between the outside members of the 13-mer and 9-mer repeats.

Opening *oriC* involves action at two types of sequence in the origin: 9 bp and 13 bp repeats. Together the 9 bp and 13 bp repeats define the limits of the 245 bp minimal origin, as indicated in **FIGURE 18.2**. An origin is activated by the sequence of events summarized in **FIGURE 18.3**, in which binding of DnaA-ATP is succeeded by association with the other proteins.

The four 9 bp consensus sequences on the right side of *oriC* provide the initial binding sites for DnaA-ATP. It binds cooperatively to form a central core around which *oriC* DNA is wrapped. DnaA then acts at three A-T-rich 13 bp tandem repeats located on the left side of *oriC*. In its active form, DnaA-ATP twists open the DNA strands at each of these sites to form an open bubble complex. All three 13 bp repeats must be opened for the reaction to proceed to the next stage. Transcription of

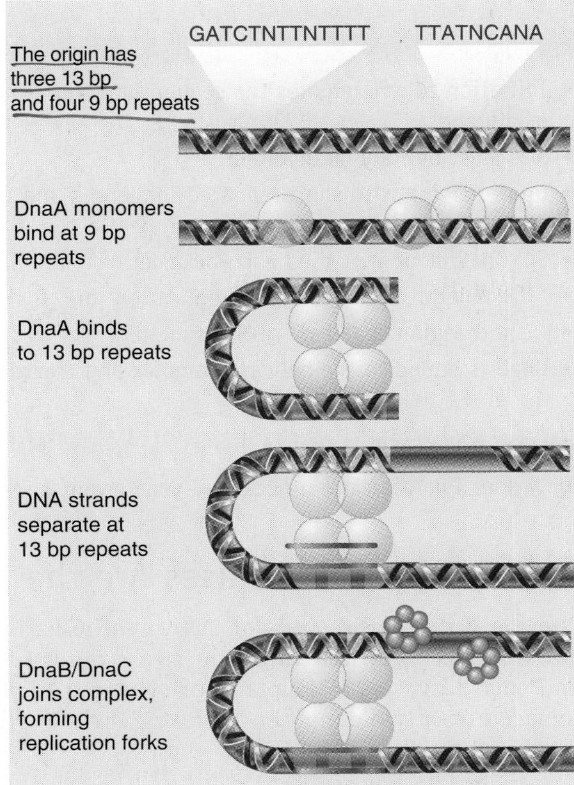
GATCTNTTNTTTT
TTATNCANA
The origin has three 13 bp and four 9 bp repeats
DnaA monomers bind at 9 bp repeats
DnaA binds to 13 bp repeats
DNA strands separate at 13 bp repeats
DnaB/DnaC joins complex, forming replication forks

FIGURE 18.3 Prepriming involves formation of a complex by sequential association of proteins, which leads to the separation of DNA strands.

either gene flanking *oriC* provides additional torsional stress to help snap apart the double-stranded DNA.

Altogether, two to four monomers of DnaA bind at the origin, and they recruit two "prepriming" complexes of the DnaB helicase bound to DnaC, so that there is one DnaB-DnaC complex for each of the two (bidirectional) replication forks. The only function of DnaC is that of a chaperone to repress the helicase activity of DnaB until it is needed. Each DnaB–DnaC complex consists of six DnaC monomers bound to a hexamer of DnaB. Note that the DnaB helicase cannot open double-stranded DNA; it can only unwind DNA that has already been opened, in this case by DnaA.

The region of strand separation in the open complex is large enough for both DnaB hexamers to bind, which initiates the two replication forks. As DnaB binds, it displaces DnaA from the 13 bp repeats and extends the length of the open region using its helicase activity.

▸ **gyrase** An enzyme that changes the number of times the two strands in a closed DNA molecule cross each other. It does this by cutting the DNA, passing DNA through the break, and resealing the DNA.

▸ **single-strand binding protein (SSB)** The protein that attaches to single-stranded DNA, thereby preventing the DNA from forming a duplex.

Some additional proteins are required to support the unwinding reaction. **Gyrase**, a type II topoisomerase, provides a swivel that allows one DNA strand to rotate around the other. Without this reaction, unwinding would generate torsional strain (overwinding) in the DNA that would resist unwinding by the helicase. The protein **SSB** (**single-strand binding protein**) stabilizes the single-stranded DNA as it is formed and modulates the helicase activity. The length of duplex DNA that usually is unwound to initiate replication is probably <60 bp. The protein HU is a general DNA-binding protein in *E. coli*. Its presence is not absolutely required to initiate replication *in vitro*, but it stimulates the reaction. HU has the capacity to bend DNA, and is involved in building the structure that leads to formation of the open complex.

Input of energy in the form of ATP is required at several stages for the prepriming reaction, and it is required for unwinding DNA. The helicase action of DnaB depends on ATP hydrolysis, and the swivel action of gyrase requires ATP hydrolysis.

▸ **primase** A type of RNA polymerase that synthesizes short segments of RNA that will be used as primers for DNA replication.

Once the prepriming complex is loaded onto the replication forks, the next step is the recruitment of the **primase**, DnaG, which is then loaded onto the DnaB hexamer. This entails release of DnaC, which allows the DnaB helicase to become active. DnaC hydrolyzes ATP in order to release DnaB. This step marks the transition from initiation to elongation.

KEY CONCEPTS

- Initiation at *oriC* requires the sequential assembly of a large protein complex on the membrane.
- *oriC* must be fully methylated.
- DnaA-ATP binds to short repeated sequences and forms an oligomeric complex that melts DNA.
- Six DnaC monomers bind each hexamer of DnaB, and this complex binds to the origin.
- A hexamer of DnaB forms the replication fork. Gyrase and SSB are also required.
- A short region of A-T-rich DNA is melted.
- DnaG is bound to the helicase complex and creates the replication forks.

CONCEPT AND REASONING CHECK

How does DnaA initiate replication, yet prevent a second round of replication?

18.3 DNA Polymerases Are the Enzymes That Make DNA

▸ **semiconservative replication** Replication that is accomplished by separation of the strands of a parental duplex, each strand then acting as a template for synthesis of a complementary strand.

There are two basic types of DNA synthesis. **FIGURE 18.4** shows the result of **semiconservative replication**. The two strands of the parental duplex are separated, and each serves as a template for synthesis of a new strand. The parental duplex is replaced with two daughter duplexes, each of which has one parental strand and one newly synthesized strand.

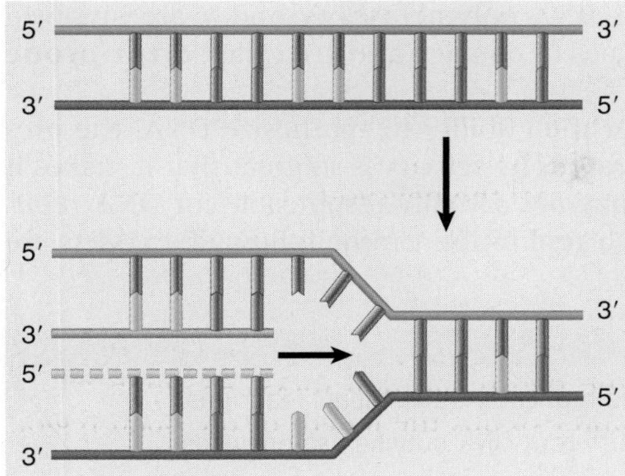

FIGURE 18.4 Semiconservative replication synthesizes two new strands of DNA.

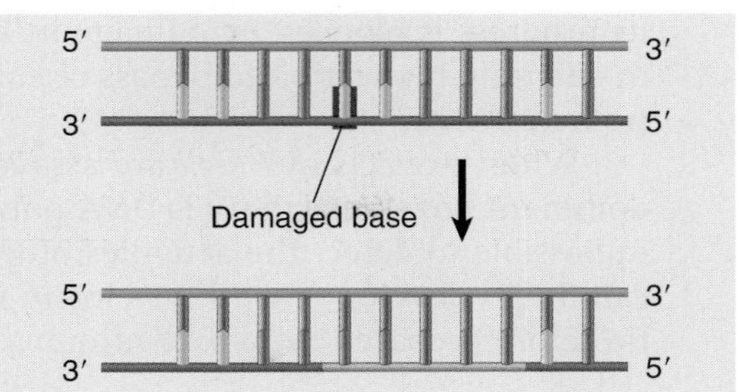

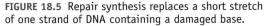

FIGURE 18.5 Repair synthesis replaces a short stretch of one strand of DNA containing a damaged base.

FIGURE 18.5 shows the consequences of a **DNA repair** reaction. One strand of DNA has been damaged. It is excised and new material is synthesized to replace it. An enzyme that can synthesize a new DNA strand on a template strand is called a **DNA polymerase** (or more properly, DNA-dependent DNA polymerase). Both prokaryotic and eukaryotic cells contain multiple DNA polymerase activities. Only a few of these enzymes actually undertake replication. The remaining enzymes are involved in repair synthesis or participate in subsidiary roles in replication.

All prokaryotic and eukaryotic DNA polymerases share the same fundamental type of synthetic activity, synthesis from 5′ to 3′ from a template that is 3′ to 5′. This means adding nucleotides one at a time to a 3′–OH end, as illustrated diagrammatically in **FIGURE 18.6**. The choice of the nucleotide to add to the chain is dictated by base pairing with the template strand.

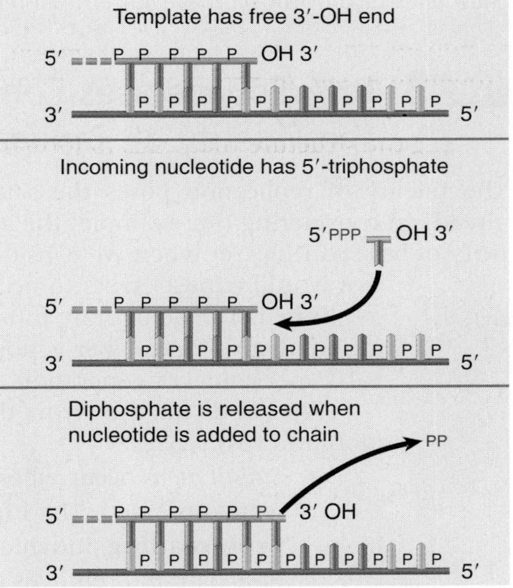

FIGURE 18.6 DNA is synthesized by adding nucleotides to the 3′–OH end of the growing chain, so that the new chain grows in the 5′⟶3′ direction. The precursor for DNA synthesis is a nucleoside triphosphate, which loses the terminal two phosphate groups in the reaction.

▸ **DNA repair** The removal and replacement of damaged DNA by the correct sequence.

▸ **DNA polymerase** An enzyme that synthesizes a daughter strand(s) of DNA (under direction from a DNA template). Any particular enzyme may be involved in repair or replication (or both).

Enzyme	Gene	Function
I	*polA*	major repair enzyme
II	*polB*	replication restart
III	*polC*	replicase
IV	*dinB*	translesion replication
V	*umuD′*$_2$*C*	translesion replication

FIGURE 18.7 Only one DNA polymerase is the replication enzyme. The others participate in repair of damaged DNA, restarting stalled replication forks, or bypassing damage in DNA.

Some DNA polymerases, such as the repair polymerases, function as independent enzymes, but others, notably the replication polymerases, are incorporated into large protein assemblies called **holoenzymes**. The DNA-synthesizing subunit is only one of several functions of the holoenzyme, which typically contains other activities concerned with fidelity.

▸ **holoenzyme** The DNA polymerase complex that is competent to initiate replication.

FIGURE 18.7 summarizes the DNA polymerases that have been characterized in *E. coli*. DNA polymerase III, a multisubunit protein, is the replication polymerase responsible for *de novo* synthesis of new strands of DNA. DNA polymerase I (coded by *polA*) is involved in the repair of damaged DNA and, in a subsidiary role, in semiconservative replication. DNA polymerase II is required to restart a replication fork when

▸ **error-prone polymerase** A DNA polymerase that incorporates noncomplementary bases into the daughter strand.

its progress is blocked by damage in DNA. DNA polymerases IV and V are involved in allowing replication to bypass certain types of damage and are called **error-prone polymerases**.

When extracts of *E. coli* are assayed for their ability to synthesize DNA, the predominant enzyme activity is DNA polymerase I. Its activity is so great that it makes it impossible to detect the activities of the enzymes actually responsible for DNA replication! To develop *in vitro* systems in which replication can be followed, extracts are therefore prepared from *polA* mutant cells.

KEY CONCEPTS

- DNA is synthesized in both semiconservative replication and repair reactions.
- A bacterium or eukaryotic cell has several different DNA polymerase enzymes.
- One bacterial DNA polymerase undertakes semiconservative replication; the others are involved in repair reactions.

CONCEPT AND REASONING CHECK

How does repair DNA synthesis differ from replication DNA synthesis?

▸ **proofreading** A mechanism for correcting errors in DNA synthesis that involves scrutiny of individual units after they have been added to the chain.

▸ **processivity** The ability of an enzyme to perform multiple catalytic cycles with a single template instead of dissociating after each cycle.

18.4 DNA Polymerases Control the Fidelity of Replication

The fidelity of replication poses the same sort of problem we have encountered already in considering (for example) the accuracy of translation. It relies on the specificity of base pairing. Yet when we consider the interactions involved in base pairing, we would expect errors to occur with a frequency of $\sim 10^{-3}$ per base pair replicated. The actual rate in bacteria seems to be $\sim 10^{-8}$ to 10^{-10}. This corresponds to ~1 error per genome per 1000 bacterial replication cycles, or $\sim 10^{-6}$ per gene per generation.

We can divide the errors that DNA polymerase makes during replication into two classes:

- *Substitutions* occur when the wrong (improperly paired) nucleotide is incorporated. The error level is determined by the efficiency of **proofreading**, in which the enzyme scrutinizes the newly formed base pair and removes the nucleotide if it is mispaired.
- *Frameshifts* occur when an extra nucleotide is inserted or omitted. Fidelity with regard to frameshifts is affected by the **processivity** of the enzyme: the tendency to remain on a single template rather than to dissociate and reassociate. This is particularly important for the replication of a homopolymeric stretch—for example, a long sequence of $dT_n{:}dA_n$, in which "replication slippage" can change the length of the homopolymeric run. As a general rule, increased processivity reduces the likelihood of such events. In multimeric DNA polymerases, processivity is usually increased by a particular subunit that is not needed for catalytic activity *per se*.

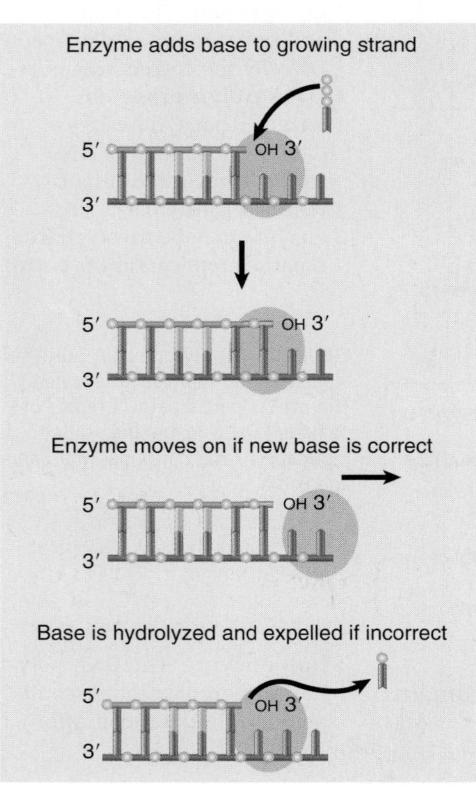
Enzyme adds base to growing strand
5′
OH 3′
3′
5′
OH 3′
3′
Enzyme moves on if new base is correct
5′
OH 3′
3′
Base is hydrolyzed and expelled if incorrect
5′
OH 3′
3′

FIGURE 18.8 Bacterial DNA polymerases scrutinize the base pair at the end of the growing chain and excise the nucleotide added in the case of a misfit.

Bacterial replication enzymes have multiple error reduction systems. All of the bacterial enzymes possess a 3′ to 5′ exonucleolytic activity that proceeds in the reverse direction from DNA synthesis. This provides a proofreading function illustrated diagrammatically in **FIGURE 18.8**. In the chain elongation step, a precursor nucleotide enters the position at the end of the growing chain. A bond is formed. The enzyme moves one base pair farther, and then is ready for the next precursor nucleotide to enter. If a mistake has been made, the DNA is structurally warped by the incorporation of the incorrect base that will cause the polymerase to pause or slow down. This will allow the enzyme to back up and remove the incorrect base (see *Section 18.5, DNA Polymerases Have a Common Structure*). In some regions errors occur more frequently than in others; that is, **mutation hotspots** occur in the DNA. This is caused by the underlying sequence context; that is, some sequences cause the polymerase to move faster or slower, which affects the ability to catch an error.

‣ **mutation hotspot** A site in the genome at which the frequency of mutation (or recombination) is very much increased, usually by at least an order of magnitude relative to neighboring sites.

As noted above in *Section 18.3, DNA Polymerases Are the Enzymes That Make DNA*, replication enzymes typically are found as multisubunit holoenzyme complexes, whereas repair DNA polymerases are typically found as single subunit enzymes. An advantage to a holoenzyme system is the availability of a specialized subunit responsible for error correction. In *E. coli* DNA polymerase III, this activity, a 3′ to 5′ exonuclease, resides in a separate subunit, the ε subunit. This subunit gives the replication enzyme a greater fidelity than the repair enzymes.

Different DNA polymerases handle the relationship between the polymerizing and proofreading activities in different ways. In some cases, the activities are part of the same protein subunit, but in others they are contained in different subunits. Each DNA polymerase has a characteristic error rate that is reduced by its proofreading activity. Proofreading typically decreases the error rate in replication from $\sim 10^{-5}$ to $\sim 10^{-7}$ per base pair replicated. Systems that recognize errors and correct them following replication then eliminate some of the errors, bringing the overall rate to $<10^{-9}$ per base pair replicated (see *Section 20.6, Controlling the Direction of Mismatch Repair*).

KEY CONCEPTS

- DNA polymerases often have a 3′ to 5′ exonuclease activity that is used to excise incorrectly paired bases.
- The fidelity of replication is improved by proofreading by a factor of ~100.

CONCEPT AND REASONING CHECK

If mutation is generally random, why are there mutation hotspots in a genome?

18.5 DNA Polymerases Have a Common Structure

FIGURE 18.9 shows that all DNA polymerases share some common structural features. The enzyme structure can be divided into several independent domains, which are described by analogy with a human right hand. DNA binds in a large cleft composed of three domains. The "palm" domain has important conserved sequence motifs that provide the catalytic active site. The "fingers" are involved in positioning the template correctly at the active site. The "thumb" binds the DNA as it exits the enzyme, and is important in processivity. The most important conserved regions of each of these three domains converge to form a continuous surface at the catalytic site. The exonuclease activity resides in an independent domain with its own catalytic site. The N-terminal domain extends into the nuclease domain. DNA polymerases fall into five families based on sequence homologies; the palm is well conserved among them, but the thumb and fingers provide analogous secondary structure elements from different sequences.

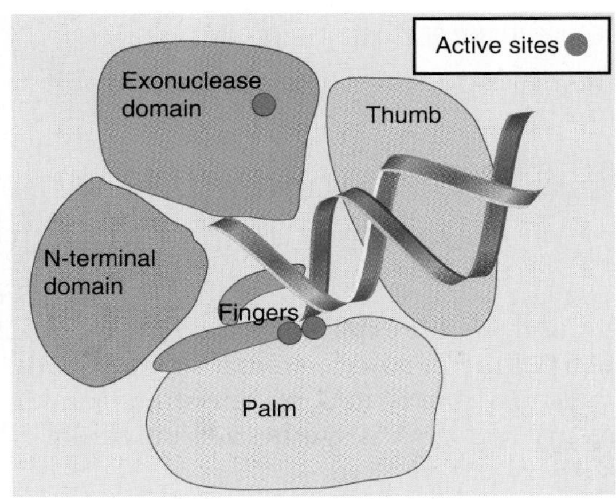

FIGURE 18.9 The common organization of DNA polymerases has a palm that contains the catalytic site, fingers that position the template, a thumb that binds DNA and is important in processivity, an exonuclease domain with its own active site, and an N-terminal domain.

The catalytic reaction in a DNA polymerase occurs at an active site in which a nucleotide triphosphate pairs

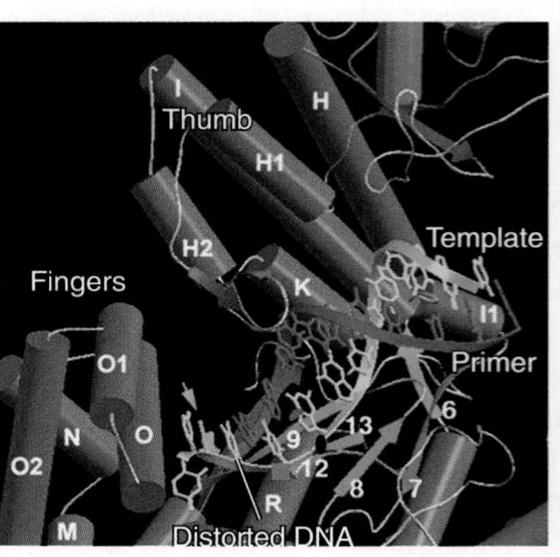

FIGURE 18.10 The crystal structure of phage T7 DNA polymerase shows that the template strand takes a sharp turn that exposes it to the incoming nucleotide. Photo courtesy of Charles Richardson and Thomas Ellenberger, Washington University School of Medicine.

with an (unpaired) single strand of DNA. The DNA lies across the palm in a groove that is created by the thumb and fingers. **FIGURE 18.10** shows the crystal structure of the T7 enzyme complexed with DNA (in the form of a primer annealed to a template strand) and an incoming nucleotide that is about to be added to the primer. The DNA is in the classic B-form duplex up to the last two base pairs at the 3′ end of the primer, which are in the more open A-form. A sharp turn in the DNA exposes the template base to the incoming nucleotide. The 3′ end of the primer (to which bases are added) is anchored by the fingers and palm. The DNA is held in position by contacts that are made principally with the phosphodiester backbone (thus enabling the polymerase to function with DNA of any sequence).

In structures of DNA polymerases of this family complexed only with DNA (that is, lacking the incoming nucleotide), the orientation of the fingers and thumb relative to the palm is more open, with the O helix (O, O1, O2; see Figure 18.10) rotated away from the palm. This suggests that an inward rotation of the O helix occurs to grasp the incoming nucleotide and create the active catalytic site. When a nucleotide binds, the fingers domain rotates 60° toward the palm, with the tops of the fingers moving by 30 Å. The thumb domain also rotates toward the palm by 8°. These changes are cyclical: they are reversed when the nucleotide is incorporated into the DNA chain, which then translocates through the enzyme to recreate an empty site.

The exonuclease activity is responsible for removing mispaired bases. The catalytic site of the exonuclease domain is distant from the active site of the catalytic domain, though. The enzyme alternates between polymerizing and editing modes, as determined by a competition between the two active sites for the 3′ primer end of the DNA. Amino acids in the active site contact the incoming base in such a way that the enzyme structure is affected by the structure of a mismatched base. When a mismatched base pair occupies the catalytic site, the fingers cannot rotate toward the palm to bind the incoming nucleotide. This leaves the 3′ end free to bind to the active site in the exonuclease domain, which is accomplished by a rotation of the DNA in the enzyme structure.

KEY CONCEPTS

- Many DNA polymerases have a large cleft composed of three domains that resemble a hand.
- DNA lies across the "palm" in a groove created by the "fingers" and "thumb."

CONCEPT AND REASONING CHECK

How does the DNA polymerase "know" that it has incorporated an incorrect base?

18.6 The Two New DNA Strands Have Different Modes of Synthesis

The antiparallel structure of the two strands of duplex DNA poses a problem for replication. As the replication fork advances, daughter strands must be synthesized on both of the exposed parental single strands. The fork template strand moves in the direction from 5′ to 3′ on one strand, and in the direction from 3′ to 5′ on the other strand. Yet DNA is synthesized only from a 5′ end toward a 3′ end (by adding new

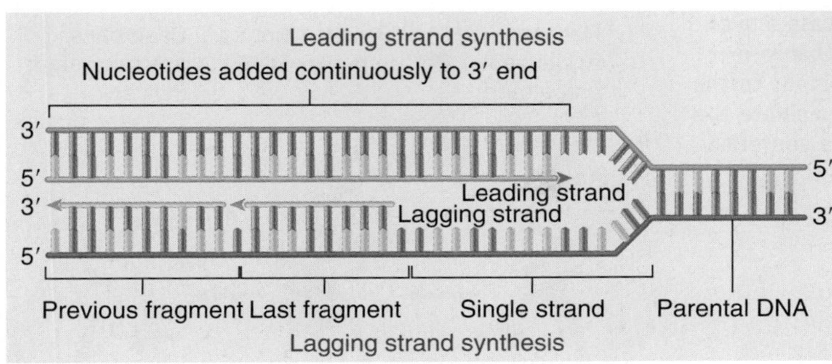

FIGURE 18.11 The leading strand is synthesized continuously, whereas the lagging strand is synthesized discontinuously.

nucleotide to the growing 3′ end) on a template that is 3′ to 5′. The problem is solved by synthesizing the new strand on the 5′ to 3′ template in a series of short fragments, each actually synthesized in the "backward" direction, that is, with the customary 5′ to 3′ polarity.

Consider the region immediately behind the replication fork, as illustrated in **FIGURE 18.11**. We describe events in terms of the different properties of each of the newly synthesized strands:

- On the **leading strand** (sometimes called the *forward strand*) DNA synthesis can proceed continuously in the 5′ to 3′ direction as the parental duplex is unwound.
- On the **lagging strand** a stretch of single-stranded parental DNA must be exposed, and then a segment is synthesized in the reverse direction (relative to fork movement). A series of these fragments are synthesized, each 5′ to 3′; they then are joined together to create an intact lagging strand.

Discontinuous replication can be followed by the fate of a very brief label of radioactivity. The label enters newly synthesized DNA in the form of short fragments, of ~1000 to 2000 bases in length. These **Okazaki fragments** are found in replicating DNA in both prokaryotes and eukaryotes. After longer periods of incubation, the label enters larger segments of DNA. The transition results from covalent linkages between Okazaki fragments.

The lagging strand is synthesized discontinuously and the leading strand is synthesized continuously. This is called **semidiscontinuous replication**.

▸ **leading strand** The strand of DNA that is synthesized continuously in the 5′ to 3′ direction.

▸ **lagging strand** The strand of DNA that must grow overall in the 3′ to 5′ direction and is synthesized discontinuously in the form of short fragments (5′–3′) that are later connected covalently.

▸ **Okazaki fragment** Short stretches of 1000 to 2000 bases produced during discontinuous replication; they are later joined into a covalently intact strand.

▸ **semidiscontinuous replication** The mode of replication in which one new strand is synthesized continuously while the other is synthesized discontinuously.

KEY CONCEPT

- The DNA polymerase advances continuously when it synthesizes the leading strand (5′–3′), but synthesizes the lagging strand by making short fragments that are subsequently joined together.

CONCEPT AND REASONING CHECK

Why can't the DNA polymerase move continuously on both strands?

18.7 Replication Requires a Helicase and Single-Strand Binding Protein

As the replication fork advances, it unwinds the duplex DNA. One of the template strands is rapidly converted to duplex DNA as the leading daughter strand is synthesized. The other remains single stranded until a sufficient length has been exposed to initiate synthesis of an Okazaki fragment complementary to the lagging strand in the backward direction. The generation and maintenance of single-stranded DNA is,

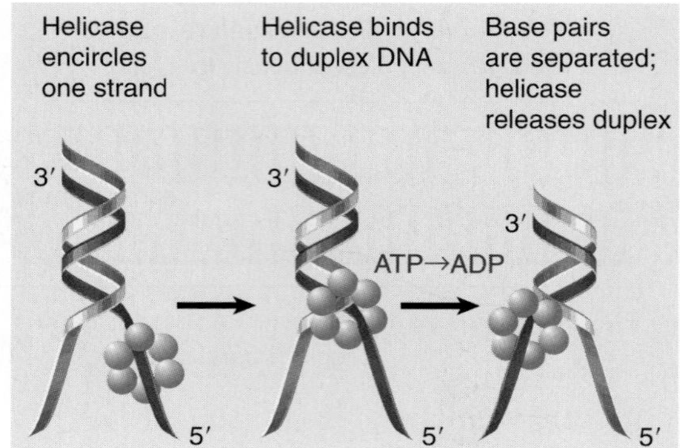
Helicase
encircles
one strand
Helicase binds
to duplex DNA
Base pairs
are separated;
helicase
releases duplex
3′
3′
3′
ATP→ADP
5′
5′
5′

FIGURE 18.12 A hexameric helicase moves along one strand of DNA. It probably changes conformation when it binds to the duplex, uses ATP hydrolysis to separate the strands, and then returns to the conformation it has when bound only to a single strand.

therefore, a crucial aspect of replication. Two types of function are needed to convert double-stranded DNA to the single-stranded state:

- A *helicase* is an enzyme that separates (or melts) the strands of DNA, usually using the hydrolysis of ATP to provide the necessary energy.
- A *single-strand binding protein* (*SSB*) binds to the single-stranded DNA, protecting it and preventing it from reforming the duplex state. The SSB binds as a monomer, but typically in a cooperative manner in which the binding of additional monomers to the existing complex is enhanced.

Helicases separate the strands of a duplex nucleic acid in a variety of situations, ranging from strand separation at the growing point of a replication fork to catalyzing migration of Holliday (recombination) junctions along DNA. There are twelve different helicases in *E. coli*. A helicase is generally multimeric. A common form of helicase is a hexamer. This typically translocates along DNA by using its multimeric structure to provide multiple DNA-binding sites.

FIGURE 18.12 shows a generalized schematic model for the action of a hexameric helicase. It is likely to have one conformation that binds to duplex DNA and another that binds to single-stranded DNA. Alternation between them drives the motor that melts the duplex and requires ATP hydrolysis—typically 1 ATP is hydrolyzed for each base pair that is unwound. A helicase usually initiates unwinding at a single-stranded region adjacent to a duplex. It may function with a particular polarity, preferring single-stranded DNA with a 3′ end (3′–5′ helicase) or with a 5′ end (5′–3′ helicase).

Under normal circumstances *in vivo*, the unwinding, coating, and replication reactions proceed in tandem. The SSB binds to DNA as the replication fork advances, keeping the two parental strands separate so that they are in the appropriate condition to act as templates. SSB is needed in stoichiometric amounts at the replication fork. It is required for more than one stage of replication.

KEY CONCEPTS

- Replication requires a helicase to separate the strands of DNA using energy provided by hydrolysis of ATP.
- A single-stranded binding protein is required to maintain the separated strands.

CONCEPT AND REASONING CHECK

Since SSB does not have an enzymatic function, why is it essential for replication?

18.8 Priming Is Required to Start DNA Synthesis

A common feature of all DNA polymerases is that they cannot initiate synthesis of a chain of DNA *de novo*, but can only elongate a chain. **FIGURE 18.13** shows the features required for initiation. Synthesis of the new strand can only start from a preexisting 3′–OH end, and the template strand must be converted to a single-stranded condition.

The 3′–OH end is called a **primer**. The primer can take various forms. Types of priming reaction are summarized in **FIGURE 18.14**:

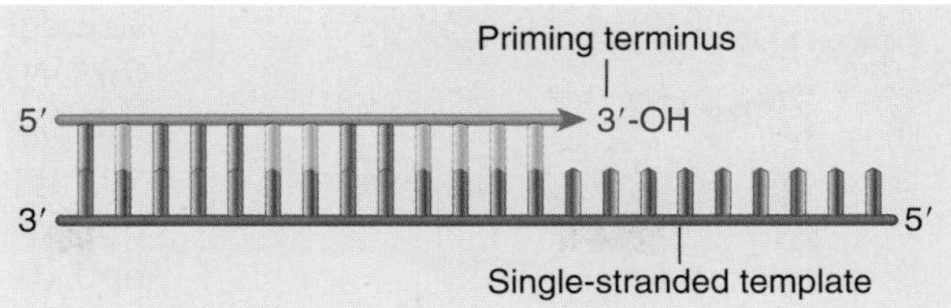

Priming terminus
5′
3′-OH
3′
5′
Single-stranded template

FIGURE 18.13 A DNA polymerase requires a 3′–OH end to initiate replication.

- A sequence of RNA is synthesized on the template, so that the free 3′–OH end of the RNA chain is extended by the DNA polymerase. This is commonly used in replication of cellular DNA and by some viruses.
- A preformed RNA (often a tRNA) pairs with the template, allowing its 3′–OH end to be used to prime DNA synthesis. This mechanism is used by retroviruses to prime reverse transcription of RNA (see Figure 21.28 in *Section 21.8, Retroviral RNA Is Converted to DNA and Integrates into the Host Genome*).
- A primer terminus is generated within duplex DNA. The most common mechanism is the introduction of a nick, as used to initiate rolling circle replication. In this case, the preexisting strand is displaced by new synthesis.
- A protein primes the reaction directly by presenting a nucleotide to the DNA polymerase. This reaction is used by certain viruses (see Figure 16.5 in *Section 16.3, Terminal Proteins Enable Initiation at the Ends of Viral DNAs*).

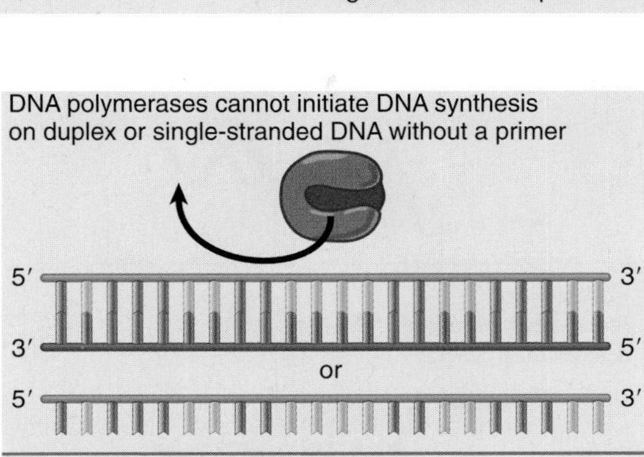

DNA polymerases cannot initiate DNA synthesis on duplex or single-stranded DNA without a primer
5′
3′
3′
5′
or
5′
3′

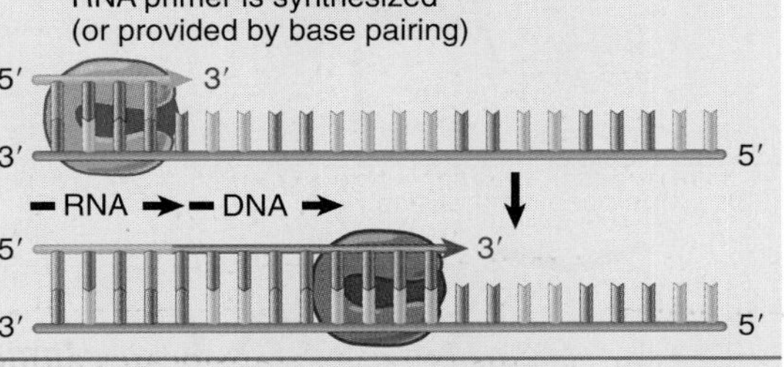

RNA primer is synthesized (or provided by base pairing)
5′
3′
3′
5′
RNA
DNA
5′
3′
3′
5′

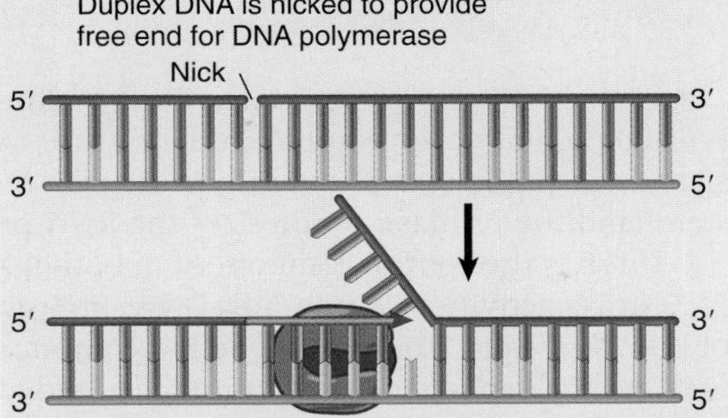

Duplex DNA is nicked to provide free end for DNA polymerase
Nick
5′
3′
3′
5′
5′
3′
3′
5′

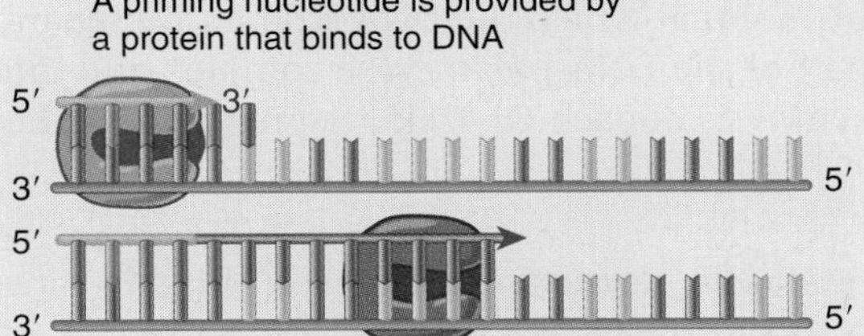

A priming nucleotide is provided by a protein that binds to DNA
5′
3′
3′
5′
5′
3′
5′

FIGURE 18.14 There are several methods for providing the free 3′–OH end that DNA polymerases require to initiate DNA synthesis.

▸ **primer** A short sequence (often of RNA) that is paired with one strand of DNA and provides a free 3′–OH end at which a DNA polymerase starts synthesis of a deoxyribonucleotide chain.

Priming activity is required to provide 3′–OH ends to start off the DNA chains on both the leading and lagging strands. The leading strand requires only one such initiation event, which occurs at the origin. There must be a series of initiation events on the lagging strand, though, because each Okazaki fragment requires its own start *de novo*. Each Okazaki fragment starts with a primer sequence of RNA ~10 bases long that provides the 3′–OH end for extension by DNA polymerase.

A primase is required to catalyze the actual priming reaction. This is provided by a special RNA polymerase activity, the product of the *dnaG* gene. The enzyme is a single polypeptide of 60 kD (much smaller than RNA polymerase). The primase is an RNA polymerase that is used only under specific circumstances, that is, to synthesize short

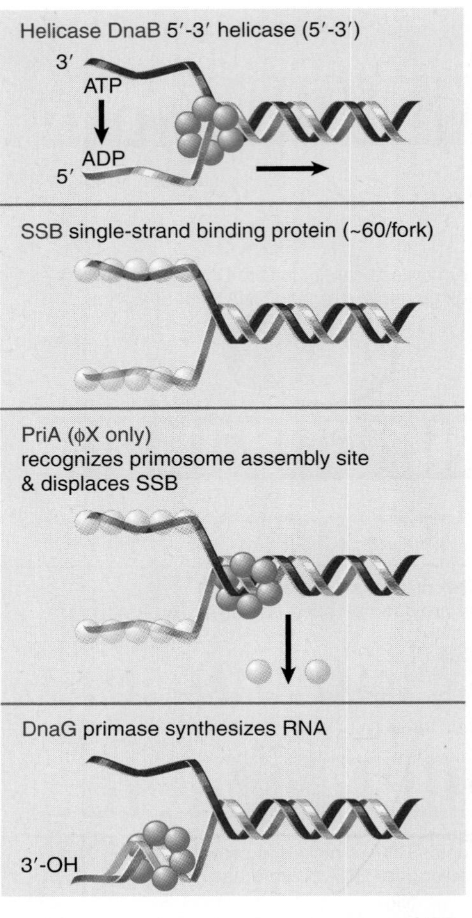

FIGURE 18.15 Initiation requires several enzymatic activities, including helicases, single-strand binding proteins, and synthesis of the primer.

stretches of RNA that are used as primers for DNA synthesis. DnaG primase associates transiently with the replication complex, and typically synthesizes an 11- to 12-base primer. Primers start with the sequence pppAG positioned opposite the sequence 3′–GTC–5′ in the template.

There are two types of priming reaction in *E. coli*:

- The *oriC* system, named for the bacterial origin, basically involves the association of the DnaG primase with the protein complex at the replication fork.
- The ϕX system, named originally for phage ϕX174, requires an initiation complex consisting of additional components, called the primosome. This system is used when damage causes the replication fork to collapse and it must be restarted (see below).

At times replicons are referred to as being of the ϕX or *oriC* type.

The types of activities involved in the initiation reaction are summarized in **FIGURE 18.15**. Although other replicons in *E. coli* may have alternatives for some of these particular proteins, the same general types of activity are required in every case. A helicase is required to generate single strands, a single-strand binding protein is required to maintain the single-stranded state, and the primase synthesizes the RNA primer.

DnaB is the central component in both ΦX and *oriC* replicons. It provides the 5′ to 3′ helicase activity that unwinds DNA. Energy for the reaction is provided by cleavage of ATP. Basically DnaB is the active component required to advance the replication fork. In *oriC* replicons, DnaB is initially loaded at the origin as part of a large complex (see *Section 18.2, Initiation: Creating the Replication Forks at the Origin*). It forms the growing point at which the DNA strands are separated as the replication fork advances. It is part of the DNA polymerase complex and interacts with the DnaG primase to initiate synthesis of each Okazaki fragment on the lagging strand.

KEY CONCEPTS

- All DNA polymerases require a 3′–OH priming end to initiate DNA synthesis.
- The priming end can be provided by an RNA primer, a nick in DNA, or a priming protein.
- For DNA replication, a special RNA polymerase called a primase synthesizes an RNA chain that provides the priming end.
- *E. coli* has two types of priming reaction, which occur at the bacterial origin (*oriC*) and the ϕX174 origin.
- Priming of replication on double-stranded DNA always requires a replicase, SSB, and primase.
- DnaB is the helicase that unwinds DNA for replication in *E. coli*.

CONCEPT AND REASONING CHECK

Why couldn't the transcription RNA polymerase be used to synthesize the primer?

18.9 DNA Polymerase Holoenzyme Consists of Subcomplexes

We can now relate the subunit structure of *E. coli* DNA polymerase III to the activities required for DNA synthesis and propose a model for its action. The replisome consists of two DNA polymerase III holoenzyme complexes and associated proteins necessary for dimerization and function. The holoenzyme is a complex of 900 kD that contains ten proteins organized into four types of subcomplex:

- There are at least two copies of the catalytic core. Each catalytic core contains the α subunit (the DNA polymerase activity), the ε subunit (the 3′–5′ proofreading exonuclease), and the θ subunit (which stimulates the exonuclease).
- There are two copies of the dimerizing subunit, τ, which link the two catalytic cores together.
- There are two copies of the **clamp**, which is responsible for holding catalytic cores on to their template strands. Each clamp consists of a homodimer of β subunits, the β ring, which binds around the DNA and ensures processivity.
- The γ complex is a group of five proteins that comprise the **clamp loader**; the clamp loader places the clamp on DNA.

▸ **clamp** A protein complex that forms a circle around the DNA; by connecting to DNA polymerase, it ensures that the enzyme action is processive.

▸ **clamp loader** A 5-subunit protein complex that is responsible for loading the ß clamp on to DNA at the replication fork.

A model for the assembly of DNA polymerase III is shown in **FIGURE 18.16**. The holoenzyme assembles on DNA in three stages:

- First the clamp loader uses hydrolysis of ATP to bind β subunits to a template-primer complex.
- Binding to DNA changes the conformation of the site on β that binds to the clamp loader, and as a result it now has a high affinity for the core polymerase. This enables core polymerase to bind, and this is the means by which the core polymerase is brought to DNA.
- A τ dimer binds to the core polymerase, and provides a dimerization function that binds a second core polymerase (associated with another β clamp). The replisome is an asymmetric dimer because it has only one clamp loader. The clamp loader is responsible for adding a pair of β dimers to each parental strand of DNA.

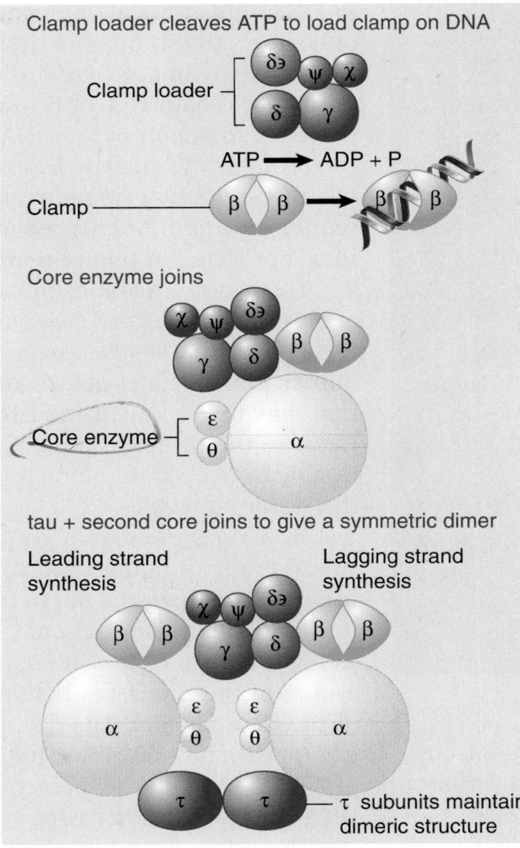

FIGURE 18.16 DNA polymerase III holoenzyme assembles in stages, generating an enzyme complex that synthesizes the DNA of both new strands.

Each of the core complexes of the holoenzyme synthesizes one of the new strands of DNA. The clamp loader is also needed for unloading the β complex from DNA; as a result, the two cores have different abilities to dissociate from DNA. This corresponds to the need to synthesize a continuous leading strand (where polymerase remains associated with the template) and a discontinuous lagging strand (where polymerase repetitively dissociates and reassociates). The clamp loader is associated with the core polymerase that synthesizes the lagging strand, and plays a key role in the ability to synthesize individual Okazaki fragments.

KEY CONCEPTS

- The *E. coli* replicase DNA polymerase III is a 900-kD complex with a dimeric structure.
- Each monomeric unit has a catalytic core, a dimerization subunit, and a processivity component.
- A clamp loader places the processivity subunits on DNA, where they form a circular clamp around the nucleic acid.
- One catalytic core is associated with each template strand.

CONCEPT AND REASONING CHECK

Why is only one clamp loader needed when there are two template strands?

18.10 The Clamp Controls Association of Core Enzyme with DNA

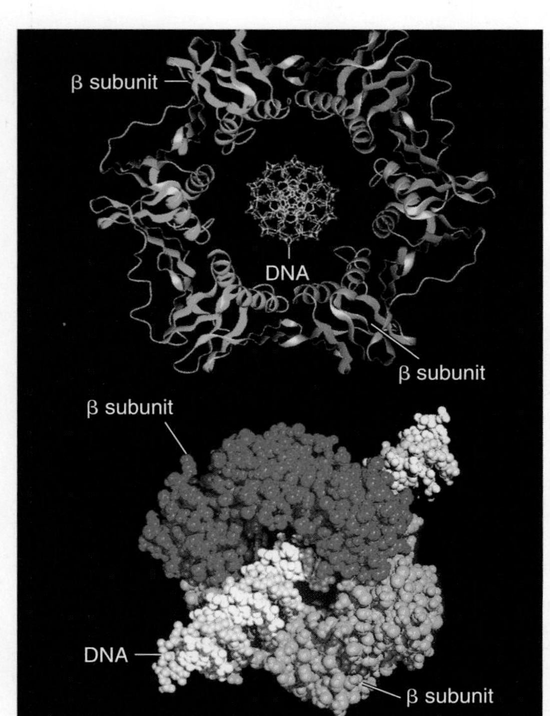

FIGURE 18.17 The β subunit of DNA polymerase III holoenzyme consists of a head-to-tail dimer (the two subunits are shown in red and orange) that forms a ring completely surrounding a DNA duplex (shown in the center). Reprinted from *Cell*, vol. 69, Kong, X. P., et al., pp. 425–437. Copyright 1992, with permission from Elsevier. [http://www.sciencedirect.com/science/journal/00928674]. Photo courtesy of John Kuriyan, University of California, Berkeley.

The β ring makes the holoenzyme highly *processive*. β is strongly bound to DNA, but can slide along a duplex molecule. The crystal structure of β shows that it forms a ring-shaped dimer. The model in **FIGURE 18.17** shows the β-ring in relationship to a DNA double helix. The ring has an external diameter of 80 Å and an internal cavity of 35 Å, almost twice the diameter of the DNA double helix (20 Å). The space between the protein ring and the DNA is filled by water. Each of the β subunits has three globular domains with similar organization (although their sequences are different). As a result, the dimer has sixfold symmetry that is reflected in twelve α-helices that line the inside of the ring.

The β-ring surrounds the duplex, providing the "sliding clamp" that allows the holoenzyme to slide along DNA. The structure explains the high processivity—the enzyme can transiently dissociate, but cannot diffuse away. The α-helices on the inside have some positive charges that may interact with the DNA via the intermediate water molecules. Because the protein clamp does not directly contact the DNA, it may be able to "ice-skate" along the DNA, making and breaking contacts via the water molecules.

How does the clamp get on to the DNA? The clamp is a circle of subunits surrounding DNA; thus its assembly or removal requires the use of an energy-dependent process by the clamp loader. The γ clamp loader is a pentameric circular structure that binds an open form of the β ring preparatory to loading it on to DNA. In effect, the ring is opened at one of the interfaces between the two β subunits by the δ subunit of the clamp loader. The clamp loader binds on top of a closed circular clamp, with its ATPase site juxtaposed to the clamp, and uses hydrolysis of ATP to provide the energy to open the ring of the clamp and insert DNA into the central cavity.

The relationship between the β clamp and the γ clamp loader is a paradigm for similar systems used by DNA polymerases ranging from bacteriophages to animal cells. The clamp is a heteromer (or possibly a dimer or trimer) that forms a ring around DNA with a set of twelve α-helices forming sixfold symmetry for the structure as a whole. The clamp loader has some subunits that hydrolyze ATP to provide energy for the reaction.

KEY CONCEPTS

- The core on the leading strand is processive because its clamp keeps it on the DNA.
- The clamp associated with the core on the lagging strand dissociates at the end of each Okazaki fragment and reassembles for the next fragment.

CONCEPT AND REASONING CHECK

How is the β-ring loaded onto the DNA?

18.11 Coordinating Synthesis of the Lagging and Leading Strands

Each new DNA strand, leading and lagging, is synthesized by an individual catalytic unit. **FIGURE 18.18** shows that the behavior of these two units is different because the new DNA strands are growing in opposite directions. One enzyme unit is moving in the same direction as the unwinding point of the replication fork and synthesizing the leading strand continuously. The other unit is moving "backward" relative to the DNA, along the exposed single strand. Only short segments of template are exposed at any one time. When synthesis of one Okazaki fragment is completed, synthesis of the next Okazaki fragment is required to start at a new location approximately in the vicinity of the growing point for the leading strand. This requires that DNA polymerase III on the lagging strand disengage from the template, move to a new location, and be reconnected to the template at a primer to start a new Okazaki fragment.

In *E. coli*, there is only a single DNA polymerase catalytic subunit used in replication, the DnaE polypeptide. Some bacteria and eukaryotes have multiple replication DNA polymerases (see *Section 18.13, Separate Eukaryote DNA Polymerases Undertake Initiation and Elongation*). The active replication complex in *E. coli* is a dimer of two DNA polymerase III holoenzymes. Each half of the dimer contains DnaE in the catalytic core as the catalytic subunit, supported by other proteins that differ on the leading and lagging strands.

The use of separate catalytic units to synthesize each new DNA strand raises a question: How is synthesis of the lagging strand coordinated with the leading strand? **FIGURE 18.19** shows that the catalytic units behave differently. As the replication complex moves along DNA, unwinding the parental strands, one catalytic unit elongates the leading strand processively in the same direction that the Y fork is moving. The other catalytic unit synthesizes an Okazaki fragment backwards, the other way, because it has to polymerize 5′to 3′. It then must dissociate from the DNA template and reassociate closer to the Y fork to start synthesis of the next Okazaki fragment where the next primer has been formed.

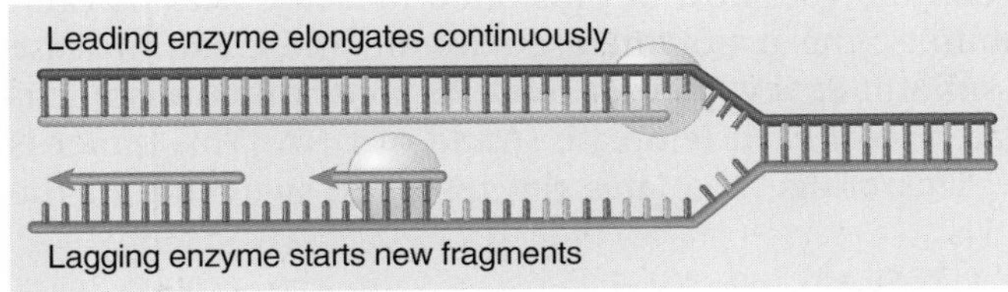

FIGURE 18.18 Leading and lagging strand polymerases move apart.

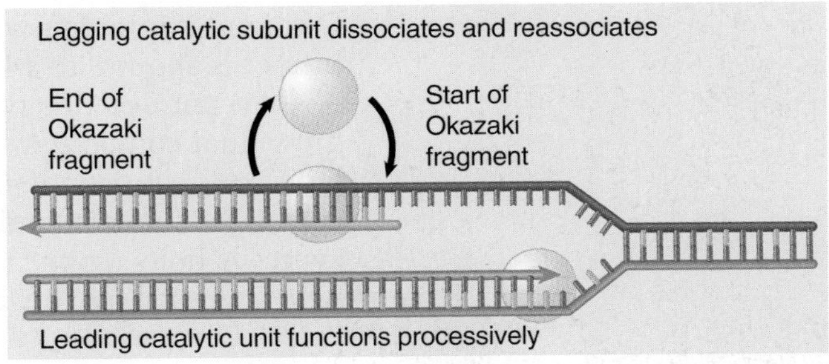

FIGURE 18.19 A replication complex contains separate catalytic units for synthesizing the leading and lagging strands.

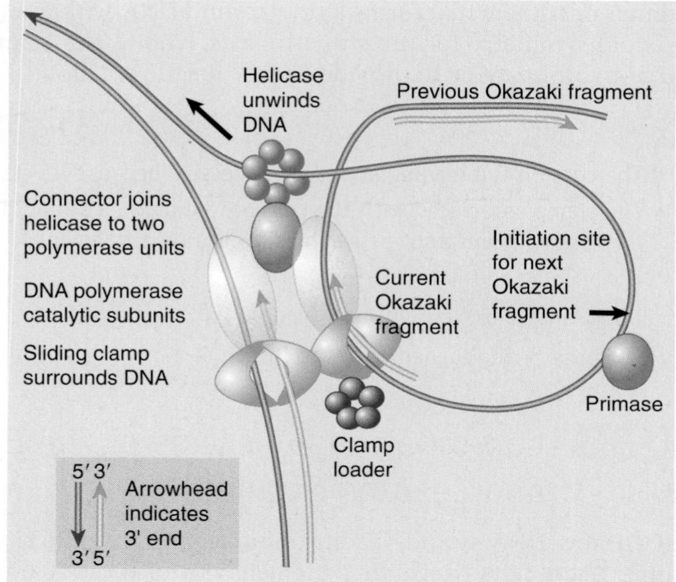

FIGURE 18.20 The helicase creating the replication fork is connected to two DNA polymerase catalytic subunits, each of which is held on to DNA by a sliding clamp. The polymerase that synthesizes the leading strand moves continuously. The polymerase that synthesizes the lagging strand dissociates at the end of an Okazaki fragment and then reassociates with a primer in the single-stranded template loop to synthesize the next fragment.

The basic principle that is established by the dimeric polymerase model is that while one polymerase subunit synthesizes the leading strand continuously, the other cyclically initiates and terminates the Okazaki fragments of the lagging strand within a large single-stranded loop formed by its template strand. **FIGURE 18.20** draws a generic model for the operation of such a replication machine. As the polymerase comes to the end of the fragment it is synthesizing (that is, as it approaches the last fragment it made), the polymerase must disengage from the template, and it must then be reattached ~2000 nucleotides downstream to the new primer just synthesized. The γ clamp loader is responsible for opening the β ring sliding clamp to release the polymerase and then reclosing it downstream. The clamp loader uses ATP hydrolysis and acts as a wrench to force the ring into an unstable configuration, which then springs apart.

The replication fork is created by a helicase—which typically forms a hexameric ring—that translocates in the 5′ to 3′ direction on the template for the lagging strand. The helicase is connected to the two DNA polymerase catalytic subunits, each of which is associated with a sliding clamp.

DnaB is a key component of the *oriC* replicon. It provides the 5′ to 3′ helicase activity that unwinds the DNA, moving the Y fork. Energy for the reaction is provided by ATP hydrolysis. Basically, DnaB is the active component of the growing point, where the DNA strands are separated as the replication fork advances. Its ability to remove SSB as it moves regulates its activity.

What is responsible for recognizing the sites for initiating synthesis of Okazaki fragments? In *oriC* replicons, the connection between priming and the replication fork is provided by the dual properties of DnaB: it is a helicase that propels the replication fork; and it interacts with the DnaG primase, at an appropriate site. The length of the primer is about 8 to 14 bases.

We can describe this model for DNA polymerase III holoenzyme in terms of the individual components of the enzyme complex, as illustrated in **FIGURE 18.21**. A catalytic core consists of three subunits, the α polymerase subunit, the ε exonuclease correctase subunit, and a third subunit θ. Seven other subunits constitute the holoenzyme. A holoenzyme is associated with each template strand of DNA. This dimer is held together by the τ bridge. This bridge regulates the template switching of the

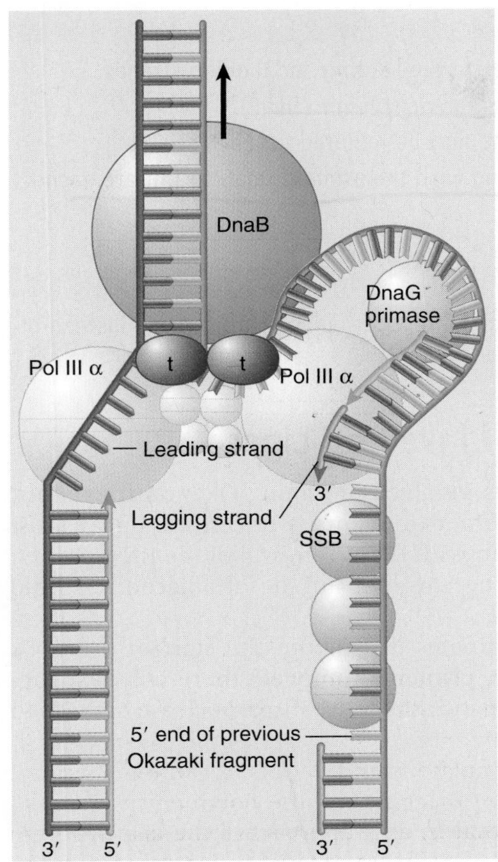

FIGURE 18.21 Each catalytic core of Pol III synthesizes a daughter strand. DnaB is responsible for forward movement at the replication fork.

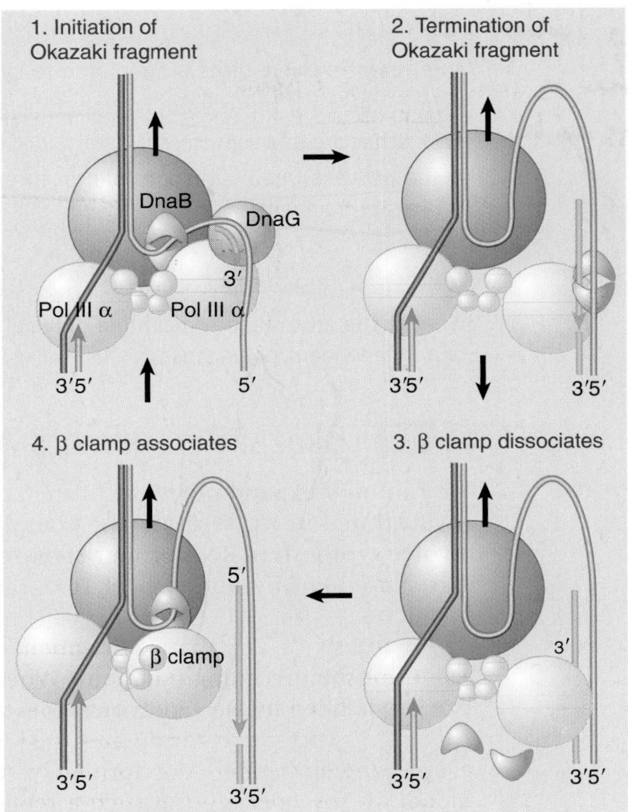

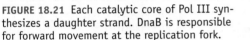

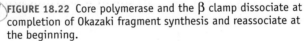

FIGURE 18.22 Core polymerase and the β clamp dissociate at completion of Okazaki fragment synthesis and reassociate at the beginning.

lagging polymerase as it comes to the end of the template and approaches the last Okazaki fragment. The polymerase moves continuously along the template for the leading strand; the template for the lagging strand is "pulled through" creating a loop in the DNA. DnaB creates the unwinding point, and translocates along the DNA in a "forward" direction.

Synthesis of the leading strand creates a loop of single-stranded DNA that provides the template for lagging strand synthesis, and this loop becomes larger as the unwinding point advances. After initiation of an Okazaki fragment, the lagging strand core complex pulls the single-stranded template through the β clamp while synthesizing the new strand. This is sometimes called the *trombone model* of replication. The single-stranded template must extend for the length of at least one Okazaki fragment before the lagging polymerase completes one fragment and is ready to begin the next.

What happens when the Okazaki fragment is completed? All of the components of the replication apparatus function processively (that is, they remain associated with the DNA), except for the primase and the β clamp. **FIGURE 18.22** shows that they dissociate when the synthesis of each fragment is completed, releasing the loop. A new β clamp is then recruited by the clamp loader to initiate the next Okazaki fragment. The lagging strand polymerase transfers from one β clamp to the next in each cycle, without dissociating from the replicating complex.

KEY CONCEPTS

- Different enzyme units are required to synthesize the leading and lagging strands.
- In *E. coli* both these units contain the same catalytic subunit (DnaE).
- In other organisms different catalytic subunits may be required for each strand.
- The helicase DnaB is responsible for interacting with the primase DnaG to initiate each Okazaki fragment.

CONCEPT AND REASONING CHECK

What is the advantage of having a polymerase dimer replicating the chromosome instead of two independent polymerases?

18.12 Okazaki Fragments Are Linked by Ligase

We can now expand our view of the actions involved in joining Okazaki fragments, as illustrated in **FIGURE 18.23**. The complete order of events is uncertain, but it must involve synthesis of RNA primer, its extension with DNA, removal of the RNA primer, its replacement by a stretch of DNA, and the covalent linking of adjacent Okazaki fragments.

Synthesis of an Okazaki fragment terminates just before the start of the RNA primer of the preceding fragment. When the primer is removed, there will be a gap. The gap is filled by DNA polymerase I, which uses its 5′ to 3′ exonuclease activity to

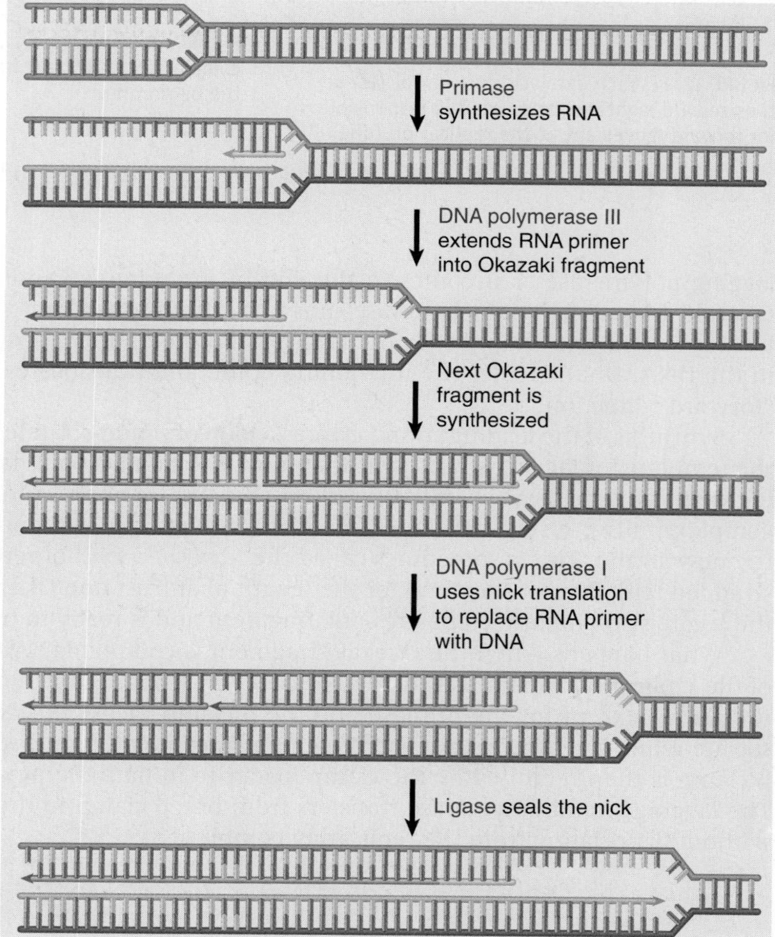
Primase
synthesizes RNA
DNA polymerase III
extends RNA primer
into Okazaki fragment
Next Okazaki
fragment is
synthesized
DNA polymerase I
uses nick translation
to replace RNA primer
with DNA
Ligase seals the nick

FIGURE 18.23 Synthesis of Okazaki fragments require priming, extension, removal of RNA primer, gap filling, and nick ligation.

remove the RNA primer while simultaneously replacing it with a DNA sequence extended from the 3′–OH end of the next Okazaki fragment. In mammalian systems (where the DNA polymerase does not have a 5′ to 3′ exonuclease activity), Okazaki fragments are connected by a two-step process. First RNase H (an enzyme that is specific for an RNA/DNA hybrid substrate) makes an endonucleolytic cleavage; then a 5′ to 3′ exonuclease called FEN1 removes the RNA.

Once the RNA has been removed and replaced, the adjacent Okazaki fragments must be linked together. The 3′–OH end of one fragment is adjacent to the 5′–phosphate end of the previous fragment. The enzyme **DNA ligase** makes a bond by using a complex with AMP. **FIGURE 18.24** shows that the AMP of the enzyme complex becomes attached to the 5′-phosphate of the nick and then a phosphodiester bond is formed with the 3′–OH terminus of the nick, releasing the enzyme and the AMP. Ligases are present in both prokaryotes and eukaryotes.

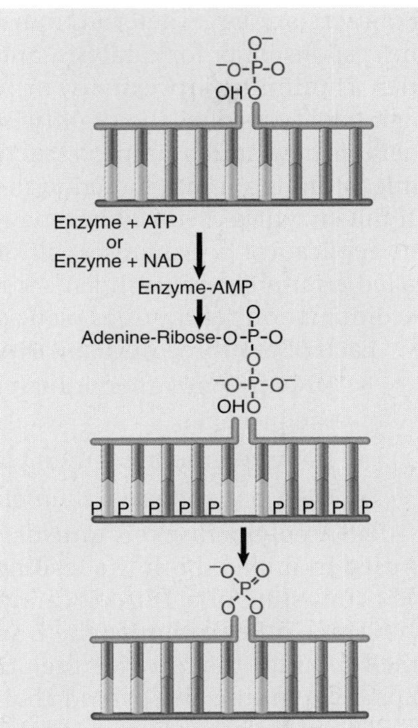

FIGURE 18.24 DNA ligase seals nicks between adjacent nucleotides by employing an enzyme-AMP intermediate.

▸ **DNA ligase** The enzyme that makes a bond between an adjacent 3′–OH and 5′–phosphate end where there is a nick in one strand of duplex DNA.

KEY CONCEPTS

- Each Okazaki fragment starts with a primer and stops before the next fragment.
- DNA polymerase I removes the primer and replaces it with DNA.
- DNA ligase makes the bond that connects the 3′ end of one Okazaki fragment to the 5′ beginning of the next fragment.

CONCEPT AND REASONING CHECK

In what processes other than replication would DNA ligase participate?

18.13 Separate Eukaryotic DNA Polymerases Undertake Initiation and Elongation

Eukaryotic replication is similar in most aspects to bacterial replication. It is semiconservative, bidirectional, and semidiscontinuous. Because of the greater amount of DNA in a eukaryote, the genome has multiple replicons. Replication takes place during S phase of the cell cycle. Replicons in euchromatin initiate before replicons in heterochromatin; replicons near active genes initiate before replicons near inactive genes. Origins of replication in eukaryotes are not well defined, except for those in yeast (called *ARS*, autonomously replicating sequences in *S. cerevisiae*). The number of replicons used in any one cycle is tightly controlled. During embryonic development more are activated than in slower growing adult cells.

Eukaryotes have a much larger number of DNA polymerases. They can be broadly divided into those required for replication and repair polymerases involved in repairing damaged DNA. Nuclear DNA replication requires DNA polymerases α, β, and ε. All the other nuclear DNA polymerases are concerned with synthesizing stretches of new DNA to replace damaged material or using damaged DNA as a template (see *Section 18.14, The Primosome Is Needed to Restart Replication,* for the error-prone DNA polymerases). **FIGURE 18.25** shows that most of the nuclear

DNA polymerase	Function	Structure
	High fidelity replicases	
α	Nuclear replication	350 kD tetramer
δ	"	250 kD tetramer
ε	"	350 kD tetramer
γ	Mitochondrial replication	200 kD dimer
	High fidelity repair	
β	Base excision repair	39 kD monomer
	Low fidelity repair	
ζ	Translesion synthesis	monomer
η	Translesion synthesis	monomer
ι	Translesion synthesis	monomer
κ	Translesion synthesis	monomer
rev 1	Translesion synthesis	monomer

FIGURE 18.25 Eukaryotic cells have many DNA polymerases. The replication enzymes operate with high fidelity. Except for the β enzyme, the repair enzymes all have low fidelity. Replication enzymes have large structures, with separate subunits for different activities. Repair enzymes have much simpler structures.

replicases are large heterotetrameric enzymes. In each case, one of the subunits has the responsibility for catalysis, and the others are concerned with ancillary functions, such as priming, processivity, or proofreading. These enzymes all replicate DNA with high fidelity, as does the slightly less complex mitochondrial enzyme. The repair polymerases have much simpler structures, which often consist of a single monomeric subunit (although it may function in the context of a complex of other repair enzymes). Of the enzymes involved in repair, only DNA polymerase β has a fidelity approaching the replication polymerases; all of the others have much greater error rates and are called *error-prone polymerases*. All mitochondrial DNA synthesis, including repair and recombination reactions as well as replication, is undertaken by DNA polymerase γ.

Each of the three nuclear DNA replication polymerases have a different function:

- DNA polymerase α initiates the synthesis of new strands of DNA by elongating the primer.
- DNA polymerase ε then elongates the leading strand.
- DNA polymerase δ then elongates the lagging strand.

DNA polymerase α is unusual because it has the ability to initiate a new strand. It is used to initiate both the leading and lagging strands. The enzyme exists as a complex consisting of a 180-kD catalytic (DNA polymerase) subunit, which is associated with three other subunits: the B subunit that appears necessary for assembly, and two small subunits that provide the primase (RNA polymerase) activity. Reflecting its dual capacity to prime and extend chains, this complex is sometimes called pol α/primase.

The pol α/primase complex binds to the initiation complex at the origin and synthesizes a short strand consisting of ~10 bases of RNA followed by 20 to 30 bases of DNA. It is then replaced by an enzyme that will extend the chain. On the leading strand, this is DNA polymerase δ; on the lagging strand, this is DNA polymerase ε. This event is called the *polymerase switch*. It involves interactions among several components of the initiation complex.

DNA polymerase δ is a highly processive enzyme that continuously synthesizes the leading strand. Its processivity results from its interaction with two other proteins, RF-C and PCNA. The roles of RF-C and PCNA are analogous to the *E. coli* γ clamp loader and β processivity unit (see *Section 18.10, The Clamp Controls Association of Core Enzyme with DNA*). RF-C is a clamp loader that catalyzes the loading of PCNA on to DNA. It binds to the 3′ end of the DNA and uses ATP hydrolysis to open the ring of PCNA so that it can encircle the DNA. The processivity of DNA polymerase δ is maintained by PCNA, which tethers DNA polymerase δ to the template. (PCNA stands for proliferating cell nuclear antigen, a historical name resulting from the fact that PCNA is enriched in dividing cells.) The crystal structure of PCNA closely resembles the *E. coli* β subunit: a trimer forms a ring that surrounds the DNA. The sequence and subunit organization are different from the dimeric β clamp; however, the function is likely to be similar. DNA polymerase ε elongates the lagging strand, in a manner similar to that of bacterial replication.

A general model suggests that a replication fork contains one complex of DNA polymerase α/primase and two additional DNA polymerase complexes. One is DNA polymerase δ and the other is DNA polymerase ε. The two complexes of DNA polymerase δ/ε behave in the same ways as the two complexes of DNA polymerase III in the *E. coli* replisome: one synthesizes the leading strand, and the other synthesizes Okazaki fragments on the lagging strand. The exonuclease MF1 removes the RNA primers of Okazaki fragments. The enzyme DNA ligase I is specifically required to seal the nicks between the completed Okazaki fragments.

KEY CONCEPTS

- A replication fork has one complex of DNA polymerase α/primase and two complexes of DNA polymerase δ and/or ε.
- The DNA polymerase α/primase complex initiates the synthesis of both DNA strands.
- DNA polymerase δ elongates the leading strand and a second DNA polymerase ε elongates the lagging strand.

CONCEPT AND REASONING CHECK

How is priming Okazaki fragments different in eukaryotes than in bacteria?

18.14 The Primosome Is Needed to Restart Replication

Damage to chromosomes that is not repaired before replication can be catastrophic and lethal. When the replication complex encounters damaged and modified bases such that it cannot place a complementary base opposite it, the polymerase stops and the replication fork collapses. A cell has two options to avoid death, **lesion bypass** or recombination.

▸ **lesion bypass** Replication by an error-prone DNA polymerase on a template that contains a damaged base. The polymerase can incorporate a noncomplementary base into the daughter strand.

Both bacteria and eukaryotes have multiple error-prone DNA polymerases that have the ability to synthesize past a lesion on the template. These enzymes have this ability because they are not constrained to follow standard base pairing rules.

FIGURE 18.26 compares an advancing replication fork with what happens when there is damage to a base in the DNA or a nick in one strand. In either case, DNA synthesis is halted, and the replication fork either is stalled or is disrupted and collapses. Replication-fork stalling appears to be quite common; estimates for the frequency in *E. coli* suggest that 18% to 50% of bacteria encounter a problem during a replication cycle.

E. coli has two error-prone DNA polymerases that can replicate through a lesion, DNA polymerases IV and V (see *Section 20.5, Error-Prone Repair*). Eukaryotes have five error-prone DNA polymerases. When used for lesion bypass during replication, these replace the replisome and are connected to the β ring temporarily to allow the lesion bypass polymerase to insert nucleotides opposite the lesion. DNA polymerase III then replaces the error-prone polymerase.

Alternatively, the situation can be rescued by a recombination event that excises and replaces the damage or provides a new duplex to replace the region containing the double-strand break. The principle of the repair event is to use the built-in redundancy

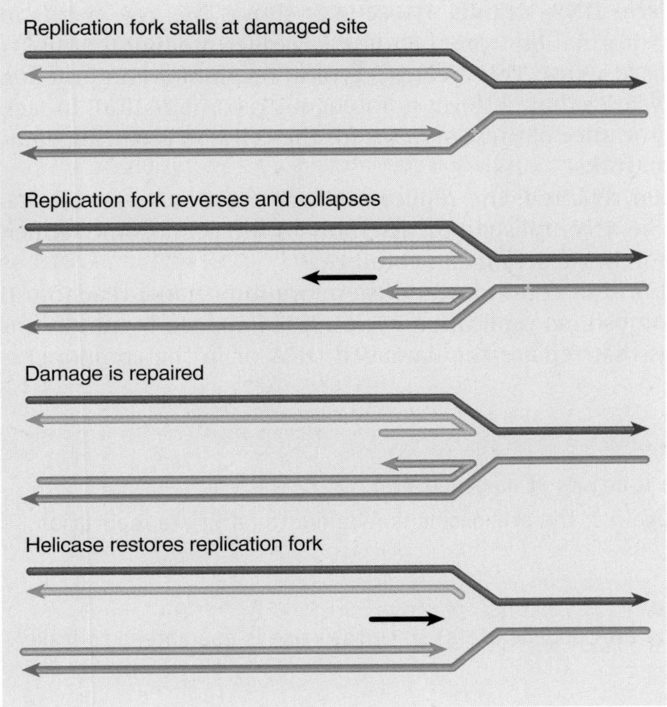

Replication fork stalls at damaged site
Replication fork reverses and collapses
Damage is repaired
Helicase restores replication fork

FIGURE 18.26 The replication fork stalls and may collapse when it reaches a damaged base or a nick in DNA. Arrowheads indicate 3′ ends.

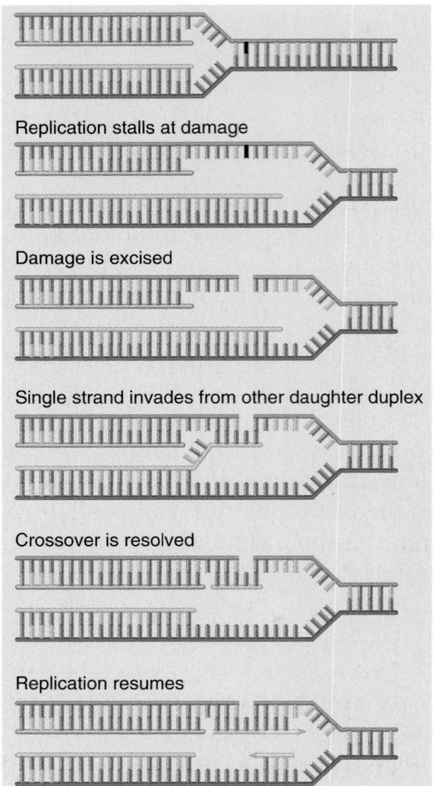

FIGURE 18.27 When replication halts at damaged DNA, the damaged sequence is excised and the complementary (newly synthesized) strand of the other daughter duplex crosses over to repair the gap. Replication can now resume, and the gaps are filled in.

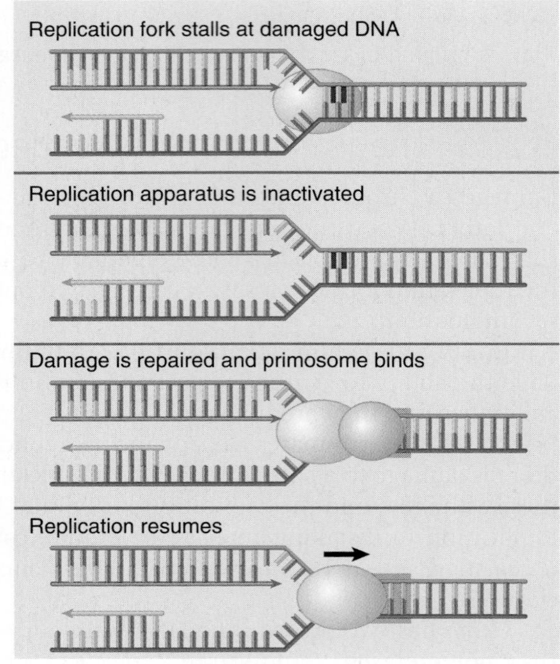

FIGURE 18.28 The primosome is required to restart a stalled replication fork after the DNA has been repaired.

of information between the two DNA strands. **FIGURE 18.27** shows the key events in such a repair event. Basically, information from the undamaged DNA daughter duplex is used to repair the damaged sequence. This creates a typical recombination junction that is resolved by the same systems that perform homologous recombination. In fact, one view is that the major importance of these systems for the cell is in repairing damaged DNA at stalled replication forks.

After the damage has been repaired, the replication fork must be restarted. **FIGURE 18.28** shows that this may be accomplished by assembly of the primosome, which in effect reloads DnaB so that helicase action can continue.

Replication fork reactivation is a common (and therefore important) reaction. It may be required in most chromosomal replication cycles. It is impeded by mutations in either the retrieval systems that replace the damaged DNA or in the components of the primosome.

KEY CONCEPTS

- A replication fork stalls when it arrives at damaged DNA.
- After the damage has been repaired, the primosome is required to reinitiate replication.

CONCEPT AND REASONING CHECK

Why do all cells have error-prone DNA polymerase(s) since their use is guaranteed to make mistakes?

18.15 Summary

The common mode of origin activation involves an initial limited melting of the double helix, followed by more general unwinding to create single strands. Several proteins act sequentially at the *E. coli* origin. Replication is initiated at *oriC* in *E. coli* when DnaA binds to a series of 9 bp repeats. This is followed by binding to a series of 13 bp repeats, where it uses hydrolysis of ATP to generate the energy to separate the DNA strands. The prepriming complex of DnaC–DnaB displaces DnaA. DnaC is released in a reaction that depends on ATP hydrolysis; DnaB is joined by the replicase enzyme, and replication is initiated by two forks that set out in opposite directions.

The availability of DnaA at the origin is an important component of the system that determines when replication cycles should initiate. Following initiation of replication, DnaA hydrolyzes its ATP under the stimulus of the β sliding clamp, thereby generating an inactive form of the protein.

Several sites that are methylated by the Dam methylase are present in the *E. coli* origin, including those of the 13-mer binding sites for DnaA. The origin remains hemimethylated and is in a sequestered state for ~10 minutes following initiation of a replication cycle. During this period it is associated with the membrane and reinitiation of replication is repressed.

DNA synthesis occurs by semidiscontinuous replication, in which the leading strand of DNA growing 5′ to 3′ is extended continuously, but the lagging strand that grows overall in the opposite 3′ to 5′ direction is made as short Okazaki fragments, each synthesized 5′ to 3′. The leading strand and each Okazaki fragment of the lagging strand initiate with an RNA primer that is extended by DNA polymerase. Bacteria and eukaryotes each possess more than one DNA polymerase activity. DNA polymerase III synthesizes both lagging and leading strands in *E. coli*. Many proteins are required for DNA polymerase III action and several constitute part of the replisome within which it functions.

The replisome contains an asymmetric dimer of DNA polymerase III; each new DNA strand is synthesized by a different core complex containing a catalytic (α) subunit. Processivity of the core complex is maintained by the β clamp, which forms a ring around DNA. The clamp is loaded onto DNA by the clamp loader complex. Clamp/clamp loader pairs with similar structural features are widely found in both prokaryotic and eukaryotic replication systems.

The looping model for the replication fork proposes that, as one half of the dimer advances to synthesize the leading strand, the other half of the dimer pulls DNA through as a single loop that provides the template for the lagging strand. The transition from completion of one Okazaki fragment to the start of the next requires the lagging strand catalytic subunit to dissociate from DNA and then reattach to a β clamp at the priming site for the next Okazaki fragment.

DnaB provides the helicase activity at a replication fork; this depends on ATP cleavage. DnaB may function by itself in oriC replicons to provide primosome activity by interacting periodically with DnaG, which provides the primase that synthesizes RNA.

CHAPTER QUESTIONS

1. The predominant DNA synthesis activity in an *E. coli* cell extract is:
   A. DNA polymerase I.
   B. DNA polymerase II.
   C. DNA polymerase III.
   D. DNA polymerase IV.
2. A mutation that greatly reduces the activity of *E. coli* DNA polymerase I would result in:
   A. quick cell death.
   B. continued growth but with a significant increase in mutation rate.
   C. continued growth but with a significant decrease in mutation rate.
   D. very little or no effect on cell growth and viability.
3. Replication slippage is reduced for DNA polymerases that have:
   A. low processivity.
   B. high processivity.
   C. low proofreading.
   D. high proofreading.
4. Proofreading improves the accuracy or fidelity of DNA replication by a factor of about:
   A. two-fold.
   B. ten-fold.
   C. 50-fold.
   D. 100-fold.
5. Which of the following genes encode the DNA primase protein for bacterial DNA replication?
   A. *dna*A
   B. *dna*B
   C. *dna*G
   D. *dna*H

Match each of the bacterial DNA polymerase enzymes from the list at the left with the correct function from the list at the right. Choices may be used more than once.

6. DNA polymerase I	**A.** translesion replication
7. DNA polymerase II	**B.** main replicase for the cell
8. DNA polymerase III	**C.** replication restart
9. DNA polymerase IV	**D.** major repair enzyme
10. DNA polymerase V	**E.** specific replication of Okazaki fragments

KEY TERMS

clamp
clamp loader
conditional lethal
DNA ligase
DNA polymerase
DNA repair
error-prone polymerase
gyrase
helicase
holoenzyme
lagging strand
leading strand
lesion bypass
licensing factor
mutation hotspot
Okazaki fragment
primase
primer
processivity
proofreading
replisome
semiconservative replication
semidiscontinuous replication
single-strand binding protein (SSB)
topoisomerase

FURTHER READING

Johnson, A. and O'Donnell, M. (2005). Cellular DNA replicases: components and dynamics at the replication fork. *Annu. Rev. Biochem.* 74, 283–315.

A review of both bacterial and eukaryotic DNA replication polymerases.

Prakash, S., Johnson, R. E., and Prakash, L. (2005). Eukaryotic translesion synthesis DNA polymerases: specificity of structure and function. *Annu. Rev. Biochem.* 74, 317–353.

Rursell, Z. F., Isoz, I., Lundström, E.-B., Johansson, E., and Kunkel, T. A. (2007). Yeast DNA polymerase ε participates in leading-strand DNA replication. *Science* 317, 127–130.

19

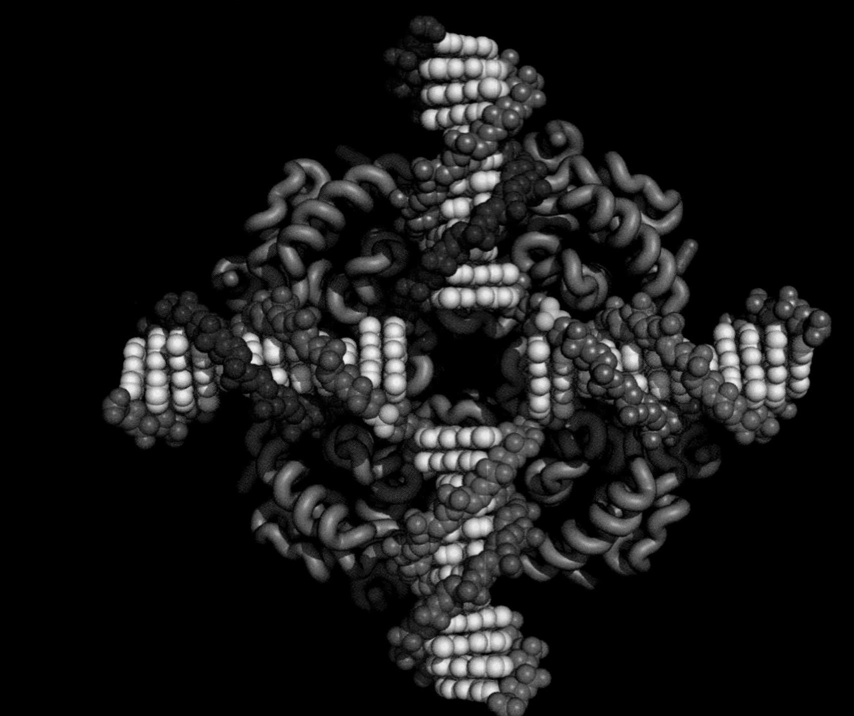

Homologous and Site-Specific Recombination

A model demonstrating the interaction of a RuvA tetramer with a DNA Holliday junction, an intermediate of recombination. Reproduced with permission from Rafferty, J. B., et al, *Science* 274 (1996): 415–421. © 1996 AAAS. Photo courtesy of David W. Rice and John B. Rafferty, University of Sheffield.

CHAPTER OUTLINE

19.1 Introduction

Genetic recombination is essential for the process of evolution. If it were not possible to exchange material between (homologous) chromosomes, the content of each individual chromosome would be irretrievably fixed (except for mutation) in its particular alleles. When mutations occurred, it would not be possible to separate favorable and unfavorable changes. By shuffling genetic information between chromosomes, recombination allows favorable and unfavorable mutations to be separated and tested as individual units in new assortments. It provides a means of escape and spreading for favorable alleles, and a means to eliminate an unfavorable allele without changing allele frequencies for all the other genes with which this allele is linked. This is the basis for natural selection.

Recombination can either be sequence specific, in which homologous chromosomes or short homologous sequences serve as the template for recombination, or be non-sequence specific and result in recombination at any site in the genome, such as occurs with some transposable elements. The sequence-specific classes of recombination will be the focus of this chapter:

- Recombination involving a reaction between homologous sequences of DNA is called **homologous recombination**. In eukaryotes, it occurs during meiosis, usually both in males (during spermatogenesis) and females (during oogenesis) in higher eukaryotes. In Chapter 2, we noted that it occurs at the "four strand" stage of meiosis and involves only two of the four strands (see *Section 2.6, Recombination Occurs by Physical Exchange of DNA*). Homologous recombination occurs between precisely corresponding sequences, so that not a single base pair is added to or lost from the recombinant chromosomes.
- Another type of event sponsors recombination between specific pairs of sequences. This **site-specific recombination** was first characterized in prokaryotes, where it is responsible for the integration of phage genomes into the bacterial chromosome. The recombination event involves specific sequences of the phage DNA and bacterial DNA, which include a short stretch of homology. The enzymes involved in this event act only on the particular pair of target sequences in an intermolecular reaction.
- Recombination can also be used for specific gene rearrangements to control gene expression. Rearrangement may create new genes, which are needed for expression in particular circumstances, as in the case of the immunoglobulins. This example of **somatic recombination** will be discussed in *Chapter 22, Immune Diversity*. Rearrangement also may be responsible for switching expression from one preexisting gene to another, as in the example of yeast mating type, where the sequence at an active locus can be replaced by a sequence from a silent locus.

Let's consider the fundamental difference between homologous and site-specific recombination reactions.

FIGURE 19.1 shows that homologous recombination occurs between two homologous DNA duplexes and can take place at any point along their length. The two chromosomes are cut at equivalent points, and then each is joined to the other to generate reciprocal recombinants. The crossover (marked by the X) is the point at which each becomes joined to the other. There is no change in the overall organization of DNA; the products have the same structure as the parents, and both parents and products are homologous.

▸ **homologous recombination** Recombination involving a reciprocal exchange of sequences of DNA, e.g., between two chromosomes that carry the same genetic loci.

▸ **site-specific recombination** Recombination that occurs between two specific sequences, as in phage integration/excision or resolution of cointegrated structures during transposition.

▸ **somatic recombination** Recombination that occurs in non-germ cells (i.e., it does not occur during meiosis); most commonly used to refer to recombination in the immune system.

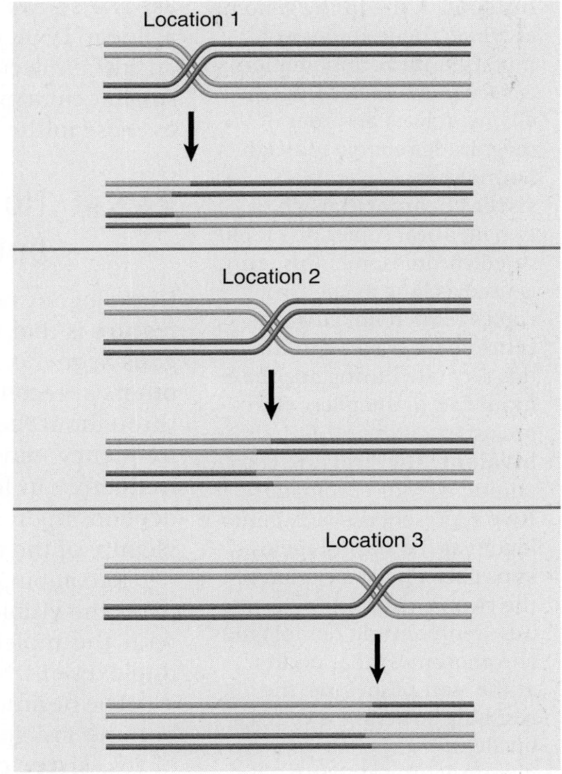

FIGURE 19.1 Homologous recombination can occur at any point along the lengths of two homologous DNAs.

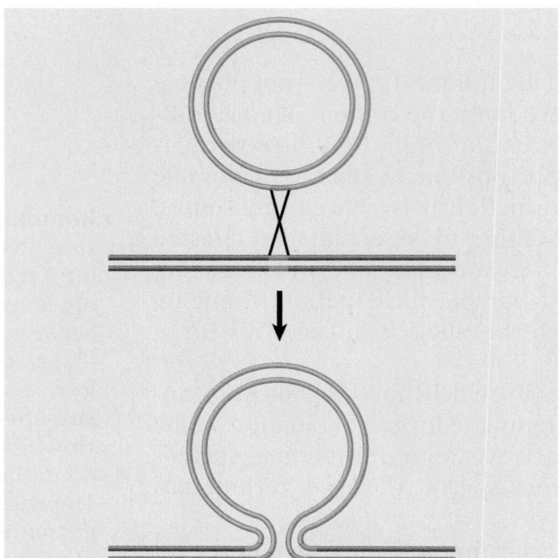

FIGURE 19.2 Site-specific recombination occurs between two specific sequences (identified in green). The other sequences in the two recombining DNAs are not homologous.

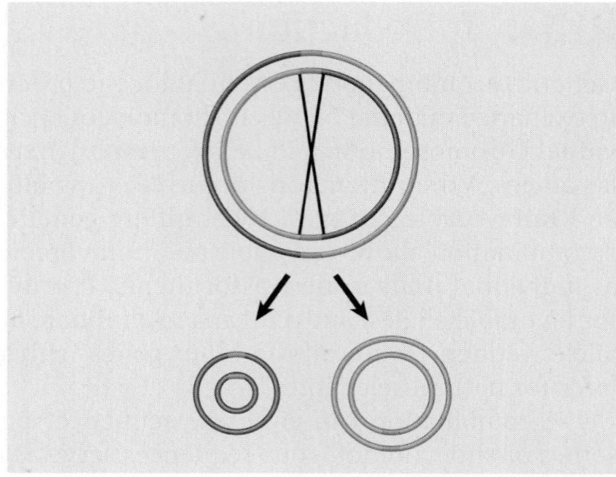

FIGURE 19.3 Site-specific recombination can be used to generate two monomeric circles from a dimeric circle.

Site-specific recombination occurs *only* between specific sites. The result of this type of recombination depends on the locations of the two recombining sites. FIGURE 19.2 shows that an intermolecular recombination between a circular DNA and a linear DNA inserts the circular DNA into the linear DNA. FIGURE 19.3 shows that an intramolecular recombination between two sites on a circular DNA releases two smaller circular DNAs. Site-specific recombination is often used to make changes such as these in the organization of DNA.

19.2 Homologous Recombination Occurs between Synapsed Chromosomes

Homologous recombination is a reaction between two duplexes of DNA. Its critical feature is that the enzymes responsible for the reaction can use any pair of homologous sequences as substrates, although some sequences appear to be favored over others at recombination "**hotspots**." The frequency of recombination is not constant throughout the genome, but is influenced by both global and local effects. The overall frequency may be different in oocytes and in sperm; recombination occurs twice as frequently in female humans as in male humans. Within the genome, its frequency depends upon chromosome structure; for example, crossing over is suppressed in the vicinity of the condensed and inactive regions of heterochromatin.

Recombination occurs during the protracted prophase of meiosis. FIGURE 19.4 compares the visible progress of chromosomes through the five stages of meiotic prophase with the molecular interactions that are involved in exchanging material between duplexes of DNA.

The beginning of meiosis is marked by the point at which individual chromosomes become visible. Each of these chromosomes has completed replication and consists of two **sister chromatids**, each of which contains a duplex DNA. The homologous chromosomes approach one another and begin to pair in one or more regions, forming **bivalents**. Pairing extends until the entire length of each chromosome is apposed with its homolog. The process is called **synapsis** or **chromosome pairing**. When the

- **hotspot** A site in the genome at which the frequency of mutation (or recombination) is very much increased, usually by at least an order of magnitude relative to neighboring sites.
- **sister chromatid** Each of two identical copies of a replicated chromosome; this term is used as long as the two copies remain linked at the centromere. Sister chromatids separate during anaphase in mitosis or anaphase II in meiosis.
- **bivalent** The structure containing all four chromatids (two representing each homologue) at the start of meiosis.
- **synapsis** The association of the two pairs of sister chromatids (representing homologous chromosomes) that occurs at the start of meiosis; the resulting structure is called a bivalent.
- **chromosome pairing** The coupling of the homologous chromosomes at the start of meiosis.

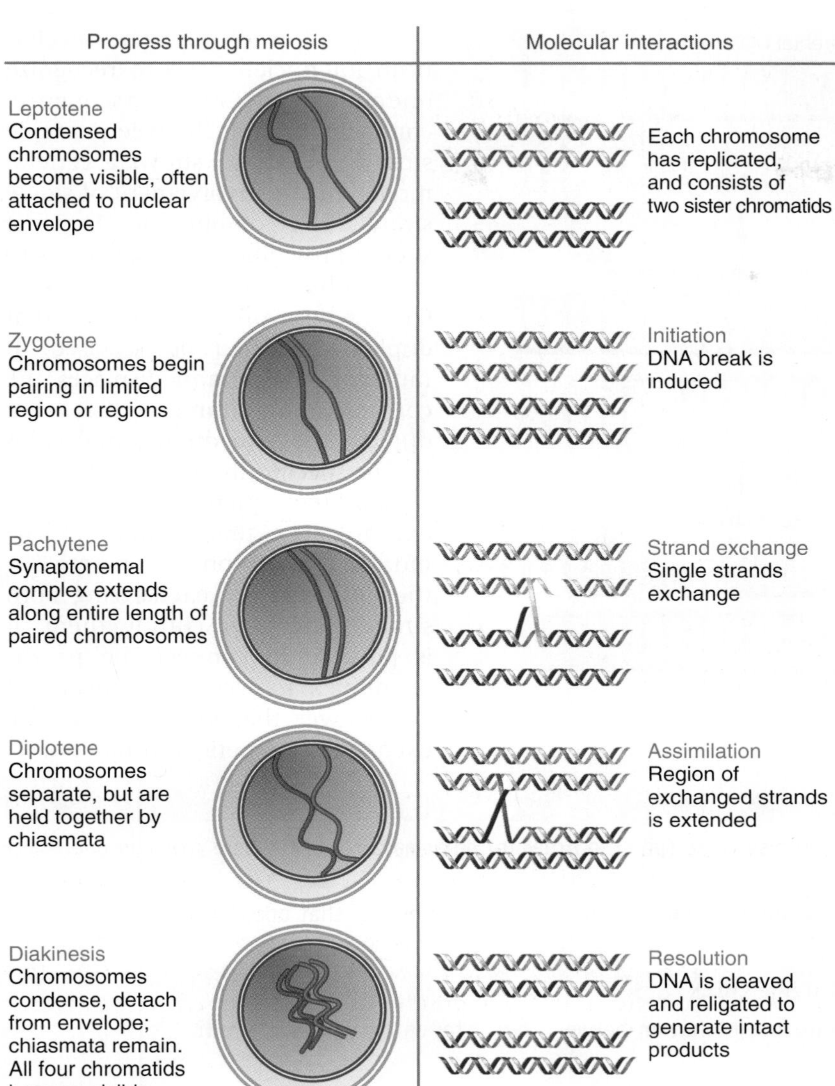
Progress through meiosis
Molecular interactions
Leptotene
Condensed chromosomes become visible, often attached to nuclear envelope
Each chromosome has replicated, and consists of two sister chromatids
Zygotene
Chromosomes begin pairing in limited region or regions
Initiation
DNA break is induced
Pachytene
Synaptonemal complex extends along entire length of paired chromosomes
Strand exchange
Single strands exchange
Diplotene
Chromosomes separate, but are held together by chiasmata
Assimilation
Region of exchanged strands is extended
Diakinesis
Chromosomes condense, detach from envelope; chiasmata remain. All four chromatids become visible
Resolution
DNA is cleaved and religated to generate intact products

FIGURE 19.4 Recombination occurs during the first meiotic prophase. The stages of prophase are defined by the appearance of the chromosomes, each of which consists of two replicas (sister chromatids), although the duplicated state becomes visible only at the end. The molecular interactions of any individual crossing-over event involve two of the four duplex DNAs.

process is completed, the chromosomes are laterally associated in the form of a **synaptonemal complex**, which has a characteristic structure in each species, although there is wide variation in the details between species.

Recombination between chromosomes involves a physical exchange of parts, usually represented as a **breakage and reunion**, in which two *nonsister* chromatids (each containing a duplex of DNA) have been broken and then linked with each other. When the chromosomes begin to separate, they remain visibly linked together at discrete sites called **chiasmata**. The number and distribution of chiasmata correspond to the frequencies and positions of genetic crossovers, and it is likely that chiasmata represent the actual sites of crossover events. The chiasmata remain visible when the chromosomes condense and all four chromatids become evident.

What is the molecular basis for these events? Each sister chromatid contains a single DNA duplex, so each bivalent contains four duplex molecules of DNA. Recombination requires a mechanism that allows the duplex DNA of one sister chromatid to interact with the duplex DNA of a sister chromatid from the other chromosome. It must be possible for this reaction to occur between any pair of corresponding sequences in the two molecules in a highly specific manner that allows material to be exchanged with precision at the level of the individual base pair.

- **synaptonemal complex** The morphological structure of synapsed chromosomes.
- **breakage and reunion** The mode of genetic recombination in which two DNA duplex molecules are broken at corresponding points and then rejoined crosswise (involving formation of a length of heteroduplex DNA around the site of joining).
- **chiasma (pl. chiasmata)** A site at which two homologous chromosomes appear to have exchanged material during meiosis.

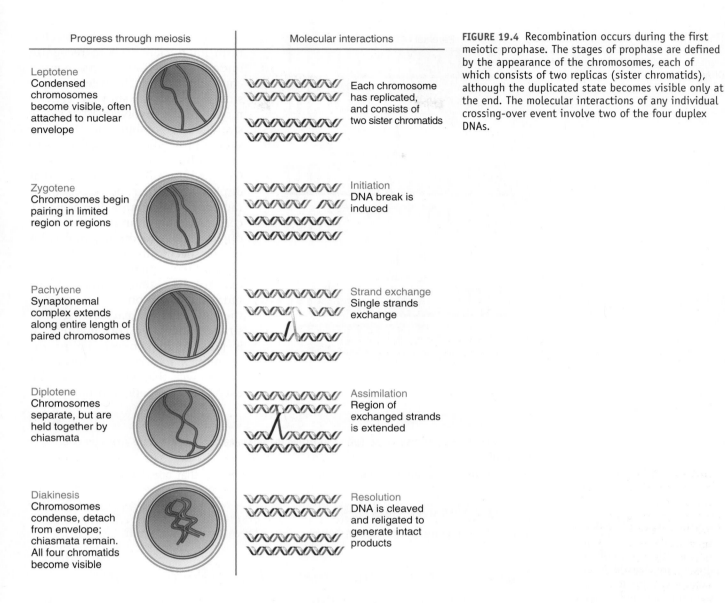

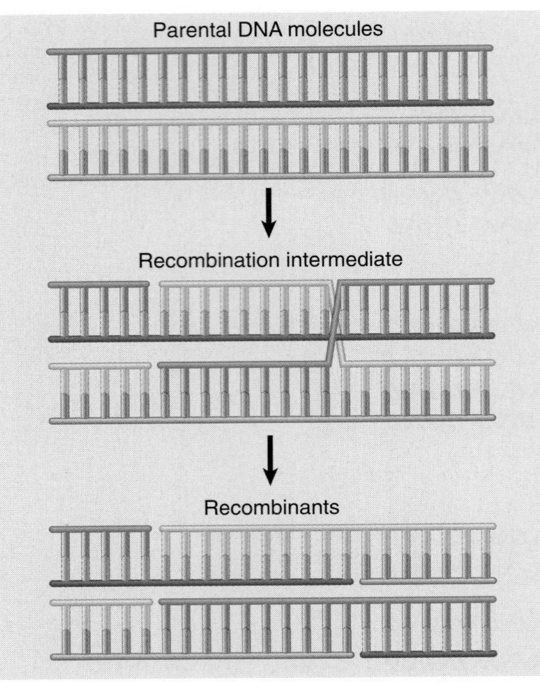

FIGURE 19.5 Recombination involves pairing between complementary strands of the two parental DNAs.

We know of only one mechanism for nucleic acids to recognize one another on the basis of sequence: complementarity between single strands. FIGURE 19.5 shows a general model for the involvement of single strands in recombination. The first step in providing single strands is to make a break in each DNA duplex. One or both of the strands of that duplex can then be released. If (at least) one strand displaces the corresponding strand in the other duplex, the two duplex molecules will be specifically connected at corresponding sequences. If the strand exchange is extended, there can be more extensive connection between the duplexes. By exchanging both strands and later cutting them, it is possible to connect the parental duplex molecules by means of a crossover that will result in the exchange of genetic information.

KEY CONCEPTS

- Chromosomes must synapse (pair) in order for chiasmata to form where crossing over occurs.
- Stages of meiosis are correlated with the molecular events that occur to DNA.

CONCEPT AND REASONING CHECK

Why is homologous recombination between *nonsister* chromatids preferred?

19.3 Double-Strand Breaks Initiate Recombination

The general model of Figure 19.4 shows that a break must be made in one duplex in order to generate a point from which single strands can unwind to participate in genetic exchange. Genetic exchange is generally initiated by the formation of a **double-strand break (DSB)**. The model is illustrated in FIGURE 19.6.

Recombination is initiated by an endonuclease that cleaves one of the partner DNA duplexes, the "recipient." The break is converted to two single-stranded overhanging regions by exonuclease action. The exonuclease nibbles away one strand on either side of the break, generating 3′ single-stranded termini. One of the free 3′ ends then invades a homologous region in the other ("donor") duplex. This is called **single-strand invasion**. This results in the formation of **heteroduplex DNA**, a region consisting of one strand from each of the parental DNA molecules. The strand invasion event also generates a **D loop**, in which one strand of the donor duplex is displaced. The D loop is extended by repair DNA synthesis, using the free 3′ end as a primer to generate double-stranded DNA.

As the D loop is displaced, it provides a complementary single-stranded sequence for the other single strand of the recipient chromosome, and these sequences anneal. Now there is heteroduplex DNA on either side of the original breakpoint. Note that even if sequences were lost from the 3′ ends of the breakpoint as well, due to accidental 3′ to 5′ exonuclease activity, this mechanism would restore the duplex by repair

- **double-strand breaks (DSB)** Breaks that occur when both strands of a DNA duplex are cleaved at the same site. Genetic recombination is initiated by such breaks. The cell also has repair systems that act on breaks that are created at other times.
- **single-strand invasion** The process in which a single strand of DNA displaces its homologous strand in a duplex.
- **heteroduplex DNA** DNA that is generated by base pairing between complementary single strands derived from the different parental duplex molecules; it occurs during genetic recombination.
- **D loop (displacement loop)** The loop of displaced DNA generated by strand invasion and extension during homologous recombination.

synthesis using the 3′ end on the left side of the gap as a primer. If information on the recipient DNA is replaced by a sequence from the donor, the resulting duplexes will be identical (homozygous) in the replaced region. This is often referred to as **gene conversion** when it results in the conversion of a heterozygous genetic locus to a homozygous state.

This molecule is a key intermediate in recombination. The reciprocal exchange of single strands creates a connection between the two DNA duplexes. The connected pair of duplexes is called a **joint molecule**. The point at which an individual strand of DNA crosses from one duplex to the other is called the **recombinant joint**.

An important feature of a recombinant joint is its ability to move along the duplex. Such mobility is called **branch migration**. FIGURE 19.7 illustrates the migration of a single strand in a duplex. The branching point can migrate in either direction as one strand is displaced by the other. Strand invasion and branch migration thus result in the structure at the bottom of Figure 19.6, a molecule with two recombinant joints, which can be at varying distances from the original breakpoint.

The joint molecule formed by strand exchange must be *resolved* into two separate duplex molecules. **Resolution** requires a further pair of nicks. We can most easily visualize the outcome by viewing the joint molecule in one plane as a **Holliday junction**. This is illustrated in FIGURE 19.8, which depicts a molecule with a single recombinant joint (for simplicity) with one duplex rotated relative to the other. The outcome of the reaction depends on which pair of strands is nicked.

Nicking one pair of strands results in **splice recombinant** DNA molecules. The duplex of one DNA parent is covalently linked to the duplex of the other DNA parent via a stretch of heteroduplex DNA. There has been a conventional recombination event between markers located on either side of the heteroduplex region.

Double-strand break made in recipient
Break is converted to 3' single-stranded overhangs
3' end invades other duplex (donor)
D-loop
Synthesis from 3' end displaces one strand of donor
Displaced strand migrates to other duplex
DNA synthesis can also occur from other 3' end
Reciprocal migration generates double crossover

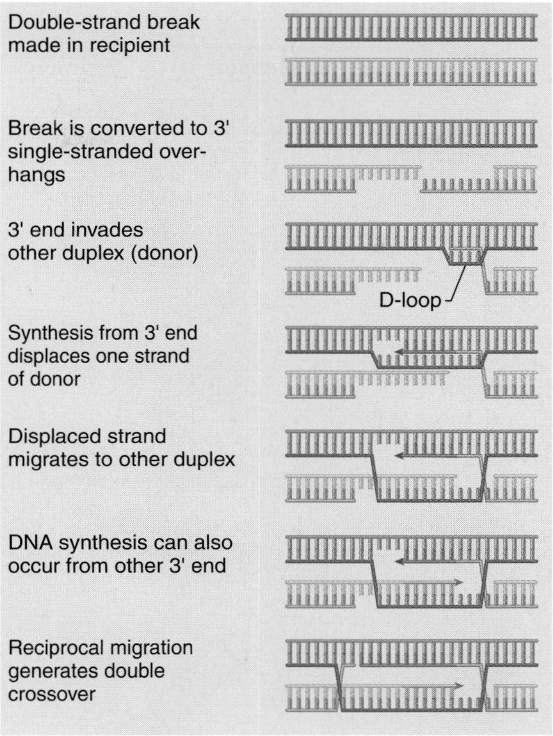

FIGURE 19.6 Recombination is initiated by a double-strand break, followed by formation of single-stranded 3′ ends, one of which invades a homologous duplex.

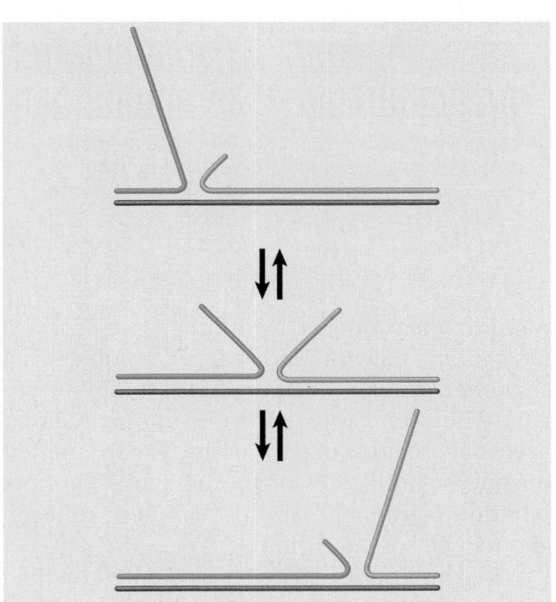

FIGURE 19.7 Branch migration can occur in either direction when an unpaired single strand displaces a paired strand.

- **gene conversion** The alteration of one strand of a heteroduplex DNA to make it complementary with the other strand at any position(s) where there were mispaired bases, or the complete replacement of genetic material at one locus by a homologous sequence.
- **joint molecule** A pair of DNA duplexes that are connected together through a reciprocal exchange of genetic material.
- **recombinant joint** The point at which two recombining molecules of duplex DNA are connected (the edge of the heteroduplex region).
- **branch migration** The ability of a DNA strand partially paired with its complement in a duplex to extend its pairing by displacing the resident strand with which it is homologous.
- **resolution** Resolution occurs by a homologous recombination reaction between the two copies of the transposon in a cointegrate. The reaction generates the donor and target replicons, each with a copy of the transposon.
- **Holliday junction** An intermediate structure in homologous recombination, for which the two duplexes of DNA are connected by the genetic material exchanged between two of the four strands, one from each duplex. A joint molecule is said to be resolved when nicks in the structure restore two separate DNA duplexes.
- **splice recombinant** DNA that results from a Holliday junction being resolved by cutting the nonexchanged strands. Both strands of DNA before the exchange point come from one chromosome; the DNA after the exchange point comes from the homologous chromosome.

FIGURE 19.8 Resolution of a Holliday junction can generate parental or recombinant duplexes, depending on which strands are nicked. Both types of product have a region of heteroduplex DNA.

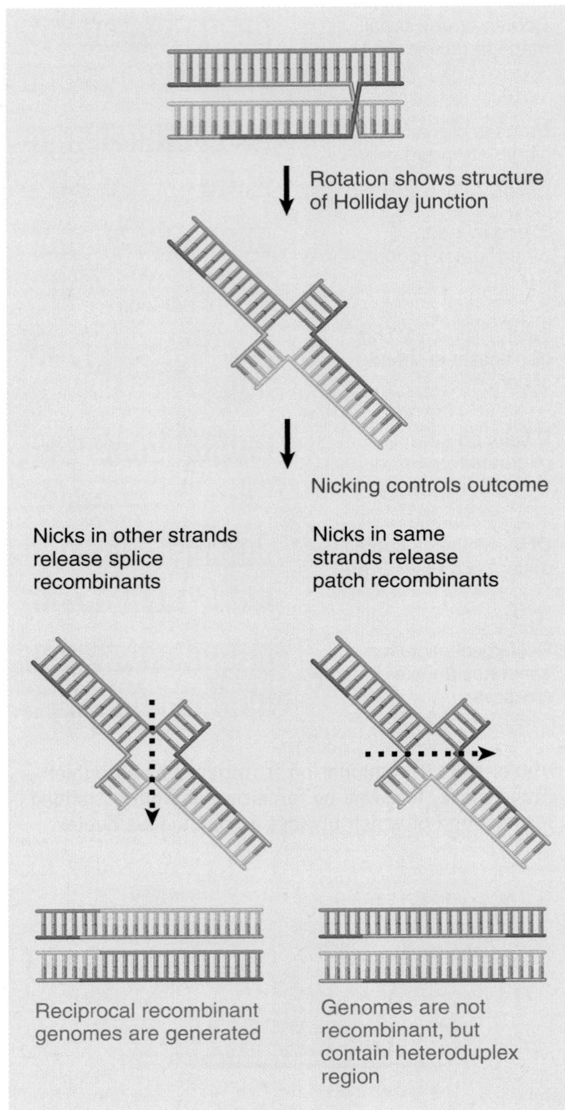

Rotation shows structure of Holliday junction
Nicking controls outcome
Nicks in other strands release splice recombinants
Nicks in same strands release patch recombinants
Reciprocal recombinant genomes are generated
Genomes are not recombinant, but contain heteroduplex region

- **patch recombinant** DNA that results from a Holliday junction being resolved by cutting the exchanged strands. The duplex is largely unchanged, except for a DNA sequence on one strand that came from the homologous chromosome.

In contrast, nicking the other pair of strands results in **patch recombinants**. This nicking releases the original parental duplexes, which remain intact with the exception that each has a residuum of the event in the form of a length of heteroduplex DNA.

These alternative resolutions of the joint molecule establish the principle that *a strand exchange between duplex DNAs always leaves behind a region of heteroduplex DNA, but the exchange may or may not be accompanied by recombination of the flanking regions.*

For a molecule with two recombinant joints, if both joints are resolved in the same way, the original noncrossover molecules will be released, each with a region of altered genetic information that is a footprint of the exchange event. If the two joints are resolved in opposite ways, a genetic crossover is produced.

This general model for homologous recombination has been validated in yeast, although there is evidence that there are other mechanisms of recombination as well. There is good evidence in yeast that double-strand breaks initiate recombination in both homologous and site-specific recombination. Double-strand breaks were initially implicated in the change of mating type, which involves the replacement of one sequence by another (see *Section 19.19, Yeast Use a Specialized Recombination Mechanism to Switch Mating Type*). Double-strand breaks also occur early in meiosis at sites that provide hotspots for recombination. Their locations are not sequence specific. They tend to occur in promoter regions and in general to coincide with more accessible regions of chromatin. The frequency of recombination declines in a gradient on one or both sides of the hotspot. The hotspot identifies the site at which recombination is initiated, and the gradient reflects the probability that the recombination events will spread from it.

Recombination in yeast is initiated by the protein Spo11, which is homologous to the catalytic subunits of a family of type II topoisomerases (discussed further in *Section 19.7, Topoisomerases Relax or Introduce Supercoils in DNA*). This suggests that Spo11 may be a topoisomerase-like enzyme that generates the initial double-strand breaks. The model for this reaction shown in FIGURE 19.9 suggests that Spo11 interacts reversibly with DNA; the break is converted into a permanent structure by an interaction with another protein that dissociates the Spo11 complex. Removal of Spo11 is then followed by nuclease action to generate the 3′ single strands, and the action of other

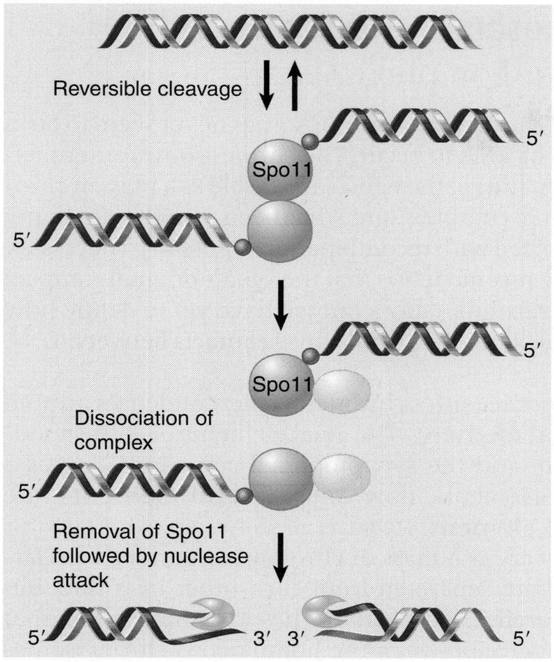

FIGURE 19.9 Spo11 is covalently joined to the 5′ ends of double-strand breaks.

proteins then enables the single-stranded ends to invade homologous duplex DNA. The system for generating the double-strand breaks that initiate recombination is generally conserved. Spo11 homologs have been identified in several multicellular eukaryotes, and a mutation in the *Drosophila* gene blocks all meiotic recombination. Many of the same factors used in meiotic recombination are also used in the repair of accidental double-strand breaks; we will discuss some of these factors in more detail in *Section 19.5, Specialized Enzymes Catalyze 5′ End Resection and Single-Strand Invasion*, and in *Chapter 20, Repair Systems*.

KEY CONCEPTS

- Recombination is initiated by making a double-strand break in one (recipient) DNA duplex.
- Exonuclease action generates 3′ single-stranded ends that invade the other (donor) duplex.
- When a single strand from one duplex displaces its counterpart in the other duplex, it creates a D loop.
- The exchange generates a stretch of heteroduplex DNA consisting of one strand from each parent.
- New DNA synthesis replaces any material that has been degraded.
- This generates a recombinant joint molecule in which the two DNA duplexes are connected by heteroduplex DNA.
- Whether recombinants are formed depends on whether the strands involved in the original exchange or the other pair of strands are nicked during resolution.

CONCEPT AND REASONING CHECK

Is it possible to tell whether a DNA gap was generated during a homologous recombination event that occurs at a heterozygous locus? Why or why not?

19.4 Recombining Chromosomes Are Connected by the Synaptonemal Complex

A basic paradox in recombination is that the parental chromosomes never seem to be in close enough contact for recombination of DNA to occur. The chromosomes enter meiosis in the form of replicated (sister chromatid) pairs, which are visible as a mass of chromatin. They pair to form the synaptonemal complex, and it has been assumed for many years that this represents some stage involved with recombination—possibly a necessary preliminary to exchange of DNA. A more recent view is that the synaptonemal complex is a consequence rather than a cause of recombination, but we have yet to define how the structure of the synaptonemal complex relates to molecular contacts between DNA molecules.

Synapsis begins when each chromosome (sister chromatid pair) condenses around a proteinaceous structure called the **axial element**. The axial elements of corresponding chromosomes then become aligned, and the synaptonemal complex forms as a tripartite structure, in which the axial elements, now called **lateral elements**, are separated from each other by a **central element**. FIGURE 19.10 shows an example.

- **axial element** A proteinaceous structure around which the chromosomes condense at the start of synapsis.
- **lateral element** A structure in the synaptonemal complex that forms when a pair of sister chromatids condenses on to an axial element.
- **central element** A structure that lies in the middle of the synaptonemal complex, along which the lateral elements of homologous chromosomes align. It is formed from Zip proteins.
- **recombination nodules (nodes)** Dense objects present on the synaptonemal complex; they may represent protein complexes involved in crossing-over.

Each chromosome at this stage appears as a mass of chromatin bounded by a lateral element. The two lateral elements are separated from each other by a fine, but dense, central element. The triplet of parallel dense strands lies in a single plane that curves and twists along its axis. The distance between the homologous chromosomes is greater than 200 nm, which is considerable in molecular terms (the diameter of DNA is 2 nm). Thus a major problem in understanding the role of the complex is that although it aligns homologous chromosomes, it is far from bringing homologous DNA molecules into contact.

The only visible link between the two sides of the synaptonemal complex is provided by spherical or cylindrical structures observed in fungi and insects. They lie across the complex and are called **nodes** or **recombination nodules**; they occur with the same frequency and distribution as the chiasmata. Their name reflects the hope that they may prove to be the sites of recombination.

From mutations that affect synaptonemal complex formation, we can identify the types of proteins that are involved in its structure. FIGURE 19.11 presents a molecular view of the synaptonemal complex. Its distinctive structural features are the result of two groups of proteins:

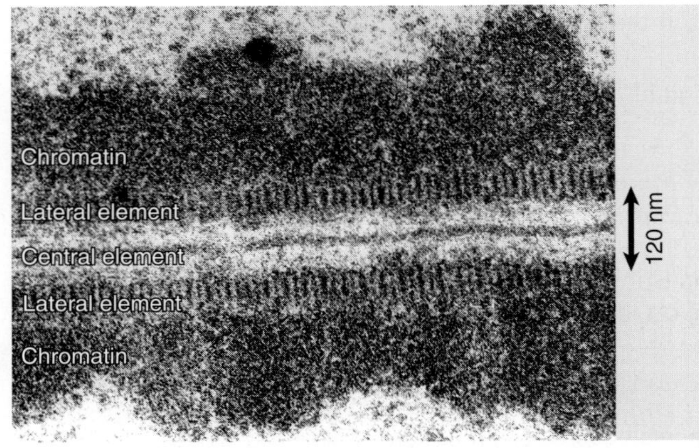

FIGURE 19.10 The synaptonemal complex brings chromosomes into juxtaposition. Reproduced from D. Von Wettstein, *Proc. Natl. Acad. Sci. USA* 68 (1971): 851–855. Photo courtesy of D. Von Wettstein, Washington State University.

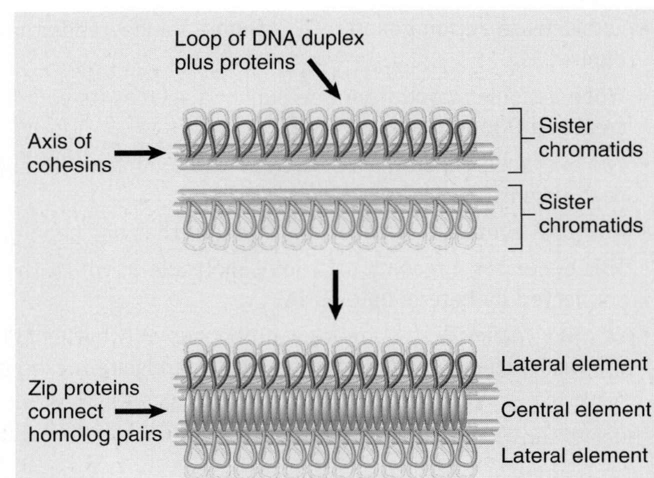

FIGURE 19.11 Each pair of sister chromatids has an axis made of cohesins. Loops of chromatin project from the axis. The synaptonemal complex is formed by linking together the axes via zip proteins.

- The *cohesins* form a single linear axis for each pair of sister chromatids from which loops of chromatin extend. This is equivalent to the lateral element of Figure 19.10. (The cohesins belong to a general group of proteins involved in connecting sister chromatids so that they segregate properly at mitosis or meiosis.)
- The *lateral elements* are connected by transverse filaments that are equivalent to the central element of Figure 19.10. These are formed from Zip proteins.

Mutations in genes for proteins that are needed for lateral elements to form are found in the genes coding for cohesins. The cohesins that are used in meiosis include Smc3 (which is also used in mitosis) and Rec8 (which is specific to meiosis and is related to the mitotic cohesin Scc1). The cohesins appear to bind to specific sites along the chromosomes in both mitosis and meiosis. They are likely to play a structural role in chromosome segregation. At meiosis, the formation of the lateral elements may be necessary for the later stages of recombination, because although these mutations do not prevent the formation of double-strand breaks, they do block formation of recombinants.

Mutation of the *ZIP1* gene allows lateral elements to form and to become aligned, but they do not become closely synapsed. The N-terminal domain of the Zip1 protein is localized in the central element, but the C-terminal domain is localized in the lateral elements. Zip1 also interacts with a number of other proteins, including Zip2-4, Mer3, and Msh4/5, collectively known as the ZMM proteins. The Zip proteins form transverse filaments that connect the lateral elements of the sister chromatid pairs, while the Mer3, Msh4, and Msh5 proteins promote recombination.

KEY CONCEPTS

- During the early part of meiosis, homologous chromosomes are paired in the synaptonemal complex.
- The mass of chromatin of each homolog is separated from the other by a proteinaceous complex.

CONCEPT AND REASONING CHECK

Based on its structure, explain how the synaptonemal complex could actually represent an obstacle to recombination.

19.5 Specialized Enzymes Catalyze 5′ End Resection and Single-Strand Invasion

The central unique feature of homologous recombination is the generation of single-stranded regions that are used to align with and invade homologous donor sequences. The conserved Mre11 complex is required for generation of single strands at meiotic breaks in eukaryotes, and this complex also dissociates Spo11 from the DNA ends by cleaving off a short oligonucleotide containing the covalently linked Spo11. The *mre11* mutation was first discovered in cells that were *me*iotic *re*combination deficient. The conserved complex contains three components: Mre11, Rad50, and Nbs1 (Xrs2 in *S. cerevisiae*), and is usually referred to as the MRN (or MRX in yeast) complex. This complex is required for the generation of the single-stranded regions, a process known as **5′ end resection**, although it is not yet clear whether MRN performs any actual end resection itself. The exonuclease Exo1 has been implicated in resection, and at least half a dozen other proteins interact with MRN and are required for meiotic DSB generation and end resection. A critical role of MRN appears to be as a "chromosomal glue" to prevent separation of DNA ends, as shown in **FIGURE 19.12**.

▸ **5′ end resection** The generation of 3′ overhanging single-stranded regions that occurs via exonucleolytic digestion of the 5′ ends at a double-strand break.

Once the single-stranded regions have been generated, the next step is to use these regions to search for and invade a homologous double-stranded donor sequence. The

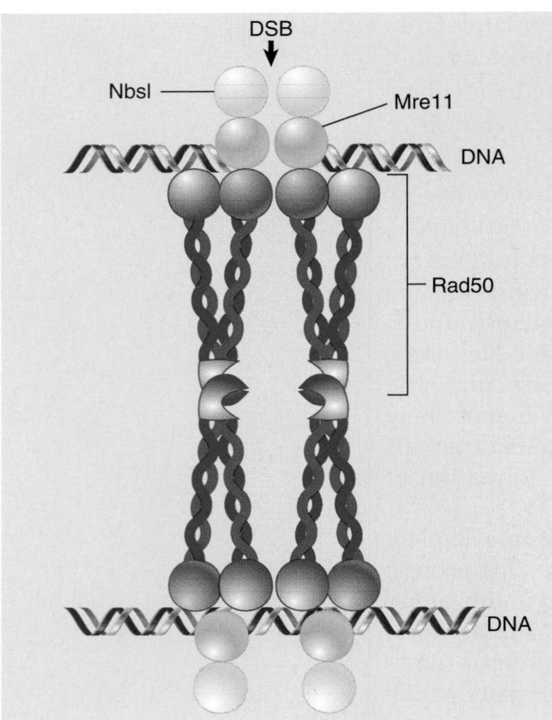

FIGURE 19.12 The MRN complex, required for 5′ end resection, also serves as a DNA bridge to prevent broken ends from separating. The "head" region of Rad50, bound to Mre11, binds DNA, while the extensive coiled coil region of Rad50 ends with a "zinc hook" that mediates interaction with another MRN complex. The precise position of Nbs1 within the complex is unknown but it interacts directly with Mre11.

E. coli protein RecA was the first example of a DNA strand-transfer protein to be discovered. It is the paradigm for a group that includes several other bacterial and archaeal proteins, and the Rad51 and Dmc1 proteins in eukaryotes. Analysis of yeast *rad51* mutants shows that this class of protein plays a central role in recombination. They accumulate double-strand breaks and fail to form normal synaptonemal complexes.

Both Rad51/Dmc1 and their prokaryotic counterpart RecA assemble onto single-stranded DNA to form helical nucleoprotein filaments in the presence of ATP. These filaments hold the single strand in an extended conformation, stretched about 50% longer than DNA in a normal duplex. They are sometimes referred to as **presynaptic filaments**, as filament formation can occur prior to any interaction with homologous donor sequences. When duplex DNA is bound, it contacts the RecA-class protein via its minor groove, leaving the major groove accessible for possible reaction with a second DNA molecule. These filaments can then promote base pairing between a single strand of DNA and its complement in a duplex molecule, creating a three-stranded intermediate.

Once the homologous duplex has been identified, the nucleoprotein filaments promote the displacement of the corresponding strand of the duplex, and lead to base pairing between the invading strand and its complementary strand in the donor. The displacement reaction can occur between DNA molecules in several configurations and has three general conditions:

- One of the DNA molecules must have a single-stranded region.
- One of the molecules must have a free 3′ end.
- The single-stranded region and the 3′ end must be located within a region that is complementary between the molecules.

presynaptic filaments Single-stranded DNA bound in a helical nucleoprotein filament with a strand transfer protein such as Rad51 or RecA.

The reaction is illustrated in FIGURE 19.13. When a linear single strand invades a duplex, it displaces the original partner to its complementary strand. The reaction can be followed most easily by making either the donor or the recipient a circular molecule. The reaction proceeds 5′ to 3′ along the strand whose partner is being displaced and replaced; that is, the reaction involves an exchange in which (at least) one of the exchanging strands has a free 3′ end.

FIGURE 19.13 RecA promotes the assimilation of invading single strands into duplex DNA so long as one of the reacting strands has a free end.

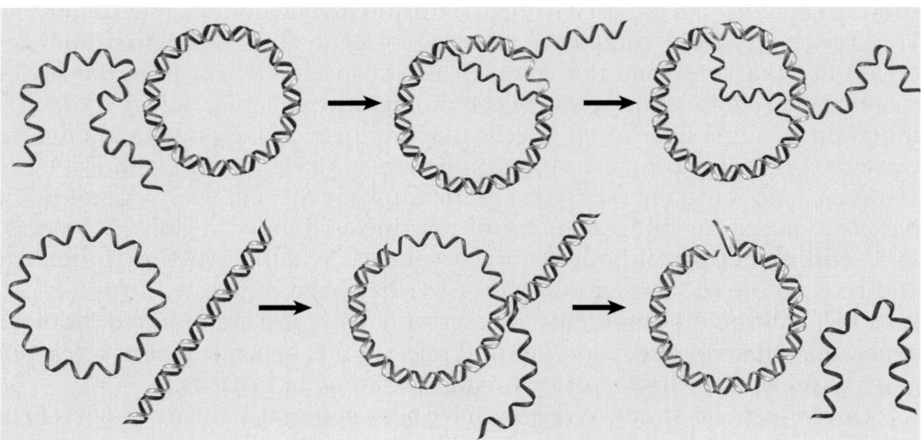

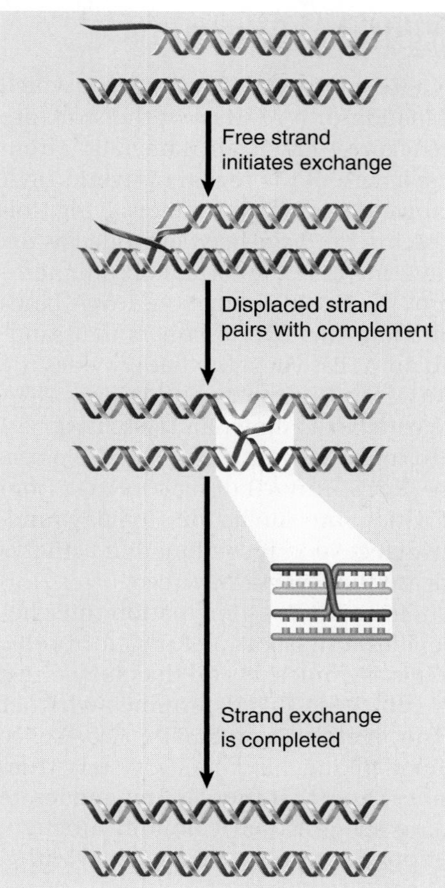
Free strand
initiates exchange
Displaced strand
pairs with complement
Strand exchange
is completed

FIGURE 19.14 RecA-mediated strand exchange between partially duplex and entirely duplex DNA generates a joint molecule with the same structure as a recombination intermediate.

The reaction between a partially duplex molecule and an entirely duplex molecule leads to the exchange of strands. An example is illustrated in **FIGURE 19.14**. Strand invasion starts with one end of the linear molecule, where the invading single strand displaces its homolog in the duplex in the customary way. When the reaction reaches the region that is duplex in both molecules, though, the invading strand unpairs from its partner, which then pairs with the other displaced strand.

At this stage, the molecule has a structure indistinguishable from the recombinant joint in Figure 19.8. The reaction sponsored *in vitro* by RecA-family proteins can generate Holliday junctions, which suggests that these enzymes can mediate reciprocal strand transfer.

KEY CONCEPTS

- MRN/MRX complexes are required for Spo11 displacement and 5′ end resection.
- RecA-type proteins form filaments with single-stranded or duplex DNA and catalyze the ability of a single-stranded DNA with a free 3′ to displace its counterpart in a DNA duplex.

CONCEPT AND REASONING CHECK

ATP binding promotes filament formation by RecA-family proteins, but ATP hydrolysis appears to promote turnover of the filament. What would be the effect of the non-hydrolyzable ATP analog ATP-γ-S on presynaptic filament formation?

19.6 The Ruv System Resolves Holliday Junctions

The critical final step in recombination is the resolution of the Holliday junction, which determines whether there is a reciprocal recombination or a reversal of the structure that leaves only a short stretch of hybrid DNA (see Figure 19.8). Branch migration from the exchange site (see Figure 19.7) determines the length of the region of hybrid DNA (with or without recombination). The proteins involved in stabilizing and resolving Holliday junctions have been elusive in eukaryotic systems, but have been identified as the products of the *ruv* genes in *E. coli.* RuvA and RuvB increase the formation of heteroduplex structures. RuvA recognizes the structure of the Holliday junction. RuvA binds to all four strands of DNA at the crossover point and forms two tetramers that sandwich the DNA. RuvB is a hexameric helicase with an ATPase activity that provides the motor for branch migration. Hexameric rings of RuvB bind around each duplex of DNA upstream of the crossover point. A diagram of the complex is shown in **FIGURE 19.15**.

The RuvAB complex can cause the branch to migrate as fast as 10 to 20 bp/sec. A similar activity is provided by another helicase, RecG. RuvAB displaces RecA from DNA during its action. The RuvAB and RecG activities both can act on Holliday junctions, but if both are mutant, *E. coli* is completely defective in recombination activity.

The third gene, *ruvC,* codes for an endonuclease that specifically recognizes Holliday junctions. It can cleave the junctions *in vitro* to resolve recombination intermediates. A common tetranucleotide sequence provides a hotspot for RuvC to resolve the Holliday junction. The tetranucleotide (ATTG) is asymmetric, and thus may direct resolution with regard to which pair of strands is nicked. This determines whether the outcome is patch recombinant formation (no overall recombination) or splice recombinant formation (recombination between flanking markers). Crystal structures of RuvC and other junction-resolving enzymes show that there is only moderate structural similarity among the group members, in spite of their common function. **FIGURE 19.16** shows the crystal structure of a bacteriophage resolvase complexed with a Holliday junction, showing the sites of cleavage.

All of this suggests that recombination uses a "resolvasome" complex that includes enzymes catalyzing branch migration as well as junction-resolving activity. It is likely that mammalian cells contain a similar complex, but no functional RuvC homologs have yet been identified in mammals.

A different system of resolvases has been characterized in yeast and mammals. Mutants in *S. cerevisiae Mus81* are defective in recombination. Mus81 is a component of an endonuclease that resolves Holliday junctions into duplex structures. This resolvase is important both in meiosis and for restarting stalled replication forks (see *Section 20.7, Recombination-Repair Systems*).

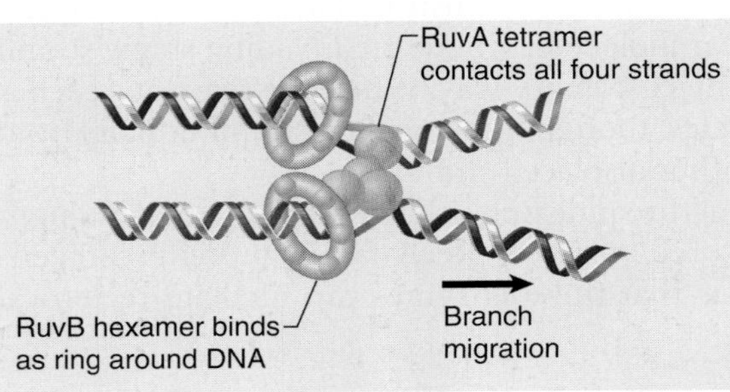
RuvA tetramer
contacts all four strands
RuvB hexamer binds
as ring around DNA
Branch
migration

FIGURE 19.15 RuvAB is an asymmetric complex that promotes branch migration of a Holliday junction.

KEY CONCEPTS

- The Ruv complex acts on recombinant junctions.
- RuvA recognizes the structure of the junction and RuvB is a helicase that catalyzes branch migration.
- RuvC cleaves junctions to resolve Holliday junctions.

CONCEPT AND REASONING CHECK

Note the symmetry of the resolvase depicted in Figure 19.16. Why is it important that resolvases function as symmetrical molecules with two active sites?

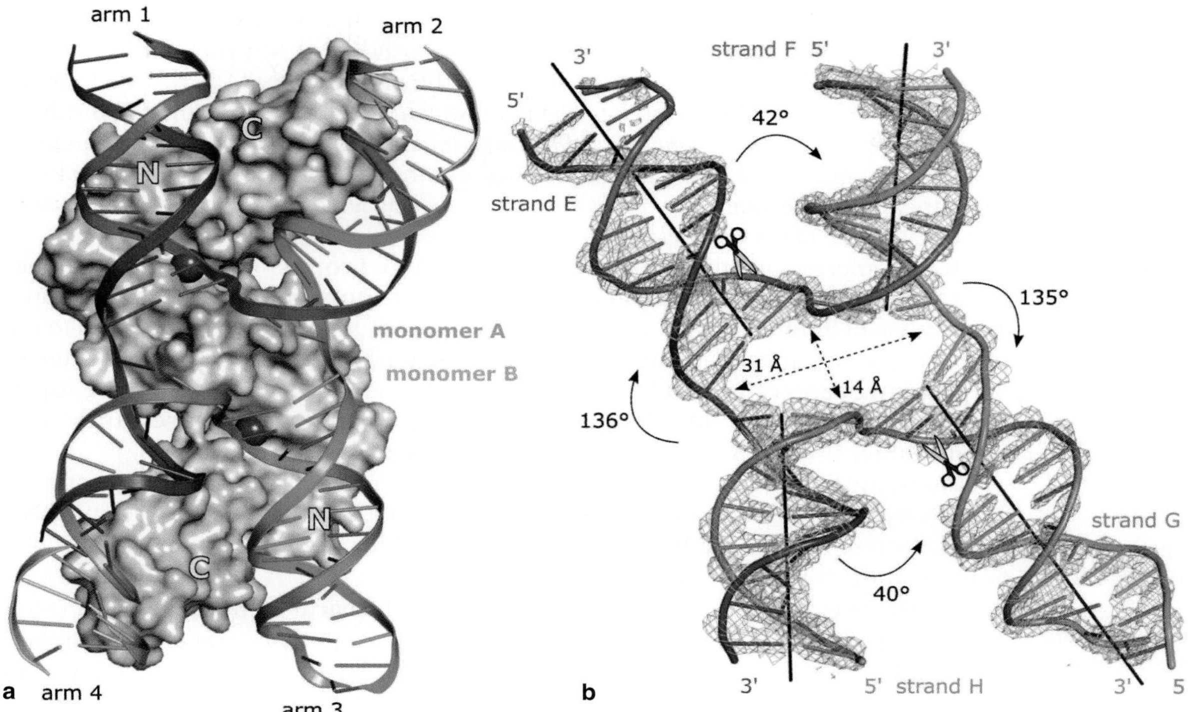

FIGURE 19.16 **(a)** Complex of T4 endo VII, with the two subunits colored differently bound to the Holliday junction (in ribbon diagrams, with each strand color-coded). Mg^{2+} ions are shown as purple spheres. **(b)** The branch point of the junction is open. The angles between DNA arms are marked. Scissors indicate the sites of cleavage. Reprinted by permission from Macmillan Publishers Ltd: *Nature* 449, Biertümpfel, C., Yang, W., and Suck, D., copyright 2007. Photo courtesy of Dietrich Suck, European Molecular Biology Laboratory.

19.7 Topoisomerases Relax or Introduce Supercoils in DNA

Topological manipulation of DNA is a central aspect of all its functional activities—recombination, replication, and transcription—as well as of the organization of higher-order structure. All synthetic activities involving double-stranded DNA require the strands to separate, and as we have seen, recombination generates topologically complex structures that must be resolved. Because the DNA strands do not simply lie side by side but are intertwined, their separation requires the strands to rotate about each other in space. The principle of **supercoiling** and the implications of over- and underwinding DNA were introduced in Chapter 1 (see *Section 1.5, Supercoiling Affects the Structure of DNA*).

▸ **supercoiling** The coiling of a closed duplex DNA in space so that it crosses over its own axis.

Changes in the topology of DNA can be caused in several ways, some of which can be detrimental to cellular processes if not properly resolved. The initiation of both replication and transcription can actually be facilitated by negative supercoiling, as this tends to promote unwinding of the two strands of DNA. However, the movement of RNA or DNA polymerases creates a region of positive supercoiling in front of the enzyme, which must be resolved before the accumulation of positive supercoils impedes the movement of the enzyme. When a circular DNA molecule is replicated, the circular products may be **catenated**, with one passed through the other, and must be separated in order for the daughter molecules to segregate to separate daughter cells. Even linear eukaryotic chromosomes can become entangled due to their lengths, and must be similarly resolved to prevent chromosome breakage during cell division. All of these situations are resolved by the actions of **topoisomerases**.

▸ **catenate** To link together two circular molecules, as in a chain.

▸ **topoisomerase** An enzyme that changes the number of times the two strands in a closed DNA molecule cross each other. It does this by cutting the DNA, passing DNA through the break, and resealing the DNA.

DNA topoisomerases are enzymes that catalyze changes in the topology of DNA by transiently breaking one or both strands of DNA, passing the unbroken strand(s) through the gap, and then resealing the gap. The ends that are generated by the break are never free, but instead are manipulated exclusively within the confines of the enzyme—in fact, they are covalently linked to the enzyme. Spo11 is related to topoisomerases and undergoes a similar covalent attachment when it forms DSBs during meiosis (see *Section 19.3, Double-Strand Breaks Initiate Recombination*). Topoisomerases act on any sequence of DNA, but some enzymes involved in site-specific recombination function in the same way as topoisomerases (see *Section 19.8, Site-Specific Recombination Resembles Topoisomerase Activity*).

- **type I topoisomerase** An enzyme that changes the topology of DNA by nicking and resealing one strand of DNA.
- **type II topoisomerase** An enzyme that changes the topology of DNA by nicking and resealing both strands of DNA.
- **gyrase** Enzymes that introduce negative supercoils into DNA.
- **reverse gyrase** Enzyme that introduces positive supercoils into DNA.

Topoisomerases are divided into two classes: **Type I topoisomerases** act by making a transient break in one strand of DNA. **Type II topoisomerases** act by introducing a transient double-strand break. Topoisomerases in general vary with regard to the types of topological change they introduce. Some topoisomerases can relax (remove) only negative supercoils from DNA; others can relax both negative and positive supercoils. Enzymes that can introduce negative supercoils are called **gyrases**; those that can introduce positive supercoils are called **reverse gyrases**.

There are four topoisomerase enzymes in *E. coli:* topoisomerases I, III, and IV, and DNA gyrase. DNA topoisomerases I and III are type I enzymes. Gyrase and DNA topoisomerase IV are type II enzymes. Each of the four enzymes is important in one or more of the following functions:

- The overall level of negative supercoiling in the bacterial nucleoid is the result of a balance between the introduction of supercoils by gyrase and their relaxation by topoisomerases I and IV. This is a crucial aspect of nucleoid structure (see *Section 23.4, The Bacterial Genome Is a Supercoiled Nucleoid*), and it affects initiation of transcription at certain promoters (see *Section 11.10, Supercoiling Is an Important Feature of Transcription*).
- The same enzymes are involved in resolving the problems created by transcription; gyrase converts the positive supercoils that are generated ahead of RNA polymerase into negative supercoils, and topoisomerases I and IV remove the negative supercoils that are left behind the enzyme. Gyrase is also critical for removing positive supercoils produced during replication.
- As replication proceeds, the daughter duplexes can become twisted around one another in a stage known as *precatenation*. The precatenanes are removed by topoisomerase IV, which also decatenates any catenated genomes that are left at the end of replication.

The enzymes in eukaryotes follow the same principles, although the detailed division of responsibilities may be different. They do not show sequence or structural similarity with the prokaryotic enzymes. Most eukaryotes contain a single topoisomerase I enzyme that is required both for replication fork movement and for relaxing supercoils generated by transcription. A topoisomerase II enzyme(s) is required to unlink chromosomes following replication. Other individual topoisomerases have been implicated in recombination and repair activities.

The common mechanism for all topoisomerases is to link one end of each broken strand to a tyrosine residue in the enzyme. A type I enzyme links to the single broken strand; a type II enzyme links to one end of each broken strand. The topoisomerases are further divided into the A and B groups according to whether the linkage is to a 5′ phosphate or 3′ phosphate. The use of the transient phosphodiester–tyrosine bond suggests a mechanism for the action of the enzyme; it transfers a phosphodiester bond(s) in DNA to the protein, manipulates the structure of one or both DNA strands, and then rejoins the bond(s) in the original strand.

The *E. coli* enzymes are all of type A and use links to 5′ phosphate. This is the general pattern for bacteria, where there are almost no type B topoisomerases. All four possible types of topoisomerase (IA, IB, IIA, and IIB) are found in eukaryotes.

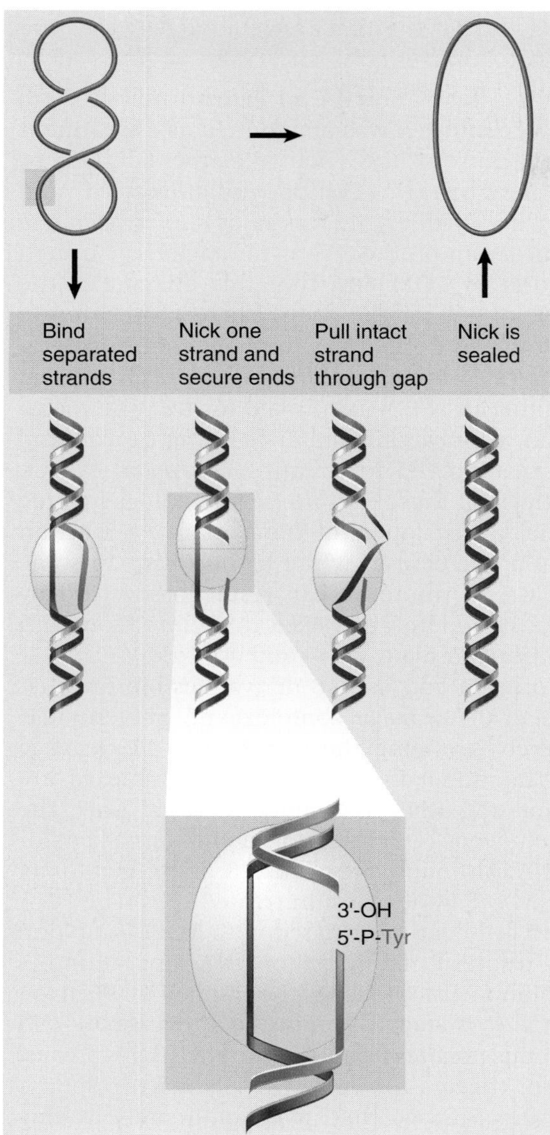
Bind separated strands
Nick one strand and secure ends
Pull intact strand through gap
Nick is sealed
3'-OH
5'-P-Tyr

FIGURE 19.17 Type I topoisomerases recognize partially unwound segments of DNA and pass one strand through a break made in the other.

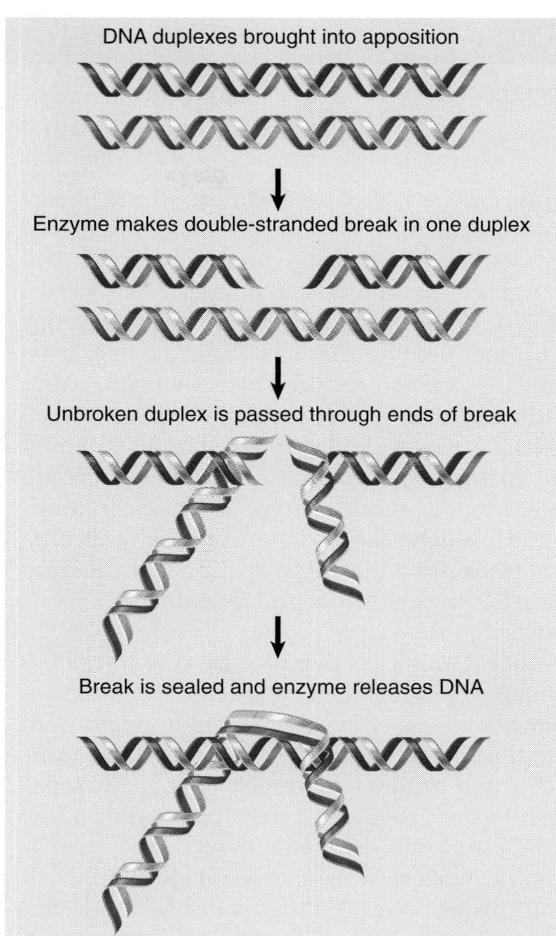
DNA duplexes brought into apposition
Enzyme makes double-stranded break in one duplex
Unbroken duplex is passed through ends of break
Break is sealed and enzyme releases DNA

FIGURE 19.18 Type II topoisomerases can pass a duplex DNA through a double-strand break in another duplex.

A model for the action of topoisomerase IA is illustrated in **FIGURE 19.17**. The enzyme binds to a region in which duplex DNA becomes separated into its single strands. The enzyme then breaks one strand, pulls the other strand through the gap, and finally seals the gap. The transfer of bonds from nucleic acid to protein explains how the enzyme can function without requiring any input of energy. There has been no irreversible hydrolysis of bonds; their energy has been conserved through the transfer reactions. The model is supported by the crystal structure of the enzyme.

Type II topoisomerases generally relax both negative and positive supercoils. The reaction requires ATP, with one ATP hydrolyzed for each catalytic event. As illustrated in **FIGURE 19.18**, the reaction is mediated by making a double-strand break in one DNA duplex. The double strand is cleaved with a 4-base stagger between the ends, and each subunit of the dimeric enzyme attaches to a protruding broken end. Another duplex

CLINICAL APPLICATIONS

Camptothecin and Topoisomerase I Inhibition

Topoisomerases are enzymes that catalyze strand breakage of DNA to relieve or introduce supercoiling into double-stranded DNA. The relaxation of supercoiled DNA is a prerequisite for essential nuclear processes including DNA replication, recombination, transcription, and chromatin condensation. There are two general categories of topoisomerases: type I enzymes that act by introducing nicks in one strand of DNA, and type II enzymes that act by ATP-dependent double-stranded cleavage. In mammals, there are four genes encoding distinct type I topoisomerases: type IA nuclear topoisomerase and mitochondrial topoisomerase, and type IB topoisomerases IIIα and IIIβ. Only the topoisomerase type IB can relax both positive and negative supercoils in double-stranded DNA.

Camptothecin and its derivatives are chemotherapy drugs that target dividing cells by specific inhibition of nuclear topoisomerase type IB (TOPI). In regions of the genome that are undergoing replication and transcription, the two strands of duplex DNA are separated, generating positively and negatively supercoiled DNA upstream and downstream, respectively. TOPI is concentrated in these regions and is able to relax both topological forms of DNA. TOPI is an essential protein, as demonstrated by the embryonic lethality observed in TOPI-deficient mice. The mechanism of DNA relaxation by TOPI proceeds by enzyme-mediated nicking of one strand of DNA, controlled rotation of the nicked strand, and religation of the DNA. TOPI nicks the DNA on one strand and then forms a transient covalent complex, referred to as TOPI cleavage complex. There is a transesterification between tyrosine 723 in the C-terminal portion of human TOPI and the phosphodiester bond of one strand of the DNA, creating an intermediate tyrosyl-phosphodiester covalent bond between the enzyme and the 3′ end of the nicked DNA. The core domain of the TOPI encircles the DNA, encompassing 14 nucleotides around the DNA cleavage site, and mediates the controlled rotation of the DNA, relaxing the supercoiling. Religation of the DNA requires the precise alignment of the 5′-hydroxyl end of the nicked DNA and the tyrosyl-phosphodiester bond of the TOPI cleavage complex. Neither TOPI cleavage nor religation requires energy and these two processes exist in equilibrium, with religation favored over cleavage.

Camptothecin is a naturally occurring substance isolated from the bark of the Chinese yew tree *Camptotheca acuminata*. It is a 5-ring heterocyclic alkaloid that was first isolated and demonstrated to have anticancer activity by the National Cancer Institute in 1966. After clinical testing, it was abandoned as a treatment for cancer because of the severity of the side effects, including hemorrhagic cystitis (inflammation and bleeding of the bladder). It was later shown that the mechanism of camptothecin's cytotoxicity is through the inhibition of TOPI and that this enzyme is the sole cellular target for the drug. Currently, there are two camptothecin derivatives, topotecan and irinotecan, that are also TOPI inhibitors being used to treat cancer. Topotecan is FDA approved for treating recurrent ovarian, cervical, and small cell lung cancer, and irinotecan is approved for treating colorectal cancer. The side effects of these camptothecin analogs include neutropenia, a disruption of bone marrow function resulting in a decrease in white blood cells. In some cases, the use of irinotecan can result in severe diarrhea. Typically, topotecan is used in conjunction with another anticancer drug, cisplatin.

Camptothecin and its two derivatives bind reversibly to the TOPI cleavage complex and prevent religation, thereby trapping the transient TOPI cleavage complex. The trapped TOPI cleavage complexes are reversed and the DNA is rapidly religated once the drugs are removed. TOPI inhibitors do not directly damage the DNA, although they cause S phase cytotoxicity and G2-M cell cycle arrest in treated cells. The DNA damage and cell death associated with TOPI inhibition occurs as a result of the process of DNA replication. As the replication fork approaches the trapped TOPI cleavage complex, it is stalled, leading to irreversible TOPI cleavage complexes and double-stranded DNA breaks. This in turn, triggers the cell's DNA damage response and arrest at the GS-M checkpoint, ultimately leading to apoptotic cell death.

The camptothecin-derived TOPI inhibitors penetrate eukaryotic cells readily and act quickly. However, they have some limitations as anticancer drugs. Camptothecin and its analogs exist in two different isomeric forms, one active and one inactive. At physiological pH conditions the inactive form is favored, resulting in greatly diminished potency of the drugs. In addition, the cellular concentration of these TOPI inhibitors is rapidly decreased due to the action of the multidrug resistance ATP-binding cassette transmembrane transport proteins that actively pump the drugs across the plasma membrane and out the cell. To date there are several non-camptothecin-derived TOPI inhibitors that are in clinical trials. These have been developed as effective TOPI inhibitors that have different chemical properties than camptothecin and its derivatives and hopefully will prove to be more stable and more effective anticancer drugs.

region is then passed through the break. The ATP is used in the following religation/release step, when the ends are rejoined and the DNA duplexes are released. This is why inhibiting the ATPase activity of the enzyme results in a "cleavable complex" that contains broken DNA.

Type II topoisomerases recognize a variety of DNA structures, most commonly sites where two double-stranded segments cross each other. The hydrolysis of ATP may be used to drive the enzyme through conformational changes that provide the force needed to push one DNA duplex through the break made in the other. As a result of the topology of supercoiled DNA, the relationship of the crossing segments allows supercoils to be removed from either positively or negatively supercoiled circles.

KEY CONCEPTS

- Topoisomerases alter supercoiling by breaking bonds in DNA, changing the conformation of the double helix in space, and remaking the bonds.
- Type I enzymes act by breaking a single strand of DNA; type II enzymes act by making double-strand breaks.
- Type I topoisomerases function by forming a covalent bond to one of the broken ends, moving one strand around the other, and then transferring the bound end to the other broken end. Bonds are conserved, and as a result no input of energy is required.
- Type II topoisomerases also form covalent bonds to the broken ends, and pass a duplex DNA region through the double-strand break. ATP is required to complete the reaction and reseal the break.

CONCEPT AND REASONING CHECK

Explain why either a single- or double-strand break can result in a change in supercoiling.

19.8 Site-Specific Recombination Resembles Topoisomerase Activity

Site-specific recombination involves a reaction between two specific sites. The lengths of target sites are short, and are typically in a range of 14 to 50 bp. In some cases the two sites have the same sequence, but in other cases they are nonhomologous. The reaction can be used to insert a free phage DNA into the bacterial chromosome or to excise an integrated phage DNA from the chromosome, and in this case the two recombining sequences are different from one another.

The enzymes that catalyze site-specific recombination are generally called **recombinases**, and more than one hundred of them are now known. Those involved in phage integration or that are related to these enzymes are also known as the **integrase** family. Prominent members of the integrase family are the prototype Int from phage lambda, Cre from phage P1, and the yeast FLP enzyme (which catalyzes a chromosomal inversion).

The classic model for site-specific recombination is illustrated by phage lambda. The conversion of lambda DNA between its different life forms involves two types of event. (The regulation of these events is described in *Chapter 14, Phage Strategies*.) The physical condition of the DNA is different in the lysogenic and lytic states:

- In the lytic state lambda DNA exists as an independent, circular molecule in the infected bacterium.
- In the lysogenic state, the phage DNA is an integral part of the bacterial chromosome (called **prophage**).

Transition between these states involves site-specific recombination:

- To enter the lysogenic condition, free lambda DNA must be inserted into the host DNA. This is called **integration**.
- To be released from lysogeny into the lytic cycle, prophage DNA must be released from the chromosome. This is called **excision**.

▸ **recombinase** Enzyme that catalyzes site-specific recombination.

▸ **prophage** A phage genome covalently integrated as a linear part of the bacterial chromosome.

▸ **integration** Insertion of a viral or another DNA sequence into a host genome as a region covalently linked on either side to the host sequences.

▸ **excision** Release of phage or episome or other sequence from the host chromosome as an autonomous DNA molecule.

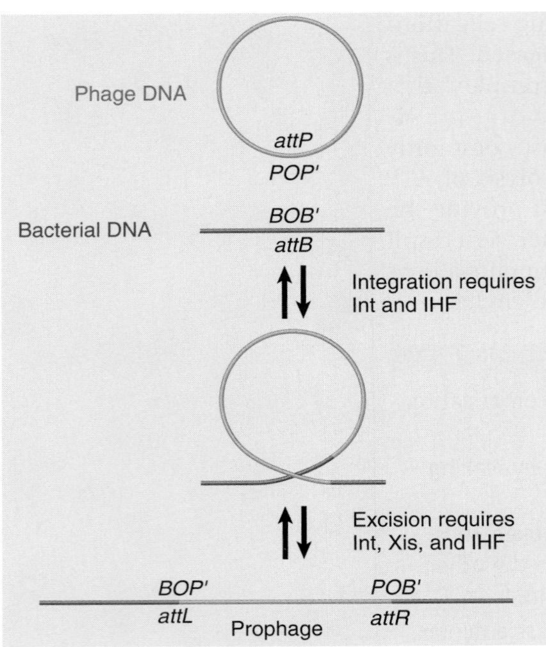

FIGURE 19.19 Circular phage DNA is converted to an integrated prophage by a reciprocal recombination between *attP* and *attB*; the prophage is excised by reciprocal recombination between *attL* and *attR*.

Integration and excision occur by recombination at specific loci on the bacterial and phage DNAs called attachment (***att***) **sites**. When describing the integration/excision reactions, the bacterial attachment site is called *attB*, consisting of the sequence components *BOB′*. The attachment site on the phage, *attP*, consists of the components *POP′*. **FIGURE 19.19** outlines the recombination reaction between these sites. The sequence *O* is common to *attB* and *attP*. It is called the **core sequence**; and the recombination event occurs within it. The flanking regions *B*, *B′* and *P*, *P′* are referred to as the *arms*; each is distinct in sequence. The phage DNA is circular, so the recombination event inserts it into the bacterial chromosome as a linear sequence. The prophage is bounded by two new *att* sites—the products of the recombination—called *attL* and *attR*.

An important consequence of the constitution of the *att* sites is that the integration and excision reactions do not involve the same pair of reacting sequences. Integration requires recognition between *attP* and *attB*, whereas excision requires recognition between *attL* and *attR*. The directional character of site-specific recombination is controlled by the identity of the recombining sites. The recombination event is reversible, but different conditions prevail for each direction of the reaction. This is an important feature in the life of the phage, because it offers a means to ensure that an integration event is not immediately reversed by an excision, and vice versa.

The difference in the pairs of sites reacting at integration and excision is reflected by a difference in the proteins that mediate the two reactions:

- Integration (*attB* × *attP*) requires the product of the phage gene *int*, which codes for the integrase Int, and a bacterial protein called integration host factor (IHF).
- Excision (*attL* × *attR*) requires the product of phage gene *xis*, in addition to Int and IHF.

Thus Int and IHF are required for both reactions. Xis plays an important role in controlling the direction; it is required for excision, but inhibits integration.

A similar system, but with somewhat simpler requirements for both sequence and protein components, is found in the bacteriophage P1. The Cre recombinase encoded by the phage catalyzes a recombination between two target sequences. Unlike phage lambda, for which the recombining sequences are different, in phage P1 they are identical. Each consists of a 34 bp-long sequence called *loxP*. The Cre recombinase is sufficient for the reaction; no accessory proteins are required. As a result of its simplicity and its efficiency, what is now known as the Cre/*lox* system has been adapted for use in eukaryotic cells, where it has become one of the standard techniques for undertaking site-specific recombination.

Enzymes such as Int and Cre use a mechanism similar to that of type I topoisomerases, in which a break is made in one DNA strand at a time. The difference is that a recombinase reconnects the ends *crosswise*, whereas a topoisomerase makes a break, manipulates the ends, and then rejoins the *original* ends. The basic principle of the system is that four molecules of the recombinase are required, one to cut each of the four strands of the two duplexes that are recombining.

FIGURE 19.20 shows the nature of the reaction catalyzed by an integrase. The enzyme is a monomeric protein that has an active site capable of cutting and ligating DNA. The reaction involves an attack by a tyrosine on a phosphodiester bond. The 3′ end of the DNA chain is linked through a phosphodiester bond to a tyrosine in the enzyme. This releases a free 5′ hydroxyl end.

Two enzyme units are bound to each of the recombination sites. At each site, only one of the units attacks the DNA. The symmetry of the system ensures that

- ***att* sites** The loci on a lambda phage and the bacterial chromosome at which recombination integrates the phage into, or excises it from, the bacterial chromosome.
- **core sequence** The segment of DNA that is common to the attachment sites on both the phage lambda and bacterial genomes. It is the location of the recombination event that allows phage lambda to integrate.

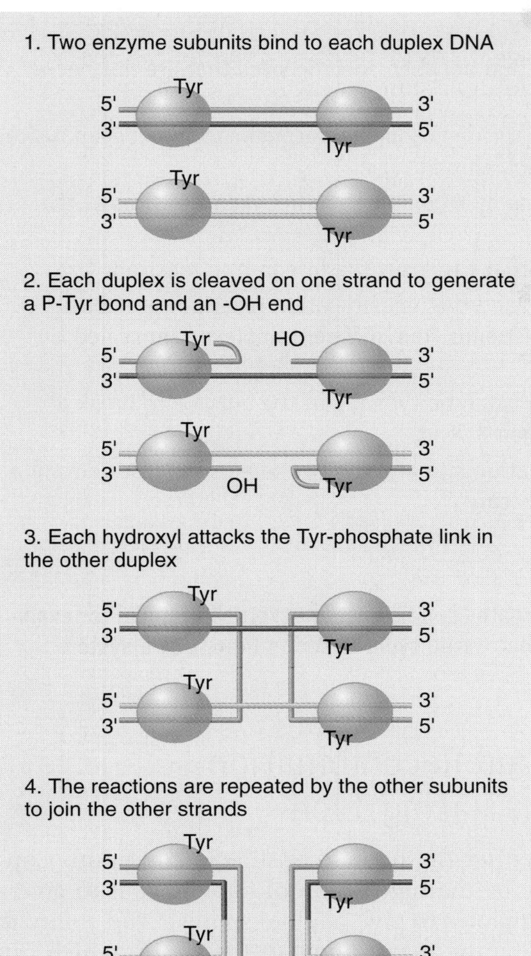
1. Two enzyme subunits bind to each duplex DNA
Tyr
5'
3'
3'
5'
Tyr
2. Each duplex is cleaved on one strand to generate a P-Tyr bond and an -OH end
HO
OH
3. Each hydroxyl attacks the Tyr-phosphate link in the other duplex
4. The reactions are repeated by the other subunits to join the other strands

FIGURE 19.20 Integrases catalyze recombination by a mechanism similar to topoisomerases. Staggered cuts are made in DNA and the 3′-phosphate end is covalently linked to a tyrosine in the enzyme. The free hydroxyl group of each strand then attacks the P-Tyr link of the other strand. The first exchange shown in the figure generates a Holliday junction. The junction is resolved by repeating the process with the other pair of strands.

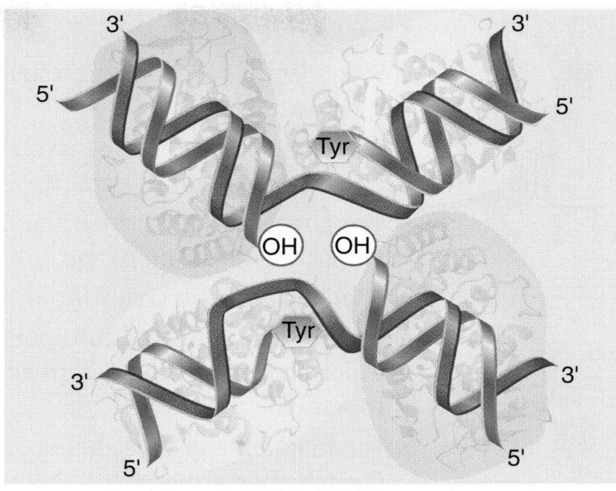
3'
5'
Tyr
OH
OH

FIGURE 19.21 A synapsed loxA recombination complex has a tetramer of Cre recombinases, with one enzyme monomer bound to each half site. Two of the four active sites are in use, acting on complementary strands of the two DNA sites.

complementary strands are broken in each recombination site. The free 5′–OH end in each site attacks the 3′–phosphotyrosine link in the other site. This generates a Holliday junction.

The structure is resolved when the other two enzyme units (which had not been involved in the first cycle of breakage and reunion) act on the other pair of complementary strands.

The successive interactions accomplish a conservative strand exchange, in which there are no deletions or additions of nucleotides at the exchange site, and there is no need for input of energy. The transient 3′–phosphotyrosine link between protein and DNA conserves the energy of the cleaved phosphodiester bond.

FIGURE 19.21 shows the reaction intermediate, based on the crystal structure. (Trapping the intermediate was made possible by using a "suicide substrate," which consists of a synthetic DNA duplex with a missing phosphodiester bond, so that the attack by the enzyme does not generate a free 5′–OH end.) The structure of the Cre-*lox* complex shows two Cre molecules, each of which is bound to a 15 bp length of DNA. The DNA is bent by ~100° at the center of symmetry. Two of these complexes assemble in an antiparallel way to form a tetrameric protein structure bound to two synapsed DNA molecules. Strand exchange takes place in a central cavity of the protein structure that contains the central six bases of the crossover region.

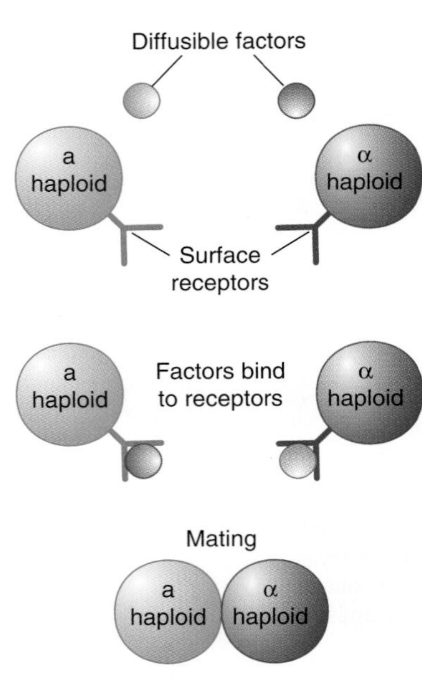

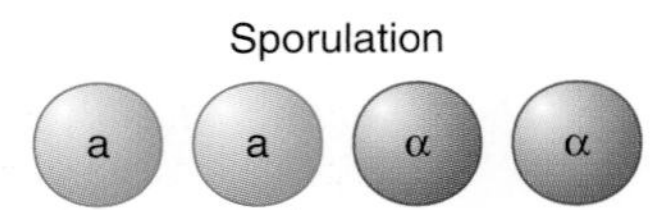

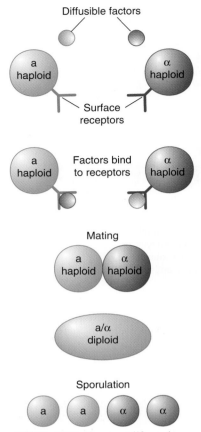

FIGURE 19.22 The yeast life cycle proceeds through mating of *MAT***a** and *MAT*α haploids to give heterozygous diploids that sporulate to give haploid spores.

KEY CONCEPTS

- Site-specific recombination involves reaction between specific sites that are not necessarily homologous.
- Phage lambda integrates into the bacterial chromosome by recombination between a site on the phage and the *att* site on the *E. coli* chromosome.
- The phage is excised from the chromosome by recombination between the sites at the end of the linear prophage.
- Phage lambda *int* codes for an integrase that catalyzes the integration reaction.
- Integrases are related to topoisomerases, and the recombination reaction resembles topoisomerase action except that nicked strands from different duplexes are sealed together.
- The reaction conserves energy by using a catalytic tyrosine in the enzyme to break a phosphodiester bond and link to the broken 3′ end.
- Two enzyme units bind to each recombination site and the two dimers synapse to form a complex in which the transfer reactions occur.

CONCEPT AND REASONING CHECK

The Cre/lox system is used extensively to create "conditional knockouts" in mice, for example to delete a gene of interest in a particular tissue type. Describe how such a system might work.

19.9 Yeast Use a Specialized Recombination Mechanism to Switch Mating Type

The yeast *S. cerevisiae* can propagate in either the haploid or diploid condition. Conversion between these states takes place by mating (fusion of haploid cells to give a diploid) and by sporulation (meiosis of diploids to give haploid spores). The ability to engage in these activities is determined by the mating type of the strain, which can be either **a** or α. Haploid cells of type **a** can mate only with haploid cells of type α to generate diploid cells of type **a**/α. The diploid cells can sporulate to regenerate haploid spores of both types. The basic yeast life cycle is shown in **FIGURE 19.22**.

Mating behavior is determined by the genetic information present at the *MAT* (mating type) locus. Cells that carry the *MAT***a** allele at this locus are type **a**; likewise, cells that carry the *MAT*α allele are type α. Recognition between cells of opposite mating type is accomplished by the secretion of pheromones: α cells secrete the small polypeptide α-factor; **a** cells secrete **a**-factor. A cell of one mating type carries a surface receptor for the pheromone of the opposite type. When an **a** cell and an α cell encounter one another, their pheromones act on their receptors to arrest the cells in the G1 phase of the cell cycle, and various morphological changes occur. In a successful mating, the cell cycle arrest is followed by cell and nuclear fusion to produce an **a**/α diploid cell.

Some yeast strains have the remarkable ability to switch their mating types. These strains carry the wild-type allele *HO* and change their mating type frequently—as often as once every generation. Strains with the recessive allele *ho* have a stable mating type, which is subject to change with a frequency of $\sim 10^{-6}$.

▸ **mating type cassette** Yeast mating type is determined by a single active locus (the active cassette) and two inactive copies of the locus (the silent cassettes). Mating type is changed when an active cassette of one type is replaced by a silent cassette of the other type.

The presence of *HO* causes the genotype of a yeast population to change. Irrespective of the initial mating type, within a very few generations there are large numbers of cells of both mating types, leading to the formation of *MAT***a**/*MAT*α diploids. Mating type switching allows sexual reproduction and the benefits of meiotic recombination in populations that may have begun as only a single mating type.

The existence of switching suggests that all cells contain the potential information needed to be either *MAT***a** or *MAT*α, but that haploids express only one type. Where does the information to change mating types come from? Two additional loci are needed for switching. *HML*α is needed for switching to give a *MAT*α type; *HMR***a** is needed for switching to give a *MAT***a** type. These loci lie on the same chromosome that carries *MAT*. *HML* is far to the left, *HMR* is far to the right.

In addition to the active **a** or α **mating type cassette** present at the *MAT* locus, yeast, therefore, also carry silent mating-type cassettes at *HML* and *HMR*, as illustrated in **FIGURE 19.23**. All cassettes carry information that codes for mating type, but only the active cassette at *MAT* is expressed. The cassettes at *HML* and *HMR* are silent because they are assembled into heterochromatin (see *Chapter 27, Epigenetic Effects Are Inherited*). Mating-type switching occurs when the active cassette is replaced by information from a silent cassette. The newly installed cassette is then expressed.

Switching is nonreciprocal; the copy at *HML* or *HMR* replaces the allele at *MAT*. We know this because a mutation at *MAT* is lost permanently when it is replaced by switching: it does not exchange with the copy that replaces it. Likewise, if the silent copy present at *HML* or *HMR* is mutated, switching introduces a mutant allele into the *MAT* locus. The mutant copy at *HML* or *HMR* remains there through an indefinite number of switches. The donor element generates a new copy at the recipient site while itself remaining inviolate.

Mating-type switching is a directed event, in which there is only one recipient *(MAT)*, but two potential donors (*HML* and *HMR*). Switching usually involves replacement of *MAT***a** by the copy at *HML*α or replacement of *MAT*α by the copy at *HMR***a**. In 80% to 90% of switches, the *MAT* allele is replaced by one of the opposite type. This is determined by the mating type of the cell. **a** cells preferentially choose *HML* as the donor; α cells preferentially choose *HMR*.

Several groups of genes are involved in establishing and switching mating type. In addition to the genes that directly determine mating type, they include genes needed to repress the silent cassettes, switch mating type, or execute the functions involved in mating.

By comparing the sequences of the two silent cassettes (*HML*α and *HMR***a**) with the sequences of the two types of active cassette (*MAT***a** and *MAT*α), we can delineate the sequences that determine mating type. The organization of the mating type loci is summarized in **FIGURE 19.24**. Each cassette contains common sequences that flank a central region that differs in the **a** and α types of cassette (called *Y***a** or *Y*α). On either side of this region, the flanking sequences are virtually identical, although they are shorter at *HMR*. The active cassette at *MAT* is transcribed from a promoter within the *Y* region. Note that the same *Y* region is present in the cassettes at *HML* or *HMR, but these cassettes are maintained in a transcriptionally silent state so that no mating type information can be expressed from these copies.*

A switch in mating type is accomplished by a gene conversion event in which the recipient site (*MAT*) acquires the sequence of the donor type (*HML* or *HMR*). This is a modified homologous recombination mechanism, which uses many of the same enzymes required for meiotic recombination or repair of DNA double-strand breaks; what is unique about mating type switching is the directionality by which homologous sequences are always transferred from the silent donor cassette to the active locus.

Switching is initiated by a formation of a site-specific double-strand break created by an endonuclease encoded by the *HO* locus. The HO endonuclease makes a staggered double-strand break just to the right of the *Y* boundary, as shown in **FIGURE 19.25**. The 24-bp

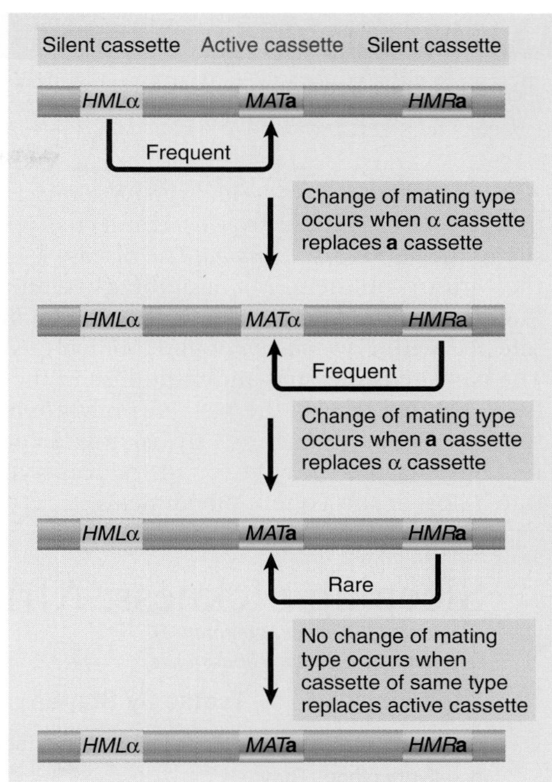

FIGURE 19.23 Changes of mating type occur when silent cassettes replace active cassettes of opposite genotype; when gene conversions occur between cassettes of the same type, the mating type remains unaltered.

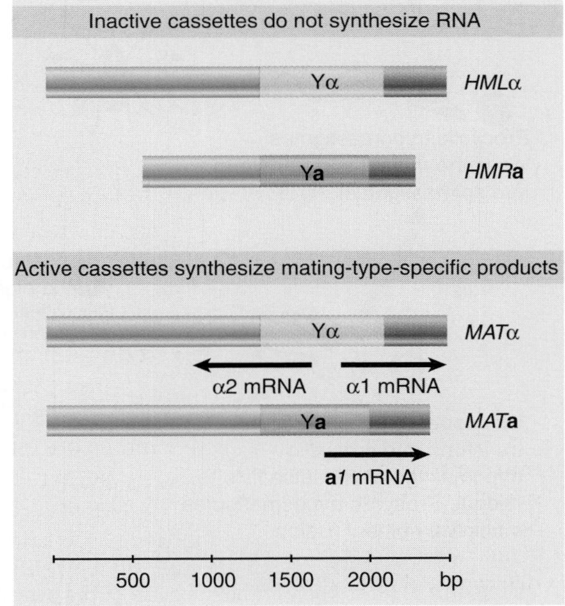

FIGURE 19.24 Silent cassettes have the same sequences as the corresponding active cassettes, except for the absence of the extreme flanking sequences in *HMR***a**. Only the *Y* region changes between **a** and α types.

MEDICAL APPLICATIONS

Trypanosomes Use Gene Switching to Evade the Host Immune System

African sleeping sickness or trypanosomiasis is caused by the bite of a tsetse fly that is infected with a parasite of the genus *Trypanosoma*. The disease is endemic on the African continent. Major outbreaks tend to occur in Sudan, Uganda, Angola, and Congo. Tsetse flies become infected with *Trypanosoma* when a blood meal is taken. The parasite is ingested and multiplies in the gut of the fly. After two weeks, the parasites migrate back to the salivary glands. The tsetse fly is now infectious and remains infectious for life as it introduces trypanosomes into a host at subsequent blood meals.

Symptoms appear about 1 to 2 weeks after a bite by an infected fly. An inflamed nodule appears at the site of the tsetse fly bite. Other symptoms take weeks to months to appear. Symptoms include fever, headache, joint pain, rash, and lymph node swelling. The parasites multiply rapidly in the blood, reaching population densities of more than 1 million parasites per milliliter of blood. This number drops rapidly (called a remission), only to be followed by another burst of population growth (relapse). This pattern continues. If untreated, the parasite invades the central nervous system, causing sleeping sickness (**FIGURE B19.1**). These infections last from weeks to years. Sleeping sickness symptoms include chronic fever, severe headaches, and sleeping during

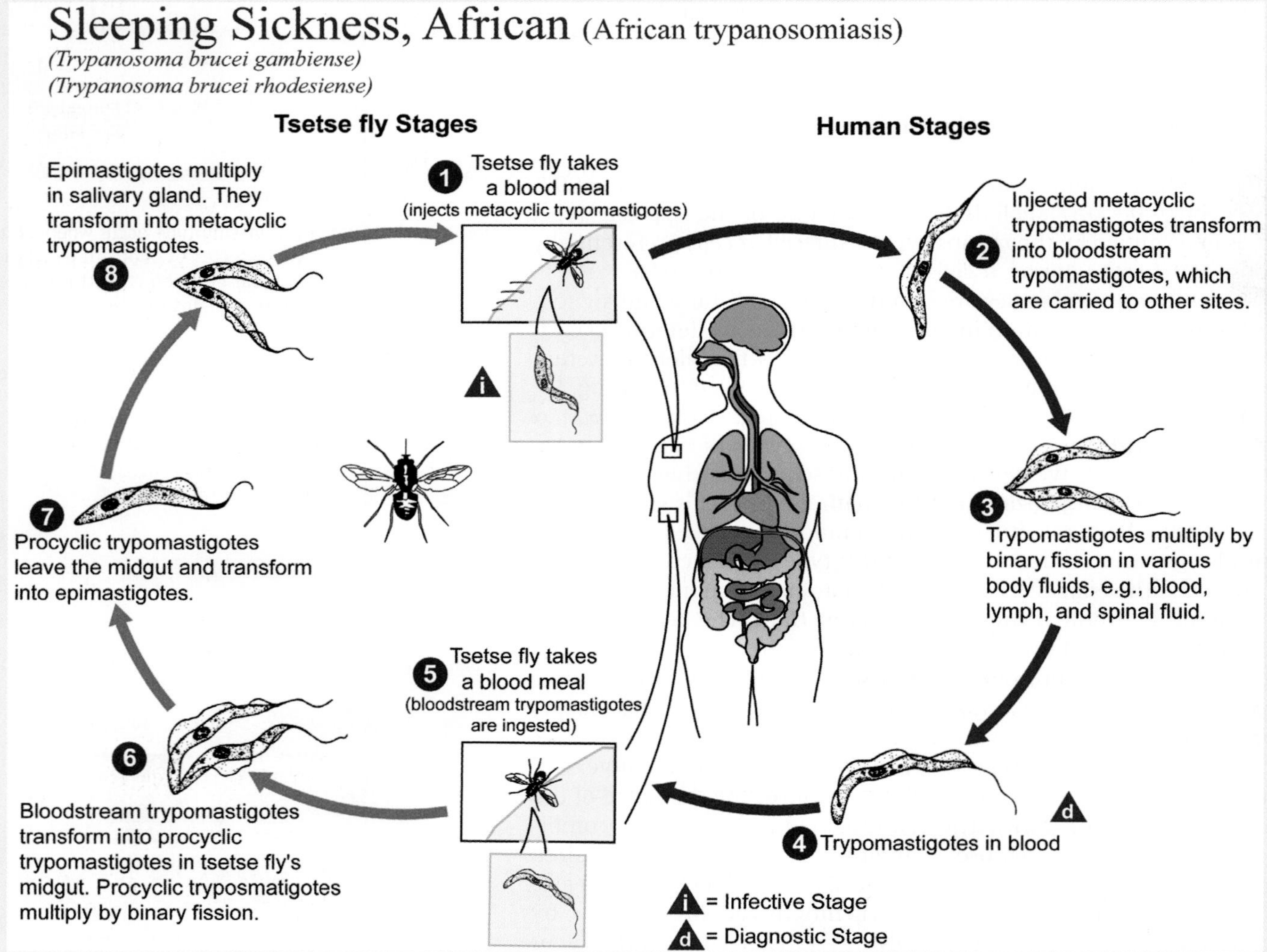
Sleeping Sickness, African (African trypanosomiasis)
(Trypanosoma brucei gambiense)
(Trypanosoma brucei rhodesiense)
Tsetse fly Stages
Human Stages
1 Tsetse fly takes a blood meal (injects metacyclic trypomastigotes)
2 Injected metacyclic trypomastigotes transform into bloodstream trypomastigotes, which are carried to other sites.
3 Trypomastigotes multiply by binary fission in various body fluids, e.g., blood, lymph, and spinal fluid.
4 Trypomastigotes in blood
5 Tsetse fly takes a blood meal (bloodstream trypomastigotes are ingested)
6 Bloodstream trypomastigotes transform into procyclic trypomastigotes in tsetse fly's midgut. Procyclic tryposmatigotes multiply by binary fission.
7 Procyclic trypomastigotes leave the midgut and transform into epimastigotes.
8 Epimastigotes multiply in salivary gland. They transform into metacyclic trypomastigotes.
i = Infective Stage
d = Diagnostic Stage

FIGURE B19.1 Illustration of the life cycle of the parasite that causes African sleeping sickness. Courtesy of Alexander J. da Silva, PhD, and Melanie Moser/CDC.

long periods of the day, as well as nighttime insomnia. Trypanosomes can invade the brain several months to years after infection, leading to coma and death.

The African trypanosomes are able to change the surface coat proteins of their outer membranes. This phenomenon is referred to as *antigenic variation*. In doing so, the parasite is always one step ahead of the immune system. *Antibodies* are highly specific proteins produced by the body in response to a foreign substance such as a parasite, and are capable of binding to it. Antibodies are produced against the parasites *v*ariant *s*urface *g*lycoproteins coat (*VSG*). The host's antibodies can neutralize and kill approximately 99% of the original parasite population within the host (remission), but a few trypanosomes switch VSGs, gaining a new antigenically distinct VSG coat (causing a relapse). The trypanosomes undergo a type of gene switching that is similar to the yeast mating-type switching mechanism. Approximately 1000 different genes encode VSGs. VSG genes represent about 10% of the total DNA of a trypanosome. The VSG genes are located either in the chromosome's interior or near the telomeres. Expressed VSG genes are always near telomeres. Only one of the VSG genes is expressed at a time; all other VSG genes are transcriptionally silent. Therefore, only 1 type of VSG molecule is present within the trypanosome surface coat, resulting in a homogenous display of identical surface antigens that is recognized by the immune system.

The mechanism of this type of antigen variation in trypanosomes involves a silent or basic copy of the VSG gene displacing a previously active VSG gene, located in a VSG expression site. The trigger for the switching is unknown. The gene that is to be expressed is copied onto a cassette and the copy is unidirectionally moved or recombined into one of potentially 20 different active *b*loodstream-form *e*xpression *s*ites (BESs) that contain *e*xpression *s*ite-*a*ssociated *g*enes (ESAGs) on the chromosome near a telomere where the gene is transcribed and subsequently translated into the VSG protein. The VSG proteins are densely packed as monolayer or protective coat around the parasite. As the antibody response against the VSG mounts, some trypanosomes switch to expression of another VSG and proliferate, causing the next wave of disease. The trypanosomes that are ingested by the tsetse fly lose their coat and revert to a basic coat type.

Drugs used to treat late-stage disease are highly toxic; there is no prophylactic chemotherapy and little or no prospect of a vaccine. A collaboration to sequence and analyze the genomes of *Trypanosoma* species that cause sleeping sickness has been under way at institutions in Africa and South America. Findings of the initial sequencing shows that the most sequenced silent VSGs are defective or pseudogenes and almost all VSGs form arrays. More than 90% of the VSGs contain 1 or more 70 bp repeats upstream of the VSG (**FIGURE B19.2**). Hopefully, this new genomic information will help scientists identify novel potential drug targets that will accelerate the development of new therapeutics.

References

Berriman, M. *et al.*, 2005. The Genome of the African Trypanosome Trypanosoma brucei. *Science* 309:416–422.

Dubois, M. E., Demick, K. P., and Mansfield, J. M. 2005. Trypanosomes Expressing a Mosaic Variant Surface Glycoprotein Coat Escape Early Detection by the Immune System. *Infection and Immunity* 73:5;2690–2697.

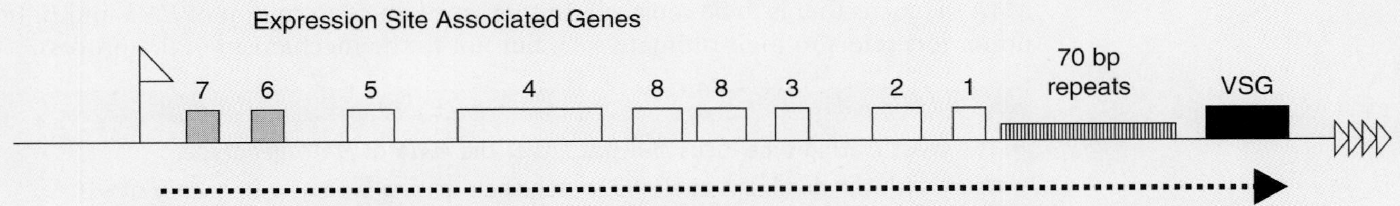

FIGURE B19.2: Organization of a bloodstream-form VSG expression site (BES) from *T. brucei.* The black box indicates the expressed VSG, located near the telomere. These sites also contain expression-site-associated genes (ESAGs) and characteristic 70-bp repeats. Adapted from Rudenko, G., *Clin. Sci.* 28 (2000): 536–540.

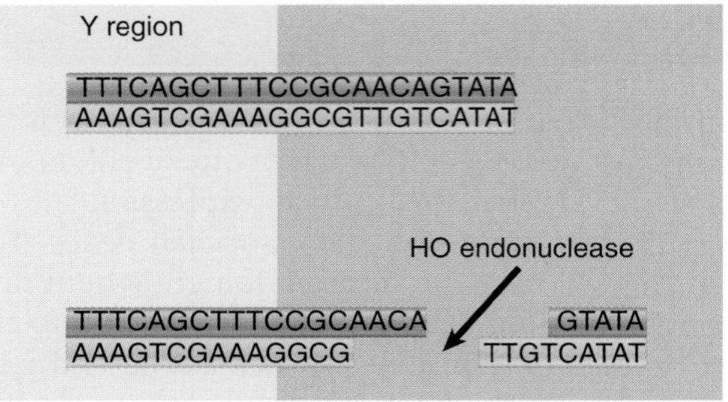

FIGURE 19.25 HO endonuclease cleaves *MAT* just to the right of the *Y* region, which generates sticky ends with a 4-base overhang.

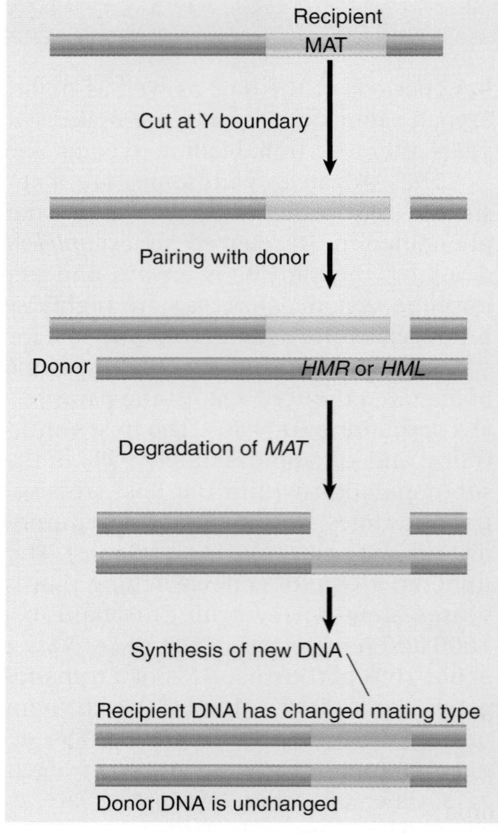

FIGURE 19.26 Cassette substitution is initiated by a double-strand break in the recipient (*MAT*) locus, and may involve pairing on either side of the *Y* region with the donor (*HMR* or *HML*) locus.

recognition site is relatively large for a nuclease, and it occurs only at the three mating-type cassettes. Only the *MAT* locus, and not the *HML* or *HMR* loci, is a target for the endonuclease. The same mechanisms that keep the silent cassettes from being transcribed also keep them inaccessible to the HO endonuclease. This inaccessibility ensures that switching is unidirectional.

The reaction triggered by the cleavage is illustrated schematically in **FIGURE 19.26** in terms of the general reaction between donor and recipient regions. In terms of the interactions of individual strands of DNA, it follows the scheme for recombination via a double-strand break drawn in Figure 19.6, and the stages following the initial cut require the enzymes involved in homologous recombination.

Like the double-strand break model for recombination, the process is initiated by *MAT*, the locus that is to be replaced. In this sense, the description of *HML* and *HMR* as donor loci refers to their ultimate role, but not to the mechanism of the process.

KEY CONCEPTS

- The yeast mating type locus *MAT* has either the *MAT***a** or *MAT*α genotype.
- Yeast with the dominant allele *HO* switch their mating type at a frequency of $\sim 10^{-6}$.
- The allele at *MAT* is called the active cassette.
- There are also two silent cassettes, *HML*α and *HMR***a**.
- Switching occurs if *MAT***a** is replaced by *HML*α or *MAT*α is replaced by *HMR***a**.
- Mating type switching is initiated by a double-strand break made at the *MAT* locus by the HO endonuclease.
- Mating type switching is achieved by a special homologous recombination event that copies information from *HML*α or *HMR***a** to the active *MAT* locus.

CONCEPT AND REASONING CHECKS

An α yeast strain is mutated so that the mating cassettes at *HML* and *HMR* are both **a** type cassettes. Can this strain still switch mating type? How many times? What sequences will be present at *HML* and *HMR* after a switching event?

19.10 Summary

Recombination involves the physical exchange of parts between corresponding DNA molecules. This results in a duplex DNA in which two regions of opposite parental origins are connected by a stretch of hybrid (heteroduplex) DNA, with one strand derived from each parent. Hybrid DNA can also be formed without recombination occurring between markers on either side.

Recombination is initiated by a double-strand break in DNA. The break is converted to a single-stranded region; a free single-stranded end then forms a heteroduplex with a donor sequence. The DNA in which the break occurs may actually incorporate the sequence of the chromosome that it invades, so the initiating DNA is called the recipient. Hotspots for recombination are sites where double-strand breaks are initiated. A gradient of gene conversion is determined by the likelihood that a sequence near the free end will be converted to a single strand; this decreases with distance from the break.

Recombination is initiated in yeast by Spo11, a topoisomerase-like enzyme that becomes linked to the free 5′ ends of DNA. The DSB is then processed by generating single-stranded DNA that can anneal with its complement in the other chromosome. Yeast mutations that block synaptonemal complex formation show that recombination is required for its formation. Formation of the synaptonemal complex may be initiated by double-strand breaks, and it may persist until recombination is completed. Mutations in components of the synaptonemal complex block its formation but do not prevent chromosome pairing, so homolog recognition is independent of recombination and synaptonemal complex formation.

The single-strand regions are generated by a complex of proteins and require the MRN/MRX complex. The single strands provide a substrate for proteins in the RecA family, which have the ability to synapse homologous DNA molecules by sponsoring a reaction in which a single strand from one molecule invades a duplex of the other molecule. Heteroduplex DNA is formed by displacing one of the original strands of the duplex. These actions create a recombination junction, which is resolved by resolvases such as the Ruv proteins. RuvA and RuvB act at a heteroduplex, and RuvC cleaves Holliday junctions.

Recombination, like replication and transcription, requires topological manipulation of DNA. Topoisomerases may relax (or introduce) supercoils in DNA, and are required to disentangle DNA molecules that have become catenated by recombination or by replication. Type I topoisomerases introduce a break in one strand of a DNA duplex; type II topoisomerases make double-strand breaks. The enzyme becomes linked to the DNA by a bond from tyrosine to either 5′ phosphate (type A enzymes) or 3′ phosphate (type B enzymes).

The enzymes involved in site-specific recombination have actions related to those of topoisomerases. Among this general class of recombinases, those concerned with phage integration form the subclass of integrases. The Cre/*lox* system uses two molecules of Cre to bind to each *lox* site, so that the recombining complex is a tetramer. This is one of the standard systems for inserting DNA into a foreign genome. Phage lambda integration requires the phage Int protein and host IHF protein and involves a precise breakage and reunion in the absence of any synthesis of DNA. Reaction in the reverse direction requires the phage protein Xis. Some integrases function by *cis*-cleavage, where the tyrosine that reacts with DNA in a half site is provided by the enzyme subunit bound to that half site; others function by *trans*-cleavage, for which a different protein subunit provides the tyrosine.

The yeast *S. cerevisiae* can propagate in either the haploid or diploid condition. Conversion between these states takes place by mating (fusion of haploid cells to give a diploid) and by sporulation (meiosis of diploids to give haploid spores). The ability to engage in these activities is determined by the mating type of the strain. The mating type is determined by the sequence of the *MAT* locus, and can be changed by a recombination event that substitutes a different sequence at this locus. The recombination event is initiated by a double-strand break—such as a homologous recombination event—but then the subsequent events ensure a unidirectional replacement of the sequence at the *MAT* locus.

Replacement is regulated so that *MAT**a*** is usually replaced by the sequence from *HMLα*, whereas *MATα* is usually replaced by the sequence from *HMR**a***. The endonuclease *HO* triggers the reaction by recognizing a unique target site at *MAT*.

CHAPTER QUESTIONS

1. What are the three requirements for RecA-mediated strand invasion?

A. ______________________

B. ______________________

C. ______________________

2. What four components are required for bacteriophage lambda integration into the bacterial chromosome, and where do these components come from?

A. ______________________

B. ______________________

C. ______________________

D. ______________________

3. Excision of an integrated lambda phage genome from the bacterial chromosome requires what three proteins?

A. ______________________

B. ______________________

C. ______________________

4. Recombination between homologous DNA sequences is called:

A. homologous recombination.

B. excision.

C. site-specific recombination.

D. transposition.

5. A classic example of site-specific recombination is:

A. HIV virus integration.

B. bacteriophage lambda integration.

C. meiotic recombination.

D. recombination repair.

6. When chromosomes begin to separate during meiosis, they can be observed to be held together at discrete sites called:

A. bivalents.

B. synaptonemal complexes.

C. chiasmata.

D. Holliday junctions.

7. Recombinant DNA molecules where the duplex of one DNA parent is covalently linked to the duplex of the other parent by a stretch of heteroduplex DNA are called:
   A. patch recombinants.
   B. splice recombinants.
   C. Holliday junctions.
   D. joint molecules.
8. What typically initiates homologous DNA recombination?
   A. a site-specific nick in one strand of duplex DNA
   B. a site-specific double-strand break in duplex DNA
   C. a random nick in one strand of duplex DNA
   D. a random double-strand break in duplex DNA
9. Which of the following enzymes remove positive supercoils that occur ahead of a replicating DNA polymerase?
   A. Topoisomerase I
   B. Topoisomerase II
   C. DNA gyrase
   D. Reverse gyrase
10. Formation of diploid *S. cerevisiae* involves mating between which pair of haploid cells?
    A. two type **a** cells
    B. two type α cells
    C. a type **a** cell and a type α cell
    D. any of the above combinations
11. What is responsible for initiation of yeast mating type switching?
    A. a double-strand break made at the *MAT* locus by the Spo11 endonuclease
    B. a single-strand nick at a silent mating cassette by the Spo11 endonuclease
    C. a single-strand nick made at the *MAT* locus by the HO endonuclease
    D. a double-strand break made at the *MAT* locus by the HO endonuclease

KEY TERMS

5′ end resection
***att* sites**
axial element
bivalent
branch migration
breakage and reunion
catenate
central element
chiasma (pl. chiasmata)
chromosome pairing
core sequence
D loop (displacement loop)
double-strand breaks (DSB)
excision
gene conversion
gyrase
heteroduplex DNA
Holliday junction
homologous recombination
hotspot
integration
joint molecule
lateral element
mating type cassette
patch recombinant
presynaptic filaments
prophage
recombinant joint
recombinase
recombination nodules (nodes)
resolution
reverse gyrase
single-strand invasion
sister chromatid
site-specific recombination
somatic recombination
splice recombinant
supercoiling
synapsis
synaptonemal complex
topoisomerase
type I topoisomerase
type II topoisomerase

FURTHER READING

Cromie G.A. and Smith, G.R. (2007). Branching out: meiotic recombination and its regulation. *Trends Cell Biol.* 17(9): 448–455.

A review of new views of the multiple pathways controlling meiotic recombination.

Haber, J.E. (1998). Mating-type gene switching in *Saccharomyces cerevisiae. Ann. Rev. Gen.* 32: 561–599.

An introduction to mating type switching in yeast.

San Filippo, J., Sung, P., and Klein, H. (2008). Mechanism of eukaryotic homologous recombination. *Ann. Rev. Biochem.* 77: 229–257.

A review of the mechanisms of homologous recombination, with a focus on the role of the Rad51 and Dmc1 recombinases.

20

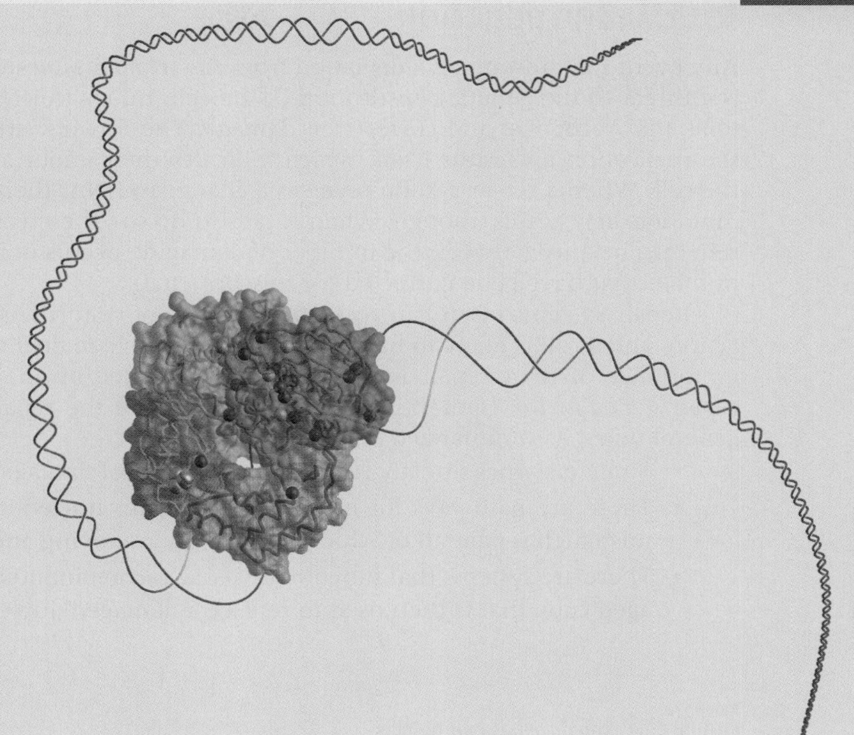

The XPD helicase, involved in both transcription and nucleotide excision repair, is shown unwinding DNA. Colors indicate different domains of XPD; the helicase domain is green. Photo courtesy of M. Pique, Li Fan, Jill Fuss, and John Tainer, The Scripps Research Institute and Lawrence Berkeley National Laboratory. More information available at Fan, L., et al., *Cell* 133 (2008): 789–800.

Repair Systems

CHAPTER OUTLINE

20.1 Introduction

Any event that introduces a deviation from the usual double-helical structure of DNA is a threat to the genetic constitution of the cell. Injury to DNA is minimized by systems that recognize and correct the damage. The repair systems are as complex as the replication apparatus itself, which indicates their importance for the survival of the cell. When a repair system reverses a change to DNA, there is no consequence. A mutation may result, though, when it fails to do so. The measured rate of mutation reflects a balance between the number of damaging events occurring in DNA and the number that have been corrected (or miscorrected).

Repair systems often can recognize a range of distortions in DNA as signals for action, and a cell is likely to have several systems able to deal with DNA damage. The importance of DNA repair in eukaryotes is indicated by the identification of >130 repair genes in the human genome. We can divide the repair systems into several general types, as summarized in **FIGURE 20.1**.

- Some enzymes directly reverse specific sorts of damage to DNA.
- There are pathways for base excision repair, nucleotide excision repair, and mismatch repair, all of which function by removing and replacing material.
- There are systems that function by using recombination to retrieve an undamaged copy that is then used to replace a damaged duplex sequence.

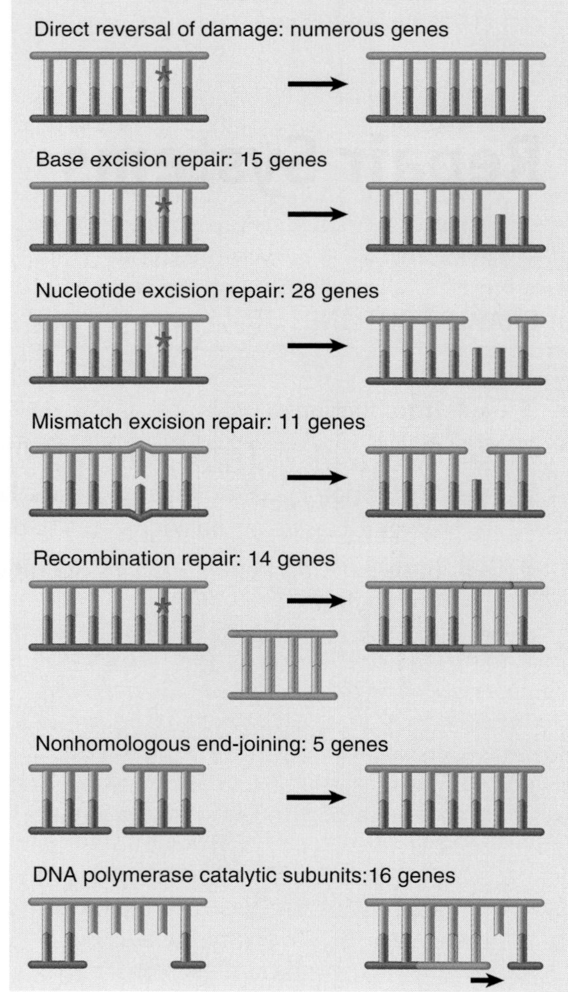

FIGURE 20.1 Repair genes can be classified into pathways that use different mechanisms to reverse or bypass damage to DNA.

- The nonhomologous end-joining pathway rejoins broken double-stranded ends.
- Several different DNA polymerases can resynthesize stretches of replacement DNA.

Direct repair is rare and involves the reversal or simple removal of the damage. One good example is **photoreactivation** of pyrimidine dimers, in which inappropriate covalent bonds between adjacent bases are reversed by a light-dependent enzyme. This system is widespread in nature, occurring in all but placental mammals, and appears to be especially important in plants. In *E. coli* it depends on the product of a single gene (*phr*) that codes for an enzyme called photolyase.

▸ **photoreactivation** A repair mechanism that uses a white light-dependent enzyme to split cyclobutane pyrimidine dimers formed by ultraviolet light.

Mismatches between the strands of DNA are one of the major targets for repair systems. **Mismatch repair** is accomplished by scrutinizing DNA for apposed bases that do not pair properly. Mismatches that arise during replication are corrected by distinguishing between the "new" and "old" strands and preferentially correcting the sequence of the newly synthesized strand. Other systems deal with mismatches generated by base conversions, such as the result of deamination. The importance of these systems is emphasized by the fact that cancer is caused in human populations by mutation of genes related to those involved in mismatch repair in yeast.

▸ **mismatch repair** Repair that corrects recently inserted bases that do not pair properly. The process preferentially corrects the sequence of the daughter strand by distinguishing the daughter strand and parental strand, sometimes on the basis of their states of methylation.

Mismatches are usually corrected by *excision repair,* which is initiated by a recognition enzyme that sees an actual damaged base or a change in the spatial path of DNA. There are two types of excision repair systems:

- *Base excision-repair (BER)* systems directly remove the damaged base and replace it in DNA. A good example is DNA uracil glycosylase, which removes uracils that are mispaired with guanines (see *Section 20.4, Base Excision Repair Systems Require Glycosylases*).
- *Nucleotide excision repair (NER)* systems excise a sequence that includes the damaged base(s); a new stretch of DNA is then synthesized to replace the excised material. **FIGURE 20.2** summarizes the main events in the operation of such a system. Such systems are common. Some recognize general damage to DNA; others act upon specific types of base damage. There are usually multiple excision repair systems in a single cell type.

Recombination-repair systems handle situations in which damage remains in a daughter molecule and replication has been forced to bypass the site, which typically creates a gap in the daughter strand. A retrieval system uses recombination to obtain another copy of the sequence from an undamaged source; the copy is then used to repair the gap.

A major feature in recombination and repair is the need to handle double-strand breaks (DSBs). DSBs initiate crossovers in homologous recombination. They can also be created by problems in replication, when they may trigger the use of recombination-repair systems. When DSBs are created by environmental damage (for example, by radiation damage) or are the result of the shortening of telomeres, they can cause mutations. In addition to recombination repair, DSBs can also be repaired by joining together nonhomologous DNA ends.

Mutations that affect the ability of *E. coli* cells to engage in DNA repair fall into groups that correspond to several repair pathways (not necessarily all independent). The major known pathways are the *uvr* excision repair system, the methyl-directed mismatch repair system, and the *recB* and *recF* recombination and recombination-repair pathways. The enzyme activities associated with these systems are endonucleases and exonucleases (important in removing damaged DNA); resolvases (endonucleases that act specifically on recombinant

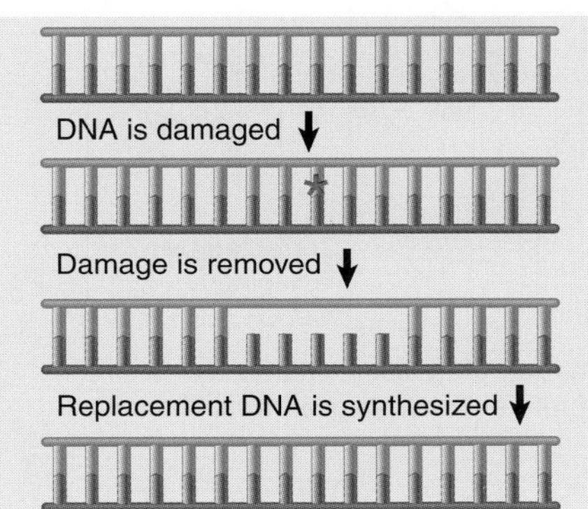
DNA is damaged
Damage is removed
Replacement DNA is synthesized

FIGURE 20.2 Excision repair directly replaces damaged DNA and then resynthesizes a replacement stretch for the damaged strand.

junctions); helicases to unwind DNA; and DNA polymerases to synthesize new DNA. Some of these enzyme activities are unique to particular repair pathways, whereas others participate in multiple pathways.

The replication apparatus devotes a lot of attention to quality control. DNA polymerases use proofreading to check the daughter strand sequence and to remove errors. Some of the repair systems are less accurate when they synthesize DNA to replace damaged material. For this reason, these systems have been known historically as *error-prone* systems.

20.2 Repair Systems Correct Damage to DNA

The types of damage that trigger repair systems can be divided into two general classes:

- *Single-base changes* affect the sequence of DNA but do not grossly distort its overall structure. They do not affect transcription or replication, when the strands of the DNA duplex are separated. Thus these changes exert their damaging effects on future generations through the consequences of the change in DNA sequence. The reason for this type of effect is the conversion of one base into another that is not properly paired with the partner base. Single-base changes may happen as the result of mutation of a base *in situ* or by replication errors. **FIGURE 20.3** shows that deamination of cytosine to uracil (spontaneously or by chemical mutagen) creates a mismatched U-G pair. **FIGURE 20.4** shows that a replication error might insert adenine instead of cytosine to create an A-G pair. Similar consequences could result from covalent addition of a small group to a base that modifies its ability to base pair. These changes may result in very minor structural distortion (as in the case of a U-G pair) or quite significant change (as in the case of an A-G pair), but the common feature is that the mismatch persists only until the next replication. Thus only limited time is available to repair the damage before it is made permanent by replication.
- *Structural distortions* may provide a physical impediment to replication or transcription. Introduction of covalent links between bases on one strand of DNA or between bases on opposite strands inhibits replication and transcription. **FIGURE 20.5** shows the example of ultraviolet (UV) irradiation, which introduces covalent bonds between two adjacent thymine bases and results in an intrastrand **pyrimidine dimer**. **FIGURE 20.6** shows that similar consequences can result from the addition of a bulky adduct to a base that distorts the structure

▸ **pyrimidine dimer** A dimer that forms when ultraviolet irradiation generates a covalent link directly between two adjacent pyrimidine bases in DNA. It blocks DNA replication and transcription.

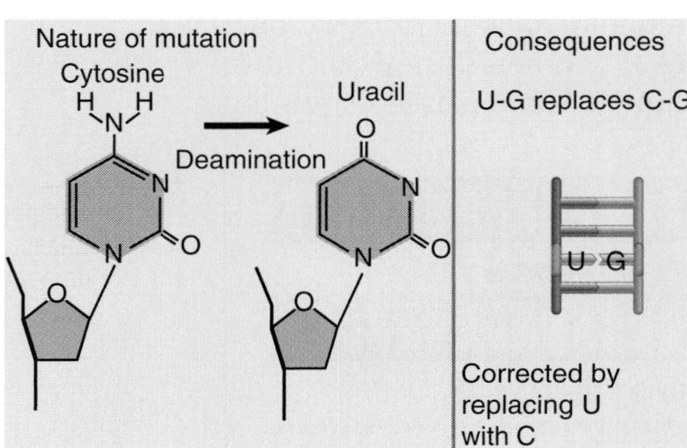

FIGURE 20.3 Deamination of cytosine creates a U-G base pair. Uracil is preferentially removed from the mismatched pair.

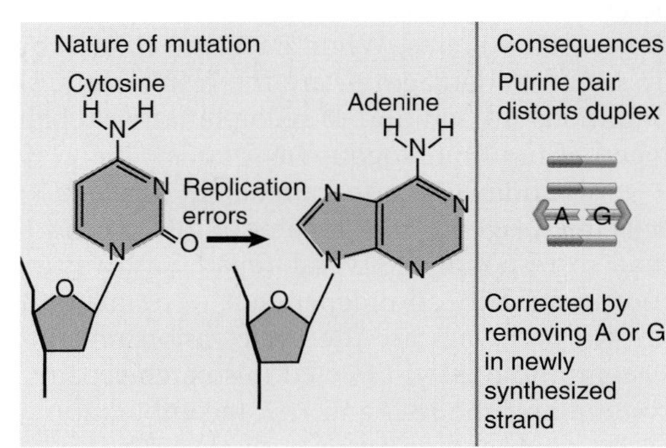

FIGURE 20.4 A replication error creates a mismatched pair that may be corrected by replacing one base; if uncorrected, a mutation is fixed in one daughter duplex.

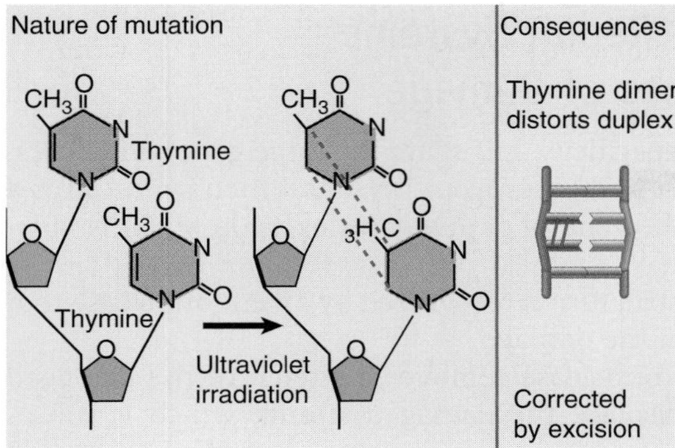

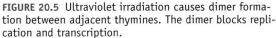

FIGURE 20.5 Ultraviolet irradiation causes dimer formation between adjacent thymines. The dimer blocks replication and transcription.

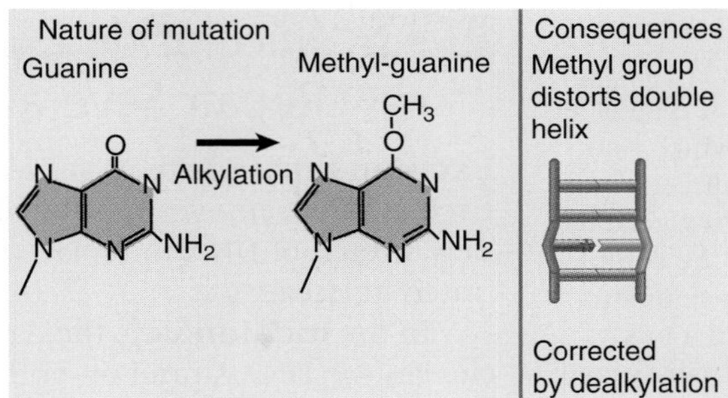

FIGURE 20.6 Methylation of a base distorts the double helix and causes mispairing at replication. Star indicates the methyl group.

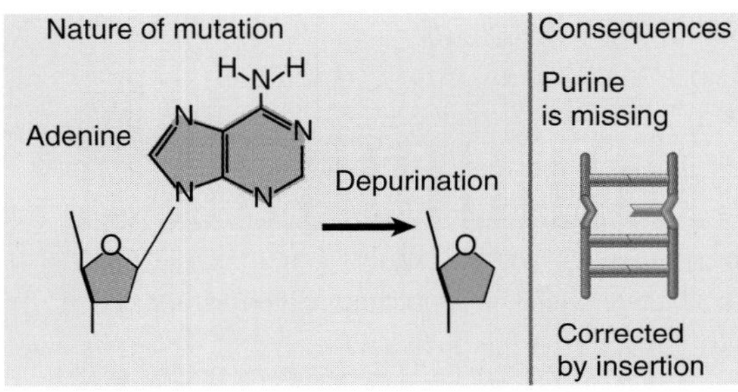

FIGURE 20.7 Depurination removes a base from DNA, blocking replication and transcription.

of the double helix. A single-strand nick or the removal of a base, as shown in FIGURE 20.7, prevents a strand from serving as a proper template for synthesis of RNA or DNA. The common feature in all these changes is that the damaged adduct remains in the DNA and continues to cause structural problems and/or induce mutations until it is removed.

When a repair system is eliminated, cells become exceedingly sensitive to agents that cause DNA damage, particularly the type of damage recognized by the missing system.

KEY CONCEPTS

- Repair systems recognize DNA sequences that do not conform to standard base pairs.
- Excision systems remove one strand of DNA at the site of damage and then replace it.
- Recombination-repair systems use recombination to replace the double-stranded region that has been damaged.
- All these systems are prone to introducing errors during the repair process.
- Photoreactivation is a nonmutagenic repair system that acts specifically on pyrimidine dimers.

CONCEPT AND REASONING CHECK

If a replication error that creates an A-G mismatch is not repaired before the next round of replication, what will be the sequences of the two daughter DNAs after replication?

20.3 Nucleotide Excision Repair Systems Repair Several Classes of Damage

- **excision repair** A type of repair system in which one strand of DNA is directly excised and then replaced by resynthesis using the complementary strand as template.
- **incision** A step in a mismatch excision-repair system in which an endonuclease recognizes the damaged area in the DNA and isolates it by cutting the DNA strand on both sides of the damage.
- **excision** The step in an excision-repair system that consists of removing a single-stranded stretch of DNA by the action of a 5′ to 3′ exonuclease.

Excision repair systems vary in their specificity, but share the same general features. Each system removes mispaired or damaged bases from DNA and then synthesizes a new stretch of DNA to replace them. The general pathway for excision repair is illustrated in **FIGURE 20.8**.

In the **incision** step, the damaged structure is recognized by an endonuclease that cleaves the DNA strand on both sides of the damage.

In the **excision** step, a 5′ to 3′ exonuclease removes a stretch of the damaged strand. Alternatively, a helicase can displace the damaged strand, which is subsequently degraded.

In the *synthesis* step, the resulting single-stranded region serves as a template for a DNA polymerase to synthesize a replacement for the excised sequence. (Synthesis of the new strand can be associated with removal of the old strand, in one coordinated action.) Finally, DNA ligase covalently links the 3′ end of the new DNA strand to the original DNA.

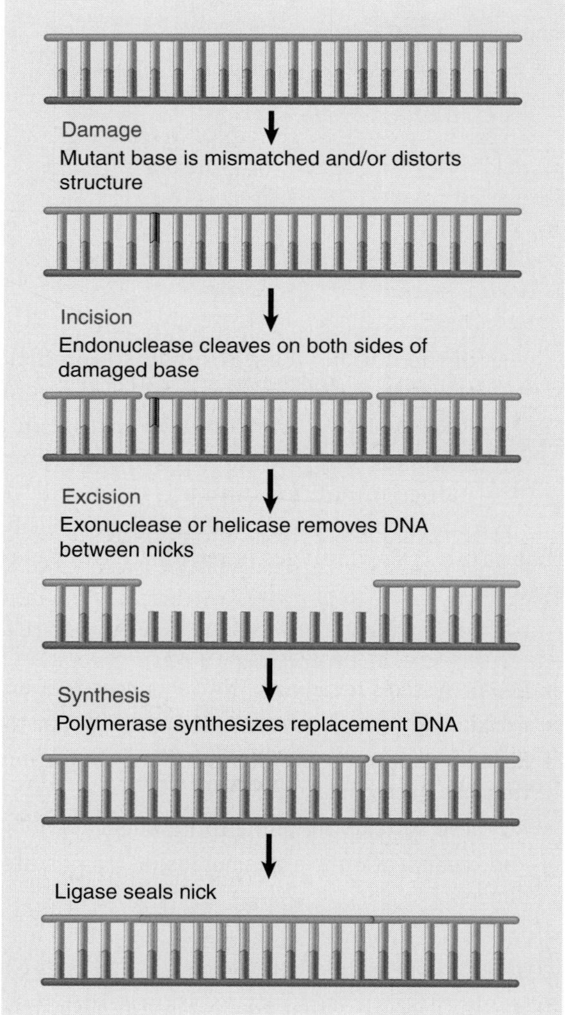

Damage
Mutant base is mismatched and/or distorts structure
Incision
Endonuclease cleaves on both sides of damaged base
Excision
Exonuclease or helicase removes DNA between nicks
Synthesis
Polymerase synthesizes replacement DNA
Ligase seals nick

FIGURE 20.8 Excision repair removes and replaces a stretch of DNA that includes the damaged base(s).

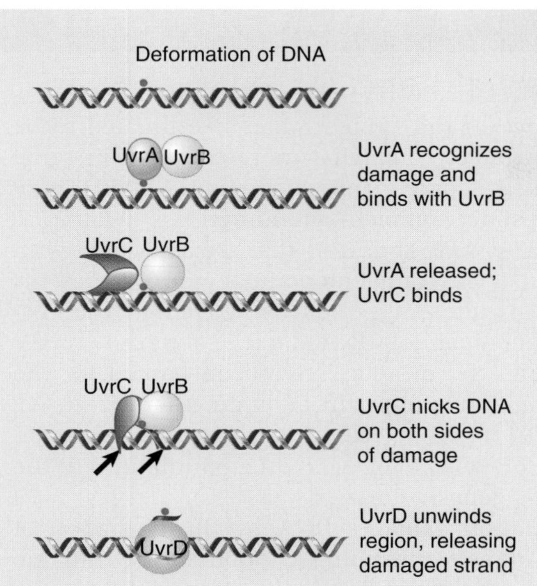
Deformation of DNA
UvrA UvrB
UvrA recognizes damage and binds with UvrB
UvrC UvrB
UvrA released; UvrC binds
UvrC UvrB
UvrC nicks DNA on both sides of damage
UvrD
UvrD unwinds region, releasing damaged strand

FIGURE 20.9 The Uvr system operates in stages in which UvrAB recognizes damage, UvrBC nicks the DNA, and UvrD unwinds the marked region.

The *E. coli uvr* system of excision repair includes three genes, *uvrA, -B,* and *-C,* which code for the components of a repair endonuclease. It functions in the stages indicated in **FIGURE 20.9**. First, a UvrAB dimer recognizes pyrimidine dimers and other bulky lesions. Next, UvrA dissociates (this requires adenosine triphosphate [ATP]), and UvrC joins UvrB. The UvrBC complex makes an incision on each side—one that is seven nucleotides from the 5′ side of the damaged site and another that is three to four nucleotides away from the 3′ side. This also requires ATP. UvrD is a helicase that helps to unwind the DNA to allow release of the single strand between the two cuts. The enzyme that excises the damaged strand is DNA polymerase I. The enzyme involved in the repair synthesis also is likely to be DNA polymerase I (although DNA polymerases II and III can substitute for it).

UvrABC repair accounts for virtually all of the excision repair events in *E. coli.* In almost all (99%) of cases, the average length of replaced DNA is ~12 nucleotides. (For this reason, the process is sometimes described as *short-patch repair*.) The remaining 1% of cases involve the replacement of stretches of DNA mostly ~1500 nucleotides long, but extending as much as >9000 nucleotides (sometimes called *long-patch repair*). We do not know why some events trigger the long-patch rather than the short-patch mode.

The general principle of excision repair in eukaryotic cells is similar to that of bacteria. Bulky lesions, such as those created by UV damage, are also recognized and repaired by a nucleotide excision repair system. The critical role of mammalian nucleotide excision repair is seen in certain human hereditary disorders. The best investigated of these is **xeroderma pigmentosum (XP)**, a recessive disease resulting in hypersensitivity to sunlight, and in particular, ultraviolet light. The deficiency results in skin disorders and cancer predisposition (for more discussion, see the Medical Applications box on pages 482 and 483).

xeroderma pigmentosum (XP) A disease caused by mutation in one of the XP genes that results in hypersensitivity to sunlight (particularly ultraviolet light), skin disorders, and cancer predisposition.

The disease is caused by a deficiency in nucleotide excision repair. XP patients cannot excise pyrimidine dimers and other bulky adducts. Mutations occur in one of eight genes called *XP-A* to *XP-G*. A protein complex that includes products of several of the *XP* genes is responsible for excision of thymine dimers. There are actually two major

MEDICAL APPLICATIONS

Xeroderma Pigmentosum

Xeroderma pigmentosum (XP) is an autosomal recessive disorder associated with extreme ultraviolet (UV) light sensitivity, increased risk of skin cancer, and neurological degeneration. The name *xeroderma pigmentosum* literally means "dry pigmented skin." XP and two other related disorders, Cockayne syndrome and trichothiodystrophy, are associated with defects in DNA nucleotide excision repair (NER). XP patients exhibit sun-induced freckles and other skin pigmentation as infants, which is rare in the general population, and this is often the basis of the clinical diagnosis. About half the patients with XP are extremely light sensitive and many exhibit sunburn and blistering with minimal sun exposure; others tan normally but still exhibit pigmentation on sun-exposed skin. With continued sun exposure the skin becomes dry and looks prematurely aged. Children with XP frequently develop precancerous lesions, and have a 1000-fold increased risk of developing either melanoma or non-melanoma skin cancer before age 20. The average age of XP patients that have skin cancer is under 10 years old, which is 50 years earlier than the average age of the appearance of skin cancers in the general population. With continued exposure to sunlight, XP patients may develop chronic UV-induced eye inflammation, corneal damage and the loss of eyelashes. About 30% of XP patients also exhibit neurological defects including progressive mental retardation, loss of deep tendon reflexes, and loss of high-frequency hearing. Patients may also develop difficulties in walking and swallowing over time.

XP is a relatively rare genetic disorder; the overall prevalence in the United States is estimated to be 1/1,000,000 but is somewhat more common in some populations. In Japan the prevalence is 1/22,000. There are seven distinct complementation groups (*XPA, XPB, XPC, XPD, XPE, XPF,* and *XPG*) that are associated with this disease. Each complementation group represents mutations in different proteins involved in the recognition and repair of DNA damage induced by UV radiation, and the proteins have been named for the complementation groups. In addition, there is an XP variant (XPV) that is defective in post-replication repair. Approximately 50% of all XP is due to mutations in the *XPA* and *XPC* genes.

UV radiation damages DNA by the formation of pyrimidine dimers, including cyclobutane pyrimidine dimers and pyrimidine-pyrimidone dimers (also known as 6,4-photoproducts). These photoproducts are normally recognized, excised, and repaired by the nucleotide excision repair (NER) pathway (see Figure 20.10). The NER pathway also removes other bulky DNA modifications induced by other carcinogens. DNA repair by the NER pathway differs depending on whether or not the damaged DNA is in a region that is actively transcribed. Actively transcribed regions of the genome are repaired more rapidly and are recognized through a distinct complex of proteins of the transcription-coupled repair (TCR) pathway. Damaged DNA is detected by the presence of stalled RNA polymerase, which, in turn, recruits DNA binding proteins CSA and CSB. In contrast, DNA damage in the non-transcribed regions of the genome is detected by the binding of proteins of the global genome repair (GGR) pathway, XPE and XPC pro-

pathways of nucleotide excision repair in eukaryotes, illustrated in **FIGURE 20.10**. The major difference between the two pathways is how the damage is initially recognized. In *global genome repair*, a protein called XPC detects the damage and initiates the repair pathway. XPC can recognize damage anywhere in the genome. On the other hand, *transcription-coupled repair*, as the name suggests, is responsible for repairing lesions that occur in the transcribed strand of active genes. In this case, the damage is recognized by RNA polymerase II itself.

Both pathways then use a common set of proteins to effect the repair. The strands of DNA are unwound for ~20 bp around the damaged site. This action is performed by the helicase activity of the transcription factor $TF_{II}H$, itself a large complex, which includes the products of two *XP* genes (*XPB* and *XPD*). Then cleavages are made on either side of the lesion by endonucleases encoded by the *XPG* and *XPH* genes. The single-stranded stretch including the damaged bases can then be replaced by new synthesis.

In cases where replication encounters a thymine dimer that has not been removed, replication requires DNA polymerase η activity in order to proceed past the dimer. This polymerase is encoded by *XPV.* This bypass mechanism allows cell division to proceed even in the presence of unrepaired damage, but this is generally a last resort as cells prefer to put a hold on cell division until all damage is repaired.

teins. After the recognition of the DNA lesions by these two distinct mechanisms, the pathways converge and the damaged DNA is unwound, excised, and repaired by common proteins. XPA protein binds the flagged, damaged DNA and recruits two additional proteins, XPB and XPD. Both of these proteins are helicases that are components of the transcription factor $TF_{II}H$ (and thus already present at sites of damage recognized by stalled RNA polymerase). Once the DNA is unwound, exonucleases are recruited to regions surrounding the damaged DNA. XPF and XPG are 5′ and 3′ exonucleases, respectively, and cut on either side of the damaged DNA to generate a fragment of approximately 30 nucleotides of single-stranded DNA. The resulting gap in the genomic DNA is repaired by *de novo* DNA synthesis. XPV is not a component of the NER pathway but is a member of the Y family of error-prone DNA polymerases that can replicate DNA containing photoproducts such as pyrimidine dimers, which cannot be replicated by the polymerases of the replication fork.

There is a very interesting genotype/phenotype correlation between mutations in the NER pathway and the sun sensitivity, cancer risk, and neurological manifestations of the XP disease. Patients with mutations in the proteins that are unique to the GGR pathway (XPC and XPE) and mutations in XPV, exhibit XP without neurological defects. Patients with mutations in the other XP proteins that are common to GGR and TCR exhibit XP with the associated neurological defects. Interestingly, mutations in CSA and CSB, the DNA binding proteins in the TCR pathway, result in the disorder Cockayne syndrome. Cockayne syndrome is characterized by sun sensitivity, mental retardation, and developmental retardation; however, there is no increased risk of skin cancer. Cockayne syndrome can also result from specific mutations in the XPB and XPD helicases. These mutations are at different sites from those that result in XP.

Trichothiodystrophy is another disease resulting from defects in proteins involved in NER; specifically, another subset of mutations in XPB and XPD which may affect their roles as components of the transcription factor $TF_{II}H$. This disorder is associated with photosensitivity, brittle hair and nails, and growth retardation but not with sun-induced pigmentation or cancer, and these patients appear to have specific transcriptional defects rather than repair deficiency. These observations of XP and other NER defective disorders has led to the suggestion that the neurological defects associated with these diseases may arise due to disruption of the TCR pathway, resulting in a blockage of transcription and the apoptosis of neural cells. Alternatively, the cancer risk associated with XP, but not the other two disorders, may arise due to genomic instability resulting from the disruption of the GGR pathway in cells.

There is no cure or any treatment for XP, other than aggressive avoidance of UV exposure. This includes limiting time spent outdoors or in the car, wearing protective clothing, using high SPF sunscreen at all times and wearing UV absorbing glasses. Patients need to be continuously monitored and treated for precancerous and cancerous skin lesions. Recently, several promising clinical trials have tested the effects of topical lotions to treat photosensitivity in XP patients. These lotions contain liposomes designed to deliver DNA repair enzymes to the site of application, replacing the absent repair functions in the skin cells and reducing the risk of skin cancer.

KEY CONCEPTS

- The Uvr system makes incisions ~12 bases apart on both sides of damaged DNA, removes the DNA between them, and resynthesizes new DNA.
- Xeroderma pigmentosum (XP) is a human disease caused by mutations in any one of several nucleotide excision repair genes.
- A complex of proteins including XP products and the transcription factor $TF_{II}H$ provides a human excision repair mechanism.
- Global genome repair recognizes damage anywhere in the genome.
- Transcriptionally active genes are preferentially repaired via transcription-coupled repair.
- Global genome repair and transcription-coupled repair differ in their mechanisms of damage recognition (XPC vs. RNA polymerase II).

CONCEPT AND REASONING CHECK

Transcription-coupled repair occurs more rapidly than global genome repair. Why?

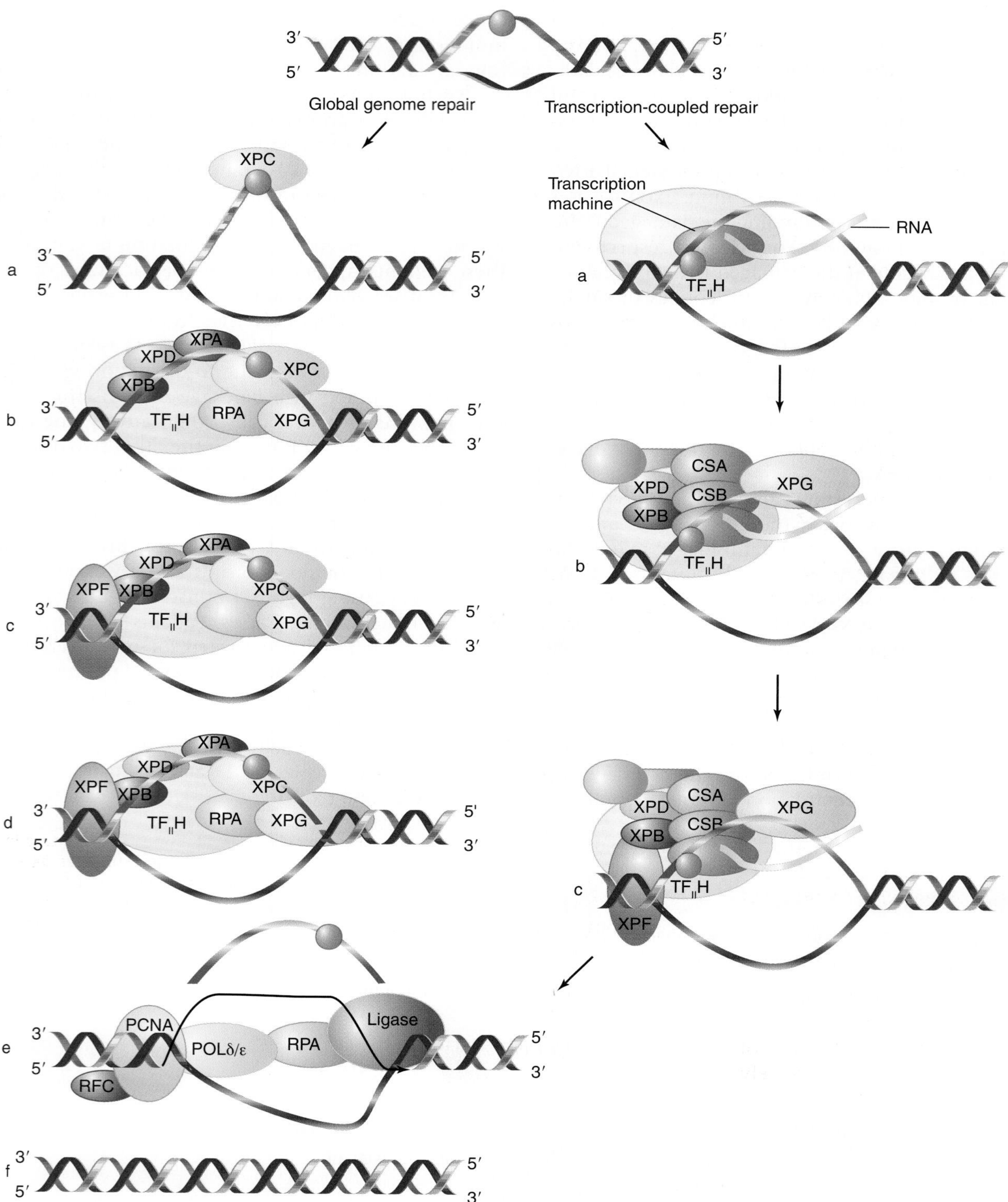

FIGURE 20.10 Nucleotide excision repair occurs via two major pathways: global genome repair, in which XPC recognizes damage anywhere in the genome, and transcription-coupled repair, in which the transcribed strand of active genes is preferentially repaired, and the damage is recognized by an elongating RNA polymerase. Adapted from Friedberg, E. C., et al., *Nature Rev. Cancer* 1 (2001): 22–23.

20.4 Base Excision Repair Systems Require Glycosylases

Base excision repair is similar to the nucleotide excision repair pathways described in the previous section. The process usually starts in a different way, however, with the removal of an *individual* damaged base. This serves as the trigger to activate the enzymes that excise and replace a stretch of DNA, including the damaged site.

Enzymes that remove bases from DNA are called **glycosylases** and **lyases**. FIGURE 20.11 shows that a glycosylase cleaves the bond between the damaged or mismatched base and the deoxyribose. FIGURE 20.12 shows that some glycosylases are also lyases that can take the reaction a stage further by using an amino (NH_2) group to attack the deoxyribose ring. This is usually followed by a reaction that introduces a nick into the polynucleotide chain. FIGURE 20.13 shows that the exact form of the pathway depends on whether the damaged base is removed by a glycosylase or lyase.

▸ **glycosylase** A repair enzyme that removes damaged bases by cleaving the bond between the base and the sugar.

▸ **lyase** A repair enzyme (usually also a glycosylase) that opens the sugar ring at the site of a damaged base.

Glycosylase action is followed by the endonuclease APE1, which cleaves the polynucleotide chain on the 5′ side. This in turn attracts a replication complex including the DNA polymerase δ/ε and ancillary components, which performs a short synthesis reaction extending for two to ten nucleotides. The displaced material is removed by the endonuclease FEN1. The enzyme ligase-1 seals the chain. This is called the *long-patch* pathway. (Note these names refer to mammalian enzymes, but the descriptions are generally applicable for all eukaryotes.)

When the initial removal involves lyase action, the endonuclease APE1 instead recruits DNA polymerase β to replace a single nucleotide. The nick is then sealed by the ligase XRCC1/ligase-3. This is called the *short-patch* pathway.

Several enzymes that remove or modify individual bases in DNA use a remarkable reaction in which a base is "flipped" out of the double helix. This type of interaction was first demonstrated for methyltransferases—enzymes that add a methyl group to cytosine in DNA. This base flipping mechanism places the base directly into the active site of the enzyme, where it can be modified and returned to its normal position in the helix, or, in the case of DNA damage, it can be immediately excised. Alkylated bases (typically in which a methyl group has been added to a base) are removed by this mechanism. A human enzyme, alkyladenine DNA glycosylase (AAG), recognizes and removes a variety of alkylated substrates,

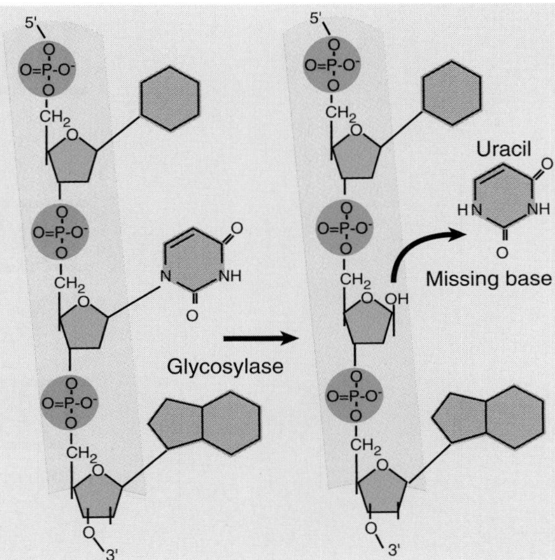

FIGURE 20.11 A glycosylase removes a base from DNA by cleaving the bond to the deoxyribose.

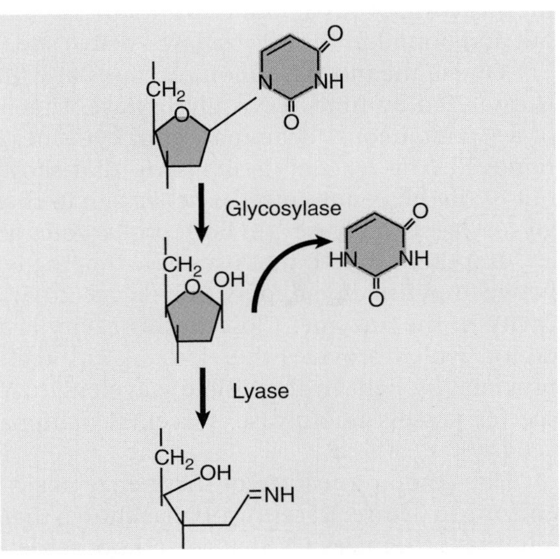

FIGURE 20.12 A glycosylase hydrolyzes the bond between base and deoxyribose (using H_2O), but a lyase takes the reaction further by opening the sugar ring (using NH_2).

FIGURE 20.13 Base removal by glycosylase or lyase action triggers mammalian excision-repair pathways.

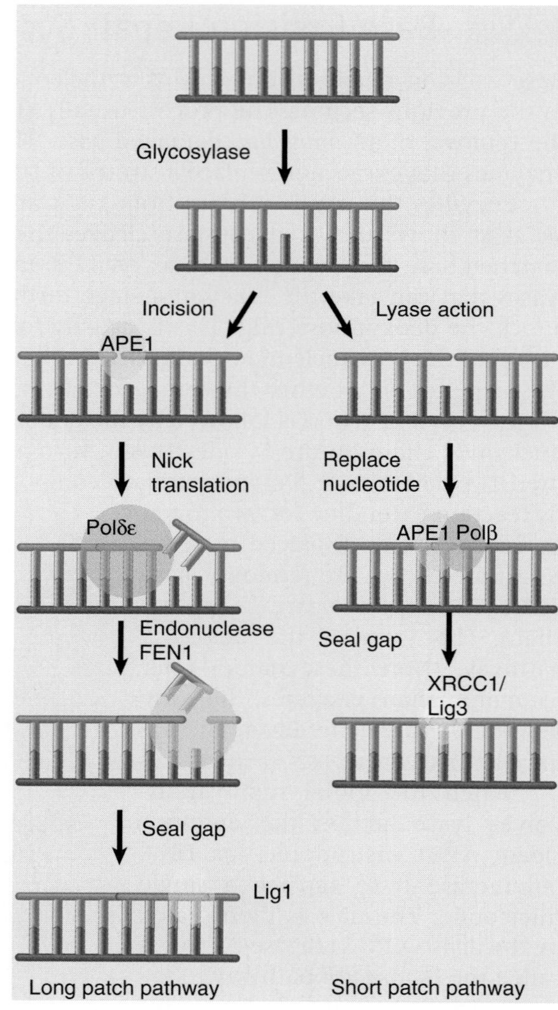

including 3-methyladenine, 7-methylguanine, and hypoxanthine. **FIGURE 20.14** shows the structure of AAG bound to a methylated adenine, in which the adenine is flipped out and bound in the glycosylase's active site.

One of the most common reactions in which a base is directly removed from DNA is catalyzed by uracil-DNA glycosylase. Uracil typically only occurs in DNA because of a (spontaneous) deamination of cytosine. It is recognized by the glycosylase and removed. The reaction is similar to that shown in Figure 20.14: the uracil is flipped out of the helix and into the active site in the glycosylase. It appears that most or all glycosylases and lyases (in both prokaryotes and eukaryotes) work in a similar way.

Another enzyme that uses base flipping is the photolyase that reverses the bonds between pyrimidine dimers (see Figure 20.5). The pyrimidine dimer is flipped into a cavity in the enzyme. Close to this cavity is an active site that contains an electron donor, which provides the electrons to break the bonds. Energy for the reaction is provided by light in the visible wavelength. While most prokaryotic and eukaryotic species possess photolyase, placental mammals (but not marsupials) have lost this activity.

The common feature of these enzymes is the flipping of the target base into the enzyme structure. Recent work has shown that Rad4, the yeast XPC homolog (the protein that recognizes UV damage and other lesions during nucleotide excision repair) uses an interesting variation on this theme. Rad4 flips out the two adenine bases that are *complementary* to the linked thymines in a pyrimidine dimer, rather than flipped out the damaged pyrimidine dimer itself. In fact, it is believed that the ease with

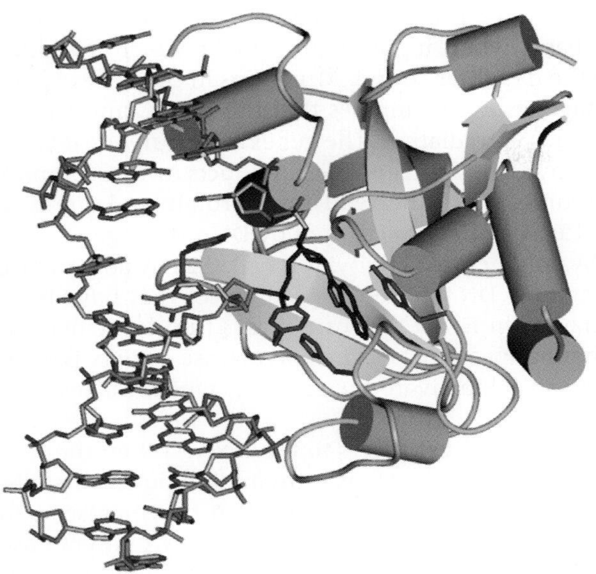

FIGURE 20.14 Crystal structure of the DNA repair enzyme alkyladenine DNA glycosylase (AAG) bound to a damaged base (3-methyladenine). The base (black) is flipped out of the DNA double helix (blue) and into AAG's active site (orange and green). Photo reproduced from Lau, A. Y., et al., *Proc. Natl. Acad. Sci. USA* 97 (2000): 13573–13578. Copyright 2000 National Academy of Sciences, USA. Photo courtesy of Tom Ellenberger, Washington University School of Medicine.

which these unpaired adenines are flipped out is actually the mechanism by which Rad4 detects the damage. Thus in this case, the target for the subsequent repair is not directly recognized by Rad4 at all, and instead the protein uses flipping as an indirect mechanism to detect the loss of a normal base-paired DNA double helix.

KEY CONCEPTS

- Base excision repair is triggered by directly removing a damaged base from DNA.
- Base removal triggers the removal and replacement of a stretch of polynucleotides.
- The nature of the base removal reaction determines which of two pathways for excision repair is activated.
- The polδ/ε pathway replaces a long polynucleotide stretch; the polβ pathway replaces a short stretch.
- Uracil and alkylated bases are recognized by glycosylases and removed directly from DNA.
- Glycosylases and photolyase act by flipping the base out of the double helix, where, depending on the reaction, it is either removed or modified and returned to the helix.

CONCEPT AND REASONING CHECK

Glycosylases flip out damaged bases, but Rad4/XPC flips out *undamaged* bases across from sites of damage. Given the roles of these different proteins, why does this difference make sense?

20.5 Error-Prone Repair

The existence of repair systems that engage in DNA synthesis raises the question of whether their quality control is comparable with that of DNA replication. As far as we know, most systems, including *uvr*-controlled excision repair, do not differ significantly from DNA replication in the frequency of mistakes. **Error-prone synthesis** of DNA, however, occurs in *E. coli* under certain circumstances.

▸ **error-prone synthesis** A repair process in which noncomplementary bases are incorporated into the daughter strand.

The error-prone pathway, also known as *translesion synthesis*, was first observed when it was found that the repair of damaged λ phage DNA is accompanied by the induction of mutations if the phage is introduced into cells that had previously been irradiated with UV. This suggests that the UV irradiation of the host has activated functions that generate mutations when repairing λ DNA. The mutagenic response also operates on the bacterial host DNA.

What is the actual error-prone activity? It is a specialized DNA polymerase that inserts random (usually incorrect) bases when it passes any site at which it cannot insert complementary base pairs in the daughter strand. Mutations in the genes *umuD* and *umuC* abolish UV-induced mutagenesis. This implies that the UmuC and UmuD proteins cause mutations to occur after UV irradiation. The genes constitute the *umuDC* operon, whose expression is induced by DNA damage. Their products form a complex UmuD'$_2$C, which consists of two subunits of a truncated UmuD protein and one subunit of UmuC. UmuD is cleaved by RecA, which is activated by DNA damage.

The UmuD'$_2$C complex has DNA polymerase activity. It is called DNA polymerase V, and is responsible for synthesizing new DNA to replace sequences that have been damaged by UV. This is the only enzyme in *E. coli* that can bypass the classic pyrimidine dimers produced by UV (or other bulky adducts). The polymerase activity is error prone. Mutations in either *umuC* or *umuD* inactivate the enzyme, which makes high doses of UV irradiation lethal.

How does an alternative DNA polymerase get access to the DNA? When the replicase (DNA polymerase III) encounters a block, such as a thymidine dimer, it stalls. It is then displaced from the replication fork and replaced by DNA polymerase V. In fact, DNA polymerase V uses some of the same ancillary proteins as DNA polymerase III. The same situation is true for DNA polymerase IV, the product of *dinB*, which is another enzyme that acts on damaged DNA.

DNA polymerases IV and V are part of a larger family of *translesion polymerases*, which includes eukaryotic DNA polymerases, whose members are specialized for repairing damaged DNA. In addition to the *dinB* and *umuCD* genes that code for DNA polymerases IV and V in *E. coli*, this family also includes the RAD30 gene coding for DNA polymerase η of *S. cerevisiae*, and the *XPV* gene described previously that encodes the human homolog. A difference between the bacterial and eukaryotic enzymes is that the latter are not error prone at thymine dimers: they accurately introduce an A-A pair opposite a T-T dimer. When they replicate through other sites of damage, however, they are more prone to introduce errors.

KEY CONCEPTS

- Damaged DNA that has not been repaired causes DNA polymerase III to stall during replication.
- DNA polymerase V (coded by *umuCD*) or DNA polymerase IV (coded by *dinB*) can synthesize a complement to the damaged strand.
- The DNA synthesized by repair DNA polymerases often has errors in its sequence.

CONCEPT AND REASONING CHECK

Why is error-prone repair useful, even if it introduces mutations?

20.6 Controlling the Direction of Mismatch Repair

▸ **mutator** A mutation or a mutated gene that increases the basal level of mutation. Such genes often code for proteins that are involved in repairing damaged DNA.

Genes whose products are involved in controlling the fidelity of DNA synthesis during either replication or repair may be identified by mutations that have a **mutator** phenotype. A mutator mutant has an increased frequency of spontaneous mutation. If identified originally by the mutator phenotype, a gene is described as *mut;* often, though, a *mut* gene is later found to be equivalent with a known replication or repair activity.

Many *mut* genes turn out to be components of *mismatch-repair systems*. Failure to remove a damaged or mispaired base before replication allows it to induce a mutation. Functions in this group include the *Dam* methylase that identifies the target for repair, and enzymes that participate directly or indirectly in the removal of particular types of damage (*MutH, -S, -L, and -Y*).

When a structural distortion is removed from DNA, the wild-type sequence is restored. In most cases, the distortion is due to the creation of a base that is not naturally found in DNA, and that is therefore recognized and removed by the repair system.

A problem arises if the target for repair is a mispaired partnership of (normal) bases created when one was mutated, or misinserted during replication. The repair system has no intrinsic means of knowing which is the wild-type base and which is the mutant! All it sees are two improperly paired bases, either of which can provide the target for excision repair.

If the mutated base is excised, the wild-type sequence is restored. If it happens to be the original (wild-type) base that is excised, though, the new (mutant) sequence becomes fixed. Often, however, the direction of excision repair is not random, but instead is biased in a way that is likely to lead to restoration of the wild-type sequence.

Some precautions are taken to direct repair in the right direction. For example, for cases such as the deamination of 5-methylcytosine to thymine, there is a special system to restore the proper sequence (see also *Section 1.13, Mutations Are Concentrated at Hotspots*). The deamination generates a G-T pair, and the system that acts on such pairs has a bias to correct them to G-C pairs (rather than to A-T pairs). The system that undertakes this reaction includes the MutL and MutS products that remove T from both G-T and C-T mismatches.

When mismatch errors occur during replication in *E. coli*, it is possible to distinguish the original strand of DNA. Immediately after replication of methylated DNA, only the original parental strand carries methyl groups. In the period during which the newly synthesized strand awaits the introduction of methyl groups, the two strands can be distinguished.

This provides the basis for a system to correct replication errors. The *dam* gene codes for a methylase whose target is the adenine in the sequence $^{\text{GATC}}_{\text{CTAG}}$ (see Figure 15.7). The hemimethylated state is used to distinguish replicated origins from nonreplicated origins. The same target sites are used by a replication-related mismatch repair system.

FIGURE 20.15 shows that DNA containing mismatched base partners is repaired preferentially by excising the strand that lacks the methylation. The excision is quite extensive; mismatches can be repaired preferentially for >1 kb around a GATC site. The result is that the newly synthesized strand is corrected to the sequence of the parental strand.

E. coli dam⁻ mutants show an increased rate of spontaneous mutation. This repair system therefore helps reduce the number of mutations caused by errors in replication. It consists of several proteins, coded by *mut* genes. MutS binds to the mismatch and is joined by MutL. MutS can use two DNA-binding sites, as illustrated in **FIGURE 20.16**. The first specifically recognizes mismatches. The second is not specific for sequence or structure, and is used to translocate along DNA until a GATC sequence is encountered. Hydrolysis of ATP is used to drive the translocation. MutS is bound to both the mismatch site and to DNA as it translocates, and as a result it creates a loop in the DNA.

Recognition of the GATC sequence causes the MutH endonuclease to bind to MutSL. The endonuclease then cleaves the unmethylated strand. This strand is then excised from the GATC site to the mismatch site. The excision can occur in either the 5′ to

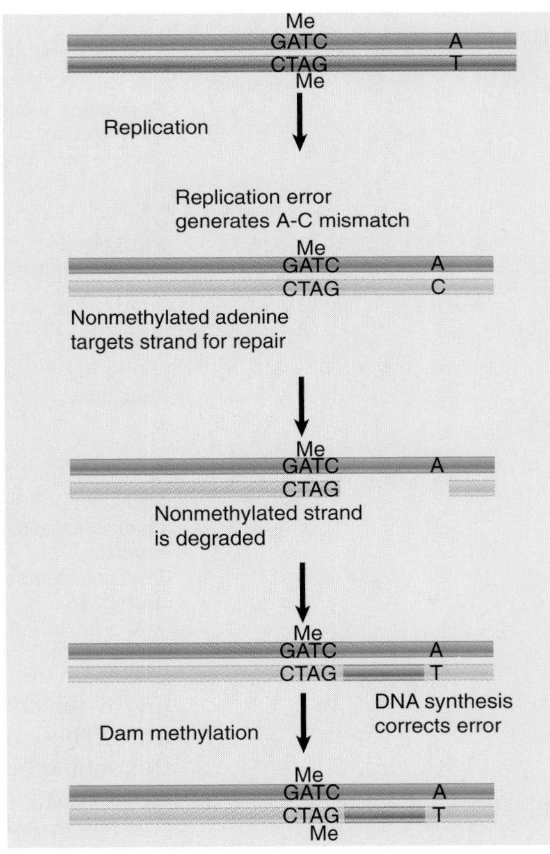

FIGURE 20.15 GATC sequences are targets for the Dam methylase after replication. During the period before this methylation occurs, the nonmethylated strand is the target for repair of mismatched bases.

3′ direction (using RecJ or exonuclease VII) or in the 3′ to 5′ direction (using exonuclease I), and is assisted by the helicase UvrD. A new DNA strand is then synthesized by DNA polymerase III.

Eukaryotic cells have systems homologous to the *E. coli mut* system. Msh2 ("MutS homolog 2") provides a scaffold for the apparatus that recognizes mismatches. Msh3 and Msh6 provide specificity factors. In addition to repairing single-base mismatches, they are responsible for repairing mismatches that arise as the result of replication slippage. The Msh2-Msh3 complex binds mismatched insertion/deletion loops, while the Msh2-Msh6 complex binds to single-base mismatches. Other proteins, including MutL homologs, are required for the repair process itself. Surprisingly, even though higher eukaryotes possess DNA methylation, eukaryotic mismatch repair systems do not use DNA methylation to select the daughter strand for repair. It is not known how eukaryotes recognize the daughter strand during mismatch repair, but MutSL homologs interact directly with the replication machinery.

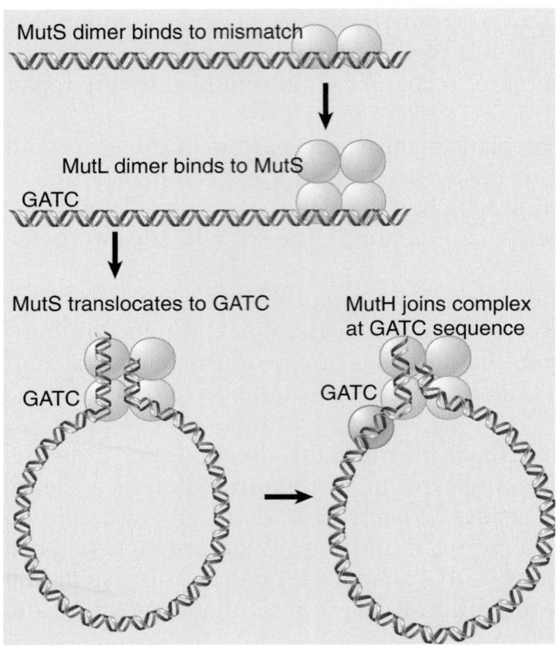
MutS dimer binds to mismatch
MutL dimer binds to MutS
GATC
MutS translocates to GATC
MutH joins complex at GATC sequence
GATC
GATC

FIGURE 20.16 MutS recognizes a mismatch and translocates to a GATC site. MutH cleaves the unmethylated strand at the GATC. Endonucleases degrade the strand from the GATC to the mismatch site.

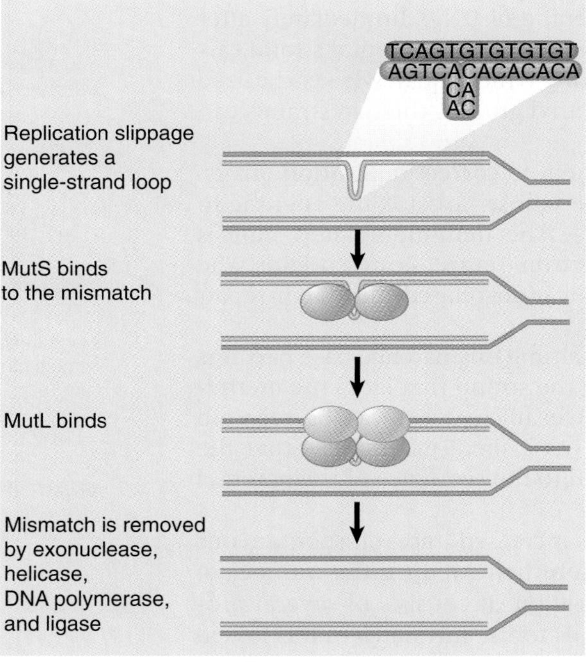
TCAGTGTGTGTGT
AGTCACACACACA
CA
AC
Replication slippage generates a single-strand loop
MutS binds to the mismatch
MutL binds
Mismatch is removed by exonuclease, helicase, DNA polymerase, and ligase

FIGURE 20.17 The MutS/MutL system initiates repair of mismatches produced by replication slippage.

The eukaryotic MutS/L system is particularly important for repairing errors caused by *replication slippage.* In a region such as a microsatellite, where a very short sequence is repeated several times, realignment be-tween the newly synthesized daughter strand and its template can lead to a stuttering in which the DNA polymerase slips backward and synthesizes extra repeating units. These units in the daughter strand are extruded as a single-stranded loop from the double helix, which is repaired by homologs of the MutS/L system, as shown in **FIGURE 20.17**.

The importance of the MutS/L system for mismatch repair is indicated by the high rate at which it is found to be defective in human cancers. Loss of this system leads to an increased mutation rate, and mutations in MutS/L components can lead to hereditary non-polyposis colorectal cancer (HNPCC).

KEY CONCEPTS

- The *mut* genes code for a mismatch repair system that deals with mismatched base pairs.
- There is a bias in the selection of which strand to replace at mismatches.
- The strand lacking methylation at a hemimethylated $\frac{\text{GATC}}{\text{CTAG}}$ is usually replaced.
- The mismatch repair system is used to remove errors in a newly synthesized strand of DNA. At G-T and C-T mismatches, the T is preferentially removed.
- Eukaryotic MutS/L systems repair mismatches and insertion/deletion loops.

CONCEPT AND REASONING CHECK

Cancer cells from patients with HNPCC show dramatic changes in the lengths of microsatellite loci. Explain.

20.7 Recombination-Repair Systems

Recombination-repair systems use activities that overlap with those involved in genetic recombination. They are also sometimes called "post-replication repair," because they function after replication. Such systems are effective in dealing with the defects produced in daughter duplexes by replication of a template that contains damaged bases. An example is illustrated in FIGURE 20.18.

Consider a structural distortion, such as a pyrimidine dimer, on one strand of a double helix. When the DNA is replicated, the dimer prevents the damaged site from acting as a template. Replication is forced to skip past it.

DNA polymerase probably proceeds up to or close to the pyrimidine dimer. The polymerase then ceases synthesis of the corresponding daughter strand. Replication restarts some distance farther along. This replication may be performed by translesion polymerases, which can replace the main DNA polymerase at such sites of unrepaired damage (see *Section 20.5, Error-Prone Repair*). A substantial gap is left in the newly synthesized strand.

The resulting daughter duplexes are different in nature. One has the parental strand containing the damaged adduct, which faces a newly synthesized strand with a lengthy gap. The other duplicate has the undamaged parental strand, which has been copied into a normal complementary strand. The retrieval system takes advantage of the normal daughter.

The gap opposite the damaged site in the first duplex is filled by utilizing the homologous single strand of DNA from the normal duplex. Following this **single-strand exchange**, the recipient duplex has a parental (damaged) strand facing a wild-type strand. The donor duplex has a normal parental strand facing a gap; the gap can be filled by repair synthesis in the usual way, generating a normal duplex. Thus the damage is confined to the original distortion (although the same recombination-repair events must be repeated after every replication cycle unless and until the damage is removed by an excision repair system).

In many cases, rather than skipping a DNA lesion, DNA polymerase instead stops replicating when it encounters DNA damage. FIGURE 20.19 shows one possible outcome when a replication fork stalls. The fork stops moving forward when it encounters the damage. The

▸ **recombination-repair** A mode of filling a gap in one strand of duplex DNA by retrieving a homologous single strand from another duplex.

▸ **single-strand exchange** A reaction in which one of the strands of a duplex of DNA leaves its former partner and instead pairs with the complementary strand in another molecule, displacing its homologue in the second duplex.

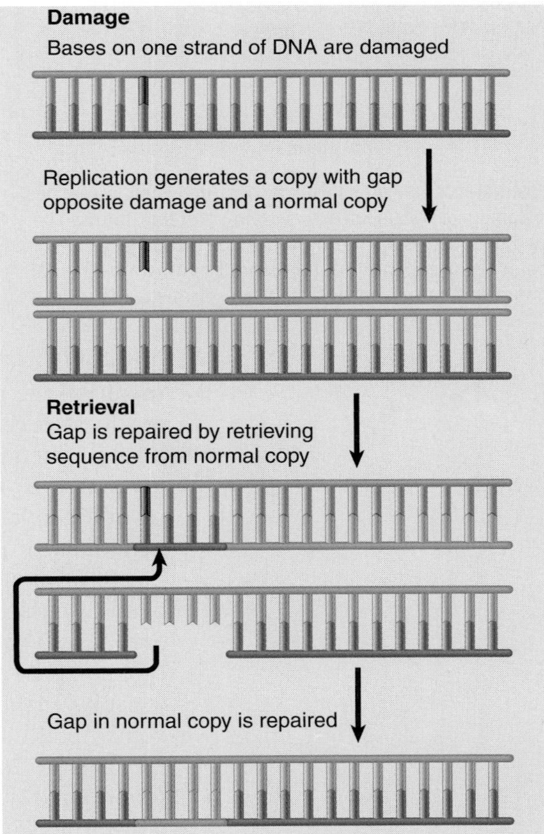
Damage
Bases on one strand of DNA are damaged
Replication generates a copy with gap opposite damage and a normal copy
Retrieval
Gap is repaired by retrieving sequence from normal copy
Gap in normal copy is repaired

FIGURE 20.18 An *E. coli* retrieval system uses a normal strand of DNA to replace the gap left in a newly synthesized strand opposite a site of unrepaired damage.

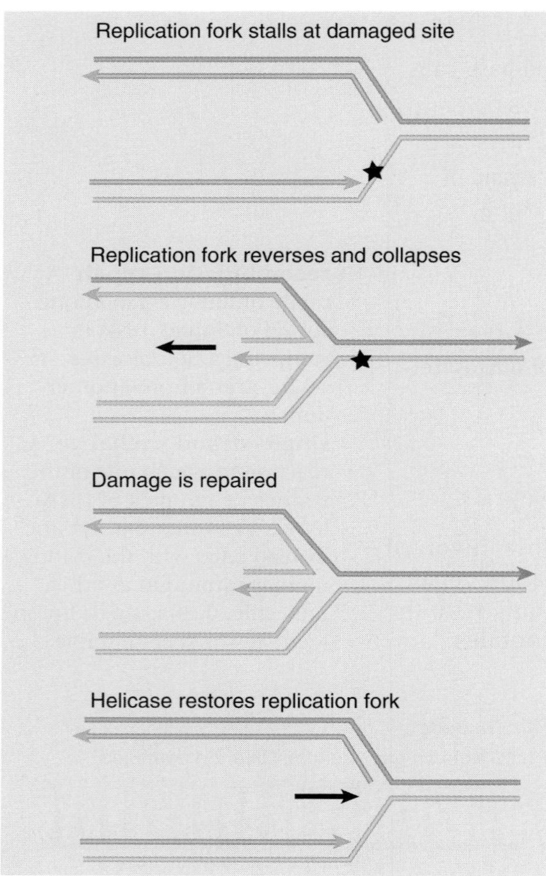
Replication fork stalls at damaged site
Replication fork reverses and collapses
Damage is repaired
Helicase restores replication fork

FIGURE 20.19 A replication fork stalls when it reaches a damaged site in DNA. Reversing the fork allows the two daughter strands to pair. After the damage has been repaired, the fork is restored by forward-branch migration catalyzed by a helicase. Arrowheads indicate 3′ ends.

replication apparatus disassembles, at least partially. This allows branch migration to occur, when the fork effectively moves backward, and the new daughter strands pair to form a duplex structure. After the damage has been repaired, a helicase rolls the fork forward to restore its structure. Then the replication apparatus can reassemble, and replication is restarted (see *Section 18.14, The Primosome Is Needed to Restart Replication*). DNA polymerase II is required for the replication restart, and is later replaced by DNA polymerase III.

The pathway for handling a stalled replication fork requires repair enzymes, and restarting stalled replication forks is thought to be a major role of the recombination-repair systems. The principal pathway for recombination-repair in *E. coli* is mediated by the products of the *rec* genes. In *E. coli* deficient in excision repair, mutation in the *recA* gene essentially abolishes all the remaining repair and recovery facilities. Attempts to replicate DNA in *uvr*$^-$ *recA*$^-$ cells produce fragments of DNA whose size corresponds with the expected distance between thymine dimers. This result implies that the dimers provide a lethal obstacle to replication in the absence of RecA function. It explains why the double mutant cannot tolerate >1 to 2 dimers in its genome (compared with the ability of a wild-type bacterium to handle as many as 50).

One *rec* pathway involves the *recBC* genes and is well characterized; the other involves *recF* and is not so well defined. They fulfill different functions *in vivo*. The *RecBC* pathway is involved in restarting stalled replication forks. The *RecF* pathway is involved in repairing the gaps in a daughter strand that are left after replicating past a pyrimidine dimer. The *RecBC* and *RecF* pathways both lead to the association of RecA with a single-stranded DNA. The ability of RecA to exchange single strands allows it to perform the retrieval step in Figure 20.18. Nuclease and polymerase activities then complete the repair action.

When a replication fork encounters a lesion in a single stand, it can result in the formation of a double-strand break (DSB). DSBs are one of the most severe types of DNA damage that can occur, particularly in eukaryotes. If a DSB on a linear chromosome is not repaired, the portion of the chromosome lacking a centromere will not be segregated at the next cell division. In addition to their occurrence during replication, DSBs can be generated in a number of other ways, including ionizing radiation, oxygen radicals generated by cellular metabolism, or action of endonucleases. The preferred mechanism for repairing DSBs is to use recombination repair, as this ensures that no critical genetic information is lost due to sequence loss at the breakpoint.

Several of the genes required for recombination repair in eukaryotes have already been discussed in the context of meiotic homologous recombination (see *Section 19.5, Specialized Enzymes Catalyze 5′ End Resection and Single-Strand Invasion*). Many eukaryotic repair genes are named *RAD* genes; they were initially characterized genetically in yeast by virtue of their sensitivity to *rad*iation. There are three general groups of repair genes in the yeast *S. cerevisiae,* identified by the *RAD3* group (involved in excision repair), the *RAD6* group (required for postreplication repair), and the *RAD52* group (concerned with recombination-like mechanisms). Homologs of these genes are present in higher eukaryotes as well.

The *RAD52* group plays essential roles in homologous recombination, and includes a large number of genes such as *RAD50, RAD51, RAD54, RAD55, RAD57,* and *RAD59.* These Rad proteins all assemble at a double-strand break. As occurs during meiotic recombination, the Mre11/Rad50/Xbs1 (MRX)-dependent exonuclease acts on the free ends to generate single-stranded tails. This single-stranded DNA serves to activate a DNA damage checkpoint, stopping cell division until the damage can be repaired.

The RecA homolog Rad51 binds to the single-stranded DNA to form a nucleoprotein filament, which is used for strand invasion of a homologous sequence. Rad52, Rad55, and Rad54 are required to form a stable Rad51 filament and also assist in the homology search and strand invasion. Following repair synthesis, the resulting structure (which resembles a Holliday junction) is resolved.

KEY CONCEPTS

- The *rec* genes of *E. coli* code for the principal recombination-repair system.
- Recombination-repair functions when replication leaves a gap in a newly synthesized strand that is opposite a damaged sequence.
- The single strand of another duplex is used to replace the gap.
- The damaged sequence is then removed and resynthesized.
- A replication fork may stall when it encounters a damaged site or a nick in DNA.
- A stalled fork may reverse by pairing between the two newly synthesized strands.
- A stalled fork may restart after repairing the damage and use a helicase to move the fork forward.
- The yeast *RAD* mutations, identified by radiation sensitive phenotypes, are in genes that code for repair systems.
- The *RAD52* group of genes are required for recombination repair.

CONCEPT AND REASONING CHECK

Explain why large gaps and double-strand breaks are preferentially repaired by recombination, but a stalled replication fork can utilize either excision repair or recombination-repair pathways.

20.8 Nonhomologous End-Joining Also Repairs Double-Strand Breaks

nonhomologous end-joining (NHEJ) The process that ligates blunt ends. It is common to many repair pathways and to certain recombination pathways (such as immunoglobulin recombination).

Repair of DSBs by homologous recombination ensures no genetic information is lost from a broken DNA end. However, in many cases a sister chromatid or homologous chromosome is not easily available to use as a template for repair. In addition, some DSBs are specifically repaired using error-prone mechanisms as intermediate in the recombination of immunoglobulin genes (see *Section 22.6, The RAG Proteins Catalyze Breakage and Reunion*). In these cases, the mechanism used to repair these breaks is called **nonhomologous end-joining (NHEJ)**, and consists of ligating the blunt ends together.

The steps involved in NHEJ are summarized in **FIGURE 20.20**. The same enzyme complex undertakes the process in both NHEJ and immune recombination. The first stage is recognition of the broken ends by a heterodimer consisting of the proteins Ku70 and Ku80. They form a scaffold that holds the ends together and allows other enzymes to act on them. A key component is the DNA-dependent protein kinase (DNA-PKcs), which is activated by DNA to phosphorylate protein targets. One of these targets is the protein Artemis, which in its activated form has both exonuclease and endonuclease activities, and can both trim overhanging ends and cleave the hairpins generated by recombination of immunoglobulin genes. The DNA polymerase activity that fills in any remaining single-stranded protrusions is not known. The actual joining of the double-stranded ends is performed by DNA ligase IV, which functions in conjunction with the protein XRCC4. Mutations in any of these components may

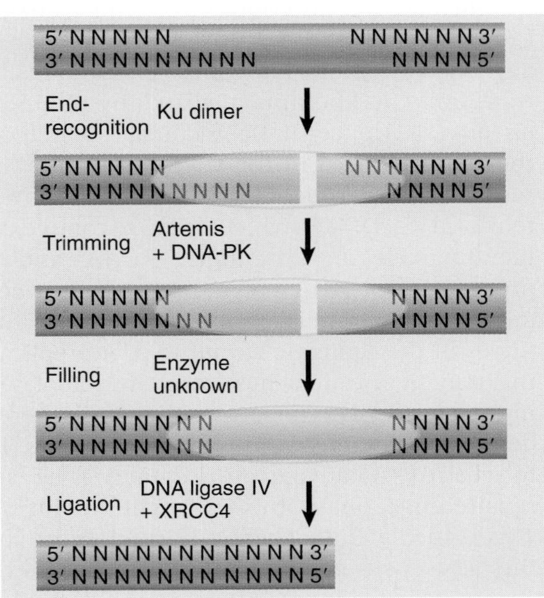

FIGURE 20.20 Nonhomologous end-joining requires recognition of the broken ends, trimming of overhanging ends and/or filling, followed by ligation.

MEDICAL APPLICATIONS

p53 Is the "Guardian of the Genome"

A key protein in the mammalian cell's response to stress in general, and to DNA damage in particular, is a protein called the *p53 transcription factor*. In normal cells, the level of activated p53 protein is very low, even though substantial amounts of p53 mRNA and protein are present. The activity of p53 is kept low by another protein called Mdm2. To function as a transcription factor, p53 must be activated first by phosphorylation and then by acetylation. Mdm2 binds to p53 and prevents phosphorylation and subsequent steps in the activation of p53 as a transcription factor. In addition, Mdm2 continuously shuttles between the nucleus and the cytoplasm, and in this process it continuously exports p53 from the nucleus for degradation by the proteasome in the cytoplasm.

When cells are treated with agents that damage DNA, the cells can arrest in at multiple points within the cell cycle, including G1, S, G2, or transitions between these phases. The DNA damage signal is sensed and transmitted, and this causes p53 to become activated by protein kinases and acetylases, overriding the inhibitory effect of Mdm2. Activation of p53 causes its release from Mdm2, which stabilizes the active p53 transcription factor and results in increased levels of the p53 protein. The activated p53 then triggers the transcription of a number of genes and the repression of others. A flow chart of how some key proteins whose genes are transcriptionally activated by p53 affect processes in the cell cycle and other cellular properties is given in **FIGURE B20-1**.

Increased transcription of the genes for p21, GADD45, and 14-3-3σ all serve to block the cell cycle at particular points, known as "checkpoints." The G1/S transition checkpoint is mediated by the increased levels of p21. Hence, if DNA damage is detected in G1, the cell is blocked in the G1/S transition. The S-phase response to DNA damage is mediated by the p21 protein and GADD45, which form a complex with proliferating cell nuclear antigen (PCNA) and reduce the processivity of DNA polymerase. The **processivity** of a DNA polymerase is the number of consecutive nucleotides in the template strand that are replicated before the polymerase detaches from the template. Decreasing the processivity of DNA polymerase therefore slows down DNA synthesis, in effect buying time for the cell to repair DNA damage. If the damage is not successfully repaired in S phase, the G2/M checkpoint is triggered. This is mediated by the 14-3-3σ protein, which blocks the G2-to-M transition. If DNA damage is detected in either the S or the G2 phase, the cell cycle arrests at the G2/M checkpoint.

DNA damage also triggers activation of another pathway, a pathway for **apoptosis,** or **programmed cell death**. When the apoptotic pathway is activated, a cascade of proteolysis is initiated that culminates in cell suicide. The proteases involved are called *caspases*. Their activation ultimately results in destruction of the cellular DNA, internal organelles, and the actin cytoskeleton, and it is accompanied by nuclear condensation and usually followed by engulfment of the cellular remnants by phagocytes. There is also evidence that p53 acts directly to increase the permeability of mitochondria, resulting in the release of cytochrome c, which triggers apoptosis. The p53 transcription factor also activates the apoptotic pathway by turning on transcription of genes that promote apoptosis. In all of its functions, p53 acts to protect the integrity of the genome with respect to nucleotide sequence and strand integrity as well as with respect to euploidy. Instability in the genome, unbalanced genomes, and damaged DNA pose a hazard to the organism. Apoptosis is a mechanism for killing such damaged—and thus dangerous—cells.

To gain an appreciation of the importance of the DNA damage checkpoint, consider what would happen if part or all of this fail-safe mechanism should malfunction. Suppose, for example, that cells lacked p21 protein but retained the rest of the mechanism. If DNA damage was properly sensed, and p53 functioned as usual to increase levels of 14-3-3σ and GADD45, cells would accumulate in G2 and be unable to undergo mitosis. However, DNA would continue to be synthesized. In addition, lacking p21 protein, the cells would have a defective G1/S checkpoint and so could embark on additional rounds of DNA synthesis. The expected result of p21 malfunction would therefore be polyploid cells, and this is indeed the result observed.

To take another example, consider the consequence of loss of p53 function. Even if DNA damage were detected, the cell would be unable to mount a response—it would be unable to buy time to repair the DNA damage. The cells could initiate new rounds of synthesis with damaged chromosomes. Such synthesis would not only result in an increased frequency of mutation but also permit gene amplification. Amplification of **oncogenes**, which are genes whose misexpression is associated with cancers, would allow cells to escape the normal controls of DNA synthesis and proliferation. Cells already in S or G2 would enter mitosis with damaged chromosomes, because not enough time for repair of lesions would have elapsed. In addition, the organism would have lost its ultimate protection against such damaged cells: apoptosis. If the description of these runaway cells reminds you of the unchecked proliferation of cancer cells, this is because cancer cells *become* cancer cells by subverting the checkpoint mechanisms. This is illustrated dramatically in **Li-Fraumeni Syndrome**, a rare hereditary disorder caused by loss of a single copy of p53. Patients with Li-Fraumeni are at high risk for numerous malignancies, including brain tumors, sarcomas, leukemias, breast cancer and others. Patients can develop multiple primary tumors, particularly at a young age, with a 50% chance of developing cancer by age 40 and a 90% chance by age 60. Recent research suggests that p53 replacement in tumors by gene therapy may yield promising treatment options not only for cancers in Li-Fraumeni patients, but also in the >60% of sporadic cancers known to have reduced or absent p53 function.

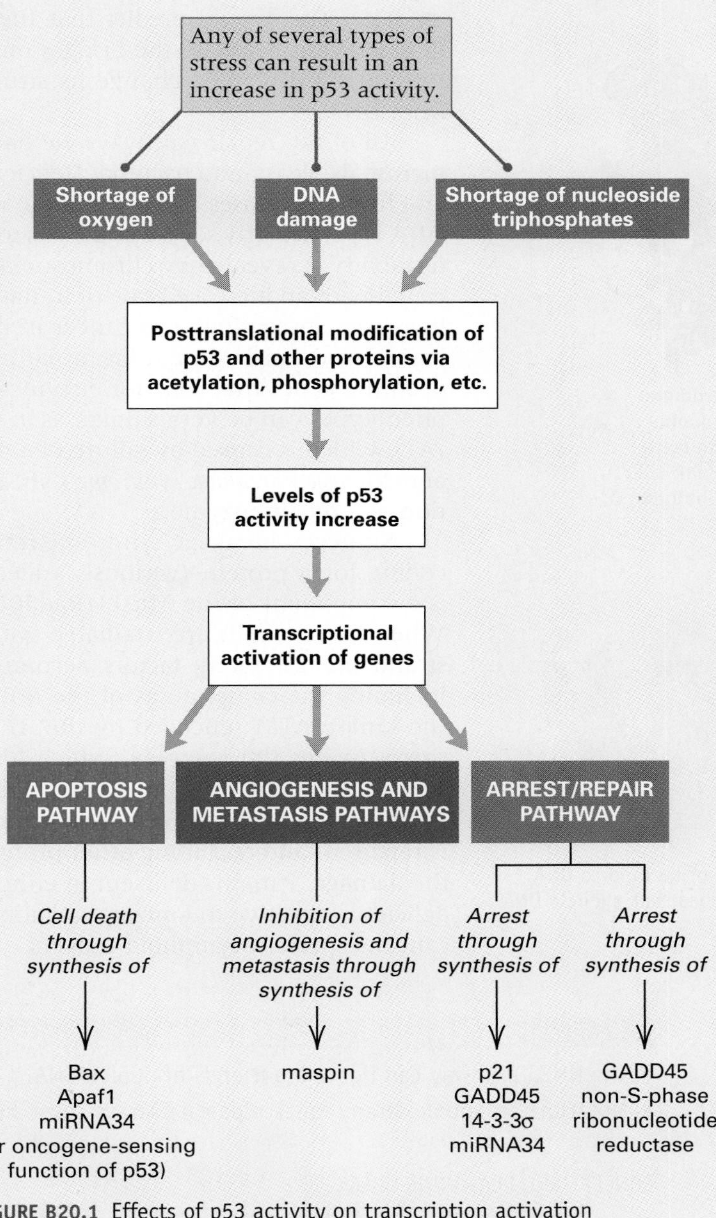

FIGURE B20.1 Effects of p53 activity on transcription activation and apoptosis.

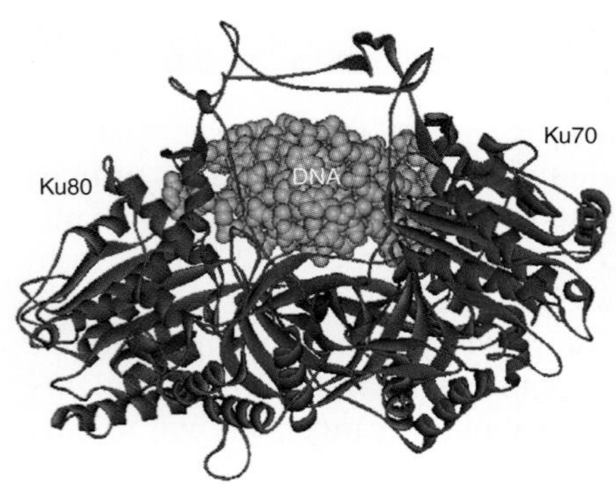

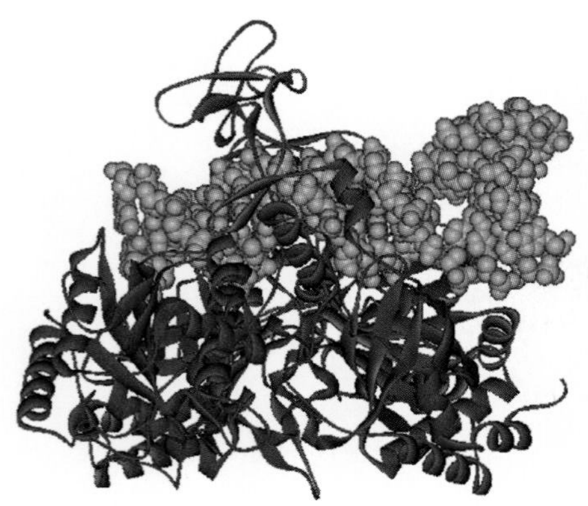

FIGURE 20.21 The Ku70-Ku80 heterodimer binds along two turns of the DNA double helix and surrounds the helix at the center of the binding site. Protein Data Bank 1JEY. Walker, J. R., Corpina, R. A., and Goldberg, J., *Nature* 412 (2001): 607–614.

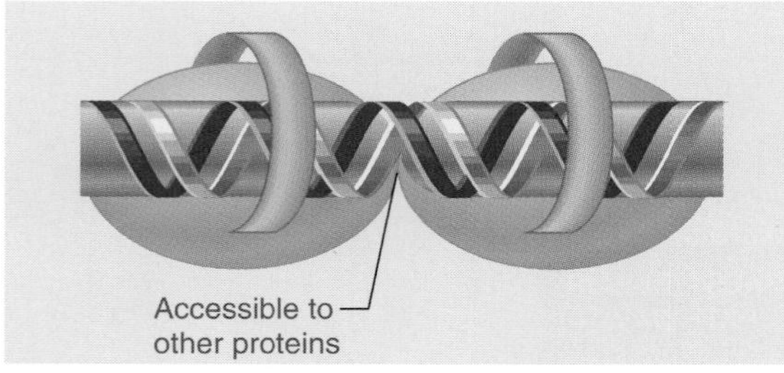

FIGURE 20.22 If two heterodimers of Ku bind to DNA, the distance between the two bridges that encircle DNA is ~12 bp.

render eukaryotic cells more sensitive to radiation. Some of the genes for these proteins are mutated in patients who have diseases due to deficiencies in DNA repair.

The Ku heterodimer is the sensor that detects DNA damage by binding to the broken ends. The crystal structure in **FIGURE 20.21** shows why it binds only to ends. The bulk of the protein extends for about two turns along one face of DNA (lower), but a narrow bridge between the subunits, located in the center of the structure, completely encircles DNA. This means that the heterodimer needs to slip onto a free end.

Ku can bring broken ends together by binding two DNA molecules. The ability of Ku heterodimers to associate with one another suggests that the reaction might take place as illustrated in **FIGURE 20.22**. This would predict that the ligase would act by binding in the region between the bridges on the individual heterodimers. Presumably Ku must change its structure in order to be released from DNA.

All of the repair pathways we have discussed are conserved in mammals, yeast, and bacteria. Deficiency in DNA repair causes several human diseases. The inability to repair double-strand breaks in DNA is particularly severe and leads to chromosomal instability. The instability is revealed by chromosomal aberrations, which are associated with an increased rate of mutation, which in turn leads to an increased susceptibility to cancer in patients with the disease. The basic cause can be mutation in pathways that control DNA repair or in the genes that code for enzymes of the repair complexes. The phenotypes can be very similar, as in the case of *Ataxia telangiectasia* (AT), which is caused by failure of a cell cycle checkpoint pathway, and *Nijmegen breakage syndrome* (NBS), which is caused by a mutation of a repair enzyme.

Nijmegen breakage syndrome results from mutations in a gene coding for a protein (variously called Nibrin, p95, or NBS1) that is a component of the Mre11/Rad50/Nbs1 (MRN) repair complex. When human cells are irradiated with agents that induce double-strand breaks, many factors accumulate at the sites of damage, including the components of the MRN complex. After irradiation, the kinase ATM (encoded by the *AT* gene) phosphorylates NBS1; this activates the complex, which localizes to sites of DNA damage. Subsequent steps involve triggering a checkpoint (a mechanism that prevents the cell cycle from proceeding until the damage is repaired) and recruiting other proteins that are required to repair the damage. Patients deficient in either ATM or NBS1 are immunodeficient, sensitive to ionizing radiation and predisposed to develop cancer, especially lymphoid cancers.

KEY CONCEPTS

- The NHEJ pathway can ligate blunt ends of duplex DNA.
- Mutations in double-strand break repair pathways cause human diseases.

CONCEPT AND REASONING CHECK

Many lymphoid cancers in patients with diseases such as NBS are linked to chromosomal translocations. Explain.

20.9 Summary

All cells contain systems that maintain the integrity of their DNA sequences in the face of damage or errors of replication and that distinguish the DNA from sequences of a foreign source.

Repair systems can recognize mispaired, altered, or missing bases in DNA, as well as other structural distortions of the double helix. Excision repair systems cleave DNA near a site of damage, remove one strand, and synthesize a new sequence to replace the excised material. The *Uvr* system provides the main excision-repair pathway in *E. coli*. The *mut* and *dam* systems are involved in correcting mismatches generated by incorporation of incorrect bases during replication and function by preferentially removing the base on the strand of DNA that is not methylated at a *dam* target sequence. Eukaryotic homologs of the *E. coli* MutSL system are involved in repairing mismatches that result from replication slippage; mutations in this pathway are common in certain types of cancer.

Repair systems can be connected with transcription in both prokaryotes and eukaryotes. Human diseases are caused by mutations in genes coding for repair activities that are associated with the transcription factor $TF_{II}H$. They have homologs in the *RAD* genes of yeast, which suggests that this repair system is widespread.

Recombination-repair systems retrieve information from a DNA duplex and use it to repair a sequence that has been damaged on both strands. The *RecBC* and *RecF* pathways both act prior to RecA, whose strand-transfer function is involved in all bacterial recombination. A major use of recombination-repair may be to recover from the situation created when a replication fork stalls. Genes in the *RAD52* group are involved in recombination repair in eukaryotes.

Nonhomologous end-joining (NHEJ) is a general mechanism for repairing broken ends in eukaryotic DNA. The Ku heterodimer brings the broken ends together so they can be ligated. Several human diseases are caused by mutations in enzymes of this pathway.

CHAPTER QUESTIONS

The enzyme activities involved in DNA repair systems include the following (insert the missing enzymes or description of activity):

1. ______________________ nicks DNA near damaged regions to initiate DNA repair.

2. Exonucleases ____.

3. Resolvases ____.

4. ______________________ unwinds DNA regions.

5. ______________________ synthesize new DNA.

6. DNA ligase ____.

Enzymes that remove bases from DNA are called

7. ____________________ and

8. ________________.

9. Which of the following repair systems is known to be highly conserved in most organisms except for placental mammals?

A. base excision
B. nucleotide excision
C. recombination repair
D. photoreactivation

10. Ultraviolet light causes which of the following in DNA?
   A. formation of thymine dimers
   B. repair of chemically damaged DNA
   C. deamination of cytosines
   D. methylation of adenines
11. Deamination of cytosine in DNA results in a:
   A. A-T base pair.
   B. G-C base pair.
   C. U-G base pair.
   D. T-U base pair.
12. Nearly all excision-repair events in *E. coli* are carried out by the:
   A. RecB system.
   B. RecF system.
   C. Uvr system.
   D. Umu system.
13. What is the average length of DNA replaced by the vast majority of excision-repair events in *E. coli*?
   A. 9 nucleotides
   B. 12 nucleotides
   C. 21 nucleotides
   D. 32 nucleotides
14. Lyases function by:
   A. cleaving the bond between the base and the deoxyribose, leading to removal of the base.
   B. cleaving the ring structure of the nucleotide, leading to its removal.
   C. nicking the phosphodiester backbone on both sides of a base, leading to its removal.
   D. opening up the deoxyribose sugar ring, leading to removal of the damaged base.
15. Umu proteins are involved in which of the following DNA repair systems?
   A. recombination repair
   B. error-prone DNA synthesis
   C. mismatch repair
   D. excision repair

KEY TERMS

error-prone synthesis
excision
excision repair
glycosylase
incision
lyase
mismatch repair
mutator
nonhomologous end-joining (NHEJ)
photoreactivation
pyrimidine dimer
recombination-repair
single-strand exchange
xeroderma pigmentosum (XP)

FURTHER READING

Baute, J. and Depicker, A. (2008). Base excision repair and its role in maintaining genome stability. *Crit. Revs. Biochem. Mol. Biol.* 43(4), 239–276.

A review of base excision repair (BER) in numerous organisms, focusing on the role of glycosylases in repair.

Bergoglio, V. and Magnaldo, T. (2006). Nucleotide excision repair and related human diseases. *Genome Dynam.* 1, 35–52.

A review of eukarytotic nucleotide exision repair (NER) with an emphasis on human diseases related to defects in NER components.

Goosen, N. and Moolenaar, G. F. (2008). Repair of UV damage in bacteria. *DNA Repair* 7(3), 353–379.

A review of numerous prokaryotic repair systems that deal with UV damage, including photolyase, BER and NER.

Green, C. M. and Lehmann, A. R. (2005). Translesion synthesis and error-prone polymerases. *Adv. Exper. Med. Biol.* 570, 199–223.

A review of error-prone pathways of repair.

Hsieh, P. and Yamane, K. (2008). DNA mismatch repair: molecular mechanism, cancer, and ageing. *Mech. Ageing Devel.* 129(7–8), 391–407.

A review of the molecular details of mismatch repair (MMR) and the connections between MMR deficiency and cancer and ageing.

Li, X. and Heyer, W. D. (2008). Homologous recombination in DNA repair and DNA damage tolerance. *Cell Res.* 18(1), 99–113.

A review of the homologous recombination (HR) pathway in the repair of double-strand breaks (DSBs) and interstrand crosslinks (ICLs).

Weterings, E. and Chen, D. J. (2008). The endless tale of non-homologous end-joining. *Cell Res.* 18(1), 114–124.

A review of all the known enzymes involved in non-homologous end-joining (NHEJ) in mammalian cells.

21

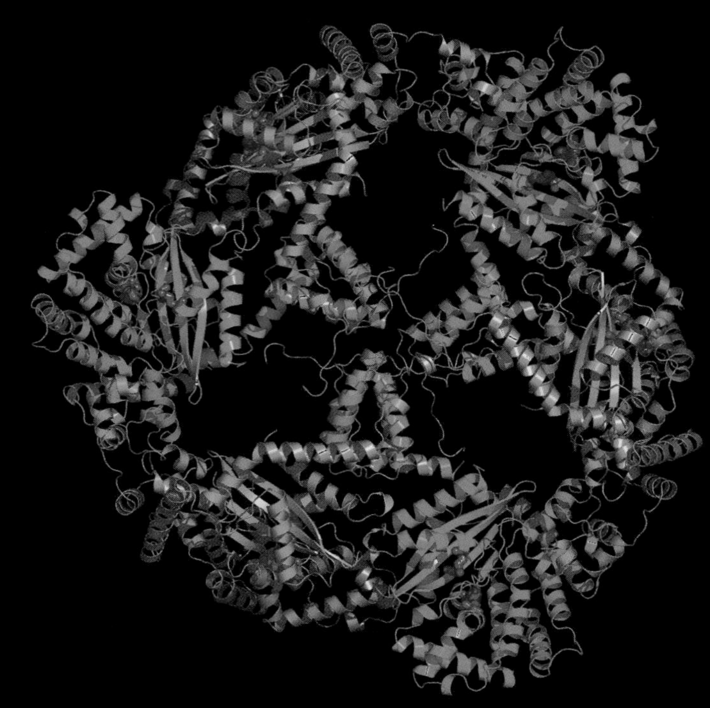

The Hermes DNA transposase, with adjacent monomers of the hexamer colored green and orange and active sites colored red. Photo courtesy of Fred Dyda, NIDDK, National Institutes of Health.

Transposons, Retroviruses, and Retrotransposons

CHAPTER OUTLINE

21.1 Introduction

Genomes evolve both by acquiring new sequences and by rearranging existing sequences.

The sudden introduction of new sequences results from the ability of vectors to carry information between genomes. Extrachromosomal elements move information horizontally by mediating the transfer of (usually rather short) lengths of genetic material. In bacteria, plasmids move by conjugation (see *Section 16.7, Conjugation Transfers Single-Stranded DNA*), whereas phages spread by infection (see *Chapter 14, Phage Strategies*). Both plasmids and phages occasionally transfer host genes along with their own replicon. Direct transfer of DNA occurs between some bacteria by means of transformation (see *Section 1.2, DNA Is the Genetic Material of Bacteria, Viruses, and Eukaryotic Cells*). In eukaryotes, some viruses, notably the retroviruses, can transfer genetic information during an infective cycle (see *Section 21.9, Retroviruses May Transduce Cellular Sequences*).

Rearrangements are sponsored by processes internal to the genome. Two of the major causes are summarized in **FIGURE 21.1**.

Unequal recombination results from mispairing by the cellular systems for homologous recombination. Nonreciprocal recombination results in duplication or rearrangement of loci (see *Section 6.7, Unequal Crossing-over Rearranges Gene Clusters*). Duplication of sequences within a genome provides a major source of new sequences. One copy of the sequence can retain its original function, whereas the other may evolve into a new function. Furthermore, significant differences between individual genomes are found at the molecular level because of polymorphic variations caused by recombination. We saw in *Section 6.12, Minisatellites Are Useful for Genetic Mapping,* that recombination between minisatellites adjusts their lengths so that every individual genome is distinct.

Another major cause of variation is provided by **transposable elements** or **transposons**: these are discrete sequences in the genome that are mobile—they are able to transport themselves to other locations within the genome. The mark of a transposon is that it does not utilize an independent form of the element (such as phage or plasmid DNA), but moves directly from one site in the genome to another. Unlike most other processes involved in genome restructuring, transposition does not rely on any relationship between the sequences at the donor and recipient sites. Transposons are restricted to moving themselves, and sometimes additional sequences, to new sites elsewhere within the same genome; they are, therefore, an internal counterpart to the vectors that can transport sequences from one genome to another. They may provide the major source of mutations in the genome.

▸ **transposon (transposable element)** A DNA sequence able to insert itself (or a copy of itself) at a new location in the genome without having any sequence relationship with the target locus.

Transposons that mobilize via DNA are found in both prokaryotes and eukaryotes. Each bacterial transposon carries gene(s) that code for the enzyme activities required for its own transposition, but it may also require ancillary functions of the genome in which it resides (such as DNA polymerase or DNA gyrase). Comparable systems exist in eukaryotes, although their enzymatic functions are not so well characterized. A genome may contain both functional and nonfunctional (defective) elements. Often the majority of elements in a eukaryotic genome are defective and have lost the ability to transpose independently, although they may still be recog-

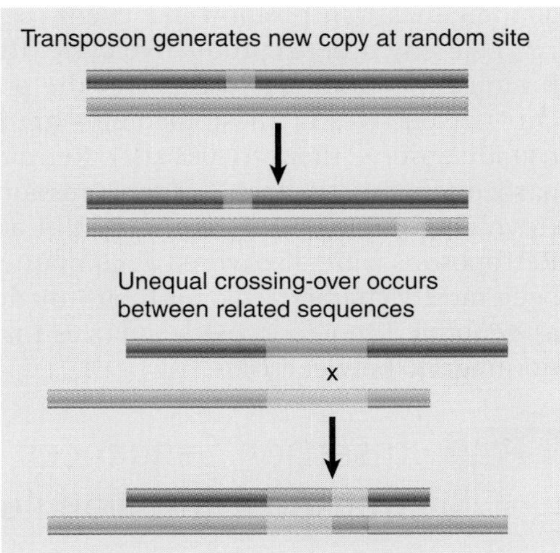
Transposon generates new copy at random site
Unequal crossing-over occurs between related sequences
x

FIGURE 21.1 A major cause of sequence change within a genome is the movement of a transposon to a new site. This may have direct consequences on gene expression. Unequal crossing-over between related sequences causes rearrangements. Copies of transposons can provide targets for such events.

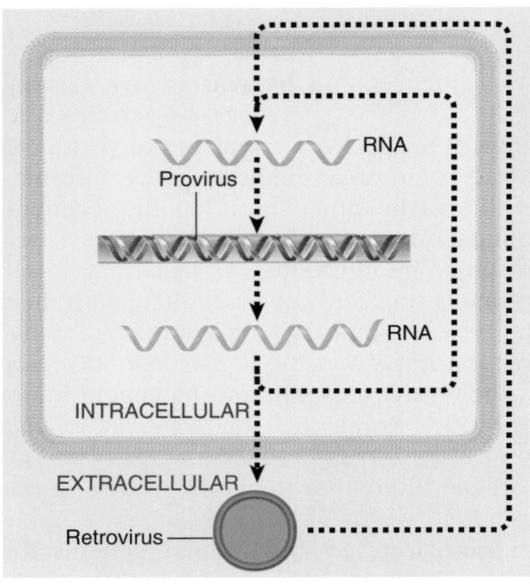

FIGURE 21.2 The reproductive cycles of retroviruses and retroposons alternate reverse transcription from RNA to DNA with transcription from DNA to RNA. Only retroviruses can generate infectious particles. Retroposons are confined to an intracellular cycle.

nized as substrates for transposition by the enzymes produced by functional transposons. A eukaryotic genome contains a large number and variety of transposons. The fly genome has >50 types of transposons, with a total of several hundred individual elements.

Transposable elements can promote rearrangements of the genome either directly or indirectly:

- The transposition event itself may cause deletions or inversions or lead to the movement of a host sequence to a new location.
- Transposons serve as substrates for cellular recombination systems by functioning as "portable regions of homology"; two copies of a transposon at different locations (even on different chromosomes) may provide sites for reciprocal recombination. Such exchanges result in deletions, insertions, inversions, or translocations.

Transposition that involves an obligatory intermediate of RNA is primarily found in eukaryotes. As a class, the elements that transpose through an RNA intermediate are called **retrotransposons** (or sometimes **retroposons**). The very simplest retrotransposons do not themselves have transposition activity, but have sequences that are recognized as substrates for transposition by active elements. Thus elements that use RNA-dependent transposition range from the retroviruses themselves (which are able to infect host cells freely), to sequences that transpose via RNA, to those that do not themselves possess the ability to transpose. They share with all transposons the diagnostic feature of generating short direct repeats of target DNA at the site of an insertion.

▸ **retrotransposon (retroposon)** A transposon that mobilizes via an RNA form; the DNA element is transcribed into RNA, and then reverse-transcribed into DNA, which is inserted at a new site in the genome. It does not have an infective (viral) form.

Even in genomes where active transposons have not been detected, footprints of ancient transposition events are found in the form of direct target repeats flanking dispersed repetitive sequences. The features of these sequences sometimes implicate an RNA sequence as the progenitor of the genomic (DNA) sequence. This suggests that the RNA must have been converted into a duplex DNA copy that was inserted into the genome by a transposition-like event.

Like any other reproductive cycle, the cycle of a retrovirus or retroposon is continuous; it is arbitrary to consider the point at which we interrupt it a "beginning." Our perspectives of these elements are biased, though, by the forms in which we usually observe them (**FIGURE 21.2**). Retroviruses were first observed as infectious virus particles that were capable of transmission between cells, and so the intracellular cycle (involving duplex DNA) is thought of as the means of reproducing the RNA virus. Retroposons were discovered as components of the genome, and the RNA forms have been mostly characterized for their functions as mRNAs. Thus we think of retroposons as genomic (duplex DNA) sequences that may transpose within a genome; they do not migrate between cells.

21.2 Insertion Sequences Are Simple Transposition Modules

Transposable elements were first identified at the molecular level in the form of spontaneous insertions in bacterial operons. Such an insertion prevents transcription and/

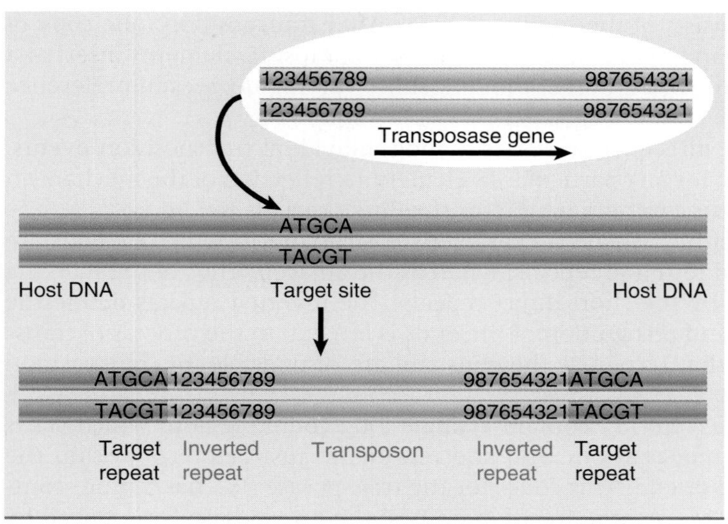

Transposon	Target repeat (bp)	Inverted repeat (bp)	Overall length (bp)	Target selection
IS1	9	23	768	random
IS2	5	41	1327	hotspots
IS4	11–13	18	1428	$AAAN_{20}TTT$
IS5	4	16	1195	hotspots
IS10R	9	22	1329	NGCTNAGCN
IS50R	9	9	1531	hotspots
IS903	9	18	1057	random

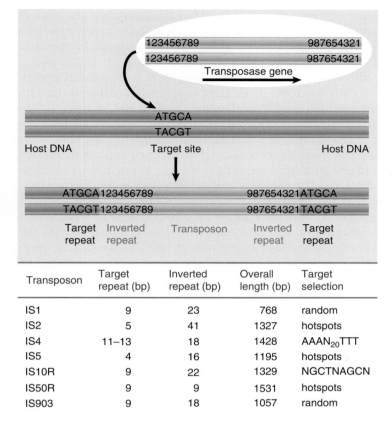

FIGURE 21.3 Transposons have inverted terminal repeats and generate direct repeats of flanking DNA at the target site. In this example, the target is a 5 bp sequence. The ends of the transposon consist of inverted repeats of 9 bp, where the numbers 1 through 9 indicate a sequence of base pairs.

or translation of the gene in which it is inserted. Many different types of transposable elements have now been characterized.

The simplest transposons are called **insertion sequences** (reflecting the way in which they were detected). Each type is given the prefix **IS**, followed by a number that identifies the type. (The original classes were numbered IS1 to IS4; later classes have numbers reflecting the history of their isolation, but not corresponding to the total number of elements so far isolated!)

▸ **insertion sequences (IS)** A small bacterial transposon that carries only the genes needed for its own transposition.

The IS elements are normal constituents of bacterial chromosomes and plasmids. A standard strain of *E. coli* is likely to contain several (<10) copies of any one of the more common IS elements. To describe an insertion into a particular site, a double colon is used; so λ::IS1 describes an IS1 element inserted into phage lambda.

The IS elements are autonomous units, each of which codes only for the proteins needed to sponsor its own transposition. Each IS element is different in sequence, but there are some common features in organization. The structure of a generic transposon before and after insertion at a target site is illustrated in **FIGURE 21.3**, which also summarizes the details of some common IS elements.

An IS element ends in short **inverted terminal repeats**; usually the two copies of the repeat are closely related rather than identical. As illustrated in the figure, the presence of the inverted terminal repeats means that the same sequence is encountered proceeding toward the element from the flanking DNA on either side of it.

▸ **inverted terminal repeats** The short related or identical sequences present in reverse orientation at the ends of some transposons.

When an IS element transposes, a sequence of host DNA at the site of insertion is duplicated. The nature of the duplication is revealed by comparing the sequence of the target site before and after an insertion has occurred. Figure 21.3 shows that at the site of insertion, the IS DNA is always flanked by very short **direct repeats**. (In this context, "direct" indicates that two copies of a sequence are repeated in the same orientation, not that the repeats are adjacent.) In the original gene (prior to insertion), however, the target site has the sequence of only one of these repeats. In the

▸ **direct repeats** Identical (or closely related) sequences present in two or more copies in the same orientation in the same molecule of DNA.

figure, the target site consists of the sequence $\frac{\text{ATGCA}}{\text{TACGT}}$. After transposition, one copy of this sequence is present on either side of the transposon. Most IS elements insert at a variety of sites within host DNA. Some, though, show varying degrees of preference for particular hotspots.

The sequence of the direct repeat varies among individual transposition events, but the length is constant for any particular IS element (a reflection of the mechanism of transposition). The most common length for the direct repeats is 9 bp.

An IS element therefore displays a characteristic structure in which its ends are identified by the inverted terminal repeats, whereas the adjacent ends of the flanking host DNA are identified by the short direct repeats. The inverted repeats define the ends of the transposon, and recognition of the ends is integral to the process of transposition. The protein(s) that recognize the ends and are responsible for transposition are called **transposases**.

▸ **Transposase** The enzyme activity involved in insertion of transposon at a new site.

All the IS elements except IS1 contain a single long coding region, which starts just inside the inverted repeat at one end and terminates just before or within the inverted repeat at the other end. This codes for the transposase. IS1 has a more complex organization, with two separate reading frames; the transposase is produced by making a frameshift during translation to allow both reading frames to be used.

The frequency of transposition varies among different elements. The overall rate of transposition is ~10^{-3} to 10^{-4} per element per generation. Insertions in individual targets occur at a level comparable with the spontaneous mutation rate, usually ~10^{-5} to 10^{-7} per generation. Reversion (by precise excision of the IS element) is usually infrequent, with a range of rates of 10^{-6} to 10^{-10} per generation, which is ~10^{3} times less frequent than insertion.

Some transposons carry drug resistance (or other) markers in addition to the functions concerned with transposition. These transposons are named **Tn** followed by a number. One class of larger transposons are called **composite elements**, because a central region carrying the drug marker(s) is flanked on either side by "arms" that consist of IS elements.

▸ **Tn** Followed by a number, it denotes bacterial transposons carrying markers that are not related to their function, e.g., drug resistance.

▸ **composite elements** Transposable elements consisting of two IS elements (can be the same or different) and the DNA sequences between the IS elements; the non-IS sequences often include gene(s) conferring antibiotic resistance.

The arms may be in either the same or (more commonly) inverted orientation. Thus a composite transposon with arms that are direct repeats has the structure

ArmL Central Region ArmR

If the arms are inverted repeats, the structure is

ArmL Central Region ArmR

The arrows indicate the orientation of the arms, which are identified as L and R according to an (arbitrary) orientation of the genetic map of the transposon from left to right. The structure of a composite transposon is illustrated in more detail in **FIGURE 21.4**, which also summarizes the properties of some common composite transposons.

Arms consist of IS modules, and each module has the usual structure ending in inverted repeats; as a result the composite transposon also ends in the same short inverted repeats. A composite transposon can have two identical IS modules, such as Tn9 (direct repeats of IS1), or two closely related, but not identical modules. Thus we can distinguish the L and R modules in Tn10 or in Tn5. Functional IS elements code for transposase activities that are responsible both for creating a target site and for recognizing the ends of the transposon, so only the ends are needed for a transposon to serve as a substrate for transposition.

A functional IS module can transpose either itself or the entire transposon. When the modules of a composite transposon are identical, presumably either module can sponsor movement of the transposon. When the modules are different, they may differ in functional ability, so transposition can depend entirely or principally on one of the modules.

What is responsible for transposing a composite transposon instead of just the individual module? This question is especially pressing in cases where both modules are functional. In the example of Tn9, where the modules are IS1 elements, presumably each is active in its own right as well as on behalf of the composite transposon. Why is the transposon preserved as a whole, instead of each insertion sequence looking out for itself?

A major force supporting the transposition of composite transposons is selection for the marker(s) carried in the central region. For example, Tn10 consists of two IS10 modules flanking a *tet*^{R} gene conferring tetracycline resistance. An IS10 module is free to move around on its own, and mobilizes an order of magnitude more frequently than Tn10. Tn10 is held together by selection for *tet*^{R}, though, so that under selective conditions, the relative frequency of intact Tn10 transposition is much increased.

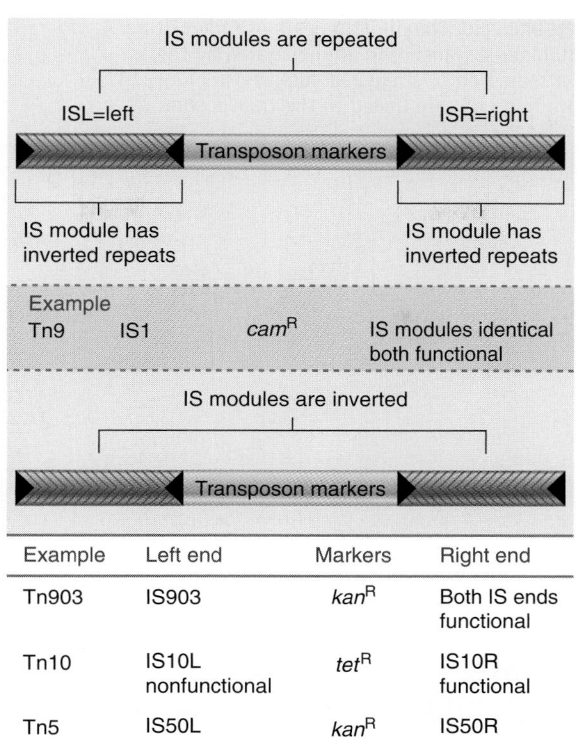

Example	Left end	Markers	Right end
Tn903	IS903	*kan*^{R}	Both IS ends functional
Tn10	IS10L nonfunctional	*tet*^{R}	IS10R functional
Tn5	IS50L nonfunctional	*kan*^{R}	IS50R functional

FIGURE 21.4 A composite transposon has a central region carrying markers (such as drug resistance) flanked by IS modules. The modules have short inverted terminal repeats. If the modules themselves are in inverted orientation, the short inverted terminal repeats at the ends of the transposon are identical.

KEY CONCEPTS

- An insertion sequence is a transposon that codes for the enzyme(s) needed for transposition flanked by short inverted terminal repeats.
- The target site at which a transposon is inserted is duplicated during the insertion process to form two repeats in direct orientation at the ends of the transposon.
- The length of the direct repeat is 5 to 9 bp and is characteristic for any particular transposon.
- Transposons can carry other genes in addition to those coding for transposition.
- Composite transposons have a central region flanked by an IS element at each end.
- Either one or both of the IS elements of a composite transposon may be able to undertake transposition.
- A composite transposon may transpose as a unit, but an active IS element at either end may also transpose independently.

CONCEPT AND REASONING CHECK

Can a transposon with intact terminal repeats but a mutant transposase gene ever transpose? Why or why not?

21.3 Transposition Occurs by Both Replicative and Nonreplicative Pathways

The insertion of a transposon into a new site is illustrated in **FIGURE 21.5**. It consists of making staggered breaks in the target DNA, joining the transposon to the protruding single-stranded ends, and filling in the gaps. The generation and filling of the staggered ends explain the occurrence of the direct repeats of target DNA at the site of

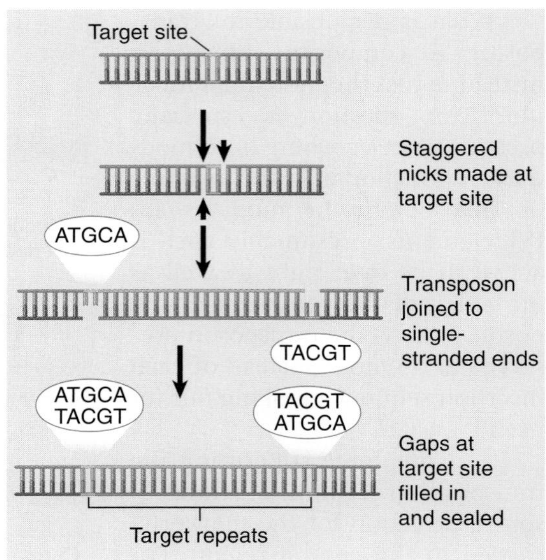
Target site
Staggered nicks made at target site
ATGCA
Transposon joined to single-stranded ends
TACGT
ATGCA
TACGT
TACGT
ATGCA
Gaps at target site filled in and sealed
Target repeats

FIGURE 21.5 The direct repeats of target DNA flanking a transposon are generated by the introduction of staggered cuts whose protruding ends are linked to the transposon.

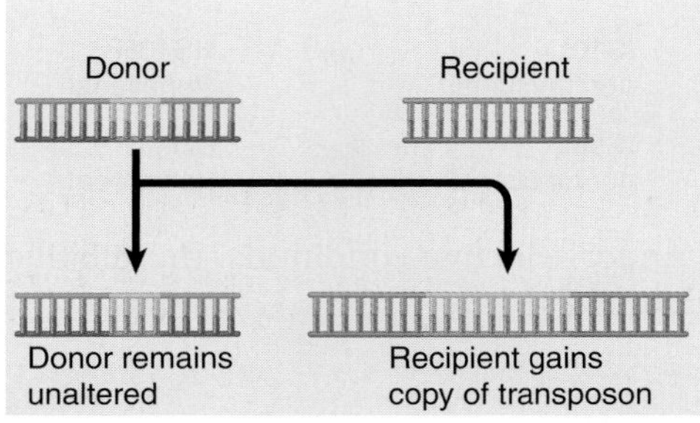
Donor
Recipient
Donor remains unaltered
Recipient gains copy of transposon

FIGURE 21.6 Replicative transposition creates a copy of the transposon, which inserts at a recipient site. The donor site remains unchanged, so both donor and recipient have a copy of the transposon.

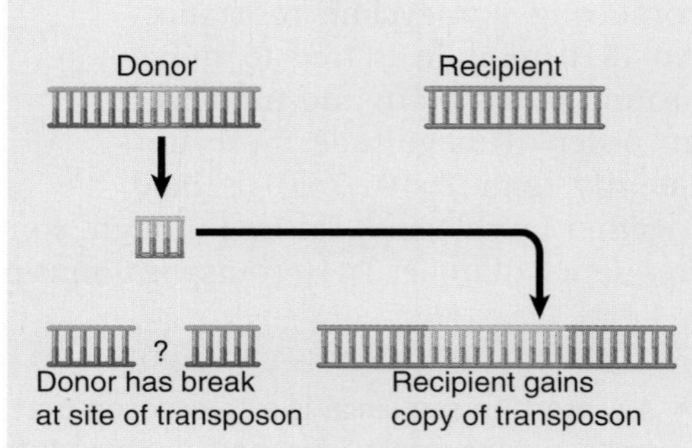
Donor
Recipient
?
Donor has break at site of transposon
Recipient gains copy of transposon

FIGURE 21.7 Nonreplicative transposition allows a transposon to move as a physical entity from a donor to a recipient site. This leaves a break at the donor site, which is lethal unless it can be repaired.

insertion. The stagger between the cuts on the two strands determines the length of the direct repeats; thus the target repeat characteristic of each transposon reflects the geometry of the enzyme involved in cutting target DNA.

The use of staggered ends is common to all means of transposition, but we can distinguish three different types of mechanism by which a transposon moves:

- In **replicative transposition**, the element is duplicated during the reaction, so that the transposing entity is a copy of the original element. FIGURE 21.6 summarizes the results of such a transposition. The transposon is copied as part of its movement. One copy remains at the original site, whereas the other inserts at the new site. Thus transposition is accompanied by an increase in the number of copies of the transposon. Replicative transposition involves two types of enzymatic activity: a transposase that acts on the ends of the original transposon, and a **resolvase** that acts on the duplicated copies.
- In **nonreplicative transposition**, the transposing element moves as a physical entity directly from one site to another and is conserved. The insertion sequences and composite transposons Tn10 and Tn5 use the mechanism shown in FIGURE 21.7, which involves the release of the transposon from the

replicative transposition The movement of a transposon by a mechanism in which first it is replicated, and then one copy is transferred to a new site.

resolvase The enzyme activity involved in site-specific recombination between two copies of a transposon that has been duplicated.

nonreplicative transposition The movement of a transposon that leaves a donor site (usually generating a double-strand break) and moves to a new site.

flanking donor DNA during transfer. This type of mechanism requires only a transposase. Another mechanism utilizes the connection of donor and target DNA sequences and shares some steps with replicative transposition. Both mechanisms of nonreplicative transposition cause the element to be inserted at the target site and lost from the donor site. What happens to the donor molecule after a nonreplicative transposition? Its survival requires that host repair systems recognize the double-strand break and repair it.

- **Conservative transposition** describes another sort of nonreplicative event, in which the element is excised from the donor site and inserted into a target site by a series of events in which every nucleotide bond is conserved. FIGURE 21.8 summarizes the result of a conservative event. This resembles the mechanism of lambda integration discussed in *Section 19.8, Site-Specific Recombination Resembles Topoisomerase Activity*, and the transposases of such elements are related to the λ integrase family. The elements that use this mechanism are large, and can mediate transfer not only of the element itself but also of donor DNA from one bacterium to another.

conservative transposition The movement of large elements that were originally classified as transposons but now are considered to be episomes. The mechanism of movement resembles that of phage excision and integration.

Some transposons use only one type of pathway for transposition, whereas others may be able to use multiple pathways. For example, the elements IS1 and IS903 use both nonreplicative and replicative pathways.

The same basic types of reaction are involved in all classes of transposition event. The ends of the transposon are disconnected from the donor DNA by cleavage reactions that generate 3′–OH ends. The exposed ends are then joined to the target DNA by transfer reactions, involving transesterification in which the 3′–OH end directly attacks the target DNA. These reactions take place within a nucleoprotein complex that contains the necessary enzymes and both ends of the transposon. Transposons differ as to whether the target DNA is recognized before or after the cleavage of the transposon itself.

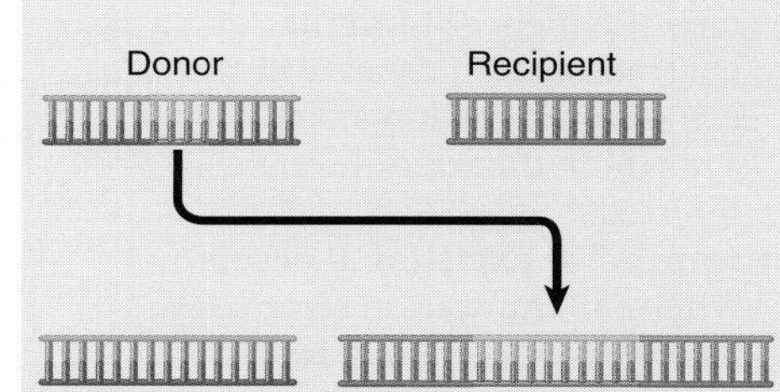
Donor
Recipient

FIGURE 21.8 Conservative transposition involves direct movement with no loss of nucleotide bonds; compare with lambda integration and excision.

The choice of target site is in effect made by the transposase. In some cases, the target is chosen virtually at random. In others, there is specificity for a consensus sequence or for some other feature in DNA. The feature can take the form of a structure in DNA, such as DNA containing an intrinsic bend, or a protein-DNA complex. In the latter case, the nature of the target complex can cause the transposon to insert at specific promoters (such as Ty1 or Ty3, which select RNA polymerase III promoters in yeast), inactive regions of the chromosome, or replicating DNA.

In addition to the "simple" intermolecular transposition that results in insertion at a new site, transposons promote other types of DNA rearrangements. Some of these events are consequences of the relationship between the multiple copies of the transposon. Others represent alternative outcomes of the transposition mechanism, and they leave clues about the nature of the underlying events.

Rearrangements of host DNA may result when a transposon inserts a copy at a second site near its original location. Host systems may undertake reciprocal recombination between the two copies of the transposon; the consequences are determined by whether the repeats are oriented in the same or inverted direction.

FIGURE 21.9 illustrates the general rule that recombination between any pair of direct repeats will delete the material between them. The intervening region is excised as a circle of DNA (which is lost from the cell); the chromosome retains a single copy of the direct repeat. A recombination between the directly repeated IS1 modules of the composite transposon Tn9 would replace the transposon with a single IS1 module.

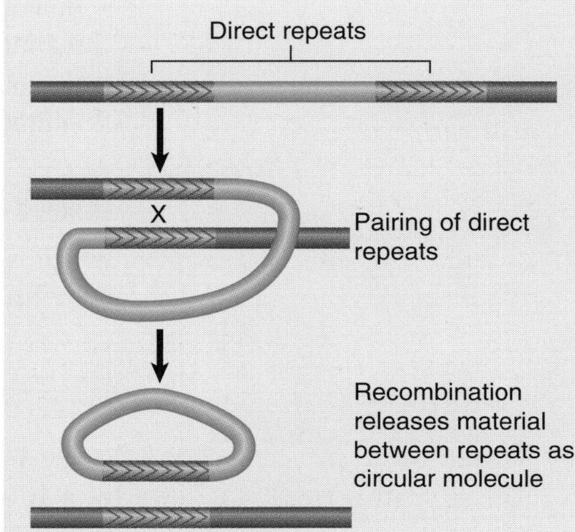
Direct repeats
X
Pairing of direct repeats
Recombination releases material between repeats as circular molecule

FIGURE 21.9 Reciprocal recombination between direct repeats excises the material between them; each product of recombination has one copy of the direct repeat.

Deletion of sequences adjacent to a transposon could therefore result from a two-stage process; transposition generates a direct repeat of a transposon, and recombination occurs between the repeats. The majority of

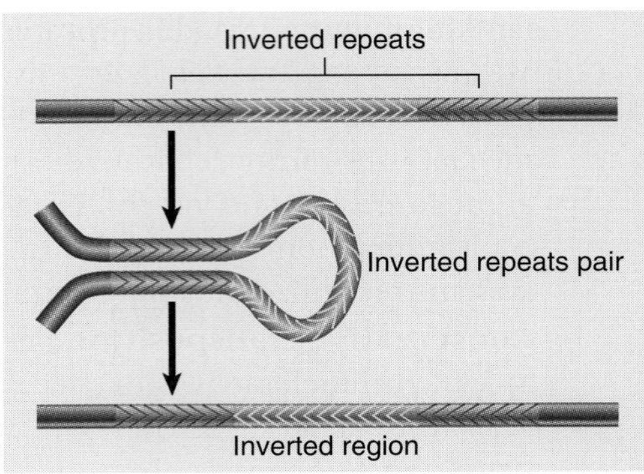

FIGURE 21.10 Reciprocal recombination between inverted repeats inverts the region between them.

deletions that arise in the vicinity of transposons, however, probably result from a variation in the pathway followed in the transposition event itself.

FIGURE 21.10 depicts the consequences of a reciprocal recombination between a pair of inverted repeats. The region between the repeats becomes inverted; the repeats themselves remain available to sponsor further inversions. A composite transposon whose modules are inverted is a stable component of the genome, although the direction of the central region with regard to the modules could be inverted by recombination.

Excision is not supported by transposons themselves, but may occur when bacterial enzymes recognize homologous regions in the transposons. This is important because the loss of a transposon may restore function at the site of insertion. **Precise excision** requires removal of the transposon plus one copy of the duplicated sequence. This is rare; it occurs at a frequency of ~10^{-6} for Tn5 and ~10^{-9} for Tn10. It probably involves a recombination between the 9 bp duplicated target sites. **Imprecise excision** leaves a remnant of the transposon, and occurs at a much higher frequency than precise excision (e.g., ~10^{-6} for Tn10). Neither type of excision relies on transposon-coded functions, but the mechanism is not known. Excision is RecA-independent and could occur by some cellular mechanism that generates spontaneous deletions between closely spaced repeated sequences.

- **precise excision** The removal of a transposon plus one of the duplicated target sequences from the chromosome. Such an event can restore function at the site where the transposon inserted.
- **imprecise excision** It occurs when the transposon removes itself from the original insertion site, but leaves behind some of its sequence.

KEY CONCEPTS

- All transposons use a common mechanism in which staggered nicks are made in target DNA, the transposon is joined to the protruding ends, and the gaps are filled.
- The order of events and exact nature of the connections between transposon and target DNA determine whether transposition is replicative or nonreplicative.
- Homologous recombination between multiple copies of a transposon causes rearrangement of host DNA.
- Homologous recombination between the repeats of a transposon may lead to precise or imprecise excision.

CONCEPT AND REASONING CHECK

Which is more likely to cause detrimental mutations in a host, a transposon that uses random insertion or one that shows site selectivity? Why?

21.4 Mechanisms of Transposition

Many mobile DNA elements transpose from one chromosomal location to another by a fundamentally similar mechanism. They include IS elements, prokaryotic and eukaryotic transposons, and certain bacteriophages. Insertion of the DNA copy of retroviral RNA uses a similar mechanism (see *Section 21.7, The Retrovirus Life Cycle Involves Transposition-Like Events*). The first stages of immunoglobulin recombination also are similar (see *Section 22.6, The RAG Proteins Catalyze Breakage and Reunion*).

CONCEPT AND REASONING CHECK

Inactive SINEs have been known to cause mutations in humans not by transposition but by unequal homologous recombination. Explain how this might occur.

21.11 Summary

Prokaryotic and eukaryotic cells contain a variety of transposons that mobilize by moving or copying DNA sequences. The transposon can be identified only as an entity within the genome; its mobility does not involve an independent form.

The archetypal transposon has inverted repeats at its termini and generates direct repeats of a short sequence at the site of insertion. The simplest types are the bacterial insertion sequences (IS), which consist essentially of the inverted terminal repeats flanking a coding frame(s) whose product(s) provide transposition activity. Composite transposons have terminal modules that consist of IS elements; one or both of the IS modules provides transposase activity, and the sequences between them (often carrying antibiotic resistance) are treated as passengers.

The generation of target repeats flanking a transposon reflects a common feature of transposition. The target site is cleaved at points that are staggered on each DNA strand by a fixed distance (often five or nine base pairs). The transposon is in effect inserted between protruding single-stranded ends generated by the staggered cuts. Target repeats are generated by filling in the single-stranded regions.

IS elements, composite transposons, and P elements mobilize by nonreplicative transposition, in which the element moves directly from a donor site to a recipient site. A single transposase enzyme undertakes the reaction. It occurs by a "cut and paste" mechanism in which the transposon is separated from flanking DNA. Cleavage of the transposon ends, nicking of the target site, and connection of the transposon ends to the staggered nicks all occur in a nucleoprotein complex containing the transposase. Loss of the transposon from the donor creates a double-strand break that must be repaired by the cell.

The best-characterized transposons in plants are the controlling elements of maize, which fall into several families. Each family contains a single type of autonomous element that is analogous to bacterial transposons in its ability to mobilize. A family also contains many different nonautonomous elements that are derived by mutations (usually deletions) of the autonomous element. The nonautonomous elements lack the ability to transpose, but display transposition activity and other abilities of the autonomous element when an autonomous element is present to provide the necessary *trans*-acting functions.

In addition to the direct consequences of insertion and excision, the maize elements may also control the activities of genes at or near the sites where they are inserted; this control may be subject to developmental regulation. Maize elements inserted into genes may be excised from the transcripts, which explains why they do not simply impede gene activity. Control of target gene expression involves a variety of molecular effects, including activation by provision of an enhancer and suppression by interference with posttranscriptional events.

Transposition of maize elements (in particular *Ac*) is nonreplicative, and probably requires only a single transposase enzyme coded by the element. Transposition occurs preferentially after replication of the element. It is likely that there are mechanisms to limit the frequency of transposition. Advantageous rearrangements of the maize genome may have been connected with the presence of the elements.

P elements in *D. melanogaster* are responsible for hybrid dysgenesis, which could be a forerunner of speciation. A cross between a male carrying P elements and a female lacking them generates hybrids that are sterile. A P element has four open reading frames, which are separated by introns. Splicing of the first three ORFs generates a 66 kD repressor and occurs in all cells. Splicing of all four ORFs to generate the 87 kD transposase occurs only in the germline by a tissue-specific splicing event.

P elements mobilize when exposed to cytoplasm lacking the repressor. The burst of transposition events inactivates the genome by random insertions. Only a complete P element can generate transposase, but defective elements can be mobilized in *trans* by the enzyme.

Reverse transcription is the unifying mechanism for reproduction of retroviruses and perpetuation of retroposons. Retroviruses have genomes of single-stranded RNA that are replicated through a double-stranded DNA intermediate. An individual retrovirus contains two copies of its genome. The genome contains the *gag, pol,* and *env* genes that are translated into polyproteins, each of which is cleaved into smaller functional proteins. The Gag and Env components are concerned with packing RNA and generating the virion; the Pol components are concerned with nucleic acid synthesis.

Reverse transcriptase is the major component of Pol, and is responsible for synthesizing a DNA (minus strand) copy of the viral (plus strand) RNA. The DNA product is longer than the RNA template; by switching template strands, reverse transcriptase copies the 3′ sequence of the RNA to the 5′ end of the DNA, and copies the 5′ sequence of the RNA to the 3′ end of the DNA. This generates the characteristic LTRs (long terminal repeats) of the DNA. A similar switch of templates occurs when the plus strand of DNA is synthesized using the minus strand as a template. Linear duplex DNA is inserted into a host genome by the integrase enzyme. Transcription of the integrated DNA from a promoter in the left LTR generates further copies of the RNA sequence.

During an infective cycle, a retrovirus may exchange part of its usual sequence for a cellular sequence; the resulting virus is usually replication-defective but can be perpetuated in the course of a joint infection with a helper virus. Many of the defective viruses have gained an RNA version (*v-onc*) of a cellular gene (*c-onc*). The *onc* sequence may be any one of a number of genes whose expression in *v-onc* form causes the cell to be transformed into a tumorigenic phenotype.

The integration event generates direct target repeats (like transposons that mobilize via DNA). An inserted provirus therefore has direct terminal repeats of the LTRs, flanked by short repeats of target DNA. Mammalian and avian genomes have endogenous (inactive) proviruses with such structures. Other elements with this organization have been found in a variety of genomes, most notably in *S. cerevisiae* (*Ty* elements) and *D. melanogaster* (*copia* elements). They may generate particles resembling viruses but do not have infectious capability. The LINE sequences of mammalian genomes are further removed from the retroviruses but retain enough similarities to suggest a common origin. LINES lack LTRs and use a different type of priming event to initiate reverse transcription, and both autonomous and nonautonomous LINEs exist. Autonomous LINEs encode functional reverse transcriptase and endonuclease activities.

The members of another class of retroposons have the hallmarks of transposition via RNA, but have no coding sequences (or at least none resembling retroviral functions). A particularly prominent family that appears to have originated from a processing event is the mammalian SINE; it includes the human Alu family. Some snRNAs, including 7SL snRNA (a component of the SRP), are related to this family. Both SINEs and LINEs comprise a large fraction of mammalian genomes.

CHAPTER QUESTIONS

1. Recombination between two transposons that are present in the same replicon in direct repeat orientation results in:
   A. degradation of the sequence between the two transposons.
   B. inversion of the sequence between the two transposons.
   C. excision and circularization of the sequence between the two transposons.
   D. any of the above are possible.
2. Which occurs least frequently: transposon insertion, imprecise transposon excision, or precise transposon excision?
   A. transposon insertion
   B. imprecise transposon excision
   C. precise transposon excision
   D. These all occur at about the same rate.
3. Which of the following maize transposons is a nonautonomous element of the same family as the activator (Ac) element?
   A. Spm
   B. Dt
   C. Mu
   D. Ds
4. The P element produces a:
   A. repressor of transposition, which is inherited paternally in the cytoplasm.
   B. activator of transposition, which is inherited paternally in the cytoplasm.
   C. repressor of transposition, which is inherited maternally in the cytoplasm.
   D. activator of transposition, which is inherited maternally in the cytoplasm.

List two activities for each of the following retroviral proteins

Gag **5.** ______________________ **6.** ______________________
Pol **7.** ______________________ **8.** ______________________
Env **9.** ______________________ **10.** ______________________

11. The enzyme responsible for generating the initial DNA copy from RNA in a retrovirus is:
    A. DNA polymerase.
    B. RNA polymerase.
    C. reverse transcriptase.
    D. integrase.
12. Translation of which retrovirus protein(s) requires ribosome frameshifting?
    A. Gag
    B. Pol
    C. Env
    D. all of the above
13. Retroviral RNA ends in:
    A. direct repeats.
    B. indirect repeats.
    C. long terminal repeats.
    D. unique sequences—no repeats.
14. Replication-defective retroviruses are most commonly generated by:
    A. recombination and rearrangement of sequences.
    B. mutation at critical sites in viral genes.
    C. deletion of a segment of the viral genome.
    D. insertion of sequences into viral genes.

15. Which of the following types of element makes up the greatest proportion of the human genome?
   A. retroviruses and retroposons
   B. LINEs
   C. SINEs
   D. DNA transposons
16. Alu elements are what type of transposable element?
   A. SINEs
   B. LINEs
   C. retroposon
   D. DNA transposon
17. About how many copies of Alu family elements are present in the haploid human genome?
   A. 20,000
   B. 50,000
   C. 160,000
   D. 300,000
18. Alu family elements are characterized by which of the following?
   A. identical elements of 170 bp with terminal inverted repeats and containing a single AluI restriction site
   B. related elements of 170 bp with terminal tandem repeats and containing a single AluI restriction site
   C. identical elements of 300 bp with terminal inverted repeats and containing a single AluI restriction site
   D. related elements of 300 bp with terminal tandem repeats and containing a single AluI restriction site

KEY TERMS

***Ac* element**
acentric fragment
Alu element
autonomous controlling element
cointegrate
composite elements
conservative transposition
controlling elements
cytotype
direct repeats
***Ds* element**
helper virus
hybrid dysgenesis
imprecise excision
insertion sequences (IS)
integrase
inverted terminal repeats
long interspersed elements (LINEs)
long terminal repeat (LTR)
minus strand DNA
nonautonomous controlling element
nonreplicative transposition
nonviral superfamily
P element
plus strand DNA
plus strand virus
precise excision
provirus
R segments
replication-defective virus
replicative transposition
resolution
resolvase
retrotransposon (retroposon)
retrovirus
reverse transcriptase
short interspersed elements (SINEs)
Tn
transducing virus
transposase
transposon (transposable element)
U3
U5
viral superfamily

FURTHER READING

Belancio, V. P., Hedges, D. J., and Deininger, P. (2008). Mammalian non-LTR retrotransposons: for better or worse, in sickness and in health. *Genome Res.* 18(3), 343–358.

A review of retrotransposons in mammals, including LINEs and SINES, with a discussion of the effects of transposition in human biology.

Castro J. P. and Carareto, C. M. (2004). *Drosophila melanogaster* P transposable elements: mechanisms of transposition and regulation. *Genetica* 121(2), 107–118.

A review of the mechanisms of P element transposition and hybrid dysgenesis.

Haniford, D. B. (2006). Transpososome dynamics and regulation in Tn10 transposition. *Crit. Rev. Biochem. Mol. Biol.* 41(6), 407–424.

A review of the nonreplicative transposition mechanism typified by the bacterial Tn10 transposon.

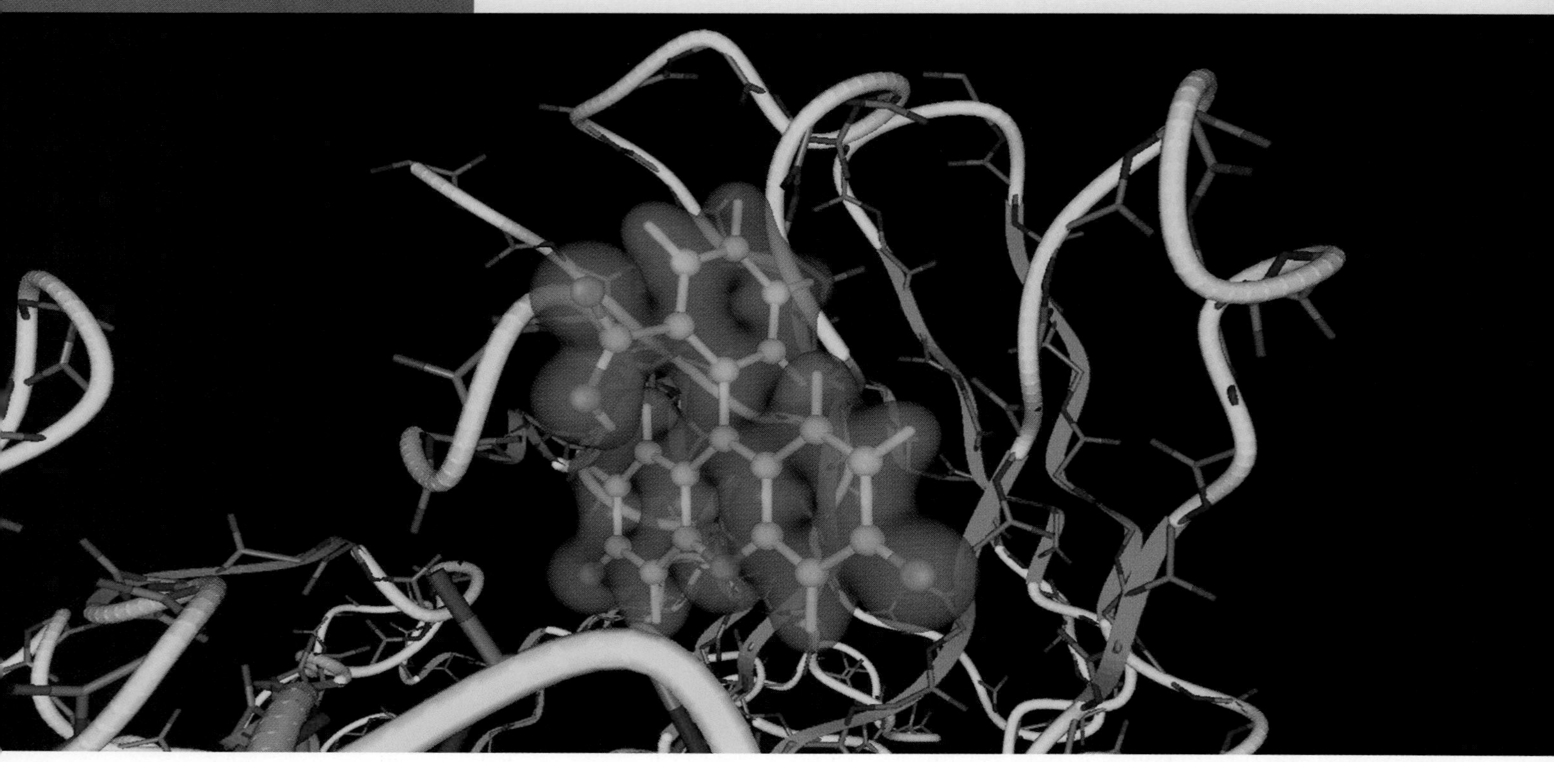

The structure of the antigen-binding site of an antibody, bound to the fluorescent molecule fluorescein (in green and orange). Photo courtesy of Kim Baldridge, University of Zürich.

Immune Diversity

CHAPTER OUTLINE

22.1 Introduction

It is an axiom of genetics that the genetic constitution created in the zygote by the combination of sperm and egg is inherited by all somatic cells of the organism. We look to differential control of gene expression, rather than to changes in DNA content, to explain the different phenotypes of particular somatic cells.

Yet there are exceptional situations in which the reorganization of certain DNA sequences is used to regulate gene expression or to create new genes. The immune system provides a striking and extensive case in which the content of the genome changes, when recombination creates active genes in lymphocytes. Other cases are represented by the substitution of one sequence for another to change the mating type of yeast or to generate new surface antigens by trypanosomes.

The **immune response** of vertebrates provides a protective system that distinguishes foreign proteins from the proteins of the organism itself. Foreign material (or part of the foreign material) is recognized as comprising an **antigen**. Often the antigen is a protein that has entered the bloodstream of the animal—for example, the coat protein of an infecting virus. Exposure to an antigen initiates production of an immune response that *specifically recognizes the antigen and destroys it.*

Immune reactions are performed by white blood cells: B and T lymphocytes and macrophages. The lymphocytes are named after the tissues that produce them. In mammals, **B cells** mature in the bone marrow, whereas **T cells** mature in the thymus. *Each class of lymphocyte uses the rearrangement of DNA as a mechanism for producing the proteins that enable it to participate in the immune response.*

The immune system has many ways to destroy an antigenic invader, but it is useful to consider them in two general classes. The type of response the immune system mounts when it encounters a foreign structure depends partly on the nature of the antigen. The response is defined according to whether it is executed principally by B cells or T cells.

The **humoral response** depends on B cells. It is mediated by the secretion of antibodies, which are **immunoglobulin** proteins. *Production of an* **antibody** *specific for a foreign molecule is the primary event responsible for recognition of an antigen.* Recognition requires the antibody to bind to a small region or structure on the antigen.

The function of antibodies is represented in FIGURE 22.1. Foreign material circulating in the bloodstream—for example, a toxin or pathogenic bacterium—has a surface that presents antigens. The antigen(s) are recognized by the antibodies, which form an antigen-antibody complex. This complex then attracts the attention of other components of the immune system.

The humoral response depends on these other components in two ways. First, B cells need signals provided by T cells to enable them to secrete antibodies. These T cells are called **helper T cells**, because they assist the B cells. Second, antigen-antibody formation is a trigger for the antigen to be destroyed. The major pathway is provided by the action of **complement**, a component whose name reflects its ability to "complement," or complete, the action

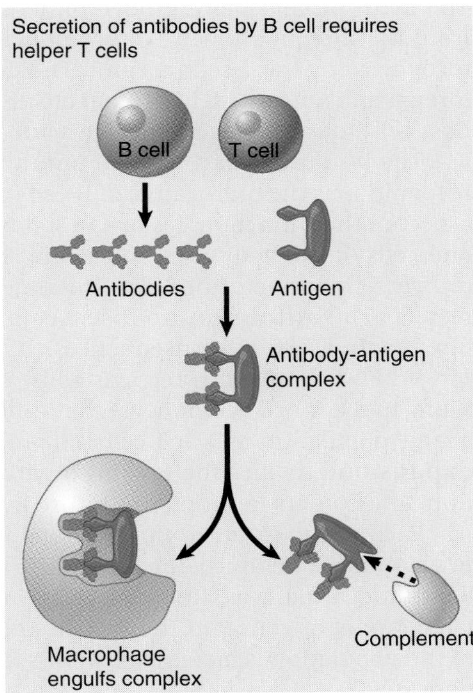

FIGURE 22.1 Humoral immunity is conferred by the binding of free antibodies to antigens to form antigen-antibody complexes that are removed from the bloodstream by macrophages or that are attacked directly by the complement proteins.

- **immune response** An organism's reaction, mediated by components of the immune system, to an antigen.
- **antigen** A molecule that can bind specifically to an antigen receptor, such as an antibody.
- **B cell** A lymphocyte that produces antibodies. Development occurs primarily in bone marrow.
- **T cells** Lymphocytes of the T (thymic) lineage; they may be subdivided into several functional types. They carry TcR and are involved in the cell-mediated immune response.
- **humoral response** An immune response that is mediated primarily by antibodies. It is defined as immunity that can be transferred from one organism to another by serum antibody.
- **immunoglobulin** A protein that is produced by B cells and that binds to a particular antigen.
- **antibody** A protein that is produced by B lymphocytes and that binds a particular antigen. They are synthesized in membrane-bound and secreted forms. Those produced during an immune response recruit effector functions to help neutralize and eliminate the pathogen.
- **helper T cell** A T lymphocyte that activates macrophages and stimulates B cell proliferation and antibody production. They usually express cell surface CD4 but not CD8.
- **complement** A set of ~20 proteins that function through a cascade of proteolytic actions to lyse infected target cells or to attract macrophages.

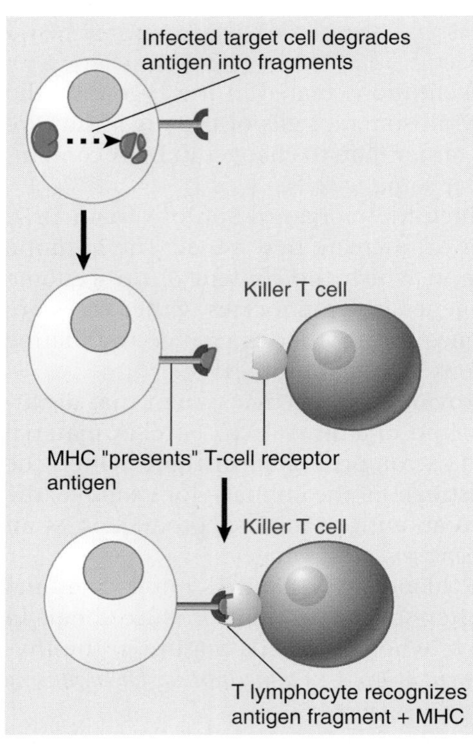

Infected target cell degrades antigen into fragments
Killer T cell
MHC "presents" antigen
T-cell receptor
Killer T cell
T lymphocyte recognizes antigen fragment + MHC

FIGURE 22.2 In cell-mediated immunity, killer T cells use the T cell receptor to recognize a fragment of the foreign antigen that is presented on the surface of the target cell by the MHC protein.

of the antibody itself. Complement consists of a set of ~20 proteins that function through a cascade of proteolytic actions. If the target antigen is part of a cell—for example, an infecting bacterium—the action of complement culminates in lysing the target cell. The action of complement also provides a means of attracting macrophages, which scavenge the target cells or their products. Alternatively, the antigen-antibody complex may be taken up directly by macrophages (scavenger cells) and destroyed.

The **cell-mediated response** is executed by a class of T lymphocytes called **cytotoxic T cells** (also called *killer T cells*). The basic function of the T cell in recognizing a target antigen is indicated in FIGURE 22.2. A cell-mediated response typically is elicited by an intracellular parasite, such as a virus that infects the body's own cells. As a result of the viral infection, fragments of foreign (viral) antigens are displayed on the surface of the cell. These fragments are recognized by the **T cell receptor (TCR)**, which is the T cell's equivalent of the antibody produced by a B cell.

- **cell-mediated response** The immune response that is mediated primarily by T lymphocytes. It is defined based on immunity that cannot be transferred from one organism to another by serum antibody.
- **cytotoxic T cell** A T lymphocyte (usually CD8+) that can be stimulated to kill cells containing intracellular pathogens, such as viruses.
- **T cell receptor (TCR)** The antigen receptor on T lymphocytes. It is clonally expressed and binds to a complex of MHC class I or class II protein and antigen-derived peptide.
- **major histocompatibility complex (MHC)** A chromosomal region containing genes that are involved in the immune response. The genes encode proteins for antigen presentation, cytokines, and complement, as well as other functions. It is highly polymorphic. Its genes and proteins are divided into three classes.
- **autoimmune disease** A pathological condition in which the immune response is directed against self antigens.
- **clonal expansion** The production of numerous daughter cells all arising from a single cell.

A crucial feature of this recognition reaction is that *the antigen must be* presented *by a cellular protein that is a member of the* **MHC (major histocompatibility complex)**. The MHC protein has a groove on its surface that binds a peptide fragment derived from the foreign antigen. The combination of peptide fragment and MHC protein is recognized by the T cell receptor. The demand that the T lymphocytes recognize both foreign antigen and MHC protein ensures that the cell-mediated response acts only on host cells that have been infected with a foreign antigen.

The purpose of each type of immune response is to attack a foreign target. Target recognition is the prerogative of B cell immunoglobulins and T cell receptors. A crucial aspect of their function lies in the ability to distinguish "self" from "nonself." Proteins and cells of the body itself must *never* be attacked. Foreign targets must be *destroyed entirely*. The property of failing to attack "self" is called *tolerance*. Loss of this ability results in an **autoimmune disease**, in which a body's immune system attacks itself, often with disastrous consequences.

When an antigen is recognized by an antibody or TCR, the recognition triggers a signal in the B or T lymphocyte that causes it to divide. Numerous cell divisions lead to a large population of B or T cells, all producing the same antibody or TCR. This **clonal expansion** provides the organism with an army of cells equipped to fight the infection, and confers immunity to future exposure to the same antigen.

Each of the three groups of proteins required for the immune response—immunoglobulins, T cell receptors, and MHC proteins—is diverse. Examining a large number of individuals, we find many variants of each protein. Each protein is coded by a large family of genes; in the case of antibodies and the T cell receptors, the diversity of the population is increased by DNA rearrangements that occur in the relevant lymphocytes.

Immunoglobulins and T cell receptors are direct counterparts, each produced by its own type of lymphocyte. The proteins are related in structure, and their genes are related in organization. The sources of variability are similar. The MHC proteins also share some common features with the antibodies, as do other lymphocyte-specific proteins. In dealing with the genetic organization of the immune system, we are therefore concerned with a series of related gene families, indeed a **superfamily** that may have evolved from some common ancestor representing a primitive immune response.

22.2 Immunoglobulin Genes Are Assembled from Their Parts in Lymphocytes

A remarkable feature of the immune response is an animal's ability to produce an appropriate antibody whenever it is exposed to a new antigen. How can the organism be prepared to produce antibody proteins, each of which is designed specifically to recognize an antigen whose structure cannot be anticipated?

For practical purposes, we usually predict that a mammal has the ability to produce 10^6 to 10^8 different antibodies. Each antibody is an immunoglobulin tetramer consisting of two identical **light chains (L)** and two identical **heavy chains (H)**. The structure of the immunoglobulin tetramer is illustrated in **FIGURE 22.3**. Light chains and heavy chains share the same general type of organization in which each protein chain consists of two principal regions: the N-terminal **variable region (V region)**, and the C-terminal **constant region (C region)**. As the names suggest, the variable regions show considerable changes in sequence from one protein to the next, whereas the constant regions show substantial homology. There are two types of light chain and ~10 types of heavy chain. Different classes of immunoglobulins, which serve different functions, are determined by the heavy chain constant region (see Figure 22.15).

Corresponding regions of the light chains and heavy chains associate to generate distinct domains in the immunoglobulin protein. The *variable (V) domain* is generated by association between the variable regions of the light chain and heavy chain. *The V domain is responsible for recognizing the antigen.* An immunoglobulin has a Y-shaped structure in which the arms of the Y are identical, and each arm has a copy of the V domain. Production of V domains of different specificities creates the ability to respond to diverse antigens. The total number of variable regions for either light- or heavy-chain proteins is measured in hundreds. *Thus the protein displays the maximum versatility in the region responsible for binding the antigen.*

The number of constant regions is vastly smaller than the number of variable regions; typically there are only one to ten C regions for any particular type of chain. The constant regions in the subunits of the immunoglobulin tetramer associate to generate several individual *C domains*. The first domain results from association of the single constant region of the light-chain (C_L) with the C_{H1} part of the heavy-chain constant region. The two copies of this domain complete the arms of the Y-shaped molecule. Association between the C regions of the heavy chains generates the remaining C domains, which vary in number depending on the type of heavy chain.

Comparing the characteristics of the variable and constant regions, we see the central dilemma in immunoglobulin gene structure. How does the genome code for a set of proteins in which any individual polypeptide chain must have one of <10 possible C regions, but can have any one of several hundred possible V regions? It turns out that the number of coding sequences for each type of region reflects its variability. There are many genes coding for V regions, but only a few genes coding for C regions.

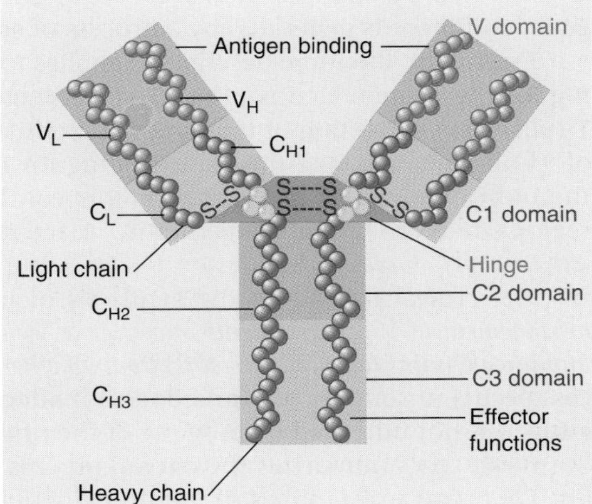

FIGURE 22.3 Heavy and light chains combine to generate an immunoglobulin with several discrete domains.

- **superfamily** A set of genes all related by presumed descent from a common ancestor, but now showing considerable variation.
- **light chain** The smaller of the two types of subunits that make up an antibody tetramer. Each antibody contains two light chains. The N-terminus of the light chain forms part of the antigen recognition site.
- **heavy chain** The larger of the two types of subunits that make up an antibody tetramer. Each antibody contains two heavy chains. The N-terminus of the heavy chain forms part of the antigen recognition site, while the C-terminus determines the antibody subclass.
- **variable region (V region)** An antigen-binding site of an immunoglobulin or T cell receptor molecule. They are composed of the variable domains of the component chains. They are coded by V gene segments and vary extensively among antigen receptors as the result of multiple, different genomic copies and of changes introduced during synthesis.
- **constant region (C region)** The part of an immunoglobulin or T cell receptor that varies least in amino acid sequence between different molecules. They are coded by C gene segments. The heavy chain regions identify the type of immunoglobulin and recruits effector functions.

FIGURE 22.4 The germline genome has three separate clusters of V gene segments separated from C gene segments. Recombination can occur in each of the clusters to create an active gene by linking a V segment to a C segment. To produce a functional immunoglobulin, a lymphocyte must successfully recombine the heavy cluster and one of the light clusters.

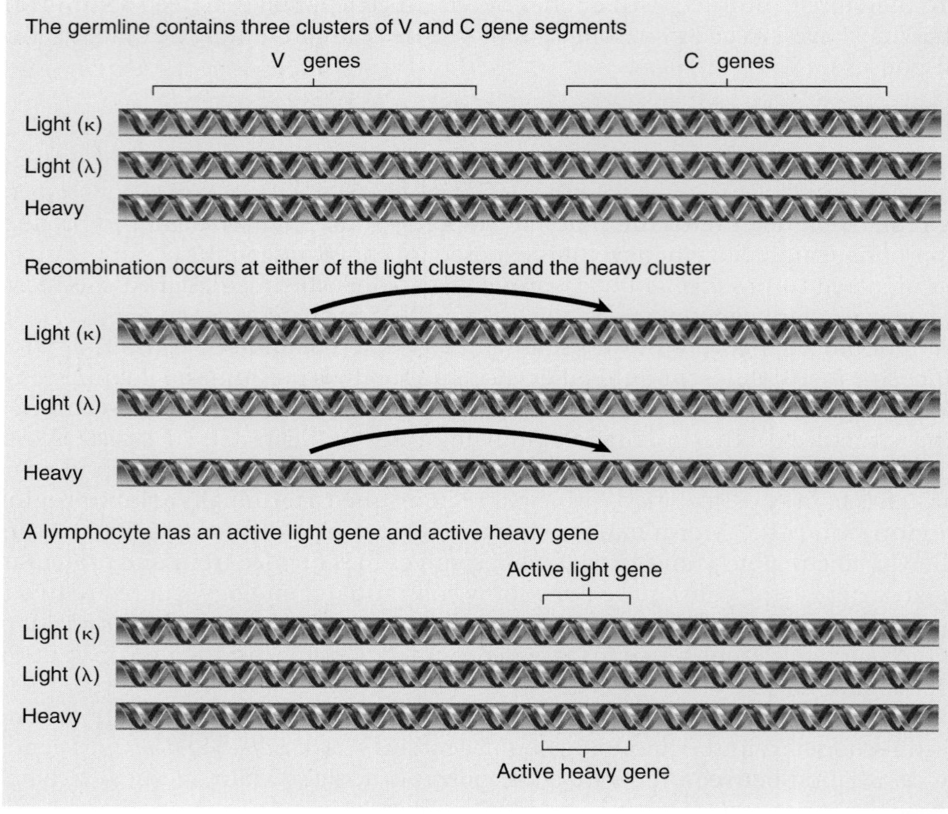

The germline contains three clusters of V and C gene segments
V genes
C genes
Light (κ)
Light (λ)
Heavy
Recombination occurs at either of the light clusters and the heavy cluster
Light (κ)
Light (λ)
Heavy
A lymphocyte has an active light gene and active heavy gene
Active light gene
Light (κ)
Light (λ)
Heavy
Active heavy gene

- **V gene** A sequence coding for the major part of the variable (N-terminal) region of an immunoglobulin chain.
- **C genes** Genes that code for the constant regions of immunoglobulin protein chains.
- **somatic recombination** The process of joining a V gene to a C gene in a lymphocyte to generate an immunoglobulin or T cell receptor.
- **somatic hypermutation** The introduction of somatic mutations in a rearranged immunoglobulin gene. The mutations can change the sequence of the corresponding antibody, especially in its antigen-binding site.

In this context, *"gene" means a sequence of DNA coding for a discrete part of the final immunoglobulin polypeptide* (heavy or light chain). Thus **V genes** code for variable regions and **C genes** code for constant regions, but *neither type of gene is expressed as an independent unit.* To construct a unit that can be expressed in the form of an authentic light or heavy chain, a V gene must be joined physically to a C gene. In this system, two "genes" code for one polypeptide. To avoid confusion, we will refer to these units as "gene segments" rather than "genes."

The sequences coding for light chains and heavy chains are assembled in the same way: *any one of many V gene segments may be joined to any one of a few C gene segments.* This **somatic recombination** occurs *in the B lymphocyte in which the antibody is expressed.* The large number of available V gene segments is responsible for a major part of the diversity of immunoglobulins. Not all diversity is coded in the genome, though; some is generated by changes that occur during the process of constructing a functional gene, and more is generated by a process of **somatic hypermutation**.

Essentially the same description applies to the formation of functional genes coding for the protein chains of the T cell receptor. Two types of receptor are found on T cells—one consisting of two types of chain called α and β, and the other consisting of γ and δ chains. Like the genes coding for immunoglobulins, the genes coding for the individual chains in T cell receptors consist of separate parts, including V and C regions, that are brought together in an active T cell (see *Section 22.10, T Cell Receptors Are Related to Immunoglobulins*).

The crucial fact about the synthesis of immunoglobulins, therefore, is that *the arrangement of V gene segments and C gene segments is different in the cells producing the immunoglobulins (or T cell receptors) from all other somatic cells or germ cells.* The entire process occurs in somatic cells and does not affect the germline; thus the response to an antigen is not inherited by progeny of the organism.

FIGURE 22.4 summarizes the overall process of creating a functional immunoglobulin. There are two families of immunoglobulin light chains, κ and λ, and one family containing all the types of heavy chain (H). Each family resides on a different chromosome and consists of its own set of both V gene segments and C gene segments. This is called the *germline pattern,* and is found in the germline and in somatic cells of all lineages other than the immune system.

In a cell expressing an antibody, though, each of its chains—one light type (either κ or λ) and one heavy type—is encoded by a single intact gene. The recombination event that brings a V gene segment to partner a C gene segment creates an active gene consisting of exons that correspond precisely with the functional domains of the protein. The introns are removed in the usual way by RNA splicing.

The principles by which functional genes are assembled are the same in each family, but there are differences in the details of the organization of the V gene segments and C gene segments, and correspondingly of the recombination reaction between them. In addition to the V gene segments and C gene segments, other short DNA sequences (including J segments and D segments) are included in the functional somatic loci.

KEY CONCEPTS

- An immunoglobulin is a tetramer of two light chains and two heavy chains.
- Light chains fall into the λ and κ families; heavy chains form a single family.
- Each chain has an N-terminal variable region (V) and a C-terminal constant region (C).
- The V domain recognizes the antigen and the C domain provides the effector response.
- V domains and C domains are separately coded by V gene segments and C gene segments, respectively.
- A gene coding for an intact immunoglobulin chain is generated by somatic recombination to join a V gene segment with a C gene segment.

CONCEPT AND REASONING CHECK

Why are there many more V gene segments than C gene segments?

22.3 Light Chains Are Assembled by a Single Recombination

A λ light chain is assembled from two parts, as illustrated in FIGURE 22.5. The V gene segment consists of the leader exon (L) separated by a single intron from the variable (V) segment. The C gene segment consists of the J segment separated by a single intron from the constant (C) exon.

The name of the **J segment** is an abbreviation for joining, because it identifies the region to which the V segment becomes connected. Thus the joining reaction does not directly involve V and C gene segments but occurs via the J segment; by the joining of "V and C gene segments" for light chains, we really mean V-JC joining.

▸ **J segments** Coding sequences in the immunoglobulin and T cell receptor loci. They are between the variable (V) and constant (C) gene segments.

The J segment is short and codes for the last few amino acids of the variable region. In the intact gene generated by recombination, the V-J segment constitutes a single exon coding for the entire variable region.

The consequences of the κ joining reaction are illustrated in FIGURE 22.6. A κ light chain also is assembled from two parts, but there is a difference in the organization of the C gene segment. A group of five J segments is spread over a region of 500 to 700 bp, separated from the C_κ exon by an intron of 2–3 kb. A V_κ segment may be joined to any one of the J segments.

Whichever J segment is used becomes the terminal part of the intact variable exon. Any J segments on the left of the recombining J segment are lost (J1 has been lost in the figure). Any J segment on the right of the recombining J segment is treated as part of the intron between the variable and constant exons (J3, J4, and J5 are included in the intron that is spliced out in the figure).

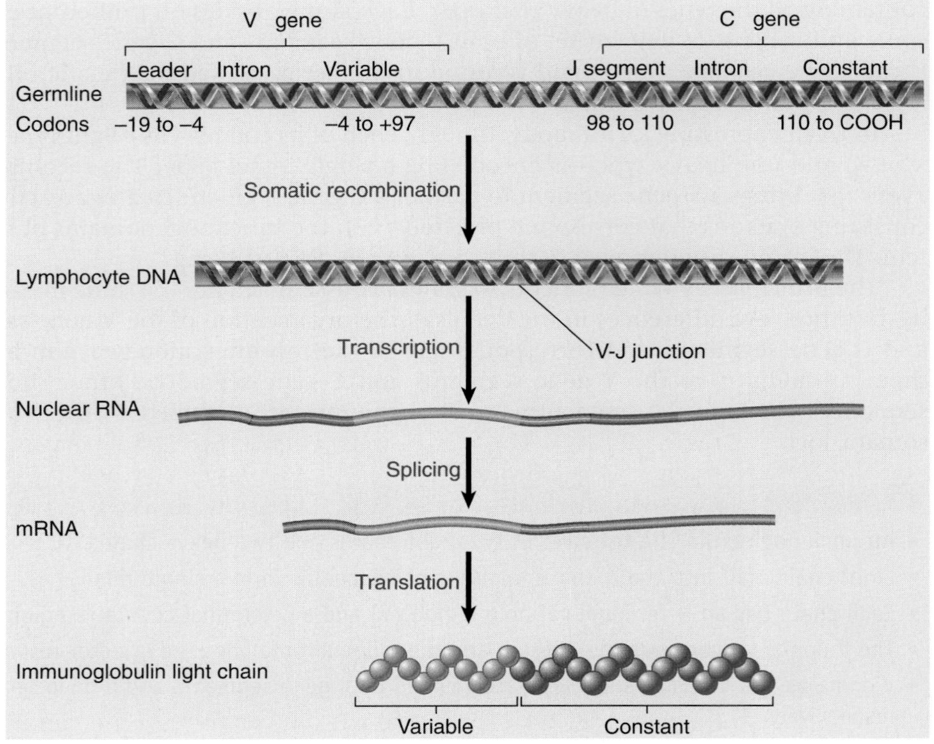

FIGURE 22.5 The lambda C gene segment is preceded by a J segment, so that V-J recombination generates a functional lambda light-chain gene. Only one V gene is shown for simplicity.

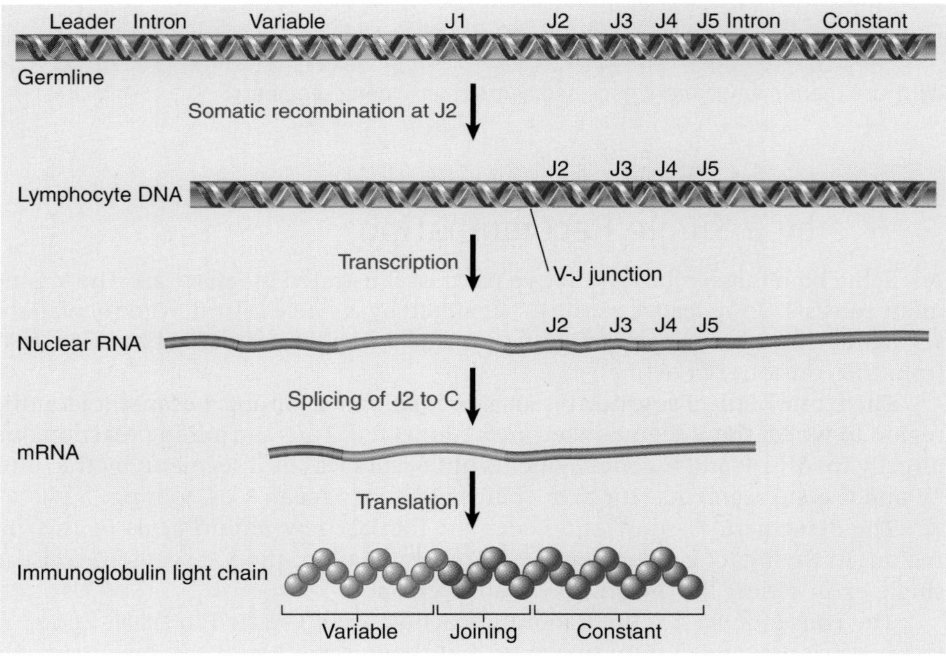

FIGURE 22.6 The kappa C gene segment is preceded by multiple J segments in the germline. V-J joining may recognize any one of the J segments, which is then spliced to the C gene segment during RNA processing. Only one V gene is shown for simplicity.

All functional J segments possess a signal at the left boundary that makes it possible to recombine with the V segment; they also possess a signal at the right boundary that can be used for splicing to the C exon. Whichever J segment is recognized in DNA V-J joining uses its splicing signal in RNA processing.

The λ locus has ~300 V_λ gene segments, along with ~6 C gene segments, each preceded by its own J segment, as shown in **FIGURE 22.7**. The λ locus in mouse is much less diverse than the λ locus in humans. The main difference is that in a mouse there

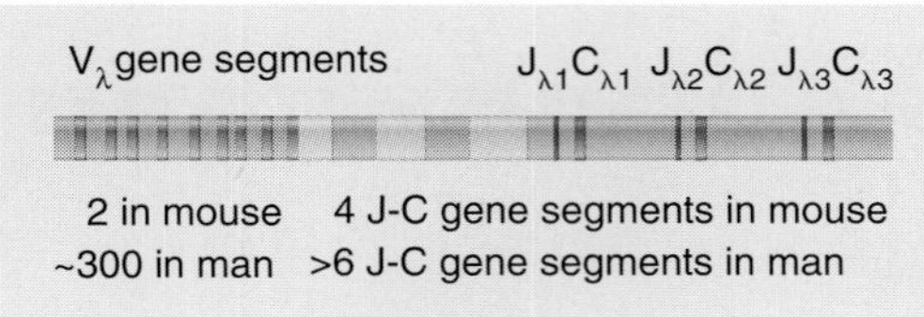

FIGURE 22.7 The lambda family consists of V gene segments linked to a small number of J-C gene segments.

FIGURE 22.8 The human and mouse kappa families consist of V gene segments linked to five J segments connected to a single C gene segment.

are only two V_λ gene segments; each is linked to two J-C regions. Of the four C_λ gene segments, one is inactive. At some time in the past, the mouse suffered a dramatic deletion of most of its germline V_λ gene segments.

FIGURE 22.8 shows that the κ locus has only one C gene segment, preceded by five J segments (one of them inactive). The V_κ gene segments occupy a large cluster on the chromosome, upstream of the constant region. The human cluster has two regions. Just preceding the C_κ gene segment, a region of 600 kb contains the five J_κ segments and 40 V_κ gene segments. A gap of 800 kb separates this region from another group of 36 V_κ gene segments.

The V_κ gene segments can be subdivided into families, which are defined by the criterion that members of a family have >80% amino acid identity. The mouse family is unusually large (~1000 genes), and there are ~18 V_κ families that vary in size from 2 to 100 members. Like other families of related genes, related V gene segments form subclusters, which are generated by duplication and divergence of individual ancestral members. Many of the V segments are inactive pseudogenes, though, and <50 are likely to be used to generate immunoglobulins.

A given lymphocyte generates *either* a κ *or* a λ light chain to associate with the heavy chain. In humans, ~60% of the light chains are κ and ~40% are λ. In mouse, 95% of B cells express the κ type of light chain, presumably because of the reduced number of λ gene segments.

KEY CONCEPTS

- A λ light chain is assembled by a single recombination between a V gene and a J-C gene segment.
- The V gene segment has a leader exon, intron, and variable-coding region.
- The J-C gene segment has a short J-coding exon, intron, and C-coding region.
- A κ light chain is assembled by a single recombination between a V gene segment and one of five J segments preceding the C gene.
- A light-chain locus can produce >1000 chains by combining 300 V gene segments with 4–5 C gene segments.

CONCEPT AND REASONING CHECK

The J segment is part of the C gene segments, but also is part of the variable region in the immunoglobulin. Explain.

22.4 Heavy Chains Are Assembled by Two Successive Recombinations

Heavy-chain construction involves an additional segment. The **D segment** (for diversity) provides an extra two to thirteen amino acids between the sequences encoded by the V segment and the J segment. A large array of D segments lies on the chromosome between the V_H segments and the four J_H segments.

▸ **D segment** An additional sequence that is found between the V and J regions of an immunoglobulin heavy chain.

V-D-J joining takes place in two stages, as illustrated in **FIGURE 22.9**. First one of the D segments recombines with a J_H segment; a V_H segment then recombines with the

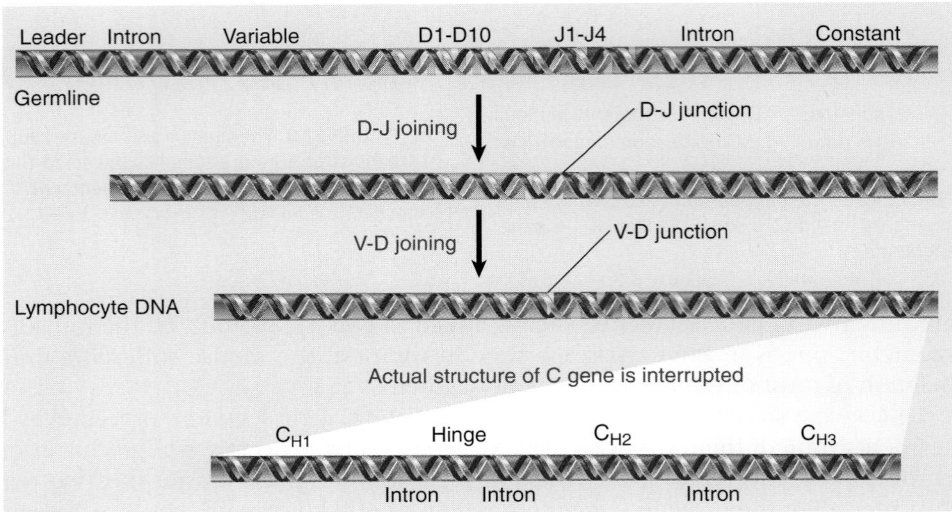

FIGURE 22.9 Heavy genes are assembled by sequential joining reactions. First a D segment is joined to a J segment, and then a V gene segment is joined to the D segment.

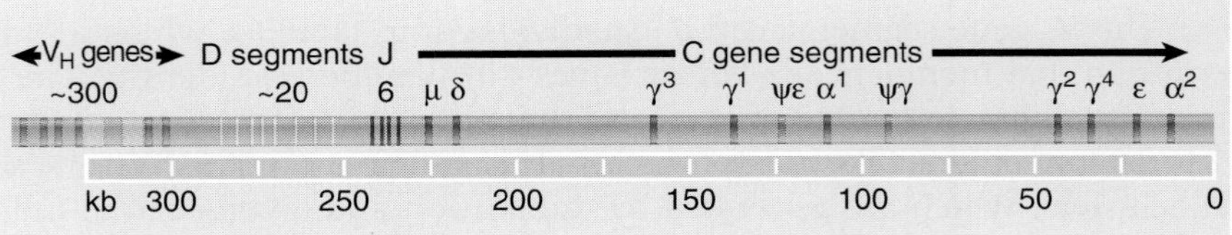

FIGURE 22.10 A single gene cluster in humans contains all the information for heavy-chain gene assembly. Note that the estimated number of active V_H genes and D segments is shown; there are more copies of each that are inactive (~300 V_H genes total and ~30 D segments).

DJ_H combined segment. The reconstruction leads to expression of the adjacent C_H segment (which consists of several exons).

The D segments are organized in a tandem array. The mouse heavy-chain locus contains twelve D segments of variable length; the human locus has ~27 D segments. Some unknown mechanism must ensure that the *same* D segment is involved in the D-J joining and V-D joining reactions. (When we discuss joining of V and C gene segments for heavy chains, we assume the process has been completed by V-D and D-J joining reactions.)

The V gene segments of all three immunoglobulin families are similar in organization. The first exon codes for a signal sequence involved in membrane attachment, and the second exon codes for the major part of the variable region itself (<100 codons long). The remainder of the variable region is provided by the D segment (in the H family only) and by a J segment (in all three families).

The structure of the constant region depends on the type of chain. For both κ and λ light chains, the constant region is encoded by a single exon (which becomes the third exon of the reconstructed, active gene). For H chains, the constant region is encoded by several exons. Corresponding with the protein chain shown in Figure 22.3, separate exons code for the regions C_{H1}, hinge, C_{H2}, and C_{H3}. Each C_H exon is ~100 codons long; the hinge is shorter. The introns usually are relatively small (~300 bp).

The single locus for heavy-chain production in humans consists of several discrete sections, as summarized in **FIGURE 22.10**. The mouse shows similar organization but different numbers of segments: there are more V_H gene segments, fewer D and J segments, and a slight difference in the number and organization of C gene segments. The 3′-most member of the V_H cluster is separated by only 20 kb from the first D segment. The D segments are spread over ~50 kb, followed by the cluster of J segments. Over the next 220 kb lie all the C_H gene segments. There are nine functional C_H gene segments and two pseudogenes. The organization suggests that a γ gene segment must have been duplicated to give the subcluster of γ-γ-ε-α, after which the entire group was then duplicated. By combining any one of ~65 functional V_H gene segments, ~27 D segments, and 6 J segments, the genome potentially can produce ~11,000 variable regions to accompany any C_H gene segment.

KEY CONCEPTS

- The units for heavy-chain recombination are a V gene, a D segment, and a J-C gene segment.
- The first recombination joins D to J-C.
- The second recombination joins V to D-J-C.
- The C segment consists of several exons.
- An H locus can produce ~11,000 chains by combining ~65 V gene segments, ~27 D gene segments, and 6 J segments.

CONCEPT AND REASONING CHECK

Based on the lymphocyte DNA structure shown at the bottom of Figure 22.9, what would be the composition of the resulting mRNA?

22.5 Immune Recombination Uses Two Types of Consensus Sequence

Assembly of light- and heavy-chain genes involves the same mechanism (although the number of different types of gene segments is different). The same consensus sequences are found at the boundaries of all germline segments that participate in joining reactions. Each consensus sequence (often called a *recombination signal sequence*) consists of a *heptamer* (7 bp element) separated by either 12 or 23 bp from a *nonamer* (9 bp element).

FIGURE 22.11 illustrates the relationship between the consensus sequences at the mouse Ig loci. At the κ locus, each V_κ gene segment is followed by a consensus sequence with a 12 bp spacer. Each J_κ segment is preceded by a consensus sequence with a 23 bp spacer. The V and J consensus sequences are inverted in orientation. At the λ locus, each V_λ gene segment is followed by a consensus sequence with a 23 bp spacer; each J_λ gene segment is preceded by a consensus of the 12 bp spacer type.

The orientation of the heptamer and nonamer sequences indicates the positions of the coding regions of the gene segments. The heptamer end of the signal sequence is adjacent to the coding region of a given gene segment, whereas the nonamer end is oriented toward the intervening sequences between gene segments.

The rule that governs the joining reaction is that *a consensus sequence with one type of spacer can be joined only to a consensus sequence with the other type of spacer.* The consensus sequences at V and J segments can lie in either order; thus (unlike the

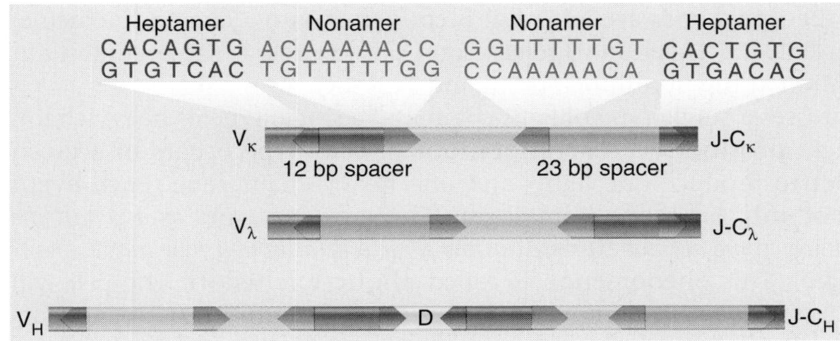

FIGURE 22.11 Consensus sequences are present in inverted orientation at each pair of recombining sites. One member of each pair has a spacing of 12 bp between its components; the other has 23 bp spacing.

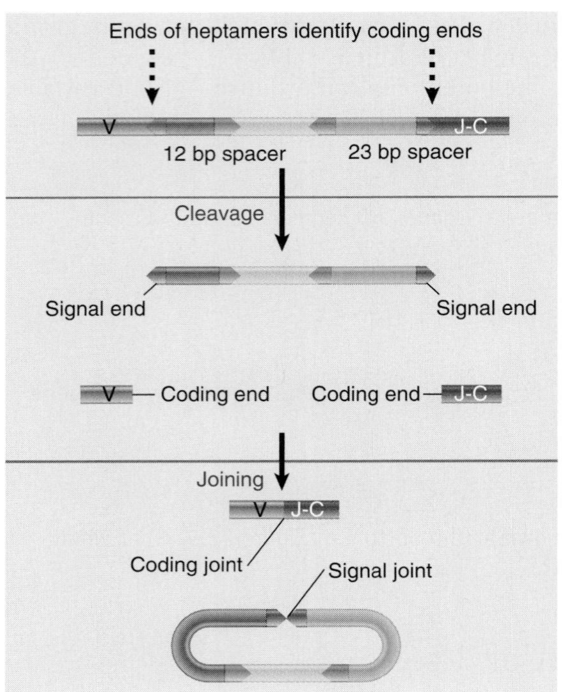

FIGURE 22.12 Breakage and reunion at consensus sequences generates immunoglobulin genes.

heptamer and nonamer sequences) the different spacers do not impart any directional information. Instead, the spacers serve to prevent one V or J gene segment from recombining with another of the same type of gene segment.

This concept is borne out by the structure of the components of the heavy gene segments. Each V_H gene segment is followed by a consensus sequence of the 23 bp spacer type. The D segments are flanked on either side by consensus sequences of the 12 bp spacer type. The J_H segments are preceded by consensus sequences of the 23 bp spacer type. Thus the V gene segment must be joined to a D segment, and the D segment must be joined to a J segment. A V gene segment cannot be joined directly to a J segment, because both possess the same type of consensus sequence.

The spacing between the components of the consensus sequences corresponds almost to one or two turns of the double helix. This may reflect a geometric relationship in the recombination reaction. For example, the recombination protein(s) may approach the DNA from one side, in the same way that RNA polymerase and repressors approach recognition elements such as promoters and operators.

The actual recombination of the components of immunoglobulin genes is accomplished by a physical rearrangement of sequences, involving breakage and reunion, but the mechanism is different from homologous recombination. The general nature of the reaction for a λ light chain is illustrated in **FIGURE 22.12**. (The reaction is similar at a heavy-chain locus, with the exception that there are two recombination events: first D-J, then V-DJ.)

Breakage and reunion occur as separate reactions. A double-strand break is made at the heptamers that lie at the ends of the coding units. This releases the entire fragment between the V gene segment and J-C gene segment; the cleaved termini of this fragment are called **signal ends**. The cleaved termini of the V and J-C loci are called **coding ends**. The two coding ends are covalently linked to form a **coding joint**; this is the connection that links the V and J segments. If the two signal ends are also connected, the excised fragment would form a circular molecule.

We have shown the V and J-C loci as organized in the same orientation. As a result, the cleavage at each consensus sequence releases the region between them as a linear fragment. If the signal ends are joined, it is converted into a circular molecule, as indicated in Figure 23.12. Deletion to release an excised circle is the predominant mode of recombination at the immunoglobulin and TCR loci.

Each B cell expresses a single type of light chain and a single type of heavy chain, because only a single **productive rearrangement** of each type occurs in a given lymphocyte in order to produce one light- and one heavy-chain gene. Each event involves the genes of only *one* of the homologous chromosomes, and as a result *the alleles on the other chromosome are not expressed in the same cell (although they may be nonproductively rearranged).* This phenomenon is called **allelic exclusion**. The cell will continue to recombine V gene segments and C gene segments until a productive rearrangement is achieved. Allelic exclusion is caused by the suppression of further rearrangement as soon as an active chain is produced.

- **signal end** It is produced during recombination of immunoglobulin and T cell receptor genes. They are at the termini of the cleaved fragment containing the recombination signal sequences. Their subsequent joining yields a signal joint.
- **coding end** It is produced during recombination of immunoglobulin and T cell receptor genes. They are at the termini of the cleaved V and (D)J coding regions. Their subsequent joining yields a coding joint.
- **coding joint** The DNA junction created by the joining of two coding ends during V(D)J recombination.
- **productive rearrangement** Occurs as a result of the recombination of V, (D), and J gene segments if all the rearranged gene segments are in the correct reading frame.
- **allelic exclusion** The expression in any particular lymphocyte of only one allele coding for the expressed immunoglobulin. This is caused by feedback from the first immunoglobulin allele to be expressed that prevents activation of a copy on the other chromosome.

KEY CONCEPTS

- The consensus sequence used for recombination is a heptamer separated by either 12 or 23 base pairs from a nonamer.
- Recombination occurs between two consensus sequences that have different spacers.
- Recombination initiates with double-strand breaks at the heptamers of two consensus sequences.
- The signal ends of the fragment between the breaks usually join to generate an excised circular fragment.
- The coding ends are covalently linked to join V to J-C (L chain), or D to J-C, and V to D-J-C (H chain).
- A productive rearrangement (resulting in an active immunoglobulin protein) prevents any further rearrangement from occurring.

CONCEPT AND REASONING CHECK

What are the different functions of the heptamer/nonamer versus the 12 and 23 bp spacers in the recombination signal sequence?

22.6 The RAG Proteins Catalyze Breakage and Reunion

The proteins RAG1 and RAG2 are necessary and sufficient to cleave DNA for V(D)J recombination. The RAG proteins together undertake the catalytic reactions of cleaving and rejoining DNA, and also provide a structural framework within which the reactions occur.

RAG1 recognizes the heptamer/nonamer signals with the appropriate 12/23 spacers and recruits RAG2 to the complex. The nonamer provides the site for initial recognition, and the heptamer directs the site of cleavage (within a span of a few base pairs).

The reactions involved in recombination are shown in **FIGURE 22.13**. The complex nicks one strand at each junction. The nick has 3′–OH and 5′–P ends. The free 3′–OH end then attacks the phosphate bond at the corresponding position *in the other strand of the duplex*. This creates a *hairpin* at the coding end, in which the 3′ end of one strand is covalently linked to the 5′ end of the other strand; it leaves a blunt double-strand break at the signal end.

This second cleavage is a transesterification reaction in which bond energies are conserved. It resembles the topoisomerase-like reactions catalyzed by the resolvase proteins of bacterial transposons. This suggests that somatic recombination of immune genes evolved from an ancestral transposon.

The hairpins at the coding ends provide the substrate for the next stage of reaction. If a single-strand break is introduced into one strand close to the hairpin, an unpairing reaction at the end generates a single-stranded overhang. Synthesis of a complement to the exposed single strand then converts the coding end to an extended duplex. This reaction explains the introduction of **P nucleotides** at coding ends; they consist of a few extra base pairs related to, but reversed in orientation from, the original coding end.

Some extra bases also may be inserted, apparently with random sequences, between the coding ends. They are called **N nucleotides**. Their insertion occurs via the activity of the enzyme deoxynucleoside transferase, active in lymphocytes, at a free 3′ coding end generated during the joining process.

Changes in sequence during recombination are, therefore, a consequence of the enzymatic mechanisms involved in breaking and rejoining the DNA. In heavy-chain recombination, base pairs are lost or inserted at the V_H-D or D-J junctions, or both. Deletion also occurs in V_λ-J_λ joining, but insertion at these joints is unusual. The changes in sequence affect the amino acid encoded at V-D junctions and D-J junctions in heavy chains or at the V-J junction in light chains.

▸ **P nucleotide** A short palindromic (inverted repeat) sequence that is generated during rearrangement of immunoglobulin and T cell receptor V, (D), and J gene segments. They are generated at coding joints when RAG proteins cleave the hairpin ends generated during rearrangement.

▸ **N nucleotide** A short nontemplated sequence that is added randomly by the enzyme at coding joints during rearrangement of immunoglobulin and T cell receptor genes. They augment the diversity of antigen receptors.

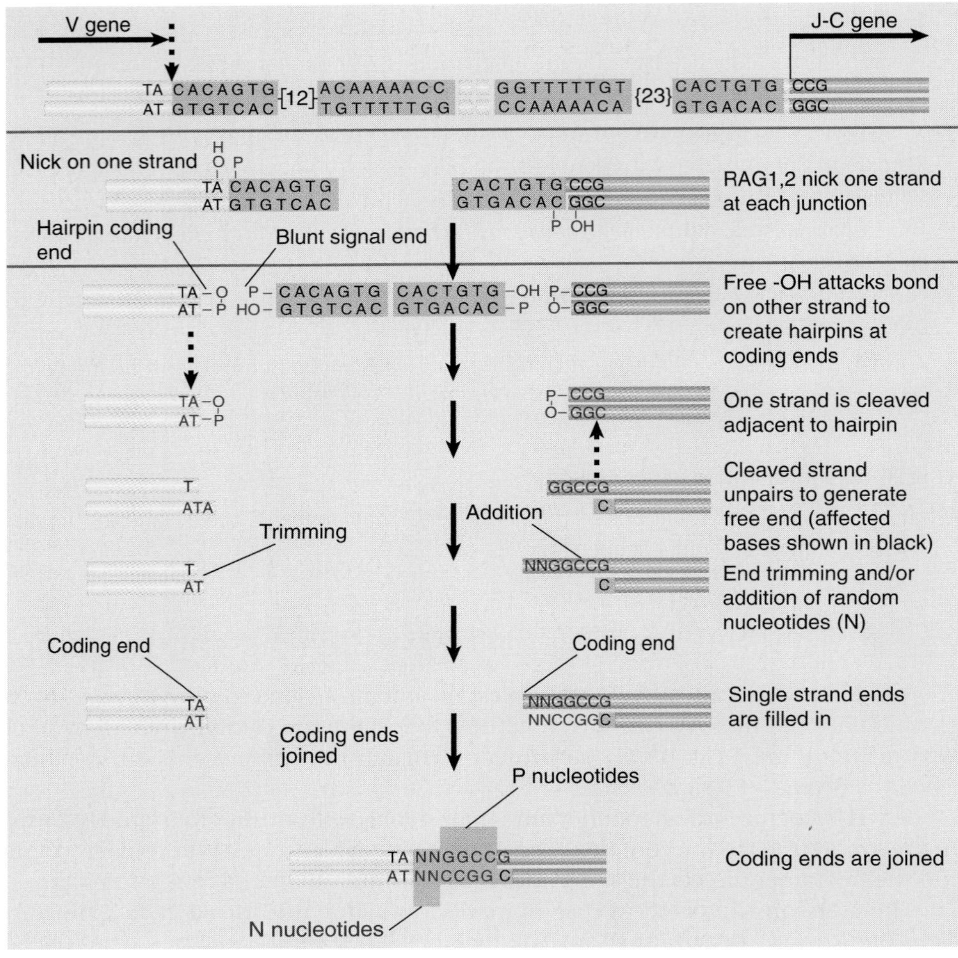

FIGURE 22.13 Processing of coding ends introduces variability at the junction.

These various mechanisms together ensure that a coding joint may have a sequence that is different from what would be predicted by a direct joining of the coding ends of the V, D, and J regions.

Changes in the sequence at the junction make it possible for a great variety of amino acids to be encoded at this site. It is interesting that the amino acid at position 96 is created by the V-J joining reaction. It forms part of the antigen-binding site and also is involved in making contacts between the light chains and the heavy chains. Thus the maximum diversity is generated at the site that contacts the target antigen.

Changes in the number of base pairs at the coding joint affect the reading frame. The joining process appears to be random with regard to reading frame, so that probably only one third of the joined sequences retain the proper frame of reading through the junctions. If the V-J region is joined so that the J segment is out of phase, translation is terminated prematurely by a nonsense codon in the incorrect frame. We may think of the formation of aberrant genes as the price the cell must pay for the increased diversity that it gains by being able to adjust the sequence at the joining site.

Similar—although even greater—diversity is generated in the joining reactions that involve the D segment of the heavy chain. The same result is seen with regard to reading frame; nonproductive genes are generated by joining events that place J and C out of phase with the preceding V gene segment.

The joining reaction that works on the coding end uses the same pathway of nonhomologous end-joining (NHEJ) that repairs double-strand breaks in cells (see *Section*

20.8, Nonhomologous End-Joining Also Repairs Double-Strand Breaks). Just as in repair, DNA-dependent protein kinase (DNA-PK) is recruited to the DNA by the Ku70 and Ku80 proteins, which bind to the DNA ends. DNA-PK phosphorylates and thereby activates the protein Artemis, which nicks the hairpin ends (it also has exonuclease and endonuclease activities that function in the NHEJ pathway). The actual ligation is performed by DNA ligase IV and also requires the protein XRCC4. Mutations in all of these proteins have been found among human patients who have diseases caused by deficiencies in DNA repair that result in increased sensitivity to radiation, and most of these mutations also result in immunodeficiency. (See the accompanying Medical Applications box for a description of SCID-A, caused by a mutation in the *ARTEMIS* gene.)

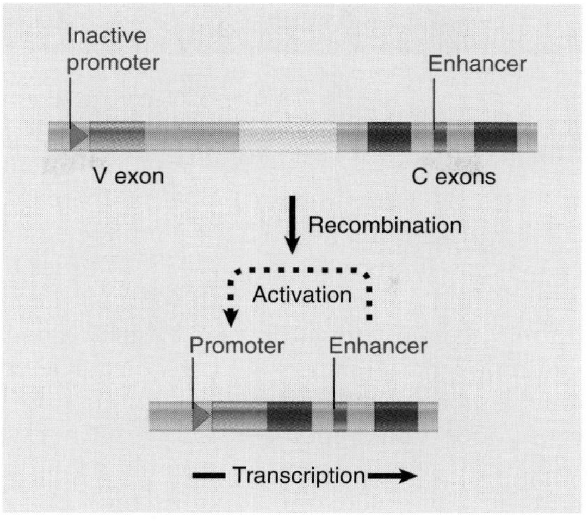

FIGURE 22.14 A V gene promoter is inactive until recombination brings it into the proximity of an enhancer in the C gene segment. The enhancer is active only in B lymphocytes.

What is the connection between joining of V and C gene segments and their transcriptional activation? Unrearranged V gene segments are not actively transcribed. When a V gene segment is joined productively to a C_κ gene segment, however, the resulting unit is transcribed. The sequence upstream of a V gene segment is not altered by the joining reaction, though, and as a result *the promoter must be the same in unrearranged, nonproductively rearranged, and productively rearranged genes.*

A promoter lies upstream of every V gene segment but is inactive. It is activated by its relocation to the C region. The effect must depend on sequences downstream. What role might they play? An enhancer located within or downstream of the C gene segment activates the promoter at the V gene segment. The enhancer is tissue specific; it is active only in B cells. Its existence suggests the model illustrated in **FIGURE 22.14**, in which the V gene segment promoter is activated when it is brought within the range of the enhancer.

KEY CONCEPTS

- The RAG proteins are necessary and sufficient for the cleavage reaction.
- RAG1 recognizes the nonamer consensus sequences for recombination. RAG2 binds to RAG1 and cleaves at the heptamer.
- The reaction resembles the topoisomerase-like resolution reaction that occurs in transposition.
- It proceeds through a hairpin intermediate at the coding end; opening of the hairpin is responsible for insertion of extra bases (P nucleotides) in the recombined gene.
- Deoxynucleoside transferase inserts additional N nucleotides at the coding end.
- The codon at the site of the V-(D)J joining reaction has an extremely variable sequence and codes for amino acid 96 in the antigen-binding site.
- The double-strand breaks at the coding joints are repaired by the same system involved in nonhomologous end-joining of damaged DNA.
- An enhancer in the C gene activates the promoter of the V gene after recombination has generated the intact immunoglobulin gene.

CONCEPT AND REASONING CHECK

Is heavy chain joining likely to produce a higher percentage of nonproductive products than light chain joining? Why or why not?

MEDICAL APPLICATIONS

ARTEMIS and SCID-A

Severe combined immunodeficiency (SCID) is an inherited immune disorder that results in severe T and B lymphocyte immunodeficiency. T and B lymphocytes are key players of the immune system. They are the second line of defense, tailoring their activities toward individual threats created by invading microbial pathogens. Children born with SCID develop bacterial, viral, and fungal infections that may cause pneumonia, meningitis, or bloodstream infections during the first six months of life. These infections are usually severe and life-threatening. If untreated, children die before they reach their first birthday. If diagnosed in time, there are effective treatments to treat the disease. SCID has been referred to as the "bubble boy disease." The disease received wide recognition in the media during the 1970s and 1980s when David Vetter, a young boy with SCID, lived in a plastic germ-free bubble for 12 years. He spent most of his life at Texas Children's Hospital.

Mutations in any of eight known genes (*IL2RG, RAGI, RAG2, ADA, CD45, IL7R, JAK3,* and *ARTEMIS*) cause SCID. Mutations in unidentified genes may also cause SCID. SCID is very rare in the U.S. population, affecting 1 in 100,000 children. Newborn screening programs are now being offered in some states. The focus of this Medical Applications box is on the role of the *ARTEMIS* gene in the development of SCID. SCID-A has a high incidence in Athabascan-speaking children. About 1 in 2000 Navajo and Apache Native Americans are born with SCID-A. The pedigrees and genetic analyses of Athabascan-speaking Navajo and Apache Native Americans allowed researchers to map the *ARTEMIS* gene for SCID-A (SCID-Athabascan) to the short arm of chromosome 10 in 1998.

ARTEMIS encodes a 77.6 kDa protein that is postulated to be a novel DNA double-stranded break repair/V(D)J recombination protein. *In vitro* studies demonstrate that the Artemis protein is phosphorylated by a DNA-dependent protein kinase (DNA-PKs) that in turn activates its hairpin-opening or nicking activity so that nucleotides can be added and/or deleted at heavy chain V-D and D-J or light chain V-J junctions during the rearrangement or recombination of V(D)J genes that are responsible for the diversity of T-cell receptor and immunoglobulin-encoding genes. Mutations in *ARTEMIS* result in SCID-A by causing a significant impairment in V(D)J coding joint formation, leading to the inability to develop pathogen-fighting T and B lymphocytes.

A founder mutation in *ARTEMIS* was discovered in 2002; it is a *nonsense mutation,* which is a point mutation in which a change in DNA replaced a codon specifying an amino acid with a stop codon, resulting in a truncated and often nonfunctional protein product. A "founder effect" occurs when a few individuals leave an original population and found a new population (e.g., colonizing an island). The founders are likely not representative of the original population. In this case, it is believed that the SCID-A defect originated in the Athabascan-speaking populations. Ancestors of the Athabascans probably migrated across the Bering Land Bridge at the end of the last ice age, about 12,000 years ago. They traditionally lived in small nomadic groups in Alaska and the Canadian Northwest territories. Dwellings were usually temporary, and they did not have any formal tribal organization. Between 700 and 1300 CE some of the Athabascans migrated to the Southwestern United States. It is believed that the Navajo and Apache are descendants from these small migratory groups of Athabascan people because their languages share Athabascan roots.

Because SCID-A occurs at a higher rate in families of Navajo Native Americans, geneticists have been studying these families for generations. The SCID-A condition is autosomal recessive. An autosomal recessive allele can be passed from one generation to the next without causing harm. But if both parents are carriers of the mutant recessive *ARTEMIS* allele, each child has a 1 in 4 chance of being born with SCID-1 (**FIGURE B22.1**). Counselors are working with affected families of the Navajo Nation to help them understand and cope with this information. Support sessions unraveled family histories that were told by grandparents about newborn children who only lived a short time. Written on death certificates were the words "severe infection." Today we know that this may be because the newborns likely suffered from SCID-A. Genetic testing results can have a powerful emotional impact on individuals, their families, and community.

22.7 Class Switching Is Caused by DNA Recombination

The *class* of immunoglobulin is defined by the type of C_H region it contains. **FIGURE 22.15** summarizes the five Ig classes. IgM (the first immunoglobulin to be produced by any B cell) and IgG (the most common immunoglobulin) possess the central ability to activate complement, which leads to destruction of invading cells. IgA is found in secretions (such as saliva), and IgE is associated with the allergic response and defense against parasites. IgD is found on the surface of B cells and may help regulate B cell function.

Treatment of SCID-A children involves a bone marrow transplant that requires suitable matched donors. An alternative approach to the treatment of SCID-A is the application of gene therapy. Gene therapy involves the use of a viral vector that carries a functional copy of the defective gene that will complement the defective *ARTEMIS* gene. Researchers have shown that the mutant allele can be complemented by a wild-type *ARTEMIS* allele in a mouse model system.

References

Li, L., *et al.*, 2002. A Founder Mutation in Artemis, an SNM1-Like Protein, Causes SCID in Athabascan-Speaking Native Americans. *J. Immunology* 168:6323–6329.

Li, L., *et al.*, 2005. Targeted Disruption of the Artemis Murine Counterpart Results in SCID and Defective V(D)J Recombination That Is Partially Corrected with Bone Marrow Transplantation. *J. Immunology* 174:2420–2428.

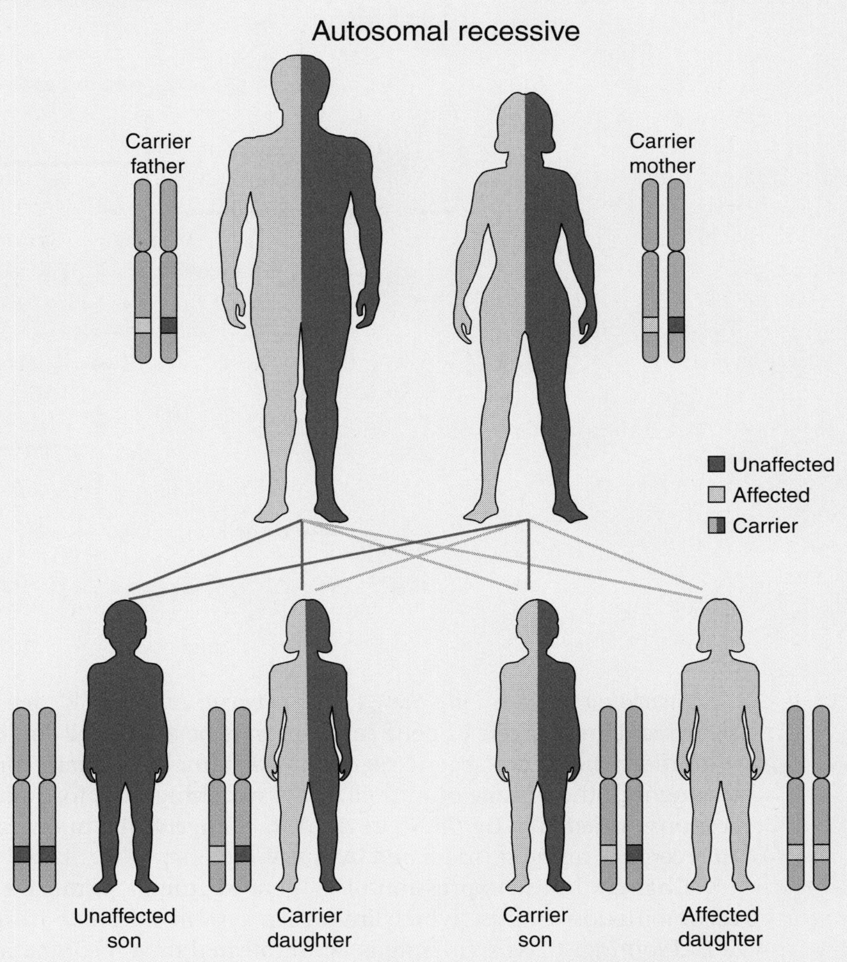

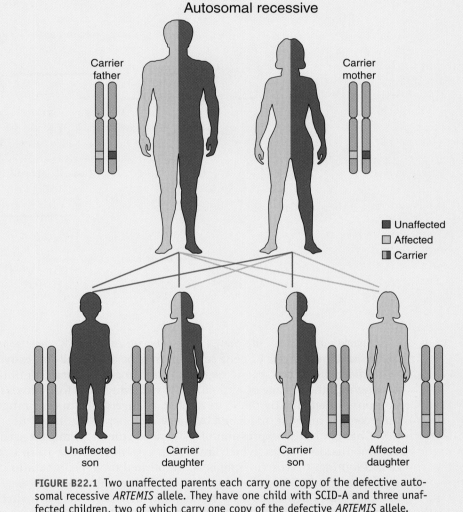

FIGURE B22.1 Two unaffected parents each carry one copy of the defective autosomal recessive *ARTEMIS* allele. They have one child with SCID-A and three unaffected children, two of which carry one copy of the defective *ARTEMIS* allele. Reproduced from U.S. National Library of Medicine. *Genetics Home Reference Handbook*, 2008. [http://ghr.nlm.nih.gov/handbook/illustrations/autorecessive]

All lymphocytes start productive life as immature cells engaged in synthesis of IgM. Cells expressing IgM have the germline arrangement of the C_H gene segment cluster shown in Figure 22.10. The V-D-J joining reaction triggers expression of the C_μ gene segment. A lymphocyte generally produces only a single class of immunoglobulin at any one time, but the class may change during the cell lineage. This change in expression is called **class switching**. It is accomplished by a substitution in the type of C_H region that is expressed. Switching can be stimulated by environmental effects; for example, the growth factor TGFβ causes switching from C_μ to C_α.

class switching A change in Ig gene organization in which the C region of the heavy chain is changed but the V region remains the same.

FIGURE 22.15 Immunoglobulin type and function are determined by the heavy chain. J is a joining protein in IgM and IgA; all other Ig types exist as tetramers.

Type	IgM	IgD	IgG	IgA	IgE
Heavy chain	μ	δ	γ	α	ε
Structure	$(\mu_2L_2)_5J$	δ_2L_2	γ_2L_2	$(\alpha_2L_2)_2J$	ε_2L_2
Proportion	5%	1%	80%	14%	<1%
Effector function	Activates complement	Development of tolerance (?)	Activates complement	Found in secretions	Allergic response

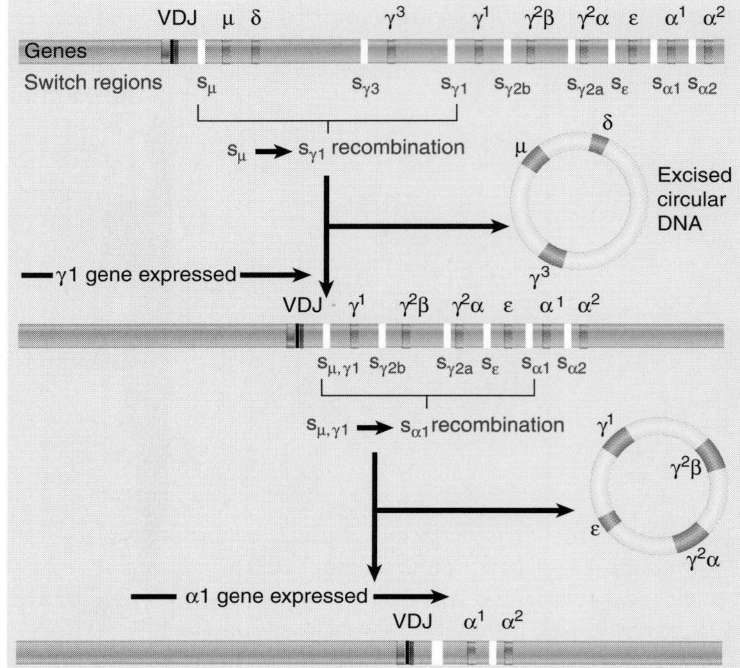

FIGURE 22.16 Class switching of heavy genes may occur by recombination between switch regions (S), deleting the material between the recombining S sites. Successive switches may occur.

Switching involves only the C_H gene segment; the same V_H gene segment continues to be expressed. Thus a given V_H gene segment may be expressed successively in combination with more than one C_H gene segment. The same light chain continues to be expressed throughout the lineage of the cell. Class switching therefore allows the type of effector response (mediated by the C_H region) to change while maintaining the same capacity to recognize antigen (mediated by the V regions).

Changes in the expression of C_H gene segments primarily occur via further DNA recombination events, which involve a system different from that concerned with V-D-J joining. Class switching is accomplished by a recombination to bring a new C_H gene segment into juxtaposition with the expressed V-D-J unit, resulting in deletion of the previously expressed C_H gene segment. The sequences of switched V-D-J-C_H units show that the sites of switching lie upstream of the C_H gene segments themselves. The switching sites are called **S regions**. **FIGURE 22.16** depicts two successive switches.

▸ **S region** A sequence involved in immunoglobulin class switching. They consist of repetitive sequences at the 5′ ends of gene segments encoding the heavy-chain constant regions.

In the first switch, expression of C_μ is succeeded by expression of $C_{\gamma1}$. The $C_{\gamma1}$ gene segment is brought into the expressed position by recombination between the sites S_μ and $S_{\gamma1}$. The S_μ site lies between V-D-J and the C_μ gene segment. The $S_{\gamma1}$ site lies upstream of the $C_{\gamma1}$ gene segment. The DNA sequence between the two switch sites is excised as a circular molecule.

The linear deletion model imposes a restriction on the heavy-gene locus: *once a class switch has been made, it becomes impossible to express any C_H gene segment that used to reside between C_μ and the new C_H gene segment.* In the example of Figure 22.16, after the S_μ to $S_{\gamma1}$ recombination event, cells expressing $C_{\gamma1}$ will then be unable to give rise to cells expressing $C_{\gamma3}$, which has been deleted.

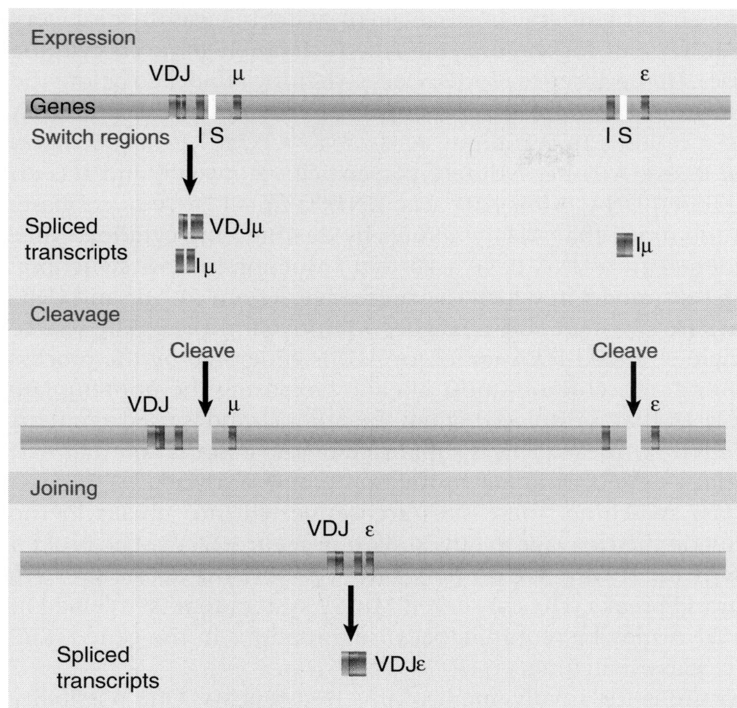

FIGURE 22.17 Class switching passes through discrete stages. The I promoters initiate transcription of noncoding transcripts. The switch regions are cleaved. Joining occurs at the cleaved regions.

It is possible, however, to undertake *another* switch to any C_H gene segment *downstream* of the expressed gene. Figure 22.16 shows a second switch to C_α expression, which is accomplished by recombination between $S_{\alpha 1}$ and the switch region $S_{\mu,\gamma 1}$ that was generated by the original switch.

We know that switch sites are not uniquely defined, because different cells expressing the same C_H gene segment prove to have recombined at different points. Switch regions vary in length (as defined by the limits of the sites involved in recombination) from 1 to 10 kb. They contain groups of short inverted repeats, with repeating units that vary from 20 to 80 nucleotides in length. The primary sequence of the switch region does not seem to be important; what matters is the presence of the inverted repeats.

An S region typically is located ~2 kb upstream of a C_H gene segment. The switching reaction releases the excised material between the switch sites as a circular DNA molecule. Two of the factors required for the joining phase of VDJ recombination, Ku70/80 and DNA-PKcs, are also required for NHEJ, suggesting that the joining reaction may use the NHEJ repair pathway. Basically, this implies that the reaction occurs by a double-strand break followed by rejoining of the cleaved ends.

We can put together the features of the reaction to propose a model for the generation of the double-strand break. The critical points are:

- transcription through the S region is required;
- the inverted repeats are crucial; and
- the break can occur at many different places within the S region.

FIGURE 22.17 shows the stages of the class-switching reaction. A promoter ("I") lies immediately upstream of each switch region. Switching requires transcription from this promoter. The promoter may respond to activators that respond to environmental conditions, such as stimulation by cytokines, thus creating a mechanism to regulate switching. The first stage in switching is therefore to activate the I promoters that are upstream of each of the switch regions that will be involved. When these promoters are activated, they generate noncoding transcripts that are spliced to join the I region with the corresponding heavy constant region.

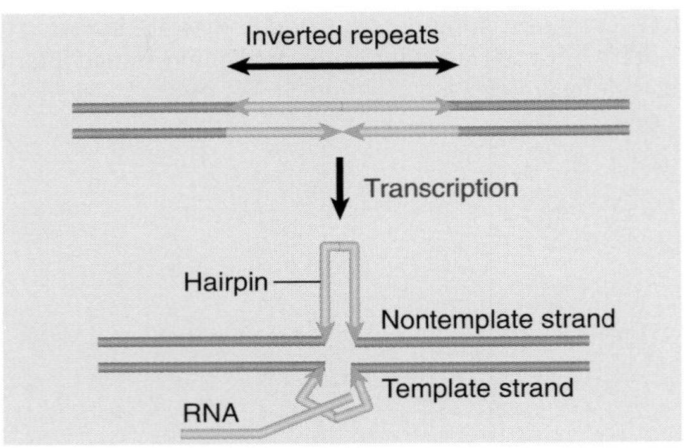

FIGURE 22.18 When transcription separates the strands of DNA, one strand may form an inverted repeat if the sequence is palindromic.

The key insight into the mechanism of switching was the discovery of the requirement for the enzyme AID (activation-induced cytidine deaminase). In the absence of AID, class switching is blocked before the nicking stage. AID is a member of a class of enzymes that act on RNA to change a cytidine to a uridine (see *Section 29.9, RNA Editing Occurs at Individual Bases*). AID has a different specificity, however, and acts on single-stranded DNA. After AID acts, UNG, a uracil DNA glycosylase, removes the uracil that AID generates by deaminating cytidine. Mice that are deficient in UNG have a 10-fold reduction in class switching. This suggests a model in which the successive actions of AID and UNG create sites from which a base has been removed in DNA. The source of the single-stranded DNA target for AID is generated by the process of noncoding transcription, most likely by exposing the nontemplate strand of DNA that is displaced when the other strand is used as a template for RNA synthesis. This is supported by the observation that AID preferentially targets cytidines in the nontemplate strand.

In order to cause class switching, these sites are converted into breaks in the nucleotide chain that provide the cleavage events shown in Figure 22.17. The broken ends are joined by the NHEJ pathway. We do not know yet how the abasic sites are converted into double-strand breaks. The conserved MutSL system that is involved in repairing mismatches in DNA may be required, because mutations in the gene *MSH2* (MutS homolog 2) reduce class switching.

One unexplained feature is the involvement of inverted repeats. One possibility is that hairpins are formed by an interaction between the inverted repeats on the displaced nontemplate strand, as shown in **FIGURE 22.18**. In conjunction with the generation of abasic sites on this strand, this might lead to breakage.

Two critical questions that remain unanswered are how the system is targeted to the appropriate regions in the heavy-chain locus, and what controls the use of switching sites.

KEY CONCEPTS

- Immunoglobulins are divided into five classes according to the type of constant region in the heavy chain.
- Class switching to change the C_H region occurs by a recombination between S regions that deletes the region between the old C_H region and the new C_H region.
- Multiple successive switch recombinations can occur.
- Switching occurs by a double-strand break followed by the nonhomologous end-joining reaction.
- The important feature of a switch region is the presence of inverted repeats.
- Switching requires activation of promoters that are upstream of the switch sites.

CONCEPT AND REASONING CHECK

Why can't a lymphocyte resume expression of an earlier immunoglobulin class, after it has switched from that class to another?

22.8 Somatic Mutation Is Induced by Cytidine Deaminase and Uracil Glycosylase

Comparisons between the sequences of expressed immunoglobulin genes and the corresponding V gene segments of the germline show that new sequences appear in the expressed population. Some of this additional diversity results from sequence changes at the V-J or V-D-J junctions that occur during the recombination process. Other changes occur upstream, however, at locations within the variable domain.

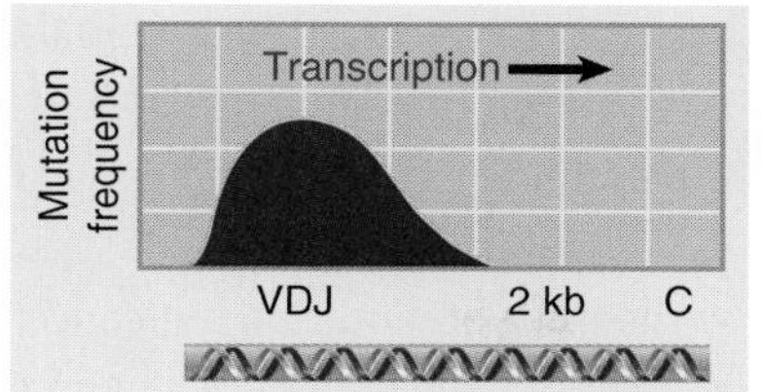

FIGURE 22.19 Somatic mutation occurs in the region surrounding the V segment and extends over the joined VDJ segments.

Two types of mechanisms can generate changes in V gene sequences after rearrangement has generated a functional immunoglobulin gene. In the mouse and human, the mechanism is the induction of **somatic mutations** at individual locations within the gene specifically in the active lymphocyte. The process is sometimes called *hypermutation*. In chicken, rabbit, and pig, a different mechanism uses gene conversion to change a segment of the expressed V gene into the corresponding sequence from a different V gene (see *Section 22.9, Avian Immunoglobulins Are Assembled from Pseudogenes*).

▸ **somatic mutation** A mutation occurring in a somatic cell, therefore affecting only its daughter cells; it is not inherited by descendants of the organism.

FIGURE 22.19 shows that sequence changes are localized around the V gene segment, extending in a region from ~150 bp downstream of the V gene promoter for ~1.5 kb. They take the form of substitutions of individual nucleotide pairs. Usually there are ~3 to ~15 substitutions, corresponding to <10 amino acid changes in the protein. They are concentrated in the antigen-binding site (thus generating the maximum diversity for recognizing new antigens). Only some of the mutations affect the amino acid sequence, since others lie in third-base coding positions as well as in nontranslated regions.

The large proportion of ineffectual mutations suggests that somatic mutation occurs more or less at random in a region including the V gene segment and extending beyond it. There is a tendency for some mutations to recur on multiple occasions. These may represent hotspots as a result of some intrinsic preference in the system.

In many cases, a single family of V gene segments is used consistently to respond to a particular antigen. Upon exposure to an antigen, presumably the V region with highest intrinsic affinity provides a starting point. Somatic mutation then increases the repertoire. Random mutations have unpredictable effects on protein function; some inactivate the protein, whereas others confer high specificity for a particular antigen. The proportion and effectiveness of the lymphocytes that respond is increased by selection among the lymphocyte population for those cells bearing antibodies in which mutation has increased the affinity for the antigen.

Somatic mutation has many of the same requirements as class switching (see *Section 22.7, Class Switching Is Caused by DNA Recombination*):

- transcription must occur in the target region;
- it requires the enzymes AID and UNG; and
- the MSH mismatch-repair system is involved.

When AID deaminates cytosine, it generates uracil, which then is removed from the DNA by UNG. The MSH repair system is then recruited to excise and replace the stretch of DNA containing the damage; this is somewhat surprising as abasic sites are usually repaired by base excision repair, not mismatch repair. One possibility is that the replacement is performed by an error-prone DNA polymerase in order to introduce mutations. Another possibility is that so many abasic sites are created that the repair systems are overwhelmed. When replication occurs, this could lead to the random insertion of bases opposite the abasic sites. We don't know yet what restricts the action of this system to the target region for hypermutation.

The difference between class switching and somatic mutation is at the end of the process, when double-strand breaks are introduced in class switching, but individual point mutations are created during somatic mutation. We do not yet know exactly where the systems diverge. One possibility is that breaks are introduced at abasic sites in class switching, but the sites are erratically repaired in somatic mutation. Another possibility is that breaks are introduced in both cases, but are repaired in an error-prone manner in somatic mutation.

KEY CONCEPTS

- Active immunoglobulin genes have V regions with sequences that are changed from the germline because of somatic mutation.
- The mutations occur as substitutions of individual bases.
- The sites of mutation are concentrated in the antigen-binding site.
- A cytidine deaminase is required for somatic mutation as well as for class switching.
- Uracil-DNA glycosylase activity influences the pattern of somatic mutations.
- Hypermutation may be initiated by the sequential action of these enzymes.

CONCEPT AND REASONING CHECK

Why is somatic mutation useful even though most of the mutants generated may be nonfunctional or produce weaker antibodies?

22.9 Avian Immunoglobulins Are Assembled from Pseudogenes

The chick immune system is the paradigm for rabbits, cows, and pigs, which rely upon using the diversity that is coded in the genome. A similar mechanism is used for both the single light-chain locus (of the λ type) and the H-chain locus. The organization of the λ locus is drawn in **FIGURE 22.20**. It has only one functional V gene segment, J segment, and C gene segment. Upstream of the functional $V_{\lambda 1}$ gene segment lie 25 V_{λ} pseudogenes, organized in either orientation. They are classified as pseudogenes because either the coding segment is deleted at one or both ends or proper signals for recombination are missing, or both. This assignment is confirmed by the fact that only the $V_{\lambda 1}$ gene segment recombines with the J-C_{λ} gene segment.

Sequences of active rearranged V_{λ}-J-C_{λ} gene segments show considerable diversity, however! A rearranged gene has one or more positions at which a cluster of changes has occurred in the sequence. A sequence identical to the new sequence can almost always be found in one of the pseudogenes (which themselves remain unchanged). The exceptional sequences that are not found in a pseudogene always represent changes at the junction between the original sequence and the altered sequence.

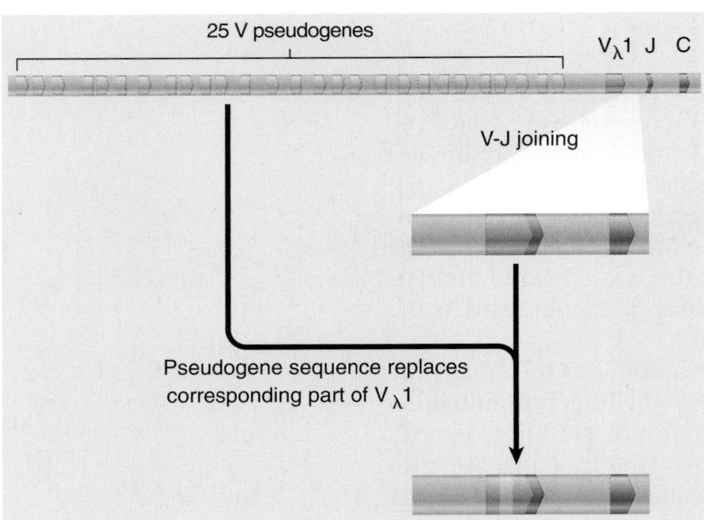

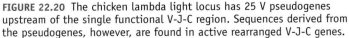

FIGURE 22.20 The chicken lambda light locus has 25 V pseudogenes upstream of the single functional V-J-C region. Sequences derived from the pseudogenes, however, are found in active rearranged V-J-C genes.

Thus a novel mechanism is employed to generate diversity. Sequences from the pseudogenes, between 10 and 120 bp in length, are *substituted* into the active $V_{\lambda 1}$ region by gene conversion. The unmodified $V_{\lambda 1}$ sequence is not expressed, even at early times during the immune response. A successful conversion event probably occurs every ten to twenty cell divisions to every rearranged $V_{\lambda 1}$ sequence. At the end of the immune maturation period, a rearranged $V_{\lambda 1}$ sequence has four to six converted segments spanning its entire length, which are derived from different donor pseudogenes. If all pseudogenes participate, this allows 2.5×10^8 possible combinations!

The enzymatic basis for copying pseudogene sequences into the expressed locus depends on enzymes involved in recombination and is related to the mechanism for somatic hypermutation that introduces diversity in mouse and human. Some of the genes involved in homologous recombination are required for the gene conversion process; for example, it is prevented by deletion of *RAD54*. Deletion of other recombination genes

(*XRCC2, XRCC3,* and *RAD51B*) has another, very interesting effect: Somatic mutation occurs at the V gene in the expressed locus. The frequency of the somatic mutation is ~10× greater than the usual rate of gene conversion.

These results show that the absence of somatic mutation in chick is not due to a deficiency in the enzymatic systems that are responsible in mouse and human. The most likely explanation for a connection between (lack of) recombination and somatic mutation is that unrepaired breaks at the locus trigger the induction of mutations. The reason somatic mutation occurs in mouse and human but not in chick may therefore lie with the details of the operation of the repair system that operates on breaks at the locus. It is more efficient in chick, so that the gene is repaired by gene conversion before mutations can be induced.

KEY CONCEPT

- An immunoglobulin gene in chicken is generated by copying sequences from one of 25 pseudogenes into the V gene at a single active locus.

CONCEPT AND REASONING CHECK

Given that V-J-C joining occurs in chick, do you think the V pseudogenes were once active V gene segments? Why or why not?

22.10 T Cell Receptors Are Related to Immunoglobulins

The lymphocyte lineage presents an example of evolutionary opportunism: a similar mechanism is used in both B cells and T cells to generate proteins that have a variable region able to provide significant diversity, whereas constant regions are more limited and account for a small range of effector functions. T cells produce either of two types of T cell receptor.

The γδ receptor is found on <5% of T lymphocytes. It is synthesized only at an early stage of T cell development. TCR αβ is found on >95% of lymphocytes. It is synthesized later in T cell development than γδ. It is synthesized by a separate lineage of cells from those involved in TCR γδ synthesis, and involves independent rearrangement events.

Like immunoglobulins, a TCR must recognize a foreign antigen of unpredictable structure. The problem of antigen recognition by B cells and T cells is resolved in the same way, and the organization of the T cell receptor genes resembles the immunoglobulin genes in the use of variable and constant regions. *Each locus is organized in the same way as the immunoglobulin genes, with separate segments that are brought together by a recombination reaction specific to the lymphocyte.* The components are the same as those found in the Ig heavy- and light-chain families. TCRα resembles the Ig light chain, whereas TCRβ resembles a heavy chain.

The organization of the TCR proteins resembles that of the immunoglobulins. The V regions have the same general internal organization in both Ig and TCR proteins. The TCR C region is related to the constant Ig regions and has a single constant domain followed by transmembrane and cytoplasmic portions. Exon-intron structure is related to protein function.

As summarized in **FIGURE 22.21**, the genomic organization of TCRα resembles that of Ig κ, with V gene segments separated from a cluster of J segments that precedes a single C gene segment. The organization of the locus is similar in both human and mouse, with some differences only in the number of V_α gene segments and J_α segments. Note the presence of δ genes in this locus as well (see below), which are not incorporated into TCRα.

The components of TCRβ resemble those of IgH. **FIGURE 22.22** shows that the organization is different, with V gene segments separated from two

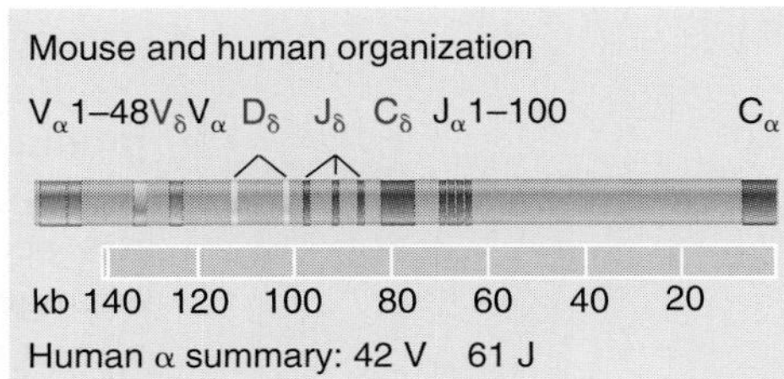

FIGURE 22.21 The human TCRα locus has interspersed α and δ segments. A V_δ segment is located within the V_α cluster. The D-J-C_δ segments lie between the V gene segments and the J-C_α segments. The mouse locus is similar, but has more V_δ segments.

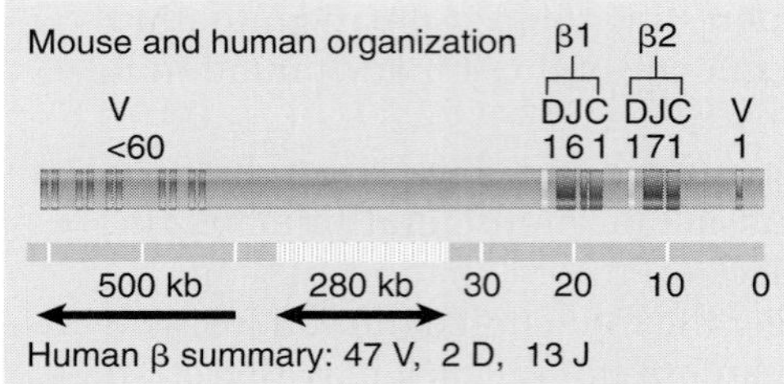

FIGURE 22.22 The TCRβ locus contains many V gene segments spread over ~500 kb that lie ~280 kb upstream of the two D-J-C clusters.

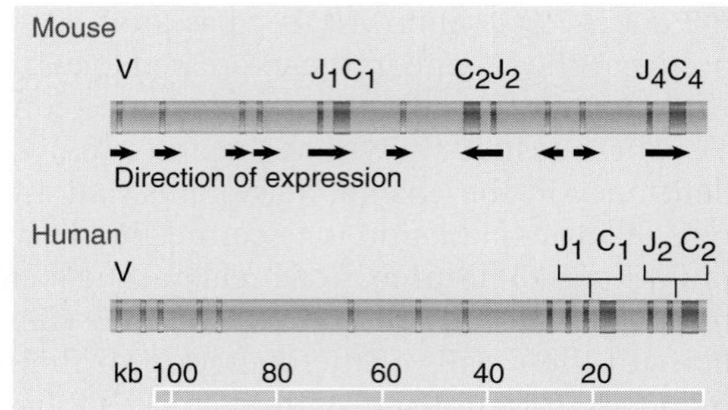

FIGURE 23.23 The TCRγ locus contains a small number of functional V gene segments (and also some pseudogenes not shown) that lie upstream of the J-C loci.

clusters each containing a D segment, several J segments, and a C gene segment. Again, the only differences between human and mouse are in the numbers of the V_β and J_β units.

Diversity is generated by the same mechanisms as in immunoglobulins. Intrinsic diversity results from the combination of a variety of V, D, J, and C segments; some additional diversity results from the introduction of new sequences at the junctions between these components (in the form of P and N nucleotides; see Figure 22.13). Some TCRβ chains incorporate two D segments, which are generated by D-D joins (directed by an appropriate organization of the nonamer and heptamer sequences). A difference between TCR and Ig is that somatic mutation does not occur at the TCR loci. Measurements of the extent of diversity show that the 10^{12} T cells in the human contain 2.5×10^7 different α chains associated with 10^6 different β chains.

The same mechanisms are likely to be involved in the reactions that recombine Ig genes in B cells and TCR genes in T cells. The recombining TCR segments are surrounded by nonamer and heptamer consensus sequences identical to those used by the Ig genes. This argues strongly that the same enzymes are involved. We do not know how the process is controlled so that Ig loci are rearranged in B cells, whereas T cell receptors are rearranged in T cells.

The organization of the γ locus resembles that of Ig λ, with V gene segments separated from a series of J-C segments. **FIGURE 22.23** shows that this locus has relatively little diversity, with ~8 functional V segments. The organization is different in human and mouse. Mouse has three functional J-C loci, but some segments are inverted in orientation. The human has multiple J segments for each C gene segment.

The δ subunit is encoded by segments that lie at the TCRα locus, as illustrated previously in Figure 22.21. The segments D_δ-D_δ-J_δ-C_δ lie between the V gene segments and the J_α-C_α segments. Both of the D segments may be incorporated into the δ chain to give the structure VDDJ. The basis for specificity in choosing V segments in α and δ rearrangement is not known. One possibility is that many of the V_α gene segments can be joined to the DDJ_δ segment, but that only some (therefore defined as V_δ) can give active proteins.

Rearrangements at the TCR loci, like those of immunoglobulin genes, may be productive or nonproductive. The β locus shows allelic exclusion in much the same way as immunoglobulin loci; rearrangement is suppressed once a productive allele has been generated. The α locus may be different; several cases of continued rearrangement suggest the possibility that substitution of V_α sequences may continue after a productive allele has been generated.

KEY CONCEPTS

- T cells use a similar mechanism of V(D)J-C joining to B cells to produce either of two types of T cell receptor.
- TCR αβ is found on >95% of T lymphocytes; TCR γδ is found on <5%.

CONCEPT AND REASONING CHECK

Why do T cell receptors need to be as diverse as immunoglobulins?

22.11 Summary

Immunoglobulins and T cell receptors are proteins that play analogous functions in the roles of B cells and T cells in the immune system. An Ig or TCR protein is generated by rearrangement of DNA in a single lymphocyte; exposure to an antigen recognized by the Ig or TCR leads to clonal expansion to generate many cells that have the same specificity as the original cell. Many different rearrangements occur early in the development of the immune system, thereby creating a large repertoire of cells of different specificities.

Each immunoglobulin protein is a tetramer containing two identical light chains and two identical heavy chains. A TCR is a dimer containing two different chains. Each polypeptide chain is expressed from a gene created by linking one of many V segments via D segments and J segments to one of a few C segments. Ig L chains (either κ or λ) have the general structure V-J-C, Ig H chains have the structure V-D-J-C, TCR α and γ have components like Ig L chains, and TCR δ and β are like Ig H chains.

Each type of chain is coded by a large cluster of V genes separated from the cluster of D, J, and C segments. The numbers of each type of segment and their organization are different for each type of chain, but the principle and mechanism of recombination appear to be the same. The same nonamer and heptamer consensus sequences are involved in each recombination; the reaction always involves joining of a consensus with 23 bp spacing to a consensus with 12 bp spacing. The cleavage reaction is catalyzed by the RAG1 and RAG2 proteins, and the joining reaction is catalyzed by the same NHEJ pathway that repairs double-strand breaks in cells. The mechanism of action of the RAG proteins is related to the action of site-specific recombination catalyzed by resolvases.

Considerable diversity is generated by joining different V, D, and J segments to a C segment; however, additional variations are introduced in the form of changes at the junctions between segments during the recombination process. Changes are also induced in immunoglobulin genes by somatic mutation, which requires the actions of cytidine deaminase and uracil glycosylase. Mutations induced by cytidine deaminase probably lead to removal of uracil by uracil glycosylase, followed by the induction of mutations at the sites where bases are missing.

Allelic exclusion ensures that a given lymphocyte synthesizes only a single Ig or TCR. A productive rearrangement inhibits the occurrence of further rearrangements. The use of the V region is fixed by the first productive rearrangement, but B cells switch use of C_H genes from the initial μ chain to one of the H chains coded farther downstream. This process involves a different type of recombination in which the sequences between the VDJ region and the new C_H gene are deleted. More than one switch occurs in C_H gene usage. Class switching requires the same cytidine deaminase that is required for somatic mutation, but its role is not known.

CHAPTER QUESTIONS

1. A set of about 20 cellular proteins that function through a cascade of proteolytic actions is:
   A. major histocompatibility complex.
   B. complement.
   C. immunoglobulins.
   D. haptens.
2. Somatic recombination to join one of many immunoglobulin V gene segments with one of a few C gene segments occurs in:
   A. B cells.
   B. T cells.
   C. germline cells.
   D. any of the above.
3. The J segment of an immunoglobulin gene is:
   A. relatively short and codes for the first few amino acids of the constant region.
   B. relatively short and codes for the last few amino acids of the constant region.
   C. relatively short and codes for the first few amino acids of the variable region.
   D. relatively short and codes for the last few amino acids of the variable region.
4. The target sequence for somatic recombination in generation of immune diversity is what type of arrangement?
   A. a heptamer-intron-nonamer
   B. a heptamer-spacer-nonamer
   C. a nonamer-intron-heptamer
   D. a nonamer-spacer-heptamer
5. Somatic recombination of immunoglobulin genes occurs by:
   A. single-strand nick and exonuclease action at the heptamers of two consensus sequences.
   B. single-strand nick and exonuclease action at the nonomers of two consensus sequences.
   C. double-strand breaks at the heptamers of two consensus sequences.
   D. double-strand breaks at the nonamers of two consensus sequences.
6. B cells that are not blocked from further rearrangement due to recombination are said to have undergone:
   A. class switching.
   B. allelic exclusion.
   C. productive rearrangement.
   D. nonproductive rearrangement.
7. _______ proteins are necessary and sufficient for the cleavage reaction. **8.** _______ recognizes the **9.** _______ sequences for **10.** _______. **11.** _______ then binds to **12.** _______ and cleaves at the **13.** _______.

KEY TERMS

adaptive (acquired) immunity
allelic exclusion
antibody
antigen
autoimmune disease
B cell
C genes
cell-mediated response
class switching
clonal expansion
coding end
coding joint
complement
constant region (C region)
cytotoxic T cell
D segment
heavy chain
helper T cell
humoral response
immune response
immunoglobulin
J segments
light chain
major histocompatibility complex (MHC)
N nucleotide
P nucleotide
productive rearrangement
S region
signal end
somatic hypermutation
somatic mutation
somatic recombination
superfamily
T cell receptor (TCR)
T cells
transmembrane region (domain)
V gene
variable region (V region)

FURTHER READING

Dudley, D. D., Chaudhuri, J., Bassing, C. H., and Alt, F. W. (2005). Mechanism and control of V(D)J recombination versus class switch recombination: similarities and differences. *Adv. Immunol.* 86, 43–112.
A review of both V(D)J recombinations and class switch recombination, comparing and contrasting the enzymes involved in each mechanism.

23

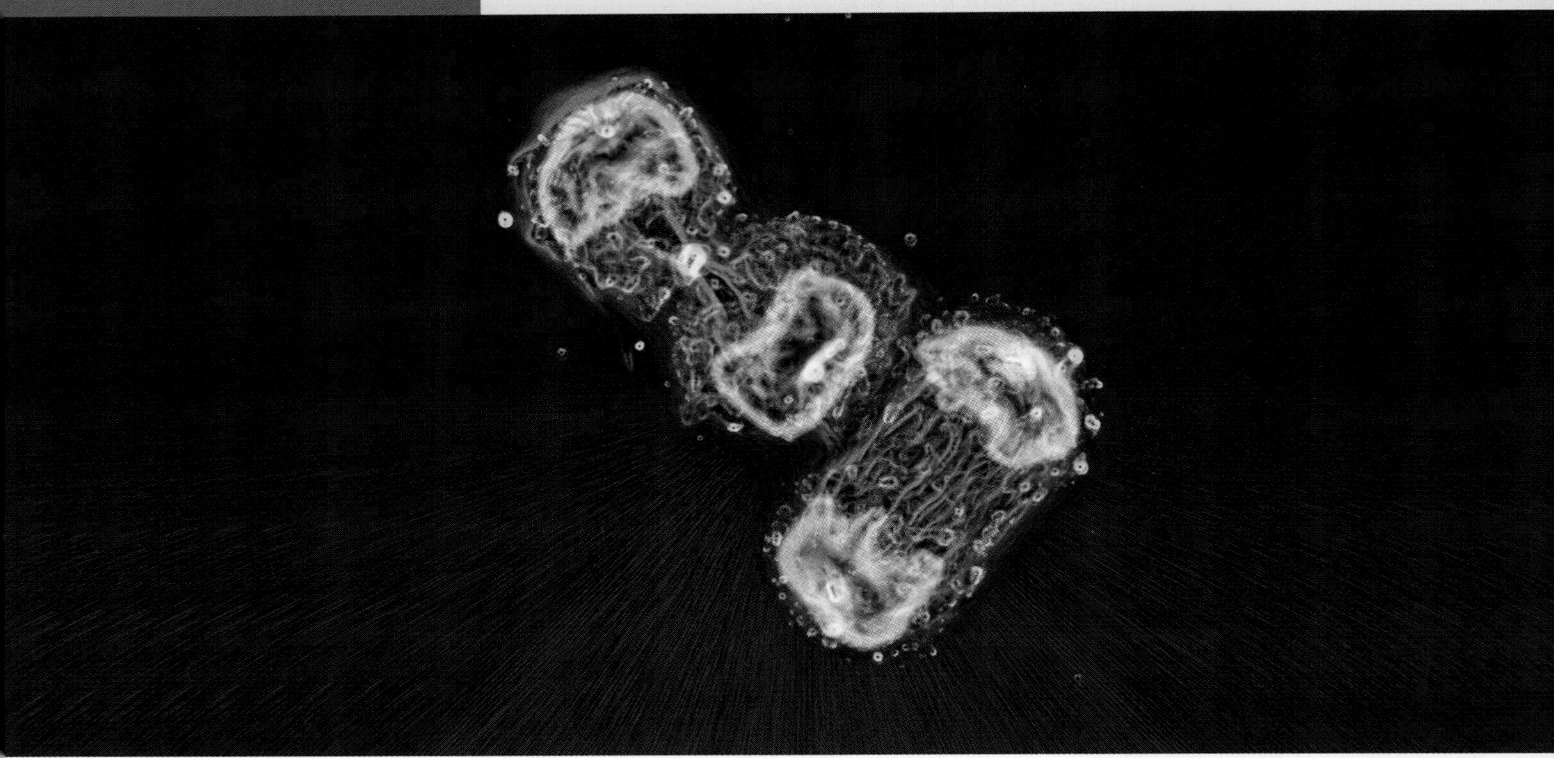

Cells undergoing mitosis. The microtubules of the spindle are stained purple and chromatin is light blue. Photo courtesy of Edward H. Hinchcliffe, Hormel Institute, University of Minnesota.

Chromosomes

CHAPTER OUTLINE

23.1 Introduction

A general principle is evident in the organization of all cellular genetic material. It exists as a compact mass that is confined to a limited volume, and its various activities, such as replication and transcription, must be accomplished within this space. The organization of this material must accommodate transitions between inactive and active states.

The condensed state of nucleic acid results from its binding to basic proteins. The positive charges of these proteins neutralize the negative charges of the nucleic acid. The structure of the nucleoprotein complex is determined by the interactions of the proteins with the DNA (or RNA).

A common problem is presented by the packaging of DNA into phages, viruses, bacterial cells, and eukaryotic nuclei. The length of the DNA as an extended molecule would vastly exceed the dimensions of the compartment that contains it. The DNA (or in the case of some viruses, the RNA) must be compressed exceedingly tightly to fit into the space available. *Thus in contrast with the customary picture of DNA as an extended double helix, structural deformation of DNA to bend or fold it into a more compact form is the rule rather than the exception.*

The magnitude of the discrepancy between the length of the nucleic acid and the size of its compartment is evident from the examples summarized in **FIGURE 23.1**. For bacteriophages and for eukaryotic viruses, the nucleic acid genome, whether single-stranded or double-stranded DNA or RNA, effectively fills the container (which can be rodlike or spherical).

For bacteria or for eukaryotic cell compartments, the discrepancy is hard to calculate exactly, because the DNA is contained in a compact area that occupies only part of the compartment. The genetic material is seen in the form of the **nucleoid** in bacteria, and as the mass of **chromatin** in eukaryotic nuclei at interphase (between divisions), or as maximally condensed **chromosomes** during mitosis.

The density of DNA in these compartments is high. In a bacterium it is ~10 mg/ml; in a eukaryotic nucleus it is ~100 mg/ml; and in the phage T4 head it is >500 mg/ml. Such a concentration in solution would be equivalent to a gel of great viscosity. We do not entirely understand the physiological implications of such high concentrations of DNA, such as the effect this has upon the ability of proteins to find their binding sites on DNA.

The packaging of chromatin is flexible; it changes during the eukaryotic cell cycle. At the time of division (mitosis or meiosis), the genetic material becomes even more tightly packaged, and individual chromosomes become recognizable. Mitotic chromosomes are thought to be five to ten times more tightly packaged than interphase chromatin.

- **nucleoid** The structure in a prokaryotic cell that contains the genome. The DNA is bound to proteins and is not enclosed by a membrane.
- **chromatin** The state of nuclear DNA and its associated proteins during the interphase (between mitoses) of the eukaryotic cell cycle.
- **chromosome** A discrete unit of the genome carrying many genes. Each consists of a very long molecule of duplex DNA and an approximately equal mass of proteins. It is visible as a morphological entity only during cell division.

Compartment	Shape	Dimensions	Type of Nucleic Acid	Length
TMV	filament	0.008 x 0.3 μm	One single-stranded RNA	2 μm = 6.4 kb
Phage fd	filament	0.006 x 0.85 μm	One single-stranded DNA	2 μm = 6.0 kb
Adenovirus	icosahedron	0.07 μm diameter	One double-stranded DNA	11 μm = 35.0 kb
Phage T4	icosahedron	0.065 x 0.10 μm	One double-stranded DNA	55 μm = 170.0 kb
E. coli	cylinder	1.7 x 0.65 μm	One double-stranded DNA	1.3 mm = 4.2×10^3 kb
Mitochondrion (human)	oblate spheroid	3.0 x 0.5 μm	~10 identical double-stranded DNAs	50 μm = 16.0 kb
Nucleus (human)	spheroid	6 μm diameter	46 chromosomes of double-stranded DNA	1.8 m = 6×10^6 kb

FIGURE 23.1 The length of nucleic acid is much greater than the dimensions of the surrounding compartment.

23.2 Viral Genomes Are Packaged into Their Coats

From the perspective of packaging the individual sequence, there is an important difference between a cellular genome and a virus. The cellular genome is essentially indefinite in size; the number and location of individual sequences can be changed by duplication, deletion, and rearrangement. Thus it requires a generalized method for packaging its DNA, one that is insensitive to the total content or distribution of sequences. By contrast, two restrictions define the needs of a virus. The amount of nucleic acid to be packaged is predetermined by the size of the genome, and it must all fit within a coat assembled from a protein or proteins coded by the viral genes.

▸ **capsid** The external protein coat of a virus particle.

A virus particle is deceptively simple in its superficial appearance. The nucleic acid genome is contained within a **capsid**, which is a symmetrical or quasi-symmetrical structure assembled from one or only a few proteins. Attached to the capsid (or incorporated into it) are other structures; these structures are assembled from distinct proteins and are necessary for infection of the host cell.

The virus particle is tightly constructed. The internal volume of the capsid is rarely much greater than the volume of the nucleic acid it must hold. The difference is usually less than twofold, and often the internal volume is barely larger than the nucleic acid. There are two types of solution to the problem of how to construct a capsid that contains nucleic acid:

- The protein shell can be assembled around the nucleic acid, thereby condensing the DNA or RNA by protein–nucleic acid interactions during the process of assembly.
- The capsid can be constructed from its component(s) in the form of an empty shell, into which the nucleic acid must be inserted, being condensed as it enters.

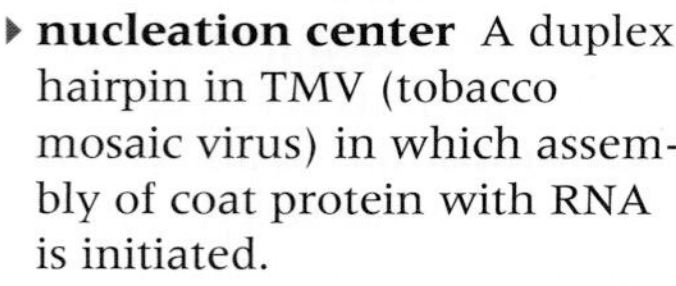

▸ **nucleation center** A duplex hairpin in TMV (tobacco mosaic virus) in which assembly of coat protein with RNA is initiated.

The capsid is assembled around the genome for single-stranded RNA viruses. The principle of assembly is that *the position of the RNA within the capsid is determined directly by its binding to the proteins of the shell.* The best-characterized example is TMV (tobacco mosaic virus). Assembly starts at a duplex hairpin that lies within the sequence of the RNA genome. From this **nucleation center**, assembly proceeds bidirectionally along the RNA until it reaches the ends. The unit of the capsid is a two-layer disk, with each layer containing 17 identical protein subunits. The disk is a circular structure, which forms a helix as it interacts with the RNA. At the nucleation center, the RNA hairpin inserts into the central hole in the disk, and the disk changes conformation into a helical structure that surrounds the RNA. Additional disks are added, with each new disk pulling a new stretch of RNA into its central hole. The RNA becomes coiled in a helical array on the inside of the protein shell, as illustrated in **FIGURE 23.2**.

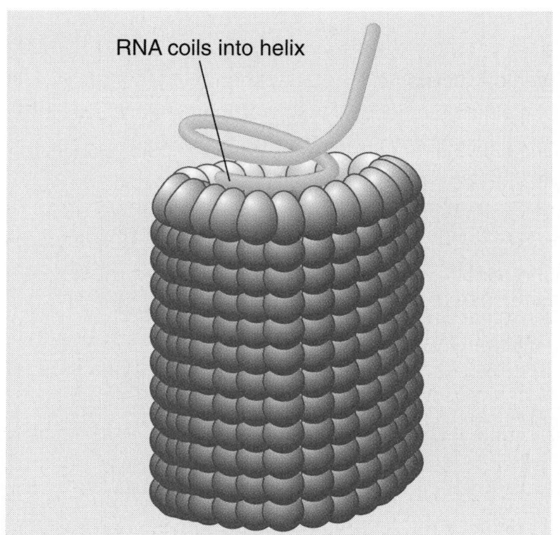

FIGURE 23.2 A helical path for TMV RNA is created by the stacking of protein subunits in the virus.

The spherical capsids of DNA viruses are assembled in a different way, as best characterized for the phages lambda and T4. In each case, an empty headshell is assembled from a small set of proteins. The duplex genome then is inserted into the head, accompanied by a structural change in the capsid.

FIGURE 23.3 summarizes the assembly of lambda. It starts with a small headshell that contains a protein "core." This is converted to an empty headshell of more distinct shape. At this point the DNA packaging begins, the headshell expands in size (though remaining the same shape), and finally the full head is sealed by the addition of the tail.

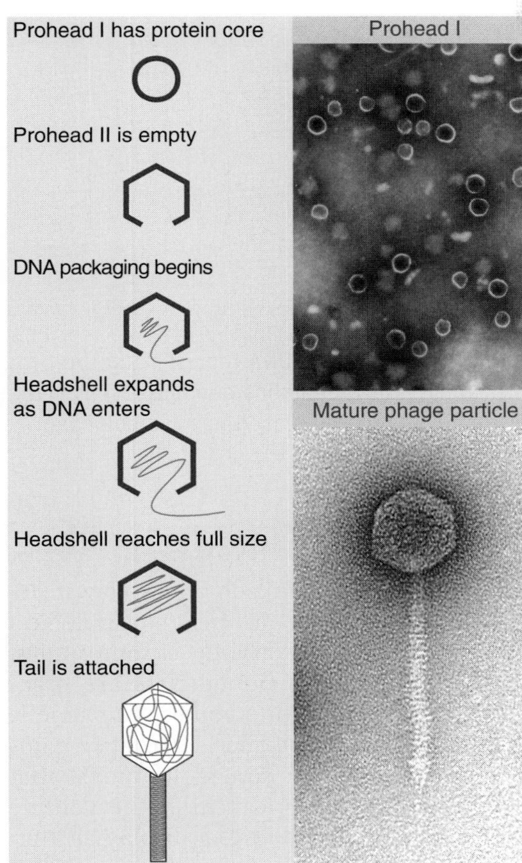
Prohead I has protein core
Prohead II is empty
DNA packaging begins
Headshell expands as DNA enters
Headshell reaches full size
Tail is attached
Prohead I
Mature phage particle

FIGURE 23.3 Maturation of phage lambda passes through several stages. The empty head changes shape and expands when it becomes filled with DNA. The electron micrographs show the particles at the start and the end of the maturation pathway. Top photo reproduced from Cue, D. and Feiss, M., *Proc. Natl. Acad. Sci. USA* 90(1993): 9290–9294. Copyright 1993 National Academy of Science, U.S.A. Photo courtesy of Michael G. Feiss, University of Iowa. Bottom photo courtesy of Robert Duda, University of Pittsburgh.

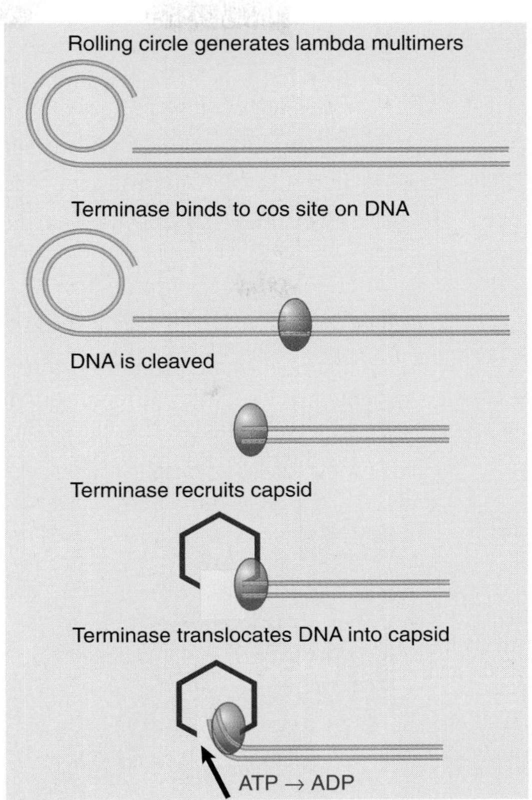
Rolling circle generates lambda multimers
Terminase binds to cos site on DNA
DNA is cleaved
Terminase recruits capsid
Terminase translocates DNA into capsid
ATP → ADP

FIGURE 23.4 Terminase protein binds to specific sites on a multimer of virus genomes generated by rolling circle replication. It cuts the DNA and binds to an empty virus capsid, and then uses energy from hydrolysis of ATP to insert the DNA into the capsid.

Inserting DNA into a phage head involves two types of reaction: *translocation* and *condensation*. Both are energetically unfavorable.

Translocation is an active process in which the DNA is driven into the head by an ATP-dependent mechanism. A common mechanism is used for many viruses that replicate by a rolling circle mechanism to generate long tails that contain multimers of the viral genome. Phage lambda inserts its genome into the empty capsid using the **terminase** enzyme. FIGURE 23.4 summarizes the process.

terminase An enzyme cleaves multimers of a viral genome and then uses hydrolysis of ATP to provide the energy to translocate the DNA into an empty viral capsid starting with the cleaved end.

Little is known about the mechanism of condensation into an empty capsid, except that the capsid contains "internal proteins" as well as DNA. One possibility is that they provide some sort of scaffolding onto which the DNA condenses. (This would be a counterpart to the use of the proteins of the shell in the plant RNA viruses like TMV.)

How specific is the packaging? It cannot depend on particular sequences, because deletions, insertions, and substitutions all fail to interfere with the assembly process. The relationship between DNA and the headshell has been investigated directly by determining which regions of the DNA can be chemically crosslinked to the proteins of the capsid. The surprising answer is that all regions of the DNA are more or less equally susceptible. This probably means that when DNA is inserted into the head, it follows a general rule for condensing, but the pattern is not determined by particular sequences.

KEY CONCEPTS

- The length of DNA that can be incorporated into a virus is limited by the structure of the headshell.
- Nucleic acid within the headshell is extremely condensed.
- Filamentous RNA viruses condense the RNA genome as they assemble the headshell around it.
- Spherical DNA viruses insert the DNA into a preassembled protein shell.

CONCEPT AND REASONING CHECK

Lamba phages are often used to construct *libraries* (collections of DNA representing the genomic material of different organisms), in which most of the lambda genome is replaced by library sequences. Is there a limit to how much DNA could be placed in a lambda carrier? Why or why not?

23.3 The Bacterial Genome Is a Supercoiled Nucleoid

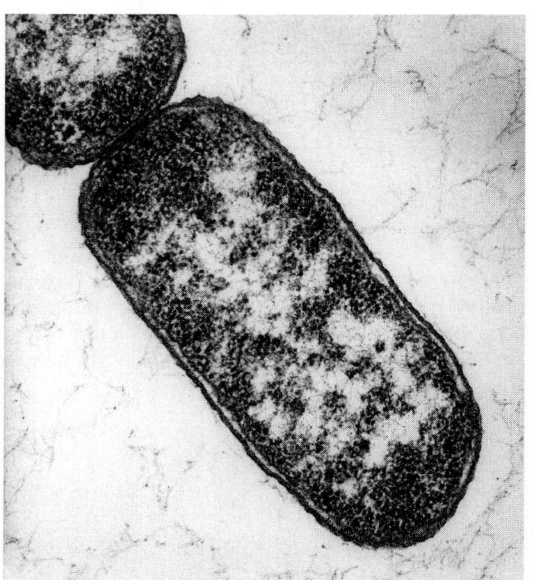

FIGURE 23.5 A thin section shows the bacterial nucleoid as a compact mass in the center of the cell. Photo courtesy of the Molecular and Cell Biology Instructional Laboratory Program, University of California, Berkeley.

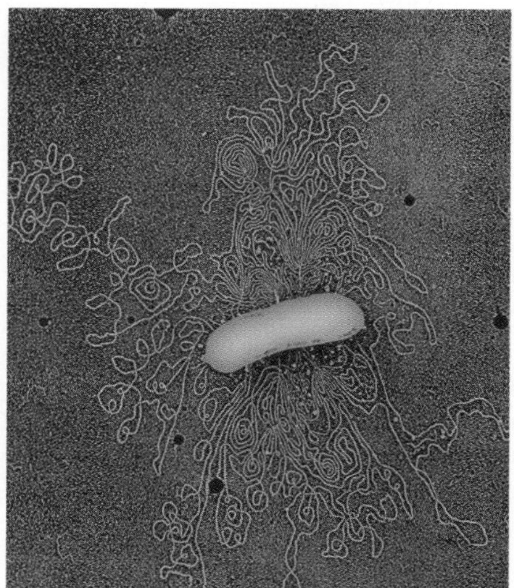

FIGURE 23.6 The nucleoid spills out of a lysed *E. coli* cell in the form of loops of a fiber. Photo © G. Murti/ Photo Researchers, Inc.

Although bacteria do not display structures with the distinct morphological features of eukaryotic chromosomes, their genomes nonetheless are organized into definite bodies. The genetic material can be seen as a fairly compact clump (or series of clumps) that occupies about a third of the volume of the cell. **FIGURE 23.5** displays a thin section through a bacterium in which this nucleoid is evident.

When *E. coli* cells are lysed, fibers are released in the form of loops attached to the broken envelope of the cell. As can be seen from **FIGURE 23.6**, the DNA of these loops is not found in the extended form of a free duplex, but instead is compacted by association with proteins. It is not known which proteins are bound to the DNA to form the nucleoid.

The isolated nucleoid consists of ~80% DNA by mass. Eukaryotic chromosomes, in comparison, contain ~50% DNA by mass. The bacterial nucleoid can be unfolded by treatment with reagents that destroy RNA or protein, indicating that both RNA and protein are essential to maintain its structure.

The DNA of the bacterial nucleoid behaves as a closed duplex structure, as judged by its response to ethidium bromide. This small molecule intercalates between base pairs to generate positive superhelical turns in "closed" circular DNA molecules, that is, molecules in which both strands have covalent integrity. (In "open" circular molecules, which contain a nick in one strand, or with linear molecules, the DNA can rotate freely in response to the intercalation, thus relieving the tension.)

In a natural closed DNA that is negatively supercoiled, the intercalation of ethidium bromide first removes the negative supercoils and then introduces positive supercoils. The amount of ethidium bromide needed to achieve zero supercoiling is a measure of the original density of negative supercoils.

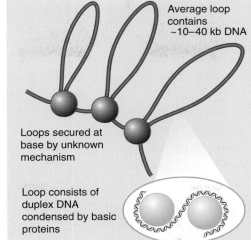

Average loop contains ~10–40 kb DNA
Loops secured at base by unknown mechanism
Loop consists of duplex DNA condensed by basic proteins

FIGURE 23.7 The bacterial genome consists of a large number of loops of duplex DNA (in the form of a fiber), each of which is secured at the base to form an independent structural domain.

Some nicks occur in the compact nucleoid during its isolation; they can also be generated by limited treatment with DNAase. This does not, however, abolish the ability of ethidium bromide to introduce positive supercoils. This capacity of the genome to retain its response to ethidium bromide in the face of nicking means that it must have many independent chromosomal **domains**, and that *the supercoiling in each domain is not affected by events in the other domains.*

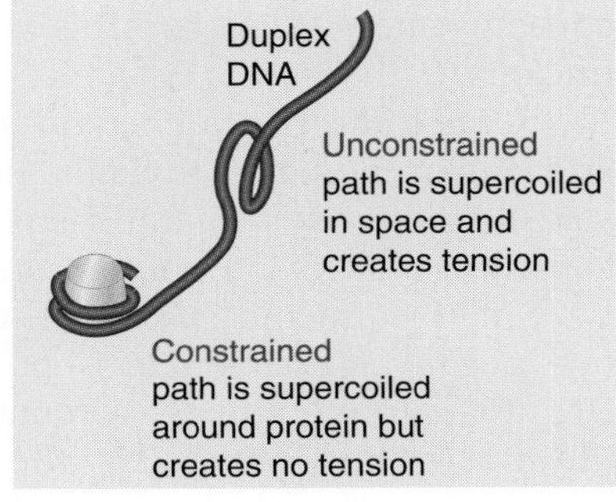

Unconstrained path is supercoiled in space and creates tension
Constrained path is supercoiled around protein but creates no tension

FIGURE 23.8 An unconstrained supercoil in the DNA path creates tension, but no tension is transmitted along DNA when a supercoil is constrained by protein binding.

▸ **domain** In reference to a chromosome, it may refer either to a discrete structural entity defined as a region within which supercoiling is independent of other regions or to an extensive region including an expressed gene that has heightened sensitivity to degradation by the enzyme DNAase I. In a protein, it is a discrete continuous part of the amino acid sequence that can be equated with a particular function.

This autonomy suggests that the structure of the bacterial chromosome has the general organization depicted diagrammatically in **FIGURE 23.7**. Each domain consists of a loop of DNA, the ends of which are secured in some (unknown) way that does not allow rotational events to propagate from one domain to another. There are estimated to be ~400 domains of ~10 kb each in the *E. coli* genome.

The existence of separate domains could permit different degrees of supercoiling to be maintained in different regions of the genome. This could be relevant in considering the different susceptibilities of particular bacterial promoters to supercoiling (see *Section 11.10, Supercoiling Is an Important Feature of Transcription*).

As shown in **FIGURE 23.8**, supercoiling in the genome can take either of two forms:

- If a supercoiled DNA is free, its path is *unconstrained*, and negative supercoils generate a state of torsional tension that is transmitted freely along the DNA within a domain. This supercoiling can be relieved by unwinding the double helix, as described in *Section 1.5, Supercoiling Affects the Structure of DNA*. The DNA is in a dynamic equilibrium between the states of tension and unwinding.
- Supercoiling can be *constrained* if proteins are bound to the DNA to hold it in a particular three-dimensional configuration. In this case, the supercoils are represented by the path the DNA follows in its fixed association with the proteins. The energy of interaction between the proteins and the supercoiled DNA stabilizes the nucleic acid, so that no tension is transmitted along the molecule.

It is estimated that about half of the total supercoils in *E. coli* are unconstrained, and the total superhelical density averages one negative superhelical turn per 100 base pairs. There is likely to be variation about an average level, but it is clear that the level of superhelicity is sufficient to exert significant effects on DNA structure, for example, in assisting melting in particular regions such as origins or promoters.

KEY CONCEPTS

- The bacterial nucleoid is ~80% DNA by mass and can be unfolded by agents that act on RNA or protein.
- The proteins that are responsible for condensing the DNA have not been identified.
- The nucleoid has ~400 independent negatively supercoiled domains.
- The average density of supercoiling is ~1 supercoil/100 bp.

CONCEPT AND REASONING CHECK

Topoisomerases can relax *unconstrained* supercoils, but not supercoils that are constrained by protein binding. What would be the effects of treating a bacterial nucleoid with increasing amounts of topoisomerase?

23.4 Eukaryotic DNA Has Loops and Domains Attached to a Scaffold

Interphase chromatin is a tangled mass occupying a large part of the nuclear volume, in contrast with the highly organized and reproducible ultrastructure of mitotic chromosomes. What controls the distribution of interphase chromatin within the nucleus?

Some indirect evidence on its nature is provided by the isolation of the genome as a single, compact body. Like the bacterial nucleoid, the genome of *Drosophila melanogaster* can be isolated and visualized as a compactly folded fiber (10 nm in diameter) consisting of DNA bound to proteins.

Supercoiling measured by the response to ethidium bromide corresponds to about one negative supercoil/200 bp. These supercoils can be removed by nicking with DNAase, although the DNA remains in the form of the 10 nm fiber. This suggests that the supercoiling is caused by the arrangement of the fiber in space, and that it represents the existing torsion.

Full relaxation of the supercoils requires one nick/85 kb, thus identifying the average length of "closed" DNA. This region could consist of a loop or domain similar to those identified in the bacterial genome. Loops can be seen directly when the majority of proteins are extracted from mitotic chromosomes. The resulting complex consists of the DNA associated with ~8% of the original protein content. As seen in FIGURE 23.9, the protein-depleted chromosomes take the form of a central **metaphase scaffold** that still resembles the general form of a mitotic chromosome, surrounded by a halo of DNA.

▸ **metaphase (or mitotic) scaffold** A proteinaceous structure in the shape of a sister chromatid pair, generated when chromosomes are depleted of histones.

The metaphase scaffold consists of a dense network of fibers. Threads of DNA emanate from the scaffold, apparently as loops of average length 10 to 30 μm (30 to 90 kb). The DNA can be digested without affecting the integrity of the scaffold, which consists of a set of specific proteins. This suggests a form of organization in which loops of DNA of ~60 kb are anchored in a central proteinaceous scaffold. In interphase nuclei, this underlying proteinaceous structure changes its organization to occupy the entire nucleus; during interphase this structure is referred to as the *matrix* rather than the scaffold.

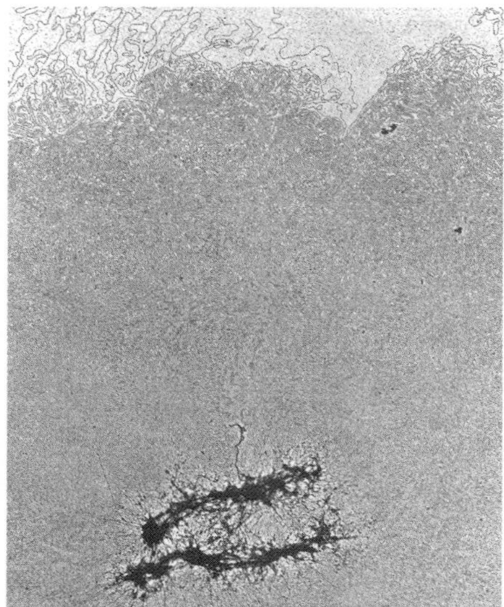

FIGURE 23.9 Histone-depleted chromosomes consist of a protein scaffold to which loops of DNA are anchored. Reprinted from *Cell*, vol. 12, Paulson, J. R., and Laemmli, U. K., *The structure of histone . . .*, pp. 817–828. Copyright 1977, with permission from Elsevier. [http://www.sciencedirect.com/science/journal/00928674]. Photo courtesy of Ulrich K. Laemmli, University of Geneva, Switzerland.

Is DNA attached to the matrix or scaffold via specific sequences? DNA sites attached to proteinaceous structures in interphase nuclei are called **MARs (matrix attachment regions)**; these sequences are also called *SARs* (*scaffold attachment regions*), as the same

▸ **matrix attachment region (MAR)** A region of DNA that attaches to the nuclear matrix. It is also known as a scaffold attachment site (SAR).

sequences appear to attach to the protein substructure in both metaphase and interphase cells. The nature of the structure in interphase cells is not clear, but there have been many suggestions that attachment to this matrix is necessary for transcription or replication. When nuclei are depleted of proteins, the DNA extrudes as loops from a residual proteinaceous structure, just as occurs in scaffold preparations. Attempts to relate the proteins found in this preparation to structural elements of intact cells have not been successful, though.

MAR sequences are usually ~70% A-T-rich, but otherwise lack any consensus sequences. However, MARs are often found close to other important sequences, such as *cis*-acting sites that regulate transcription, and a recognition site for topoisomerase II is usually present in the MAR. It is therefore possible that a MAR serves more than one function by providing a site for attachment to the matrix and also containing other sites at which topological changes in DNA are effected.

The nuclear matrix and chromosome scaffold different proteins, although there are some common components. Topoisomerase II is a prominent component of the chromosome scaffold, and is a constituent of the nuclear matrix. This suggests that the control of topology is important in both cases.

KEY CONCEPTS

- DNA of interphase chromatin is negatively supercoiled into independent domains of ~85 kb.
- Metaphase chromosomes have a protein scaffold to which the loops of supercoiled DNA are attached.
- DNA is attached to the nuclear matrix at specific sequences called MARs or SARs.
- The MARs are A-T-rich but do not have any specific consensus sequence.

CONCEPT AND REASONING CHECK

Even though linear DNA can't be supercoiled, linear eukaryotic chromosomes have ~1 negative supercoil/200 bp. Explain.

23.5 Chromatin Is Divided into Euchromatin and Heterochromatin

Each chromosome contains a single, very long duplex of DNA, folded into a fiber that runs continuously throughout the chromosome. *Thus in accounting for interphase chromatin and mitotic chromosome structure, we have to explain the packaging of a single, exceedingly long molecule of DNA into a form in which it can be transcribed and replicated, and can become cyclically more and less condensed.*

Individual eukaryotic chromosomes come into the limelight for a brief period, during the act of cell division. At that time, each can be seen as a compact unit. **FIGURE 23.10** is an electron micrograph of a sister chromatid pair captured at metaphase. (The sister chromatids are daughter chromosomes produced by the previous replication event and still joined together at this stage of mitosis.) Each consists of a fiber with a diameter of ~30 nm and a nubbly appearance. The DNA is five to ten times more condensed in mitotic chromosomes than in interphase chromatin.

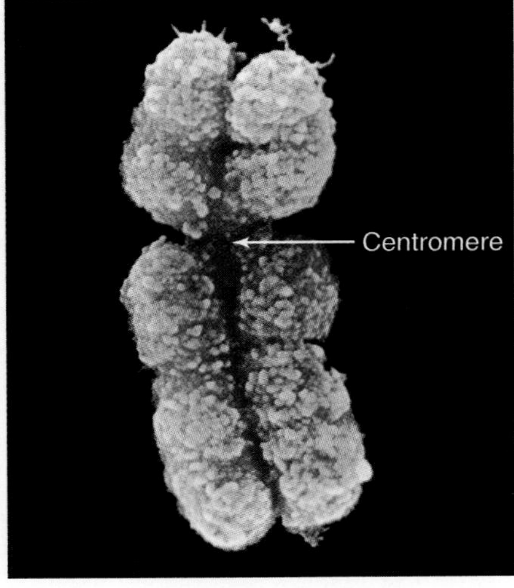

FIGURE 23.10 The sister chromatids of a mitotic pair each consist of a fiber (~30 nm in diameter) compactly folded into the chromosome. © Biophoto Associates/Photo Researchers, Inc.

During most of the life cycle of the eukaryotic cell, however, its genetic material occupies

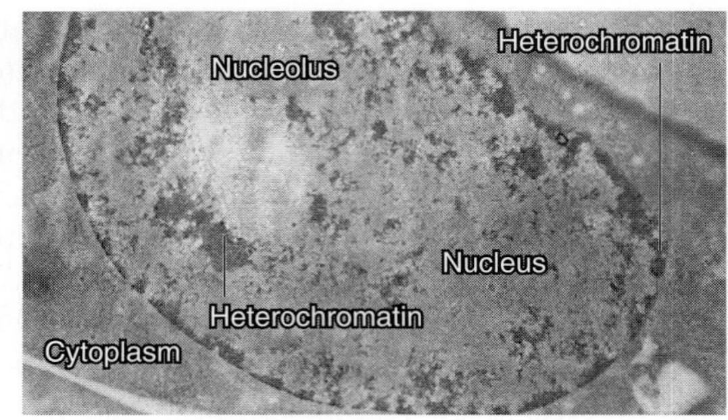

FIGURE 23.11 A thin section through a nucleus stained with feulgen shows heterochromatin as compact regions clustered near the nucleolus and nuclear membrane. Photo courtesy of Edmund Puvion, Centre National De La Recherche Scientifique.

an area of the nucleus in which individual chromosomes cannot be distinguished. The structure of the interphase chromatin does not change visibly between divisions. No disruption is evident during the period of replication, when the amount of chromatin doubles. Interphase chromatin consists of a fiber that is similar or identical to that of the mitotic chromosomes.

Chromatin can be divided into two types of material, which can be seen in the nuclear section of **FIGURE 23.11**:

- In most regions, the fibers are much less densely packed than in the mitotic chromosome. This material is called **euchromatin**. It has a relatively dispersed appearance, stains lightly with DNA-specific dyes, and occupies most of the nuclear region in Figure 23.11.
- Some regions of chromatin are very densely packed with fibers, displaying a density comparable to that of the chromosome at mitosis. This material is called **heterochromatin**, and stains strongly with DNA-specific dyes. It is found at a number of locations within the chromatin, including the centromeres. It passes through the cell cycle with relatively little change in its degree of condensation. It forms a series of discrete clumps in Figure 23.11, but in some cell types the various heterochromatic regions aggregate into a densely staining **chromocenter**. (This description applies to regions that are always heterochromatic, which are called *constitutive heterochromatin*. In addition, there is another sort of heterochromatin, called *facultative heterochromatin*, in which regions of euchromatin are converted to a heterochromatic state).

- **euchromatin** It comprises most of the genome in the interphase nucleus, is less tightly coiled than heterochromatin, and contains most of the active or potentially active single copy genes.
- **heterochromatin** Regions of the genome that are highly condensed, are not transcribed, and are late-replicating. It is divided into two types: constitutive and facultative.
- **chromocenter** An aggregate of heterochromatin from different chromosomes.

The same fibers run continuously between euchromatin and heterochromatin, showing that these states represent different degrees of condensation of the genetic material. In the same way, euchromatic regions exist in different states of condensation during interphase and during mitosis. Thus the genetic material is organized in a manner that permits alternative states to be maintained side by side in chromatin, and allows cyclical changes to occur in the packaging of euchromatin between interphase and mitosis. We discuss the molecular basis for these states in *Chapter 26, Eukaryotic Transcription Regulation* and *Chapter 27, Epigenetic Effects Are Inherited*. Constitutive heterochromatin has some striking characteristic features:

- It is permanently condensed.
- It often consists of multiple repeats of a few sequences of DNA that are not transcribed or are transcribed at very low levels.
- It replicates late in S phase and has a reduced frequency of genetic recombination; this is probably a result of its condensed state.
- The density of genes in this region is very much reduced compared with euchromatin, and genes that are translocated into or near it are often inactivated. The one dramatic exception to this is the ribosomal DNA in the nucleolus, which has the general compacted appearance and behavior of heterochromatin (such as late replication), yet is engaged in very active transcription.

Although most active genes are contained within euchromatin, only a subset of euchromatic genes are transcribed at any time. Thus location in euchromatin is *necessary* for gene expression (at least for RNA Polymerase II), but is not *sufficient* for it.

KEY CONCEPTS

- Individual chromosomes can be seen only during mitosis.
- During interphase, the general mass of chromatin is in the form of euchromatin, which is less tightly packed than mitotic chromosomes.
- Regions of heterochromatin remain densely packed throughout interphase.

CONCEPT AND REASONING CHECK

Why are genes not transcribed in heterochromatin? What do you predict happens to transcription during mitosis?

23.6 Chromosomes Have Banding Patterns

Because of the diffuse state of interphase chromatin, it is very difficult to determine the specificity of its organization. We can, however, ask whether the structure of the mitotic chromosome is ordered. Do particular sequences always lie at particular sites, or is the folding of the fiber into the overall structure a more random event?

At the level of the chromosome, each member of the complement has a different and reproducible ultrastructure. When subjected to certain treatments and then stained with the chemical dye Giemsa, chromosomes generate a series of reproducible **G-bands**. FIGURE 23.12 shows a Giemsa-stained set of human chromosomes.

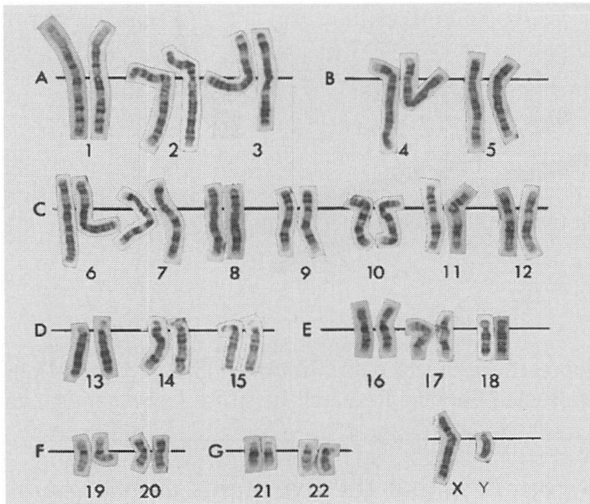

FIGURE 23.12 G-banding generates a characteristic lateral series of bands in each member of the chromosome set. Photo courtesy of Lisa Shaffer, Washington State University–Spokane.

▸ **G-bands** Bands generated on eukaryotic chromosomes by staining techniques that appear as a series of lateral striations. They are used for karyotyping (identifying chromosomes and chromosomal regions by the banding pattern).

G-banding allows each chromosome to be identified by its characteristic banding pattern. This pattern allows translocations from one chromosome to another to be identified by comparison with the original diploid set. FIGURE 23.13 shows a diagram of the bands of the human X chromosome. The bands are large structures, each ~10^7 bp of DNA, which could include many hundreds of genes. This figure also shows the nomenclature used to identify genetic positions on individual chromosomes. A given location is indicated by its position on the long (q) or short (p) arm, then by the region of the arm, band, and subband(s). For example, *CFTR*, the gene that is mutated in cystic fibrosis, is located at 7q31.2; in other words, on the long arm of chromosome 7, region 3, band 1, subband 2.

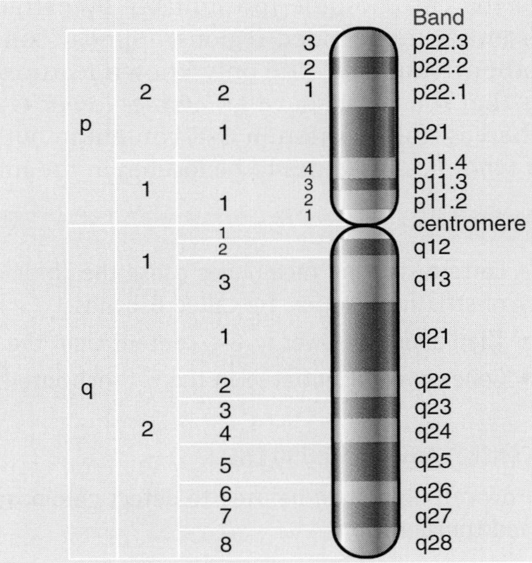

FIGURE 23.13 The human X chromosome can be divided into distinct regions by its banding pattern. The short arm is *p* and the long arm is *q;* each arm is divided into larger regions that are further subdivided. This map shows a low resolution structure; at higher resolution, some bands are further subdivided into smaller bands and interbands, e.g., *p*21 is divided into *p*21.1, *p*21.2, and *p*21.3.

The banding technique is of enormous practical use, but the mechanism of banding remains a mystery. All that

METHODS AND TECHNIQUES

FISH, Chromosome Painting, and Spectral Karyotyping

Classical staining techniques do not provide sufficient sensitivity to detect translocations, deletions, or insertions that involve small segments within a chromosome. However, investigators have taken advantage of lessons learned from molecular biology to devise a very sensitive technique for detecting even very small chromosomal changes in a sample fixed to a microscope slide.

This technique, called **fluorescent *in situ* hybridization** (**FISH**), takes advantage of the fact that a DNA probe with an attached fluorescent dye will bind to a specific DNA sequence within a denatured chromosome (**FIGURE B23.1**). The fluorescent probes can be prepared by nick translation or by the polymerase chain reaction.

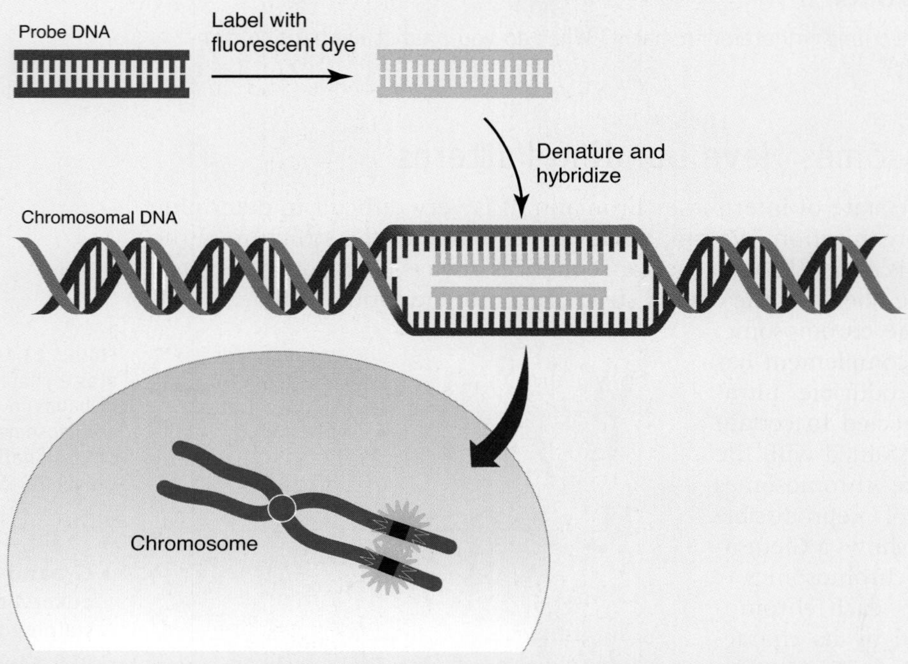

FIGURE B23.1 Fluorescence *in situ* hybridization (FISH). Adapted from an illustration by Darryl Leja, National Human Genome Research Institute (www.genome.gov).

is certain is that the dye stains *untreated* chromosomes more or less uniformly. Thus the generation of bands depends on a variety of treatments that change the response of the chromosome (presumably by extracting the component that binds the stain from the nonbanded regions). Similar bands can be generated by an assortment of other treatments. The only known features that distinguish bands from **interbands** is that the bands have an *average* lower G-C content than the interbands (although there is still variation in G-C content in both bands and interbands), and that there is a tendency for genes to be located in the interband regions.

▸ **interbands** The relatively dispersed regions of polytene chromosomes that lie between the bands.

KEY CONCEPTS

- Certain staining techniques cause the chromosomes to have the appearance of a series of striations, which are called G-bands.
- The bands are lower in G-C content than the interbands.
- Genes are concentrated in the G-C-rich interbands.

CONCEPT AND REASONING CHECK

How can G-banding be used to detect chromosomal deletions, inversions, and translocations?

FISH has a wide variety of applications, which include detecting aneuploidy, identifying chromosomal aberrations, and locating genes and other DNA segments on a chromosome. It can also be used to locate DNA segments during interphase when the chromosome is not visible.

A variation of FISH, called **chromosome painting**, uses a fluorescent dye bound to DNA pieces that bind all along a particular chromosome. The major shortcoming of FISH and chromosome painting is that they cannot be used to study all chromosomes at the same time, because there are not enough fluorescent dyes with sufficient color differences to mark all 23 chromosomes in a unique color.

This problem was solved by labeling the painting probes for each chromosome with a different assortment of fluorescent dyes, a technique called **spectral karyotyping**. When the fluorescent probes hybridize to a chromosome, each kind of chromosome is labeled with a different assortment of fluorescent dye combinations. Stained chromosomes are then viewed through a series of filters, each of which transmits only light emitted by a single fluorescent dye. Alternatively, an interferometer determines the full spectrum of light emitted by the stained chromosome. In either case, a computer provides a composite picture that shows different chromosome pairs as if they were stained in different colors (**FIGURE B23.2**).

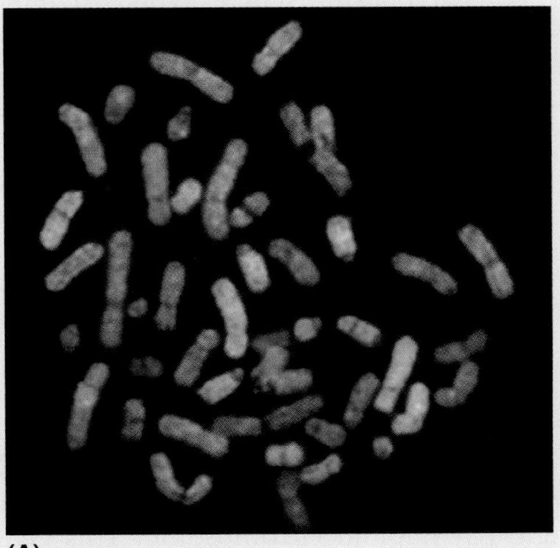

(A)

(B)

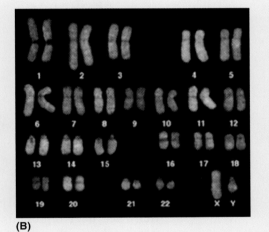

1 2 3 4 5 6 7 8 9 10 11 12 13 14 15 16 17 18 19 20 21 22 X Y

FIGURE B23.2 Spectral karyotyping (SKY), an application of chromosome painting. **(A)** A chromosome spread hybridized with SKY fluorescent probes. **(B)** The same labeled chromosomes as in (A), sorted by size to show the karyotype. Photos courtesy of Johannes Wienberg, Ludwig-Maximilians-University, and Thomas Ried, National Institutes of Health.

23.7 Polytene Chromosomes Form Bands That Expand at Sites of Gene Expression

Much of our understanding of chromosome structure and the structural changes associated with transcription comes from the study of an interesting type of chromosome found in dipteran (two-winged) insects. The interphase nuclei of specific tissues of the larvae of dipteran flies contain chromosomes that are greatly enlarged relative to their usual condition. They possess both increased diameter and greater length. **FIGURE 23.14** shows an example of a chromosome set from the salivary gland of *D. melanogaster*. The members of this set are called **polytene** chromosomes.

Each member of the polytene set consists of a visible series of **bands** (more properly, but rarely, described as **chromomeres**). The bands range in size from the largest, with a breadth of ~0.5 μm, to the smallest, at ~0.05 μm. The bands contain most of the mass of DNA and stain intensely with appropriate reagents. The regions between them stain more lightly and are called **interbands**. There are ~5000 bands in the *D. melanogaster* set.

The centromeres of all four chromosomes of *D. melanogaster* aggregate to form a chromocenter that consists largely of heterochromatin (in the male it includes the entire

- **polytene chromosomes** Chromosomes that are generated by successive replications of a chromosome set without separation of the replicas.
- **bands** They are visible on polytene chromosomes as dense regions that contain the majority of DNA; they include active genes.
- **chromomeres** Densely staining granules visible in chromosomes under certain conditions, especially early in meiosis, when a chromosome may appear to consist of a series of chromomeres.

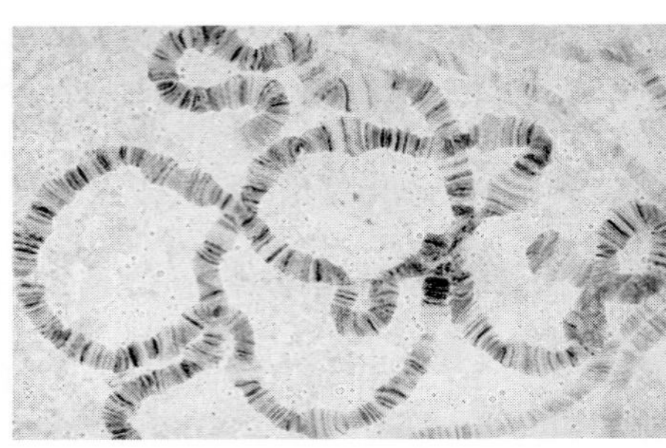

FIGURE 23.14 The polytene chromosomes of *D. melanogaster* form an alternating series of bands and interbands. Photo courtesy of José Bonner, Indiana University.

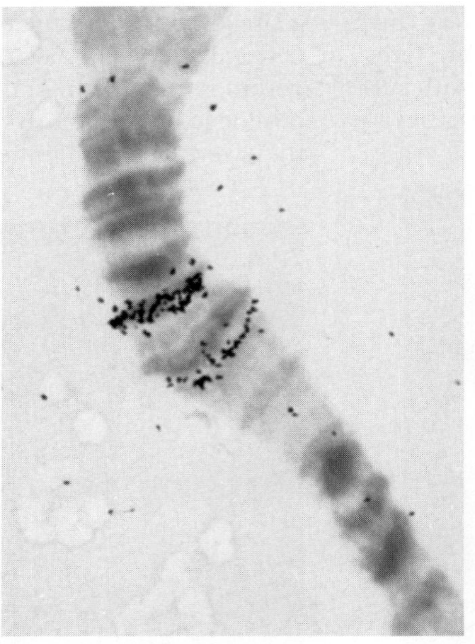

FIGURE 23.15 A magnified view of bands 87A and 87C shows their hybridization *in situ* with labeled RNA extracted from heat-shocked cells. Photo courtesy of José Bonner, Indiana University.

Y chromosome). Allowing for this, ~75% of the haploid DNA set is organized into alternating bands and interbands. The length of the chromosome set is ~2000 μm. The DNA in extended form would stretch for ~40,000 μm, so this gives a ratio of the extended length to the actual length of ~20. Typically, this ratio is about 1000–2000 for interphase chromatin, and up to ~7000 for highly condensed mitotic chromosomes! This demonstrates vividly the extension of the genetic material of polytene chromosomes relative to the usual condensation states chromatin.

What is the structure of these giant chromosomes? Each is produced by the successive replications of a synapsed diploid pair. The replicas do not separate, but instead remain attached to each other in their extended state. At the start of the process, each synapsed pair has a DNA content of 2C (where C represents the DNA content of the individual chromosome). This amount then doubles up to nine times, at its maximum giving a content of 1024C. The number of doublings is different in the various tissues of the *D. melanogaster* larva. This process is known as *endoreduplication*.

Each chromosome can be visualized as a large number of parallel fibers running longitudinally that are tightly condensed in the bands and less so in the interbands. It is likely that each fiber represents a single (C) haploid chromosome. This gives rise to the name *polytene*: the degree of polyteny is the number of haploid chromosomes contained in the giant chromosome.

The banding pattern is characteristic for each strain of *Drosophila*. The constant number and linear arrangement of the bands was first noted in the 1930s, when it was realized that they form a *cytological map* of the chromosomes. Rearrangements—such as deletions, inversions, or duplications—result in alterations of the order of bands.

The linear array of bands can be equated with the linear array of genes. Thus genetic rearrangements, as seen in a linkage map, can be correlated with structural rearrangements of the cytological map. Ultimately, a particular mutation can be located in a particular band. The total number of genes in *D. melanogaster* exceeds the number of bands, so there are probably multiple genes in most or all bands.

The positions of particular genes on the cytological map can be determined directly by the technique of ***in situ* hybridization**. This method is described in the accompanying *Methods Box*. Although fluorescent probes are currently preferred, when the method was originally developed a radioactive probe representing the gene of interest was used, and the position or positions of the corresponding genes was detected by autoradiography. An example is shown in **FIGURE 23.15**.

One of the intriguing features of the polytene chromosomes is that sites of active transcription can be visualized. Some of the bands pass transiently through an expanded state in which they appear like a **puff** on the chromosome, when chromosomal material is extruded from the axis. The dramatic puffing of a heat shock gene locus upon heat shock is shown in **FIGURE 23.16**. In this example, the correlation between puffing and transcription is illustrated by staining the chromosomes for RNA polymerase II, which is concentrated in the puffs.

- ***in situ* hybridization** Hybridization performed by denaturing the DNA of cells squashed on a microscope slide so that reaction is possible with an added single-stranded RNA or DNA; the added preparation is radioactively labeled and its hybridization is followed by autoradiography.
- **puff** An expansion of a band of a polytene chromosome associated with the synthesis of RNA at some locus in the band.

What is the nature of the puff? It consists of a region in which the chromosome fibers unwind from their usual state of packing in the band. The fibers remain continuous with those in the chromosome axis. Puffs usually emanate from single bands, although when they are very large, the swelling may be so extensive as to obscure the underlying array of bands.

The puffs are *sites where RNA is being synthesized.* The accepted view of puffing has been that expansion of the band is a consequence of the need to relax its structure in order to synthesize RNA. Puffing has therefore been viewed as a consequence of transcription. A puff can be generated by a single active gene. The sites of puffing differ from ordinary bands in accumulating additional proteins, which include RNA polymerase II (as shown in Figure 23.16) and other proteins associated with transcription. These observations suggest that, in order to be transcribed, the genetic material is dispersed from its usual, more tightly packed state. The question to keep in mind is whether this dispersion at the gross level of the chromosome mimics the events that occur at the molecular level within the mass of ordinary interphase euchromatin.

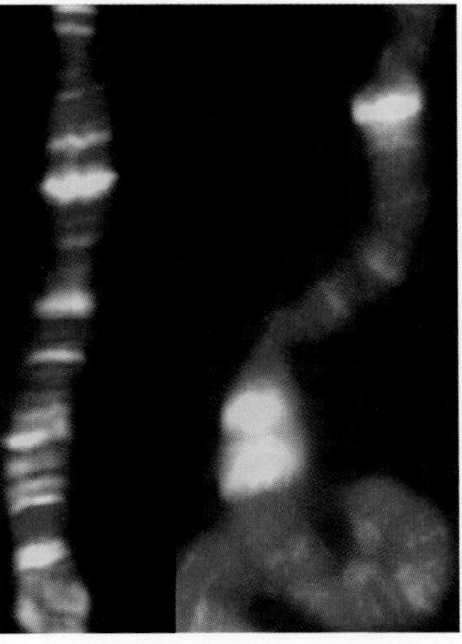

FIGURE 23.16 Heat-shock-induced puffing at major heat shock loci 87A and C. Displayed is a small segment of chromosome 3 before (left) and after (right) heat shock. Chromosomes are stained for DNA (blue) and for Pol II (yellow). HS, heat shock. Photo courtesty of Victor G. Corces, Emory University.

KEY CONCEPTS

- Polytene chromosomes of dipterans have a series of bands that can be used as a cytological map.
- Bands that are sites of gene expression on polytene chromosomes expand to give "puffs."

CONCEPT AND REASONING CHECK

How do the bands on polytene chromosomes differ from the G-bands described in the previous section?

23.8 Centromeres Often Contain Repetitive DNA

During mitosis, the sister chromatids move to opposite poles of the cell. Their movement depends on the attachment of the chromosome to microtubules, which are connected at their other end to the poles. (The microtubules comprise a cellular filamentous system, reorganized at mitosis into the **spindle**, which connects the chromosomes to the poles of the cell.) The sites in the two regions where microtubule ends are organized—in the vicinity of the centrioles at the poles and at the chromosomes—are called **MTOCs (microtubule organizing centers)**.

FIGURE 23.17 illustrates the separation of sister chromatids as mitosis proceeds from metaphase to telophase. The region of the chromosome that is responsible for its segregation at mitosis and meiosis is called the **centromere**. The centromeric region on

- **spindle** A structure made up of microtubules that guides the movements of the chromosomes during mitosis.
- **microtubule organizing center (MTOC)** A region from which microtubules emanate. In animal cells the centrosome is the major microtubule organizing center.
- **centromere** A constricted region of a chromosome that includes the site of attachment (the kinetochore) to the mitotic or meiotic spindle. It consists of unique DNA sequences and proteins not found anywhere else in the chromosome.

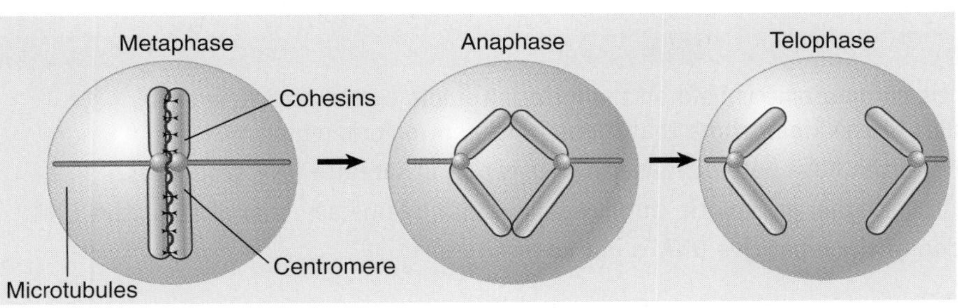

FIGURE 23.17 Chromosomes are pulled to the poles via microtubules that attach at the centromeres. The sister chromatids are held together until anaphase by glue proteins (cohesins). The centromere is shown here in the middle of the chromosome (metacentric), but can be located anywhere along its length, including close to the end (acrocentric) and at the end (telocentric).

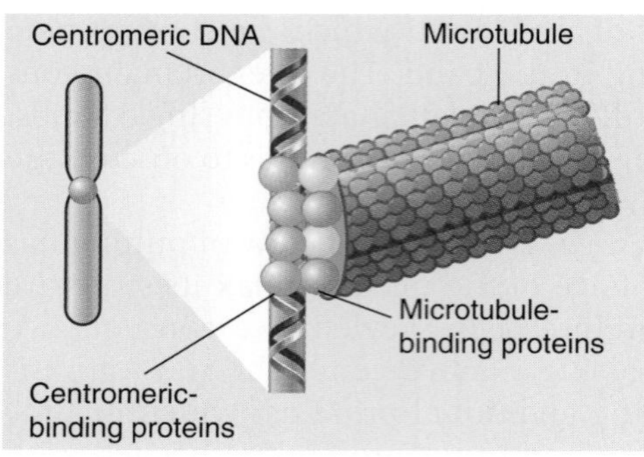

FIGURE 23.18 The centromere is identified by a DNA sequence that binds specific proteins. These proteins do not themselves bind to microtubules, but establish the site at which the microtubule-binding proteins in turn bind.

each sister chromatid is pulled by microtubules to the opposite pole. Opposing this motive force, "glue" proteins called *cohesins* hold the sister chromatids together. Initially the sister chromatids separate at their centromeres, and then they are released completely from one another during anaphase when the cohesins are degraded. The centromere is pulled toward the pole during mitosis. The centromere is the site at which the sister chromatids are held together prior to the separation of the individual chromosomes, which shows as a constricted region connecting all four chromosome arms, as in the photo of Figure 23.10.

The centromere is essential for segregation, as shown by the behavior of chromosomes that have been broken. A single break generates one piece that retains the centromere, and another, an **acentric fragment**, that lacks it. The acentric fragment does not become attached to the mitotic spindle, and as a result it fails to be included in either of the daughter nuclei.

▸ **acentric fragment** A fragment of a chromosome (generated by breakage) that lacks a centromere and is lost at cell division.

The regions flanking the centromere often are rich in repetitive satellite DNA sequences and frequently display a considerable amount of heterochromatin. The region of the chromosome at which the centromere forms is defined by DNA sequences. The centromeric DNA binds specific proteins that are responsible for establishing the structure that attaches the chromosome to the microtubules, which is called the **kinetochore**. It is a darkly staining fibrous object of ~400 nm. The kinetochore provides the MTOC on a chromosome. **FIGURE 23.18** shows the hierarchy of organization that connects centromeric DNA to the microtubules. Proteins bound to the centromeric DNA bind other proteins that bind to microtubules.

▸ **kinetochore** A small organelle associated with the surface of the centromere that attaches a chromosome to the microtubules of the mitotic spindle. Each mitotic chromosome contains two "sisters" that are positioned on opposite sides of its centromere and face in opposite directions.

The length of DNA required for centromeric function is often quite long. (The short, discrete elements of *Saccharomyces cerevisiae*, discussed in the next section, may be an exception to the general rule.) In those cases where we can equate specific DNA sequences with the centromeric region, they usually include repetitive sequences. The centromeres of the yeast *S. pombe* consist of 40 to 100 kb regions made up largely or entirely of repetitious DNA. Attempts to localize centromeric functions in *Drosophila* chromosomes suggest that they are dispersed in a large region, consisting of 200 to 600 kb. The large size of this type of centromere suggests that it is likely to contain several separate specialized functions, such as sequences required for kinetochore assembly and sister chromatid pairing.

The primary motif of the heterochromatin of primate centromeres is α satellite DNA, which consists of tandem arrays of a 170 bp repeating unit. There is significant variation between individual repeats, although those at any centromere tend to be more closely related to one another than to members of the family in other locations. It is clear that the sequences required for centromeric function reside within the blocks of α satellite DNA, but it is not clear whether the α satellite sequences themselves provide this function, or whether other critical sequences are embedded within the α satellite arrays.

KEY CONCEPTS

- A eukaryotic chromosome is held on the mitotic spindle by the attachment of microtubules to the kinetochore that forms in its centromeric region.
- Centromeres often have heterochromatin that is rich in satellite DNA sequences.
- Centromeres in higher eukaryotic chromosomes contain large amounts of repetitive DNA.
- The function of the repetitive DNA is not known.

CONCEPT AND REASONING CHECK

What would happen to a chromosome that, as a result of a translocation, contains two centromeres?

23.9 *S. cerevisiae* Centromeres Have Short Protein-Binding DNA Sequences

S. cerevisiae chromosomes do not display large, visible kinetochores comparable to those of higher eukaryotes, but otherwise use the same mechanisms for mitotic and meiotic divisions. A *CEN* fragment is defined as the minimal sequence that can support accurate segregation of a plasmid at mitosis. Every chromosome has a *CEN* region, and *CEN* fragments from different chromosomes are interchangeable. This indicates that centromeres only serve *to attach the chromosome to the spindle, and play no role in distinguishing one chromosome from another.*

The sequences required for centromeric function fall within a stretch of ~120 bp. The centromeric region is packaged into a nuclease-resistant structure and binds a single microtubule. We may therefore look to the *S. cerevisiae* centromeric region to identify proteins that bind centromeric DNA and proteins that connect the chromosome to the spindle.

There are three types of sequence element in the *CEN* region, summarized in **FIGURE 23.19**:

- The cell cycle-dependent element (*CDE*)-*I* is a sequence of 9 bp that is conserved with minor variations at the left boundary of all centromeres.
- *CDE-II* is a >90% A-T-rich sequence of 80–90 bp found in all centromeres; its function could depend on its length rather than exact sequence. Its base composition may cause some characteristic distortions of the DNA double helical structure.
- *CDE-III* is an 11 bp sequence highly conserved at the right boundary of all centromeres. Sequences on either side of the element are less well conserved, but may also be needed for centromeric function. (*CDE-III* could be longer than 11 bp if it turns out that the flanking sequences are essential.)

Mutations in *CDE-I* or *CDE-II* reduce, but do not inactivate, centromere function; point mutations in the central CCG of *CDE-III* completely inactivate the centromere.

A large protein complex assembles at the *CDE* sequences, and connects the chromosome to microtubules. The structure is summarized in **FIGURE 23.20**.

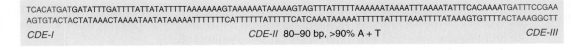

TCACATGATGATATTTGATTTTATTATATTTTTAAAAAAAGTAAAAAATAAAAAGTAGTTTATTTTTAAAAAATAAAATTTAAAATATTTCACAAAATGATTTCCGAA
AGTGTACTACTATAAACTAAAATAATATAAAAATTTTTTTCATTTTTTATTTTTCATCAAATAAAAATTTTTTATTTTAAATTTTATAAAGTGTTTTACTAAAGGCTT
CDE-I *CDE-II* 80–90 bp, >90% A + T *CDE-III*

FIGURE 23.19 Three conserved regions can be identified by the sequence homologies between yeast *CEN* elements.

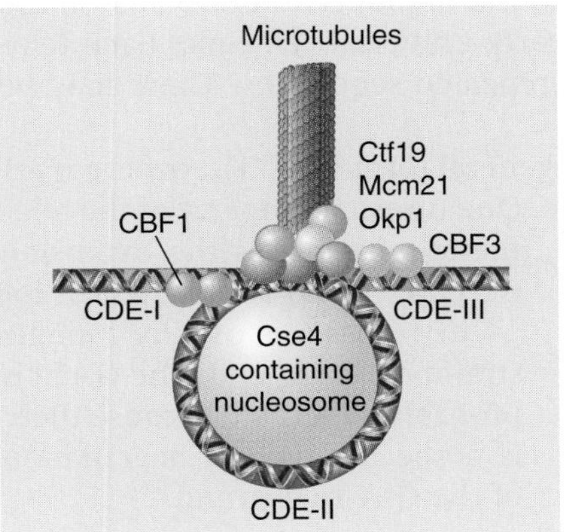

FIGURE 23.20 The DNA at *CDE-II* is wound around an alternative nucleosome including Cse4, *CDE-III* is bound by the CBF3 complex, and *CDE-I* is bound by a CBF1 homodimer. These proteins are connected by the group of Ctf19, Mcm21, and Okp1 proteins, and numerous other factors serve to link this complex to a microtubule.

The *CEN* region recruits three DNA-binding factors, Cbf1, CBF3 (an essential four-protein complex) and Mif2. In addition, a specialized chromatin structure is built by binding the *CDE-II* region to a protein called Cse4, a variant of one of the histone proteins that comprise the basic subunits of chromatin (see *Section 24.4, Histone Variants Produce Alternative Nucleosomes*). A protein called Scm3 is required for proper association of Cse4 with *CEN*. Inclusion of histone variants related to Cse4 are a universal aspect of centromere construction in all species.

The proteins bound at *CDE-I*, *-II*, and *-III* interact with another group of proteins (Ctf19, Mcm21, Okp1), which in turn link the centromeric complex to the kinetochore proteins (~70 individual kinetochore proteins have been identified in yeast) and to the microtubule.

The overall model suggests that the complex is localized at the centromere by a protein structure that resembles the normal building block of chromatin (the nucleosome). The bending of DNA at this structure allows proteins bound to the flanking elements to become part of a single complex. The DNA-binding components of the complex form a scaffold for assembly of the kinetochore, linking the centromere to the microtubule. The construction of kinetochores follows a similar pattern and uses related components, in a wide variety of organisms.

KEY CONCEPTS

- *CEN* elements are identified in *S. cerevisiae* by the ability to allow a plasmid to segregate accurately at mitosis.
- *CEN* elements consist of the short conserved sequences *CDE-I* and *CDE-III* that flank the A-T-rich region *CDE-II*.
- A specialized protein complex that is an alternative to the usual chromatin structure is formed at *CDE-II*.
- The proteins that bind *CEN* serve as an assembly platform for the kinetochore and provide the connection to microtubules.

CONCEPT AND REASONING CHECK

Why might the length of the *CDE-II* region, but not its precise sequence, have importance for *CEN* function?

23.10 Telomeres Have Simple Repeating Sequences

▸ **telomere** The natural end of a chromosome; the DNA sequence consists of a simple repeating unit with a protruding single-stranded end.

Another essential feature of all eukaryotic chromosomes is the **telomere**, which "seals" the ends of chromosomes. We know that the telomere must be a special structure, because chromosome ends generated by breakage are "sticky" and tend to react with other chromosomes, whereas natural ends are stable. The cell has to be able to distinguish the telomere from free ends caused by damage.

Telomeric sequences have been characterized from a wide range of lower and higher eukaryotes. The same type of sequence is found in plants and humans, so the construction of the telomere seems to follow a nearly universal principle. Each telomere consists of a long series of short, tandemly repeated sequences. There may be 100 to 1000 repeats, depending on the organism.

All telomeric sequences can be written in the general form $C_n(A/T)_m$, where $n > 1$ and m is 1 to 4. **FIGURE 23.21** shows a generic example. One unusual property of the telomeric sequence is the extension of the G-T-rich strand, which for 14 to 16 bases is usually a single strand with a 3′ end. The G-tail is probably generated because there is a specific limited degradation of the C-A-rich strand.

CCCCAACCCCAACCCCAACCCCAACCCCAACCCCAA
GGGGTTGGGGTTGGGGTTGGGGTTGGGGTTGGGGTT

↓

CCCCAACCCCAACCCCAA 5′
GGGGTTGGGGTTGGGGTTGGGGTTGGGGTTGGGGTT3′

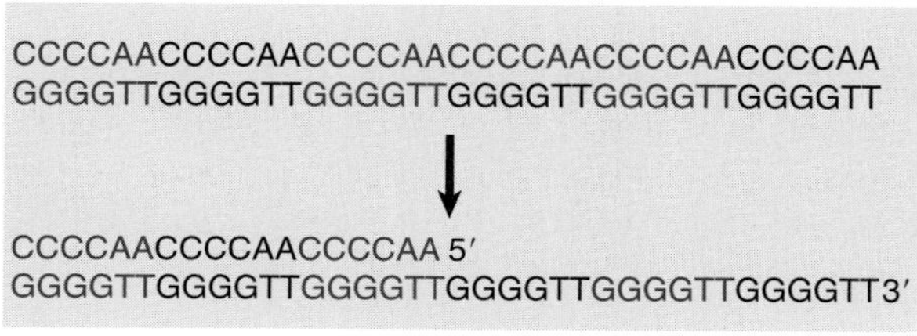

FIGURE 23.21 A typical telomere has a simple repeating structure with a G-T-rich strand that extends beyond the C-A-rich strand. The G-tail is generated by a limited degradation of the C-A-rich strand.

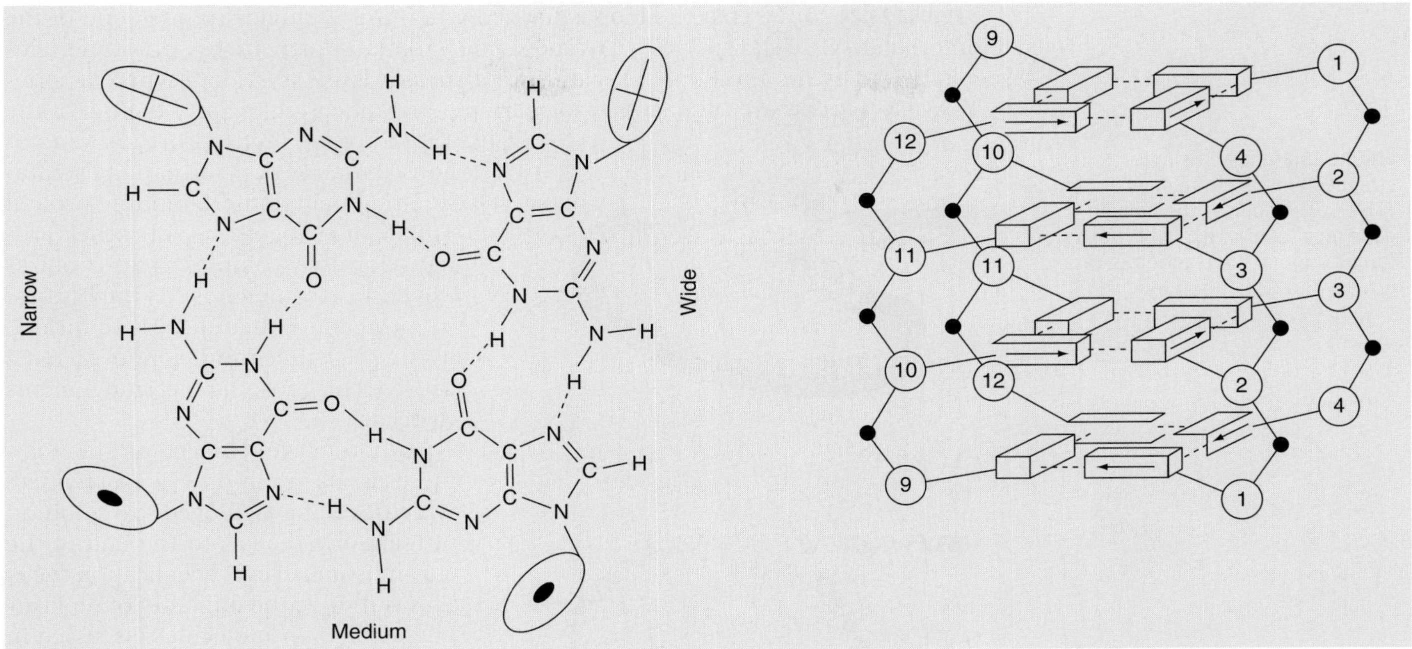

FIGURE 23.22 The crystal structure of a short repeating sequence from the human telomere forms three stacked G quartets. The top quartet contains the first G from each repeating unit. This is stacked above a quartet that contains the second G and the third G.

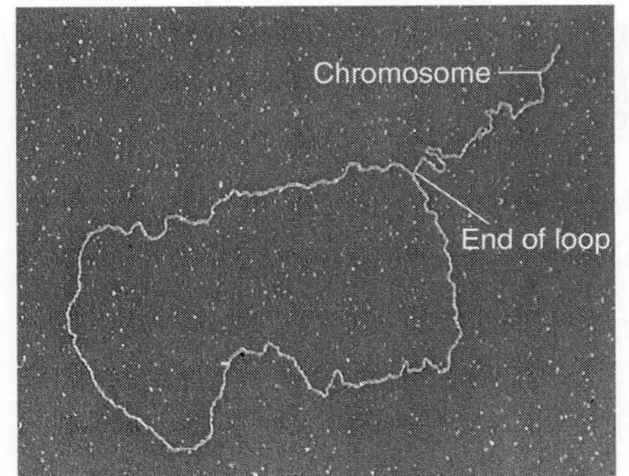

FIGURE 23.23 A loop forms at the end of chromosomal DNA. Photo Courtesy of Jack Griffith, University of North Carolina at Chapel Hill.

The G-tail has the capacity to form a unique structure. Guanine bases have an unusual ability to associate with one another. The single-stranded G-rich tail of the telomere can form "quartets" of G residues. Each quartet contains four guanines that hydrogen bond with one another to form a planar structure. Each guanine comes from the corresponding position in a successive TTAGGG repeating unit. **FIGURE 23.22** shows a quartet representing an association between the first guanine in each repeating unit. It is stacked on top of another quartet that has the same organization, but is formed from the second guanine in each repeating unit. A series of quartets can be stacked like this in a helical manner. It is suggested that this unusual structure may help protect telomere ends *in vivo*, though it is not yet known if this structure is required for telomere function.

In addition to the possible role of the G-tail, another mechanism has been discovered that is believed to promote the stability of chromosome ends. **FIGURE 23.23** shows that a loop of DNA forms at the telomere. The absence of any free end may be the crucial feature that stabilizes the end of the chromosome. The average length of the loop in animal cells is 5 to 10 kb.

FIGURE 23.24 shows that the loop is formed when the 3′ single-stranded end of the telomere $(TTAGGG)_n$ displaces the same sequence in an upstream region of the telomere. This converts the duplex region into a structure like a D-loop, where a series of TTAGGG repeats are displaced to form a single-stranded region, and the tail of the telomere is paired with the homologous strand. Loop formation is catalyzed by the telomere-binding protein Trf2, which together with other proteins forms a complex that stabilizes the chromosome ends. Its importance in protecting the ends is indicated by the fact that the deletion of TRF2 causes chromosome rearrangements to occur.

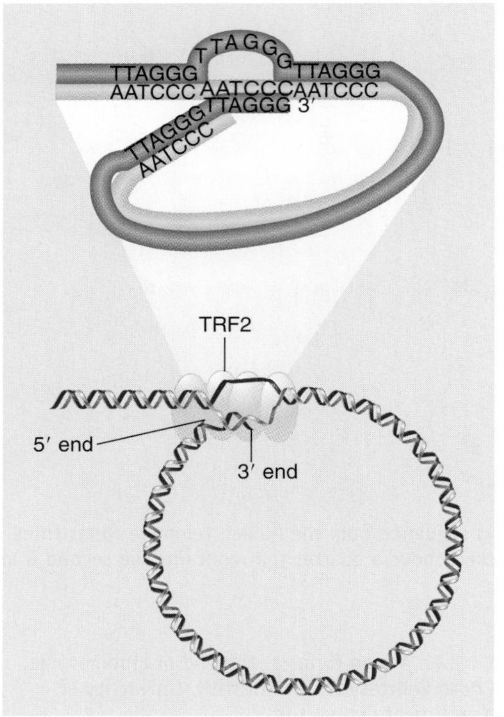

FIGURE 23.24 The 3′ single-stranded end of the telomere $(TTAGGG)_n$ displaces the homologous repeats from duplex DNA to form a t-loop. The reaction is catalyzed by Trf2.

Another key aspect of the telomere is its ability to be extended, which can be achieved by addition of telomeric repeats to the end of the chromosome in every replication cycle. This can solve the difficulty of replicating linear DNA molecules discussed in *Section 16.2, The Ends of Linear DNA Are a Problem for Replication*. The addition of repeats by *de novo* synthesis would counteract the loss of repeats resulting from failure to replicate up to the end of the chromosome. Extension and shortening would be in dynamic equilibrium.

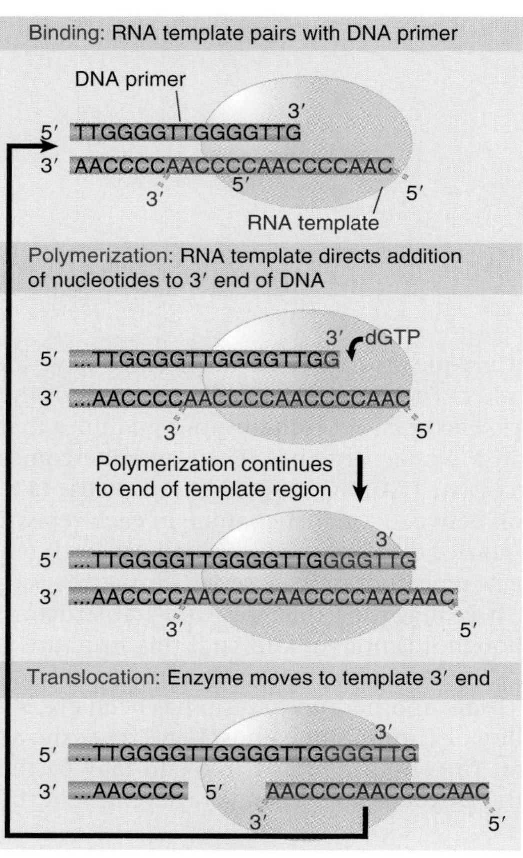

FIGURE 23.25 Telomerase positions itself by base pairing between the RNA template and the protruding single-stranded DNA primer. It adds G and T bases one at a time to the primer, as directed by the template. The cycle starts again when one repeating unit has been added.

▸ **telomerase** The ribonucleoprotein enzyme that creates repeating units of one strand at the telomere by adding individual bases to the DNA 3′ end, as directed by an RNA sequence in the RNA component of the enzyme.

A **telomerase** enzyme extends the C-A-rich strand. Telomerase uses the 3′–OH of the G+T telomeric strand as a primer for synthesis of tandem TTGGGG repeats. Only dGTP and dTTP are needed for the activity. The telomerase is a large ribonucleoprotein that consists of a *templating RNA* and a protein with polymerase activity. The RNA component (ranging from ~150 nt to >1300 nt, depending on species) includes a sequence of 15 to 22 bases that is identical to two repeats of the C-rich repeating sequence. This RNA provides the template for synthesizing the G-rich repeating sequence. The protein component of the telomerase is a catalytic subunit that can act only upon the RNA template provided by the nucleic acid component.

FIGURE 23.25 shows the action of telomerase. The enzyme progresses discontinuously: the template RNA is positioned on the DNA primer, several nucleotides are added to the primer, and then the enzyme translocates to begin again. The telomerase is a specialized example of a reverse transcriptase, an enzyme that synthesizes a DNA sequence using an RNA template. We do not know how the complementary (C-A-rich) strand of the telomere is assembled, but we may speculate that it could be synthesized by using the 3′–OH of a terminal G-T hairpin as a primer for DNA synthesis.

Telomerase synthesizes the individual repeats that are added to the chromosome ends, but does not itself control the number of repeats. Other proteins are involved in determining the length of the telomere, such as the Est1 and Est3 proteins in yeast. These proteins may bind telomerase and also influence the length of the telomere by controlling the access of telomerase to its substrate. Proteins that bind telomeres in mammalian cells have been found, but less is known about their functions. As a result of the actions of these types of proteins, each organism has a characteristic range of telomere lengths. They are long in mammals (typically 5 to 15 kb in humans) and short in yeast (typically ~300 bp in *S. cerevisiae*).

KEY CONCEPTS

- The telomere is required for the stability of the chromosome end.
- A telomere consists of a simple repeat where a C + A-rich strand has the sequence $C_{>1}(A/T)_{1-4}$.
- The protein Trf2 catalyzes a reaction in which the 3′ repeating unit of the G + T-rich strand forms a loop by displacing its homolog in an upstream region of the telomere.
- Telomerase uses the 3′–OH of the G + T telomeric strand to prime synthesis of tandem TTGGGG repeats.
- The RNA component of telomerase has a sequence that pairs with the C + A-rich repeats.
- One of the protein subunits is a reverse transcriptase that uses the RNA as a template to synthesize the G + T-rich sequence.

CONCEPT AND REASONING CHECK

Most somatic cells in mammals do not contain active telomerase. What is likely to happen as these cells continue to divide?

23.11 Summary

The genetic material of all organisms and viruses takes the form of tightly packaged nucleoprotein. Some virus genomes are inserted into preformed viral particles, whereas others assemble a protein coat around the nucleic acid. The bacterial genome forms a dense nucleoid, with ~20% protein by mass, but details of the interaction of the proteins with DNA are not known. The DNA is organized into ~400 domains that maintain independent supercoiling, with a density of unrestrained supercoils corresponding to ~1/100 to 200 bp. In eukaryotes, interphase chromatin and metaphase chromosomes both appear to be organized into large loops. Each loop may be an independently supercoiled domain. The bases of the loops are connected to a metaphase scaffold or to the nuclear matrix by specific DNA sites.

Transcriptionally active sequences reside within the euchromatin that comprises the majority of interphase chromatin. The regions of heterochromatin are packaged ~5 to 10× more compactly, and are generally transcriptionally inert. All chromatin becomes densely packaged during cell division, when the individual chromosomes can be distinguished. The existence of a reproducible ultrastructure in chromosomes is indicated by the production of G-bands by treatment with Giemsa stain. The bands are very large regions (~10^7 bp) that can be used to map chromosomal translocations or other large changes in structure.

Polytene chromosomes of insects have unusually extended structures. Polytene chromosomes of *D. melanogaster* are divided into ~5000 bands. These bands vary in size by an order of magnitude, with an average of ~25 kb. Transcriptionally active regions can be visualized in even more unfolded ("puffed") structures, in which material is extruded from the axis of the chromosome. This may resemble the changes that occur on a smaller scale when a sequence in euchromatin is transcribed.

The centromeric region contains the assembly site for the kinetochore, which is responsible for attaching a chromosome to the mitotic spindle. The centromere often is surrounded by heterochromatin. Centromeric sequences have been identified in the yeast *S. cerevisiae,* where they consist of short conserved elements. The elements *CDE-I* and *CDE-III* bind Cbf1 and the CBF3 complex, respectively, and a long A-T-rich region called *CDE-II* binds Cse4 to form a specialized structure in chromatin. Another group of proteins that binds to this assembly provides the connection to microtubules.

Telomeres make the ends of chromosomes stable. Almost all known telomeres consist of multiple repeats in which one strand has the general sequence $C_n(A/T)_m$, where $n > 1$ and $m = 1$ to 4. The other strand, $G_n(T/A)_m$, has a single protruding end that provides a template for addition of individual bases in defined order. The enzyme telomerase is a ribonucleoprotein whose RNA component provides the template for synthesizing the G-rich strand. This overcomes the problem of the inability to replicate at the very end of a duplex. The telomere stabilizes the chromosome end because the overhanging single strand $G_n(T/A)_m$ displaces its homolog in earlier repeating units in the telomere to form a loop, so there are no free ends.

CHAPTER QUESTIONS

What are the two ways that bacteriophage capsids containing DNA or RNA are assembled, and what shape of capsid is associated with each?

1. ______________________ **2.** shape:______________________

3. ______________________ **4.** shape:______________________

What are the four common features of constitutive heterochromatin?

5. ______________________ **6.** ______________________

7. ______________________ **8.** ______________________

9. The density of DNA in a eukaryotic nucleus is about:

A. 1 mg/ml.
B. 10 mg/ml.
C. 100 mg/ml.
D. 500 mg/ml.

10. Which of the following viruses has a filamentous shape?

A. adenovirus
B. bacteriophage T4
C. tobacco mosaic virus
D. HIV

11. For spherical capsids of DNA viruses such as lambda phage, assembly occurs:

A. by capsid coat proteins assembling around the linear DNA.
B. by capsid coat proteins assembling around a tightly wound DNA.
C. by a coordinated DNA and capsid protein monomer assembly process.
D. by capsid proteins forming a shell into which the DNA is inserted.

12. Centromeric regions in chromosomes contain:

A. euchromatin.
B. constitutive heterochromatin.
C. facultative heterochromatin.
D. acentric fragments.

13. Matrix attachment regions (MARs) are DNA sequences that attach to the nuclear matrix, and are characterized by:
   A. A-T rich regions with a specific consensus sequence.
   B. G-C rich regions with a specific consensus sequence.
   C. A-T rich regions without any specific consensus sequence.
   D. G-C rich regions without any specific consensus sequence.
14. Giemsa staining of eukaryotic chromosomes allows visualization of dense bands with less dense interbands. This is due to:
   A. higher G-C content in the interbands.
   B. higher G-C content in the bands.
   C. higher purine content in the interbands.
   D. higher purine content in the bands.
15. Bands in an expanded state observed on chromosomes are sites of:
   A. DNA replication.
   B. DNA recombination.
   C. gene expression.
   D. protein binding.
16. Telomeres consist of:
   A. simple sequence repeats of a C+A rich strand that interact with proteins.
   B. simple sequence repeats of a G+C rich strand that interact with proteins.
   C. simple sequence repeats of a C+T rich strand that interact with proteins.
   D. simple sequence repeats of a T+A rich strand that interact with proteins.

KEY TERMS

acentric fragment
bands
capsid
centromere
chromatin
chromocenter
chromomeres
chromosome
domain
euchromatin
G-bands
heterochromatin
***in situ* hybridization**
interbands
kinetochore
matrix attachment region (MAR)
metaphase (or mitotic) scaffold
microtubule organizing center (MTOC)
nucleation center
nucleoid
polytene chromosomes
puff
spindle
telomerase
telomere
terminase

FURTHER READING

Bailey, S. M. and J. P. Murname (2006). Telomeres, chromosome instability and cancer. *Nucleic Acids Research* 34(8):2408–2417.
An overview of telomere structure and function, including the role of telomeres in maintaining chromosome stability.

Bloom, K. (2007). Centromere dynamics. *Current Opinion in Genetics and Development* 17(2):151–156.
A review of eukaryotic kinetochore structure and function, focusing on the conserved role of the centromeric H3 variant.

Chattopadhyay, S. and L. Pavithra (2007). MARs and MARBPs: key modulators of gene regulation and disease manifestation. *Subcellular Biochemistry* 41:213–230.
A review of MARs/SARs and their known interacting proteins, with an emphasis on the role MARs/SARs play in gene regulation and cancer.

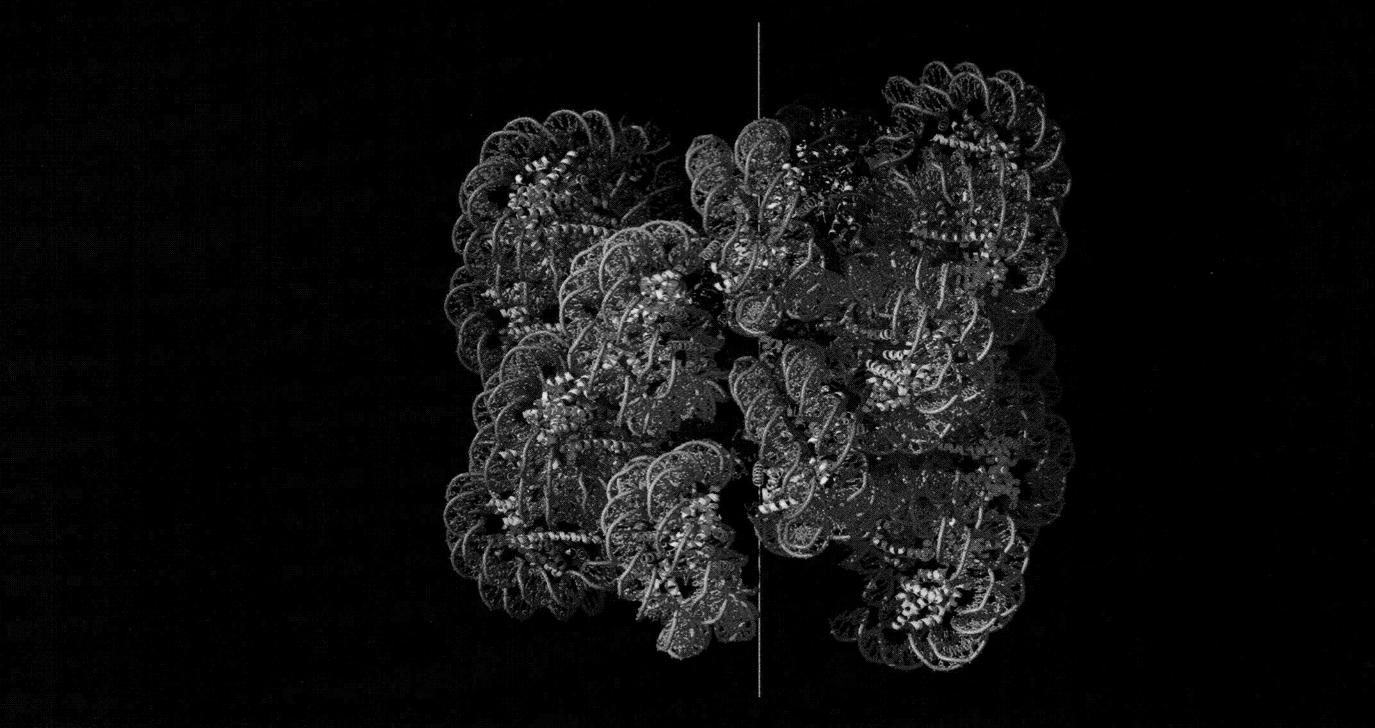

Chromatin is a compacted structure made up of DNA (blue and purple) wrapped around histone proteins. Reproduced from Robinson, P. J. J., et al., *Proc. Natl. Acad. Sci USA* 103 (2006): 6506–6511. Copyright 2006 National Academy of Sciences, USA. Photo courtesy of Daniela Rhodes, MRC Laboratory of Molecular Biology, Cambridge, UK.

Eukaryotic Gene Expression

24

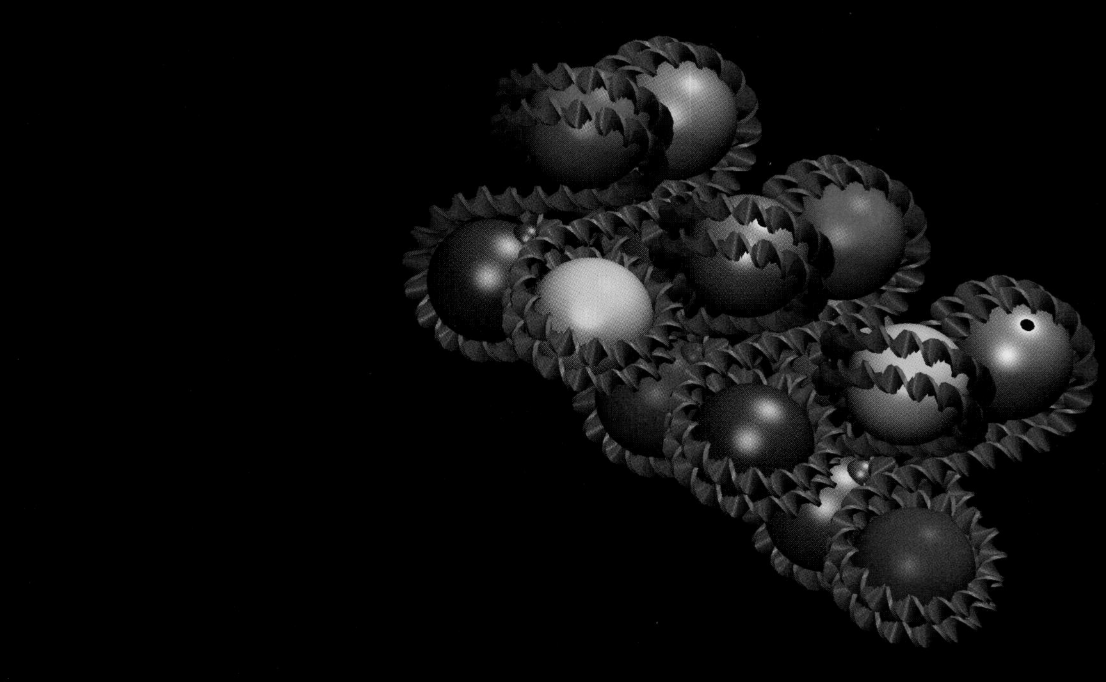

A three-dimensional model of chromatin showing one possible arrangement for the nucleosomes in the 30 nm fiber. Photo courtesy of Thomas Bishop, Tulane University, using VDNA plug-in for VMD [http://www.ks.uiuc.edu/research/vmd/plugins/vdna/].

Chromatin

CHAPTER OUTLINE

24.1 Introduction

- **nucleosome** The basic structural subunit of chromatin, consisting of ~200 bp of DNA and an octamer of histone proteins.
- **histones** Conserved DNA-binding proteins that form the basic subunit of chromatin in eukaryotes. H2A, H2B, H3, and H4 form an octameric core around which DNA coils to form a nucleosome. Linker histones are external to the nucleosome.
- **histone tails** Flexible amino- or carboxy-terminal regions of the core histones that extend beyond the surface of the nucleosome; histone tails are sites of extensive post-translational modification.
- **10 nm fiber** A linear array of nucleosomes, generated by unfolding from the natural condition of chromatin.
- **30 nm fiber** A coil of nucleosomes. It is the basic level of organization of nucleosomes in chromatin.
- **nonhistone** Any structural protein found in a chromosome except one of the histones.

Chromatin has a compact organization in which most DNA sequences are structurally inaccessible and functionally inactive. Within this mass is the minority of active sequences. What is the general structure of chromatin, and what is the difference between active and inactive sequences? The high overall packing ratio of the genetic material immediately suggests that DNA cannot be directly packaged into the final structure of chromatin. There must be *hierarchies* of organization.

The fundamental subunit of chromatin has the same basic structure in all eukaryotes. The **nucleosome** contains ~200 bp of DNA, organized by an octamer of small, basic proteins into a beadlike structure. The proteins are called **histones**. They form an interior core; the DNA lies on the surface of the particle. Additional regions of the histones, known as the **histone tails**, extend from the surface. Nucleosomes are an invariant component of euchromatin and heterochromatin in the interphase nucleus and of mitotic chromosomes. The nucleosome provides the first level of organization, compacting the DNA ~6-fold over the length of naked DNA, resulting in a fiber ~10 nm in diameter. Its components and structure are well characterized.

The second level of organization is the coiling of the **10 nm fiber** into a helical array to constitute the fiber of diameter ~30 nm that is found in both interphase chromatin and mitotic chromosomes (see Figure 23.10). This compacts the DNA ~40-fold.

This **30 nm fiber** is then further folded and compacted into interphase chromatin or into mitotic chromosomes. This results in ~1000-fold compaction in euchromatin, cyclically interchangeable with packing into mitotic chromosomes to achieve ~10,000-fold compaction. Heterochromatin is generally as compact as mitotic chromatin.

We need to work through these levels of organization to characterize the events involved in cyclical packaging, replication, and transcription. Association with additional proteins, or modification of existing chromosomal proteins, is involved in changing the structure of chromatin. Both replication and transcription require unwinding of DNA, and thus must involve an unfolding of the structure that allows the relevant enzymes to manipulate the DNA. This is likely to involve changes in all levels of organization.

The mass of chromatin contains up to twice as much protein as DNA. Approximately half of the protein mass is accounted for by the nucleosomes. The mass of RNA is <10% of the mass of DNA. Much of the RNA consists of nascent transcripts still associated with the template DNA.

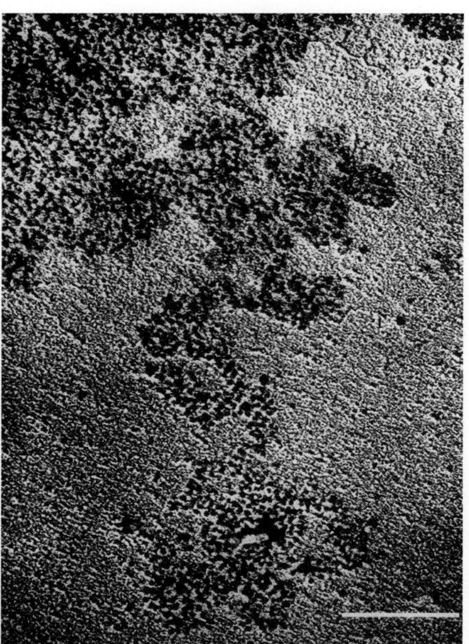

FIGURE 24.1 Chromatin spilling out of lysed nuclei consists of a compactly organized series of particles. The bar is 100 nm. Reproduced from *Cell*, Vol. 4, Oudet, P., *et al.*, Electron Microscopic . . . , pp. 281–300. Copyright 1975, with permission from Elsevier [http://www.sciencedirect.com/science/journal/00928674]. Photo courtesy of Pierre Chambon.

The **nonhistones** include all the proteins of chromatin except the histones. They are more variable between tissues and species, and they comprise a smaller proportion of the mass than the histones. They also comprise a much larger number of proteins, so that any individual protein is present in amounts much smaller than any histone.

24.2 The Nucleosome Is the Subunit of All Chromatin

When interphase nuclei are suspended in a solution of low ionic strength, they swell and rupture to release fibers of chromatin. **FIGURE 24.1** shows a lysed nucleus in which fibers are streaming out. In some regions, the fibers consist of tightly packed material, but in regions that have become stretched, they can be seen to consist of discrete particles. These are the nucleosomes. In especially extended regions, individual nucleosomes are visibly connected by a fine thread, which is a free duplex of DNA. A continuous duplex thread of DNA runs through the series of particles.

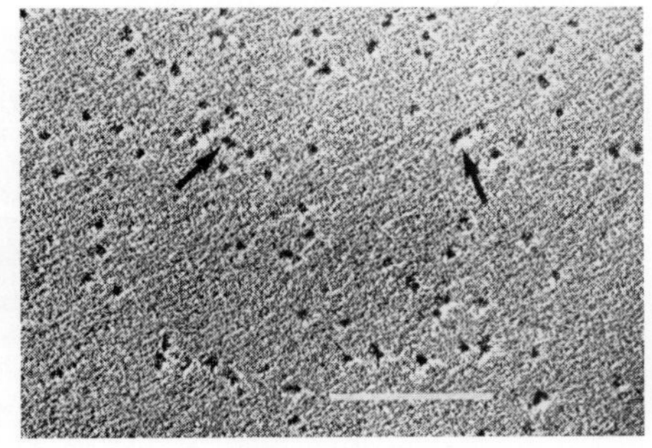

FIGURE 24.2 Individual nucleosomes are released by digestion of chromatin with micrococcal nuclease. The bar is 100 nm. Reproduced from *Cell*, vol. 4, Oudet, P., *et al.*, Electron microscopic . . . , pp. 281–300. Copyright 1975, with permission from Elsevier [http://www.sciencedirect.com/science/journal/00928674]. Photo courtesy of Pierre Chambon.

Individual nucleosomes can be obtained by treating chromatin with the endonuclease **micrococcal nuclease (MNase)**, which cuts the DNA thread at the junction between nucleosomes, a region known as **linker DNA**. Ongoing digestion with MNase releases groups of particles, and eventually single nucleosomes. Individual nucleosomes can be seen in FIGURE 24.2 as compact particles. When chromatin is digested with MNase, the DNA is cleaved into integral multiples of a unit length. Fractionation by gel electrophoresis reveals the "ladder" presented in FIGURE 24.3. Such ladders extend for ~10 steps, and the unit length, determined by the increments between successive steps, is ~200 bp.

The ladder is generated by groups of nucleosomes. The fastest-migrating band contains single nucleosomes (monomers), whereas the higher bands correspond to dimers, trimers, and so on, still connected by linker DNA. The monomeric nucleosome contains DNA of the ~200 bp unit length, the nucleosome dimer contains DNA of twice the unit length, and so on. >95% of the DNA of chromatin can be recovered in the form of the 200 bp ladder, indicating that almost all DNA must be organized in nucleosomes.

The length of DNA present in the nucleosome can vary from the "typical" value of 200 bp. The chromatin of any particular cell type has a characteristic average value (±5 bp). The average most often is between 180 and 200, but there are extremes as low as 154 bp (in a fungus) or as high as 260 bp (in a sea urchin sperm). The average value may be different in individual tissues of the adult organism, and there can be differences between different parts of the genome in a single cell type.

What is the nature of the nucleosome itself? The nucleosome contains ~200 bp of DNA associated with a **histone octamer** that consists of two copies each of H2A, H2B, H3, and H4. These are known as the **core histones**. Their association is illustrated diagrammatically in FIGURE 24.4.

Histones H3 and H4 are among the most conserved proteins known. This suggests that their functions are identical in all eukaryotes. H2A and H2B are also conserved in all eukaryotes, but show appreciable species-specific variation in sequence, particularly in the histone tails.

- **micrococcal nuclease (MNase)** An endonuclease that cleaves DNA; in chromatin, DNA is cleaved preferentially between nucleosomes.
- **linker DNA** Non-nucleosomal DNA present between nucleosomes.
- **histone octamer** The complex of 2 copies each of the four different core histones (H2A, H2B, H3, and H4); DNA wraps around the histone octamer to form the nucleosome.
- **core histone** One of the four types of histone (H2A, H2B, H3, and H4 and their variants) found in the core particle derived from the nucleosome. (This excludes linker histones.)

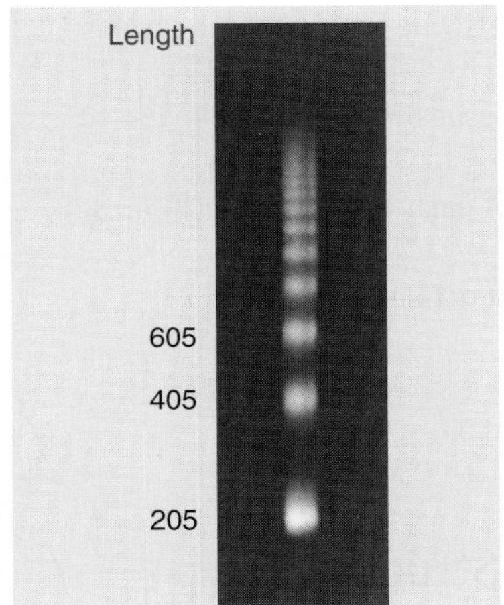

FIGURE 24.3 Micrococcal nuclease digests chromatin in nuclei into a multimeric series of DNA bands that can be separated by gel electrophoresis. Photo courtesy of Markus Noll, Universität Zürich.

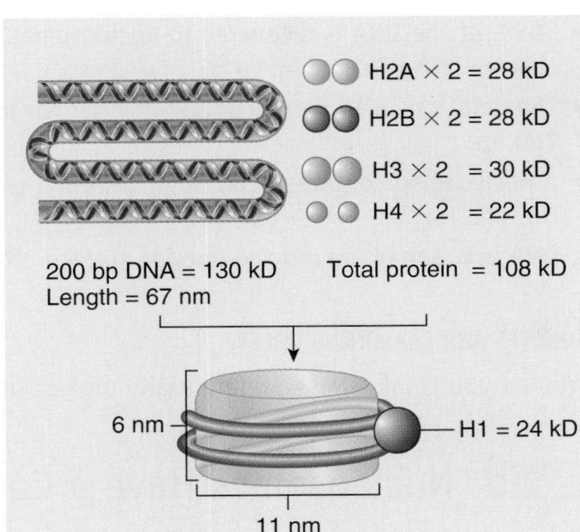

FIGURE 24.4 The nucleosome consists of approximately equal masses of DNA and histones (including H1). The predicted mass of the nucleosome is 262 kD.

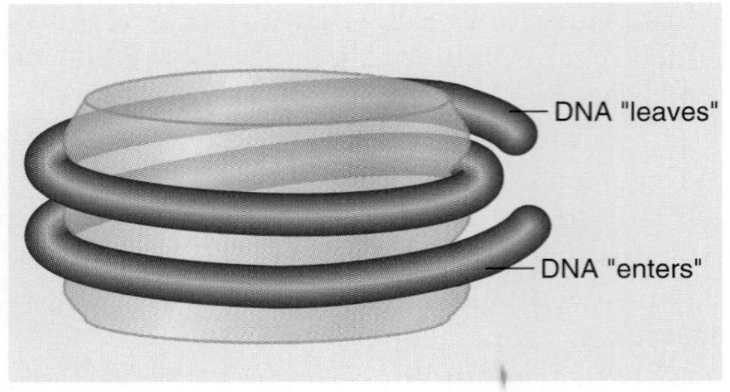

FIGURE 24.5 The nucleosome is roughly cylindrical, with DNA organized into 1¾ turns around the surface.

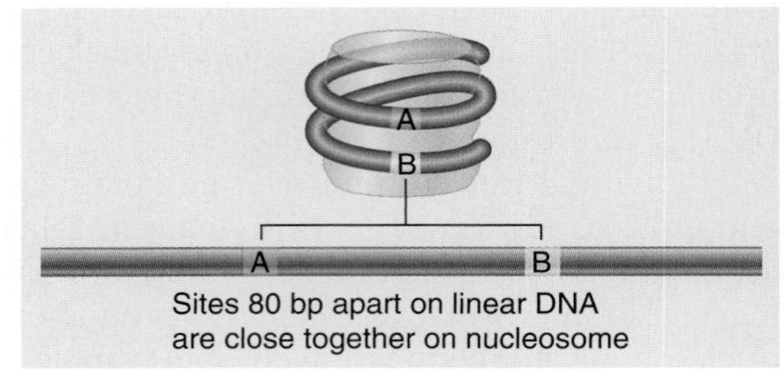

FIGURE 24.6 Sequences on the DNA that lie on different turns around the nucleosome may be close together.

▸ **linker histones** A family of histones (such as histone H1) that are not components of the nucleosome core; linker histones bind nucleosomes and/or linker DNA and promote 30 nm fiber formation.

The family of **linker histones**, typified by histone H1, comprises a set of closely related proteins that show appreciable variation both between tissues and between species. The role of H1 is different from that of the core histones. It is present in half the amount of a core histone and can be extracted more readily from chromatin. H1 can be removed without affecting the structure of the nucleosome, which suggests that its location is external to the particle.

The shape of the nucleosome corresponds to a flat disk or cylinder of diameter 11 nm and height 6 nm. The length of the DNA is roughly twice the ~34 nm circumference of the particle. The DNA follows a symmetrical path around the octamer. **FIGURE 24.5** shows the DNA path diagrammatically as a helical coil that makes ~1.75 turns around the cylindrical octamer. Note that the DNA "enters" and "leaves" the nucleosome at points relatively close to one another. In addition, the two gyres of DNA lie close to one another throughout the turns. The height of the cylinder is 6 nm, of which 4 nm is occupied by the two turns of DNA (each of diameter 2 nm).

The pattern of the two turns has a possible functional consequence. One turn around the nucleosome takes ~80 bp of DNA, so two points separated by 80 bp in the free double helix may actually be close on the nucleosome surface, as illustrated in **FIGURE 24.6**.

KEY CONCEPTS

- Micrococcal nuclease cleaves linker DNA and releases individual nucleosomes from chromatin.
- >95% of the DNA is recovered in nucleosomes or multimers when micrococcal nuclease cleaves DNA of chromatin.
- The length of DNA per nucleosome varies for individual tissues in a range from 154 to 260 bp.
- A nucleosome contains ~200 bp of DNA and two copies of each core histone (H2A, H2B, H3, and H4).
- DNA is wrapped around the outside surface of the protein octamer.

CONCEPT AND REASONING CHECK

Why do you think MNase preferentially digests linker DNA?

24.3 Nucleosomes Have a Common Structure

Although the length of linker DNA may vary in nucleosomes of different sources, all nucleosomes nevertheless reveal a common structure. The association of DNA with the histone octamer forms a **core particle** containing 146 bp of DNA, which can be revealed by extended digestion by micrococcal nuclease. The initial reaction

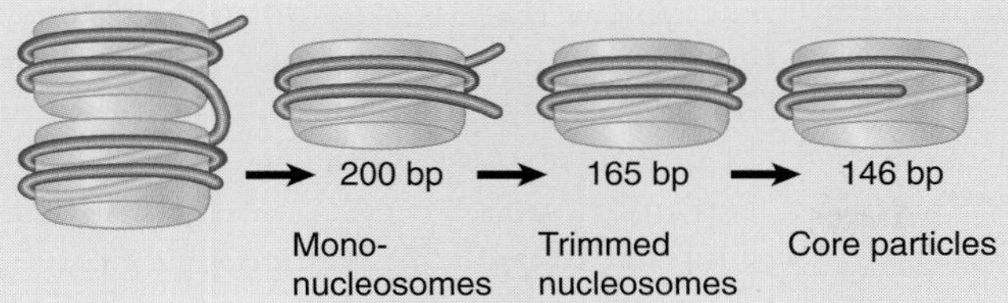

FIGURE 24.7 Micrococcal nuclease initially cleaves between nucleosomes. Mononucleosomes typically have ~200 bp DNA. End-trimming reduces the length of DNA first to ~165 bp, and then generates core particles with 146 bp.

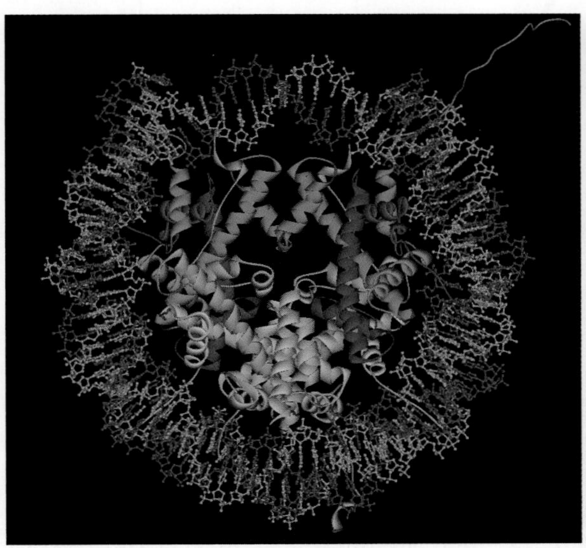

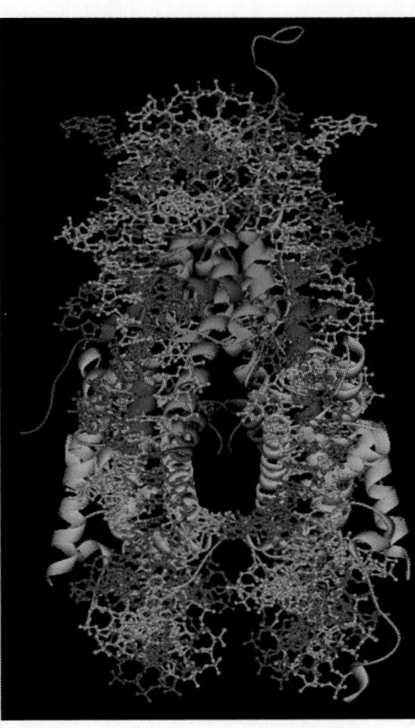

FIGURE 24.8 The crystal structure of the histone core octamer is represented in a ribbon model, including the 146-bp DNA phosphodiester backbones (orange and blue) and eight histone protein main chains (green: H3; purple: H4; turquoise: H2A; yellow: H2B). Protein Data Bank 1AOI. Luger, K., et al., *Nature* 389(1997):251–260.

of the enzyme is to cut between nucleosomes, but if it is allowed to continue after monomers have been generated, then it proceeds to digest some of the DNA of the individual nucleosome. This occurs by a reaction in which DNA is "trimmed" from the ends of the nucleosome, as shown in **FIGURE 24.7**.

Nucleosomal DNA is therefore divided into two regions: *core DNA*, which has an invariant length of 146 bp and is relatively resistant to digestion by nucleases, and *linker DNA*, which comprises the rest of the repeating unit. Its length varies from as little as 8 bp to as much as 114 bp per nucleosome.

Core particles have properties similar to those of the nucleosomes themselves, although they are smaller. Their shape and size are similar to those of nucleosomes; this suggests that the essential geometry of the particle is established by the interactions between DNA and the protein octamer in the core particle. Core particles are more readily obtained as a homogeneous population, and as a result they are often used for structural studies in preference to nucleosome preparations. Studies using reconstituted core particles have revealed detailed structural information about the organization of the nucleosome.

The core histones tend to form two types of complexes. H3 and H4 form a very stable tetramer in solution ($H3_2$–$H4_2$). H2A and H2B most typically form a dimer (H2A–H2B). The $H3_2$–$H4_2$ tetramer can organize DNA *in vitro* into particles that display some of the properties of the core particle.

The crystal structure of the nucleosome core particle (at 2.8 Å resolution) shows the interactions between histones and DNA, as shown in **FIGURE 24.8**. This structure shows clearly that histones are not organized as individual globular proteins, but that each is interdigitated with its partner H3 with H4, and H2A with H2B.

The $H3_2$–$H4_2$ tetramer accounts for the diameter of the octamer. It forms the shape of a horseshoe. The H2A–H2B pairs fit in as two dimers, each binding to the opposite

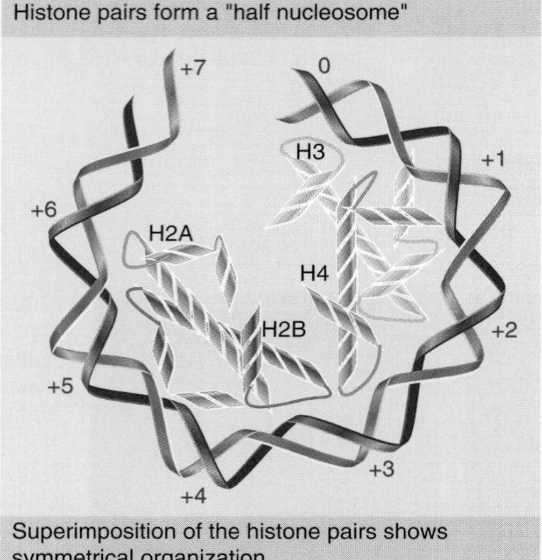

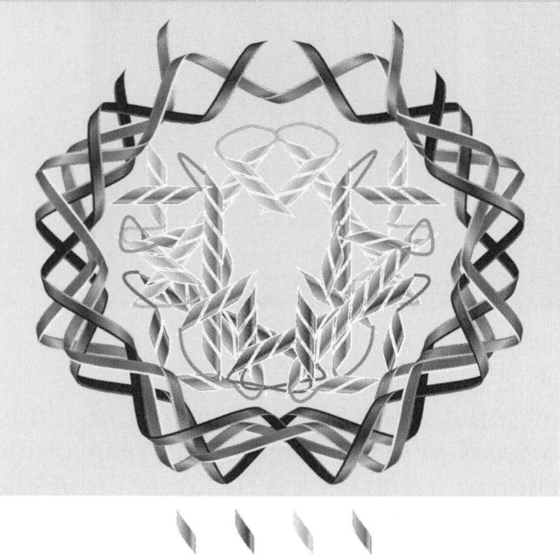

FIGURE 24.9 Histone positions in a top view show H3-H4 and H2A-H2B pairs in a half nucleosome; the symmetrical organization can be seen in the superimposition of both halves in the bottom view.

face of the $H3_2$–$H4_2$ tetramer. This is easy to see in the side view of the nucleosome in the right panel of Figure 24.8. Here the responsibilities of the $H3_2$–$H4_2$ tetramer and of the separate H2A–H2B dimers can be distinguished. The protein forms a sort of spool, with a superhelical path that corresponds to the binding site for DNA, which is wound ~1¾ turns in a nucleosome. The nucleosome displays twofold symmetry, about a vertical axis in this figure.

A simplified view of the positions of the histones is summarized in **FIGURE 24.9**. The upper view shows the position of one histone of each type relative to one turn around the nucleosome (numbered from 0 to +7). All four core histones show a similar type of structure in which three α helices are connected by two loops This is called the **histone fold**. These regions interact to form crescent-shaped heterodimers; each heterodimer binds 2.5 turns of the DNA double helix. (H2A–H2B binds at +3.5 – +6; H3–H4 binds at +0.5 – +3 for the circumference that is illustrated.) Binding is mostly to the phosphodiester backbones (consistent with the need to package any DNA irrespective of sequence). The $H3_2$–$H4_2$ tetramer is formed by interactions between the two H3 subunits, as can be seen in the lower part of the figure.

▸ **histone fold** A motif found in all four core histones in which three α-helices are connected by two loops.

Each of the core histones has a histone fold domain that contributes to the central protein mass of the nucleosome, sometimes referred to as the *globular core*. Each histone also has a flexible N-terminal tail (and, in the case of H2A, a C-terminal tail), which contains sites for covalent modification that are important in chromatin function (we will discuss these modifications in more detail in *Chapter 26, Eukaryotic Transcription Regulation*). The positions of the tails, which account for about one quarter of the protein mass, are not so well defined, as indicated in **FIGURE 24.10**. The tails extend from both faces of the nucleosome, however, and the tails of H3 and H2B can be seen to pass between the turns of the DNA superhelix, as shown in **FIGURE 24.11**.

The interaction of histone H1 with the nucleosome is poorly understood. H1 is retained on nucleosome monomers that have at least 165 bp of DNA, but does not bind to the 146 bp core particle. This suggests that H1 could be located in the region of the linker DNA immediately adjacent to the core DNA. If H1 is located at the linker, it could "seal" the DNA in the nucleosome by binding at the point where the nucleic acid enters and leaves (see Figure 24.5).

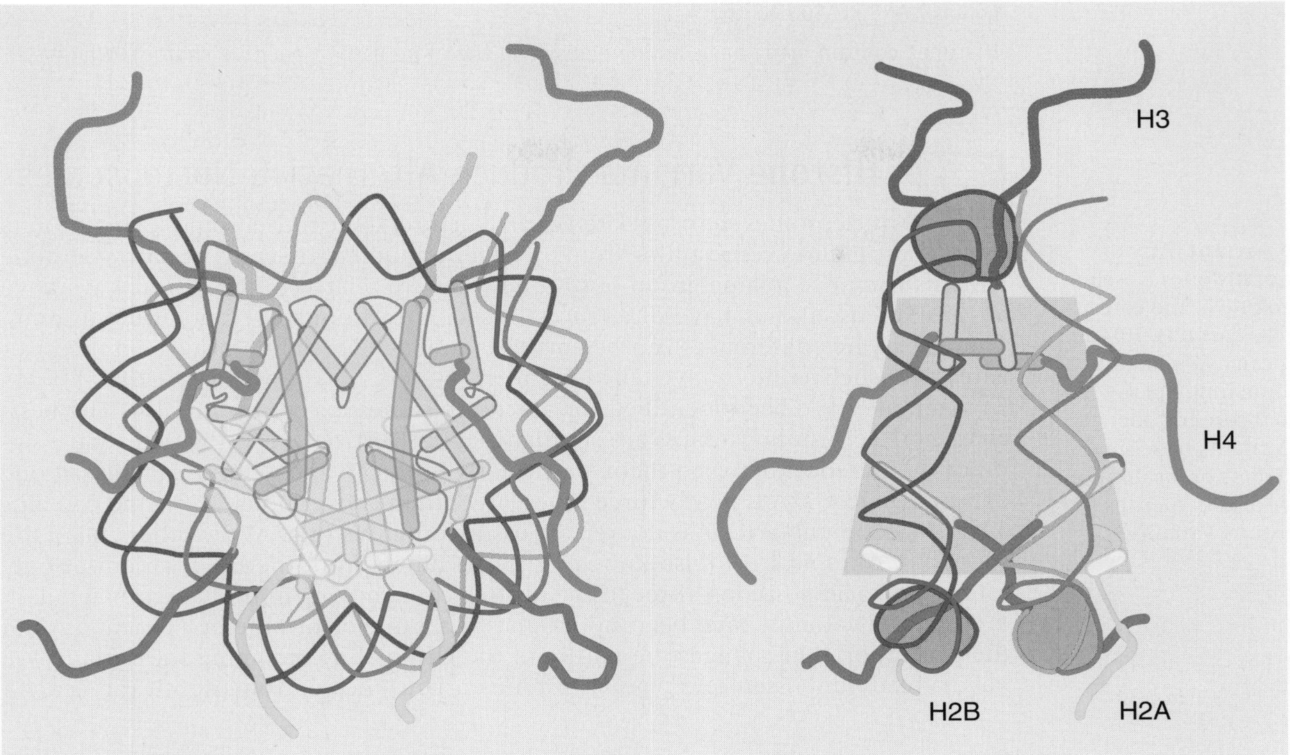

FIGURE 24.10 The histone fold domains of the histones are located in the core of the nucleosome. The N- and C-terminal tails, which carry many sites for modification, are flexible and their positions cannot be determined by crystallography.

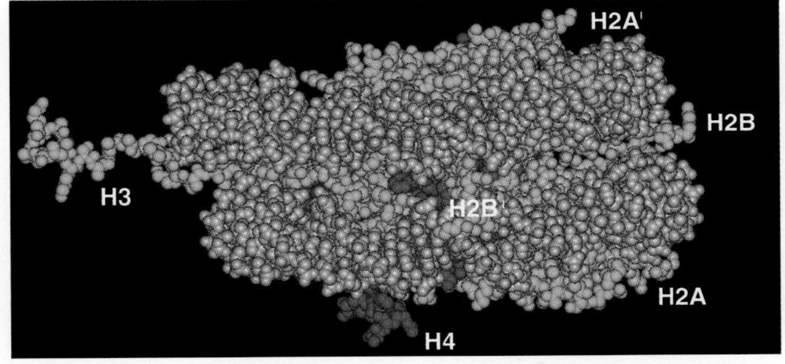

FIGURE 24.11 The histone tails are disordered and exit from both faces of the nucleosome and between turns of the DNA. Note this figure shows only the first few amino acids of the tails, as the complete tails were not present in the crystal structure. Protein Data Bank 1OAI. Luger, K., et al., *Nature* 389(1997):251–260.

KEY CONCEPTS

- Nucleosomal DNA is divided into the core DNA and linker DNA depending on its susceptibility to micrococcal nuclease.
- The core DNA is the length of 146 bp that is found on the core particles produced by prolonged digestion with micrococcal nuclease.
- Linker DNA is the region of 8 to 114 bp that is susceptible to early cleavage by the enzyme.
- The histone octamer has a kernel of an $H3_2$–$H4_2$ tetramer associated with two H2A–H2B dimers.
- Each histone is extensively interdigitated with its partner.
- All core histones have the structural motif of the histone fold. N-terminal tails extend out of the nucleosome.
- H1 is associated with linker DNA and may lie at the point where DNA enters and leaves the nucleosome.

CONCEPT AND REASONING CHECK

Histones contain many basic amino acids (and therefore carry a positive charge). Why is this so?

24.4 Histone Variants Produce Alternative Nucleosomes

▸ **histone variant** Any of a number of histones closely related to one of the core histones (H2A, H2B, H3, or H4) that can assemble into a nucleosome in the place of the related core histone; many histone variants have specialized functions or localization. There are also numerous linker histone variants.

While all nucleosomes share a related core structure, some nucleosomes exhibit subtle or dramatic differences resulting from the incorporation of **histone variants**. Histone variants comprise a large group of histones that are related to the histones we have already discussed, but have differences in sequence from the "canonical" histones. These sequence differences can be small (as few as four amino acid differences) or extensive (such as alternative tail sequences).

Variants have been identified for all core histones except histone H4. The best-characterized histone variants are summarized in **FIGURE 24.12**. Most variants have significant differences between them, particularly in the N- and C-terminal tails. At one extreme, macroH2A is nearly three times larger than conventional H2A, and contains a large C-terminal tail that is not related to any other histone. At the other end of the spectrum, canonical H3 (also known as H3.1) differs from the H3.3 variant at only four amino acid positions, three in the histone core and one in the N-terminal tail.

Histone variants have been implicated in a number of different functions, and their incorporation changes the nature of the chromatin containing the variant. We have already discussed one type of histone variant, the centromeric H3 (or CenH3)

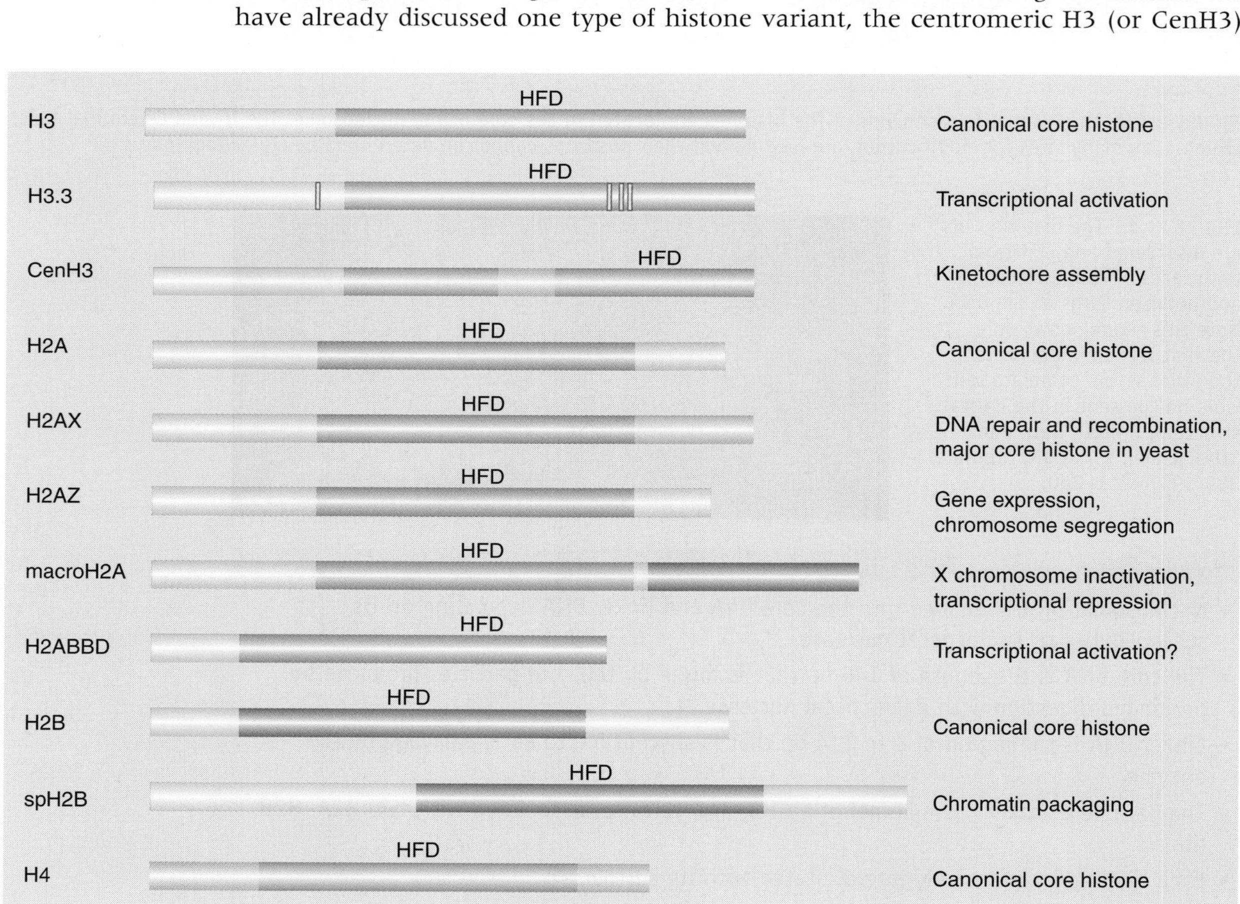

FIGURE 24.12 The major core histones contain a conserved histone-fold domain. In the histone H3.3 variant, the residues that differ from the major histone H3 (also known as H3.1) are highlighted in yellow. The centromeric histone CenH3 has a unique N terminus, which does not resemble other core histones. Most H2A variants contain alternative C-termini, except H2ABBD, which contains a distinct N terminus. The sperm-specific spH2B has a long N-terminus. Proposed functions of the variants are listed. Reprinted by permission from Macmillan Publishers Ltd: *Nat. Rev. Mol. Cell Biol.* 6, Sarma, K., and Reinberg, D., copyright 2005.

histone, known as Cse4 in yeast). CenH3 histones are incorporated into specialized nucleosomes present at centromeres in all eukaryotes (see *Section 23.9,* S. cerevisiae *Centromeres Have Short Protein-Binding DNA Sequences*). In yeast, it has been shown that these centromeric nucleosomes consist of Cse4, H4, and a non-histone protein Scm3, which replaces H2A/H2B dimers. In *Drosophila,* the centromeric chromatin appears to consist of "hemisomes" containing one copy each of CenH3, H4, H2A, and H2B. It is not known whether any centromeric chromatin in higher eukaryotes contains an Scm3-like protein at a subset of centromeric nucleosomes.

The other major H3 variant is histone H3.3. In multicellular eukaryotes this variant is a minority component of the total H3 in the cell, but in yeast, the major H3 is actually of the H3.3 type. H3.3 is expressed throughout the cell cycle, in contrast to most histones that are expressed during S phase, when new chromatin assembly is required during DNA replication. As a result, H3.3 is available for assembly at any time in the cell cycle, and is incorporated at sites of active transcription, where nucleosomes become disrupted. Because of this, H3.3 is often referred to as a "replacement" histone, in contrast to the "replicative" histone H3.1.

The H2A variants are the largest and most diverse family of core histone variants, and have been implicated in a variety of distinct functions. One of the best studied is the variant H2AX. H2AX is normally present in only 10% to 15% of the nucleosomes in multicellular eukaryotes, though again (like H3.3) this subtype is the major H2A present in yeast. This variant has a C-terminal tail that is distinct from the canonical H2A, which is characterized by a SQEL/Y motif at the end. This motif is the target of phosphorylation by ATM/ATR kinases, activated by DNA damage, and this histone variant is involved in DNA repair, particularly repair of double strand breaks (see *Chapter 20, Repair Systems*). H2AX phosphorylated at the SQEL/Y motif is referred to as "γ-H2AX," and is required to stabilize binding of various repair factors at DNA breaks and to maintain checkpoint arrest. γ-H2AX appears within moments at broken DNA ends, as can be seen in **FIGURE 24.13**, which shows foci of γ-H2AX forming along the path of double-strand breaks induced by a laser.

Other H2A variants have different roles. The H2AZ variant, which has ~60% sequence identity with canonical H2A, has been shown to be important in several processes, such as gene activation, heterochromatin-euchromatin boundary formation, and cell cycle progression. The vertebrate-specific macroH2A is named for its extremely long C-terminal tail, which contains a leucine-zipper dimerization motif that may mediate chromatin compaction by facilitating inter-nucleosome interactions. Mammalian macroH2A is enriched in the inactive X chromosome in females, which is assembled into a silent, heterochromatic state. In contrast, the mammalian H2ABbd variant is *excluded* from the inactive X, and forms a less stable nucleosome

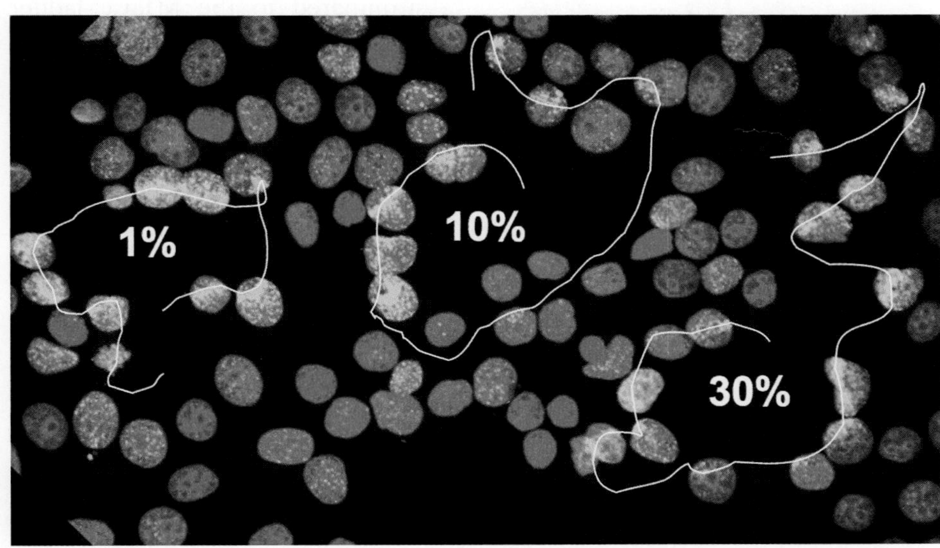

FIGURE 24.13 γ-H2AX is detected by an antibody (yellow) and appears along the path traced by a laser that produces double-strand breaks. The percentages refer to the relative laser energy used in each transit. Nuclei are red. © Rogakou et al., 1999. Originally published in *The Journal of Cell Biology*, 146: 905–915. Photo courtesy of William M. Bonner. National Cancer Institute, NIH.

than canonical H2A; perhaps this histone is designed to be more easily displaced in transcriptionally active regions of euchromatin.

Still other variants are expressed in limited tissues, such as spH2B, present in sperm and required for chromatin compaction. The presence and distribution of histone variants shows that individual chromatin regions, entire chromosomes, or even specific tissues can have unique "flavors" of chromatin specialized for different function. In addition, the histone variants, like the canonical histones, are subject to numerous covalent modifications (see *Section 26.8, Histone Modification Regulates Chromatin Function*), adding levels of complexity to the roles chromatin plays in nuclear processes.

KEY CONCEPTS

- All core histones except H4 are members of families of related variants.
- Histone variants can be closely related or highly divergent from canonical histones.
- Different variants serve different functions in the cell.

CONCEPT AND REASONING CHECK

Why would a less stable nucleosome (such as one containing H2ABbd) be more likely to be present in euchromatin than in heterochromatin?

24.5 DNA Structure Varies on the Nucleosomal Surface

So far we have focused on the protein components of the nucleosome. The DNA wrapped around these proteins is in an unusual conformation. The exposure of DNA on the surface of the nucleosome explains why it is accessible to cleavage by certain nucleases. The reaction with nucleases that attack single strands has been especially informative. The enzymes DNase I and DNase II make single-strand nicks in DNA; they cleave a bond in one strand, but the other strand remains intact. When DNA is free in solution, it is nicked (relatively) at random. The DNA on nucleosomes also can be nicked by the enzymes, but in this case the nicking occurs only at regular intervals. When the points of cutting are determined by using radioactively end-labeled DNA and then DNA is denatured and electrophoresed, a ladder of the sort displayed in **FIGURE 24.14** is obtained.

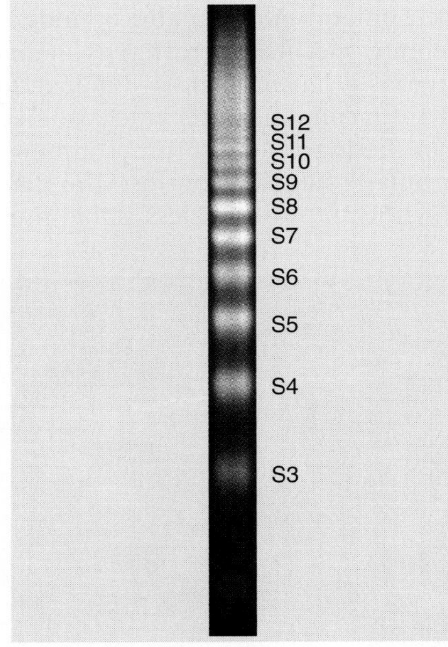

FIGURE 24.14 Sites for nicking lie at regular intervals along core DNA, as seen in a DNase I digest of nuclei. Cleavage sites are numbered as S1 through S13 (where S1 is ~10 bases from the labeled 5′ end, S2 is ~20 bases from it, and so on). Photo courtesy of Leonard C. Lutter, Henry Ford Hospital, Detroit, MI.

Compared to the MNase ladder shown in Figure 24.3, DNase digestion generates products with a much shorter interval: 10 to 11 bases. The ladder extends for the full distance of core DNA. The same cutting pattern is obtained by cleaving with a hydroxyl radical, which argues that the pattern reflects the structure of the DNA itself rather than any sequence preference.

When DNA is immobilized on a flat surface, sites are cut at a regular distance apart. This reflects the recurrence of the exposed site with the helical periodicity of B-form DNA, as shown in **FIGURE 24.15**. The cutting periodicity (the spacing between cleavage points) is a reflection of the structural

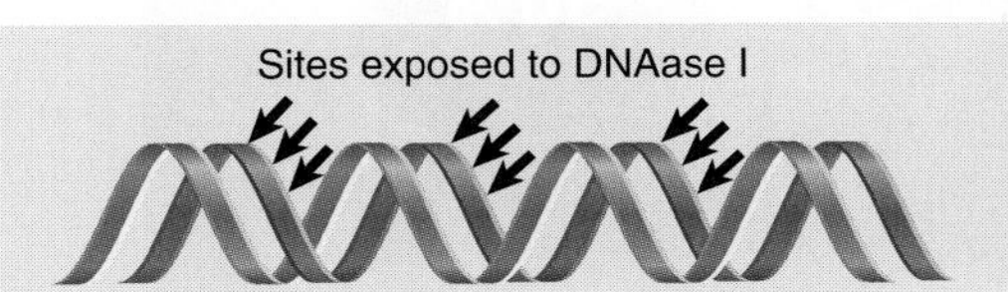

FIGURE 24.15 The most exposed positions on DNA recur with a periodicity that reflects the structure of the double helix (for clarity, sites are shown for only one strand.)

periodicity (the number of base pairs per turn of the double helix). Thus the distance between the sites corresponds to the number of base pairs per turn (also known as the *repeat length*). Measurements of this type suggest that the average repeat length for double-helical B-type DNA is 10.5 bp/turn. On the nucleosome, however, the average repeat length varies from ~10.0 at the ends, to ~10.7 in the middle. This means that there is variation in the structural periodicity of core DNA. The DNA has more bp/turn than its solution value in the middle, but has fewer bp/turn at the ends. The average periodicity over the entire nucleosome is only 10.17 bp/turn, which is significantly less than the 10.5 bp/turn of DNA in solution.

The crystal structure of the core particle shows that DNA is wound into a *solenoidal* (spring-shaped) supercoil, with 1.65 turns wound around the histone octamer. The structure of DNA is distorted. The central 129 bp are in the form of B-DNA, but with a substantial curvature that is needed to form the superhelix. The major groove is smoothly bent, but the minor groove has abrupt kinks. These conformational changes may explain why the central part of nucleosomal DNA is not usually a target for binding by regulatory proteins, which typically bind to the terminal parts of the core DNA or to the linker sequences.

Some insights into the structure of nucleosomal DNA emerge when we compare predictions for supercoiling in the path that DNA follows with actual measurements of supercoiling of nucleosomal DNA. Much work on the structure of sets of nucleosomes has been carried out with the virus SV40. The DNA of SV40 is a circular molecule of 5200 bp. In both the viral particle and infected nucleus, it is packaged into a series of nucleosomes, which together are called a *minichromosome*.

When isolated, the minichromosome is compacted ~7-fold over naked DNA (essentially the same as the ~6 of the nucleosome itself). The degree of supercoiling constrained by the individual nucleosomes of the minichromosome can be measured as illustrated in **FIGURE 24.16**. First, the free supercoils of the minichromosome itself are relaxed, so that the nucleosomes form a circular string with a superhelical density of 0. Next, the histone octamers are extracted. This releases the DNA to follow a free path. Every supercoil that was present but constrained in the minichromosome will appear in the deproteinized DNA as –1 turn. Now the total number of supercoils in the SV40 DNA is measured.

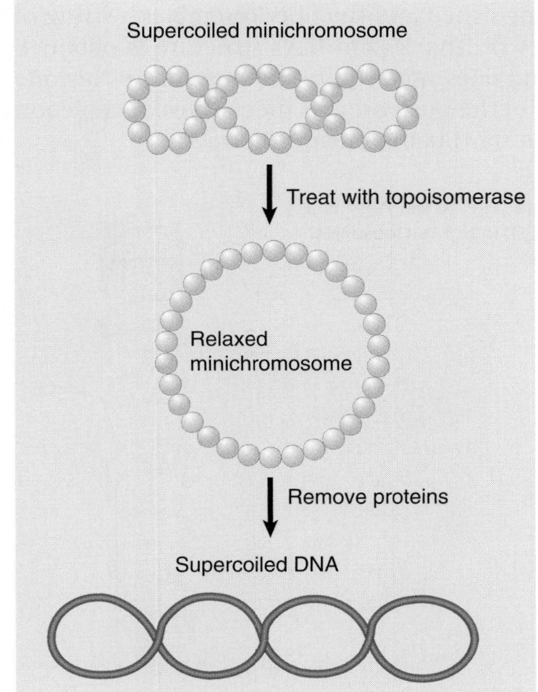

FIGURE 24.16 The supercoils of the SV40 minichromosome can be relaxed to generate a circular structure, whose loss of histones then generates supercoils in the free DNA.

The observed value is close to the number of nucleosomes. Thus the DNA follows a path on the nucleosomal surface that generates ~1 negative supercoiled turn when the restraining protein is removed. The path that DNA follows on the nucleosome, though, corresponds to –1.67 superhelical turns (see Figure 24.5). This discrepancy is sometimes called the *linking number paradox.*

The discrepancy is explained by the difference between the 10.17 average bp/turn of nucleosomal DNA and the 10.5 bp/turn of free DNA. In a nucleosome of 200 bp, there are 200/10.17 = 19.67 turns. When DNA is released from the nucleosome, it now has 200/10.5 = 19.0 turns. The path of the less tightly wound DNA on the nucleosome absorbs –0.67 turns, which explains the discrepancy between the physical path of –1.67 and the measurement of –1.0 superhelical turns. In effect, some of the torsional strain in nucleosomal DNA goes into increasing the number of bp/turn; only the remainder is measured as a supercoil.

KEY CONCEPTS

- DNA is wrapped 1.65 times around the histone octamer.
- The structure of the DNA is altered so that it has an increased number of base pairs/turn in the middle, but a decreased number at the ends.
- ~0.6 negative turns of DNA are absorbed by the change in bp/turn from 10.5 in solution to an average of 10.2 on the nucleosomal surface, which explains the linking-number paradox.

CONCEPT AND REASONING CHECK

Explain the different digestion patterns generated by MNase and DNase on nucleosomal DNA.

24.6 The Path of Nucleosomes in the Chromatin Fiber

When chromatin is examined with the electron microscope, two types of fibers are seen: the 10 nm fiber and 30 nm fiber. They are described by the approximate diameter of the thread (that of the 30 nm fiber actually varies from ~25–30 nm).

The 10 nm fiber is a continuous string of nucleosomes and represents the least compacted level of chromatin structure. In fact, in a stretched-out 10 nm fiber, linker DNA and nucleosomes can be easily distinguished and the fiber resembles a string of beads, as seen in the example of **FIGURE 24.17**. The 10 nm fiber structure is obtained under conditions of low ionic strength and does not require the presence of histone H1. This means that it is a function strictly of the nucleosomes themselves. A depiction of the continuous series of nucleosomes in this fiber is shown in **FIGURE 24.18**.

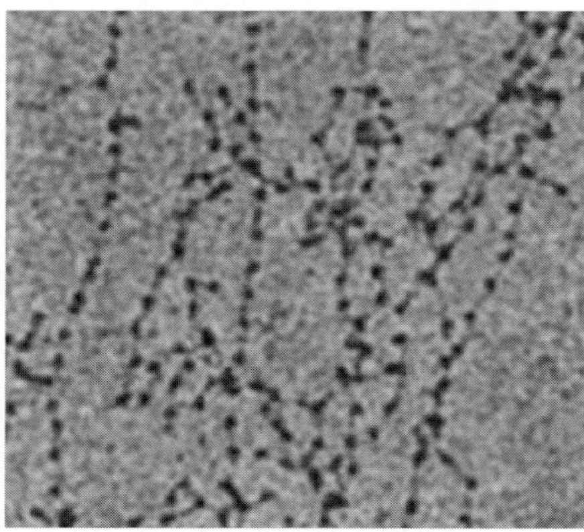

FIGURE 24.17 The 10 nm fiber in partially unwound state can be seen to consist of a string of nucleosomes. Photo courtesy of Barbara Hamkalo, University of California, Irvine.

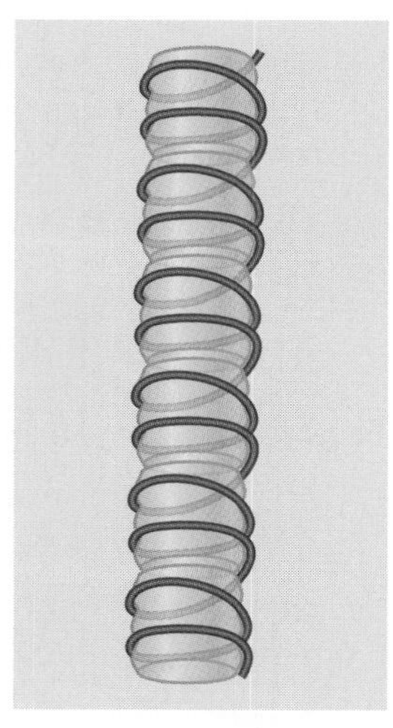

FIGURE 24.18 The 10 nm fiber is a continuous string of nucleosomes.

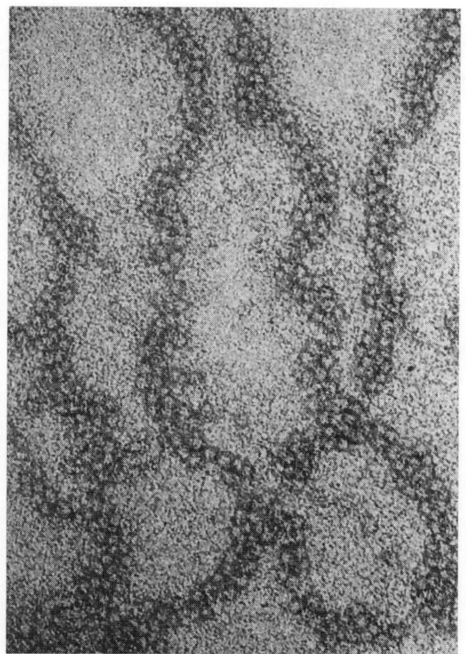

FIGURE 24.19 The 30 nm fiber has a coiled structure. Photo courtesy of Barbara Hamkalo, University of California, Irvine.

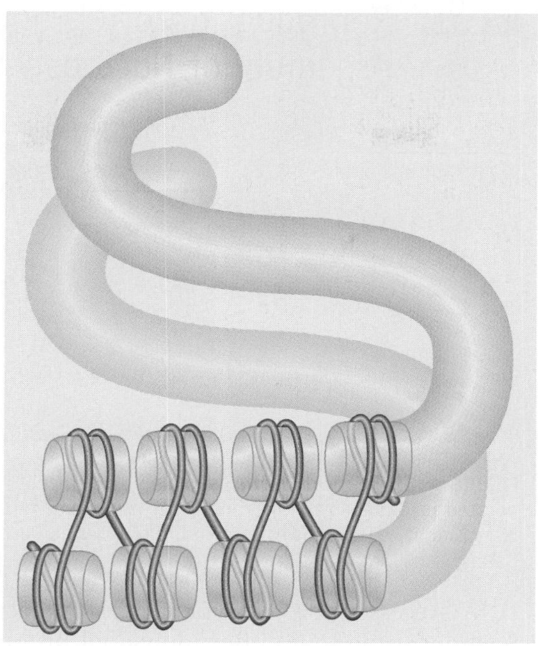

FIGURE 24.20 The 30 nm fiber is a helical ribbon consisting of two parallel rows of nucleosomes coiled into a solenoid.

When chromatin is visualized in conditions of greater ionic strength, the 30 nm fiber is obtained. An example is given in **FIGURE 24.19**. The fiber can be seen to have an underlying coiled structure. It has ~6 nucleosomes for every turn, which corresponds to a compaction of ~40-fold (that is, each μm along the axis of the fiber contains 40 μm of DNA). The formation of this fiber requires the histone tails, which are involved in internucleosomal contacts, and is facilitated by the presence of a linker histone such as H1. This fiber is the basic constituent of both interphase chromatin and mitotic chromosomes.

The most likely arrangement for packing nucleosomes into the fiber is a coiled or solenoidal structure, in which the nucleosomes turn in a helical array that is coiled around a central cavity. Recent data reveal that the 30 nm fiber is most likely what is known as a *two-start* helix, which consists of a double row of nucleosomes, as depicted in **FIGURE 24.20**.

The 30 nm and 10 nm fibers can be reversibly converted by changing the ionic strength. This suggests that the linear array of nucleosomes in the 10 nm fiber is coiled into the 30 nm structure at higher ionic strength and in the presence of H1.

KEY CONCEPTS

- 10 nm chromatin fibers are unfolded from 30 nm fibers and consist of a string of nucleosomes.
- 30 nm fibers have 10 nm fibers coiled into two parallel rows of nucleosomes.
- Histone H1 promotes the formation of the 30 nm fiber.

CONCEPT AND REASONING CHECK

Why might histone tails be important for formation of the 30 nm fiber?

24.7 Reproduction of Chromatin Requires Assembly of Nucleosomes

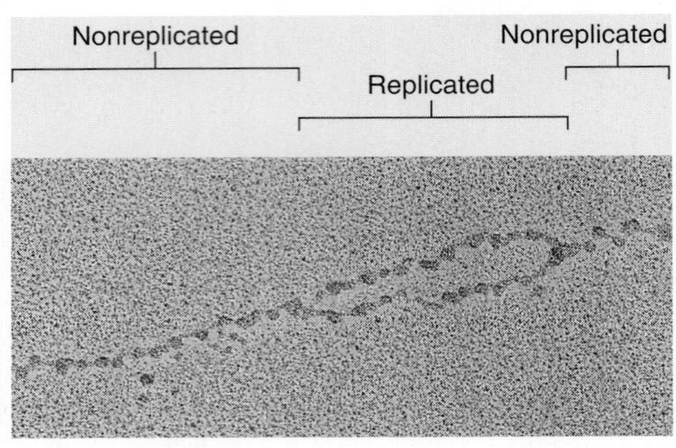

FIGURE 24.21 Replicated DNA is immediately incorporated into nucleosomes. Photo courtesy of Steven L. McKnight, UT Southwestern Medical Center at Dallas.

Replication separates the strands of DNA and, therefore, must inevitably disrupt the structure of the nucleosome. However, this disruption is confined to the immediate vicinity of the replication fork. Once DNA has been replicated, nucleosomes are quickly generated on both the duplicates. This point is illustrated by the electron micrograph of **FIGURE 24.21**, which shows a recently replicated stretch of DNA that is already packaged into nucleosomes on both daughter duplex segments.

Accessory proteins are involved in assisting histones to associate with DNA. Accessory proteins may act as "molecular chaperones" that bind to the histones in order to present either individual histones or complexes ($H3_2$–$H4_2$ or H2A–H2B) to the DNA in a controlled manner. This could be necessary because the histones, as basic proteins, have a general high affinity for DNA. *Such interactions allow histones to form nucleosomes without becoming trapped in other kinetic intermediates (that is, other complexes resulting from indiscreet binding of histones to DNA).*

Numerous histone chaperones have been identified. Chromatin assembly factor (CAF)-1 and Anti-silencing function 1 (ASF1) are two chaperones that function at the replication fork. CAF-1 is a conserved 3-subunit complex that is directly recruited to the replication fork by proliferating cell nuclear antigen (PCNA), the processivity factor for DNA polymerase. ASF1 interacts with the replicative helicase that unwinds the replication fork. Furthermore, CAF-1 and ASF1 interact with each other. These interactions provide the link between replication and nucleosome assembly, ensuring that nucleosomes are assembled as soon as DNA has been replicated.

CAF-1 acts stoichiometrically, and functions by binding only to newly synthesized H3 and H4. New nucleosomes form by first assembling the $H3_2$–$H4_2$ tetramer onto the DNA, and then adding the H2A–H2B dimers. ASF1 appears to play an important role in transfer of parental nucleosomes from ahead of the replication fork to the newly synthesized region behind the fork, although ASF1 can bind and assemble newly synthesized histones as well.

The pattern of disassembly and reassembly has been difficult to characterize in detail, but a working model is illustrated in **FIGURE 24.22**. The replication fork displaces histone octamers, which then dissociate into $H3_2$–$H4_2$ tetramers and H2A–H2B dimers. The "old" tetramers may be transferred directly to the newly synthesized regions with the assistance of chaperones. The "old" dimers enter a pool that also includes "new" tetramers and dimers, which are assembled from newly synthesized histones. Nucleosomes assemble ~600 bp behind the replication fork. Assembly is initiated when $H3_2$-$H4_2$ tetramers bind to each of the daughter duplexes, assisted by CAF-1 or ASF1. Two H2A-H2B dimers then bind to each $H3_2$–$H4_2$ tetramer to complete the histone octamer. The assembly of tetramers and dimers is random with respect to "old" and "new" subunits.

During S phase (the period of DNA replication) in a eukaryotic cell, the duplication of chromatin requires synthesis of sufficient histone proteins to package an entire genome—basically the same quantity of histones must be synthesized that are already contained in nucleosomes. The synthesis of most histone mRNAs is controlled as part of the cell cycle, and increases enormously in S phase. The pathway for assembling

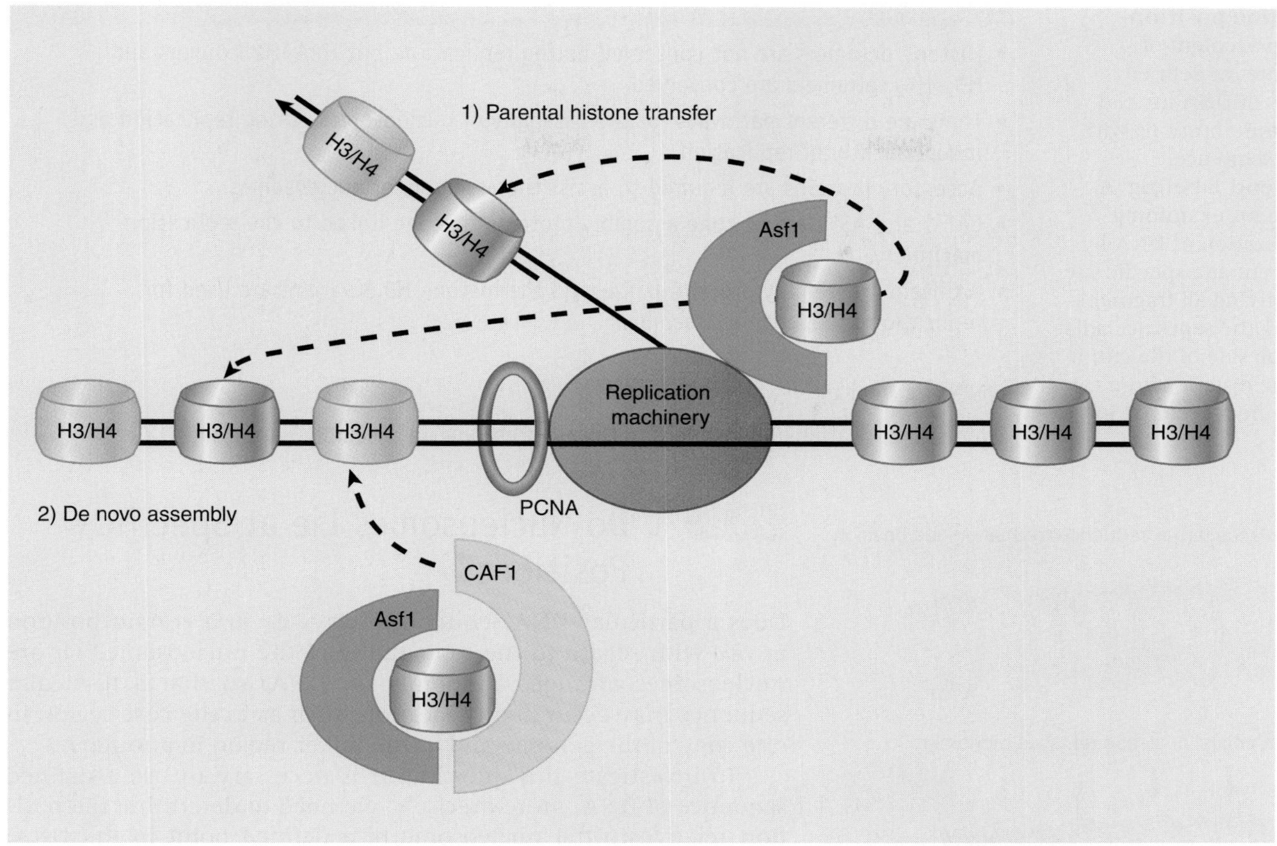

FIGURE 24.22 Replication fork passage displaces histone octamers from DNA. They disassemble into H3–H4 tetramers and H2A–H2B dimers. H3–H4 tetramers are directly transferred behind the replication forks. Newly synthesized histones are assembled into H3–H4 tetramers and H2A–H2B dimers. The old and new tetramers and dimers are assembled with the aid of histone chaperones into new nucleosomes immediately behind the replication fork. Adapted from Rocha, W., and Verreault, A., *FEBS Lett.* 582(2008):1938–1949.

chromatin from this equal mix of old and new histones during S phase is called the *replication-coupled (RC) pathway*.

Another pathway, called the *replication-independent (RI) pathway*, exists for assembling nucleosomes during other phases of cell cycle, when DNA is not being synthesized. This may become necessary as the result of damage to DNA or because nucleosomes are displaced during transcription. The assembly process must necessarily have some differences from the replication-coupled pathway, because it cannot be linked to the replication apparatus. The replication-independent pathway uses the histone H3.3 variant, as discussed in *Section 24.4, Histone Variants Produce Alternative Nucleosomes*. H3.3 slowly replaces H3 in differentiating cells that do not have replication cycles. This happens as the result of assembly of new histone octamers to replace those that have been displaced from DNA for whatever reason.

CAF-1 is not involved in replication-independent assembly. (There are organisms such as yeast and *Arabidopsis* for which CAF-1 is not essential, implying that alternative assembly processes may also be used in replication-coupled assembly.) A factor involved in replication-independent assembly is called HIRA. HIRA functions as a chaperone to assist the incorporation of histones into nucleosomes. This pathway appears to be generally responsible for replication-independent assembly; for example, HIRA is required for the decondensation of the sperm nucleus, when protamines are replaced by histones, in order to generate chromatin that is competent to be replicated following fertilization.

- **nucleosome positioning** The placement of nucleosomes at defined sequences of DNA instead of at random locations with regard to sequence.
- **indirect end labeling** A technique for examining the organization of DNA by making a cut at a specific site and identifying all fragments containing the sequence adjacent to one side of the cut; it reveals the distance from the cut to the next break(s) in DNA.

KEY CONCEPTS

- Histone octamers are not conserved during replication, but H2A–H2B dimers and $H3_2$–$H4_2$ tetramers are conserved.
- There are different pathways for the assembly of nucleosomes during replication and independently of replication.
- Accessory proteins are required to assist the assembly of nucleosomes.
- CAF-1 and ASF1 are histone assembly proteins that are linked to the replication machinery.
- A different assembly protein, HIRA, and the histone H3.3 variant are used for replication-independent assembly.

CONCEPT AND REASONING CHECK

Why is a replication-independent pathway of chromatin assembly important?

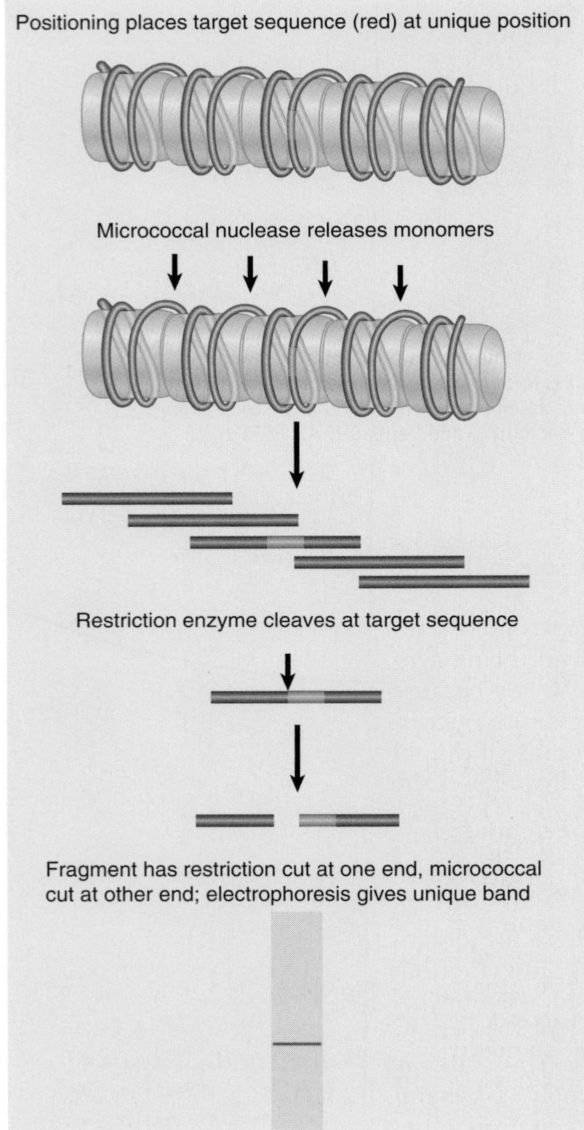

FIGURE 24.23 Nucleosome positioning places restriction sites at unique positions relative to the linker sites cleaved by micrococcal nuclease.

24.8 Do Nucleosomes Lie at Specific Positions?

Does a particular DNA sequence always lie in a certain position *in vivo* with regard to the topography of the nucleosome? Or are nucleosomes arranged randomly on DNA, so that a particular sequence may occur at any location, such as in the core region in one copy of the genome and in the linker region in another?

To investigate this question, it is necessary to use a defined sequence of DNA; more precisely, we need to determine the position relative to the nucleosome of a defined point in the DNA. **FIGURE 24.23** illustrates the principle of a procedure used to achieve this.

Suppose the DNA sequence is organized into nucleosomes in only one particular configuration, so that each site on the DNA always is located at a particular position on the nucleosome. This type of organization is called **nucleosome positioning** (or sometimes *nucleosome phasing*). In a series of positioned nucleosomes, the linker regions of DNA comprise unique sites.

Consider the consequences for just a single nucleosome. Cleavage with micrococcal nuclease generates a monomeric fragment that constitutes a *specific sequence.* If the DNA is isolated and cleaved with a restriction enzyme that has only one target site in this fragment, it should be cut at a unique point. This produces two fragments, each of unique size.

The products of the micrococcal/restriction double digest are separated by gel electrophoresis. A probe representing the sequence on one side of the restriction site is used to identify the corresponding fragment in the double digest. This technique is called **indirect end labeling** (because it is not possible to label the end of the nucleosomal DNA fragment itself, so it must be detected indirectly with a probe).

Reversing the argument, the identification of a single sharp band demonstrates that the position of the restriction site is uniquely defined with respect to the end of the nucleosomal DNA (as defined by the micrococcal nuclease cut). Thus the nucleosome has a unique sequence of DNA.

What happens if the nucleosomes do *not* lie at a single position? Now the linkers consist of *different* DNA sequences in each copy of the genome. Thus the restriction site lies at a different position each time; in fact, it lies at all possible locations relative to the ends of the monomeric nucleosomal DNA. **FIGURE 24.24** shows that the double cleavage then generates a broad smear, ranging from the smallest detectable fragment (~20 bases) to the length of the monomeric DNA.

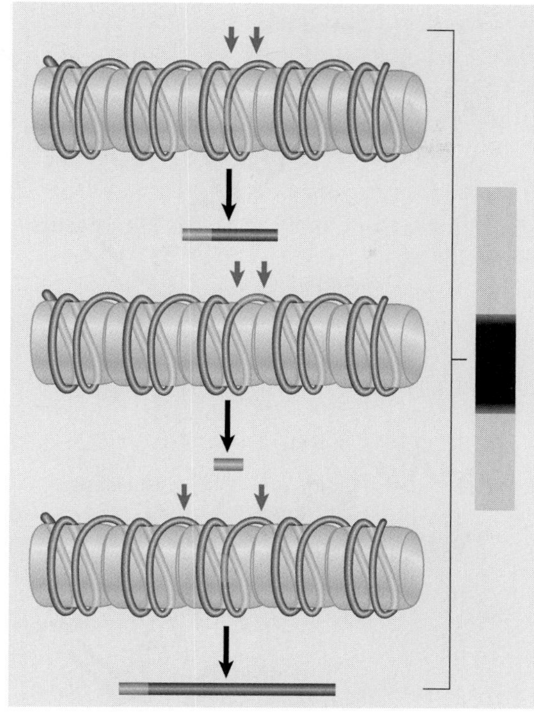

FIGURE 24.24 In the absence of nucleosome positioning, a restriction site lies at all possible locations in different copies of the genome. Fragments of all possible sizes are produced when a restriction enzyme cuts at a target site (red) and micrococcal nuclease cuts at the junctions between nucleosomes (green).

Nucleosome positioning might be accomplished in either of two ways:

- Intrinsic mechanisms: *Nucleosomes are deposited specifically at particular DNA sequences, or excluded by specific sequences.* This modifies our view of the nucleosome as a subunit able to form between any sequence of DNA and a histone octamer.
- Extrinsic mechanisms: *The first nucleosome in a region is preferentially assembled at a particular site.* A preferential starting point for nucleosome positioning can result either from the exclusion of a nucleosome from a particular region (due to competition with another protein binding that region), or by specific deposition of a nucleosome at a particular site. The excluded region or the positioned nucleosome provides a *boundary* that restricts the positions available to the adjacent nucleosome. A series of nucleosomes may then be assembled sequentially, with a defined repeat length.

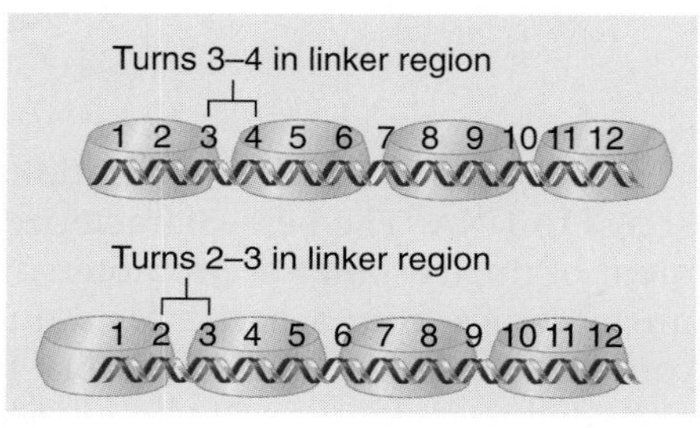

FIGURE 24.25 Translational positioning describes the linear position of DNA relative to the histone octamer. Displacement of the DNA by 10 bp changes the sequences that are in the more exposed linker regions, but does not alter which face of DNA is protected by the histone surface and which is exposed to the exterior. DNA is really coiled around the nucleosomes, and is shown in linear form only for convenience.

It is now clear that the deposition of histone octamers on DNA is not random with regard to sequence. The pattern is intrinsic in some cases, determined by structural features in DNA. It is extrinsic in other cases, resulting from the interactions of other proteins with the DNA and/or histones.

Certain structural features of DNA affect placement of histone octamers. Some sequences of DNA have intrinsic tendencies to bend in one direction rather than another. For example, AT dinucleotides bend easily, and thus AT-rich sequences are easier to wrap tightly in a nucleosome. A-T-rich regions locate so that the minor groove faces in toward the octamer, whereas G-C-rich regions are arranged so that the minor groove points out. Poly(A) stretches, in contrast, stiffen the DNA and assemble poorly into nucleosomes. It is not yet possible to sum all of the relevant structural effects and thus entirely predict the location of a particular DNA sequence with regard to the nucleosome. Sequences that cause DNA to take up more extreme structures may have effects such as the exclusion of nucleosomes, and thus could cause boundary effects. Specific sequences, such as a portion of the 5S rDNA and certain simple sequence satellites, can robustly position nucleosomes. Recent genome-wide studies have begun to reveal patterns of intrinsic positioning, including the prevalence of nucleosome-excluding sequences in critical promoter regions.

▸ **translational positioning** The location of a histone octamer at successive turns of the double helix, which determines which sequences are located in linker regions.

The location of DNA on nucleosomes can be described in two ways. **FIGURE 24.25** shows that **translational positioning** describes the position of DNA with regard to

FIGURE 24.26 Rotational positioning describes the exposure of DNA on the surface of the nucleosome. Any movement that differs from the helical repeat (~10.2 bp/turn) displaces DNA with reference to the histone surface. Nucleotides on the inside are more protected against nucleases than nucleotides on the outside.

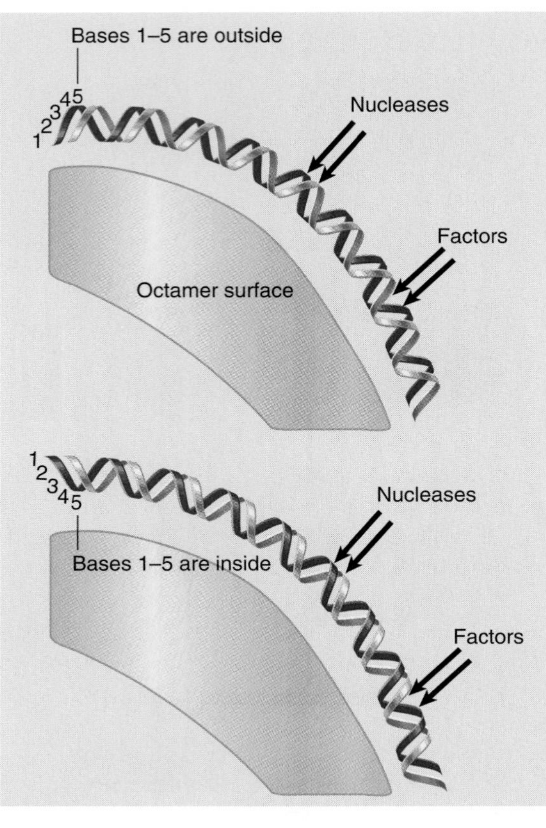

▸ **rotational positioning** The location of the histone octamer relative to turns of the double helix, which determines which face of DNA is exposed on the nucleosome surface.

the boundaries of the nucleosome. In particular, it determines which sequences are found in the linker regions. Shifting the DNA by 10 bp brings the next turn into a linker region. Thus translational positioning determines which regions are more accessible (at least as judged by sensitivity to micrococcal nuclease).

Because DNA lies on the outside of the histone octamer, one face of any particular sequence is obscured by the histones, whereas the other face is accessible. Depending upon its positioning with regard to the nucleosome, a site in DNA that must be recognized by a regulator protein could be inaccessible or available. The **rotational positioning** of the double helix with regard to the histone octamer can therefore be important, as shown in FIGURE 24.26. If the DNA is moved by a partial number of turns (imagine the DNA as rotating relative to the protein surface), there is a change in the exposure of sequence to the outside.

Both translational and rotational positioning can be important in controlling access to DNA. The best-characterized cases of positioning involve the specific placement of nucleosomes at promoters. Translational positioning and/or the exclusion of nucleosomes from a particular sequence may be necessary to allow a transcription complex to form. Some regulatory factors can bind to DNA only if a nucleosome is excluded to make the DNA freely accessible, and this creates a boundary for translational positioning. In other cases, regulatory factors can bind to DNA on the surface of the nucleosome, but rotational positioning is important to ensure that the face of DNA with the appropriate contact points is exposed.

We discuss the connection between nucleosomal organization and transcription in *Chapter 26, Eukaryotic Transcription Regulation,* but note for now that promoters (and some other structures) often have short regions that exclude nucleosomes. These regions typically form a boundary next to which nucleosome positions are restricted. A survey of an extensive region in the *Saccharomyces cerevisiae* genome (mapping 2278 nucleosomes over 482 kb of DNA) showed that in fact 60% of the nucleosomes have specific positions as the result of boundary effects, most often from promoters.

KEY CONCEPTS

- Nucleosomes may form at specific positions as the result of either the local structure of DNA or proteins that interact with specific sequences.
- A common cause of nucleosome positioning is when proteins binding to DNA establish a boundary.
- Positioning may affect which regions of DNA are in the linker and which face of DNA is exposed on the nucleosome surface.

CONCEPT AND REASONING CHECK

How can a boundary effect be created by either extrinsic or intrinsic mechanisms?

24.9 DNase Hypersensitive Sites Reflect Changes in Chromatin Structure

In addition to the pattern of nucleosomal or nucleosome-free regions, active or potentially active regions exhibit structural changes that occur at specific sites associated with initiation of transcription. These changes were first detected by the effects of digestion with very low concentrations of the enzyme DNase I.

When chromatin is digested with DNase I, the first effect is the introduction of breaks in the duplex at specific, **hypersensitive sites**. Susceptibility to DNase I reflects the accessibility of DNA in chromatin, so we take these sites to represent chromatin regions in which the DNA is particularly exposed because it is not organized in the usual nucleosomal structure. A typical hypersensitive site is 100× more sensitive to enzyme attack than bulk chromatin. These sites are also hypersensitive to other nucleases and chemical agents.

▸ **hypersensitive site** A short region of chromatin detected by its extreme sensitivity to cleavage by DNase I and other nucleases; it comprises an area from which nucleosomes are excluded.

Hypersensitive sites are created by the local structure of chromatin, which may be tissue specific. Hypersensitive sites are found associated with promoters, other elements that regulate transcription, origins of replication, centromeres, and sites with other structural significance. Most hypersensitive sites are related to gene expression. Every active gene has a site, or sometimes more than one site, in the region of the promoter. Most hypersensitive sites are found only in chromatin of cells in which the associated gene is being (or will be) expressed; they do not occur when the gene is inactive. These site(s) generally appear before transcription begins, and the DNA sequences contained within these hypersensitive sites are required for gene expression.

A particularly well-characterized nuclease-sensitive region lies on the SV40 minichromosome. A short segment near the origin of replication, just upstream of the promoter for the late transcription unit, is cleaved preferentially by DNase I, micrococcal nuclease, and other nucleases (including restriction enzymes).

The state of the SV40 minichromosome can be visualized by electron microscopy. In up to 20% of the samples, a "gap" is visible in the nucleosomal organization, as evident in **FIGURE 24.27**. The gap is a region of ~120 nm in length (about 350 bp), surrounded on either side by nucleosomes. The visible gap corresponds with the nuclease-sensitive region. This shows directly that increased sensitivity to nucleases is associated with the exclusion of nucleosomes.

What is the structure of a hypersensitive site? Its preferential accessibility to nucleases indicates that it is not protected by histone octamers, but this does not necessarily imply that it is free of protein. A region of free DNA might be vulnerable to damage, and in any case, how would it be able to exclude nucleosomes? A hypersensitive site results from the binding of specific regulatory proteins that exclude nucleosomes. Indeed, many hypersensitive sites exhibit a central protected region, not susceptible to digestion, which indicates the binding of such regulatory proteins.

In addition to detecting hypersensitive sites, DNase I digestion can also be used to assess the relative accessibility of a genomic region. A region of the genome that contains an active gene may have an altered structure in addition to any hypersensitive sites. Increased DNase I sensitivity defines a *chromosomal domain*, which is a region of altered structure including at least one active transcription unit, and sometimes extending farther. (Note that use of the term *domain* does not imply any necessary connection with the structural domains identified by the loops of chromatin or chromosomes.) Active genes are preferentially degraded by DNase I.

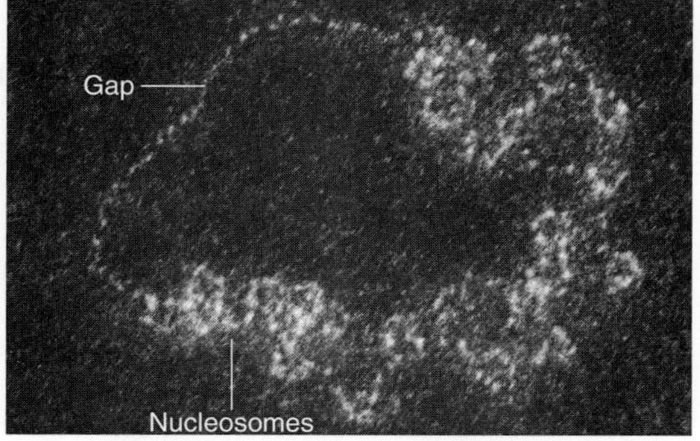

FIGURE 24.27 The SV40 minichromosome has a nucleosome gap. Photo courtesy of Moshe Yaniv, Pasteur Institute.

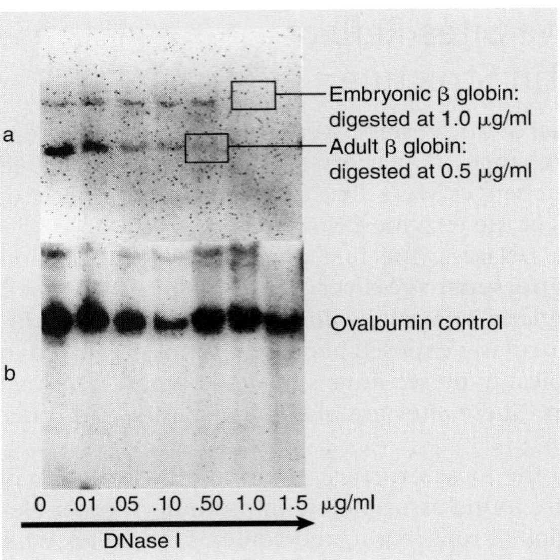

FIGURE 24.28 In adult erythroid cells, the adult β-globin gene is highly sensitive to DNase I digestion; the embryonic β-globin gene is partially sensitive (probably due to spreading effects), but ovalbumin is not sensitive. Photos courtesy of Harold Weintraub, Fred Hutchinson Cancer Research Center. Used with permission of Mark T. Groudine.

The fate of individual genes can be followed by quantitating the amount of DNA that survives to react with a specific probe. **FIGURE 24.28** shows what happens to β-globin genes and an ovalbumin gene in chromatin extracted from chicken red blood cells (in which globin genes are expressed and the ovalbumin gene is inactive). The restriction fragments representing the β-globin genes are rapidly lost, whereas those representing the ovalbumin gene show little degradation. (The ovalbumin gene in fact is digested at the same rate as the bulk of DNA.)

Thus the bulk of chromatin is relatively resistant to DNase I and contains nonexpressed genes (as well as other sequences). A gene becomes relatively susceptible to the enzyme specifically in the tissue(s) in which it is expressed. It is important to note that this DNase I sensitivity is not simply a function of high levels of transcription, which can deplete nucleosomes (discussed in Chapter 26), as even rarely expressed genes show increased DNase I sensitivity. In addition, the region of high sensitivity to DNase I extends over a considerable distance, and is not limited to the promoter or transcribed region.

KEY CONCEPTS

- Hypersensitive sites are found at the promoters of expressed genes.
- They are generated by the binding of transcription factors that exclude histone octamers.
- A domain containing a transcribed gene is defined by increased sensitivity to degradation by DNase I.

CONCEPT AND REASONING CHECK

What is the difference between a DNase I hypersensitive site and a DNase I-sensitive domain?

24.10 An LCR May Control a Domain

Every gene is controlled by its promoter, and many genes also respond to enhancers (containing similar control elements but located farther away); see *Chapter 25, Eukaryotic Transcription*. These local controls are not sufficient for all genes, though. In some cases, a gene lies within a domain of several genes, all of which are influenced by regulatory elements that act on the whole domain.

The best-characterized example of a regulated gene cluster is provided by the mammalian β-globin genes. Recall from Figure 6.5 that the α- and β-globin genes in mammals each exist as clusters of related genes that are expressed at different times during embryonic and adult development. These genes are provided with a large number of regulatory elements, which have been analyzed in detail. In the case of the adult human β-globin gene, regulatory sequences are located both 5′ and 3′ to the gene. The regulatory sequences include both positive and negative elements in the promoter region, as well as additional positive elements within and downstream of the gene.

However, all of these control regions are not sufficient for proper expression of the human β-globin gene, so some further regulatory sequence is required. Regions that provide the additional regulatory function are identified by DNase I hypersensitive sites that are found at the ends of the cluster. The map of FIGURE 24.29 shows that the 20 kb upstream of the ε gene contains a group of five sites, and that there is a single site 30 kb downstream of the β gene.

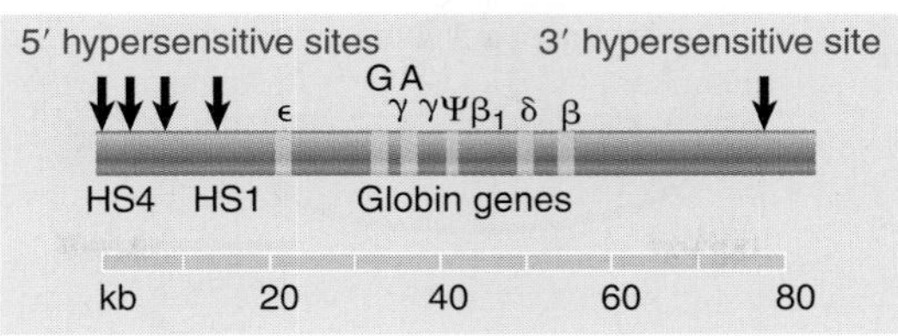

FIGURE 24.29 A globin domain is marked by hypersensitive sites at either end. The group of sites at the 5′ side constitutes the LCR and is essential for the function of all genes in the cluster.

The 5′ regulatory sites are the primary regulators, and the cluster of hypersensitive sites is called the **locus control region (LCR)**. (We do not know if the 3′ site has any function, but a 3′ hypersensitive site in the chicken β-globin locus acts as an *insulator*, described in the next section.) The LCR is absolutely required for expression of each of the globin genes in the cluster. Each gene is then further regulated by its own specific controls. Some of these controls are autonomous; for example, expression of the ε gene appears intrinsic to this locus in conjunction with the LCR. Other controls appear to rely upon position in the cluster, which provides a suggestion that *gene order* in a cluster is important for regulation.

▸ **locus control region (LCR)** The region that is required for the expression of several genes in a domain.

The entire region containing the globin genes, and extending well beyond them, constitutes a chromosomal **domain**. It shows increased sensitivity to digestion by DNase I. Deletion of the 5′ LCR restores normal resistance to DNase over the whole region. It is not clear whether the LCR is required to activate the individual promoters, or whether it acts to increase the rate of transcription from the promoters. The exact nature of the sequential interactions between the LCR and the individual promoters has not yet been fully defined, but it has recently become clear that the LCR contacts individual promoters directly, forming loops when these promoters are active.

▸ **domain** In reference to a chromosome, it may refer either to a discrete structural entity defined as a region within which supercoiling is independent of other regions or to an extensive region including an expressed gene that has heightened sensitivity to degradation by the enzyme DNase I.

Does this model apply to other gene clusters? The α-globin locus has a similar organization of genes that are expressed at different times, with a group of hypersensitive sites at one end of the cluster, and increased sensitivity to DNase I throughout the region. Only a small number of other cases are known in which an LCR controls a group of genes.

One of these cases involves an LCR that controls genes on more than one chromosome. The T_H2 LCR coordinately regulates a group of genes that are spread out over 120 kb on chromosome 11 by interacting with their promoters. It also interacts with the promoter of the IFNγ gene on chromosome 10. The two types of interaction are alternatives that comprise two different cell fates; that is, in one group of cells the LCR causes expression of the genes on chromosome 11, whereas in the other group it causes the gene on chromosome 10 to be expressed.

KEY CONCEPTS

- An LCR is located at the 5′ end of a domain and consists of several hypersensitive sites.
- LCRs regulate gene clusters.

CONCEPT AND REASONING CHECK

Loss of the β-globin LCR in hematopoietic cells results in decreased DNase I sensitivity across the entire globin domain. What would be the effect of this loss in liver cells?

24.11 Insulators Define Independent Domains

Insulators are a class of elements that can prevent the passage of activating or inactivating effects. They have either or both of two key properties:

▸ **insulator** A sequence that prevents an activating or inactivating effect passing from one side to the other.

- When an insulator is placed between an enhancer and a promoter, *it prevents the enhancer from activating the promoter.* The blocking effect is shown in FIGURE 24.30. This may explain how the action of a given enhancer is limited to a particular promoter.

FIGURE 24.30 An enhancer activates a promoter in its vicinity, but may be blocked from doing so by an insulator located between them.

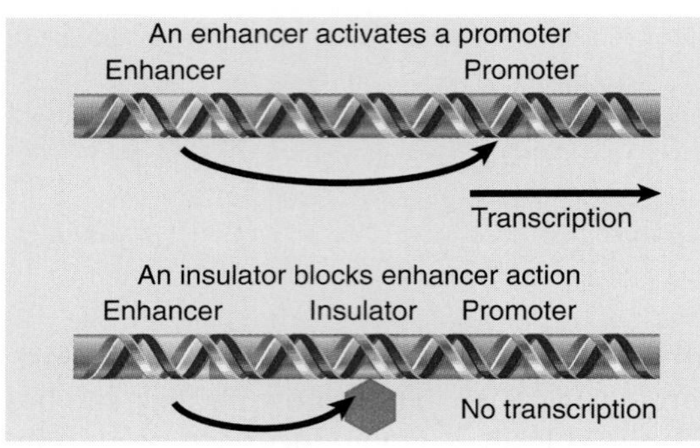

- When an insulator is placed between an active gene and heterochromatin, *it provides a barrier that protects the gene against the inactivating effect that spreads from the heterochromatin.* (Heterochromatin is a region of chromatin that is inactive as the result of its higher order structure; see *Section 27.2, Heterochromatin Propagates from a Nucleation Event.*) The barrier effect is shown in **FIGURE 24.31**.

Some insulators possess both these properties, but others have only one, or the blocking and barrier functions can be separated. Although both actions are likely to be mediated by changing chromatin structure, they may involve different effects. In either case, however, the insulator defines a limit for long-range effects.

What is the purpose of an insulator? A major function may be to counteract the indiscriminate actions of enhancers on promoters. Most enhancers will activate any promoter in the vicinity. An insulator can restrict an enhancer by blocking the effects from passing beyond a certain point, so that it can act only on a specific promoter. Similarly, when a gene is located near heterochromatin, an insulator can prevent it from being inadvertently inactivated by the spread of the heterochromatin. Insulators therefore function as elements for increasing the precision of gene regulation.

Insulators were discovered during the analysis of the region of the *Drosophila melanogaster* genome summarized in **FIGURE 24.32**. Two genes for the protein Hsp (heat shock protein) 70 lie within an 18 kb region that constitutes band 87A7. Special structures, called *scs* and *scs′* (specialized chromatin structures), are found at the ends of the band. Each consists of a region that is highly resistant to degradation by DNase I, and each is flanked on either side by hypersensitive sites that are spaced at about 100 bp. The cleavage pattern at these sites is altered when the genes are turned on by heat shock.

The *scs* elements insulate the *hsp70* genes from the effects of surrounding regions. If we take *scs* units and place them on either side of a *reporter* gene, the gene can function anywhere it is placed in the genome. See the accompanying *Methods Box* for a description of the experiments that allowed the *scs* and *scs′* elements to be identified as the first insulators.

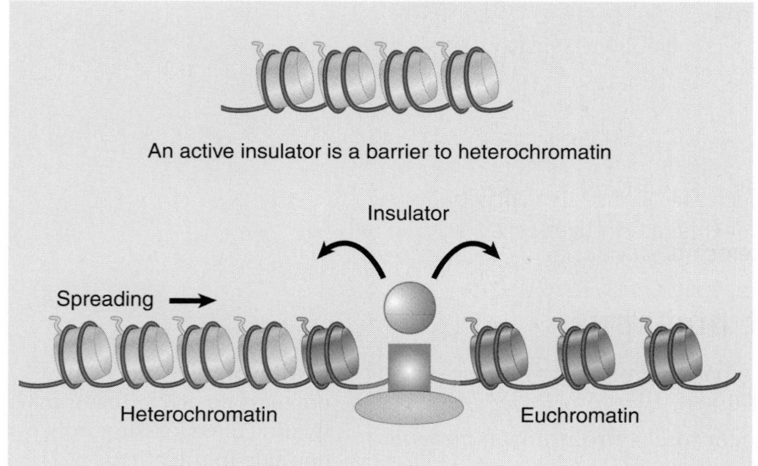

FIGURE 24.31 Heterochromatin may spread from a center and then block any promoters that it covers. An insulator may be a barrier to propagation of heterochromatin that allows the promoter to remain active.

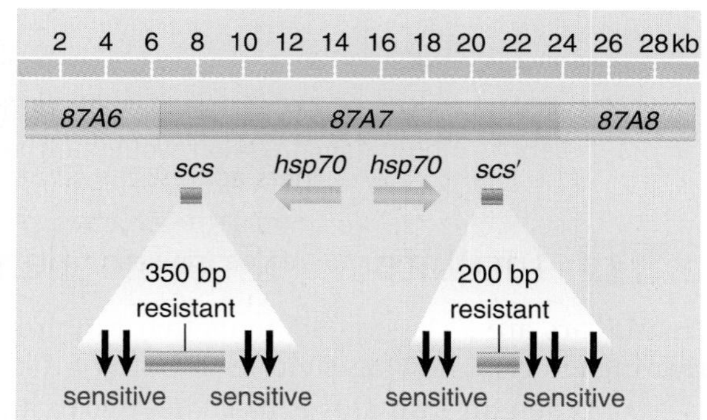

FIGURE 24.32 Specialized chromatin structures that include hypersensitive sites mark the ends of a domain in the *D. melanogaster* genome and insulate genes between them from the effects of surrounding sequences.

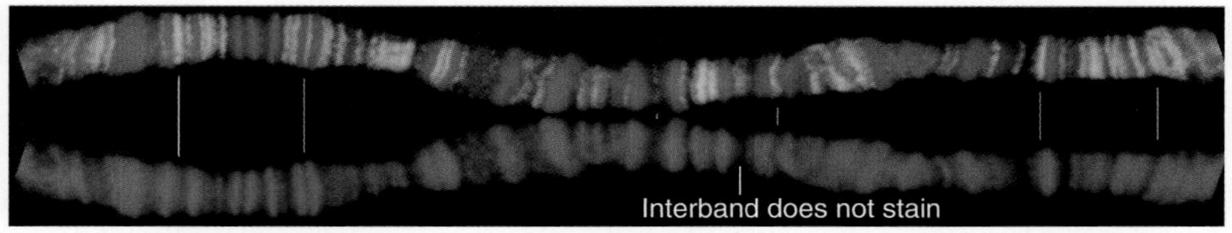

FIGURE 24.33 A protein that binds to the insulator *scs′* is localized at interbands in *Drosophila* polytene chromosomes. Red staining identifies the DNA (the bands) on both the upper and lower samples; green staining identifies BEAF32 (often at interbands) on the upper sample. Yellow shows coincidence of the two labels (meaning that BEAF32 is in a band). Reproduced from *Cell*, vol. 81, Zhao, K., Hart, C. M., and Laemmli, U. K., Visualization of chromosomal . . . , pp. 879–889. Copyright 1995, with permission from Elsevier [http://www.sciencedirect.com/science/journal/00928674]. Photo courtesy of Ulrich K. Laemmli, University of Geneva, Switzerland.

The *scs* and *scs′* elements themselves do not play either positive or negative roles in controlling gene expression, but just restrict effects from passing from one region to the next. If adjacent regions have repressive effects, however, the *scs* elements would be needed to block the spread of such effects, and therefore would be essential for gene expression. In this case, deletion of such elements could eliminate the expression of the adjacent gene(s).

The *scs* and *scs′* elements have different structures, and each appears to have a different basis for its insulator activity. The key sequence in the *scs* element is a stretch of 24 bp that binds the product of the *zw5* gene. The insulator property of *scs′* resides in a series of CGATA repeats. The repeats bind a group of related proteins called BEAF-32. The protein shows discrete localization within the nucleus, but the most remarkable data derive from its localization on polytene chromosomes. **FIGURE 24.33** shows that an anti-BEAF-32 antibody stains ~50% of the interbands of polytene chromosomes. This suggests that there are many insulators in the genome, and that BEAF-32 is a common part of the insulating apparatus. This result also implies that the band is a functional unit, and that interbands often have insulators that block the propagation of activating or inactivating effects.

The chicken β-globin domain is flanked by insulators. As in mammals, the chicken β-globin cluster is regulated by an LCR. The furthest upstream hypersensitive site of the LCR (HS4) is an insulator that marks the 5′ end of the functional domain. This restricts the LCR to acting only on the globin genes in the domain. It also prevents silencing of the locus by a region of condensed chromatin located upstream of this insulator. A 3′ hypersensitive site downstream of the cluster corresponds to a second insulator, which isolates the β-globin locus from a cluster of olfactory receptor genes. These two insulators combine to define a transcriptionally independent domain, in which the LCR regulates promoters within the domain, and signals from beyond the insulators are blocked.

KEY CONCEPTS

- Insulators are able to block passage of any activating or inactivating effects from enhancers, silencers, and LCRs.
- Insulators can provide barriers against the spread of heterochromatin.
- Insulators are specialized chromatin structures that have hypersensitive sites. Two insulators can protect the region between them from all external effects.

CONCEPT AND REASONING CHECK

Why are two insulators needed to define a domain?

METHODS AND TECHNIQUES

Position Effect Variegation (PEV) and the Discovery of Insulators

Genes near the breakpoints of chromosomal rearrangement become repositioned in the genome and flanked by new neighboring genes or regulatory sequences. In many cases, the repositioning of a gene affects its level of expression or, in some cases, its ability to function; this is called **position effect**. Such effects have been studied extensively in *Drosophila* and yeast. In *Drosophila*, the most common type of position effect results in a mottled (mosaic) phenotype that is observed as interspersed patches of wild-type cells, in which the wild-type allele is expressed, and mutant cells, in which the wild-type allele is inactivated. The phenotype is said to show *variegation,* and the phenomenon is called **position-effect variegation (PEV)**.

PEV can result from a chromosome aberration that moves a wild-type gene from a position in euchromatin to a new position in or near heterochromatin (**FIGURE B24.1**). **FIGURE B24.2** illustrates some of the patterns of wild-type (red) and mutant (white) facets that are observed in male flies that carry a rearranged X chromosome in which an inversion repositions the wild-type *white* (w^+) allele into heterochromatin. The same types of patterns are found in females heterozygous for the rearranged X chromosome and an X chromosome carrying the *w* allele. The patterns of w^+ expression coincide with the clonal lineages in the eye; that is, all of the red cells in a particular patch derive from a single ancestral cell in the embryo in which the w^+ allele was activated.

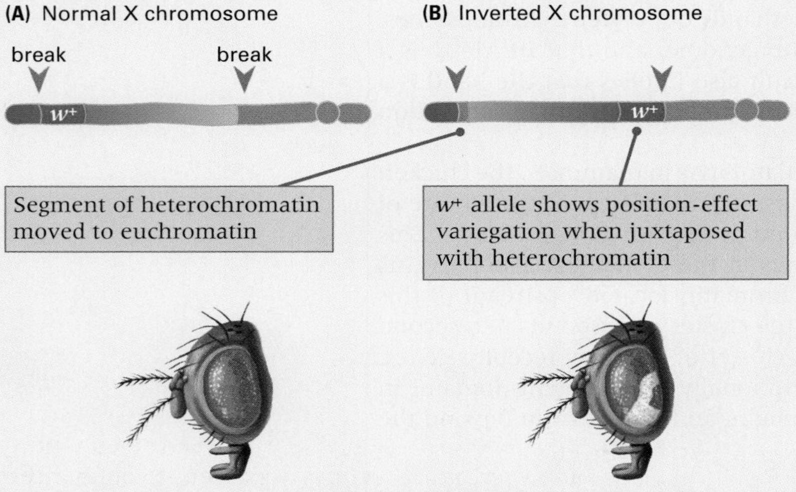

Although the mechanism of PEV is not understood in detail, it is thought to result from the unusual chromatin structure of heterochromatin interfering with gene activation. The determination of gene expression or nonexpression is thought to take place when the boundary between condensed heterochromatin and euchromatin is established. Where heterochromatin is juxtaposed with euchromatin, the chromatin condensation characteristic of heterochromatin may spread into the adjacent euchromatin, inactivating euchromatic genes in the cell and all of its descendants.

The phenomenon of position effects has proven to be a useful system for scientists studying various aspects of chromosome organization. For example, genes encoding products involved in heterochromatin formation have been identified in screens for factors that alter levels of PEV when deleted or overexpressed. Deletion of a gene that promotes heterochromatin spreading, such that encoding the structural heterochromatin protein 1 (HP1), would result in *more* expression of a gene located close to heterochromatin.

The first insulators to be functionally characterized, *scs* and *scs′*, were investigated using PEV as a tool. Researchers placed *scs* and *scs′* elements on both sides of a *white* reporter gene and then randomly integrated these reporters into the *Drosophila* genome. When the *white* gene *lacking* flanking insulators was used, the resulting transgenic flies exhibited a wide array of expression levels of the *white* gene, including silencing of the gene, reflecting its integration into or near heterochromatin, as well as high expression resulting from integration near strong enhancers. In contrast, when the *white* reporter was flanked by insulators, both these

FIGURE B24.1 Position-effect variegation (PEV) is often observed when an inversion or other chromosome rearrangement repositions a gene normally in euchromatin to a new location in or near heterochromatin. In this example, an inversion in the X chromosome of *Drosophila melanogaster* repositions the wild-type allele of the *white* gene near heterochromatin. PEV of the w^+ allele is observed as mottled red and white eyes.

24.12 What Constitutes a Regulatory Domain?

If we put together the various types of structures that have been found in different systems, we think about the possible nature of a chromosomal domain. The basic feature of a regulatory domain is that regulatory elements act only on transcription units in the same domain. A domain might contain more than one transcription unit and/or enhancer. **FIGURE 24.34** summarizes the structures that might be involved in defining a domain.

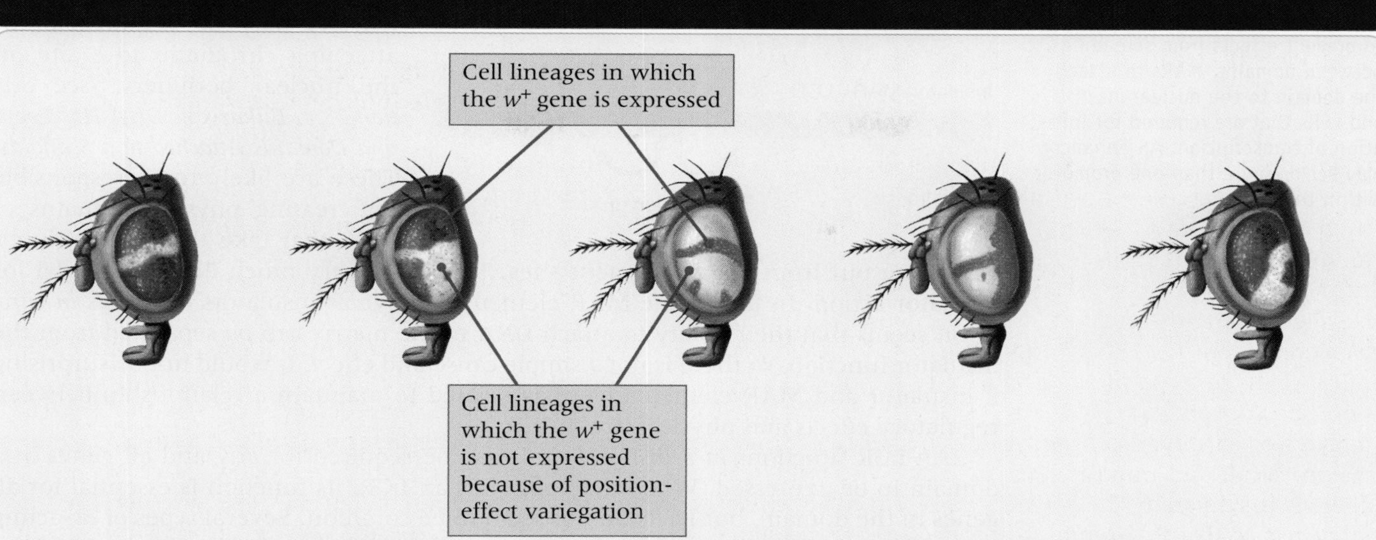

FIGURE B24.2 Patterns of red and white sectors in the eye of *Drosophila melanogaster* resulting from position-effect variegation. Each group of contiguous facets of the same color derives from a single cell in development. Such large patches of red are atypical. More often, one observes numerous very small patches of red or a mixture of many small and a few large patches.

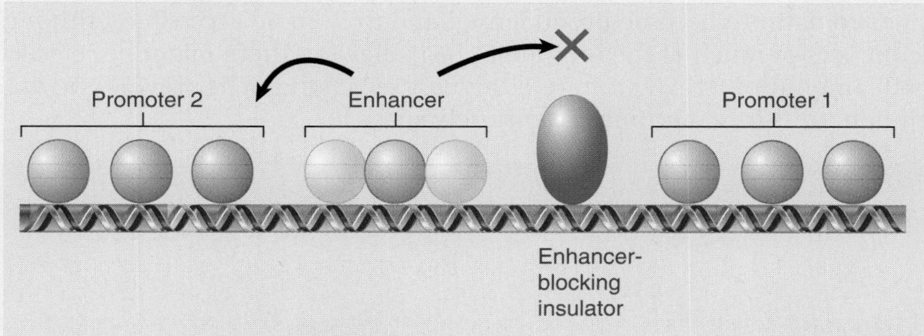

FIGURE B24.3 An enhancer-blocking insulator interferes with enhancer–promoter communication in a position-dependent manner. An enhancer-blocking insulator blocks transcriptional activation only when it lies between a promoter and an enhancer (as in the case of promoter 1); in other situations (such as for promoter 2), activation is not blocked. A transcriptional repressor, by contrast, would reduce the level of transcription from both promoters when placed in the same position. Adapted from Gaszner, M., and Felsenfeld, G., *Nat. Rev. Genet.* 7(2006): 703–713.

negative and positive position effects were blocked, and the transgenic flies showed remarkably uniform levels of *white* expression no matter where the gene had integrated. The discovery that *positive* position effects were also prevented led to a now-commonly used assay for insulator function: the "enhancer-blocking" assay. This assay provides a simple test for identifying insulators, as shown in **FIGURE B24-3**: when placed between an enhancer and a promoter, an insulator will prevent the enhancer from activating the promoter without actually repressing the function of the enhancer (note in the figure the enhancer can still act on promoters in the "other direction"). This assay, and variations of PEV assays, are still the only way to positively identify insulator elements.

An insulator stops activating or repressing effects from passing. In its simplest form, an insulator blocks either type of effect from passing across it, but there can be more complex relationships in which the insulator blocks only one type of effect and/or acts directionally. We assume that insulators act by affecting higher-order chromatin structure, but we do not know the details and varieties of such effects.

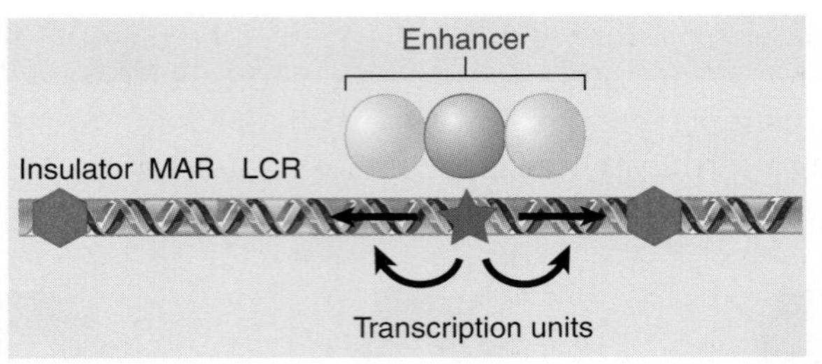

FIGURE 24.34 Domains may possess three types of sites: insulators to prevent effects from spreading between domains, MARs to attach the domain to the nuclear matrix, and LCRs that are required for initiation of transcription. An enhancer may act on more than one promoter within the domain.

A **matrix attachment site (MAR)** may be responsible for attaching chromatin to a site on the nuclear periphery (see *Section 23.4, Eukaryotic DNA Has Loops and Domains Attached to a Scaffold*). These are likely to be responsible for creating physical domains of DNA that take the form of loops extending out from the attachment sites. This looks very much like one model for insulator action. In fact, some MAR elements behave as insulators in assays *in vitro*, but it seems that their ability to attach DNA to the matrix can be separated from the insulator function, so there is not a simple cause and effect. It would not be surprising if insulator and MAR elements were associated to maintain a relationship between regulatory effects and physical structure.

An LCR functions at a distance and may be required for any and all genes in a domain to be expressed. When a domain has an LCR, its function is essential for all genes in the domain, but LCRs do not seem to be common. Several types of *cis*-acting structures could be required for function. As defined originally, the property of the LCR rests with an enhancer-like hypersensitive site that is needed for the full activity of promoter(s) within the domain.

The organization of domains may help to explain the large size of the genome. A certain amount of space could be required for such a structure to operate, for example, to allow chromatin to become decondensed and to become accessible. Although the exact sequences of much of the unit might be irrelevant, there might be selection for the overall amount of DNA within it, or at least selection might prevent the various transcription units from becoming too closely spaced.

KEY CONCEPT

- A domain may have an insulator, an LCR, a matrix attachment site, and transcription unit(s).

CONCEPT AND REASONING CHECK

Does a domain require all of the elements shown in Figure 24.34?

24.13 Summary

All eukaryotic chromatin consists of nucleosomes. A nucleosome contains a characteristic length of DNA, usually ~200 bp, which is wrapped around an octamer containing two copies each of histones H2A, H2B, H3, and H4. A single H1 protein may associate with a nucleosome. Virtually all genomic DNA is organized into nucleosomes. Treatment with micrococcal nuclease shows that the DNA packaged into each nucleosome can be divided operationally into two regions. The linker region is digested rapidly by the nuclease; the core region of 146 bp is resistant to digestion. Histones H3 and H4 are the most highly conserved, and an $H3_2$–$H4_2$ tetramer accounts for the diameter of the particle. The H2A and H2B histones are organized as two H2A–H2B dimers. Octamers are assembled by the successive addition of two H2A–H2B dimers to the $H3_2$–$H4_2$ tetramer.

The path of DNA around the histone octamer creates –1.65 supercoils. The DNA "enters" and "leaves" the nucleosome in the same vicinity, and could be "sealed" by histone H1. Removal of the core histones releases –1.0 supercoils. The difference can be largely explained by a change in the helical pitch of DNA, from an average of 10.2 bp/turn in nucleosomal form to 10.5 bp/turn when free in solution. There is variation in the structure of DNA from a periodicity of 10.0 bp/turn at the nucleosome ends to 10.7 bp/turn in the center. There are kinks in the path of DNA on the nucleosome.

Nucleosomes are organized into a fiber of 30 nm diameter that has six nucleosomes per turn and a compaction level of 40-fold. Removal of H1 allows this fiber to unfold into a 10 nm fiber that consists of a linear string of nucleosomes. The 30 nm fiber probably consists of the 10 nm fiber wound into a 2-start solenoid. The 30 nm fiber is the basic constituent of both euchromatin and heterochromatin; nonhistone proteins are responsible for further organization of the fiber into chromatin or chromosome ultrastructure.

There are two pathways for nucleosome assembly. In the replication-coupled pathway, the PCNA processivity subunit of the replisome recruits CAF-1, which is a nucleosome assembly factor. CAF-1 assists the deposition of $H3_2$–$H4_2$ tetramers onto the daughter duplexes resulting from replication. The tetramers may be produced either by disruption of existing nucleosomes by the replication fork or as the result of assembly from newly synthesized histones. Similar sources provide the H2A–H2B dimers that then assemble with the $H3_2$–$H4_2$ tetramer to complete the nucleosome. The $H3_2$–$H4_2$ tetramer and the H2A–H2B dimers assemble at random, so the new nucleosomes may include both preexisting and newly synthesized histones. HIRA assembles nucleosomes outside of S phase, and ASF1 acts both during and outside replication to assemble chromatin.

Two types of changes in sensitivity to nucleases are associated with gene activity. Chromatin capable of being transcribed has a generally increased sensitivity to DNase I, reflecting a change in structure over an extensive region that can be defined as a domain containing active or potentially active genes. Hypersensitive sites in DNA occur at discrete locations, and are identified by greatly increased sensitivity to DNase I. A hypersensitive site consists of a sequence of ~200 bp from which nucleosomes are excluded by the presence of other proteins. A hypersensitive site forms a boundary that may cause adjacent nucleosomes to be restricted in position. Nucleosome positioning may be important in controlling access of regulatory proteins to DNA.

Hypersensitive sites occur at several types of regulators. Those that regulate transcription include promoters, enhancers, and LCRs. Other sites include origins for replication and centromeres. A promoter or enhancer acts on a single gene, but an LCR contains a group of hypersensitive sites and may regulate a domain containing several genes.

An insulator blocks the transmission of activating or inactivating effects in chromatin. An insulator that is located between an enhancer and a promoter prevents the enhancer from activating the promoter. Two insulators define the region between them as a regulatory domain; regulatory interactions within the domain are limited to it, and the domain is insulated from outside effects. Most insulators block regulatory effects from passing in either direction, but some are directional. Insulators usually can block both activating effects (enhancer–promoter interactions) and inactivating effects (mediated by the spread of heterochromatin), but some are limited to one or the other. Insulators are thought to act via changing higher-order chromatin structure, but the details are not certain.

CHAPTER QUESTIONS

Insulators have either or both of two key properties. When an insulator is placed between a(n) **1.** __________ and a promoter, it **2.** __________ the enhancer from activating the **3.** __________. When an insulator is placed between an active **4.** __________ and heterochromatin, it provides a **5.** __________ that protects the gene against the **6.** __________ effect that spreads from the **7.** __________.

8. An average nucleosome contains about:
A. 100 bp of DNA.
B. 200 bp of DNA.
C. 300 bp of DNA.
D. 500 bp of DNA.

9. Which histone is not part of the core particle?
A. H1
B. H2A
C. H3
D. H4

10. Which two histones are among the most conserved of all known proteins?
A. H1 and H2B
B. H2A and H3
C. H3 and H4
D. H2B and H4

11. One turn of the DNA double helix around a nucleosome takes about?
A. 50 bp of DNA
B. 80 bp of DNA
C. 140 bp of DNA
D. 200 bp of DNA

12. When nucleosomes are treated with DNase I and the products are separated in a polyacrylamide gel, a ladder of bands is observed, with the average distance between bands being about:
A. 6 base pairs.
B. 10 base pairs.
C. 12.5 base pairs.
D. 16 base pairs.

13. A chromosomal domain containing a transcribed gene is defined by:
A. increased sensitivity to micrococcal nuclease.
B. decreased sensitivity to micrococcal nuclease.
C. increased sensitivity to DNase I.
D. decreased sensitivity to DNase I.

KEY TERMS

10 nm fiber
30 nm fiber
core histone
domain
histone fold
histone octamer
histone tails
histone variant
histones
hypersensitive site
indirect end labeling
insulator
linker DNA
linker histones
locus control region (LCR)
micrococcal nuclease (MNase)
nonhistone
nucleosome
nucleosome positioning
rotational positioning
translational positioning

FURTHER READING

Boulard, M., Bouvet, P., Kundu, T. K., and Dmitrov, S. 2007. Histone variant nucleosomes: structure, function and implication in disease. *Subcell. Biochem.* 41, 91–109.

A review of core histone variants, the structural/functional consequences of histone variant incorporation into nucleosomes, and links to human disease.

Chodaparambil, J. V., Edayathumangalam, R. S., Bao, Y., Park, Y. J., and Luger, K. 2006. Nucleosome structure and function. *Ernst Schering Res. Found. Workshop* 57, 29–46.

A review of nucleosome structure and dynamics.

Eitoku, M., Sato, L., Senda, T., and Horikoshi, M. 2008. Histone chaperones: 30 years from isolation to elucidation of the mechanisms of nucleosome assembly and disassembly. *Cell Mol. Life Sci.* 65, 414–444.

An overview of the best-studied histone chaperones, including insights from structural studies.

Gaszner M. and Felsenfeld., G. 2006. Insulators: exploiting transcriptional and epigenetic mechanisms. *Nature Revs. Gen.* 7, 703–713.

A review of insulator function for both enhancer blocking and heterochromatin barrier activities.

Izzo, A., Kamieniarz, K., and Schneider, R. 2008. The histone H1 family: specific members, specific functions? *Biological Chemistry* 389, 333–343.

A review of the linker histone H1 and the H1 variants, including a review of the differences in specificity and function of different H1 family members.

Palstra, R. J., de Laat, W., and Grosvel, F. 2008. Beta-globin regulation and long-range interactions. *Adv. Gen.* 61, 107–142.

A review of the role of the beta-globin LCR in controlling long-range, developmentally-regulated activation of globin genes in the beta-globin locus.

25

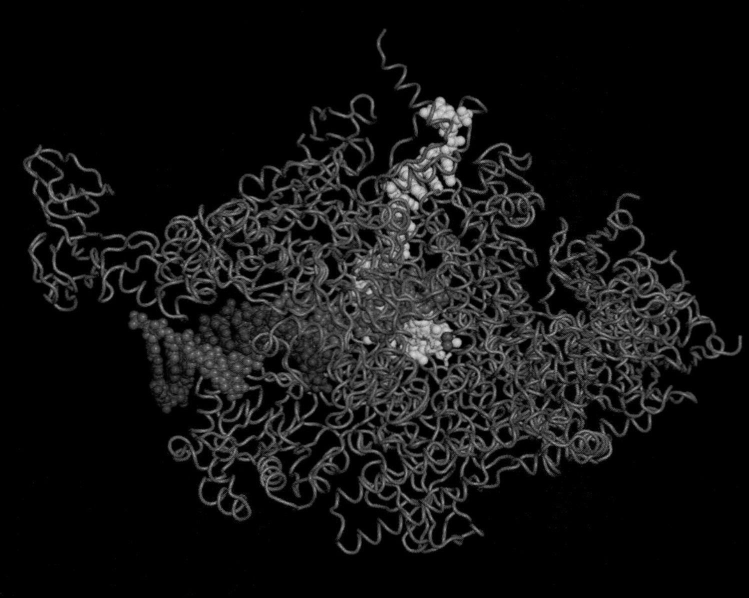

Eukaryotic Transcription

A structure of RNA polymerase in the act of transcription, showing the polymerase itself (green), the DNA template (red and blue), and the RNA transcript (yellow). The incoming nucleotide is shown in cyan. Photo courtesy of Irina Artsimovitch, Ohio State University.

CHAPTER OUTLINE

25.1 Introduction

Initiation of transcription on a chromatin template that is already opened requires the enzyme RNA polymerase to bind at the promoter and transcription factors to bind to enhancers. *In vitro* transcription on a DNA template requires a different subset of transcription factors than are needed to transcribe a chromatin template (we will examine how chromatin is opened in Chapter 26). Any protein that is needed for the initiation of transcription, but that is not itself part of RNA polymerase, is defined as a transcription factor. Many transcription factors act by recognizing *cis*-acting sites on DNA. Binding to DNA, however, is not the only means of action for a transcription factor. A factor may recognize another factor, may recognize RNA polymerase, or may be incorporated into an initiation complex only in the presence of several other proteins. The ultimate test for membership in the transcription apparatus is functional: a protein must be needed for transcription to occur at a specific promoter or set of promoters.

A significant difference between the transcription of eukaryotic and prokaryotic RNAs is that in bacteria, transcription takes place on a DNA template, whereas in eukaryotes, transcription takes place on a chromatin template. Chromatin changes everything and must be taken into account at every step. The chromatin must be in an open structure and accessible before RNA polymerase can bind.

A second major difference is that the bacterial RNA polymerase, with its sigma factor subunit, can read the DNA sequence to find and bind to its promoter. A eukaryotic RNA polymerase cannot read the DNA. Initiation at eukaryotic promoters, therefore, involves a large number of factors that must prebind to a variety of *cis*-acting elements before the RNA polymerase can bind. These factors are called **basal transcription factors**. The RNA polymerase then binds to this basal transcription factor/DNA complex. This binding region is defined as the **core promoter**, the region containing all the binding sites necessary for RNA polymerase to bind and function. RNA polymerase itself binds around the **startpoint** of transcription, but does not directly contact the extended upstream region of the promoter. By contrast, the bacterial promoters discussed in *Chapter 11, Bacterial Transcription*, are largely defined in terms of the binding site for RNA polymerase in the immediate vicinity of the startpoint.

While bacteria have a single RNA polymerase that transcribes all three major classes of genes, transcription in eukaryotic cells is divided into three classes. Each class is transcribed by a different RNA polymerase:

- RNA polymerase I transcribes 18S/28S rRNA.
- RNA polymerase II transcribes mRNA and a few small RNAs.
- RNA polymerase III transcribes tRNA, 5S ribosomal RNA, and other small RNAs.

This is the picture that we have of the major classes of genes. As we have seen in *Chapter 13, Regulatory RNA*, recent discoveries by whole genome tiling arrays have uncovered a new world of antisense transcripts, intergenic transcripts, and heterochromatin transcripts. We do not yet know anything about the promoters for these classes or their regulation, but we do know that many (possibly most) of these transcripts are produced by RNA polymerase II.

Basal transcription factors are needed for initiation but most are not required subsequently. For the three eukaryotic RNA polymerases, the transcription factors, rather than the RNA polymerases themselves, are responsible for recognizing the promoter DNA sequence. For all eukaryotic RNA polymerases, the basal transcription factors create a structure at the promoter to provide the target that is recognized by the RNA polymerase. For RNA polymerases I and III, these factors are relatively simple, but for RNA polymerase II they form a sizeable group. The basal factors join with RNA polymerase II to form a complex surrounding the startpoint, and they determine the site of initiation. The basal factors together with RNA polymerase constitute the basal transcription apparatus.

- **basal transcription factors** Transcription factors required by RNA polymerase II to form the initiation complex at all RNA polymerase II promoters. Factors are identified as $TF_{II}X$, where X is a letter.
- **core promoter** The shortest sequence at which an RNA polymerase can initiate transcription (typically at a much lower level than that displayed by a promoter containing additional elements). For RNA polymerase II it is the minimal sequence at which the basal transcription apparatus can assemble, and it includes three sequence elements: the Inr, the TATA box and the DPE. It is typically ~40 bp long.
- **startpoint** The position on DNA corresponding to the first base incorporated into RNA.

The promoters for RNA polymerases I and II are (mostly) upstream of the startpoint, but a large number of promoters for RNA polymerase III lie downstream (within the transcription unit) of the startpoint. Each promoter contains characteristic sets of short conserved sequences that are recognized by the appropriate class of basal transcription factors. RNA polymerases I and III each recognize a relatively restricted set of promoters, and rely upon a small number of accessory factors.

Promoters utilized by RNA polymerase II show much more variation in sequence, and have a modular organization. All RNA polymerase II promoters have sequence elements close to the startpoint that are bound by the basal apparatus and the polymerase to establish the site of initiation. Other sequences farther upstream (or downstream), called **enhancer** sequences, determine whether the promoter is expressed, and if expressed, whether this occurs in all cell types or is cell type-specific. An enhancer is another type of site involved in transcription and is identified by sequences that stimulate initiation, but that are located a variable distance from the core promoter. Enhancer elements are often targets for tissue-specific or temporal regulation. Some enhancers bind transcription factors that function by short-range interactions and are located near the promoter, whereas others can be located thousands of base pairs away. FIGURE 25.1 illustrates the general properties of promoters and enhancers. A regulatory site that binds more negative regulators than positive regulators to control transcription is called a **silencer**.

▸ **enhancer** A *cis*-acting sequence that increases the utilization of (most) eukaryotic promoters, and can function in either orientation and in any location (upstream or downstream) relative to the promoter.

▸ **silencer** A short sequence of DNA that can inactivate expression of a gene in its vicinity.

▸ **housekeeping genes** Genes that are (theoretically) expressed in all cells because they provide basic functions needed for sustenance of all cell types.

Promoters that are constitutively expressed and needed in all cells (their genes are sometimes called **housekeeping genes**) have upstream sequence elements that are recognized by ubiquitous activators. No one element/factor combination is an essential component of the promoter, which suggests that initiation by RNA polymerase II may be regulated in many different ways. Promoters that are expressed only in certain times or places have sequence elements that require activators that are available only at those times or places.

The components of an enhancer or silencer resemble those of the promoter, in that they consist of a variety of modular elements that can bind positive regulators or negative regulators in a closely packed array. Enhancers do not need to be near the promoter. They can be upstream, inside a gene, or beyond the end of a gene. Proteins bound at enhancer elements interact with proteins bound at promoter elements, very often through intermediates called **coactivators**.

▸ **coactivator** Factors required for transcription that do not bind DNA, but are required for (DNA-binding) activators to interact with the basal transcription factors.

Eukaryotic transcription is most often under positive regulation: A transcription factor is provided under tissue-specific control to activate a promoter or set of promoters that contain a common target sequence. This is a multistep process that first involves opening the chromatin and then binding the basal transcription factors, and then binding the polymerase. Regulation by specific repression of a target promoter is less common.

A eukaryotic transcription unit generally contains a single gene, and termination occurs beyond the end of the coding region. Termination lacks the regulatory impor-

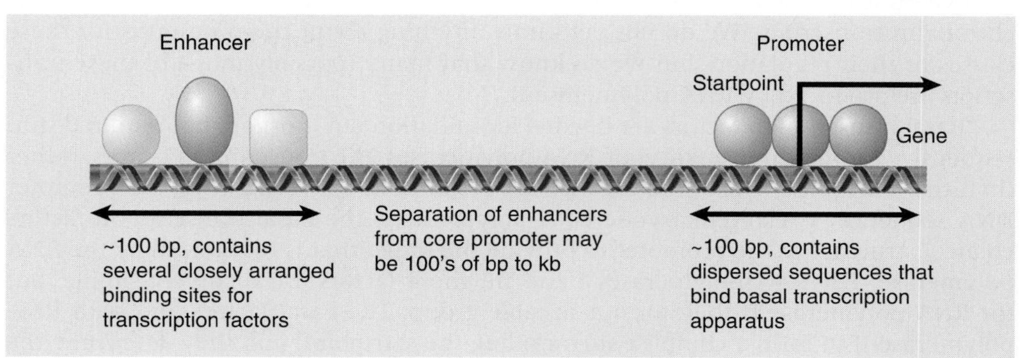

FIGURE 25.1 A typical gene transcribed by RNA polymerase II has a promoter that extends upstream from the site where transcription is initiated. The promoter contains several short (~10 bp) sequence elements that bind transcription factors, dispersed over ~100 bp. An enhancer containing a more closely packed array of elements that also bind transcription factors may be located several hundred bp to several kb distant. (DNA may be coiled or otherwise rearranged so that transcription factors at the promoter and at the enhancer interact to form a large protein complex.)

tance that applies in prokaryotic systems. RNA polymerases I and III terminate at discrete sequences in defined reactions, but the mode of termination by RNA polymerase II is not clear. The significant event in generating the 3′ end of an mRNA, however, is not the termination event itself, but instead results from a cleavage reaction in the primary transcript (see *Chapter 28, RNA Splicing and Processing*).

KEY CONCEPT

- Chromatin must be opened before RNA polymerase can bind the promoter.

CONCEPT AND REASONING CHECK

What are the two major differences between bacterial and eukaryotic transcription?

25.2 Eukaryotic RNA Polymerases Consist of Many Subunits

The three eukaryotic RNA polymerases have different locations in the nucleus that correspond with the different genes that they transcribe.

The most prominent activity is the enzyme RNA polymerase I, which resides in the nucleolus and is responsible for transcribing the genes coding for the 18S and 28S rRNA. It accounts for most cellular RNA synthesis (in terms of quantity).

The other major enzyme is RNA polymerase II, which is located in the nucleoplasm (the part of the nucleus excluding the nucleolus). It represents most of the remaining cellular activity and is responsible for synthesizing most of the **heterogeneous nuclear RNA (hnRNA)**, the precursor for most mRNA and a lot more. The classical definition was that hnRNA includes everything but rRNA and tRNA in the nucleus (again, classically, mRNA is only found in the cytoplasm). With modern molecular tools, we can now look a little closer at hnRNA and find many low abundance RNAs that are very important, plus a lot that we are just now starting to understand. The mRNA is the least abundant of the three major RNAs, accounting for just 2%–5% of the cytoplasmic RNA.

▸ **heterogeneous nuclear RNA (hnRNA)** RNA that comprises transcripts of nuclear genes made primarily by RNA polymerase II; it has a wide size distribution and variable stability.

▸ **carboxy-terminal domain (CTD)** The domain of eukaryotic RNA polymerase II that is phosphorylated at initiation and is involved in coordinating several activities with transcription.

RNA polymerase III is a minor enzyme in terms of activity, but it produces a collection of stable, essential RNAs. This nucleoplasmic enzyme synthesizes the 5S rRNA, tRNAs and other small RNAs that constitute over a quarter of the cytoplasmic RNAs.

All eukaryotic RNA polymerases are large proteins, appearing as aggregates of ~500 kD. They typically have ~12 subunits. The purified enzyme can undertake template-dependent transcription of RNA, but is not able to initiate selectively at promoters. The general constitution of a eukaryotic RNA polymerase II enzyme as typified in *Saccharomyces cerevisiae* is illustrated in **FIGURE 25.2**. The two largest subunits are homologous to the β and β′ subunits of bacterial RNA polymerase. Three of the remaining subunits are common to all the RNA polymerases; that is, they are also components of RNA polymerases I and III. Note that there is no subunit related to the bacterial sigma factor. Its function is contained in the basal transcription factors.

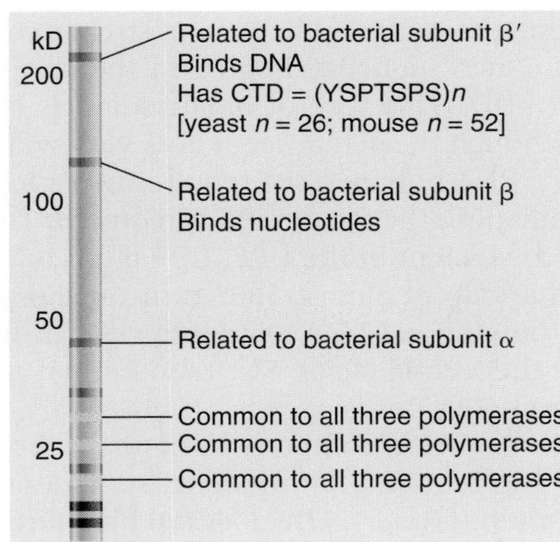

FIGURE 25.2 Some subunits are common to all classes of eukaryotic RNA polymerases and some are related to bacterial RNA polymerase. This figure is an illustration of purified RNA and polymerase II subunits run on an SIDS gel.

The largest subunit in RNA polymerase II has a **carboxy-terminal domain (CTD)**, which consists of multiple repeats of a consensus sequence of seven amino acids. The sequence is unique to RNA polymerase II. There are ~26 repeats in yeast and ~50 in mammals. The number of repeats is important because deletions that remove (typically) more than half of the repeats are lethal (in yeast). The CTD can be highly phosphorylated on serine or threonine residues. The CTD is involved in regulating the initiation reaction (see *Section 25.8, Initiation Is Followed by Promoter Clearance and Elongation*), transcription elongation, and all aspects of mRNA processing, even export of mRNA to the cytoplasm.

The RNA polymerases of mitochondria and chloroplasts are smaller, and they resemble bacterial RNA polymerase rather than any of the nuclear enzymes (since they evolved from eubacteria). Of course, the organelle genomes are much smaller, the resident polymerase needs to transcribe relatively few genes, and the control of transcription is likely to be very much simpler (if it exists at all).

KEY CONCEPTS

- RNA polymerase I synthesizes rRNA in the nucleolus.
- RNA polymerase II synthesizes mRNA in the nucleoplasm.
- RNA polymerase III synthesizes small RNAs in the nucleoplasm.
- All eukaryotic RNA polymerases have ~12 subunits and are aggregates of ~500 kD.
- Some subunits are common to all three RNA polymerases.
- The largest subunit in RNA polymerase II has a CTD (carboxy-terminal domain) consisting of multiple repeats of a heptamer.

CONCEPT AND REASONING CHECK

Explain the apparent paradox that ribosomal genes number only a few hundreds, yet they account for about 75% of the total cytoplasmic RNA.

25.3 RNA Polymerase I Has a Bipartite Promoter

▸ **nontranscribed spacer** The region between transcription units in a tandem gene cluster.

RNA polymerase I transcribes from a single type of promoter only the genes for ribosomal RNA. The precursor transcript includes the sequences of both large 28S and small 18S rRNAs, which are later processed by cleavages and modifications. There are many copies of the transcription unit. They alternate with **nontranscribed spacers** and are organized in a cluster as discussed in *Section 6.7, Genes for rRNA Form Tandem Repeats Including an Invariant Transcription Unit*. The organization of the promoter, and the events involved in initiation, are illustrated in **FIGURE 25.3**. RNA polymerase I exists as a holoenzyme that contains additional factors required for initiation, and is recruited by its transcription factors directly as a giant complex to the promoter.

The promoter consists of two separate regions. The core promoter surrounds the startpoint, extending from –45 to 20, and is sufficient for transcription to initiate. It is generally G-C-rich (unusual for a promoter), except for the only conserved sequence element, a short A-T-rich sequence around the startpoint. The core promoter's efficiency, however, is very much increased by the upstream promoter element (UPE, sometimes also called the upstream control element, or UCE). The UPE is another G-C-rich sequence related to the core promoter sequence, and extends from –180 to –107. This type of organization is common to pol I promoters in many species, although the actual sequences vary widely.

RNA polymerase I requires two ancillary transcription factors. For high frequency initiation, the factor UBF is required. This is a single polypeptide that binds to a G-C-rich element in the UPE. UBF binds to the minor groove of DNA and wraps the DNA in a loop of almost 360° turn on the protein surface, with the result that the core promoter and UPE come into close proximity. This enables UBF to stimulate binding of the second factor, SL1 (also known as TIF-IB, Rib1 in different species), to the core promoter.

▸ **TATA-binding protein (TBP)** The subunit of transcription factor TFIID that binds to the TAT box in the promoter and is positioned at the promoters that do not contain a TATA box by other factors.

One of its components of SL1 is the **TATA-binding protein** (**TBP**), a factor that also is required for initiation by RNA polymerases II and III (see *Section 25.6, TBP Is a Universal Factor*). TBP does not bind directly to G-C-rich DNA, and DNA binding is the

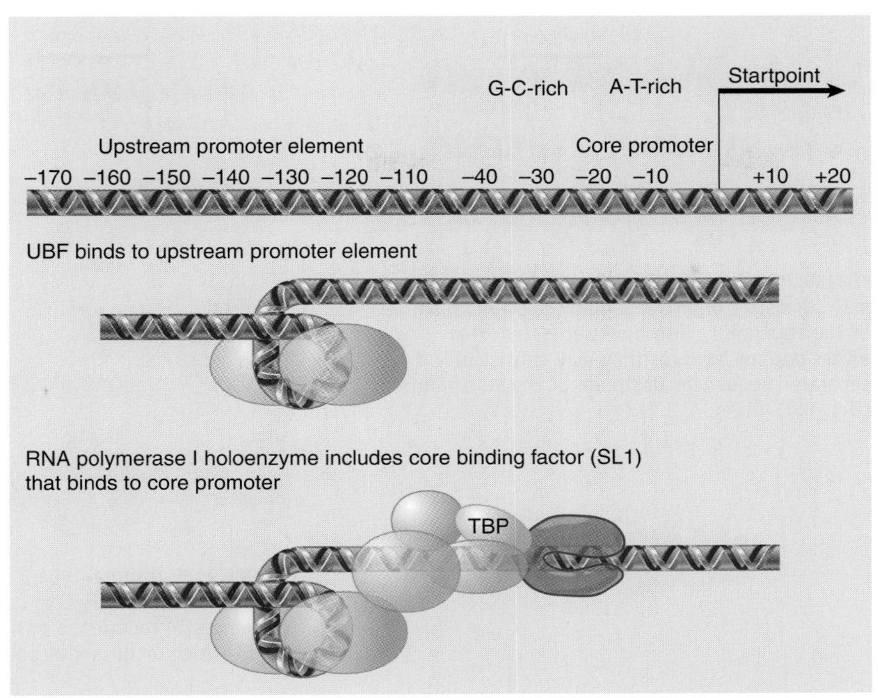

FIGURE 25.3 Transcription units for RNA polymerase I have a core promoter separated by ~70 bp from the upstream promoter element. UBF binding to the UPE increases the ability of core-binding factor to bind to the core promoter. Core-binding factor (SL1) positions RNA polymerase I at the startpoint.

responsibility of the other components of SL1. It is likely that TBP interacts with RNA polymerase, probably with a common subunit or a feature that has been conserved among polymerases. SL1 enables RNA polymerase I to initiate from the promoter at a low basal frequency. SL1 has primary responsibility for ensuring that the RNA polymerase is properly localized at the startpoint. It consists of four proteins, one of which (TBP) is a component of "positioning factors" that are also required by RNA polymerases II and III. Its exact mode of action is different in each of the positioning factors; at the promoter for RNA polymerase I, it does not bind DNA, whereas at the promoter for RNA polymerase II it is the principal means for locating the factor on DNA.

KEY CONCEPTS

- The RNA polymerase I promoter consists of a core promoter and an upstream promoter element (UPE).
- The factor UBF1 wraps DNA around a protein structure to bring the core and UPE into proximity.
- SL1 includes the factor TBP that is involved in initiation by all three RNA polymerases.
- RNA polymerase I binds to the UBF1-SL1 complex at the core promoter.

CONCEPT AND REASONING CHECK

Why would there be a conserved A-T rich element in the core promoter?

25.4 RNA Polymerase III Uses Both Downstream and Upstream Promoters

Recognition of promoters by RNA polymerase III strikingly illustrates the relative roles of transcription factors and the polymerase enzyme. The promoters fall into two general classes that are recognized in different ways by different groups of factors. The promoters for 5S and tRNA genes are internal; they lie downstream of the startpoint.

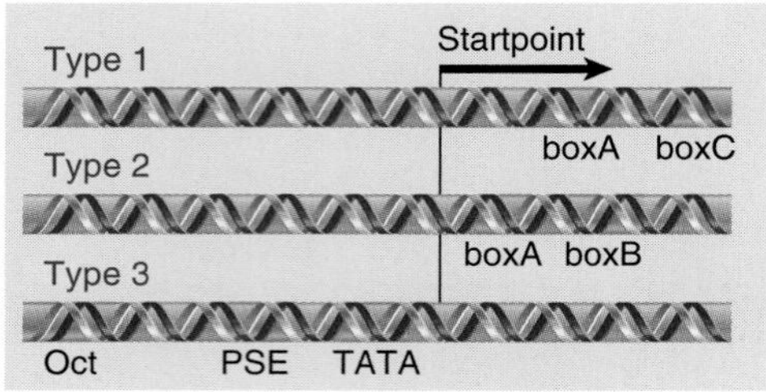

FIGURE 25.4 Promoters for RNA polymerase III may consist of bipartite sequences downstream of the startpoint, with *boxA* separated from either *boxC* or *boxB*, or they may consist of separated sequences upstream of the startpoint (Oct, PSE, TATA).

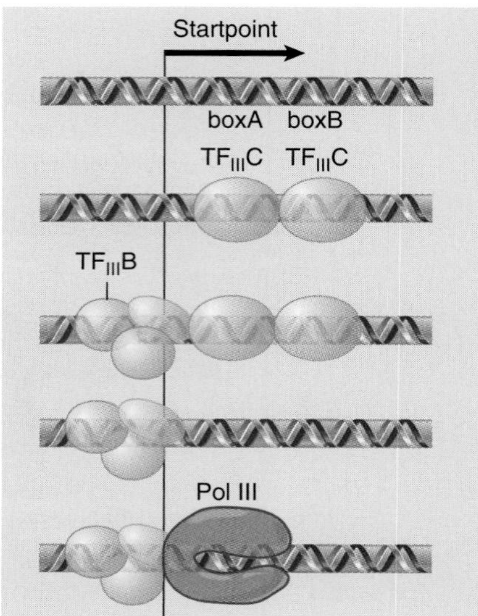

FIGURE 25.5 Internal type 2 pol III promoters use binding of $TF_{III}C$ to *boxA* and *boxB* sequences to recruit the positioning factor $TF_{III}B$, which recruits RNA polymerase III.

▸ **assembly factors** Proteins that are required for formation of a macromolecular structure but are not themselves part of that structure.

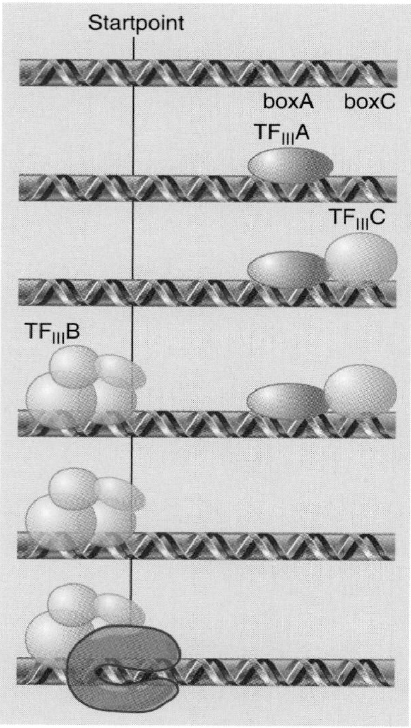

FIGURE 25.6 Internal type 1 pol III promoters use the assembly factors $TF_{III}A$ and $TF_{III}C$, at *boxA* and *boxC*, to recruit the positioning factor $TF_{III}B$, which recruits RNA polymerase III.

The promoters for snRNA (small nuclear RNA) genes lie upstream of the startpoint in the more conventional manner of other promoters. In both cases, the individual elements that are necessary for promoter function consist exclusively of sequences recognized by transcription factors, which in turn direct the binding of RNA polymerase.

The structures of three types of promoter for RNA polymerase III are summarized in **FIGURE 25.4**. There are two types of internal promoter. Each contains a bipartite structure, in which two short sequence elements are separated by a variable sequence. The 5S ribosomal gene Type 1 promoter consists of a *boxA* sequence separated from a *boxC* sequence, and the tRNA type 2 promoter consists of a *boxA* sequence separated from a *boxB* sequence. A common group of type 3 promoters coding for other small RNAs have three sequence elements that are all located upstream of the startpoint.

The detailed interactions are different at the two types of internal promoter, but the principle is the same. $TF_{III}C$ binds downstream of the startpoint, either independently (tRNA type 2 promoters) or in conjunction with $TF_{III}A$ (5S type 1 promoters). The presence of $TF_{III}C$ enables the positioning factor $TF_{III}B$ to bind at the startpoint. RNA polymerase is then recruited.

FIGURE 25.5 summarizes the stages of reaction at type 2 internal promoters used for tRNA genes. The distance between *boxA* and *boxB* can vary since many tRNA genes contain a small intron. $TF_{III}C$ binds to both *boxA* and *boxB*. This enables $TF_{III}B$ to bind at the startpoint. At this point RNA polymerase III can bind.

The difference at type 1 internal promoters (for 5S genes) is that $TF_{III}A$ must bind at *boxA* to enable $TF_{III}C$ to bind at *boxC*. $TF_{III}A$ is a 5S sequence-specific binding factor that binds to the promoter and to the 5S RNA as a chaperone and gene regulator. **FIGURE 25.6** shows that once $TF_{III}C$ has bound, events follow the same course as at type 2 promoters, with $TF_{III}B$ (which contains the ubiquitous TBP) binding at the startpoint, and RNA polymerase III joining the complex. Type 1 promoters are found only in the genes for 5S rRNA.

$TF_{III}A$ and $TF_{III}C$ are **assembly factors**, whose sole role is to assist the binding of the positioning factor $TF_{III}B$ at the correct location. Once $TF_{III}B$ has bound, $TF_{III}A$ and $TF_{III}C$ can be removed from the promoter without affecting the initiation reaction. $TF_{III}B$ remains bound in the vicinity of the startpoint, and its presence is sufficient to allow RNA polymerase III to identify and bind at the startpoint. Thus $TF_{III}B$ is the only true initiation factor required by RNA polymerase III. This sequence of events explains how the promoter boxes downstream can cause RNA polymerase to bind at the startpoint, farther upstream. Although the ability to transcribe these genes is conferred by the internal promoter, changes in the region immediately upstream of the startpoint can alter the efficiency of transcription.

The upstream region has a conventional role in the third class of polymerase III promoters. In the example shown in Figure 25.5, there are three upstream elements. These elements are also found in promoters for snRNA genes that are transcribed by RNA polymerase II. (Genes for some snRNAs are transcribed by RNA polymerase II, whereas others are transcribed by RNA polymerase III.) The upstream elements function in a similar manner in promoters for both RNA polymerases II and III.

Initiation at an upstream promoter for RNA polymerase III can occur on a short region that immediately precedes the startpoint and contains only the TATA element. Efficiency of transcription, however, is much increased by the presence of the enhancer PSE (proximal sequence element) and OCT (named because it has an 8 base pair binding sequence) elements. The factors that bind at these elements interact cooperatively. The TATA element confers specificity for the type of polymerase (II or III) that is recognized by an snRNA promoter. It is bound by a factor that includes TBP, which actually recognizes the sequence in DNA. TBP is associated with other proteins, which are specific for the type of promoter. The function of TBP and its associated proteins is to position the RNA polymerase correctly at the startpoint. We discuss this in more detail for RNA polymerase II (see *Section 25.6, TBP Is a Universal Factor*).

The factors work in the same way for both types of promoters for RNA polymerase III. The factors bind at the promoter before RNA polymerase itself can bind. They form a **preinitiation complex** that directs binding of the RNA polymerase. RNA polymerase III does not itself recognize the promoter sequence, but binds adjacent to factors that are themselves bound just upstream of the startpoint. In all cases, the chromatin must be modified and in an open configuration.

▸ **preinitiation complex** The assembly of transcription factors at the promoter before RNA polymerase binds in eukaryotic transcription.

KEY CONCEPTS

- RNA polymerase III has two types of promoters.
- Internal promoters have short consensus sequences located within the transcription unit and cause initiation to occur at a fixed distance upstream.
- Upstream promoters contain three short consensus sequences upstream of the startpoint that are bound by transcription factors.
- $TF_{III}A$ and $TF_{III}C$ bind to the consensus sequences and enable $TF_{III}B$ to bind at the startpoint.
- $TF_{III}B$ has TBP as one subunit and enables RNA polymerase to bind.

CONCEPT AND REASONING CHECK

The tRNA and 5S gene promoters are inside the gene. How can the polymerase find the startpoint at the beginning of the gene?

25.5 The Startpoint for RNA Polymerase II

The basic organization of the apparatus for transcribing protein-coding genes was revealed by the discovery that purified RNA polymerase II can catalyze synthesis of mRNA, but cannot initiate transcription unless an additional extract is added. The purification of this extract led to the definition of the general transcription factors, or *basal transcription factors*—a group of proteins that are needed for initiation by RNA polymerase II at all promoters. RNA polymerase II in conjunction with these factors constitutes the basal transcription apparatus that is needed to transcribe any promoter. The general factors are described as $TF_{II}X$, where "X" is a letter that identifies

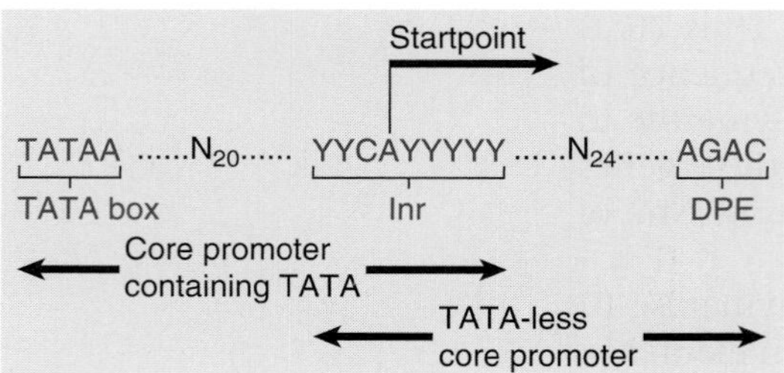

FIGURE 25.7 The minimal pol II promoter may have a TATA box ~25 bp upstream of the Inr. The TATA box has the consensus sequence of TATAA. The Inr has pyrimidines (Y) surrounding the CA at the startpoint. The DPE is downstream of the startpoint. The sequence shows the coding strand.

the individual factor. The subunits of RNA polymerase II and the general transcription factors are conserved among eukaryotes.

Our starting point for considering promoter organization is to define the *core promoter* as the shortest sequence at which RNA polymerase II can initiate transcription. A core promoter can in principle be expressed in any cell. It is the minimum sequence that enables the general transcription factors to assemble at the startpoint. These factors are involved in the mechanics of binding to DNA and enable RNA polymerase II to initiate transcription. A core promoter functions at only a low efficiency. Other proteins, called *activators*, another class of transcription factors, are required for a proper level of function (see *Section 25.9, Enhancers Contain Bidirectional Elements That Assist Initiation*). The activators are not described systematically, but have casual names reflecting their histories of identification.

We may expect any sequence components involved in the binding of RNA polymerase and general transcription factors to be conserved at most or all promoters. As with bacterial promoters, when promoters for RNA polymerase II are compared, homologies in the regions near the startpoint are restricted to rather short sequences. These elements correspond with the sequences implicated in promoter function by mutation. **FIGURE 25.7** shows the construction of a typical pol II core promoter.

At the startpoint, there is no extensive homology of sequence, but there is a tendency for the first base of mRNA to be A, flanked on either side by pyrimidines. (This description is also valid for the CAT start sequence of bacterial promoters.) This region is called the **initiator** (**Inr**), and may be described in the general form Py_2CAPy_5, where Py stands for any pyrimidine. The Inr is contained between positions –3 and 5.

Many promoters have a sequence called the **TATA box**, usually located ~25 bp upstream of the startpoint in higher eukaryotes. It constitutes the only upstream promoter element that has a relatively fixed location with respect to the startpoint. The core sequence is TATAA, usually followed by three more A-T base pairs. The TATA box tends to be surrounded by G-C-rich sequences, which could be a factor in its function. It is almost identical with the –10 TATA box sequence found in bacterial promoters; in fact, it could pass for one except for the difference in its location at –25 instead of –10 (except in yeast, where the TATA box is more typically found at –90).

Promoters that do not contain a TATA element are called **TATA-less promoters**. Surveys of promoter sequences suggest that 50% or more of promoters may be TATA-less. When a promoter does not contain a TATA box, it often contains another element, the **DPE** (**downstream promoter element**), which is located at +28 to 32.

Most core promoters consist either of a TATA box plus Inr, or of an Inr plus DPE, although other combinations exist as well.

- **initiator (Inr)** The sequence of a pol II promoter between –3 and +5 and has the general sequence Py2CAPy5. It is the simplest possible pol II promoter.
- **TATA box** A conserved A-T-rich octamer found about 25 bp before the startpoint of each eukaryotic RNA polymerase II transcription unit; it is involved in positioning the enzyme for correct initiation.
- **TATA-less promoter** It does not have a TATA box in the sequence upstream of its startpoint.
- **downstream promoter element (DPE)** A common component of RNA polymerase II promoters that do not contain a TATA box.

KEY CONCEPTS

- RNA polymerase II requires general transcription factors (called $TF_{II}X$) to initiate transcription.
- RNA polymerase II promoters have a short conserved sequence Py_2CAPy_5 (the initiator Inr) at the startpoint.
- The TATA box is a common component of RNA polymerase II promoters and consists of an A-T-rich octamer located ~25 bp upstream of the startpoint.
- The DPE is a common component of RNA polymerase II promoters that do not contain a TATA box.
- A core promoter for RNA polymerase II includes the Inr and commonly either a TATA box or a DPE.

CONCEPT AND REASONING CHECK

Why might RNA polymerase II promoters be more variable than RNA polymerase I promoters?

25.6 TBP Is a Universal Factor

Before transcription initiation can begin, the chromatin has to be modified and remodeled to the open configuration, and any nucleosome octamer positioned over the promoter has to be moved or removed at all classes of eukaryote promoters (we will examine this aspect of transcription control more closely in Chapter 26). Then it is possible for a positioning factor to bind to the promoter. Each class of RNA polymerase is assisted by a positioning factor that contains TBP associated with other components. The name TBP has an interesting history. It was initially so named because it was a protein that bound to the TATA box in RNA polymerase II genes. It was subsequently discovered to also be part of the positioning factors for both RNA polymerase I (see *Section 25.3, RNA Polymerase I Has a Bipartite Promoter*) and RNA polymerase III (see *Section 25.4, RNA Polymerase III Uses Both Downstream and Upstream Promoters*). For these latter two RNA polymerases, TBP does not recognize the TATA box sequence; thus, the name is misleading. In addition, many RNA polymerase II promoters lack TATA boxes but still require the presence of TBP.

▸ **$TF_{II}D$** The transcription factor that binds to the TATA sequence upstream of the startpoint of promoters for RNA polymerase II. It consists of TBP (TATA binding protein) and the TAF subunits that bind to TBP.

▸ **TAFs** The subunits of TFIID that assist TBP in binding to DNA. They also provide points of contact for other components of the transcription apparatus.

The positioning factor for RNA polymerase I is SL1 and for RNA polymerase III is $TF_{III}B$. For RNA polymerase II, the positioning factor is **$TF_{II}D$**, which consists of TBP associated with ~11 other subunits called **TAFs** (for TBP-associated factors). Some TAFs are stoichiometric with TBP; others are present in lesser amounts. $TF_{II}Ds$ containing different TAFs could recognize different promoters. Some TAFs are tissue-specific. The total mass of $TF_{II}D$ typically is ~800 kD. The TAFs in $TF_{II}D$ were originally named in the form $TAF_{II}00$, for example, where the number "00" gives the molecular mass of the subunit. Recently, the RNA polymerase II TAFs have been renamed TAF1, TAF2, etc.; in this nomenclature TAF1 is the largest TAF, TAF2 is the next largest, and homologous TAFs in different species thus have the same names.

FIGURE 25.8 shows that the positioning factor recognizes the promoter in a different way in each case. At promoters for RNA polymerase III, $TF_{III}B$ binds adjacent to $TF_{III}C$. At promoters for RNA polymerase I, SL1 binds in conjunction with UBF. $TF_{II}D$ is solely responsible for recognizing promoters for RNA polymerase II. At a promoter that has a TATA element, TBP binds specifically to the TATA box; but, at TATA-less promoters, it may be incorporated by association with other factors that bind to DNA first. Whatever its means of entry into the initiation complex, it has the common purpose of interaction with the RNA polymerase.

TBP has the unusual property of binding to DNA in the minor groove. (The vast majority of DNA-binding proteins bind in the major groove.) The crystal structure of TBP suggests a detailed model for its binding to DNA. **FIGURE 25.9** shows that it surrounds one face of DNA, forming a "saddle" around a stretch of the minor groove, which is bent to fit into this saddle. In effect, the inner surface of TBP binds to DNA, and the larger outer surface is available to extend contacts to other proteins. The DNA-binding site consists of a C-terminal domain that is conserved between species, and the variable N-terminal tail is exposed to interact with other proteins. It is a measure of the conservation of mechanism in transcriptional initiation that the DNA-binding sequence of TBP is 80% conserved between yeast and humans.

Binding of TBP may be inconsistent with the presence of nucleosome octamers. Nucleosomes form preferentially by placing A-T-rich sequences with the minor grooves facing inward; as a result, they could prevent binding of TBP. This may explain why the presence of a nucleosome at the promoter prevents initiation of transcription.

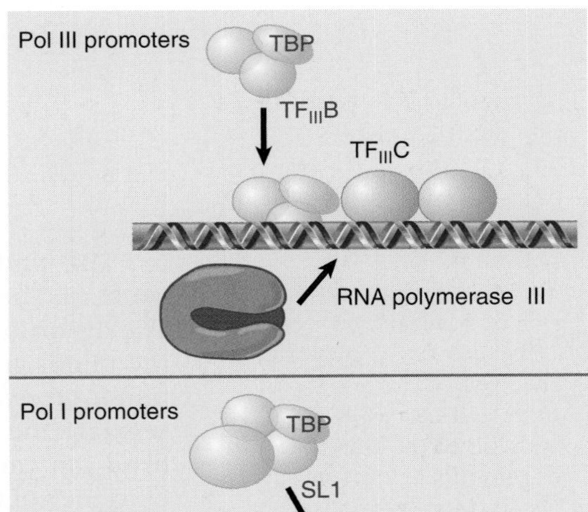

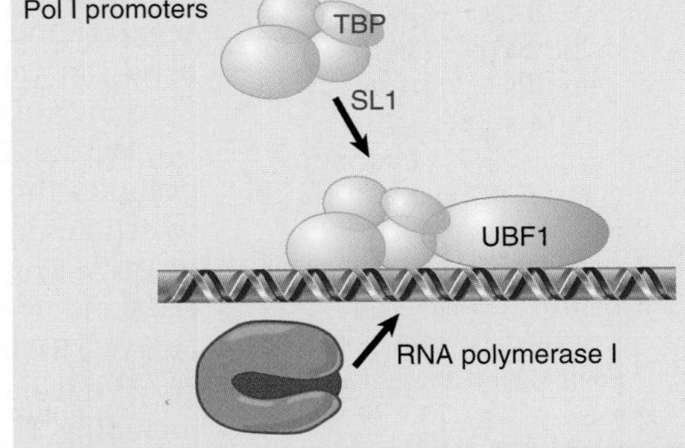

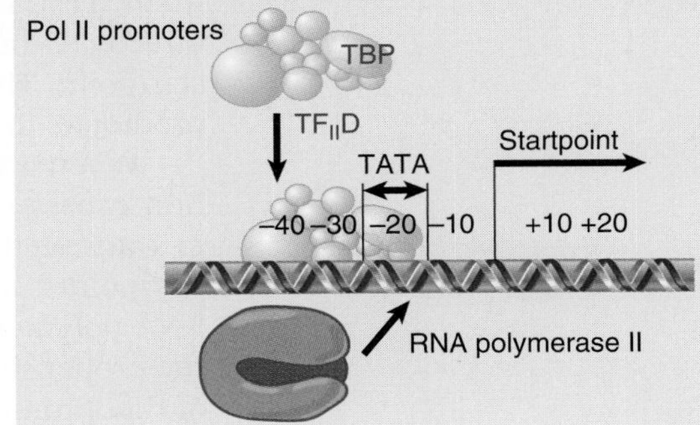

FIGURE 25.8 RNA polymerases are positioned at all promoters by a factor that contains TBP.

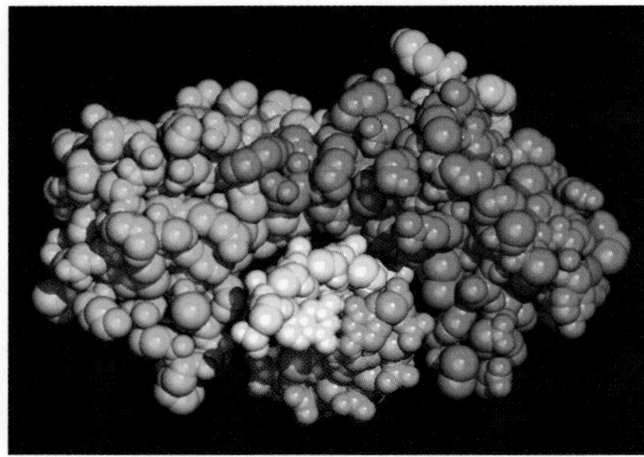

FIGURE 25.9 A view in cross-section shows that TBP surrounds DNA from the side of the narrow groove. TBP consists of two related (40% identical) conserved domains, which are shown in light and dark blue. The N-terminal region varies extensively and is shown in green. The two strands of the DNA double helix are in light and dark gray. Photo courtesy of Stephen K. Burley.

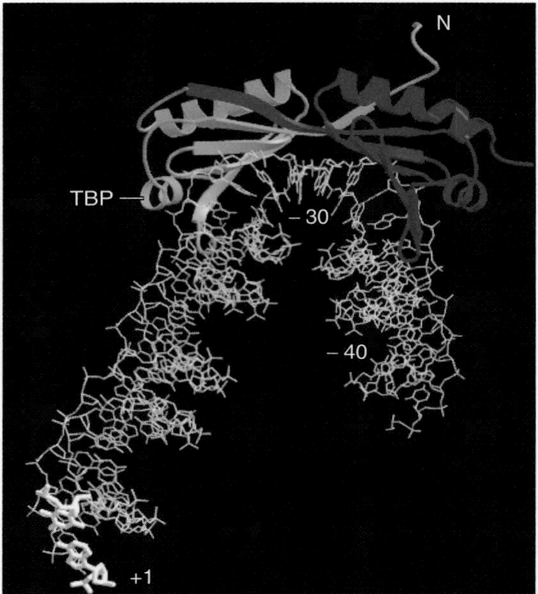

FIGURE 25.10 The cocrystal structure of TBP with DNA from −40 to the startpoint shows a bend at the TATA box that widens the narrow groove where TBP binds. Photo courtesy of Stephen K. Burley.

TBP binds to the minor groove and bends the DNA by ~80°, as illustrated in **FIGURE 25.10**. The TATA box bends toward the major groove, widening the minor groove. The distortion is restricted to the 8 bp of the TATA box; at each end of the sequence, the minor groove has its usual width of ~5 Å, but at the center of the sequence the minor groove is >9 Å. This is a deformation of the structure, but does not actually separate the strands of DNA because base pairing is maintained. The extent of the bend can vary with the exact sequence of the TATA box, and is correlated with the efficiency of the promoter.

Within $TF_{II}D$ as a free protein complex, the factor TAF1 binds to TBP, where it occupies the concave DNA-binding surface. In fact, the structure of the binding site, which lies in the N-terminal domain of TAF1, mimics the surface of the minor groove in DNA. This molecular mimicry allows TAF1 to control the ability of TBP to bind to DNA; the N-terminal domain of TAF1 must be displaced from the DNA-binding surface of TBP in order for $TF_{II}D$ to bind to DNA.

What happens at TATA-less promoters? The same general transcription factors, including $TF_{II}D$ are needed. The Inr, when present, can provide the positioning element; $TF_{II}D$ binds to it via the ability of one or more of the TAFs to recognize the Inr directly. Other TAFs in $TF_{II}D$ also recognize the DPE element downstream from the startpoint. The positioning of TBP at TATA-less promoters is more like that at internal promoters for RNA polymerase III.

When a TATA box is present, it determines the location of the startpoint. Its deletion causes the site of initiation to become erratic, although any overall reduction in transcription is relatively small. Indeed, some TATA-less promoters lack unique startpoints, so initiation occurs within a cluster of startpoints. The TATA box aligns the RNA polymerase via the interaction with $TF_{II}D$ and other factors so that it initiates at the proper site. Binding of TBP to TATA is the predominant feature in the recognition of that promoter.

KEY CONCEPTS

- TBP is a component of the positioning factor that is required for each type of RNA polymerase to bind its promoter.
- The factor for RNA polymerase II is $TF_{II}D$, which consists of TBP and ~11 TAFs, with a total mass ~800 kD.
- TBP binds to the TATA box in the minor groove of DNA.
- It forms a saddle around the DNA and bends it by ~80°.

CONCEPT AND REASONING CHECK

Why does initiation at a TATA-less promoter still require the TATA binding factor?

25.7 The Basal Apparatus Assembles at the Promoter

In a cell, gene promoters can be found in three types of chromatin with respect to activity. The first is an inactive gene in closed chromatin. The second is a potentially active gene in open chromatin, called a poised gene. This class may assemble the basal apparatus, but cannot proceed to transcribe the gene without a second signal to start transcription. Heat shock genes are poised so that they can be activated immediately upon a rise in temperature. The third class (which we will examine below) is an active gene in open chromatin. Initiation requires the basal transcription factors to act in a defined order to build a complex that is joined by RNA polymerase. The series of events is summarized in **FIGURE 25.11**. Once a polymerase is bound, its activity then is controlled by enhancer-binding transcription factors.

A promoter for RNA polymerase II often consists of two types of region. The core promoter contains the startpoint itself, typically identified by the Inr, and often includes either the TATA box or DPE close by. The efficiency and specificity with which a promoter is recognized, however, depend upon short sequences farther upstream, which are recognized by a different group of transcription factors, sometimes called *activators*. In general, the target sequences are 100 bp upstream of the startpoint, but sometimes they are more distant. Binding of activators at these sites may influence the formation of the initiation complex at (probably) any one of several stages. Promoters are organized on a principle of "mix and match." A variety of elements can contribute to promoter function, but none is essential for all promoters.

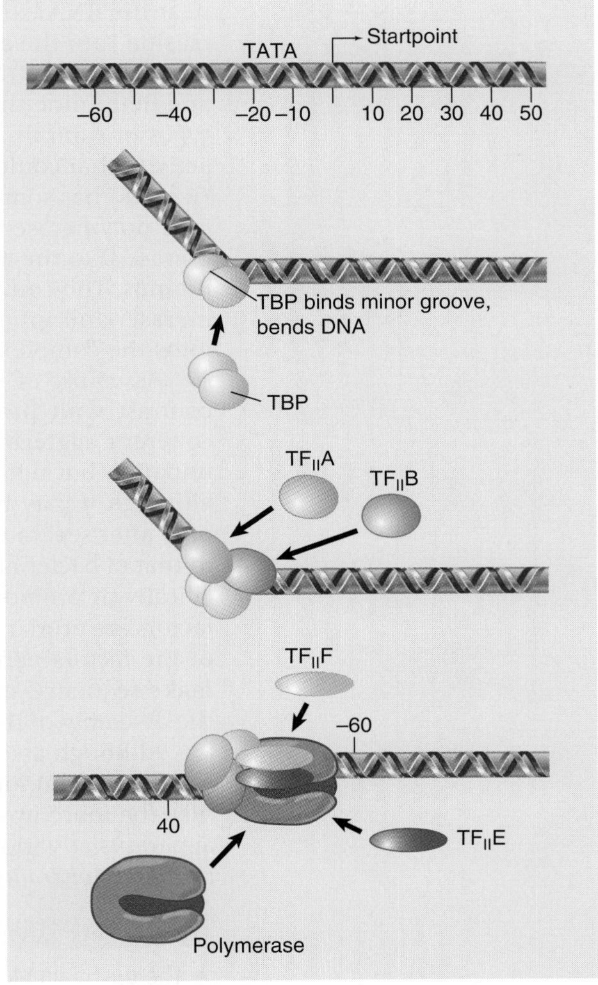

FIGURE 25.11 An initiation complex assembles at promoters for RNA polymerase II by an ordered sequence of association of transcription factors. Note that only TBP (not the entire $TF_{II}D$ complex) is shown for simplicity.

The first step in activating a promoter in open chromatin is initiated when $TF_{II}D$ binds the TATA box. This may be enhanced by upstream elements acting through a coactivator ($TF_{II}D$ also recognizes the Inr sequence at the startpoint.) When $TF_{II}A$ joins the complex, $TF_{II}D$ becomes able to protect a region extending farther upstream. $TF_{II}A$ may activate TBP by relieving the repression that is caused by the TAF1.

$TF_{II}B$ binds downstream of the TATA box, adjacent to TBP, extending contacts along one face of the DNA. It makes contacts in the minor groove downstream of the TATA box, and contacts the major groove upstream of the TATA box in a region called the BRE. In archaea, the homolog of $TF_{II}B$ actually makes

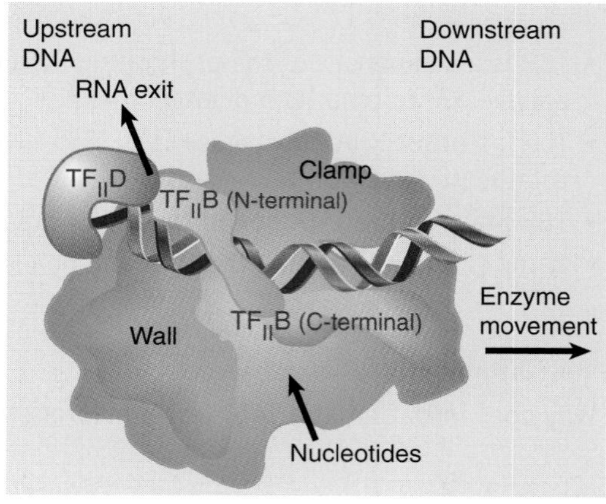

FIGURE 25.12 $TF_{II}B$ binds to DNA and contacts RNA polymerase near the RNA exit site and at the active center, and orients it on DNA. Compare with Figure 25.13, which shows the polymerase structure engaged in transcription.

sequence-specific contacts with the promoter in the BRE region. The schematic of **FIGURE 25.12** shows the relationship between $TF_{II}B$, $TF_{II}D$, and RNA polymerase II. $TF_{II}B$ binds adjacent to $TF_{II}D$, and its N-terminal region contacts RNA polymerase near the RNA exit site. Its C-terminal region extends across the enzyme, with a protrusion into the active site. $TF_{II}B$ determines the path of the DNA where it contacts the factors $TF_{II}E$, $TF_{II}F$, and $TF_{II}H$, which may align them in the basal factor complex and determine the startpoint. The factor $TF_{II}F$ is a heterotetramer consisting of two types of subunit. The larger subunit (RAP74) has an ATP-dependent DNA helicase activity that could be involved in melting the DNA at initiation. The smaller subunit (RAP38) has some homology to the regions of bacterial sigma factor that contact the core polymerase; it binds tightly to RNA polymerase II. $TF_{II}F$ may bring RNA polymerase II to the assembling transcription complex and provide the means by which it binds. The complex of TBP and TAFs may interact with the CTD tail of RNA polymerase, and interaction with $TF_{II}B$ may also be important when $TF_{II}F$/polymerase joins the complex.

Assembly of the RNA polymerase II initiation complex provides an interesting contrast with prokaryotic transcription. Bacterial RNA polymerase is essentially a coherent aggregate with intrinsic ability to bind DNA; the sigma factor, needed for initiation but not for elongation, becomes part of the enzyme before DNA is bound, although it may be later released. RNA polymerase II can bind to the promoter, but only after separate transcription factors have bound. The factors play a role analogous to that of bacterial sigma factor—to allow the basic polymerase to recognize DNA specifically at promoter sequences—but have evolved more independence. Indeed, the factors are primarily responsible for the specificity of promoter recognition. Only some of the factors participate in protein–DNA contacts (and only TBP and certain TAFs make sequence-specific contacts); thus protein–protein interactions are important in the assembly of the complex.

Although assembly can take place just at the core promoter *in vitro*, this reaction is not sufficient for transcription *in vivo*, where interactions with activators that recognize the more upstream elements are required. The activators interact with the basal apparatus at various stages during its assembly (see *Section 26.4, Activators Interact with the Basal Apparatus*).

KEY CONCEPTS

- The upstream elements and the factors that bind to them increase the frequency of initiation.
- Binding of $TF_{II}D$ to the TATA box or Inr is the first step in initiation.
- Other transcription factors bind to the complex in a defined order, extending the length of the protected region on DNA.
- When RNA polymerase II binds to the complex, it initiates transcription.

CONCEPT AND REASONING CHECK

Why are there so many basal transcription factors for RNA polymerase II promoters compared to RNA polymerase I or III promoters?

25.8 Initiation Is Followed by Promoter Clearance and Elongation

Some final steps are needed to release the RNA polymerase from the promoter once the first nucleotide bonds have been formed. This step is called *promoter clearance* and is the key regulated step in determining if a poised gene or an active gene will be transcribed. This step is controlled by enhancers. (Remember, the key step in bacterial transcription is conversion of the closed complex to the open complex; see *Section 11.3, The Transcription Reaction Has Three Stages.*)

The transcription factors that bind enhancers usually do not directly contact elements at the promoter to control it, but rather bind to a coactivator that binds to the promoter elements. The coactivator Mediator is one of the most common coactivators. This is a very large multisubunit protein complex, conserved from yeast to humans, that integrates signals from many transcription factors. Both poised and active genes require the interaction of the transcription factors bound to enhancers with the promoter.

The last factors to join the initiation complex are $TF_{II}E$ and $TF_{II}H$. They act at the later stages of initiation. $TF_{II}H$ is the only general transcription factor that has multiple independent enzymatic activities. Its several activities include an ATPase, helicases of both polarities, and a kinase activity that can phosphorylate the CTD tail of RNA polymerase II. $TF_{II}H$ is an exceptional factor that may also play a role in elongation. Its interaction with DNA downstream of the startpoint is required for RNA polymerase to escape from the promoter. $TF_{II}H$ is also involved in repair of damage to DNA (see *Section 20.3, Nucleotide Excision Repair Systems Repair Several Classes of Damage*).

Phosphorylation of the CTD tail is needed to release RNA polymerase II from the promoter and transcription factors so that it can make the transition to the elongating form as shown in **FIGURE 25.13**. The phosphorylation pattern on the CTD is dynamic during the elongation process, controlled and catalyzed by multiple protein kinases and phosphatases. Most of the basal transcription factors are released from the promoter at this stage.

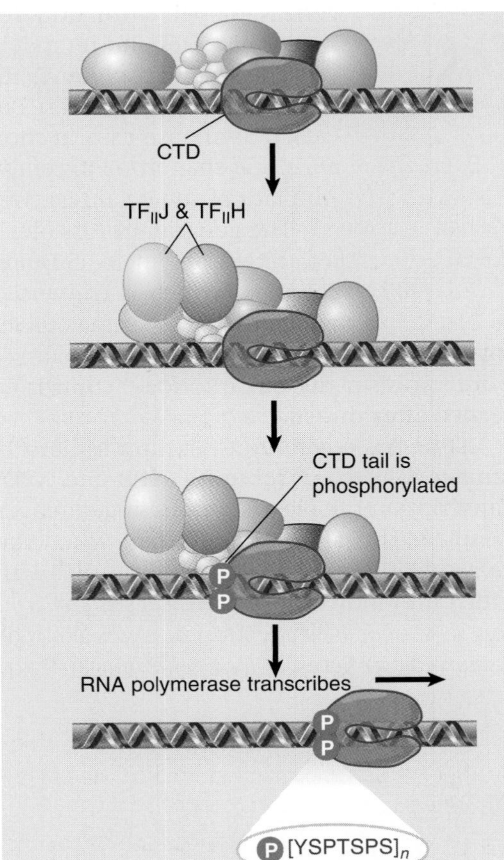

FIGURE 25.13 Phosphorylation of the CTD by the kinase activity of $TF_{II}H$ may be needed to release RNA polymerase to start transcription.

The CTD is involved, directly or indirectly, in processing mRNA while it is being synthesized and after it has been released by RNA polymerase II. Each site of phosphorylation on the CTD serves as a recognition or anchor point for other proteins to dock with the polymerase. **FIGURE 25.14** summarizes processing reactions in which the CTD is involved. The capping enzyme (guanylyl transferase), which adds the G residue to the 5′ end of newly

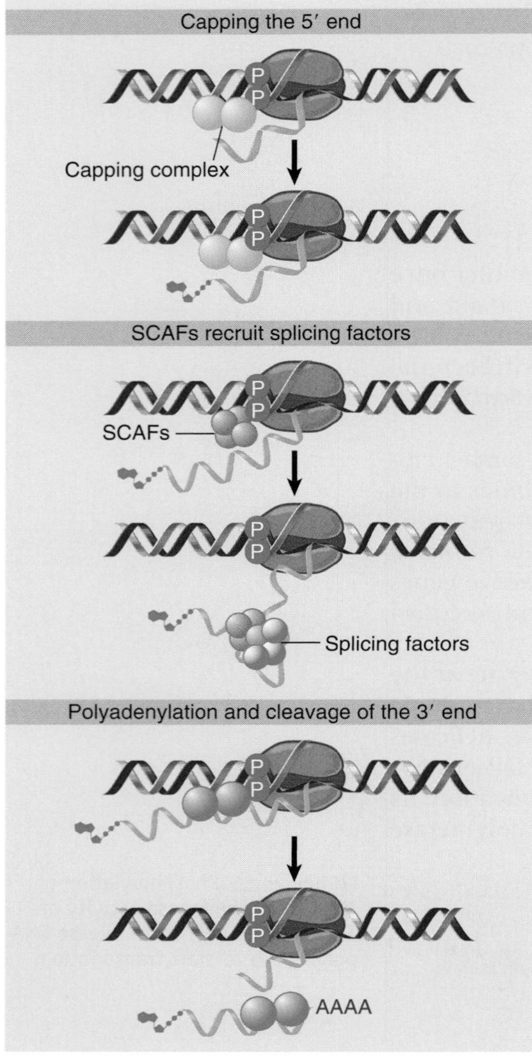

FIGURE 25.14 The CTD is important in recruiting enzymes that modify RNA.

synthesized mRNA, binds to the phosphorylated CTD. This may be important in enabling it to modify (and thus protect) the 5′ end as soon as it is synthesized. A set of proteins called SCAFs bind to the CTD, and they may in turn bind to splicing factors. This may be a means of coordinating transcription and splicing. Some components of the cleavage/polyadenylation apparatus used after transcription termination also bind to the CTD. Oddly enough, they do so at the time of initiation, so that RNA polymerase is ready for the 3′ end processing reactions as soon as it sets out. Export from the nucleus through the nuclear pore is also coordinated by the CTD. All of this suggests that the CTD may be a general focus for connecting other processes with transcription. In the cases of capping and splicing, the CTD functions indirectly to promote formation of the protein complexes that undertake the reactions. In the case of 3′ end generation, it may participate directly in the reaction.

The key event in determining whether a gene will be expressed is *promoter clearance*, release from the promoter. Once that has occurred and initiation factors are released, there is a transition to the elongation phase. The transcription complex now consists of the RNA polymerase II, the basal factors $TF_{II}E$ and $TF_{II}H$, and elongation factors like $TF_{II}S$ to prevent inappropriate pausing and all of the enzymes and factors bound to the CTD. This complex now has to transcribe a chromatin template, through nucleosomes. The whole gene may be in open chromatin, especially if it is not too large, or only the area around the promoter. Some genes, like the Muscular Dystrophy gene (*DMD*), can be megabases in size and require many hours to transcribe. There is a model in which the first polymerase to leave the promoter acts as a pathfinder polymerase. Its major function is to ensure that the entire gene is in open chromatin. It carries with it enzyme complexes to modify the histones and remodel the chromatin.

The second problem is that even open chromatin contains nucleosomes that all polymerases must traverse. The most recent model has each polymerase using a chromatin remodeling complex together with a histone chaperone to remove an H2A/H2B dimer, leaving a hexamer (in place of the octamer), which is easier to temporarily displace.

The general process of initiation is similar to that catalyzed by bacterial RNA polymerase. Binding of RNA polymerase generates a closed complex, which is converted at a later stage to an open complex in which the DNA strands have been separated. In the bacterial reaction, formation of the open complex completes the necessary structural change to DNA; a difference in the eukaryotic reaction is that further unwinding of the template is needed after this stage.

$TF_{II}H$ has a common function in both initiating transcription and repairing DNA damage. The same subunits (XPB and XPD) participate in the initial opening of the transcription bubble and in melting damaged DNA for repair, in different forms of the complex. This allows rapid repair when the RNA polymerase stalls due to damaged DNA on the template strand. Subunits with the name XP are coded for by genes in which mutations cause the disease *xeroderma pigmentosum*, which causes a predisposition to cancer (see the *Medical Applications* box in *Section 20.3, Nucleotide Excision Repair Systems Repair Several Classes of Damage*).

KEY CONCEPTS

- $TF_{II}E$ and $TF_{II}H$ are required to melt DNA to allow polymerase movement.
- Phosphorylation of the CTD is required for elongation to begin.
- Further phosphorylation of the CTD is required at some promoters to end abortive initiation.
- The CTD coordinates processing of RNA with transcription.
- Transcribed genes are preferentially repaired when DNA damage occurs.
- $TF_{II}H$ provides the link to a complex of repair enzymes.

CONCEPT AND REASONING CHECK

Why is the CTD important for transcription?

25.9 Enhancers Contain Bidirectional Elements That Assist Initiation

We have largely considered the promoter as an isolated region responsible for binding RNA polymerase. Eukaryotic promoters do not necessarily function alone, though. In most cases, the activity of a promoter is enormously increased by the presence of an enhancer located at a variable distance from the core promoter. Some enhancers function through long-range interactions of tens of kilobases, other enhancers function through short-range interactions and may lie quite close to the core promoter.

The concept that the enhancer is distinct from the promoter reflects two characteristics. The position of the enhancer relative to the promoter need not be fixed, but can vary substantially. **FIGURE 25.15** shows that it can be upstream, downstream, or within a gene (typically in introns). In addition, it can function in either orientation (that is, it can be inverted) relative to the promoter. Manipulations of DNA show that an enhancer can stimulate any promoter placed in its vicinity, even tens of kilobases away in either direction.

Like the promoter, an enhancer (or its alter ego, a silencer) is a modular element constructed of short DNA sequence elements that bind various types of transcription

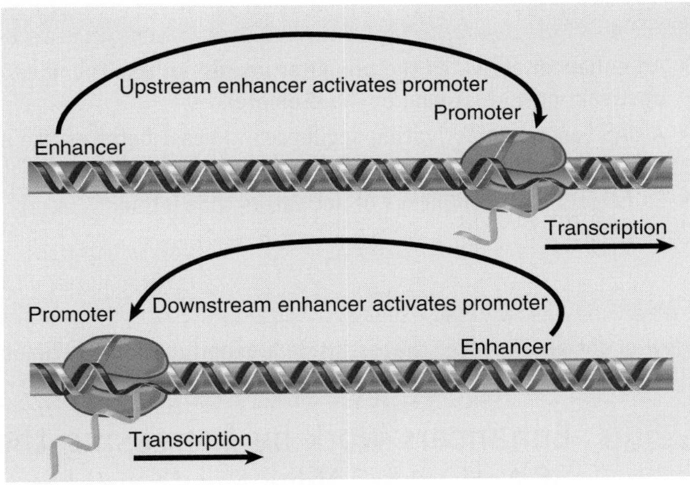

FIGURE 25.15 An enhancer can activate a promoter from upstream or downstream locations, and its sequence can be inverted relative to the promoter.

factors. Enhancers can be simple or complex depending on the number of binding elements and the type of transcription factors they bind.

One way to divide up the world of enhancer-binding transcription factors is to consider positive and negative factors. Transcription factors can be positive and stimulate transcription: **activators**, or negative and repress transcription: **repressors**. At any given time in a cell, determined by its developmental history, that cell will contain a mixture of transcription factors that can bind to an enhancer. If more activators bind than repressors, the element will be an enhancer. If more repressors bind than activators, the element will be a silencer.

- **activator** A protein that stimulates the expression of a gene, typically by interacting with a promoter to stimulate RNA polymerase. In eukaryotes, the sequence to which it binds in the promoter is called an enhancer.
- **repressor** A protein that inhibits expression of a gene. It may act to prevent transcription by binding to an enhancer or silencer.

A second way to examine the transcription factors that bind enhancers is by function. The first class we will consider is called *true activators*; that is, they function by making contact at the promoter, either directly by themselves, or, more commonly, through coactivators like Mediator. This class functions equally well on a DNA template or a chromatin template. There are two additional classes of activators that have a completely different mechanism of activation. The second class includes those that function by recruiting chromatin modification enzymes and chromatin remodeling complexes. The third class includes architectural modifiers. Their sole function is to change the structure of the DNA, typically to bend it. This can then result in bringing together two transcription factors separated by a short distance to synergize. In the next section, we will examine more closely how the different classes of activators and repressors work together in an enhancer, and in the next chapter we will examine transcription regulation in more detail.

- **upstream activating sequence (UAS)** The equivalent in yeast of the enhancer in higher eukaryotes; a UAS cannot function downstream of the promoter.

Elements analogous to enhancers, called **upstream activating sequences** (**UAS**), are found in yeast. They can function in either orientation at variable distances upstream of the promoter, but cannot function when located downstream. They have a regulatory role: the UAS is bound by the regulatory protein(s) that activates the genes downstream.

Reconstruction experiments in which the enhancer sequence is removed from the DNA and then is inserted elsewhere show that normal transcription can be sustained as long as it is present anywhere on the DNA molecule. If a β-globin gene is placed on a DNA molecule that contains an enhancer, its transcription is increased *in vivo* more than 200-fold, even when the enhancer is several kb upstream or downstream of the startpoint, in either orientation. We have yet to discover at what distance the enhancer fails to work.

KEY CONCEPTS

- An enhancer activates the promoter nearest to itself, and can be any distance either upstream or downstream of the promoter.
- A UAS (upstream activating sequence) in yeast behaves like an enhancer, but works only upstream of the promoter.
- Enhancers form complexes of activators that interact directly or indirectly with the promoter.

CONCEPT AND REASONING CHECK

How might an enhancer that is inside a gene function to stimulate transcription?

25.10 Enhancers Work by Increasing the Concentration of Activators Near the Promoter

Enhancers function by binding combinations of transcription factors, either positive or negative, that control the promoter and, by extension, gene expression. The promoter is the site where, in open chromatin, basal transcription factors prebind so that RNA

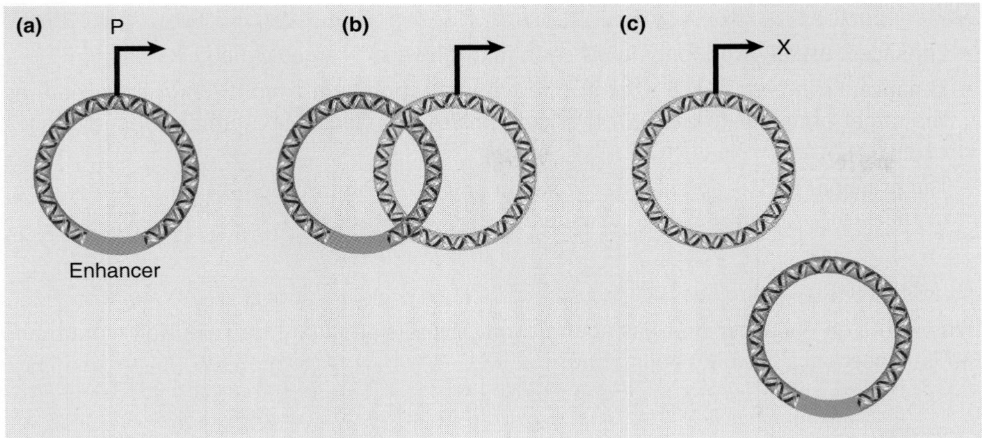

FIGURE 25.16 An enhancer may function by bringing proteins into the vicinity of the promoter. An enhancer and promoter on separate circular DNAs do not interact as in (c), but can interact when the two molecules are catenated as in (b).

polymerase can find the promoter. How can an enhancer stimulate initiation at a promoter that can be located any distance away on either side of it?

Enhancer function involves interaction with the basal apparatus at the core promoter element. Enhancers are modular, like promoters. Some elements are found in both long range enhancers and enhancers near promoters. Some individual elements found near promoters share with distal enhancers the ability to function at variable distance and in either orientation. Thus the distinction between long-range enhancers and short-range enhancers is blurred.

The essential role of the enhancer may be to increase the concentration of activator in the vicinity of the promoter (vicinity in this sense being a relative term) in *cis*. Numerous experiments have demonstrated that the level of gene expression (that is, the rate of transcription) is proportional to the net number of activator binding sites. The more activators bound at an enhancer site, the higher the level of expression.

The *Xenopus laevis* ribosomal RNA enhancer is able to stimulate transcription from its RNA polymerase I promoter. This stimulation is relatively independent of location and is able to function from a circle. There is, however, no stimulation when the enhancer and promoter are on separated circles. Yet, when the enhancer is placed on a circle of DNA that is catenated (interlocked) with a circle that contains the promoter, initiation is almost as effective as when the enhancer and promoter are on the same circular molecule as shown in **FIGURE 25.16** (even though, in this case, the enhancer is acting on its promoter in *trans*). Again, this suggests that the critical feature is localization of the protein bound at the enhancer, which increases the enhancer's chance of contacting a protein bound at the promoter.

If proteins bound at an enhancer several kb distant from a promoter interact directly with proteins bound in the vicinity of the startpoint, the organization of DNA must be flexible enough to allow the enhancer and promoter to be closely located. This requires the intervening DNA to be extruded as a large "loop." Such loops have been directly observed in the case of the bacterial enhancer.

What limits the activity of an enhancer? Typically it works upon the nearest promoter. There are situations in which an enhancer is located between two promoters, but activates only one of them on the basis of specific protein–protein contacts between the complexes bound at the two elements. The action of an enhancer may be limited by an insulator—an element in DNA that prevents the enhancer from acting on promoters beyond the insulator (see *Section 24.11, Insulators Define Independent Domains*).

KEY CONCEPTS

- Enhancers usually work only in *cis* configuration with a target promoter.
- Enhancers can be made to work in *trans* configuration by linking the DNA that contains the target promoter to the DNA that contains the enhancer via a protein bridge or by catenating the two molecules.
- The principle is that an enhancer works in any situation in which it is constrained to be in the same proximity as the promoter.

CONCEPT AND REASONING CHECK

Why would an enhancer that has bound two activators stimulate transcription more than an enhancer that has only bound one?

25.11 Summary

Of the three eukaryotic RNA polymerases, RNA polymerase I transcribes rDNA and accounts for the majority of activity, RNA polymerase II transcribes structural genes for mRNA and has the greatest diversity of products, and RNA polymerase III transcribes small RNAs. The enzymes have similar structures, with two large subunits and many smaller subunits; there are some common subunits among the enzymes.

None of the three RNA polymerases recognize their promoters directly. A unifying principle is that transcription factors have primary responsibility for recognizing the characteristic sequence elements of any particular promoter, and they serve in turn to bind the RNA polymerase and to position it correctly at the startpoint. At each type of promoter, the initiation complex is assembled by a series of reactions in which individual factors join (or leave) the complex. The factor TBP is required for initiation by all three RNA polymerases. In each case it provides one subunit of a transcription factor that binds in the vicinity of the startpoint.

An RNA polymerase II promoter consists of a number of short-sequence elements in the region upstream of the startpoint. Each element is bound by one or more transcription factors. The basal apparatus, which consists of the TF_{II} factors, assembles at the startpoint and enables RNA polymerase to bind. The TATA box (if there is one) near the startpoint and the initiator region immediately at the startpoint are responsible for selection of the exact startpoint at promoters for RNA polymerase II. TBP binds directly to the TATA box when there is one; in TATA-less promoters it is located near the startpoint by binding to the Inr or to the DPE downstream. After binding of $TF_{II}D$, the other general transcription factors for RNA polymerase II assemble the basal transcription apparatus at the promoter. Other elements in the promoter, located upstream of the TATA box, bind activators that interact with the basal apparatus. The activators and basal factors are released when RNA polymerase begins elongation.

The CTD of RNA polymerase II is phosphorylated during the initiation reaction. It provides a point of contact for proteins that modify the RNA transcript, including the 5′ capping enzyme, splicing factors, the 3′ processing complex and mRNA export from the nucleus.

Promoters may be stimulated by enhancers, sequences that can act at great distances and in either orientation on either side of a gene. Enhancers also consist of sets of elements, although they are more compactly organized. Some elements are found both close to promoters and in distant enhancers. Enhancers probably function by assembling a protein complex that interacts with the proteins bound at the promoter, requiring that DNA between is "looped out."

CHAPTER QUESTIONS

1. List three ways the C-terminal domain of the largest RNA polymerase II subunit may be involved in posttranscriptional RNA modifications:
   A. ______________________________
   B. ______________________________
   C. ______________________________
2. Where in the cell is each type of RNA transcribed? Choices may be used more than once or not at all.

Ribosomal RNA	**A.** cytoplasm
Small RNA	**B.** nucleolus
Messenger RNA	**C.** nucleoplasm

3. Genes transcribed from which of the following RNA polymerases often have promoters located downstream of the transcription start sites?
   A. RNA polymerase I
   B. RNA polymerase II
   C. RNA polymerase III
   D. All of the above
4. Which enzyme is responsible for synthesis of heterogeneous nuclear RNA (hnRNA)?
   A. RNA polymerase I
   B. RNA polymerase II
   C. RNA polymerase III
   D. RNA polymerase IV
5. Which RNA polymerase utilizes bipartite promoters?
   A. RNA polymerase I
   B. RNA polymerase II
   C. RNA polymerase III
   D. RNA polymerase IV
6. The first base of a transcript synthesized by RNA polymerase II tends to be:
   A. an A surrounded by pyrimidines.
   B. an A surrounded by purines.
   C. a C surrounded by pyrimidines.
   D. a C surrounded by purines.
7. Which of the following promoter elements are required for RNA polymerase II transcription?
   A. A TATA box, Inr element, and DPE element
   B. A TATA box and DPE element
   C. A TATA box and Inr element or TATA box and DPE element
   D. A TATA box and Inr element or Inr element and DPE element
8. What is the first factor to bind to an RNA polymerase II-transcribed promoter containing a TATA box?
   A. $TF_{II}A$
   B. $TF_{II}D$
   C. $TF_{II}H$
   D. RNA polymerase II

9. What is thought to be responsible for releasing RNA polymerase II from its associated transcription factors to allow transition to elongation?
   - **A.** methylation of the C-terminal domain of a large RNA polymerase II subunit
   - **B.** phosphorylation of the C-terminal domain of a large RNA polymerase II subunit
   - **C.** methylation of the N-terminal domain of a large RNA polymerase II subunit
   - **D.** phosphorylation of the N-terminal domain of a large RNA polymerase II subunit

10. Enhancers function to activate or stimulate:
    - **A.** the nearest promoter in *cis* to it.
    - **B.** the nearest promoter in *trans* to it.
    - **C.** the nearest promoter in *cis* or *trans* to it, either upstream or downstream.
    - **D.** the nearest promoter in *cis* or *trans* to it, upstream of the promoter.

KEY TERMS

activator
assembly factors
basal transcription factors
carboxy-terminal domain (CTD)
coactivator
core promoter
downstream promoter element (DPE)
enhancer
heterogeneous nuclear RNA (hnRNA)
housekeeping genes
initiator (Inr)
nontranscribed spacer
preinitiation complex
repressor
silencer
startpoint
TAFs
TATA-binding protein (TBP)
TATA box
TATA-less promoter
$TF_{II}D$
upstream activating sequence (UAS)

FURTHER READING

Ares, M. Jr. and Proudfoot, N. J. (2005). The Spanish connection: transcription and mRNA processing get even closer. *Cell* 120, 163–166.
A good review of how RNA polymerase II transcription and the processing of the transcript are coupled.

Grummt, I. (2003). Life on a small planet of its own: regulation of RNA polymerase I transcription in the nucleolus. *Genes Dev.* 17, 1691–1702.
A good review of RNA polymerase I transcription.

Schramm, L. and Hernandez, N. (2003). Recruitment of RNA polymerase III to its target promoters. *Genes Dev.* 16, 2593–2620.
A good review of RNA polymerase III transcription.

Shilatifard, A., Conway, R. C., and Conway, J. W. (2003). The RNA polymerase II elongation complex. *Annu. Rev. Biochem.* 72, 693–715.

Smale, S. T. and Kadonaga, J. T. (2003). The RNA polymerase II core promoter. *Annu. Rev. Biochem.* 72, 449–479.

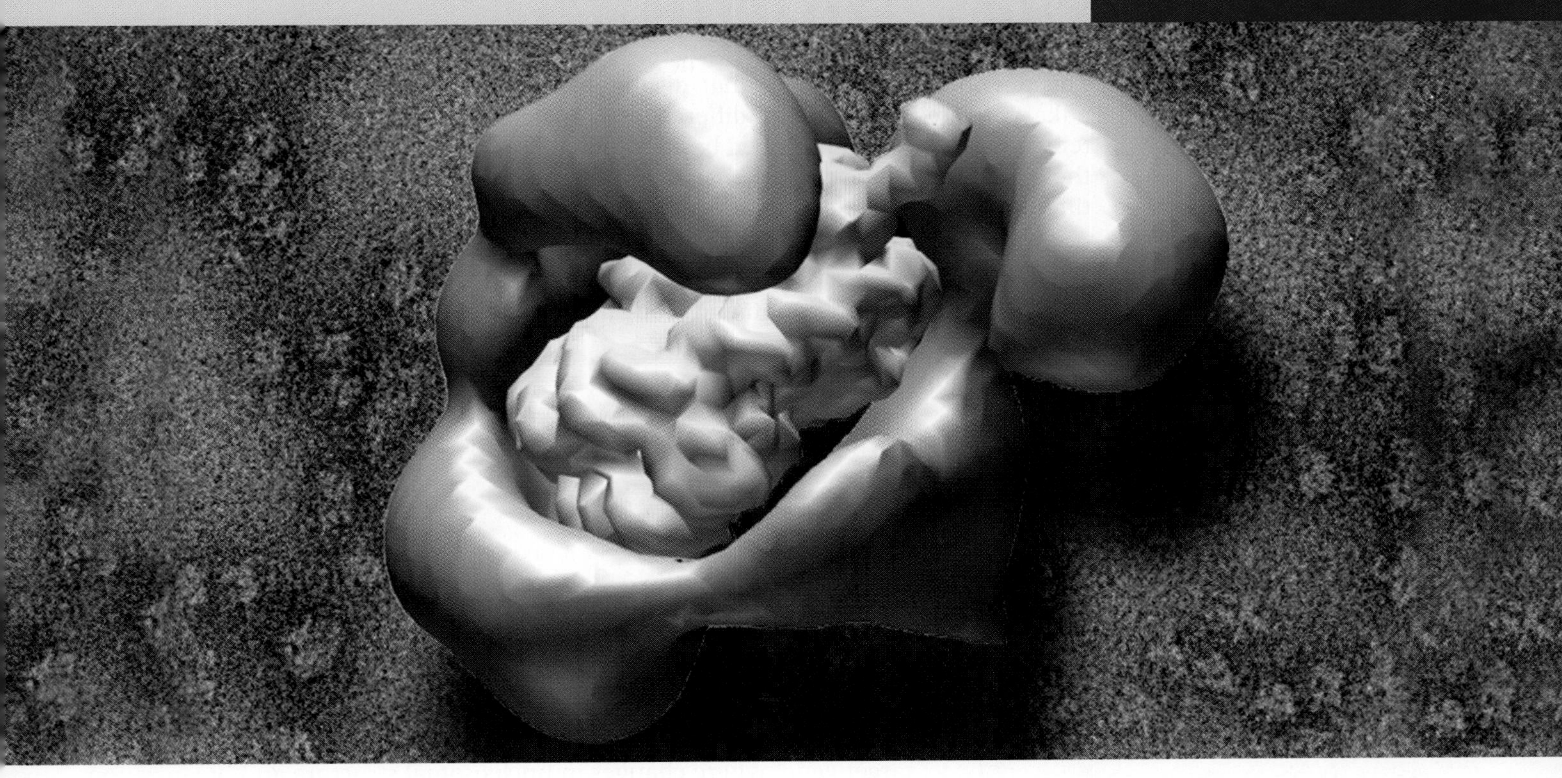

A model of the ATP-dependent chromatin remodeling complex RSC bound to a nucleosome. Photo courtesy of Andres Leschziner, Harvard University.

Eukaryotic Transcription Regulation

CHAPTER OUTLINE

26.1 Introduction

The phenotypic differences that distinguish the various kinds of cells in a higher eukaryote are largely due to differences in the expression of genes that code for proteins, that is, those transcribed by RNA polymerase II. In principle, the expression of these genes might be regulated at any one of several stages. In **FIGURE 26.1**, we can distinguish (at least) six potential control points, which form the following series:

Activation of gene structure: open chromatin
↓
Initiation of transcription and elongation
↓
Processing the transcript
↓
Transport to cytoplasm from the nucleus
↓
Translation of mRNA
↓
Degradation and turnover of mRNA

Control of transcription initiation: used for most genes
Local structure of the gene is changed
General transcription apparatus binds to promoter
RNA is modified and processed:
can control expression of alternative products from gene
AAAA
mRNA is exported from nucleus to cytoplasm:
AAAA
Nucleus
Cytoplasm
mRNA is translated and degraded:

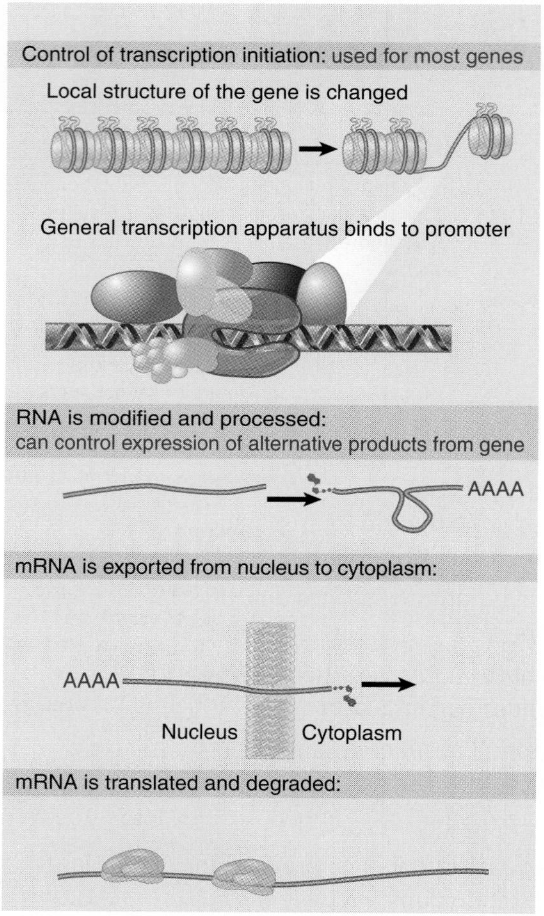

FIGURE 26.1 Gene expression is controlled principally at the initiation of transcription. Control of processing may be used to determine which form of a gene is represented in mRNA. The mRNA may be regulated during transport to the cytoplasm, during translation, and by degradation.

The determination of whether a gene is expressed depends on the structure of chromatin both locally (at the promoter) and in the surrounding domain. Chromatin structure correspondingly can be regulated by individual activation events or by changes that affect a wide chromosomal region. The most localized events concern an individual target gene, where changes in nucleosomal structure and organization occur in the immediate vicinity of the promoter. More general changes may affect regions as large as a whole chromosome. Activation of a gene requires changes in the state of chromatin. The essential issue is how the transcription factors gain access to the promoter DNA.

Local chromatin structure is an integral part of controlling gene expression. Genes may exist in either of two structural conditions. Genes are found in an "active" state only in the cells in which they are expressed. The change of structure precedes the act of transcription and indicates that the gene is "transcribable." This suggests that acquisition of the "active" structure must be the first step in gene expression. Active genes are found in domains of euchromatin with a preferential susceptibility to nucleases, and hypersensitive sites are created at promoters before a gene is activated (see *Section 24.9, DNase Hypersensitive Sites Reflect Changes in Chromatin Structure*).

There is an intimate and continuing connection between initiation of transcription and chromatin structure. Some activators of gene transcription directly modify histones; in particular, acetylation of histones is associated with gene activation. Conversely, some repressors of transcription function by deacetylating histones. Thus a reversible change in histone structure in the vicinity of the promoter is involved in the control of gene expression. These changes influence the association of histone octamers with DNA, and are responsible for controlling the presence and structure of nucleosomes at specific sites. This is an important aspect of the mechanism by which a gene is maintained in an active or inactive state.

The mechanisms by which regions of chromatin are maintained in an inactive (silent) state are related to the means by which an individual promoter is repressed. The proteins involved in the formation of heterochromatin act on chromatin via the histones, and modifications of the histones are an important feature in the interaction. Once established, such changes in chromatin can persist through cell divisions, creating an **epigenetic** state in which the properties of a gene are determined by the self-perpetuating structure of chromatin. The name *epigenetic* reflects the fact that a gene may have an inherited condition (it may be active or inactive) that does not depend on its sequence (see *Chapter 27, Epigenetic Effects Are Inherited*).

Once transcription begins, the primary transcript is modified by capping at the 5′ end, and in general also is modified by polyadenylation at the 3′ end. Introns must be excised from the transcripts of interrupted genes. The mature RNA must then be exported from the nucleus to the cytoplasm. Regulation of gene expression at the level of nuclear RNA processing might involve any or all of these stages, but the one for which we have most evidence concerns changes in splicing; some genes are expressed by means of alternative splicing patterns whose regulation controls the type of protein product (see *Section 28.11, Alternative Splicing Involves Differential Use of Splice Junctions*).

The translation of an mRNA in the cytoplasm can be specifically controlled, as can the turnover rate of the mRNA. While translation level control is uncommon in adult somatic cells, it does occur in some embryonic situations. This can also involve the localization of the mRNA to specific sites where it is expressed and/or the blocking of initiation of translation by specific protein factors. Different mRNAs may have different intrinsic half-lives determined by specific sequence elements.

Regulation of tissue-specific gene transcription lies at the heart of eukaryotic differentiation. It is also important for control of metabolic and catabolic pathways. A regulatory transcription factor serves to provide common control of a large number of target genes, and we seek to answer two questions about this mode of regulation: How does the transcription factor identify its group of target genes? and How is the activity of the transcription factor itself regulated in response to intrinsic or extrinsic signals?

▸ **epigenetic** Changes that influence the phenotype without altering the genotype. They consist of changes in the properties of a cell that are inherited, but that do not represent a change in genetic information.

26.2 Mechanism of Action of Activators and Repressors

Initiation of transcription involves many protein–protein interactions between transcription factors bound at enhancers with the basal apparatus that assembles at the promoter, including RNA polymerase. We can divide these transcription factors into two opposing classes, positive **activators** and negative **repressors**.

We have seen in Chapter 12 that **positive control** in bacteria entails a regulator that aids the RNA polymerase in the transition from the closed complex to the open complex. Transcription factors like the *E. coli* CRP typically bind close to the promoter to allow the CTD of the α subunit of RNA polymerase to make direct physical contact. This usually occurs in a gene having a poor promoter sequence. The activator functions to overcome the inability of the RNA polymerase to open the promoter. Positive control in eukaryotes is quite different. We can identify three classes of activators that differ by function.

The first class is the **true activators** (see *Section 25.9, Enhancers Contain Bidirectional Elements That Assist Initiation*). These are the classical transcription factors that function by making direct physical contact with the basal apparatus at the promoter (see *Section 26.3, Independent Domains Bind DNA and Activate Transcription*) either directly, or indirectly through a coactivator. These transcription factors function on DNA or chromatin templates.

The activity of a true activator may be regulated in any one of several ways, as illustrated schematically in **FIGURE 26.2**:

- A factor is tissue-specific because it is synthesized only in a particular type of cell. This is typical of factors that regulate development, such as homeodomain proteins.
- The activity of a factor may be directly controlled by modification. HSF (Heat Shock transcription Factor) is converted to the active form by phosphorylation.
- A factor is activated or inactivated by binding a ligand. The steroid receptors are prime examples. Ligand binding may influence the localization of the protein

▸ **activator** A protein that stimulates the expression of a gene, typically by acting at a promoter to stimulate RNA polymerase. In eukaryotes, the sequence to which it binds in the promoter is called a response element.

▸ **repressor** A protein that inhibits expression of a gene. It may act to prevent transcription by binding to an operator site in DNA or to prevent translation by binding to RNA.

▸ **positive control** The default state of genes that are under positive control is that they cannot be expressed unless a positive regulator is bound.

▸ **true activator** A positive transcription faction that functions by making contact, direct or indirect, with the basal apparatus to activate transcription.

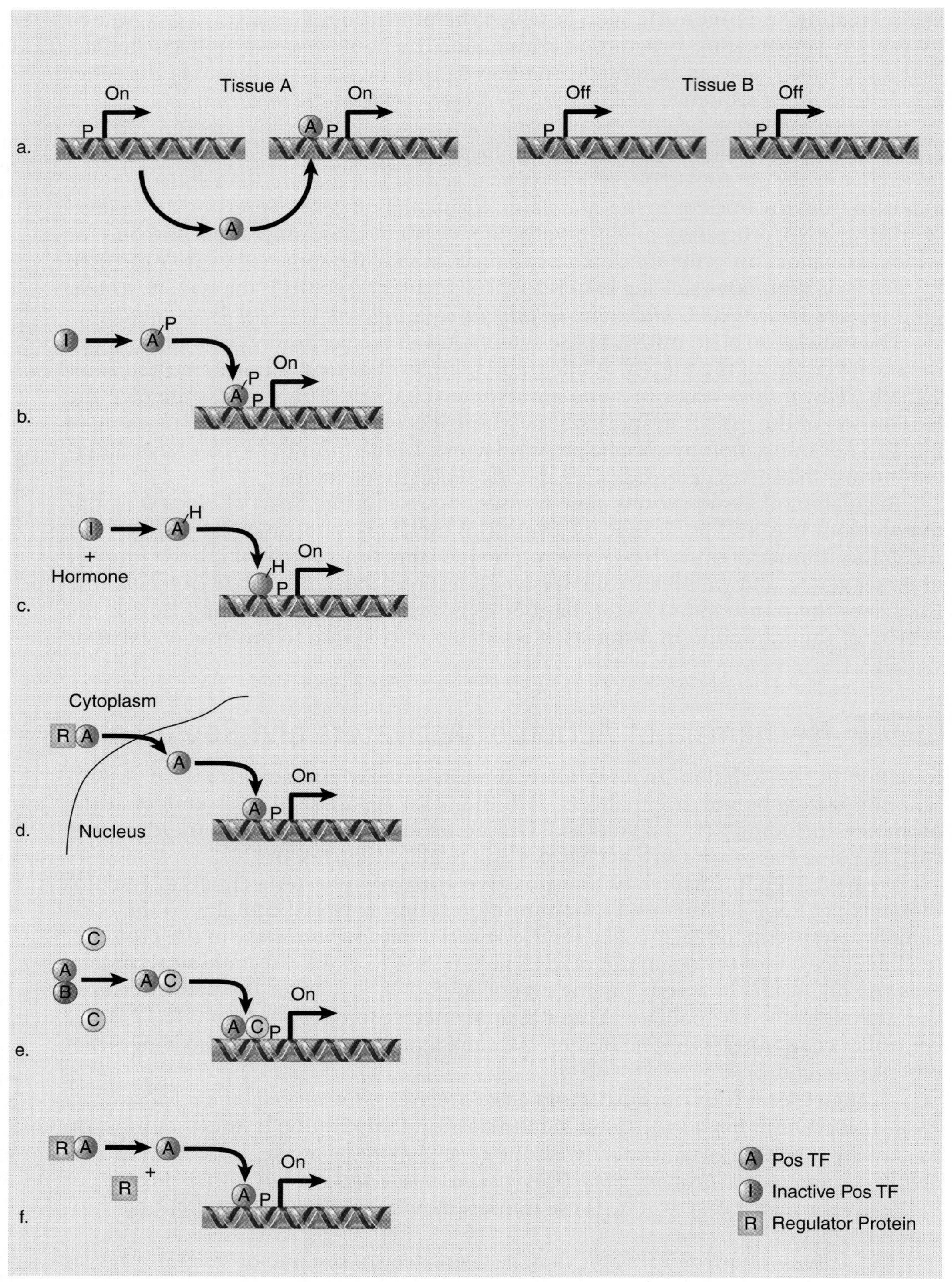

FIGURE 26.2 The activity of a positive regulatory transcription factor may be controlled by (a) the synthesis of new protein, (b) by covalent modification such as phosphorylation, (c) by ligand or hormone binding, (d) by counteracting the binding of inhibitors that sequester the regulator in the cytoplasm, (e) by the ability to select the correct binding partner for activation, and (f) by cleavage from an inactive precursor.

(causing transport from cytoplasm to nucleus), as well as determine its ability to bind to DNA.

- Availability of a factor may vary; for example, the factor NF-κB (which activates immunoglobulin κ genes in B lymphocytes) is present in many cell types. It is sequestered or masked in the cytoplasm, however, by the inhibitory protein I-κB. In B lymphocytes, NF-κB is released from I-κB and moves to the nucleus, where it activates transcription.
- A dimeric factor may have alternative partners. One partner may cause it to be inactive; synthesis of the active partner may displace the inactive partner. Such situations may be amplified into networks in which various alternative partners pair with one another, especially among the HLH proteins.
- The factor may be cleaved from an inactive precursor. One activator is produced as a protein bound to the nuclear envelope and endoplasmic reticulum. The absence of sterols (such as cholesterol) causes the cytosolic domain to be cleaved; it then translocates to the nucleus and provides the active form of the activator.

The second class includes the **antirepressors**. When one of these activators is bound to its enhancer, it recruits the histone modifier enzymes and/or the chromatin remodeler complexes to convert the chromatin from the closed state to the open state. This class has no activity on a DNA template; it only functions on chromatin templates (described below in *Section 26.6, Chromatin Remodeling Is an Active Process*).

▸ **antirepressor** A positive regulator that functions in opening chromatin.

The third class includes **architectural proteins** such as Yin-Yang; these proteins function to bend the DNA, either bringing bound proteins together to facilitate forming a cooperative complex, or bending the DNA the other way to prevent complex formation, as shown in **FIGURE 26.3**. Note that a strand of DNA may thus be bent in two different directions depending on whether the regulator binds to the top or to the bottom. This is a difference of one half of a turn of the helix, which is 5 base pairs (10.5 bp per turn).

▸ **architectural protein** A protein that when bound to DNA, can alter its structure, e.g., introduce a bend. They may have no other function.

We have seen several examples of **negative control** in bacteria in the *lac* operon and in the *trp* operon in Chapter 12. Repression can occur in bacteria when the repressor prevents the RNA polymerase from converting from the closed complex to the open complex or binds to the promoter sequence to prevent polymerase from binding. There are many more mechanisms by which repressors act in eukaryotes, which are illustrated in **FIGURE 26.4**.

▸ **negative control** The default state of genes that are under negative control is to be expressed. A specific intervention is required to turn them off.

- One mechanism of action by which a eukaryote repressor can prevent gene expression is to sequester an activator in the cytoplasm. Eukaryote proteins are synthesized in the cytoplasm. Proteins that function in the nucleus have a domain that directs their transport through the nuclear membrane. A repressor can bind to that domain and mask it.
- Several variations of that mechanism are possible. One that takes place in the nucleus is that the repressor can bind to an activator that is already bound to an enhancer and mask its activation domain, preventing it from functioning (such as the Gal80 repressor; see *Section 26.14, The Yeast* GAL *Genes: A Model for Activation and Repression*).

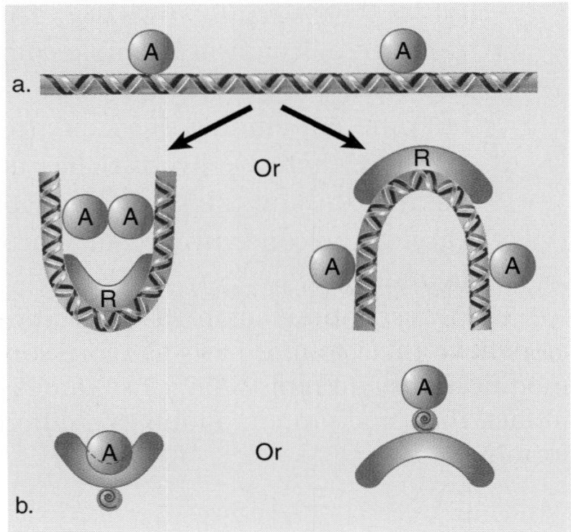

FIGURE 26.3 Architectural proteins control the structure of DNA and thus control whether bound proteins can contact each other. A = activator; R = architectural protein.

FIGURE 26.4 A repressor may control transcription by sequestering an activator in the cytoplasm, by binding an activator and masking its activation domain, by being held in the cytoplasm until it is needed, or by competition with an activator for a binding site.

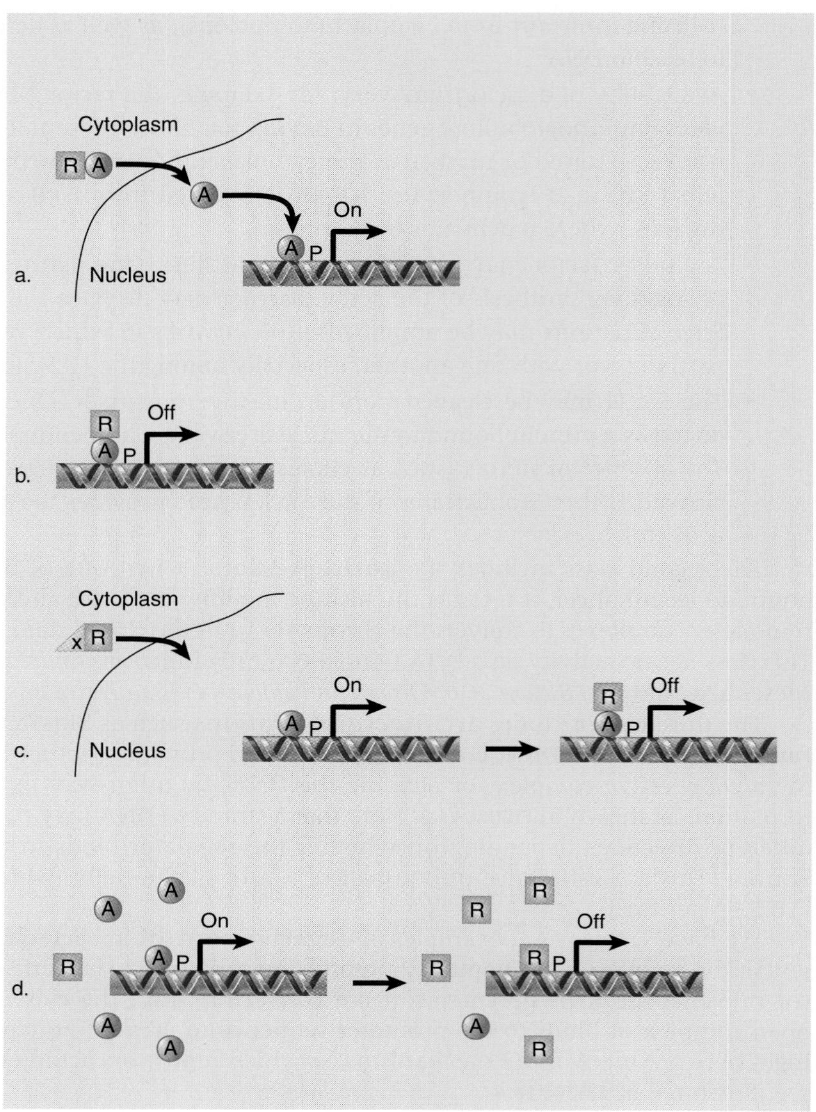

- Alternatively, the repressor can be masked and held in the cytoplasm until it is released to enter the nucleus.
- A fourth mechanism is simple competition for an enhancer, where either the repressor and activator have the same binding site sequence or overlapping but different binding site sequences. This is a very versatile mechanism for a cell since there are two variables at work here. One is strength of factor binding to DNA and the second variable is factor concentration. By only slightly varying the concentration of a factor, a cell can dramatically alter its developmental path.

The transcription factors that recruit the histone modifiers and chromatin remodelers have their counterparts as repressors that recruit the complexes that undo the modifications and remodeling. The same is true for the architectural proteins, where, in fact, the same protein bound to a different site prevents activator complexes from forming.

KEY CONCEPTS

- Activators determine the frequency of transcription.
- Activators work by making protein–protein contacts with the basal factors.
- Activators may work via coactivators.
- Activators are regulated in many different ways.
- Some components of the transcriptional apparatus work by changing chromatin structure.
- Repression is achieved by affecting chromatin structure or by binding to and masking activators.

CONCEPT AND REASONING CHECKS

- How is positive control in bacteria different from positive control in eukaryotes? Why do you think it is different?
- How is negative control in bacteria different from negative control in eukaryotes? Why do you think it is different?

26.3 Independent Domains Bind DNA and Activate Transcription

We know the most about the activator class of transcription factors. Activators require protein domains with multiple functions:

- They recognize specific DNA target sequences located in enhancers that affect a particular target gene.
- Having bound to DNA, an activator exercises its function by binding to components of the basal transcription apparatus.
- Many require a dimerization domain to form complexes with other proteins.

Can we characterize domains in the activator that are responsible for these activities? Often an activator has a separate domain that binds DNA and a separate domain that activates transcription. Each domain behaves as a separate module that functions independently when it is linked to a domain of the other type. The geometry of the overall transcription complex must allow the activating domain to contact the basal apparatus irrespective of the exact location and orientation of the DNA-binding domain.

Enhancer elements near the promoter may still be an appreciable distance from the startpoint, and in many cases may be oriented in either direction. Enhancers may even be farther away and always show orientation independence. This organization has implications for both the DNA and proteins. The DNA may be looped or condensed in some way to allow the formation of the transcription complex. In addition, the domains of the activator may be connected in a flexible way, as illustrated diagrammatically in **FIGURE 26.5**. The main point here is that the DNA-binding and activating domains are independent, and are connected in a way that allows the activating domain to interact with the basal apparatus irrespective of the orientation and exact location of the DNA-binding domain.

Binding to DNA is necessary for activating transcription, but does activation depend on the particular DNA-binding domain? This question has been answered by making hybrid proteins that consist of the DNA-binding domain of one activator linked to the activation domain of another activator. The hybrid functions in transcription at sites dictated by its DNA-binding domain, but in a way determined by its activation domain.

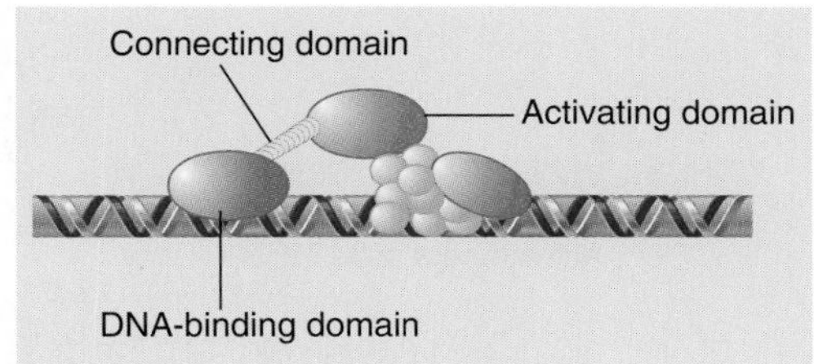

FIGURE 26.5 DNA-binding and activating functions in a transcription factor may comprise independent domains of the protein.

This result fits the modular view of transcription activators. The function of the DNA-binding domain is to bring the activation domain to the basal apparatus at the promoter. Precisely how or where it is bound to DNA is irrelevant, but once it is there, the activation domain can play its role. This explains why the exact locations of DNA-binding sites can vary. The ability of the two types of module to function in hybrid proteins suggests that each domain of the protein folds independently into an active structure that is not influenced by the rest of the protein.

KEY CONCEPTS

- DNA-binding and transcription-activation activities are carried by independent domains of an activator.
- The role of the DNA-binding domain is to bring the transcription-activation domain into the vicinity of the promoter.

CONCEPT AND REASONING CHECK

What would happen when one subunit of a heterodimeric activator is in a cell that contains a partner subunit with a mutation in the DNA-binding domain?

- **coactivator** Factors required for transcription that do not bind DNA, but are required for (DNA-binding) activators to interact with the basal transcription factors.
- **basal apparatus** The complex of transcription factors that assembles at the promoter before RNA polymerase is bound.

26.4 Activators Interact with the Basal Apparatus

The true activator class of transcription factors may work directly when it consists of a DNA-binding domain linked to a transcription-activating domain, as illustrated in Figure 25.4. In other cases, the activator does not itself have a transcription-activating domain (or contains only a weak activation domain), but binds another protein—a coactivator—that has the transcription-activating activity. **FIGURE 26.6** shows the action of such an activator. We may regard **coactivators** as transcription factors whose specificity is conferred by the ability to bind to DNA-binding transcription factors instead of directly to DNA. A particular activator may require a specific coactivator.

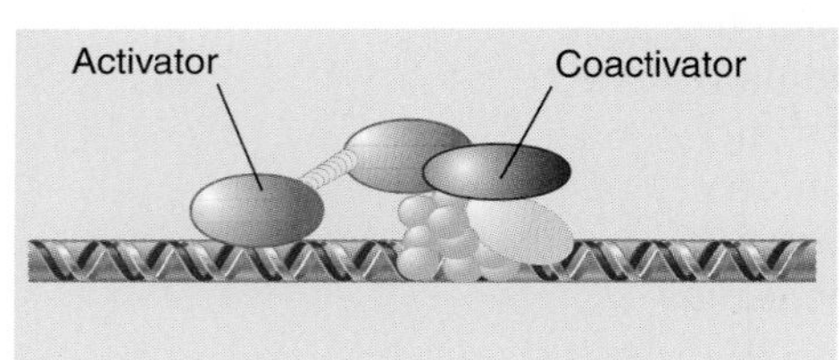

FIGURE 26.6 An activator may bind a coactivator that contacts the basal apparatus.

Although the protein components are organized differently, the mechanism is the same. An activator that contacts the basal apparatus directly has an activation domain covalently connected to the DNA-binding domain. When an activator works through a coactivator, the connections involve noncovalent binding between protein subunits (compare Figure 26.4 and Figure 26.5). The same interactions are responsible for activation, irrespective of whether the various domains are present in the same protein subunit or divided into multiple protein subunits. In addition, many coactivators also contain additional enzymatic activities that promote transcription activation, such as activities that modify chromatin structure (see *Section 26.9, Histone Acetylation Is Associated with Transcription Activation*).

An activation domain works by making protein–protein contacts with general transcription factors that promote assembly of the **basal apparatus**. Contact with the basal apparatus may be made with any one of several basal factors, but typically occurs with $TF_{II}D$, $TF_{II}B$, or $TF_{II}A$. All of these factors participate in early stages of assembly of the basal apparatus (see Figure 25.11). **FIGURE 26.7** illustrates the situation when such a contact is made. The major effect of the activators is to influence the assembly of the basal apparatus.

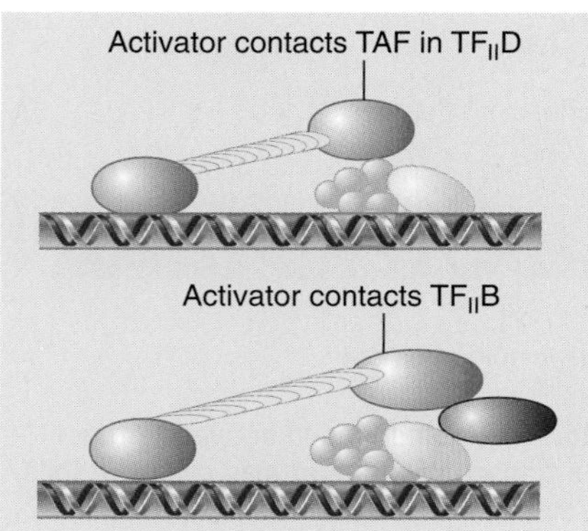

FIGURE 26.7 Activators may work at different stages of initiation by contacting the TAFs of $TF_{II}D$ or by contacting $TF_{II}B$.

$TF_{II}D$ may be the most common target for activators, which may contact any one of several TAFs. In fact, a major role of the TAFs is to provide the connection from the basal apparatus to activators. This explains why TBP alone can support basal-level transcription, whereas the TAFs of $TF_{II}D$ are required for the higher levels of transcription that are stimulated by activators. Different TAFs in $TF_{II}D$ may provide surfaces that interact with different activators. Some activators interact only with individual TAFs; others interact with multiple TAFs. We assume that the interaction assists the binding of $TF_{II}D$ to the TATA box, assists the binding of other basal apparatus components around the $TF_{II}D$-TATA box complex, or controls the phosphorylation of the CTD. In any case, the interaction stabilizes the basal transcription complex, speeds the process of initiation, and thereby increases use of the promoter.

How does an activator stimulate transcription? We can imagine two general types of model:

- The recruitment model argues that its sole effect is to increase the binding of RNA polymerase to the promoter.
- An alternative model is to suppose that it induces some change in the transcriptional complex, for example, in the conformation of enzymes such as protein kinases, which increases its efficiency.

When we add up all the components required for efficient transcription—basal factors, RNA polymerase, activators, and coactivators—we get a very large apparatus that consists of >40 proteins. Is it feasible for this apparatus to assemble step by step at the promoter? Some activators, coactivators, and basal factors may assemble stepwise at the promoter, but then they may be joined by a very large complex consisting of RNA polymerase preassembled with further activators and coactivators, as illustrated in **FIGURE 26.8**.

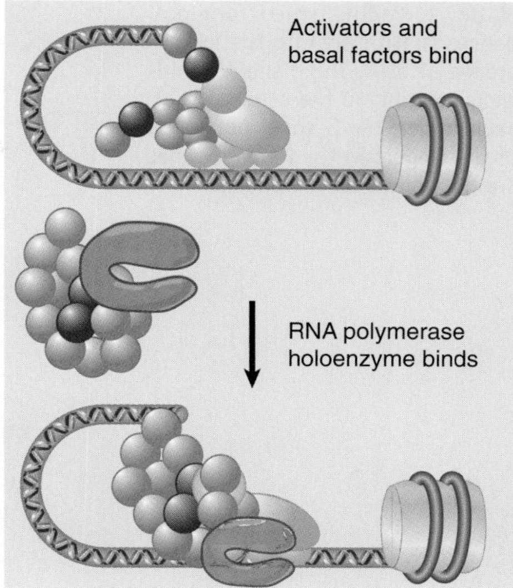

FIGURE 26.8 RNA polymerase exists as a holoenzyme containing many activators.

Several forms of RNA polymerase in which the enzyme is associated with various transcription factors have been found. The most prominent "holoenzyme complex" in yeast (defined as being capable of initiating transcription without additional components) consists of RNA polymerase associated with a 20-subunit complex called **Mediator**. Mediator is necessary for transcription of most yeast genes. Homologous complexes are required for the transcription of most higher eukaryotic genes. Mediator undergoes a conformational change when it interacts with the CTD of RNA polymerase. It can transmit either activating or repressing effects from upstream components to the RNA polymerase. It is probably released when a polymerase starts elongation.

▸ **Mediator** A large protein complex associated with yeast bacterial RNA polymerase II. It contains factors that are necessary for transcription from many or most promoters.

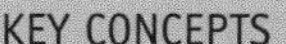

KEY CONCEPTS

- The principle that governs the function of all activators is that a DNA-binding domain determines specificity for the target promoter or enhancer.
- The DNA-binding domain is responsible for localizing a transcription-activating domain in the proximity of the basal apparatus.
- An activator that works directly has a DNA-binding domain and an activating domain.
- An activator that does not have an activating domain may work by binding a coactivator that has an activating domain.
- Several factors in the basal apparatus are targets with which activators or coactivators interact.
- RNA polymerase may be associated with various alternative sets of transcription factors in the form of a holoenzyme complex.

CONCEPT AND REASONING CHECK

Do transcription activators "turn on the gene"? If not, what does?

26.5 There Are Many Types of DNA-Binding Domains

It is common for an activator to have a modular structure in which different domains are responsible for binding to DNA and for activating transcription. Factors are often classified according to the type of DNA-binding domain. In general, a relatively short motif in this domain is responsible for binding to DNA:

- The **zinc finger** motif comprises a DNA-binding domain. It was originally recognized in factor $TF_{III}A$, which is required for RNA polymerase III to transcribe 5S rRNA genes. This motif has since been identified in numerous other tran-

▸ **zinc finger** A DNA-binding motif that typifies a class of transcription factor.

FIGURE 26.9 Zinc fingers form α helices that insert into the major groove of DNA. The β sheets sandwich the zinc on the other side of each finger; the β-sheet portions of the fingers in the bottom panel are not shown for simplicity.

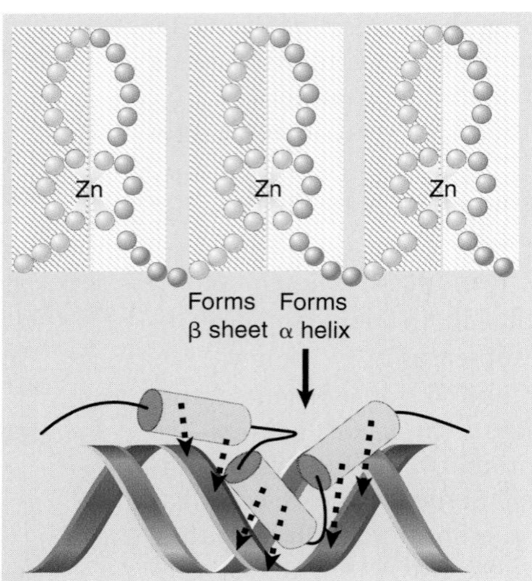

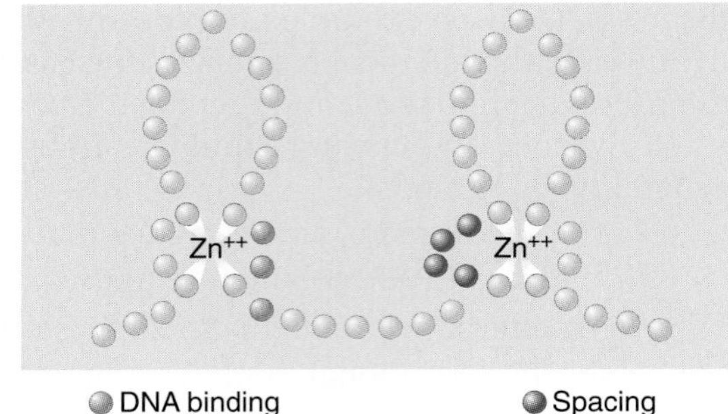

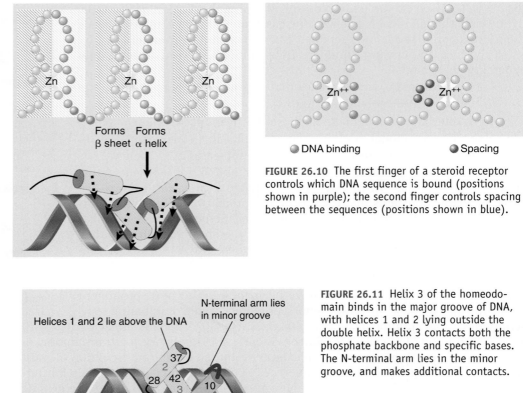

FIGURE 26.10 The first finger of a steroid receptor controls which DNA sequence is bound (positions shown in purple); the second finger controls spacing between the sequences (positions shown in blue).

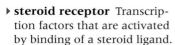

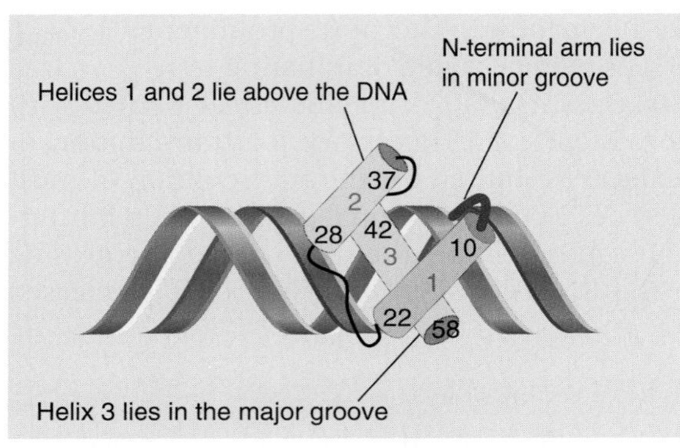

FIGURE 26.11 Helix 3 of the homeodomain binds in the major groove of DNA, with helices 1 and 2 lying outside the double helix. Helix 3 contacts both the phosphate backbone and specific bases. The N-terminal arm lies in the minor groove, and makes additional contacts.

scription factors (and presumed transcription factors). Proteins often contain multiple zinc fingers, such as the three shown in FIGURE 26.9.

- The **steroid receptors** are defined as a group by a functional relationship: each receptor is activated by binding a particular steroid, such as glucocorticoid binding to the glucocorticoid receptor. Together with other receptors, such as the thyroid hormone receptor or the retinoic acid receptor, the steroid receptors are members of the superfamily of ligand-activated activators with the same general modus operandi: the protein factor is inactive until it binds a small ligand, as shown in FIGURE 26.10.
- The **helix-turn-helix** motif was originally identified as the DNA-binding domain of phage repressors. One helix lies in the major groove of DNA; the other lies at an angle across DNA. A related form of the motif is present in the **homeodomain**, a sequence first characterized in several proteins encoded by genes involved in developmental regulation in *Drosophila,* shown in FIGURE 26.11. It is also present in the comparable mammalian genes.
- The amphipathic **helix-loop-helix (HLH)** motif has been identified in some developmental regulators and in genes coding for eukaryotic DNA-binding proteins. Each amphipathic helix presents a face of hydrophobic residues on one side and charged residues on the other side. The length of the connecting loop varies from 12 to 28 amino acids. The motif enables proteins to dimerize, and a basic region near this motif contacts DNA as seen in FIGURE 26.12.

▸ **steroid receptor** Transcription factors that are activated by binding of a steroid ligand.
▸ **helix-turn-helix** The motif that describes an arrangement of two α-helices that form a site that binds to DNA, one fitting into the major groove of DNA and the other lying across it.
▸ **homeodomain** A DNA-binding motif that typifies a class of transcription factors.
▸ **helix-loop-helix (HLH)** The motif that is responsible for dimerization of a class of transcription factors called HLH proteins. A bHLH protein has a basic DNA-binding sequence close to the dimerization motif.

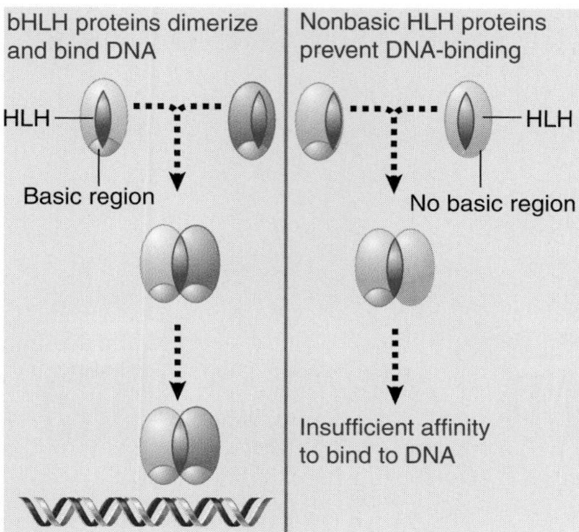

FIGURE 26.12 An HLH dimer in which both subunits are of the bHLH type can bind DNA, but a dimer in which one subunit lacks the basic region cannot bind DNA.

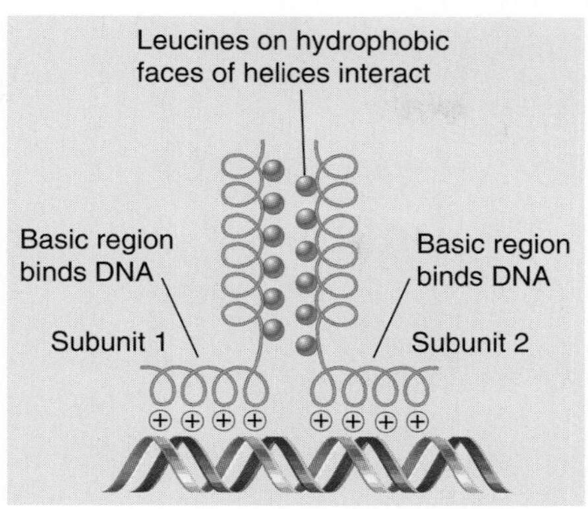

FIGURE 26.13 The basic regions of the bZIP motif are held together by the dimerization at the adjacent zipper region when the hydrophobic faces of two leucine zippers interact in parallel orientation.

- **Leucine zippers** consist of a stretch of amino acids with a leucine residue in every seventh position. A leucine zipper in one polypeptide interacts with a zipper in another polypeptide to form a dimer. Adjacent to each zipper is a stretch of positively charged residue that is involved in binding to DNA; this is known as the **bZIP (basic zipper)** structural motif shown in FIGURE 26.13.

▸ **leucine zipper** A dimerization motif that is found in a class of transcription factors.

▸ **bZIP (basic zipper)** A bZIP protein has a basic DNA-binding region adjacent to a leucine zipper dimerization motif.

KEY CONCEPTS

- Activators are classified according to the type of DNA-binding domain.
- Members of the same group have sequence variations of a specific motif that confer specificity for individual DNA target sites.

CONCEPT AND REASONING CHECK

Why do eukaryotic organisms have so many different sequence motifs for binding DNA?

26.6 Chromatin Remodeling Is an Active Process

Transcriptional activators face a challenge when trying to bind to their recognition sites in eukaryotic chromatin. FIGURE 26.14 illustrates two general states that can exist at a eukaryotic promoter. In the inactive state, nucleosomes are present, and they prevent basal factors and RNA polymerase from binding. In the active state, the basal apparatus occupies the promoter, and histone octamers cannot bind to it. Each type of state is stable. In order to convert a promoter from the inactive state to the

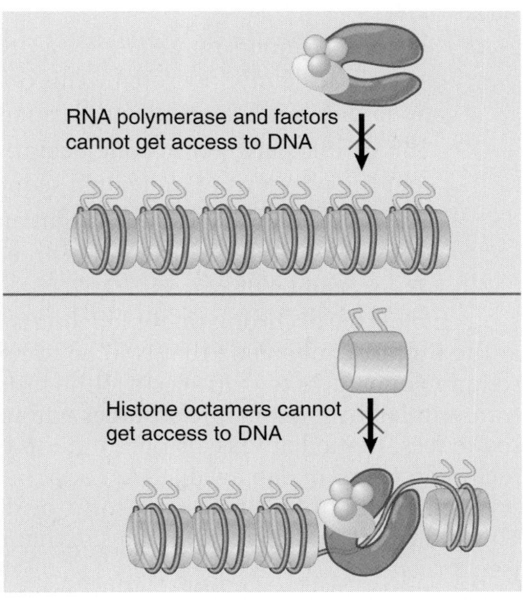

FIGURE 26.14 If nucleosomes form at a promoter, transcription factors (and RNA polymerase) cannot bind. If transcription factors (and RNA polymerase) bind to the promoter to establish a stable complex for initiation, histones are excluded.

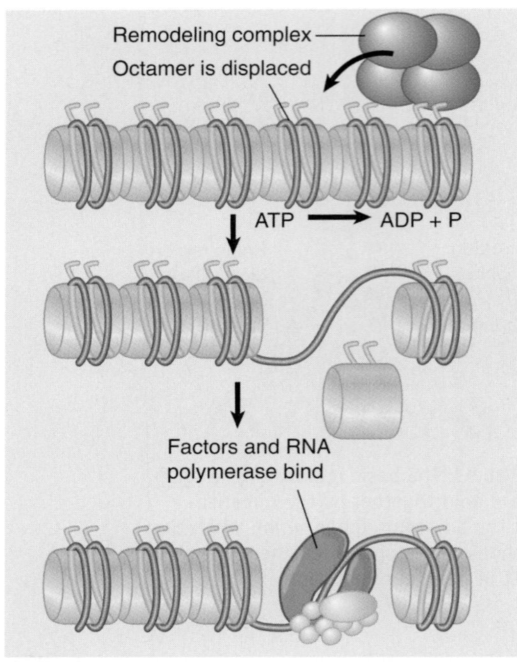

FIGURE 26.15 The dynamic model for transcription of chromatin relies upon factors that can use energy provided by hydrolysis of ATP to displace nucleosomes from specific DNA sequences.

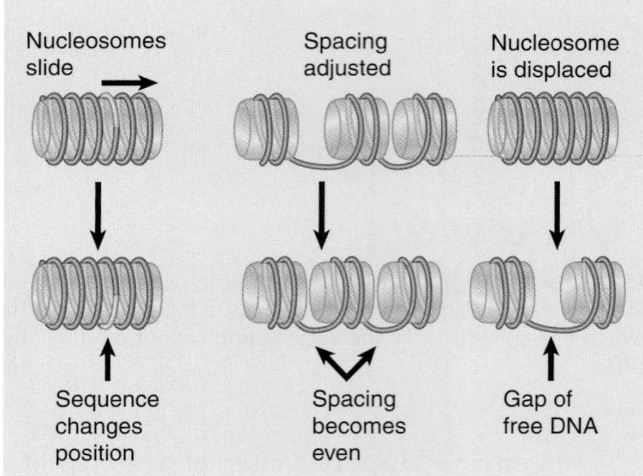

FIGURE 26.16 Remodeling complexes can cause nucleosomes to slide along DNA, can displace nucleosomes from DNA, or can reorganize the spacing between nucleosomes.

active state, the chromatin structure must be perturbed in order to allow binding of the basal factors.

chromatin remodeling The energy-dependent displacement or reorganization of nucleosomes that occurs in conjunction with activation of genes for transcription.

The general process of inducing changes in chromatin structure is called **chromatin remodeling**. This consists of mechanisms for displacing histones that depend on the input of energy. Many protein–protein and protein–DNA contacts need to be disrupted to release histones from chromatin. There is no free ride: energy must be provided to disrupt these contacts. **FIGURE 26.15** illustrates the principle of a dynamic model by a factor that hydrolyzes ATP. When the histone octamer is released from DNA, other proteins (in this case transcription factors and RNA polymerase) can bind.

There are several alternative outcomes of chromatin remodeling, summarized in **FIGURE 26.16**:

- Histone octamers may *slide* along DNA, changing the relationship between the nucleic acid and the protein. This can alter both the rotational and the translational position of a particular sequence on the nucleosome.
- The *spacing* between histone octamers may be changed, again with the result that the positions of individual sequences are altered relative to protein.
- The most extensive change is that an octamer(s) may be *displaced entirely* from DNA to generate a nucleosome-free gap. Alternatively, one or both H2A-H2B dimers can be displaced.

A major role of chromatin remodeling is to change the organization of nucleosomes at the promoter of a gene that is to be transcribed. This is required to allow the transcription apparatus to gain access to the promoter. Remodeling can also act to prevent transcription by moving nucleosomes onto, rather than away from, essential promoter sequences. Remodeling is also required to enable other manipulations of chromatin, including repair of damaged DNA.

Remodeling often takes the form of displacing one or more histone octamers. This can result in the creation of a site that is hypersensitive to cleavage with DNase I (see *Section 24.9, DNase Hypersensitive Sites Reflect Changes in Chromatin Structure*). Sometimes there are less dramatic changes in the positioning of a single nucleosome. Thus

Type of Complex	SWI/SNF	ISWI	CHD	INO80/SWRI
Yeast	SWI/SNF RSC	ISW1a, ISWb ISW2	CHDI	INO80 SWRI
Fly	dSWI/SNF (brahma)	NURF CHRAC ACF	JMIZ	Tip60
Human	hSWI/SNF	RSF hACF/WCFR hCHRAC WICH	NuRD	INO80 SRCAP
Frog		WICH CHRAC ACF	Mi-2	

FIGURE 26.17 Remodeling complexes can be classified by their ATPase subunits. This table is not exhaustive but gives some examples of each class of remodeler.

changes in chromatin structure can extend from altering the positions of nucleosomes to removing them altogether.

Chromatin remodeling is undertaken by **ATP-dependent chromatin remodeling complexes**, which use ATP hydrolysis to provide the energy for remodeling. The heart of the remodeling complex is its *ATPase subunit*. The ATPase subunits of all remodeling complexes are related members of a large *superfamily* of proteins, which is divided into *subfamilies* of more closely related members. Remodeling complexes are classified according to the subfamily of ATPase that they contain as their catalytic subunit. There are many subfamilies, but the four major subfamilies (SWI/SNF, ISWI, CHD, and INO80/SWR1) are shown in **FIGURE 26.17**. The first remodeling complex described was the SWI/SNF ("switch sniff") complex in yeast, which has homologs in all eukaryotes. The chromatin remodeling superfamily is large and diverse, and most species have multiple complexes in different subfamilies. Yeast has two SWI/SNF-related complexes and three ISWI complexes. Eight different ISWI complexes have been identified thus far in mammals. Remodeling complexes range from small heterodimeric complexes (the ATPase subunit plus a single partner) to massive complexes of 10 or more subunits. Each type of complex may undertake a different range of remodeling activities.

▸ **ATP-dependent chromatin remodeling complex** A complex of one or more proteins associated with an ATPase of the SWI2/SNF2 superfamily that uses the energy of ATP hydrolysis to alter or displace nucleosomes.

SWI/SNF is the prototypic remodeling complex. Its name reflects the fact that many of its subunits are encoded by genes originally identified by *swi* or *snf* mutations in *Saccharomyces cerevisiae*. (*swi* mutants cannot switch mating type, and *snf*—sucrose nonfermenting—mutants cannot use sucrose as a carbon source.) Mutations in these loci are pleiotropic, and the range of defects is similar to those shown by mutants that have lost part of the carboxyl-terminal domain (CTD) of RNA polymerase II. Early hints that these genes might be linked to chromatin came from evidence that these mutations show genetic interactions with mutations in genes that code for components of chromatin: *SIN1*, which codes for a nonhistone chromatin protein, and *SIN2*, which codes for histone H3. The *SWI* and *SNF* genes are required for expression of a variety of individual loci (~120, or 2%, of *S. cerevisiae* genes require SWI/SNF for normal expression). Expression of these loci may require the SWI/SNF complex to remodel chromatin at their promoters.

SWI/SNF acts catalytically *in vitro*, and there are only ~150 complexes per yeast cell. All of the genes encoding the SWI/SNF subunits are nonessential, which implies that yeast must also have other ways of remodeling chromatin. The related RSC (remodels the structure of chromatin) complex is more abundant and also is essential. It acts at ~700 target loci.

Different subfamilies of remodeling complexes have distinct modes of remodeling, reflecting differences in their ATPase subunits as well as effects of other proteins in individual remodeling complexes. SWI/SNF complexes can remodel chromatin *in vitro* without overall loss of histones or can displace histone octamers. These reactions likely pass through the same intermediate in which the structure of the target nucleosome is altered, leading either to reformation of a (remodeled) nucleosome on the original DNA or to displacement of the histone octamer to a different DNA molecule. In contrast, the ISWI family primarily affects nucleosome positioning *without* displacing octamers, in a sliding reaction in which the octamer moves along DNA. The activity of ISWI requires the histone H4 tail as well as binding to linker DNA.

There are many contacts between DNA and a histone octamer; fourteen are identified in the crystal structure. All of these contacts must be broken for an octamer to be released or for it to move to a new position. How is this achieved? The ATPase subunits are distantly related to helicases (enzymes that unwind double-stranded nucleic acids), but remodeling complexes do not have any unwinding activity. Present thinking is that remodeling complexes in the SWI/NSF and ISWI classes use the hydrolysis of ATP to *twist* DNA on the nucleosomal surface. This twisting creates a mechanical force that allows a small region of DNA to be released from the surface and then repositioned. This mechanism creates transient loops of DNA on the surface of the octamer; these loops are themselves accessible to interact with other factors, or they can propagate along the nucleosome, ultimately resulting in nucleosome sliding.

Different remodeling complexes have different roles in the cell. SWI/SNF complexes are generally involved in transcriptional activation, whereas some ISWI complexes act as repressors, using their remodeling activity to slide nucleosomes *onto* promoter regions to prevent transcription. Members of the CHD (chromodomain helicase DNA-binding) family have also been implicated in repression, particularly the Mi-2/NuRD complexes, which contain both chromatin remodeling and histone deacetylase activities. Remodelers in the SWR1/INO80 class have a unique activity: in addition to their normal remodeling capabilities, some members of this class also have *histone exchange* capability, in which individual histones (usually H2A/H2B dimers) can be replaced in a nucleosome, typically with a histone variant (see *Section 24.4, Histone Variants Produce Alternative Nucleosomes*).

KEY CONCEPTS

- There are numerous chromatin remodeling complexes that use energy provided by hydrolysis of ATP.
- All remodeling complexes contain a related ATPase catalytic subunit, and are grouped into subfamilies containing more closely related ATPase subunits.
- Remodeling complexes can alter, slide, or displace nucleosomes.
- Some remodeling complexes can exchange one histone for another in a nucleosome.

CONCEPT AND REASONING CHECK

How can remodeling complexes act as either activators or repressors?

26.7 Nucleosome Organization or Content May Be Changed at the Promoter

How are remodeling complexes targeted to specific sites on chromatin? They do not themselves contain subunits that bind specific DNA sequences. This suggests the model shown in **FIGURE 26.18**, in which they are recruited by activators or (sometimes) by repressors.

The interaction between transcription factors and remodeling complexes gives a key insight into their modus operandi. The transcription factor Swi5 activates the *HO* gene in yeast, a gene involved in mating-type switching. (Note that despite its name Swi5 is not a member of the SWI/SNF complex.) Swi5 enters nuclei toward the end of mitosis and binds to the *HO* promoter. It then recruits SWI/SNF to the promoter.

Swi5 is then released, leaving SWI/SNF at the promoter. This means that a transcription factor can activate a promoter by a "hit and run" mechanism, in which its function is fulfilled once the remodeling complex has bound.

The involvement of remodeling complexes in gene activation was discovered because the complexes are necessary to enable certain transcription factors to activate their target genes. One of the first examples was the GAGA factor, which activates the *Drosophila hsp70* promoter. Binding of GAGA to four $(CT)_n$-rich sites near the promoter disrupts the nucleosomes, creates a hypersensitive region, and causes the adjacent nucleosomes to be rearranged so that they occupy preferential instead of random positions. Disruption is an energy-dependent process that requires the NURF remodeling complex, a complex in the ISWI subfamily. The organization of nucleosomes is altered so as to create a boundary that determines the positions of the adjacent nucleosomes. During this process, GAGA binds to its target sites and DNA, and its presence fixes the remodeled state.

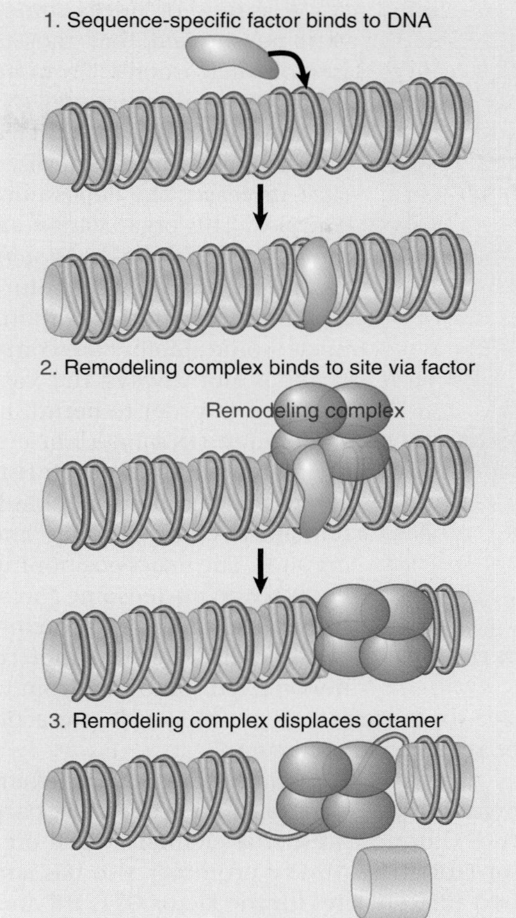

FIGURE 26.18 A remodeling complex binds to chromatin via an activator (or repressor).

The *PHO* system was one of the first in which it was shown that a change in nucleosome organization is involved in gene activation. At the *PHO5* promoter, the bHLH activator Pho4 responds to phosphate starvation by inducing the disruption of four precisely positioned nucleosomes, as depicted in **FIGURE 26.19**. This event is independent of transcription (it occurs in a TATA$^-$ mutant) and independent of replication. There are two binding sites for Pho4 (and another activator, Pho2) at the promoter. One is located between nucleosomes, which can be bound by the isolated DNA-binding domain of Pho4, and the other lies within a nucleosome, which cannot be recognized. Disruption of the nucleosome to allow DNA binding at the second site is necessary for gene activation. This action requires the presence of the transcription-activating domain, and appears to involve at least two remodelers: SWI/SNF and INO80. In addition, chromatin disassembly at *PHO5* also requires a histone chaperone, Asf1, which may assist in nucleosome removal or act as a recipient of displaced histones.

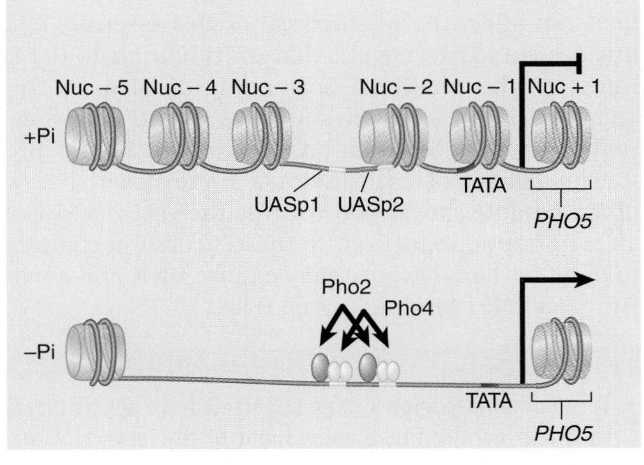

FIGURE 26.19 Nucleosomes are displaced from promoters during activation. The *PHO5* promoter contains nucleosomes positioned over the TATA box and one of the binding sites for the Pho4 and Pho2 activators. When *PHO5* is induced by phosphate starvation (-Pi), promoter nucleosomes are displaced.

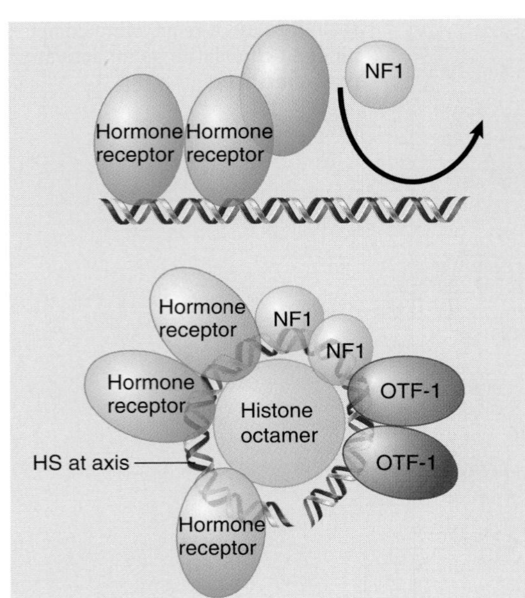

FIGURE 26.20 Hormone receptor and NF1 cannot bind simultaneously to the MMTV promoter in the form of linear DNA, but can bind when the DNA is presented on a nucleosomal surface.

A survey of nucleosome positions in a large region of the yeast genome showed that most sites that bind transcription factors are free of nucleosomes. Promoters for RNA polymerase II typically have a nucleosome-free region (NFR) ~200 bp upstream of the startpoint, which is flanked by positioned nucleosomes on either side. These positioned nucleosomes typically contain the histone variant H2AZ (called Htz1 in yeast); the deposition of H2AZ requires the SWR1 remodeling complex. This organization appears to be present in many human promoters as well. It has been suggested that H2AZ-containing nucleosomes are more easily evicted during transcription activation, thus "poising" promoters for activation; however, the actual effects of H2AZ on nucleosome stability are controversial.

It is not always the case, however, that nucleosomes must be excluded in order to permit initiation of transcription. Some activators can bind to DNA on a nucleosomal surface. Nucleosomes appear to be precisely positioned at some steroid hormone response elements in such a way that receptors can bind. Receptor binding may alter the interaction of DNA with histones, and may even lead to exposure of new binding sites. The exact positioning of nucleosomes could be required either because the nucleosome "presents" DNA in a particular rotational phase or because there are protein–protein interactions between the activators and histones or other components of chromatin. Thus we have now moved some way from viewing chromatin exclusively as a repressive structure to considering which interactions between activators and chromatin can be required for activation.

The MMTV promoter presents an example of the need for specific nucleosomal organization. It contains an array of six partly palindromic sites which constitute the HRE (hormone response element). Each site is bound by one dimer of hormone receptor (HR). The MMTV promoter also has a single binding site for the factor NF1, and two adjacent sites for the factor OTF. HR and NF1 cannot bind simultaneously to their sites in free DNA. **FIGURE 26.20** shows how the nucleosomal structure controls binding of the factors.

The HR protects its binding sites at the promoter when hormone is added, but does not affect the micrococcal nuclease-sensitive sites that mark either side of the nucleosome. This suggests that HR is binding to the DNA on the nucleosomal surface; however, the rotational positioning of DNA on the nucleosome prior to hormone addition allows access to only two of the four sites. Binding to the other two sites requires a change in rotational positioning on the nucleosome. This can be detected by the appearance of a sensitive site at the axis of dyad symmetry (which is in the center of the binding sites that constitute the HRE). NF1 can be detected on the nucleosome after hormone induction, so these structural changes may be necessary to allow NF1 to bind, perhaps because they expose DNA and abolish the steric hindrance by which HR blocks NF1 binding to free DNA.

KEY CONCEPTS

- A remodeling complex does not itself have specificity for any particular target site, but must be recruited by a component of the transcription apparatus.
- Remodeling complexes are recruited to promoters by sequence-specific activators.
- The factor may be released once the remodeling complex has bound.
- Transcription activation often involves nucleosome displacement at the promoter.
- Promoters contain nucleosome-free regions flanked by nucleosomes containing the H2A variant H2AZ (Htz1 in yeast).
- The MMTV promoter requires a change in rotational positioning of a nucleosome to allow an activator to bind to DNA on the nucleosome.

CONCEPT AND REASONING CHECK

Why does it make sense that chromatin remodelers do not recognize promoters directly, but rather are recruited by site-specific factors?

26.8 Histone Modification Regulates Chromatin Function

All of the histones are subject to numerous covalent modifications, most of which occur in the histone tails. The histone tails consist of 15–30 amino acids at the N-termini of all four core histones and the C-termini of H2A and H2B. The N-terminal tails of H2B and H3 pass between the turns of DNA (see Figure 24.11 in *Section 24.3, Nucleosomes Have a Common Structure*). All of the histones can be modified at several sites by methylation, acetylation, or phosphorylation, as shown in **FIGURE 26.21**. Other modifications occur as well, such as mono-ubiquitylation or sumoylation, and many of their functions are yet to be characterized.

Acetylation and methylation occur on the free epsilon (ε) amino group of lysine. As seen in **FIGURE 26.22**, acetylation neutralizes the positive charge that resides on the NH_3^+ form of the group. Methylation also occurs on arginine. Phosphorylation occurs on the hydroxyl group of serine and threonine. This introduces a negative charge in the form of the phosphate group. Lysine can be mono-, di-, or trimethylated (all still positively charged), and arginine can be mono- or dimethylated (symmetrically or asymmetrically).

These modifications are transient. They can change the charge of the protein molecule, and as a result they are potentially able to change the functional properties of the octamers. Modification of histones is associated with structural changes that occur in chromatin at replication and transcription, and specific modifications also facilitate DNA repair. Modifications at specific positions on specific histones define different functional states of chromatin. Newly synthesized core histones carry specific patterns

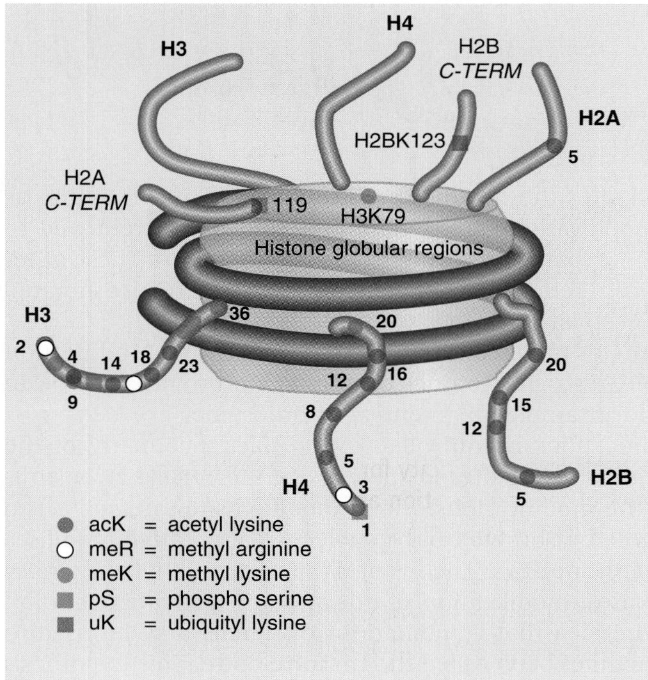

FIGURE 26.21 The histone tails can be acetylated, methylated, phosphorylated, and ubiquitylated at numerous sites. Not all possible modifications are shown. Adapted from *The Scientist* 17(2003): p. 27.

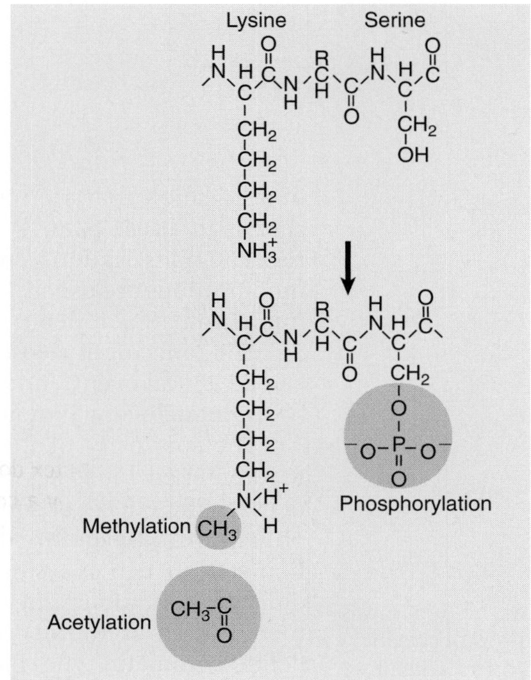

FIGURE 26.22 Acetylation of lysine or phosphorylation of serine reduces the overall positive charge of a protein. Methylated lysine retains a positive charge.

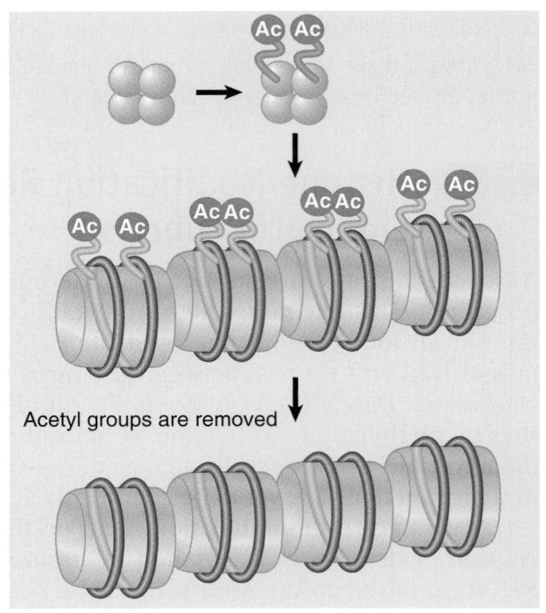

FIGURE 26.23 Acetylation at replication occurs on histones before they are incorporated into nucleosomes.

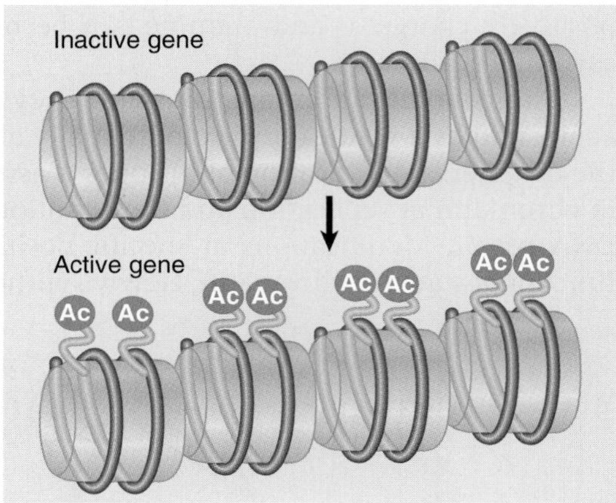

FIGURE 26.24 Acetylation associated with gene activation occurs by directly modifying histones in nucleosomes.

of acetylation that are removed after the histones are assembled into chromatin, as shown in **FIGURE 26.23**. Other modifications are dynamically added and removed to regulate transcription, replication, repair, and chromosome condensation. These other modifications are usually added and removed from nucleosomes that are incorporated into chromatin, as depicted for acetylation in **FIGURE 26.24**.

The range of nucleosomes that is targeted for modification can vary. Modification can be a local event, for example, restricted to nucleosomes at a promoter. It also can be a general event, extending, for example, to an entire chromosome.

The specificity of the modifications is controlled by the fact that many of the modifying enzymes have individual target sites in specific histones. **FIGURE 26.25** summarizes the effects of some of the modifications. Many modified sites are subject to only a single type of modification, but others can be subject to alternative modification states (such as lysine 9 of histone H3, which is acetylated or methylated under different conditions). In some cases, modification of one site may activate or inhibit modification of another site. The idea that combinations of signals may be used to define chromatin types has sometimes been called the **histone code**. This hypothesis proposes that the *collective impact* of multiple modifications at particular sites defines the function of a chromatin domain. These modifications are not restricted to a single

▸ **histone code** The hypothesis that combinations of specific modifications on specific histone residues act cooperatively to define chromatin function.

Histone	Site	Modification	Function
H3	K-4	Methylation	Activation
H3	K-9	Methylation	Chromatin condensation
	K-9	Methylation	Repression
	K-9	Acetylation	Activation
H3	S-10	Phosphorylation	Activation
H3	K-14	Acetylation	Prevents methylation at Lys-9
H3	K-79	Methylation	Telomeric silencing and activation
H4	R-3	Methylation	
H4	K-5	Acetylation	Assembly
H4	K-12	Acetylation	Assembly
H4	K-16	Acetylation	Nucleosome assembly
	K-16	Acetylation	Fly X activation

FIGURE 26.25 Most modified sites in histones have a single, specific type of modification, but some sites can have more than one type of modification. Individual functions can be associated with some of the modifications.

histone; the code is derived from all the modifications within a nucleosome or even nearby nucleosomes.

The changes in charge caused by some histone modifications can directly alter the structure of chromatin, but a major function of histone modification lies in the *creation of binding sites* for the attachment of nonhistone proteins that change the properties of chromatin. In recent years, a number of protein domains have been identified that bind to specifically modified histone tails.

The **bromodomain** is found in a variety of proteins that interact with chromatin, including components of some chromatin remodeling complexes. Bromodomains recognize acetylated lysine, and different bromodomain-containing proteins recognize different acetylated targets. The bromodomain itself recognizes only a very short sequence of four amino acids, including the acetylated lysine, so specificity for target recognition must depend on interactions involving other regions. Besides remodeling complexes, the bromodomain is found in a range of proteins that interact with chromatin, including components of the transcription apparatus and some of the enzymes that actually acetylate histones, discussed in the next section.

▸ **bromodomain** A 110-amino acid domain that binds to acetylated lysines in histones.

Methylated lysines (and arginines) are recognized by a number of different domains, which not only can recognize specific modified sites but also can distinguish between mono-, di- or trimethylated lysines. The **chromodomain** is a common protein motif of 60 amino acids present in a number of chromatin-associated proteins. Some chromodomain proteins are important for targeting proteins to heterochromatin by recognizing specific "silencing" modifications (such as H3 lysine 9 methylation) associated with heterochromatin. This will be discussed further in *Section 27.3, Heterochromatin Depends on Interactions with Histones* and *Section 27.4, Polycomb and Trithorax Are Antagonistic Repressors and Activators*. A number of other methyl lysine binding domains have been identified, such as the PHD (plant homeodomain), Tudor, and WD40 domains; the number of different motifs designed to recognize particular methylated sites emphasizes the importance and complexity of histone modifications.

▸ **chromodomain** ~60 amino acid domains that recognize methylated lysines in histones; some chromodomains have different functions such as RNA binding.

The idea that *combinations* of modifications are critical, as proposed in the histone code hypothesis, has been reinforced by recent discoveries of proteins or complexes that can recognize multiple sites of modification. For example, some proteins have tandem bromodomains or chromodomains, with particular spacing, which could promote binding to histones that are acetylated or methylated at two specific sites. There are also cases in which modification at one site can prevent a protein from recognizing its target modification at another site. It is clear that the effects of a single modification may not always be predictable, and the context of other modifications must be accounted for in order to assign a function to a region of chromatin.

KEY CONCEPTS

- Histones are modified by methylation, acetylation, and phosphorylation (and other modifications).
- The bromodomain is found in a variety of proteins that interact with chromatin; it is used to recognize acetylated sites on histones.
- The chromodomain is found in several chromatin proteins that have either activating or repressing effects on gene expression.

CONCEPT AND REASONING CHECK

Some proteins contain multiple domains that bind modified histones. What do you think might be the purpose of (for example) two adjacent bromodomains in one protein?

26.9 Histone Acetylation Is Associated with Transcription Activation

All of the core histones are dynamically acetylated on lysine residues in the tails (and occasionally within the globular core). As described in the previous section, certain patterns of acetylation are associated with newly synthesized histones that are deposited during DNA synthesis in S phase. This specific acetylation pattern is then erased after histones are incorporated into nucleosomes.

Outside of S phase, acetylation of histones in chromatin is generally correlated with the state of gene expression. The correlation was first noticed because histone acetylation is increased in a domain containing active genes, and acetylated chromatin is more sensitive to DNase I. We now know that this occurs largely because of acetylation of the nucleosomes (on specific lysines) in the vicinity of the promoter when a gene is activated.

In addition to events at individual promoters, global changes in acetylation occur on sex chromosomes. This is part of the mechanism by which the activities of genes on the X chromosome are altered to compensate for the presence of two X chromosomes in one sex but only one X chromosome (in addition to the Y chromosome) in other sex (see *Section 27.5, X Chromosomes Undergo Global Changes*). The inactive X chromosome in female mammals has underacetylated histones. The super-active X chromosome in *Drosophila* males has increased acetylation of H4. This suggests that the presence of acetyl groups may be a prerequisite for a less condensed, active structure. In male *Drosophila*, the X chromosome is acetylated specifically at K16 of histone H4. The enzyme responsible for this acetylation is called MOF; MOF is recruited to the chromosome as part of a large protein complex. This "dosage compensation" complex is responsible for introducing general changes in the X chromosome that enable it to be more highly expressed. The increased acetylation is only one of its activities.

Acetylation is reversible. Each direction of the reaction is catalyzed by a specific type of enzyme. Enzymes that can acetylate lysine residues in histones are called **lysine (K) acetyltransferases** or **KATs** (these were previously known as **histone acetyltransferases** or **HATs**). The acetyl groups are removed by **histone deacetylases** or **HDACs**. There are two groups of KAT enzymes: those in group A act on histones in chromatin and are involved with the control of transcription; those in group B act on newly synthesized histones in the cytosol, and are involved with nucleosome assembly.

Two inhibitors have been useful in analyzing acetylation. Trichostatin and butyric acid inhibit histone deacetylases, and cause acetylated nucleosomes to accumulate. The use of these inhibitors has supported the general view that acetylation is associated with gene expression; in fact, the ability of butyric acid to cause changes in chromatin resembling those found upon gene activation was one of the first indications of the connection between acetylation and gene activity.

- **lysine (K) acetyltransferase (KAT)** An enzyme (typically present in large complexes) that acetylates lysine residues in histones (or other proteins). Previously known as histone acetyltransferase (HAT).
- **histone acetyltransferase (HAT)** An enzyme that modifies histones by addition of acetyl groups; some transcriptional coactivators have this activity. Also known as **lysine (K) acetyltransferase (KAT)**.
- **histone deacetylase (HDAC)** Enzyme that removes acetyl groups from histones; may be associated with repressors of transcription.

The breakthrough in analyzing the role of histone acetylation was provided by the characterization of the acetylating and deacetylating enzymes, and their association with other proteins that are involved in specific events of activation and repression. A basic change in our view of histone acetylation was caused by the discovery that previously identified activators of transcription turned out to also have KAT activity.

The connection was established when the catalytic subunit of a group A KAT was identified as a homolog of the yeast regulator protein Gcn5 (see the accompanying Methods and Techniques box for the history of this discovery). It then was shown that yeast Gcn5 itself has KAT activity, with histones H3 and H2B as its preferred substrates *in vivo*. Gcn5 had previously been identified as part of an adaptor complex required for the function of certain enhancers and their target promoters. It is now known that Gcn5's KAT activity is required for activation of a number of target genes.

Gcn5 was the prototypic KAT that opened the way to the identification of a large family of related acetyltransferase complexes conserved from yeast to mammals. In yeast, Gcn5 is the catalytic KAT subunit of the 1.8 MDa Spt-Ada-Gcn5-acetyltransferase (SAGA) complex, which contains several proteins that are involved in transcription. Among these proteins are several $TAF_{II}s$. In addition, the Taf1 subunit of $TF_{II}D$ is itself an acetyltransferase. There are some functional overlaps between $TF_{II}D$ and SAGA, most notably that yeast can survive the loss of either Taf1 or Gcn5, but cannot tolerate the deletion of both. This suggests that an acetylase activity is essential for gene expression, but can be provided by either $TF_{II}D$ or SAGA. As might be expected from the size of the SAGA complex, acetylation is only one of its functions. The SAGA complex has histone H2B deubiquitylation activity (dynamic H2B ubiquitylation/deubiquitylation is also associated with transcription), and also contains subunits possessing bromodomains and chromodomains, allowing this complex to interact with acetylated and methylated histones.

One of the first general activators to be characterized as KAT was p300/CREB-binding protein (CBP). (Actually, p300 and CBP are different proteins, but they are so closely related that they are often referred to as a single type of activity.) p300/CBP is a coactivator that links an activator to the basal apparatus (see Figure 26.6). p300/CBP interacts with various activators, including hormone receptors, AP-1 (c-Jun and c-Fos), and MyoD. p300/CBP acetylates multiple histone targets, with a preference for the H4 tail. p300/CBP interacts with another coactivator, PCAF, which is related to Gcn5 and preferentially acetylates H3 in nucleosomes. p300/CBP and PCAF form a complex that functions in transcriptional activation. In some cases yet another KAT can be involved, such as the hormone receptor coactivator ACTR, which is itself a KAT that acts on H3 and H4. One explanation for the presence of multiple KAT activities in a coactivating complex is that each KAT has a different specificity, and that multiple different acetylation events are required for activation. This enables us to redraw our picture for the action of coactivators as shown in **FIGURE 26.26**, where RNA polymerase II is bound at a hypersensitive site and coactivators are acetylating histones on the nucleosomes in the vicinity.

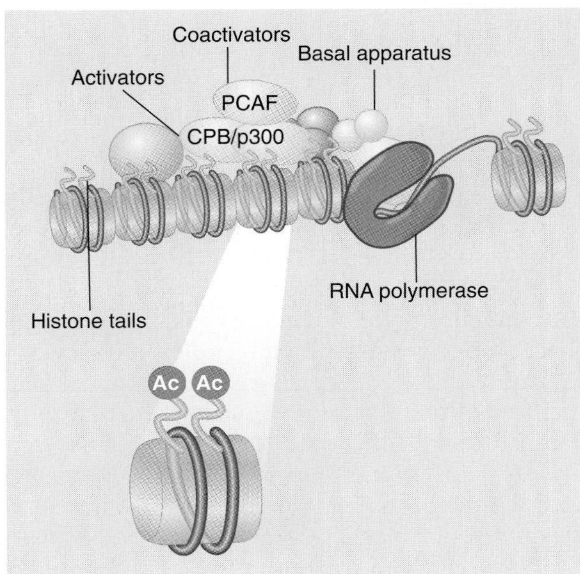

FIGURE 26.26 Coactivators may have KAT activities that acetylate the tails of nucleosomal histones.

As acetylation is linked to activation, deacetylation is linked to transcriptional repression. Where site-specific activators recruit coactivators with

METHODS AND TECHNIQUES

A Tale of Two Nuclei—*Tetrahymena thermophila*

Tetrahymena thermophila is a ciliated protist, a unicellular eukaryote that has been the subject of intense genetic studies in part because of its unique nuclear organization. Typically, the DNA in the nucleus of a eukaryotic cell has two roles: it is the stored genetic material that is transmitted to the next generation and it is the blueprint for the gene expression of the cell. The most striking genetic feature of *Tetrahymena* is the compartmentalization of DNA into two nuclei with distinct functions: one functions as the "germline nucleus" to be used for gamete production and reproduction, and the second functions as the "somatic nucleus" to be used for transcription and gene expression. The germline nucleus is the smaller of the two and is referred to as the *micronucleus*. This nucleus is diploid and contains five pairs of chromosomes that undergo normal mitosis and meiosis. The somatic nucleus, called the *macronucleus* for its larger size, is polyploid and has an estimated 200 to 300 chromosomes. *Tetrahymena* can divide vegetatively, through mitosis of both the micronucleus and macronucleus, followed by cell division. However, under conditions of starvation, the cell undergoes meiosis and the micronucleus produces haploid gamete pronuclei. A mating pair of *Tetrahymena* cells undergoes sexual reproduction when they conjugate, and there is reciprocal exchange of the gamete pronuclei. When the two gametes fuse, they generate a diploid zygotic nucleus, which in turn undergoes mitosis to give rise to two daughter nuclei, one that becomes the new micronucleus and the second that becomes the new macronucleus. The old macronucleus is destroyed, and the two fused cells separate and continue to grow vegetatively. These life cycles are illustrated in **FIGURE B26.1**.

The newly formed macronucleus is derived from a diploid zygotic nucleus that undergoes extensive genomic reorganization and amplification. The chromosomes are fragmented in a site-specific fashion to generate approximately 200 to 300 chromosomes, ranging in size from 21 to 3000 kilobases. During the process of chromosome breakage, ~10%–15% of the genome is eliminated in the macronucleus. In addition to this fragmentation, there is an amplification of the chromosomes resulting in approximately 50 copies of each chromosome. Telomeres are added to the newly formed chromosomes at their ends. Interestingly, these chromosomes lack centromeres so that the copies of homologous chromosomes are randomly distributed during mitosis.

The unique genomic organization of the *Tetrahymena* macronucleus and its high level of transcriptional activity have facilitated the discovery of many novel structures and mechanisms of chromosome organization and transcription, which in time have been demonstrated to be universal in all eukaryotes. For instance, telomeres and telomerase, as well as self-splicing RNAs, were first described in *Tetrahymena*. The role of chromatin remodeling, specifically the link between histone acetylation and transcription activation, was first made by the biochemical isolation of histone acetyltransferase (HAT) activity [now referred to as lysine acetyltransferase, or KAT, activity] from *Tetrahymena* macronuclei. In the mid-1990s, David Allis and his colleagues used a novel activity gel assay to isolate the first HAT from *Tetrahymena* macronuclei. Micronuclei were chosen as a likely source of HAT activity, as they are highly transcribed and enriched in histone acetylation. In these experiments, histone proteins were incorporated into a polyacrylamide gel prior to polymerization to create a "histone-containing gel." Then, crude extracts from macronuclei were subjected to electrophoresis through this histone-containing gel, separating the many different proteins. The presence of HAT activity was determined by the ability of a single polypeptide to incorporate ^{3}H [tritium]-acetate into the histones contained in the gel. Following gel electrophoresis, the gel was soaked in ^{3}H-acetate, rinsed, dried, and subjected to fluorography to detect tritiated histones. This revealed the presence of a 55-kilodalton polypeptide (p55) that was capable of specifically acetylating histone proteins. Using this assay, Allis and his colleagues were able to purify a single polypeptide with histone acetyltransferase activity. Using the peptide sequence from this polypeptide, they designed oligonucleotide primers and cloned the *Tetrahymena* p55 gene that encodes HAT activity. The sequence of this gene revealed a predicted amino acid sequence of p55 that is homologous (with 60% similarity) to the yeast Gcn5 protein, a protein known to be important for transcriptional activation in yeast. This was the first direct evidence that histone modification and the associated chromatin remodeling is a mechanism of transcriptional activation.

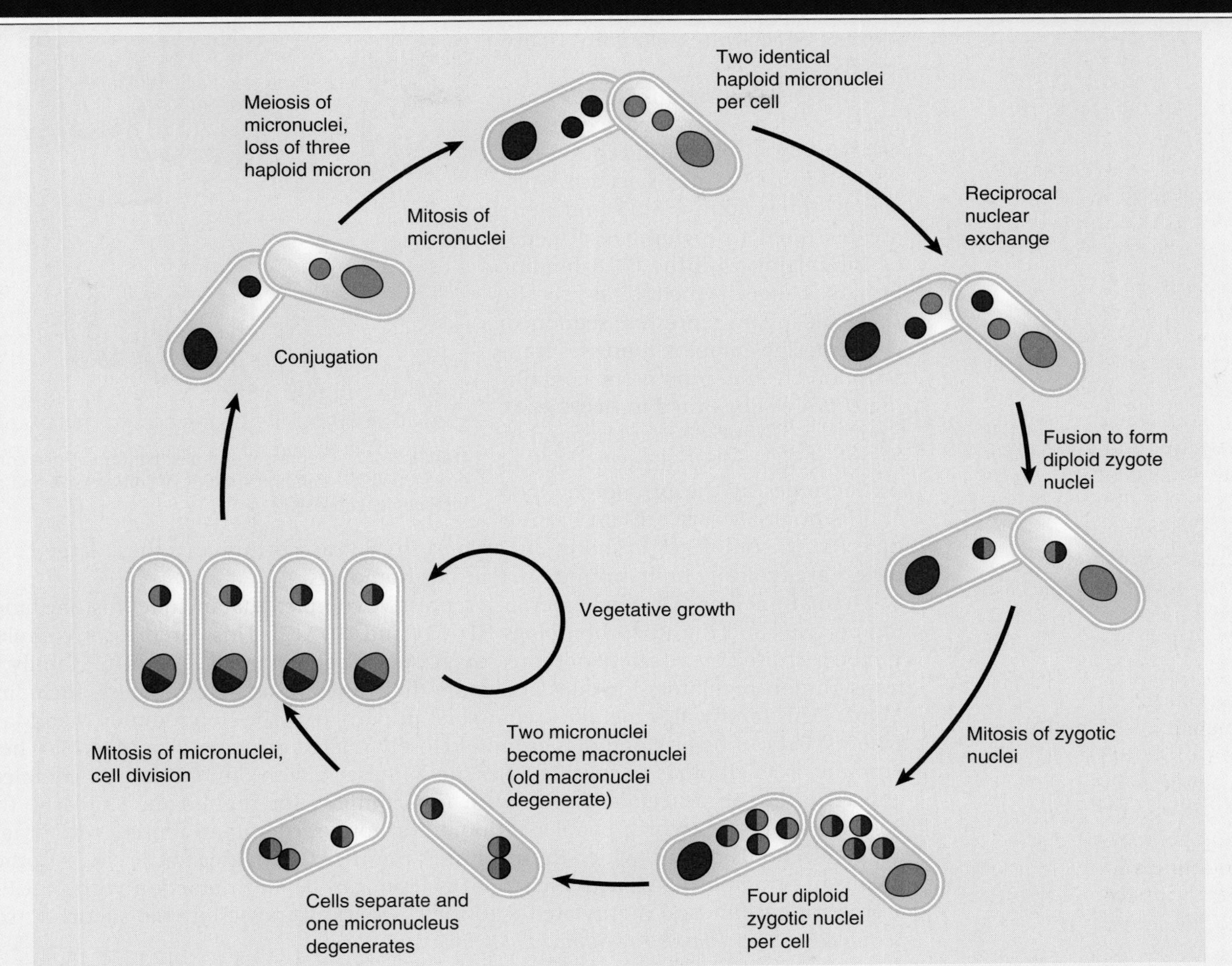

FIGURE B26.1 Life cycle of *Tetrahymena thermophila*. Vegetative cells of *T. thermophila* contain one micronucleus and one macronucleus. Conjugation begins with pairing of complementary mating types. The micronucleus undergoes meiosis to produce four haploid nuclei (three of which degenerate), followed by mitosis of the remaining haploid micronucleus. The conjugating cells then reciprocally exchange single micronuclei, which fuse to form diploid zygote nuclei. The zygote nuclei undergo two mitotic divisions to form four diploid nuclei. Two of these develop into new macronuclei (the old macronucleus degenerates), during which time chromosome fragmentation and sequence elimination occur. Cells then separate and one micronucleus degenerates. The remaining micronucleus divides mitotically and subsequent cell divisions produce four daughter cells, which can reproduce vegetatively or can enter another mating cycle.

KAT activity, site-specific repressor proteins can recruit corepressor complexes, which often contain HDAC activity.

In yeast, mutations in *SIN3* and *RPD3* result in increased expression of a variety of genes, indicating that Sin3 and Rpd3 proteins act as repressors of transcription. Sin3 and Rpd3 are recruited to a number of genes by interacting with the DNA-binding protein Ume6, which binds to the *URS1* (upstream repressive sequence) element. The complex represses transcription at the promoters containing *URS1*, as illustrated in **FIGURE 26.27**. Rpd3 is a histone deacetylase, and its recruitment leads to deacetylation of nucleosomes at the promoter. Rpd3 and its homologs are present in multiple HDAC complexes found in eukaryotes from yeast to humans; these large complexes are typically built around Sin3 and its homologs.

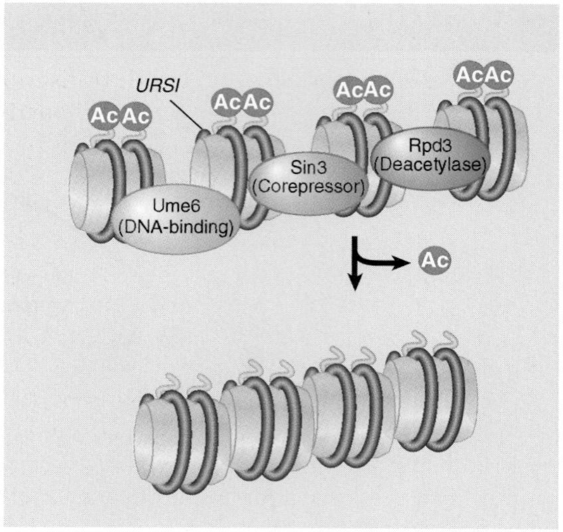

FIGURE 26.27 A repressor complex contains three components: a DNA-binding subunit, a corepressor, and a histone deacetylase.

In mammalian cells, Sin3 is part of a repressive complex that includes histone binding proteins and the Rpd3 homologs HDAC1 and HDAC2. This corepressor complex can be recruited by a variety of repressors to specific gene targets. The bHLH family of transcription regulators includes activators that function as heterodimers, including MyoD. This family also includes repressors, in particular the heterodimer Mad:Max, where Mad can be any one of a group of closely related proteins. The Mad:Max heterodimer (which binds to specific DNA sites) interacts with Sin3/HDAC1/2 complex, and requires the deacetylase activity of this complex for repression. Similarly, the SMRT corepressor (which enables retinoid hormone receptors to repress certain target genes) binds mSin3, which in turn brings the HDAC activities to the site. Another means of bringing HDAC activities to a DNA site can be an interaction with MeCP2, a protein that binds to methylated cytosines, a mark of transcriptional silencing (see *Section 27.6, CpG Islands Are Subject to Methylation*).

Absence of histone acetylation is also a feature of heterochromatin. This is true of both constitutive heterochromatin (typically involving regions of centromeres or telomeres) and facultative heterochromatin (regions that are inactivated in one cell although they may be active in another). Typically the N-terminal tails of histones H3 and H4 are not acetylated in heterochromatic regions (see *Section 27.3, Heterochromatin Depends on Interactions with Histones*).

KEY CONCEPTS

- Newly synthesized histones are acetylated at specific sites, then deacetylated after incorporation into nucleosomes.
- Histone acetylation is associated with activation of gene expression.
- Transcription activators are associated with histone acetylase activities in large complexes.
- Histone acetyltransferases vary in their target specificity.
- Deacetylation is associated with repression of gene activity.
- Deacetylases are present in complexes with repressor activity.

CONCEPT AND REASONING CHECK

What would be the effect on transcription of adding an HDAC inhibitor to a cell?

26.10 Methylation of Histones and DNA Is Connected

DNA methylation is associated with transcriptional inactivity, whereas histone methylation can be linked to either active or inactive regions, depending on the specific site of methylation. There are numerous sites of lysine methylation in the tail and core of histone H3 (a few of which occur only in some species), and a single lysine in the tail of H4. In addition, three arginines in H3 and one in H4 are also methylated.

Di- or trimethylation of H3K4 is associated with transcriptional activation, and trimethylated H3K4 occurs around the start sites of active genes. In contrast, H3 methylated at K9 or K27 is a feature of transcriptionally silent regions of chromatin, including heterochromatin and smaller regions that are known to not be expressed. Whole genome studies have begun to uncover general patterns of modifications linked to different transcriptional states, as shown in **FIGURE 26.28**.

Histone lysine methylation is catalyzed by lysine methyltransferases (KMTs), most of which contain a conserved region called the SET domain. Like acetylation, methylation is reversible, and two different families of lysine demethylases (KDMs) have been identified: the LSD1 (lysine-specific demethylase 1, also known as KDM1) family and the Jumonji family. Different classes of enzymes demethylate arginines.

In silent or heterochromatic regions, the methylation of H3 at K9 is linked to DNA methylation. The enzyme that targets this lysine is a SET-domain containing enzyme called Suv39h1. Deacetylation of H3K9 by HDACs must occur before this lysine can be methylated. H3K9 methylation then recruits a protein called HP1 (heterochromatin protein 1), which binds H3K9me via its chromodomain. HP1 then targets the activity of DNA methyltransferases (DNMTs). Most of the methylation sites in DNA are CpG islands (see *Section 27.6, CpG Islands Are Subject to Methylation*). CpG sequences in heterochromatin are usually methylated. Conversely, it is necessary for the CpG islands located in promoter regions to be unmethylated in order for a gene to be expressed.

Methylation of DNA and methylation of histones is connected in a mutually reinforcing circuit. In addition to the recruitment of DNMTs via HP1 binding to H3K4me, DNA methylation can in turn result in histone methylation. Some histone

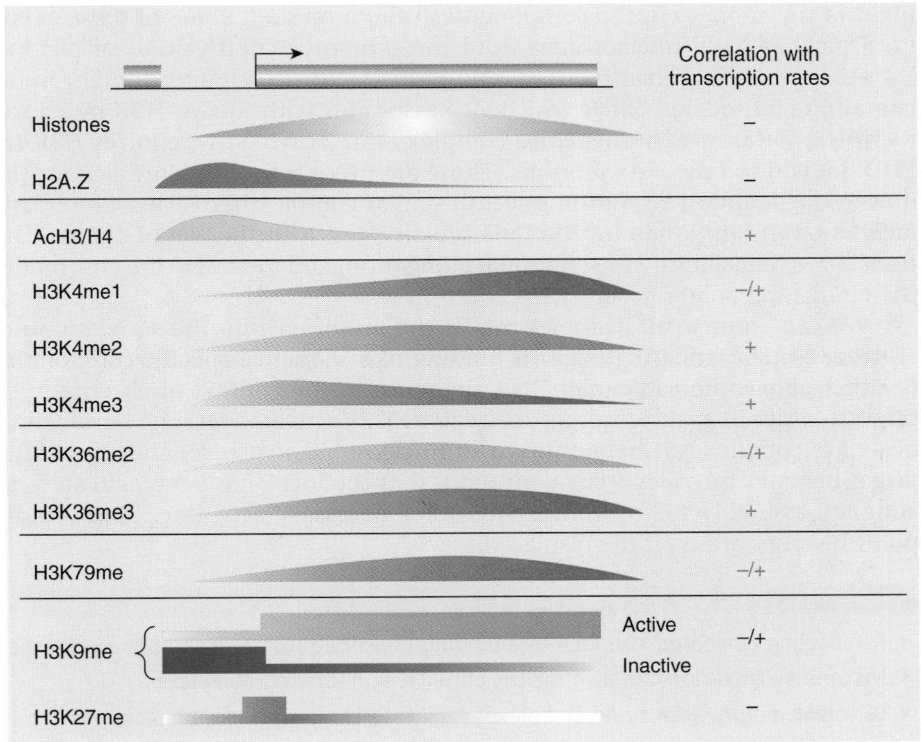

FIGURE 26.28 The distribution of histones and their modifications are mapped on an arbitrary gene relative to its promoter. The curves represent the patterns that are determined via genome-wide approaches. The location of the histone variant H2A.Z is also shown. With the exception of the data on K9 and K27 methylation, most of the data are based on yeast genes. Reprinted from *Cell*, vol. 128, Li, B., Carey, M., and Workman, J. L., pp. 707–719. Copyright 2007, with permission from Elsevier [http://www.sciencedirect.com/science/journal/00928674].

methyltransferase complexes (as well as some HDAC complexes) contain binding domains that recognize the methylated CpG doublet, so the DNA methylation reinforces the circuit by providing a target for the histone deacetylases and methyltransferases to bind. The important point is that one type of modification can be the trigger for another. These systems are widespread, as can be seen by evidence for these connections in fungi, plants, and animal cells, and for regulating transcription at promoters used by both RNA polymerases I and II, as well as maintaining heterochromatin in an inert state.

KEY CONCEPTS

- Methylation of both DNA and specific sites on histones is a feature of inactive chromatin.
- The SET domain is part of the catalytic site of protein methyltransferases.
- The two types of methylation event are connected.

CONCEPT AND REASONING CHECK

HP1 recognizes H3 methylated on lysine 9, but NOT H3 methylated on lysine 4. Why is this specificity critical?

26.11 Promoter Activation Involves Multiple Changes to Chromatin

How are histone-modifying enzymes such as acetyltransferases (or deacetylases) recruited to their specific targets? As we have seen with remodeling complexes, the process is likely to be indirect. A sequence-specific activator (or repressor) may interact with a component of the acetylase (or deacetylase) complex to recruit it to a promoter.

There can also be direct interactions between remodeling complexes and histone-modifying complexes. Binding by the SWI/SNF remodeling complex may lead in turn to binding by the SAGA acetyltransferase complex. Acetylation of histones may then stabilize the association with the SWI/SNF complex, making a mutual reinforcement of the changes in the components at the promoter. Some of these events result in displacement of nucleosomes from the promoter. Methylation of histone H3 on K4 also results in recruitment of numerous factors, including the chromodomain-containing remodeler Chd1, which also associates with SAGA. H3K4me also directly recruits another acetyltransferase complex, NuA3, which recognizes H3K4me via a PHD domain in one of its subunits. These are just a few of the interactions that occur during transcription activation in yeast; similar complex networks of interactions also facilitate transcription in multicellular eukaryotes. A further set of dynamic modifications serve to facilitate transcriptional elongation, and to "reset" the chromatin behind the elongating polymerase.

We can connect all of the events at the promoter into the series summarized in **FIGURE 26.29**. The initiating event is binding of a sequence-specific component, which is either able to find its target DNA sequence in the context of chromatin or which binds to a site in a nucleosome-free region. This activator recruits remodeling and/or acetylase complexes. Changes occur in nucleosome structure, and the acetylation of target histones provides a covalent mark that the locus has been activated. Initiation complex assembly follows (after any other necessary activators bind), and at some point histones are typically displaced.

KEY CONCEPTS

- Remodeling complexes can facilitate binding of acetyltransferase complexes, and vice versa.
- Histone methylation can also recruit chromatin modifying complexes.
- Different modifications and complexes facilitate transcription elongation.

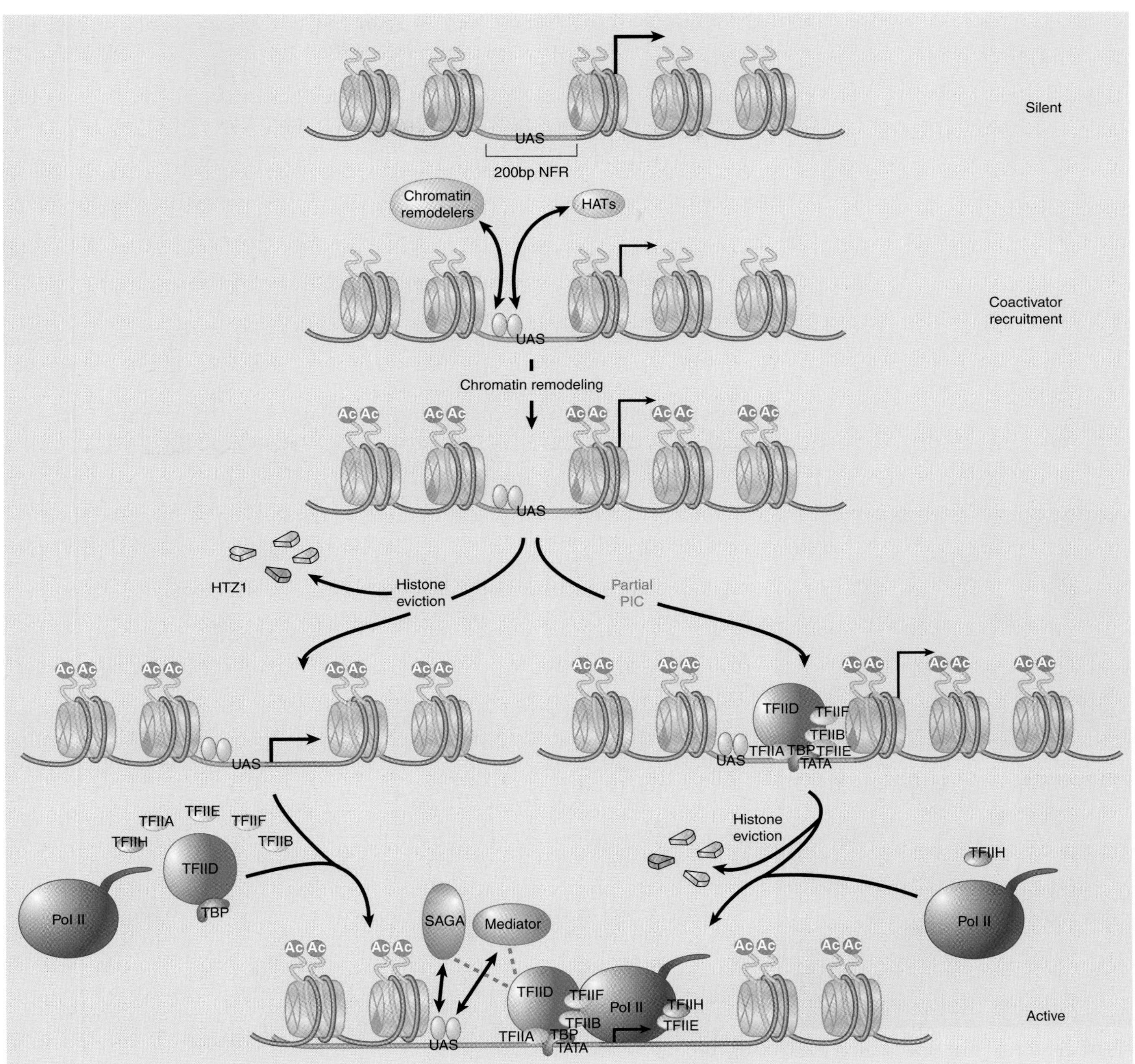

FIGURE 26.29 Htz1-containing nucleosomes flank a 200 bp NFR on both sides of a promoter. Upon targeting to the upstream-activation sequence (UAS), activators recruit various coactivators (such as Swi/Snf or SAGA). This recruitment further increases the binding of activators, particularly for those bound within nucleosomal regions. More importantly, histones are acetylated at promoter-proximal regions, and these nucleosomes become much more mobile. In one model (left), a combination of acetylation and chromatin remodeling directly results in the loss of Htz1-containing nucleosome, thereby exposing the entire core promoter to the GTFs and Pol II. SAGA and mediator then facilitate PIC formation through direct interactions. In the other model (right), which represents the remodeled state, partial PICs could be assembled at the core promoter without loss of Htz1. It is the binding of Pol II and TFIIH that leads to the displacement of Htz1-containing nucleosomes and the full assembly of PIC. Reprinted from *Cell*, vol. 128, Li, B., Carey, M., and Workman, J. L., pp. 707–719. Copyright 2007, with permission from Elsevier [http://www.sciencedirect.com/science/journal/00928674].

CONCEPT AND REASONING CHECK

How does acetylation facilitate chromatin remodeling?

26.12 Histone Phosphorylation Affects Chromatin Structure

All histones can be phosphorylated *in vivo* in different contexts. Histones are phosphorylated in three circumstances:

- cyclically during the cell cycle,
- in association with chromatin remodeling during transcription, and
- during DNA repair.

It has been known for a long time that the linker histone H1 is phosphorylated at mitosis, and more recently it was discovered that H1 is an extremely good substrate for the Cdc2 kinase that controls cell division. This led to speculation that the phosphorylation might be connected with the condensation of chromatin, but so far no direct effect of this phosphorylation event has been demonstrated, and we do not know whether it plays a role in cell division.

Phosphorylation of serine 10 of histone H3 is linked to transcriptional activation (where it promotes acetylation of K14 in the same tail), as well to chromosome condensation and mitotic progression. In *Drosophila melanogaster*, loss of a kinase that phosphorylates histone H3S10 (JIL-1) has devastating effects on chromatin structure. **FIGURE 26.30** compares the usual extended structure of the polytene chromosome (upper photograph) with the structure that is found in a null mutant that has no JIL-1 kinase (lower photograph). The absence of JIL-1 is lethal, but the chromosomes can be visualized in the larvae before they die.

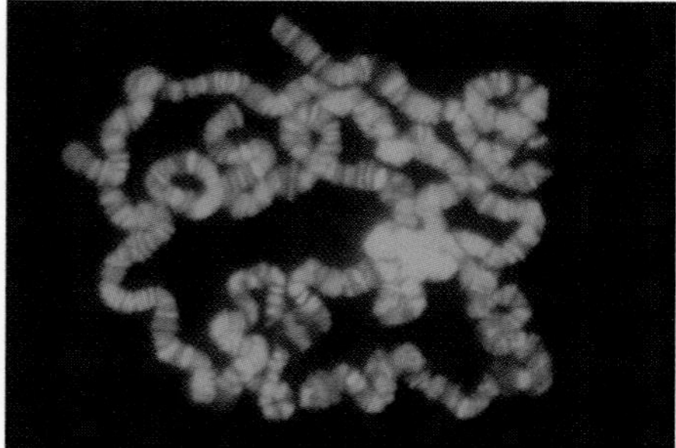
Loss of JIL-1 causes condensation

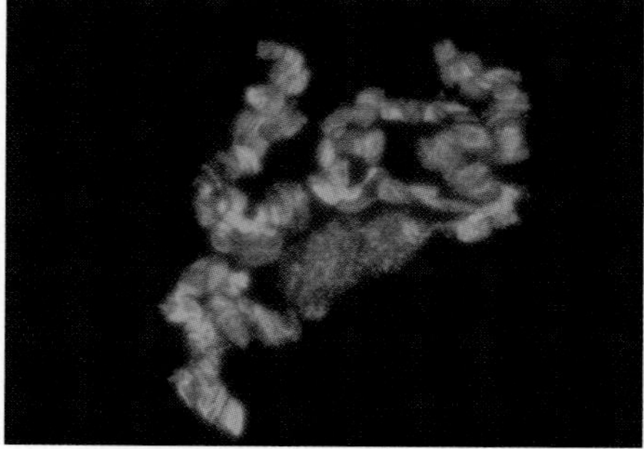

FIGURE 26.30 Flies that have no JIL-1 kinase have abnormal polytene chromosomes that are condensed instead of extended. Photos courtesy of Jorgen Johansen and Kristen M. Johansen, Iowa State University.

This suggests that H3 phosphorylation is required to generate the more extended chromosome structure of euchromatic regions. Evidence supporting the idea that JIL-1 acts directly on chromatin is that it associates with the complex of proteins that binds to the X chromosome to increase its gene expression in males (see *Section 27.5, X Chromosomes Undergo Global Changes*). It is not clear how this role of H3 phosphorylation is related to the requirement for H3 phosphorylation to initiate chromosome condensation in at least some species (including mammals and the ciliate *Tetrahymena*).

This leaves us with somewhat conflicting impressions of the roles of histone phosphorylation. Where it is important in the cell cycle, it is likely to be as a signal for condensation. Its effect in transcription and repair appears to be the opposite, where it contributes to open chromatin structures compatible with transcription activation and repair processes. (Histone phosphorylation during repair is discussed in *Section 24.4, Histone Variants Produce Alternative Nucleosomes*.)

It is possible, of course, that phosphorylation of different histones, or even of different amino acid residues in one histone, has opposite effects on chromatin structure.

KEY CONCEPT

- Histone phosphorylation is linked to transcription, repair, chromosome condensation, and cell cycle progression.

CONCEPT AND REASONING CHECK

What is the likely effect of JIL-1 mutation on transcription in polytene chromosomes?

26.13 How Is a Gene Turned On?

Multicellular eukaryotes typically begin life through the fertilization of an egg by a sperm. In both of these haploid gametes, but especially the sperm, the chromosomes are in super-condensed modified chromatin. Males of some species use positively charged *polyamines* like spermines and spermidines to replace the histones in sperm chromatin; others include sperm-specific histone variants. Once the process of fusion of the two haploid nuclei is complete in the egg, genes are then activated in a cascade of regulatory events. The general question of how a gene in closed chromatin is turned on can be broken down into (at least) two parts. How do we identify and target for activation an individual gene that is wrapped up in condensed chromatin? And, when we begin to modify the histones and remodel the chromatin, how do we prevent that from spreading to genes we do not wish to turn on?

First of all, we can imagine that replication is one mechanism by which closed chromatin can be disrupted in order to allow DNA-binding sequences to become accessible. Replication opens higher order chromatin structure by temporarily displacing histone octamers. The occupation of enhancer DNA sites on daughter strands subsequently can be viewed as competition. Chromatin can be opened if transcription factors are present in high enough concentration, as shown in **FIGURE 26.31**. If transcription factor concentration is low, then nucleosomes can bind and condense the region. This occurs in *Xenopus* embryos as oocyte 5S ribosomal genes are repressed in the embryo after fertilization.

Secondly, it is clear that some transcription factors can bind to their DNA target sequence in closed chromatin. The DNA exposed on the surface of the histone octamer is potentially accessible. These transcription factors can then recruit the histone modifiers and chromatin remodelers to begin the process of opening the gene region and clearing the promoter (see Figure 26.18). Recently described examples of antisense transcription through a gene region can facilitate this process (see Figure 13.14).

Chromatin modification typically originates from a point source, an enhancer, and spreads, usually unidirectionally, but not always. In those cases where modification spreads in a unidirectional fashion, we can ask why it is not spread bidirectionally. In any event, the next question is what prevents chromatin modification from spreading into distant gene regions.

Activation is limited by boundaries called **insulators** or **boundary elements** (see *Section 24.11, Insulators Define Independent Domains*). Very few of these have been described in detail. In one sense, they are very much like enhancers. They are modular compact sequence sets that bind specific proteins. Insulators can also function within complex loci to separate multiple temporal and tissue-specific enhancers so that only

- **insulator** A sequence that prevents an activating or inactivating effect passing from one side to the other.
- **boundary element** A DNA sequence element bound by proteins that prevent the spread of open or closed chromatin.

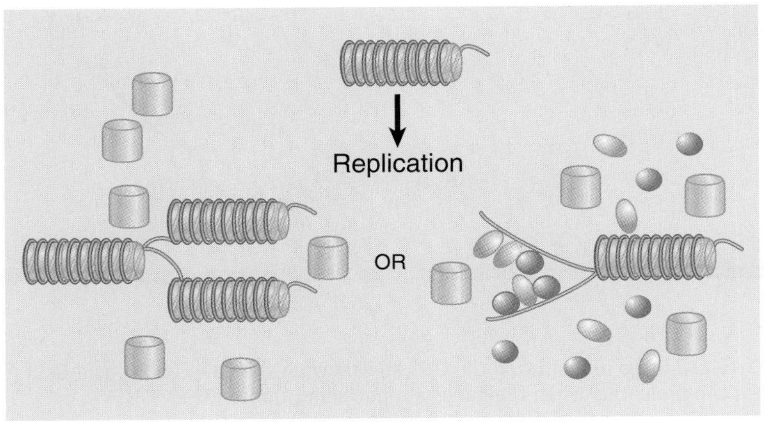

FIGURE 26.31 When replication disrupts chromatin structure, after the Y fork has passed, either chromatin can reform or transcription factors can bind and prevent chromatin formation.

one can function at a time. Boundary elements are also required to prevent the heterochromatin at centromeres and telomeres from spreading into euchromatin.

KEY CONCEPTS

- Some transcription factors may compete with histones for DNA after passage of a replication fork.
- Some transcription factors can recognize their targets in closed chromatin to initiate activation.
- The genome is divided into domains by boundary elements (insulators).
- Insulators can block the spreading of chromatin modifications from one domain to another.

CONCEPT AND REASONING CHECK

How could a transcription factor bind to closed chromatin?

26.14 The Yeast *GAL* Genes: A Model for Activation and Repression

Yeast, like bacteria, need to be able to rapidly respond to their environment (see *Section 12.3, The* lac *Operon Is Negative Inducible*). In the yeast *Saccharomyces cerevisiae* the *GAL* genes serve a similar function to the *lac* operon in *E. coli*. In an emergency, when there is little or no glucose as an energy source and only galactose (or in *E. coli*, lactose) is available, then the cell will survive because it can catabolize the alternate sugar to generate ATP. The *GAL* system in *S. cerevisiae* has been a model system to investigate gene regulation in eukaryotes for many years. We will focus on two of the genes, *GAL1* and *GAL10*, shown in **FIGURE 26.32**. Like most eukaryotic genes, the *GAL* genes are monocistronic. This pair of genes are divergently transcribed and regulated from a central control region called the **UAS (Upstream Activating Sequence)**. Like the *lac* operon in *E. coli*, the *GAL* genes are induced by their substrate, galactose. For the same reason as in *E. coli*, the *GAL* genes are also under a second level of control: catabolite repression. They cannot be activated by the substrate galactose when there is a sufficient supply of glucose, the preferred energy source.

UAS (Upstream Activating Sequence) The equivalent in yeast of the enhancer in higher eukaryotes and is bound by transcriptional activator proteins.

The *GAL* genes are under four different levels of control. The first level of control is chromatin structure. Mutations in any of the subunits of SWI/SNF and in the acetyl-

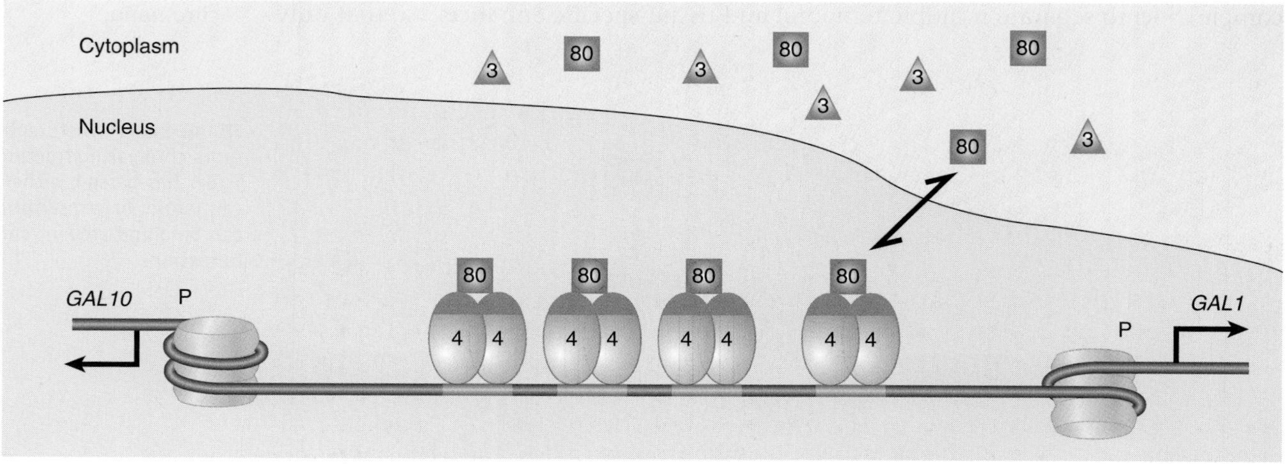

Cytoplasm
Nucleus
GAL10
P
GAL1
P

FIGURE 26.32 The yeast *GAL1/GAL10* locus highlighting the UAS and showing the Gal4, Gal80, and Gal3 regulatory proteins. Nucleosomes are positioned at the promoters when the genes are not being transcribed.

transferase complex SAGA will result in reduced expression of the *GAL* genes. Second, in the upstream control region there are both general enhancer and Mig1 repressor binding sites. The third level is the *GAL* gene-specific, galactose induction mechanism. The fourth level is catabolite (glucose) repression.

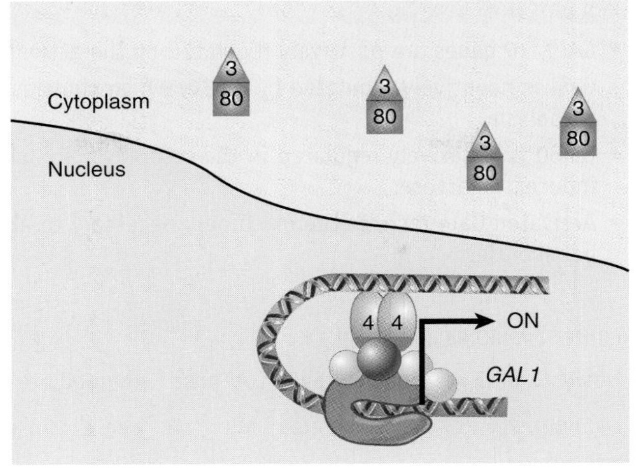

FIGURE 26.33 The yeast *GAL1* gene as it is being activated. Gal3 is holding Gal80 in the cytoplasm, allowing Gal4 to recruit the transcription machinery and activate transcription.

GAL1 is an unusual gene, in that it lacks the typical nucleosome-free region present at the start sites of most yeast genes. Instead, the start site is contained in a well-positioned nucleosome, whereas the ~170 bp UAS region is held in a nucleosome-free state, which may be partly dependent on the chromatin remodeler SWI/SNF. This DNA region has an unusual base composition, short phased AT repeats every 10 base pairs, which causes the DNA to bend. Nucleosomes containing the histone variant H2AZ (Htz in yeast) are positioned over the promoters of both *GAL1* and *GAL10*, presumably aided in their positioning by the bent DNA.

The *GAL* genes are ultimately controlled by the positive regulator Gal4, which binds as a dimer to four binding sites in the control region as shown in Figure 26.32 and **FIGURE 26.33**. Gal4 in turn is regulated by Gal80, a negative regulator that binds to Gal4 and masks its activation domain, preventing it from activating transcription. Gal80 normally shuttles back and forth between the cytoplasm and the nucleus, reentering the nucleus because of a nuclear localization domain. Gal80 in turn is regulated in the cytoplasm by the negative regulator Gal3, which is itself controlled by the inducer galactose. Gal3, when it is activated by a phosphorylated derivative of galactose in conjunction with NADP, has the ability to bind to Gal80. When it does, Gal3 masks the nuclear localization signal of Gal80, preventing it from shuttling back into the nucleus. Gal3 is thus a negative regulator of a negative regulator, which makes it a positive regulator of Gal4. This depletes the nuclear level of Gal80, unmasking Gal4 and allowing activation of the genes. NADP is thought to be a "second messenger" metabolic sensor.

Unmasked Gal4 is now able to recruit an H2B histone ubiquitylation factor (Rad6), which then stimulates histone methylation of histone H3. Next, the SAGA acetyltransferase complex both deubiquitylates H2B and acetylates histone H3, ultimately resulting in the eviction of the poised nucleosomes from the two promoters. The removal is facilitated by the chaperones Hsp90/70. This allows the recruitment of TBP/$TF_{II}D$, which then recruits RNA polymerase II and the coactivator complex Mediator. The elongation control factor $TF_{II}S$ is also recruited, which actually plays a role in initiation for at least some genes.

Although catabolite repression in eukaryotes is used for the same purpose as in *E. coli* (which uses cAMP as a positive coregulator), it has a completely different mechanism. Glucose repression of the yeast *GAL* genes is multifaceted. Glucose activates the glucose-dependent repressor Mig1 at the *GAL* loci, which interacts with the Cyc8-Tup1 corepressor, known to recruit histone deacetylases. Transcription of several other genes, including *GAL3* and the galactose permease transporter, is also similarly repressed. Overcoming repression requires only low glucose levels. This activates a protein kinase (Snf1), which inactivates Mig1 by phosphorylation. Mig1 phosphorylation disrupts its interaction with the Cyc8-Tup1 corepressor and also leads to its export from the nucleus, relieving glucose repression.

KEY CONCEPTS

- *GAL1/10* genes are positively regulated by the activator Gal4.
- Gal4 is negatively regulated by Gal80, which shuttles between the nucleus and the cytoplasm.
- Gal80 is negatively regulated in the cytoplasm by Gal3, which is activated by the inducer, galactose.
- Activated Gal4 recruits the machinery necessary to alter the chromatin and recruit RNA polymerase.

CONCEPT AND REASONING CHECK

How can Gal3, a negative regulator, positively regulate *GAL1* and *GAL10*?

26.15 Summary

Transcription factors include basal factors, activators, and coactivators. Basal factors interact with RNA polymerase at the startpoint within the promoter. Activators bind specific short DNA sequence elements located near promoters or in enhancers. Activators function by making protein–protein interactions with the basal apparatus. Some activators interact directly with the basal apparatus; others require coactivators to mediate the interaction. Activators often have a modular construction, in which there are independent domains responsible for binding to DNA and activating transcription. The main function of the DNA-binding domain may be to tether the activating domain in the vicinity of the initiation complex. Some response elements are present in many genes and are recognized by ubiquitous factors; others are present in a few genes and are recognized by tissue-specific factors.

Near the promoters for RNA polymerase II are a variety of short *cis*-acting elements, each of which is recognized by a *trans*-acting factor. The *cis*-acting elements can be located upstream of the TATA box and may be present in either orientation and at a variety of distances with regard to the startpoint or downstream within an intron. These elements are recognized by activators or repressors that interact with the basal transcription complex to determine the efficiency with which the promoter is used. Some activators interact directly with components of the basal apparatus; others interact via intermediaries called coactivators. The targets in the basal apparatus are the TAFs of $TF_{II}D$, or $TF_{II}B$ or $TF_{II}A$. The interaction stimulates assembly of the basal apparatus.

Several groups of transcription factors have been identified by sequence homology. The homeodomain is a 60-amino acid sequence that regulates development in insects, worm, and humans. It is related to the prokaryotic helix-turn-helix motif and is the DNA-binding motif for these transcription factors.

Another motif involved in DNA binding is the zinc finger, which is found in proteins that bind DNA or RNA (or sometimes both). A zinc finger has cysteine and histidine residues that bind zinc. One type of finger is found in multiple repeats in some transcription factors; another is found in single or double repeats in others.

The leucine zipper contains a stretch of amino acids rich in leucine that are involved in dimerization of transcription factors. An adjacent basic region is responsible for binding to DNA in the bZIP transcription factors.

Steroid receptors were the first members identified of a group of transcription factors in which the protein is activated by binding a small hydrophobic hormone. The activated factor becomes localized in the nucleus and binds to its specific response element, where it activates transcription. The DNA-binding domain has zinc fingers.

HLH (helix-loop-helix) proteins have amphipathic helices that are responsible for dimerization, which are adjacent to basic regions that bind to DNA. bHLH proteins have a basic region that binds to DNA and fall into two groups: ubiquitously expressed and tissue-specific. An active protein is usually a heterodimer between two subunits, one from each group. When a dimer has one subunit that does not have the basic region, it fails to bind DNA, so such subunits can prevent gene expression. Combinatorial associations of subunits form regulatory networks.

Many transcription factors function as dimers, and it is common for there to be multiple members of a family that form homodimers and heterodimers. This creates the potential for complex combinations to govern gene expression. In some cases, a family includes inhibitory members, whose participation in dimer formation prevents the partner from activating transcription.

Genes whose control regions are organized in nucleosomes usually are not expressed. In the absence of specific regulatory proteins, promoters and other regulatory regions are organized by histone octamers into a state in which they cannot be activated. This may explain the need for nucleosomes to be precisely positioned in the vicinity of a promoter, so that essential regulatory sites are appropriately exposed. Some transcription factors have the capacity to recognize DNA on the nucleosomal surface, and a particular positioning of DNA may be required for initiation of transcription.

Chromatin remodeling complexes have the ability to slide or displace histone octamers by a mechanism that involves hydrolysis of ATP. Remodeling complexes range from small to extremely large and are classified according to the type of the ATPase subunit. Common types are SWI/SNF, ISWI, CHD, and SWR1/INO80. A typical form of this chromatin remodeling is to displace one or more histone octamers from specific sequences of DNA, creating a boundary that results in the precise or preferential positioning of adjacent nucleosomes. Chromatin remodeling may also involve changes in the positions of nucleosomes, sometimes involving sliding of histone octamers along DNA.

Extensive covalent modifications occur on histone tails, all of which are reversible. Acetylation of histones occurs at both replication and transcription and facilitates formation of a less compact chromatin structure. Some coactivators, which connect transcription factors to the basal apparatus, have histone acetylase activity. Conversely, repressors may be associated with deacetylases. The modifying enzymes are usually specific for particular amino acids in particular histones. Some histone modifications may be exclusive or synergistic with others.

Large activating (or repressing) complexes often contain several activities that undertake different modifications of chromatin. Some common motifs found in proteins that modify chromatin are the chromodomain (which binds methylated lysine), the bromodomain (which targets acetylated lysine), and the SET domain (which is part of the active sites of histone methyltransferases).

CHAPTER QUESTIONS

1. Eukaryotic gene expression is controlled principally at:
   A. initiation of transcription.
   B. elongation of transcription.
   C. initiation of translation.
   D. RNA processing.
2. True activators are transcription factors that bind to:
   A. other proteins to enhance transcription.
   B. promoters.
   C. enhancers.
   D. promoters and enhancers.
3. A DNA-binding domain important in genes involved in developmental regulation in *Drosophila*:
   A. homeodomains
   B. steroid receptors
   C. leucine zippers
   D. zinc fingers

4. Leucine zippers have a stretch of amino acids with a leucine residue in every:
   A. fifth position.
   B. seventh position.
   C. tenth position.
   D. twelfth position.
5. The leucine zipper is a(n):
   A. positively charged helix that binds as a monomer in the major groove.
   B. negatively charged helix that binds as a monomer in the major groove.
   C. amphipathic helix that dimerizes, with a basic region that binds DNA.
   D. amphipathic helix that dimerizes, with an acidic region that binds DNA.
6. Changes in chromatin independent of DNA sequence that persist through multiple cell divisions are called:
   A. epigenetic effects.
   B. remodeling effects.
   C. translational effects.
   D. prions.
7. Promoters of protein-coding genes in eukaryotic chromatin have a nucleosome-free region of about:
   A. 75 bp upstream of the startpoint.
   B. 75 bp downstream of the startpoint.
   C. 200 bp upstream of the startpoint.
   D. 200 bp with the startpoint about in the middle.
8. Methylation of histones occurs most often on which of the following pairs of amino acids?
   A. lysine and asparagine
   B. asparagine and glutamine
   C. lysine and arginine
   D. lysine and glutamine
9. Bromodomains bind:
   A. methylated lysine.
   B. methylated arginine.
   C. acetylated lysine.
   D. phosphorylated serine.
10. Which histone variant is associated with nucleosomes that flank nucleosome-free regions in promoters?
   A. H2AX
   B. H2AZ
   C. macroH2A
   D. H3.3

KEY TERMS

activator
antirepressor
architectural protein
ATP-dependent chromatin remodeling complex
basal apparatus
boundary element
bromodomain
bZIP
chromatin remodeling
chromodomain
coactivator
epigenetic
helix-loop-helix
helix-turn-helix
histone acetyltransferase (HAT)
histone code
histone deacetylase (HDAC)
homeodomain
insulator
leucine zipper
lysine (K) acetyltransferase (KAT)
mediator
negative control
positive control
repressor
steroid receptor
true activator
UAS (Upstream Activating Sequence)
zinc finger

FURTHER READING

Cairns, B. (2005). Chromatin remodeling complexes: strength in diversity, precision through specialization. *Curr. Opin. Gen. Devel.* 15, 185–190.

A review of the diverse family of chromatin remodelers and their roles in transcription and other functions.

Lee, K. K. and Workman, J. L. (2007). Histone acetyltransferase complexes: one size doesn't fit all. *Nature Revs. Molec. Cell. Biol.* 8, 284–295.

A review of a variety of HAT (KAT) complexes and their roles in multiple DNA transactions, including transcription, chromosome decondensation, and others.

Li, B., M. Carey, and Workman, J. L. (2007). The role of chromatin during transcription. *Cell* 128, 707–719.

A review of many aspects of chromatin modification and remodeling and their relationship to transcriptional activation.

Peng, G. and Hopper, J. E. (2002). Gene activation by interaction of an inhibitor with a cytoplasmic signaling protein. *Proc. Natl. Acad. Sci. USA* 99, 8548–8553.

A report describing the *GAL* gene system regulation.

Ruthenburg, A. J., Li, H., Patel, D. J., and Allis, C. D. (2007). Multivalent engagement of chromatin modifications by linked binding modules. *Nature Revs. Molec. Cell. Biol.* 8, 983–994.

A review of histone modifications and their recognition, with an emphasis on models for recognition of combinations of histone modifications.

Schnitzler, G. R. (2008). Control of nucleosome positions by DNA sequence and remodeling machines. *Cell Biochem. Biophys*. 51, 67–80.

A review of the interplay between intrinsic DNA sequence and chromatin remodeling in the establishment of activating vs. repressing nucleosome positions.

Science. (2008). 319, 1781–1799.

A collection of reviews on eukaryotic gene regulation.

27

A prion protein in cellular (noninfectious) form. The inner ribbon diagram shows the secondary structure motifs of the protein backbone. The transparent overlay depicts the actual protein surface, with the charged surfaces color-coded in blue (negative) and red (positive). Photo courtesy of Martin Stumpe, Max-Planck-Institute for Biophysical Chemistry, Germany [http://www.proto-vision.com].

Epigenetic Effects Are Inherited

CHAPTER OUTLINE

27.1 Introduction

Epigenetic inheritance describes the inheritance of different functional states, which may have different phenotypic consequences, without any change in the *sequence* of DNA. This means that two individuals with the *same* DNA sequence at a locus that can exhibit different epigenetic patterns may show *different* phenotypes. The basic cause of this phenomenon is the existence of a self-perpetuating structure in one of the individuals that does not depend on DNA sequence. Several different types of structure have the ability to sustain epigenetic effects:

- A proteinaceous structure that assembles on DNA.
- A covalent modification of DNA (such as methylation of a base).
- A protein aggregate that controls the conformation of new subunits as they are synthesized.

In each case the epigenetic state results from a difference in function (typically inactivation) that is determined by the structure.

Self-perpetuating structures that assemble on DNA usually have a repressive effect by forming heterochromatic regions that prevent the expression of genes within them. Their perpetuation depends on the ability of proteins in a heterochromatic region to remain bound to those regions after replication, and then to recruit more protein subunits to sustain the complex. If individual subunits are distributed at random to each daughter duplex at replication, the two daughters will continue to be marked by the protein, although its density will be reduced to half of the level before replication. **FIGURE 27.1** shows that the existence of epigenetic effects forces us to take the view that a protein responsible for such a situation must have some sort of self-templating or self-assembling capacity to restore the original complex.

It can be the state of protein modification, rather than the presence of the protein *per se*, that is responsible for an epigenetic effect. Usually the tails of histones H3 and H4 are not acetylated in constitutive heterochromatin (and instead are methylated at specific sites). If centromeric heterochromatin is inappropriately acetylated, though, silenced genes may become active. The effect may be perpetuated through mitosis and meiosis, which suggests that an epigenetic effect has been created by changing the state of histone acetylation.

In the case of DNA methylation, a methylated DNA sequence may fail to be transcribed, whereas the nonmethylated sequence will be expressed. **FIGURE 27.2** shows

▸ **epigenetic** Changes that influence the phenotype without altering the genotype. They consist of changes in the properties of a cell that are inherited, but that do not represent a change in genetic information.

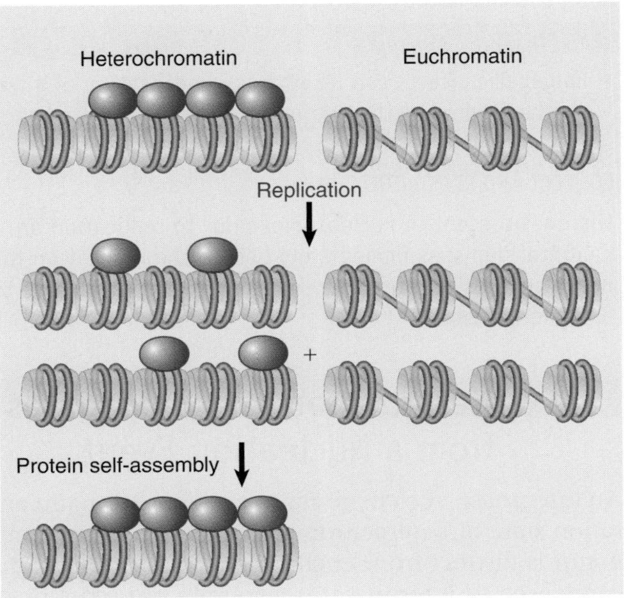

FIGURE 27.1 Heterochromatin is created by proteins that associate with histones. Perpetuation through division requires that the proteins associate with each daughter duplex and then recruit new subunits to reassemble the repressive complexes.

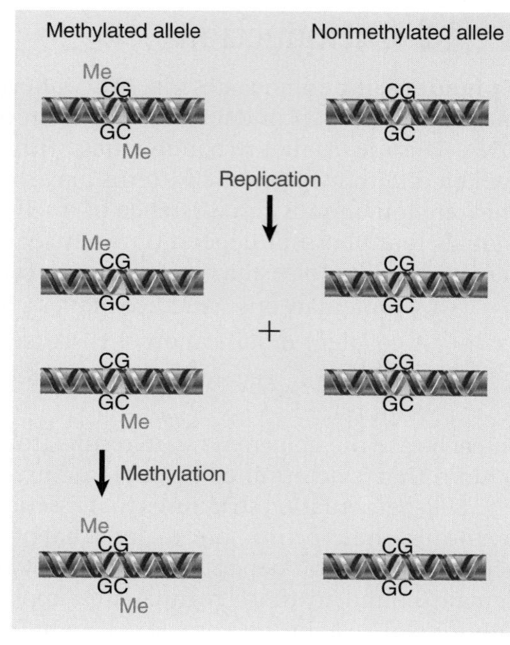

FIGURE 27.2 Replication of a methylated site produces hemimethylated DNA, in which only the parental strand is methylated. A perpetuation methylase recognizes hemimethylated sites and adds a methyl group to the base on the daughter strand. This restores the original situation, in which the site is methylated on both strands. A nonmethylated site remains nonmethylated after replication.

how this situation is inherited. One allele has a sequence that is methylated on both strands of DNA, whereas the other allele has an unmethylated sequence. Replication of the methylated allele creates hemimethylated daughters that are restored to the methylated state by a constitutively active DNA **methyltransferase** enzyme, sometimes called a "maintenance methylase." Replication does not affect the state of the nonmethylated allele. If the state of methylation affects transcription, the two alleles differ in their state of gene expression, even though their sequences are identical. As we will see, DNA methylation, histone modifications, and heterochromatin assembly can all mutually reinforce an epigenetically silenced state, contributing to stability of the silenced state over multiple cell divisions.

▸ **methyltransferase** An enzyme that adds a methyl group to a substrate, which can be a small molecule, a protein, or a nucleic acid.

▸ **prion** A proteinaceous infectious agent that behaves as an inheritable trait, although it contains no nucleic acid. Examples are PrPSc, the agent of scrapie in sheep and bovine spongiform encephalopathy, and Psi, which confers an inherited state in yeast.

Independent protein aggregates that cause epigenetic effects (called **prions**) work by sequestering the protein in a form in which its normal function cannot be displayed. Once the protein aggregate has formed, it forces newly synthesized protein subunits to join it in the inactive conformation.

KEY CONCEPT

- Epigenetic effects can result from modification of a nucleic acid after it has been synthesized or by the perpetuation of protein structures.

CONCEPT AND REASONING CHECK

Histones present in nucleosomes prior to replication are randomly distributed to both daughter duplexes immediately behind the replication fork, and are mixed together with newly synthesized histones. How might this contribute to the maintenance of a particular epigenetic state?

27.2 Heterochromatin Propagates from a Nucleation Event

An interphase nucleus contains both euchromatin and heterochromatin. The condensation state of heterochromatin is close to that of mitotic chromosomes. Heterochromatin is distinct from euchromatin in a number of ways. Heterochromatin remains condensed in interphase, is generally transcriptionally repressed, replicates late in S phase, and may be localized to the nuclear periphery. Centromeric heterochromatin typically consists of satellite DNAs; however, the formation of heterochromatin is not rigorously defined by sequence. When a gene is transferred, either by a chromosomal translocation or by transfection and integration, into a position adjacent to heterochromatin, it may become inactive as the result of its new location, implying that it has become heterochromatic.

Such inactivation is the result of an epigenetic effect. It may differ between individual cells in an animal, and results in the phenomenon of **position effect variegation (PEV)**, in which genetically identical cells have different phenotypes (PEV was introduced in the *Medical Applications* box in Chapter 24). This has been well characterized in *Drosophila*. FIGURE 27.3 shows an example of position effect variegation in the fly eye. Some of the regions in the eye lack color, whereas others are red. This occurs because the *white* gene (required to develop red pigment) is inactivated by adjacent heterochromatin in some cells, but remains active in others.

▸ **position effect variegation (PEV)** Silencing of gene expression that occurs as the result of proximity to heterochromatin.

The explanation for this effect is shown in FIGURE 27.4. Inactivation spreads from heterochromatin into the adjacent region for a variable distance. In some cells it goes far enough to inactivate a nearby gene, but in others it does not. This takes place at a certain point in embryonic development, and after that point the state of the gene is stably inherited by the progeny cells. Cells descended from an ancestor in which the gene was inactivated form patches corresponding to the phenotype of loss-of-function (in the case of *white*, absence of color).

The closer a gene lies to heterochromatin, the higher the probability that it will be inactivated. This results from the fact that the formation of heterochromatin is a two-stage process: a *nucleation* event occurs at a specific sequence (triggered by binding of a protein that recognizes this sequence), and then the inactive structure *propagates* along the chromatin fiber. The distance for which the inactive structure extends is not precisely determined and may be stochastic, being influenced by parameters such as the quantities of limiting protein components. One factor that may affect the

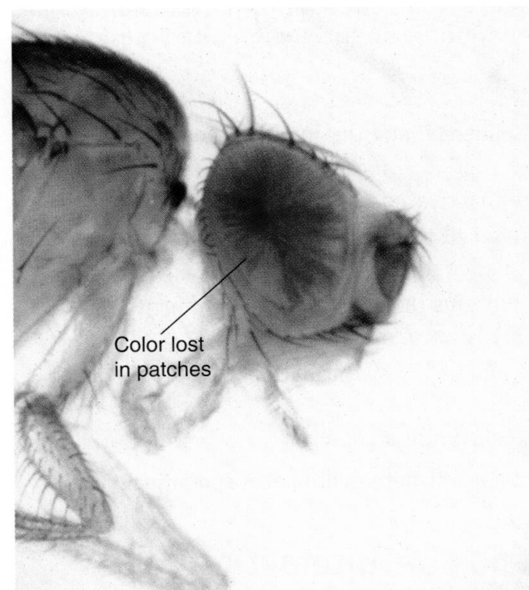

FIGURE 27.3 Position effect variegation in eye color results when the *white* gene is integrated near heterochromatin. Cells in which *white* is inactive give patches of white eye, whereas cells in which *white* is active give red patches. The severity of the effect is determined by the closeness of the integrated gene to heterochromatin. Photo courtesy of Steven Henikoff, Fred Hutchinson Cancer Research Center.

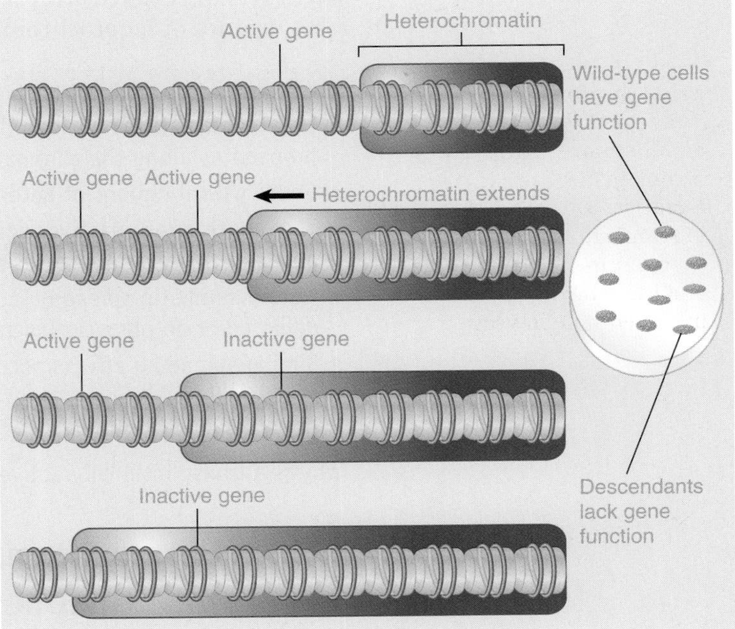

FIGURE 27.4 Extension of heterochromatin inactivates genes. The probability that a gene will be inactivated depends on its distance from the heterochromatin region.

spreading process is the activation of promoters in the region; an active promoter may inhibit spreading. In addition, insulators can protect a transcriptionally active region by preventing heterochromatin spreading (see *Section 24.11, Insulators Define Independent Domains*).

Genes that are closer to heterochromatin are more likely to be inactivated, and will, therefore, be inactive in a greater proportion of cells. This model predicts that the boundaries of a heterochromatic region might be terminated by exhausting the supply of one of the proteins that is required. In fact, many proteins that regulate or contribute to heterochromatin formation were originally identified in *Drosophila* using screens for dosage-dependent effects of these proteins on heterochromatin spreading.

▸ **telomeric silencing** The repression of gene activity that occurs in the vicinity of a telomere.

The effect of **telomeric silencing** in yeast is analogous to position effect variegation in *Drosophila;* genes translocated to a telomeric location show the same sort of variable loss of activity. This results from a spreading effect that propagates from the telomeres. In this case, the binding of the Rap1 protein to telomeric repeats triggers the nucleation event, which results in the recruitment of heterochromatin proteins, as described in *Section 27.3, Heterochromatin Depends on Interactions with Histones.*

In addition to the telomeres, there are two other sites at which heterochromatin is nucleated in yeast. Yeast mating type is determined by the activity of a single active locus (*MAT*), but the genome contains two other copies of the mating type sequences (*HML* and *HMR*), which are maintained in an inactive form (this system was introduced in *Section 19.9, Yeast Use a Specialized Recombination Mechanism to Switch Mating Type*). The silent loci *HML* and *HMR* nucleate heterochromatin via binding of several proteins (rather than the single protein, Rap1, required at telomeres), which then lead to propagation of heterochromatin similar to that at telomeres. Heterochromatin in yeast exhibits features typical of heterochromatin in other species, such as transcriptional inactivity and self-perpetuating protein structures superimposed on nucleosomes (which are generally deacetylated). The only notable difference between yeast heterochromatin and that of most other species is that histone methylation in yeast is not associated with silencing, whereas specific sites of histone methylation are a key feature of heterochromatin formation in most multicellular eukaryotes.

KEY CONCEPTS

- Heterochromatin is nucleated at a specific sequence, and the inactive structure propagates along the chromatin fiber.
- Genes within regions of heterochromatin are inactivated.
- The length of an inactive region can vary from cell to cell; as a result, inactivation of genes in this vicinity causes position effect variegation.
- Heterochromatin spreading continues until proteins required for heterochromatin are depleted or an obstacle (such as an insulator) is encountered.
- Similar spreading effects occur at telomeres and at the silent mating type loci in yeast.

CONCEPT AND REASONING CHECK

Why is a transcriptionally active gene likely to prevent heterochromatin spreading?

27.3 Heterochromatin Depends on Interactions with Histones

Inactivation of chromatin occurs by the addition of proteins to the nucleosomal fiber. The inactivation may be due to a variety of effects, including the condensation of chromatin to make it inaccessible to the apparatus needed for gene expression, the addition of proteins that directly block access to regulatory sites, or the presence of proteins that directly inhibit transcription.

Two systems that have been characterized at the molecular level involve HP1 in mammals and the SIR complex in yeast. Although many of the proteins involved in each system are not evolutionarily related, the general mechanism of reaction is similar: the points of contact in chromatin are the N-terminal tails of the histones.

Our insights into the molecular mechanisms for regulating the formation of heterochromatin originated with mutants that affect position effect variegation. Some 30 genes have been identified in *Drosophila*. They are named systematically as *Su(var)* for genes whose products act to suppress variegation and *E(var)* for genes whose products enhance variegation. Remember that the genes were named for the behavior of the *mutant* loci; thus *Su(var)* mutations lie in genes whose products are needed for the formation of heterochromatin. They include enzymes that act on chromatin, such as histone deacetylases, and proteins that are localized to heterochromatin. In contrast, *E(var)* mutations lie in genes whose products are needed to activate gene expression. They include members of the SWI/SNF complex (see *Section 26.6, Chromatin Remodeling Is an Active Process*).

HP1 (heterochromatin protein 1) is one of the most important Su(var) proteins. It was originally identified as a protein that is localized to heterochromatin by staining polytene chromosomes with an antibody directed against the protein. It was later shown to be the product of the gene *Su(var)2-5*. HP1 is now called HP1α, because two related proteins, HP1β and HP1γ, have since been found.

HP1 contains a chromodomain near the N-terminus, and another domain that is related to it, called the chromo-shadow domain, at the C-terminus. The HP1 chromodomain binds to histone H3 that is trimethylated at lysine 9 (H3K9me3).

Mutation of a deacetylase that acts on the H3 N-terminus prevents the methylation at K9. This leads to a model for initiating formation of heterochromatin shown in **FIGURE 27.5**. First the deacetylase acts to remove acetyl groups from the H3 tail, which then allows the SUV39H1 methyltransferase (also known as KMT1A) to methylate H3K9 to create the signal to which HP1 will bind. This is a trigger for forming inactive chromatin. **FIGURE 27.6** shows that the inactive region may then be extended by the ability of further HP1 molecules to interact with one another.

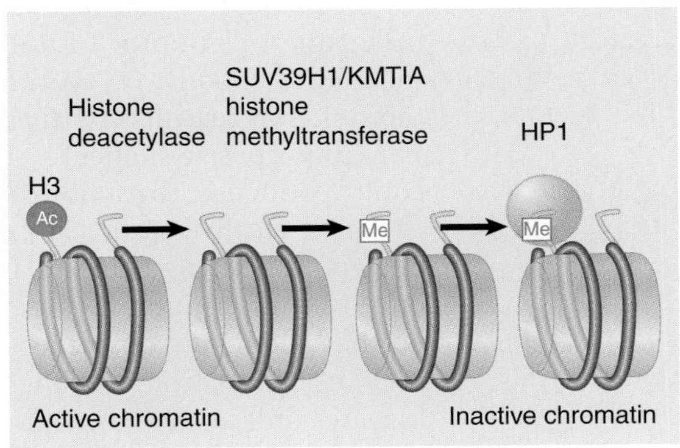

FIGURE 27.5 SUV39H1/KMTIA is a histone methyltransferase that acts on K9 of histone H3. HP1 binds to the methylated histone.

Heterochromatin formation at telomeres and silent mating type loci in yeast relies on an overlapping set of genes, known as *silent information regulators* (*SIR* genes). Mutations in *SIR2*, *SIR3*, or *SIR4* cause *HML* and *HMR* to become activated, and also relieve the inactivation of genes that have been integrated near telomeric heterochromatin. The products of these loci therefore function to maintain the inactive state of both types of heterochromatin.

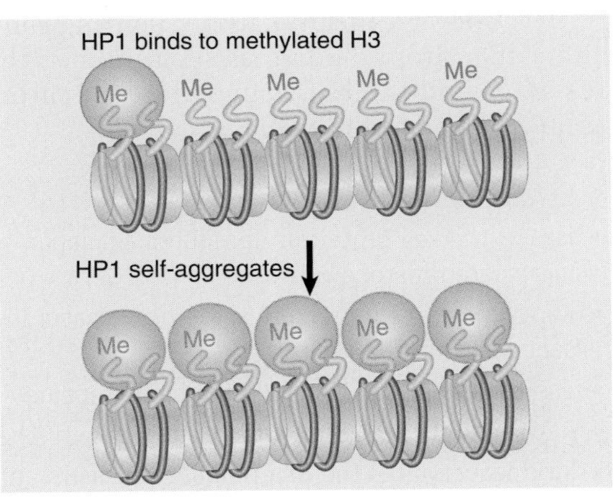

FIGURE 27.6 Binding of HP1 to methylated histone H3 forms a trigger for silencing because further molecules of HP1 aggregate on the nucleosome chain.

FIGURE 27.7 shows a model for actions of these proteins. At telomeres, the process is initiated by a sequence-specific DNA-binding protein, Rap1, which binds to the $C_{1-3}A$ repeats at the telomeres. The proteins Sir3 and Sir4 interact with Rap1 and also with one another

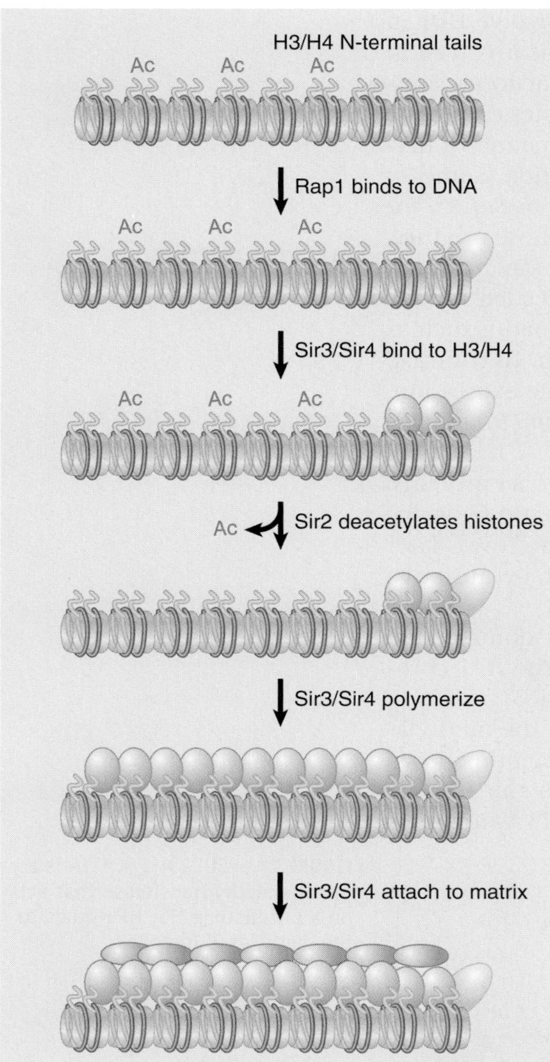

FIGURE 27.7 Formation of heterochromatin is initiated when Rap1 binds to DNA. Sir3/4 bind to Rap1 and also to histones H3/H4. Sir2 deacetylates histones. The SIR complex polymerizes along chromatin and may connect telomeres to the nuclear matrix.

(they may function as a heteromultimer). Sir3/Sir4 interact with the N-terminal tails of the histones H3 and H4, with a preference for unacetylated tails. Another SIR protein, Sir2, is a deacetylase, and its activity is necessary to maintain binding of the Sir3/4 complex to chromatin.

Rap1 has the crucial role of identifying the DNA sequences at which heterochromatin forms. It recruits Sir3/Sir4, and they interact directly with the histones H3/H4. Once Sir3/Sir4 have bound to histones H3/H4, the complex may polymerize further and spread along the chromatin fiber. This may inactivate the region, either because coating with Sir3/Sir4 itself has an inhibitory effect, or because binding to histones H3/H4 induces some further change in structure. The C-terminus of Sir3 has a similarity to nuclear lamin proteins (constituents of the nuclear matrix) and may be responsible for tethering heterochromatin to the nuclear periphery.

A similar series of events forms the silenced regions at *HMR* and *HML*. In this case, three sequence-specific factors are involved in triggering formation of the complex: Rap1, Abf1 (a transcription factor), and ORC (the origin replication complex). Another SIR protein, Sir1, binds to ORC and then recruits Sir2, -3, and -4 to form the repressive structure.

Formation of heterochromatin in the yeast *S. pombe* utilizes an RNAi-dependent pathway (see *Section 13.8, Eukaryotes Contain Regulator RNAs*). This pathway is initiated by the production of siRNA molecules resulting from transcription of centromeric repeats. These siRNAs result in formation of RNA-induced transcriptional gene silencing (RITS) complex. The siRNA components are responsible for localizing the complex at centromeres. The siRNA complex promotes methylation of histone H3K9 by the Clr4 methyltransferase (also known as KMT1, a homolog of *Drosophila Su(Var)3-9*). H3K9 methylation recruits the *S. pombe* homolog of HP1, Swi6.

How does a silencing complex repress chromatin activity? It could condense chromatin so that regulator proteins cannot find their targets. The simplest case would be to suppose that the presence of a silencing complex is mutually incompatible with the presence of transcription factors and RNA polymerase. The cause could be that silencing complexes block remodeling (and thus indirectly prevent factors from binding) or that they directly obscure the binding sites on DNA for the transcription factors. The situation may not be this simple, though, because transcription factors and RNA polymerase can be found at promoters in silenced chromatin. This could mean that the silencing complex prevents the factors from working rather than from binding as such. In fact, there may be competition between gene activators and the repressing effects of chromatin, so that activation of a promoter inhibits spread of the silencing complex.

KEY CONCEPTS

- HP1 is the key protein in forming mammalian heterochromatin, and acts by binding to methylated histone H3.
- Rap1 initiates formation of heterochromatin in yeast by binding to specific target sequences in DNA.
- The targets of Rap1 include telomeric repeats and silencers at *HML* and *HMR*.
- Rap1 recruits Sir3/Sir4, which interact with the N-terminal tails of H3 and H4.
- Sir2 deacetylates the N-terminal tails of H3 and H4 and promotes spreading of Sir3/Sir4.
- RNAi pathways promote heterochromatin formation at centromeres.

CONCEPT AND REASONING CHECK

What is the role of sequence-specific binding proteins in heterochromatin formation?

27.4 Polycomb and Trithorax Are Antagonistic Repressors and Activators

Regions of constitutive heterochromatin, such as at telomeres and centromeres, provide one example of specific repression by chromatin structure. Another insight into chromatin silencing is provided by the genetics of homeotic genes (which affect the identity of body segments) in *Drosophila*. These studies have led to the identification of a protein complex that may *maintain* certain genes in a repressed state. The gene *Pc* (*polycomb*) is the prototype for a class of genes called the *Pc* group (*Pc-G*). *Pc* mutants show transformations of cell type that are equivalent to gain-of-function mutations in homeotic genes, because in *Pc* mutants these genes are expressed in tissues in which they are normally repressed.

The Pc-G proteins are not conventional repressors. They are not responsible for determining the *initial* pattern of expression of the genes on which they act. In the absence of Pc-G proteins, these genes are initially repressed as usual, but later in development the repression is lost without Pc-G group functions. This suggests that *the Pc-G proteins in some way recognize the state of repression when it is established, and they then act to perpetuate it through cell division of the daughter cells*. In other words, Pc-G proteins are necessary for the *maintenance* phase of repression, but not for the initial *establishment* phase. **FIGURE 27.8** shows a model in which Pc-G proteins bind in conjunction with a repressor, but the Pc-G proteins remain bound after the repressor is no longer available. This is necessary to maintain repression, so that if Pc-G proteins are absent, the gene becomes activated.

The Pc proteins function in large complexes, known as PRCs (Polycomb-repressive complexes). These complexes contain proteins with a variety of functions and enzymatic activities, such as histone deacetylase, methyltransferase, and ubiquitin ligase activities. Pc itself has a chromodomain that binds to methylated H3. These properties reveal a connection between chromatin remodeling and repression by PRCs.

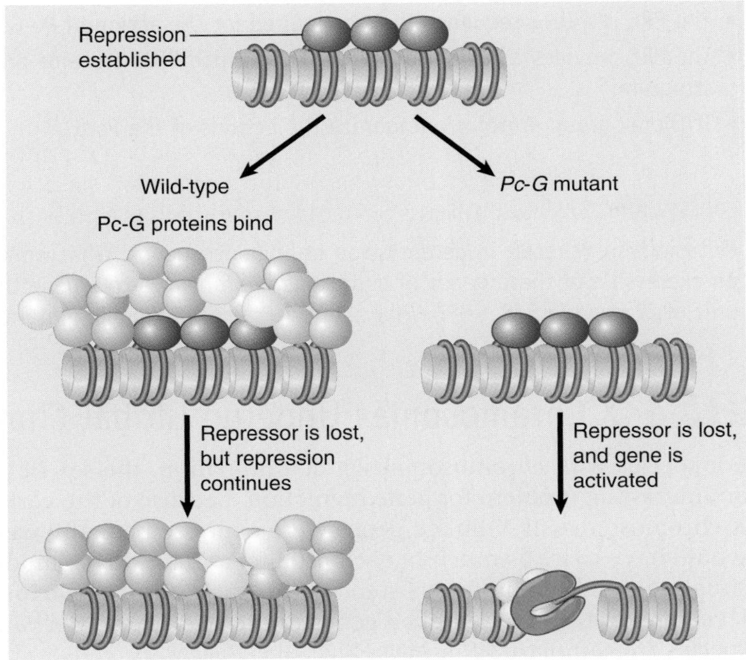

FIGURE 27.8 Pc-G proteins do not initiate repression, but are responsible for maintaining it.

A region of DNA that is sufficient to enable the response to the *Pc-G* genes is called a PRE (*Polycomb* response element). It can be defined operationally by the property that it maintains repression in its vicinity throughout development. The assay for a PRE is to insert it close to a reporter gene that is controlled by an enhancer that is repressed in early development, and then to determine whether the reporter becomes expressed subsequently in the descendants. An effective PRE will prevent such re-expression.

The PRE is a complex structure that measures ~10 kb. Several proteins, including Pho, Phol and GAGA factor (GAF), with DNA-binding activity for sites within the PRE have been identified, but there could be others. When a locus is repressed by Pc-G, however, the Pc-G proteins occupy a much larger region of DNA than the PRE itself. Pc is found locally over a few kilobases of DNA surrounding a PRE. This suggests that the PRE may provide a nucleation center, from which a structural state depending on Pc-G proteins may propagate. This model is supported by the observation of effects related to position effect variegation (see Figure 27.4); that is, a gene near a locus whose repression is maintained by Pc-G may become heritably inactivated in some cells but not others.

A working model for Pc-G binding at a PRE is suggested by the properties of the individual proteins. First Pho and Phol bind to specific sequences within the PRE. A methyltransferase KMT6 (also known as E(z)) is recruited by Pho/Phol; it then methylates histone H3 at lysine 27. This creates the binding site for the PRC, because the chromodomain of Pc binds to methylated H3K27. The Polycomb complex induces a more compact structure in chromatin; each PRC1 complex causes about three nucleosomes to become less accessible.

The *trithorax* group (*trxG*) of proteins have the opposite effect to the Pc-G proteins: they act to maintain genes in an active state. There may be some similarities in the actions of the two groups: mutations in some loci prevent both Pc-G and trxG from functioning, suggesting that they could rely on common components. trxG proteins are quite diverse, some comprise subunits of chromatin remodeling enzymes such as SWI/SNF, whereas others also possess important histone modification activities, such as histone **demethylases** which could oppose the activities of Pc-G proteins.

▸ **demethylase** A name for an enzyme that removes a methyl group, typically from DNA, RNA, or protein.

KEY CONCEPTS

- Polycomb group proteins (Pc-G) perpetuate a state of repression through cell divisions.
- The PRE is a DNA sequence that is required for the action of Pc-G.
- The PRE provides a nucleation center from which Pc-G proteins propagate an inactive structure.
- Trithorax group proteins antagonize the actions of the Pc-G.

CONCEPT AND REASONING CHECK

Mutations in *Pc* result in derepression of homeotic genes. What would be the likely effect on expression of these genes of mutations in *brahma*, a trxG gene that is the ATPase subunit of *Drosophila* SWI/SNF? What about a *Pc brahma* double mutant?

27.5 X Chromosomes Undergo Global Changes

For species with chromosomal sex determination, the sex of the individual presents an interesting problem for gene regulation, because of the variation in the number of X chromosomes. If X-linked genes were expressed equally well in each sex, females would have twice as much of each product as males. The importance of avoiding this situation is shown by the existence of **dosage compensation**, which equalizes the level of expression of X-linked genes in the two sexes. Mechanisms used in different species are summarized in **FIGURE 27.9**:

▸ **dosage compensation** Mechanisms employed to compensate for the discrepancy between the presence of two X chromosomes in one sex but only one X chromosome in the other sex.

- In mammals, one of the two female X chromosomes is inactivated. The result is that females have only one active X chromosome, which is the same situation found in males. The active X chromosome of females and the single X chromosome of males are expressed at the same level.
- In *Drosophila*, the expression of the single male X chromosome is doubled relative to the expression of each female X chromosome.
- In *Caenorhabditis elegans*, the expression of each female X chromosome is halved relative to the expression of the single male X chromosome.

	Mammals	Flies	Worms
	Inactivate one ♀ X	Double expression ♂ X	Halve expression two ♀ X
X			
X			
X			
Y			

FIGURE 27.9 Different means of dosage compensation are used to equalize X chromosome expression in male and female.

The common feature in all these mechanisms of dosage compensation is that *the entire chromosome is the target for regulation*. A global change occurs that quantitatively affects all of the promoters on the chromosome. We know most about the inactivation of the X chromosome in mammalian females, where the entire chromosome becomes heterochromatic.

The twin properties of heterochromatin are its condensed state and its associated transcriptional inactivity. Heterochromatin can be divided into two types:

- **Constitutive heterochromatin** contains specific sequences that have no coding function. In general these include satellite DNAs, often found at the centromeres, and telomeric repeats. These regions are invariably heterochromatic because of their intrinsic sequence.
- **Facultative heterochromatin** takes the form of chromosome segments or entire chromosomes that are inactive in one cell lineage, although they can be expressed in other lineages. The example *par excellence* is the mammalian X chromosome. The inactive X chromosome in females is perpetuated in a heterochromatic state, whereas the active X chromosome is euchromatic. Thus *identical DNA sequences are involved in both states*. Once the inactive state has been established, it is inherited by descendant cells. This is an example of epigenetic inheritance, because it does not depend on the DNA sequence.

▸ **constitutive heterochromatin** The inert state of permanently nonexpressed sequences, usually satellite DNA.

▸ **facultative heterochromatin** The inert state of sequences that also exist in active copies, for example, one mammalian X chromosome in females.

▸ **single X hypothesis** The theory that describes the inactivation of one X chromosome in female mammals.

Our basic view of the situation of the female mammalian X chromosomes was formed by the **single X hypothesis** in 1961. Female mice that are heterozygous for X-linked coat color mutations have a variegated phenotype in which some areas of the coat are wild type, but others are mutant. **FIGURE 27.10** shows that this can be explained *if one of the two X chromosomes is inactivated at random in each cell of a small precursor population*. Cells in which the X chromosome carrying the wild-type gene is inactivated give rise to progeny that express only the mutant allele on the active chromosome. Cells derived from a precursor where the other chromosome was inactivated have an active wild-type gene. In the case of coat color, cells descended from a particular precursor stay together and thus form a patch of the same color, creating the pattern of visible variegation. In other cases, individual cells in a population will express one or the other of X-linked alleles; for example, in heterozygotes for the X-linked locus *G6PD*, any particular red blood cell will express only one of the two allelic forms.

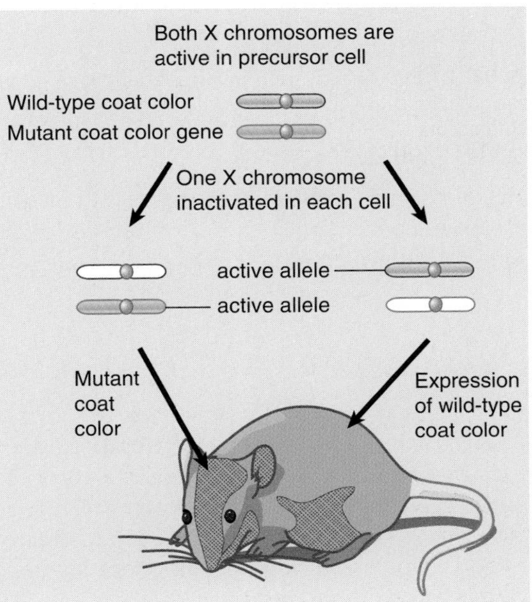

FIGURE 27.10 X-linked variegation is caused by the random inactivation of one X chromosome in each precursor cell. Cells in which the + allele is on the active chromosome have wild phenotype; cells in which the – allele is on the active chromosome have mutant phenotype.

n – 1 rule The rule that states that only one X chromosome is active in female mammalian cells; any others are inactivated.

Inactivation of the X chromosome in females is governed by the **n – 1 rule**: However many X chromosomes are present, all but one will be inactivated. In normal females there are of course two X chromosomes, but in rare cases where nondisjunction has generated a 3X or greater genotype, only one X chromosome remains active. This suggests a general model in which a specific event is limited to one X chromosome and protects it from an inactivation mechanism that applies to all the others.

A single locus on the X chromosome is sufficient for inactivation. When a translocation occurs between the X chromosome and an autosome, this locus is present on only one of the reciprocal products, and only that product can be inactivated. By comparing different translocations, it was possible to map this locus, which is called the *Xic* (X-inactivation center). A cloned region of 450 kb contains all the properties of the *Xic*. When this sequence is inserted as a transgene onto an autosome, the autosome can become subject to inactivation.

Xic is a *cis*-acting locus that contains the information necessary to both count X chromosomes and inactivate all copies but one. Inactivation spreads from *Xic* along the entire X chromosome. When *Xic* is present on an X chromosome-autosome translocation, inactivation spreads into the autosomal regions (although the effect is not always complete).

Xic contains a gene, called *Xist*, which is stably expressed only on the *inactive* X chromosome. The behavior of this gene is effectively the opposite from all other loci on the chromosome, which are turned off. Deletion of *Xist* prevents an X chromosome from being inactivated. It does not, however, interfere with the counting mechanism (because other X chromosomes can be inactivated). Thus we can distinguish two features of *Xic*: an unidentified element(s) required for counting and the *Xist* gene required for inactivation.

FIGURE 27.11 illustrates the role of *Xist* RNA in X-inactivation. *Xist* codes for a noncoding RNA that lacks open reading frames. The *Xist* RNA "coats" the X chromosome from which it is synthesized; this suggests that it has a structural role. Prior to X-inactivation, it is synthesized by both female X chromosomes. Following inactivation, the RNA is found only on the inactive X chromosome. The transcription rate remains the same before and after inactivation, so the transition depends on posttranscriptional events. An antisense RNA, *Tsix*, plays a role in regulating the stability of *Xist*.

Accumulation of *Xist* on the future inactive X results in exclusion of transcription machinery (such as RNA polymerase II), and leads to the recruitment of Polycomb repressor complexes (PRC1 and PRC2), which trigger a series of chromosome-wide histone modifications (H2AK119 ubiquitination, H3K27 methylation, and H4K20 methylation, and H4 deacetylation). Late in the process, an inactive X-specific his-

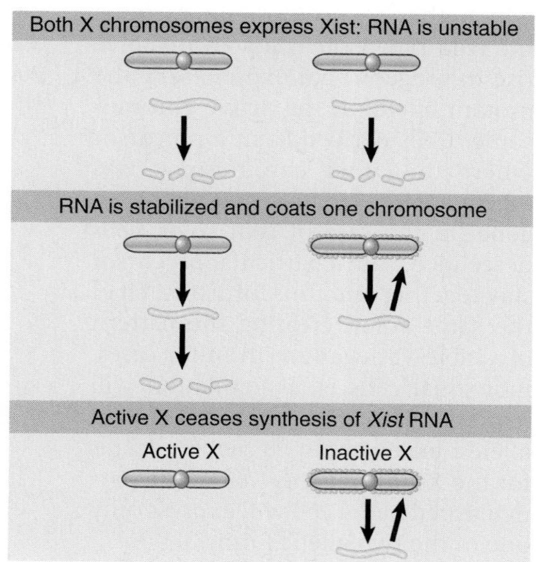
Both X chromosomes express Xist: RNA is unstable
RNA is stabilized and coats one chromosome
Active X ceases synthesis of Xist RNA
Active X
Inactive X

FIGURE 27.11 X-inactivation involves stabilization of *Xist* RNA, which coats the inactive chromosome.

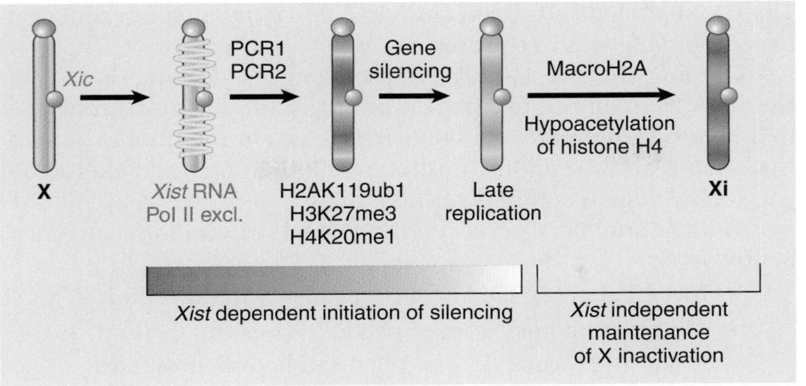

FIGURE 27.12 *Xist* RNA produced from the *Xic* locus accumulates on the future inactive X (Xi). This excludes transcription machinery, such as RNA polymerase II (Pol II). Polycomb group complexes are recruited to the *Xist*-covered chromosome and establish chromosome-wide histone modifications. Histone macroH2A becomes enriched on the Xi and promoters of genes on the Xi are methylated. In this phase X inactivation is irreversible and *Xist* is not required for maintenance of the silent state. Adapted from Wutz, A., and Gribnau, J., *Curr. Opin. Genet. Dev.* 17 (2007): 387–393.

tone variant, macroH2A, is incorporated into the chromatin, and promoter DNA is methylated. These changes are summarized in **FIGURE 27.12**. (The repressive effects of promoter methylation are discussed in the following sections.) At this point, the heterochromatic state of the inactive X is stable, and *Xist* is not required to maintain the silent state of the chromosome.

KEY CONCEPTS

- One of the two female X chromosomes is inactivated at random in each cell during embryogenesis of placental mammals.
- In exceptional cases where there are >2 X chromosomes, all but one are inactivated.
- The *Xic* (X inactivation center) is a cis-acting region on the X chromosome that is necessary and sufficient to ensure that only one X chromosome remains active.
- *Xic* includes the *Xist* gene, which codes for an RNA that is found only on inactive X chromosomes.
- *Xist* recruits Polycomb complexes, which modify histones on the inactive X.
- The mechanism that is responsible for preventing *Xist* RNA from accumulating on the active chromosome is unknown.

CONCEPT AND REASONING CHECK

Dosage compensation in *Drosophila* also depends on noncoding RNAs, which recruit a dosage compensation complex to the single X chromosome in males. This complex contains an acetyltransferase. Explain how this promotes dosage compensation in *Drosophila*.

27.6 CpG Islands Are Subject to Methylation

Methylation of DNA is a critical parameter that affects transcription. The typical relationship is that methylation in the vicinity of a promoter inhibits transcription and demethylation is required for gene expression. The reality is a little more complex than that, as some areas are methylated in active genes. Critical regions for methylation in vertebrates usually occur in **CpG islands** that are found in the 5′ region of the gene. These islands are detected by the presence of an increased density of the dinucleotide sequence, CpG. (CpG = 5′–CG-3′)

CpG islands Stretches of 1–2 kb in mammalian genomes that are enriched in CpG dinucleotides; frequently found in promoter regions of genes.

The CpG dinucleotide occurs in vertebrate DNA at only ~20% of the frequency that would be expected from the proportion of G-C base pairs. This may be because many CpG dinucleotides are methylated on C, and spontaneous deamination of methyl-C converts it to T. This introduces a mutation that can result in the loss of the CpG dinucleotide, as it is impossible for the damage repair system to determine if the T or the G should be replaced (although some repair systems are biased to remove the T in this context). In certain regions, however, the density of CpG doublets reaches the predicted value; in fact, it is increased by 10× relative to the rest of the genome.

The CpG doublets in these regions can be methylated or unmethylated, which affects expression of genes in the area.

These CpG-rich islands have an average G-C content of ~60%, compared with the 40% average in bulk DNA. They take the form of stretches of DNA typically 1 to 2 kb long. There are ~45,000 such islands in the human genome. In several cases, CpG-rich islands begin just upstream of a promoter and extend downstream into the transcribed region before petering out.

Methylation of a CpG island can affect transcription. One of two mechanisms can be involved:

- Methylation of a binding site for some factor may prevent that factor from binding.
- Methylation may cause specific repressors to bind to the DNA, which promotes the formation of a region of heterochromatin.

- **maintenance methyltransferase** Adds a methyl group to a target site that is already hemimethylated.
- **hemimethylated site** A palindromic sequence that is methylated on only one strand of DNA.
- **fully methylated** A site that is a palindromic sequence that is methylated on both strands of DNA.
- ***de novo* methyltransferase** Adds a methyl group to an unmethylated target sequence on DNA.

Repression is caused by either of two types of protein that bind to methylated CpG sequences. The protein MeCP1 requires the presence of several methyl groups to bind to DNA, whereas MeCP2 and a family of related proteins can bind to a single methylated CpG base pair. This explains why a methylation-free zone is required for initiation of transcription. The absence of methyl groups is associated with gene expression. It is important to note that DNA methylation in not used for controlling gene expression in every species. In the case of *Drosophila melanogaster* (and other Dipteran insects), there is very little methylation of DNA, and neither the yeast *S. cerevisiae* nor the nematode *C. elegans* have any DNA methylation. The other differences between inactive and active chromatin appear to be the same as in species that display methylation. Thus, in these organisms, any role that methylation has in vertebrates is replaced by some other mechanism.

There are two types of DNA methyltransferase, whose actions are distinguished by the state of the methylated DNA, as shown in **FIGURE 27.13**. A **maintenance (or "perpetuation") methyltransferase** acts constitutively *only on* **hemimethylated sites** to convert them to **fully methylated** sites (see Figure 27.2). Its existence means that any methylated site is perpetuated after replication. Maintenance methylation is virtually 100% efficient. There is one maintenance methyltransferase (Dnmt1) in mouse, and it is essential: mouse embryos in which its gene has been disrupted do not survive past early embryogenesis. The absence of Dnmt1 causes widespread demethylation at promoters, and we assume this is lethal because of the uncontrolled gene expression.

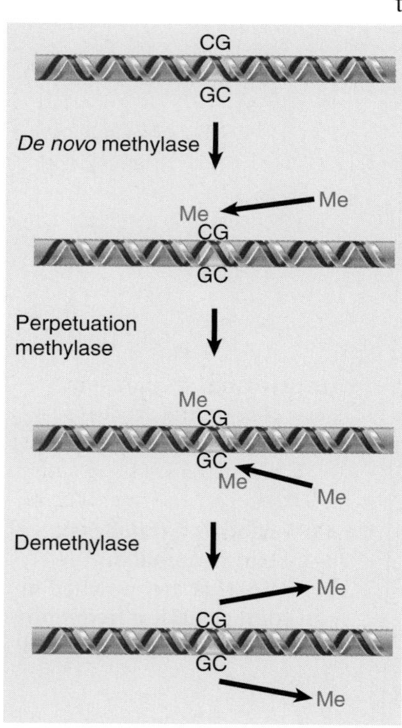

FIGURE 27.13 The state of methylation is controlled by three types of enzyme. *De novo* and perpetuation methyltransferases are known, but DNA demethylases have not been identified.

To modify DNA at a new position requires the action of a ***de novo* methyltransferase**, which recognizes DNA by virtue of a specific sequence or is recruited to a specific site in DNA. It acts *only* on nonmethylated DNA, to add a methyl group to one strand. There are two *de novo* methyltransferases (Dnmt3A and Dnmt3B) in mouse; they have different target sites, and both are essential for development. Mutations in Dnmt3B prevent methylation of satellite DNA, which causes centromere instability at the cellular level. Mutations in the corresponding human gene cause a disease called ICF (immunodeficiency, centromere instability, facial anomalies). The importance of methylation is emphasized by another human disease, Rett syndrome, which is caused by mutation of the gene for the protein MeCP2 that binds methylated CpG sequences. Patients with Rett syndrome exhibit autism-like symptoms that appear to be the result of a failure of normal gene silencing in the brain.

DNA methylation is closely linked to histone modifications associated with heterochromatin formation, and in fact DNA methylation and heterochromatin are mutually reinforcing, as depicted in **FIGURE 27.14** (also see *Section 26.10, Methylation of Histones and DNA Is Connected*). Recall that HP1 is recruited to regions in which histone H3 has been methylated at lysine 9, a modification involved in heterochromatin formation. It turns out that HP1 can interact with Dnmt1, which can promote DNA methylation in the vicinity of HP1 binding. Furthermore, Dnmt1 can directly interact with the methyltransferase responsible for H3K9 methylation, creating a positive feedback loop to ensure continued DNA and histone methylation. These interactions (and other similar networks of interactions) contribute to the stability of epigenetic states, allowing a heterochromatin region to be maintained through many cell divisions.

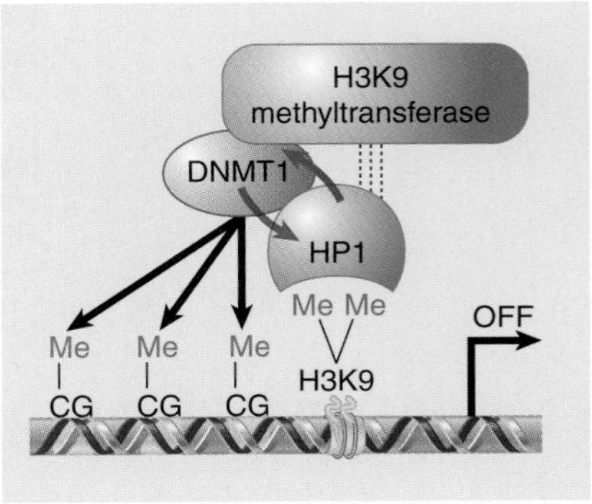

KEY CONCEPTS

- Demethylation at the 5′ end of the gene is necessary for transcription.
- CpG islands surround the promoters of some genes.
- There are ~45,000 CpG islands in the human genome.
- Methylation of a CpG island prevents activation of a promoter within it.
- Repression is caused by proteins that bind to methylated CpG doublets.
- Replication converts a fully methylated site to a hemimethylated site.
- Hemimethylated sites are converted to fully methylated sites by a maintenance methyltransferase.
- DNA and histone methylation are mutually reinforcing.

CONCEPT AND REASONING CHECK

Some methyl-DNA binding complexes also contain histone deacetylases. Explain how this would also reinforce the stability of a silenced region.

FIGURE 27.14 Mammalian HP1 is recruited to regions where lysine 9 of histone H3 (H3K9) has been methylated by a histone methyltransferase. HP1 then binds to DNMT1 and potentiates its DNA methyltransferase activity (blue arrow), thereby enhancing cytosine methylation (meCG) on nearby DNA. DNMT1 could in turn assist HP1 loading onto chromatin (red arrow). Furthermore, association of DNMT1 with the histone methyltransferase could allow a positive feedback loop to stabilize inactive chromatin.

27.7 DNA Methylation Is Responsible for Imprinting

The pattern of methylation of germ cells is established in each sex during gametogenesis by a two-stage process. First the existing pattern is erased by a genome-wide demethylation, and then the pattern specific for each sex is imposed.

All allelic differences are lost when primordial germ cells develop in the embryo; irrespective of sex, the previous patterns of methylation are erased, and a typical gene is then unmethylated. In males, the pattern develops in two stages. The methylation pattern that is characteristic of mature sperm is established in the spermatocyte, but further changes are made in this pattern after fertilization. In females, the maternal pattern is imposed during oogenesis, when oocytes mature through meiosis after birth.

Systematic changes occur in early embryogenesis. Some sites will continue to be methylated, whereas others will be specifically unmethylated in cells in which a gene is expressed. From the pattern of changes, we may infer that individual sequence-specific demethylation events occur during somatic development of the organism as particular genes are activated.

The specific pattern of methyl groups in germ cells is responsible for the phenomenon of **imprinting**, which describes a difference in behavior between the alleles inherited from each parent. The expression of certain genes in mouse embryos depends upon the sex of the parent from which they were inherited. For example, the allele coding for IGF-II (insulin-like growth factor II) that is inherited from the father is expressed, but the allele that is inherited from the mother is not expressed. The *IGF-II* gene in oocytes is methylated, but the *IGF-II* gene in sperm is not methylated, so that the two alleles behave differently in the zygote. The dependence on sex is reversed (that is, the maternal copy is expressed) for IGF-IIR, the receptor for IGF-II. The allele that is silenced is referred to as the *imprinted* allele.

▸ **imprinting** A change in a gene that occurs during passage through the sperm or egg with the result that the paternal and maternal alleles have different properties in the very early embryo. This is caused by methylation of DNA.

This sex-specific mode of inheritance requires that the pattern of methylation be established specifically during each gametogenesis. The fate of a hypothetical locus in a mouse is illustrated in **FIGURE 27.15**. In the early embryo, the paternal allele is nonmethylated and expressed, and the maternal allele is methylated and silent. What happens when this mouse itself forms gametes? If it is a male, the allele contributed to the sperm must be nonmethylated, irrespective of whether or not it was originally methylated. Thus when the maternal allele finds itself in a sperm, it must

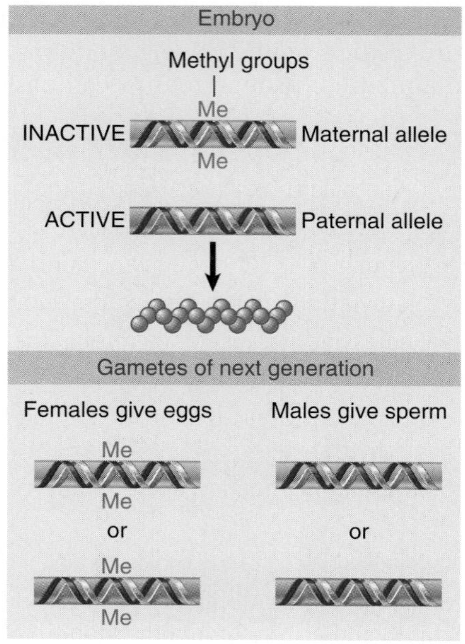

FIGURE 27.15 The typical pattern for imprinting is that a methylated locus is inactive. If this is the maternal allele, only the paternal allele is active and will, therefore, be essential to produce a functional product of the locus. The methylation pattern is reset when gametes are formed, so that all sperm have the paternal type and all oocytes have the maternal type.

be demethylated. If the mouse is a female, the allele contributed to the egg must be methylated; if it was originally the paternal allele, methyl groups must be added.

The consequence of imprinting is that an embryo is *hemizygous* for any imprinted gene. Thus in the case of a heterozygous cross where the allele of one parent has an inactivating mutation, the embryo will survive if the wild-type allele comes from the parent in which this allele is active, but will die if the wild-type allele is the imprinted (silenced) allele. Imprinted alleles exhibit classic epigenetic inheritance: although the paternal and maternal alleles have identical sequences, they display different properties, depending on which parent provided them. These properties are inherited through meiosis and the subsequent somatic mitoses.

Imprinted genes are sometimes clustered. More than half of the 17 known imprinted genes in mouse are contained in two particular regions, each containing both maternally and paternally expressed genes. This suggests the possibility that imprinting mechanisms may function over long distances. Some insights into this possibility come from deletions in the human population that cause the Prader–Willi and Angelman diseases. Many cases are caused by the same 4-Mb deletion, but the syndromes are different, depending on which parent contributed the deletion. The reason is that the deleted region includes an "imprint center" that acts at a distance to control sex-specific imprinting of genes in the surrounding area. This large region contains a number of genes, some of which are paternally imprinted and at least one that is maternally imprinted. The basic effect of the deletion is to prevent a father from resetting the paternal mode to a chromosome inherited from his mother. The result is that these genes remain in maternal mode, so that the paternal as well as maternal alleles are silent in the offspring, resulting in Prader-Willi. The inverse effect is found in Angelman's syndrome.

Imprinting is determined by the state of methylation of a *cis*-acting site near a target gene or genes. These regulatory sites are known as differentially methylated domains (DMDs) or imprinting control regions (ICRs). Deletion of these sites removes imprinting, and the target loci then behave the same in both maternal and paternal genomes.

The behavior of a region containing two genes, *Igf2* and *H19*, illustrates the ways in which methylation can control gene activity. **FIGURE 27.16** shows that these two genes react oppositely to the state of methylation at the ICR located between them. The ICR is methylated on the paternal allele. *H19* shows the typical response of inactivation. Note,

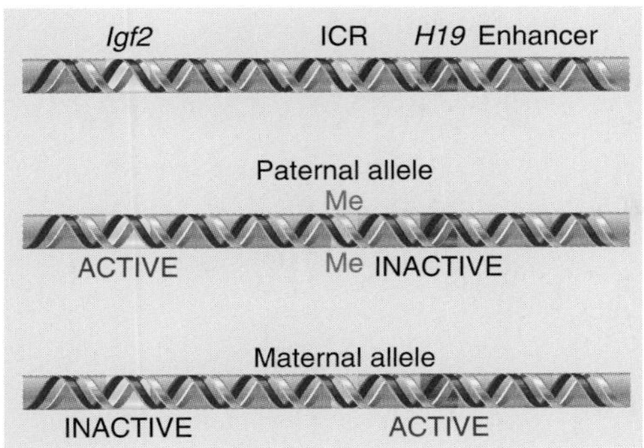

FIGURE 27.16 *ICR* is methylated on the paternal allele, where *Igf2* is active and *H19* is inactive. *ICR* is unmethylated on the maternal allele, where *Igf2* is inactive and *H19* is active.

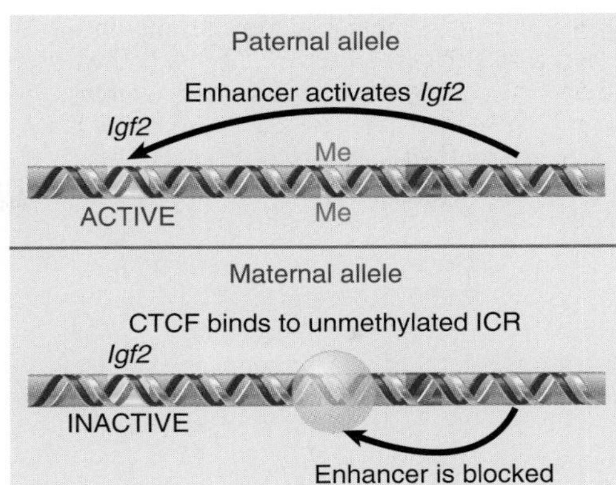

FIGURE 27.17 The *ICR* is an insulator that prevents an enhancer from activating *Igf2*. The insulator functions only when it binds CTCF to unmethylated DNA.

however, that *Igf2* is expressed. The reverse situation is found on a maternal allele, where the ICR is not methylated. *H19* now becomes expressed, but *Igf2* is inactivated.

The control of *Igf2* is exercised by an insulator function of the ICR. **FIGURE 27.17** shows that when the ICR is unmethylated, it binds the protein CTCF. This creates a functional insulator that blocks an enhancer from activating the *Igf2* promoter. This is an unusual effect in which methylation indirectly activates a gene by blocking an insulator. The regulation of *H19* shows the more usual mechanism of control in which methylation creates an inactive imprinted state.

KEY CONCEPTS

- Paternal and maternal alleles may have different patterns of methylation at fertilization.
- Methylation is usually associated with inactivation of the gene.
- When genes are differentially imprinted, survival of the embryo may require that the functional allele be provided by the parent with the unmethylated allele.
- Survival of heterozygotes for imprinted genes is different, depending on the direction of the cross.
- Imprinted genes occur in clusters and may depend on a local control site where *de novo* methylation occurs unless specifically prevented.
- Imprinted genes are controlled by methylation of *cis*-acting sites.

CONCEPT AND REASONING CHECK

In placental mammals, X-inactivation is random, but in marsupials, the paternal X chromosome is always inactivated, analogous to imprinting an entire chromosome. What does this mean for the expression of X-linked traits in marsupials?

27.8 Yeast Prions Show Unusual Inheritance

One of the clearest cases of the dependence of epigenetic inheritance on the condition of a protein is provided by the behavior of **proteinaceous infectious agents, or prions**. They have been characterized in two circumstances: by genetic effects in yeast, and as the causative agents of neurological diseases in mammals, including humans. A striking epigenetic effect is found in yeast, where two different states can be inherited that map to a single genetic locus, *although the sequence of the gene is the same in both states*. A number of prions have been identified in yeast, one of the best characterized is [*PSI*+]. The two different states are [*psi*$^-$] (non-infectious) and [*PSI*$^+$] (prion). A switch in condition occurs at a low frequency as the result of a spontaneous transition between the states.

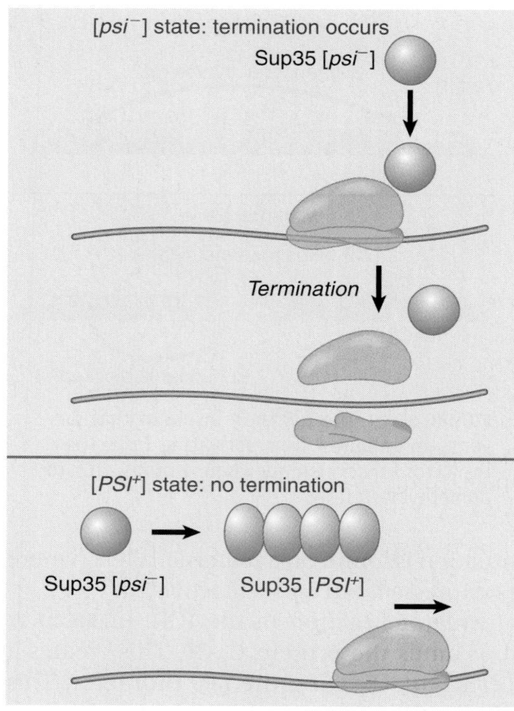
[psi⁻] state: termination occurs
Sup35 [psi⁻]
Termination
[PSI⁺] state: no termination
Sup35 [psi⁻]
Sup35 [PSI⁺]

FIGURE 27.18 The state of the Sup35 protein determines whether termination of translation occurs.

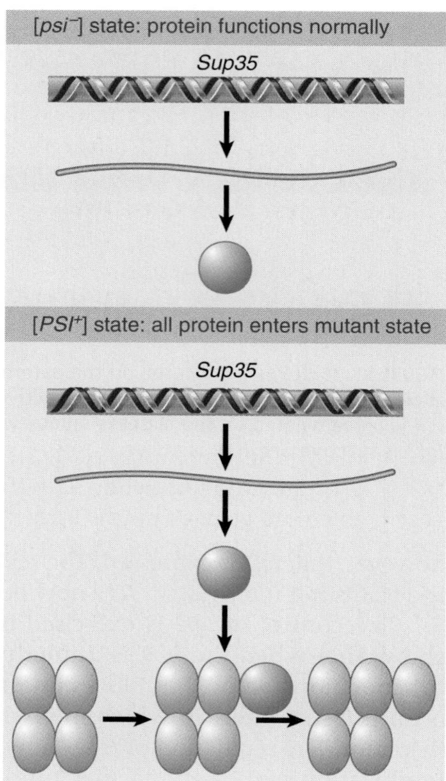
[psi⁻] state: protein functions normally
Sup35
[PSI⁺] state: all protein enters mutant state
Sup35

FIGURE 27.19 Newly synthesized Sup35 protein is converted into the [*PSI*⁺] state by the presence of preexisting [*PSI*⁺] protein.

The [*psi*] genotype maps to the locus *SUP35*, which codes for a translation termination factor. **FIGURE 27.18** summarizes the effects of the Sup35 protein in yeast. In wild-type cells, which are characterized as [*psi*⁻], the gene is active, and Sup35 protein terminates protein synthesis. In cells of the mutant [*PSI*⁺] type, the factor does not function, which causes a failure to terminate protein synthesis properly.

[*PSI*⁺] strains have unusual genetic properties. When a [*psi*⁻] strain is crossed with a [*PSI*⁺] strain, *all of the progeny are* [*PSI*⁺]. This is a pattern of inheritance that would be expected of an extrachromosomal agent, but the [*PSI*⁺] trait cannot be mapped to any such nucleic acid. The [*PSI*⁺] trait is *metastable*, which means that although it is inherited by most progeny, it is lost at a higher rate than is consistent with mutation. Similar behavior is shown also by the locus *URE2*, which codes for a protein required for nitrogen-mediated repression of certain catabolic enzymes. When a yeast strain is converted into an alternative state, called [*URE3*], the Ure2 protein is no longer functional.

The [*PSI*⁺] state is determined by the conformation of the Sup35 protein. In a wild-type [*psi*⁻] cell, the protein displays its normal function. In a [*PSI*⁺] cell, though, the protein is present in an alternative conformation in which its normal function has been lost. To explain the unilateral dominance of [*PSI*⁺] over [*psi*⁻] in genetic crosses, we must suppose that *the presence of protein in the* [*PSI*⁺] *state causes all the protein in the cell to enter this state*. This requires an interaction between the [*PSI*⁺] protein and newly synthesized protein, which probably reflects the generation of an oligomeric state in which the [*PSI*⁺] protein has a nucleating role, as illustrated in **FIGURE 27.19**.

A feature common to both the Sup35 and Ure2 proteins is that each consists of two domains that function independently. The C-terminal domain is sufficient for the activity of the protein. The N-terminal domain is sufficient for formation of the structures that make the protein inactive. Thus yeast in which the N-terminal domain of Sup35 has been deleted cannot acquire the [*PSI*$^+$] state, and the presence of a [*PSI*$^+$] N-terminal domain is sufficient to maintain Sup35 protein in the [*PSI*$^+$] condition. The critical feature of the N-terminal domain is that it is rich in glutamine and asparagine residues.

Loss of function in the [*PSI*$^+$] state is due to the sequestration of the protein in an oligomeric complex. Sup35 protein in [*PSI*$^+$] cells is clustered in discrete foci, whereas the protein in [*psi*$^-$] cells is diffused in the cytosol. Sup35 protein from [*PSI*$^+$] cells forms **amyloid fibers** *in vitro;* these have a characteristic high content of β-sheet structures.

The involvement of protein conformation (rather than covalent modification) is suggested by the effects of conditions that affect protein structure. Denaturing treatments cause loss of the [*PSI*$^+$] state. In particular, the chaperone Hsp104 is involved in inheritance of [*PSI*$^+$]. Deletion of *HSP104* prevents maintenance of the [*PSI*$^+$] state, and it is believed that Hsp104 actually disaggregates amyloid filaments to create growth points for new filaments.

Using the ability of Sup35 to form the inactive structure *in vitro*, it is possible to provide biochemical proof that the inheritance of the [*PSI*+] state is entirely dependent on the structure of the protein. **FIGURE 27.20** illustrates a striking experiment in which the protein was converted to the inactive form *in vitro*, put into liposomes (where, in effect, the protein is surrounded by an artificial membrane), and then introduced directly into cells by fusing the liposomes with [*psi*$^-$] yeast. The yeast cells were converted to [*PSI*$^+$]!

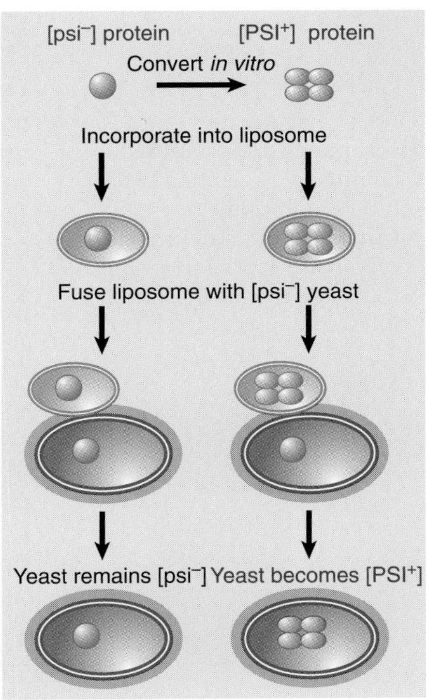

FIGURE 27.20 Purified protein can convert the [*psi*$^-$] state of yeast to [*PSI*$^+$].

The ability of yeast to form the [*PSI*$^+$] prion state depends on the genetic background. The yeast must be [*PIN*$^+$] in order for the [*PSI*$^+$] state to form. The [*PIN*$^+$] condition itself is an epigenetic state: the prion form of the protein Rnq1. Rnq1 and other prion-forming proteins share the characteristic of Sup35 of having Gln/Asn-rich domains. Overexpression of these domains in yeast stimulates formation of the [*PSI*$^+$] state. This suggests that there is a common model for the formation of the prion state that involves aggregation of the Gln/Asn domains into self-propagating amyloid structure.

amyloid fibers Insoluble fibrous protein polymers with a cross β-sheet structure, generated by prions or other dysfunctional protein aggregations (such as in Alzheimers).

How does the presence of one Gln/Asn protein influence the formation of prions by another? We know that the formation of Sup35 prions is specific to Sup35 protein; that is, it does not occur by cross-aggregation with other proteins. This suggests that the yeast cell may contain soluble proteins that antagonize prion formation. These proteins are not specific for any one prion. As a result, the introduction of any Gln/Asn domain protein that interacts with these proteins will reduce the concentration. This will allow other Gln/Asn proteins to aggregate more easily.

KEY CONCEPTS

- The Sup35 protein in its wild-type soluble form [*psi*-] is a termination factor for translation.
- It can also exist in an alternative form of oligomeric aggregates [*PSI*+], in which it is not active in protein synthesis.
- The presence of the oligomeric form causes newly synthesized protein to acquire the inactive structure.
- Conversion between the two forms is influenced by chaperones.
- A number of other prion-forming proteins have been identified in yeast; all share Gln/Asn-rich domains.

CONCEPT AND REASONING CHECK

Hsp104 promotes prion formation, but *overexpression* of Hsp104 interferes with prion inheritance. Why might this be?

27.9 Prions Cause Diseases in Mammals

▸ **scrapie** A disease caused by an infective agent made of protein (a prion).

▸ **kuru** A human neurological disease caused by prions. It may be caused by eating infected brains.

Prion diseases have been found in sheep, elk, cows, and humans. The basic phenotype is an ataxia—a neurodegenerative disorder that is manifested by an inability to remain upright. The name of the disease in sheep, **scrapie**, reflects the phenotype: The sheep rub against walls in order to stay upright. Scrapie can be perpetuated by inoculating sheep with tissue extracts from infected animals. The human disease **kuru** was discovered in New Guinea, where it was perpetuated by cannibalism, in particular the eating of brains. Related diseases in Western populations with a pattern of genetic transmission include Gerstmann–Straussler syndrome and the related Creutzfeldt–Jakob disease (CJD), which occurs sporadically. Most recently, a disease resembling CJD appears to have been transmitted by consumption of meat from cows suffering from "mad cow" disease.

When tissue from scrapie-infected sheep is inoculated into mice, the disease occurs in a period ranging from 75 to 150 days. The active component is a protease-resistant protein. The protein is encoded by a gene that is normally expressed in the brain. The form of the protein in normal brain, called PrP^{C}, is sensitive to proteases. Its conversion to the resistant form, called Prp^{Sc}, is associated with occurrence of the disease. As with the yeast prions, this is a case of epigenetic inheritance in which there is no change in genetic information (because normal and diseased cells have the same *PrP* gene sequence), but the PrP^{Sc} form of the protein is the infectious agent (whereas PrP^{C} is harmless). The PrP^{Sc} form has a high content of β sheets, which form an amyloid fibrillous structure that is absent from the PrP^{C} form.

The assay for infectivity in mice allows the dependence on protein sequence to be tested. **FIGURE 27.21** illustrates the results of some critical experiments. In the normal situation, PrP^{Sc} protein extracted from an infected mouse will induce disease (and ultimately kill) when it is injected into a recipient mouse. If the *PrP* gene is deleted, a mouse becomes resistant to infection. This experiment demonstrates two things. First, the endogenous protein is necessary for an infection, because it provides the raw material that is converted into the infectious agent. Second, the cause of disease is not the removal of the PrP^{C} form of the protein, because a mouse with no PrP^{C} survives normally: the disease is caused by a gain-of-function in PrP^{Sc}.

The existence of species barriers allows hybrid proteins to be constructed to delineate the features required for infectivity. The original preparations of scrapie were perpetuated in several types of animal, but these cannot always be transferred readily.

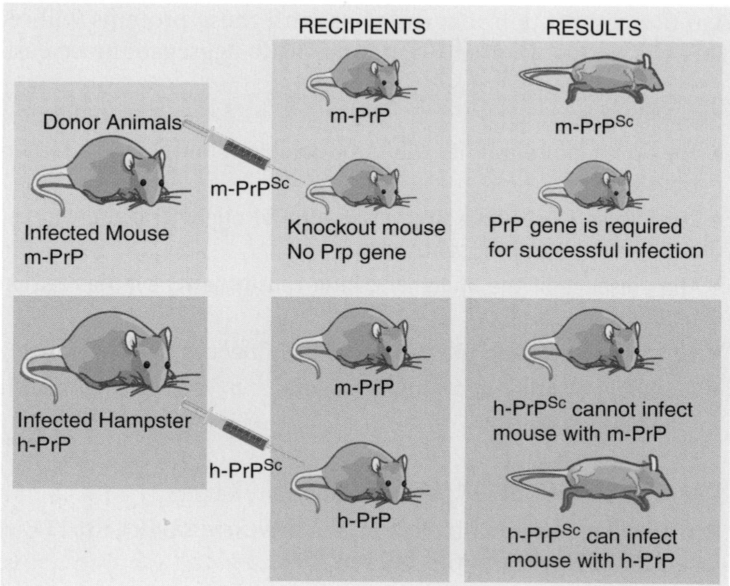

FIGURE 27.21 A Prp^{Sc} protein can only infect an animal that has the same type of endogenous PrP^{C} protein.

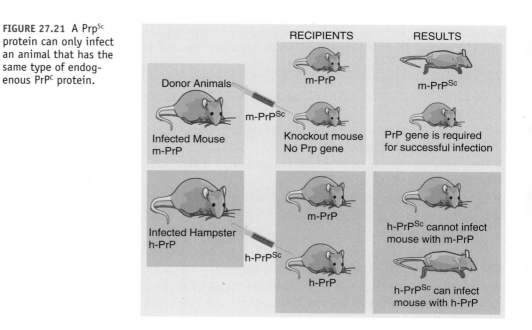

For example, mice are resistant to infection from prions of hamsters. This means that hamster-PrP^{Sc} cannot convert mouse-PrP^{C} to PrP^{Sc}. The situation changes, though, if the mouse *PrP* gene is replaced by a hamster *PrP* gene. A mouse with a hamster *PrP* gene is sensitive to infection by hamster PrP^{Sc}. This suggests that the conversion of cellular PrP^{C} protein into the Sc state requires that the PrP^{Sc} and PrP^{C} proteins have matched sequences.

There are different "strains" of PrP^{Sc}, which are distinguished by characteristic incubation periods upon inoculation into mice. This implies that the protein is not restricted solely to alternative states of PrP^{C} and PrP^{Sc}, but rather that there may be multiple Sc states. These differences must depend on some self-propagating property of the protein other than its sequence. If conformation is the feature that distinguishes PrP^{Sc} from PrP^{C}, then there must be multiple conformations, each of which has a self-templating property when it converts PrP^{C}.

The probability of conversion from PrP^{C} to PrP^{Sc} is affected by the sequence of PrP. Gerstmann–Straussler syndrome in humans is caused by a single amino acid change in PrP. This is inherited as a dominant trait. If the same change is made in the mouse PrP gene, mice develop the disease. This suggests that the mutant protein has an increased probability of spontaneous conversion into the Sc state. Similarly, the sequence of the PrP gene determines the susceptibility of sheep to develop the disease spontaneously; the combination of amino acids at three positions (codons 136, 154, and 171) determines susceptibility.

The prion offers an extreme case of epigenetic inheritance, in which the infectious agent is a protein that can adopt multiple conformations, each of which has a self-templating property. This property is likely to involve the state of aggregation of the protein.

KEY CONCEPTS

- The protein responsible for scrapie exists in two forms: the wild-type noninfectious form PrP^{C}, which is susceptible to proteases, and the disease-causing form PrP^{Sc}, which is resistant to proteases.
- The neurological disease can be transmitted to mice by injecting the purified PrP^{Sc} protein into mice.
- The recipient mouse must have a copy of the PrP gene coding for the mouse protein.
- The PrP^{Sc} protein can perpetuate itself by causing the newly synthesized PrP protein to take up the PrP^{Sc} form instead of the PrP^{C} form.
- Multiple strains of PrP^{Sc} may have different conformations of the protein.

CONCEPT AND REASONING CHECK

Would injection of a [*PSI*+] prion from yeast induce PrP^{Sc} in mice? Why or why not?

27.10 Summary

The formation of heterochromatin occurs by proteins that bind to specific chromosomal regions (such as telomeres) and that interact with histones. The formation of an inactive structure may propagate along the chromatin thread from an initiation center. Similar events occur in silencing of the inactive yeast mating type loci. Repressive structures that are required to maintain the inactive states of particular genes are formed by the Pc-G protein complex in *Drosophila*. They share with heterochromatin the property of propagating from an initiation center.

Formation of heterochromatin may be initiated at certain sites and then propagated for a distance that is not precisely determined. When a heterochromatic state has been established, it is inherited through subsequent cell divisions. This gives rise to a pattern of epigenetic inheritance, in which two identical sequences of DNA may be associated with different protein structures, and therefore have different abilities to be expressed. This explains the occurrence of position effect variegation in *Drosophila*.

Modification of histone tails is a trigger for chromatin reorganization. Acetylation is generally associated with gene activation. Histone acetylases are found in activating complexes, whereas histone deacetylases are found in inactivating complexes. Histone methylation at specific sites in the histone N-termini is associated with gene inactivation. Some histone modifications may be exclusive or synergistic with others.

Inactive chromatin at yeast telomeres and silent mating type loci appears to have a common cause, and involves the interaction of silent information regulator (SIR) proteins with the N-terminal tails of histones H3 and H4. Formation of the inactive complex may be initiated by binding of one protein to a specific sequence of DNA; the other components may then polymerize in a cooperative manner along the chromosome.

Inactivation of one X chromosome in female placental mammals occurs at random. The *Xic* locus is necessary and sufficient to count the number of X chromosomes. The n – 1 rule ensures that all but one X chromosome are inactivated. *Xic* contains the gene *Xist*, which codes for an RNA that is expressed only on the inactive X chromosome. Stabilization of *Xist* RNA is the mechanism by which the inactive X chromosome is distinguished; it is then inactivated by the activities of Polycomb complexes, heterochromatin formation, and DNA methylation.

Methylation of DNA is inherited epigenetically. Replication of DNA creates hemimethylated products, and a maintenance methylase restores the fully methylated state. Some methylation events depend on parental origin. Sperm and eggs contain specific and different patterns of methylation, with the result that paternal and maternal alleles are differently expressed in the embryo. This is responsible for imprinting, in which the nonmethylated allele inherited from one parent is essential because it is the only active allele; the allele inherited from the other parent is silent. Patterns of methylation are reset during gamete formation in every generation.

Prions are proteinaceous infectious agents that are responsible for the disease of scrapie in sheep and for related diseases in human beings. The infectious agent is a variant of a normal cellular protein. The PrP^{Sc} form has an altered conformation that is self-templating: The normal PrP^{C} form does not usually take up this conformation, but does so in the presence of PrP^{Sc}. A similar effect is responsible for inheritance of the [*PSI*] element in yeast.

CHAPTER QUESTIONS

List five characteristics of heterochromatin:

1. ______________________________

2. ______________________________

3. ______________________________

4. ______________________________

5. ______________________________

Explain briefly the characteristics of constitutive heterochromatin and facultative heterochromatin.

6. Constitutive ______________________________

7. Facultative ______________________________

Dosage compensation equalizes the level of expression of X-linked genes in the two sexes to avoid double expression of X chromosome-encoded genes in females. List three ways this has been shown to occur (in different organisms):

8. ______________________________

9. ______________________________

10. ______________________________

What are two general mechanisms by which epigenetic effects can be inherited, or self-perpetuated?

11. ______________________________

12. ______________________________

List two animal diseases known to be caused by prions:

13. ______________ **14.** ______________

15. Hemimethylated DNA is methylated on:

A. only the newly synthesized DNA strand.
B. only the parental DNA strand.
C. only some scattered methylation sites on both DNA strands.
D. only methylated sites associated with heterochromatin.

16. Position effect variegation describes:

A. phenotypically identical cells with different genetic elements.
B. phenotypically identical cells with similar genetic elements.
C. genetically identical cells with similar phenotypes.
D. genetically identical cells with different phenotypes.

17. The HP1 protein plays a key role in formation of heterochromatin in mammals by:

A. binding to methylated histone H3.
B. binding to methylated histone H5.
C. binding to acetylated histone H1.
D. binding to acetylated histone H2B.

18. A *de novo* methylase recognizes:

A. nonmethylated sites and adds a methyl group to the base on one strand.
B. hemimethylated sites and adds a methyl group to the base on one strand.
C. methylated sites and adds methyl groups to additional bases on both strands.
D. methylated sites and adds methyl groups to additional bases on the parental strand.

19. Prion diseases in animals cause:

A. muscular degeneration.
B. neurodegenerative disorder.
C. digestive system disorder.
D. blindness.

KEY TERMS

amyloid fibers
constitutive heterochromatin
CpG islands
***de novo* methyltransferase**
demethylase
dosage compensation
epigenetic
facultative heterochromatin
fully methylated
hemimethylated site
imprinting
kuru
maintenance methyltransferase
methyltransferase
n – 1 rule
position effect variegation (PEV)
prion
scrapie
single X hypothesis
telomeric silencing

FURTHER READING

Horn, P. J. and Peterson, C. L. (2006). Heterochromatin assembly: a new twist on an old model. *Chromo. Res.* 14, 83–94.

A review of heterochromatin assembly and maintenance focused on data obtained from studies of fission yeast.

Horsthemke, B. and Buiting, K. (2008). Genomic imprinting and imprinting defects in humans. *Adv. Gen.* 61, 225–246.

A review of imprinting in placental mammals and defects in imprinting that lead to human disease.

Kim, J. K., Samaranayake, M. and Pradhan, S. (2008). Epigenetic mechanisms in mammals. *Cell. Mol. Life Sci.* DOI 10.1007/s00018-008-8432-4.

A comprehensive review of mammalian epigenetics, which includes discussion of DNA methylation, histone modification, insulators, X inactivation, imprinting, and developmental epigenetics (e.g., PcG and trxG).

Payer, B. and Lee, J. T. (2008). X chromosome dosage compensation: how mammals keep the balance. *Ann. Rev. Gen.* 42, 733–772.

A review of the mechanism and evolution of X chromosome inactivation.

Schwartz, Y. B. and Pirotta, V. (2008). Polycomb complexes and epigenetic states. *Curr. Opin. Cell Biol.* 20, 266–273.

A review of Polycomb Group (PcG) complexes and the regulation of PcG target genes.

Shkundina, I. S. and Ter-Avanesyan, M. D. (2007). Prions. *Biochemistry (Moscow)* 72, 1519–1536.

A review of prions focusing on lessons learned from the [*PSI*+] prion of yeast.

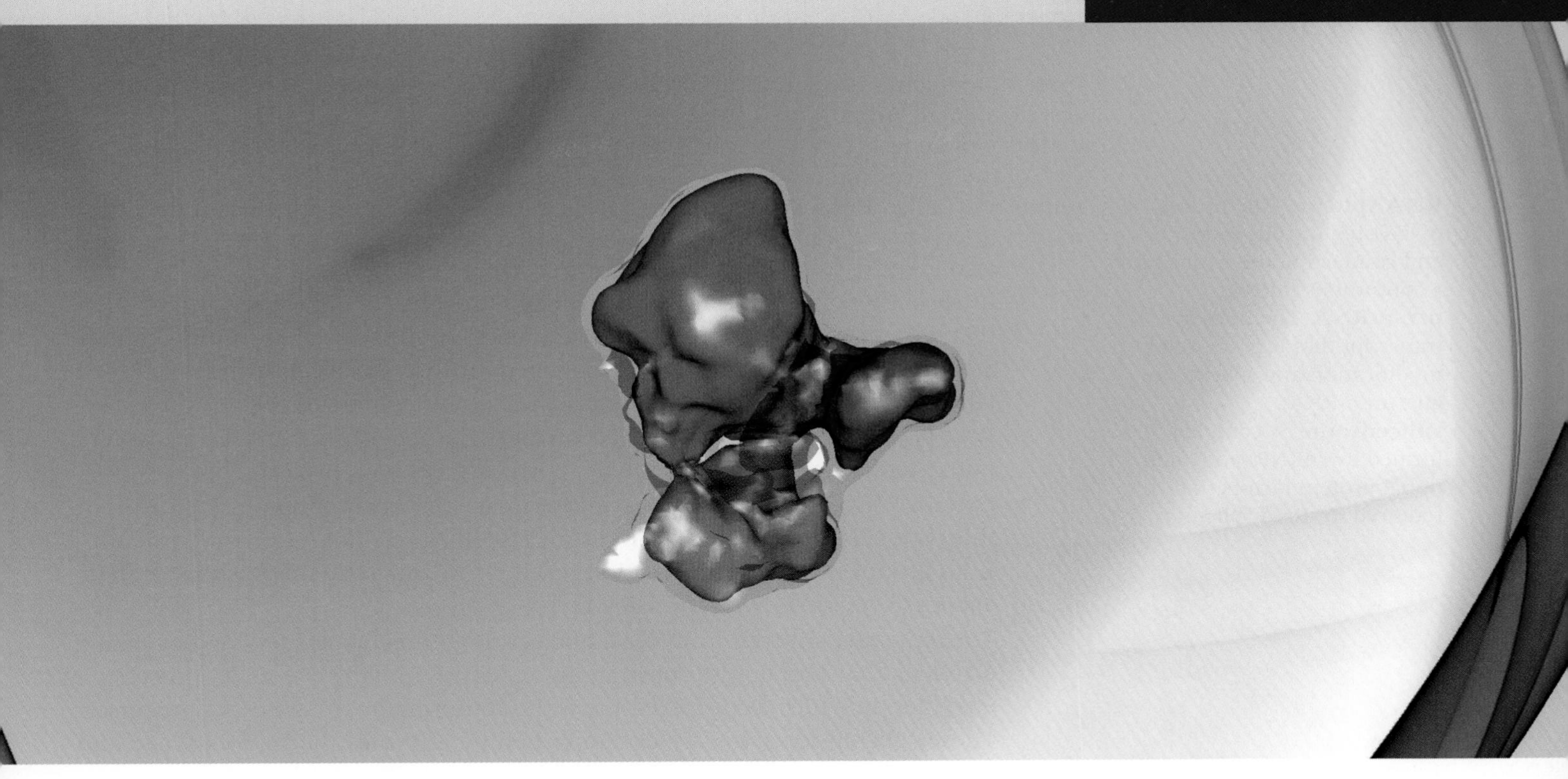

RNA Splicing and Processing

A three-dimensional structure of the mammalian spliceosomal C complex. Spliceosome structure courtesy of Nikolaus Grigorieff, Brandeis University. Background photo © Bella D/ ShutterStock, Inc.

CHAPTER OUTLINE

28.1 Introduction

Interrupted genes are found in all major taxonomic groups of organisms. They represent a small proportion of the genes of the unicellular eukaryotes, but the vast majority of genes in multicellular eukaryotic genomes. Removal of introns is a major part of the processing of RNAs in all eukaryotes. The process by which the introns are removed is called **RNA splicing**, and it occurs in the nucleus, together with the other modifications that are made to newly synthesized RNAs.

- **RNA splicing** The process of excising introns from RNA and connecting the exons into a continuous mRNA.
- **pre-mRNA** The nuclear transcript that is processed by modification and splicing to give an mRNA.
- **spliceosome** A complex formed by snRNPs and additional protein factors that is required for RNA splicing.

We can identify several types of splicing systems:

- Introns are removed from the nuclear **pre-mRNAs** of eukaryotes by a system that recognizes only short consensus sequences conserved at exon–intron boundaries and within the intron. This reaction requires a large splicing apparatus, which takes the form of an array of proteins and ribonucleoproteins that functions as a large particulate complex (the **spliceosome**). The mechanism of splicing involves transesterifications, and the catalytic center includes both RNA and proteins.
- Certain RNAs have the ability to excise their introns autonomously. Introns of this type fall into two groups, as distinguished by secondary/tertiary structure. Both groups use transesterification reactions in which the RNA is the catalytic agent (see *Chapter 29, Catalytic RNA*).
- The removal of introns from yeast nuclear tRNA precursors involves enzymatic activities that handle the substrate in a way resembling the tRNA processing enzymes, in which a critical feature is the conformation of the tRNA precursor. These splicing reactions are accomplished by enzymes that use cleavage and ligation.

The process of expressing an interrupted gene is reviewed in **FIGURE 28.1**. The transcript is capped at the 5′ end (see *Section 7.9, The 5′ End of Eukaryotic mRNA Is Capped*), has the introns removed, and is polyadenylated at the 3′ end (see *Section 7.10, The 3′*

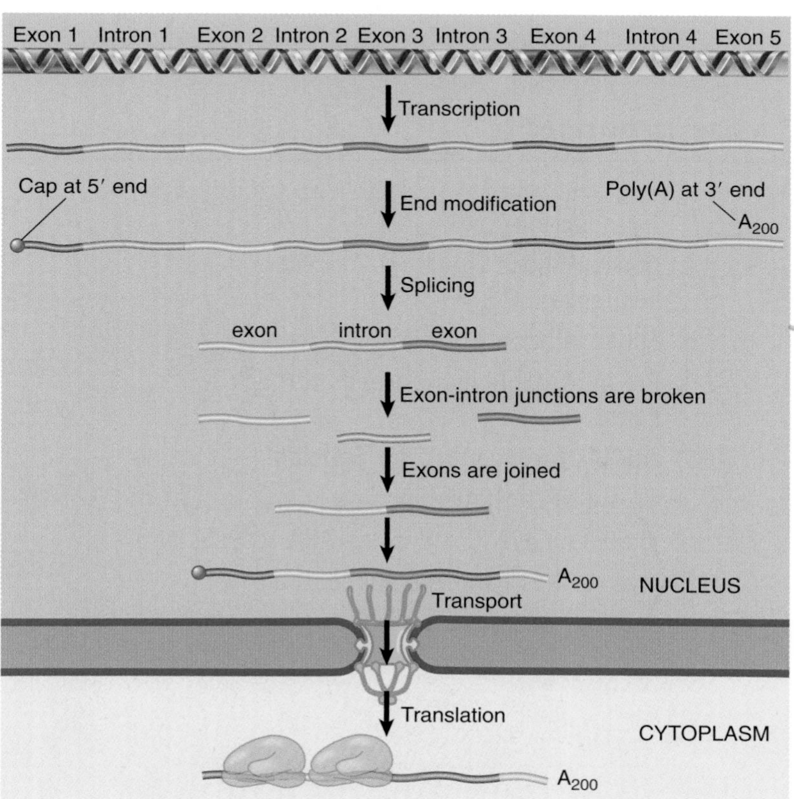

FIGURE 28.1 RNA is modified in the nucleus by additions to the 5′ and 3′ ends and by splicing to remove the introns. The splicing event requires breakage of the exon–intron junctions and joining of the ends of the exons. Mature mRNA is transported through nuclear pores to the cytoplasm, where it is translated.

Terminus of Eukaryotic mRNA Is Polyadenylated). The processed mRNA is then transported through nuclear pores to the cytoplasm, where it is available to be translated.

When the pre-RNA is synthesized, it becomes bound by proteins to form a ribonucleoprotein particle. Taking its name from its broad size distribution, the RNA was originally called **heterogeneous nuclear RNA (hnRNA)**, and the particle is called **hnRNP**. Some of the proteins may have a structural role in packaging the hnRNA; several are known to shuttle between the nucleus and cytoplasm, and to play roles in exporting the RNA or otherwise controlling its activity.

▸ **heterogeneous nuclear RNA (hnRNA)** RNA that comprises transcripts of nuclear genes made by RNA polymerase II; it has a wide size distribution and low stability.

▸ **hnRNP** The ribonucleoprotein form of hnRNA (heterogeneous nuclear RNA), in which the hnRNA is complexed with proteins. Pre-mRNAs are not exported until processing is complete; thus they are found only in the nucleus.

28.2 Nuclear Splice Junctions Are Short Sequences

To focus on the molecular events involved in nuclear intron splicing, we must consider the nature of the splice sites, the boundaries at both ends of each intron that include the sites of breakage and reunion.

By comparing the nucleotide sequence of mRNA with that of the original gene, the junctions between exons and introns can be assigned. There is no extensive homology or complementarity between the two ends of an intron. The junctions, however, have well conserved, though rather short, consensus sequences. A specific end can be assigned to every intron by aligning the exon–intron junctions to conform to the consensus sequence given in **FIGURE 28.2**.

The subscripts in the figure indicate the percentage of the specified base at each consensus position. High conservation is found only *immediately within the intron* at the presumed junctions, although the internal branch point sequence is conserved within major taxonomic groups. This identifies the sequence of a generic (using the consensus branch point sequence for animals) intron as:

GU . . . UACUAAC . . . AG

Because the intron defined in this way starts with the dinucleotide GU and ends with the dinucleotide AG, the junctions are often described as conforming to the **GU-AG rule**. (Of course, the coding strand sequence of DNA has GT-AG.)

▸ **GU-AG rule** The rule that describes the presence of these constant dinucleotides at the first two and last two positions of introns of nuclear genes.

Note that the two sites have different sequences and so they define the ends of the intron *directionally*. They are named proceeding from left to right along the intron as the *5′ splice site* (sometimes called the *left* or *donor site*) and the *3′ splice site* (also called the *right* or *acceptor site*). The consensus sequences are implicated as the sites recognized in splicing by point mutations that prevent splicing *in vivo* and *in vitro*.

GU-AG sequences constitute the vast majority (>98%) of splicing junctions in the human genome. A small number of introns (<1%) use the related junctions GC-AG. In addition, there is a minor class of introns marked by the ends AU-AC (0.1%).

KEY CONCEPTS

- The 5′ splice site at the 5′ (left) end of the intron includes the consensus sequence GU.
- The 3′ splice site at the 3′ (right) end of the intron includes the consensus sequence AG.
- The GU-AG rule describes the requirement for these constant dinucleotides at the first two and last two positions of introns in pre-mRNAs.

CONCEPT AND REASONING CHECK

What would be the effect of a mutation that altered the GU or AG sequence of an intron splice site?

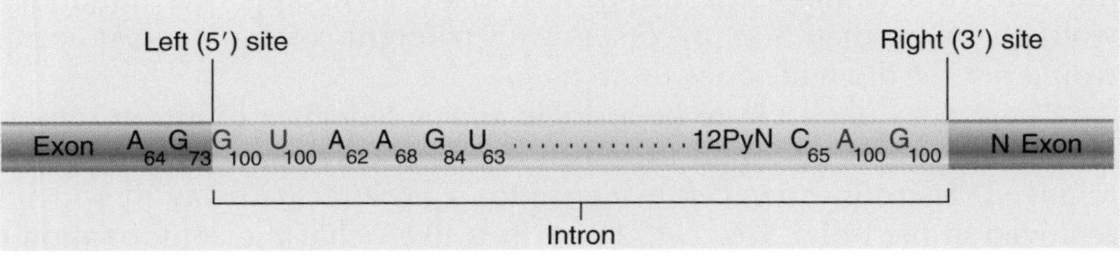

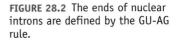

FIGURE 28.2 The ends of nuclear introns are defined by the GU-AG rule.

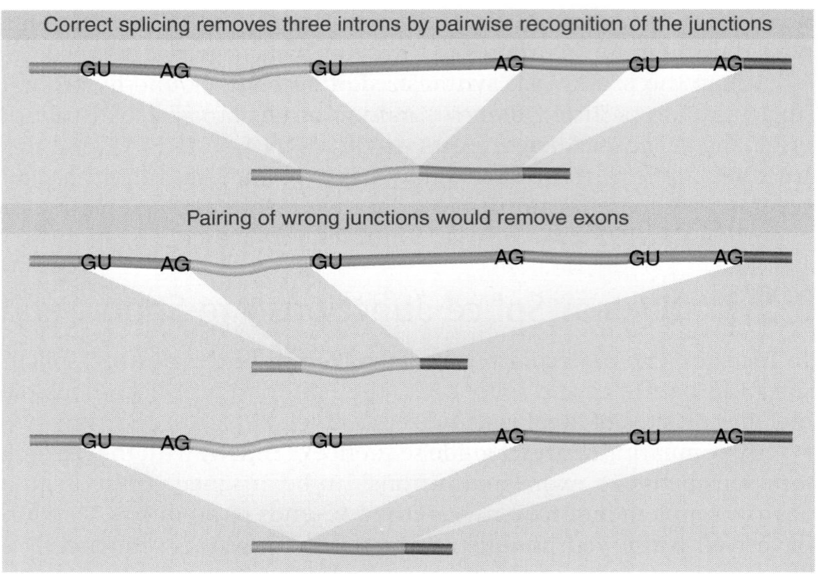

FIGURE 28.3 Splicing junctions are recognized only in the correct pairwise combinations.

28.3 Splice Junctions Are Read in Pairs

A typical mammalian gene has many introns. The basic problem of pre-mRNA splicing results from the simplicity of the splice sites and is illustrated in **FIGURE 28.3**: What ensures that the correct pairs of sites are spliced together? In other words, how are splice junctions identified and brought together? The corresponding GU-AG pairs must often be connected across great distances, since some introns are >10 kb long. The recognition and splicing processes involve interactions between mRNA sequences and other RNAs and proteins as part of the spliceosome. These interactions are detailed in the next few sections.

Experiments using hybrid RNA precursors show that any 5′ splice site can, in principle, be connected to any 3′ splice site. Such experiments have two general conclusions:

- *Splice sites are generic.* They do not have specificity for individual RNA precursors, and individual precursors do not convey specific information (such as secondary structure) that is needed for splicing.
- *The apparatus for splicing is not tissue-specific.* An RNA can usually be properly spliced by any cell, regardless of whether it is usually synthesized in that cell. (We discuss exceptions in which there are tissue-specific alternative splicing patterns in *Section 28.11, Alternative Splicing Involves Differential Use of Splice Junctions.*)

Here is a paradox: it is likely that all 5′ splice sites and all 3′ splice sites are similarly recognized by the splicing apparatus. *In principle, any 5′ splice site may be joined with any 3′ splice site.* In the usual circumstances, though, splicing occurs only between the 5′ and 3′ sites of the *same* intron. *What rules ensure that recognition of splice sites is restricted so that only the 5′ and 3′ sites of the same intron are spliced?*

Are introns removed from a particular RNA in a specific *order*? Using RNA blotting, we can identify nuclear RNAs that represent intermediates from which some introns have been removed. **FIGURE 28.4** shows a blot of the precursors to ovomucoid mRNA. There is a discrete series of bands, which suggests that splicing occurs via definite pathways. (If the seven introns were removed in an entirely random order, there would be more than 300 precursors with different combinations of introns, and we would not see discrete bands.)

There does not seem to be a single *unique* sequence of intron removal, because intermediates can be found in which different combinations of introns have been removed. There is, however, evidence for a *preferred* pathway in which introns are removed in the order 5/6, 7/4, 2/1, 3. It is likely that the conformation of the RNA influences the accessibility of the splice sites. As particular introns are removed, the conformation changes, and new pairs of splice sites become available. The ability of the precursor to remove its introns in more than one order, though, suggests that alternative conformations are available at each stage. One important conclusion of this analysis is that *the reaction does not proceed sequentially along the precursor in the 5′ to 3′ direction*.

A simple and obvious model to control recognition of splice sites would be for the splicing apparatus to act in a processive manner. Having recognized a 5′ site, the apparatus might scan the RNA in the appropriate direction until it meets the next 3′ site. This would restrict splicing to adjacent sites. This model, however, is excluded by experiments that show that splicing can occur in *trans* as an intermolecular reaction under special circumstances (see *Section 28.12,* trans-*Splicing Reactions Use Small RNAs*) or in RNA molecules in which part of the nucleotide chain is replaced by a chemical linker. This means that there cannot be a requirement for strict scanning along the RNA from the 5′ splice site to the 3′ splice site. Another problem with the scanning model is that it cannot explain the existence of alternative splicing patterns, where (for example) a common 5′ site is spliced to more than one 3′ site under different circumstances (see *Section 28.11, Alternative Splicing Involves Differential Use of Splice Junctions*).

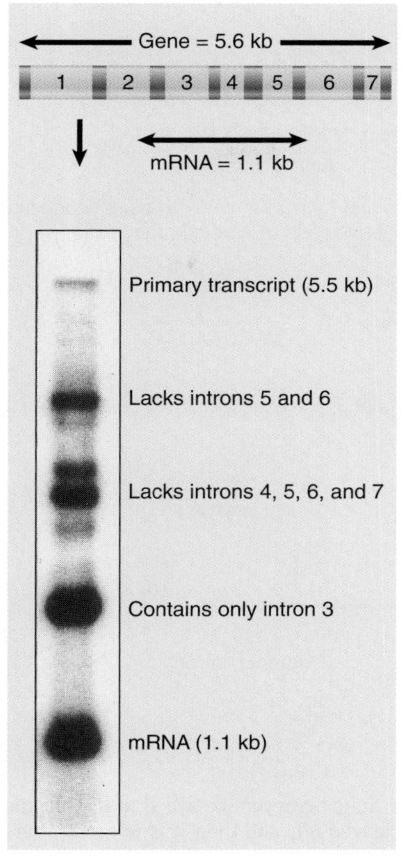

FIGURE 28.4 Northern blotting of nuclear RNA with an ovomucoid probe identifies discrete precursors to mRNA. The contents of the more prominent bands are indicated. Photo courtesy of Bert W. O'Malley, Baylor College of Medicine.

An alternate model is that there is a sequence downstream from the 5′ site that influences or determines the choice of the 3′ site. This is explored in the next section.

KEY CONCEPTS

- Splicing depends only on recognition of pairs of splice junctions.
- All 5′ splice sites are functionally equivalent, and all 3′ splice sites are functionally equivalent.

CONCEPT AND REASONING CHECK

What evidence is there that the splicing apparatus does *not* scan from a recognized 5′ splice site to the next 3′ splice site?

28.4 Pre-mRNA Splicing Proceeds through a Lariat

In addition to the 5′ and 3′ splice sites, the splicing apparatus recognizes a sequence downstream from the 5′ site called the **branch site**, so-called because of the branched intermediate formed during the splicing reaction.

The stages of splicing are illustrated in the pathway of **FIGURE 28.5**. We discuss the reaction in terms of the individual RNA species that can be identified, but remember that *in vivo* the species containing exons are not released as free molecules, but remain held together by the splicing apparatus.

The first step is to make a cut at the 5′ splice site, separating the left exon and the right intron–exon molecule. The left exon takes the form of a linear molecule. The right intron–exon molecule forms a branched structure called the **lariat**, in which the

- **branch site** A short sequence just before the end of an intron at which the lariat intermediate is formed in splicing by joining the 5′ nucleotide of the intron to the 2′ position of an adenosine.
- **lariat** An intermediate in RNA splicing in which a circular structure with a tail is created by a 5′ to 2′ bond.

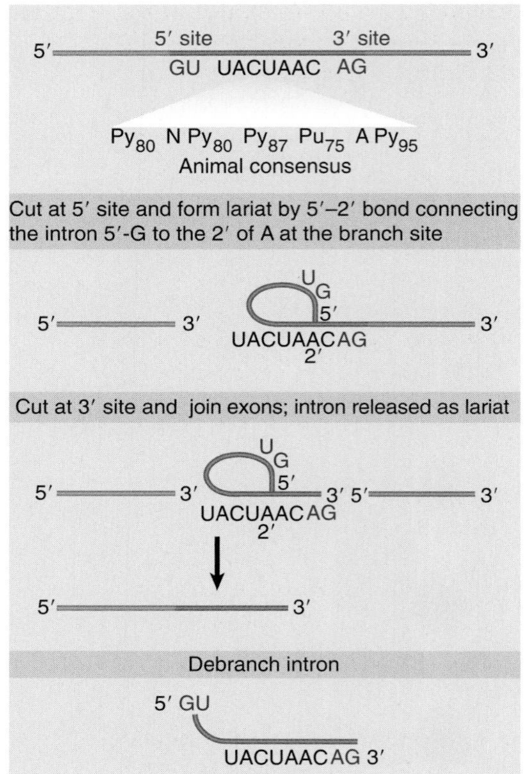

FIGURE 28.5 Splicing occurs in two stages. First the 5′ exon is cleaved off, and then it is joined to the 3′ exon.

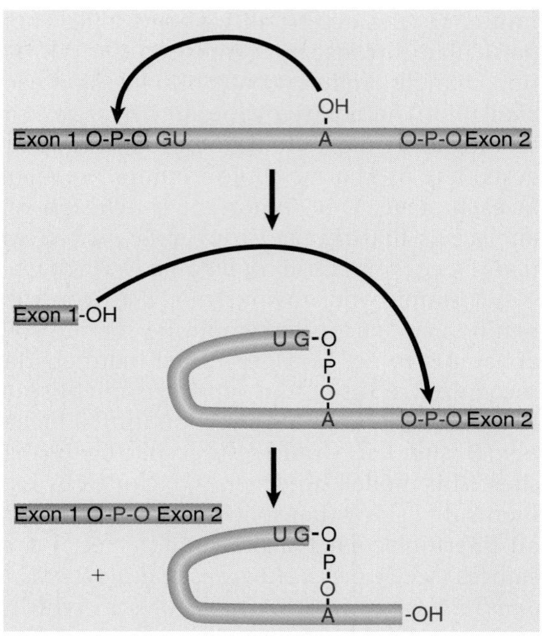

FIGURE 28.6 Nuclear splicing occurs by two transesterification reactions in which an OH group attacks a phosphodiester bond.

5′ terminus generated at the end of the intron becomes linked by a 5′–2′ bond to a base within the intron. The target base is an A in the branch site.

Cutting at the 3′ splice site releases the free intron in lariat form; the right exon is ligated (spliced) to the left exon. The cleavage and ligation reactions are shown separately in the figure for illustrative purposes but actually occur as one coordinated transfer.

The lariat is then "debranched" to give a linear excised intron, which is rapidly degraded.

The sequences needed for splicing are the short consensus sequences at the 5′ and 3′ splice sites and at the branch site.

The branch site plays an important role in identifying the 3′ splice site. The branch site in yeast is highly conserved and has the consensus sequence UACUAAC. The branch site in multicellular eukaryotes is not well conserved, but has a bias for either purines or pyrimidines at each position and retains the target A nucleotide (see Figure 28.6). The branch site lies 18 to 40 nucleotides upstream of the 3′ splice site. *Its role is to identify the nearest 3′ splice site as the target for connection to the 5′ splice site*; it is always the 3′ consensus sequence nearest to the 3′ side of the branch that becomes the target for splicing. This can be explained by the fact that an interaction occurs between protein complexes that bind to these two sites.

The bond that forms the lariat goes from the 5′ position of the invariant G from the 5′ end of the intron to the 2′ position of the invariant A in the branch site. This corresponds to the third A residue in the yeast UACUAAC box.

▸ **transesterification** A reaction that breaks and makes chemical bonds in a coordinated transfer so that no energy is required.

The chemical reactions proceed by **transesterification**: in effect, a bond is transferred from one location to another. **FIGURE 28.6** shows that the first step is a nucleophilic attack by the 2′–OH of the invariant A of the UACUAAC consensus sequence of the 5′ splice site. In the second step, the free 3′–OH of the exon that was released by the first reaction now attacks the bond at the 3′ splice site. Note that the number of phosphodiester bonds is conserved. There were originally two 5′–3′ bonds at the exon–intron splice sites; one has been replaced by the 5′–3′ bond between the exons, and the other has been replaced by the 5′–2′ bond that forms the lariat.

KEY CONCEPTS

- Splicing requires the 5′ and 3′ splice sites and a branch site just upstream of the 3′ splice site.
- The branch sequence is conserved in yeast but less well conserved in multicellular eukaryotes.
- A lariat is formed when the intron is cleaved at the 5′ splice site, and the 5′ end is joined to a 2′ position at an A at the branch site in the intron.
- The intron is released as a lariat when it is cleaved at the 3′ splice site, and the left and right exons are then ligated together.
- The reactions occur by transesterifications, in which a bond is transferred from one location to another.

CONCEPT AND REASONING CHECK

Describe the sequence of chemical events that occur in RNA splicing.

28.5 snRNAs Are Required for Splicing

The 5′ and 3′ splice sites and the branch sequence are recognized by components of the splicing apparatus that assemble to form a large complex. This complex brings together the 5′ and 3′ splice sites before any reaction occurs, which explains why a deficiency in any one of the sites may prevent the reaction from initiating. The complex assembles sequentially on the pre-mRNA, and passes through several "presplicing complexes" before forming the final, active complex, which is called the spliceosome. Splicing occurs only after all the components have assembled.

The splicing apparatus contains both proteins and RNAs (in addition to the pre-mRNA). The RNAs take the form of small molecules that exist as ribonucleoprotein particles. Both the nucleus and cytoplasm of eukaryotic cells contain many discrete small RNA species, which typically range in size from 100 to 300 bases in multicellular eukaryotes. Those restricted to the nucleus are called **small nuclear RNAs (snRNA)**; those found in the cytoplasm are called **small cytoplasmic RNAs (scRNA)**. In their natural state, they exist as ribonucleoprotein particles (*snRNP* and *scRNP*). Colloquially, they are sometimes known as **snurps** and **scyrps**. There is also a class of small RNAs found in the nucleolus, called snoRNAs, which are involved in processing ribosomal RNA (see *Section 28.15, Small RNAs Are Required for rRNA Processing*).

The spliceosome is a large particle, greater in mass than the ribosome, and contains five snRNPs as well as many additional proteins. The snRNPs involved in splicing are U1, U2, U5, U4, and U6. They are named according to the snRNAs that are present. Each snRNP contains a single snRNA and several (<20) proteins. The U4 and U6 snRNPs are usually found as a single (U4/U6) particle. A common structural core for each snRNP consists of a group of eight proteins, all of which are recognized by an autoimmune antiserum called **anti-Sm**; conserved sequences in the proteins form the target for the antibodies. The other proteins in each snRNP are unique to it. The Sm proteins bind to the conserved sequence PuAU$_{3-6}$GPu, which is present in all snRNAs except U6. The U6 snRNP instead contains a set of Sm-like (Lsm) proteins. The Sm proteins must be involved in the autoimmune reaction, although their relationship to the phenotype of the autoimmune disease is not clear.

FIGURE 28.7 summarizes the components of the spliceosome. The five snRNAs account for more than a quarter of the mass; together with their 41 associated proteins, they account for almost half of the mass. Some 70 other proteins found in the spliceosome are described as **splicing factors**. They include proteins required for

- **small nuclear RNA (snRNA; snurps)** One of many small RNA species confined to the nucleus; several of them are involved in splicing or other RNA processing reactions. Snurps are the ribonucleoprotein particles that include a specific snRNA and its protein partners.
- **small cytoplasmic RNAs (scRNA; scyrps)** RNAs that are present in the cytoplasm (and sometimes are also found in the nucleus). Scyrps are the ribonucleoprotein particles that include an scRNA and its associated proteins.
- **anti-Sm** An autoimmune antiserum that defines the Sm domain that is common to a group of proteins found in snRNPs that are involved in RNA splicing.
- **splicing factor** A protein component of the spliceosome that is not part of one of the snRNPs.

FIGURE 28.7 The spliceosome is ~12 MDa. Five snRNPs account for almost half of the mass. The remaining proteins include known splicing factors, as well as proteins that are involved in other stages of gene expression.

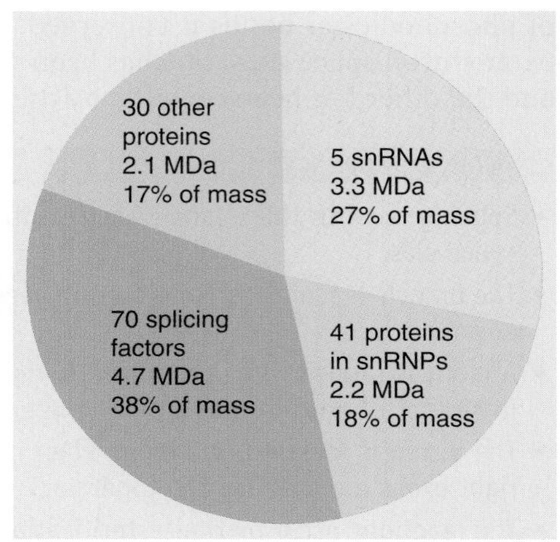

assembly of the spliceosome, proteins required for it to bind to the RNA substrate, and proteins involved in the catalytic process. In addition to these proteins, another ~30 proteins associated with the spliceosome are believed to be acting at other stages of gene expression, which suggests that the spliceosome may serve as a coordinating apparatus. Some of these proteins bind to the C-terminal domain (CTD) of RNA polymerase II, which coordinates RNA splicing during transcription.

Some of the proteins in the snRNPs may be directly involved in splicing; others may be required in structural roles, or just for assembly or interactions between the snRNP particles. About one third of the proteins involved in splicing are components of the snRNPs. Increasing evidence for a direct role of RNA in the splicing reaction suggests that relatively few of the splicing factors play a direct role in catalysis; most are involved in structural or assembly roles.

KEY CONCEPTS

- The five snRNPs involved in splicing are U1, U2, U5, U4, and U6.
- Together with some additional proteins, the snRNPs form the spliceosome.
- All the snRNPs except U6 contain a conserved sequence that binds the Sm proteins that are recognized by antibodies generated in autoimmune disease.

CONCEPT AND REASONING CHECK

What are the probable respective roles of RNAs and proteins in RNA splicing?

28.6 U1 snRNP Initiates Splicing

Splicing can be broadly divided into two stages:

- First, the consensus sequences at the 5′ splice site, branch sequence, and adjacent pyrimidine tract are recognized. A complex assembles that contains all of the splicing components.
- Next, the cleavage and ligation reactions change the structure of the substrate RNA. Components of the complex are released or reorganized as the splicing reactions proceed.

The important point is that all of the splicing components are assembled and have ensured that the splice sites are available before any irreversible change is made to the RNA.

Recognition of the consensus sequences involves both RNAs and proteins. Certain snRNAs have sequences that are complementary to the mRNA consensus sequences or to one another, and base pairing between snRNA and pre-mRNA, or between snRNAs, plays an important role in splicing.

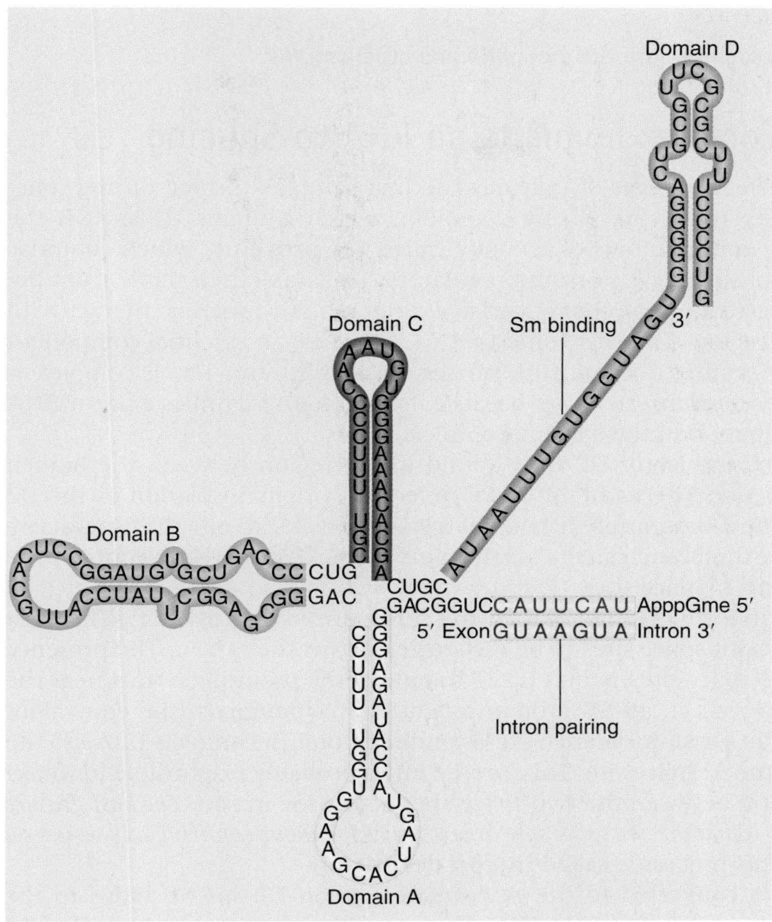

FIGURE 28.8 U1 snRNA has a base-paired structure that creates several domains. The 5′ end remains single stranded and can base pair with the 5′ splicing site.

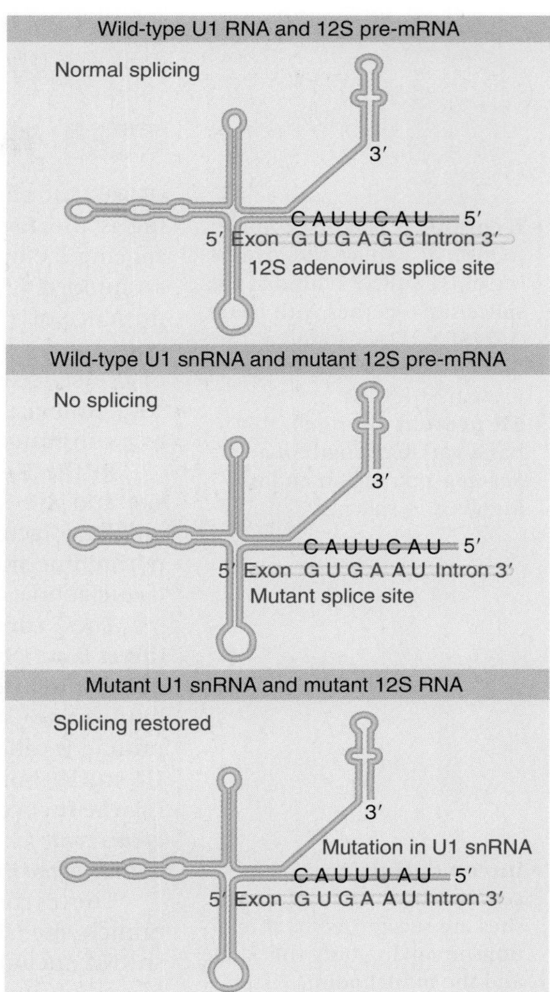

FIGURE 28.9 Mutations that abolish function of the 5′ splice site can be suppressed by compensating mutations in U1 snRNA that restore base pairing.

Binding of U1 snRNP to the 5′ splice site is the first step in splicing. The human U1 snRNP contains eight proteins as well as the RNA. The secondary structure of the U1 snRNA is shown in **FIGURE 28.8**. It contains several domains. The Sm-binding site is required for interaction with the common snRNP proteins. Domains identified by the individual stem-loop structures provide binding sites for proteins that are unique to U1 snRNP.

U1 snRNA base pairs with the 5′splice site by means of a single-stranded region at its 5′ terminus, which usually includes a stretch of four to six bases that is complementary with the splice site. **FIGURE 28.9** describes an experiment that directly demonstrated the need for this base pairing. The wild-type sequence of the splice site of the 12S adenovirus pre-mRNA pairs at five out of six positions with U1 snRNA. A mutant in the 12S RNA that cannot be spliced has two sequence changes; the GG residues at positions 5 and 6 in the intron are changed to AU. Splicing can be restored by a compensating change in the U1 snRNA that allows pairing.

KEY CONCEPT

- U1 snRNP initiates splicing by binding to the 5′ splice site by means of an RNA–RNA pairing reaction.

CONCEPT AND REASONING CHECK

How does U1 snRNP recognize both the pre-mRNA and other snRNAs?

28.7 The E Complex Commits an RNA to Splicing

▸ **E complex** The first complex to form at a splice site, consisting of U1 snRNP bound at the splice site together with factor ASF/SF2, U2AF bound at the branch site, and the bridging protein SF1/BBP.

▸ **SR protein** A protein that has a variable length of a Ser-Arg-rich region and is involved in splicing.

FIGURE 28.10 shows the early stages of splicing. The first complex formed during splicing is the **E complex** (early presplicing complex), which contains U1 snRNP, the splicing factor U2AF, and members of a family called **SR proteins**, which comprise an important group of splicing factors and regulators. They take their name from the presence of a Ser-Arg-rich region that is variable in length. SR proteins interact with one another via this region. They also bind to RNA. They are an essential component of the spliceosome, forming a framework on the RNA substrate. The E complex is sometimes called the *commitment complex*, because its formation identifies a pre-mRNA as a substrate for formation of the splicing complex.

In the E complex, the factor U2AF is bound to the region between the branch site and the 3′ splice site. The name of U2AF reflects its original isolation as the U2 auxiliary factor. In most organisms, it has a large subunit (U2AF65) that contacts a pyrimidine tract near the branch site; a small subunit (U2AF35) directly contacts the dinucleotide AG at the 3′ splice site.

The E complex can form in either of the ways summarized in **FIGURE 28.11**. The most direct reaction is for both splice sites to be recognized across the intron. The presence of U1 snRNP at the 5′ splice site enables U2AF to bind at the pyrimidine tract near the branch site. A splicing factor, an SR protein called SF1 in mammals (the equivalent protein is called BBP in yeast, for branch point binding protein), connects U2AF to the U1 snRNP bound at the 5′ splice site. This interaction is probably responsible for making the first connection between the two splice sites across the intron. *The basic feature of this route for splicing is that the two splice sites are recognized without requiring any sequences outside of the intron.* This process is called **intron definition**.

▸ **intron definition** The process in which a pair of splicing sites are recognized by interactions involving only the 5′ site and the branchpoint/3′ site.

▸ **A complex** The second splicing complex, formed by the binding of U2 snRNP to the E complex.

The E complex is converted to the **A complex** when U2 snRNP binds to the branch site. Both U1 snRNP and U2AF/Mud2 are needed for U2 binding. The U2 snRNA includes sequences complementary to the branch site; a sequence near the 5′ end of the snRNA base pairs with the branch sequence in the intron. In yeast this typically involves formation of a duplex with the UACUAAC box (see Figure 28.13). Several proteins of the U2 snRNP are bound to the substrate RNA just upstream of the branch site. The addition of U2 snRNP to the E complex generates the A presplic-

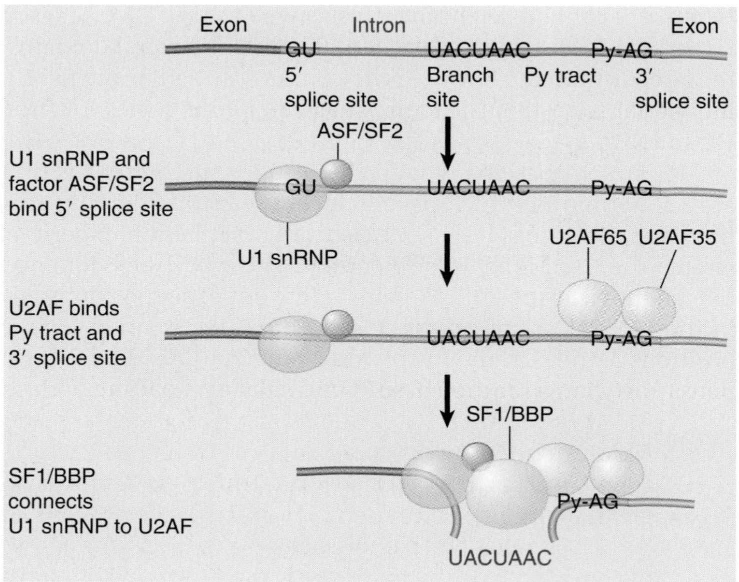

FIGURE 28.10 The commitment (E) complex forms by the successive addition of U1 snRNP to the 5′ splice site, U2AF to the pyrimidine tract/3′ splice site, and the bridging protein SF1/BBP.

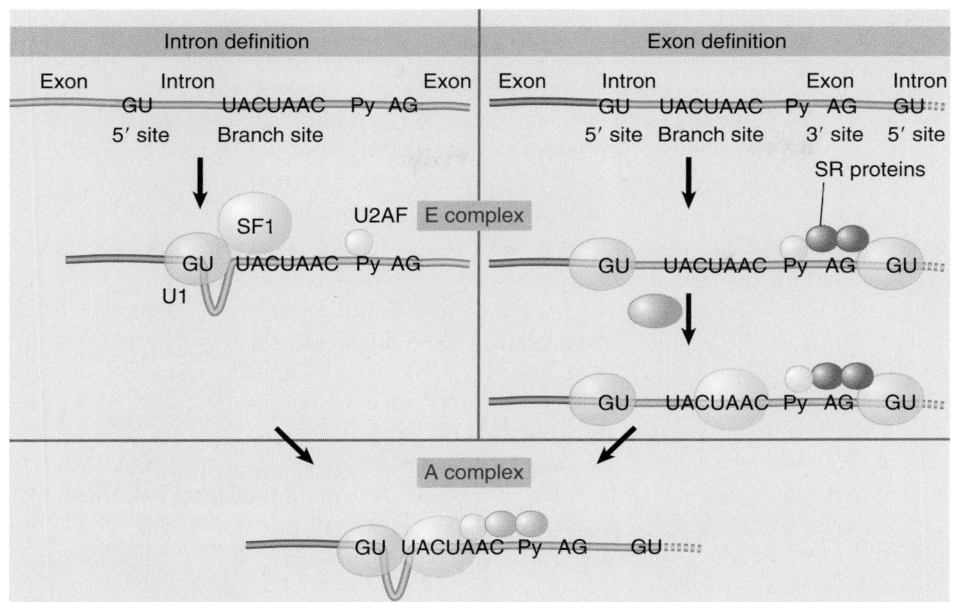

FIGURE 28.11 There may be multiple routes for initial recognition of 5′ and 3′ splice sites.

ing complex. The binding of U2 snRNP requires ATP hydrolysis, and commits a pre-mRNA to the splicing pathway.

An alternative route to form the spliceosome may be followed when the introns are long and the splice sites are weak. As shown on the right side of Figure 28.11, the 5′ splice site is recognized by U1 snRNA in the usual way. However, the 3′ splice site is recognized as part of a complex that forms across the next exon in which the next 5′ splice site is also bound by U1 snRNA. This U1 snRNA is connected by SR proteins to the U2AF at the pyrimidine tract. When U2 snRNP joins to generate the A complex, there is a rearrangement in which the correct (leftmost) 5′ splice site displaces the downstream 5′ splice site in the complex. The important feature of this route for splicing is that sequences downstream of the intron itself are required. Usually these sequences include the next 5′-splice site. This process is called **exon definition**. This mechanism is not universal; neither SR proteins nor exon definition are found in *S. cerevisiae*.

exon definition The process in which a pair of splicing sites are recognized by interactions involving the 5′ site of the intron and also the 5′ site of the next intron downstream.

KEY CONCEPTS

- The direct way of forming an E complex is for U1 snRNP to bind at the 5′ splice site and U2AF to bind at a pyrimidine tract between the branch site and the 3′ splice site. This is intron definition.
- Another possibility is for the complex to form between U2AF at the pyrimidine tract and U1 snRNP at a downstream 5′ splice site. This is exon definition.

CONCEPT AND REASONING CHECK

Describe the formation of the E and A presplicing complexes.

28.8 Five snRNPs Form the Spliceosome

snRNPs and other factors involved in splicing associate with the presplicing complexes in a defined order. FIGURE 28.12 shows the components of the complexes that can be identified as the reaction proceeds.

The B1 complex is formed when a trimer containing the U5 and U4/U6 snRNPs binds to the A complex containing U1 and U2 snRNPs. It is converted to the B2 complex after U1 is released. The dissociation of U1 is necessary to allow other components to come into juxtaposition with the 5′-splice site, most notably U6 snRNA. At

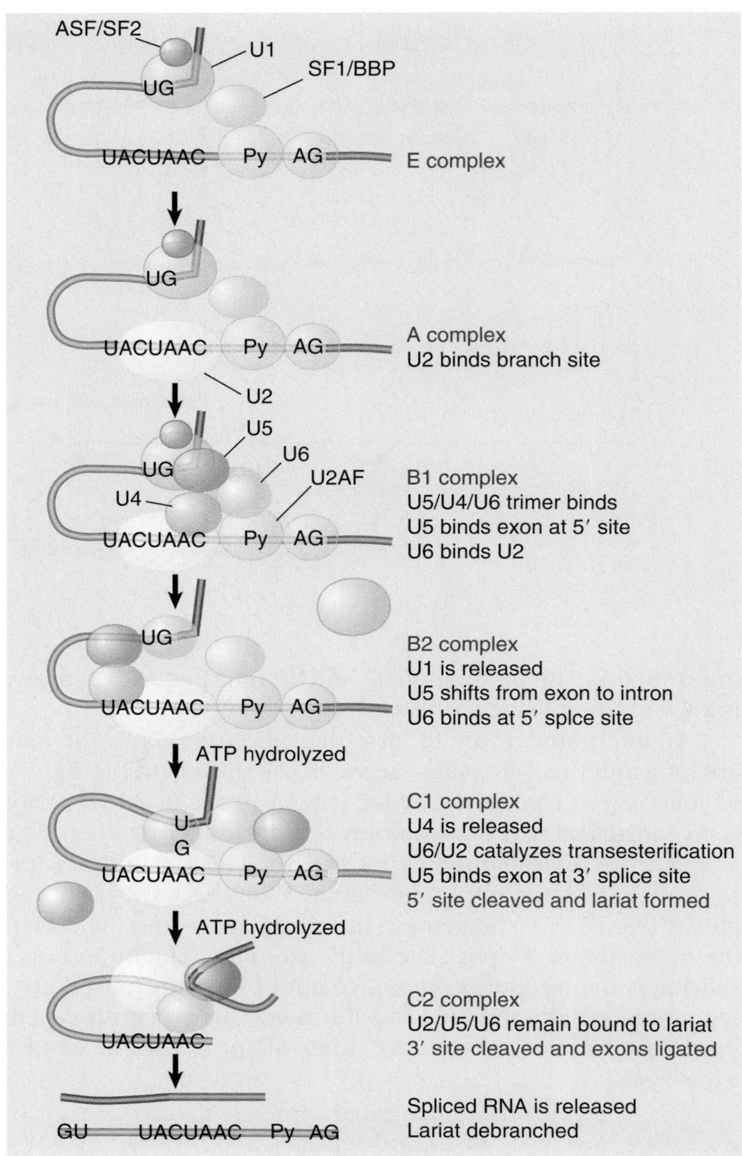
ASF/SF2
U1
SF1/BBP
UG
UACUAAC
Py
AG
E complex
A complex
U2 binds branch site
U2
U5
U6
U2AF
U4
B1 complex
U5/U4/U6 trimer binds
U5 binds exon at 5′ site
U6 binds U2
B2 complex
U1 is released
U5 shifts from exon to intron
U6 binds at 5′ splce site
ATP hydrolyzed
C1 complex
U4 is released
U6/U2 catalyzes transesterification
U5 binds exon at 3′ splice site
5′ site cleaved and lariat formed
ATP hydrolyzed
C2 complex
U2/U5/U6 remain bound to lariat
3′ site cleaved and exons ligated
Spliced RNA is released
Lariat debranched
GU UACUAAC Py AG

FIGURE 28.12 The splicing reaction proceeds through discrete stages in which spliceosome formation involves the interaction of components that recognize the consensus sequences.

this point U5 snRNA changes its position; initially it is close to exon sequences at the 5′-splice site, but it must shift its position to bind to the 3′-splice site for the catalytic reaction to occur.

The catalytic reaction is triggered by the release of U4; this requires hydrolysis of ATP. The role of U4 snRNA may be to sequester U6 snRNA until it is needed. **FIGURE 28.13** shows the changes that occur in the base pairing interactions between snRNAs during splicing. In the U6/U4 snRNP, a continuous length of 26 bases of U6 is paired with two separated regions of U4. When U4 dissociates, the region in U6 that is released becomes free to take up another structure. The first part of it pairs with U2; the second part forms an intramolecular hairpin. The interaction between U4 and U6 is mutually incompatible with the interaction between U2 and U6, so the release of U4 allows the spliceosome to proceed.

Although for clarity the figure shows the RNA substrate in extended form, the 5′-splice site is actually close to the U6 sequence immediately on the 5′ side of the stretch bound to U2. This sequence in U6 snRNA pairs with sequences in the intron just downstream of the conserved GU at the 5′ splice site (mutations that enhance such pairing improve the efficiency of splicing).

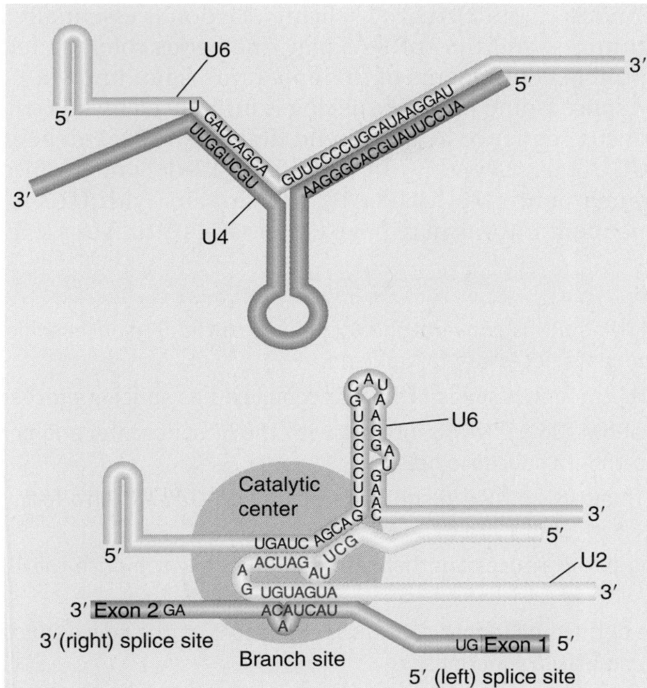

FIGURE 28.13 U6-U4 pairing is incompatible with U6-U2 pairing. When U6 joins the spliceosome it is paired with U4. Release of U4 allows a conformational change in U6; one part of the released sequence forms a hairpin (dark gray), and the other part (black) pairs with U2. An adjacent region of U2 is already paired with the branch site, which brings U6 into juxtaposition with the branch. Note that the substrate RNA is reversed from the usual orientation and is shown 3′ to 5′.

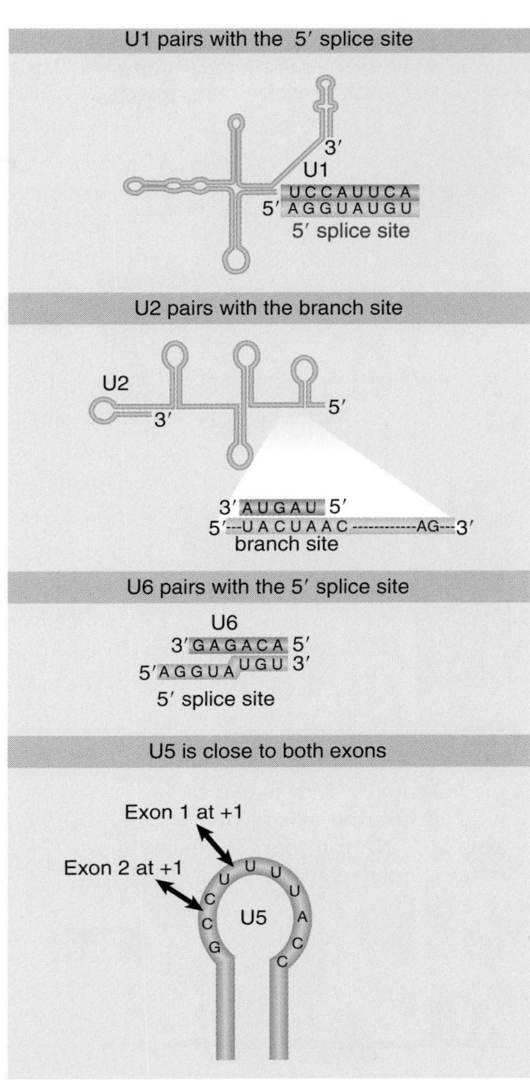

FIGURE 28.14 Splicing utilizes a series of base pairing reactions between snRNAs and splice sites.

As discussed above, several pairing reactions between snRNAs and the substrate pre-mRNA occur in the course of splicing. They are summarized in **FIGURE 28.14**. The snRNPs have sequences that pair with the pre-mRNA substrate and with one another. They also have single-stranded regions in loops that are in close proximity to sequences in the substrate, and that play an important role, as judged by the ability of mutations in the loops to prevent splicing.

The base pairing between U2 and the branch point, and between U2 and U6, creates a structure that resembles the active center of group II self-splicing introns (see Figure 28.19). This suggests the possibility that the catalytic component could comprise an RNA structure generated by the U2-U6 interaction. U6 is paired with the 5′ splice site, and crosslinking experiments show that a loop in U5 snRNA is immediately adjacent to the first base positions in both exons. Although we can define the proximities of the substrate (5′ splice site and branch site) and snRNPs (U2 and U6) at the catalytic center (as shown in Figure 28.13), the components that undertake the transesterifications have not been directly identified.

The important conclusion suggested by these results is that *the snRNA components of the splicing apparatus interact both among themselves and with the substrate pre-mRNA by means of base pairing interactions, and these interactions allow for changes in structure that may bring reacting groups into apposition and may even create catalytic centers*. Furthermore, the conformational changes in the snRNAs are reversible; for example, U6 snRNA is not used up in a splicing reaction, and at completion must be released from U2 so that it can reform the duplex structure with U4 to undertake another cycle of splicing.

A small proportion of introns is spliced by an alternative apparatus, called the U12 spliceosome, consisting of U11 and U12 (related to U1 and U2, respectively), a U5 variant, and the $U4_{atac}$ and $U6_{atac}$ snRNAs. The splicing reaction is essentially similar to that at U2-dependent introns, and the snRNAs play analogous roles. Whether there are differences in the protein components of this apparatus is not known.

The specific type of spliceosome used for splicing is influenced by sequences in the intron. A strong consensus sequence at the 5′ end defines the U12-dependent type of intron: $5'^{G}_{A}$UAUCCUUU. . . . . .PyA^{G}_{C} 3′. In addition, U12-dependent introns have a highly conserved branch point, UCCUUPuAPy, which pairs with U12. Both U12-dependent and U2-dependent introns may have either GU-AG or AU-AC termini.

KEY CONCEPTS

- Binding of U5 and U4/U6 snRNPs converts the A complex to the B1 spliceosome, which contains all the components necessary for splicing.
- The spliceosome passes through a series of further complexes as splicing proceeds.
- Release of U1 snRNP allows U6 snRNA to interact with the 5′ splice site, and converts the B1 spliceosome to the B2 spliceosome.
- When U4 dissociates from U6 snRNP, U6 snRNA can pair with U2 snRNA to form the catalytic active site.
- An alternative splicing pathway uses another set of snRNPs that comprise the U12 spliceosome.
- The target introns are defined by longer consensus sequences at the splice junctions, but usually include the same GU-AG junctions.
- Some introns have the splice junctions AU-AC, including some that are U2-dependent and some that are U12-dependent.

CONCEPT AND REASONING CHECK

What is the relationship between the U2, U4, and U6 snRNPs?

▸ **EJC (exon junction complex)** A protein complex that assembles at exon–exon junctions during splicing and assists in RNA transport, localization, and degradation.

28.9 Splicing Is Connected to Export of mRNA

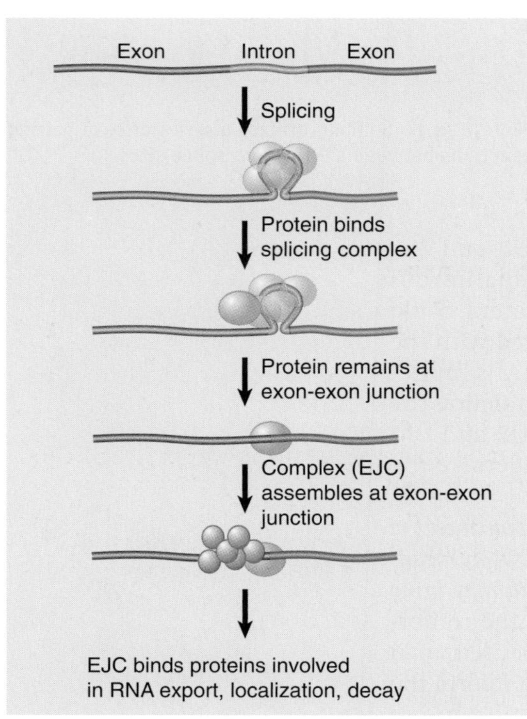

FIGURE 28.15 The EJC (exon junction complex) binds to RNA by recognizing the splicing complex.

After it has been synthesized and processed, mRNA is exported from the nucleus to the cytoplasm in the form of a ribonucleoprotein complex. One means for ensuring that transport occurs only after the completion of splicing may be that introns can prevent export of mRNA because they are associated with the splicing apparatus. The spliceosome also may provide the initial point of contact for the export apparatus. **FIGURE 28.15** shows a model in which a protein complex binds to the RNA via the splicing apparatus. The complex consists of >9 proteins and is called the **EJC (exon junction complex)**.

The EJC is involved in several functions of spliced mRNAs. Some of the proteins of the EJC are directly involved in these functions, and others recruit additional proteins for particular functions. The first contact in assembling the EJC is made with one of the splicing factors. After splicing, the EJC remains attached to the mRNA just upstream of the exon–exon junction. The EJC is not associated with RNAs transcribed from genes that lack introns, so it is uniquely involved with spliced products.

If introns are deleted from a gene, its RNA product is exported much more slowly to the cytoplasm. This suggests that the intron may provide a signal for attachment of the export apparatus. We can now account for this phenomenon in terms of a series of protein interactions, as shown in **FIGURE 28.16**. The EJC includes a group of proteins called the REF family (the best characterized member is called Aly). The REF proteins in turn interact with a transport protein (variously called TAP and Mex), which has direct responsibility for interaction with the nuclear pore.

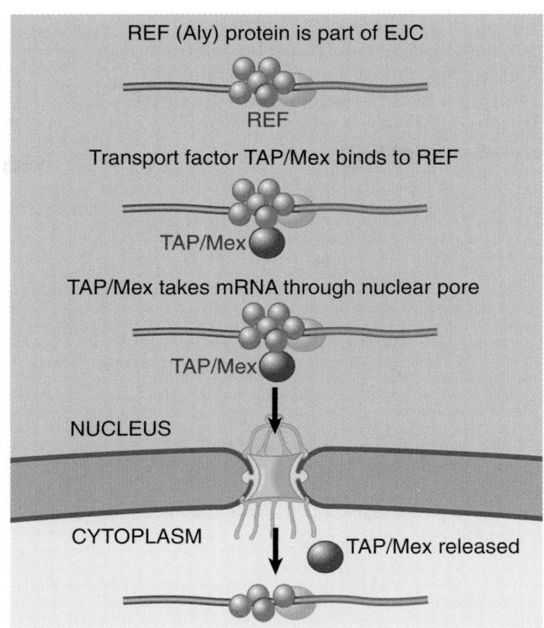

FIGURE 28.16 A REF protein binds to a splicing factor and remains with the spliced RNA product. REF binds to an export factor that binds to the nuclear pore.

A similar system may be used to identify a spliced RNA so that nonsense mutations prior to the last exon trigger its degradation in the cytoplasm (see *Section 7.13, Nonsense Mutations Trigger a Surveillance System*).

KEY CONCEPTS

- The REF proteins bind to splicing junctions by associating with the spliceosome.
- After splicing, they remain attached to the RNA at the exon–exon junction.
- They interact with the transport protein TAP/Mex that exports the RNA through the nuclear pore.

CONCEPT AND REASONING CHECK

Explain how the presence of introns in a gene may aid in export of the mRNA product to the cytoplasm.

28.10 Group II Introns Autosplice via Lariat Formation

Two groups of introns that are quite separate from the introns in nuclear protein-coding genes are found in organelles and in bacteria. Group I and group II introns are classified according to their internal organization. Each can be folded into a typical type of secondary structure. Group I introns are also found in the nucleus of unicellular eukaryotes.

The group I and group II introns have the remarkable ability to excise themselves from an RNA. This is called **autosplicing**. Group I introns are more common than group II introns. There is little relationship between the two classes, but in each case the RNA can perform the splicing reaction *in vitro* by itself, without requiring enzymatic activities provided by proteins; however, proteins are almost certainly required *in vivo* to assist with folding (see *Chapter 29, Catalytic RNA*).

autosplicing (self-splicing) The ability of an intron to excise itself from an RNA by a catalytic action that depends only on the sequence of RNA in the intron.

FIGURE 28.17 shows that three classes of introns are excised by two successive trans-esterifications (shown previously for nuclear introns in Figure 28.5). In the first reaction, the 5′ exon–intron junction is attacked by a free hydroxyl group (provided by an internal 2′–OH position in nuclear and group II introns, and by a free guanine nucleotide in group I introns). In the second reaction, the free 3′–OH at the end of the released exon in turn attacks the 3′ intron–exon junction.

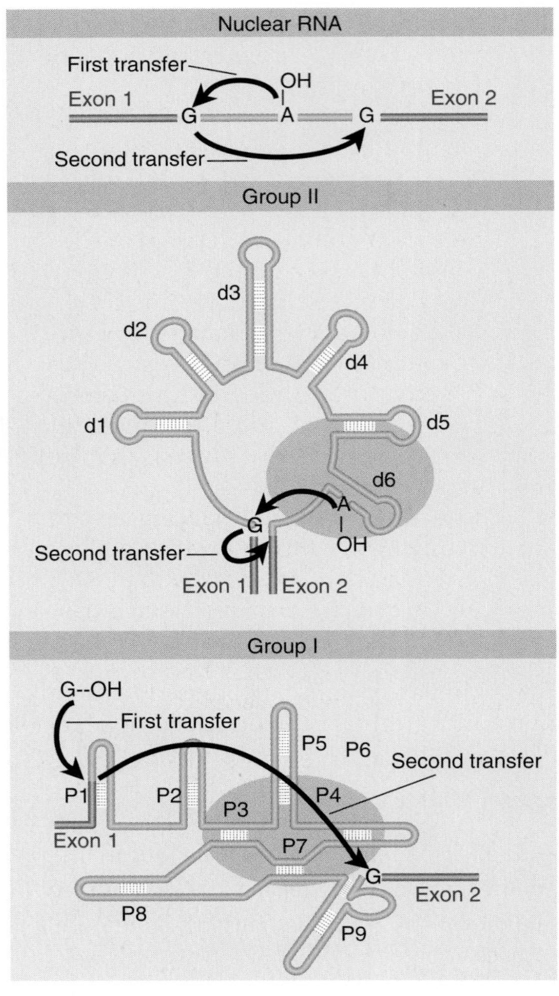

FIGURE 28.17 Three classes of splicing reactions proceed by two transesterifications. First, a free -OH group attacks the exon 1-intron junction. Second, the -OH created at the end of exon 1 attacks the intron-exon 2 junction.

There are parallels between group II introns and pre-mRNA splicing. Group II mitochondrial introns are excised by the same mechanism as nuclear pre-mRNAs via a lariat that is held together by a 5′–2′ bond. An example of a lariat produced by splicing a group II intron is shown in **FIGURE 28.18**. When an isolated group II RNA is incubated *in vitro* in the absence of additional components, it is able to perform the splicing reaction. This means that the two transesterification reactions shown in Figure 28.17 can be performed by the group II intron RNA sequence itself. The number of phos-

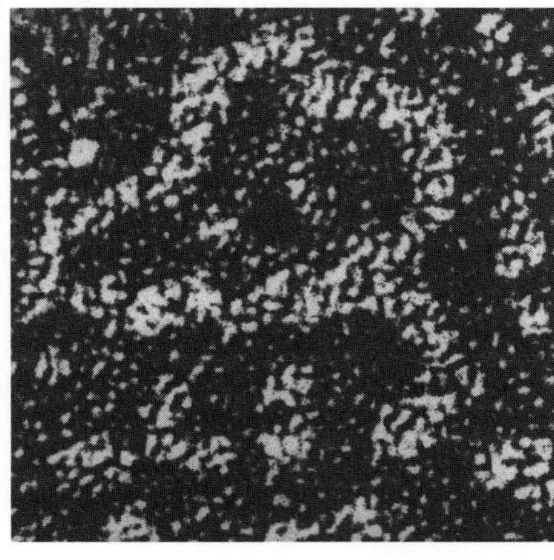

FIGURE 28.18 Splicing releases a mitochondrial group II intron in the form of a stable lariat. Reproduced from Van derVeen, R., *et al.*, *EMBO J.* 6(1987): 1079–1084. Photo courtesy of Leslie A. Grivell, European Molecular Biology Organisation.

phodiester bonds is conserved in the reaction, and as a result an external supply of energy is not required.

A group II intron forms into a secondary structure that contains several domains formed by base-paired stems and single-stranded loops. Domain 5 is separated by two bases from domain 6, which contains an A residue that donates the 2′–OH group for the first transesterification. This constitutes a catalytic domain in the RNA. **FIGURE 28.19** compares this secondary structure with the structure formed by the combination of U6 with U2 and of U2 with the branch site. The similarity suggests that U6 may have a catalytic role.

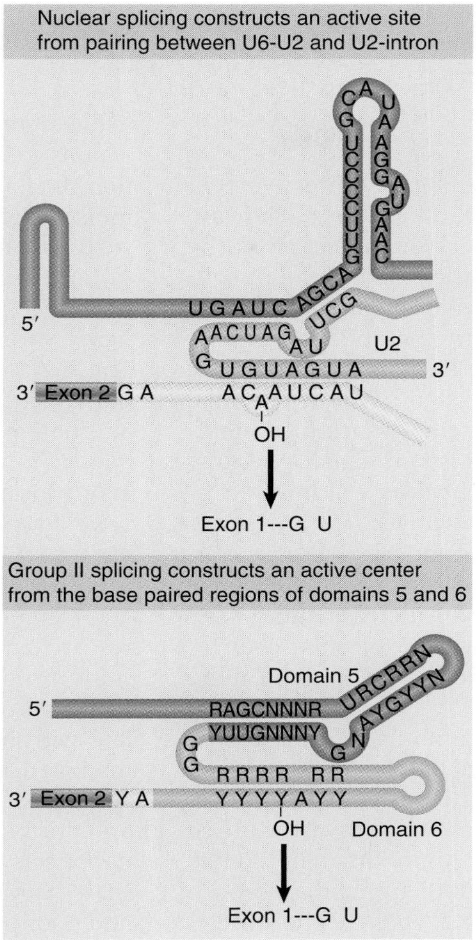

The features of group II splicing suggest that splicing evolved from an autocatalytic reaction undertaken by an individual RNA molecule, in which it accomplished a controlled deletion of an internal sequence. It is likely that such a reaction would require the RNA to fold into a specific conformation, or series of conformations, and would occur exclusively in *cis* conformation.

The ability of group II introns to remove themselves by an autocatalytic splicing event stands in great contrast to the requirement for a complex apparatus to splice nuclear introns. We may regard the snRNAs of the spliceosome as compensating for the lack of sequence information in the intron, and providing the information required to form particular structures in RNA. The system involving snRNAs may have evolved from the original autocatalytic system. These snRNAs act in *trans* upon the substrate pre-mRNA; we might imagine that the ability of U1 to pair with the 5′ splice site, or of U2 to pair with the branch sequence, replaced a similar reaction that required the relevant sequence to be carried by the intron. So the snRNAs may undergo reactions, both with the pre-mRNA substrate and with one another, that have substituted for the series of conformational changes that occur in RNAs that splice by group II mechanisms. In effect, these changes have relieved the substrate pre-mRNA of the obligation to carry the sequences needed to sponsor the reaction. As the splicing apparatus has become more complex (and as the number of potential substrates has increased), proteins have played a more important role. Also, the loss of autosplicing has allowed the possibility of alternative splicing patterns for greater flexibility in products from a single gene (see the next section).

FIGURE 28.19 Nuclear splicing and group II splicing involve the formation of similar secondary structures. The sequences are more specific in nuclear splicing; group II splicing uses positions that may be occupied by either purine (R) or pyrimidine (Y).

KEY CONCEPTS

- Group II introns excise themselves from RNA by an autocatalytic splicing event.
- The splice junctions and mechanism of splicing of group II introns are similar to splicing of nuclear introns.
- A group II intron folds into a secondary structure that generates a catalytic site resembling the structure of U6-U2 bound to a nuclear intron.

CONCEPT AND REASONING CHECK

Compare the roles of RNA sequences in both autosplicing and the splicing of nuclear RNAs.

MEDICAL APPLICATIONS

Alternative Splicing and Cancer

One in every two men, and one in every three women, will develop cancer at some time in their lives. There are over 200 different types of cancer that affect all of the different body tissues. *Cancer* is the result of the uncontrolled growth of a single cell that eventually forms a clone of tumor cells that have the potential to *metastasize* to other sites of the body. Each of us is composed of approximately 10^{14} cells, all of which have the potential to become cancerous; however the chance of an individual cell becoming cancerous is very small. We now know that for the change of a healthy cell into a cancerous cell to occur, a series of changes slowly releases the cell from the multiple checks and balances that control its normal growth. Over the years, researchers have identified many factors that may increase the risk of developing certain types of cancers. Molecular changes in cells at the DNA and RNA levels make them more likely to become cancerous.

Breast cancer is the most commonly diagnosed type of cancer in women. It was estimated that in 2006, there would be 212,920 new cases of invasive breast cancer in women and 61,980 new cases of noninvasive breast cancer in the United States. Approximately 40,970 women were expected to die in 2006 from breast cancer. Changes in certain genes (*BRCA1*, *BRCA2*, and others) make women more susceptible to breast cancer. Recent research implies that the processing of *CD44* pre-mRNA plays a role in breast cancer. The *CD44* gene is located on the short arm of chromosome 11. It codes for a transmembrane cell-surface glycoprotein that is present in a variety of cell types including the epidermis, central nervous system, lung, liver, and pancreas. It participates in a variety of cellular functions including cell-cell interactions, release of cytokines, cell adhesion, and migration. It is a receptor for hyaluronic acid (a natural skin moisturizer) and can also interact with other ligands such as osteopontin (a glycoprotein that is abundant in bone mineral matrix), collagens (major component of connective tissue) and metalloproteinases. There are many variant isoforms of CD44 found in normal cells. The *CD44* gene contains in-frame constant and alternative variable exons. These isoforms are coded for by alternative splicing in two different regions containing exons of the gene. The *CD44* gene contains 10 constant exons located at the 5′ and 3′ ends of the gene and 10 variable exons located between the constant exons (**FIGURE B28.1**). Some of these variations diversify the biological function(s) in the final protein molecule.

Tremendous interest in CD44 was generated when researchers discovered that CD44 variants have metastatic properties. Research findings showed that the inclusion of *CD44* variable exons v4 through v6 correlate with both tumorigenesis and metastasis of several malignancies, including breast cancer. *CD44* exons v4 and v5 contain several copies of cytosine/adenine-rich and purine-rich sequences that are known to be strong enhancers of transcription. Overexpression of abnormal CD44 variants has been associated with breast cancer and poor prognosis.

The number of diseases associated with missplicing is increasing. Missplicing can be caused by either mutations in regulatory sequences such as splice sites, mutations in enhancer/silencer sequences or alterations in *trans*-acting factors such as *splicing factors*. Given the correlation between the overexpression of the CD44 variants and breast tumors, a group of German researchers

28.11 Alternative Splicing Involves Differential Use of Splice Junctions

▸ **alternative splicing** The production of different RNA products from a single product by changes in the usage of splicing junctions.

When an interrupted gene is transcribed into an RNA that gives rise to a single type of spliced mRNA, there is no ambiguity in assignment of exons and introns. The RNAs of many genes, however, follow patterns of **alternative splicing**, which results in a single gene giving rise to more than one mRNA sequence. In some cases, the ultimate pattern of expression is dictated by the primary transcript, because the use of different startpoints or the generation of alternative 3′ ends alters the pattern of splicing. In other cases, a single primary transcript is spliced in more than one way, and internal exons are substituted, added, or deleted. In some cases, the multiple products all are made in the same cell, but in others the process is regulated so that particular splicing patterns occur only under particular conditions. For example, in humans, different versions of a protein may be expressed in different tissues or at different developmental stages of the same tissue.

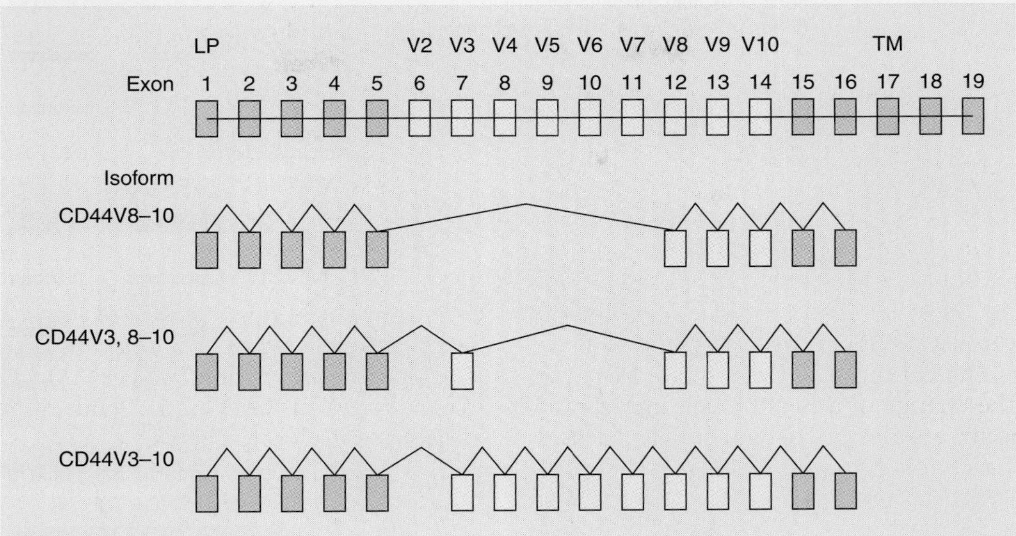

FIGURE B28.1 Top of the figure shows a schematic of the constant and variable exons of the normal human CD44 gene. Isoforms showing alternative splicing are shown below. Reproduced with permission, from Piepkorn, M., et al., 1997, *Biochem. J.*, vol. 327, pp. 499–506. © The Biochemical Society.

working on a model of breast cancer development performed experiments to determine the cause of *CD44* missplicing in breast cancers. To do this, they tested whether the expression of splicing activators or factors Tra2-α and Tra2-β expression change in mastectomy tissue removed from primary invasive adenocarcinomas of the breast compared to normal tissue. All tissue specimens showed a significant induction of Tra2-β in the invasive breast cancer at both the RNA and protein levels. This was not observed in normal breast tissue. It was also shown that Tra2-β was bound to the enhancers found in the *CD44* exon v4 and exon v5 sequences, inducing missplicing. Therefore the induction of the Tra2-β splicing factor might explain the observed alternative splicing of CD44 isoforms associated with tumor progression and metastasis during breast cancer development.

Reference

Watermann, DO, *et al.* (2006). Splicing Factor Tra2-β1 Is Specifically Induced in Breast Cancer and Regulates Alternative Splicing of the CD44 Gene. *Cancer Research* 66(9):4774–4780.

FIGURE 28.20 shows examples of alternative splicing in which one splice site remains constant, but the other varies. The large T/small t antigens of SV40 and the products of the adenovirus E1A region are generated by connecting a varying 5′ site to a constant 3′ site. In the T/t antigens, the 5′ site used for the T antigen removes a termination codon that is present in the t antigen mRNA, so that the T antigen is larger than the t antigen. In the case of the E1A transcripts, one of the 5′ sites connects to the last exon in a different reading frame, again making a significant change in the C-terminal part of the protein. In these examples, all the relevant splicing events take place in every cell in which the gene is expressed, so that all the protein products are made.

There are differences in the ratios of T/t antigens in different cell types. The relative usage of the alternative splice sites is determined by the splicing factor ASF/SF2 (a component of the E complex). When a pre-mRNA has more than one 5′ splice site preceding a single 3′ splice site, increased concentrations of ASF/SF2 promote use of the 5′ site nearest to the 3′ site at the expense of the other site. This effect of ASF/SF2 can be counteracted by another splicing factor, SF5. As a general rule, alternative

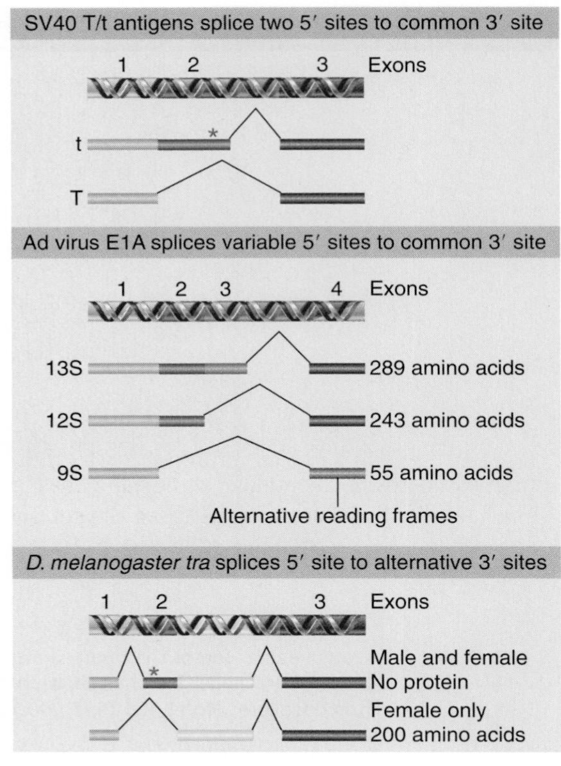

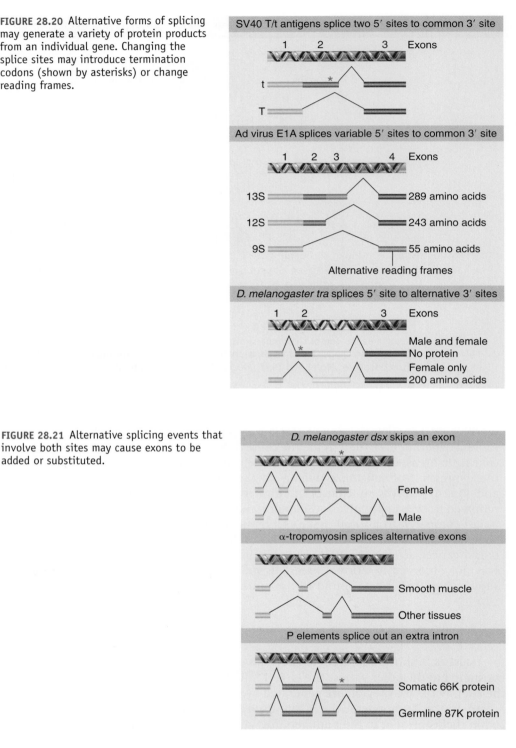

FIGURE 28.20 Alternative forms of splicing may generate a variety of protein products from an individual gene. Changing the splice sites may introduce termination codons (shown by asterisks) or change reading frames.

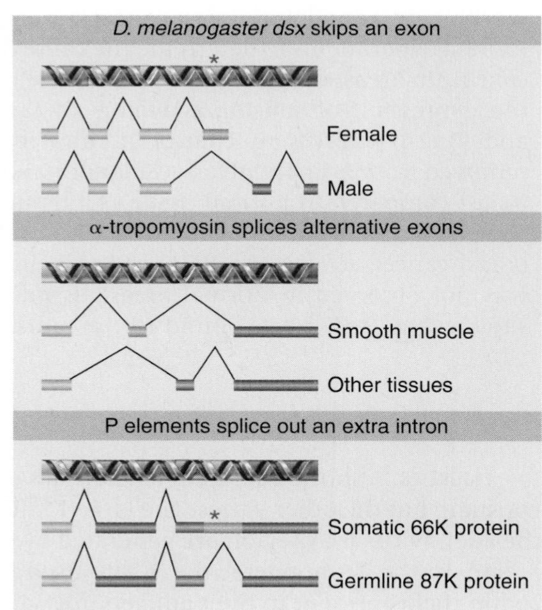

FIGURE 28.21 Alternative splicing events that involve both sites may cause exons to be added or substituted.

splicing involving different 5′ sites may be influenced by proteins in spliceosome assembly that either stimulate or repress the usage of one of the possible sites.

FIGURE 28.21 shows examples of cases in which splice sites are used to add or to substitute exons or introns, again with the consequence that different protein products are generated. In the *Drosophila doublesex* (*dsx*) gene, females splice the 5′ site of intron 3 to the 3′ site of that intron; as a result, translation terminates at the end of exon 4. Males splice the 5′ site of intron 3 directly to the 3′ site of intron 4, thus omitting exon 4 from the mRNA and allowing translation to continue through exon 6. The result of the alternative splicing is that different proteins are produced in each sex: the male product blocks female sexual differentiation, whereas the female product represses expression of male-specific genes. Alternative splicing of *dsx* RNA is controlled by competition between 3′ splice sites, which relies on the binding of splicing factors to those sites.

KEY CONCEPTS

- Specific exons may be excluded or included in the RNA product by using or failing to use a pair of splicing junctions.
- Exons may be extended by changing one of the splice junctions to use an alternative junction.
- Sex determination in *Drosophila* involves sex-specific alternative splicing events.

CONCEPT AND REASONING CHECK

Initial estimates of the number of human genes, based on the number of mRNAs produced in human cells, were much higher than the current count of human genes. How is this possible?

28.12 *trans*-Splicing Reactions Use Small RNAs

In both mechanistic and evolutionary terms, splicing has been viewed as an *intramolecular* reaction, essentially amounting to a controlled deletion of the intron sequences at the level of RNA. In genetic terms, splicing occurs only in *cis*. This means that *only sequences on the same molecule of RNA can be spliced together*. The upper part of **FIGURE 28.22** shows the normal situation. The introns can be removed from each RNA molecule, allowing the exons of that RNA molecule to be spliced together, but there is no *intermolecular* splicing of exons between different RNA molecules. However, *trans*-splicing can occur in the situation in which different introns have complementary sequences.

Although *trans*-splicing is rare, it occurs *in vivo* in some special situations. One is revealed by the presence of a common 35-base leader sequence at the beginning of numerous mRNAs in trypanosomes. The leader sequence is not located upstream of the individual transcription units, however. Instead it is transcribed into an independent

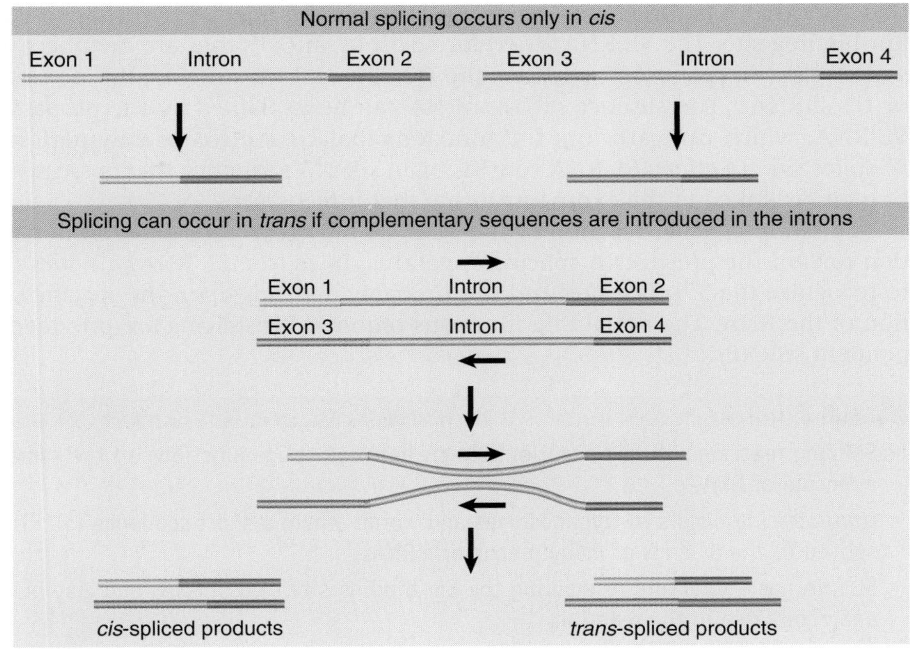

FIGURE 28.22 Splicing usually occurs only in *cis* between exons carried on the same physical RNA molecule, but *trans*-splicing can occur when special constructs are made that support base pairing between introns.

FIGURE 28.23 The SL RNA provides an exon that is connected to the first exon of an mRNA by *trans*-splicing. The reaction involves the same interactions as nuclear *cis*-splicing, but generates a Y-shaped RNA instead of a lariat.

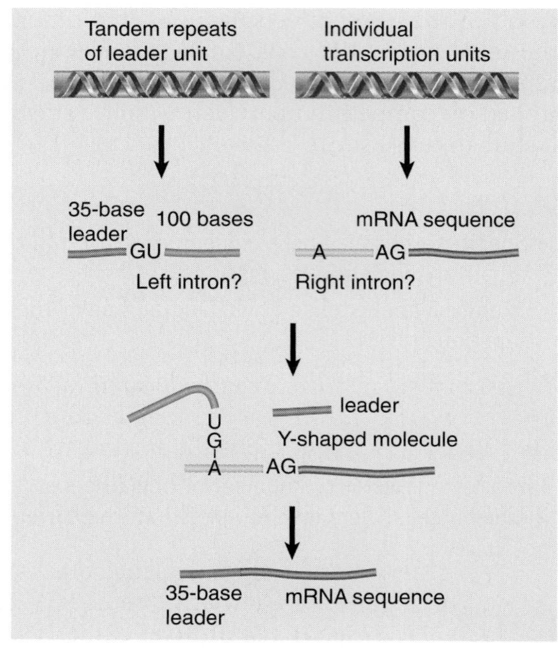

RNA, carrying additional sequences at its 3′ end, from a repetitive unit located elsewhere in the genome. FIGURE 28.23 shows that this RNA carries the 35-base leader sequence followed by a 5′-splice site sequence. The sequences coding for the mRNAs carry a 3′-splice site just preceding the sequence found in the mature mRNA. A similar situation is found in *C. elegans*, which contains two types of **SL RNA (spliced leader RNA)**.

SL RNA (spliced leader RNA) A small RNA that donates an exon in the *trans*-splicing reaction of trypanosomes and nematodes.

When the leader and the mRNA are connected by a *trans*-splicing reaction, the 3′ region of the leader RNA and the 5′ region of the mRNA in effect comprise the 5′ and 3′ halves of an intron. When splicing occurs, a 5′–2′ link forms by the usual reaction between the GU of the 5′ intron and the branch sequence near the AG of the 3′ intron. The two parts of the intron are not covalently linked, and therefore generate a Y-shaped molecule instead of a lariat.

The SL RNAs found in several species of trypanosomes and also in the nematode (*C. elegans*) have some common features. They fold into a common secondary structure that has three stem-loops and a single-stranded region that resembles the Sm-binding site. The SL RNAs therefore exist as snRNPs that are members of the Sm snRNP class. Trypanosomes possess the U2, U4, and U6 snRNAs, but do not have U1 or U5 snRNAs. The absence of U1 snRNA can be explained by the properties of the SL RNA, which can carry out the functions that U1 snRNA usually performs at the 5′-splice site. In effect, SL RNA consists of an snRNA sequence that possesses U1 function and is linked to the exon–intron site that it recognizes.

The *trans*-splicing reaction of the SL RNA may represent an evolutionary transition toward the pre-mRNA splicing apparatus. In *cis*, the SL RNA provides the ability to recognize the 5′ splice site, and this probably depends upon the specific conformation of the RNA. The remaining functions required for splicing are provided by independent snRNPs.

KEY CONCEPTS

- Splicing reactions usually occur only in *cis* between splice junctions on the same molecule of RNA.
- *trans*-splicing occurs in trypanosomes and worms where a short sequence (SL RNA) is spliced to the 5′ ends of many precursor mRNAs.
- SL RNA has a structure resembling the Sm-binding site of U snRNAs and may play an analogous role in the reaction.

CONCEPT AND REASONING CHECK

Compare the processes of self-splicing, *trans*-splicing, and pre-mRNA splicing.

28.13 Yeast tRNA Splicing Involves Cutting and Rejoining

Most splicing reactions depend on short consensus sequences and occur by transesterification reactions in which breaking and formation of bonds is coordinated. The splicing of tRNA genes is achieved by a different mechanism that relies upon separate cleavage and ligation reactions.

Some 59 of the 272 nuclear tRNA genes in the yeast *S. cerevisiae* are interrupted. Each has a single intron that is located just one nucleotide beyond the 3′ side of the anticodon. The introns vary in length from 14 to 60 bp. Those in related tRNA genes are related in sequence, but the introns in tRNA genes representing different amino acids are unrelated. *There is no consensus sequence that could be recognized by the splicing enzymes.* This is also true of interrupted nuclear tRNA genes of plants, amphibians, and mammals.

All the introns include a sequence that is complementary to the anticodon of the tRNA. This creates an alternative conformation for the anticodon arm in which the anticodon is base paired to form an extension of the usual arm. An example is shown in **FIGURE 28.24**. Only the anticodon arm is affected—the rest of the molecule retains its usual structure.

The exact sequence and size of the intron is not important. Most mutations in the intron do not prevent splicing. *Splicing of tRNA depends principally on recognition of a common secondary structure in tRNA rather than a common sequence of the intron.* Regions in various parts of the molecule are important, including the stretch between the acceptor arm and D arm, in the TψC arm, and especially the anticodon arm. This is reminiscent of the structural demands placed on tRNA for translation (see *Chapter 8, Translation*).

The intron sequence is not entirely irrelevant, however. Pairing between a base in the intron loop and an unpaired base in the stem is required for splicing. Mutations at other positions that influence this pairing (for example, to generate alternative patterns for pairing) influence splicing. The rules that govern availability of tRNA precursors for splicing resemble the rules that govern recognition by aminoacyl-tRNA synthetases (see *Section 9.8, tRNAs Are Charged with Amino Acids by Synthetases*).

The reaction occurs in two stages that are catalyzed by different enzymes:

- The first step does not require ATP. In this step, a phosphodiester bond is cleaved by an atypical nuclease reaction that is catalyzed by an endonuclease.
- The second step requires ATP and involves bond formation; it is a ligation reaction, and the responsible enzyme activity is described as an **RNA ligase**.

▸ **RNA ligase** An enzyme that functions in tRNA splicing to make a phosphodiester bond between the two exon sequences that are generated by cleavage of the intron.

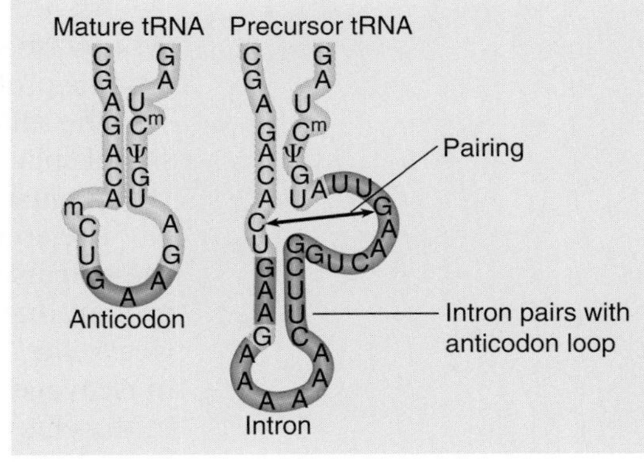

FIGURE 28.24 The intron in yeast $tRNA^{Phe}$ base pairs with the anticodon to change the structure of the anticodon arm. Pairing between an excluded base in the stem and the intron loop in the precursor may be required for splicing.

The overall tRNA splicing reaction is summarized in **FIGURE 28.25**. The products of cleavage are a linear intron and two half-tRNA molecules. These intermediates have unique ends. Each 5′ terminus ends in a hydroxyl group; each 3′ terminus ends in a 2′, 3′–cyclic phosphate group. (All other known RNA splicing enzymes cleave on the other side of the phosphate bond.)

The two half-tRNAs base pair to form a tRNA-like structure. When ATP is added, the second reaction occurs. Both of the unusual ends generated by the endonuclease must be altered.

The cyclic phosphate group is opened to generate a 2′–phosphate terminus. This reaction requires cyclic phosphodiesterase activity. The product has a 2′–phosphate group and a 3′–OH group.

The 5′–OH group generated by the nuclease must be phosphorylated to give a 5′–phosphate. This generates a site in which the 3′–OH is

FIGURE 28.25 Splicing of tRNA requires separate nuclease and ligase activities. The exon–intron boundaries are cleaved by the nuclease to generate 2′ to 3′ cyclic phosphate and 5′ OH termini. The cyclic phosphate is opened to generate 3′–OH and 2′ phosphate groups. The 5′–OH is phosphorylated. After releasing the intron, the tRNA half molecules fold into a tRNA-like structure that now has a 3′–OH, 5′–P break. This is sealed by a ligase.

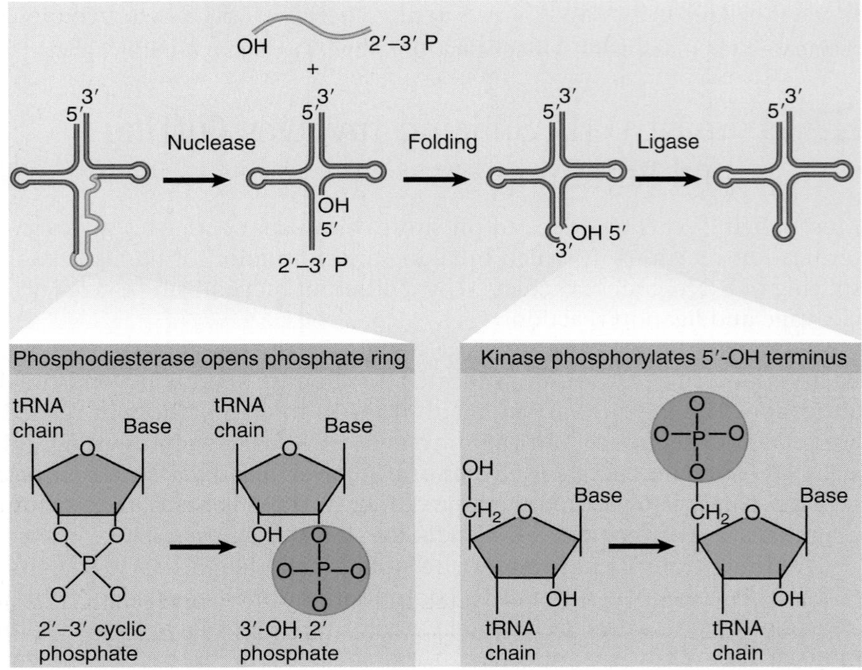

FIGURE 28.26 The 3′ and 5′ cleavages in *S. cerevisiae* pre-tRNA are catalyzed by different subunits of the endonuclease. Another subunit may determine location of the cleavage sites by measuring distance from the mature structure. The AI base pair is also important.

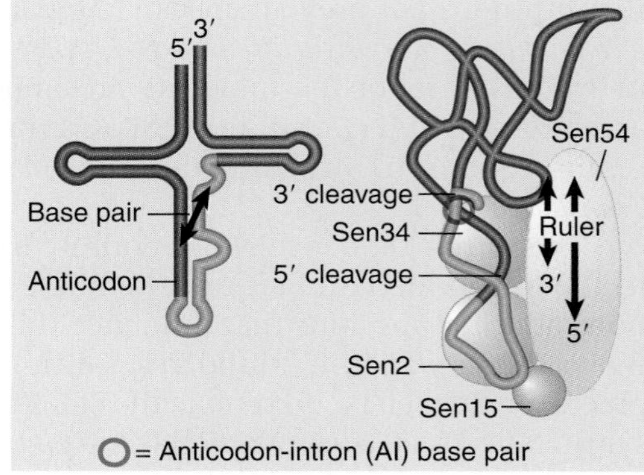

next to the 5′–phosphate. The two polynucleotide chains are then covalently bonded by ligase activity.

All three activities—phosphodiesterase, polynucleotide kinase, and adenylate synthetase (which provides the ligase function)—are different functional domains of a single protein. They act sequentially to join the two tRNA halves.

The spliced molecule is now uninterrupted, with a 5′–3′ phosphate linkage at the site of splicing, but it also has a 2′–phosphate group marking the event. This surplus group must be removed by a phosphatase.

The endonuclease is responsible for the specificity of intron recognition. It cleaves the precursor at both ends of the intron. The yeast endonuclease is a heterotetrameric protein that functions as shown in FIGURE 28.26. The related subunits Sen34 and Sen2 cleave the 3′ and 5′ splice sites, respectively. Subunit Sen54 may determine the sites of cleavage by "measuring" distance from a point in the tRNA structure. This point is in the elbow of the (mature) L-shaped structure.

An interesting insight into the evolution of tRNA splicing is provided by the endonucleases of archaeans. These are homodimers or homotetramers, in which each subunit has an active site (although only two of the sites function in the tetramer) that cleaves one of the splice sites. The subunit has sequences related to the sequences of the active sites in the Sen34 and Sen2 subunits of the yeast enzyme. The archaeal enzymes recognize their substrates in a different way, however. Instead of measuring distance from particular sequences, they recognize a structural feature called the bulge-helix-bulge. **FIGURE 28.27** shows that cleavage occurs in the two bulges.

The existence of tRNA splicing in both archaea and eukaryotes indicates that its origin must have preceded their evolutionary separation. If interrupted tRNA genes originated by insertion of the intron into tRNAs, this must have been a very ancient event.

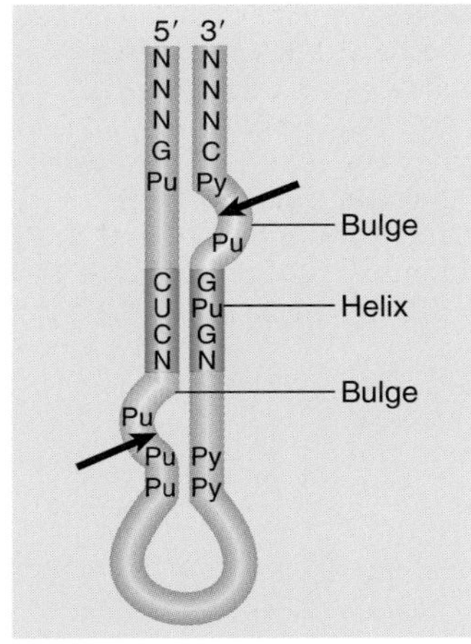

FIGURE 28.27 Archaeal tRNA splicing endonuclease cleaves each strand at a bulge in a bulge-helix-bulge motif.

KEY CONCEPTS

- tRNA splicing occurs by successive cleavage and ligation reactions.
- An endonuclease cleaves the tRNA precursors at both ends of the intron.
- Release of the intron generates two half-tRNAs that pair to form the mature structure.
- The halves have the unusual ends 5′ hydroxyl and 2′–3′ cyclic phosphate.
- The 5′-OH end is phosphorylated by a polynucleotide kinase, the cyclic phosphate group is opened by phosphodiesterase to generate a 2′-phosphate terminus and 3′-OH group, exon ends are joined by an RNA ligase, and the 2′-phosphate is removed by a phosphatase.
- The yeast endonuclease is a heterotetramer with two (related) catalytic subunits.
- It uses a measuring mechanism to determine the sites of cleavage by their positions relative to a point in the tRNA structure.
- The archaeal nuclease has a simpler structure and recognizes a bulge-helix-bulge structural motif in the substrate.

CONCEPT AND REASONING CHECK

Compare tRNA splicing and pre-mRNA splicing.

28.14 The 3′ Ends of mRNAs Are Generated by Cleavage and Polyadenylation

It is not clear whether RNA polymerase II actually engages in a termination event at a specific site. It is possible that its termination is only loosely specified. In some transcription units, termination occurs >1000 bp downstream of the site corresponding to the mature 3′ end of the mRNA (which is generated by cleavage at a specific sequence). Instead of using specific terminator sequences, the enzyme ceases RNA synthesis within multiple sites located in rather long "terminator regions." The nature of the individual termination sites is largely unknown.

The 3′ ends of mRNAs are generated by cleavage followed by polyadenylation (see *Section 7.10, The 3′ Terminus of Eukaryotic mRNA Is Polyadenylated*). The reactions are coupled by enzymes that are present in a single multiprotein complex.

Generation of the 3′ end is illustrated in **FIGURE 28.28**. RNA polymerase transcribes past the site corresponding to the 3′ end, and sequences in the

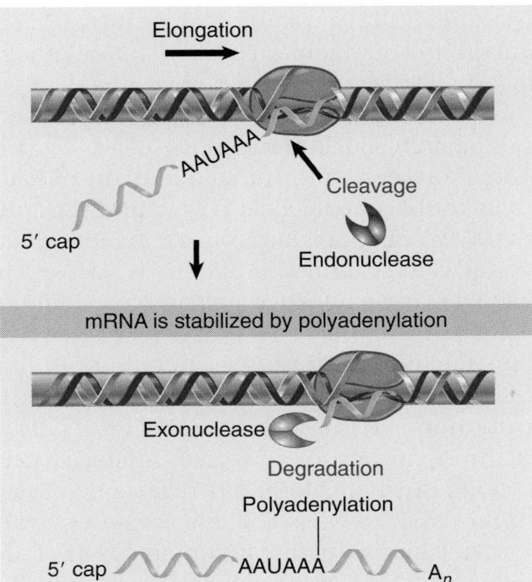

FIGURE 28.28 The sequence AAUAAA is necessary for cleavage to generate a 3′ end for polyadenylation.

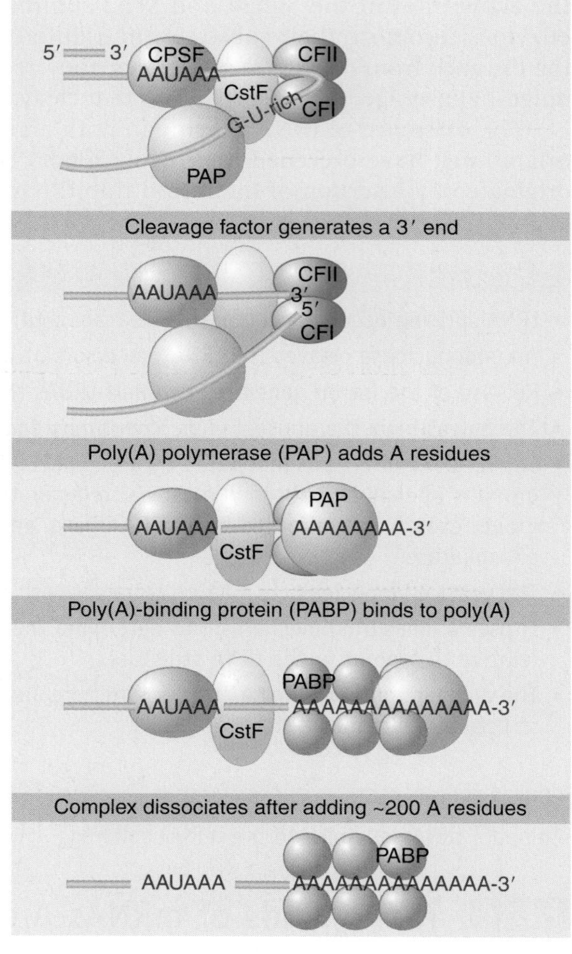

FIGURE 28.29 The 3′ processing complex consists of several activities. CPSF and CstF each consist of several subunits; the other components are monomeric. The total mass is >900 kD.

RNA are recognized as targets for an endonucleolytic cut followed by polyadenylation. A single processing complex undertakes both the cutting and polyadenylation. The polyadenylation stabilizes the mRNA against degradation from the 3′ end and is required for translation of all mRNAs except for those of the major histones. Its 5′ end is already stabilized by the cap. RNA polymerase continues transcription after the cleavage, but the 5′ end that is generated by the cleavage is unprotected. As a result, the rest of the transcript is rapidly degraded. This makes it difficult to determine what is happening beyond the point of cleavage.

A common feature of mRNAs in multicellular eukaryotes (but not in yeast, although their mRNAs are polyadenylated) is the presence of the highly conserved sequence AAUAAA in the region from 11 to 30 nucleotides upstream of the site of poly(A) addition. Deletion or mutation of the AAUAAA hexamer prevents generation of the polyadenylated 3′ end. The signal is needed for both cleavage and polyadenylation.

The formation and functions of the complex that undertakes 3′ processing are illustrated in **FIGURE 28.29**. Generation of the proper 3′ terminal structure requires an *endonuclease* (consisting of the components CFI and CFII) to cleave the RNA, a **poly(A) polymerase (PAP)** to synthesize the poly(A) tail, and a *specificity component* (CPSF) that recognizes the AAUAAA sequence and directs the other activities. A stimulatory factor, CstF, binds to a G-U-rich sequence that is downstream from the cleavage site itself.

▸ **poly(A) polymerase (PAP)** The enzyme that adds the stretch of polyadenylic acid to the 3′ end of eukaryotic mRNA. It does not use a template.

The specificity factor contains four subunits, which together bind specifically to RNA containing the sequence AAUAAA. The individual subunits are proteins that have common RNA-binding motifs, but that by themselves bind nonspecifically to

RNA. Protein–protein interactions between the subunits may be needed to generate the specific AAUAAA-binding site. CPSF binds strongly to AAUAAA only when CstF is also present to bind to the G-U-rich site.

The specificity factor is needed for both the cleavage and polyadenylation reactions. It exists in a complex with the endonuclease and poly(A) polymerase, and this complex usually undertakes cleavage followed by polyadenylation in a tightly coupled manner.

The two components CFI and CFII (cleavage factors I and II), together with specificity factor, are necessary and sufficient for the endonucleolytic cleavage.

The poly(A) polymerase has a nonspecific catalytic activity. When it is combined with the other components, the synthetic reaction becomes specific for RNA containing the sequence AAUAAA. The polyadenylation reaction passes through two stages. First, a rather short oligo(A) sequence (~10 residues) is added to the 3′ end. This reaction is absolutely dependent on the AAUAAA sequence, and poly(A) polymerase performs it under the direction of the specificity factor. In the second phase, the oligo(A) tail is extended to the full ~200 residue length. This reaction requires another stimulatory factor that recognizes the oligo(A) tail and directs poly(A) polymerase specifically to extend the 3′ end of a poly(A) sequence.

The poly(A) polymerase by itself adds A residues individually to the 3′ position. Its intrinsic mode of action is distributive; it dissociates after each nucleotide has been added. In the presence of CPSF and PABP (poly(A)-binding protein), however, it functions processively to extend an individual poly(A) chain. The PABP is a 33 kD protein that binds to the poly(A) stretch. Copies of PABP accumulate as poly(A) is extended. The length of poly(A) is controlled by the PABP, which in some way limits the action of poly(A) polymerase to ~200 additions of A residues. The limit may represent the accumulation of a critical mass of PABP on the poly(A) chain. PABP binds to the translation initiation factor eIF4G, thus generating a closed loop in which a protein complex contains both the 5′ and 3′ ends of the mRNA.

KEY CONCEPTS

- The sequence AAUAAA is a signal for cleavage to generate a 3′ end of mRNA that is polyadenylated.
- The reaction requires a protein complex that contains a specificity factor, an endonuclease, and poly(A) polymerase.
- The specificity factor and endonuclease cleave RNA downstream of AAUAAA.
- The specificity factor and poly(A) polymerase add ~200 A residues processively to the 3′ end.

CONCEPT AND REASONING CHECK

What would be the effect of deletion of the AATAAA consensus sequence at the 3′ end of the coding strand of a gene?

28.15 Small RNAs Are Required for rRNA Processing

The major eukaryotic rRNAs are synthesized as part of a single primary transcript that is processed by cleavage and trimming events to generate the mature products. The precursor contains the sequences of the 18S, 5.8S, and 28S rRNAs. In multicellular eukaryotes, the precursor is named for its sedimentation rate as 45S RNA. In bacteria, the precursor contains the sequences of 16S and 23S rRNAs (and sometimes also the 5S RNA). The ribosomal precursor RNA is modified by the addition of many methyl groups.

Processing and modification of rRNA requires a class of small RNAs called **snoRNAs (small nucleolar RNAs)**. There are 71 snoRNAs in the yeast (*S. cerevisiae*) genome. They are associated with the protein fibrillarin, which is an abundant

▸ **small nucleolar RNA (snoRNA)** A small nuclear RNA that is localized in the nucleolus.

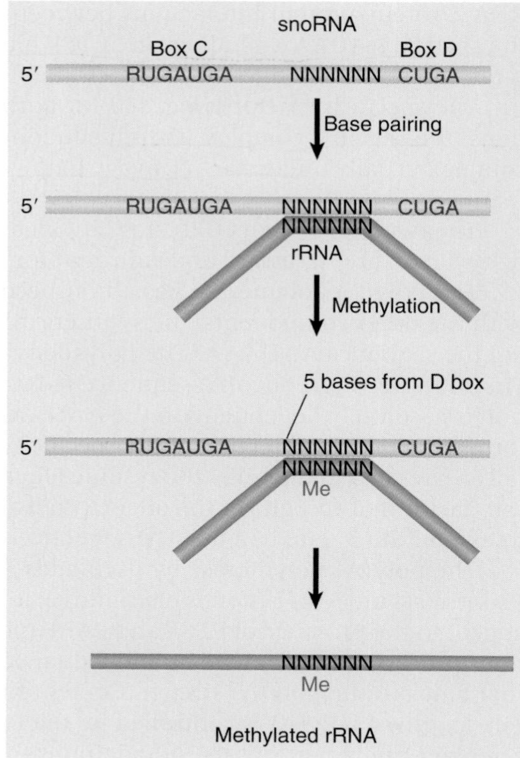

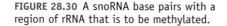

FIGURE 28.30 A snoRNA base pairs with a region of rRNA that is to be methylated.

component of the nucleolus (the region of the nucleus where the rRNA genes are transcribed). Some snoRNAs are required for cleavage of the precursor to rRNA; one example is U3 snoRNA, which is required for the first cleavage event in both yeast and *Xenopus*. We do not know what role the snoRNA plays in cleavage. It could be required to pair with the rRNA sequence to form a secondary structure that is recognized by an endonuclease.

Two groups of snoRNAs are required for the modifications that are made to bases in the rRNA. The members of each group are identified by very short conserved sequences and common features of secondary structure.

The C/D group of snoRNAs is required for adding a methyl group to the 2′ position of ribose. There are >100 2′–O–methyl groups at conserved locations in vertebrate rRNAs. This group takes its name from two short conserved sequence motifs called boxes C and D. Each snoRNA contains a sequence near the D box that is complementary to a region of the 18S or 28S rRNA that is methylated. Loss of a particular snoRNA prevents methylation in the rRNA region to which it is complementary.

FIGURE 28.30 suggests that the snoRNA base pairs with the rRNA to create the duplex region that is recognized as a substrate for methylation. Methylation occurs within the region of complementarity at a position that is fixed five bases on the 5′ side of the D box. It is likely that each methylation event is specified by a different snoRNA; ~40 snoRNAs have been characterized so far. The methylase(s) has not been characterized; it may be that the snoRNA itself provides part of the methylase activity.

Another group of snoRNAs is involved in the synthesis of pseudouridine (ψ). There are 43 ψ residues in yeast rRNAs and ~100 in vertebrate rRNAs. The synthesis of pseudouridine involves the reaction shown in **FIGURE 28.31**, in which the N1 bond from uridylic acid to ribose is broken, the base is rotated, and C5 is rejoined to the sugar.

Pseudouridine formation in rRNA requires the H/ACA group of ~20 snoRNAs. They are named for the presence of an ACA triplet three nucleotides from the 3′ end and a partially conserved sequence (the H box) that lies between two stem-loop hairpin structures. Each of these snoRNAs has a sequence complementary to rRNA within

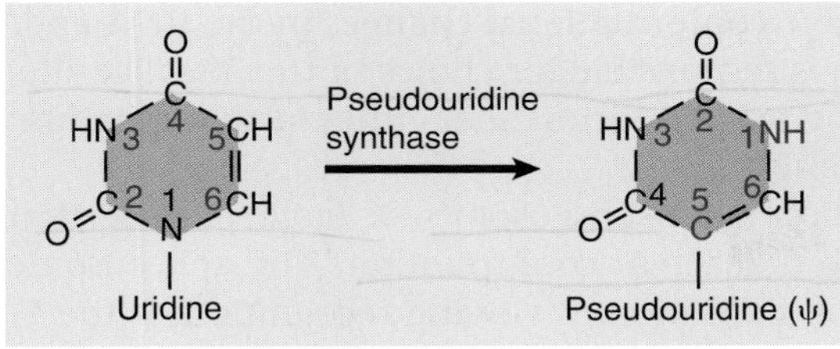

FIGURE 28.31 Uridine is converted to pseudouridine by replacing the N1-sugar bond with a C5-sugar bond and rotating the base relative to the sugar.

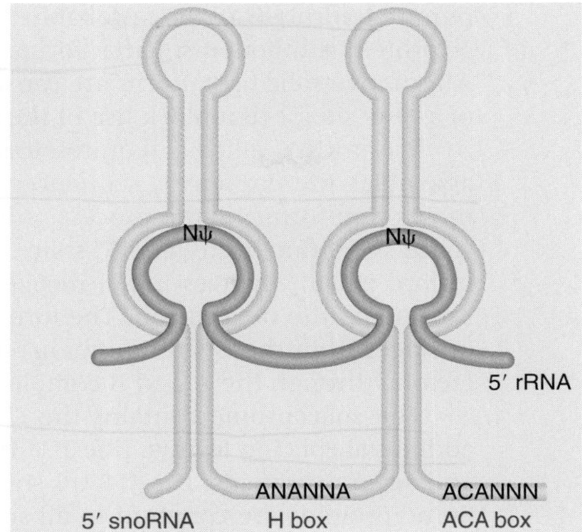

FIGURE 28.32 H/ACA snoRNAs have two short conserved sequences and two hairpin structures, each of which has regions in the stem that are complementary to rRNA. Pseudouridine is formed by converting an unpaired uridine within the complementary region of the rRNA.

the stem of each hairpin. **FIGURE 28.32** shows the structure that would be produced by pairing with the rRNA. Within each pairing region, there are two unpaired bases, one of which is a uridine that is converted to pseudouridine. The enzymatic activity that catalyzes the reaction has not been identified.

KEY CONCEPTS

- The C/D group of snoRNAs is required for modifying the 2′ position of ribose with a methyl group.
- The H/ACA group of snoRNAs is required for converting uridine to pseudouridine.
- In each case, the snoRNA base pairs with a sequence of rRNA that contains the target base to generate a typical structure that is the substrate for modification.

CONCEPT AND REASONING CHECK

What is the role of snoRNAs in the processing of rRNAs?

28.16 Summary

RNA splicing accomplishes the removal of introns and the joining of exons into a mature RNA sequence. There are at least four types of reaction, including those for eukaryotic nuclear introns, group I and group II introns, and tRNA introns. Each reaction changes the organization within an individual RNA molecule, and is therefore a *cis*-acting event.

Pre-mRNA splicing follows preferred, but not obligatory, pathways. Only very short consensus sequences are necessary; the rest of the intron sequence appears to be irrelevant. All 5′ splice sites are probably equivalent, as are all 3′ splice sites. The required sequences are given by the GU-AG rule, which describes the ends of the intron. The UACUAAC branch site of yeast, or a less well conserved consensus in mammalian introns, is also required. The reaction with the 5′ splice site involves formation of a lariat that joins the GU end of the intron via a 5′–2′ linkage to the A at position 6 of the branch site. The 3′–OH end of the exon then attacks the 3′ splice site, so that the exons are ligated and the intron is released as a lariat. Both reactions are transesterifications in which bonds are conserved. Several stages of the reaction require hydrolysis of ATP, probably to drive conformational changes in the RNA and/or protein components. Lariat formation is responsible for choice of the 3′ splice site. Alternative splicing patterns are caused by protein factors that either facilitate the use of a new site or that block use of the default site.

Pre-mRNA splicing requires formation of a spliceosome—a large particle that assembles the consensus sequences into a reactive conformation. The spliceosome most often forms by the process of intron definition, involving recognition of the 5′ splice site, branch site, and 3′ splice site. An alternative pathway involves exon definition, which involves initial recognition of the 5′ splice sites of both the substrate intron and the next intron. The formation of the spliceosome passes through a series of stages from the E (commitment) complex, which contains U1 snRNP and splicing factors, through the A and B complexes as additional components are added.

The spliceosome contains the U1, U2, U4/U6, and U5 snRNPs, as well as some additional splicing factors. The U1, U2, and U5 snRNPs each contain a single snRNA and several proteins; the U4/U6 snRNP contains two snRNAs and several proteins. Some proteins are common to all snRNP particles. The snRNPs recognize consensus sequences. U1 snRNA base pairs with the 5′ splice site, U2 snRNA base pairs with the branch sequence, and U5 snRNP acts at the 5′ splice site. When U4 releases U6, the U6 snRNA base pairs with U2, and this may create the catalytic center for splicing. An alternative set of snRNPs provides analogous functions for splicing the U12-dependent subclass of introns. The snRNA molecules may have catalyst-like roles in splicing and other processing reactions.

Splicing is usually intramolecular, but *trans*-splicing (intermolecular splicing) occurs in trypanosomes and nematodes. It involves a reaction between a small SL RNA and the pre-mRNA. The SL RNA resembles U1 snRNA and may combine the role of providing the exon and the functions of U1. In nematodes there are two types of SL RNA, one used for splicing to the 5′ end of an mRNA, and the other for splicing to an internal site.

Group II introns share with nuclear introns the use of a lariat as an intermediate, but are able to perform the reaction as a self-catalyzed property of the RNA. These introns follow the GU-AG rule, but form a characteristic secondary structure that holds the reacting splice sites in the appropriate apposition.

Yeast tRNA splicing includes separate endonuclease and ligase reactions. The endonuclease recognizes the secondary (or tertiary) structure of the precursor and cleaves both ends of the intron. The two half-tRNAs released by loss of the intron are ligated in the presence of ATP.

The termination capacity of RNA polymerase II has not been characterized, and 3′ ends of its transcripts are generated by cleavage. The sequence AAUAAA, located 11 to 30 bases upstream of the cleavage site, provides the signal for both cleavage and polyadenylation. An endonuclease and the poly(A) polymerase are associated in a complex with other factors that confer specificity for the AAUAAA signal.

In the nucleolus, two groups of snoRNAs are responsible for pairing with rRNAs at sites that are modified; group C/D snoRNAs indicate target sites for methylation, and group ACA snoRNAs identify sites where uridine is converted to pseudouridine.

CHAPTER QUESTIONS

1. Which class(es) of introns have the ability to excise from the transcript autonomously, without any proteins (at least *in vitro*)?

A. nuclear introns

B. group I introns

C. group II introns

D. tRNA introns

2. The first two bases and the last two bases of nuclear introns are (almost) always:
   A. GC and AT.
   B. GU and AG.
   C. CU and AT.
   D. CG and GA.
3. Which base in the intron branch sequence is completely conserved?
   A. A
   B. G
   C. T
   D. C
4. Which two snRNPs are usually found as a single particle?
   A. U1 and U2
   B. U4 and U5
   C. U4 and U6
   D. U2 and U5
5. The U1 snRNA base pairs with:
   A. a sequence spanning the first exon–intron splicing site.
   B. the 3′ splice site of the intron.
   C. a sequence spanning the intron–second exon splicing site.
   D. the branch sequence within the intron.
6. The U2 snRNA base pairs with:
   A. a sequence spanning the first exon–intron boundary.
   B. the 3′ splice site of the intron.
   C. a sequence spanning the intron–second exon boundary.
   D. the branch sequence within the intron.
7. Removal of introns is connected to export of the mRNA from the nucleus to the cytoplasm by:
   A. binding of specific export proteins to the 5′ end of the processed mRNA.
   B. binding of specific export proteins to the first exon–exon junction.
   C. binding of specific export proteins to multiple exon–exon junctions.
   D. binding of specific export proteins to the 3′ end of the processed mRNA.
8. Which type of intron does not excise via a lariat mechanism?
   A. nuclear introns
   B. group I introns
   C. group II introns
   D. more than one of the above
9. In most cases, eukaryotic transcription actually terminates at:
   A. the mRNA cleavage site.
   B. a point upstream of the mRNA cleavage site.
   C. a point downstream of the mRNA cleavage site.
   D. a termination codon.
10. The polyadenylation signal in primary mRNA transcripts is:
   A. A-rich.
   B. G-rich.
   C. C-rich.
   D. U-rich.

KEY TERMS

A complex
alternative splicing
anti-Sm
autosplicing (self-splicing)
branch site
E complex
EJC (exon junction complex)
exon definition
GU-AG rule
heterogeneous nuclear RNA (hnRNA)
hnRNP
intron definition
lariat
poly(A) polymerase (PAP)
pre-RNA
RNA ligase
RNA splicing
SL RNA (spliced leader RNA)
small cytoplasmic RNAs (scRNA; scyrps)
small nuclear RNA (snRNA; snurps)
small nucleolar RNA (snoRNA)
spliceosome
splicing factor
SR protein
transesterification

FURTHER READING

Brow, D. A. (2002). Allosteric cascade of spliceosome activation. *Annu. Rev. Genet.* 36, 333–360.
A review of the steps in the assembly of the spliceosome and its removal of an intron.

Dreyfuss, G., Kim, V. N., and Kataoka, N. (2002). Messenger-RNA-binding proteins and the messages they carry. *Nat. Rev. Mol. Cell Biol.* 3, 195–205.
A review of the functions of proteins that bind to eukaryotic mRNAs.

Krämer, A. (1996). The structure and function of proteins involved in mammalian pre-mRNA splicing. *Annu. Rev. Biochem.* 65, 367–409.
A comprehensive review of the snRNPs and protein factors functioning in the mammalian spliceosome.

Reed, R. and Hurt, E. (2002). A conserved mRNA export machinery coupled to pre-mRNA splicing. *Cell* 108, 523–531.
A review of the mechanism of eukaryotic mRNA transport from the nucleus to the cytoplasm and its links to pre-mRNA splicing and nonsense-mediated mRNA decay.

Wahle, E. and Keller, W. (1992). The biochemistry of 3′-end cleavage and polyadenylation of messenger RNA precursors. *Annu. Rev. Biochem.* 61, 419–440.
A review of polyadenylation of animal mRNAs.

Zhou, Z., Licklider, L. J., Gygi, S. P., and Reed, R. (2002). Comprehensive proteomic analysis of the human spliceosome. *Nature* 419, 182–185.
The identification of the ~145 proteins that make up the human spliceosomal complex, many of which have yet to be characterized. Many appear to play roles in gene expression other than RNA splicing.

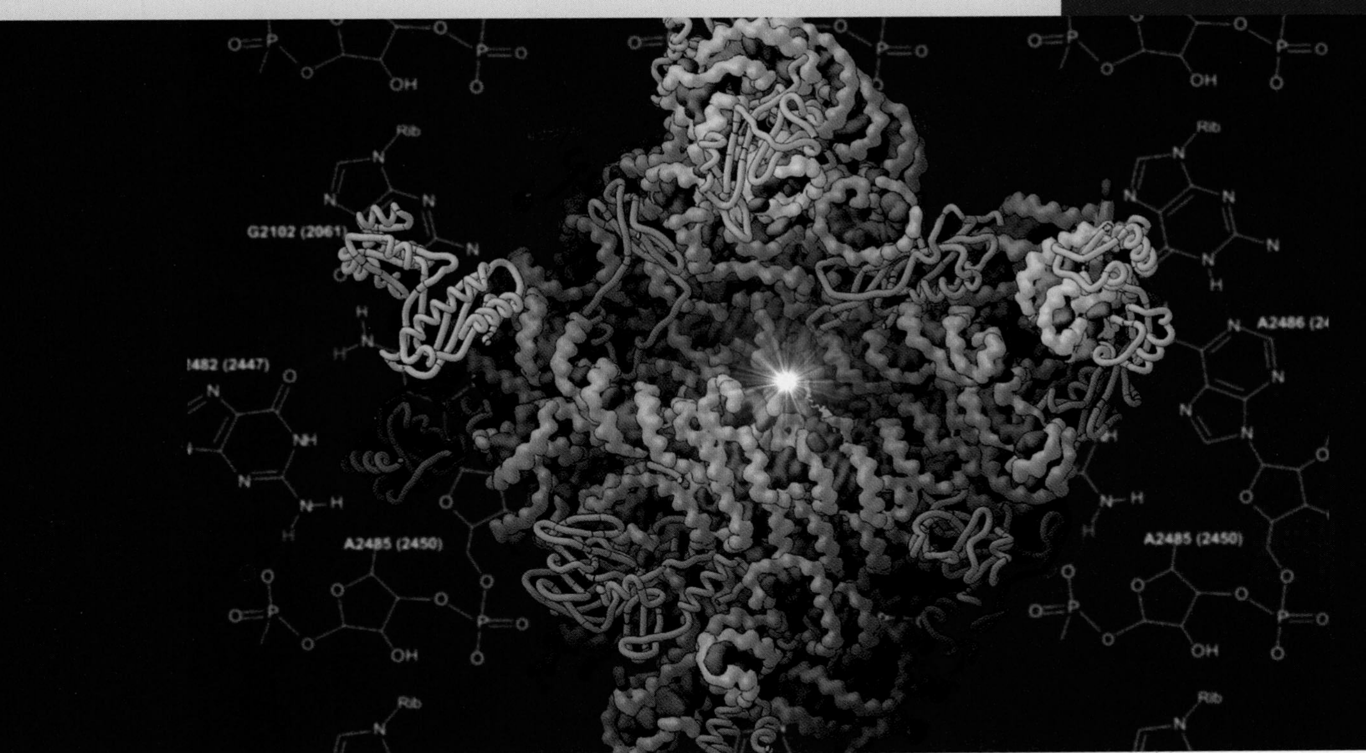

The RNA/protein architecture of the large ribosomal subunit with the active site highlighted. The background shows a schematic diagram of the peptidyl transferase active site of the ribosome. Photo courtesy of Nenad Ban, Institute for Molecular Biology & Biophysics, Zurich.

29 Catalytic RNA

CHAPTER OUTLINE

29.1 Introduction

▸ **ribozyme** An RNA that has catalytic activity.

Several types of catalytic reaction are now attributed to RNA. **Ribozyme** has become a general term used to describe an RNA with catalytic activity, and it is possible to characterize the enzymatic activity in the same way as a more conventional (protein) enzyme. Some RNA catalytic activities are directed against separate substrates, whereas others are intramolecular (which limits the catalytic action to a single cycle). This has led to the hypothesis that early, perhaps pre-cell genetic systems were based on RNA serving as both a genetic information storage and carrier molecule and a provider of catalytic functions, preceding proteins in these functions.

Introns of the group I and group II classes possess the ability to splice themselves out of the RNAs that contain them. Engineering of group I introns has generated RNA molecules that have several other catalytic activities related to the original activity.

The enzyme ribonuclease P is a ribonucleoprotein that contains a single RNA molecule bound to a protein. The RNA possesses the ability to catalyze cleavage in a tRNA substrate, whereas the protein component plays an indirect role, probably to maintain the structure of the catalytic RNA.

The common theme of these reactions is that the RNA can perform an intramolecular or intermolecular reaction that involves cleavage or joining of phosphodiester bonds *in vitro*. Although the specificity of the reaction and the basic catalytic activity is provided by RNA, proteins associated with the RNA may be needed in order for the reaction to occur efficiently *in vivo*.

▸ **RNA editing** A change of sequence at the level of RNA following transcription.

RNA splicing is not the only means by which changes can be introduced in the informational content of RNA. In the process of **RNA editing**, individual bases are changed or added at particular positions within an mRNA. The insertion of bases (most commonly uridine residues) occurs for RNAs of several genes in the mitochondria of certain unicellular eukaryotes; like splicing, it involves the breakage and reunion of bonds between nucleotides, but also requires a template for encoding the information of the new sequence.

29.2 Group I Introns Undertake Self-Splicing by Transesterification

▸ **autosplicing (self-splicing)** The ability of an intron to excise itself from an RNA by a catalytic action that depends only on the sequence of RNA in the intron.

Group I introns are found in diverse taxonomic groups. They occur in the genes coding for rRNA in the nuclei of the unicellular or syncytial (multinucleated) eukaryotes *Tetrahymena thermophila* (a ciliate) and *Physarum polycephalum* (a slime mold). They are common in the genes of fungal mitochondria. They are present in three genes of phage T4 and also are found in bacteria. Group I introns have an intrinsic ability to splice themselves. This is called **autosplicing** or **self-splicing**. (This property also is found in the group II introns discussed in *Section 28.10, Group II Introns Autosplice via Lariat Formation*.)

Self-splicing was discovered as a property of the transcripts of the rRNA genes in *T. thermophila*. The genes for the two major rRNAs follow the usual organization, in which both are expressed as part of a common transcription unit. The product is a 35S precursor RNA with the sequence of the small rRNA in the 5′ part, and the sequence of the larger (26S) rRNA toward the 3′ end.

In some strains of *T. thermophila*, the sequence coding for 26S rRNA is interrupted by a single, short intron. When the 35S precursor RNA is incubated *in vitro*, splicing occurs as an autonomous reaction. The intron is excised from the precursor and accumulates as a linear fragment of 400 bases, which is subsequently converted to a circular RNA. These events are summarized in **FIGURE 29.1**.

The reaction requires only a monovalent cation, a divalent cation, and a guanine nucleotide cofactor. No other base can be substituted for G, but a triphosphate is not needed: GTP, GDP, GMP, and guanosine itself all can be used, so there is no net energy requirement. The guanine nucleotide must have a 3′–OH group.

FIGURE 29.2 shows that three transfer reactions occur. In the first transfer, the guanine nucleotide behaves as a cofactor that provides a free 3′–OH group that attacks the 5′ end of the intron. This reaction creates the G–intron link and generates a 3′–OH group at the end of the exon. The second transfer involves a similar chemical reaction in which this 3′–OH then attacks the second exon. The two transfers are connected; no free exons have been observed, so their ligation may occur as part of the same reaction that releases the intron. The intron is released as a linear molecule, but the third transfer reaction converts it to a circle.

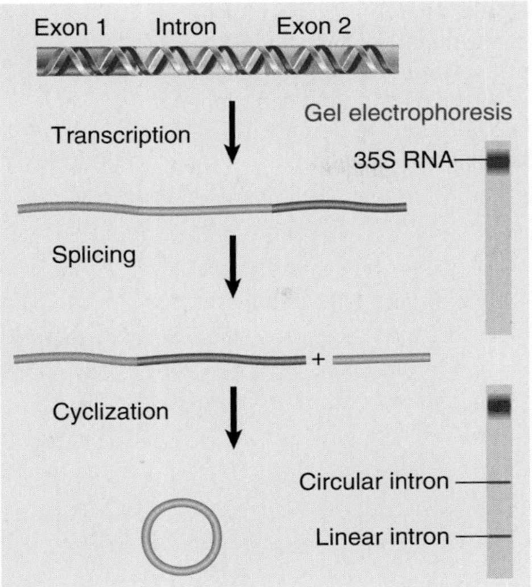

FIGURE 29.1 Splicing of the *Tetrahymena* 35S rRNA precursor can be followed by gel electrophoresis. The removal of the intron is revealed by the appearance of a rapidly moving small band. When the intron becomes circular, it electrophoreses more slowly, as seen by a higher band.

Each stage of the self-splicing reaction occurs by a transesterification, in which one phosphate ester is converted directly into another without any intermediary hydrolysis. Bonds are exchanged directly, and energy is conserved, so the reaction does not require input of energy from hydrolysis of ATP or GTP.

If each of the consecutive transesterification reactions involves no net change of energy, why does the splicing reaction proceed to completion instead of coming to equilibrium between spliced product and nonspliced precursor? The concentration of GTP is high relative to that of RNA, and therefore drives the reaction forward; a change in secondary structure in the RNA prevents the reverse reaction.

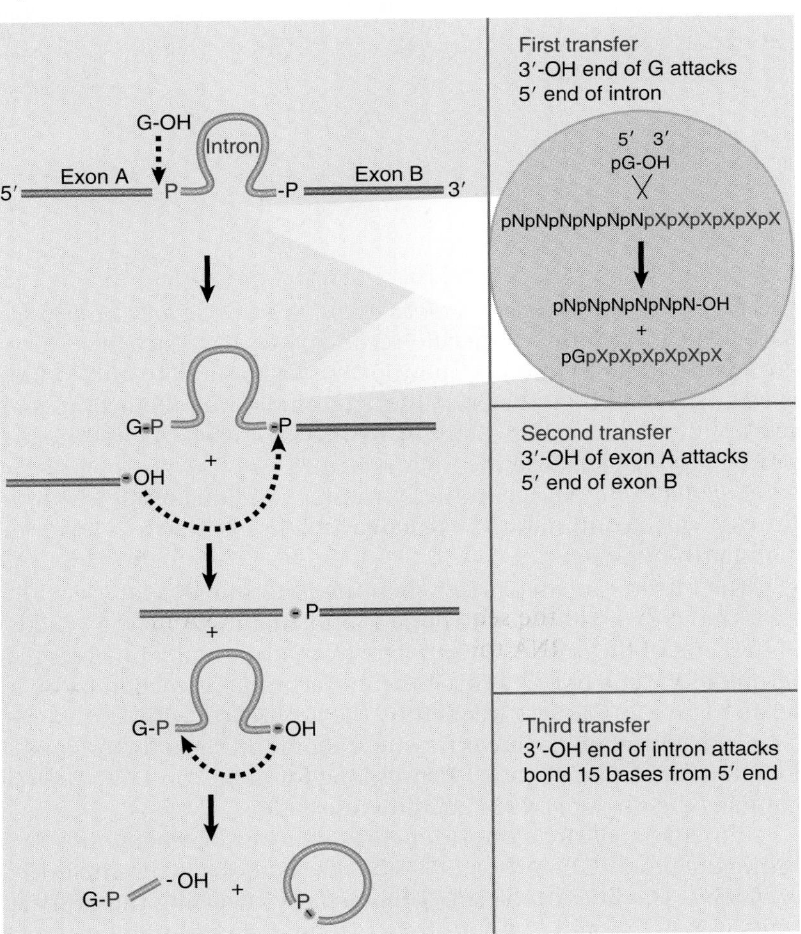

FIGURE 29.2 Self-splicing occurs by transesterification reactions in which bonds are exchanged directly. The bonds that have been generated at each stage are indicated by the shaded boxes.

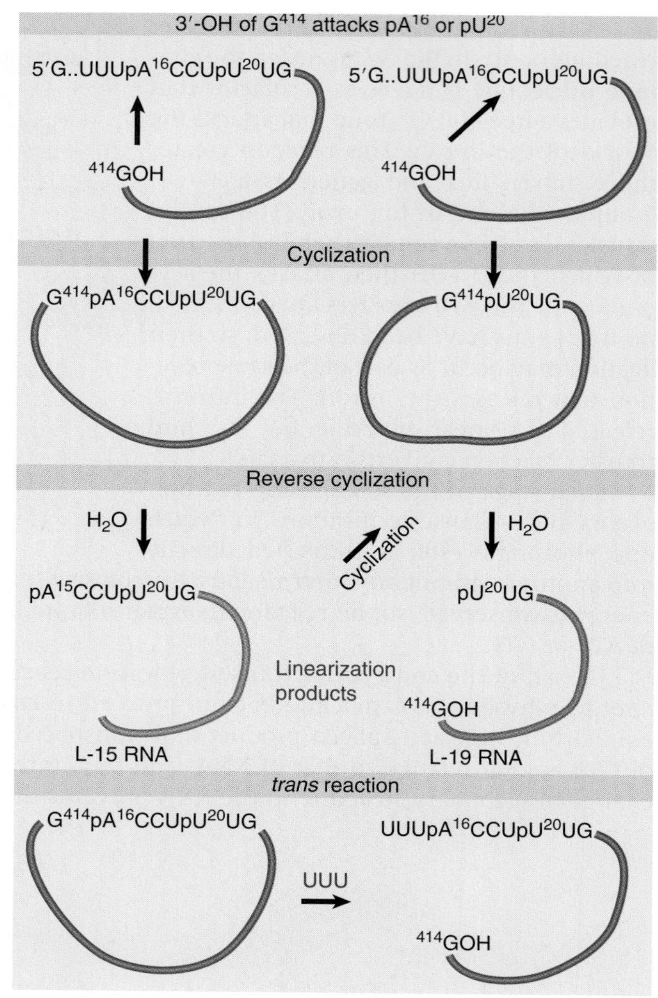

FIGURE 29.3 The excised intron can form circles by using either of two internal sites for reaction with the 5′ end, and can reopen the circles by reaction with water or oligonucleotides.

The in vitro *system does not include any proteins, so the ability to splice is intrinsic to the RNA.* The RNA forms a specific secondary/tertiary structure in which the relevant groups are brought into juxtaposition so that a guanine nucleotide can be bound to a specific site, and then the bond breakage and reunion reactions shown in Figure 29.2 can occur. Although it is catalyzed by the RNA itself, the reaction is assisted *in vivo* by proteins, which stabilize the RNA structure.

The ability to engage in these transfer reactions resides with the sequence of the intron, which continues to be reactive after its excision as a linear molecule. **FIGURE 29.3** summarizes its activities.

The intron can circularize when the 3′ terminal G attacks either of two positions near the 5′ end. The internal bond is broken and the new 5′ end is transferred to the 3′–OH end of the intron. The *primary cyclization* usually involves reaction between the terminal G^{414} and the A^{16}. This is the most common reaction (shown as the third transfer in Figure 29.2). Less frequently, the G^{414} reacts with U^{20}. Each reaction generates a circular intron and a linear fragment that represents the original 5′ region (15 bases long for attack on A^{16}, and 19 bases long for attack on U^{20}). The released 5′ fragment contains the original added guanine nucleotide.

Either type of circle can regenerate a linear molecule *in vitro* by specifically hydrolyzing the bond (G^{414}–A^{16} or G^{414}–U^{20}) that had closed the circle. This is called a *reverse cyclization*. The linear molecule generated by reversing the primary cyclization at A^{16} remains reactive and can perform a secondary cyclization by attacking U^{20}.

The final product of the spontaneous reactions following release of the intron is the L-19 RNA, a linear molecule generated by reversing the shorter circular form. This molecule has an enzymatic activity that allows it to catalyze the extension of short oligonucleotides (not shown in this figure, but see Figure 29.7).

The reactivity of the released intron extends beyond merely reversing the cyclization reaction. Addition of the oligonucleotide UUU reopens the primary circle by reacting with the G^{414}–A^{16} bond. The UUU (which resembles the 3′ end of the 15-mer released by the primary cyclization) becomes the 5′ end of the linear molecule that is formed. This is an *intermolecular* reaction, and thus demonstrates the ability to connect two different RNA molecules.

KEY CONCEPTS

- The only factors required for autosplicing *in vitro* by group I introns are a monovalent cation, a divalent cation, and a guanine nucleotide.
- Splicing occurs by two transesterifications, without requiring input of energy.
- The 3′-OH end of the guanine cofactor attacks the 5′ end of the intron in the first transesterification.
- The 3′-OH end generated at the end of the first exon attacks the junction between the intron and second exon in the second transesterification.
- The intron is released as a linear molecule that circularizes when its 3′-OH terminus attacks a bond at one of two internal positions.
- The G^{414}–A^{16} internal bond of the *Tetrahymena* intron can also be attacked by other nucleotides in a *trans*-splicing reaction.

CONCEPT AND REASONING CHECK

What is the evidence that group I introns are self-splicing?

29.3 Group I Introns Form a Characteristic Secondary Structure

All group I introns can be organized into a characteristic secondary structure with nine helices (P1–P9). **FIGURE 29.4** shows a model for the secondary structure of the *Tetrahymena* intron.

The group I splicing reaction depends on the formation of secondary structures between pairs of consensus sequences within the intron. The principle established by this work is that *sequences distant from the splice junctions themselves are required to form the active site that makes self-splicing possible.*

Two of the base-paired regions are generated by pairing between conserved sequence elements that are common to group I introns. P4 is constructed from the sequences *P* and *Q*; P7 is formed from sequences *R* and *S*. The other base-paired regions vary in sequence in individual introns. Mutational analysis identifies an intron "core" containing P3, P4, P6, and P7, which represents the minimal region that can undertake a catalytic reaction.

Some of the pairing reactions are directly involved in bringing the splice junctions into a conformation that supports the enzymatic reaction. P1

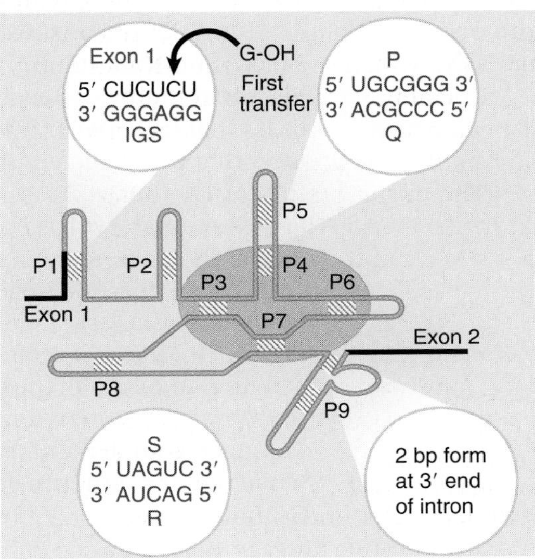

FIGURE 29.4 Group I introns have a common secondary structure that is formed by nine base-paired regions. The sequences of regions P4 and P7 are conserved, and identify the individual sequence elements P, Q, R, and S. P1 is created by pairing between the end of the left exon and the IGS of the intron; a region between P7 and P9 pairs with the 3′ end of the intron.

includes the 3′ end of the left exon. The sequence within the intron that pairs with the exon is called the IGS, or internal guide sequence. A very short sequence—sometimes as short as two bases—between P7 and P9 base pairs with the sequence that immediately precedes the reactive G (position 414 in *Tetrahymena*) at the 3′ end of the intron.

KEY CONCEPTS

- Group I introns form a secondary structure with nine duplex regions.
- The cores of regions P3, P4, P6, and P7 have catalytic activity.
- Regions P4 and P7 are both formed by pairing between conserved consensus sequences.
- A sequence adjacent to P7 base pairs with the sequence that contains the reactive G.

CONCEPT AND REASONING CHECK

What would be the effect of substitution mutations that weaken *P-Q* or *R-S* base pairing?

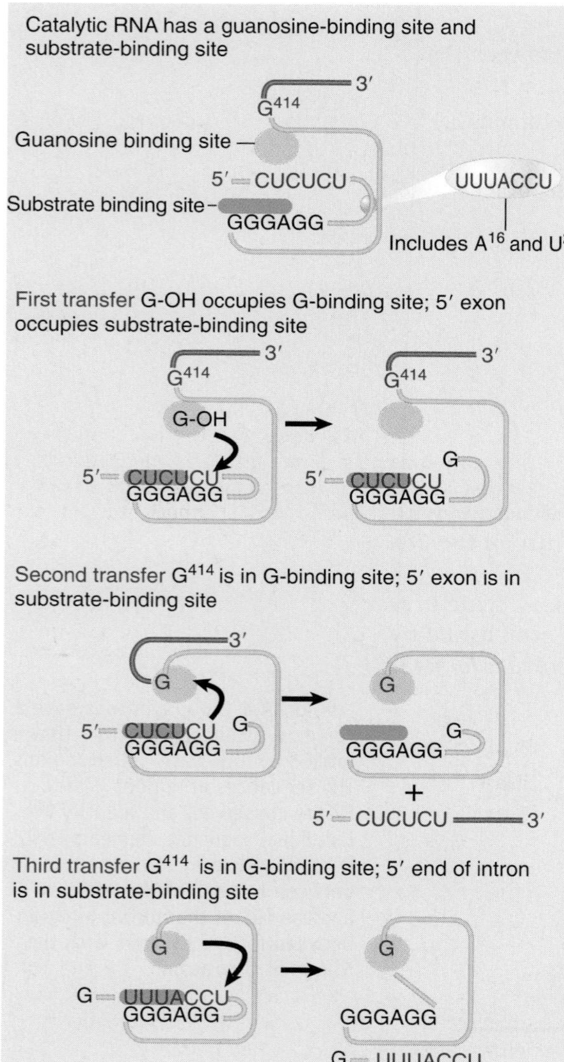

FIGURE 29.5 Excision of the group I intron in *Tetrahymena* rRNA occurs by successive reactions between the occupants of the guanosine-binding site and the substrate-binding site. The left exon is pink, and the right exon is purple.

29.4 Ribozymes Have Various Catalytic Activities

The catalytic activity of group I introns was discovered by virtue of their ability to autosplice, but they also are able to undertake other catalytic reactions *in vitro*. All of these reactions are based on transesterifications. We will review these reactions in terms of their relationship to the splicing reaction itself.

The catalytic activity of a group I intron is conferred by its ability to generate particular secondary and tertiary structures that create active sites equivalent to the active sites of a conventional (protein) enzyme. **FIGURE 29.5** illustrates the splicing reaction in terms of these sites (this is the same series of reactions shown previously in Figure 29.2).

The substrate-binding site is formed from the P1 helix, in which the 3′ end of the first intron base pairs with the IGS in an intermolecular reaction. A guanosine-binding site is formed by sequences in P7. This site may be occupied either by a free guanosine nucleotide or by the G residue in position 414. In the first transfer reaction, it is used by a free guanosine nucleotide; it is subsequently occupied by G^{414}. The second transfer releases the joined exons. The third transfer creates the circular intron.

Binding to the substrate involves a change of conformation. Before substrate binding, the 5′ end of the IGS is close to P2 and P8; after binding, when it forms the P1 helix, it is close to conserved bases that lie between P4 and P5. The reaction is visualized by contacts that are detected in the secondary structure in **FIGURE 29.6**. In the tertiary structure, the two sites alternatively contacted by P1 are 37 Å apart, which implies a substantial movement in the position of P1.

The L-19 RNA is generated by opening the circular intron (shown as the last stage of the intramolecular rearrangements in Figure 29.3). It still retains enzymatic abilities, which resemble the activities involved in the original splicing reaction. We may consider ribozyme function in terms of the ability to bind an intramolecular sequence complementary to the IGS in the substrate-binding site while binding either the terminal G^{414} or a free G-nucleotide in the G-binding site.

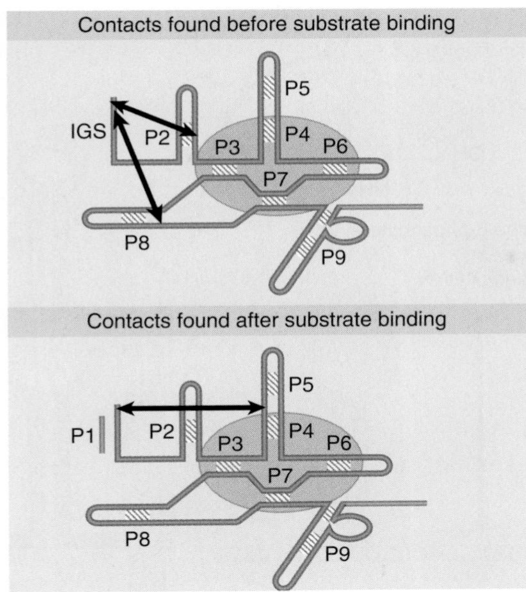

FIGURE 29.6 The position of the IGS in the tertiary structure changes when P1 is formed by substrate binding.

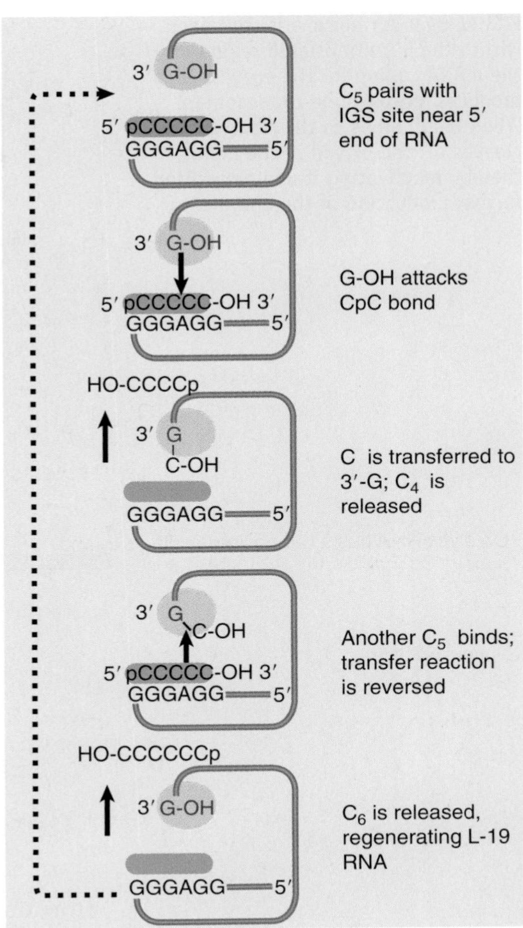

FIGURE 29.7 The L-19 linear RNA can bind C in the substrate-binding site; the reactive G-OH 3′ end is located in the G-binding site, and catalyzes transfer reactions that convert two C_5 oligonucleotides into a C_4 and a C_6 oligonucleotide.

FIGURE 29.7 illustrates the mechanism by which the oligonucleotide C_5 is extended to generate a C_6 chain. The C_5 oligonucleotide binds in the substrate-binding site, whereas G^{414} occupies the G-binding site. By transesterification reactions, a C is transferred from C_5 to the 3′–terminal G, and then back to a new C_5 molecule. Further transfer reactions lead to the accumulation of longer cytosine oligonucleotides. The reaction is a true catalysis, because the L-19 RNA remains unchanged and is available to catalyze multiple cycles. The ribozyme is behaving as a nucleotidyl transferase.

The reactions catalyzed by RNA can be characterized in the same way as classical enzymatic reactions in terms of Michaelis–Menten kinetics. FIGURE 29.8 analyzes the reactions catalyzed by RNA. K_M (the Michaelis constant) is a measure of substrate concentration at which half of the maximum velocity of the reaction will take place; a lower K_M is an indication that the catalyst and substrate have a higher affinity for each other. The K_M values for RNA-catalyzed reactions are low, and therefore imply that the RNA can bind its substrate with high specificity. The turnover numbers are low, which reflects a low catalytic rate. In effect, the RNA molecules behave in the same general manner as traditionally defined for enzymes, although they are relatively slow compared to protein catalysts (where a typical range of turnover numbers is 10^3 to 10^6).

A powerful extension of the activities of ribozymes has been made with the discovery that they can be regulated by ligands (see *Section 13.6, A Riboswitch in the 5′ UTR Region Can Control Translation of the mRNA*). FIGURE 29.9 summarizes the regulation of a **riboswitch**. The small metabolite GlcN6P binds to a ribozyme and activates its ability to cleave the RNA in an intramolecular reaction. The purpose of the system is to regulate production of GlcN6P; the ribozyme is located in the 5′ untranslated region of the mRNA that codes for the enzyme involved in producing GlcN6P, and the cleavage prevents translation.

Enzyme	Substrate	K_M (mM)	Turnover (/min)
19-base virusoid	24-base RNA	0.0006	0.5
L-19 Intron	CCCCCC	0.04	1.7
RNase P RNA	pre-tRNA	0.00003	0.4
RNase P complete	pre-tRNA	0.00003	29
RNase T1	GpA	0.05	5,700
β galactosidase	lactose	4.0	12,500

FIGURE 29.8 Reactions catalyzed by RNA have the same features as those catalyzed by proteins, although the rate is slower. The K_M gives the concentration of substrate required for half-maximum velocity; this is an inverse measure of the affinity of the enzyme for substrate. The turnover number gives the number of substrate molecules transformed in unit time by a single catalytic site.

▸ **riboswitch** A catalytic RNA whose activity responds to a small ligand.

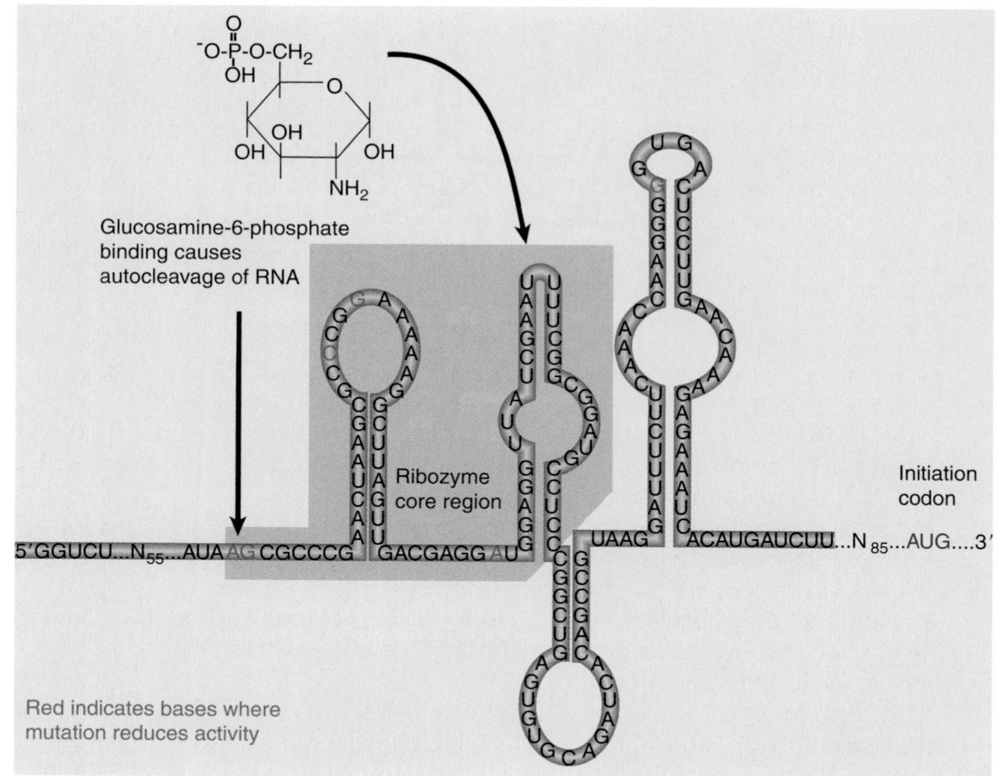

Glucosamine-6-phosphate binding causes autocleavage of RNA
Ribozyme core region
Initiation codon
5′GGUCU...N55...AUAAGCGCCCG
GACGAGGAUG
UAAG
ACAUGAUCUU...N85...AUG....3′
Red indicates bases where mutation reduces activity

FIGURE 29.9 A ribozyme is contained within the 5′ untranslated region of the mRNA coding for the enzyme that produces glucosamine-6-phosphate. When Glc6P binds to the ribozyme, it cleaves off the 5′ end of the mRNA, thereby inactivating it and preventing further production of the enzyme.

How does RNA provide a catalytic center? Its ability seems reasonable if we think of an active center as a surface that exposes a series of active groups in a fixed relationship. In a protein, the active groups are provided by the side chains of the amino acids, which have appreciable variety, including positive and negative ionic groups and hydrophobic groups. In an RNA, the available moieties are more restricted, consisting primarily of the exposed groups of bases. Short regions are held in a particular structure by the secondary/tertiary conformation of the molecule, providing a surface of active groups able to maintain an environment in which bonds can be broken and formed in another molecule. It seems inevitable that the interaction between the RNA catalyst and the RNA substrate will rely on base pairing to create the environment. Divalent cations (typically Mg^{2+}) play an important role in structure, typically by being present at the active site where they coordinate the positions of the various groups. They play a direct role in the endonucleolytic activity of virusoid ribozymes (see *Section 29.8, Viroids Have Catalytic Activity*).

KEY CONCEPTS

- By changing the substrate binding-site of a group I intron, it is possible to introduce alternative sequences that interact with the reactive G.
- The reactions follow classical enzyme kinetics with a low catalytic rate.

CONCEPT AND REASONING CHECK

Describe how an RNA can act as a catalytic active site like that of a protein enzyme.

29.5 Some Group I Introns Code for Endonucleases That Sponsor Mobility

Certain introns of both the group I and group II classes contain open reading frames that are translated into proteins. Expression of the proteins allows the intron (either in its original DNA form or as a DNA copy of the RNA) to be *mobile*; it is able to insert itself into a new genomic site. Introns of both groups I and II are extremely widespread, being found in both prokaryotes and eukaryotes. Group I introns migrate by DNA-mediated mechanisms, whereas group II introns migrate by RNA-mediated mechanisms.

Intron mobility was first detected by crosses in which the alleles for the relevant gene differ with regard to the presence of the intron. Polymorphisms for the presence or absence of introns are common in fungal mitochondria. This is consistent with the view that these introns originated by insertion into the gene. Some light on the process that could be involved is cast by an analysis of recombination in crosses involving the large rRNA gene of the yeast mitochondrion.

This gene has a group I intron that contains a coding sequence. The intron is present in some strains of yeast (called ω^+) but absent in others (ω^-). Genetic crosses between ω^+ and ω^- are *polar*; the progeny are usually ω^+.

If we think of the ω^+ strain as a donor and the ω^- strain as a recipient, we form the view that in $\omega^+ \times \omega^-$ crosses, a new copy of the intron is generated in the ω^- genome. As a result, the progeny are all ω^+.

Mutations can occur in either parent to abolish the polarity. Mutants show normal segregation, with equal numbers of ω^+ and ω^- progeny. The mutations indicate the nature of the process. Mutations in the ω^- strain occur close to the site where the intron would be inserted. Mutations in the ω^+ strain lie in the reading frame of the intron and prevent production of the protein. This suggests the model of **FIGURE 29.10**, in which the protein coded by the intron in an ω^+ strain recognizes the site where the intron should be inserted in an ω^- strain and causes it to be preferentially inherited.

What is the action of the protein? The product of the ω intron is an endonuclease *that recognizes the ω^- gene as a target for a double-strand break*. The endonuclease recognizes an 18 bp target sequence that contains the site where the intron is inserted. The target sequence is cleaved on each strand of DNA two bases to the 3′ side of the insertion site, so the cleavage sites are 4 bp apart and generate overhanging single strands.

This type of cleavage is related to the mechanism of transposons when they migrate to new sites (see *Chapter 21, Transposons, Retroviruses, and Retrotransposons*). The double-strand break probably initiates a gene conversion process in which the sequence of the ω^+ gene is copied to replace the sequence of the ω^- gene. The reaction involves transposition by a duplicative mechanism, and occurs solely at

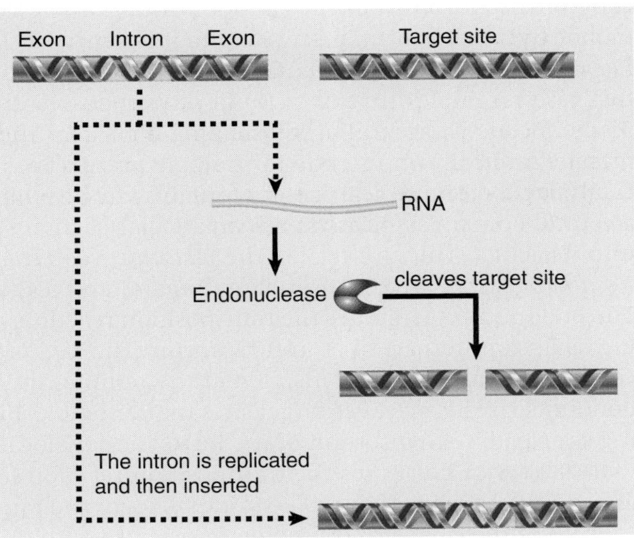

FIGURE 29.10 An intron codes for an endonuclease that makes a double-strand break in DNA. The sequence of the intron is duplicated and then inserted at the break.

the level of DNA. Insertion of the intron interrupts the sequence recognized by the endonuclease, thus ensuring stability.

Many group I introns code for endonucleases that make them mobile. Similar introns often carry quite different endonucleases. The dissociation between the intron sequence and the endonuclease sequence is emphasized by the fact that the same endonuclease sequences are found in inteins (sequences that code for self-splicing proteins; see *Section 29.11, Protein Splicing Is Autocatalytic*).

The variation in the endonucleases means that there is no homology between the sequences of their target sites. The target sites are among the longest and therefore the most specific known for any endonucleases (with a range of 14 to 40 bp). The specificity ensures that the intron perpetuates itself only by insertion into a single target site and not elsewhere in the genome. This is called **intron homing**.

▸ **intron homing** The ability of certain introns to insert themselves into a target DNA. The reaction is specific for a single target sequence.

Introns carrying sequences that code for endonucleases are found in a variety of bacteria and unicellular eukaryotes. These results strengthen the view that introns carrying coding sequences originated as independent elements.

KEY CONCEPTS

- Mobile introns are able to insert themselves into new sites.
- Mobile group I introns code for an endonuclease that makes a double-strand break at a target site.
- The intron transposes into the site of the double-strand break by a DNA-mediated replicative mechanism.

CONCEPT AND REASONING CHECK

How are mobile group I introns similar to transposons? How are they different?

29.6 Some Group II Introns Code for Reverse Transcriptases

Most of the open reading frames contained in group II introns encode regions that are related to reverse transcriptases. Introns of this type are found in organelles of unicellular eukaryotes and also in some bacteria. The reverse transcriptase activity is specific for the intron, and is involved in homing. The reverse transcriptase generates a DNA copy of the intron from the pre-mRNA, and therefore allows the intron to become mobile by a mechanism resembling that of retroviruses (see *Section 21.7, The Retrovirus Life Cycle Involves Transposition-Like Events*). The type of retrotransposition involved in this case resembles that of a group of retroposons that lack LTRs, and generate the 3′–OH needed for priming by making a nick in the target. The best characterized mobile group II introns code for a single protein in a region of the intron beyond its catalytic core. The typical protein contains an N-terminal reverse transcriptase activity, a central domain associated with an ancillary activity that assists folding of the intron into its active structure (called the *maturase*; see *Section 29.7, Some Autosplicing Introns Require Maturases*), a DNA-binding domain, and a C-terminal endonuclease domain. The endonuclease initiates the transposition reaction, and plays the same role in homing as its counterpart in a group I intron. The reverse transcriptase generates a DNA copy of the intron that is inserted at the homing site. At much lower frequency, the endonuclease also cleaves target sites that resemble, but are not identical to, the homing site, leading to insertion of the intron at new locations.

FIGURE 29.11 illustrates the transposition reaction for a typical group II intron. The endonuclease makes a double-strand break at the target site. A 3′ end is generated at the site of the break, and provides a primer for the reverse transcriptase. The intron RNA provides the template for the synthesis of cDNA. Because the RNA includes exon sequences on either side of the intron, the cDNA product is longer than the region of the intron itself, so that it can span the double-strand break, allowing the cDNA to repair the break. The result is the insertion of the intron.

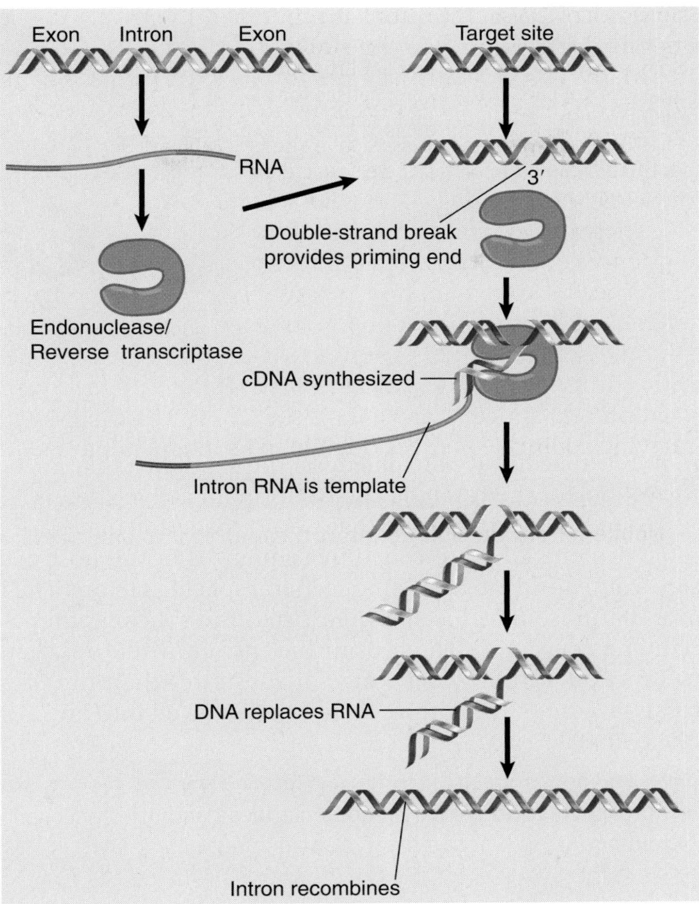
Exon
Intron
Exon
Target site
RNA
3′
Double-strand break provides priming end
Endonuclease/ Reverse transcriptase
cDNA synthesized
Intron RNA is template
DNA replaces RNA
Intron recombines

FIGURE 29.11 Reverse transcriptase coded by an intron allows a copy of the RNA to be inserted at a target site generated by a double-strand break.

KEY CONCEPT

- Some group II introns code for a reverse transcriptase that generates a DNA copy of the RNA sequence that transposes by a retroposon-like mechanism.

CONCEPT AND REASONING CHECK

How are mobile group II introns similar to retroposons? How are they different?

29.7 Some Autosplicing Introns Require Maturases

Both types of autosplicing intron may code for proteins involved in their perpetuation and homing, such as endonucleases in group I and reverse transcriptases in group II. In addition, both types of intron may code for **maturase** activities that are required to assist the splicing reaction.

▸ **maturase** A protein encoded by a group I or group II intron that is needed to assist the RNA to form the active conformation that is required for self-splicing.

The maturase activity is part of the product of the single open reading frame coded by the intron. In the example of introns that code for homing endonucleases, the single protein product has both endonuclease and maturase activity. The activities are provided by different active sites in the protein, each coded by a separate domain. The endonuclease site binds to DNA, but the maturase site binds to the intron RNA. **FIGURE 29.12** shows the structure of one such protein bound to DNA. A characteristic feature of the endonuclease is the presence of parallel α helices, which contain the hallmark LAGLIDADG amino acid sequences of the endonuclease, leading to the two catalytic amino acids. The maturase activity is located some distance away on the surface of the protein.

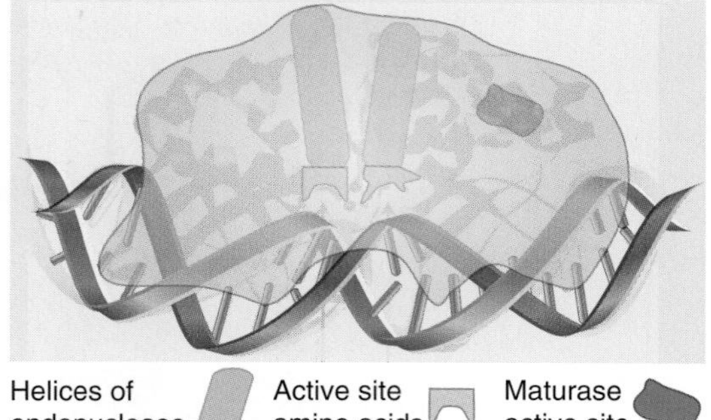

FIGURE 29.12 A homing intron codes for an endonuclease of the LAGLIDADG family that also has maturase activity. The LAGLIDADG sequences are part of the two α helices that terminate in the catalytic amino acids close to the DNA duplex. The maturase active site is identified by an arginine residue elsewhere on the surface of the protein.

Introns that code for maturases may be unable to splice themselves effectively in the absence of the protein activity. The maturase is basically a splicing factor that is required specifically for splicing of the sequence that encodes it. It functions to assist the folding of the catalytic core to form an active site.

Some group II introns that do not code for maturase activities may use comparable proteins that are coded by sequences in the host genome. This suggests a possible route for the evolution of general splicing factors. The factor may have originated as a maturase that specifically assisted the splicing of a particular intron. The coding sequence became isolated from the intron in the host genome, and then it evolved to function with a wider range of substrates than the original intron sequence. The catalytic core of the intron could have evolved into an snRNA.

KEY CONCEPT

- Autosplicing introns may require maturase activities encoded within the intron to assist folding into the active catalytic structure.

CONCEPT AND REASONING CHECK

What is a maturase and what is its role in self-splicing of introns?

29.8 Viroids Have Catalytic Activity

Another example of the ability of RNA to function as an endonuclease is provided by some small plant RNAs (~350 bases) that undertake a self-cleavage reaction. As with the case of the *Tetrahymena* group I intron, however, it is possible to engineer constructs that can function on external substrates.

These small plant RNAs fall into two general groups: viroids and virusoids. The **viroids** are infectious RNA molecules that function independently without encapsidation by any protein coat. The **virusoids** are similar in organization but are encapsidated by plant viruses, being packaged together with a viral genome. The virusoids cannot replicate independently, but require assistance from the virus. The virusoids are sometimes called *satellite RNAs*.

- **viroid** A small infectious nucleic acid that does not have a protein coat.
- **virusoid** A small infectious nucleic acid that is encapsidated by a plant virus together with its own genome.

Viroids and virusoids both replicate via rolling circles (see Figure 16.7). The strand of RNA that is packaged into the virus is called the "plus" strand. The complementary strand, generated during replication of the RNA, is called the "minus" strand. Multimers of both plus and minus strands are found. Both types of monomer are generated by cleaving the tail of a rolling circle; circular plus-strand monomers are generated by ligating the ends of the linear monomer.

Both plus and minus strands of viroids and virusoids undergo self-cleavage *in vitro*. The cleavage reaction is promoted by divalent metal cations; it generates 5′–OH and 2′–3′–cyclic phosphodiester termini. Some of the RNAs cleave *in vitro* under physiological conditions. Others do so only after a cycle of heating and cooling; this suggests that the isolated RNA has an inappropriate conformation, but can generate an active conformation when it is denatured and renatured.

The viroids and virusoids that undergo self-cleavage form a "hammerhead" secondary structure at the cleavage site, as shown in the upper part of **FIGURE 29.13**. The sequence of this structure is sufficient for cleavage. When the surrounding sequences are deleted, the need for a heating–cooling cycle is eliminated, and the small RNA self-cleaves spontaneously. This suggests that the sequences beyond the hammerhead usually interfere with its formation.

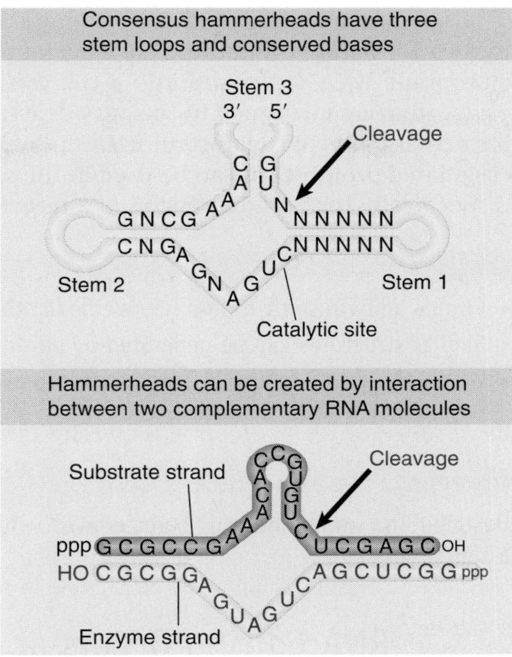

FIGURE 29.13 Self-cleavage sites of viroids and virusoids have a consensus sequence and form a hammerhead secondary structure by intramolecular pairing. Hammerheads can also be generated by pairing between a substrate strand and an "enzyme" strand.

The active site is a sequence of only 58 nucleotides. The hammerhead contains three stem-loop regions whose position and size are constant, and 13 conserved nucleotides, mostly in the regions connecting the center of the structure. The conserved bases and duplex stems generate an RNA with the intrinsic ability to cleave.

An active hammerhead can also be generated by pairing an RNA representing one side of the structure with an RNA representing the other side. The lower part of Figure 29.13 shows an example of a hammerhead generated by hybridizing a 19 base molecule with a 24 base molecule. The hybrid mimics the hammerhead structure, with the omission of loops I and III. When the 19 base RNA is added to the 24 base RNA, cleavage occurs at the appropriate position in the hammerhead.

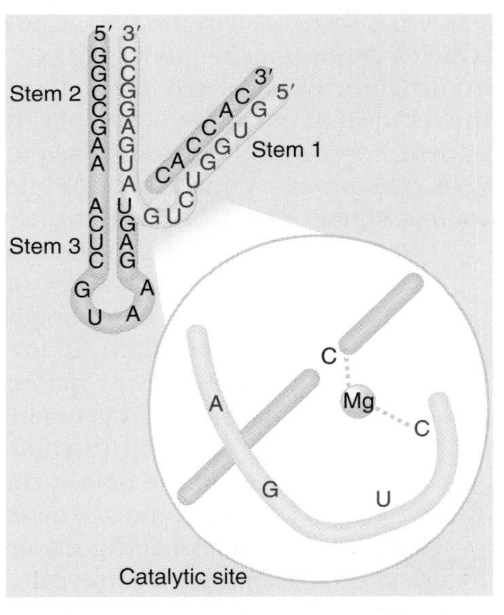

FIGURE 29.14 A hammerhead ribozyme forms a V-shaped tertiary structure in which stem 2 is stacked upon stem 3. The catalytic center lies between stems 2 and 3 and stem 1. It contains a magnesium ion that initiates the hydrolytic reaction.

We may regard the top (24 base) strand of this hybrid as comprising the "substrate" and the bottom (19 base) strand as the "enzyme." When the 19 base RNA is mixed with an excess of the 24 base RNA, multiple copies of the 24 base RNA are cleaved. This suggests that there is a cycle of 19 base–24 base pairing, cleavage, dissociation of the cleaved fragments from the 19 base RNA, and pairing of the 19 base RNA with a new 24 base substrate. The 19 base RNA is therefore a ribozyme with endonuclease activity. The parameters of the reaction are similar to those of other RNA-catalyzed reactions.

The crystal structure of a hammerhead shows that it forms a compact V-shape in which the catalytic center lies in a turn, as indicated diagrammatically in **FIGURE 29.14**. An Mg^{2+} ion located in the catalytic site plays a crucial role in the reaction. It is positioned by the target cytidine and by the cytidine at the base of stem 1; it may also be connected to the adjacent uridine. It extracts a proton from the 2′–OH of the target cytidine, and then directly attacks the labile phosphodiester bond. Mutations in the hammerhead sequence that affect the transition state of the cleavage reaction occur in both the active site and other locations, suggesting that there may be a substantial rearrangement of structure prior to cleavage.

It is possible to design many enzyme-substrate combinations that can form hammerhead structures, and these have been used to demonstrate that introduction of the appropriate RNA molecules into a cell can allow the enzymatic reaction to occur *in vivo*. A ribozyme designed in this way essentially provides a highly specific restriction-like activity directed against an RNA target. By placing the ribozyme under control of a regulated promoter, it can be used in the same way as, for example, antisense RNAs, to specifically turn off expression of a target gene under defined circumstances.

KEY CONCEPTS

- Viroids and virusoids form a hammerhead structure that has a self-cleaving activity.
- Similar structures can be generated by pairing a substrate strand with an enzyme strand.
- When an enzyme strand is introduced into a cell, it can pair with a substrate strand target that is then cleaved.

CONCEPT AND REASONING CHECK

Describe how an engineered "hammerhead" ribozyme may be used to inhibit expression of a target gene.

29.9 RNA Editing Occurs at Individual Bases

A prime axiom of molecular biology is that the sequence of an mRNA can only represent what is encoded in the DNA. The central dogma describes a linear relationship in which a continuous sequence of DNA is transcribed into a sequence of mRNA that is in turn directly translated into polypeptide. The occurrence of interrupted genes and the removal of introns by RNA splicing introduces an additional step into the process of gene expression: The coding sequences (exons) in DNA must be reconnected in RNA. The process remains one of information transfer, though, in which the actual coding sequence in DNA remains unchanged.

Changes in the information coded by DNA occur in some exceptional circumstances, most notably in the generation of new sequences coding for immunoglobulins in mammals and birds. These changes occur specifically in the somatic cells (B lymphocytes) in which immunoglobulins are synthesized (see Chapter 22, *Immune Diversity*). New information is generated in the DNA of an individual during the process of reconstructing an immunoglobulin gene, and information coded in the DNA is changed by somatic mutation. The information in DNA continues to be faithfully transcribed into RNA.

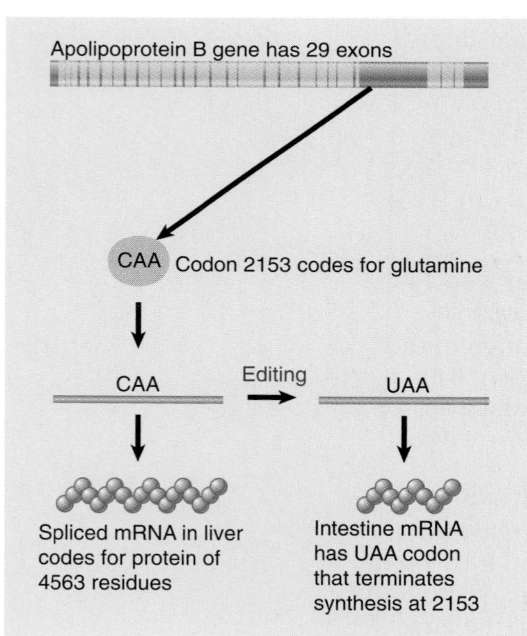

FIGURE 29.15 The sequence of the apo-B gene is the same in intestine and liver, but the sequence of the mRNA is modified by a base change that creates a termination codon in intestine.

RNA editing is a process in which *information is changed at the level of mRNA*. It is revealed by situations in which the coding sequence in an mRNA differs from the sequence of DNA from which it was transcribed. RNA editing occurs in two different situations, each with different causes. In mammalian cells there are cases in which a substitution occurs in an individual base in mRNA, causing a change in the sequence of the polypeptide that is encoded. In trypanosome mitochondria, more widespread changes occur in transcripts of several genes, when bases are systematically added or deleted.

FIGURE 29.15 summarizes the sequences of the apolipoprotein-B gene and mRNA in mammalian intestine and liver cells. The genome contains a single (interrupted) gene whose sequence is identical in all tissues, with a coding region of 4563 codons. In liver cells, this gene is transcribed into an mRNA that is translated into a protein of 512 kD representing the full coding sequence.

A shorter form of the protein, ~250 kD, is synthesized in intestine cells. This protein consists of the N-terminal half of the full-length protein. It is translated from an mRNA whose sequence is identical with that of liver cells except for a change from C to U at codon 2153. This substitution changes the codon CAA for glutamine into the ochre termination codon UAA.

What is responsible for this substitution? No alternative gene or exon to encode the new sequence is found in the genome, and no change in the pattern of splicing has been detected. We are forced to conclude that a direct change to the sequence of the transcript has been made.

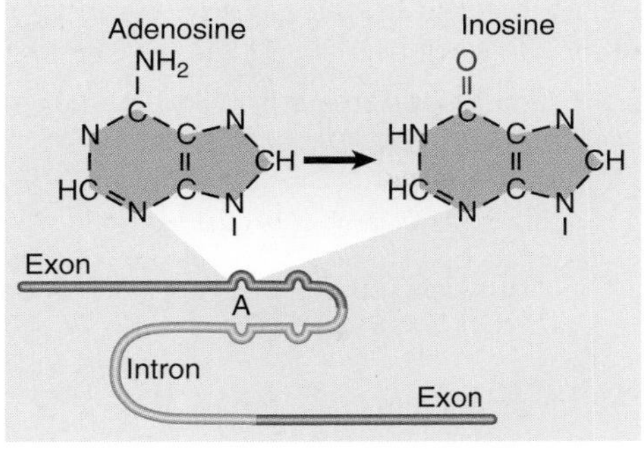

FIGURE 29.16 Editing of mRNA occurs when a deaminase acts on an adenine in an imperfectly paired RNA duplex region.

The editing event in apo-B causes C_{2153} to be changed to U. Similar events in rat brain glutamate receptors change an A to I (inosine). These events are *deaminations* in which the amino group on the nucleotide ring is removed. Such events are catalyzed by enzymes called *cytidine deaminases* and *adenosine deaminases*, respectively. This type of editing appears to occur largely in cells of the nervous system. There are 16 (potential) targets for cytidine deaminase in *D. melanogaster*, and all are genes involved in neurotransmission. In many cases, the editing event changes an amino acid at a functionally important position in the protein.

What controls the specificity of an editing reaction? Enzymes that undertake deamination often have broad specificity—for example, the best-characterized adenosine deaminase acts on any A residue in a duplex RNA region. Editing enzymes are related to the general deaminases, but have other regions or additional subunits that control their specificity. In the case of apo-B editing, the catalytic subunit of an editing complex is related to bacterial cytidine deaminase, but has an additional RNA-binding region that helps to recognize the specific target site for editing. A special adenosine deaminase enzyme recognizes the target sites in the glutamate receptor RNA, and similar events occur in a serotonin receptor RNA.

The complex may recognize a particular region of secondary structure in a manner analogous to tRNA-modifying enzymes, or it could directly recognize a nucleotide sequence. The development of an *in vitro* system for the apo-B editing event suggests that a relatively small sequence (~26 bases) surrounding the editing site provides a sufficient target. **FIGURE 29.16** shows that in GluR-B RNA, a base-paired region that is necessary for recognition of the target site is formed between the edited region in the exon and a complementary sequence in the downstream intron. A pattern of mispairing within the duplex region is necessary for specific recognition. So different editing systems may have different types of requirement for sequence specificity in their substrates.

KEY CONCEPT

- Apolipoprotein-B and glutamate receptors have site-specific deaminations catalyzed by cytidine and adenosine deaminases that change the coding sequence.

CONCEPT AND REASONING CHECK

The product of a gene may be different from the predicted product based on the DNA sequence of the gene for several reasons, including RNA editing. What are the other reasons, and how can RNA editing be demonstrated to be the reason for this difference for a particular gene?

29.10 RNA Editing Can Be Directed by Guide RNAs

Another type of editing is revealed by dramatic changes in sequence in the products of several genes of trypanosome mitochondria. In the first case to be discovered, the sequence of the cytochrome oxidase subunit II protein has a frameshift relative to the

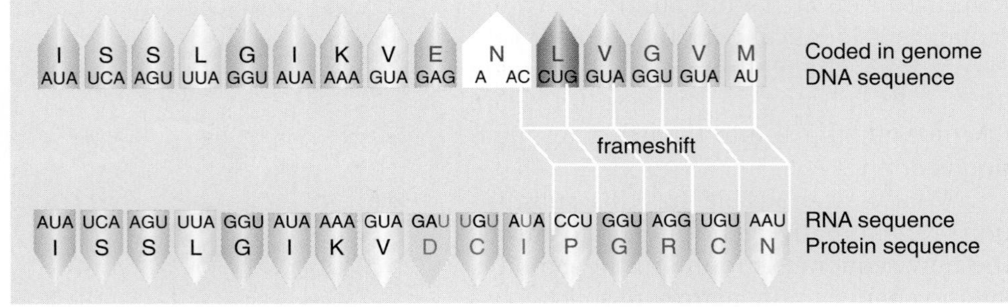

FIGURE 29.17 The mRNA for the trypanosome *coxII* gene has a frameshift relative to the DNA; the correct reading frame is created by the insertion of four uridines.

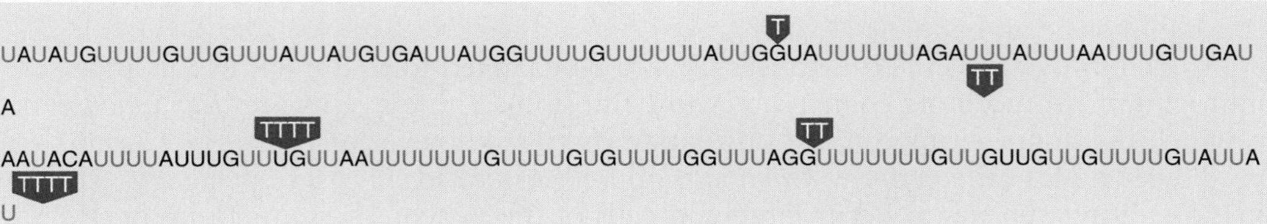

FIGURE 29.18 Part of the mRNA sequence of *T. brucei coxIII* shows many uridines that are not coded in the DNA (shown in red) or that are removed from the RNA (shown as T).

sequence of the *coxII* gene. The sequences of the gene and protein given in **FIGURE 29.17** are conserved in several trypanosome species. How does this gene function?

The *coxII* mRNA has an insert of an additional four nucleotides (all uridines) around the site of frameshift. The insertion changes the reading frame; it inserts an extra amino acid and changes the amino acids downstream. No second gene with this sequence can be discovered, and we are forced to conclude that the extra bases are inserted during or after transcription. A similar discrepancy between mRNA and genomic sequences (involving the addition of G residues in the mRNA) is found in genes of the SV5 and measles paramyxoviruses.

Similar editing of RNA sequences occurs for other genes, and includes deletions as well as additions of uridine. The extraordinary case of the *coxIII* gene of *Trypanosoma brucei* is summarized in **FIGURE 29.18**.

More than half of the residues in the mRNA consist of uridines that are not encoded by the gene. Comparison between the genomic DNA and the mRNA shows that no stretch longer than seven nucleotides is represented in the mRNA without alteration, and runs of uridine up to seven bases long are inserted.

▸ **guide RNA** A small RNA whose sequence is complementary to the sequence of an RNA that has been edited. It is used as a template for changing the sequence of the pre-edited RNA by inserting or deleting nucleotides.

What provides the information for the specific insertion of uridines? A **guide RNA** contains a sequence that is complementary to the correctly edited mRNA. **FIGURE 29.19** shows a model for its action in the cytochrome b gene of *Leishmania*.

The sequence at the top of the figure shows the original transcript, or pre-edited RNA. Gaps show where bases will be inserted in the editing process. Eight uridines must be inserted into this region to create the final mRNA sequence.

The guide RNA is complementary to the mRNA for a significant distance, including and surrounding the edited region. Typically the complementarity is more extensive on the 3′ side of the edited region and is rather short on the 5′ side. Pairing between the guide RNA and the pre-edited RNA leaves gaps where unpaired A residues in the guide RNA do not find complements in the pre-edited RNA. The guide RNA provides a template that allows the missing U residues to be inserted at these positions. When the reaction is completed, the guide RNA separates from the mRNA, which becomes available for translation.

Specification of the final edited sequence can be quite complex. In this example, a lengthy stretch of the transcript is edited by the insertion of a total of 39 U residues, which appears to require two guide RNAs that act at adjacent sites. The first guide RNA pairs at the 3′–most site, and the edited sequence then becomes a substrate for further editing by the next guide RNA.

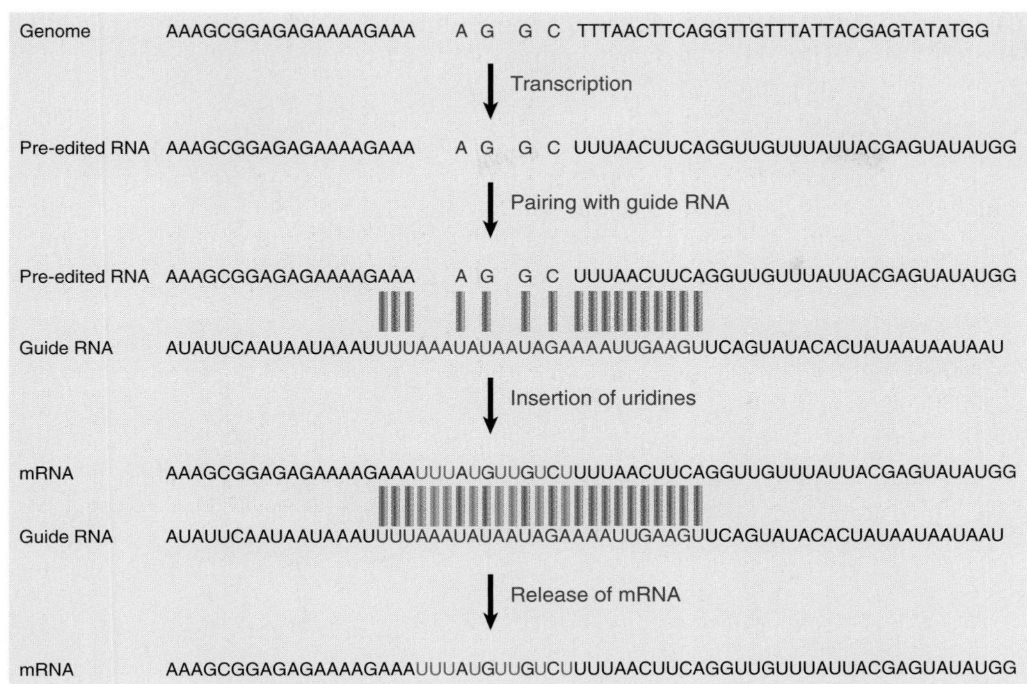

FIGURE 29.19 Pre-edited RNA base pairs with a guide RNA on both sides of the region to be edited. The guide RNA provides a template for the insertion of uridines. The mRNA produced by the insertions is complementary to the guide RNA.

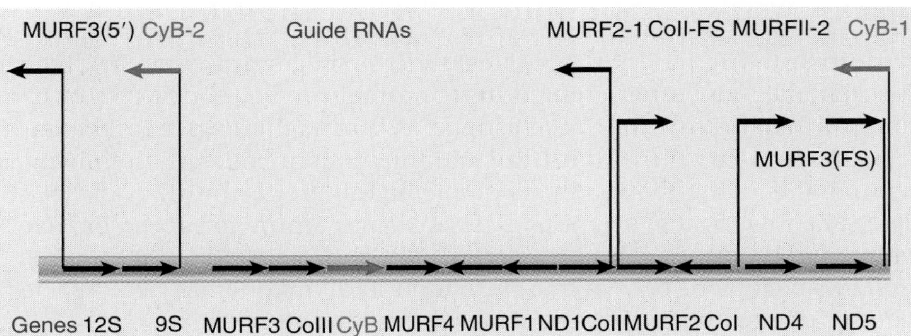

FIGURE 29.20 The *Leishmania* genome contains genes coding for pre-edited RNAs interspersed with units that code for the guide RNAs required to generate the correct mRNA sequences. Some genes have multiple guide RNAs. CyB is the gene for pre-edited cytochrome b, and CyB-1 and CyB-2 are genes for the guide RNAs involved in its editing.

The guide RNAs are encoded as independent transcription units. **FIGURE 29.20** shows a map of the relevant region of the *Leishmania* mitochondrial DNA. It includes the gene for cytochrome b, which encodes the pre-edited sequence, and two regions that encode guide RNAs. Genes for the major coding regions and for their guide RNAs are interspersed.

The characterization of intermediates that are partially edited suggests that the reaction proceeds along the pre-edited RNA in the 3′–5′ direction. The guide RNA determines the specificity of uridine insertions by its pairing with the pre-edited RNA.

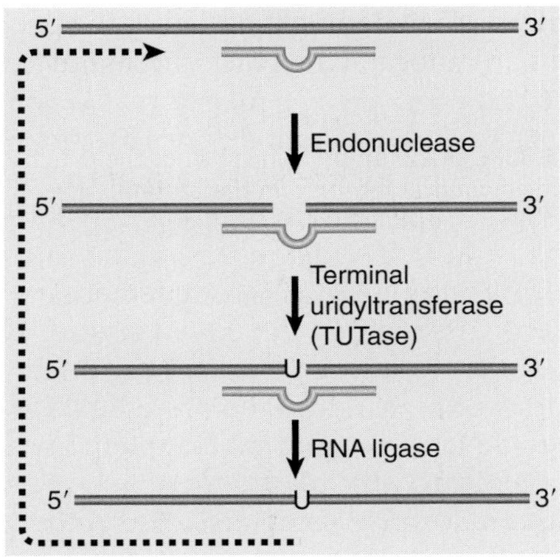

FIGURE 29.21 Addition or deletion of U residues occurs by cleavage of the RNA, removal or addition of the U, and ligation of the ends. The reactions are catalyzed by a complex of enzymes under the direction of guide RNA.

Editing of uridines is catalyzed by a 20S enzyme complex that contains an endonuclease, a terminal uridyltransferase (TUTase), and an RNA ligase, as illustrated in **FIGURE 29.21**. It binds the guide RNA and uses it to pair with the pre-edited mRNA. The substrate RNA is cleaved at a site that is presumably identified by the absence of pairing with the guide RNA, a uridine is inserted or deleted to base pair with the guide RNA, and then the substrate RNA is ligated.

Uracil triphosphate (UTP) provides the source for the uridyl residue. It is added by the TUTase activity; it is not clear whether this activity, or a separate exonuclease, is responsible for deletion.

The structures of partially edited molecules suggest that the U residues are added one at a time rather than in groups. It is possible that the reaction proceeds through successive cycles in which U residues are added, tested for complementarity with the guide RNA, retained if acceptable, and removed if unacceptable. If so, the construction of the correct edited sequence occurs gradually. We do not know whether the same types of reaction are involved in editing reactions that add C residues.

KEY CONCEPTS

- Extensive RNA editing in trypanosome mitochondria occurs by insertions or deletions of uridine.
- The substrate RNA base pairs with a guide RNA on both sides of the region to be edited.
- The guide RNA provides the template for addition (or less often, deletion) of uridines.
- Editing is catalyzed by a complex of endonuclease, terminal uridyltransferase activity, and RNA ligase.

CONCEPT AND REASONING CHECK

Describe the process by which editing of the *Leishmania* cytochrome b RNA occurs.

29.11 Protein Splicing Is Autocatalytic

- **protein splicing** The autocatalytic process by which an intein is removed from a protein and the exteins on either side become connected by a standard peptide bond.
- **extein** A sequence that remains in the mature protein that is produced by processing a precursor via protein splicing.
- **intein** The part that is removed from a protein that is processed by protein splicing.

Protein splicing has the same effect as RNA splicing: a sequence that is represented in a gene fails to be represented in its protein product. The parts of the protein are named by analogy with RNA splicing: **exteins** are the sequences that are represented in the mature protein, and **inteins** are the sequences that are removed. The mechanism for removing the intein is completely different from that of RNA splicing. FIGURE 29.22 shows that the gene is translated into a protein precursor that contains the intein, and then the intein is excised from the protein. About a hundred examples of protein splicing are known and are found in all taxonomic groups. The typical gene whose product undergoes protein splicing has a single intein.

The first intein was discovered in an archaeal DNA polymerase gene in the form of an intervening sequence in the gene that shows no similarity to conventional introns. It was then demonstrated that the purified protein can splice this sequence out of itself in an autocatalytic reaction. The reaction does not require input of energy and occurs

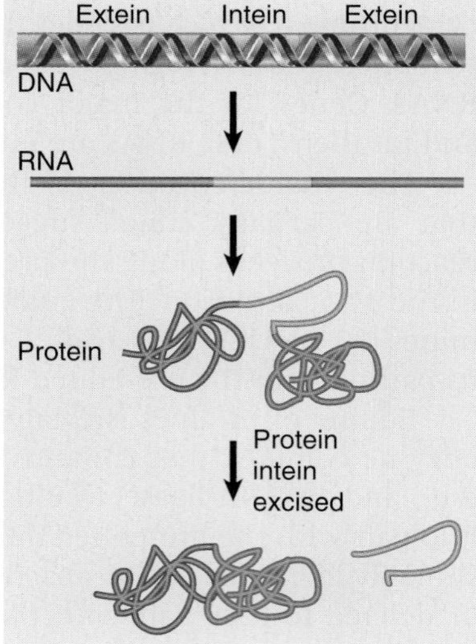

FIGURE 29.22 In protein splicing, the exteins are connected by removing the intein from the protein.

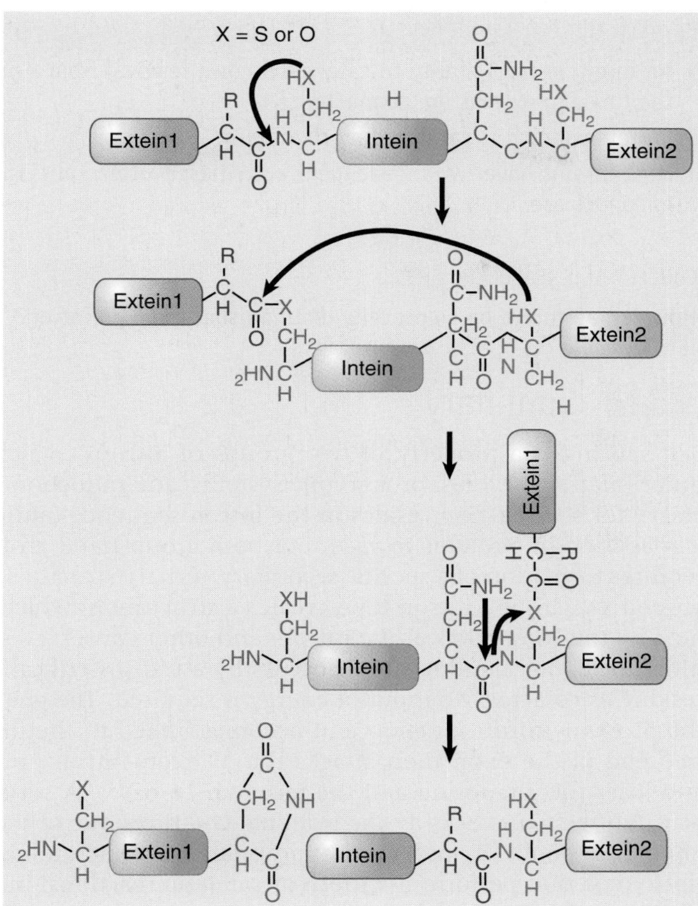

FIGURE 29.23 Bonds are rearranged through a series of transesterifications involving the –OH groups of serine or threonine or the –SH group of cysteine until the exteins are connected by a peptide bond and the intein is released with a circularized C-terminus.

through the series of bond rearrangements shown in **FIGURE 29.23**. The reaction is a function of the intein, although its efficiency can be influenced by the exteins.

The first reaction is an attack by an –OH or –SH side chain of the first amino acid in the intein on the peptide bond that connects it to the first extein. This transfers the extein from the amino-terminal group of the intein to an N-O or N-S acyl connection. This bond is then attacked by the –OH or –SH side chain of the first amino acid in the second extein. The result is to transfer extein-1 to the side chain of the amino-terminal acid of extein-2. Finally, the C-terminal asparagine of the intein cyclizes, and the terminal –NH of extein-2 attacks the acyl bond to replace it with a conventional peptide bond. Each of these reactions can occur spontaneously at very low rates, but the intein catalyzes these reactions in a coordinated manner and rapidly enough to achieve protein splicing.

Inteins have characteristic features. They are found as in-frame insertions into coding sequences and are recognized as such because of the existence of homologous genes that lack the insertion. They have an N-terminal serine or cysteine (to provide the –XH side chain) and a C-terminal asparagine. A typical intein has a sequence of ~150 amino acids at the N-terminal end and ~50 amino acids at the C-terminal end that are involved in catalyzing the protein splicing reaction. The sequence in the center of the intein can have other functions.

An extraordinary feature of many inteins is that they have homing endonuclease activity. A homing endonuclease cleaves a target DNA to create a site into which the DNA sequence coding for the intein can be inserted (see Figure 29.10 in *Section 29.5, Some Group I Introns Code for Endonucleases That Sponsor Mobility*). The protein splicing and homing endonuclease activities of an intein are independent.

KEY CONCEPTS

- An intein has the ability to catalyze its own removal from a protein in such a way that the flanking exteins are connected.
- Protein splicing is catalyzed by the intein.
- Most inteins have two independent activities: protein splicing and a homing endonuclease.

CONCEPT AND REASONING CHECK

How can an intein be empirically distinguished from an intron?

29.12 Summary

Self-splicing is a property of two groups of introns, which are widely dispersed in unicellular eukaryotes, prokaryotic systems, and mitochondria. The information necessary for the reaction resides in the intron sequence (although the reaction is actually assisted by proteins *in vivo*). For both group I and group II introns, the reaction requires formation of a specific secondary/tertiary structure involving short consensus sequences. Group I intron RNA creates a structure in which the substrate sequence is held by the IGS region of the intron, and other conserved sequences generate a guanine nucleotide binding site. It occurs by a transesterification involving a guanosine residue as cofactor. No input of energy is required. The guanosine breaks the bond at the 5′ exon–intron junction and becomes linked to the intron; the hydroxyl at the free end of the exon then attacks the 3′ exon–intron junction. The intron cyclizes and loses the guanosine and the terminal 15 bases. A series of related reactions can be catalyzed via attacks by the terminal G-OH residue of the intron on internal phosphodiester bonds. By providing appropriate substrates, it has been possible to engineer ribozymes that perform a variety of catalytic reactions, including nucleotidyl transferase activities.

Some group I and group II mitochondrial introns have open reading frames. The proteins encoded by group I introns are endonucleases that make double-stranded cleavages in target sites in DNA; the cleavage initiates a gene conversion process in which the sequence of the intron itself is copied into the target site. The proteins coded by group II introns include an endonuclease activity that initiates the transposition process, and a reverse transcriptase that enables an RNA copy of the intron to be copied into the target site. These types of intron probably originated by insertion events. The proteins encoded by both groups of introns may include maturase activities that assist splicing of the intron by stabilizing the formation of the secondary/tertiary structure of the active site.

Virusoid RNAs can undertake self-cleavage at a "hammerhead" structure. Hammerhead structures can form between a substrate RNA and a ribozyme RNA, which allows cleavage to be directed at highly specific sequences. These reactions support the view that RNA can form specific active sites that have catalytic activity.

RNA editing changes the sequence of an RNA after or during its transcription. The changes are required to create the final coding sequence. Substitutions of individual bases occur in mammalian systems; they take the form of deaminations in which C is converted to U, or A is converted to I. A catalytic subunit related to cytidine deaminase or adenosine deaminase functions as part of a larger complex that has specificity for a particular target sequence.

Additions and deletions (most often of uridine) occur in trypanosome mitochondria and in paramyxoviruses. Extensive editing reactions occur in trypanosomes in which as many as half of the bases in an mRNA are derived from editing. The editing reaction uses a template consisting of a guide RNA that is complementary to the mRNA sequence. The reaction is catalyzed by an enzyme complex that includes an endonuclease, terminal uridyltransferase, and RNA ligase, using free nucleotides as the source for additions, or releasing cleaved nucleotides following deletion.

Protein splicing is an autocatalytic reaction that occurs by bond transfer reactions without requiring the input of energy. The intein catalyzes its own splicing out of the flanking exteins. Many inteins have a homing endonuclease activity that is independent of the protein splicing activity.

CHAPTER QUESTIONS

1. Which class of introns has been modified to carry out catalytic activities other than the original activity?
   A. group I introns
   B. group II introns
   C. nuclear introns
   D. both group I and group II introns
2. What base is most commonly inserted into transcripts of unicellular eukaryote mitochondria by RNA editing?
   A. A
   B. U
   C. G
   D. C
3. Which class of introns is found in bacteria and rRNA genes in unicellular eukaryotes?
   A. group I introns
   B. group II introns
   C. nuclear introns
   D. both group I and group II introns
4. Group I introns, after excision from the precursor RNA:
   A. form lariat structures.
   B. form Y-shaped structures.
   C. form circular structures.
   D. remain linear.
5. As a required cofactor for group I intron splicing, a guanine nucleotide with a ________ group must be present.
   A. 5′ phosphate
   B. 3′ phosphate
   C. 5′ OH
   D. 3′ OH
6. A subset of which class(es) of introns contain(s) mobile introns that encode endonucleases that facilitate mobility of the intron to new sites?
   A. group I introns
   B. group II introns
   C. nuclear introns
   D. group I and group II introns
7. The required sequence elements needed for group II intron splicing are:
   A. short conserved sequences at the very ends of the intron.
   B. conserved sequence elements in the center of the intron.
   C. a characteristic secondary structure based on conserved internal inverted repeats.
   D. no conserved sequence elements or secondary structure.

8. Small infectious plant RNAs that function independently without encapsidation are called:
   A. viruses.
   B. viroids.
   C. virusoids.
   D. satellite RNAs.
9. Editing of apolipoprotein-B transcripts in mammalian intestine and liver involves:
   A. conversion of a U to a C.
   B. conversion of a C to a U.
   C. insertion of several U bases.
   D. deletion of several C bases.
10. RNA editing in trypanosome mitochondria involves:
   A. conversion of a U to a C.
   B. conversion of a C to a U.
   C. insertion or deletion of several U bases.
   D. insertion or deletion of several C bases.

KEY TERMS

autosplicing (self-splicing)
extein
guide RNA
intein
intron homing
maturase
protein splicing
RNA editing
riboswitch
ribozyme
viroid
virusoid

FURTHER READING

Cech, T. R. (1990). Self-splicing of group I introns. *Annu. Rev. Biochem.* 59, 543–568.
A review of the mechanism of this process.

Doherty, E. A. and Doudna, J. A. (2000). Ribozyme structures and mechanisms. *Annu. Rev. Biochem.* 69, 597–615.
A comparison of the structures of four ribozymes.

Lambowitz, A. M. and Zimmerly, S. (2004). Mobile group II introns. *Annu. Rev. Genet.* 38, 1–35.
A review of the molecular mechanisms for the mobility of group II introns and the evolutionary relationship between them, eukaryotic retroposons, and eukaryotic spliceosomal introns.

Liu, X.-Q. (2000). Protein-splicing intein: genetic mobility, origin, and evolution. *Annu. Rev. Genet.* 34, 61–76.
How the structure and splicing mechanism of inteins suggest the origin of these elements.

Paulus, H. (2000). Protein splicing and related forms of protein autoprocessing. *Annu. Rev. Biochem.* 69, 447–496.
A review of the process of protein splicing and a comparison to protein autoprocessing.

Winkler, W. C., Nahvi, A., Roth, A., Collins, J. A., and Breaker, R. R. (2004). Control of gene expression by a natural metabolite-responsive ribozyme. *Nature* 428, 281–286.
A report of the discovery of a new class of ribozyme that inhibits expression of an enzyme by cleaving its mRNA. Through a negative feedback system, the ribozyme is activated by a metabolic product of the enzyme.

30

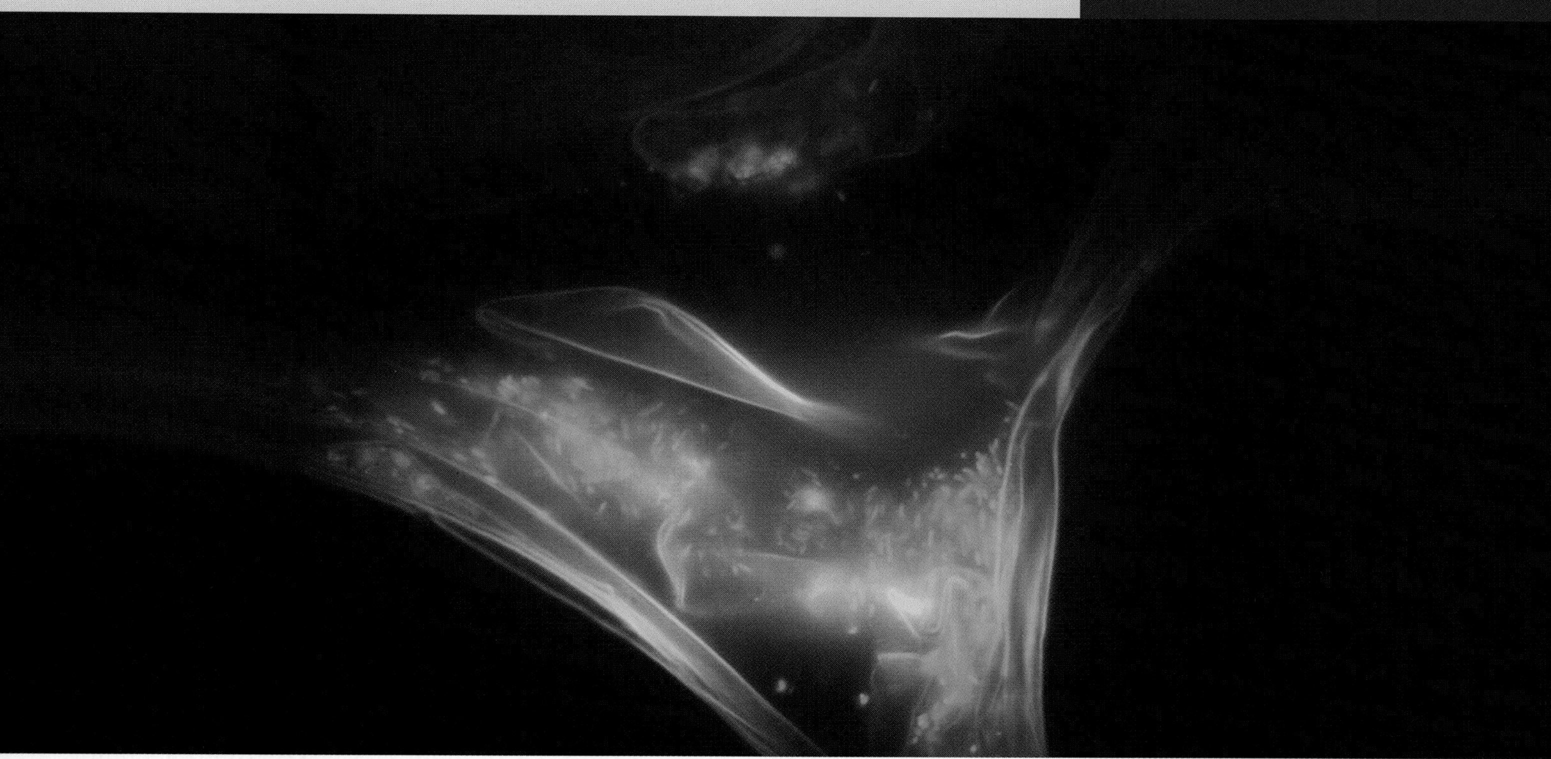

Wheat roots colonized with the bacterium *Azospirillum lipoferum* expressing the green fluorescent protein (GFP) reporter gene. With kind permission from Springer Science+Business Media: *Biol. Fertil. Soils*, "Mitigation of salt stress in wheat seedlings...," vol. 40, 2004, Bacilio, M., et al., figure 1c. Photo courtesy of Yoav Banshan, The Northwestern Center for Biological Research, Mexico.

Genetic Engineering

CHAPTER OUTLINE

- **restriction endonuclease** An enzyme that recognizes a specific short sequence of DNA and cleaves the duplex (sometimes at the target site, sometimes elsewhere, depending on the type of enzyme).
- **cloning vector** DNA (often derived from a plasmid or a bacteriophage genome) that can be used to propagate an incorporated DNA sequence in a host cell; vectors contain selectable markers and replication origins to allow identification and maintenance of the vector in the host.

30.1 Introduction

The field of molecular biology was dramatically advanced by the development of tools and methods that allow the direct manipulation of DNA both *in vitro* and *in vivo* in numerous organisms. Two essential items in the molecular biologist's toolkit are **restriction endonucleases**, which allow DNA to be cut into precise pieces, and **cloning vectors**, such as plasmids or phages used to "carry" inserted foreign DNA fragments for the purposes of producing more material or a protein product. *Genetic engineering* was originally used as a term to describe the range of manipulations of DNA that become possible with the ability to clone a gene by placing its DNA into another context in which it could be propagated. From this beginning, when recombinant DNA was used as a tool to analyze gene structure and expression, we moved to the ability to change the DNA content of bacteria and eukaryotic cells by directly introducing cloned DNA that could become part of the genome. Then, by changing the genetic content in conjunction with the ability to develop an animal from an embryonic cell, it became possible to generate mice with deletions or additions of specific genes that are inherited via the germline. We now use genetic engineering to describe a range of activities including the manipulation of DNA, the introduction of changes into specific somatic cells within an animal, and even changes in the germline itself.

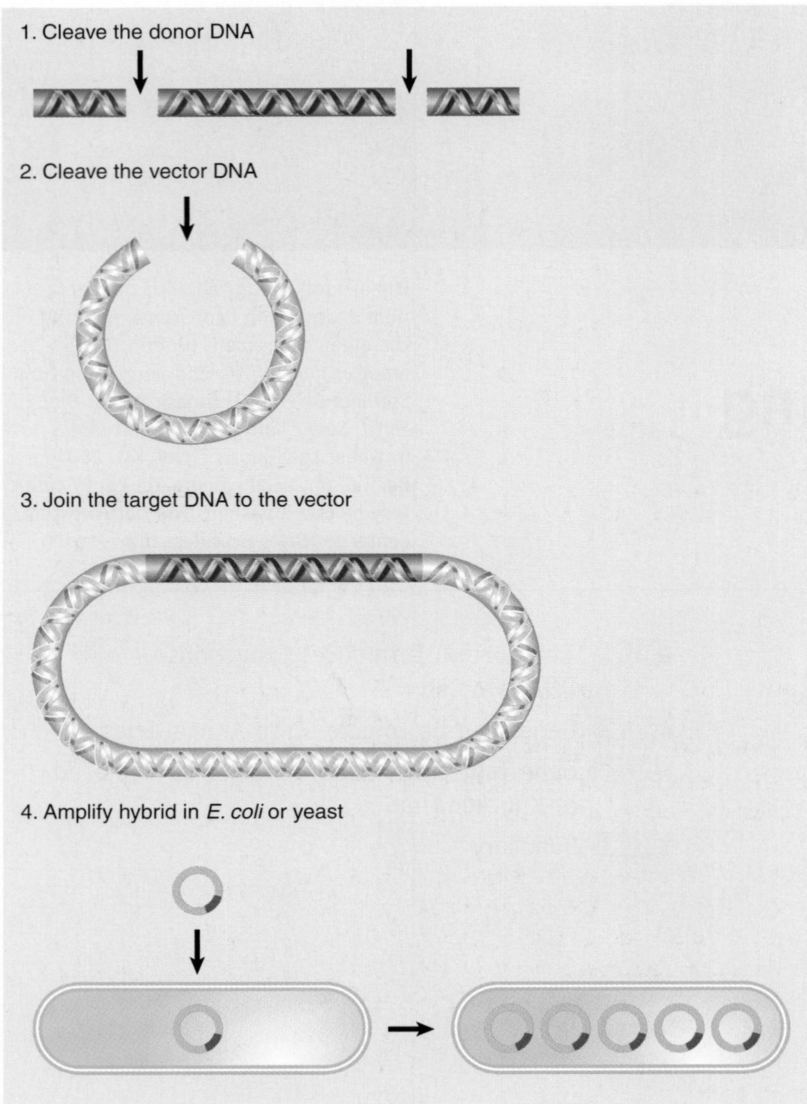
1. Cleave the donor DNA
2. Cleave the vector DNA
3. Join the target DNA to the vector
4. Amplify hybrid in E. coli or yeast

FIGURE 30.1 A donor DNA is generated by cleavage so that its ends can be joined to the ends of a cloning vector. The hybrid molecule is then introduced into a bacterium or yeast in which it can replicate to produce many more copies.

FIGURE 30.1 summarizes the steps in cloning a sequence of DNA. The basic principle is that the DNA of interest is inserted into a cloning vector, a vehicle that carries it and amplifies it within the target cell. In principle, the reaction is accomplished by cutting the donor DNA to release a fragment containing a gene or some other sequence, cutting a vector DNA (shown in the figure as a circular plasmid), and then joining the ends crosswise to generate a hybrid DNA molecule. The vector is chosen to carry an origin of replication that will let it perpetuate itself in an appropriate host, most often *E. coli* or *S. cerevisiae*. After amplification, hosts carrying the hybrid vector are selected, and the DNA can be isolated.

FIGURE 30.2 gives an overview of the stages used in changing the genome of a target cell. First the DNA that we want to introduce must be cloned as shown in the previous figure. Then we need a delivery method to introduce the cloned DNA into the cell, and it must either be placed directly in the nucleus (if the host cell is eukaryotic) or be transferred there. Once in the nucleus, the cloned DNA must be integrated into the host chromosome, typically by a recombination event. Depending on the system used, the DNA of the cloning vector may or may not be integrated with the desired donor DNA sequence.

There is now quite a wide range of methods for introducing DNA into target cells. Bacteria and simple eukaryotes like yeast can be induced to take up DNA easily, using chemical

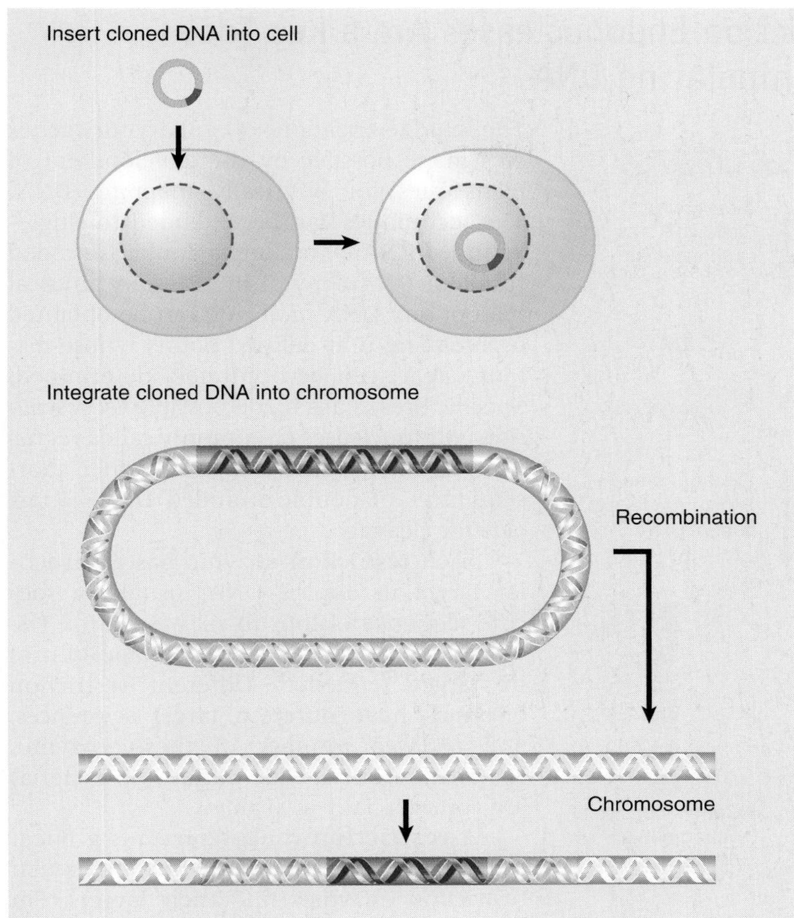

FIGURE 30.2 Introducing new DNA into the genome requires cloning the donor sequence, delivery of the cloned DNA into the cell, and integration into the genome.

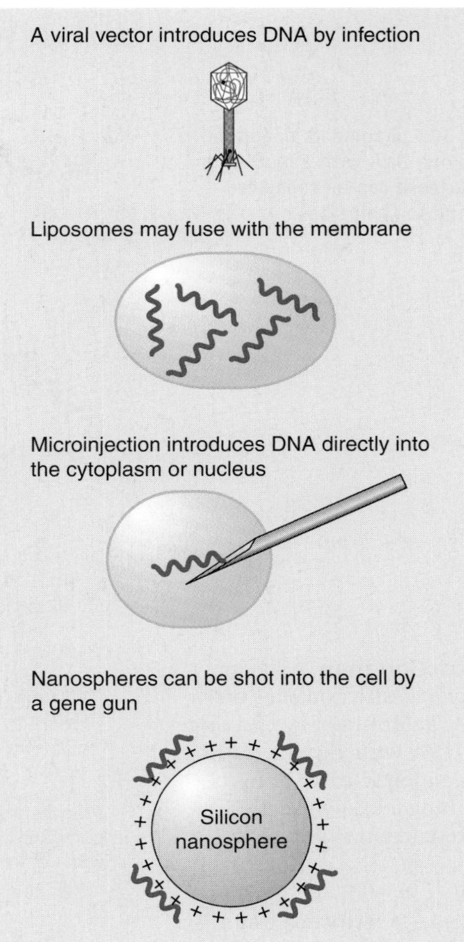

A viral vector introduces DNA by infection
Liposomes may fuse with the membrane
Microinjection introduces DNA directly into the cytoplasm or nucleus
Nanospheres can be shot into the cell by a gene gun
Silicon nanosphere

FIGURE 30.3 DNA can be released into target cells by methods that pass it across the membrane naturally, such as by means of a viral vector (in the same way as a viral infection) or by encapsulating it in a liposome (which fuses with the membrane). Or it can be passed manually, by microinjection, or by coating it on the exterior of nanoparticles that are shot into the cell by a gene gun that punctures the membrane at very high velocity.

treatments that permeabilize the cell membranes. These treated cells are then *transformed* by incubation with the DNA. However, many types of cells cannot be transformed so easily, and other methods must be used, as summarized in FIGURE 30.3. Some types of cloning vectors use natural methods of infection to pass the DNA into the cell, such as a viral vector that uses the viral infective process to enter the cell. *Liposomes* are small spheres made from artificial membranes, which can contain DNA or other biological materials. Liposomes can fuse with plasma membranes and release their contents into the cell. *Microinjection* uses a very fine needle to puncture the cell membrane. A solution containing DNA can be introduced into the cytoplasm, or directly into the nucleus in the case where the nucleus is large enough to pick out as a target (such as an egg). The thick cell walls of plants are an impediment to many transfer methods, and the "gene gun" was invented as a means for overcoming this obstacle. A gene gun shoots very small particles into the cell by propelling them through the wall at high velocity. The particles can consist of gold or nanospheres coated with DNA. This method now has been adapted for use with a variety of species, including mammalian cells.

30.2 Restriction Endonucleases Are a Key Tool in Manipulating DNA

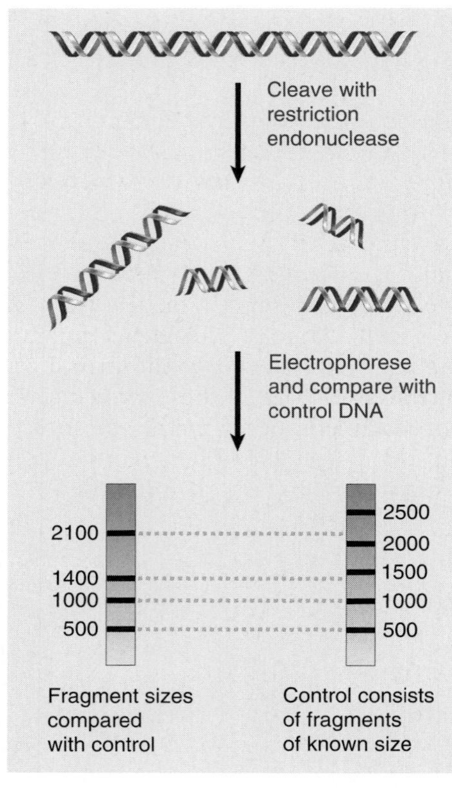

FIGURE 30.4 Fragments generated by cleaving DNA with a restriction endonuclease can be separated according to their sizes.

The characterization of eukaryotic genes was made possible by the development of techniques for physically mapping DNA. The techniques can be extended to single-stranded RNA by making a double-stranded (duplex) DNA copy of the RNA. A physical map of any DNA molecule can be obtained by breaking it at defined points whose distance apart can be accurately determined. Specific breaks are made possible by restriction endonucleases (commonly called *restriction enzymes*), which recognize rather short sequences of double-stranded DNA as targets for cleavage.

Each restriction enzyme has a particular target in duplex DNA, usually a specific sequence of four to six base pairs. The enzyme cuts the DNA at every location of its target sequence. Different restriction enzymes have different target sequences, and a large number of these enzymes (obtained from a wide variety of bacteria) are commercially available.

▸ **restriction map** A linear array of restriction sites on DNA, determined by cleaving the DNA with various restriction endonucleases or by scanning a known sequence for restriction sites.

A **restriction map** represents a linear sequence of the sites at which particular restriction enzymes find their targets. For short distances, the distance along such maps is measured directly in base pairs (bp). Longer distances are given in kilobases (kb), which correspond to kilobase (10^3) pairs in DNA or to kilobases in RNA. At the level of the chromosome, the scale of a map is in megabase pairs (1 Mb = 10^6 bp).

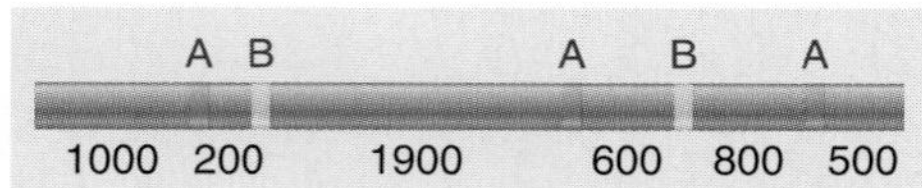

FIGURE 30.5 A restriction map is a linear sequence of sites separated by defined distances on DNA. The map identifies the sites cleaved by enzymes A and B, as defined by the individual fragments produced by the single and double digests.

When a DNA molecule is cut with a restriction enzyme, it is cleaved into distinct fragments. These fragments can be separated on the basis of their size by gel electrophoresis, as shown in **FIGURE 30.4**. The cleaved DNA is placed in a gel made of agarose or polyacrylamide. When an electric current is passed through the gel, each fragment moves through it at a rate that is determined by its size. This movement produces a series of bands in the gel, each of which corresponds to a fragment of a particular size, with smaller fragments moving farther in the gel.

By determining the restriction fragments of a section of DNA, we can generate a map of the original molecule in the form shown in **FIGURE 30.5**. The map shows the positions at which particular restriction enzymes cut DNA, with the distances between the sites measured in base pairs.

Cleavage by restriction enzymes (also called *digestion*) results in the generation of defined fragments of DNA that can be used in cloning, as described in the next section (*Section 30.3, Cloning Vectors Are Used to Amplify Donor DNA*).

KEY CONCEPTS

- Restriction endonucleases can be used to cleave DNA into defined fragments.
- A map can be generated by using the overlaps between the fragments generated by different restriction enzymes.

CONCEPT AND REASONING CHECK

What would be the advantages of the naturally produced restriction endonucleases in bacterial cells?

30.3 Cloning Vectors Are Used to Amplify Donor DNA

The basic principle of gene **cloning** is to insert the DNA of interest into a cloning vector, which then replicates to produce many more copies. The key issues are how we identify the donor DNA, what means we use to insert it into the vector, how we select the hybrid molecule that results from the insertion, and the method of amplification.

▸ **cloning** Propagation of a DNA sequence by incorporating it into a hybrid construct that can be replicated in a host cell.

The ability to identify donor DNA and insert it into a vector has benefited enormously from genome sequencing (see the accompanying *Methods and Techniques* box). Instead of using various indirect techniques, we can work directly on the basis of the sequence. One of the early cloning methods is well adapted to this development. In this method, the donor DNA is cleaved with a restriction enzyme that introduces double-strand breaks by making cuts in the two strands of DNA. If these two cuts are staggered, this generates single-stranded ends on each side of the break, as shown in **FIGURE 30.6**. Depending on the pattern of cutting, the protruding ends may be 3′ or 5′ ends. Restriction enzymes recognize specific sequences that are palindromic (that

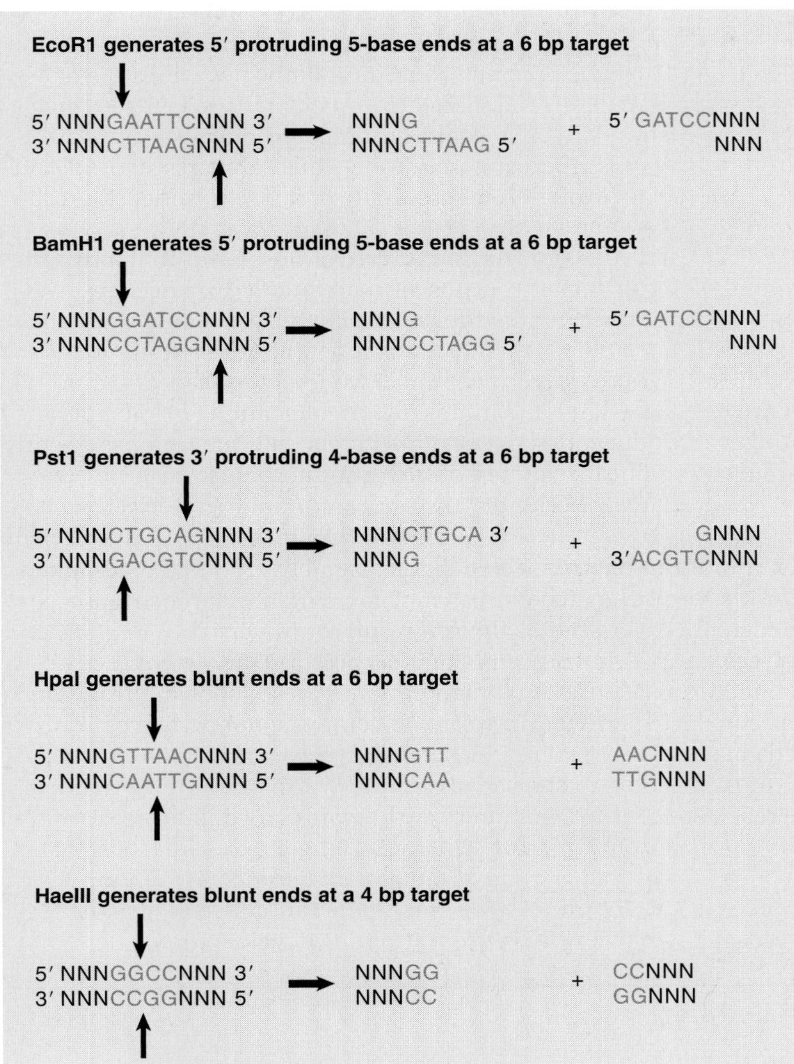

FIGURE 30.6 Each restriction enzyme cleaves a specific target sequence (usually 4–6 bp long). The sites of cleavage on the two strands may generate 5′ protruding ends, 3′ protruding ends, or blunt ends.

METHODS AND TECHNIQUES

$1000 Genome

We all know the label "one size fits all" is not exactly accurate. Although most people can wear an item so labeled, there are always some for whom it is too small or too large. The same holds true for medicine, but it isn't as easy to predict who will respond to a treatment and who won't. Hence, the idea of "personalized" medicine, in which therapies can be targeted to individuals based on their genetic makeup or even the changes that have taken place in the DNA of tumor cells, is very attractive.

But we're not there yet. The human genome is composed of 3 billion bases. The rough draft of the genome, based on sequencing a mixture of samples from a number of people, took many years at an estimated cost of $300 million. For individual human sequencing to be viable in the clinic, the technology is going to have to cost about $1000 and take only a few days to complete.

The method of DNA sequencing hasn't changed much since Frederick Sanger and colleagues developed a technique in 1977 called *dideoxy sequencing*. This method requires many identical copies of the DNA, a primer that is complementary to a short stretch of the DNA, DNA polymerase, deoxynucleotides (dNTPs: dATP, dCTP, dGTP, and dTTP), and dideoxynucleotides (ddNTPs). Dideoxynucleotides are modified nucleotides that can be incorporated into the growing DNA strand but lack the hydroxyl group needed to attach the next nucleotide. Thus their incorporation terminates the synthesis reaction. The ddNTPs are added at much lower concentrations than the normal nucleotides so that they are incorporated at low rate, often only after synthesis has proceeded normally for a strand length of up to several hundred nucleotides.

Originally, four separate reactions were necessary, with a single different ddNTP added to each one. The reason for this was that the strands were labeled with radioisotopes and could not be distinguished from each other on the basis of the label. Thus, the reactions were loaded into adjacent lanes on a denaturing acrylamide gel and separated by electrophoresis at a resolution that distinguished between strands differing by a length of one nucleotide. The gel was transferred to a solid support, dried, and exposed to a film. The results were read from top to bottom, with a band appearing in the ddATP lane indicating that the strand terminated with an adenine, the next band appearing in the ddTTP lane indicating that the next base was a thymine, and so on.

Two recent modifications have aided in the automation and scaling up of the procedure. The incorporation of a different fluorescent label for each ddNTP allows a single reaction to be run that is read as the strands are hit with a laser and pass by an optical sensor. The information as to which ddNTP terminated the fragment is fed directly into a computer. The second modification is the replacement of large slabs of polyacrylamide gels with very thin, long glass capillary tubes filled with gel. These tubes can dissipate heat more rapidly, allowing the electrophoresis to be run at a higher voltage, greatly reducing the time required for separation. A schematic illustrating this process is shown in **FIGURE B30.1**.

To get from a single sequencing reaction to the sequence of the entire human genome, two approaches were taken initially. In the publically funded human genome project, sets of human DNA inserted into bacterial artificial chromosomes (BACs) were positioned on a physical map of the genome that had been derived by a variety of techniques. Then small sets of overlapping BACs were chosen for a given sequencing run. Their DNA inserts were randomly cleaved into smaller fragments and cloned into vectors from which single-stranded DNA could be prepared and used in a dideoxy sequencing reaction. Thus, this research group knew roughly where their sequence output fell on the genome map.

By contrast, the private effort by the company Celera skipped the step of mapping BAC clones and simply fragmented the entire genome for sequencing. They assembled their sequence without having any idea where a given sequence was likely to fall.

Why might this be problematic? Assembling the individual sequences into larger runs that ultimately give a linear sequence of each chromosome is no easy matter, largely because there are confounding elements such as long repeats in the genome. Computer algorithms played an important role in the assembly process, but the question of whether the whole genome approach used by Celera actually works is still controversial.

As you can imagine, the efforts to derive the initial draft of the human genome were labor-intensive and expensive. How can we get to the point where it is feasible to routinely sequence everyone's genome? In an effort to spur advances toward this goal, the Archon X Prize for Genomics will award $10 million to the first group that can sequence 100 genomes in 10 days at a cost of less than $10,000 per genome.

A number of "next-generation" sequencing technologies are currently under development. These aim to eliminate the need for time-consuming gel separation and reliance on human labor. Sequencing-by-synthesis and sequencing through nanopores are two of the many technologies being explored right now.

Sequencing-by-synthesis relies on the detection and identification of each nucleotide as it is added to a growing strand. In one such application, the primer is tethered to a glass surface and the complementary DNA to be sequenced anneals to the primer. Sequencing proceeds by adding polymerase and fluorescently labeled nucleotides individually, washing away any unused dNTPs. After illuminating with a laser, the nucleotide that has been incorporated into the DNA strand can be detected. Other versions use nucleotides with reversible termination, so that only one nucleotide can be incorporated at a time even if there is a stretch of homopolymeric DNA (e.g., a run of adenines). Still another version, called *pyrosequencing,* detects the release of pyrophosphate from the newly added base. Although these technologies are still under development, they have the advantage that massively parallel reactions can be run.

A completely different approach aims to detect individual nucleotides as a DNA sequence is run through a silicone nanopore. Tiny transistors are used to control a current passing through the pore. As a nucleotide passes through the pore, it disturbs the current in a manner unique to its chemical structure. This technology would have the advantage of reading DNA by simply using electronics, with no chemistry or optical detection required. Nevertheless, there are many kinks to work out before it becomes feasible.

Unlike many breakthroughs, which seem to lie forever on the horizon, the $1000 genome is clearly achievable in the near future. It does not rely on understanding the complexity of the human body, which routinely foils drug development efforts. In fact, many in the community are urging policymakers to act now to ensure that this information is not exploited or misused when it becomes available. And even though the medical advances that are able to make use of the knowledge of how we differ may lag behind, the availability of large numbers of individual genomes is likely to spur advances in our understanding of disease processes and differing susceptibilities.

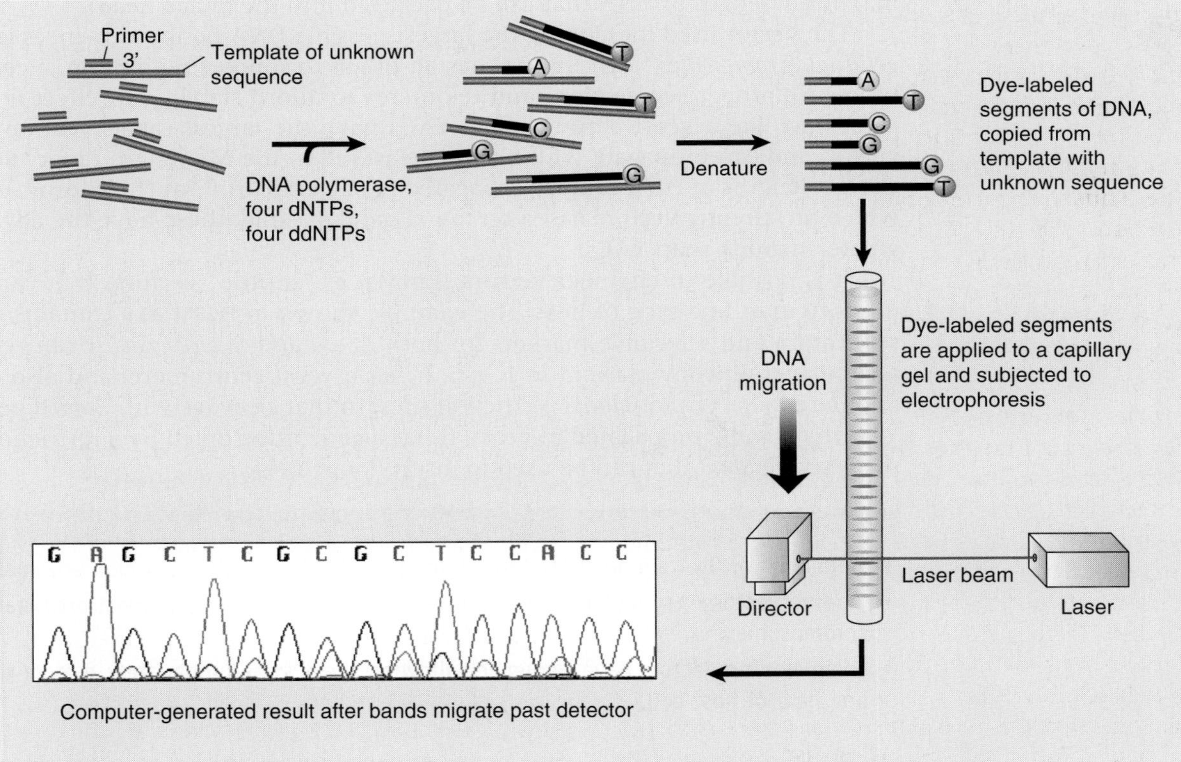

FIGURE B30.1 DideoxyNTP sequencing using fluorescent tags. Inset photo courtesy of Jan Kieleczawa.

is, the sequence is the same read 5′ to 3′ on either strand). As a result, the single-stranded ends on either side of the break are complementary. These are often referred to as "sticky" ends, because they are available for base pairing with a complementary overhang. Some restriction enzymes cleave at the same point on both strands, and therefore generate "blunt" ends.

The vector is cleaved with the same restriction enzyme(s) as the donor DNA. **FIGURE 30.7** shows that the complementary single-stranded ends can then hybridize to join the donor and vector DNAs. By using different enzymes to cleave each end of the donor sequence, we can ensure that the insertion occurs in only one orientation. Ligase, the same enzyme used to seal DNA nicks during replication and repair, is used to covalently join the vector and donor DNAs. Ligase will also join blunt ends together, although less efficiently without the stabilizing effect of hybridized single-stranded ends.

The design of the cloning vector is determined by the nature of the experiment. **FIGURE 30.8** summarizes the properties of the most common classes of cloning vectors. Cloning vectors are usually propagated in bacteria or yeast. Plasmids can be easily engineered to carry sequences of donor DNA in bacteria, but plasmids with large inserts (greater than 10 kb) become difficult to isolate and to transform. Bacteriophages can be used in bacteria but have the disadvantage that only a limited amount of DNA can be packaged into the viral coat (although more than can be carried in plasmid). The advantages of plasmids and phages are combined in the **cosmid**, which propagates like a plasmid but uses the packaging mechanism of phage lambda to deliver the DNA to the bacterial cells. Lambda DNA ends in *cos* sites that are recognized by the packaging system, which cuts them and inserts the DNA into the phage head. By incorporating *cos* sites into a cloning vector, the lambda system can be used to package the DNA and transfer it with high efficiency to the bacteria. It is limited to containing 47 kb (the maximum length of DNA that can be packaged into the phage head).

▸ **cosmid** Cloning vector derived from a bacterial plasmid by incorporating the *cos* sites of phage lambda, which make the plasmid DNA a substrate for the lambda packaging system.

The vector used for cloning the largest possible DNA donor sequences is the *yeast artificial chromosome* (YAC). A YAC has an origin to support replication, a centromere to ensure proper segregation, and telomeres to afford stability. In effect it is propagated just like a yeast chromosome. YACs have the largest capacity of any cloning vector, and can propagate with inserts measured in the Mb length range. Somewhat shorter lengths of DNA can be propagated in bacterial artificial chromosomes (BACs), which are cloning vectors based on the *E. coli* F plasmid; these have the advantage of greater stability than YACs.

It is possible to engineer vectors (known as "shuttle" vectors) that can be used in more than one type of host. The example shown in **FIGURE 30.9** contains origins of replication and selectable markers for both *E. coli* and *S. cerevisiae*. It can replicate as a circular multicopy plasmid in *E. coli*. It has a yeast centromere, and also has yeast telomeres adjacent to *Bam*HI restriction sites, so that cleavage with *Bam*HI generates a YAC that can be propagated in yeast. It has a gene interrupted by a restriction site for the enzyme *Sma*I that can be used to insert donor DNA.

KEY CONCEPTS

- Fragments of DNA are generated for cloning by cleavage with restriction enzymes.
- Cloning vectors may be bacterial plasmids, phages, or cosmids, or yeast artificial chromosomes.
- Some cloning vectors have sequences allowing them to be propagated in more than one type of host cell.

CONCEPT AND REASONING CHECK

If you digest a vector and donor DNA with restriction enzymes that make blunt ends, do you have to use the same enzyme for both vector and donor DNA? Why or why not?

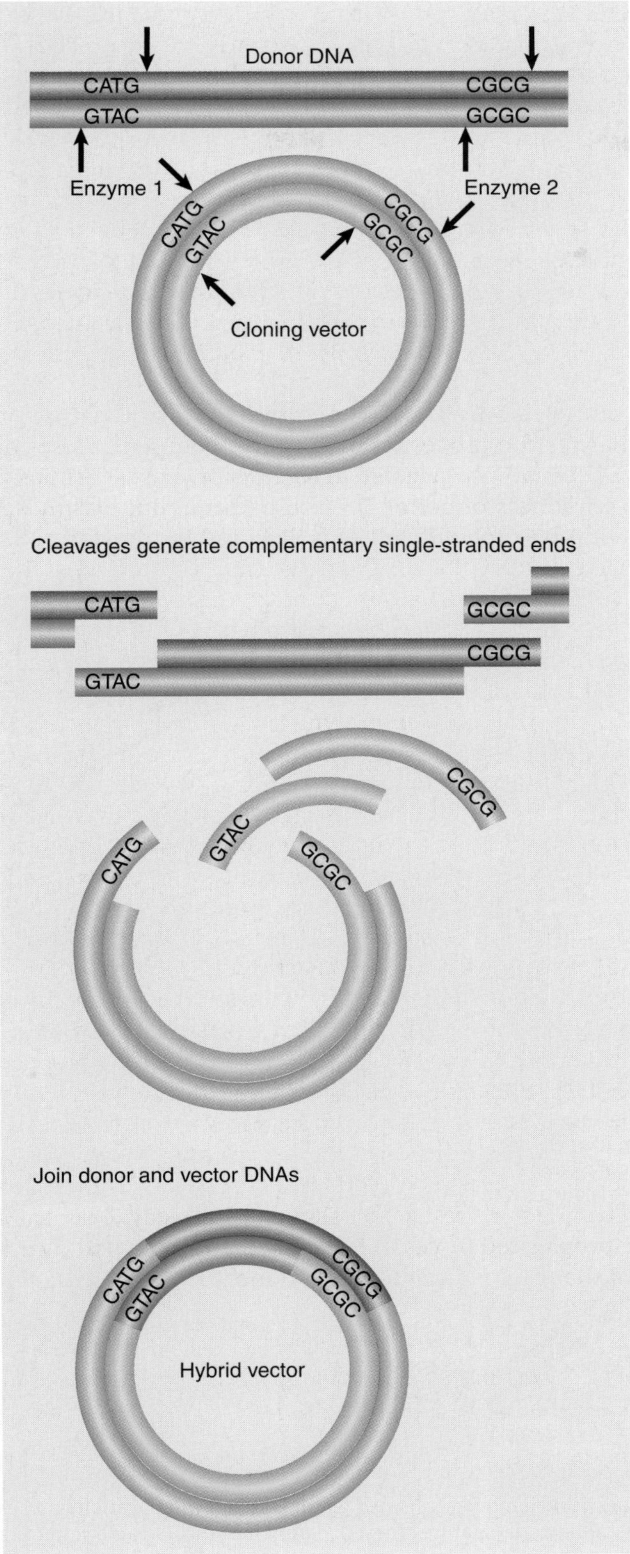

FIGURE 30.7 Restriction enzymes can generate complementary single-stranded ends in donor and vector DNA by making staggered breaks in their target sequences. Crosswise joining of the fragments generates a hybrid DNA molecule.

Vector	Features	Isolation of DNA	DNA limit
Plasmid	High copy number	Physical	10 kb
Phage	Infects bacteria	Via phage packaging	20 kb
Cosmid	High copy number	Via phage packaging	48 kb
BAC	Based on F plasmid	Physical	300 kb
YAC	Origin + centromere+ telomere	Physical	>1 Mb

FIGURE 30.8 Cloning vectors may be based on plasmids or phages or may mimic eukaryotic chromosomes.

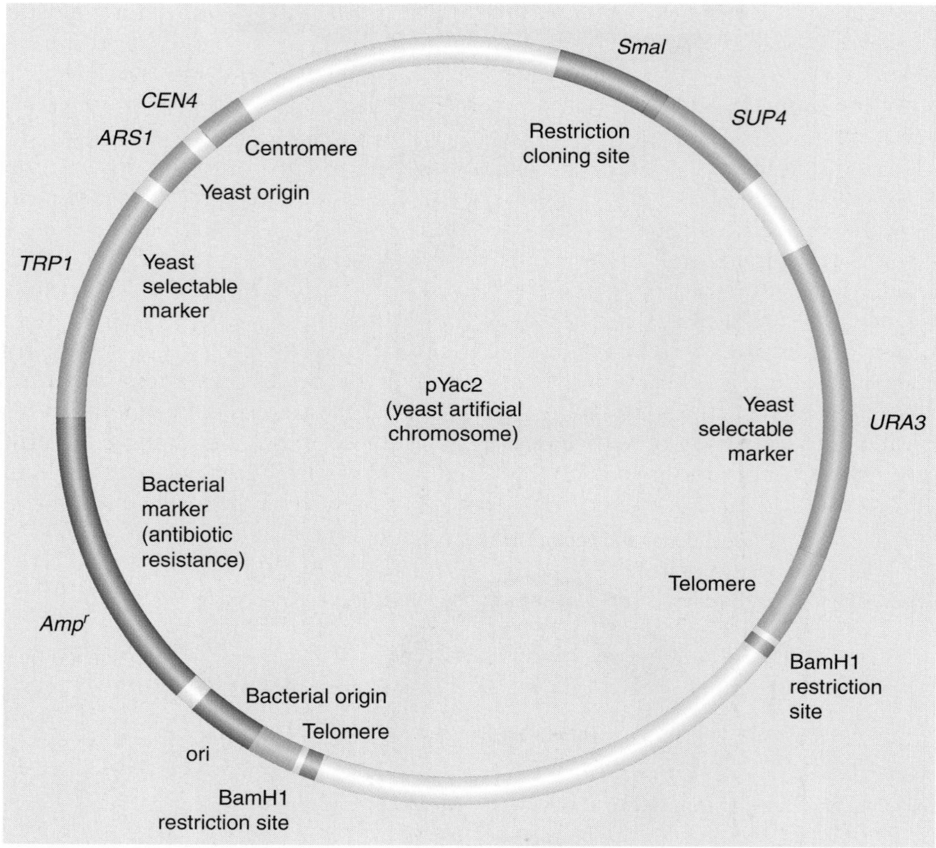

FIGURE 30.9 pYac2 is a cloning vector with features to allow replication and selection in both bacteria and yeast. Bacterial features (described in blue) include an origin of replication and antibiotic resistance gene. Yeast features (described in red) include an origin, centromere, two selectable markers, and telomeres.

30.4 Cloning Vectors Can Be Specialized for Different Purposes

Cloning vectors are now customarily designed for particular purposes. Aside from the issue of which type of host cell they use, the main questions are whether they are intended simply to amplify the cloned sequence, to investigate the properties of a promoter (or other regulatory elements), or to express the protein product of a cloned gene.

One feature that all vectors must contain is a *selectable marker* to allow identification of cells that have taken up the vector, and ideally an additional mechanism to distinguish the hybrid vector, carrying the donor DNA, from the original vector used for cloning. (Not all copies of the cloning vector will actually get an insertion.) One approach is to use a cloning vector that carries two resistance genes to two different antibiotics, as shown in **FIGURE 30.10**. The site where the donor DNA is inserted lies within one of the resistance genes, and the insertion therefore inactivates the gene. The other resistance gene is unaffected by the insertion. The original cloning vector therefore makes its host resistant to both antibiotics, but the hybrid vector is resistant to only one. So we can perpetuate the cloning vector itself by selecting bacteria that are resistant to either of the antibiotics. (We need to apply selection because otherwise the vector is lost from some bacteria.) After insertion of donor DNA, we can identify bacteria carrying the hybrid vector because they are resistant to one of the antibiotics but sensitive to the other. Another popular method, known as "blue-white selection," is to insert the donor DNA into the middle of the *lacZ* gene, which encodes the enzyme β-galactosidase (β-gal), and clone into *lacZ*⁻ cells. β-gal converts a substrate called Xgal into a blue product; vectors that have received an insert will therefore produce white colonies, whereas "empty" vectors will result in blue colonies.

When our main purpose is to examine the regulation of a gene, usually the crucial issue is how initiation of transcription is controlled at the promoter. In many cases, it would be very difficult to assay the product of the gene. The technique for dealing with this is to connect the promoter to a **reporter gene** whose product can be easily assayed. Usually the function of the promoter is not affected by sequences in the coding region, so from the perspective of regulation, it is quite irrelevant what gene the promoter is connected to. This allows us freedom to design constructs in which the reporter codes for a gene that can be easily assayed in the circumstances in question.

▸ **reporter gene** Any gene whose product (RNA or protein) can be easily detected; reporter genes are often used as a substitute for a gene of interest.

The type of reporter gene that is most appropriate depends on whether we are interested in quantitating the efficiency of the promoter (and, for example, determining the effects of mutations in it or the activities of transcription factors that bind to it) or determining its tissue-specific pattern of expression. **FIGURE 30.11** summarizes a

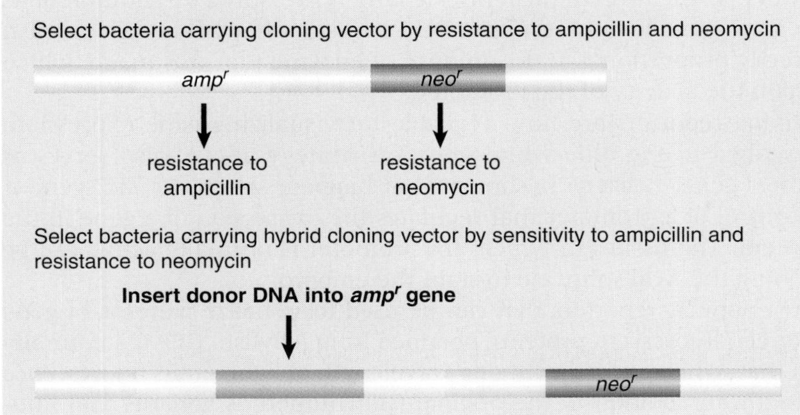

FIGURE 30.10 Bacteria carrying a cloning vector can be selected by resistance to an antibiotic marker carried by the vector. Bacteria carrying a hybrid cloning vector with an insertion of donor DNA can be distinguished by loss of activity of the vector gene into which the insertion was made.

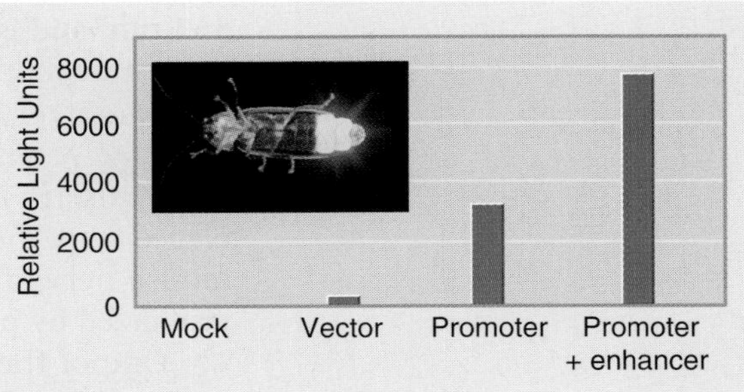

FIGURE 30.11 Luciferase (derived from fireflies such as the one shown here) is a popular reporter gene. The graph shows the results from mammalian cells transfected with a luciferase vector driven by a minimal promoter or the promoter plus a putative enhancer. The levels of luciferase activity correlate with the activities of the promoters. Photo © Cathy Keifer/Dreamstime.com.

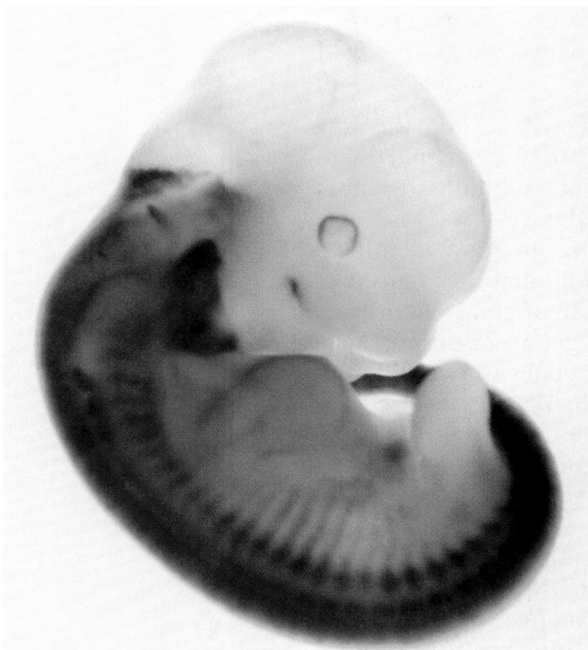

FIGURE 30.12 Expression of a *lacZ* gene can be followed in the mouse by staining for β-galactosidase (in blue). In this example, *lacZ* was expressed under the control of a promoter of a mouse gene that is expressed in the nervous system. The corresponding tissues can be visualized by blue staining. Photo courtesy of Robb Krumlauf, Stowers Institute for Medical Research.

(a)

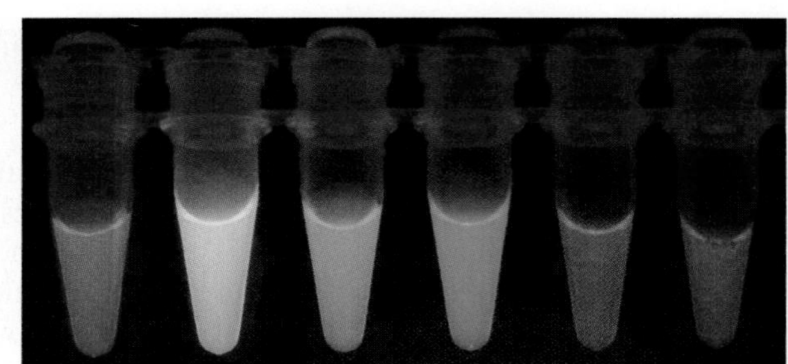

(b)

FIGURE 30.13 (**a**) Since the discovery of GFP, derivatives that fluoresce in different colors have been engineered. Photo courtesy of Joachim Goedhart, Molecular Cytology, SILS, University of Amsterdam. (**b**) A live transgenic mouse expressing human rhodopsin (a protein expressed in the retina of the eye) fused to GFP. Reproduced from *Vision Res.*, vol. 45, Wensel, T. G., et al., Rhodopsin-EGFP knock-ins..., pp. 3445–3453. Copyright 2005, with permission from Elsevier [http://www.sciencedirect.com/science/journal/00426989]. Photo courtesy of Theodore G. Wensel, Baylor College of Medicine.

common system for assaying promoter activity. A cloning vector is created that has a eukaryotic promoter linked to the coding region of *luciferase*, a gene that encodes the enzyme responsible for bioluminescence in the firefly. Usually an end termination signal is added to ensure the proper generation of the mRNA. The hybrid vector is introduced into target cells, and the cells are grown and subjected to any appropriate experimental treatments. The level of luciferase activity is measured by addition of its substrate luciferin. Luciferase activity results in light emission that can be measured at 562 nm, and is directly proportional to the amount of enzyme that was made, which in turn depends upon the activity of the promoter.

Some very striking reporters are now available for visualizing gene expression. The *lacZ* gene, described in the blue-white selection strategy above, also serves as a very useful reporter gene. **FIGURE 30.12** shows what happens when the *lacZ* gene is placed under the control of a promoter that regulates the expression of a gene in the mouse nervous system. The tissues in which this promoter is normally active can be visualized by providing the Xgal substrate to stain the embryo.

One of the most popular reporters that can be used to visualize patterns of gene expression is GFP (green fluorescent protein), obtained from jellyfish. GFP is a naturally fluorescent protein that, when excited with one wavelength of light, emits fluorescence in another wavelength. In addition to the original GFP, numerous variants that fluoresce in different colors, such as yellow (YFP), cyan (CFP) and blue (BFP), have been developed. GFP and its variants can be used as reporter genes on their own, or they can be used to generate *fusion proteins* in which a protein of interest is fused to GFP and can thus be visualized in living tissues, as is shown in the example in **FIGURE 30.13**.

When the purpose of cloning is to express a gene in large amounts, the vector is engineered to have a promoter and a translational start sequence adjacent to the site of insertion. Then any open reading frame can be inserted into the vector and expressed without further modification. These "expression vectors" can have a number of useful features, such as inducible promoters and *epitope tags*. Epitope tags are short sequences that are fused in the same reading frame as the open reading frame to be expressed. These tags encode peptides that are recognized by commercially available antibodies, and provide an easy means for identifying or purifying the tagged protein.

KEY CONCEPTS

- Amplification of a cloned sequence requires a selective technique to distinguish hybrid vectors from the original vector.
- Reporter genes can be used to measure promoter activity or tissue-specific expression.

CONCEPT AND REASONING CHECK

Why must epitope tags or GFP fusions be fused in the same reading frame as the protein of interest?

30.5 Transfection Introduces Exogenous DNA into Cells

The challenge in delivering DNA to a eukaryotic cell is that eukaryotes lack natural DNA uptake systems of which we can take advantage. Delivery methods for animal cells fall into two general classes: those based on addition of DNA to cells in a form that relies on some sort of spontaneous uptake mechanism and those based on mechanical delivery of the DNA, such as microinjection (with a syringe) or electroporation (the use of electrical pulses to create transient holes in the membrane). With plant cells, the thick cell wall poses an obstacle to DNA uptake, and two types of method are used: the Ti plasmid of the bacterium *A. tumefaciens*, a natural pathogen of plants, can be used as the basis for cloning vectors (see *Section 16.8, The Bacterial Ti Plasmid Transfers Genes into Plant Cells*), or mechanical methods are employed, such as the gene gun method mentioned earlier.

▸ **transfection** The acquisition of new genetic material by eukaryotic cells via incorporation of added DNA.

The first procedure that was developed for introducing exogenous donor DNA into recipient cells is called **transfection**. Early transfection experiments involved addition of preparations of metaphase chromosomes to cell suspensions. The chromosomes were taken up rather inefficiently by the cells and gave rise to unstable variants at a low frequency. Intact chromosomes rarely survive the procedure; the recipient cell usually gains a fragment of a donor chromosome (which is unstable because it lacks a centromere). Rare cases of stable lines result from integration of donor material into a resident chromosome.

FIGURE 30.14 shows that similar results are obtained when purified DNA is added to a recipient cell preparation. However, with purified DNA it is possible to add particular sequences instead of relying on random fragmentation of chromosomes. This greatly increases the efficiency of the process. Transfection with

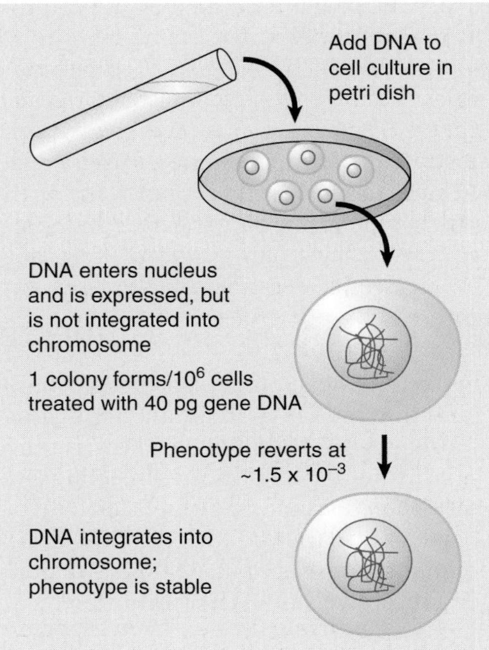
Add DNA to
cell culture in
petri dish
DNA enters nucleus
and is expressed, but
is not integrated into
chromosome
1 colony forms/10^6 cells
treated with 40 pg gene DNA
Phenotype reverts at
~1.5×10^{-3}
DNA integrates into
chromosome;
phenotype is stable

FIGURE 30.14 DNA added to a cell culture can enter the cell and become expressed in the nucleus, initially as an unstable transfectant. The cell becomes a stable transfectant when the DNA is integrated into a random chromosomal site.

DNA yields stable as well as unstable lines, with the former relatively predominant. (These experiments are directly analogous to those performed in bacterial transformation but are described as transfection because of the historical use of "transformation" to describe changes that allow unrestrained growth of eukaryotic cells.) Transfection efficiency can also be dramatically improved by use of liposomes or other methods to deliver the DNA to the cells.

▸ **transient transfectants (unstable transfectants)** Eukaryotic cells that have foreign DNA in an unstable—i.e., extrachromosomal—form.

Transient transfectants (sometimes called unstable transfectants) reflect the survival of the transfected DNA in extrachromosomal form; *stable* lines result from integration into the genome. The transfected DNA can be expressed in both cases. Because the frequency of transfection is low, the technique is restricted to genes whose transfer produces a selectable phenotype. The classic example was to introduce the thymidine kinase gene into cells that lack the activity, so that when plated on a medium on which thymidine kinase is necessary for survival, only the transfectants can survive. Some antibiotics, such as neomycin, can also be used for selection in cultured mammalian cells. Another approach is to screen transfected cells for the transformed (tumorigenic) phenotype, because the transfected cells can be identified by the change in their morphological properties. This type of protocol has led to the isolation of several cellular *onc* genes involved in cancer.

Integration to form a stable transfectant occurs by the insertion of the donor DNA into a random site in a host chromosome. The material that is integrated may contain multiple copies of the donor sequence, and if two different types of DNA are "cotransfected" together, the material usually contains both sequences. For example, when cells are transfected with a DNA preparation containing both an isolated selectable gene and DNA that does not result in a detectable phenotype, both DNA donors are present after selection for the selectable marker gene. This is a useful observation, because it allows unselectable markers to be introduced routinely by *cotransfection* with a selectable marker.

An individual stable transfectant has only a single site of integration; but the site is different in each case. The sites at which exogenous material becomes integrated usually do not appear to have any sequence relationship to the transfected DNA. The integration event involves a nonhomologous recombination between the mass of added DNA and a random site in the genome.

KEY CONCEPTS

- DNA that is transfected into a eukaryotic cell forms a large repeating unit of many head-to-tail tandem repeats.
- The transfected unit is unstable unless it becomes integrated into a host chromosome.
- Genes carried by the transfected DNA can be expressed.

CONCEPT AND REASONING CHECK

Individual stable transfectants can have very different levels of expression of a transfected gene, even though the same DNA was used for transfection. Why is this?

30.6 Genes Can Be Injected into Animal Eggs

An exciting development of transfection techniques is their application to introduce genes into animals. An animal that gains new genetic information from the addition of foreign DNA is described as **transgenic**.

▸ **transgenic animals** Animals created by introducing DNA prepared in test tubes into the germline. The DNA may be inserted into the genome or exist in an extrachromosomal structure.

The approach of directly injecting DNA can be used with mouse eggs, as shown in **FIGURE 30.15**. Plasmids carrying the gene of interest are injected into the nucleus of the oocyte or into the pronucleus of the fertilized egg. The egg is implanted into a pseudopregnant mouse, a mouse that has mated with a vasectomized (sterilized) male to trigger a receptive state. After birth, the recipient mouse can be examined to see whether

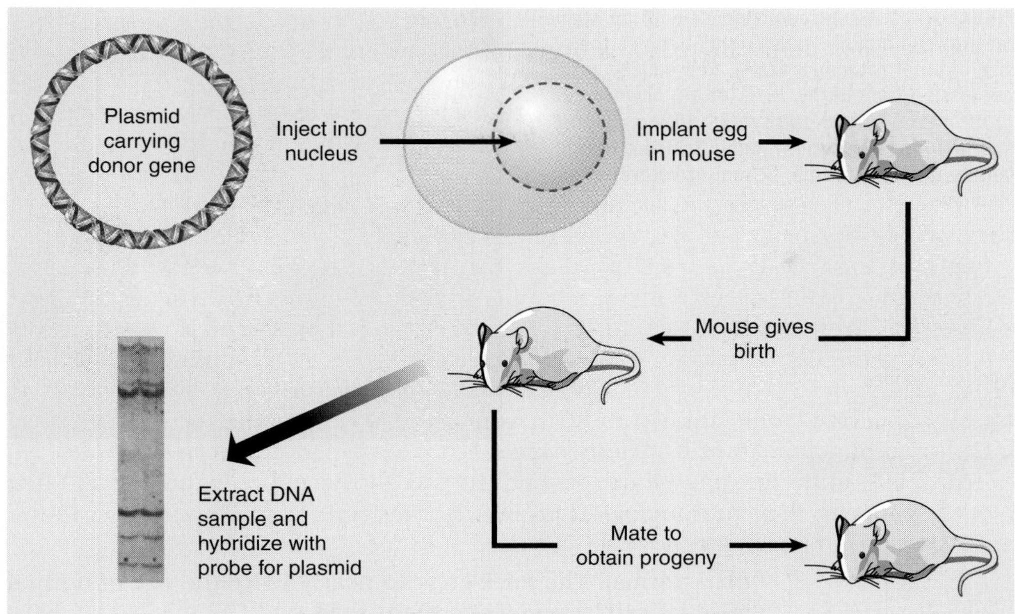

FIGURE 30.15 Transfection can introduce DNA directly into the germline of animals. Southern blot inset reproduced from Chambon, P., *Sci. Am.* 244 (1981): 60–71. Used with permission of Pierre Chambon, Institute of Genetics and Molecular and Cellular Biology, College of France.

it has gained the foreign DNA, and, if so, whether it is expressed. Typically, a minority (~15%) of the injected mice carry the transfected sequence. Usually, multiple copies of the plasmid appear to have been integrated in a tandem array into a single chromosomal site. The number of copies varies from 1–150, and they are inherited by the progeny of the injected mouse.

An important issue that can be addressed by experiments with transgenic animals concerns the independence of genes and the effects of the region within which they reside. If we take a gene, including the flanking sequences that contain its known regulatory elements, can it be expressed independently of its location in the genome? In other words, do the regulatory elements function independently, or is gene expression also controlled by other effects, such as location in an appropriate chromosomal domain? In general, there is a reasonable facsimile of proper control: the transfected genes are generally expressed in appropriate cells and at the usual time.

However, the absolute levels of expression of transgenes can vary dramatically. In the progeny of the injected mice, expression of the donor gene is extremely variable; it may be extinguished entirely, reduced somewhat, or even increased. The level of gene expression does not correlate with the number of tandemly integrated genes, and it is likely that only some of the genes are active.

What is responsible for the variation in gene expression? One factor is that the site of integration is important. A gene may be highly expressed if it integrates within an active chromatin domain, but not if it integrates in or near a heterochromatic region. Another possibility is the occurrence of epigenetic modification; for example, changes in the pattern of methylation might be responsible for changes in activity. Alternatively, the genes that happened to be active in the parents may have been deleted or amplified in the progeny.

A particularly striking example of the effects of an injected gene is provided by a strain of transgenic mice derived from eggs injected with a fusion consisting of the MT (metallthionein) promoter linked to the rat growth hormone structural gene. Growth hormone levels in some of the transgenic mice were several hundred times greater

FIGURE 30.16 A transgenic mouse with an active rat growth hormone gene (right) is twice the size of a normal mouse (left). Reproduced with permission from Batey, R. T., et al., *Science* 222 (1983): 809–814. © 1983 AAAS. Photo adapted and provided courtesy of Ralph Brinster, University of Pennsylvania, School of Veterinary Medicine.

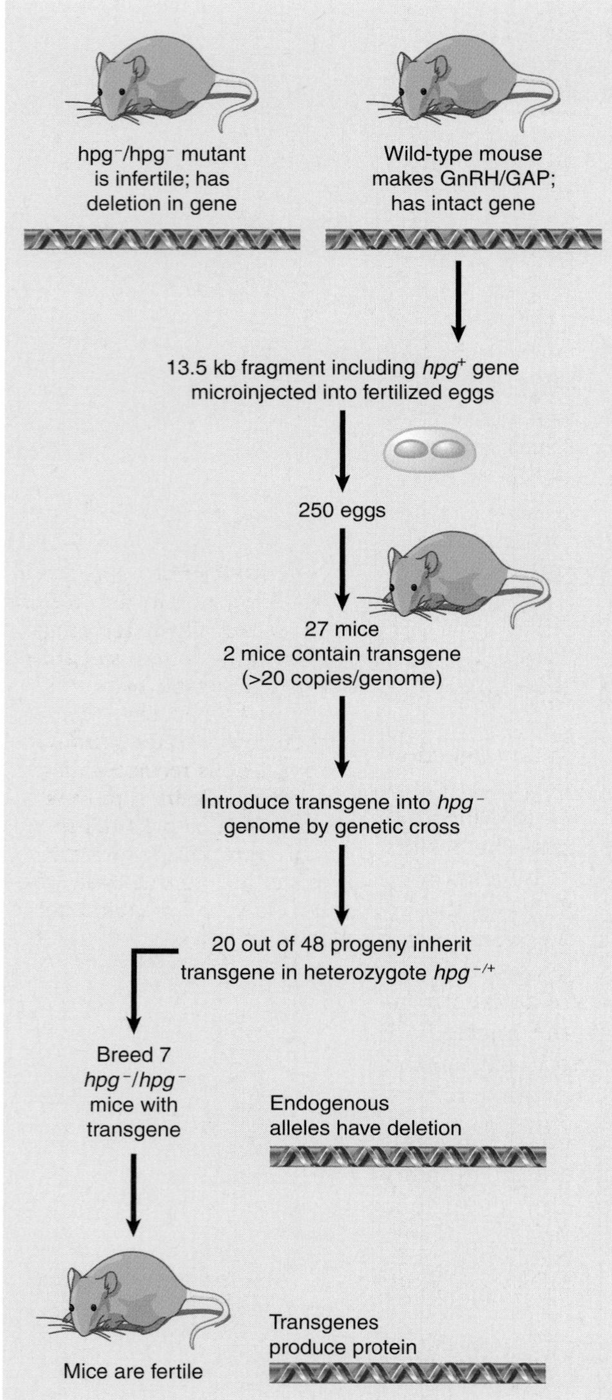

FIGURE 30.17 Hypogonadism can be averted in the progeny of *hpg* mice by introducing a transgene that has the wild-type sequence.

than normal. The mice grew to nearly twice the size of normal mice, as can be seen from **FIGURE 30.16**.

Defective genes can also be replaced by functional genes using transgenic techniques. One example is the cure of the defect in the *hypogonadal* mouse. The *hpg* mouse has a deletion that removes the distal part of the gene coding for the precursor to GnRH (gonadotropin-releasing hormone) and GnRH-associated peptide (GAP). As a result, the mouse is infertile. When an intact *hpg* gene is introduced into the mouse by transgenic techniques, it is expressed in the appropriate tissues. **FIGURE 30.17** summarizes experiments to introduce a transgene into a line of $hpg^{-/-}$ homozygous mutant mice. The resulting progeny are normal. This provides a striking demonstration that expression of a transgene under normal regulatory control can be indistinguishable from the behavior of the normal allele.

While promising, there are impediments to using such techniques to cure human genetic defects. The transgene must be introduced into the germline of the *preceding* generation, the ability to express a transgene is not predictable, and an adequate level of expression of a transgene may be obtained in only a small minority of the transgenic individuals. Also, the large number of transgenes that may be introduced into the germline, and their erratic expression, could pose problems in cases in which overexpression of the transgene is harmful.

In the *hpg* murine experiments, for example, only 2 out of 250 eggs injected with intact *hpg* genes gave rise to transgenic mice. (If this were to be attempted in human embryos, such a high rate of failure would create an ethical dilemma.) Each transgenic animal contained 720 copies of the transgene. Only 20 of the 48 offspring of the transgenic mice retained the transgenic trait. When inherited by their offspring, however, the transgene(s) could substitute for the lack of endogenous *hpg* genes. Gene replacement via a transgene is therefore effective only under restricted conditions.

An alternative approach is to use a retroviral vector carrying the donor gene. A single proviral copy inserts at a chromosomal site, and infection by retrovirus is an efficient way to introduce donor sequences to a cell. It is possible also to treat cells at different stages of development, and thus to target a particular somatic tissue; however, it is difficult to infect germ cells, so this method is not useful for creating heritable changes in the genome.

KEY CONCEPTS

- DNA that is injected into animal eggs can integrate into the genome.
- Usually a large array of tandem repeats integrates at a single site.
- Expression of the DNA is variable and may be affected by the site of integration and other epigenetic effects.

CONCEPT AND REASONING CHECK

How might you use insulators to improve methods for generating transgenic animals? (See *Section 24.11, Insulators Define Independent Domains*.)

30.7 Gene Targeting Allows Genes to Be Replaced or Knocked Out

The transfection and injection methods discussed in the previous sections allow DNA to be *added* to cells or animals, but in order to understand the function of a gene, it is most useful to be able to *remove* the gene or its function and observe the resulting phenotype. The most powerful techniques for changing the genome use *gene targeting* to delete or replace genes by homologous recombination. Gene deletions are usually referred to as **knockouts**, while replacement of a gene with an alternative mutated version is called a **knock-in**.

▸ **gene knockout** A process in which a gene function is eliminated, usually by replacing most of the coding sequence with a selectable marker *in vitro* and transferring the altered gene to the genome by homologous recombination.

▸ **gene knock-in** A process similar to a knockout, but more subtle mutations are made, such as nucleotide substitutions or the addition of epitope tags.

In mammals, the target is usually the genome of an embryonic stem (ES) cell, which is then used to generate a mouse with the mutation. ES cells are derived from the mouse blastocyst (an early stage of development, which precedes implantation of the egg in the uterus). **FIGURE 30.18** illustrates the general approach.

ES cells are transfected with DNA in the usual way (most often by microinjection or electroporation). By using a donor that carries an additional sequence such as a drug resistance marker or some particular enzyme, it is possible to select ES cells that have obtained an integrated transgene carrying any particular donor trait. This results in a population of ES cells in which there is a high proportion carrying the marker.

These ES cells are then injected into a recipient blastocyst. The ability of the ES cells to participate in normal development of the blastocyst forms the basis of the technique. The blastocyst is implanted into a foster mother, and in due course develops into a *chimeric* mouse. Some of the tissues of the chimeric mice are derived from the cells of the recipient blastocyst; other tissues are derived from the injected ES cells. The proportions of tissues in the adult mouse that are derived from cells in the recipient blastocyst and from injected ES cells varies widely in individual progeny; if a visible marker (such as coat color gene) is used, areas of tissue representing each type of cell can be seen.

To determine whether the ES cells contributed to the germline, the chimeric mouse is crossed with a mouse that lacks the donor trait. Any progeny that have the trait must be derived from germ cells that have descended from the injected ES cells. By this means, it is known that an entire mouse has been generated from an original ES cell!

When a donor DNA is introduced into the cell, it may insert into the genome by either nonhomologous or homologous recombination. Homologous recombination is relatively rare, probably representing $<1\%$ of all recombination events, and thus occurring at a frequency of $\sim 10^{-7}$. However, by designing the donor DNA appropriately, we can use selective techniques to identify those cells in which homologous recombination has occurred.

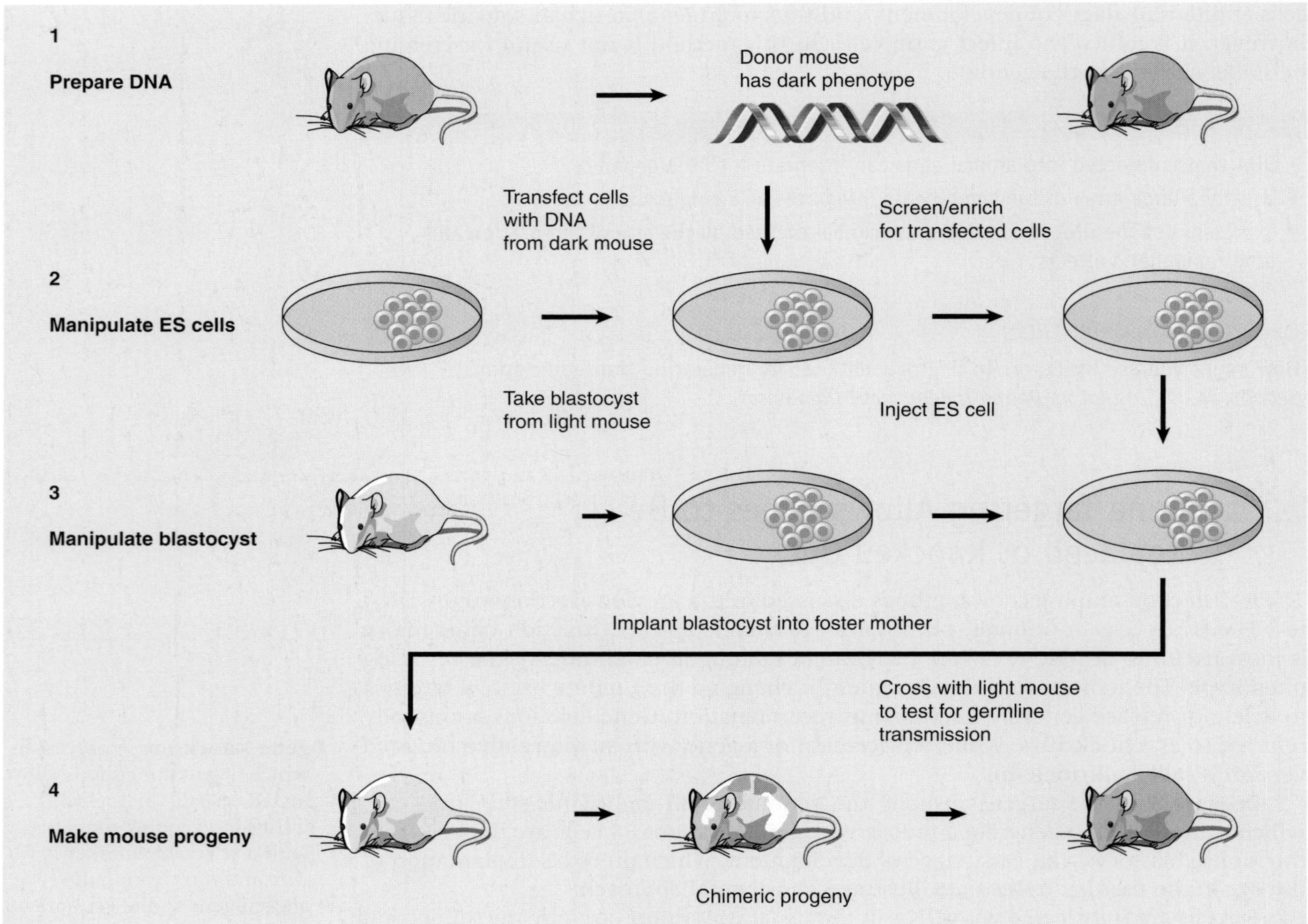

FIGURE 30.18 ES cells can be used to generate mouse chimeras, which breed true for the transfected DNA when the ES cell contributes to the germline.

FIGURE 30.19 illustrates the knockout technique that is used to disrupt endogenous genes. The basis for the technique is the design of a knockout construct with two different markers, designed to allow nonhomologous and homologous recombination events in the ES cells to be distinguished. The donor DNA is homologous to a target gene, but has two key modifications. First, the gene is inactivated by interrupting or replacing an exon with a gene encoding a selectable marker (most often the *neo*^{R} gene that confers resistance to the drug G418 is used). Second, a *counter-selectable* marker (a gene that can be selected *against*) is added on one side of the gene; for example, the *TK* gene of the herpes virus.

When this knockout construct is introduced into an ES cell, homologous and nonhomologous recombination will result in different outcomes. Nonhomologous recombination inserts the entire construct, including the flanking *TK* gene. These cells are resistant to neomycin, and they also express thymidine kinase, which makes them *sensitive* to the drug gancyclovir (thymidine kinase phosphorylates gancyclovir, which makes it toxic). In contrast, homologous recombination involves two exchanges within the sequence of the donor gene, resulting in the loss of the flanking *TK* gene. Cells in which homologous recombination has occurred, therefore, gain neomycin resistance in the same way as cells that have nonhomologous recombination, but they do *not* have thymidine kinase activity, and so are resistant to gancyclovir. So plating the cells in the presence of neomycin plus gancyclovir specifically selects those in which homologous recombination has replaced the endogenous gene with the donor gene.

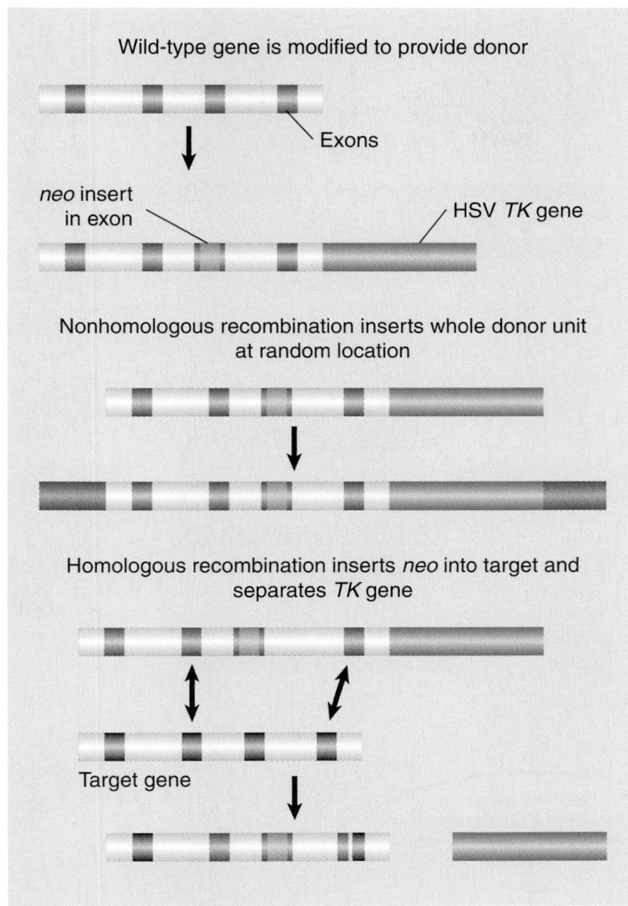
Wild-type gene is modified to provide donor
Exons
neo insert in exon
HSV TK gene
Nonhomologous recombination inserts whole donor unit at random location
Homologous recombination inserts neo into target and separates TK gene
Target gene

FIGURE 30.19 A transgene containing *neo* within an exon and *TK* downstream can be selected by resistance to G418 and loss of *TK* activity.

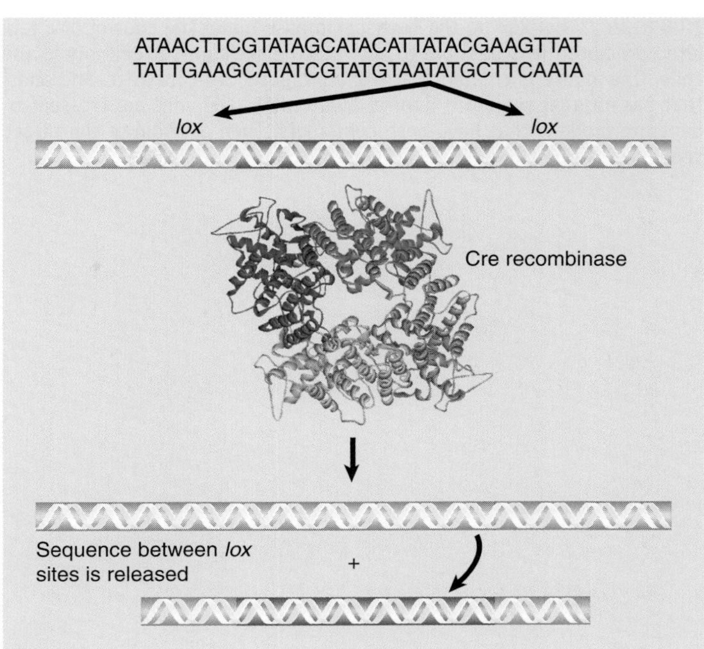
ATAACTTCGTATAGCATACATTATACGAAGTTAT
TATTGAAGCATATCGTATGTAATATGCTTCAATA
lox
lox
Cre recombinase
Sequence between lox sites is released
+

FIGURE 30.20 The Cre recombinase catalyzes a site-specific recombination between two identical *lox* sites, releasing the DNA between them. Structure from Protein Data Bank 1OUQ. Ennifar, E., et al., *Nucleic Acids Res.* 31 (2003): 5449–5460.

The presence of the *neo*^{R} gene in an exon of the donor gene disrupts translation, and thereby creates a null allele. A particular target gene can therefore be knocked out by this means; and once a mouse with one null allele has been obtained, it can be bred to generate the homozygote. This is a powerful technique for investigating whether a particular gene is essential, and what functions in the animal are perturbed by its loss. Sometimes phenotypes can even be observed in the heterozygote.

A major extension of ability to manipulate a target genome has been made possible by using the phage Cre/*lox* system to engineer site-specific recombination in a eukaryotic cell. The Cre enzyme catalyzes a site-specific recombination reaction between two *lox* sites, which are identical 34 bp sequences (see *Section 19.8, Site-Specific Recombination Resembles Topoisomerase Activity*). **FIGURE 30.20** shows that the consequence of the reaction is to excise the stretch of DNA between the two *lox* sites.

The great utility of the Cre/*lox* system is that it requires no additional components and works when the Cre enzyme is produced in any cell that has a pair of *lox* sites. **FIGURE 30.21** shows that we can control the reaction to make it work in a particular cell by placing the *cre* gene under the control of a regulated promoter. The procedure starts by making two mice. One mouse has the *cre* gene, typically controlled by a promoter that can be turned on specifically in a certain cell or under certain conditions. The other mouse has a target sequence flanked by *lox* sites. When we cross the two mice, the progeny have both elements of the system; and the system can be turned on by controlling the promoter of the *cre* gene. This allows the sequence between the *lox* sites to be excised in a controlled way.

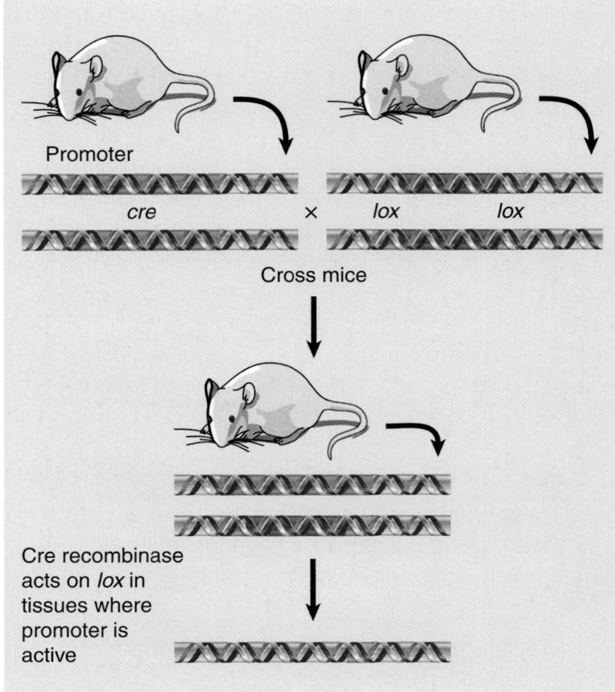

FIGURE 30.21 By placing the Cre recombinase under the control of a regulated promoter, it is possible to activate the excision system only in specific cells. One mouse is created that has a promoter-*cre* construct, and another that has a target sequence flanked by *lox* sites. The mice are crossed to generate progeny that have both constructs. Then excision of the target sequence can be triggered by activating the promoter.

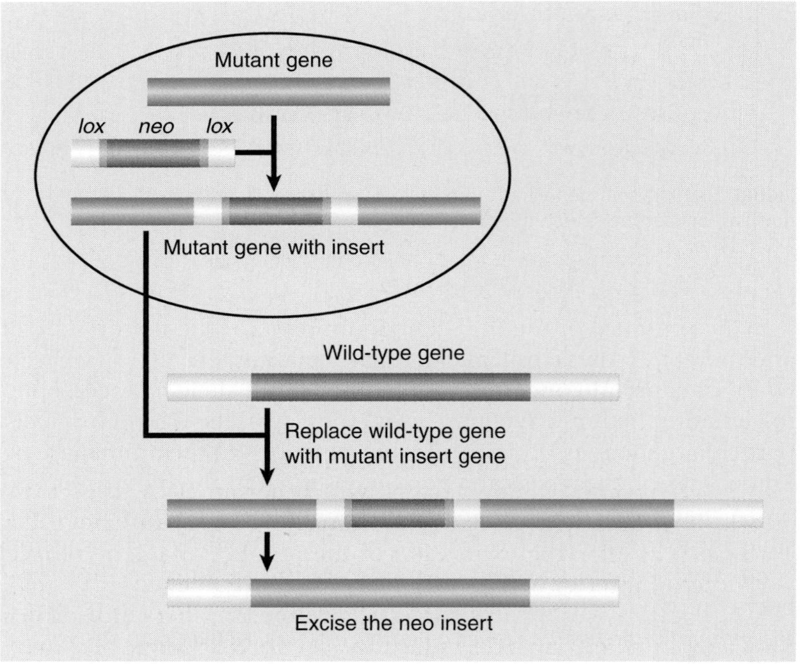

FIGURE 30.22 An endogenous gene is replaced in the same way as when a knockout is made (see Figure 30.19), but the neomycin gene is flanked by *lox* sites. After the gene replacement has been made using the selective procedure, the neomycin gene can be removed by activating Cre, leaving an active insert.

The Cre/*lox* system can be combined with the knockout technology to give us even more control over the genome. Inducible knockouts can be made by flanking the neo^R gene (or any other gene that is used similarly in a selective procedure) with *lox* sites. After the knockout has been made, the target gene can be reactivated by causing Cre to excise the neo^R gene in some particular circumstance (such as in a specific tissue).

FIGURE 30.22 shows a modification of this procedure that allows a knock-in to be created. Basically we use a construct in which some mutant version of the target gene is used to replace the endogenous gene, replying on the usual selective procedures. Then when the inserted gene is reactivated by excising the neo^R sequence, we have in effect replaced the original gene with a different version.

A useful variant of this method is to introduce a wild-type copy of the gene of interest in which the gene itself (or one of its exons) is flanked by *lox* sites. This results in a normal animal that can be crossed to a mouse containing *Cre* under control of a tissue-specific or otherwise regulated promoter. The offspring of this cross are *conditional knockouts*, in which the function of the gene is lost only in cells that express *Cre*. This is particularly useful for studying genes that are essential for embryonic development; genes in this class would be lethal in homozygous embryos and thus are very difficult to study.

With these techniques, we are able to investigate the functions and regulatory features of genes in whole animals. The ability to introduce DNA into the genome allows us to make changes in it, to add new genes that have had particular modifications introduced *in vitro*, or to inactivate existing genes. So it becomes possible to delineate the features responsible for tissue-specific gene expression. Ultimately, we may expect routinely to replace defective genes in the genome in a targeted manner.

KEY CONCEPTS

- ES (embryonic stem) cells that are injected into a mouse blastocyst generate descendant cells that become part of a chimeric adult mouse.
- When the ES cells contribute to the germline, the next generation of mice may be derived from the ES cell.
- Genes can be added to the mouse germline by transfecting them into ES cells before the cells are added to the blastocyst.
- An endogenous gene can be replaced by a transfected gene using homologous recombination.
- The occurrence of successful homologous recombination can be detected by using two selectable markers, one of which is incorporated with the integrated gene, the other of which is lost when recombination occurs.
- The Cre/*lox* system is widely used to make inducible knockouts and knock-ins.

CONCEPT AND REASONING CHECK

What is an advantage to being able to create a conditional knockout in a selected tissue or at a particular time in development?

30.8 Summary

DNA can be manipulated and propagated using the techniques of cloning. These include digestion by restriction endonucleases, which cut DNA at specific sequences, and insertion into cloning vectors, which permit DNA to be maintained and amplified in host cells such as bacteria. Cloning vectors can have specialized functions as well, such as allowing expression of the product of a gene of interest, or fusion of a promoter of interest to an easily assayed reporter gene.

New sequences of DNA may be introduced into a cultured cell by transfection or into an animal egg by microinjection. The foreign sequences may become integrated into the genome, often as large tandem arrays. The array appears to be inherited as a unit in a cultured cell. The sites of integration appear to be random. A transgenic animal arises when the integration event occurs into a genome that enters the germ cell lineage. Often a transgene responds to tissue and temporal regulation in a manner that resembles the endogenous gene. Under conditions that promote homologous recombination, an inactive sequence can be used to replace a functional gene, thus creating a knockout, or deletion, of the target locus. Extensions of this technique can be used to make conditional knockouts, where the activity of the gene can be turned on or off (e.g., by Cre-dependent recombination), and knock-ins, where a donor gene specifically replaces a target gene. Transgenic mice can be obtained by injecting recipient blastocysts with ES cells that carry transfected DNA.

CHAPTER QUESTIONS

1. Restriction enzymes that generate sticky ends:
   - **A.** make a single cut in the DNA backbone.
   - **B.** make two cuts in the DNA backbone in the same strand.
   - **C.** make two staggered cuts in the DNA backbone.
   - **D.** make two cuts in the DNA backbone directly across the double helix from each other.
2. Recombinant plasmids are usually introduced into bacterial cells by:
   - **A.** gene guns.
   - **B.** transformation of chemically treated cells.
   - **C.** transfection using liposomes.
   - **D.** microinjection.
3. The best vector for cloning very large (>1 Mb) fragments of DNA is a:
   - **A.** plasmid.
   - **B.** bacteriophage.
   - **C.** cosmid.
   - **D.** YAC.
4. Reporter genes are:
   - **A.** genes that produce products that are easy to detect and/or quantify.
   - **B.** genes that are essential in all species.
   - **C.** genes that never occur in nature.
   - **D.** all of the above.
5. Transfection:
   - **A.** is the most common means of introducing DNA into plants.
   - **B.** can result in stable integration of added DNA into the genome.
   - **C.** only results in transient maintenance of DNA.
   - **D.** none of the above.
6. A knockout construct does NOT usually contain:
   - **A.** regions homologous to an endogenous gene.
   - **B.** a selectable marker.
   - **C.** a marker for counterselection.
   - **D.** a reporter gene.
7. Put the following events in the generation of a knockout mouse in their correct order:
   - **A.** breed heterozygotes to generate homozygous null offspring
   - **B.** inject knockout construct into ES cells
   - **C.** implant blastocyst into foster mother and obtain offspring
   - **D.** breed chimeric offspring to obtain heterozygotes from ES-derived germ cells
   - **E.** make knockout construct to target gene of interest
   - **F.** introduce ES cells to blastocyst
   - **G.** select for ES cells that have integrated knockout construct via homologous recombination

KEY TERMS

cloning
cloning vector
cosmid
gene knock-in
gene knockout
reporter gene
restriction endonuclease
restriction map
transfection
transgenic animals
transient transfectants (unstable transfectants)

Glossary

10-nm fiber—A linear array of nucleosomes, generated by unfolding from the natural condition of chromatin.

3′ UTR—The untranslated sequence downstream from the coding region of an mRNA.

30-nm fiber—A coiled coil of nucleosomes. It is the basic level of organization of nucleosomes in chromatin.

-35 box—The consensus sequence centered about 35 bp before the startpoint of a bacterial gene. It is involved in initial recognition by RNA polymerase.

5′ UTR—The untranslated sequence upstream from the coding region of an mRNA.

5′-end resection—The generation of 3′ overhanging single-stranded regions that occurs via exonucleolytic digestion of the 5′ ends at a double strand break.

7S RNA—A small RNA component of the signal recognition particle (SRP), transcribed by RNA polymerase III. Also known as 7SL RNA.

A complex—The second splicing complex, formed by the binding of U2 snRNP to the E complex.

A site—The site of the ribosome that an aminoacyl-tRNA enters to base pair with the codon.

abortive initiation—It describes a process in which RNA polymerase starts transcription but terminates before it has left the promoter. It then reinitiates. Several cycles may occur before the elongation stage begins.

abundance—The average number of mRNA molecules per cell.

abundant mRNA—Consists of a small number of individual species, each present in a large number of copies per cell.

***Ac* element**—Activator element; an autonomous transposable element in maize.

acceptor arm—A short duplex on tRNA that terminates in the CCA sequence to which an amino acid is linked.

acentric fragment—A fragment of a chromosome (generated by breakage) that lacks a centromere and is lost at cell division.

acridines—Mutagens that act on DNA to cause the insertion or deletion of a single base pair. They were useful in defining the triplet nature of the genetic code.

activator—A protein that stimulates the expression of a gene, typically by acting at a promoter to stimulate RNA polymerase. In eukaryotes, the sequence to which it binds in the promoter is called a response element.

adaptive (acquired) immunity—The response mediated by lymphocytes that are activated by their specific interaction with antigen. The response develops over several days as lymphocytes with antigen-specific receptors are stimulated to proliferate and become effector cells. It is responsible for immunological memory.

alarmone—A small molecule in bacteria that is produced as a result of stress and that acts to alter the state of gene expression. The unusual nucleotides ppGpp and pppGpp are examples.

allele—One of several alternative forms of a gene occupying a given locus on a chromosome.

allelic exclusion—The expression in any particular lymphocyte of only one allele coding for the expressed immunoglobulin. This is caused by feedback from the first immunoglobulin allele to be expressed that prevents activation of a copy on the other chromosome.

allolactose—A byproduct of the LacZ enzyme, the true inducer of the *lac* operon.

allosteric control—The ability of a protein to change its conformation (and, therefore, activity) at one site as the result of binding a small molecule to a second site located elsewhere on the protein.

alternative splicing—The production of different RNA products from a single product by changes in the usage of splicing junctions.

Alu domain—The parts of the 7S RNA of the SRP that are related to Alu RNA.

Alu element—One of a set of dispersed, related sequences, each ~300 bp long, in the human genome (members of the **SINE** family). The individual members have Alu cleavage sites at each end.

amber codon—The triplet UAG, one of the three termination codons that end polypeptide translation.

aminoacyl-tRNA—A tRNA linked to an amino acid. The COOH group of the amino acid is linked to the 3′- or 2′-OH group of the terminal base of the tRNA.

aminoacyl-tRNA synthetases—Enzymes responsible for covalently linking amino acids to the 2′- or 3′-OH position of tRNA.

amyloid fibers—Insoluble fibrous protein polymers with a cross β-sheet structure, generated by prions or other dysfunctional protein aggregations (such as in Alzheimer's disease).

anchor sequence—A segment of a transmembrane protein that resides in the membrane.

annealing—The renaturation of a duplex structure from single strands that were obtained by denaturing duplex DNA.

antibody—A protein that is produced by B lymphocytes and that binds a particular antigen. They are synthesized in membrane-bound and secreted forms. Those produced during an immune response recruit effector functions to help neutralize and eliminate the pathogen.

anticodon—A trinucleotide sequence in tRNA that is complementary to the codon in mRNA and enables the tRNA to place the appropriate amino acid in response to the codon.

anticodon arm—A stem-loop structure in tRNA that exposes the anticodon triplet at one end.

antigen—A molecule that can bind specifically to an antigen receptor, such as an antibody.

antiparallel—Strands of the double helix are organized in opposite orientation, so that the 5′ end of one strand is aligned with the 3′ end of the other strand.

antirepressor—A positive regulator that functions in opening chromatin.

antisense RNA—RNA that has a complementary sequence to an RNA that is its target.

antisense strand (template strand)—The DNA strand that is complementary to the sense strand and acts as the template for synthesis of mRNA.

anti-Sm—An autoimmune antiserum that defines the Sm domain that is common to a group of proteins found in snRNPs that are involved in RNA splicing.

antitermination—A mechanism of transcriptional control in which termination is prevented at a specific terminator site, allowing RNA polymerase to read into the genes beyond it.

antitermination protein—Protein that allows RNA polymerase to transcribe through certain terminator sites.

anucleate cells—Bacteria that lack a nucleoid but are of similar shape to wild-type bacteria.

architectural protein—A protein that when bound to DNA, can alter its structure, e.g., introduces a bend. They may have no other function.

arm—One of the four (or in some cases five) stem-loop structures that make up the secondary structure of tRNA.

ARS—An origin for replication in yeast. The common feature among different examples of these sequences is a conserved 11-bp sequence called the A domain.

assembly factors—Proteins that are required for formation of a macromolecular structure but are not themselves part of that structure.

ATP-dependent chromatin remodeling complex—A complex of one or more proteins associated with an ATPase of the SWI2/SNF2 superfamily that uses the energy of ATP hydrolysis to alter or displace nucleosomes.

***att* sites**—The loci on a lambda phage and the bacterial chromosome at which recombination integrates the phage into, or excises it from, the bacterial chromosome.

attenuation—The regulation of bacterial operons by controlling termination of transcription at a site located before the first structural gene.

attenuator—A terminator sequence at which attenuation occurs.

autoimmune disease—A pathological condition in which the immune response is directed against self antigens.

autonomous controlling element—An active transposon in maize with the ability to transpose.

autoregulation—A site or mutation that affects the properties only of its own molecule of DNA, often indicating that a site does not code for a diffusible product.

autosplicing (self-splicing)—The ability of an intron to excise itself from an RNA by a catalytic action that depends only on the sequence of RNA in the intron.

axial element—A proteinaceous structure around which the chromosomes condense at the start of synapsis.

B cell—A lymphocyte that produces antibodies. Development occurs primarily in bone marrow.

back mutation—A mutation that reverses the effect of a mutation that had inactivated a gene; thus, it restores the original sequence or function of the gene product.

bacteriophage—A bacterial virus.

Bam islands—A series of short, repeated sequences found in the nontranscribed spacer of *Xenopus* rDNA genes.

bands—They are on polytene chromosomes visible as dense regions that contain the majority of DNA; they include active genes.

basal apparatus—The complex of transcription factors that assembles at the promoter before RNA polymerase is bound.

basal transcription factors—Transcription factors required by RNA polymerase II to form the initiation complex at all RNA polymerase II promoters. Factors are identified as TFIIX, where X is a letter.

bidirectional replication—A system in which an origin generates two replication forks that proceed away from the origin in opposite directions.

bivalent—The structure containing all four chromatids (two representing each homologue) at the start of meiosis.

boundary (insulator) element—A DNA sequence element bound by proteins that prevent the spread of open or closed chromatin.

branch migration—The ability of a DNA strand partially paired with its complement in a duplex to extend its pairing by displacing the resident strand with which it is homologous.

branch site—A short sequence just before the end of an intron at which the lariat intermediate is formed in splicing by joining the 5′ nucleotide of the intron to the 2′ position of an adenosine.

breakage and reunion—The mode of genetic recombination in which two DNA duplex molecules are broken at corresponding points and then rejoined crosswise (involving formation of a length of heteroduplex DNA around the site of joining).

bromodomain—A 110–amino acid domain that binds to acetylated lysines in histones.

bZIP—A bZIP (basic zipper) protein has a basic DNA-binding region adjacent to a leucine zipper dimerization motif.

C genes—Genes that code for the constant regions of immunoglobulin protein chains.

cAMP (cyclic AMP)—The coregulator of CRP, it has an internal 3′–5′ phosphodiester bond. Its concentration is inverse to the concentration of glucose.

cap—The structure at the 5′ end of eukaryotic mRNA, and is introduced after transcription by linking the terminal phosphate of 5′ GTP to the terminal base of the mRNA.

cap 0—A cap at the 5′ end of mRNA that has only a methyl group on 7-guanine.

cap 1—A cap at the 5′ end of mRNA that has methyl groups on the terminal 7-guanine and the 2′-O position of the next base.

cap 2—A cap that has three methyl groups (7-guanine, 2′-O position of next base, and N6 adenine) at the 5′ end of mRNA.

capsid—The external protein coat of a virus particle.

carboxy terminal domain (CTD)—The domain of eukaryotic RNA polymerase II that is phosphorylated at initiation and is involved in coordinating several activities with transcription.

carrier protein—General term for a protein that carries other molecules within the cell, between cellular compartments, or around the body.

cascade—A sequence of events, each of which is stimulated by the previous one. In transcriptional regulation, as seen in sporulation and phage lytic development, it means that regulation is divided into stages, and at each stage, one of the genes that is expressed codes for a regulator needed to express the genes of the next stage.

catabolite regulation—The ability of the glucose to prevent the expression of a number of genes. In bacteria this is a positive control system; in eukaryotes, it is completely different.

catenate—To link together two circular molecules, as in a chain.

cDNA—A single-stranded DNA complementary to an RNA, synthesized from it by reverse transcription *in vitro*.

cell-mediated response—The immune response that is mediated primarily by T lymphocytes. It is defined based on immunity that cannot be transferred from one organism to another by serum antibody.

central dogma—Information cannot be transferred from protein to protein or protein to nucleic acid but can be transferred between nucleic acids and from nucleic acid to protein.

central element—A structure that lies in the middle of the synaptonemal complex, along which the lateral elements of homologous chromosomes align. It is formed from Zip proteins.

centromere—A constricted region of a chromosome that includes the site of attachment (the kinetochore) to the mitotic or meiotic spindle. It consists of unique DNA sequences and proteins not found anywhere else in the chromosome.

chaperone—A class of proteins that bind to incompletely folded or assembled proteins in order to assist their folding or prevent them from aggregating.

checkpoint—A biochemical control mechanism that prevents the cell from progressing from one stage to the next unless specific goals and requirements have been met.

chemical proofreading—A proofreading mechanism in which the correction event occurs after the addition of an incorrect subunit to a polymeric chain, by means of reversing the addition reaction.

chiasma (pl. chiasmata)—A site at which two homologous chromosomes synapse during meiosis.

chromatin—The complex of nuclear DNA and associated proteins that compacts DNA in the nucleus.

chromatin remodeling—The energy-dependent displacement or reorganization of nucleosomes that occurs in conjunction with activation of genes for transcription.

chromocenter—An aggregate of heterochromatin from different chromosomes.

chromodomain—~60 amino acid domains that recognize methylated lysines in histones; some chromodomains have different functions such as RNA binding.

chromomeres—Densely staining granules visible in chromosomes under certain conditions, especially early in meiosis, when a chromosome may appear to consist of a series of chromomeres.

chromosomal walk—A technique for locating a gene by using the mostly closely linked markers as a probe for a genetic library.

chromosome—A discrete unit of the genome carrying many genes. Each consists of a very long molecule of duplex DNA and an approximately equal mass of proteins. It is visible as a morphological entity only during cell division.

chromosome pairing—The coupling of the homologous chromosomes at the start of meiosis.

chromosome scaffold—A proteinaceous structure in the shape of a sister chromatid pair, generated when chromosomes are depleted of histones.

***cis*-acting sequence**—A site that affects the activity only of sequences on its own molecule of DNA (or RNA); this property usually implies that the site does not code for protein.

***cis*-dominant**—A site or mutation that affects the properties only of its own molecule of DNA, often indicating that a site does not code for a diffusible product.

cistron—The genetic unit defined by the complementation test; it is equivalent to a gene.

clamp—A protein complex that forms a circle around the DNA; by connecting to DNA polymerase, it ensures that the enzyme action is processive.

clamp loader—A five-subunit protein complex that is responsible for loading the β clamp onto DNA at the replication fork.

class switching—A change in Ig gene organization in which the C region of the heavy chain is changed but the V region remains the same.

clonal expansion—The production of numerous daughter cells all arising from a single cell.

cloning—Propagation of a DNA sequence by incorporating it into a hybrid construct that can be replicated in a host cell.

cloning vector—DNA (often derived from a plasmid or a bacteriophage genome) that can be used to propagate an incorporated DNA sequence in a host cell; vectors contain selectable markers and replication origins to allow identification and maintenance of the vector in the host.

closed (blocked) reading frame—A reading frame that cannot be translated into protein because of the occurrence of termination codons.

closed complex—The stage of initiation of transcription before RNA polymerase causes the two strands of DNA to separate to form the "transcription bubble." The DNA is double stranded.

cloverleaf—The structure of tRNA drawn in two dimensions, forming four distinct arm-loops.

coactivator—Factors required for transcription that do not bind DNA but are required for (DNA-binding) activators to interact with the basal transcription factors.

coding end—It is produced during recombination of immunoglobulin and T cell receptor genes. They are at the termini of the cleaved V and (D)J coding regions. Their subsequent joining yields a coding joint.

coding joint—The DNA junction created by the joining of two coding ends during V(D)J recombination.

coding region—A part of a gene that codes for a polypeptide sequence.

coding strand (sense strand)—The DNA strand that has the same sequence as the mRNA and is related by the genetic code to the polypeptide sequence that it represents.

codon—A triplet of nucleotides that codes for an amino acid or a termination signal.

codon bias—A higher usage of one codon in genes to encode amino acids for which there are several synonymous codons.

codon usage—A description of the relative abundance of tRNAs for each codon.

cognate tRNAs—tRNAs recognized by a particular aminoacyl-tRNA synthetase. All are charged with the same amino acid.

cointegrate—A structure that is produced by fusion of two replicons, one originally possessing a transposon and the other lacking it; the cointegrate has copies of the transposon present at both junctions of the replicons, oriented as direct repeats.

colinearity—The relationship that describes the 1:1 correspondence of a sequence of triplet nucleotides to a sequence of amino acids.

compatibility group—A group of plasmids that contains members unable to coexist in the same bacterial cell.

complement—A set of ~20 proteins that function through a cascade of proteolytic actions to lyse infected target cells, or to attract macrophages.

complementary—Base pairs that match up in the pairing reactions in double helical nucleic acids (A with T in DNA or with U in RNA, and C with G).

complementation test—A test that determines whether two mutations are alleles of the same gene. It is accomplished by crossing two different recessive mutations that have the same phenotype and determining whether the wild-type phenotype can be produced. If so, the mutations are said to complement each other and are probably not mutations in the same gene.

complex mRNA —See **scarce mRNA**.

composite elements—Transposable elements consisting of two IS elements (can be the same or different) and the DNA sequences between the IS elements; the non-IS sequences often include gene(s) conferring antibiotic resistance.

concerted evolution (coincidental evolution)—The ability of two or more related genes to evolve together as though constituting a single locus.

conditional lethal—A mutation that is lethal under one set of conditions but not lethal under a second set of conditions, such as temperature.

conjugation—A process in which two cells come in contact and transfer genetic material. In bacteria, DNA is transferred from a donor to a recipient cell. In protozoa, DNA passes from each cell to the other.

consensus sequence—An idealized sequence in which each position represents the base most often found when many actual sequences are compared.

conservative transposition—The movement of large elements that were originally classified as transposons but now are considered to be episomes. The mechanism of movement resembles that of phage excision and integration.

conserved sequences—Sequences in which many examples of a particular nucleic acid or protein are compared and the same individual bases or amino acids are always found at particular locations.

constant region (C region)—The part of an immunoglobulin or T cell receptor that varies least in amino acid sequence between different molecules. They are are coded by C gene segments. The heavy chain regions identify the type of immunoglobulin and recruits effector functions.

constitutive expression—This describes a gene system where the gene is expressed continuously.

constitutive gene—See **housekeeping genes**.

constitutive heterochromatin—The inert state of permanently nonexpressed sequences, usually satellite DNA.

context (sequence context)—The fact that neighboring sequences may change the efficiency with which a codon is recognized by its aminoacyl-tRNA or is used to terminate polypeptide translation.

controlling elements—Transposable units in maize originally identified solely by their genetic properties. They may be autonomous (able to transpose independently) or nonautonomous (able to transpose only in the presence of an autonomous element).

copy number—The number of copies of a plasmid that is maintained in a bacterium (relative to the number of copies of the origin of the bacterial chromosome).

core enzyme—The complex of RNA polymerase subunits needed for elongation. It does not include additional subunits or factors that may be needed for initiation or termination.

core histone—One of the four types of histone (H2A, H2B, H3, and H4 and their variants) found in the core particle derived from the nucleosome. (This excludes linker histones.)

core promoter—The shortest sequence at which an RNA polymerase can initiate transcription (typically at a much lower level than that displayed by a promoter containing additional elements). For RNA polymerase II it is the minimal sequence at which the basal transcription apparatus can assemble, and it includes three sequence elements: the Inr, the TATA box, and the DPE. It is typically ~40 bp long.

core sequence—The segment of DNA that is common to the attachment sites on both the phage lambda and bacterial genomes. It is the location of the recombination event that allows phage lambda to integrate.

corepressor—A small molecule that triggers repression of transcription by binding to a regulator protein.

cosmid—Cloning vector derived from a bacterial plasmid by incorporating the *cos* sites of phage lambda, which make the plasmid DNA a substrate for the lambda packaging system.

cotranslational translocation—The movement of a protein across a membrane as the protein is being synthesized. The term is usually restricted to cases in which the ribosome binds to the channel. It may be restricted to the endoplasmic reticulum.

coupled transcription/translation—The phenomena in bacteria where translation of the mRNA occurs simultaneously with its transcription.

CpG islands—Stretches of 1–2 kb in mammalian genomes that are enriched in CpG dinucleotides; frequently found in promoter regions of genes.

crossover control—Control that limits the number of recombination events between meiotic chromosomes to one to two crossovers per pair of homologs.

crossover fixation—A possible consequence of unequal crossing over that allows a mutation in one member of a tandem cluster to spread through the whole cluster (or to be eliminated).

crown gall disease—A tumor that can be induced in many plants by infection with the bacterium *Agrobacterium tumefaciens*.

CRP—A positive regulator protein activated by cyclic AMP. It is needed for RNA polymerase to initiate transcription of many operons of *E. coli*.

cryptic satellite—A satellite DNA sequence not identified as such by a separate peak on a density gradient; that is, it remains present in main band DNA.

ctDNA (cpDNA)—chloroplast DNA.

C-terminal domain—See **carboxy terminal domain**.

CUT—Cryptic unstable transcripts generated by promoters located at the 3′ end of genes.

C-value—The total amount of DNA in the genome (per haploid set of chromosomes).

C-value paradox—The lack of relationship between the DNA content (C-value) of an organism and its coding potential.

cytotoxic T cell—A T lymphocyte (usually CD8+) that can be stimulated to kill cells containing intracellular pathogens, such as viruses.

cytotype—A cytoplasmic condition that affects P element activity. The effect of cytotype is due to the presence or absence of a repressor of transposition, which is provided by the mother to the egg.

D arm—The arm of tRNA that has a high content of the base dihydrouridine.

D loop—A region of the animal mitochondrial DNA molecule that is variable in size and sequence and contains the origin of replication.

D loop (displacement loop)—The loop of displaced DNA generated by strand invasion and extension during homologous recombination.

D segment—An additional sequence that is found between the V and J regions of an immunoglobulin heavy chain.

***de novo* methyltransferase**—Adds a methyl group to an unmethylated target sequence on DNA.

deacylated tRNA—tRNA that has no amino acid or polypeptide chain attached because it has completed its role in protein synthesis and is ready to be released from the ribosome.

degradosome—A complex of bacterial enzymes, including RNAase and helicase activities, that may be involved in degrading mRNA.

delayed early genes—Genes in phage lambda that are equivalent to the middle genes of other phages. They cannot be transcribed until regulator protein(s) coded by the immediate early genes have been synthesized.

deletion—The removal of a sequence of DNA, the regions on either side being joined together except in the case of a terminal deletion at the end of a chromosome.

demethylase—A casual name for an enzyme that removes a methyl group, typically from DNA, RNA, or protein.

denaturation—A molecule's conversion from the physiological conformation to some other (inactive) conformation. In DNA, this involves the separation of the two strands due to breaking of hydrogen bonds between bases.

Dicer—An endonuclease that processes double-stranded RNA to 21 to 23 nucleotide RNA molecules.

direct repeats—Identical (or closely related) sequences present in two or more copies in the same orientation in the same molecule of DNA.

divergence—The corrected percent difference in nucleotide sequence between two related DNA sequences or in amino acid sequences between two proteins.

DNA fingerprinting—A technique for analyzing the differences between individuals of the fragments generated by using restriction enzymes to cleave regions that contain short repeated sequences or by PCR. The lengths of the repeated regions are unique to every individual, and, as a result, the presence of a particular subset in any two individuals can be used to define their common inheritance (e.g., a parent-child relationship).

DNA ligase—The enzyme that makes a bond between an adjacent 3′-OH and 5′-phosphate end where there is a nick in one strand of duplex DNA.

DNA polymerase—An enzyme that synthesizes a daughter strand(s) of DNA (under direction from a DNA template). Any particular enzyme may be involved in repair or replication (or both).

DNA repair—The removal and replacement of damaged DNA by the correct sequence.

DNase—An enzyme that degrades DNA.

domain—In reference to a chromosome, it may refer either to a discrete structural entity defined as a region within which supercoiling is independent of other regions or to an extensive region including an expressed gene that has heightened sensitivity to degradation by the enzyme DNase I. In a protein, it is a discrete continuous part of the amino acid sequence that can be equated with a particular function.

dominant negative—A mutation that results in a mutant gene product that prevents the function of the wild-type gene product, causing loss or reduction of gene activity in cells containing both the mutant and wild-type alleles. The most common cause is that the gene codes for a homomultimeric protein whose function is lost if only one of the subunits is a mutant.

dosage compensation—Mechanisms employed to compensate for the discrepancy between the presence of two X chromosomes in one sex but only one X chromosome in the other sex.

double-strand breaks (DSB)—Breaks that when both strands of a DNA duplex are cleaved at the same site. Genetic recombination is initiated by such breaks. The cell also has repair systems that act on breaks that are created at other times.

doubling time—The period (usually measured in minutes) that it takes for a bacterial cell to reproduce.

down mutation—A mutation in a promoter that decreases the rate of transcription.

downstream—Sequences proceeding farther in the direction of expression.

downstream promoter element (DPE)—A common component of RNA polymerase II promoters that do not contain a TATA box.

***Ds* element**—Dissociation element; a non-autonomous transposable element in maize, related to the autonomous Activator (*Ac*) element.

E complex—The first complex to form at a splice site, consisting of U1 snRNP bound at the splice site together with factor ASF/SF2, U2AF bound at the branch site, and the bridging protein SF1/BBP.

early genes—Genes that are transcribed before the replication of phage DNA. They code for regulators and other proteins needed for later stages of infection.

early infection—The part of the phage lytic cycle between entry and replication of the phage DNA. During this time, the phage synthesizes the enzymes needed to replicate its DNA.

EF-Tu—The elongation factor that binds aminoacyl-tRNA and places it into the A site of a bacterial ribosome.

EJC (exon junction complex)—A protein complex that assembles at exon-exon junctions during splicing and assists in RNA transport, localization, and degradation.

elongation—The stage in a macromolecular synthesis reaction (replication, transcription, or translation) in which the nucleotide or polypeptide chain is extended by the addition of individual subunits.

elongation factors—Proteins that associate with ribosomes cyclically during the addition of each amino acid to the polypeptide chain.

endonuclease—An enzyme that cleaves bonds within a nucleic acid chain; it may be specific for RNA or for single-stranded or double-stranded DNA.

endoplasmic reticulum—A network of membranes extending from the nuclear membrane responsible for the synthesis, modification, and transport of proteins.

enhancer—A *cis*-acting sequence that increases the utilization of (most) eukaryotic promoters and can function in either orientation and in any location (upstream or downstream) relative to the promoter.

epigenetic—Changes that influence the phenotype without altering the genotype. They consist of changes in the properties of a cell that are inherited that do not represent a change in genetic information.

episome—A plasmid able to integrate into bacterial DNA.

equilibrium density-gradient centrifugation—A gradient method used to separate macromolecules on the basis of differences in their density. For DNA, it is prepared from a heavy soluble compound such as CsCl.

error-prone polymerase—A DNA polymerase that incorporates noncomplementary bases into the daughter strand.

error-prone synthesis—A repair process in which noncomplementary bases are incorporated into the daughter strand.

euchromatin—Regions that constitute most of the genome in the interphase nucleus, are less tightly coiled than heterochromatin and contain most of the active or potentially active single copy genes.

excision—1. Release of phage or episome or other sequence from the host chromosome as an autonomous DNA molecule. 2. The step in an excision-repair system that consists of removing a single-stranded stretch of DNA by the action of a 5′ to 3′ exonuclease.

excision repair—A type of repair system in which one strand of DNA is directly excised and then replaced by resynthesis using the complementary strand as template.

exon—Any segment of an interrupted gene that is represented in the mature RNA product.

exon definition—The process in which a pair of splicing sites are recognized by interactions involving the 5′ site of the intron and the 5′ site of the next intron downstream.

exon shuffling—The hypothesis that genes have evolved by the recombination of various exons coding for functional protein domains.

exon trapping—Inserting a genomic fragment into a vector whose function depends on the provision of splicing junctions by the fragment.

exonuclease—An enzyme that cleaves nucleotides one at a time from the end of a polynucleotide chain; it may be specific for either the 5′ or 3′ end of DNA or RNA.

exosome—A complex of several exonucleases involved in degrading mRNA.

expressed sequence tag (EST)—A short sequenced fragment of a cDNA sequence that can be used to identify an actively expressed gene.

extein—A sequence that remains in the mature protein that is produced by processing a precursor via protein splicing.

extra arm—The arm of tRNA that shows length variation among different tRNAs.

extranuclear genes—Genes that reside outside the nucleus in organelles such as mitochondria and chloroplasts.

F plasmid—An episome that can be free or integrated in *E. coli* and that can sponsor conjugation in either form.

facultative heterochromatin—The inert state of sequences that also exist in active copies, e.g., one mammalian X chromosome in females.

fixation—The process by which a new allele replaces the allele that was previously predominant in a population.

forward mutation—A mutation that inactivates a functional gene.

frameshift—A mutation caused by deletions or insertions that are not a multiple of three base pairs. They change the frame in which triplets are translated into polypeptide.

fully methylated—A site that is a palindromic sequence that is methylated on both strands of DNA.

gain-of-function mutation—A mutation that causes an increase in the normal gene activity. It sometimes represents acquisition of certain abnormal properties. It is often, but not always, dominant.

G-bands—Bands generated on eukaryotic chromosomes by staining techniques that appear as a series of lateral striations. They are used for karyotyping (identifying chromosomes and chromosomal regions by the banding pattern).

gene cluster—A group of adjacent genes that are identical or related.

gene conversion—The alteration of one strand of a heteroduplex DNA to make it complementary with the other strand at any position(s) where there were mispaired bases, or the complete replacement of genetic material at one locus by a homologous sequence.

gene expression—The process by which the information in a sequence of DNA in a gene is used to produce an RNA or polypeptide, involving transcription and (for polypeptides) translation.

gene family—A set of genes within a genome that code for related or identical proteins or RNAs. The members were derived by duplication of an ancestral gene followed by accumulation of changes in sequence between the copies. Most often the members are related but not identical.

gene knock-in—A process similar to a knockout, but more subtle mutations are made, such as nucleotide substitutions or the addition of epitope tags.

gene knockout—A process in which a gene function is eliminated, usually by replacing most of the coding sequence with a selectable marker *in vitro* and transferring the altered gene to the genome by homologous recombination.

genetic code—The correspondence between triplets in DNA (or RNA) and amino acids in polypeptide.

genetic drift—The chance fluctuation (without selective pressure) of the frequencies of alleles in a population.

genetic map—See **linkage map**.

genetic recombination—A process by which separate DNA molecules are joined into a single molecule, due to such processes as crossing over or transposition.

genome—The complete set of sequences in the genetic material of an organism. It includes the sequence of each chromosome plus any DNA in organelles.

glycosylase—A repair enzyme that removes damaged bases by cleaving the bond between the base and the sugar.

Golgi apparatus—A stack of flattened, membrane-enclosed compartments involved in the modification and sorting of lipids and proteins.

gratuitous inducer—Inducers that resemble authentic inducers of transcription but are not substrates for the induced enzymes.

growing point—See **replication fork**.

GU-AG rule—The rule that describes the presence of these constant dinucleotides at the first two and last two positions of introns of nuclear genes.

guide RNA—A small RNA whose sequence is complementary to the sequence of an RNA that has been edited. It is used as a template for changing the sequence of the pre-edited RNA by inserting or deleting nucleotides.

gyrase—An enzyme that changes the number of times the two strands in a closed DNA molecule cross each other. It does this by cutting the DNA, passing DNA through the break, and resealing the DNA.

hairpin—A "stem and loop" structure formed by intramolecular base pairing, particularly in RNA.

haplotype—The particular combination of alleles in a defined region of some chromosome-in effect, the genotype in miniature. Originally used to described combinations of MHC alleles, it now may be used to describe particular combinations of RFLPs, SNPs, or other markers.

Hb anti-Lepore—A fusion gene produced by unequal crossing over that has the N-terminal part of β globin and the C-terminal part of δ globin.

Hb Kenya—A fusion gene produced by unequal crossing over between the between Aγ and β globin genes.

Hb Lepore—An unusual globin protein that results from unequal crossing over between the β and δ genes. The genes become fused together to produce a single β-like chain that consists of the N-terminal sequence of δ joined to the C-terminal sequence of β.

HbH disease—A condition in which there is a disproportionate amount of the abnormal tetramer β4 relative to the amount of normal hemoglobin (α2β2).

heavy chain—The larger of the two types of subunits that make up an antibody tetramer. Each antibody contains two heavy chains. The N-terminus of the heavy chain forms part of the antigen recognition site, while the C-terminus determines the antibody subclass.

helicase—A enzyme that uses energy provided by ATP hydrolysis to separate the strands of a nucleic acid duplex.

helix-loop-helix—The motif that is responsible for dimerization of a class of transcription factors called HLH proteins. A bHLH protein has a basic DNA-binding sequence close to the dimerization motif.

helix-turn-helix—The motif that describes an arrangement of two α-helices that form a site that binds to DNA, one fitting into the major groove of DNA and other lying across it.

helper T cell—A T lymphocyte that activates macrophages and stimulates B cell proliferation and antibody production. They usually express cell surface CD4 but not CD8.

helper virus—A virus that provides functions absent from a defective virus, enabling the latter to complete the infective cycle during a mixed infection with the helper virus.

hemimethylated DNA—DNA that is methylated on one strand of a target sequence that has a cytosine on each strand.

hemimethylated site—A palindromic sequence that is methylated on only one strand of DNA.

heterochromatin—Regions of the genome that are highly condensed, are not transcribed, and are late-replicating. It is divided into two types: constitutive and facultative.

heteroduplex DNA—DNA that is generated by base pairing between complementary single strands derived from the different parental duplex molecules; it occurs during genetic recombination.

heterogeneous nuclear RNA (hnRNA)—RNA that constitutes transcripts of nuclear genes made primarily by RNA polymerase II; it has a wide size distribution and variable stability.

heteromultimer—A molecular complex (such as a protein) composed of different subunits.

heteroplasmy—Having more than one allelic variant in a mitochondrion.

Hfr—A bacterium that has an integrated F plasmid within its chromosome. Hfr stands for high frequency recombination, referring to the fact that chromosomal genes are transferred from an Hfr cell to an F^- cell much more frequently than from an F^+ cell.

histone acetyltransferase (HAT)—An enzyme that modifies histones by addition of acetyl groups; some transcriptional coactivators have this activity. Also known as **lysine acetyltransferase (KAT)**.

histone code—The hypothesis that combinations of specific modifications on specific histone residues act cooperatively to define chromatin function.

histone deacetylase (HDAC)—Enzyme that removes acetyl groups from histones; may be associated with repressors of transcription.

histone fold—A motif found in all four core histones in which three α-helices are connected by two loops.

histone octamer—The complex of two copies each of the four different core histones (H2A, H2B, H3, and H4); DNA wraps around the histone octamer to form the **nucleosome**.

histone tails—Flexible amino- or carboxy-terminal regions of the core histones that extend beyond the surface of the nucleosome; histone tails are sites of extensive post-translational modification.

histone variant—Any of a number of histones closely related to one of the core histones (H2A, H2B, H3 or H4) that can assemble into a nucleosome in the place of the related core histone; many histone variants have specialized functions or localization. There are also numerous linker histone variants.

histones—Conserved DNA-binding proteins that form the basic subunit of chromatin in eukaryotes. H2A, H2B, H3, and H4 form an octameric core around which DNA coils to form a nucleosome. Linker histones are external to the nucleosome.

hnRNP—The ribonucleoprotein form of hnRNA (heterogeneous nuclear RNA), in which the hnRNA is complexed with proteins. Pre-mRNAs are not exported until processing is complete; thus, they are found only in the nucleus.

Holliday junction—An intermediate structure in homologous recombination, for which the two duplexes of DNA are connected by the genetic material exchanged between two of the four strands, one from each duplex. A joint molecule is said to be resolved when nicks in the structure restore two separate DNA duplexes.

holoenzyme—1. The DNA polymerase complex that is competent to initiate replication. 2. The RNA polymerase form that is competent to initiate transcription. It consists of the four subunits of the core enzyme ($\alpha 2\beta\beta'$) and σ factor.

homeodomain—A DNA-binding motif that typifies a class of transcription factors.

homologous genes (homologs)—Related genes in the same species, such as alleles on homologous chromosomes, or multiple genes in the same genome sharing common ancestry.

homologous recombination—Recombination involving a reciprocal exchange of sequences of DNA, e.g., between two chromosomes that carry the same genetic loci.

homomultimer—A molecular complex (such as a protein) in which the subunits are identical.

horizontal transfer—The transfer of DNA from one cell to another by a process other than cell division, such as bacterial conjugation.

hotspot—A site in the genome at which the frequency of mutation (or recombination) is very much increased, usually by at least an order of magnitude relative to neighboring sites.

housekeeping genes—Genes that are (theoretically) expressed in all cells because they provide basic functions needed for sustenance of all cell types.

humoral response—An immune response that is mediated primarily by antibodies. It is defined as immunity that can be transferred from one organism to another by serum antibody.

hybrid dysgenesis—The inability of certain strains of *D. melanogaster* to interbreed, because the hybrids are sterile (although otherwise they may be phenotypically normal).

hybridization—The pairing of complementary RNA and DNA strands to give an RNA-DNA hybrid.

hydrops fetalis—A fatal disease resulting from the absence of the hemoglobin α gene.

hypermutation—The introduction of somatic mutations in a rearranged immunoglobulin gene. The mutations can change the sequence of the corresponding antibody, especially in its antigen-binding site.

hypersensitive site—A short region of chromatin detected by its extreme sensitivity to cleavage by DNase I and other nucleases; it comprises an area from which nucleosomes are excluded.

idling reaction—This results in the production of pppGpp and ppGpp by ribosomes when an uncharged tRNA is present in the A site; this triggers the stringent response.

IF-1—A bacterial initiation factor that stabilizes the initiation complex for polypeptide translation.

IF-2—A bacterial initiation factor that binds the initiator tRNA to the initiation complex for polypeptide translation.

IF-3—A bacterial initiation factor required for 30S ribosomal subunits to bind to initiation sites in mRNA. It also prevents 30S subunits from binding to 50S ribosomal subunits.

immediate early genes—Genes in phage lambda that are equivalent to the early class of other phages. They are transcribed immediately upon infection by the host RNA polymerase.

immune response—An organism's reaction, mediated by components of the immune system, to an antigen.

immunity—In phages, the ability of a prophage to prevent another phage of the same type from infecting a cell. In plasmids, the ability of a plasmid to prevent another of the same type from becoming established in a cell. It can also refer to the ability of certain transposons to prevent others of the same type from transposing to the same DNA molecule.

immunity region—A segment of the phage genome that enables a prophage to inhibit additional phage of the same type from infecting the bacterium. This region has a gene that encodes for the repressor, as well as the sites to which the repressor binds.

immunoglobulin—A protein that is produced by B cells and that binds to a particular antigen.

imprecise excision—It occurs when the transposon removes itself from the original insertion site but leaves behind some of its sequence.

imprinting—A change in a gene that occurs during passage through the sperm or egg with the result that the paternal and maternal alleles have different properties in the very early embryo. This is caused by methylation of DNA.

***in situ* hybridization**—Hybridization performed by denaturing the DNA of cells squashed on a microscope slide so that reaction is possible with an added single-stranded RNA or DNA; the added preparation is radioactively labeled and its hybridization is followed by autoradiography.

incision—A step in a mismatch excision repair system in which an endonuclease recognizes the damaged area in the DNA and isolates it by cutting the DNA strand on both sides of the damage.

indirect end labeling—A technique for examining the organization of DNA by making a cut at a specific site and identifying all fragments containing the sequence adjacent to one side of the cut; it reveals the distance from the cut to the next break(s) in DNA.

induced mutations—Mutations that result from the action of a mutagen. The mutagen may act directly on the bases in DNA or it may act indirectly to trigger a pathway that leads to a change in DNA sequence.

inducer—A small molecule that triggers gene transcription by binding to a regulator protein.

inducible gene—A gene that is turned on by the presence of its substrate.

induction—The ability to synthesize certain enzymes only when their substrates are present; applied to gene expression, it refers to switching on transcription as a result of interaction of the inducer with the regulator protein.

induction of phage—A phage's entry into the lytic (infective) cycle as a result of destruction of the lysogenic repressor, which leads to excision of free phage DNA from the bacterial chromosome.

initiation—The stages of translation up to synthesis of the first peptide bond of the polypeptide.

initiation codon—A special codon (usually AUG) used to start synthesis of a polypeptide.

initiation factors (IFs)—Proteins that associate with the small subunit of the ribosome specifically at the stage of initiation of polypeptide translation.

initiator (Inr)—The sequence of a pol II promoter between –3 and +5 and has the general sequence Py_2CAPy_5. It is the simplest possible pol II promoter.

insertion sequence (IS)—A small bacterial transposon that carries only the genes needed for its own transposition.

insulator—A sequence that prevents an activating or inactivating effect passing from one side to the other.

integral membrane protein—A protein that requires disruption of the lipid bilayer in order to be released from the membrane.

integrase—An enzyme that is responsible for a site-specific recombination that inserts one molecule of DNA into another.

integration—Insertion of a viral or another DNA sequence into a host genome as a region covalently linked on either side to the host sequences.

intein—The part that is removed from a protein that is processed by protein splicing.

interactome—The complete set of protein complexes/protein-protein interactions present in a cell, tissue, or organism.

interallelic complementation—The change in the properties of a heteromultimeric protein brought about by the interaction of subunits coded by two different mutant alleles; the mixed protein may be more or less active than the protein consisting of subunits only of one or the other type.

interbands—The relatively dispersed regions of polytene chromosomes that lie between the bands.

intercistronic region—In a polycistronic mRNA, the distance between the termination codon of one gene and the initiation codon of the next gene.

interrupted gene—A gene in which the coding sequence is not continuous due to the presence of introns.

intron—A segment of DNA that is transcribed, but later removed from within the transcript by splicing together the sequences (exons) on either side of it.

intron definition—The process in which a pair of splicing sites are recognized by interactions involving only the 5′ site and the branchpoint/3′ site.

intron homing—The ability of certain introns to insert themselves into a target DNA. The reaction is specific for a single target sequence.

introns early model—The hypothesis that the earliest genes contained introns and some genes subsequently lost them.

introns late model—The hypothesis that the earliest genes did not contain introns and that introns were subsequently added to some genes.

invariant sequences—Base positions in tRNA that have the same nucleotide in virtually all (>95%) tRNAs.

inverted terminal repeats—The short related or identical sequences present in reverse orientation at the ends of some transposons.

IRES (internal ribosome entry site)—A eukaryotic messenger RNA sequence that allows a ribosome to initiate polypeptide translation without migrating from the 5′ end.

isoaccepting tRNAs—See **cognate tRNAs**.

J segments—Coding sequences in the immunoglobulin and T cell receptor loci. They are between the variable (V) and constant (C) gene segments.

joint molecule—A pair of DNA duplexes that are connected together through a reciprocal exchange of genetic material.

kinetic proofreading—A proofreading mechanism that depends on incorrect events proceeding more slowly than correct events, so that incorrect events are reversed before a subunit is added to a polymeric chain.

kinetochore—A small organelle associated with the surface of the centromere that attaches a chromosome to the microtubules of the mitotic spindle. Each mitotic chromosome contains two "sisters" that are positioned on opposite sides of its centromere and face in opposite directions.

kuru—A human neurological disease caused by prions. It may be caused by eating infected brains.

***lac* repressor**—A negative gene regulator encoded by the *lacI* gene that turns off the *lac* operon.

lagging strand—The strand of DNA that must grow overall in the 3′ to 5′ direction and is synthesized discontinuously in the form of short fragments (5′–3′) that are later connected covalently.

large subunit—The subunit of the ribosome (50S in bacteria, 60S in eukaryotes) that has the peptidyl transferase active site that synthesizes the peptide bond.

lariat—An intermediate in RNA splicing in which a circular structure with a tail is created by a 5′ to 2′ bond.

late gene—A gene transcribed when phage DNA is being replicated. It codes for components of the phage particle.

late infection—The part of the phage lytic cycle from DNA replication to lysis of the cell. During this time, the DNA is replicated and structural components of the phage particle are synthesized.

lateral element—A structure in the synaptonemal complex that forms when a pair of sister chromatids condenses on to an axial element.

leader—In a protein, it is a short N-terminal sequence responsible for initiating passage into or through a membrane; in mRNA, it is the untranslated sequence at the 5′ end that precedes the initiation codon.

leading strand—The strand of DNA that is synthesized continuously in the 5′ to 3′ direction.

lesion bypass—Replication by an error prone DNA polymerase on a template that contains a damaged base. The polymerase can incorporate a noncomplementary base into the daughter strand.

leucine zipper—A dimerization motif that is found in a class of transcription factors.

leucine-rich region—A motif found in the extracellular domains of some surface receptor proteins in animal and plant cells.

licensing factor—A factor located in the nucleus and necessary for replication; it is inactivated or destroyed after one round of replication. New factors must be provided for further rounds of replication to occur.

light chain—The smaller of the two types of subunits that make up an antibody tetramer. Each antibody contains two light chains. The N-terminus of the light chain forms part of the antigen recognition site.

linkage—The tendency of genes to be inherited together as a result of their location on the same chromosome; measured by percent recombination between loci.

linkage map—A map of the positions of loci or other genetic markers on a chromosome obtained by measuring recombination frequencies between markers.

linker DNA—Non-nucleosomal DNA present between nucleosomes.

linker histones—A family of histones (such as histone H1) that are not components of the nucleosome core; linker histone bind nucleosomes and/or linker DNA and promote 30-nm fiber formation.

locus—The position on a chromosome at which the gene for a particular trait resides; it may be occupied by any one of the alleles for the gene.

locus control region (LCR)—The region that is required for the expression of several genes in a domain.

long interspersed elements (LINEs)—Long interspersed nuclear elements; a major class of retrotransposons that occupy ~21% of the human genome (see **retrotransposon**).

long terminal repeat (LTR)—The sequence that is repeated at each end of the provirus (integrated retroviral sequence).

loop—A single-stranded region at the end of a hairpin in RNA (or single-stranded DNA); it corresponds to the sequence between inverted repeats in duplex DNA.

loss-of-function mutation—A mutation that eliminates or reduces the activity of a gene. It is often, but not always, recessive.

luxury gene—A gene coding for a specialized function, synthesized (usually) in large amounts in particular cell types.

lyase—A repair enzyme (usually also a glycosylase) that opens the sugar ring at the site of a damaged base.

lysine (K) acetyltransferase (KAT)—An enzyme (typically present in large complexes) that acetylates lysine residues in histones (or other proteins). Previously known as histone acetyltransferase (HAT).

lysis—The death of bacteria at the end of a phage infective cycle when they burst open to release the progeny of an infecting phage (because phage enzymes disrupt the bacterium's cytoplasmic membrane or cell wall). The same term also applies to eukaryotic cells; for example, when infected cells are attacked by the immune system.

lysogeny—The ability of a phage to survive in a bacterium as a stable prophage component of the bacterial genome.

lytic infection—Infection of a bacterium by a phage that ends in the destruction of the bacterium with release of progeny phage.

maintenance methyltransferase—Adds a methyl group to a target site that is already hemimethylated.

major groove—A fissure running the length of the DNA double helix that is 22 Å across.

major histocompatibility complex (MHC)—A chromosomal region containing genes that are involved in the immune response. The genes encode proteins for antigen presentation, cytokines, and complement, as well as other functions. It is highly polymorphic. Its genes and proteins are divided into three classes.

maternal inheritance—The preferential survival in the progeny of genetic markers provided by one parent.

mating type cassette—Yeast mating type is determined by a single active locus (the active cassette) and two inactive copies of the locus (the silent cassettes). Mating type is changed when an active cassette of one type is replaced by a silent cassette of the other type.

matrix attachment region—A region of DNA that attaches to the nuclear matrix. It is also known as a scaffold attachment site (SAR).

maturase—A protein coded by a group I or group II intron that is needed to assist the RNA to form the active conformation that is required for self-splicing.

mature transcript—A modified RNA transcript. Modification may include the removal of intron sequences and alterations to the 5′ and 3′ ends.

mediator—A large protein complex associated with yeast bacterial RNA polymerase II. It contains factors that are necessary for transcription from many or most promoters.

melting temperature—The midpoint of the temperature range over which the strands of DNA separate.

messenger RNA (mRNA)—The intermediate that represents one strand of a gene coding for polypeptide. Its coding region is related to the polypeptide sequence by the triplet genetic code.

metaphase (or mitotic) scaffold—A proteinaceous structure in the shape of a sister chromatid pair, generated when chromosomes are depleted of histones.

methyltransferase—An enzyme that adds a methyl group to a substrate, which can be a small molecule, a protein, or a nucleic acid.

microarray—An arrayed series of thousands of tiny DNA oligonucleotide samples imprinted on a small chip. mRNAs can be hybridized to microarrays to assess the amount and level of gene expression.

micrococcal nuclease (MNase)—An endonuclease that cleaves DNA; in chromatin, DNA is cleaved preferentially between nucleosomes.

microsatellite—DNAs consisting of repetitions of extremely short (typically <10 bp) units.

microtubule organizing center (MTOC)—A region from which microtubules emanate. In animal cells the centrosome is the major microtubule organizing center.

middle genes—Phage genes that are regulated by the proteins coded by early genes. Some proteins coded by them catalyze replication of the phage DNA; others regulate the expression of a later set of genes.

minicell—An anucleate bacterial (*E. coli*) cell produced by a division that generates a cytoplasm without a nucleus.

minisatellite—DNAs consisting of ~10 copies of a short repeating sequence. The length of the repeating unit is measured in 10s of base pairs. The number of repeats varies between individual genomes.

minor groove—A fissure running the length of the DNA double helix that is 12 Å across.

minus strand DNA—The single-stranded DNA sequence that is complementary to the viral RNA genome of a plus strand virus.

miRNA—Very short RNAs that may regulate gene expression.

mismatch repair—Repair that corrects recently inserted bases that do not pair properly. The process preferentially corrects the sequence of the daughter strand by distinguishing the daughter strand and parental strand, sometimes on the basis of their states of methylation.

missense suppressor—A suppressor that codes for a tRNA that has been mutated to recognize a different codon. By inserting a different amino acid at a mutant codon, the tRNA suppresses the effect of the original mutation.

modification—All changes made to the nucleotides of DNA or RNA after their initial incorporation into the polynucleotide chain.

molecular clock—An approximately constant rate of evolution that occurs in DNA sequences, such as by the genetic drift of neutral mutations.

monocistronic mRNA—mRNA that codes for one polypeptide.

mtDNA—mitochondrial DNA.

multicopy replication control—Occurs when the control system allows the plasmid to exist in more than one copy per individual bacterial cell.

multiforked chromosome—A bacterial chromosome that has more than one set of replication forks, because a second initiation has occurred before the first cycle of replication has been completed.

mutagens—Substances that increase the rate of mutation by inducing changes in DNA sequence, directly or indirectly.

mutation hotspot—A site in the genome at which the frequency of mutation (or recombination) is very much increased, usually by at least an order of magnitude relative to neighboring sites.

mutator—A mutation or a mutated gene that increases the basal level of mutation. Such genes often code for proteins that are involved in repairing damaged DNA.

N nucleotide—A short non-templated sequence that is added randomly by the enzyme at coding joints during rearrangement of immunoglobulin and T cell receptor genes. They augment the diversity of antigen receptors.

n-1 rule—The rule that states that only one X chromosome is active in female mammalian cells; any others are inactivated.

nascent polypeptide—A protein that has not yet completed its synthesis; the polypeptide chain is still attached to the ribosome via a tRNA.

nascent RNA—A ribonucleotide chain that is still being synthesized, so that its 3′ end is paired with DNA where RNA polymerase is elongating.

ncRNA—RNA that does not contain an open reading frame.

negative complementation—It occurs when interallelic complementation allows a mutant subunit to suppress the activity of a wild-type subunit in a multimeric protein.

negative control—This describes a gene system that is on and a regulator is required to turn it off.

negative inducible—A control circuit in which an active repressor is inactivated by the substrate of the operon.

negative repressible—A control circuit in which an inactive repressor is activated by the product of the operon.

nested genes—A gene located within an intron of another gene.

neutral mutation—A mutation that has no significant effect on evolutionary fitness and usually has no effect on the phenotype.

neutral substitutions—Substitutions in a protein that cause changes in amino acids that do not affect activity.

N-formyl-methionyl-tRNA—The aminoacyl-tRNA that initiates bacterial polypeptide translation. The amino group of the methionine is formylated.

nonallelic genes—Two (or more) copies of the same gene that are present at different locations in the genome (contrasted with alleles, which are copies of the same gene derived from different parents and present at the same location on the homologous chromosomes).

nonautonomous controlling element—A transposon in maize that encodes a non-functional transposase; it can transpose only in the presence of a *trans*-acting autonomous member of the same family.

nonhistone—Any structural protein found in a chromosome except one of the histones.

non-homologous end-joining (NHEJ)—The process that ligates blunt ends. It is common to many repair pathways and to certain recombination pathways (such as immunoglobulin recombination).

non-Mendelian inheritance—a pattern of inheritance that does not follow that expected by Mendelian principles (each parent contributing a single allele to offspring). Extranuclear genes show a non-Mendelian inheritance pattern.

nonproductive rearrangement—Occurs as a result of the recombination of V, (D), J gene segments if the rearranged gene segments are not in the correct reading frame. It occurs when nucleotide addition or subtraction disrupts the reading frame or when a functional protein is not produced.

nonrepetitive DNA—DNA that is unique (present only once) in a genome.

nonreplicative transposition—The movement of a transposon that leaves a donor site (usually generating a double-strand break) and moves to a new site.

nonsense-mediated mRNA decay—A pathway that degrades an mRNA that has a nonsense mutation prior to the last exon.

nonsense suppressor—A gene coding for a mutant tRNA that is able to respond to one or more of the termination codons and insert an amino acid at that site.

nontranscribed spacer—The region between transcription units in a tandem gene cluster.

nonviral superfamily—Transposons originated independently of retroviruses.

nuclear localization signal (NLS)—Short sequence of amino acids that targets a polypeptide to the nucleus via the nuclear pore.

nuclear pore—Pore in the nuclear membrane that allows transport of RNA, proteins, and other molecules in and out of the nucleus.

nucleation center—A duplex hairpin in TMV (tobacco mosaic virus) in which assembly of coat protein with RNA is initiated.

nucleoid—The structure in a prokaryotic cell that contains the genome. The DNA is bound to proteins and is not enclosed by a membrane.

nucleolar organizer—The region of a chromosome carrying genes coding for rRNA.

nucleolus—A discrete region of the nucleus where ribosomes are produced.

nucleoside—A molecule consisting of a purine or pyrimidine base linked to the 1′ carbon of a pentose sugar.

nucleosome—The basic structural subunit of chromatin, consisting of ~200 bp of DNA and an octamer of histone proteins.

nucleosome positioning—The placement of nucleosomes at defined sequences of DNA instead of at random locations with regard to sequence.

nucleotide—A molecule consisting of a purine or pyrimidine base linked to the 1′ carbon of a pentose sugar and a phosphate group linked to either the 5′ or 3′ carbon of the sugar.

null mutation—A mutation that completely eliminates the function of a gene.

nut—An acronym for N utilization site, the sequence of DNA that is recognized by the N antitermination factor.

ochre codon—The triplet UAA, one of the three termination codons that end polypeptide translation.

Okazaki fragment—Short stretches of 1000 to 2000 bases produced during discontinuous replication; they are later joined into a covalently intact strand.

one gene:one enzyme hypothesis—Beadle and Tatum's hypothesis that a gene is responsible for the production of a single enzyme.

one gene:one polypeptide hypothesis—A modified version of the not generally correct one gene:one enzyme hypothesis; the hypothesis that a gene is responsible for the production of a single polypeptide.

opal codon—The triplet UGA, one of the three termination codons that end polypeptide translation. It has evolved to code for an amino acid in a small number of organisms or organelles.

open complex—The stage of initiation of transcription when RNA polymerase causes the two strands of DNA to separate to form the "transcription bubble."

open reading frame—A sequence of DNA consisting of triplets that can be translated into amino acids starting with an initiation codon and ending with a termination codon.

operator—The site on DNA at which a repressor protein binds to prevent transcription from initiating at the adjacent promoter.

operon—A unit of bacterial gene expression and regulation, including structural genes and control elements in DNA recognized by regulator gene product(s).

opine—A derivative of arginine that is synthesized by plant cells infected with crown gall disease.

ORC—Origin recognition complex, found in eukaryotes, a multiprotein complex that binds to the replication origin, the ARS, and remains associated with it throughout the cell cycle.

origin—A sequence of DNA at which replication is initiated.

orthologous genes (orthologs)—Related genes in different species.

overlapping gene—A gene in which part of the sequence is found within part of the sequence of another gene.

overwound—B-form DNA that has more than 10 base pairs per turn of the helix.

P element—A type of transposon in *D. melanogaster*.

P nucleotide—A short palindromic (inverted repeat) sequence that is generated during rearrangement of immunoglobulin and T cell receptor V, (D), and J gene segments. They are generated at coding joints when RAG proteins cleave the hairpin ends generated during rearrangement.

P site—The site in the ribosome that is occupied by peptidyl-tRNA, the tRNA carrying the nascent polypeptide chain, still paired with the codon to which it bound in the A site.

packing ratio—The ratio of the length of DNA to the unit length of the fiber containing it.

palindrome—A symmetrical sequence that reads the same forward and backward.

paralogous genes (paralogs)—Genes that share a common ancestry due to gene duplication.

patch recombinant—DNA that results from a Holliday junction being resolved by cutting the exchanged strands. The duplex is largely unchanged, except for a DNA sequence on one strand that came from the homologous chromosome.

pathogenicity islands—DNA segments that are present in pathogenic bacterial genomes but absent in their nonpathogenic relatives.

peptidyl transferase—The activity of the large ribosomal subunit that synthesizes a peptide bond when an amino acid is added to a growing polypeptide chain. The actual catalytic activity is a property of the rRNA.

peptidyl-tRNA—The tRNA to which the nascent polypeptide chain has been transferred following peptide bond synthesis during polypeptide translation.

periplasm (periplasmic space)—The region between the inner and outer membranes in the bacterial cell envelope.

periseptal annulus—A ringlike area where inner and outer membranes appear fused. Formed around the circumference of the bacterium, it determines the location of the septum.

peroxins—The protein components of the peroxisome.

peroxisome—An organelle in the cytoplasm enclosed by a single membrane. It contains oxidizing enzymes and is involved in fatty acid metabolism.

phage—An abbreviation of bacteriophage or bacterial virus.

photoreactivation—A repair mechanism that uses a white light-dependent enzyme to split cyclobutane pyrimidine dimers formed by ultraviolet light.

pilin—The subunit that is polymerized into the pilus in bacteria.

pilus (pl. **pili**)—A surface appendage on a bacterium that allows the bacterium to attach to other bacterial cells. It appears as a short, thin, flexible rod. During conjugation, it is used to transfer DNA from one bacterium to another.

plasmid—Circular, extrachromosomal DNA. It is autonomous and can replicate itself.

plastids—A group of related organelles found in plants and algae that contain multiple copies of a circular plastid genome. Different plastids have specialized functions, such as chloroplasts, which contain the machinery of photosynthesis, or chromoplasts, which are used for pigment storage. All plastid types develop from undifferentiated protoplastids.

plus strand DNA—The strand of the duplex sequence representing a retrovirus that has the same sequence as that of the RNA.

plus strand virus—A virus with a single-stranded nucleic acid genome whose sequence directly codes for the protein products.

point mutation—A change in the sequence of DNA involving a single base pair.

poly(A)—A stretch of ~200 bases of adenylic acid that is added to the 3′ end of mRNA following its synthesis.

poly(A) binding protein (PABP)—The protein that binds to the 3′ stretch of poly(A) on a eukaryotic mRNA.

poly(A) polymerase (PAP)—The enzyme that adds the stretch of polyadenylic acid to the 3′ end of eukaryotic mRNA. It does not use a template.

poly(A)+ mRNA—mRNA that has a 3′ terminal stretch of poly(A).

polycistronic mRNA—mRNA that includes coding regions representing more than one gene.

polymorphism—The simultaneous occurrence in the population of alleles showing variations at a given position.

polynucleotide—A chain of nucleotides, such as DNA or RNA.

polyribosome (polysome)—An mRNA that is simultaneously being translated by several ribosomes.

polytene chromosomes—Chromosomes that are generated by successive replications of a chromosome set without separation of the replicas.

position effect variegation (PEV)—Silencing of gene expression that occurs as the result of proximity to heterochromatin.

positive control—This describes a system in which a gene is not expressed unless some action turns it on.

positive inducible—A control circuit in which an inactive positive regulator is converted into an active regulator by the substrate of the operon.

positive repressible—A control circuit in which an active positive regulator is inactivated by the product of the operon.

postmeiotic segregation—The segregation of two strands of a duplex DNA that bear different information (created by heteroduplex formation during meiosis) when a subsequent replication allows the strands to separate.

postreplication complex—A protein-DNA complex in *S. cerevisiae* that consists of the ORC complex bound to the origin.

posttranslational translocation—The movement of a protein across a membrane after the synthesis of the protein is completed and it has been released from the ribosome.

ppGpp—Guanosine tetraphosphate, with diphosphate groups attached to both the 5′ and 3′ positions, is an alarmone.

pppGpp—Guanosine pentaphosphate, with a triphosphate group attached to the 5′ position and a diphosphate gropup attached to the 3′ position, is an alarmone.

precise excision—The removal of a transposon plus one of the duplicated target sequences from the chromosome. Such an event can restore function at the site where the transposon inserted.

preinitiation complex—The assembly of transcription factors at the promoter before RNA polymerase binds in eukaryotic transcription.

preprotein—A protein to be imported into an organelle or secreted from bacteria before its signal sequence has been removed.

pre-mRNA—The nuclear transcript that is processed by modification and splicing to give an mRNA.

prereplication complex—A protein-DNA complex at the origin in *S. cerevisiae* that is required for DNA replication. The complex contains the ORC complex, Cdc6, and the MCM proteins.

presynaptic filaments—Single-stranded DNA bound in a helical nucleoprotein filament with a strand transfer protein such as Rad51 or RecA.

primary transcript—The original unmodified RNA product corresponding to a transcription unit.

primase—A type of RNA polymerase that synthesizes short segments of RNA that will be used as primers for DNA replication.

primer—A short sequence (often of RNA) that is paired with one strand of DNA and provides a free 3′-OH end at which a DNA polymerase starts synthesis of a deoxyribonucleotide chain.

prion—A proteinaceous infectious agent that behaves as an inheritable trait, although it contains no nucleic acid. One example is PrPSc, the agent of scrapie in sheep and bovine spongiform encephalopathy.

processed pseudogene—An inactive gene copy that lacks introns, contrasted with the interrupted structure of the active gene. Such genes originate by reverse transcription of mRNA and insertion of a duplex copy into the genome.

processivity—The ability of an enzyme to perform multiple catalytic cycles with a single template instead of dissociating after each cycle.

productive rearrangement—Occurs as a result of the recombination of V, (D), J gene segments if all the rearranged gene segments are in the correct reading frame.

programmed frameshifting—Frameshifting that is required for expression of the polypeptide sequences encoded beyond a specific site at which a +1 or –1 frameshift occurs at some typical frequency.

promoter—A region of DNA where RNA polymerase binds to initiate transcription.

proofreading—A mechanism for correcting errors in DNA synthesis that involves scrutiny of individual units after they have been added to the chain.

prophage—A phage genome covalently integrated as a linear part of the bacterial chromosome.

protein sorting—The direction of different types of proteins for transport into or between specific organelles.

protein splicing—The autocatalytic process by which an intein is removed from a protein and the exteins on either side become connected by a standard peptide bond.

protein translocation—The movement of a protein across a membrane. This occurs across the membranes of organelles in eukaryotes, or across the plasma membrane in bacteria. Each membrane across which proteins are moved has a channel specialized for the purpose.

proteome—The complete set of proteins that is expressed by the entire genome. Sometimes the term is used to describe the complement of proteins expressed by a cell at any one time.

provirus—A duplex sequence of DNA integrated into a eukaryotic genome that represents the sequence of the RNA genome of a retrovirus.

pseudogenes—Inactive but stable components of the genome derived by mutation of an ancestral active gene. Usually they are inactive because of mutations that block transcription or translation or both.

puff—An expansion of a band of a polytene chromosome associated with the synthesis of RNA at some locus in the band.

purine—A double-ringed nitrogenous base, such as adenine or guanine.

pyrimidine—A single-ringed nitrogenous base, such as cytosine, thymine, or uracil.

pyrimidine dimer—A dimer that forms when ultraviolet irradiation generates a covalent link directly between two adjacent pyrimidine bases in DNA. It blocks DNA replication and transcription.

R segments—The sequences that are repeated at the ends of a retroviral RNA. They are called R-U5 and U3-R.

rDNA—Genes encoding ribosomal RNA (rRNA).

reading frame—One of three possible ways of reading a nucleotide sequence. Each divides the sequence into a series of successive triplets.

readthrough—This occurs at transcription or translation when RNA polymerase or the ribosome, respectively, ignores a termination signal because of a mutation of the template or the behavior of an accessory factor.

rec mutations—Mutations of *E. coli* that cannot undertake general recombination.

recoding—Events that occur when the meaning of a codon or series of codons is changed from that predicted by the genetic code. It may involve altered interactions between aminoacyl-tRNA and mRNA that are influenced by the ribosome.

recognition helix—One of the two helices of the helix-turn-helix motif that makes contacts with DNA that are specific for particular bases. This determines the specificity of the DNA sequence that is bound.

recombinant joint—The point at which two recombining molecules of duplex DNA are connected (the edge of the heteroduplex region).

recombinase—Enzyme that catalyzes site-specific recombination.

recombination nodules (nodes)—Dense objects present on the synaptonemal complex; they may represent protein complexes involved in crossing-over.

recombination-repair—A mode of filling a gap in one strand of duplex DNA by retrieving a homologous single strand from another duplex.

redundancy—The concept that two or more genes may fulfill the same function, so that no single one of them is essential.

regulator gene—A gene that codes for a product (typically protein) that controls the expression of other genes (usually at the level of transcription).

relaxase—An enzyme that cuts one strand of DNA and binds to the free 5′ end.

relaxed mutants—In *E. coli*, these do not display the stringent response to starvation for amino acids (or other nutritional deprivation).

release factor (RF)—A protein required to terminate polypeptide translation to cause release of the completed polypeptide chain and the ribosome from mRNA.

renaturation—The reassociation of denatured complementary single strands of a DNA double helix.

repair—When a DNA strand has been damaged, it is excised and replaced by the synthesis of a new stretch through repair synthesis. This can also take place by recombination reactions, when the duplex region containing the damaged strand is replaced by an undamaged region from another copy of the genome.

repetitive DNA—DNA that is present in many (related or identical) copies in a genome.

replacement sites—Sites in a coding region at which mutations have altered the amino acid that is encoded.

replication-defective virus—A virus that cannot perpetuate an infective cycle because some of the necessary genes are absent (replaced by host DNA in a transducing virus) or mutated.

replication eye—A region in which DNA has been replicated within a longer, unreplicated region.

replication fork—The point at which strands of parental duplex DNA are separated so that replication can proceed. A complex of proteins including DNA polymerase is found there.

replicative transposition—The movement of a transposon by a mechanism in which first it is replicated, and then one copy is transferred to a new site.

replicon—A unit of the genome in which DNA is replicated. Each contains an origin for initiation of replication.

replisome—The multiprotein structure that assembles at the bacterial replication fork to undertake synthesis of DNA. It contains DNA polymerase and other enzymes.

reporter gene—Any gene whose product (RNA or protein) can be easily detected; reporter genes are often used as a substitute for a gene of interest.

repressible gene—A gene that is turned off by its product.

repression—The ability to prevent synthesis of certain enzymes when their products are present; more generally, it refers to inhibition of transcription (or translation) by binding of repressor protein to a specific site on DNA (or mRNA).

repressor—A protein that inhibits expression of a gene. It may act to prevent transcription by binding to an operator site in DNA or to prevent translation by binding to RNA.

resolution—Resolution occurs by a homologous recombination reaction between the two copies of the transposon in a cointegrate. The reaction generates the donor and target replicons, each with a copy of the transposon.

resolvase—The enzyme activity involved in site-specific recombination between two copies of a transposon that has been duplicated.

restriction endonuclease—An enzyme that recognizes a specific short sequence of DNA and cleaves the duplex (sometimes at the target site, sometimes elsewhere, depending on the type of enzyme).

restriction fragment length polymorphism (RFLP)—Inherited differences in sites for restriction enzymes (for example, caused by base changes in the target site) that result in differences in the lengths of the fragments produced by cleavage with the relevant restriction enzyme. They are used for genetic mapping to link the genome directly to a conventional genetic marker.

restriction map—A linear array of restriction sites on DNA, determined by cleaving the DNA with various restriction endonucleases or by scanning a known sequence for restriction sites.

retrotransposon (retroposon)—A transposon that mobilizes via an RNA form; the DNA element is transcribed into RNA, and then reverse-transcribed into DNA, which is inserted at a new site in the genome. It does not have an infective (viral) form.

retrovirus—An RNA virus with the ability to convert its sequence into DNA by reverse transcription.

reverse gyrase—Enzyme that introduces positive supercoils into DNA.

reverse transcriptase—An enzyme that uses single stranded RNA as a template to synthesize a complementary DNA strand.

reverse transcription—Synthesis of DNA on a template of RNA. It is accomplished by the enzyme reverse transcriptase.

reverse transcription polymerase chain reaction (RT-PCR)—A technique for the detection and quantification of expression of a gene by reverse transcription and amplification of RNAs from a cell sample.

reverse translocation—Transfer of proteins from the endoplasmic reticulum to the cytosol, usually for degradation.

revertants—Reversions of a mutant cell or organism to the wild-type phenotype.

RF1—The bacterial release factor that recognizes UAA and UAG as signals to terminate polypeptide translation.

RF2—The bacterial release factor that recognizes UAA and UGA as signals to terminate polypeptide translation.

RF3—A polypeptide translation termination factor related to the elongation factor EF-G. It functions to release the factors RF1 or RF2 from the ribosome when they act to terminate polypeptide translation.

rho dependent termination—Transcriptional termination by bacterial RNA polymerase in the presence of the rho factor.

rho factor—A protein involved in assisting *E. coli* RNA polymerase to terminate transcription at certain terminators (called rho-dependent terminators).

ribonucleoprotein—A complex of RNA with proteins.

ribosomal RNAs (rRNAs)—A major component of the ribosome.

ribosome—A large assembly of RNA and proteins that synthesizes proteins under direction from an mRNA template.

ribosome-binding site—A sequence on bacterial mRNA that includes an initiation codon that is bound by a 30S subunit in the initiation phase of polypeptide translation.

riboswitch—A catalytic RNA whose activity responds to a metabolite product or another small ligand.

ribozyme—An RNA that has catalytic activity.

RISC—RNA-induced silencing complex, a ribonucleoprotein particle composed of a short single-stranded siRNA and a nuclease that cleaves mRNAs complementary to the siRNA. It receives siRNA from Dicer and delivers it to the mRNA.

RNA editing—A change of sequence at the level of RNA following transcription.

RNA interference (RNAi)—A process by which short 21 to 23 nucleotide antisense RNAs, derived from longer double-stranded RNAs, can modulate expression of mRNA by translation inhibition or degradation.

RNA ligase—An enzyme that functions in tRNA splicing to make a phosphodiester bond between the two exon sequences that are generated by cleavage of the intron.

RNA polymerase—An enzyme that synthesizes RNA using a DNA template (formally described as DNA-dependent RNA polymerases).

RNA processing—Modifications to RNA transcripts of genes. This may include alterations to the 3′ and 5′ ends and the removal of introns.

RNA silencing—The ability of an RNA, especially ncRNA, to alter chromatin structure in order to prevent gene transcription.

RNA splicing—The process of excising introns from RNA and connecting the exons into a continuous mRNA.

RNase—An enzyme that degrades RNA.

rolling circle—A mode of replication in which a replication fork proceeds around a circular template for an indefinite number of revolutions; the DNA strand newly synthesized in each revolution displaces the strand synthesized in the previous revolution, giving a tail containing a linear series of sequences complementary to the circular template strand.

rotational positioning—The location of the histone octamer relative to turns of the double helix, which determines which face of DNA is exposed on the nucleosome surface.

rut—An acronym for rho utilization site, the sequence of RNA that is recognized by the rho termination factor.

S domain—The sequence of 7S RNA of the SRP that is not related to Alu RNA.

S phase—The restricted part of the eukaryotic cell cycle during which synthesis of DNA occurs.

S region—A sequence involved in immunoglobulin class switching. They consist of repetitive sequences at the 5′ ends of gene segments encoding the heavy-chain constant regions.

satellite DNA—DNA that consists of many tandem repeats (identical or related) of a short basic repeating unit.

scarce mRNA—mRNA that consists of a large number of individual mRNA species, each present in very few copies per cell. This accounts for most of the sequence complexity in RNA.

scrapie—A disease caused by an infective agent made of protein (a prion).

Sec61 complex—A heterotrimeric complex of proteins that forms a membrane-spanning pore, the translocon. Known as **SecY** in prokaryotes.

second-site reversion—A second mutation suppressing the effect of a first mutation.

sector—A patch of cells made up of a single altered cell and its progeny.

selfish DNA—DNA sequences that do not contribute to the phenotype of the organism but have self-perpetuation within the genome as their sole function.

semiconservative replication—replication accomplished by separation of the strands of a parental duplex, with each strand then acting as a template for synthesis of a complementary strand.

semiconserved (semi-invariant)—A position where comparison of many individual sequences finds the same type of base (pyrimidine or purine) always present.

semidiscontinuous replication—The mode of replication in which one new strand is synthesized continuously while the other is synthesized discontinuously.

septal ring—A complex of several proteins coded by *fts* genes of *E. coli* that forms at the mid-point of the cell. It gives rise to the septum at cell division. The first of the proteins to be incorporated is FtsZ, which gave rise to the original name of the Z-ring.

septum—The structure that forms in the center of a dividing bacterium, providing the site at which the daughter bacteria will separate. The same term is used to describe the cell wall that forms between plant cells at the end of mitosis.

sequence context—See **context**.

Shine-Dalgarno sequence—The polypurine sequence AGGAGG centered about 10 bp before the AUG initiation codon on bacterial mRNA. It is complementary to the sequence at the 3′ end of 16S rRNA.

short interspersed elements (SINEs)—Short interspersed nuclear elements; a major class of short (<500 bp) nonautonomous retrotransposons that occupy ~13% of the human genome (see **retrotransposon**).

sigma factor—The subunit of bacterial RNA polymerase needed for initiation; it is the major influence on selection of promoters.

signal end—It is produced during recombination of immunoglobulin and T cell receptor genes. They are at the termini of the cleaved fragment containing the recombination signal sequences. Their subsequent joining yields a signal joint.

signal patch—A protein targeting signal consisting of amino acids that can be distant in the primary sequence of the protein but are clustered together in the folded protein to form a functional signal.

signal peptidase—An enzyme within the membrane of the ER that specifically removes the signal sequences from proteins as they are translocated. Analogous activities are present in bacteria, archaea, and in each organelle in a eukaryotic cell into which proteins are targeted and translocated by means of removable targeting sequences. It is one component of a larger protein complex.

signal recognition particle (SRP)—A ribonucleoprotein complex that recognizes signal sequences during translation and guides the ribosome to the translocation channel. Particles from different organisms may have different compositions, but all contain related proteins and RNAs.

signal sequence—A short region of a protein that directs it to the endoplasmic reticulum for cotranslational translocation.

silencer—A short sequence of DNA that can inactivate expression of a gene in its vicinity.

silent mutation—A mutation that does not change the sequence of a polypeptide because it produces synonymous codons.

silent sites—Sites in a coding region at which mutations have not changed the amino acid that is encoded.

simple sequence DNA—Short repeating units of DNA sequence.

single-copy replication control—A control system in which there is only one copy of a replicon per unit bacterium. The bacterial chromosome and some plasmids have this type of regulation.

single nucleotide polymorphism (SNP)—A polymorphism (variation in sequence between individuals) caused by a change in a single nucleotide. This is responsible for most of the genetic variation between individuals.

single-strand binding protein (SSB)—The protein that attaches to single-stranded DNA, thereby preventing the DNA from forming a duplex.

single-strand exchange—A reaction in which one of the strands of a duplex of DNA leaves its former partner and instead pairs with the complementary strand in another molecule, displacing its homologue in the second duplex.

single-strand invasion—The process in which a single strand of DNA displaces its homologous strand in a duplex.

single-strand passage—A reaction catalyzed by type I topoisomerase in which one section of single-stranded DNA is passed through another strand.

single X hypothesis—The theory that describes the inactivation of one X chromosome in female mammals.

siRNA—Short interfering RNA, a miRNA that prevents gene expression.

sister chromatid—Each of two identical copies of a replicated chromosome; this term is used as long as the two copies remain linked at the centromere. Sister chromatids separate during anaphase in mitosis or anaphase II in meiosis.

site-specific recombination—Recombination that occurs between two specific sequences, as in phage integration/excision or resolution of cointegrate structures during transposition.

SL RNA (spliced leader RNA)—A small RNA that donates an exon in the *trans*-splicing reaction of trypanosomes and nematodes.

small cytoplasmic RNAs (scRNA; scyrps)—RNAs that are present in the cytoplasm (and sometimes are also found in the nucleus). Scyrps are the ribonucleoprotein particles that include a specific snRNA and its protein partners.

small nuclear RNA (snRNA; snurps)—One of many small RNA species confined to the nucleus; several of them are involved in splicing or other RNA processing reactions. Snurps are the ribonucleoprotein particles that include a specific snRNA and its protein partners.

small nucleolar RNA (snoRNA)—A small nuclear RNA that is localized in the nucleolus.

small subunit—The subunit of the ribosome (30S in bacteria, 40S in eukaryotes) that binds the mRNA.

somatic mutation—A mutation occurring in a somatic cell, therefore affecting only its daughter cells; it is not inherited by descendants of the organism.

somatic recombination—Recombination that occurs in non-germ cells (i.e., it does not occur during meiosis); mostly commonly used to refer to recombination in the immune system.

SOS box—The DNA sequence (operator) of ~20 bp recognized by LexA repressor protein.

SOS response—In *E. coli*, the coordinate induction of many enzymes, including repair activities, in response to irradiation or other damage to DNA; it results from activation of protease activity by RecA to cleave LexA repressor.

spindle—A structure made up of microtubules that guides the movements of the chromosomes during mitosis.

splice recombinant—DNA that results from a Holliday junction being resolved by cutting the nonexchanged strands. Both strands of DNA before the exchange point come from one chromosome; the DNA after the exchange point comes from the homologous chromosome.

spliceosome—A complex formed by snRNPs and additional protein factors that is required for RNA splicing.

splicing—The process of excising introns from RNA and connecting the exons into a continuous mRNA.

splicing factor—A protein component of the spliceosome that is not part of one of the snRNPs.

spontaneous mutations—Mutations occurring in the absence of any added reagent to increase the mutation rate, as the result of errors in replication (or other events involved in the reproduction of DNA) or by random changes to the chemical structure of bases.

sporulation—The generation of a spore by a bacterium (by morphological conversion) or by a yeast (as the product of meiosis).

SR protein—A protein that has a variable length of a Ser-Arg-rich region and is involved in splicing.

sRNA—A small bacterial RNA that functions as a regulator of gene expression.

SRP receptor—A dimeric protein in the ER membrane that recognizes the signal recognition particle (SRP) and is involved in the insertion of the nascent polypeptide into the ER translocon.

startpoint—The position on DNA corresponding to the first base incorporated into RNA.

stem—The base-paired segment of a hairpin structure in RNA.

steroid receptor—Transcription factors that are activated by binding of a steroid ligand.

stop codon—One of three triplets (UAG, UAA, or UGA) that cause polypeptide translation to terminate. They are also known historically as nonsense codons. The UAA codon is called ochre and the UAG codon is called amber, after the names of the nonsense mutations by which they were originally identified.

strand displacement—A mode of replication of some viruses in which a new DNA strand grows by displacing the previous (homologous) strand of the duplex.

stringent factor—The protein RelA, which is associated with ribosomes. It synthesizes ppGpp and pppGpp when an uncharged tRNA enters the A site.

stringent response—The ability of a bacterium to shut down synthesis of tRNA and ribosomes in a poor-growth medium.

stRNA—Short temporal RNA, a form of miRNA in eukaryotes that modulates mRNA expression during development.

structural gene—A gene that codes for any RNA or polypeptide product other than a regulator.

supercoiling—The coiling of a closed duplex DNA in space so that it crosses over its own axis.

superfamily—A set of genes all related by presumed descent from a common ancestor but now showing considerable variation.

suppression mutation—A second event eliminates the effects of a mutation without reversing the original change in DNA.

suppressor—A second mutation that compensates for or alters the effects of a primary mutation.

surveillance systems—Systems that check nucleic acids for errors. The term is used in several different contexts. One example is the system that degrades mRNAs that have nonsense mutations. Another is the set of systems that react to damage in the double helix. The common feature is that the system recognizes an invalid sequence or structure and triggers a response.

synapsis—The association of the two pairs of sister chromatids (representing homologous chromosomes) that occurs at the start of meiosis; the resulting structure is called a bivalent.

synaptonemal complex—The morphological structure of synapsed chromosomes.

synonym codons—Codons that have the same meaning (specifying the same amino acid, or specifying termination of translation) in the genetic code.

synteny—A relationship between chromosomal regions of different species where homologous genes occur in the same order.

synthetic genetic array analysis (SGA)—An automated technique in budding yeast whereby a mutant is crossed to an array of approximately 5000 deletion mutants to determine if the mutations interact to cause a synthetic lethal phenotype.

synthetic lethality—It occurs when two mutations that by themselves are viable cause lethality when combined.

T cell receptor (TCR)—The antigen receptor on T lymphocytes. It is clonally expressed and binds to a complex of MHC class I or class II protein and antigen-derived peptide.

T cells—Lymphocytes of the T (thymic) lineage; they may be subdivided into several functional types. They carry TcR and are involved in the cell-mediated immune response.

TAFs—The subunits of TFIID that assist TBP in binding to DNA. They also provide points of contact for other components of the transcription apparatus.

TATA-binding protein (TBP)—The subunit of transcription factor TFIID that binds to the TAT box in the promoter and is positioned at the promoters that do not contain a TATA box by other factors.

TATA box—A conserved A-T-rich octamer found about 25 bp before the startpoint of each eukaryotic RNA polymerase II transcription unit; it is involved in positioning the enzyme for correct initiation.

TATA-less promoter—A promoter lacking a TATA box in the sequence upstream of its startpoint.

T-DNA—The segment of the Ti plasmid of *Agrobacterium tumefaciens* that is transferred to the plant cell nucleus during infection. It carries genes that transform the plant cell.

telomerase—The ribonucleoprotein enzyme that creates repeating units of one strand at the telomere by adding individual bases to the DNA 3′ end, as directed by an RNA sequence in the RNA component of the enzyme.

telomere—The natural end of a chromosome; the DNA sequence consists of a simple repeating unit with a protruding single-stranded end.

telomeric silencing—The repression of gene activity that occurs in the vicinity of a telomere.

temperate phage—A phage that can enter a lysogenic cycle within the host (can become a prophage integrated into the host genome).

template strand—See **antisense strand**.

terminal protein—A protein that allows replication of a linear phage genome to start at the very end. It attaches to the 5′ end of the genome through a covalent bond, is associated with a DNA polymerase, and contains a cytosine residue that serves as a primer.

terminase—An enzyme cleaves multimers of a viral genome and then uses hydrolysis of ATP to provide the energy to translocate the DNA into an empty viral capsid starting with the cleaved end.

termination—A separate reaction that ends a macromolecular synthesis reaction (replication, transcription, or translation), by stopping the addition of subunits, and (typically) causing disassembly of the synthetic apparatus.

termination codon—One of the three codons (UAA, UAG, UGA) that signal the termination of translation of a polypeptide.

terminator—A sequence of DNA that causes RNA polymerase to terminate transcription.

ternary complex—The complex in initiation of transcription that consists of RNA polymerase and DNA as well as a dinucleotide that represents the first two bases in the RNA product.

tetrad—The four (haploid) spores that result from meiosis in yeast.

$TF_{II}D$—The transcription factor that binds to the TATA sequence upstream of the startpoint of promoters for RNA polymerase II. It consists of TBP (TATA binding protein) and the TAF subunits that bind to TBP.

thalassemia—A disease of red blood cells resulting from lack of either α or ß globin.

third-base degeneracy—The lesser effect on codon meaning of the nucleotide present in the third (3′) codon position.

Ti plasmid—An episome of the bacterium *Agrobacterium tumefaciens* that carries the genes responsible for the induction of crown gall disease in infected plants.

tight binding—The binding of RNA polymerase to DNA in the formation of an open complex (when the strands of DNA have separated).

TIM complex—A complex that resides in the inner membrane of mitochondria and is responsible for transporting proteins from the intermembrane space into the interior of the organelle.

Tn—Followed by a number, it denotes bacterial transposons carrying markers that are not related to their function, e.g., drug resistance.

TOM complex—A complex that resides in the outer membrane of the mitochondrion and is responsible for importing proteins from the cytosol into the space between the membranes.

topoisomerase—An enzyme that changes the number of times the two strands in a closed DNA molecule cross each other. It does this by cutting the DNA, passing DNA through the break, and resealing the DNA.

trailer (3′ UTR)—An untranslated sequence at the 3′ end of an mRNA following the termination codon.

***trans*-acting sequence**—DNA sequence coding for a product that can function on any copy of its target DNA. This implies that it is a diffusible protein or RNA.

transcription—Synthesis of RNA on a DNA template.

transcription unit—The sequence between sites of initiation and termination by RNA polymerase; it may include more than one gene.

transcriptome—The complete set of RNAs present in a cell, tissue, or organism. Its complexity is due mostly to mRNAs, but it also includes noncoding RNAs.

transducing virus—A virus that carries part of the host genome in place of part of its own sequence. The best known examples are retroviruses in eukaryotes and DNA phages in *E. coli*.

transesterification—A reaction that breaks and makes chemical bonds in a coordinated transfer so that no energy is required.

transfection—The acquisition of new genetic material by eukaryotic cells via incorporation of added DNA.

transfer region—A segment on the F plasmid that is required for bacterial conjugation.

transfer RNA (tRNA)—The intermediate in protein synthesis that interprets the genetic code. Each molecule can be linked to an amino acid. It has an anticodon sequence that is complementary to a triplet codon representing the amino acid.

transformation—In bacteria, it is the acquisition of new genetic material by incorporation of added DNA.

transforming principle—DNA that is taken up by a bacterium and whose expression then changes the properties of the recipient cell.

transgenic animals—Animals created by introducing DNA prepared in test tubes into the germline. The DNA may be inserted into the genome or exist in an extrachromosomal structure.

transient transfectants (unstable transfectants)—Eukaryotic cells that have foreign DNA in an unstable—i.e., extrachromosomal—form.

transition—A mutation in which one pyrimidine is replaced by the other, or in which one purine is replaced by the other.

translation—Synthesis of protein on an mRNA template.

translational positioning—The location of a histone octamer at successive turns of the double helix, which determines which sequences are located in linker regions.

translocation—1. The movement of the ribosome one codon along mRNA after the addition of each amino acid to the polypeptide chain. 2. The reciprocal or nonrecriprocal exchange of chromosomal material between nonhomologous chromosomes.

translocon—An integral membrane protein that provides a channel for displacement of polypeptide segments across the membrane; it is usually required in movement of a protein across a lipid bilayer.

transmembrane region (domain)—The part of a protein that spans the membrane bilayer. It is hydrophobic and in many cases contains approximately 20 amino acids that form an α-helix.

transplantation antigen—Protein coded by a major histocompatibility locus, present on all mammalian cells, involved in interactions between lymphocytes.

transposase—The enzyme activity involved in insertion of transposon at a new site.

transposon (transposable element)—A DNA sequence able to insert itself (or a copy of itself) at a new location in the genome without having any sequence relationship with the target locus.

transversion—A mutation in which a purine is replaced by a pyrimidine or vice versa.

$tRNA_f^{Met}$—The special RNA used to initiate polypeptide translation in bacteria. It mostly uses AUG but can also respond to GUG and CUG.

$tRNA_m^{Met}$—The bacterial tRNA that inserts methionine at internal AUG codons.

true activator—A positive transcription faction that functions by making contact, direct or indirect, with the basal apparatus to activate transcription.

true reversion—A mutation that restores the original sequence of the DNA.

Ty—It stands for transposon yeast, the first transposable element to be identified in yeast.

type I topoisomerase—An enzyme that changes the topology of DNA by nicking and resealing one strand of DNA.

type II topoisomerase—An enzyme that changes the topology of DNA by nicking and resealing both strands of DNA.

U3—The repeated sequence at the 3′ end of a retroviral RNA.

U5—The repeated sequence at the 5′ end of a retroviral RNA.

UAS (spstream activating sequence)—The equivalent in yeast of the enhancer in higher eukaryotes and is bound by the Gal4 transcriptional activator proteins.

underwound—B-form DNA that has fewer than 10 base pairs per turn of the helix.

unequal crossing over (nonreciprocal recombination)—It results from an error in pairing and crossing-over in which nonequivalent sites are involved in a recombination event. It produces one recombinant with a deletion of material and one with a duplication.

unidentified reading frame—An open reading frame with an as yet undetermined function.

uninducible—A mutant in which the affected gene(s) cannot be expressed.

unit cell—The state of an *E. coli* bacterium generated by a new division. It is 1.7 μm long and has a single replication origin.

up mutation—A mutation in a promoter that increases the rate of transcription.

upstream—Sequences in the opposite direction from expression.

upstream activating sequence (UAS)—The equivalent in yeast of the enhancer in higher eukaryotes; a UAS cannot function downstream of the promoter.

V gene—A sequence coding for the major part of the variable (N-terminal) region of an immunoglobulin chain.

variable number tandem repeat (VNTR)—Very short repeated sequences, including microsatellites and minisatellites.

variable region (V region)—An antigen-binding site of an immunoglobulin or T cell receptor molecule. They are composed of the variable domains of the component chains. They are coded by V gene segments and vary extensively among antigen receptors as the result of multiple, different genomic copies and of changes introduced during synthesis.

variegation—It is produced by a change in genotype during somatic development.

vector—A plasmid or phage chromosome that is used to perpetuate a cloned DNA segment.

vegetative phase—The period of normal growth and division of a bacterium. For a bacterium that can sporulate, this contrasts with the sporulation phase, when spores are being formed.

viral superfamily—Transposons that are related to retroviruses. They are defined by sequences that code for reverse transcriptase or integrase.

viroid—A small infectious nucleic acid that does not have a protein coat.

virulent mutations—Phage mutants that are unable to establish lysogeny.

virulent phage—A bacteriophage that can only follow the lytic cycle.

virusoid—A small infectious nucleic acid that is encapsidated by a plant virus together with its own genome.

wobble hypothesis—The ability of a tRNA to recognize more than one codon by unusual (non-G-C, non-A-T) pairing with the third base of a codon.

xeroderma pigmentosum (XP)—A disease caused by mutation in one of the *XP* genes that results in hypersensitivity to sunlight (particularly ultraviolet light), skin disorders, and cancer predisposition.

zinc finger—A DNA-binding motif that typifies a class of transcription factor.

zoo blot—The use of Southern blotting to test the ability of a DNA probe from one species to hybridize with the DNA from the genomes of a variety of other species.

Z-ring—See **septal ring.**

Appendix

Answers to Even-Numbered End-of-Chapter Questions

Chapter 1

2. D; **4**. A; **6**. A; **8**. D; **10**. A

Chapter 2

2. B; **4**. B; **6**. C; **8**. C; **10**. C

Chapter 3

2. C; **4**. A; **6**. B; **8**. A; **10**. D

Chapter 4

2. B; **4**. D; **6**. C; **8**. C; **10**. B

Chapter 5

2. C; **4**. C; **6**. B; **8**. B; **10**. B

Chapter 6

2. A; **4**. C; **6**. C; **8**. A; **10**. A

Chapter 7

2. B; **4**. B; **6**. B; **8**. B; **10**. C

Chapter 8

2. B; **4**. A; **6**. B; **8**. D; **10**. D

Chapter 9

2. C; **4**. C; **6**. D; **8**. D; **10**. D

Chapter 10

2. matrix, inner membrane, intermembrane space, outer membrane.
4. E, B, D, A, F, C
6. A; **8**. C
10. Many proteins contain regions of charged or polar amino acids compared with the highly hydrophobic lipid bilayer membrane. Hydrophobic and hydrophilic elements generally repel each other. These proteins require a passage or pore through which they can move across the membrane. This pore is composed of integral membrane proteins.

Chapter 11

2. C; **4**. A; **6**. D; **8**. B; **10**. A.

Chapter 12

2. D; **4**. D; **6**. D; **8**. A; **10**. A

Chapter 13

2. C; **4**. A; **6**. C; **8**. D; **10**. B.

Chapter 14

2. A; **4**. B; **6**. D; **8**. C; **10**.A

Chapter 15

2. A; **4**. D; **6**. A; **8**. A; **10**. B

Chapter 16

2. B; **4**. D; **6**. C; **8**. D; **10**. B

Chapter 17

2. C; **4**. A; **6**. D; **8**. B; **10**. A

Chapter 18

2. B; **4**. D; **6**. D; **8**. B; **10**. A

Chapter 19

2. bacterial protein IHF (integration host factor); phage protein Int (integrase); *attB* sequence in the bacterial chromosome; *attP* sequence in the phage genome
4. A; **6**. C; **8**. D; **10**. C

Chapter 20

2. removes DNA between nicks around the damaged site; **4**. helicases; **6**. seals nick in DNA to reestablish a continuous phosphodiester backbone. **8**. lyases
10. A; **12**. C; **14**. D

Chapter 21

2. C; **4**. A; **6**. any of the following: matrix protein; capsid protein; nucleocapsid protein; **8**. any of the following: protease; reverse transcriptase; integrase; **10**. surface protein (spikes to allow virion to interact with host)
12. C; **14**. A; **16**. B; **18**. D

Chapter 22

2. A; **4** B; **6**. D; **8**. RAG1; **10**. cleavage; **12**. RAG1

Chapter 23

2. filaments; **4**. spherical or icosahedron; **6**. It often consists of multiple repeats that are not transcribed. **8**. It replicates late in S phase and has a reduced incidence of genetic recombination.
10. C; **12**. B; **14**. A; **16**. A

Chapter 24

2. prevents; **4**. gene; **6**. inactivating; **8**. B; **10**. C; **12**. B

Chapter 25

2. BCC; **4**. B; **6**. A; **8**. B; **10**. C

Chapter 26

2. C; **4**. B; **6**. A; **8**. C; **10**. B

Chapter 27

2. It remains condensed in interphase.
4. It replicates late in S phase.
6. Constitutive heterochromatin contains specific DNA sequences that have no coding function and includes satellite DNAs located at the centromeres.
8. In mammals, one of the two female X chromosomes is completely inactivated.
10. In *C. elegans*, gene expression from each female X chromosome is only half relative to expression from the single X chromosome.
12. A self-perpetuating state or conformation of a protein may be established and maintained.
14. Scrapie (other answers can be Gerstmann-Staussler syndrome and Creutzfeldt-Jakob disease [CJD])
16. D; **18**. A

Chapter 28

2. B; **4**. C; **6**.D; **8**. B; **10**. A

Chapter 29

2. B; **4**. C; **6**. D; **8**. B; **10**. C

Chapter 30

2. B; **4**. A; **6**. D

Index

A

B

C

D

E

H

I

J

K

L

M

N

O

P

Q

R

S

U

V

W

X

Y

Z